Lecture Notes in Networks and Systems

Volume 577

The series "Lecture Notes in Networks and Systems" publishes the latest developments in Networks and Systems—quickly, informally and with high quality. Original research reported in proceedings and post-proceedings represents the core of LNNS.

Volumes published in LNNS embrace all aspects and subfields of, as well as new challenges in, Networks and Systems.

The series contains proceedings and edited volumes in systems and networks, spanning the areas of Cyber-Physical Systems, Autonomous Systems, Sensor Networks, Control Systems, Energy Systems, Automotive Systems, Biological Systems, Vehicular Networking and Connected Vehicles, Aerospace Systems, Automation, Manufacturing, Smart Grids, Nonlinear Systems, Power Systems, Robotics, Social Systems, Economic Systems and other. Of particular value to both the contributors and the readership are the short publication timeframe and the world-wide distribution and exposure which enable both a wide and rapid dissemination of research output.

The series covers the theory, applications, and perspectives on the state of the art and future developments relevant to systems and networks, decision making, control, complex processes and related areas, as embedded in the fields of interdisciplinary and applied sciences, engineering, computer science, physics, economics, social, and life sciences, as well as the paradigms and methodologies behind them.

Indexed by SCOPUS, INSPEC, WTI Frankfurt eG, zbMATH, SCImago.

All books published in the series are submitted for consideration in Web of Science.

For proposals from Asia please contact Aninda Bose (aninda.bose@springer.com).

Ivan Petrovic · Emanuele Menegatti ·
Ivan Marković
Editors

Intelligent Autonomous Systems 17

Proceedings of the 17th International Conference IAS-17

Springer

Editors
Ivan Petrovic
Faculty of Electrical Engineering
University of Zagreb
Zagreb, Croatia

Emanuele Menegatti
Department of Information Engineering
University of Padua
Padua, Italy

Ivan Marković
Faculty of Electrical Engineering
University of Zagreb
Zagreb, Croatia

ISSN 2367-3370 ISSN 2367-3389 (electronic)
Lecture Notes in Networks and Systems
ISBN 978-3-031-22215-3 ISBN 978-3-031-22216-0 (eBook)
https://doi.org/10.1007/978-3-031-22216-0

This Springer imprint is published by the registered company Springer Nature Switzerland AG
The registered company address is: Gewerbestrasse 11, 6330 Cham, Switzerland

Program Committee

Aamir Ahmad	University of Stuttgart
Luis Almeida	University of Porto
Francesco Amigoni	Politecnico di Milano
Qi An	The University of Tokyo
Shinya Aoi	Kyoto University
Sven Behnke	University of Bonn
Saso Blazic	University of Ljubljana
Stjepan Bogdan	University of Zagreb, Faculty of Electrical Engineering and Computing
Davide Brugali	Università degli Studi di Bergamo
Zheng Chen	Shanghai Jiao Tong University
Grazia Cicirelli	Institute of Intelligent Systems for Automation
Davide Antonio Cucci	Politecnico di Milano
Antonio D'Angelo	Department of Mathematics and Computer Science, University of Udine
Tiziana D'Orazio	Institute of Intelligent Systems for Automation, National Research Council
Miha Deniša	Jožef Stefan Institute
Zheng Fang	Northeastern University
Damir Filko	Faculty of Electrical Engineering, Computer Science and Information Technology Osijek
Alberto Finzi	Universita' di Napoli Federico II
Andrej Gams	Jožef Stefan Institute
Stefano Ghidoni	University of Padova
Giuseppina Gini	Politecnico di Milano
Edward Grant	North Carolina State University
Giorgio Grisetti	University "Sapienza" of Rome
Ales Hace	UM FERI
Zhu Haiyue	Singapore Institute of Manufacturing Technology
Tom Henderson	University of Utah
Yasuhisa Hirata	Tohoku University

Tianjiang Hu	NUDT
Han Pang Huang	National Taiwan University
Shuhei Ikemoto	Kyushu Institute of Technology
Tetsunari Inamura	National Institute of Informatics
Giovanni Indiveri	Universita' degli Studi di Genova
Yonghoon Ji	JAIST
Gregor Klančar	Faculty of Electrical Engineering, University of Ljubljana
Stefan Klink	Institute AIFB, University of Karlsruhe
Toshiyuki Kondo	Tokyo University of Agriculture and Technology
Zdenko Kovacic	University of Zagreb Faculty of Electrical Engineering and Computing
Miroslav Kulich	CTU in Prague
Yoshinori Kuno	Saitama University
Daisuke Kurabayashi	Tokyo Institute of Technology
Bakir Lacevic	Faculty of Electrical Engineering, University of Sarajevo
Joo-Ho Lee	Ritsumeikan University
Sukhan Lee	Sungkyunkwan University
Kristijan Lenac	University of Trieste
Xiangpeng Li	Soochow University
Yusuke Maeda	Yokohama National University
Ester Martinez-Martin	University ES
Hisataka Maruyama	Nagoya University
Matteo Matteucci	Politecnico di Milano
Emanuele Menegatti	University of Padua
Nikola Miskovic	University of Zagreb, Faculty of Electrical Engineering and Computing
Ikuo Mizuuchi	Tokyo University of Agriculture and Technology
Surya Girinatha Nurzaman	Bio-Inspired Robotics Lab., ETH Zürich
Emmanuel Karlo Nyarko	Faculty of Electrical Engineering, Computer Science and Information Technology Osijek
Dinko Osmankovic	ETF Sarajevo
Jun Ota	The University of Tokyo
Christian Pek	KTH Royal Institute of Technology
Tadej Petric	Jozef Stefan Institute
Tamara Petrovic	University of Zagreb Faculty of Electrical Engineering and Computing
Alberto Pretto	IT+Robotics Srl
Libor Přeučil	CTU in Prague
Chenkun Qi	Shanghai Jiao Tong University
Monica Reggiani	University of Padua
Paolo Rocco	Polytechnic University of Milan
Arne Roennau	Forschungszentrum Informatik
Andre Rosendo	Shanghai Tech University

Marija Seder	University of Zagreb, Faculty of Electrical Engineering and Computing
Yantao Shen	University of Nevada, Reno
Nobutaka Shimada	Ritsumeikan University
Shouhei Shirafuji	The University of Tokyo
Domenico G. Sorrenti	Universita' degli Studi di Milano—Bicocca
Sutthiphong Srigrarom	National University of Singapore
Ivo Stancic	Faculty of Electrical Engineering, Mechanical Engineering, and Naval Architecture
Gerald Steinbauer	Graz University of Technology
Todor Stoyanov	AASS Research Center, Örebro University
Marko Svaco	University of Zagreb, Faculty of Mechanical Engineering and Naval Architecture
Adnan Tahirovic	University of Sarajevo, Faculty of Electrical Engineering
Masaki Takahashi	Department of System Design Engineering, Keio University
Takashi Tsubouchi	University of Tsukuba
Jingchuan Wang	Shanghai Jiao Tong University
Kundong Wang	Shanghai Jiao Tong University
Qining Wang	Peking University
Atsushi Yamashita	The University of Tokyo
Wenwei Yu	Chiba University

Preface

This volume contains the papers presented at the 17th International Conference on Intelligent Autonomous Systems (IAS-17), which was held from June 13–16, 2022, in Zagreb, Croatia.

IAS has been held every two years since 1986, allowing researchers to become acquainted with the latest accomplishments and innovations in advanced intelligent autonomous systems, which are increasingly applied in areas ranging from industrial applications to professional service and household domains. New technologies and application domains push forward the need for research and development resulting in new challenges to be overcome in order to apply intelligent autonomous systems in a reliable and user-independent way. Recent advances in the areas of artificial intelligence, machine learning, and adaptive control enable autonomous systems with improved robustness and flexibility. Contributions on basic research as well as on relevant applications are included in this volume.

In our view, 62 papers, authored by 196 authors from 19 countries, are a testimony to the appeal of the conference considering travel restrictions imposed by the COVID-19 pandemic. Most papers were evaluated by three or more independent expert reviewers and by a member of the International Program Committee. We are very proud to have welcomed distinguished plenary speakers: Professor Danica Kragi´c Jensfelt (KTH Royal Institute of Technology, Sweden), Professor Dana Kuli´c (Monash University, Australia), Professor Wolfram Burgard (Technical University of Nuremberg, Germany), and Professor Takayuki Kanda (Kyoto University, Japan).

We would like to thank all those who contributed to the success of IAS-17. First, we would like to express our thanks to the IAS Society for entrusting us with the organization of IAS-17 and for their help. Furthermore, our sincere thanks are due to all people whose cooperation and hardwork made this conference possible. First and foremost, we would like to thank the members of the International Program Committee and Advisory Committee for their outstanding work and our sponsors whose support is particularly appreciated. Our special thanks go to the authors for submitting their work to IAS-17 and to the reviewers for their time and effort in evaluating the submissions. The results of their joint work are visible in this volume.

We are looking forward to seeing you at IAS-18 in 2023 in Suwon, South Korea!

July 2022

Ivan Petrovic
General Chair

Emanuele Menegatti
General Co-chair

Ivan Marković
Program Chair

Contents

Perception

Advanced and Intelligent Control Design for Underwater Robots

Autonomous Robots in Warehouse Automation

Collaborative Robots

Evaluation of Software Solutions for Risk Assessment Focusing on Human-Robot Collaboration

Luisa Hornung(✉), Christian Wurll, and Björn Hein

Hochschule Karlsruhe - University of Applied Sciences, 76133 Karlsruhe, Germany
luisa.hornung@h-ka.de

Abstract. There is still limited support for performing human-robot collaboration (HRC) risk assessments, so users must rely on their expertise and experience. Therefore, risk assessment is challenging for inexperienced users without this knowledge. In this paper, we present a study that identifies evaluation criteria for risk assessment software based on expert opinions and evaluate the software solution against the criteria. In addition, the software solution is evaluated for its applicability to HRC applications. The results show that the available software solutions takes different approaches and only partially meets the experts' requirements. The specifics of HRC applications are only rudimentary addressed by the software solutions, but can be mapped with the available functions and the necessary knowledge.

Keywords: Risk assessment · Software solutions · Human-robot collaboration · Evaluation criteria

1 Introduction

For strengthening small and medium-sized enterprises (SMEs), flexible human-machine interaction systems take important key roles, as human-robot collaboration (HRC) has the potential to automate areas where the use of robots was not feasible so far. Despite the many advantages offered by HRC, the major breakthrough of cyber-physical systems has not yet been achieved. One of the main reasons for the inhibited use of collaborative robot applications are the high requirements for the safety of the systems and the resulting uncertainties regarding safety aspects to be fulfilled [1–4].

In particular, the risk assessment, the execution of which is prescribed by the Machinery Directive 2006/42/EC (MD) [5] and which therefore must be conducted for each HRC application [6], represents a major hurdle for many users. Conducting a reliable and high-quality risk assessment requires extensive experience and expertise on the part of the user [1,7] and is highly dependent

I. Petrovic et al. (Eds.): IAS 2022, LNNS 577, pp. 3–14, 2023.
https://doi.org/10.1007/978-3-031-22216-0_1

on subjective assessments. Due to the lack of protective fences in an HRC application, the comparatively new technology and limited existing assistance and support in the area of safety, the risk assessment is thus very complex and proves to be extremely difficult especially without an expert and/or experience [7].

Particularly the target group of HRC, SMEs, often do not have the necessary expert knowledge and also the required resources to build up the knowledge to be able to perform a risk assessment to the full extent.

In order to improve clarity and objectivity in risk assessment for HRC applications, and to support and enable non-experts to conduct a risk assessment for HRC, the development of an HRC-specific approach or standard procedure is crucial and necessary. For the successful applicability of a standard procedure, it is advisable to build on the existing expertise and experience and thus counteract the barriers to the use of HRC.

This paper provides a first step towards a standard procedure based on current practice and thus integrates the proven experience and expertise into the development, that simplifies the risk assessment procedure of HRC applications. In this paper, we present a study that examines software solutions by MD. The study pursued multiple aims: (1) to compare the approach of commercially available software solutions, (2) to analyse whether the software solutions meet the requirements of experts, (3) to evaluate them for applicability to HRC applications and (4) to identify possible gaps in relation to HRC. The study aims to answer the following Research Questions (RQ):

> ***RQ 1:*** *Does the state of the art of software for conducting risk assessments comply with the requirements of the users?*
> ***RQ 2:*** *Is the state of the art able to address the specifics of HRC applications?*

The paper is structured as follows: following an overview of the state of the art and research in Sect. 2, the procedure of the study is explained in Sect. 3. The results are presented in Sect. 4 and discussed in Sect. 5. The paper ends with a Conclusion in Sect. 6.

2 Risk Assessment

Machinery Directive 2006/42/EC: The MD requires that a risk assessment must be carried out before placing a machine on the market (reason 23) and the machine must be constructed and designed according to the results of the risk assessment [5]. Annex 1 specifies an iterative process for risk assessment, requiring the following aspects to be included: limits of the machine, all hazards emanating from the machine and the associated hazardous situations, estimation and evaluation of the risks and their mitigation or elimination [5]. The MD is an EU directive and must be transposed into national law by the EU member states [8].

EN ISO 12100: The international standard presents a general overview and guidance for designing a safe machine in accordance with its intended use. In EN

ISO 12100, the process of risk assessment is described as a sequence of logical steps for the systematic analysis and evaluation of risks posed by the machine [9]. EN ISO 12100 specifies the following sequence: (1) limits of the machine, (2) identification of hazards, (3) risk estimation for each hazard, (4) risk evaluation, (5) risk reduction. The risk assessment includes steps (1)–(4). The standard lists all aspects to be considered and the annexes provide examples of hazard consequences, causes and situations. EN ISO 12100 is a harmonised standard of the MD [10].

Some research work already address the simplification of risk assessments for HRC applications, but are limited to specific aspects and, in contrast to software solutions, do not consider the entire process. For example, [11–15] use (semi) formal methods for identification of hazards. The HRC applications are described with the help of formal models on the basis of which the hazard identification is performed. References [16,17] focus on rule- and task-based approaches to identify hazards and [18–22] developed simulation-based approaches to automatically identify critical situations for humans. As a research result of *Cover* [23], a method for identifying hazards is made available on their website.[1] The developed procedure is accessible and can also be used by non-experts, but does not follow the classic procedure according to EN ISO 12100. These methods simplify the risk assessment for HRC applications, but still need to be embedded in the whole process.

3 Study Procedure

The aim of the study was to analyze, compare and evaluate different commercial available software for performing risk assessments according to defined criteria and the applicability to HRC applications. The study was carried out according to a consistent procedure and evaluation scheme.

3.1 Selection of Software Solutions Included in the Study

An extensive Google search was performed to identify the subjects of the study. The keyword search included all combinations of the following keywords (in German as well as in English) (Fig. 1). Stop criterion of the keyword search: with the corresponding keyword combination, no new findings were obtained up to the third page level. Only software solutions for risk assessment according to MD were included in the study and no (commercially or freely available) Microsoft Office templates.

A total of 13 software solutions were identified, eleven of which could be included in the study (one software provider did not grant access to the software and no feedback was received from another). Table 1 lists the identified software solutions in the order in which the study was conducted by name and provider. The *version* indicates which version of the software solution the study was conducted with, *tested* specifies whether the software solution was considered in

[1] www.safearoundrobots.com.

$$\{\text{software}\}\text{AND}\begin{Bmatrix}\text{risk assessment}\\ \text{OR}\\ \text{none}\end{Bmatrix}\text{AND}\begin{Bmatrix}\text{CE marking}\\ \text{OR}\\ \text{CE conformidity}\\ \text{OR}\\ \text{CE assessment}\\ \text{none}\end{Bmatrix}\text{AND}\begin{Bmatrix}\text{machinery directive}\\ \text{OR}\\ \text{2006/42/EC}\\ \text{OR}\\ \text{none}\end{Bmatrix}$$

Fig. 1. Equation Google search

the study, and *type* denotes whether the demo or full version was concerned. *Code* represents a triple of {*version, type, software*} that uniquely specifies the software solution.

Table 1. Identified research objects

Code	Software	Software provider	Version	Evaluated	Type
I	DOCUFY Machine Safety	DOCUFY GmbH	6.0.12	Yes	Demo
II	CE-CON	CE-CON Safety Inc.	3.8.6	Yes	Demo
III	Secutify Solo	Secutify Solo	V1.0	Yes	Demo
IV	SafetyToolBox	pgx software solutions	5.4.3.1	Yes	Full
V	Safexpert	IBF Solutions GmbH	8.5	Yes	Demo
VI	CSafe	CSafe	2016.01	Yes	Demo
VII	GESIMA	Beuth	1.0.6.5	Yes	Demo
VIII	CE-Ready	PC-Force GmbH	n.a.	Yes	Full
IX	WEKA Manager CE	Weka	3.7	Yes	Demo
X	COMPLIANCE Risk Software	DD IT Solutions Ltd	1.0.6.0	Yes	Demo
XI	CEM4	Certifico s.r.l	CE4 4.20.12	Yes	Demo
XII	Quentic	Quentic	n.a.	No	n.a.
XIII	FAM	FAM e.V.		No	n.a.

3.2 Study Realization

The study was conducted in the period from 30.11.2020 to 10.03.2021 in the order 1–11 according to Table 1. In order to ensure a standardized study procedure and thus the comparability of the results, a study procedure was defined in advance and documents were prepared for the standardized documentation of the results. The evaluation of each software solution was conducted in three phases:

Introduction Phase: The goal was the initial introduction to the software solution, familiarization with the functionality of the software solution and the corresponding risk assessment process according to MD, ensuring that a basic understanding of the software solution was available in the *Conduction Phase.*

After the download and installation, all functions of the respective software solution were extensively tested. The risk assessment process was run through in rudimentary form, the reaction of the system to possible incorrect entries or

small changes in the input was tested and the documentation (if available) for the software solution was analyzed in parallel. The phase lasted between 1.5 and 11 h, depending on the complexity of the software solution (6.25 h on average).
Conduction Phase: The risk assessments were performed for each software solution for two different HRC use cases. To ensure comparability of results, the same inputs were entered for each software solution (e.g., same phrases for machine limits, same input of hazards, same mitigation measures, and same risk evaluation, etc.). The phase lasted between 2 and 7 h per software (3.22 h on average).
Evaluation Phase: In the *Evaluation Phase* the software solution was evaluated according to predefined evaluation criteria.

3.3 Definition of Evaluation Criteria

An evaluation scheme based on expert opinions provides the basis for analyzing whether the software solutions meet the requirements of experts. For determine the requirements of experts and thus evaluation criteria, a two-stage descriptive study was conducted. First, expert interviews were conducted as a basis to obtain core statements and identify the important areas (in general and HRC-specific). The results from the interviews were evaluated in the second step through the use of an expert survey. A total of 44 experts participated in the study (expert interviews: 16 participants; survey: 28 participants). Figure 2 presents the distribution of the experts according to their profession and their years of experience in the field of risk assessment. The evaluation criteria are based on the experts' requirements for the software solutions, the individual approach to the risk assessment, and the aspects to be considered in the risk assessment. An expert is defined as a person who has already conducted several risk assessments and has experience in robotics.

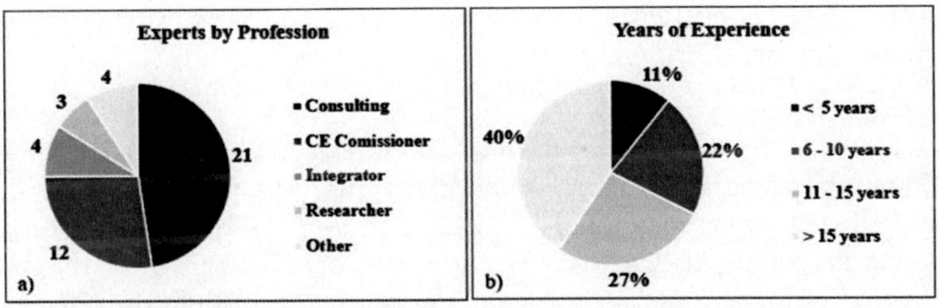

Fig. 2. Distribution of experts by (a) profession, (b) years of experience

Expert Interviews The expert interviews were conducted as qualitative, structured interviews with a total of 20 questions in presence, by telephone or in writing. The interviews were recorded and transcribed. Qualitative content analysis

Table 2. Evaluation scheme

Evaluation criteria	x_{C_n}	w_{C_n}	$P_{C_{max}}/P_{CS_n}$
1 Structured, table-based, comprehensible format	**30**	**1**	**5.00**
1.1 Software is structured and tabular	30		2.68
1.2 Structure corresponds to the procedure from DIN ISO 12100	26		2.32
2 Clear layout	**30**	**1**	**5.00**
2.1 Color coding (e.g. of tasks to be processed)	30		0.98
2.2 Display of not processed hazards	30		0.98
2.3 Display of the status of processing	30		0.98
2.4 Hazards are listed clearly recognizable and separable	23		0.75
2.5 Navigation bar available	20		0.65
2.6 Hierarchical structure/tree structure view	20		0.65
3 Flexibility	**28**	**0.93**	**4.65**
3.1 Adding of hazards is possible			5.00
4 The procedure is defined by directives and standards	**26**	**0.87**	**4.35**
4.1 Software fulfills the requirements of the MD	26		1.86
4.2 Software fulfills the requirements of DIN EN ISO 12100	26		1.86
4.3 CE conformity assessment procedure can be carried out	18		1.29
5 A process/tasks are depictable with the software	**21**	**0.70**	**3.50**
5.1 Software includes tasks/process flow	21		1.69
5.2 Software includes operation modes	21		1.69
5.3 Software supports modular structure	20		1.61
6 Completeness	**19**	**0.63**	**3.15**
6.1 All hazard consequences and causes from EN ISO 12100 Annex B are included	19		1.48
6.2 All hazard consequences from EN ISO 12100	18		1.41
6.3 All hazards causes from EN ISO 10218-2	14		1.09
6.4 All hazard consequences from EN ISO 12100 and all hazard causes from EN ISO 10218-2 are included	13		1.02
7 Integration of standards	**18**	**0.6**	**3.00**
7.1 All required standards are included in the software	18		5.00
8 Traceability/personalization	**17**	**0.57**	**2.85**
8.1 Several people can work traceably on one projects	17		5.00

based on Mayring [24] was used for data analysis and to extract the core statements. This step involved two researchers independently reviewing the interview transcripts, separately coding the expert statements, and then comparing and merging them in discussion. The requirements identified in the first stage represent the first level of evaluation criteria in Table 2 and serve as the basis of the question development in the expert survey.

Expert Survey For the second part of the descriptive study, an online expert survey was performed applying a standardized questionnaire. The questionnaire development, the structure of the questionnaire and the question wording were carried out according to the recommendations of [25–27]. For a reliable and reproducible evaluation of the software solutions, the evaluation criteria must be formulated as objectively as possible, thus ensuring others reach the same conclusions. For this reason, the questions were formulated objectively, with verifiable theses and as concretely as possible. The categories identified in the expert interviews were subdivided into objectively assessable subcategories and inserted into the questionnaire as binary questions (important/not important). Both German (46%) and non-German Europeans (54%) participated in the survey.

3.4 Evaluation System and Weighting

The evaluation system is presented in Table 2 and includes a total of eight categories. The categories are based on the descriptive study and each category is given a value (P_{C_n}) and a weighting (w_{C_n}). The evaluation scheme is used to evaluate each software solution, i.e. if one of the categories applies to a particular software solution, the corresponding value is assigned to it. The sum results in the final value, which indicates how strongly the software solutions meet the requirements of the experts (the higher the value, the more likely the software solution meets the requirements). For each category, a maximum value of $P_{max} = 5$ (without weighting) is determined. The categories are not to be considered equal, as there are requirements of the experts for the software solutions that are regarded as more important or less important compared to other requirements. For this reason, the categories are weighted depending on the number of experts' nominations.

The weighting of each category is calculated from the number of nominations of the category C_n in relation to the total highest number of nominations of a category. The calculation of the weighting factor is presented in Eq. 1 with w_{C_n} = *weighting factor*, x_{C_n} = *number of nominations of category* C_n and $a = 30$ (highest number of nominations). The number of nominations relates to both the *Expert Interviews* and the *Expert Survey.*

$$w_{C_n} = \frac{x_{C_n}}{a} \tag{1}$$

Each category is divided into a varying number of subcategories. For categories 3, 7 and 8, the evaluation value is set to P_{max} if the subcategory applies to the corresponding software solution and 0 if the subcategory does not apply. For all other categories P_{max} is distributed among the subcategories depending on the number of nominations of the subcategory. The value of each subcategory P_{CS_n} = *value of subcategory* S_n is calculated according to Eq. 2 with x_{S_n} = *number of nominations of subcategory* S_n and q_{C_n} = *total number of subcategories of category* C_n.

$$P_{CS_n} = 5 \cdot \frac{x_{S_n}}{\sum_{m=1}^{q_{C_n}} x_{S_m}} \tag{2}$$

The value of the category P_{Cn} = *value of category* C_n is calculated according to Eq. 3 and the total value of the software solution $P_{Software}$ is calculated from the sum of the achieved value in each category (Eq. 4). The maximum possible value is $P_{Software_{max}} = 31.50$.

$$P_{C_n} = \sum_{m=1}^{q_C} P_{CS_m} \cdot w_{C_n} \tag{3}$$

$$P_{Software} = \sum_{n=1}^{9} P_{Cn} \tag{4}$$

4 Evaluation Results

4.1 Expert Requirements

The results of the evaluation of the software solution according to the experts' requirements are summarized in Table 3. The requirements analysis focused on risk assessment for robotic applications and solely experts with a background in robotics were surveyed, i.e. the results may differ for an other machine sector.

Software solution *V* ($P_{Software_V}$ = *26.67*) fulfills the requirements of the experts most effectively, followed by *IX* ($P_{Software_{IX}}$ = *25.39*) and *I* ($P_{Software_I}$ = *23.81*). *IV* ($P_{Software_{IV}}$ = *9.56*) is the least compliant with the experts' requirements.

Software solutions *II*, *IV* and *VIII* have no structured tabular layout and the structure of *III*, *VI* and *X* do not correspond to the procedure of EN ISO 12100. All software solutions display hazards which have not been processed. However, there is no color coding in *I*, *III* or *VI*, and no feature to indicate the status of processing in *I*, *III* and *IV*. In contrast to the other software solutions examined, the hazards are not clearly and separably listed in *I*, *IV* and *X*. In addition, at *III*, *IV* and *IX* there is no navigation bar and *I* and *III* have no hierarchical or tree structure. Adding additional hazards is not possible with *IV*, *VI*, *VII* and *VIII*. Software solutions *III*, *IV*, *VI* and *X* do not comply with the MD requirements, therefore performing the CE conformity assessment procedure with the software solution is not possible. *III* and *IV*, however, meet the requirements of EN ISO 12100.

I, *III*, *VI*, *IX* and *XI* support both task and operation mode definition, in *IV* it is only possible to define tasks and in *V* there is no extra function for task description, but there is the possibility to include the tasks and operation modes in the life phases section. A modular structure of the risk assessment is supported by *I*, *II*, *X* and *XI*. In *II*, *III* and *XI* all hazard consequences of EN ISO 12100 are predefined and in *I*, *IV*, *V*, *VI*, *VIII* and *IX* the hazard causes are additionally provided. The hazard consequences of EN ISO 10218-2 are integrated solely in *VI*. Software solution *V* is the only software solution where the standards are available for reference, i.e. the corresponding standards do not have to be purchased additionally and are always available up-to-date. Only with *I*, *V*, *VIII* and *IX* is traceability and personalization possible.

4.2 Suitability for HRC

Due to the fence-less and partly close collaboration between robot and human in HRC applications, there are special requirements that have to be considered: (a) **EN ISO 10218-1 and -2**: compliance with the C-standards for industrial robots, (b) **ISO/TS 15066**: compliance with the technical specification for collaborative robot systems, (c) **Forms of interaction**: collaboration, cooperation, coexistence can be applied in the same HRC application in different process steps and impact the selection of protection principles, (d) **Consideration of tasks/process flow**: in cooperation and collaboration humans are actively

Table 3. Assessment of the software solutions according to the requirements of the experts

Category	1	2	3	4	5	6	7	8	$P_{Software_n}$	Rank
V	5.00	5.00	4.65	4.35	0.00	1.82	3.00	2.85	26.67	1
IX	5.00	4.35	4.65	4.35	2.37	1.82	0.00	2.85	25.39	2
I	5.00	1.63	4.65	4.35	3.50	1.82	0.00	2.85	23.81	3
XI	5.00	5.00	4.65	4.35	3.50	0.89	0.00	0.00	23.39	4
II	4.64	5.00	4.65	4.35	1.13	0.89	0.00	0.00	18.34	5
VIII	4.64	5.00	0.00	4.35	0.00	1.82	0.00	2.85	16.34	6
VII	5.00	5.00	0.00	4.35	0.00	0.00	0.00	0.00	14.35	7
III	5.36	1.73	4.65	1.62	2.37	0.89	0.00	0.00	13.93	8
X	5.36	4.24	4.65	0.00	1.13	0.00	0.00	0.00	12.70	9
VI	5.36	4.02	0.00	0.00	2.37	3.15	0.00	0.00	12.22	10
IV	4.64	2.61	0.00	1.62	1.19	1.82	0.00	0.00	9.56	11

involved in the work process, which can cause contacts in certain process steps, (e) **Biomechanical limits**: for each possible contact situation between robot and human, compliance with the biomechanical limit values must be ensured, and (f) **Contact scenarios**: a distinction is made between quasi-static and transient contacts. An overview of the results is given in Table 4.

In *I, II, IV, V, VI* and *IX* references to EN ISO 10218-1 and -2 are included, ISO/TS 15066 is referenced solely in *V*. Tasks or the representation of the process flow as tasks and their assignment to hazards is only possible in *I, III, IV, VI, IX* and *XI*. None of the software solutions explicitly includes either the interaction forms, the biomechanical limit values, or the distinction of contact scenarios. There is also no information in the documentation or info fields.

Several of the aspects specified can be realized with the help of the functions available in the software solution. An inclusion of the interaction forms and their linkage with the tasks and hazards can be realized e.g. via the operation mode or life phase function by defining the interaction form instead of an operation mode resp. life phase. This is possible using all software solutions except *I, IV* and *X*. In the explicit definition of contact situations, the process step in which the contact can take place, as well as the location where the contact takes place (hazard zone) and the type of contact must be defined. Specifying hazard zones is possible in all software solutions except *IV* and *X* and adding hazards is also possible in all software solutions except *VI, VII* and *VIII*. Thus, considering the task function, the definition of contact situations is possible in *I, III, IX* and *XI*.

5 Discussion

With regard to *RQ1*, none of the software solutions examined meets the experts' requirements 100%, while *V* (85%), *IX* (81%) and *I* (76%) achieve high levels

Table 4. HRC suitability of the software solutions

	I	II	III	IV	V	VI	VII	VIII	IX	X	XI
EN ISO 10218-1 and -2	x	x		x	x	x			x		
ISO/TS 15066					x						
Forms of interaction		(x)	(x)		(x)	(x)	(x)	(x)	(x)		(x)
Tasks/process flow	x		x	x		x			x		x
Biomechanical limits											
Contact scenarios	(x)		(x)						(x)		(x)

of agreement. From the results, it appears that the available software solution has different focuses and supports different approaches to performing risk assessments. For example, *I*, *II*, *X* and *XI* provide procedures based on dividing the machine into individual modules, while other software solutions do not use modularization. This diversity is a necessary criterion, since there is a multiplicity of different machines, with which certain procedures and representations have proven to be effective.

One reason for the heterogeneity of the approaches is that the standards allow a high degree of leeway for individuality in the performance of risk assessments and only provide a general structure. On the one hand, this gives users the freedom to develop their own procedures according to their individual needs and preferences, but on the other hand, it complicates the entire process and poses challenges, especially for inexperienced new users. It is also evident from the fact that many users prefer self-created MS Office templates for documentation [7] that software solutions cannot necessarily fulfill all of the user's requirements.

Referring to *RQ2*, the state of the art of risk assessment software solutions does not sufficiently address the specifics of HRC applications. General functions, such as task definition or integration of C-standards are partially available, but specifics such as *inclusion of biomechanical limits*, *forms of interaction* and *differentiation of contact scenarios* are not covered in functionality or documentation. The representation of the specifics is partially possible through existing functions. An experienced user is able to find ways and solutions to perform and present the risk assessment for HRC applications as transparent and detailed as necessary, but for an inexperienced user it is very difficult and can be a barrier.

6 Conclusion

This paper describes a first stage for the development of a standard procedure based on current practice. The work identified evaluation criteria for risk assessment software based on expert opinions and evaluated the state of the art. The results show the different approaches to risk assessments and illustrate the individuality in performing it. For HRC applications, the software solutions offer few support and assistance and a detailed, transparent and successful risk assessment can only be performed through experience and a deeper understanding of the risk assessment process. For inexperienced users, the diversity of the software solution combined with a lack of prior knowledge and assistance can present an insurmountable obstacle. This can lead to HRC applications or machines being placed on the market that are inadequately safeguarded and thus pose a major risk to workers. There is therefore a need for research to create more clarity and standardization and to provide more assistance to HRC users. In order to provide a comprehensive overview of current practice, the next step is to explore the expertise and experience of those conducting risk assessments in the context of a field study.

References

1. Huck, T.P., Münch, N., Hornung, L., Ledermann, C., Wurll, C.: Risk assessment tools for industrial human-robot collaboration: novel approaches and practical needs. Saf. Sci. **141**, 105288 (2021)
2. Saenz, J., Elkmann, N., Gibaru, O., Neto, P.: Survey of methods for design of collaborative robotics applications - why safety is a barrier to more widespread robotics uptake. In: Proceedings of the 2018 4th International Conference on Mechatronics and Robotics Engineering, pp. 95–101 (2018)
3. Hornung, L.: Ergebnisse der Umfrage Mensch-Roboter-Kollaboration (2021). https://doi.org/10.13140/RG.2.2.29402.26560
4. Valori, M., et al.: Validating safety in human-robot collaboration: standards and new perspectives. Robotics **10**, 65 (2021)
5. European Union: Directive 2006/42/EC of the European Parliament and of the Council of 17 May 2006 on machinery, and amending Directive 95/16/EC (recast). Off. J. Eur. Union (2006)
6. ISO: ISO 10218-2:2011: Robots and Robotic Devices - Safety Requirements for Industrial Robots - Part 2: Robot Systems and Integration. pub-ISO (2011)
7. Hornung, L.: Ergebnisse der Umfrage Risikobeurteilung für Mensch-Roboter-Kollaboration (HRC) (2021). https://doi.org/10.13140/RG.2.2.27292.00644
8. TÜV SÜD: MASCHINENRICHTLINIE 2006/42/EG. https://www.tuvsud.com/de-de/indust-re/maschinensicherheit-info/maschinenrichtlinie. Accessed 11.02.2022
9. ISO: ISO 12100:2011: Safety of Machinery - General Principles for Design - Risk Assessment and Risk Reduction (2011)
10. European Commission: Commission communication in the framework of the implementation of the Directive 2006/42/EC of the European Parliament and of the Council on machinery, and amending Directive 95/16/EC. Off. J. Eur. Union (2018)
11. Guiochet, J.: Hazard analysis of human-robot interactions with HAZOP-UML. Saf. Sci. **86**, 225–237 (2016)
12. Adriaensen, A., Pintelon, L., Costantino, F., Di Gravio, G., Patriarca, R.: An STPA safety analysis case study of a collaborative robot application. IFAC-PapersOnLine **54**, 534–539 (2021)
13. Zhou, J., Hänninen, K., Lundqvist, K., Provenzano, L.: An ontological approach to hazard identification for safety-critical systems. In: 2017 Second International Conference on Reliability Systems Engineering (ICRSE), pp. 1–7 (2017)

14. Askarpour, M., Mandrioli, D., Rossi, D., Vicentini, F.: SAFER-HRC: Safety Analysis through Formal vERification in Human-Robot Collaboration. In: International Conference on Computer Safety, Reliability, and Security (2016)
15. Vicentini, F., Askarpour, M., Rossi, M., Mandrioli, D.: Safety assessment of collaborative robotics through automated formal verification. IEEE Trans. Robot. **36**, 42–61 (2019)
16. Awad, R., Fechter, M., van Heerden, J.: Integrated risk assessment and safety consideration during design of HRC workplaces. In: 2017 22nd IEEE International Conference on Emerging Technologies and Factory Automation (ETFA), pp. 1–10 (2017)
17. Marvel, J.A., Falco, J., Marstio, I.: Characterizing task-based human-robot collaboration safety in manufacturing. IEEE Trans. Syst. Man Cybern. Syst. **45**, 260–275 (2014)
18. Ganglbauer, M., Ikeda, M., Plasch, M., Pichler, A.: Human in the loop online estimation of robotic speed limits for safe human robot collaboration. In: 30th International Conference on Flexible Automation and Intelligent Manufacturing (FAIM2021), vol. 51, pp. 88–94 (2020)
19. Chitturi, N., Ganglbauer, M., Plasch, M., Kulha, P., Pichler, A.: HolisafeHRC: Holistic Safety Concepts in Human-Robot Collaboration (2019)
20. Huck, T.P., Ledermann, C., Kröger, T.: Simulation-based testing for early safety-validation of robot systems. In: 2020 IEEE Symposium on Product Compliance Engineering-(SPCE Portland), pp. 1–6 (2020)
21. Huck, T.P., Ledermann, C., Kröger, T.: Virtual adversarial humans finding hazards in robot workplaces. In: 2021 IEEE International Conference on Robotics and Automation (ICRA) (2021)
22. Poot, L., Johansen, K., Gopinath, V.: Supporting risk assessment of human-robot collaborative production layouts: a proposed design automation framework. Procedia Manuf. **25**, 543–548 (2018)
23. Fraunhofer Institute for Factory Operation and Automation IFF: COVR - being safe around collaborative and versatile robots in shared spaces (2021)
24. Mayring, P.: Qualitative Inhaltsanalyse - Grundlagen und Techniken, p. 152. Julius Beltz, 12, überarb. Aufl. (2015)
25. Tausendpfund, M.: Quantitative Methoden in der Politikwissenschaft - Eine Einführung, vol. 1, p. 356. Springer VS, Wiesbaden (2018)
26. Jacob, R., Heinz, A., Décieux, J.P.: Umfrage: Einführung in die Methoden der Umfrageforschung, p. 295. Oldenbourg Wissenschaftsverlag (2013)
27. Porst, R.: Praxis der Umfrageforschung, vol. 2, p. 211. VS Verlag für Sozialwissenschaften, Wiesbaden (2000)

Goal Inference via Corrective Path Demonstration for Human-Robot Collaboration

Fumiya Ohnishi[1](✉), Yosuke Kawasaki[2], and Masaki Takahashi[1]

[1] Department of System Design Engineering, Keio University, 3-14-1 Hiyoshi, Kohoku-ku, Yokohama 223-8522, Japan
fumiya-onishi@keio.jp, takahashi@sd.keio.ac.jp

[2] Graduate School of Science and Technology, Keio University, 3-14-1 Hiyoshi, Kohoku-ku, Yokohama 223-8522, Japan
y-kawasaki@keio.jp

Abstract. Recently, collaborative robots, such as collaborative delivery robots, have been expected to improve the work efficiency of users. For natural human-robot collaboration, it is necessary to infer the appropriate goal position to transport instruments, where the user's convenience and the surrounding environment are considered. In conventional research, the goal is inferred by demonstrating the user's desired positions, but position demonstration requires many trials to obtain the inference model, which is burdensome for the user. Therefore, we focus on the user's correction of the robot position and generate multiple position samples from the user's corrective path. In addition, these position samples are weighted based on the implicit intention of the correction to learn both the desired and undesired positions. Consequently, the robot improves goal inference in fewer trials. The effectiveness of the proposed method was evaluated by experiment that simulated human-robot collaborative environments.

Keywords: Learning from demonstration · Human-robot collaboration · Goal inference

1 Introduction

Recently, there has been growing demand for robots that can work together with humans in coexisting environments. For example, in hospitals, instead of the medical staff, collaborative delivery robots are expected to transport medical instruments and improve work efficiency [1]. In the research on autonomous mobile robot navigation in human-robot coexisting environment [2–4], the navigation goal is set by the robot designer. However, because the user works while placing the instruments in and out of the robot, the robot needs to infer the appropriate goal position to transport the instruments, where the user's convenience and the surrounding environment are considered. For example, we assume a situation in which a collaborative delivery robot transports medical instruments to a nurse (user) who takes care of a patient in a hospital room. In this situation, as shown in Fig. 1 (a), the robot position where the user can comfortably work with the patient depends on the user's preference. In addition, as shown in Fig. 1 (b), the desired

I. Petrovic et al. (Eds.): IAS 2022, LNNS 577, pp. 15–28, 2023.
https://doi.org/10.1007/978-3-031-22216-0_2

robot position of the user varies depending on the surrounding environment, such as the presence or absence of a visitor. The robot should infer these positions as the navigation goals. However, in general, it is difficult to accurately describe such preferences and complex human-robot coexisting environments in advance [5, 6]. To address this difficulty, some researchers have utilized user corrections to improve robot motions. Bajcsy et al. [7] proposed that the robot should utilize the user's corrections to improve its motion because the user's corrections are intentionally added when the robot moves incorrectly. In addition, Kollmitz et al. [4] proposed human-aware navigation considering the user's preferred distance from the robot, which is learned by physically pushing the robot to correct its position. They confirmed that this demonstration method was intuitive for the users.

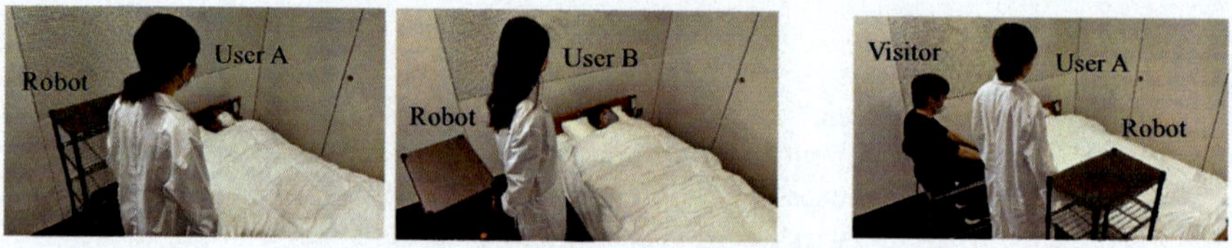

(a) Different user preferences (b) When a visitor comes

Fig. 1. Example situations and robot positions. The desired robot position of the user varies depending on the user's preference and the surrounding environment.

Based on the above, in this study, we focus on the inference of the user's desired robot position (goal), which varies depending on the user's preference and the surrounding environment. In addition, we consider that if the robot encounters an environment that is beyond its prior knowledge and infers a different goal from the desired one, the user can correct the robot position. Then, using this correction, the robot improves the goal inference sequentially. In this way, we aim to improve the robot's inference of the user's desired goal through collaboration with the user and achieve user-friendly human-robot collaboration.

Related work on goal inference includes designing inference rules for known surrounding environments [8] and inference based on personal space and affordances [9]. However, it is difficult to describe the rules and affordances when considering various user preferences and surrounding environments. To address this issue, Kawasaki et al. [10] inferred the user's desired goal from the labels and positions of surrounding objects by demonstrating the user's desired goal position. They proposed bottom-up action modeling, which was confirmed to capture the tendency of the arrangement position derived from the rules and the user preferences, and successfully infer the goal position even in unknown environments. Therefore, this method is useful for addressing the various user preferences and surrounding environments assumed in our study. However, in their study, the inference model was obtained after acquiring many position demonstrations through many trials in advance. Because requiring many trials is burdensome for the user, it is necessary to sequentially improve the inference with fewer trials.

Therefore, we focus on the user's corrective path of the robot position. Subsequently, by generating multiple position samples from the corrective path, the robot acquires many position demonstrations sequentially in fewer trials. Furthermore, these position samples are weighted based on the implicit intention of the correction to learn both the desired and undesired positions, and the inferences of desired and undesired positions are integrated. Consequently, the robot sequentially improves goal inference in fewer trials, and achieves natural human-robot collaboration.

2 Overview of the Proposed System

2.1 Assumed Environment

As described above, we assume that a collaborative delivery robot transports medical instruments to a nurse (user) who takes care of a patient in a hospital room. We also assume that the nurse visits several hospital rooms every day. Therefore, the arrangement of objects in the room changes depending on the day and the layout of the room. It is assumed that the robot can acquire the labels and positions of objects from onboard or environmental sensors and object detection methods such as Mask R-CNN [11]. The robot then infers the goal (the user's desired position) based on the labels and positions of the surrounding objects. We assume that the robot can move in all directions and its footprint is considered as circular shape; therefore, this study does not consider the orientation of the robot. In addition, the changes in the object positions during goal inference are not considered.

2.2 Flow of the Proposed System

The flow of the proposed system is illustrated in Fig. 2. First, the robot acquires its surrounding environment and infers the goal. Next, the robot moves to the goal, and the user physically pushes or pulls the robot if necessary to correct its position. Then, the robot learns from the corrective path as a demonstration and sequentially improves the data used for goal inference. The above process is repeated every time the user performs a task. Learning from corrective path demonstration consists of two steps: position samples generation and learning relations with surrounding objects probabilistically. Goal inference comprises three steps: partial probability map calculation, integration, and goal inference. Here, we define a probability map as the map generated from the probabilistically learned data.

3 Learning from Corrective Path Demonstration

3.1 Position Samples Generation

In this section, we describe the generation of multiple position samples from a corrective path. Figure 3 presents an overview of this process. First, n positions are evenly sampled from the corrective path. n is calculated using the following equation:

$$n = \max\left(\mathrm{round}\left(\frac{l_{\mathrm{correction}}}{\Delta l_{\mathrm{correction}}}\right), 1\right) \tag{1}$$

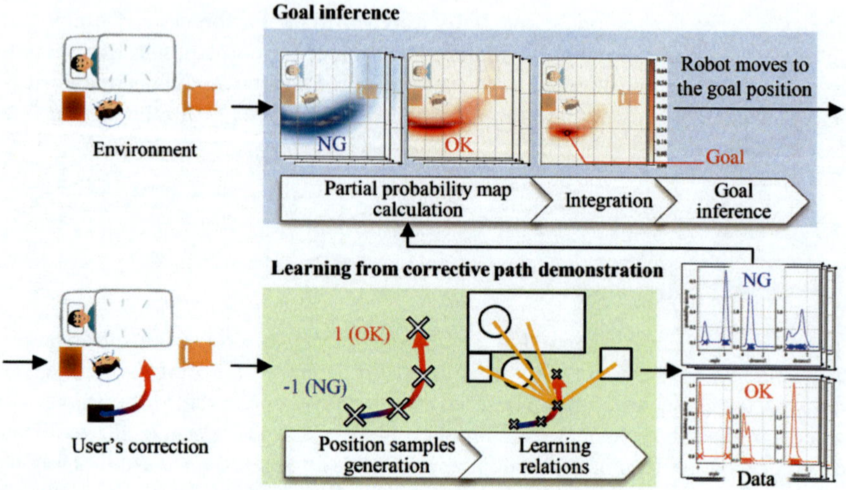

Fig. 2. Flow of the proposed system

where $l_{\text{correction}}$ is the length of the corrective path and $\Delta l_{\text{correction}}$ is the sampling interval. In the case of $n \geq 2$, the positions are sampled at equal intervals, including the starting and the end points, and in the case of $n = 1$, the end point is adopted as a position sample. In addition, if no correction is given, we assume that $l_{\text{correction}} = 0$ and the inferred goal is adopted as a position sample.

Next, each position sample p_i $(1 \leq i \leq n)$ is weighted based on the implicit intentions of the corrective path. We interpret the starting point of the corrective path as undesired, the end point as desired, and the midpoint as a gradual change. Therefore, we set the weight of the starting point as -1.0 and that of the end point as 1.0, and linearly interpolated weights are given to the position samples. Then, the position samples with positive weights are considered to be desired (OK), and those with negative weights are undesired (NG). In addition, the weights of the NG position samples are taken as absolute values and are positive. The weights of the position samples thus obtained are used in the next section to consider the degrees of OK and NG.

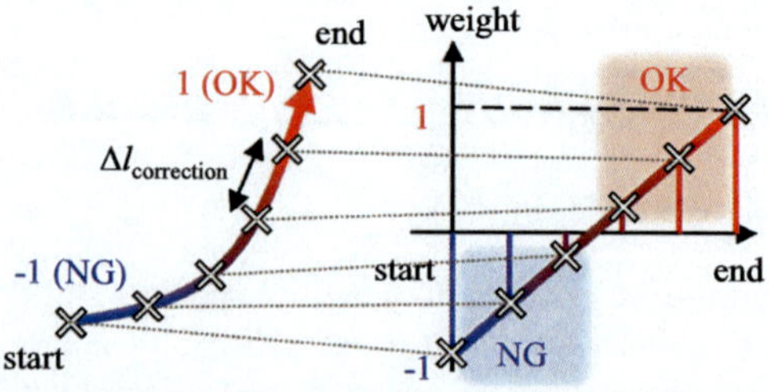

Fig. 3. Position samples generation. Generate multiple position samples from a corrective path and each position sample is weighted based on the implicit intention of correction.

3.2 Learning Relations with Surrounding Objects Probabilistically

In this section, we describe the probabilistic learning of the spatial relations between the position samples and surrounding objects. In this study, based on the idea of bottom-up action modeling [10], we interpret the spatial relations between all surrounding objects as the integration of partial spatial relations. First, as shown in Fig. 4, the position samples p and surrounding objects S are factorized into partial relations consisting of three objects. The partial spatial relation is then described as a spatial feature, $\mathbf{T} = [\theta, d_1, d_2]$, as shown in Fig. 5. Here, θ is the angle formed by o_B, o_A, and o_C counterclockwise. d_1 represents the characteristic distance between o_A and o_B, and d_2 is the characteristic distance between o_A and o_C. Taking the position sample p_i (Robot) as o_A and the two surrounding objects S_j, S_k as o_B and o_C, the spatial feature in partial relation, $\mathbf{T}_c = [\theta_c, d_{1,c}, d_{2,c}]$, is obtained, where c is the label of the partial relation (Robot-S_j-S_k). Note that S is sorted in alphabetical order and always $j < k$ to treat the same object group as the same label. In addition, the obtained samples of spatial feature are weighted the same as the position sample p_i. As a result, new feature samples are obtained for each trial and sequentially added to the data.

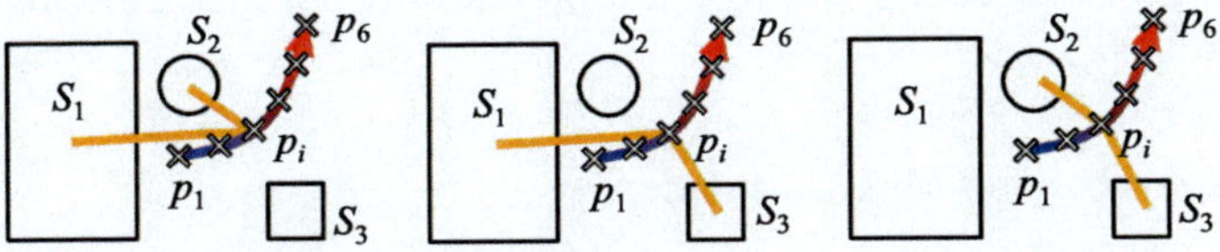

(a) Robot-S_1-S_2 (b) Robot-S_1-S_3 (c) Robot-S_2-S_3

Fig. 4. Factorization into partial relations

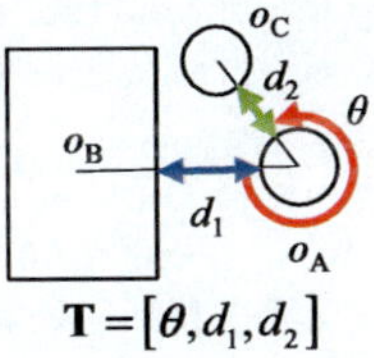

Fig. 5. Spatial factorization

Next, we learn the probability distribution of the obtained data. We use weighted kernel density estimation to estimate the probability distribution with fewer weighted samples. Using weighted kernel density estimation, the probability density function of sample x is formulated using Eq. (2)

$$P(x) = \sum_{i=1}^{N} w_i K(x, x_i) \tag{2}$$

where x is the spatial feature θ_c, $d_{1,c}$, $d_{2,c}$, x_i is the value of the i-th feature sample, and w_i is its weight and is utilized after normalization so that the sum is 1. N is the number of feature samples, and K is a kernel function. We use the Gaussian kernel for d_1 and d_2, and the von Mises' density [12] for θ, referring to the selection of the conventional method [10]. The bandwidth h of the kernel function is calculated as follows:

$$h = (h_i - h_f)N^s + h_f \tag{3}$$

where h_i and h_f are the bandwidth in the case of $N = 1$ and $N \to \infty$, respectively. h_i and h_f are set separately for the distance and the angle, that is, h_i^d, h_f^d, h_i^θ, h_f^θ. s is a factor representing the decay rate, and $s = -1.0$ was adopted in this study. Equation (3) expresses the uncertainty of the estimation when the number of samples is small: Note that Scott's rule [13] and Silverman's rule [14] are known methods for selecting the bandwidth based on the number of samples. However, in the case of one-dimensional data, the factor of these methods is $s = -0.2$. Because this value was insufficient in this study, Eq. (3) was used. The learned data are utilized in the goal inference process.

4 Goal Inference

In this section, we describe the goal inference process. The pipeline of the goal inference is shown in Fig. 6.

4.1 Partial Probability Map Calculation

First, partial probability maps, which are probability maps of partial relations, are calculated. To calculate partial probability maps, we set up a square grid with a side G [m] and resolution r [m/pix] in the robot configuration space and select a goal candidate (x, y) by searching the entire grid. Next, the spatial relation between the goal candidate and the surrounding objects is factorized into partial relations using the method described in Sect. 3.2. Then, the spatial feature of partial relation c at (x, y), $\mathbf{T}_c(x, y) = [\theta_c, d_{1,c}, d_{2,c}]$, is calculated. Finally, the value of the map at (x, y) is calculated using the product of the probability densities, which are obtained by substituting θ_c, $d_{1,c}$, and $d_{2,c}$ into the estimated probability density function. By performing the above calculations for each of the OK and NG data and all the goal candidates, we obtain the partial probability maps ${M_c}^{\mathrm{OK}}$ and ${M_c}^{\mathrm{NG}}$, where c is the label of the partial relation. The above calculation is also performed for all partial relations.

4.2 Integration and Goal Inference

First, the partial probability maps ${M_c}^{\mathrm{OK}}$ and ${M_c}^{\mathrm{NG}}$ are integrated using the following equation to obtain the partial probability map ${M_c}^{\mathrm{OKNG}}$, which considers both OK and NG:

$$M_c^{\mathrm{OKNG}} = M_c^{\mathrm{OK}} - M_c^{\mathrm{NG}} \tag{4}$$

In ${M_c}^{\mathrm{OKNG}}$, positive values represent the degree of OK and negative values represent the degree of NG. Next, we obtain a partial probability map M_c by setting all negative values of $M_c{}^{\mathrm{OKNG}}$ to 0. Here, M_c represents the degree of OK considering NG. The above calculation is performed for all the partial relations.

Next, we integrate each partial probability map M_c. Following the conventional method [10], M_c is weighted and integrated based on the importance of each partial relation. The weight of probability map M is calculated as follows:

$$w_M = \frac{\alpha}{H_M - H_{\min} + \varepsilon} \tag{5}$$

where $H_{\min}$ is the minimum value of each M_c's entropy and ε is a small constant to prevent the denominator from becoming 0. α is a normalization term, such that the sum of w_M is 1. As a result, a map with less variation in values has more weight. Then, by summing each M_c multiplied by w_{M_c}, we obtain a probability map Ψ for the overall spatial relation.

Finally, we infer the goal using the probability map Ψ. Here, the goal must not be generated where the robot physically collides with the obstacles. Therefore, Ψ is passed through an obstacle filter that sets the grid value to 0 in the area where the robot collides with the obstacles. The position where this map takes the maximum value is inferred as the goal $\hat{\mathbf{p}}$ [m].

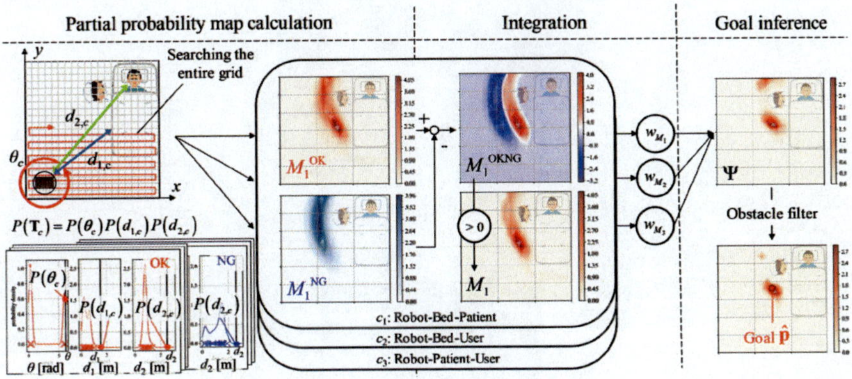

Fig. 6. Pipeline of the goal inference

5 Experiment

In this section, we verify that the goal inference is improved in fewer trials by multiple position samples generation from the corrective path and considering both OK and NG of implicit intention. (The experiment was approved by the Bioethics Committee of the Faculty of Science and Technology, Keio University. Approval No. 2021-107.)

5.1 Experimental Setting

The assumed environment described in Sect. 2.1 was simulated in our laboratory room and the method was verified with experiment, as shown in Fig. 8. We assumed that the room had the objects shown in Fig. 7. Furthermore, as the robot configuration space, we assumed a square area of 3.0 m on one side of the bed in the hospital room, as shown in Fig. 9. That is, we set $G = 3.0$ [m]. Note that we asked demonstrators to assume that there was a wall on the head side of the bed. Furthermore, the objects' positions including the robot were set by a moderator.

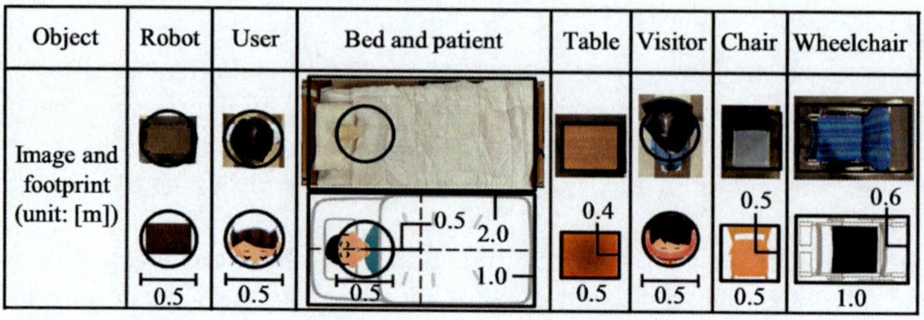

Fig. 7. Assumed objects (real images and illustrations)

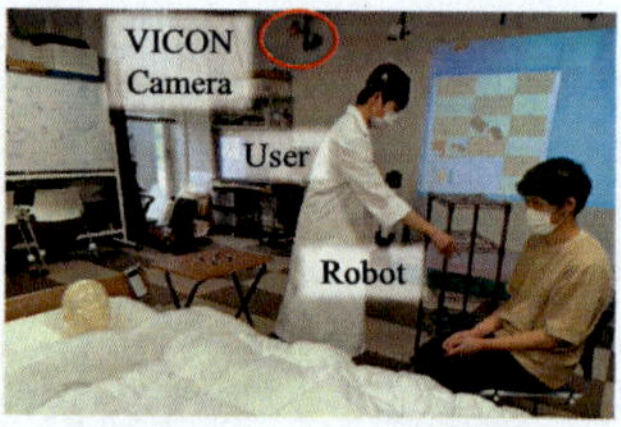

Fig. 8. Experimental scene

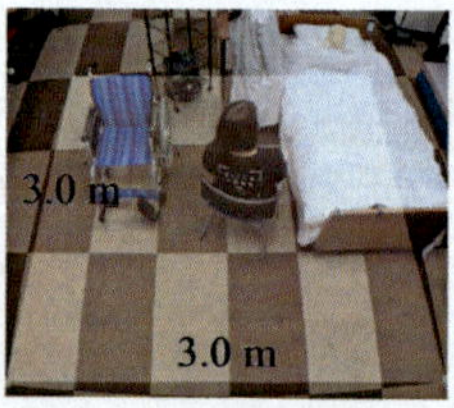

Fig. 9. Example scene

The proposed method and comparison methods were verified with four demonstrators. The experimental flow consisted of the following five steps: In step 1, we randomly generated an environment with different object arrangements and types, excluding the robot and the user. In step 2, the demonstrator (user) moved to the working position. Here, we obtained the objects' positions and labels by motion capture system VICON [15] (Bonita B10). In step 3, the robot inferred the goal and was moved. In step 4, the demonstrator corrected the robot position to the desired position as required. The corrective path was obtained by VICON, too. In step 5, the robot learned from the corrective

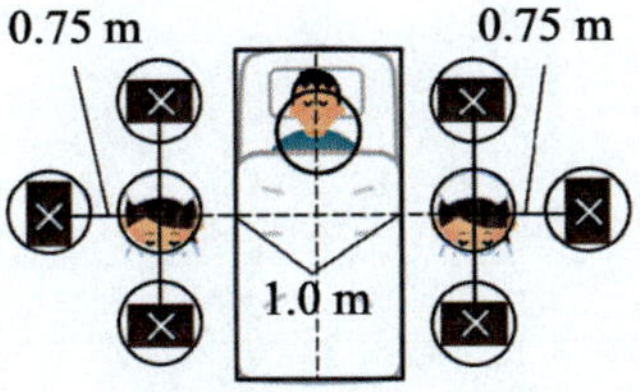

Fig. 10. Initial data

path. Note that steps 3–5 were repeated for each comparison method to compare the performance. We also asked the demonstrators that the corrected positions be equal for each method for comparison. These trials were repeated 50 times.

The proposed method and the comparison methods are shown in Fig. 11. The comparison method 1 (Comp. 1) corresponds to the conventional method [10] and generates an OK position sample from the end point. The comparison method 2 (Comp. 2) generates multiple position samples from the path but does not consider NG. The proposed method generates multiple position samples and considers both OK and NG. In addition, the evaluation indices were the correction count and accumulated error of inference. The accumulated error was defined as the accumulated distance between the inferred goal $\hat{\mathbf{p}}$ [m] and corrected position $\mathbf{p}$ [m] in each trial. We empirically adjusted these parameters: $\Delta l_{\text{correction}} = 0.15$ [m], ${h_i}^d = 1.50$ [m], ${h_f}^d = 0.10$ [m], ${h_i}^\theta = 1.00$ [rad], ${h_f}^\theta = 0.04$ [rad], $r = 0.15$ [m/pix] and $\varepsilon = 0.1$. Additionally, the initial data for goal inference were the position demonstrations shown in Fig. 10 and were stored as OK position samples with a weight of 1.0. The specification of computer we used was Intel Core i7-9750H @ 2.6 GHz CPU, 16 GB RAM, and RTX 2070 MaxQ 8 GB GPU.

5.2 Result

Table 1 shows the correction count for each demonstrator; Fig. 12 shows the graphs of the accumulated error for each demonstrator, where the horizontal axis represents the number of trials and the vertical axis represents the accumulated error after each trial. The mean and standard deviation of the computation time required by each method for goal inference across all trials for all demonstrators were 1.12 ± 0.581 [s], 1.56 ± 0.824 [s] and 2.10 ± 1.13 [s] for Comp. 1, Comp. 2, and the proposed method, respectively.

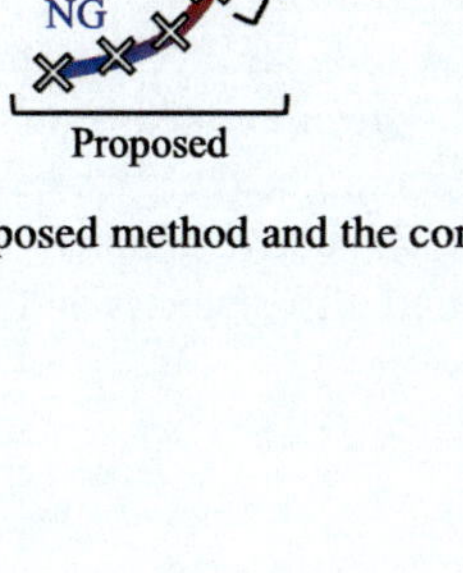

Fig. 11. The proposed method and the comparison methods

Table 1. Correction count

Method	Demonstrator				Ave
	1	2	3	4	
Comp. 1	32	27	29	19	26.8
Comp. 2	29	25	29	19	25.5
Proposed	23	19	27	17	21.5

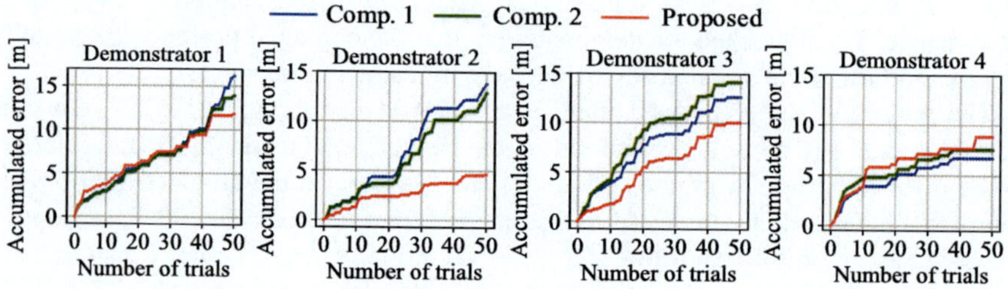

Fig. 12. Accumulated error

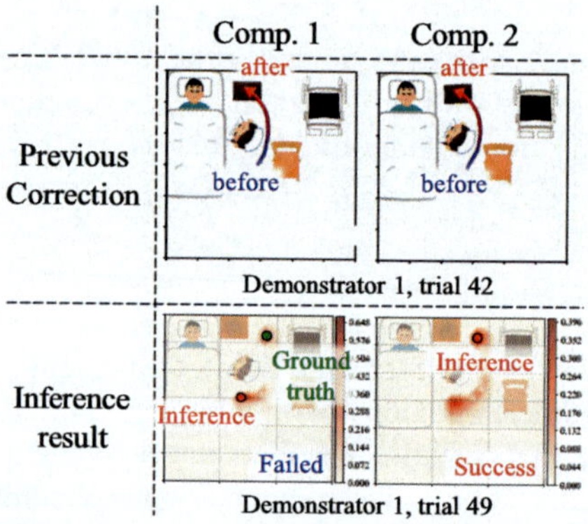

Fig. 13. The specific scene where Comp. 2 outweighs Comp. 1

Table 1 shows that the average correction count for each demonstrator decreased in the order of Comp. 1, Comp. 2, and the proposed method, confirming the effectiveness of multiple position samples generation and consideration of NG. Figure 12 shows that in demonstrators 1, 2 and 3, the accumulated error of the proposed method tended to be the lowest. However, in demonstrator 4, only the proposed method made significant inference errors in some cases, and the final accumulated error was the largest. In addition, for demonstrators 1 and 2, the final accumulated error was Comp. 2 < Comp. 1, while for 3 and 4 it was Comp. 2 > Comp. 1. The above confirms the reduction of correction count by multiple samples generation, and effectiveness and limitation of considering NG. In

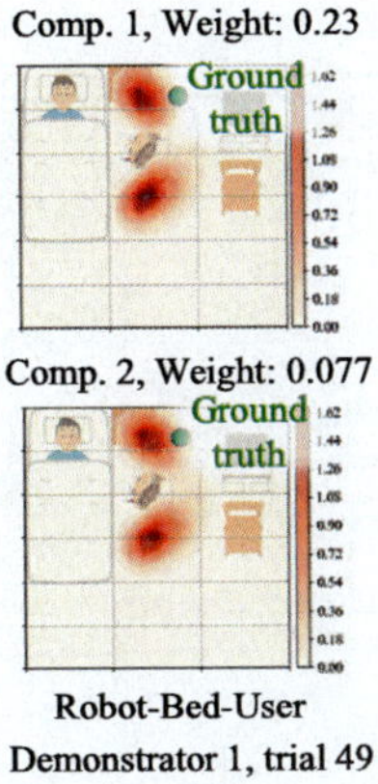

Fig. 14. Comparison of the partial relation importance

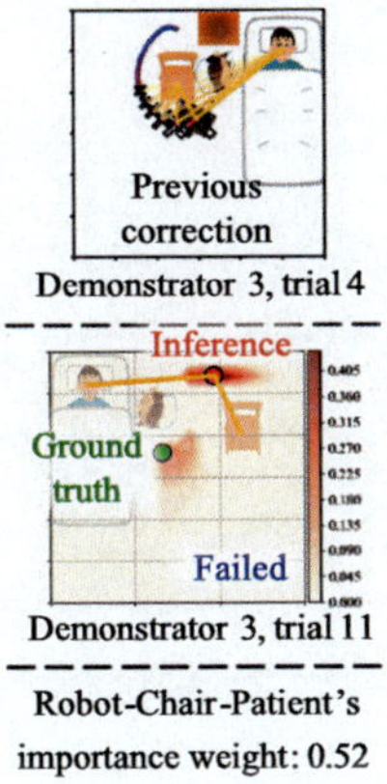

Fig. 15. The specific scene where Comp. 2 incorrectly inferred

the following paragraphs, we visualize specific scenarios and discuss the characteristics of multiple position samples generation and considering NG.

First, as shown in Fig. 13, Comp. 2 successfully inferred the goal at the end point of the previous corrective path, whereas Comp. 1 incorrectly inferred at the starting point. Following the long corrective path shown in Fig. 13, the Robot-Bed-User's importance was reduced in Comp. 2 compared to Comp. 1, as shown in Fig. 14, preventing wrong inference. These results confirm that the robot can efficiently learn spatial relations by multiple position samples generation. However, as shown in Fig. 15, it was also observed that Comp. 2 learned incorrect spatial relations from the corrective path and made incorrect inference when the data was scarce.

Second, as shown in Fig. 16, the proposed method successfully inferred the goal at the end point of the previous corrective path, whereas Comp. 2 incorrectly inferred at the starting point. The probability map of the proposed method shows that the degree of OK

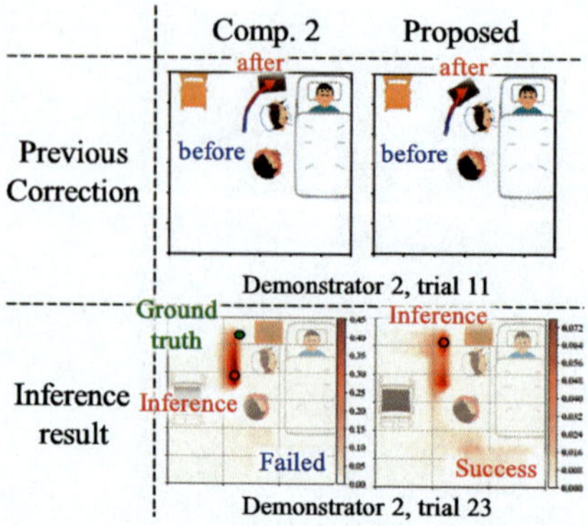

Fig. 16. The specific scene where the proposed method outweighs Comp. 2

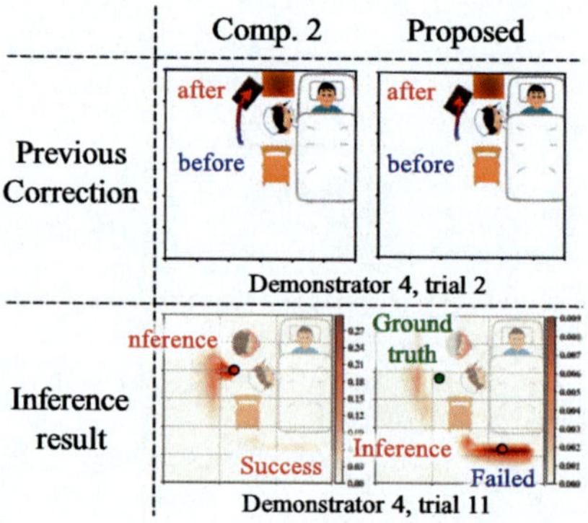

Fig. 17. The specific scene where the proposed method is inferior to Comp. 2

around the starting point decreases. This indicates that the proposed method effectively reflected the user's correction in the goal inference and improved the inference efficiently by consideration of NG.

Finally, Fig. 17 shows that the proposed method considering NG incorrectly inferred the goal by canceling out the degree of OK around the user's desired position. Figure 17 also shows that the area around the starting point of the previous corrective path was assumed to be NG, and the grid values of this area were canceled out. The same tendency was observed in the 45th trials of demonstrator 4 and in the 3rd trial of demonstrator 1, where the accumulated error was increased by the proposed method, as shown in Fig. 12. The reasons for the ineffectiveness of NG in this case are that the current goal inference process sometimes fails to address the variation in the user's desired goal depending on the surrounding environment, and the proposed method currently takes NG into account too strongly. Therefore, to effectively consider NG and improve the inference, we intend to investigate the process of integrating partial probability maps and appropriately compressing or truncating the previous data.

6 Conclusion

In this study, for natural human-robot collaboration, we proposed a method to improve the robot's goal inference with fewer trials by utilizing the user's corrective path of the robot position. The robot generated multiple position samples from a corrective path. These samples were then weighted based on the implicit intention of the correction to learn both the desired and undesired positions, and the inferences of desired and undesired positions were integrated. The effectiveness was verified in terms of the correction count and accumulated error of inference by experiment that simulated human-robot collaborative environments. As a result, we confirmed that the proposed method tends to improve inference in fewer trials. However, in some cases, the proposed method is ineffective; therefore, we intend to improve the process to further generalize its effectiveness. In the future, we would like to consider changes in object positions during goal inference and the reachability of robots, to assume more realistic environments.

Acknowledgments. This work was supported by Core Research for Evolutional Science and Technology (CREST) of the Japan Science and Technology Agency (JST) [grant number JPMJCR19A1].

References

1. Sierra Marín, S.D., et al.: Expectations and perceptions of healthcare professionals for robot deployment in hospital environments during the covid-19 pandemic. Front. Robot. AI **8**, 102 (2021)
2. Sisbot, E.A., Marin-Urias, L.F., Alami, R., Simeon, T.: A human aware mobile robot motion planner. IEEE Trans. Robot. **23**(5), 874–883 (2007)
3. Che, Y., Okamura, A.M., Sadigh, D.: Efficient and trustworthy social navigation via explicit and implicit robot-human communication. IEEE Trans. Robot. **36**(3), 692–707 (2020)
4. Kollmitz, M., Koller, T., Boedecker, J., Burgard, W.: Learning human-aware robot navigation from physical interaction via inverse reinforcement learning. In: IEEE/RSJ International Conference on Intelligent Robots and Systems (IROS), pp. 11025–11031 (2020)
5. Abdo, N., Stachniss, C., Spinello, L., Burgard, W.: Robot, organize my shelves! Tidying up objects by predicting user preferences. In: IEEE International Conference on Robotics and Automation (ICRA), pp. 1557–1564 (2015)
6. Asama, H.: Robotics in Heisei: a review: trends and future prospects of robot technology. J. Jpn. Soc. Precis. Eng. **86**(1), 23–27 (2020) (Japanese)
7. Bajcsy, A., Losey, D.P., O'Malley, M.K., Dragan, A.D.: Learning robot objectives from physical human interaction. In: Conference on Robot Learning (CoRL) (2017)
8. Cosgun, A., Christensen, H.I.: Context-aware robot navigation using interactively built semantic maps. Paladyn J. Behav. Robot. **9**(1), 254–276 (2018)
9. Lindner, F.: A conceptual model of personal space for human-aware robot activity placement. In: IEEE/RSJ International Conference on Intelligent Robots and Systems (IROS), pp. 5770–5775 (2015)
10. Kawasaki, Y., Takahashi, M.: Bottom-up action modeling via spatial factorization for serving food. Adv. Robot. **35**(12), 756–770 (2021)
11. He, K., Gkioxari, G., Dollar, P., Girshick, R.: Mask R-CNN. In: Proceedings of the IEEE International Conference on Computer Vision (ICCV), pp. 2961–2969 (2017)

12. Taylor, C.C.: Automatic bandwidth selection for circular density estimation. Comput. Stat. Data Anal. **52**(7), 3493–3500 (2008)
13. Scott, D.W.: Multivariate Density Estimation: Theory, Practice, and Visualization. Wiley, New York (1992)
14. Silverman, B.W.: Density Estimation for Statistics and Data Analysis. Chapman and Hall, London (1986)
15. Vicon Motion Systems Ltd.: VICON. http://vicon.com (as of Apr 2022)

Skeleton-Based Action and Gesture Recognition for Human-Robot Collaboration

Matteo Terreran(✉), Margherita Lazzaretto, and Stefano Ghidoni

Department of Information Engineering, University of Padova, Padova, Italy
matteo.terreran@unipd.it, margherita.lazzaretto@studenti.unipd.it, stefano.ghidoni@unipd.it

Abstract. Human action recognition plays a major role in enabling an effective and safe collaboration between humans and robots. Considering for example a collaborative assembly task, the human worker can use gestures to communicate with the robot while the robot can exploit the recognized actions to anticipate the next steps in the assembly process, improving safety and the overall productivity. In this work, we propose a novel framework for human action recognition based on 3D pose estimation and ensemble techniques. In such framework, we first estimate the 3D coordinates of the human hands and body joints by means of OpenPose and RGB-D data. The estimated joints are then fed to a set of graph convolutional networks derived from Shift-GCN, one network for each set of joints (i.e., body, left hand and right hand). Finally, using an ensemble approach we average the output scores of all the networks to predict the final human action. The proposed framework was evaluated on a dedicated dataset, named IAS-Lab Collaborative HAR dataset, which includes both actions and gestures commonly used in human-robot collaboration tasks. The experimental results demonstrated how the ensemble of the different action recognition models helps improving the accuracy and the robustness of the overall system.

Keywords: Human action recognition · Gesture recognition · 3D pose estimation · Ensemble learning · Human-robot collaboration

1 Introduction

Human-robot collaboration (HRC) aims to a close and direct collaboration between robots and humans to reach higher productivity and ergonomics thanks to the synergy between human intelligence and robot mechanical power [1–3]. In such scenario, the robot should always be aware of the position and the intentions of the human worker, in order to foresee possible dangerous situations and

I. Petrovic et al. (Eds.): IAS 2022, LNNS 577, pp. 29–45, 2023.
https://doi.org/10.1007/978-3-031-22216-0_3

guarantee the safety of the human partner. Moreover, by knowing which step of the assembly the person is working on, the robot can plan its actions accordingly, such as working in a different part of the workspace or preparing parts and tools for the next step in the assembly process.

Human action recognition has been widely investigated in the literature to provide such awareness to the robot [4–6]. Generally, the actions to be recognised are steps in an assembly sequence (e.g. pick a part, place a part, screw) or general actions such as walking, interacting or standing still; thus all actions that involve various parts of the body and can be easily distinguished. Some works focus instead on gesture recognition, where the actions to be classified are usually very small movements of few parts of the body, such as hands, used to communicate information to the robot (e.g. go left, go right, stop). Given this difference, the problems of action and gesture recognition are generally tackled separately with dedicated approaches and setups, for example a camera framing only the human hands in case of gesture recognition.

In this work, we address the problem of human action recognition in collaborative scenarios by proposing a general framework capable of recognizing both hands gestures and full-body gestures (i.e., general actions involving the whole body). Our framework relies on an ensemble of different skeleton-based classifiers, each one trained to recognize actions on a subset of skeleton joints (e.g., body joints, hands joints); the use of dedicated classifiers allows to be robust to partial skeleton inputs with missing joints, and to handle together body actions and hand gestures. Moreover, using skeletons as an intermediate representation between the classifiers and the raw RGB-D data, our system is able to generalize to different scenarios and collaborative tasks. Indeed, in such a manner, the action classifiers can leverage on mid-level features describing only people movements: picking up a piece to be assembled or grabbing up a hammer from a toolbox can both be considered a "pick" action despite the particular object being picked up.

The proposed system has been evaluated on a dataset acquired on purpose in our laboratory, namely the IAS-LAB Collaborative HAR dataset,[1] which includes RGB-D videos of several subjects executing typical actions which occur in a human-robot collaboration task. Unlike other works in the literature, in our dataset we did not restrict the action set to a specific application (e.g., take part A, place part B, take hammer from the toolbox), but we tried to generalize the most recurrent actions and gestures in the literature in order to create a common basis for several collaborative tasks.

Summarizing, the work presents 3 main contributions: (i) a unified framework for human action and gesture recognition in a human-robot collaboration scenario; (ii) an experimental comparison of different ensembling techniques to improve the overall accuracy and robustness of the system; (iii) a novel RGB-D dataset for action recognition in a human-robot collaboration scenario, including both general actions and hand gestures, to further drive research in this field. The remainder of the paper is organized as follows. Section 2 reviews the

[1] Available at http://robotics.dei.unipd.it/.

works related to action recognition, with a focus on human-robot collaboration scenarios. In Sect. 3 the main elements of our system are described in details. In Sect. 4 we present the action recognition dataset acquired in our laboratory, used to thoroughly evaluate the system proposed in Sect. 5. Finally, in Sect. 6, conclusions are drawn and future directions of research identified.

2 Related Works

Human action recognition (HAR) is generally defined as the process of identifying and analyzing the movements of one or several parts of the human body, applied in a large set of scenarios such as public events [7], video surveillance and home monitoring [8,9], human-robot interaction [5,10] and safety monitoring in industry [11]. According to the sensory modalities and sources adopted, HAR systems can be classified into two main categories: contact-based and vision-based systems. Contact-based approaches require a physical interaction of the user with sensors and acquisition devices such as accelerometers, multi-touch screens, body-mounted sensors or wearable sensors like sensorized gloves [9,10,12]. Vision-based approaches rely instead on images or video sequences to recognize activities, using either a single RGB-D camera or a network of cameras to be robust to occlusions. Unlike contact-based activity recognition systems, vision-based systems are considered "non-intrusive" since they do not require ordinary users to wear several and uncomfortable devices on different parts of their body, making them an easier solution to be used in real scenarios.

A main challenge in vision-based action recognition tasks consists in handling both a spatial and a temporal dimension, since an action is generally seen as a sequence of consecutive movements in time. Many approaches have been proposed in the literature to deal with it, such as LSTMs [13,14], 3D-CNNs [15,16] and multi-stream 2D-CNNs [17,18]. In 3D-CNNs the temporal information is exploited by considering an input sequence of RGB frames, then elaborated using 3D convolution kernels. In multi-stream CNNs instead, spatial and temporal information are analyzed by two different branches in the network, taking as input RGB frames and optical flow information respectively. Recently, many pose estimation models such as OpenPose [19] achieved very high performance and many researchers started addressing action classification using the human body pose as input of graph convolutional networks (GCNs) [20–22]. Body pose allows to encode spatial and temporal information in a more compact way than images, leading GCN models such as the Shift-GCN architecture [20] to outperform other approaches on many popular action recognition datasets [23,24].

Human action recognition is largely applied in many human-robot interaction scenarios, such as social robotics and manufacturing industry. Usually, the set of actions to be recognized includes hand gestures and face expressions, in order to provide a fast and easy interaction with the robots [4,25]. For example, in [4] authors propose six gestures to allow the human to communicate with a collaborative robot; the gestures (i.e., *on*, *off*, *right*, *left*, *up*, *down*) are recognised by fusing together three different modalities, namely speech command (with a CNN), hand motion (with a LSTM), and body motion.

Considering an industrial context, the set of actions of interest can also include either general actions (e.g., walking, standing) or very specific actions and gestures, depending on the main purpose of the action recognition. For example, action recognition could be used to guarantee human safety in human-robot collaboration by monitoring what people do while inside the robot workspace. In [5], authors monitor the people moving close to a robotic arm and recognize actions such as *passing*, *observing*, *dangerously observing*, *interacting*. The recognition is done with a multi-modal approach: a 3D-CNN is used to learn features from sequences of RGB frames, while signals collected with a haptic sensors are employed to detect collisions between human and robot by means of a 1D-CNN.

When a task involves close collaboration between humans and robots, human action recognition could be used to improve productivity, by monitoring the different stages of the collaboration. In such case the set of actions to be recognized generally includes the different operations that the human needs to fulfill the overall assembly task. For example, in [10] the set of actions includes *grab a tool* from a toolbox, *insert a screw*, *tight the screw*, *put back the tool* in the toolbox. In such work, authors proposed an action recognition classifier based on a CNN trained on a combination of skeleton features and signals from EMG and IMU sensors. In [6] similar activities are considered, such as *taking a product or a component*, *move a product*, *grab a tool*, *put on screws*, *hold a product*, *tighten the screws*, *check product* and *place product*. To classify such actions the authors focus on hands information only, proposing a system that combines hands' pose and images cropped around the hands. In [17] the assembly actions include instead *cleaning*, *hammering*, *polishing*, *smearing*, *installing*, *screwing* and *marking*, all recognised using a two-stream CNN.

Despite sharing some common high-level actions and gestures, many of the human-robot collaboration systems presented in this section consider a very specific set of actions related to a particular task; in many cases authors acquire also an ad-hoc dataset for their application which, however, is either not published or proves to be too specific to be used by other researchers. Moreover, action recognition is usually addressed focusing either on body information or hands information. To the best of our knowledge no work has attempted to recognise actions using body and hands information together.

In this work we address both such problems, namely the lack of a general dataset and the recognition of actions involving either the body or the hands. On one hand, we propose a general framework to recognize both body actions and hands gestures in a collaborative scenario; on the other hand the system is evaluated on a novel action recognition dataset acquired on purpose, including common actions of a collaborative task so as to be generalisable to various applications.

3 Methods

In this section, we provide a detailed description of the main parts of our system. A schematic representation of the proposed pipeline is shown in Fig. 1, highlighting the main steps involved. Our system takes as input a sequence of RGB-D frames and predicts the corresponding action performed according to a given set of actions of interest. In the first stage, we estimate the pose of the person in a sequence of RGB frames by means of OpenPose [19]; in particular, we focus on estimating the pose information of both the body and the hands, since many collaborative gestures could be executed by using the hands only. OpenPose predicts a set of 2D joints describing the pose of the person in the image, which is then projected in 3D space thanks to the information provided by the sequence of depth images. The estimated 3D joints are then fed to a set of graph convolutional networks derived from Shift-GCN [20], a state-of-the-art action recognition architecture. Indeed, for the final action recognition stage, we rely on an ensemble of classifiers, each one trained to recognize actions from a different set of joints (e.g., body, left hand and right hand). The final output of the system is a weighted average of the output scores of all the classifiers, which according to our experiments in Sect. 5 proves to be more accurate and robust with respect to a single model trained on body and hands poses together.

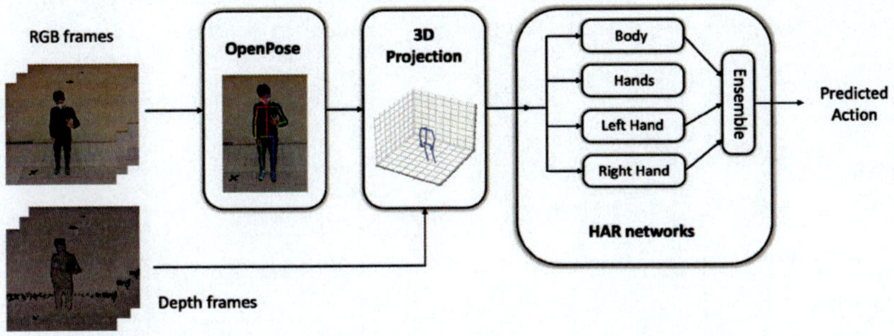

Fig. 1. Overview of the proposed action recognition system. In the first stages, 3D skeletons including body and hand joints are estimated from sequences of RGB-D frames by means of OpenPose architecture. In the last stage, the classification step, actions are predicted using an ensemble of skeleton-based action recognition models, each one trained to recognize actions from a particular set of 3D joints.

3.1 Pose Estimation

To develop our general framework we focused on skeleton-based action recognition models, which are in general more accurate and robust than other approaches. Indeed, skeletons present many advantages with respect to raw RGB-D data, since they represent a lightweight information describing both spatial and temporal information in a compact form (i.e., joints' coordinates). Moreover, skeletons provide a robust representation of the human movements free of any disturbances such as external objects, lighting and aesthetic differences of people such as clothes or skin colour; this is important especially for collaborative scenarios, where both the human and the robot are moving and the human worker should interact with many objects and tools.

For pose estimation in our framework we rely on the OpenPose architecture [19], which provides different pretrained models for multi-person pose estimation, allowing to estimate in real-time either 15, 18 or 25 body keypoints, 42 hand keypoints and 70 face keypoints. Although very accurate in general, OpenPose predictions could be missing of some keypoints when the subject is far away from the camera or partially occluded; this is especially true for the hands' keypoints, which can be easily occluded during interactions with objects.

Considering such limitations, in our framework we consider as input of the HAR models only a subset of the most significant joints for the body and hands skeletons provided by OpenPose. In particular, as shown in Fig. 2, our choice for the hand joints is limited to: the wrist, three joints each for the thumb, index and middle finger, and two joints for the ring finger; for a total of 12 joints for each hand. The rest of the joints are omitted, being more difficult to be estimated and less useful for the action recognition process. For what concerns the body, some keypoints such as the ones corresponding to eyes, ears and feet are not considered either, leading to a 15 joints body model comprising head, neck, shoulders, elbows, wrists, pelvis, hips, knees and ankles.

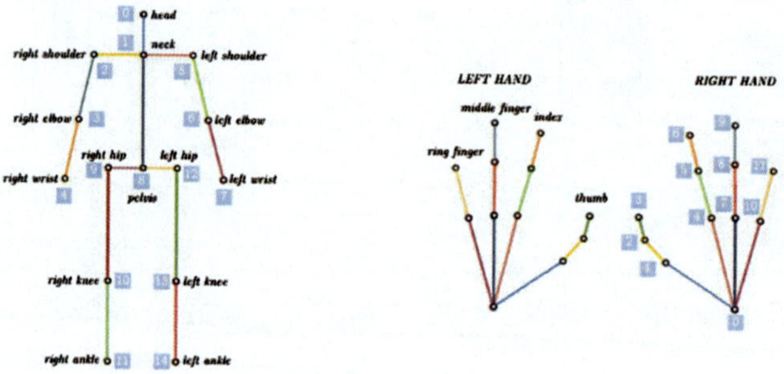

Fig. 2. Body and hand joints considered in the proposed framework, together with their associated ID. We consider a total of 15 joints describing the body, and 12 joints to describe each one of the hands.

3.2 3D Pose Estimation

The output of OpenPose is a 2D skeleton describing the pose of each person in the input image. From this information we compute a 3D skeleton by means of re-projection, in order to have a more standard and general representation to be used as input for the action classifiers. Given a calibrated RGB-D camera and its extrinsic parameters, it is possible to transform the depth image by aligning it to the RGB image, obtaining a direct mapping for each pixel; for each 2D keypoint (x_P, y_P) estimated by OpenPose in the input image, we use the information in the aligned depth image to compute the keypoints coordinates in the 3D space using re-projection and the intrinsics parameters of the camera.

The depth information acquired with a RGB-D sensors is usually not accurate around borders and for small objects, like the fingers of the hand in our case. Therefore, to improve accuracy and robustness of the 3D keypoints, we do not consider the raw depth value, but we take instead the median value in a 5×5 window centered at coordinates (x_P, y_P) in the depth image.

3.3 Action and Gesture Recognition

The action recognition module in our framework is based on the Shift-GCN architecture [20], a graph convolutional network which achieved state-of-the-art performance on the NTU RGB+D dataset [23]. Compared to other architectures, Shift-GCN has the advantage of being lightweight and requiring a lower computational cost, thanks to the introduction of the graph shift convolution operations. This was one of the main aspects that guided our choice, since in human-robot collaboration scenarios we would like to have a fast recognition of the human actions to obtain a smooth and responsive collaboration.

The Shift-GCN architecture takes as input a sequence of skeletons (i.e., a sequence of 3D joints coordinates), and is composed of 10 blocks, each one including a spatial graph shift convolution operation and an adaptive temporal shift operation. The final output layer contains 60 nodes, since it has been originally proposed to classify the actions in the NTU RGB+D dataset.

Unlike the NTU dataset, where actions were classified based only on the body pose information, in this work we aim to develop a general framework capable of recognizing both actions and gestures which can occur during the collaboration between a human worker and a robot; in such scenario, many possible actions and gestures involve only the use of hands while the body of the worker stands still (e.g., "stop" and "confirm" signals). This leads us to propose different action recognition networks based on the Shift-GCN architecture, each one designed to recognize collaborative actions from a particular set of joints. In particular, we consider the following set of joints:

- **wholebody**, which includes the 39 joints shown in Fig. 2 describing the pose of both the body and the hands;
- **body**, including the 15 joints which describe the pose of the body;
- **hands**, including the 24 joints which describe both hands together;
- **single hand**, including the 12 joints of a hand (i.e., wrist and fingers' joints).

As in the original Shift-GCN architecture, the input sequences of skeleton joints undergo some pre-processing operations before being fed to the network: a translation of the joint coordinate reference frame to a central joint of the skeleton, and a joints coordinate normalization. These operations allow to consider all the body movements with respect to the body itself, which makes them a better input for the network and easier to generalize to different scenarios. For our networks, we choose as origin of the new reference frame the joint corresponding to the neck (i.e., Joint 1 in Fig. 2); the $z-$axis of the new reference frame is taken parallel to the segment connecting the pelvis (i.e., Joint 8) and the neck joints, while the $x-$axis is considered parallel to segment connecting the shoulder joints (i.e., Joints 2 and 5). For the networks considering only the hands joints we use the same convention for the reference frame's axes, but placing its origin on the wrist joint (i.e., Joint 0) of each hand or the right hand wrist when both hands are considered. In such a manner, also the hands movements are expressed with respect to a local reference frame maintaining their relative orientation with respect to the rest of the body.

3.4 Ensemble Averaging of the Classifiers Predictions

As the last step of our proposed framework, all the action recognition models' outputs are combined together by means of ensemble techniques to compute the final prediction. Ensemble is a machine learning technique that combines several base models in order to produce one optimal predictive model, improving the overall accuracy and robustness.

We propose two main approaches to combine the information from the action recognition models: an ensemble of the *body* and *hands* models, and an ensemble of the *body* model with both single hand models (i.e., *left_hand* and *right_hand* models). In both approaches, the information from the models is combined at the score-level, namely the output of the *softmax* activation function in the last layer of the networks. Considering for example the first approach (i.e., *body* + *hand* models), we compute the final score as a weighted sum of the score of each model. The predicted action l_{pred} is then obtained by taking the argmax of the final score,

$$l_{pred} = arg\,max\left(\alpha_b\,\mathbf{o}_b + \alpha_h\,\mathbf{o}_h\right), \quad with\ \alpha_b + \alpha_h = 1 \tag{1}$$

where N is the number of actions of interest, $\mathbf{o}_b \in [0,1]^N$ is the output score of the *body* network, $\mathbf{o}_h \in [0,1]^N$ is the output score of the *hands* network and α_o, α_h are the corresponding weights.

In the second approach (i.e., *body* + *left_hand* + *right_hand* models) we take into account also the fact that some actions or gestures can be made using only one hand (e.g., confirm, left, right, stop) while the other one remains still or even not visible. The final score is still computed as a weighed sum of the models' score, but considering a weight $\alpha_{ih} = 0$ if the hand $i, i \in \{left, right\}$ is the only one not visible or in a rest position. We assume in this case that the hand whose action is labelled as "rest" falls in a group of actions in which only one of the two hands is actively moving, while the other one stands still; the actual action is therefore related to the moving hand and the other one is irrelevant in terms of action recognition.

4 IAS-Lab Collaborative HAR Dataset

As highlighted in Sect. 2, there exist different actions that are common to many human-robot applications. In order to obtain a general framework suitable for different collaborative settings, we trained our models on a set of actions and gestures that generalize those commonly found in the literature. For example, in many works we have actions such as *grab a tool*, *pick a piece* or *pick the hammer* which can be all summarized as a *pick* action if we only focus on the human movements; the specific object to be picked can be then detected from a dedicated object detector. The actions that we selected are reported in Table 1, categorized in four main groups. The first group refers to general movements that a person can do in the robot workspace, where *WALK* includes all the movements of a worker moving around the workspace and *REST* indicates that the human operator is not working. The second group includes the most common assembly actions performed by the human worker, while the third group includes all the signals that the human can do to request or pass objects to the robot during the collaboration. Finally, the fourth group includes all the gestures used to communicate an instruction to the robot, including directions of movements and signals of confirmation or halting.

Table 1. Actions and gestures in the IAS-Lab Collaborative HAR dataset

Group	Actions
Spatial movements	WALK, REST
Assembly actions	PICK, PLACE, SCREW, INSERT/JOIN, HAMMER
Collaborative gestures	HAND TO, REQUIRE
Communication gestures	STOP, OK/CONFIRM, UP, DOWN, FORWARD, BACKWARD, LEFT, RIGHT, POINT

The set of actions to be recognised was chosen in order to include movements that are as generic as possible, and that can in general be distinguished without pairing them to objects or tools and without setting a world coordinates system to locate the human operator in the workspace. It can be noticed that many actions require an active use of hands and that the hands movements are potentially discriminative in the recognition of the performed action, especially for what concerns gestures and assembly actions.

Once defined the set of actions in Table 1, we collected a novel dataset of people performing such actions in our laboratory. We asked to 6 different subjects to perform 5 times each the 18 actions selected, for a total of $18 \times 6 \times 5 = 540$ samples. Each sample is a sequence of RGB-D frames of about 5 s, acquired with a Intel Realsense L515 camera.[2] The camera was placed at a distance of about 2.5 m from the subjects, in order to frame the whole subject body during all the actions. Some samples from the acquired dataset are shown in Fig. 3. During the dataset collection, subjects were only told the name of the action to be executed, without receiving any further instruction on how to perform it. This allowed to increase the variability of the dataset, since many subjects performed the same actions in different manners as shown in Fig. 4. Indeed, we aim to recognise actions and gestures executed as naturally as possible, without creating a rigid

[2] https://www.intelrealsense.com/lidar-camera-l515/.

dictionary of movements and having to give many specific instructions to users. This can help achieving an easier, more immediate and natural communication between people and robots.

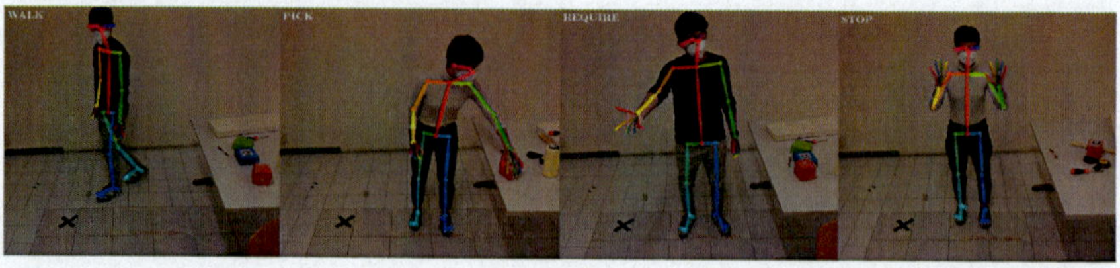

Fig. 3. Some samples from the IAS-Lab Collaborative HAR dataset, together with the skeletons outputs estimated by means of OpenPose.

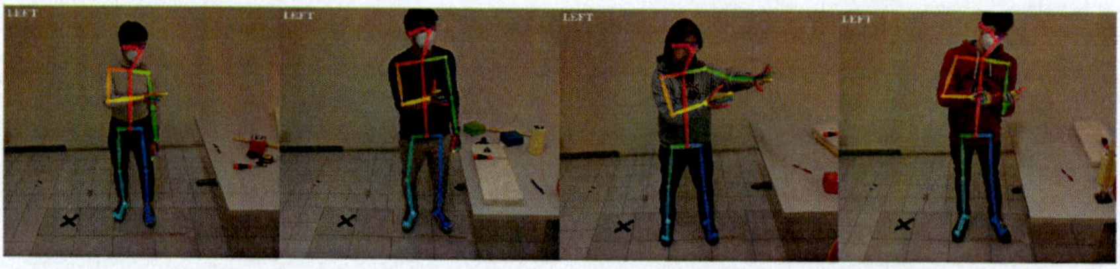

Fig. 4. Examples of variability in the IAS-Lab Collaborative HAR dataset. Each subject performed the actions requested in a different manner.

5 Experimental Results

The proposed framework has been evaluated using the IAS-Lab Collaborative HAR dataset described in Sect. 4. Such dataset is very small with respect to the dataset commonly used to train deep learning models, and hence not suitable to train from scratch the action recognition networks proposed in our framework. Therefore, the action recognition models have all been trained on a large collection of data such as the NTU RGB+D dataset and then fine-tuned on the IAS-Lab Collaborative dataset. This allows to exploit the larger dataset to learn several features encoding the human movements, which are then specialized for the human-robot collaboration scenario. The models described in the following have all been implemented starting from the official implementation of the Shift-GCN architecture.[3] In all the experiments the models have been trained using the hyper-parameters suggested by the Shift-GCN authors, using a NVIDIA® Titan RTX 2080 GPU.

[3] https://github.com/kchengiva/Shift-GCN.

5.1 Pre-training on the NTU RGB+D Dataset

The NTU RGB+D dataset [23] is a large dataset containing RGB-D frames and pose annotations of several subjects performing 60 actions, acquired with a multi-view setup composed of 3 cameras. Pose annotations are provided as skeleton models composed of 25 joints describing only the body pose. However, they do not include hands joints and the body joints slightly differ from the ones estimated by the OpenPose architecture used in our framework. For these reason, since our framework has been designed to exploit both body and hands pose information, we recreated all the pose annotations for the NTU RGB+D dataset using the OpenPose network to predict both body and hands joints. For each RGB frame in the NTU RGB+D dataset, we run the OpenPose pose estimator to predict the 2D poses of the people in the images. Then, by using the associated depth frame, the corresponding 3D pose is computed by means of re-projection as described in Sect. 3.

The authors of the NTU RGB+D dataset proposed two benchmarks: a cross-subject benchmark where all the subjects are split into training and testing groups, and a cross-view benchmark where data from different cameras are used as train and test data. However, in this work, we are interested in using the NTU dataset primarily as a tool to train our models on a large collection of actions; for this reason, we did not follow any of the proposed divisions but tried to use a training set with the largest number of images. We took as training set all the data from the cameras 2 and 3, adding also the data from camera 1 reserved for training in the cross-subject benchmark; the remaining data, namely the test data in the cross-subject benchmark acquired with the camera 1, are used as a small validation set to monitor the training and avoid overfitting.

With such division of the dataset, we trained the models described in Sect. 3 on the sequences of 3D skeletons derived from OpenPose outputs, considering the subset of the most important joints shown in Fig. 2. In particular, we trained a separate model for each set of joints (e.g., *body* joints, *left hand* and *right hand* joints) and a *whole_body* model considering all the available joints (i.e., *body* and *hand* joints). The results obtained for each trained model on the NTU dataset are reported in Table 2. We evaluated the models in terms of accuracy, considering a *Top1 accuracy* and *Top5 accuracy*. The former represents the percentage of correctly predicted actions in the test set, while the Top5 accuracy is the percentage of actions whose correct prediction falls in the five highest softmax scores estimated by the network.

As reported in Table 2, the best results are obtained with the model trained only on the 15 selected body joints, while the models trained on the hands' joints achieve very low accuracy. These results were somehow expected considering the actions of the NTU dataset, which includes daily actions (e.g., drinking, eating, reading) where the use of hands is very limited. Many of these actions differ mainly in body posture, while the hands represent only marginal information which is not sufficient for a model trained only on hands joints to distinguish such a large set of actions.

Table 2. Experimental results on the NTU RGB+D dataset for each proposed model. Results are provided in terms of accuracy using the skeleton sequences extracted with OpenPose.

Model	#Joints	Top1 %	Top5 %
Wholebody	39	59.44	81.34
Body	15	**90.40**	**97.68**
Hands	24	36.25	67.63
Left hand	12	21.12	49.38
Right hand	12	31.90	62.68

Bold font indicates the best result obtained for each metric among all models

Even the model trained using both body and hands joints achieves lower accuracy than the model with only body information. This result seems to suggest that adding hand information is even detrimental to the model, leading it to confuse more actions than using body information alone: considering the 39 input joints, only 15 joints describe the body pose, while more than half represent hand information which does not provide enough information about the actions to be recognized.

Note that the Shift-GCN architecture achieves a Top1 accuracy of 96.5% on the original NTU dataset with body pose annotations, while our "body" model performs slightly worse with a Top1 accuracy of 90.4%. A detailed and direct comparison between the two results is not possible, since we used slightly different train and test sets. However, the drop in performance could be partly due to our new annotations and how partial inputs to the network are handled: if some of the required joints are missing in the input, the entire input skeleton is discarded. We found this happening on several occasions when using the skeletons estimated by OpenPose, leading us to discard whole sequences when too many skeletons were missing.

5.2 Fine-Tuning on the IAS-Lab Collaborative HAR Dataset

Using the NTU RGB+D dataset, we were able to train several action recognition models for recognizing a large number of daily activities. Given the large size of the dataset, these models were able to learn several low and mid-level features to be exploited for our action recognition system in collaborative scenarios. We fine-tuned all the models on the IAS-Lab Collaborative HAR dataset described in Sect. 4, substituting the final layer of 60 nodes with a layer of 18 nodes to match the size of the new set of actions. In order to preserve the low and mid-level features already learnt, during fine-tuning we froze all the layers' weights except the last ones. In particular, denoting with $\ell i, i \in [1, 10]$ the 10 blocks of the Shift-GCN architecture, for each model we froze all the blocks except for the final ones reported in Table 3. We allowed more blocks to be retrained for models that achieved low accuracy on the NTU dataset (e.g., hand models), since in general low accuracy denotes that poor mid- and high-level features have been learned and more layers' weights should be updated.

For fine-tuning the models on the IAS-Lab dataset we followed a cross-subject benchmark, where the first 5 subjects are used for training and the sixth subject is used for validation. All the fine-tuned models have been evaluated in terms of Top1 and Top3 accuracy, reported in Table 3. Since the number of actions in this case is much smaller than in the NTU dataset, we chose to measure a Top3 accuracy instead of the Top5 of the previous case.

Table 3. Experimental results on the IAS-Lab Collaborative HAR dataset for each proposed model. Results are provided in terms of accuracy using the models pre-trained on the NTU RGB+D dataset.

Model	#Joints	Valid inputs	Top1 %	Top3 %	Learnable blocks
Wholebody	39	45%	44.00	54.00	$\ell7$, $\ell8$, $\ell9$, $\ell10$
Body	15	100%	**62.22**	**86.67**	$\ell9$, $\ell10$
Hands	24	45%	50.00	64.00	$\ell8$, $\ell9$, $\ell10$
Left hand	12	75%	59.42	73.91	$\ell6$, $\ell7$, $\ell8$, $\ell9$, $\ell10$
Right hand	12	70%	59.42	71.01	$\ell6$, $\ell7$, $\ell8$, $\ell9$, $\ell10$

Bold font indicates the best result obtained for each metric among all models

Among the results reported in Table 3 the best performances are still obtained with the model trained on the body joints which, however, achieves just 62.22% on Top1 accuracy. Interestingly, we obtain better performance for the models trained on the hands joints compared to the NTU dataset, especially for the models considering each hands separately. In many sequences of the IAS-Lab dataset the subjects' actions were done with one hand, while the other one was in a resting position. These represent ambiguous situations for a model that recognizes actions using information from both hands, reducing its overall accuracy.

However, no model can correctly classify all actions in the test set data, highlighting the complexity of the assigned recognition task. As described in Sect. 4, the IAS-Lab dataset collects several typical actions of a human-robot collaboration scenario, including both general actions (e.g., *pick*, *place*) and hands gestures (e.g., *confirm*, *stop*). In the former case, the actions to be recognized are large movements involving many body parts, such as walking or hammering. Hand gestures, instead, are commonly used to communicate with the robot and involve only movements of the worker's hands while the rest of the body remains mostly stationary. Both the *body* model and the *single hands* models are very accurate in classifying only one of the two types of actions, but fail on the other one, not having enough information. For example, as shown in Fig. 5, the *body* model predicts very accurately actions like *place*, *hammer* or *walk*, but struggles to recognize all the gestures based on hands movements (e.g., *hand to*).

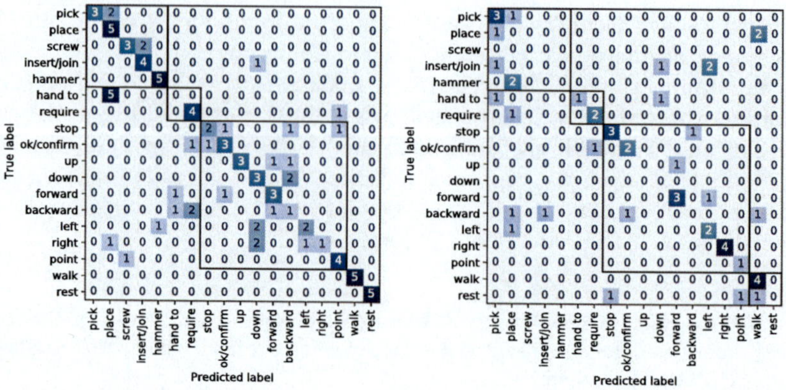

Fig. 5. Confusion matrices for the models fine-tuned on the IAS-Lab Collaborative HAR dataset. On the left, the confusion matrix for the *body* model. On the right, the confusion matrix for the *hands* model.

Intuitively, a model that uses body and hand information together should be the best way to recognize all the given actions, but in our experiments we found the opposite to be true, with the *wholebody* model achieving the lowest accuracy among the results reported in Table 3. This is mainly due to the fact that in several sequences the skeletons found were not complete with all joints, leading to discarding such sequences from the model training set and thus limiting the model's ability to learn to recognize all actions. The percentage of valid sequences available for each models has been reported in Table 3, highlighting how the available data were particularly limited to train the *wholebody* model. Although more accurate in theory, in practice using a model based on the body and hands together proved to be less robust and difficult to train; this observation led us to investigate alternative approaches for combining body and hand information together, such as ensemble techniques.

5.3 Ensemble Results Using Body and Hands Models

Body posture and hand posture are complementary information that, if properly combined, can greatly aid in the classification of actions. As highlighted in the previous experiment, combining this information when training a network proved to be inefficient: on the one hand, more sequences are required to learn the various relationships that may exist between body and hands during the actions of interest; on the other hand, it requires that in each frame all joints are always detected, which could be not true in case of occlusions.

For these reasons, we developed our action recognition system by combining information at the score level by means of ensemble techniques: we run in parallel the models fine-tuned on the IAS-Lab dataset, and combine together their score predictions (i.e. the predicted probability for each actions) by means of a weighted sum. In particular, we investigated two main approaches: an ensemble of the *body* and *hands* models, and an ensemble of the *body* model with the models for each hand. In the former case we found that best results are obtained when weighting equally the contributions, that is using weights $\alpha_b = \alpha_h = 0.5$ in Eq. (1). When using instead a separate model for each hand we found that less importance should be given to the scores of the *left_hand* model, probably due to the fact that the majority of the subjects in the dataset were right-handed and tended use their right hand to perform the requested actions; the final weights we selected for this approach are $\alpha_b = \alpha_{rh} = 0.358$ and $\alpha_{lh} = 0.284$.

Table 4. Experimental results on the IAS-Lab Collaborative HAR dataset using different ensembles of the fine-tuned models.

Model	Top1%	Top3%
Body + hands	68.00	**90.00**
Body + left hand + right hand	**76.54**	87.65

Bold font indicates the best result obtained for each metric among all models

The results obtained using the two ensemble strategies proposed are reported in Table 4. Both strategies lead to an improvement of the Top1 accuracy with respect to the previous results, showing how in general combining body and hands information helps classifying actions. The best result is obtained with the ensemble of the *body* model with the single models for each hand, achieving a huge improvement with respect to the results obtained by each model in Table 3. Although better on the Top3 accuracy, the *body + hands* ensemble achieves a lower Top1 accuracy, mainly due to the many one-handed actions in the dataset, since the second hand not performing the "real" action is misleading for the model. Moreover, such ensemble is also limited by the low number of valid input sequences available, as reported in Table 3. The *body + single hands* ensemble approach represents instead a more robust solution for implementing our action recognition system, since it relies on the base models with the lowest number of joints required: a sequence is considered a valid input if at least the body or a hand is detected, reducing the number of overall frames and sequences discarded.

6 Conclusions

In this work, we propose a unified framework for action recognition in human-robot collaboration scenarios. Our framework can recognize many body-actions and hand-gestures which are often used in several collaborative tasks, thus representing a general solution to various possible real applications. The system has been evaluated on a novel dataset including general actions of human-robot collaboration scenarios, which could be used as a benchmark to further drive research in this field. In our experiments, we demonstrated how ensemble techniques help to achieve higher accuracy when averaging the predictions of several skeleton-based classifiers, each one trained to recognize actions from a different set of joints. Best results have been obtained using an ensemble of classifiers for body joints and single hands joints, which also represents a robust solution to handle possible missing joints in the estimated skeletons. As future research directions, we will apply the proposed framework in a real human-robot collaboration task to monitor human movements during an assembly process. Moreover, we will investigate the use of a multi-camera setup to improve the 3D pose estimation step, reducing inaccuracies and possible missing joints due to occlusions.

Acknowledgment. The research leading to these results has received funding from the European Unions Horizon 2020 research and innovation program under grant agreement No. 101006732. Part of this work was supported by MIUR (Italian Minister for Education) under the initiative "Departments of Excellence" (Law 232/2016).

References

1. Villani, V., Pini, F., Leali, F., Secchi, C.: Survey on human-robot collaboration in industrial settings: safety, intuitive interfaces and applications. Mechatronics **55**, 248–266 (2018)
2. Matheson, E., Minto, R., Zampieri, E.G., Faccio, M., Rosati, G.: Human-robot collaboration in manufacturing applications: a review. Robotics **8**(4), 100 (2019)
3. Kim, W., Peternel, L., Lorenzini, M., Babič, J., Ajoudani, A.: A human-robot collaboration framework for improving ergonomics during dexterous operation of power tools. Robot. Comput.-Integr. Manuf. **68**, 102084 (2021)
4. Liu, H., Fang, T., Zhou, T., Wang, L.: Towards robust human-robot collaborative manufacturing: multimodal fusion. IEEE Access **6**, 74762–74771 (2018)
5. Mohammadi Amin, F., Rezayati, M., van de Venn, H.W., Karimpour, H.: A mixed-perception approach for safe human-robot collaboration in industrial automation. Sensors **20**(21), 6347 (2020)
6. Kobayashi, T., Aoki, Y., Shimizu, S., Kusano, K., Okumura, S.: Fine-grained action recognition in assembly work scenes by drawing attention to the hands. In: 2019 15th International Conference on Signal-Image Technology & Internet-Based Systems (SITIS), pp. 440–446. IEEE (2019)
7. Liu, K., Zhu, M., Fu, H., Ma, H., Chua, T.S.: Enhancing anomaly detection in surveillance videos with transfer learning from action recognition. In: Proceedings of the 28th ACM International Conference on Multimedia, pp. 4664–4668 (2020)
8. Prati, A., Shan, C., Wang, K.I.K.: Sensors, vision and networks: from video surveillance to activity recognition and health monitoring. J. Ambient Intell. Smart Environ. **11**(1), 5–22 (2019)
9. Ranieri, C.M., MacLeod, S., Dragone, M., Vargas, P.A., Romero, R.A.F.: Activity recognition for ambient assisted living with videos, inertial units and ambient sensors. Sensors **21**(3), 768 (2021)
10. Al-Amin, M., Tao, W., Doell, D., Lingard, R., Yin, Z., Leu, M.C., Qin, R.: Action recognition in manufacturing assembly using multimodal sensor fusion. Procedia Manuf. **39**, 158–167 (2019)
11. Bo, W., Fuqi, M., Rong, J., Peng, L., Xuzhu, D.: Skeleton-based violation action recognition method for safety supervision in the operation field of distribution network based on graph convolutional network. CSEE J. Power Energy Syst. (2021)

12. Chen, C., Jafari, R., Kehtarnavaz, N.: UTD-MHAD: a multimodal dataset for human action recognition utilizing a depth camera and a wearable inertial sensor. In: 2015 IEEE International Conference on Image Processing (ICIP), pp. 168–172. IEEE (2015)
13. Yu, J., Gao, H., Yang, W., Jiang, Y., Chin, W., Kubota, N., Ju, Z.: A discriminative deep model with feature fusion and temporal attention for human action recognition. IEEE Access **8**, 43243–43255 (2020)
14. Ullah, A., Ahmad, J., Muhammad, K., Sajjad, M., Baik, S.W.: Action recognition in video sequences using deep bi-directional LSTM with CNN features. IEEE Access **6**, 1155–1166 (2017)
15. Feichtenhofer, C.: X3D: expanding architectures for efficient video recognition. In: Proceedings of the IEEE/CVF Conference on Computer Vision and Pattern Recognition, pp. 203–213 (2020)
16. Wen, X., Chen, H., Hong, Q.: Human assembly task recognition in human-robot collaboration based on 3D CNN. In: 2019 IEEE 9th Annual International Conference on CYBER Technology in Automation, Control, and Intelligent Systems (CYBER), pp. 1230–1234. IEEE (2019)
17. Xiong, Q., Zhang, J., Wang, P., Liu, D., Gao, R.X.: Transferable two-stream convolutional neural network for human action recognition. J. Manuf. Syst. **56**, 605–614 (2020)
18. Simonyan, K., Zisserman, A.: Two-stream convolutional networks for action recognition in videos. arXiv preprint arXiv:1406.2199 (2014)
19. Cao, Z., Hidalgo, G., Simon, T., Wei, S.E., Sheikh, Y.: OpenPose: realtime multi-person 2D pose estimation using part affinity fields. IEEE Trans. Pattern Anal. Mach. Intell. **43**(1), 172–186 (2019)
20. Cheng, K., Zhang, Y., He, X., Chen, W., Cheng, J., Lu, H.: Skeleton-based action recognition with shift graph convolutional network. In: Proceedings of the IEEE/CVF Conference on Computer Vision and Pattern Recognition (2020)
21. Liu, Z., Zhang, H., Chen, Z., Wang, Z., Ouyang, W.: Disentangling and unifying graph convolutions for skeleton-based action recognition. In: Proceedings of the IEEE/CVF Conference on Computer Vision and Pattern Recognition, pp. 143–152 (2020)
22. Chen, Y., Zhang, Z., Yuan, C., Li, B., Deng, Y., Hu, W.: Channel-wise topology refinement graph convolution for skeleton-based action recognition. In: Proceedings of the IEEE/CVF International Conference on Computer Vision (2021)
23. Shahroudy, A., Liu, J., Ng, T.T., Wang, G.: NTU RGB+D: a large scale dataset for 3D human activity analysis. In: Proceedings of the IEEE Conference on Computer Vision and Pattern Recognition, pp. 1010–1019 (2016)
24. Wang, J., Nie, X., Xia, Y., Wu, Y., Zhu, S.C.: Cross-view action modeling, learning and recognition. In: Proceedings of the IEEE Conference on Computer Vision and Pattern Recognition, pp. 2649–2656 (2014)
25. Martins, G.S., Santos, L., Dias, J.: The GrowMeUp project and the applicability of action recognition techniques. In: Third Workshop on Recognition and Action for Scene Understanding (REACTS), Ruiz de Aloza (2015)

Benchmarking Variable-Stiffness Grippers

Daniel Cardin-Catalan(✉), Angel P. del Pobil, and Antonio Morales

Robotic Intelligence Laboratory, Universitat Jaume I, Castellón de la Plana, Spain
cardin@uji.es

Abstract. Designs for grippers using variable-stiffness principles have become common in recent years. Researchers have developed various tests and setups to measure and validate the properties of their designs. However, there are no clear standards or benchmarks to analyse, categorise and compare this type of grippers. This paper proposes a set of benchmarks to evaluate and categorise variable-stiffness grippers. After reviewing the tests commonly applied to evaluate existing variable-stiffness and generic grippers, we propose four tests to measure different properties of a gripper. The tests are independent of one another and allow to classify each gripper in a three-category taxonomy. In order to validate the benchmark tests, three simple variable-stiffness grippers have been built and analysed with the benchmark. The results show that the grippers can be easily analysed with our benchmarks.

Keywords: Variable stiffness · Benchmarking · Grippers

1 Introduction

Grippers made up of rigid parts have been the dominant solution for robotic manipulation over the years, thanks to the stiffness of the fingers and pincers. This permits the use of grippers in typical industrial applications to perform repetitive tasks with precision. They are still an initial solution in many fields [1]. However, their lack of adaptation to contacts in applications where softness and flexibility are required make them an unsuitable solution. Soft robotics encompasses the section of robotics created from soft materials or those that interact with soft or unknown objects [2–5]. Rigid and soft grippers have their pros and cons. Rigid grippers were originally designed for industrial applications in which task specifications required high precision and the ability to exert intense forces, but at the cost of low mechanical flexibility. Both kinds of grippers haves their pros and cons. Rigid grippers are precise and strong, but not so reliable in unexpected and uncontrolled environments. Soft grippers are the exact opposite, they are adaptable and able to perform in uncontrolled events, but they lack grasping force and are difficult to control.

Grippers which integrate variable-stiffness principles have emerged and have been a rising research topic during the last years [6–15]. This type of grippers

I. Petrovic et al. (Eds.): IAS 2022, LNNS 577, pp. 46–61, 2023.
https://doi.org/10.1007/978-3-031-22216-0_4

offer the controllable duality for working in a rigid or soft manner. Variable-stiffness grippers rely on different methods to change their stiffness. Over recent years, some researches have developed different approaches to perform a variation in the mechanical properties of the grippers. Some have applied materials with stiffness that varies according to the temperature to which they are exposed [7,8,10] (see Fig. 1a). Other grippers rely on magnets or springs as external elements to change their grasping force [9] (Fig. 1b) or the gripper overall stiffness. Others, such as the universal gripper [16] or derivatives [6], work thanks to the interference of granular material within a sealed plastic cover when negative air-pressure is applied (Fig. 1c). Finally, certain grippers use co-actuation as a means of adjusting their stiffness [12] (Fig. 1d).

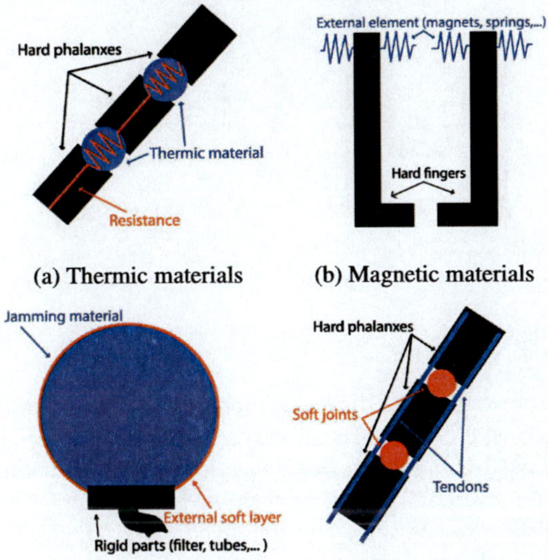

(a) Thermic materials (b) Magnetic materials

(c) Jamming principle (d) Co-actuation

Fig. 1. Schemes of stiffness tuning methods

Though there is active research on this kind of grippers, there are no standard methods to evaluate, benchmark and compare the grippers and their specific properties.

In this paper, we propose four benchmark tests for analysing and categorising the variable-stiffness grippers. Each test is centred in a specific property of the variable-stiffness grippers, such as finger force, grasping force, gripper payload and object adaptation. After the tests, the analysed grippers are organised in categories depending on the results obtained in the benchmark.

Figure 2 shows the setup for the different tests, with two different designed grippers on it. The figure shows a structure of aluminium rods that can perform all the proposed and designed tests for the benchmark.

The paper is organised as follows. Section 2 reviews the different tests that has been used to evaluate different variable-stiffness grippers. In Sect. 3 we propose four benchmark tests for variable-stiffness grippers. In order to validate and demonstrate the use of the tests, we apply them on three simple grippers. They are described in Sect. 4. Finally, the results of applying the tests are summarises and discussed in Sect. 5.

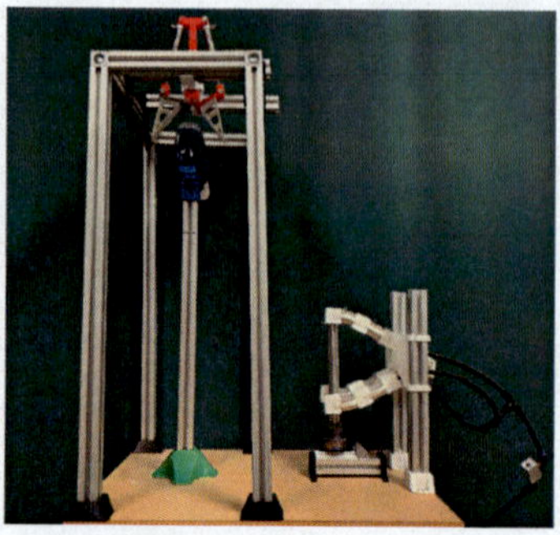

Fig. 2. The set up for the benchmark tests

2 State of the Art

Typically, works on variable-stiffness grippers [6,17–23] tend to analyse how gripper force varies when the stiffness of the gripper or fingers is changed. This same approach is used to measure the finger force in the various stiffness states of the gripper. The approach usually relies on the use of a force-torque sensor attached to the finger to measure the deployed force. This is done to test the force of the fingers in the different stiffness states.

A standardised benchmark test has been proposed to measure the finger force at the gripper's distinct stiffness or impedance states. Figure 3 presents some examples of systems used to evaluate variable stiffness. The figure follows the same colour code for all the examples: the fingers are coloured in blue and the force-torque sensor used for the readings is represented in orange. In Fig. 3a the sensor is moved using a threaded rod [18], so the finger is also moved and a force is provided. In Fig. 3b the finger is directly actuated over the sensor [6,21]. In Fig. 3c and d the finger is actuated, and a tendon moves towards a pulley system until reaching the force torque sensor to get the reading [17,22].

Apart from the previously mentioned methods, other possible tests have been considered and adapted for our benchmark. There is a total gripper maximum weight test, in which the variable-stiffness gripper grasps increasingly heavier objects until ultimately failing. This test permits the calculation of the difference in grasping stiffness, so that it can be added, after some modifications to make it repeatable, into the benchmark.

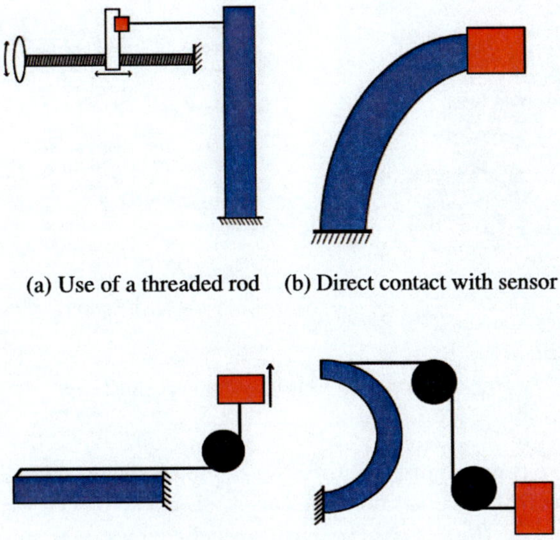

(a) Use of a threaded rod (b) Direct contact with sensor

(c) Through one pulley (d) Through various pulleys

Fig. 3. Schemes for measuring the finger force with a F/T sensor

3 Test Description

After considering some of the characteristics of the variable-stiffness grippers, presented in Sect. 1, and some of the methods applied by researchers to measure their performance, as explained in Sect. 2, a new set of protocol tests has been created and implemented to analyse and categorise variable-stiffness grippers.

A benchmark must be able to analyse and categorise all kind of variable-stiffness grippers and sort them into different categories. This categories will help future researchers choose the desired and optimal gripper, depending on the task that they are performing. To fulfill these specifications, the following four tests have been designed and implemented.

3.1 Adaptation Test

The aim of the object grasping test is to empirically measure how many objects the gripper can grasp, depending on the changes in finger stiffness.

This test also estimates the extent to which the gripper design is efficient enough to grasp a wide range of objects, even without changing the stiffness. Also, if more objects are grasped in one state or another, it would show which states are better for manipulating different types of objects.

To grasp the objects from the top, the gripper must be attached in a fixed position with the fingers pointing downwards. In this position, the fingers should

(a) Kitchen items from the YCB Benchmark

Object	Orientation	Grasping position
Pitcher	Handle face up	Handle
Bleach cleanser	Label face up	Middle of bottle
Glass cleaner	Label face up	Middle of bottle
Plastic wine glass	As in fig. 4a	Top of glass
Metal bowl	As in fig. 4a	Top of bowl
Metal mug	As in fig. 4a	Top of mug
Abrasive sponge	As in fig. 4a	Middle of object
Cooking skillet	As in fig. 4a	Handle
Metal plate	Standing perpendicular	Edge of plate
Eating utensils	As in fig. 4a	Handle
Spatula	As in fig. 4a	Handle
White table cloth	As in fig. 4a	Middle of object

(b) Grasping positions

Fig. 4. Grasping test objects and positions

be able to move freely. Figure 2 shows the required set-up with the two gripper prototypes in the testing position. The target object is placed over a stand with adaptable height, so that it can be easily reached by the gripper fingers. The gripper must be placed above the object, where it can perform an enveloping grasp by simply activating it.

The objects selected for the adaptation analysis have been chosen from the kitchen subset of the YCB Benchmark set [24], thanks to the variety of objects included in this set. The items are shown in Fig. 4a and the grasping configurations are described in Fig. 4b.

For each of the stiffness states, the gripper must follow these steps:

1. The object is placed over the stand and is set up so that the gripper can easily grasp it when activated.
2. The gripper is activated in the softest state.
3. The object is grasped.
4. The stiffness is changed to the determined state (if the determined state is the softest one, nothing changes).
5. The stand is removed.
6. Hold for 5 s.
7. The object is released.

The grasp is considered successful if the object continues to be firmly held as of the moment when the gripper is activated. Each case is repeated five times for each object and stiffness state. If the grasp fails during one of these attempts, all five attempts are considered to have failed. This is done to create the worst possible scenario. If one grasp fails, this means that, in a real world situation, there will be some failures, unacceptable in real business.

3.2 Finger Force Test

The aim of this test is to measure the finger force in all of the stiffness states. It is based on the Finger Strength test proposed by the National Institute of Standards and Technology (NIST) [25]. However, a modification has been introduced to consider specific properties of the variable-stiffness mechanism.

A force-torque sensor (Schunk model FTN-Mini-45) is used to obtain the force data. It is attached to a linear guide on one side and to a rod on the other, with a 3D-printed piece. The gripper must be attached in a fixed position that is perpendicular to the ground and can move freely to adapt to the rod, so that the readings are as accurate as possible. This set-up can be observed in Fig. 5.

The linear guide is allowed to move freely along the rod so that it can be easily adjusted to the finger length.

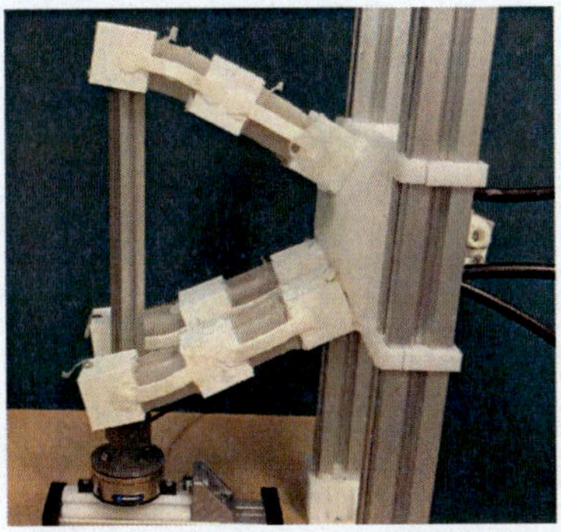

Fig. 5. Finger force set up

The gripper is mounted to a support consisting of two aluminium profiles parallel to each other. This configuration allows the fingertip to perpendicularly contact the end of the rod. In the initial position, the fingertip touches the rod without exerting any force. After the initial adjustments, the finger and rod positions must be fixed at the beginning of the tests to prevent false readings due to the displacement of the gripper or the rod. In the software, the sensor readings must be zeroed to prevent false readings.

The gripper must follow the following steps:

1. Recording of the sensor readings begins.
2. The gripper is activated and the rod is pushed.
3. Hold for 5 s.
4. The gripper is turned off.
5. Stop recording the sensor readings.

The result of the measurements is a graph with the evolution of the force during the recording of the whole experiment.

3.3 Grasp Strength Test

This part of the benchmark testing is carried out to determine two characteristics of the gripper's overall grasping strength. Firstly, the inner strength imposed on the grasped objects during the grasping cycle. Secondly, how the dynamic change of stiffness and grasping force during the grasp procedure can affect the grasped object.

To collect these data, a special artifact is used. This artifact is based on a sensorised cylinder proposed by the National Institute of Standards and Technology (NIST) [25], but modified for application to our systems.

The artifact consists of two semi-cylinders with a force/torque sensor accommodated in its interior. It is, therefore, an essential piece for the test. The artifact is shown in Fig. 6a and consists of three parts:

- Two PLA semi-cylinders. The groove permits the accommodation of the sensor and the cables. A scheme of the parts is shown in Fig. 6b.
- A force-torque sensor (Schunk model FTN-Mini-45), identical to the one in the finger force test.

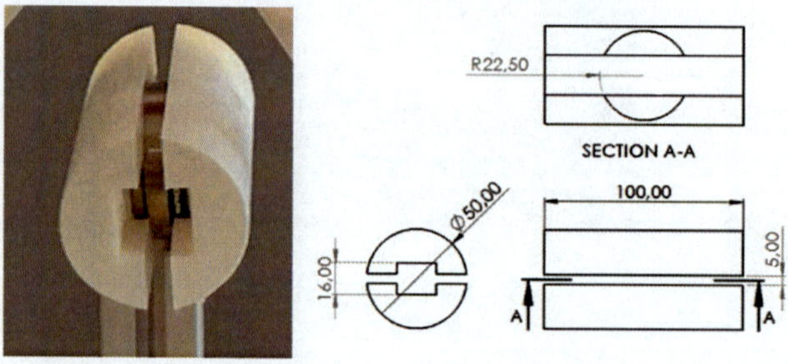

(a) Cylinder artifact (b) Cylinder scheme

Fig. 6. Grasping force cylinder

The test consists of activating the gripper to grasp the cylinder during a short period of time and recording the sensor readings for all of the different gripper stiffness states:

1. The artifact is placed over the stand and is set up so that the gripper can easily grasp it.
2. Recording of the sensor readings begins.
3. The gripper is activated and the cylinder is grasped.
4. Hold the cylinder for 5 s.
5. The cylinder is released.
6. Stop recording the sensor readings.

3.4 Gripper Payload Test

The fourth test measures the gripper payload, using the maximum weight that the gripper is capable of holding. This should be measured after verifying that the grasping of the objects can be carried out without failing. All of the different stiffness states are used.

It is important to determine the maximum weight that a gripper is capable of lifting as well as how the variations in stiffness can influence this value. With this data, it may be established if the fingers can lift more or less weight in the different stiffness states. If so, the change in the stiffness of the gripper is revealed.

The test consists of a basket welded with a cylinder for improved grasping, which has an approximate weight of 340 g.

Some bottles are gradually added to the basket in order to increase the grasped payload. The entire system is shown in Fig. 7a. The gripper will perform the grasping action to grasp the cylinder from above. The cylinder measures 60 mm in diameter per 180 mm length.

(a) Basket used for payload test with bottles

(b) Grasping the basket

Fig. 7. Payload test basket and grasping method

Initially, the grasped basket is empty. If it is successfully grasped and does not fall, one plastic bottle is added. The bottle is filled with water to reach a total weight of 335 g. This process is repeated until failure. The annotated payload will be the one taken just before the grasping failure. There are a total of 12 bottles of water with 330 mL, each measured a priori to ensure that its individual weight is approximately 335 g, with an error of 5 g. The test is repeated five times. The resulting payload will be the lowest score of the five repetitions. The gripper must follow these steps:

1. The basket is manually placed behind the gripper so that it may be grasped.
2. The grasp is performed, as in Fig. 7b.
3. The researcher's hand releases the basket.
4. Hold for 5 s.
5. The grasp is released.

The grasp is considered successful if, during the five seconds of the test, the basket does not escape from the gripper grasp.

4 Experiments

For the benchmark testing, three different grippers will undergo all four tests. The results obtained from these tests will be analysed for an improved gripper characterisation. The grippers are described below.

4.1 Variable-Stiffness Gripper (VSGrip)

This gripper, designed in [26], shown in Fig. 8, is a 3-fingered gripper. The mechanism to change the stiffness in done by an empty rectangular silicone shell filled with grounded coffee. When the void is applied into the silicone shell it passes from a soft to a rigid state, depending on the quantity of negative pressure. It has its main inspiration in the Universal Gripper stiffness change [16].

The mechanism to change the stiffness in done by an empty rectangular silicone shell filled with grounded coffee. When the void is applied into the silicone shell it passes from a soft to a rigid state, depending on the quantity of negative pressure inside the core it becomes more or less rigid.

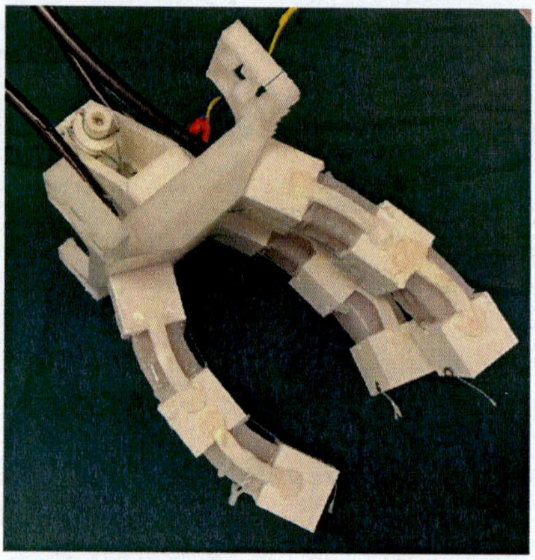

Fig. 8. Variable-stiffness gripper

4.2 Piston Gripper (PistGrip)

This is a 2-fingered gripper. Scheme is shown in Fig. 9. The main finger activation is carried out by simultaneously activating two pistons, one per finger. Therefore, the fingers encircle the object at the grasping time. The fingers have been made according to the Festo® FinGripper design. The finger is a rectangle with inner nerves that are 3D-printed in a flexible material so that they can adapt passively to the object while grasping. The mechanism required to change the stiffness consists of changing the air pressure in the activation. By altering the activation, the grasping force and movement will change accordingly.

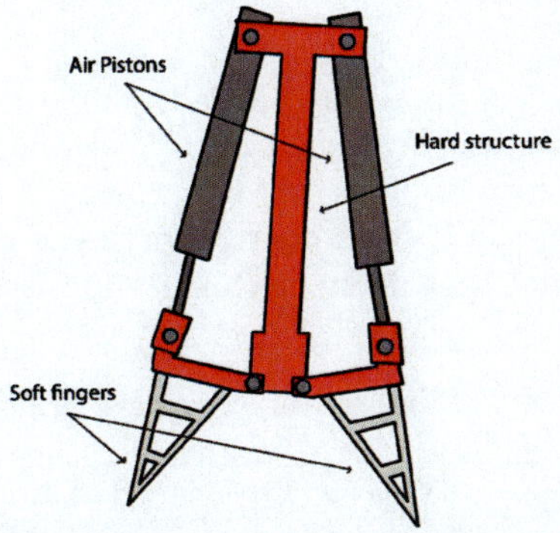

Fig. 9. Piston gripper scheme

4.3 Parallel Gripper (ParGrip)

This is a 2-finger gripper. It is integrally made of PLA, a rigid 3D-printable plastic. Its scheme is shown in Fig. 10. The inner mechanism is moved with gears and provides a parallel grasp. The gripper uses the whole finger surface when grasping rectangular objects. The mechanism to change the stiffness is carried out by changing the activation power, so that the grasping force and movement will change accordingly.

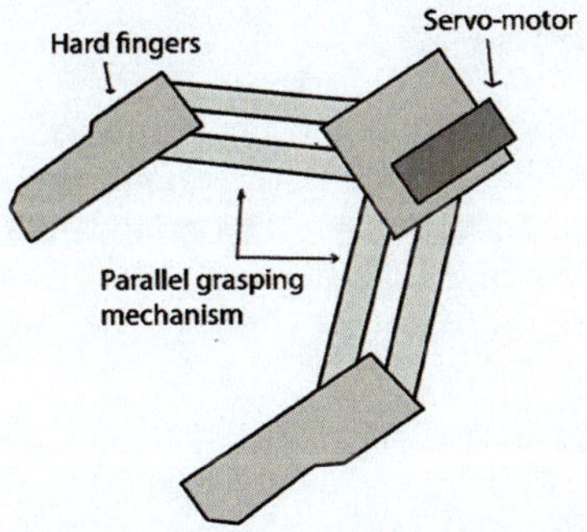

Fig. 10. Parallel gripper scheme

These two last grippers may not be considered variable-stiffness grippers, depending on the context. In some research works, they are considered to be of variable stiffness. In other situations, however, they are considered to be variable-impedance grippers, since the fingers do not change the stiffness on their own, but rather, the grasping stiffness or overall force changes.

4.4 Stiffness States

Four different stiffness states are defined for all grippers in order to conduct the experiments. The different states will help to analyse the stiffness changes of the gripper. If a gripper has a continuous stiffness change, 4 discrete states must be defined. This depends on the specific characteristics of the gripper's actuation. Table 1 presents the values that define each state for each gripper.

Table 1. Stiffness states for each gripper

	VSGrip	PistGrip	ParGrip
State 1	0 bar	2 bar	6 V
State 2	1 bar	3 bar	7 V
State 3	3 bar	4 bar	8 V
State 4	5 bar	5 bar	9 V

5 Results and Discussions

The four proposed tests have been applied to the three grippers described in the previous section. In this section we present the results along with a discussion.

5.1 Object Grasping Test

Figure 11 shows the results of the object grasping test. The results of this test reveal the adaptability of the gripper in the previously mentioned categories.

These results allow us to determine the adaptability of a specific gripper. The performance of VSGrip and PistGrip is good, except with the more difficult objects, and is consistently better than that of ParGrip.

5.2 Finger Force Test

Figure 12 shows the results of the finger force test. This test measures the variation of the force deployed by an individual finger.

Results of VSGrip show that even without reaching the highest activation state, the finger can exert a force of 4 N, as measured by the sensor. This is due to the soft materials, that provide resilience to the finger as well as the finger design.

With this design, it acts more like a more complex rigid gripper. For the other grippers, our results show that the finger force varies when changing the actuation. Thus, this test can help us to establish the initial stiffness state required before performing a grasp, saving time and resources, and helping to optimise the grasp force.

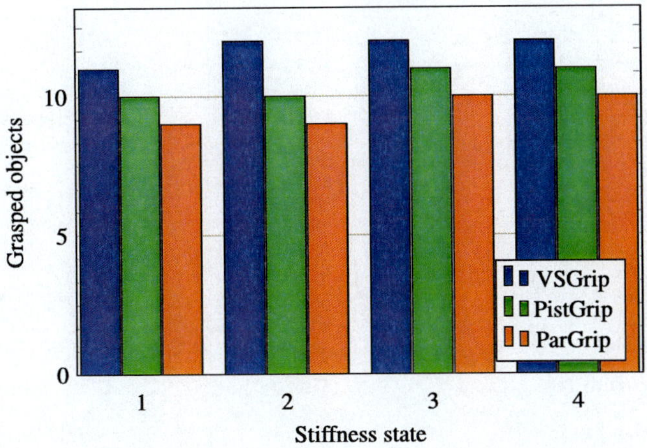

Fig. 11. Results of the object grasping test

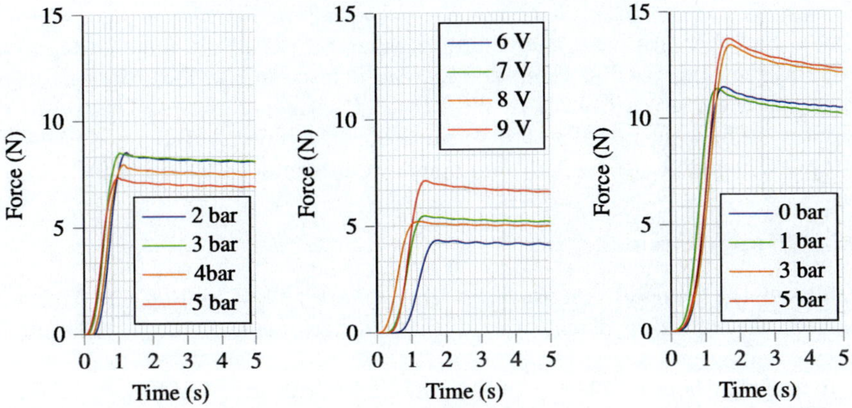

(a) PistGrip (b) ParGrip (c) VSGrip

Fig. 12. Results of the finger force test

5.3 Grasp Strength Test

Results of the grasp strength test are shown in Fig. 13.

The results of VSGrip indicate that the stiffness change of the fingers will not cause an inner force interference in the object when grasping.

Thus, the gripper's stiffness can be changed depending on the object being grasped without obstructing the manipulated object. This allows us to grasp a wide variety of objects with the same grasp plan, by simply changing the stiffness once the object has been secured by the gripper.

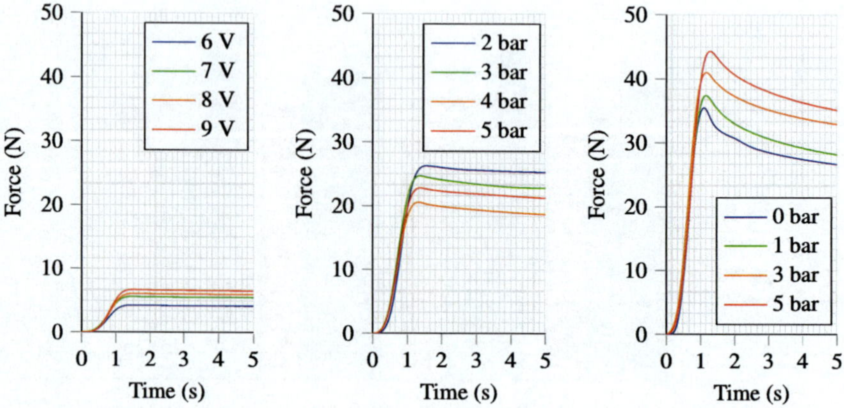

(a) ParGrip (b) PistGrip (c) VSGrip

Fig. 13. Results of the grasp strength test

For the other grippers, the results demonstrate that the finger design can alter the grasping results. From the increased force in the ParGrip with rigid fingers, to the decreased force in the PistGrip with soft fingers. This protocol, therefore, is relevant in order to determine how the object will behave when the stiffness is changed, even when the fingers are in contact with an object.

5.4 Gripper Payload Test

Results of the payload test for each gripper are shown in Fig. 14. This test analyses the variation of the weight that the gripper can lift in all stiffness states.

In general, these results reveal that the fingers can withstand more weight as the stiffness state is increased. In the case of VSGrip, since the gripper has two actuation systems, the test was carried out by deactivating the DC motor when the basket was grasped. Then, the weight of the objects was only supported by the stiffness of the fingers.

6 Conclusions

In this paper a new set of benchmark protocols has been proposed to analyse and categorise variable-stiffness grippers. Its applicability has been determined. To test the utility of each benchmark, three variable-stiffness grippers were designed and built for subsequent analysis by protocol.

The benchmark set consists of four different tests, each of which measures a distinct gripper characteristic and subsequently sorts it into one of a variety of

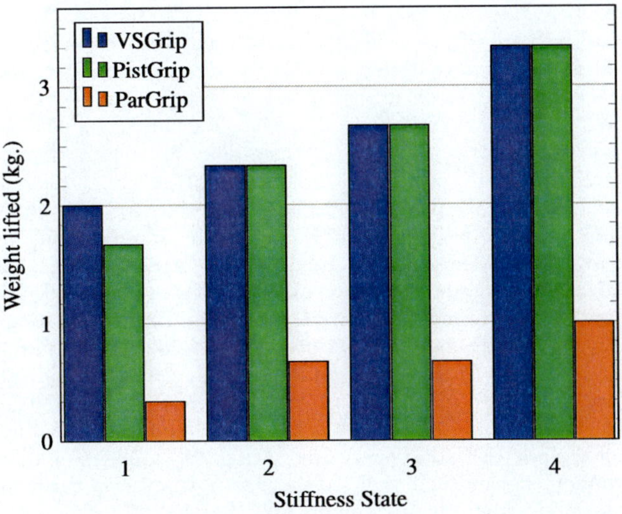

Fig. 14. Results of payload test

categories. Each test is independent from the others and can be combined without any restriction. All three grippers were subjected to the designed protocols. The results revealed informative values for all of them.

In future research, a variable-stiffness gripper will be designed and optimised for different tasks, using the benchmark protocols. This will allow us to determine how the benchmark testing can change and optimise gripper design depending on the task.

Acknowledgements. This paper describes research conducted at UJI Robotic Intelligence Laboratory. Support for this laboratory is provided in part by Universidad Jaume I (UJI-B2021-42, UJI-B2021-27), by Generalitat Valenciana (PROMETEO/2020/034) and by Ministerio de Ciencia y Teconologia (PDC2021-121011-I00).

References

1. Rodić, A., Miloradović, B., Popić, S., Spasojević, S., Karan, B.: Development of modular compliant anthropomorphic robot hand. In: Mechanisms and Machine Science, vol. 16, pp. 205–219. Springer, Cham (2014)
2. Rus, D., Tolley, M.T.: Design, fabrication and control of soft robots. Nature **521**(7553), 467–475 (2015)
3. Pfeifer, R., Lungarella, M., Iida, F.: The challenges ahead for bio-inspired "soft" robotics. Commun. ACM **55**(11), 76 (2012)
4. Rich, S.I., Wood, R.J., Majidi, C.: Untethered soft robotics. Nat. Electron. **1**(2), 102–112 (2018)
5. Pfeifer, R., Lungarella, M., Iida, F.: Self-organization, embodiment, and biologically inspired robotics (2007)

6. Mizushima, K., Oku, T., Suzuki, Y., Tsuji, T., Watanabe, T.: Multi-fingered robotic hand based on hybrid mechanism of tendon-driven and jamming transition. In: 2018 IEEE International Conference on Soft Robotics, RoboSoft 2018, Apr 2018, pp. 376–381. IEEE (2018)
7. Hao, Y., Wang, T., Fang, X., Yang, K., Mao, L., Guan, J., Wen, L.: A variable stiffness soft robotic gripper with low-melting-point alloy. In: Chinese Control Conference, CCC, July 2017, pp. 6781–6786. IEEE (2017)
8. Yang, Y., Chen, Y.: 3D printing of smart materials for robotics with variable stiffness and position feedback. In: IEEE/ASME International Conference on Advanced Intelligent Mechatronics, July 2017, pp. 418–423. IEEE (2017)
9. Memar, A.H., Mastronarde, N., Esfahani, E.T.: Design of a novel variable stiffness gripper using permanent magnets. In: Proceedings - IEEE International Conference on Robotics and Automation, May 2017, pp. 2818–2823. IEEE (2017)
10. Firouzeh, A., Paik, J.: An under-actuated origami gripper with adjustable stiffness joints for multiple grasp modes. Smart Mater. Struct. **26**(5), 055035 (2017)
11. Manti, M., Cacucciolo, V., Cianchetti, M.: Stiffening in soft robotics: a review of the state of the art. IEEE Robot. Autom. Mag. **23**(3), 93–106 (2016)
12. Cardin-Catalan, D., del Pobil, A.P., Morales, A.: Analysis of variable-stiffness soft finger joints. Adv. Intell. Syst. Comput. **867**, 334–345 (2019)
13. Shahid, Z., Glatman, A.L., Ryu, S.C.: Design of a soft composite finger with adjustable joint stiffness. Soft Robot. 1–11 (2019)
14. Firouzeh, A., Paik, J.: Grasp mode and compliance control of an underactuated origami gripper using adjustable stiffness joints. IEEE/ASME Trans. Mechatron. **22**(5), 2165–2173 (2017)
15. Li, J., Sun, M., Wu, Z., Yin, H.: Design, analysis, and grasping experiments of a novel soft hand: hybrid actuator using shape memory alloy actuators, motors, and electromagnets. Soft Robot. **7**(3), 396–407 (2020). PMID: 31905330
16. Amend, J.R., Brown, E., Rodenberg, N., Jaeger, H.M., Lipson, H.: A positive pressure universal gripper based on the jamming of granular material. IEEE Trans. Robot. **28**(2), 341–350 (2012)
17. Chen, Q., Wei, Y., Ren, T., Li, Y., Yang, Y., Chen, Y., Yan, C.: A novel, variable stiffness robotic gripper based on integrated soft actuating and particle jamming. Soft Robot. **3**(3), 134–143 (2016)
18. Yang, Y., Chen, Y., Wei, Y., Li, Y.: Novel design and three-dimensional printing of variable stiffness robotic grippers. J. Mech. Robot. **8**(6), 061010 (2016)
19. Ham, K.B., Han, J., Park, Y.J.: Soft gripper using variable stiffness mechanism and its application. Int. J. Precis. Eng. Manuf. **19**(4), 487–494 (2018)
20. Li, X., Chen, W., Lin, W., Low, K.H.: A variable stiffness robotic gripper based on structure-controlled principle. IEEE Trans. Autom. Sci. Eng. **15**(3), 1104–1113 (2018)
21. Fei, Y., Wang, J., Pang, W.: A novel fabric-based versatile and stiffness-tunable soft gripper integrating soft pneumatic fingers and wrist. Soft Robot. **6**(1), 29–31 (2018)
22. Jiang, Y., Chen, D., Liu, C., Li, J.: Chain-like granular jamming: a novel stiffness-programmable mechanism for soft robotics. Soft Robot. 1–15 (2018)
23. Ge, Q., Gu, G., Wang, D., Zhang, B., Zhang, N., Yuan, C., Hingorani, H., Ding, N., Zhang, Y.F.: Fast-response, stiffness-tunable soft actuator by hybrid multimaterial 3D printing. Adv. Funct. Mater. **1806698**, 1806698 (2019)
24. Calli, B., Walsman, A., Singh, A., Srinivasa, S., Abbeel, P., Dollar, A.M.: Benchmarking in manipulation research: using the Yale-CMU-Berkeley object and model set. IEEE Robot. Autom. Mag. **22**(3), 36–52 (2015)

25. Falco, J., Wyk, K.V., Messina, E.: Performance Metrics and Test Methods for Robotic Hands
26. Cardin-Catalan, D., Ceppetelli, S., del Pobil, A.P., Morales, A.: Design and analysis of a variable-stiffness robotic gripper. Alex. Eng. J. **61**(2), 1235–1248 (2022)

Check for updates

Evaluation of Safe Reinforcement Learning with CoMirror Algorithm in a Non-Markovian Reward Problem

Megumi Miyashita(✉), Shiro Yano, and Toshiyuki Kondo

Tokyo University of Agriculture and Technology, 2-24-16 Naka-cho, Koganei-shi, Tokyo, Japan
miyashita@livingsyslab.org, shiroyano@ieee.org, t_kondo@cc.tuat.ac.jp

Abstract. In reinforcement learning, an agent in an environment improves the skill depending on a reward, which is the feedback from an environment. For practical, reinforcement learning has several important challenges. First, reinforcement learning algorithms often use assumptions for an environment such as Markov decision processes; however, the environment in the real world often cannot be represented by these assumptions. Especially we focus on the environment with non-Markovian rewards, which allows the reward to depend on past experiences. To handle non-Markovian rewards, researchers have used a reward machine, which decomposes the original task into the sub-tasks. In those works, they assume that the sub-tasks are usually represented by a Markov decision process. Second, safety is also one of the challenges in reinforcement learning. G-CoMDS is a safe reinforcement learning algorithm based on CoMirror algorithm, an algorithm for constrained optimization problems. We have developed G-CoMDS algorithm to learn safely under environments without a Markov decision process. Therefore, the promising approach in complex situations would be decomposing the original task as the reward machine does, then solving the sub-tasks with G-CoMDS. In this paper, we provide additional experimental results and discussions of G-CoMDS, as a preliminary step of combining G-CoMDS with a reward machine. We evaluate G-CoMDS and existing reinforcement learning algorithm in the mobile robot simulation with a kind of non-Markovian rewards. The experimental result shows that G-CoMDS has the effect of suppressing the cost spike and slightly exceeds the performance of the existing safe reinforcement learning algorithm.

Keywords: Reinforcement learning · Constrained optimization · Mirror descent algorithm

1 Introduction

In reinforcement learning (RL), an agent without prior knowledge trains its skill to solve a task by interacting with a dynamic environment. The agent chooses and takes an action based on a policy, which is a strategy of the action. In

I. Petrovic et al. (Eds.): IAS 2022, LNNS 577, pp. 62–72, 2023.
https://doi.org/10.1007/978-3-031-22216-0_5

response to the action, a state of the environment transitions to the next state. Then the agent observes the state and a reward, which is a feedback from the environment. By repeating this procedure, the agent learns the optimal policy to maximize the expected cumulative reward.

RL problems can also be thought of as optimization problems. Optimization problems typically consider a known objective function and aim to search for the optimal decision variable maximizing or minimizing an objective function. However, we can sometimes obtain only the value of an objective function corresponding to a given decision variable. Such problems are called black-box optimization (BBO) problems. In particular, RL problems with a parameterized policy can be assumed as BBO problems by assigning a parameterized policy to a decision variable, and the expected cumulative reward to the value of an objective function. Assuming RL problems as BBO problems enable to relax assumptions for an environment. A popular assumption for an environment is a Markov decision process (MDP), in which the next state and reward depend on only the current state and action.

Many RL algorithms assume that an environment follows a MDP. However, in the real world, the environment might not be represented as a normal MDP. In this paper, we consider "Markovian action-state transition and non-Markovian reward function" setting which is one of the well-studied relaxation of the MDP setting. In a environment with non-Markovian rewards, the reward can depend on the past state and action. As one existing approach to handle non-Markovian rewards, there is a reward machine (RM) [1]. The RM can decompose the original task into the sub-tasks. Icarte et al. [1] assume both state transitions of the original task and the sub-tasks are Markov, hence they assumed that the environment follows MDP. However, a task sometimes has non-Markovian state transitions as well as non-Markovian rewards. In such a task, the promising approach would be (1) to decompose the original task as RM does and (2) to solve the sub-tasks with BBO. Particularly, this paper focus on the environment that returns the reward, which depends on all past states and actions, at the end of one trial, as looking toward a simple RM. Examples of this reward in the real world include running an errand, in other words, a task to reach checkpoints in the predefined order. For a task designer, it is more reasonable to give all rewards for a certain period at once than to give a reward for each interaction between the agent and the environment.

Moreover, safety is also one of the challenges in RL to apply in the real world. As the agent learns its skill by trial-and-error, the agent with a physical body might face dangerous situations, such as harming humans or objects by collision. Hence researchers have proposed safe RL algorithms in several directions [2]. We focus on approaches based on constrained optimization, by adding a constraint on a value called cost, which means the degree of danger. Gaussian CoMirror Descent Search (G-CoMDS) [3] is a safe RL algorithm without MDP based on CoMirror algorithm [4]. CoMirror algorithm is an extension of mirror descent algorithm to handle constrained optimization problems. G-CoMDS can solve BBO problems as well as RL problems, hence we consider it as a candidate of BBO algorithm to combine to RM.

From the background, in this paper, we provide additional experiments and discussion for G-CoMDS, as a preliminary step of combining G-CoMDS with a RM. The advantages of G-CoMDS are (1) handling constraints with a simple implementation; (2) applying for an environment not assumed MDP. We show the algorithms of G-CoMDS then evaluate it with a robot simulation with a kind of non-Markovian rewards.

2 Related Work

Safe RL algorithms are divided into several classes. One category of them is based on constrained optimization. Thomas et al. [5] proposed projected natural actor critic (PNAC) by extending NAC, which is a RL algorithm based on natural gradient. PNAC updates a parameterized policy while projecting into the predefined safe area over the parameter space. Achiam et al. [6] derived constrained policy optimization (CPO) based on trust region policy optimization (TRPO) [7]. CPO guarantees the satisfaction of the constraints approximately at each iteration and monotonic improvement theoretically. These algorithms are based on MDP assumption, hence they are effective in environments that follow MDP. On the other hand, CoMDS can solve RL problems in environments that do not follow MDP, and BBO problems.

A non-Markovian reward decision process (NMRDP) [8] is a decision process with a non-Markovian reward, which is determined depending on the past states. NMRDP and non-Markovian rewards have been considered in the field of temporal logic, which is a logic that changes depending on time. In recent years, the algorithms for them have also developed in RL. Icarte et al. [1] assumed known reward structure such as where subgoal is. They introduced RM, which is an extension of a finite automaton to output a reward, as a description of an environment. Furthermore, based on a classical RL algorithm Q-learning [9], they proposed Q-Learning for RM (QRM) as an algorithm for solving RM. Gaon and Brafman [10] were interested in unknown reward structure. They proposed four RL algorithms that combine RL algorithms Q-Learning and R-MAX [11] with two automaton learning algorithms L∗ [12] and Evidence Driven State Merging (EDSM) [13]. Q-learning is a model-free RL algorithm, which learns from experience without a transition probability. On the other hand, R-MAX is a model-based RL algorithm, which estimates a transition probability before learning. In the RL algorithm with EDSM, EDSM learns a reward structure offline by merging state spaces after some trials according to the RL algorithm. In the RL algorithm with L∗, L∗ provides a reward structure online while the RL algorithm learns a task. Neider et al. [14] pointed out that prior full knowledge about reward structure provides fast convergence but is unrealistic, while convergence is slow without prior knowledge. To deal trade-off between convergence and prior knowledge, they proposed AdvisoRL as an RL algorithm in which humans provide advice. Wen and Topcu [15] proposed a safe RL algorithm based on a cross-entropy method and demonstrated an experimental evaluation on an environment with non-Markovian rewards and costs. These studies deal with problem settings similar to our study, however, each approach is different. It is beneficial to have various approaches because different approaches are likely to have different strengths depending on problems and assumptions.

3 Preliminaries and Problem Setting

In RL, the agent aims to find the policy to maximize the expected cumulative reward, as mentioned in Sect. 1. This section provides the problem setting as preliminaries for Sect. 4.

First, let us denote a state by $\boldsymbol{s} \in \mathcal{S}$ and an action by $\boldsymbol{a} \in \mathcal{A}$. A policy $\pi : \mathcal{S} \to \mathcal{A}$ is a function that maps a state into an action. The function is usually represented by the probability distribution $\pi_{\boldsymbol{\theta}} (\boldsymbol{a} \mid \boldsymbol{s})$ or the deterministic function $\boldsymbol{a} = \pi_{\boldsymbol{\theta}} (\boldsymbol{s})$, both varying from the policy parameter $\boldsymbol{\theta}$.

Second, we consider the interaction between the agent and the environment. Let $\boldsymbol{S}_t$ and $\boldsymbol{A}_t$ be a state and an action at the time step t, respectively. When the agent executes the action $\boldsymbol{A}_t = \boldsymbol{a}$ following the policy in the state $\boldsymbol{S}_t = \boldsymbol{s}$, an environment transitions the next state $\boldsymbol{S}_{t+1} = \boldsymbol{s}'$ and returns a reward $R_t \in \mathbb{R}$ according to a reward function $r : \mathcal{S} \times \mathcal{A} \to \mathbb{R}$. Now, $\sum_{t=0}^{T} R_t$ denotes the cumulative reward after the time step $t = 0$. The policy is evaluated by the expected cumulative reward.

Third, we introduce a concept of an *episode* and a *trajectory*. An episode is the period from an initial state to a terminal state. A trajectory is the sequence of the states and actions in an episode, which is denoted as $\boldsymbol{\xi} := (\boldsymbol{S}_0, \boldsymbol{A}_0, \boldsymbol{S}_1, \boldsymbol{A}_1 \ldots, \boldsymbol{A}_{T-1}, \boldsymbol{S}_T)$. The trajectory varies from the policy function $\pi_{\boldsymbol{\theta}}$. In other words, the trajectory is parameterized by $\boldsymbol{\theta}$, hence the probability distribution over the trajectory can be represented as $p(\boldsymbol{\xi} \mid \boldsymbol{\theta})$. Similarly, because the cumulative reward is determined from the trajectory, we can represent the cumulative reward as a function of $\boldsymbol{\xi}$, as $j(\boldsymbol{\xi}) := \sum_{t=0}^{T} R_t$.

Moreover, the expectation of $j(\boldsymbol{\xi})$ can be written as $J(\boldsymbol{\theta}) = \int_{\boldsymbol{\xi}} p(\boldsymbol{\xi} \mid \boldsymbol{\theta}) j(\boldsymbol{\xi}) d\boldsymbol{\xi}$ because the trajectory follows the distribution $p(\boldsymbol{\xi} \mid \boldsymbol{\theta})$. Under the standard problem settings of the RL algorithms, it is assumed that the agent does not have the mathematical formulation of $p(\boldsymbol{\xi} \mid \boldsymbol{\theta})$. Accordingly, the agent does not know the mathematical formulation of $J(\boldsymbol{\theta})$ explicitly too. Thus, it sounds difficult to find the optimal $\boldsymbol{\theta}$. However, such a problem is known as the BBO problem, zeroth order optimization problem and so on, and there exists some algorithms for them. Our fundamental problem setting in this paper is to search the optimal distribution $p(\boldsymbol{\theta})$ regarding Eq. (1), not the policy parameter $\boldsymbol{\theta}$ directly.

$$\mathcal{J} = \int_{\boldsymbol{\theta}} p(\boldsymbol{\theta}) J(\boldsymbol{\theta}) d\boldsymbol{\theta}. \tag{1}$$

We further introduce constraint condition in this problem setting as noted in the next paragraph.

In RL with constraints, the environment returns a cost $C_t \in \mathbb{R}$ as well as the reward R_t. The cost represents the degree of constraint violation and follows a cost function $c : \mathcal{S} \times \mathcal{A} \rightarrow \mathbb{R}$. Here, we introduce the expected cumulative cost $D(\boldsymbol{\theta}) = \int_{\boldsymbol{\xi}} d(\boldsymbol{\xi}) p(\boldsymbol{\xi} \mid \boldsymbol{\theta}) d\boldsymbol{\xi}$, where $d(\boldsymbol{\xi}) := \sum_{t=0}^{T} C_t$. As the agent searches $p(\boldsymbol{\theta})$ as noted above, we define the evaluative criteria for $p(\boldsymbol{\theta})$ as Eq. (2).

$$\mathcal{D} = \int_{\boldsymbol{\theta}} p(\boldsymbol{\theta}) D(\boldsymbol{\theta}) d\boldsymbol{\theta} \tag{2}$$

In this paper, we consider the following constrained RL problem:

$$p^*(\boldsymbol{\theta}) = \underset{p(\boldsymbol{\theta})}{\arg\max}\, \mathcal{J} \quad \text{s.t.} \quad \mathcal{D} \leq \epsilon \tag{3}$$

The threshold $\epsilon \in \mathbb{R}$ can be designed to fit the task. Note that $D(\boldsymbol{\theta}) \leq \epsilon$ is not always satisfied even if we choose $p(\boldsymbol{\theta})$ which satisfies $\mathcal{D} \leq \epsilon$. The reason is that $\boldsymbol{\theta}$ is sampled stochastically, and $\mathcal{D}$ is the expectation of $D(\boldsymbol{\theta})$.

4 CoMirror Descent Search

Miyashita et al. [3] proposed CoMirror Descent Search (CoMDS) to solve the constrained RL problem shown in Eq. (3). Let the policy parameters at k-th iteration be $\boldsymbol{\theta}_{k,1}, \ldots, \boldsymbol{\theta}_{k,L}$, where $l = 1, \ldots, L$ are the indexes of the sampled candidates (rollouts). Then CoMDS calculates discretized distribution $\boldsymbol{q}_k := [p(\boldsymbol{\theta}_{k,1}), \ldots, p(\boldsymbol{\theta}_{k,L})]$ and updates $\boldsymbol{q}_k$ to $\boldsymbol{q}_k + 1$.

The update equation in k-th iteration is as follows:

$$\boldsymbol{q}_{k+1} = \underset{\boldsymbol{q} \in \mathcal{Q}}{\arg\max} \left\{ \langle \boldsymbol{E}_k, \boldsymbol{q} \rangle - \eta B_\phi(\boldsymbol{q}, \boldsymbol{q}_k) \right\}, \tag{4}$$

$$\text{where } \boldsymbol{E}_k = \begin{cases} \boldsymbol{J}_k & \text{if } \frac{1}{L}\sum_{l=1}^{L} \mathbb{1}[D_{k,l} \leq \epsilon] \geq \delta \qquad (5) \\ \boldsymbol{D}_k & \text{otherwise,} \qquad (6) \end{cases}$$

where η is a learning rate, and B_ϕ is the Bregman divergence associated with ϕ. The first term $\langle \boldsymbol{E}_k, \boldsymbol{q} \rangle$ in Eq. (4) approximates $\mathcal{J}$ or $\mathcal{D}$ using $\boldsymbol{q}_k$. The second term encourages new parameter $\boldsymbol{q}$ nearby the old one $\boldsymbol{q}_k$. Then η balances both two effects.

In Eqs. (5) and (6), $\boldsymbol{J}_k := [J(\boldsymbol{\theta}_{k,1}), \ldots, J(\boldsymbol{\theta}_{k,L})]$ is the vector of expected cumulative reward, and $\boldsymbol{D}_k = [D_{k,1}, \ldots, D_{k,L}] := [D(\boldsymbol{\theta}_{k,1}), \ldots, D(\boldsymbol{\theta}_{k,L})]$ is the vector of expected cumulative cost. In Eq. (5), $\mathbb{1}[D_{k,l} \leq \epsilon]$ becomes 1, only if $D_{k,l} \leq \epsilon$ is satisfied, where ϵ is the threshold of the expected cumulative cost. Thus, Eqs. (5) and (6) mean that if the policy satisfies the constraint with probability $\delta \in [0, 1]$, the gradient becomes $\boldsymbol{J}_k$, otherwise $\boldsymbol{D}_k$. Note that the parameter ϵ is defined by user depending on the task specification.

In the following experiments, we introduced the Gaussian CoMDS (G-CoMDS), which is the simplified version of the CoMDS. The G-CoMDS introduces an additional assumption that the policy parameter follows a Gaussian distribution of mean $\boldsymbol{\mu}$ and variance $\boldsymbol{\Sigma}$, and J is interpreted as $j(\boldsymbol{\xi})$ by using an online mirror descent technique [16]. Algorithm 1 shows G-CoMDS algorithm.

Algorithm 1 Gaussian CoMirror Descent Search

initialize
  Mean of policy distribution $\boldsymbol{\mu}_1$, variance of policy distribution $\boldsymbol{\Sigma}$,
  parameter ϵ, and $\delta \in [0,1]$ arbitrarily.
for iteration $k \in \{1, \ldots, K\}$ **do**
  for sampling $l \in \{1, \ldots, L\}$ **do**
    Sample policy parameter $\boldsymbol{\theta}_l \sim \mathcal{N}(\boldsymbol{\theta} \mid \boldsymbol{\mu}_k, \boldsymbol{\Sigma})$.
    Sample trajectory $\boldsymbol{\xi} \sim p(\boldsymbol{\xi} \mid \boldsymbol{\theta}_l)$.
    Evaluate policy parameter $J_{k,l} = j(\boldsymbol{\xi})$, and $C_{k,l} = c(\boldsymbol{\xi})$.
  end for
  Update policy distribution with Eq. (4)
  Estimate the mean $\boldsymbol{\mu}_{k+1}$ from $\boldsymbol{q}_{k+1}$.
end for

5 Experiments

5.1 Experiment with Non-Markovian Reward and Cost

We demonstrated a task based on `Safexp-PointGoal1-v0` in Safety Gym [17] (Fig. 1). In `Safexp-PointGoal1-v0`, a mobile robot learns to reach a goal while avoiding dangerous objects. In a field, one goal, eight dangerous objects, and one vase are placed randomly. When the robot reaches the goal, the next goal appears at a random place.

Fig. 1. The task that a red mobile robot learns to reach a goal (green cylinder) while avoiding dangerous objects (blue circles). The robot observes the goal, the dangerous objects, and a vase (light blue rectangular) with a lidar. The lidar observations are visualized over the robot in a ring shape. Approaching the goal provides a reward, and touching the dangerous objects provides a cost. Touching the vase has no effect.

In the original task, the reward and the cost can be observed at each step. In our task, we modified the reward and the cost to non-Markovian by giving the cumulative reward and the cumulative cost at only the end of the episode.

A state and action are the same as the original task. The state consists of the following 60-dimensional sensors: each 16-dimensional lidar for observing the goal, the dangerous objects, and the vase; and three dimensions for each of acceleration, velocity, gyro, and geomagnetic sensor. The action consists of two control commands toward a front-back direction and a rotation direction.

We compared our algorithm G-CoMDS which does not assume MDP environment and the existing algorithm CPO which assumes MDP environment. The comparison was held from the following perspectives: (1) the possibilities of learning and its performances in the environment with non-Markovian reward and cost and; (2) the performances in the environment with non-Markovian reward and cost compared to that with Markovian. Moreover, G-MDS, an unconstrained algorithm related to G-CoMDS, was added in the comparison as a reference. We set ϵ, the value of cost violating a constraint, as 45 and 90 in both CPO and G-CoMDS except for Markovian reward and cost.

The policy is represented as a fully connected neural network with two hidden layers with 100 units, respectively. In addition, CPO needs two value functions for the reward and the cost, hence we represented them in the same structure as the policy network.

The hyper-parameters of all algorithms were tuned by using Optuna [18]. The parameters for G-MDS were searched to maximize the reward. On the other hand, the parameters for G-CoMDS and CPO were searched as a multi-objective optimization problem, maximizing the reward and minimizing the cost.

5.2 Result

The result is shown in Fig. 2. Figure 2a shows the reward of G-MDS was higher than G-CoMDS. Figure 2b shows that the cost were similar between G-MDS and G-CoMDS at the end of the learning. However, only the cost of G-MDS spiked around time step 1.0×10^6. As Fig. 2a and c illustrate, G-MDS showed the highest reward, followed in order by G-CoMDS ($\epsilon = 90$), G-CoMDS ($\epsilon = 45$), CPO ($\epsilon = 90$), and CPO ($\epsilon = 45$). Figure 2d illustrates the cost of CPO is higher than that of G-CoMDS. Figure 3 shows the performance of CPO in Markovian reward and cost was significantly better than that of non-Markovian. On the other hand, the performance of G-CoMDS was almost unchanged between Markovian reward and cost, and non-Markovian.

5.3 Discussion

First, we discuss the result of G-MDS and G-CoMDS. The similar cost between G-MDS and G-CoMDS at the end of learning may have been caused by the task structure. The task structure in this context means the compatibility between the reward and the cost; i.e. the behavior increasing the reward decreases the cost simultaneously. On the other hand, the reward of G-MDS was higher than that of G-CoMDS. In other word, unexpectedly, G-MDS performed better than

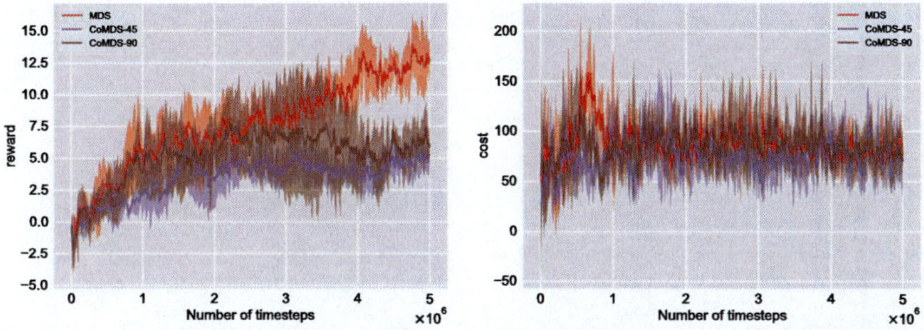

(a) Reward (MDS variant algorithms) (b) Cost (MDS variant algorithms)

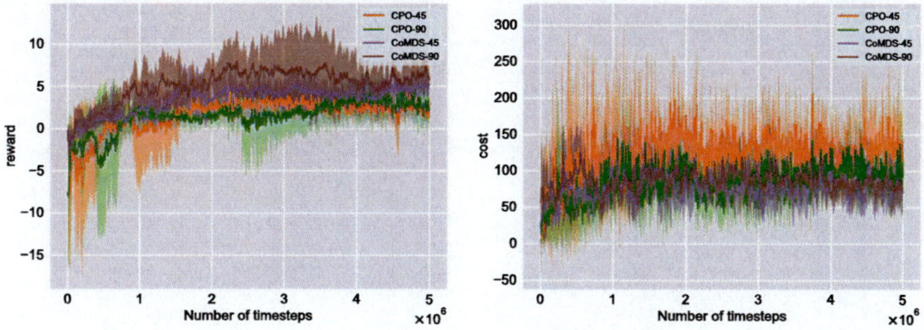

(c) Reward (algorithms for constraint) (d) Cost (algorithms for constraint)

Fig. 2. The performances of the RL algorithms in the environment with non-Markovian reward and cost. The green color shows G-MDS, the red color shows G-CoMDS ($\epsilon = 45$), the purple color shows G-CoMDS ($\epsilon = 90$), the blue color shows CPO ($\epsilon = 45$), and the orange color shows CPO ($\epsilon = 90$). These experiments were executed five times in the different seeds. The solid line represents the average, and the light-colored area means the standard deviation.

G-CoMDS. It seems to arise from that G-MDS performs various searches without persisting in constraints. At least for this task, we might obtain a better solution by choosing the algorithm that does not persist in constraints strongly or interfere with exploration.

Second, we discuss the result of CPO and G-CoMDS. From the result, the performance of G-CoMDS was slightly higher than that of CPO. We suppose the result is as G-CoMDS does not assume MDP for the environment. In addition, assumptions of MDP seem to be important for CPO because the performance of Markovian reward and cost is significantly better than that of non-Markovian in CPO. In the algorithms assumed MDP such as CPO, the performance may degrade as the assumptions are reduced, e.g., the non-Markovian state transition. Therefore, we expect the algorithms based on BBO such as G-CoMDS are stably used even in the real world, in which the fulfillment of MDP assumptions is uncertain. In addition, regarding the number of parameters, CoMDS

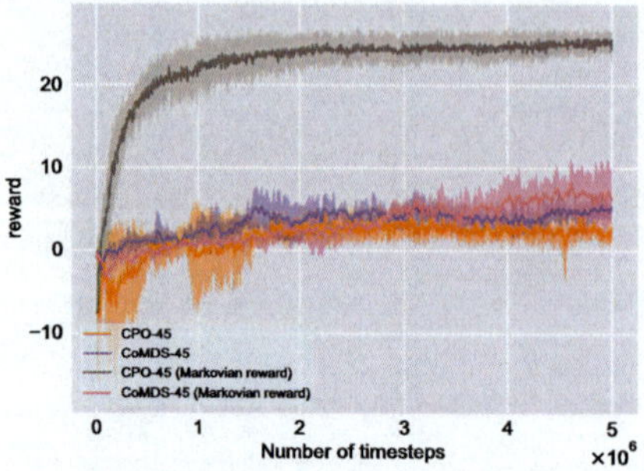

(a) Reward ($\epsilon = 45$)

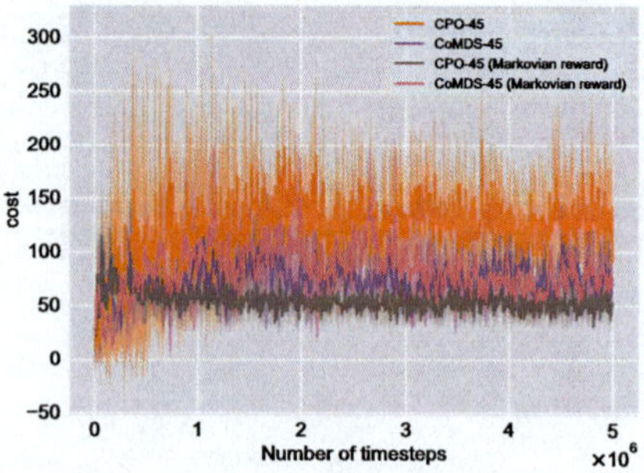

(b) Cost ($\epsilon = 45$)

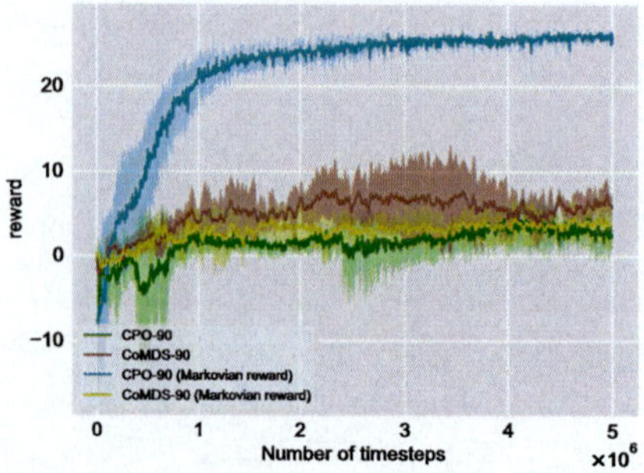

(c) Reward ($\epsilon = 90$)

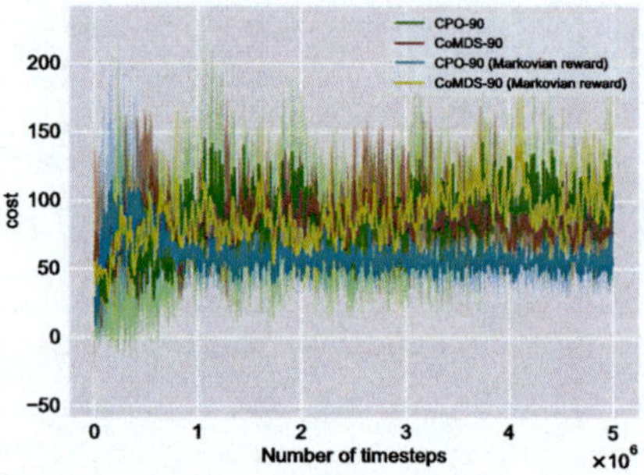

(d) Cost ($\epsilon = 90$)

Fig. 3. The performances of the RL algorithms in the environment with non-Markovian reward and cost, and that of Markovian. For the environment with Markovian reward and cost, the gray color shows CPO ($\epsilon = 45$), the pink color shows G-CoMDS ($\epsilon = 45$), the light blue color shows CPO ($\epsilon = 90$), and the light green color shows G-CoMDS ($\epsilon = 90$).

only requires the parameters for the policy function. On the other hand, CPO requires the parameter of the value function as well as the policy function. Thus, CoMDS has the advantage of having fewer parameters.

6 Conclusion

In this paper, we provided additional experiments and discussions for G-CoMDS [3], a safe RL algorithm proposed by Miyashita et al. [3]. In the experiment, we evaluated G-CoMDS by the robot simulation with a non-Markovian reward. The result shows that G-CoMDS has the effect of suppressing the cost spike and slightly exceeds CPO performance.

In the proposed method, the higher the dimension of the policy parameter, the larger the computational effort. In other words, because the dimension of the policy parameter is depending on the dimension of state, that of action, and the representation of the policy, the computational effort is also depending on them. However, because it would be possible to parallelize sampling during each iteration, the sum of the computational effort itself will keep, but the parallelization of a large number of sampling reduces the actual calculation time.

In the future, we will address the following topics. First, we hope that solving the sub-tasks in RM by a BBO algorithm such as CoMDS will provide effective learning. We will work on the implementation of a combining RM and BBO. Second, when we handle an environment that does not follow MDP, it is necessary to choose an appropriate algorithm while understanding non-Markovian elements such as a non-Markovian reward and a non-Markovian state transition. We also need to select an algorithm with an understanding of task characteristics. Therefore, we plan to investigate and improve our algorithm while clarifying the positioning of our algorithm in the existing algorithms.

Acknowledgement. This work was partially supported by JSPS KAKENHI (Grant numbers JP26120005, 17KK0064, JP18K19732, and JP17K12737).

References

1. Icarte, R.T., Klassen, T., Valenzano, R., McIlraith, S.: Using reward machines for high-level task specification and decomposition in reinforcement learning. In: International Conference on Machine Learning, PMLR, pp. 2107–2116 (2018)
2. García, J., Fernández, F.: A comprehensive survey on safe reinforcement learning. J. Mach. Learn. Res. **16**(1), 1437–1480 (2015)
3. Miyashita, M., Kondo, T., Yano, S.: Reinforcement learning with constraint based on mirror descent algorithm. Results Control Optim. **4**, 100048 (2021)
4. Beck, A., Ben-Tal, A., Guttmann-Beck, N., Tetruashvili, L.: The comirror algorithm for solving nonsmooth constrained convex problems. Oper. Res. Lett. **38**(6), 493–498 (2010)
5. Thomas, P.S., Dabney, W.C., Giguere, S., Mahadevan, S.: Projected natural actor-critic. In: Advances in Neural Information Processing Systems, pp. 2337–2345 (2013)
6. Achiam, J., Held, D., Tamar, A., Abbeel, P.: Constrained policy optimization. In: Proceedings of the 34th International Conference on Machine Learning, JMLR. org, vol. 70, pp. 22–31 (2017)
7. Schulman, J., Levine, S., Abbeel, P., Jordan, M., Moritz, P.: Trust region policy optimization. In: International Conference on Machine Learning, pp. 1889–1897 (2015)
8. Bacchus, F., Boutilier, C., Grove, A.: Rewarding behaviors. In: Proceedings of the Thirteenth National Conference on Artificial Intelligence - Volume 2. AAAI'96, Portland, Oregon, pp. 1160–1167. AAAI Press (1996)
9. Watkins, C.J., Dayan, P.: Q-learning. Mach. Learn. **8**(3–4), 279–292 (1992)
10. Gaon, M., Brafman, R.: Reinforcement learning with non-Markovian rewards. In: Proceedings of the AAAI Conference on Artificial Intelligence, vol. 34, pp. 3980–3987 (2020)
11. Brafman, R.I., Tennenholtz, M.: R-max - a general polynomial time algorithm for near-optimal reinforcement learning. J. Mach. Learn. Res. **3**, 213–231 (2002)

12. Angluin, D.: Learning regular sets from queries and counterexamples. Inf. Comput. **75**(2), 87–106 (1987)
13. Lang, K.J., Pearlmutter, B.A., Price, R.A.: Results of the Abbadingo one DFA learning competition and a new evidence-driven state merging algorithm. In: International Colloquium on Grammatical Inference, pp. 1–12. Springer (1998)
14. Neider, D., Gaglione, J.R., Gavran, I., Topcu, U., Wu, B., Xu, Z.: Advice-guided reinforcement learning in a non-Markovian environment. In: Proceedings of the AAAI Conference on Artificial Intelligence, vol. 35, pp. 9073–9080 (2021)
15. Wen, M., Topcu, U.: Constrained cross-entropy method for safe reinforcement learning. IEEE Trans. Autom. Control 1 (2020)
16. Miyashita, M., Yano, S., Kondo, T.: Mirror descent search and its acceleration. Robot. Auton. Syst. **106**, 107–116 (2018)
17. Ray, A., Achiam, J., Amodei, D.: Benchmarking Safe Exploration in Deep Reinforcement Learning (2019)
18. Akiba, T., Sano, S., Yanase, T., Ohta, T., Koyama, M.: Optuna: a next-generation hyperparameter optimization framework. In: Proceedings of the 25th ACM SIGKDD International Conference on Knowledge Discovery & Data Mining, pp. 2623–2631 (2019)

Validation of *Shared Intelligence* Approach for Teleoperating Telepresence Robots Through Inaccurate Interfaces

Gloria Beraldo[1,2](✉), Luca Tonin[1], Amedeo Cesta[2], Emanuele Menegatti[1], and José del R. Millán[3]

[1] Department of Information Engineering, University of Padua, Padua, Italy
Gloria.Beraldo@dei.unipd.it, Gloria.Beraldo@istc.cnr.it, Luca.Tonin@dei.unipd.it, Emanuele.Menegatti@dei.unipd.it

[2] Institute of Cognitive Sciences and Technologies, National Research Council of Italy (CNR-ISTC), Rome, Italy
Amedeo.Cesta@istc.cnr.it

[3] Department of Electrical and Computer Engineering and the Department of Neurology, University of Texas at Austin, Austin, TX 78712, USA
jose.millan@austin.utexas.edu

Abstract. Telepresence robots can support people with special needs (e.g., who cannot move) to remotely interact with people and the environment at a distance. In this application, people can communicate with the robot via alternative channels of communication, such as brain-machine interfaces, that are less accurate than the traditional mediums and allow the user to send limited classes of commands to the robot. To overcome these limitations, *shared intelligence* approaches are born to fuse the user's inputs with a sort of intelligence on the robot, that interprets the user's commands with respect to the environment and provides the robot deliberative ability in choosing the next action to implement. In this paper, we investigate how a *shared intelligence* system could be affected by the kind of inaccurate user's input interfaces. For this purpose, we compare a brain-machine interface vs. a more reactive keyboard endowed with the same percentage of noise. Overall, the results revealed comparable navigation performance in the two conditions except for the accuracy (e.g., number of target positions reached), indicating similar assistance derived by the system. However, differences between the two modalities are found by correlating the performance of the system to the navigation situation, suggesting a different user's inclination (more in control vs. robot's autonomy) with respect to the interface's responsiveness and the target to reach, and encouraging to adapt the *shared intelligence* system according to the real-time user's ability and the surrounding environment.

Authors Emanuele Menegatti and José del R. Millán equally contribute to this paper.

I. Petrovic et al. (Eds.): IAS 2022, LNNS 577, pp. 73–84, 2023.
https://doi.org/10.1007/978-3-031-22216-0_6

Keywords: Robot teleoperation · Brain-machine interfaces · Neurorobotics · Behavior based architecture

1 Introduction

In the last years, in response to the COVID pandemic, telepresence robots have regained importance as a medium to allow remote interaction. Telepresence robots are robotic platforms generally endowed with a display and camera, driven by a remote operator to virtually interact with people. For their nature, telepresence robots are an inclusive tool promoting social interaction, assistance and support, and can increase the quality of life for people with special needs, for instance, who cannot move and/or are affected by serious disabilities [1], that often communicate with low throughput human-machine interfaces. For instance, considering people with motor, sensory and cognitive impairments, in the last decades, it has spread the tendency of using neurophysiological signals to measure and understand human intention and convert them into control signals for the robot, setting the birth of a new inter-disciplinary field, the neurorobotics [2]. In this scenario, brain-machine interfaces (BMIs) have allowed healthy and disabled people to successfully drive different categories of robots, including telepresence robots [3]. Nevertheless, despite the recent advancements in signals processing, BMIs are still too noise interfaces due to the non-stationary of neural signals, that can be translated only into few classes of user's commands. As a consequence, the use of these communication channels to control telepresence robots requires high workload for the user making their application unfeasible outside a laboratory environment for a long-term period. Furthermore, the tendency of investigating new methods to correctly decode user's intentions from brain signals has led to neglect of the last advancements in robotics and human-robot interaction. Indeed, traditionally, the robot has a passive role and a limited autonomy that consists of directly implementing the user's commands. This aspect increases the risk of achieving undesired behaviors due to the noise and intrinsic low bitrate of BMIs.

Given these considerations, it is emerging the necessity of developing advanced robotic semi-autonomous solutions, to interpret the user's commands with respect to the context and maximize the efficacy and the assistance provided by the robot to the user [4]. Among them, on the one side, *shared control* approaches allow the robot reactively adjust motion variables (e.g., speed, acceleration, direction) from a user's motion command (e.g., to avoid collision) [5], on the other *shared autonomy* algorithms make the robot implement pre-defined procedures in autonomy under the human's supervision [6]. Although the latest studies have demonstrated the effectiveness of these shared approaches in augmenting the human-robot interaction than direct teleoperation, in both these approaches, the robot is once again very subordinated to the human limiting its functionalities in the *perception* and *act* phases.

Recently, we have proposed a third level of interaction, called *shared intelligence*, where a sort of robot's intelligence intervenes at the *decision process* and it is equally fused with the user's commands to establish the next robot's action [7]. The first tests have already demonstrated that our system supports the users in teleoperating the robot and allows them to achieve a real-time navigation performance via BMI comparable to one through keyboard thanks to the active role assigned to the robot. In this paper, we provide an extended validation of our *shared intelligence* approach by investigating whether the navigation performance is affected by the time taken by the user to interact with the robot in the case of inaccurate user's input interfaces. Indeed, since any BMI is characterised by low bit rate, we hypothesise that the intrinsic delay in the communication could influence the performances of the system and the interaction when the user cannot correct the robot's motion in time. Thus, the rationale behind this study is to assess the performance of our *shared intelligence* system with imprecise user's input interfaces for disabled people supporting the same classes of commands and characterised by different interaction times, but "noisy" in the same way. To verify our hypothesis, we have evaluated our system with two imprecise user's input interfaces: the brain-computer interface vs. a keyboard affected by noise where the user's commands were artificially modified according to the subject-dependent BMI performance to simulate the limited user's capability in controlling the robot. In other words, we simulated a BMI with a keyboard without delay (e.g., the users can immediately correct the robot just by pressing a key), that is only characterised by uncertainty in the communication. Moreover, coherently with the kind of commands deliverable via BMIs, in both cases, the user can interact with the robot from only two-classes of commands that make the robot turn left or right.

The remaining paper is organized as follows. Section 2 is dedicated to explain the key features of our *shared intelligence* system. Then, Sect. 3 reports the methods exploited to verify our hypothesis, including the details of the examined user's interfaces, our telepresence platform, the involved subjects and the experimental setup. Finally, the preliminary results in both modalities are reported in Sects. 4 and 5 concludes the paper.

2 Shared Intelligence

This section aims to describe the main features behind our *shared intelligence* system. As we previously anticipated, the main novelty of our approach is to concurrently mix the user's inputs delivered via a specific interface and the robot's intelligence based on its awareness of the situation, giving the robot the opportunity of taking the user over if needed. In our system, the fusion of the two contributions is managed by a set of *policies*, where each of them is in charge of managing a specific source of information affecting the robot's action. Specifically, in the current version of the system, we include the *obstacle-avoidance policy* for representing the robot's awareness of obstacles, the *user input policy* for considering the effect of the user's commands, the *distance policy* for including the preferable distance spanned by the robot while moving from an iteration to the next, and finally, the *direction policy* evaluating the preferable direction the robot should keep. According to the examined information, each *policy* returns a probabilistic grid that stores the probability of the most probable cell towards which the robot should move. Then, all the probabilistic grids in output from the four *policies* are equally fused by simply multiplying them in the same way (e.g., same weight) resulting in a new probability distribution Σ_p:

$$\Sigma_p = \prod_{k=1}^{4} p_k \tag{1}$$

p_k is modelled with a specific function according to the source of information handled by the *policy* (e.g., the *obstacle-avoidance policy* is inverse proportional to the distance between the robot and the closest obstacle). Please refer to [7] for further details about the implementation of each *policy* p_k.

Finally, the most probable cell in Σ_p, that we call *subgoal*, is provided in input to the navigation system that, in turn, plans the best plan to move the robot towards it. An illustrative representation of the pipeline behind our system is shown in Fig. 1.

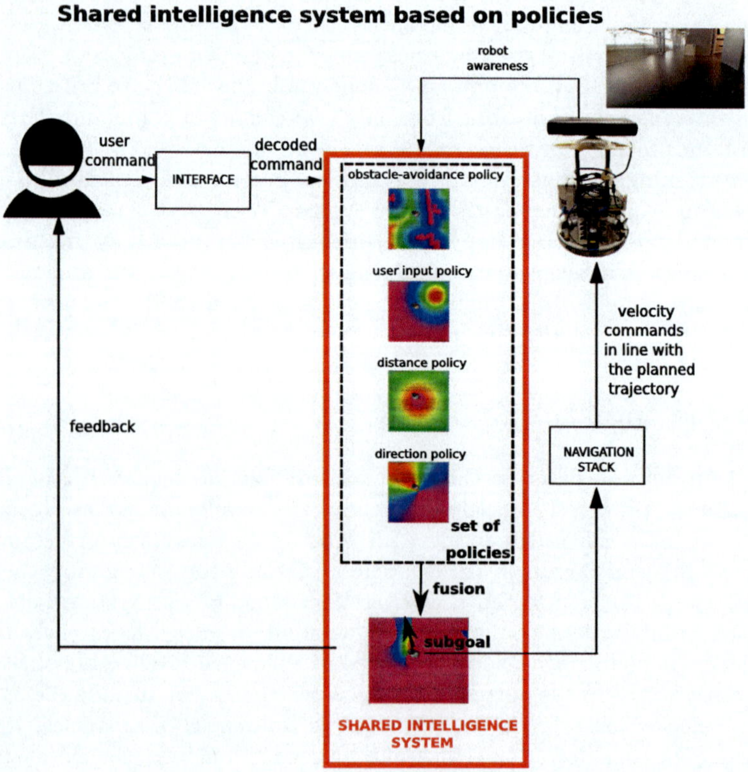

Fig. 1. Schematic representation of our *shared intelligence* system based on *policies.*

Considering the trade-off between the necessity of updating the robot's awareness to take into account the incoming information, from the environments and the user, and the related computational burden, our system computes the *subgoal* with a certain frequency that is determined by the occurrence of the following events: (a) a new user's command is sent, (b) the robot reaches the *subgoal*, and (c) a failure happens in the robot's navigation.

3 Methods

3.1 Participants

We recruited thirteen healthy people (S1–S13, 28.92 ± 5.18, years old, 5 females). Among them, four participants have already taken part to other BMI experiments, but nobody had previous experience in controlling a telepresence robot via BMI. All the participants were required to sign a written informed consent in accordance with the principles of the Declaration of Helsinki. The study was approved by the Cantonal Committee of Vaud (Switzerland) for ethics in human research under the protocol number PB 2017-00295.

3.2 User's Interfaces

In this work, we have examined two kinds of imperfect user's interfaces for driving the telepresence robot: (a) a 2-class motor imagery based BMI (both hands vs. both feet motor imagination that are mapped respectively to turn left and right the robot); (b) a 2-class keyboard affected by errors that we called *noisy keyboard*. In the former, the two turning commands were detected by the BMI system according to a classifier calibrated on each participant. In the latter (*noisy keyboard*), a certain percentage of noise was introduced in the user commands, delivered through a keyboard, according to the individual performance achieved in the BMI task. Indeed, since we used the same performance achieved by each user in the BMI in the case of *noisy keyboard*, the two modalities theoretically differentiate only in terms of the typology of the interface and hence the time required to the user to deliver a new command.

Specifically, in that modality, some user commands were artificially missed (not delivered to the robot) or swapped (left command become right and viceversa). In other words, the *noisy keyboard* simulates the limited participants' ability in controlling the robot, achieved from the BMI performance, with a more reactive interface such as a keyboard.

As regards the brain-machine system, we recorded the EEG signals with an active 16-channel amplifier 512 Hz sampling rate (g.USBamp, Guger Technologies, Graz, Austria), band-pass filtered within 0.1 and 100 Hz and notch-filtered 50 Hz. We considered the following electrodes configuration Fz, FC3, FC1, FCz, FC2, FC4, C3, C1, Cz, C2, C4, CP3, CP1, CPz, CP2, CP4 according to the international 10–20 system layout and then, we pre-processed the acquired signals by using a Laplacian spatial filter. Afterwards, we continuously computed the power spectral density (PSD) via Welch's periodogram algorithm 2 Hz resolution, in the frequency range from 4 to 48 Hz, in 1 s sliding window every 62.5 ms. The most discriminative features (channel-frequency pairs, subject-specific) were provided in input for training a Gaussian classifier on the subject-specific dataset achieved during the *calibration phase* [8].

In the *evaluation phase*, features were converted into a probability distribution over the two MI classes (imagination of both hands vs. both feet) every 62.5 ms. We rejected the probabilities below 0.55. Then, we integrated the raw posterior probabilities over time to accumulate evidence of the user's intention based on an exponential smoothing [9], and, finally, a new command was delivered to the system when reaching a threshold. The integrated posterior probabilities were also communicated to the user through a visual feedback.

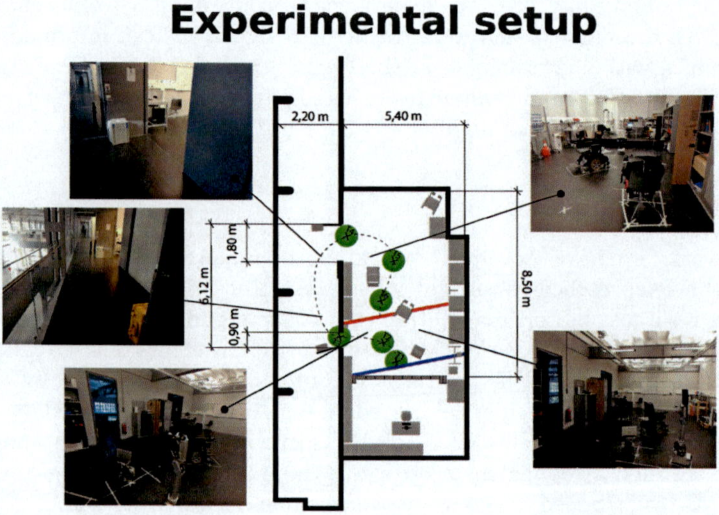

Fig. 2. Representation of the experimental setup for the navigation task, including free areas, obstacles, passage through two doors and a crossroad. The task starts from the blue line. Then, we identified seven targets positions represented with green circles and the red line (e.g., finish line) the user should reach in the shortest possible time.

3.3 Telepresence Platform

We exploited Robotino by Festo AG as telepresence platform (see Fig. 1). We mounted a laser range finder (Hokuyo URG-04LX-UG01, range 240°, distance 5600 mm) at the height of 53.5 mm and a 3D camera (Microsoft Kinect v1, field of view 43° vertical × 57° horizontal) at the height of 490 mm for obstacle detection. Furthermore, we provided the RGB streaming from the camera as feedback of the field of view of the robot to the user. The robot was equipped with an additional laptop (Dell XPS15 9570, Intel Core i7-8750H 2.20 GHz, 16 GB RAM) to overcome the limited computational power of the embedded PC, that communicated via Ethernet interface.

3.4 Experiment

The experiments were split into two sessions (e.g., days): the first was dedicated to acquire the data to train the subject-specific BMI decoder, the second to robot's teleoperation based on *shared intelligence* via both the examined input interfaces mentioned in Sect. 3.2. Specifically, each participant was involved in three *calibration runs* (60 trials, 30 per class) and three-five *evaluation runs* (40–60 trials, 20–40 per class) to test the BMI performances. Participants were sitting in front of a monitor and instructed by symbolic cues about the motor imagery task to be performed. A positive visual feedback was provided to participants during the *calibration*, while it changed coherently with the user's performance in doing the mental task during the *evaluation*. During the robot teleoperations, we asked participants to drive the mobile robot in an office environment that has typical indoor navigation situations such as narrow passages, doorways, furniture as obstacles and open spaces.[1] Participants teleoperated the robot without looking it directly but relying on the video stream from its camera. Furthermore, subjects were instructed to reach seven targets positions (T1–T7) (see Fig. 2) in order and in the shortest possible time. A target was reached with a confidence interval of 0.5 m and within the established timeout (e.g., such a timeout is equal to 240 s for targets T1 and T6; 300 s for T2–T5). When a target was missed, the navigation task continued from the next one.

Each participant completed six navigation runs (546 targets in total): four via BMI and two via *noisy keyboard* as input modalities. We compared both modalities to an ideal trajectory (dashed line in Fig. 2) achieved in a separated experiment where the robot has been continuously controlled via joystick (14 repetitions).

To compare the two conditions, we consider the typical navigation metrics in these applications such as (1) the *accuracy* (i.e., the percentage of the targets correctly reached), (2) the *Fréchet distance* (i.e., a traditional measure of similarity between the trajectories in the two modalities and the ideal one. It measures the smallest of the maximum pairwise distances [10]), (3) *path length* (i.e., the length of the robot's path by summing the distances between every sequential pair of points in the path), (4) *time* (i.e., the time taken to reach each target) and (5) *number of commands* (i.e., the number of user's intervention).

4 Results

First of all, we report the results related to the *BMI evaluation runs* where the robot was not involved in the experiment. In this phase, on average, participants

[1] An illustrative video is available at https://www.youtube.com/watch?v=FYUFP8prY6w.

achieved an accuracy of 69.30 ± 10.60% and they were able to send a command within 4.91 ± 1.11 s.

Then, we focused on analysing the data collected during the navigation runs, and we compared the two examined modalities (e.g., *shared intelligence* via BMI vs. *noisy keyboard*) to verify the presence of significant differences. Figure 3 (a)–(b) represent the heat maps of trajectories followed by all participants in the two cases. The color palette varies from blue (less common trajectories) to yellow (most common). The maps have a 5 cm resolution and they report the start and the finish lines (T7) in white, targets T1–6 with white circles markers. Furthermore, the average trajectories are displayed with a dashed white lines, while the ideal with the dashed black ones.

From a qualitative evaluation of the heat maps, overall, the trajectories appear similar and coherent with the ideal, even if those achieved through the BMI deviate more than the ones via the *noisy keyboard*. In particular, there is a major variance in the trajectories in the BMI modality for targets T2, T4 and T5. Coherently with our expectations, the average navigation accuracy were remarkably different in the two conditions: 79.94 ± 8.32% in the case of BMI, 98.9% ± 1.55% in the case of *noisy keyboard* (e.g., mean ± SD). Nevertheless, by considering the distribution of the examined navigation metrics, we did not find any statistical difference[2] between the two conditions (see Fig. 3(c)–(f)).

Then, to have a more detailed overview of how the kind of user's interface can affect the navigation performance of our *shared intelligence system*, we analysed the examined metrics per each target (e.g., navigation situation). The results are shown in Fig. 4.

The analysis of the accuracy (per each target) identified the targets T3 and T4, namely the doors, as the most demanding. For these targets, also in the *noisy keyboard* modality, participants achieved accuracy of less than 100%. These results indeed are related to the ability of the user in sending the correct turning command at the right time that might be challenging in the case of noisy teleoperation especially with low/medium performance. For these two targets we found significant difference between the two conditions (Wilcoxon rank-sum test, T3: $p = 0.001 < 0.01$; T4 $= p$: $0.0001 < 0.001$) as well as for the targets T5 ($p = 0.026 < 0.05$) and T6 ($p = 0.0122 < 0.05$). These results suggest that in the case of *noisy keyboard* the user was able to promptly correct the robot when its behavior was in contrast with his/her intent. This outcome is supported also by evaluating the Fréchet distance, the length of the path and the time required in Fig. 4(b)–(d): these metrics take smaller values in the case of *noisy keyboard* than *BMI* navigation.

Furthermore, coherently with the previous qualitative observations from the heat maps, the trajectories are significantly closer to the ideal one and shorter in the *noisy keyboard* than *BMI* condition for targets T2 (Fréchet distance: $p = 0.0023 < 0.01$; Path length: $p = 0.0221 < 0.05$), T4 ($p = 0.0067 < 0.01$; Path length: $p = 0.0084 < 0.01$) and T5 ($p = 0.0002 < 0.001$; Path length: $p = 0.0005$

[2] For statistical analyses, we applied non-parametric tests because data were not normal distributed (*Kolmogorov-Smirnov* test $p \geq 0.05$).

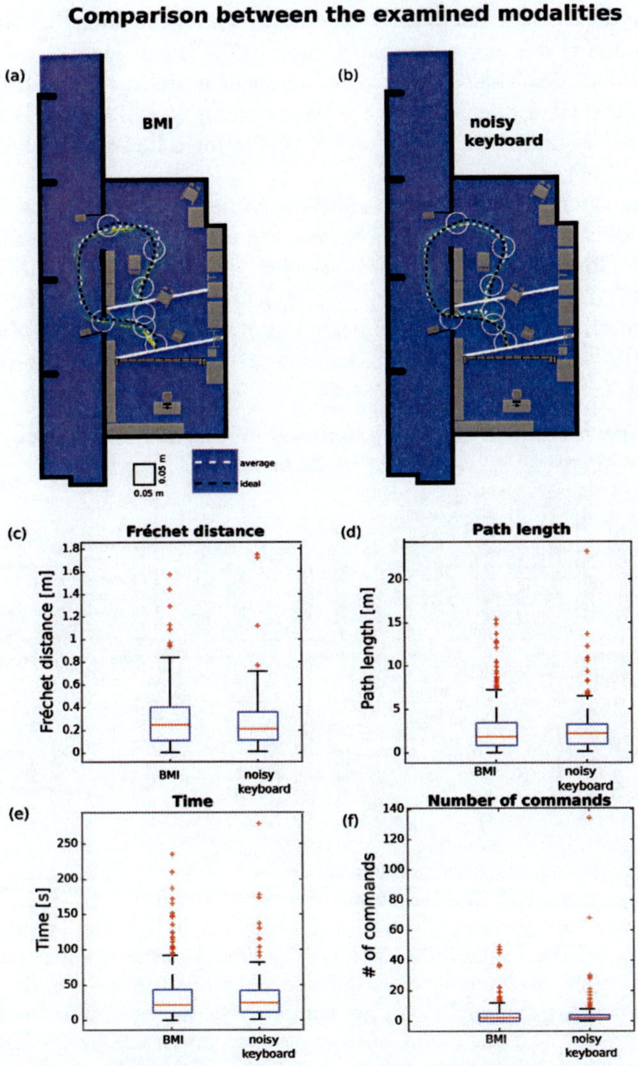

Fig. 3. Comparison of the performance of our *shared intelligence system* through the two examined modalities: the BMI and the *noisy keyboard.* (a)–(b) Heat map of the trajectories performed by the robot in the two conditions. Maps resolution is 5 cm. The color palette varies from blue (low) to yellow (in common with the most). The start and the finish line (T7) are reported in white, targets T1–6 are represented with white circles. The average trajectory for the BMI and *noisy keyboard* navigation are shown in white dashed line, while the ideal in the black dashed line. (c) Boxplot of the distribution of the Fréchet distance in m to the ideal trajectory in the two modalities. (d) Boxplot of the distribution of the length of the path in m in the two modalities. (e) Boxplot of the distribution of the time spent in seconds in the two modalities. (f) Boxplot of the distribution of the number of commands delivered in the two modalities. The box edges signify the 75th (top) and 25(th) (bottom) percentiles and the horizontal line represents the median of the corresponding distribution. The whiskers extend to the largest and smallest nonoutlier values. Outliers are marked with red crosses.

< 0.001). Target T3 represents an exception in which the length of the path was significantly higher for *noisy keyboard* than *BMI* (Path length: $p = 0.0419 < 0.05$), due probably to the temptation of the user in delivering many commands in few time (the time needed with the keyboard is less than BMI) that makes the robot turn around itself for the effect of the introduced noise and the strict passage (see also Fig. 4).

Regarding the time spent, the results are in line with the length of the path: the longer trajectory is routed by the robot, the more is the duration for reaching the target. We found significant differences for targets T3 ($p = 0.015 < 0.05$), T4 ($p = 0.009 < 0.01$) and T5 ($p = 0.002 < 0.01$). However, the ratio between times in the two conditions (*shared intelligence* based on *BMI* and *shared intelligence* based on *noisy keyboard*) is 115.97%, meaning that the two conditions differ very little in terms of time.

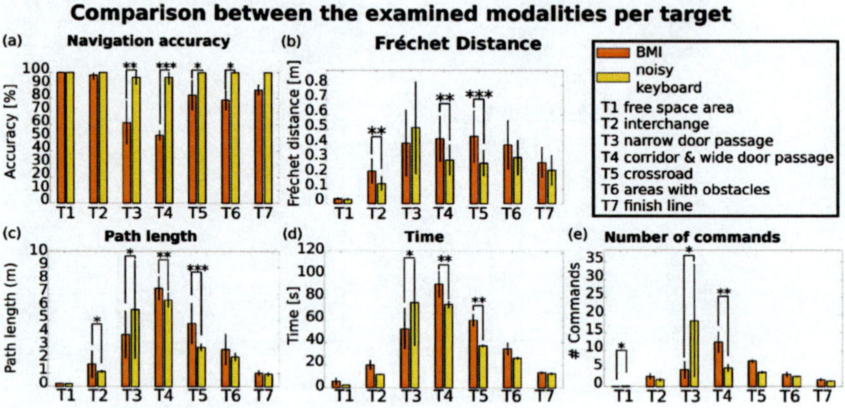

Fig. 4. (a) Navigation accuracy per modalities per each target, corresponding to the percentage of run in which the target was successfully reached. (b) Fréechet distance to the ideal trajectory per modalities to reach each target. (c) The length of the path involved per modalities to reach each target. (d) Time in seconds spent per modalities to reach each target. (e) Number of commands per modalities delivered to reach each target. In the computation of (b)–(e), we compute the corresponding metrics only for the reached targets. The mean and the standard deviation are shown for all metrics. Statistically significant differences are shown with two-sided Wilcoxon rank-sum tests, (*): p <0.05; (**): p < 0.01, (***): p <0.001.

Finally, the number of commands delivered by the user is generally higher for *BMI*, due apparently to wrong commands sent especially while reaching targets T2, T4 ($p = 0.0052 < 0.01$), T5 in which we have already highlighted the presence of extra routes by evaluating the heat maps and lengths of path. Targets T1 (free space) and T3 (narrow door passage) are exceptions in which the users delivered significantly more commands in *noisy keyboard* than *BMI* navigation (T1: $p = 0.0267 < 0.05$; T3: $p = 0.0132 < 0.05$), as we have already commented, probably for the tendency of the user in sending many inputs in a short time to correct the robot behavior as soon as possible.

5 Conclusion

In this paper, we focus on telepresence robots controlled by inaccurate interfaces, and we compare the performance of our *shared intelligence* system with two kinds of inaccurate user's interfaces affected by the same percentage of noise, but with different bit-rates: (a) a 2-classes brain-machine interface vs. (b) a 2-classes *noisy keyboard*. Our purpose consists of verifying how the responsiveness of the user's interface can affect the navigation based on a *shared intelligence* system. With respect to other common approaches in the literature, such a system equally fuses different information with the same weight, giving the same importance to the user and the robot in the *decision process*.

In our experiments, the navigation performances appear different in the two modalities, respectively equal to 79.95% in *BMI* modality and 98.9% in the *noisy keyboard*. We assume that both conditions are affected by the same participants' ability in controlling a noisy interface (we modified the keyboard by introducing the percentage of errors achieved during the *BMI evaluation*). From this point of view, it is worth mentioning that also in the *BMI evaluation* where the robot was not involved, participants achieved low performance – about 70% of accuracy on average – probably due to the lack of experience in controlling BMI acquired in few training sessions. However, literature reports evidence that MI-BMI users are able to improve their performances by adapting to the system and learn the right strategy of performing the mental task by repeating it for many sessions [11,12]. Based on these findings, we speculate that the performance will be higher in *BMI evaluation* and in the navigation based on both modalities if the subjects would be trained for more days.

Therefore, we can deduce that such differences in accuracy strongly depend on the time required to deliver a new command in the two interfaces. More precisely, since in our system we do not use any procedure designed ad-hoc to trigger challenging targets, the performances are related to the user's ability in sending the corresponding turning commands in the correct time window that is surely influenced by the responsiveness of the interaction interface. Beyond this consideration about the accuracy, overall, the considered navigation metrics (Fréchet distance, path length, time and number of commands) revealed no statistical differences between the two conditions, suggesting that our system provides the same assistance from both user's interfaces.

Nevertheless, by correlating the performances with respect to the navigation situation (e.g., per target), some significant differences between the conditions appear. From this point of view, in accordance with our hypothesis, it has clearly emerged the different responsiveness of the two interfaces and the different user's inclinations in interacting with the robot (e.g., more in control vs. more robot's autonomy) that coherently impact the accuracy, Fréchet distance, path length, time and number of commands. On the one side, in the *noisy keyboard*, participants were able to more promptly correct the undesired behaviors than BMI, reducing the time to reach targets, the path length and the deviation from the ideal trajectory (e.g., Fréchet distance). On the other, it appeared that they had a major inclination for delivering a higher number of commands in a short time than BMI, with the consequence of not allowing the robot to properly re-arrange its context awareness and, causing undesired robot's rotations, especially in the free space area T1 and in the doorway passage T3.

From these preliminary results, future works will deeply investigate how to adapt the contribution of the *shared intelligence* considering the online user's performance (e.g., tendency of sending a lot of wrong commands or not, user's ability in controlling the robot) and the surrounding environments (e.g., navigation situation) also examining other inaccurate user's interfaces and robotic platforms.

Acknowledgments. This research was partially supported by M.I.U.R. under the initiative "Departments of Excellence" (Law 232/2016) and by Fondazione Ing. Aldo Gini. G.B. and A.C. are supported by "SI-Robotics: SocIal ROBOTICS for active and healthy ageing" project (Italian M.I.U.R., PON - Ricerca e Innovazione 2014–2020 - G.A. ARS01 01120).

References

1. Zhang, G., Hansen, J.P.: Telepresence robots for people with special needs: a systematic review. Int. J. Hum. Comput. Interact. 1–17 (2022)
2. Arbib, M.A., Metta, G., van der Smagt, P.: Neurorobotics: from vision to action. In: Springer Handbook of Robotics, pp. 1453–1480 (2008)
3. Tonin, L., Millán, J.D.R.: Noninvasive brain–machine interfaces for robotic devices. Annu. Rev. Control Robot. Auton. Syst. **4** (2020)
4. Tonin, L., Leeb, R., Tavella, M., Perdikis, S., Millán, J.D.R.: The role of shared-control in BCI-based telepresence. In: 2010 IEEE International Conference on Systems, Man and Cybernetics, pp. 1462–1466. IEEE (2010)
5. Abbink, D.A., et al.: A topology of shared control systems-finding common ground in diversity. IEEE Trans. Hum. Mach. Syst. **48**(5), 509–525 (2018)
6. Schilling, M., et al.: Towards a multidimensional perspective on shared autonomy. In: Proceedings of the 2016 AAAI Fall Symposium Series (2016)
7. Beraldo, G., Tonin, L., Millán, J.D.R., Menegatti, E.: Shared intelligence for robot teleoperation via BMI. IEEE Trans. Hum. Mach. Syst. (2022)
8. Leeb, R., et al.: Transferring brain–computer interfaces beyond the laboratory: successful application control for motor-disabled users. Artif. Intell. Med. **59**(2), 121–132 (2013)
9. Gardner, E.S., Jr.: Exponential smoothing: the state of the art—part II. Int. J. Forecast. **22**(4), 637–666 (2006)
10. Eiter, T., Mannila, H.: Computing discrete Fréchet distance (1994)
11. Carmena, J.M.: Advances in neuroprosthetic learning and control. PLoS Biol. **11**(5) (2013)
12. Perdikis, S., Tonin, L., Saeedi, S., Schneider, C., Millán, J.D.R.: The Cybathlon BCI race: successful longitudinal mutual learning with two tetraplegic users. PLoS Biol. **16**(5), e2003787 (2018)

Uncertainty Estimation for Safe Human-Robot Collaboration Using Conservation Measures

Woo-Jeong Baek(✉), Christoph Ledermann, and Torsten Kröger

Karlsruhe Institute of Technology (KIT) Institute for Anthropomatics and Robotics - Intelligent Process Automation and Robotics (IAR-IPR), Karlsruhe, Germany
baek@kit.edu, christoph.ledermann@kit.edu, torsten@kit.edu

Abstract. We present an online and data-driven *uncertainty* quantification method to enable the development of safe human-robot collaboration applications. Safety and risk assessment of systems are strongly correlated with the accuracy of measurements: Distinctive parameters are often not directly accessible via known models and must therefore be measured. However, measurements generally suffer from uncertainties due to the limited performance of sensors, even unknown environmental disturbances, or humans. In this work, we quantify these *measurement uncertainties* by making use of *conservation measures* which are quantitative, system specific properties that are constant over time, space, or other state space dimensions. The key idea of our method lies in the immediate data evaluation of incoming data during run-time referring to conservation equations. In particular, we estimate violations of a-priori known, domain specific conservation properties and consider them as the consequence of measurement uncertainties. We validate our method on a use case in the context of human-robot collaboration, thereby highlighting the importance of our contribution for the successful development of safe robot systems under real-world conditions, e.g., in industrial environments. In addition, we show how obtained uncertainty values can be directly mapped on arbitrary safety limits (e.g., ISO 13849) which allows to monitor the compliance with safety standards during run-time.

1 Introduction

Human-Robot Collaboration (HRC) systems have gained much attention in recent years: while enabling human workers to fully focus on more complex tasks, simple and repetitive ones are supposed to be performed by the robot. However, realizing a shared work space is bound to assuring a *safe* environment throughout the run time. Existing literature in this domain as [1–3] focus

The authors would like to thank Tamim Asfour for his guidance and fruitful discussions throughout this work.

I. Petrovic et al. (Eds.): IAS 2022, LNNS 577, pp. 85–102, 2023.
https://doi.org/10.1007/978-3-031-22216-0_7

on hazard identification methods in simulation or based on pre-defined system models. The central assumption behind these approaches is that the measurement uncertainty in real-world scenarios is negligibly small such that findings obtained through models can be transferred to real-world applications. We, in contrast, argue that ignoring the existence of measurement uncertainties might lead to misinterpretations: dangerous situations could be classified as safe due to underestimated probabilities for undesired shifts. Specifically in HRC systems, the robot parameters are usually accessible via control options while human parameters such as the position must be determined via measurements. Here, the actual human position might be closer to the robot as provided by the corresponding measurement device, such that the distance between human and robot could be underestimated. Depending on the robot velocity, the human might not be able to safely avoid the collision. Reversely, knowing the actual amount of measurement uncertainties would allow to prevent risks by adjusting parameters accordingly. For instance, an automatic reduction of the robot velocity could be initiated when exceeding a pre-defined threshold in the human localization uncertainty.

In this work, we present an online and data-driven method to quantify measurement uncertainties by leveraging domain-specific knowledge. In particular, we propose to make use of so-called *conservation properties* of the system which are measures that are constant over state space parameters. In the context of human joint position detection, one conservation property is given by the non-changing distance between human joints. Formulating equations representing these conservation properties, we estimate violations thereof by means of incoming data. As we view the origin of such violations as the consequence of measurement uncertainties, we quantify latter ones by treating the data statistically.

The scientific novelty of our framework is given by its online and data-driven character: By immediately analyzing incoming data with respect to formulated conservation equations and mapping the obtained uncertainty on arbitrary safety requirements (e.g., ISO 13849), we provide the possibility to monitor the compliance with safety standards online. In order to demonstrate the applicability our approach, we perform experiments for human-pose detection. The tolerated uncertainty is calculated in the context of standard ISO 13849, which used for safety verification purposes in industrial robotic applications. After presenting state-of-the-art methods for uncertainty determination in HRC, we introduce our methodology in Sect. 3. To validate our approach, we apply it on data collected with the safety scanner SICK3000 and compare our findings with the uncertainty stated in the corresponding sensor data sheet. Further, we discuss how the uncertainty quantification performs without taking conservation equations into account. In the last step, we demonstrate how the obtained uncertainty value can be mapped on safety requirements and thereby discuss the importance and limitations of our approach for safe HRC (Fig. 1).

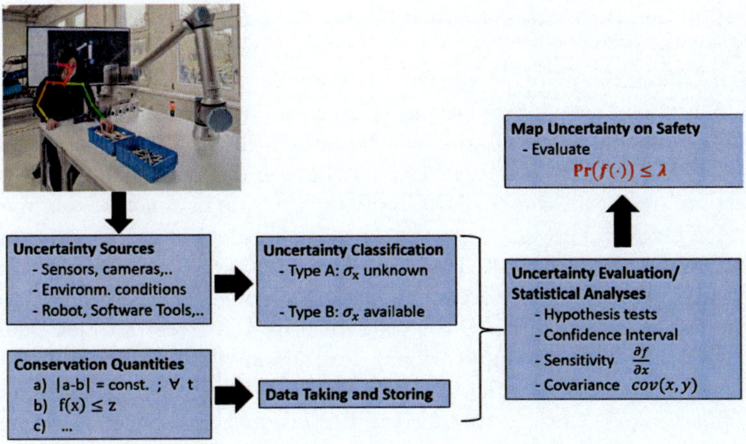

Fig. 1. Combining the knowledge of uncertainty sources and conservation properties of a robot system with measurement data allows to effectively quantify uncertainties. Treating obtained data with statistical tools, existing dependencies are explored. Results are discussed in the context of safety compliance by referring to user-defined limits (e.g., ISO standards).

2 Uncertainty and Safety in HRC

One of the main challenges in attaining safe HRC systems lies in steadily satisfying pre-defined limits. However, as highlighted in [4], current safety engineering approaches are inadequate for flexible HRC. The authors state that existing safety assurance methods are well suited for traditional automation systems, but cannot be applied for intelligent or adaptive collaborative setups. In this work, we address this issue by presenting an online uncertainty quantification method. We claim that providing an uncertainty value throughout run-time is a crucial step towards flexible and safe HRC. For instance, the study of Bartneck et al. in [5] highlights the necessity of an interpretable value for human beings to communicate the risk level. Contributions in the domain of safe HRC as [1–3,6] present hazard identification methods by modeling the system beforehand. While Askarpour et al. in [1] present a formal verification method build upon a logic language to detect possible risks of the system based on its model, Inam et al. search for potential hazards in [3] and generate a list of such by referring to a simulation setup. Similarly, Araiza et al. in [7] present a simulation-based verification method by developing a test generation approach based on different system variables. However, the authors of these contributions assume employed models to be accurate as they form the basis for suggested search strategies. In fact, uncertainties arising from the measurement process in real world environments are not taken into account. The survey in [8] discusses additional methods addressing the development of safe HRC systems, reaching from motion planning approaches to the consideration of psychological aspects. However, in contrast to our contribution, above works strongly focus on developing active search methods for possible risks.

In terms of human-involved systems, existing approaches as [9–12] address the question how the knowledge on human injuries can be embedded in robot control frameworks to avoid critical harms on the human being. To do so, these works refer to parameters as the critical impact force and stress on different human body parts to evaluate whether HRC systems are safe or must be adapted with respect to control and design strategies to meet pre-defined safety requirements. One basic assumption of these contributions however is the accuracy of critical parameters which are decisive for decision-making. Therefore, uncertainties due to the measurement process the critical force and stress are not taken into account. In terms of robot applications, several contributions focus on the consideration of uncertainties in motion planning. In [13], Thrun et al. introduce a variety of techniques to capture uncertainties originating from robot kinematics and to incorporate them in control frameworks. By presenting different representations of uncertainties, it is discussed how manipulation tasks can be accomplished in the face of uncertainties. Highlighting the need of online uncertainty tracking, several contributions as [14–17] suggest to treat robot manipulation tasks in probabilistic manner. Specifically, the authors present methods where robot tasks are modeled in the belief space. Here, each state is represented by a tuple of a state and the corresponding control input. By estimating subsequent states with a Bayesian framework, where the knowledge about the system state is combined with measurement updates, above works address the development of planning algorithms by taking robot uncertainties into account. Although suggested methods treat parameters in probabilistic manner, uncertainties of measurement updates originating from sensor devices are not explored in detail. To be specific, the authors note that the formulation of measurement updates considering possible uncertainties is challenging for unstructured environments: Uncertainties due to sensors and measurement devices as well as their correlations with each other are either completely neglected or modeled as Gauss distributions. This assumption however is not applicable in all cases: In particular, the accumulated uncertainty resulting from the interplay of different sources might show a more complex behavior. In the context of safety, such incorrect assumptions could affect the decision making process in dangerous manner such that accurately quantifying uncertainties becomes highly crucial.

Generally, the contribution of Giancola et al. in [18] shows the highest similarity to our work: By presenting methods to determine the uncertainty behavior of three camera devices, the influence of environmental disturbances and system parameters is discussed in detail. Obtained results on correlations allow the user to adjust system parameters accordingly and to thus keep the uncertainty on a desired level.

Motivated by the fact that in existing literature, scarce attention is devoted to explicitly quantifying measurement uncertainties, we develop a framework which enables their online quantification during run-time. Furthermore, our method allows to map the uncertainty values directly on arbitrary safety-limits (e.g.,

ISO 13849). We adapt terminologies and methods introduced in [19]. A detailed overview of definitions and analysis techniques will be given in Sect. 3. We apply our method on human position detection in a real-world environment to estimate the probability for the occurrence of dangerous failures per hour (PFDH), a distinctive measure stated in ISO 13849 [20]. We thereby address following research questions: (1) to which degree does our method provide accurate uncertainty results?, and (2) how can we quantify uncertainties of black box tools in the context of safety? To the best of our knowledge, this work is the first contribution in this field that (1) tackles measurement uncertainty quantification and propagation to evaluate robotic applications with respect to functional safety requirements, and (2) discusses how measurement uncertainties could contribute to the development of flexible and safe HRC systems.

3 Methodology

3.1 Problem Statement

Generally, developing a safe system is related to satisfying an equation of the form

$$Pr(f(\cdot)) \leq \lambda, \tag{1}$$

where $Pr(\cdot)$ denotes the probability, $f(\cdot)$ a user-defined constraint and λ an arbitrary limit (e.g., ISO 13849 [20]). At the same time, safety is achieved by minimizing the risks of the system. According to ISO 12100 [21], the term *risk* can be written as

$$risk_{system} = \sum_i severity(i) \cdot Pr(i). \tag{2}$$

Here, i stands for an incident and *Pr(i)* for the probability of its occurrence. In our work, we assume the severity, which can be for example biomechanical limits of the human's bones, as a constant and known quantity. Thus, the risk minimization can be achieved by minimizing the probability:

$$\min(risk_{system}) \equiv \min(Pr(i)). \tag{3}$$

However, the probability for the occurrence of a dangerous outcome originates from the incomplete knowledge of the system behavior, specifically from the range of possible deviations from the expected outcome. As indicated in Sect. 2, the *measurement uncertainty* u, which we will refer to with the term *uncertainty* for the remainder of this work, covers these kind of deviations, i.e.:

$$Pr(i) \propto u(i). \tag{4}$$

Reversely, the complete knowledge on the system with its possible deviations $u(\cdot)$ from the expected behavior would allow to predict and thus incorporate measures to avoid all dangerous situations beforehand.

Therefore, the development of a safe system requires a thorough analysis of its uncertainties. One main challenge in the uncertainty analysis is the identification of its dependencies on various parameters as time, space and environmental conditions:

$$u_{system} = f(u_1(s,t,\ldots), u_2(s,t,\ldots),\ldots, u_n(s,t,\ldots)), \tag{5}$$

where the system uncertainty u_{system} is a function of the system components' uncertainties $u_j; j \in [1;n]$.

In particular, the system uncertainty is always related to a specific property which we denote as *attribute a*. For instance, in safety-critical applications, such an attribute might be the distance between human and robot, representing a critical quantity. The obtained result on the uncertainty would reflect the possible deviation on the distance. Generally, the goal is to quantify the total uncertainty which is induced by the system on an arbitrary, user-selected attribute. Hence, in our analysis, we provide a result of the form

$$a \pm u_{system,a} \tag{6}$$

3.2 Classification of Uncertainties

To approach the issue of uncertainty estimation, we first need a proper classification thereof. As stated in Sect. 2, we apply the terminology introduced in [19]. According to this guide, the so-called *input quantities* $x_{a,i}$ on attribute a are classified regarding two evaluation methods:

- Frequency based uncertainty evaluation based on series of observations of the input quantities $x_{a,i}$ and subsequent statistical analyses (Type A)
- according to a-priori knowledge on the uncertainty behavior of input quantity $x_{a,i}$, that is provided by a theoretical model or the manufacturer (Type B).

In the following, we denote the uncertainty of input quantity $x_{a,i}$ with $u_{x_{a,i}}$.

3.3 Uncertainty Quantification

To properly determine the uncertainty on the attribute, its functional relationship with input quantities $x_{a,i}$

$$a = g(x_{a,1}, x_{a,2}, \ldots x_{a,m}); \quad m \leq n \tag{7}$$

must be known beforehand. Here, the input quantities represent the measurands of components c_j. Specifically, the variable $x_{a,i}$ stands for the input quantity of component c_i, where $i \in [1;m]$ with $m \leq n$. Above relationship which shows the dependency of attribute a on the system components c_j with $j \in [1;n]$ is required to study the uncertainty propagation: The influence of an input quantity on the attribute determines in which sense its measurement is affected by the considered system component. Assuming the independence of measurement tools, the propagation results in a so-called *combined standard uncertainty* $u_C(a)$ of

$$u_C(a) = \sqrt{\sum_j \left| \frac{\partial a}{\partial x_{a,j}} \right| u(c_j)} \tag{8}$$

for attribute a. However, the explicit form of Eq. (7) might not be available. In particular, the input quantities directly depend on the measurement uncertainty of employed devices such that the amount of $\frac{\partial a}{\partial x_{a,j}}$ is not directly quantifiable on the basis of theoretical knowledge. Hence, we aim to determine it via measurements. To do so, we leverage the knowledge of certain conservation properties which allows us to quantify the uncertainty originating from a measurement device. We apply common statistical tools including bootstrapping, hypothesis testing and correlation analyses. Since these analyses highly depend on the specific form of Eq. (7), we discuss the uncertainty quantification on a specified use case in HRC (Fig. 2).

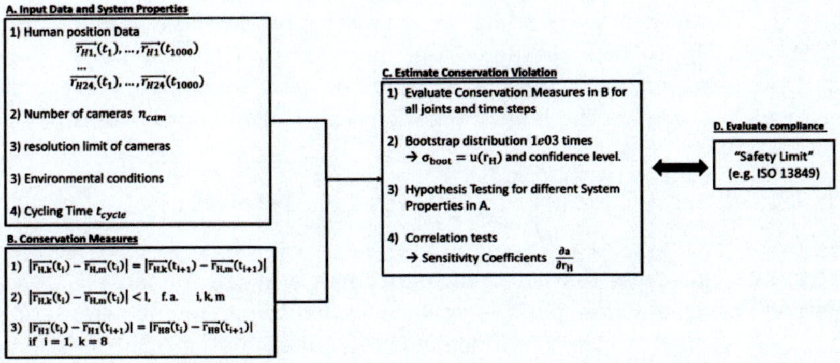

Fig. 2. We leverage the knowledge of non-changing parameters in the state space for uncertainty quantification. Regarding violations on the conservation of these non-changing properties by means of incoming data, we determine the uncertainty online. In the last step, we map the obtained uncertainty result on an arbitrary safety limit.

3.4 Use Case: Human Robot Collaboration

In Sect. 2, we outlined the relationship between uncertainties and safety assurance. Functional safety of HRC setups depends on relative measures such as *distance, velocity* and *force* between human and robot. To allow for a reliable assessment whether a situation is critical or safe, these parameters must be measured with a high degree of accuracy. On the other hand, the accuracy is limited by measurement uncertainties which are often not understood in detail. Therefore, estimating their influence on above distinctive measures is an essential step towards safe HRC. Specifying Eq. (1) for the distance d between human and robot yields

$$Pr(|\vec{r}_H - \vec{r}_R| \leq d) \leq \lambda_{HRC}\,; \qquad \forall t \tag{9}$$

where $\vec{r}_H, \vec{r}_R$ stand for the positions in 3D-Cartesian space, respectively. The safety assessment generally is achieved by evaluating Eq. (9). It is obvious that this evaluation necessitates the knowledge of the robot and human position. While $\vec{r}_R$ is easily accessible via control options, $\vec{r}_H$ is challenging to capture. Since an accurate model representing the human behavior does not exist, its position is determined through measurements resulting in

$$\tilde{\vec{r}}_H = \vec{r}_H \pm \vec{u}_H. \tag{10}$$

As a consequence, Eq. (9) changes to

$$Pr(|(\vec{r}_H \pm \vec{u}_H) - \vec{r}_R| \leq d) \leq \lambda_{HRC}\,; \qquad \forall t. \tag{11}$$

Thus, to evaluate Eq. (11), the knowledge of $\vec{u}_H$ is required. In particular, its minimization is desired to allow for a high confidence level for the compliance with λ_{HRC}. In contrast to $\vec{u}_R$ which we can derive on the basis of manufacturer specifications, $\vec{u}_H$ reflects the detection uncertainty of the human movement which depends on employed tools. We target the determination of the combined standard uncertainty on the human position which can be formulated by

$$u_C(\vec{r_H}) = \sqrt{\left|\frac{\partial r_H}{\partial det}\right| u(det) + \left|\frac{\partial r_H}{\partial \rho}\right| u(\rho) + \ldots}. \tag{12}$$

Here, *det* represents the detection performance of the human position and ρ arbitrary environmental parameter such as lightning, temperature etc. We assume that applied devices are independent from each other, that is, the measurements of environmental parameters do not stand in any relation to the human position detection. As will be specified in Sect. 4, we use the SICK safety scanner S3000 [22] in addition to OpenPose [23] for human localization. Consequently, the first term in above equation is split as follows:

$$\left|\frac{\partial r_H}{\partial det}\right| u(det) = \left|\frac{\partial r_H}{\partial det_{OP}}\right| u(det_{OP}) \cdot \left|\frac{\partial r_H}{\partial det_{LS}}\right| u(det_{LS}). \tag{13}$$

The safety scanner uncertainty $u(det_{LS})$ and such due to the behavior of environmental parameters $u(\rho)$ are assumed to be known and thus considered as Type B uncertainties. In contrast, $u(det_{OP})$ arising from OpenPose is not known due to a missing understanding of neural network uncertainties to date. To this end, we categorize it as Type A uncertainty and conduct analyses as explained in the following.

The key idea of this work lies in the *conservation based uncertainty estimation*, where conservation properties reflect characteristics of the system which can not change over state space parameters. Formulating conservation equations and evaluating them by means of incoming data during run-time allows to study the amount of violations. To be specific, conservation equations represent conditions on user-defined parameters which are known to stay constant. Given the knowledge that violations thereof can not occur, we view their occurrence as a direct consequence of the measurement uncertainty of the applied tool. Since the conservation equations are evaluated online, we obtain a distribution which reflects the amount of violations. Applying statistical techniques, we determine the measurement uncertainty and its confidence level.

In the use case of HRC, we focus on the determination of $\frac{\partial r_H}{\partial det,OP}$. One basic assumption of our work is the general functionality of the applied tool, that is, we do not focus on out-of-distribution events. Instead, we aim to quantify deviations within the scope of functionality and determine a confidence metric referring to measurement data. Here, we leverage the conservation of the constant distance between two human joints throughout time, that is

$$|\vec{r}_{H,j}(t) - \vec{r}_{H,k}(t)| = const.; \quad \forall t; \; j \neq k. \tag{14}$$

In our analysis, we assume that all joints of the human body are detected with the same performance. For each of our datasets, we calculate the distance between two joint pairs of the human body and obtain distributions. After collecting data of ten time steps, we conduct following procedure for the uncertainty determination and the mapping of such on the user-chosen safety limit:

1. **Bootstrapping:** We perform random sampling with replacement from our dataset ten thousand times. Each data point is assigned equal probability such that one data point can be selected several times. For each distribution, the mean value is calculated. According to the Central Limit Theorem, these mean values follow a Gaussian distribution which enables the uncertainty estimation on a confidence interval.
2. **Hypothesis testing:** Hypothesis tests are performed on the bootstrapped distributions. We state which dependencies are indicated, that is, whether the uncertainty behavior shows tendencies regarding parameters of interest.
3. **Covariance:** In case a dependency is indicated, we analyse the corresponding covariance. Leveraging the knowledge of this measure allows the user to adapt the system accordingly by reducing the uncertainty.
4. **Mapping on customizable safety limit:** As stated above, we view uncertainties as the main cause for the probability of occurrence for dangerous situations. Thus, we directly map the uncertainty on any safety limit: For instance, standard ISO 13849 states a maximum rate of 10^{-6} for the occurrence of dangerous situations per hour.

Generally, our method is applicable on arbitrary systems where conservation equations can be formulated and evaluated online according to Algorithm 1. However, for most well-understood devices, the uncertainty information is accessible via manufacturer specifications. Therefore, our method becomes particularly interesting for cases, where the uncertainties are not known. We view OpenPose as one measurement tool for the detection of human joint positions with a complex and non-obvious uncertainty behavior. This motivates us to apply our method on OpenPose to explore its uncertainty.

Algorithm 1 Conservation Based Uncertainty Estimation

Input: measurement data $x_{a,i}$; conservation Eq. $f_C(\cdot)$;
relationship betw. attribute and data $a(x_{a,i})$; confidence level σ; parameter for correl. analysis ξ; safety limit λ; time steps t.
Output: total combined uncertainty $u_C(a)$;
safety limit check (bool)

```
for i ≤ t do
    dev[i] ← f_C(x_{a,i}, ..., x_{n,i})
end for
for z ≤ 10.000 do
    b[z] ← bootstrap dev[]
end for
From b[] compute u_C(a) for user-defined σ
Test H_0 for given p-value and parameter ξ
if H_0 rejected then
    cov(u_C, ξ)
    print cov(u_C, ξ)
end if
r ← u_C · l_bio
if r ≤ λ then
    return 1
else
    return 0
end if
```

4 Experiments

To validate our method, we refer to the SICK safety scanner. Since the data sheet of the safety scanner provides an uncertainty value, it offers the possibility to assess the accuracy of our approach. In the second step, we address the uncertainty analysis of OpenPose. As will be detailed in the following, we make use of datasets that provide ground truth information. Treating OpenPose as a measurement device for human joint localization, we formulate conservation equations upon domain specific knowledge and quantify the uncertainty by applying our method. We consider the human body model shown in Fig. 3 and perform our analyses by means of following datasets.

4.1 Datasets

1. Dataset 1: Action Recognition NTU RGB+D is a large scale data set which contains markerless human movement data. The published ground truth data for the joint positions was obtained with Microsoft Kinect V2. Each skeleton is represented by 25 joints. As the uncertainty of Microsoft Kinect V2 has been studied in detail in [18], we consider findings in our analysis respectively. Thus, we treat the ground truth values as distributions to represent according deviations.

2. Dataset 2: We have gathered RGB+D data of six moving humans in a HRC setup. To obtain the ground truth for the conservation property in Eq. (14), we measured the distance between joints manually. Since this method is quite inaccurate, we consider an uncertainty of 5% arising from the manual measurement process. In addition, we refer to the positioning uncertainty of UR10e robot provided by Universal Robots (UR) to calculate the uncertainty on the d.

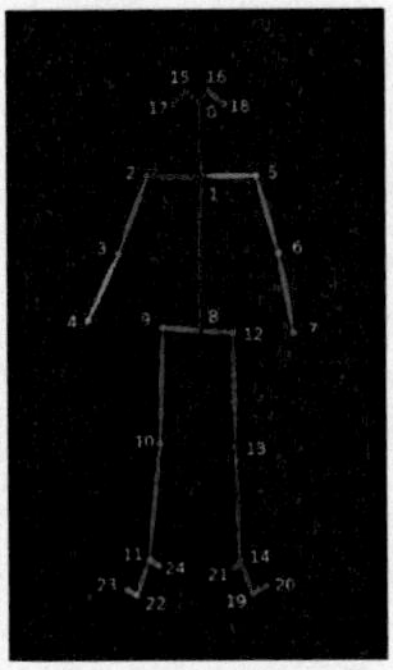

Fig. 3. We apply the model of Body25 to detect the key points with OpenPose.

4.2 Conservation Properties and Type B Uncertainties

We define the distance between the human joints as conservation properties as formulated in Eq. (14) and apply OpenPose to above datasets. In both cases, we estimate deviations with respect to the conservation equations online to explore the uncertainty of OpenPose. As a next step, we combine our results with the a-priori known uncertainty behavior of applied devices (Type B uncertainties). In case of dataset 1, we refer to the uncertainty of Microsoft Kinect V2. As presented in [18], the uncertainty behavior can be modeled by a linear function

$$u_{Kinect}(r_K) = 8 \cdot 10^{-4} \cdot r_K - 1 \cdot 10^{-4}, \tag{15}$$

where $u_{Kinect}(r_K)$ stands for the Microsoft Kinect V2 uncertainty and r_K for the distance to the camera. For dataset 2, in contrast, we refer to UR10e specifications given by

$$u_{UR10e} = 0.1 \cdot 10^{-3}\,\text{m}. \tag{16}$$

In particular, we consider this value to determine the distance uncertainty between human and robot for each time step. We use a safety laser scanner as an additional device to track the human feet. Since its uncertainty is provided by the manufacturer SICK, it is treated as Type B uncertainty. Applying our method on static data collected by the safety scanner allows us to assess its accuracy as will be discussed below.

4.3 Validation, Evaluation and Results

Prior to applying our method on the data, we conduct a validation thereof. Since the safety scanner uncertainty is provided by SICK, it enables us to compare the results obtained through our approach. We record static situations, that is, we perform our data collection in an environment with static obstacles and non-changing environmental conditions. The measurement data consists of data points collected at a scanning range of $4.0\,m$ covering an angular range of 275°, where each scanning interval amounts to 0.3850°. We follow the steps presented in Sect. 3.4 for uncertainty calculation. The bootstrapping procedure is carried out 10.000 times for each angular setting. Example histograms depicting the bootstrapped distributions are shown in Fig. 4. According to our results, the systematic uncertainty of the SICK safety scanner S3000 amounts to $u_{LS,exp} = 0.0038\,m$ at a scanning range of $4.0\,m$ which equals the relative uncertainty of

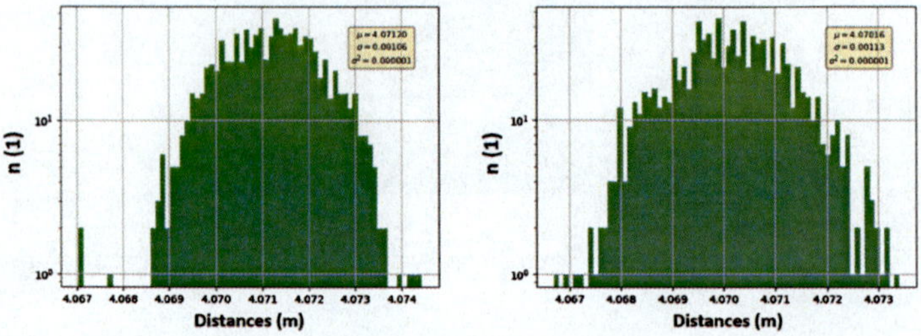

Fig. 4. We estimate the uncertainty of SICK safety scanner 3000 to validate our approach. We assume a Gaussian behavior of the uncertainty and perform our calculation referring to 100 measurements of static situations.

$$u_{LS,exp} = 0.095\% \tag{17}$$

on a 95% confidence level. The data sheet provided by SICK states an uncertainty of $u_{LS,theo} = 0.0050\,m$ at a scanning range of $5.5\,m$ which refers to a relative uncertainty of

$$u_{LS,theo} = 0.091\%. \tag{18}$$

Thus, our method yields a result which differs approximately 4.0% from the manufacturer specifications. In order to examine to which extent our method improves the uncertainty quantification, we compare this result with the baseline study: Hence, we determine the uncertainty by analysing the spread of obtained data from the laser scanner. As we recorded static situations, we do not expect any fluctuations in the measured values. However, without incorporating any conservation equations in our analysis, we obtain an uncertainty of $u_{LS,baseline} = 0.0054\,m$ for the same data set. Hence, the baseline method without conservation properties yields a relative uncertainty of

$$u_{LS,baseline} = 0.14\%, \tag{19}$$

which equals a difference of 30% from the SICK data sheet.

To analyse the uncertainty of OpenPose, we evaluate conservation equations Eq. (14), more specifically deviations thereof, online. In analogy to the safety scanner analysis, the resulting distribution allows to determine the uncertainty on a certain confidence level. However, we are additionally interested in the relationship between the uncertainty and the human velocity. We therefore target the exploration of this correlation by performing statistical tests with respect to following null hypothesis:

H0: The uncertainty does not increase with higher velocities in the human movement.

Our analysis focuses on certain human joints for which we assume no significant variations during the movements. To investigate dataset 1, we refer to uncertainty of Microsoft Kinect V2 stated in Eq. (15). Consequently, we obtain distributions for the ground truth positions. We apply OpenPose making use of the body-25 key point model and compare results with the ground truth data. Regarding dataset 2, we additionally consider the uncertainty of Intel RealSense D435 as well as the detection performance of SICK S3000 (Type B uncertainties). Finally, we determine the total uncertainty on the distance between human and robot:

$$u_C(d_{HR}) = \sqrt{\left|\frac{\partial d_{HR}}{\partial r_H}\right| \cdot u_C(r_H) + \left|\frac{\partial d_{HR}}{\partial r_R}\right| \cdot u(r_R)}, \tag{20}$$

where d_{HR} is given by

$$d_{HR} = \sqrt{\sum_{p=1}^{3}(r_{H,p} - r_{R,p})^2} \tag{21}$$

since we conduct analyses for 3D human pose estimation. We obtain following expression for the human-robot distance uncertainty:

$$u_C(d_{HR}) = \frac{\sum_{p=1}^{3}(r_{H,p} - r_{R,p})}{d_{HR}} \cdot [u_c(r_H) + u(r_R)], \tag{22}$$

where we use the manufacturer specifications for $u(r_R)$ defined in Eq. (16). Furthermore, we evaluate Eq. (12) online during run-time. To explore the uncertainty behavior of OpenPose in dataset 2, we subtract the relative uncertainty of $\leq 2\%$ stated in the data sheet of Intel RealSense D435 from the total uncertainty. It is noted that the uncertainty arising from the calibration process of the cameras and the triangulation into 3D space is negligible.

4.4 Results

In our analysis, we focus on the joint pairs of the upper body part. As can be seen in Fig. 5, the lower body part is often not detected, especially in the direct

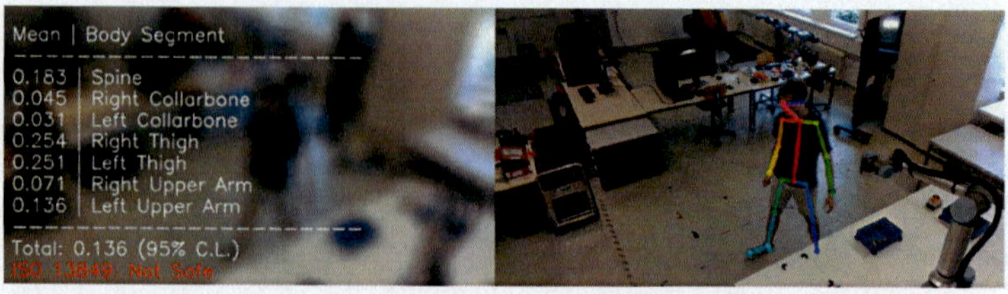

Fig. 5. Our method allows to estimate the uncertainties on user-selected joint pairs online. During run-time, our method provides uncertainty values and an evaluation on an arbitrary safety limit (e.g., ISO 13849).

surrounding of the robot. We refer to joint pairs where minimal variations are expected. While the segment between the elbow and the hand is assumed to underlie more movements and thus more fluctuations, the part between the neck and hip is expected to show a more constant behavior. Two example histograms illustrating the bootstrapped distributions of the uncertainty are shown in Fig. 6.

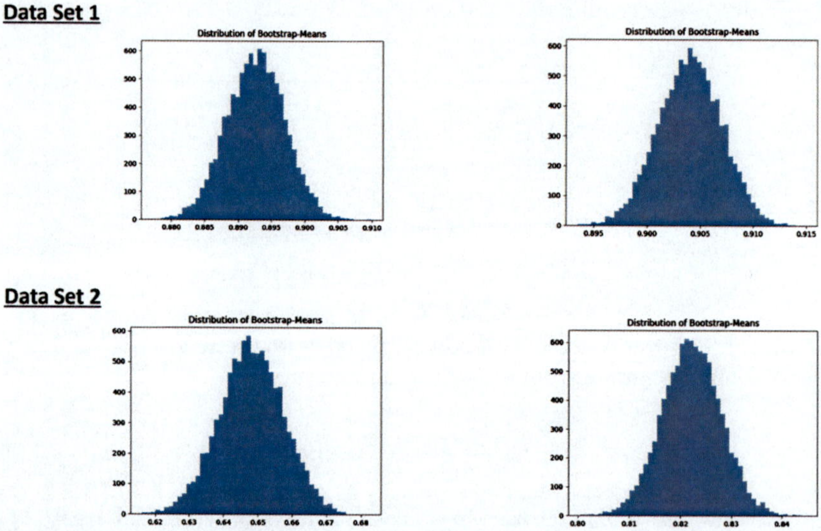

Fig. 6. Bootstrapping the deviations from the ground truth value of the OpenPose output 10.000 times yields Gaussian distributions. We show two example plots for each dataset.

Our method yields following results for the 3D spatial detection uncertainty of OpenPose:

$$u_{D1}(det_{OP}) = 0.008\% \tag{23}$$

$$u_{D2}(det_{OP}) = 0.015\%. \tag{24}$$

Accordingly, the uncertainty estimated in dataset 2 is double as high as determined on dataset 1. It is very likely that this discrepancy is due to the lower statistical significance of dataset 2. While dataset 1 consists of a large data amount, we refer to data of six different human beings in dataset 2. Furthermore, the uncertainty of $u_{FR} = 5\%$ which we assumed to originate from the manual measurement of the ground truth is one possible origin for above discrepancy. Our final result which we obtain through averaging both values is given by

$$u_{tot}(det_{OP}) = 0.009\% \tag{25}$$

on a 95% confidence level. Regarding the hypothesis test, above defined null hypothesis is rejected for both data collections with a p-value of $p = 0.9816$. Conducting correlation analyses with the human velocity leads to an averaged correlation of 0.32.

4.5 Mapping on Safety Limit

Our approach allows to map the uncertainty on any user-defined limit. Since we are interested in robotic applications, we refer to the safety limit in ISO 13849 stating a maximum probability for the occurrence of dangerous failures per hour PFD_H of

$$\text{PFD}_\text{H,max} = 10^{-6}/h. \tag{26}$$

As we consider uncertainties as the main origin for the evolution of this probability, we directly map our results on this quantity. According to ISO 13849, one dangerous failure in 10^6 h is tolerated. Above calculated uncertainty of 0.009% applies to one dangerous failure per 10^4 h, assuming that slight uncertainties in the human joint position detection could yield critical situations. Figure 5 shows a static depiction of our methodology.

4.6 Limitations

Our experiments and validation demonstrated that our conservation based uncertainty quantification method yields reasonably accurate results. Nevertheless, our framework requires the formalization of conservation equations by the user which could cause challenges in applying it to more complex systems. For instance, identifying conservation measures which offer the possibility to be monitored during run-time might not always be straight-forward. In contrast to above presented scenarios, where the length between the human body joints could be evaluated directly by incoming data, conservation properties could be separated from the monitored data. In addition, more complex conservation equations might cause computational burden on our framework. As a consequence, the online uncertainty quantification could be affected. However, these possible shortcomings have not been studied within the frame of this contribution. We therefore suggest to perform more detailed studies thereof in future.

4.7 Discussion

In order to validate the accuracy of our method, we applied it on static data recorded with the safety scanner SICK 3000. It was found that our result lies in the same order of magnitude as stated by SICK. In fact, a discrepancy of 4% could be verified. However, the data sheet corresponds to a larger scanning range which could explain the difference. On the other hand, the statistical significance of our dataset might be one origin for the deviation. As a next step, we performed our analysis on data gathered by the use of OpenPose on two datasets. We applied OpenPose first on Action Recognition NTU RGB+D, considering the uncertainty behavior of Microsoft Kinect V2. In addition, the human pose detection was performed on our data collection. Since we made use of three Intel RealSense D435 devices, we considered their uncertainties in our calculations. For the analysis, we took the frame rates into account. The time-invariant, constant distance between the human joints was defined as conservation measure. Conducting statistical analyses, a correlation of 0.32 between the uncertainty and the relative velocity between human and robot could be identified. Furthermore, we showed that OpenPose is not suitable for safe HRC: According to our results, the uncertainty of OpenPose exceeds the limit required in ISO 13849 by two orders of magnitude. In the context of future works, we suggest the incorporation of measurement uncertainties in the robot control framework. Uncertainties estimated online could be used to allow for the generation of safe trajectories. As the main purpose of our experiments was to demonstrate the applicability of our method, we performed analyses only for short video sequences (max. $30\,s$). For a more thorough uncertainty evaluation with a higher statistical significance, we suggest to take more data and use cases into account.

5 Conclusion and Outlook

We presented an online uncertainty calculation method that combines a-priori knowledge and system specific data collected at run-time. The key idea of our approach lies in the quantification of measurement uncertainties based on system-specific conservation properties. Our method allows the online calculation of uncertainties on a user-selected confidence level for customizable attributes. Throughout this work, we referred to fundamentals for uncertainty calculation provided in [19]. To do so, we assumed the independence of measurement quantities. In future, we aim to consider more complex relationships between the parameters. Therefore, we will extend the presented work by Monte Carlo simulations to cover functional behaviors of different uncertainty sources which can be not modeled analytically. In the last Section, we explored the uncertainty behavior of OpenPose which is a tool for human position detection. By identifying violations of pre-defined conservation properties, we quantified uncertainties of the spatial human detection performance. In contrast to sensor devices with manufacturer specifications providing uncertainty information, OpenPose lacks on uncertainty values such that quantifying them is necessary to enable evaluations with respect to safety requirements. Furthermore, the incorporation of uncertainties into robot control was not addressed within the scope of this work. Assuming the real-time character of robot control, implementing uncertainties arising from the human movement into control algorithms could offer new possibilities in realizing safe and efficient HRC.

References

1. Askarpour, M., Mandrioli, D., Rossi, M., Vicentini, F.: SAFER-HRC: safety analysis through formal verification in human-robot collaboration. In: International Conference on Computer Safety, Reliability and Security 2016 (2016)
2. Wadekar, P., Gopinath, V., Johansen, K.: Safe layout design and evaluation of a human-robot collaborative application cell through risk assessment - a computer aided approach. Procedia Manuf. **25**, 602–611 (2018)
3. Inam, R., Raizer, K., Hata, A., Souza, R., Fersman, E., Chao, E., Wang, S.: Risk assessment for human-robot collaboration in an automated warehouse scenario. In: 2018 IEEE 23rd International Conference on Emerging Technologies and Factory Automation (ETFA), pp. 743–751 (2018)
4. Hanna, A., Bengtsson, K., Götvall, P.-L., Ekström, M.: Towards safe human robot collaboration - risk assessment of intelligent automation. In: 2020 25th IEEE International Conference on Emerging Technologies and Factory Automation (ETFA), vol. 1, pp. 424–431 (2020)
5. Bartneck, C., Moltchanova, E.: Expressing uncertainty in human-robot interaction. PLoS ONE **15**, e0235361 (2020)
6. Gopinath, V., Ore, F., Johansen, K.: Safe assembly cell layout through risk assessment – an application with hand guided industrial robot. Procedia CIRP **63**, 430–435 (2017)
7. Araiza-Illan, D., Western, D., Pipe, A.G., Eder, K.: Systematic and realistic testing in simulation of control code for robots in collaborative human-robot interactions. In: Annual Conference Towards Autonomous Robotic Systems, pp. 20–32. Springer (2016)
8. Lasota, P.A., Fong, T., Shah, J.A.: A survey of methods for safe human-robot interaction. In: Foundations and Trends in Robotics (2014)
9. Haddadin, S., Haddadin, S., Khoury, A., Rokahr, T., Parusel, S., Burgkart, R., Bicchi, A., Albu-Schäffer, A.: On making robots understand safety: embedding injury knowledge into control. Int. J. Robot. Res. **31**(13), 1578–1602 (2012)
10. Haddadin, S., Albu-Schaffer, A., De Luca, A., Hirzinger, G.: Collision detection and reaction: a contribution to safe physical human-robot interaction. In: 2008 IEEE/RSJ International Conference on Intelligent Robots and Systems, pp. 3356–3363. IEEE (2008)
11. Nokata, M., Ikuta, K., Ishii, H.: Safety-optimizing method of human-care robot design and control. In: Proceedings 2002 IEEE International Conference on Robotics and Automation (Cat. No. 02CH37292), vol. 2, pp. 1991–1996. IEEE (2002)
12. Ikuta, K., Ishii, H., Nokata, M.: Safety evaluation method of design and control for human-care robots. Int. J. Robot. Res. **22**(5), 281–297 (2003)
13. Parsons, S.: Probabilistic robotics by Sebastian Thrun, Wolfram Burgard and Dieter Fox, MIT Press, 647 pp., 55.00, ISBN 0-262-20162-3. Knowl. Eng. Rev. **21**(3), 287–289 (2006)

14. Nguyen, H., Pham, Q.-C.: A probabilistic framework for tracking uncertainties in robotic manipulation. arXiv arXiv:1901.00969 (2019)
15. Wirnshofer, F., Schmitt, P.S., Feiten, W., Wichert, G.V., Burgard, W.: Robust, compliant assembly via optimal belief space planning. In: 2018 IEEE International Conference on Robotics and Automation (ICRA), pp. 5436–5443. IEEE (2018)
16. Liu, C., Tomizuka, M.: Safe exploration: addressing various uncertainty levels in human robot interactions. In: 2015 American Control Conference (ACC), pp. 465–470 (2015)
17. Rahman, S., Waslander, S.: Uncertainty constrained differential dynamic programming in belief space for vision based robots. IEEE Robot. Autom. Lett. **PP**, 1 (2021)
18. Giancola, S., Valenti, M., Sala, R.: A Survey on 3D Cameras: Metrological Comparison of Time-of-Flight, Structured-Light and Active Stereoscopy Technologies, 1st edn. Springer Publishing Company, Incorporated (2018)
19. Joint Committee for Guides in Metrology (JCGM): Evaluation of Measurement Data - Guide to the Expression of Uncertainty in Measurement (2009)
20. ISO 13849-2:2015 Safety of Machinery - Safety-Related Parts of Control Systems - Part 1: General Principles for Design. International Organization for Standardization (2015)
21. ISO 12100:2010 Safety of Machinery - General Principles for Design - Risk Assessment and Risk Reduction. International Organization for Standardization (2010)
22. Data Sheet "ICK Safety Laser Scanner S3000". SICK AG (2021)
23. Cao, Z., Hidalgo Martinez, G., Simon, T., Wei, S., Sheikh, Y.A.: OpenPose: realtime multi-person 2D pose estimation using part affinity fields. IEEE Trans. Pattern Anal. Mach. Intell. (2019)

Post-facto Misrecognition Filter Based on Resumable Interruptions for Coping with Real World Uncertainty in the Development of Reactive Robotic Behaviors

Guilherme de Campos Affonso(✉), Kei Okada, and Masayuki Inaba

The University of Tokyo, 7-3-1, Hongo, Bunkyo-ku, Tokyo, Japan
affonso@jsk.imi.i.u-tokyo.ac.jp

Abstract. In this paper we propose a resumable interruption framework for robotic applications which allows to "filter" misrecognition signals after their occurrence. Handling misrecognition is essential for deploying reactive systems into the real world, since being over-reactive to detection errors can lead to livelocks and stagnation. For example, constantly interrupting and resuming a picking task due to misrecognition can make the robot alternate between pre-grasping and grasping motions, without ever achieving the task. Our solution is based on resumable interruptions, continuing interrupted procedures from the exact preemption point if similar execution requests are received shortly after a cancellation order. This acts as a post-facto misrecognition filter, which stabilizes execution and ensures task completion. Compared with standard filtering, the post-facto approach allows to deliver signals faster and recover from misrecognition longer. The proposed system is verified through real robot experiments in dynamic and static environments.

Keywords: Robotics · Reactive robotics · Misrecognition handling · Robot manipulation · Dynamic environments

1 Introduction

The development of recognition algorithms and elaborated control structures played a key role during the advance of robots into unstructured environments. Now that robots are attempting to ingress into more dynamic social environments, with humans-in-the-loop who can interact with the robot and its surroundings at any time, more reactive systems are needed.

Reactive systems, however, will also promptly react to detection errors, configuring a challenge for the design and deployment of reactive robotic behaviors in the real world. In this work we address this problem by proposing a resumable

I. Petrovic et al. (Eds.): IAS 2022, LNNS 577, pp. 103–121, 2023.
https://doi.org/10.1007/978-3-031-22216-0_8

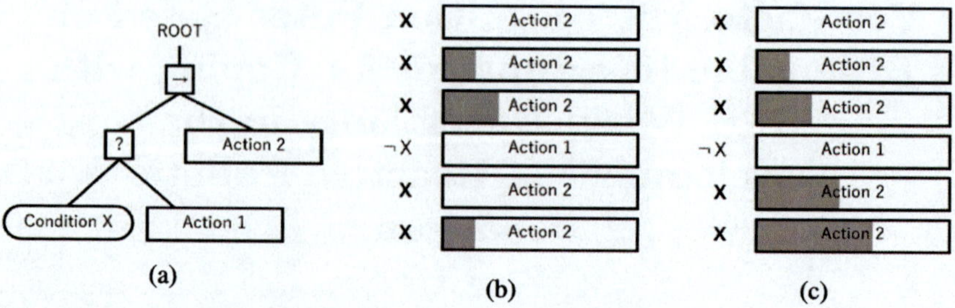

Fig. 1. Overview of the proposed framework, which uses resumable interruptions for coping with detection errors. (a) Simple task formulation using a behavior tree. In this formulation the condition X is evaluated at each time frame, switching between the execution of Action 2 if X is true and Action 1 if it is false. (b) Hypothetical runtime results using a standard implementation that does not support resumptions. Here, a misrecognition result in the fourth step signalizing that X is false would cause the cancellation of Action 2, which is then re-evaluated from start at the next correct recognition result in the fifth step. (c) Hypothetical runtime results using the proposed method. Here, a misrecognition result in the fourth step causes Action 2 to enter a standby phase instead, which is then resumed from the exact preemption point at the next correct recognition result in the fifth step.

interruption framework, which can "filter" detection errors after their occurrence by autonomously keeping track of the current progress of each action and resuming execution from the exact preemption point. The proposed framework is illustrated in Fig. 1, highlighting how resumable interruptions can be used to seamlessly overcome detection errors when assuming recognition nodes with reasonable accuracy.

A generalized reactive robotic system is defined by its ability to promptly detect and correctly handle environmental changes, and is composed of three main elements:

1. a control architecture based on the continuous monitoring of the environment;
2. robust recognition algorithms with real-time capabilities;
3. action modules capable of being interrupted and started at any execution point.

When considering each of these components, we can state that

i reactive control structures are largely available, including behavior trees [9,14] and reinforcement learning models [2,8,13,16] as the most recent developments;
ii breakthrough advance in machine learning technologies have provided robust and real-time recognition algorithms;
iii implementing actions which can handle a wide range of environmental states remains an extensive task.

The last point is contrasted with motion planning algorithms such as Inverse Kinematics and SLAM, which are developed for single motions and can deal with a wide range of initial conditions. However, during the development of more complex robotic procedures it is common to assume that the input of an action is the specific environmental state caused by the execution of the previous action. This assumption allows to adopt simplified manipulation and recognition models which greatly reduce the development time. However, it is not possible to assume any particular environmental state when dealing with dynamic conditions, since an action can be canceled and started at any time due to environmental changes. For example, when considering the task of pilling up boxes, we could assume a situation where there are no boxes pilled and then place each new box at a fixed xy coordinate and consecutively higher z value. Even when disregarding possible human intervention, if the robot is required to stop execution to deal with a higher priority event and is then asked to resume the pilling task, the assumption that there are no pilled boxes will not hold. In such cases, correct execution relies on the introduction of new recognition models or state variables, which make it possible to select optimal resumption behaviors by inferring or programmatically retaining the current progress of the preempted action. If during the resumption procedure the robot does not re-evaluate any motions that have already been executed, we say that the handling has zero overhead.

Although in this case ensuring resumption from an appropriate point is critical for task completion, there are many situations where restarting the action from start yields valid (but not optimal) results. For example, if a robot engaged in a picking task by executing a pre-grasp and a grasp motion is interrupted, repeating the pre-grasp motion will likely not affect task completion. In general terms, zero resumption overhead is non-critical for actions that do not rely on specific environmental states prior to or during their execution. In such cases, it is possible to deploy and benefit from simplified representation models.

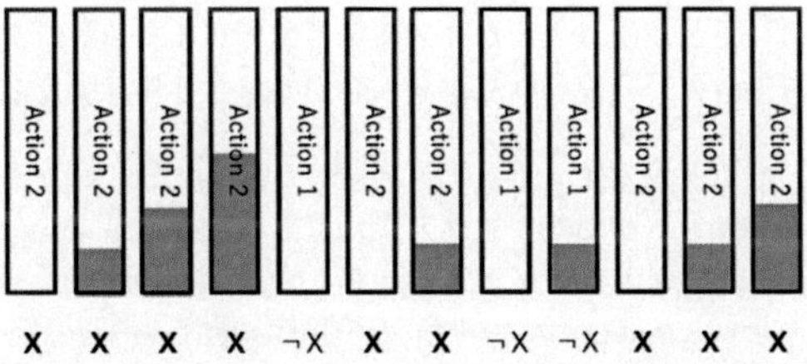

Fig. 2. Example of chattering behavior, where the program enters a livelock state which stagnates further task execution. Assuming the formulation in Fig. 1a, occasional misrecognition results signalizing that X is false keep interrupting and discarding the current progress of Action 2. Since reactive frameworks are also over-reactive to noise, chattering behaviors can occur even when deploying relatively high-reliance detection models.

However, this is not valid when considering the deployment of reactive systems in real world scenarios. When dealing with potential detection errors caused by real world uncertainties, any overhead can lead to an oscillation between two robot motions, in a situation similar to a livelock state (Fig. 2). This is because such reactive systems are also over-reactive to noise – a recognition algorithm with an accuracy of 80% will only fail 20% of the time if a single result is asserted, but almost certainly yield at least one incorrect result if the recognition time is much shorter than the actuation time. In the previous picking task example, this means that even if no other higher priority request is considered, the grasp motion could still be interrupted due to fluctuations in recognition results. In such case, the next correct recognition result would then cause the robot to repeat the picking action from start, which could be interrupted again in a loop between grasp and pre-grasp motions. In this paper we refer to this phenomenon as "chattering behavior".

When working with real world applications, it is vital to avoid chattering behavior by either:

1. enforcing zero resumption overhead;
2. deploying high reliance recognition algorithms and filtering strategies;
3. ignoring possible interruptions at predefined task points through selective reactivity.

In this work we adopt a variation of the first and second approaches, proposing a resumable interruption framework that ensures zero overhead when resuming from detection errors. The proposed framework works by autonomously keeping track of the current progress of each action module or control substructure, and then resume execution from the exact preemption point when similar execution requests are received within a short time interval. By doing this, the proposed system acts as a post-facto misrecognition filter, allowing to ignore detection errors while remaining reactive to environmental changes.

2 Past Work

2.1 Control Structures for Reactive Task Definition

The reactivity of a control structure is determined by the frequency in which it performs environmental checks – the more constant the checks, the earlier the framework can detect and adapt to new conditions. Such kind of continuous checking approach marks a departure from deliberative planning and the classic *sense-plan-act* [19,20] methodology, promoting the blending of the sensing and acting phases rather than their segregation.

Although any control framework could be made more reactive by increasing the frequency of environmental checks, architectures that inherently support checking as a part of the task structure are beneficial. In general, tick-driven switching structures which evaluate the environment condition at every time frame (every tick) and derive which action should be executed as a function of the given environment state are an example of inherently reactive frameworks.

Behavior Trees [5,9,14] are perhaps the most popular example of such tick-driven switching structures. Originally developed by gaming industries as a more reactive and modular alternative to finite state machines for modeling the behavior of non-playable characters, behavior trees work by transversing a human-defined structure according to a predefined syntax and the runtime environmental state at each time frame. The structure is articulated as an oriented tree, where internal nodes indicate how the tree should be transversed, and leaf nodes are either condition or action nodes to be executed when transversed. The placement of each node within the tree structure determines a fixed priority, which is used to represent progression order and pre-condition relations between actions.

In [6] a human design approach based on the expansion of state machines is also observed. In this framework, clear interruption and resumption callbacks are defined for each action module and administered by a monitoring node that is constantly open for new requests within a multi-task scenario. Differently from behavior trees, the emphasis of this study is placed on the handling of the side effects of each motion, which relies on the task state and is therefore hard to be defined in a purely stateless framework.

Temporal Logic formulation approaches [10,18,27,28] represent a rule-based formulation of tick-driven switching structures. Here, instead of directly designing the task structure and hierarchy, the user is responsible for defining a set of rules general to the given environment and task, which are then used by the logic engine to compute which actions should be executed under each situation. By coupling this approach with real time recognition algorithms, it is possible to attain highly reactive results. When compared with approaches relying on human design transitions, the temporal logic formulation has the advantage of being able to provide provably correct results, and also explanations for task failure under unexpected execution scenarios [4,11,26]. However, such general rules tend to be less intuitive to visualize and harder to define.

Another approach is given by data-driven methods, typically using machine learning algorithms [2,8,13,16]. Here, sensory data gathered in simulation or real world environments is used to learn a switching model for determining the optimal action to be executed at each time. Such frameworks have the benefits of being end-to-end and useful for defining adaptive systems. However, they still face limitations regarding higher-level interactions, logical congruence in multi-step scenarios, and handling of environmental states which are not directly derived from one-to-one sensor interactions.

Compared to temporal logic and data-driven approaches, human design methods have the advantage of being able to selectively inhibit the reactivity of certain control substructures in order to mitigate the effect of detection errors. Temporal logic formulations are fundamentally inconsistent with this methodology, and in learning approaches it could be hard to design value functions capable of ignoring otherwise relevant information. Alternately, suppressing certain transitions while exposing others is sufficient for selectively inhibiting reactivity in state machine based formulations. Similar approaches for including non-reactive structures in behavior trees have also been verified [1,15].

2.2 Robust Recognition Using Neural Networks and Data Filtering Algorithms

It was not until recent breakthrough discoveries in machine learning technology that robust and time-efficient detection mechanisms able to withstand execution in unstructured and dynamic environments emerged. Until then, recognition algorithms would often replace the detection problem with a more concrete query on a restricted environment – such as HSV filtering, shape detection, point clustering, and others. Although fast and accurate under the designed conditions, this approach imposed significant limitations on the acting environment, being unsuited for application in social scenarios and everyday usage.

However, when compared with more analytical approaches, machine learning algorithms are inherently probabilistic and prone to data fluctuation [7,12]. Therefore, data filtering mechanisms are still relevant for the real world usage of such technologies within reactive systems.

Data filtering is a standardized approach for handling real world uncertainty, and is widely used in robotics and autonomous driving applications [22–24]. Classical approaches include the Kalman and Particle filters, as well as more generalized sliding window methods. More recent approaches include anomaly detection through neural networks [21,25]. Although highly beneficial, such approaches tend to smooth out the input data, which may lead to an oversight of peak signals and an overall delay in detecting environmental changes.

2.3 Attaining Zero Resumption Overhead

Zero resumption overhead means the robot can resume operation without repeating any procedures whose goals have already been attained after handling an interruption signal. As discussed above, having zero resumption overhead allows for the stable execution of robotic behaviors even when considering detection errors. Zero resumption overhead can be attained by either using planning algorithms capable of providing optimal strategies for a wide range of environmental conditions, or by storing and verifying the current progress of a preempted procedure through simulated memory.

The first point is the main objective in the field of manipulation planning [3,17,29], where classic motion planners are extended to consider both robot and environmental states during the composition of more generalized motion sequences and manipulation procedures. When considering the previously discussed example of a picking task, using such manipulation planning frameworks would make it possible to replace the hand-designed pre-grasp and grasp motions with a single planning request. This request would then be able to evaluate the environmental and robot conditions to reach an optimal strategy with no overhead. By doing this, it would be possible to avoid eventual chattering behaviors in the task. However, at the time of this writing, there is still no widely available manipulation planning approach capable of handling a variety of action formulations.

The development of memory-enabled control frameworks is generally hard to be attained in purely switching structures, which are fundamentally stateless and, therefore, unable to retain information on past states. However, it is possible to emulate memory through manually assigned variables, or to attain zero resumption overhead by exclusively deploying single-motion modules with sufficiently capable planners. Unfortunately, both approaches require significantly more complex structures.

3 Post-facto Misrecognition Filter Based on Resumable Interruptions

In this work we propose using resumable interruptions at both action module and control substructure level to ensure zero resumption overhead and avoid chattering behavior when deploying reactive frameworks in real world scenarios. Compared with standard filtering strategies, the usage of resumable interruptions as a post-facto filtering strategy has increased detection coverage and lower reaction time, as shown in Fig. 3 and later discussed in this section.

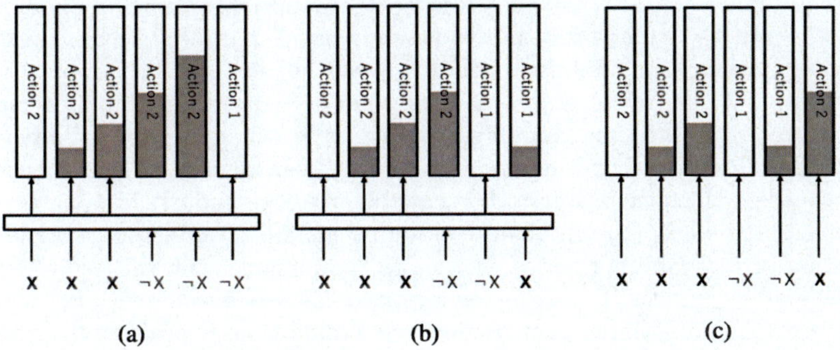

Fig. 3. Pre and post-facto filtering approaches for dealing with uncertain input. Here we assume the task definition given in Fig. 1a, and an uncertain input received at the fifth execution step. (a) A more conservative filtering approach with larger window size is good for avoiding even prolonged misrecognition results, but it takes longer to communicate environmental changes to the program. (b) A more sensitive filtering approach with smaller window size can rapidly communicate environmental changes to the program, but is unable to filter more prolonged misrecognition results. (c) Post-facto filtering approaches can attain maximal reactivity while also allowing for the resumption from prolonged misrecognition results, according to the timeout and checking function parameters.

At action module level, this framework works by stashing the current progress of preempted actions as a programming language level continuation, and then recalling this continuation if the cancellation request is later judged to be caused by flawed detection. Under the assumption that the recognition module works at a reasonable frequency, has an acceptable hit rate, and is not subject to any localized bias such as occlusions at a certain execution point, a correct recognition result will shortly follow a misrecognition signal within a few iterations cycle. Therefore, it is possible to judge that a certain cancellation request was caused by misrecognition if a new execution request is received shortly after a cancellation command, and is consistent with task execution – *i.e.* describes an environmental state which matches the robot's inner assumptions. The timeout for resuming stashed continuations and the function for checking congruence between old requests, new requests, and the given robot state are the two meta parameters of the proposed system, open to user customization and tuning.

At control substructure level, the introduction of these two meta parameters allows to enhance previous memory node formulations in behavior trees [15]. Instead of blindly retaining the last state in denial of any new environmental change, adding the ability to "forget" previous states through timeouts and check functions allows to build systems that remain sensitive to important events while resilient to detection error. Under the same assumptions, such systems behave similarly to the action module level formulation. The mirroring of the resumption capabilities in both action module level and control substructure level also allows for a more free task design, in which the granularity of modules is defined by user requirements rather than the presence or absence of runtime functionalities.

Compared to similar strategies to attain zero resumption overhead through simulated memory and minimal module size, the inclusion of action module level continuations in the proposed framework benefits from simplified models and reduced planning time. Simplified models are possible because (i) implementation of additional recognition nodes or manual management of global variables is not needed, since the framework is capable of autonomously keeping track of the preempted state; (ii) initial state assumptions are accepted, allowing for the definition of non-minimal nodes with more comfortable size; (iii) multi-motion modules allow for easier sharing of localized data between motions, without requiring a global variable pool. Reduced planning time is possible because the action module level can also retain planning results of single motions. When considering only structural memory, the minimal tracking unit is determined by module size. This means that even when considering single motion modules a new planning result will need to be recalculated at each new evaluation request.

Compared to classic data filtering algorithms, the proposed post-facto methodology ensures maximal misrecognition detection coverage and minimal reaction time to real events. This is because in filtering algorithms there is a tradeoff between coverage and reactivity dictated by the filter sensitivity and window size – large window sizes are good at detecting misrecognition data, but also have a delay in detecting abrupt environmental changes (Fig. 3a); smaller window sizes react to data more rapidly, but are also more prone to communicating erroneous data points to the program (Fig. 3b). In post-facto methods, however, the "act first think later" approach ensures maximal reactivity at all times (the signal is always passed to the program), while detection sensitivity can be freely manipulated by adjusting the timeout and checking function parameters independently (Fig. 3c). It is also worth noting that pre and post-facto filtering strategies are not mutually exclusive. It is valid and encouraged to deploy both pre and post-facto filtering methods for more stable execution.

Although enabling to resume procedures after arbitrary time intervals, excessively elongated timeout values can lead to chattering behavior and unexpected errors. Chattering behavior is caused by having to revert any potential side effects which have occurred during the interruption state. Unexpected errors may occur when such potential side effects are incongruent with the continuation plan. Ideally, heuristics determining if an action can be safely resumed or not should be included in the check function parameter.

4 Experiments

4.1 Setup

In this work we perform a total of four experiments (Table 1) to verify the ability of the proposed system to adapt to real world uncertainties and detection noise during the implementation of reactive frameworks. All experiments are performed using the PR2 robot during the pouring task illustrated in Fig. 4. The recognition setup is shown in Fig. 5, mixing a high reliance model for object manipulation and a high frequency model for detecting environmental changes in reactive setups. Both models are trained using real data from the robot observations in the laboratory. A side camera is added to cover up for partial occlusion and recognition bias in the models, which have made it difficult for the robot to recognize the cup during manipulation at the pouring task. Small plastic balls are used instead of actual liquids to avoid damaging the robot.

Table 1. Experiment conditions

Exp	Reactive framework	Misrecognition filtering	Human intervention
Exp. 1	No	No	No
Exp. 2	Yes	No	No
Exp. 3	Yes	Yes	No
Exp. 4	Yes	Yes	Yes

In the first experiment, the task is implemented using a deliberative approach with no reactivity. In this implementation the robot verifies the position of the bottle and the cup beforehand, and then preforms the whole manipulation phase completely blind to any environmental changes. This experiment is designed as a baseline for execution stability when there is no human intervention and the environment remains static.

In the second experiment, the picking and the pouring actions are rewritten to include concurrent environment monitoring in a reactive formulation. In this

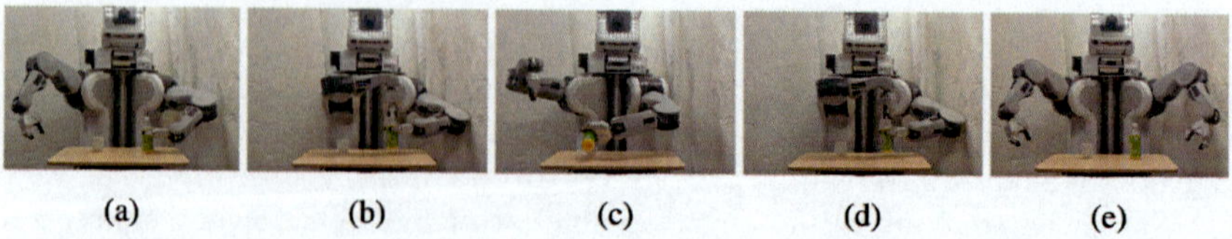

Fig. 4. Pouring task by consecutively (a) picking up the bottle, (b) opening it, (c) pouring its contents into a cup, (d) closing it, and (e) placing it on the table.

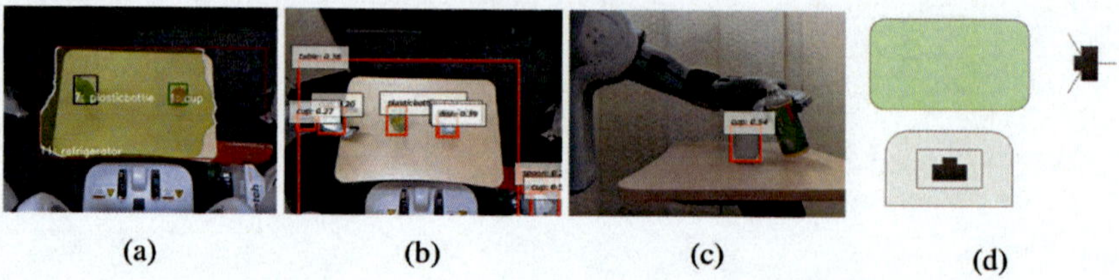

Fig. 5. Recognition system setup including (a) a higher reliance lower frequency RCNN detection model running on the head camera (Kinect V2) on approximately 4 Hz; (b) a lower reliance higher frequency SSD model running on the head camera (Kinect V2) on approximately 10 Hz; (c) a lower reliance higher frequency SSD model running on the side camera (Realsense d435) on approximately 10 Hz; (d) relative position of the robot (bottom), workspace (middle), and side camera (right).

setup, the robot is instructed to re-plan each motion whenever a significantly different recognition result is received. The robot is also required to lift the bottle and stop pouring if the pouring task is interrupted. An additional *search* action in which the robot looks to both sides searching for an object is also implemented and activated whenever the target object remains undetectable for more than five seconds. No filtering is applied to the data, and the current progress of each action is not tracked. Although both the picking and the pouring actions can be started from any execution point, there is noticeable resumption overhead in both actions, which occur as a natural implication of the standard implementation methodology of robotic actions through state progression. This experiment is designed to verify the impacts of adopting reactive frameworks with no filtering on practical scenarios.

In the third experiment, the above reactive formulation is rewritten as a behavior tree task (Fig. 13), which is then extended by the proposed system to enable resumptions at a modular level. Actions for opening the bottle, closing the bottle, and placing the bottle on the table are implemented using non-reactive frameworks, since they do not depend on recognition results. This experiment verifies the ability of the proposed system to filter detection errors within a static environment, ensuring stable execution comparable to the one presented by the first experimental setup.

Finally, in the fourth experiment the user actively intervenes with the task, moving and hiding the bottle and cup during the picking and pouring action.

This experiment tests the ability of the proposed system to correctly distinguish between detection errors and actual environmental changes when asserting if a preempted module should be resumed or not.

Considering the frame rate and reliability of our recognition system, in the third and fourth experiments we have set a large resumption timeout and evaluated the validity of new goals mainly by their resemblance with interrupted ones. Incoming requests were considered consistent with interrupted ones when the difference between target coordinates was no greater than 75 mm. Timeout was set to 30 s in the picking task and 10 s in the pouring task, which has proven to be far over the required value.

4.2 Results

Results of the first experiment are shown in Figs. 6 and 7. As expected, the robot successfully and stably executed both the picking and the pouring actions in static environments.

Results of the second experiment are shown in Figs. 8 and 9. Here, introducing a reactive framework has led to unstable execution and chattering behavior, making the robot unable to accomplish the task. In Fig. 8, detection errors in time $t = 2$ s and $t = 8$ s have caused the robot to stagnate while attempting different pre-grasp motions and ultimately knock down the bottle when trying to move from grasp to pre-grasp motion. In Fig. 9, mistakenly interpreting the robot's gripper as a cup in times $t = 2$ s, $t = 8.5$ s, and $t = 15.5$ s has lead to chattering behavior, alternating between tilting the bottle forwards and backwards without finishing the pour.

(a) $t = 0$ s (b) $t = 2$ s (c) $t = 4$ s (d) $t = 6$ s (e) $t = 10$ s

Fig. 6. Picking action in the first experimental setup. Execution was stable and smooth, since there was no recognition involved.

(a) $t = 0$ s (b) $t = 2$ s (c) $t = 4$ s (d) $t = 6$ s (e) $t = 8$ s (f) $t = 10$ s (g) $t = 12$ s (h) $t = 14$ s

Fig. 7. Pouring action in the first experimental setup. Execution was stable and smooth.

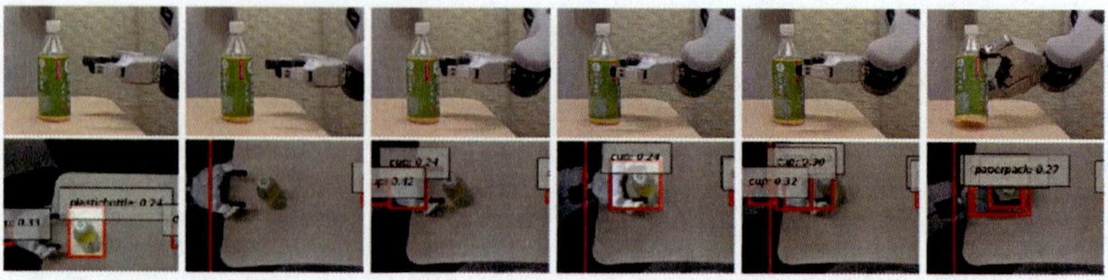

(a) $t = 0\,\mathrm{s}$ (b) $t = 2\,\mathrm{s}$ (c) $t = 4\,\mathrm{s}$ (d) $t = 6\,\mathrm{s}$ (e) $t = 8\,\mathrm{s}$ (f) $t = 12\,\mathrm{s}$

Fig. 8. Picking action in the second experimental setup. SSD recognition results are displayed below each image, highlighting the highest accuracy detection. Here, adopting a reactive system has caused stagnation around a misrecognition signal in $t = 2\,\mathrm{s}$ and manipulation errors due to a misrecognition in $t = 8\,\mathrm{s}$.

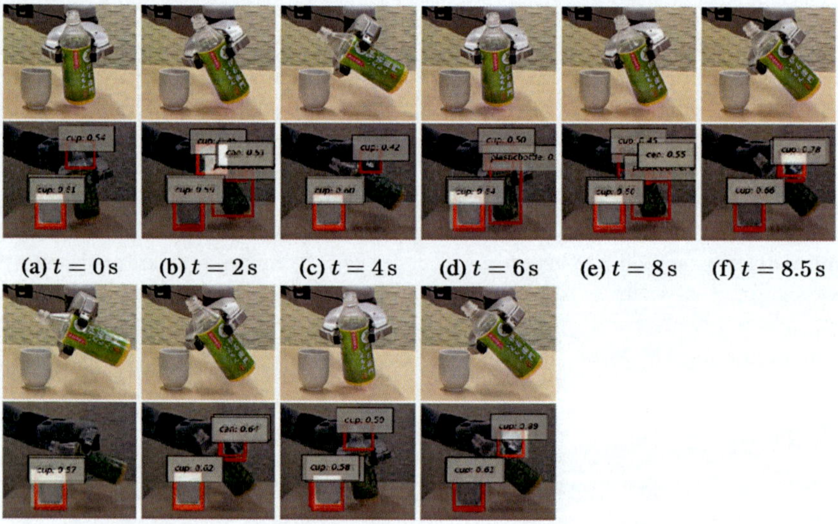

(a) $t = 0\,\mathrm{s}$ (b) $t = 2\,\mathrm{s}$ (c) $t = 4\,\mathrm{s}$ (d) $t = 6\,\mathrm{s}$ (e) $t = 8\,\mathrm{s}$ (f) $t = 8.5\,\mathrm{s}$

(g) $t = 10\,\mathrm{s}$ (h) $t = 12\,\mathrm{s}$ (i) $t = 14\,\mathrm{s}$ (j) $t = 15.5\,\mathrm{s}$

Fig. 9. Pouring action in the second experimental setup. SSD recognition results are displayed below each image, highlighting the highest accuracy detection. Here, adopting a reactive system has lead to chattering behavior as misrecognition results in $t = 2\,\mathrm{s}$, $t = 8.5\,\mathrm{s}$, and $t = 15.5\,\mathrm{s}$ prevent the robot from completing the action.

Results of the third experiment are shown in Figs. 10, 11 and Table 2. In this experiment, the proposed system is able to successfully filter misrecognition signals, attaining stable execution. In Fig. 10, three misrecognition signals at times $t = 1.00\,\mathrm{s}$, $t = 1.50\,\mathrm{s}$, and $t = 2.25\,\mathrm{s}$ cause the task to be interrupted. Since the bottle position is verified to be sufficiently close to its previous position by the RCNN node, execution is resumed in all three cases. In Fig. 10, a total of 13 misrecognition results were received, including at times $t = 0.0\,\mathrm{s}$ and $t = 15.0\,\mathrm{s}$.

(a) $t = 0.00\,\text{s}$ (b) $t = 0.25\,\text{s}$ (c) $t = 0.50\,\text{s}$ (d) $t = 0.75\,\text{s}$ (e) $t = 1.00\,\text{s}$ (f) $t = 1.25\,\text{s}$

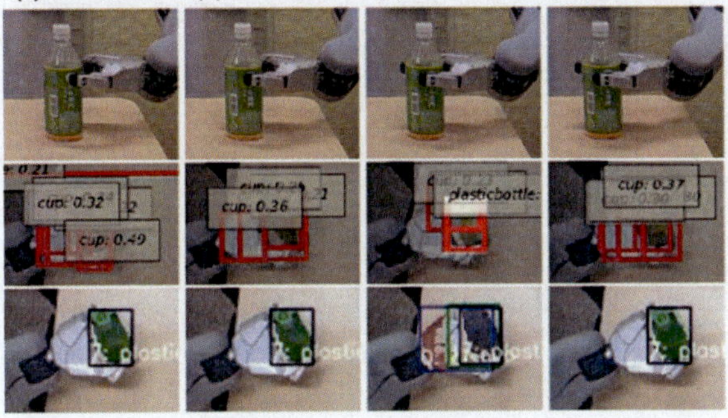

(g) $t = 1.50\,\text{s}$ (h) $t = 1.75\,\text{s}$ (i) $t = 2.00\,\text{s}$ (j) $t = 2.25\,\text{s}$

Fig. 10. Picking action in the third experimental setup. SSD and RCNN recognition results are displayed below each image, in this order. At $t = 1.00\,\text{s}$, $t = 1.50\,\text{s}$, $t = 1.75\,\text{s}$, and $t = 2.25\,\text{s}$ no SSD detections are received, compelling to robot to abort the picking procedure and engage in the search action. Since the RCNN result is still accurate, the proposed framework can seamlessly resume execution from the preempted point, ensuring stability and leading to task completion.

Preempting the pouring action at several points causes execution to be considerably slower than that of the first experiment, but the proposed system is able to correctly filter the detection errors and ultimately achieve task completion. Details of each resumption request are displayed in Table 2, in which signals regarding the picking action are shown in the top half and signals regarding the pouring action are shown in the bottom half. Separate data from two different executions (one for the picking action and another for the pouring action) are selected to highlight scenarios with multiple misrecognition signals. Data shows that the timeout settings of 30 s and 10 s were far larger than required, with the largest time interval being of about 2 s.

Finally, results of the fourth experiment are shown in Fig. 12 and Table 3. In this experiment, the proposed system is able to differentiate between misrecognition signals and actual environmental changes, allowing for the robot to correctly handle user interventions in a dynamic environment. Among the 11 resumption

(a) $t = 0.0\,\text{s}$ (b) $t = 2.5\,\text{s}$ (c) $t = 5.0\,\text{s}$ (d) $t = 7.5\,\text{s}$ (e) $t = 10.0\,\text{s}$ (f) $t = 12.5\,\text{s}$

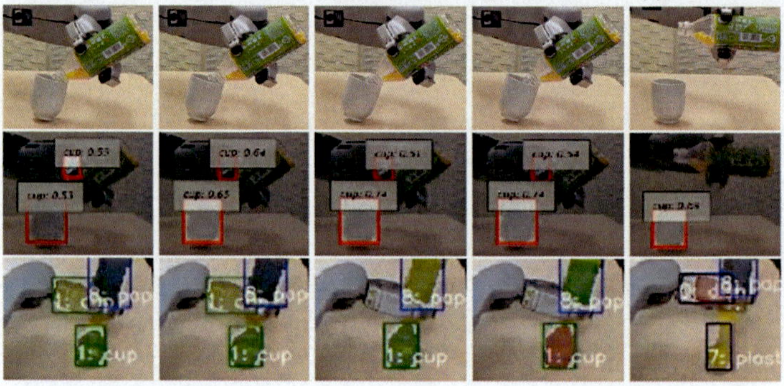

(g) $t = 15.0\,\text{s}$ (h) $t = 17.5\,\text{s}$ (i) $t = 20.0\,\text{s}$ (j) $t = 22.5\,\text{s}$ (k) $t = 25.0\,\text{s}$

Fig. 11. Pouring action in the third experimental setup. SSD and RCNN recognition results are displayed below each image, in this order. SSD detection errors such as the ones at $t = 0.0\,\text{s}$ and $t = 15.0\,\text{s}$ indicating that the cup has been moved compel the robot to stop the pouring task and tilt the bottle up. However, reasserting the RCNN detection results and verifying that the cup remains in the same place allows the robot to resume the action, avoiding chattering behavior and ensuring completion.

requests received by the robot, the proposed system mistakenly interprets one of them (fourth line in Table 3). The request is received after the human has taken away the bottle and, almost 15 s later, returned it to a similar spot. The robot then resumes the grasping action, closing the gripper. Since nothing is grasped, the robot then retries the picking action, this time successfully grasping the bottle. This reveals that the timeout value of 30 s was inconsistent with the given task and recognition framework. The robot also mistakenly tilts the bottle up once during the pouring task. This is due to a considerable delay between preemption and resumption requests (last line in Table 3), caused by perturbations in the low frequency RCNN detection model (Fig. 13).

Table 2. Evaluation of new requests and potential resumptions during picking (top) and pouring (bottom) actions on the third experiment.

Timestamp	Time difference	Coordinate difference	Resume
0.874	0.226988 s	3.82027 mm	Yes
1.244	0.297037 s	16.0534 mm	Yes
2.124	0.33567 s	15.8564 mm	Yes
12.031	0.219711 s	1.69375 mm	Yes
14.175	1.64306 s	3.55315 mm	Yes
15.591	1.31932 s	0.556701 mm	Yes
16.947	0.434762 s	2.64317 mm	Yes
18.404	1.00952 s	3.24859 mm	Yes
19.846	0.793067 s	3.48632 mm	Yes
22.004	2.07167 s	2.02815 mm	Yes
30.910	0.318637 s	5.39584 mm	Yes
32.836	0.382704 s	4.07519 mm	Yes
34.221	0.547311 s	57.6142 mm	Yes
46.029	0.37714 s	58.6319 mm	Yes
47.881	0.38899 s	19.1482 mm	Yes
50.128	0.353826 s	6.98617 mm	Yes

Table 3. Evaluation of new requests and potential resumptions during picking (top) and pouring (bottom) actions on the fourth experiment.

Timestamp	Time difference	Coordinate difference	Resume
3.969	2.04038 s	136.584 mm	No
14.165	0.256114 s	20.316 mm	Yes
14.460	0.231192 s	2.10993 mm	Yes
30.194	*14.7257 s*	*60.2746 mm*	*Yes*
43.490	0.180732 s	11.4077 mm	Yes
43.947	0.398434 s	9.50462 mm	Yes
44.545	0.196417 s	3.2379 mm	Yes
88.425	0.275922 s	2.11846 mm	Yes
91.459	1.40879 s	150.395 mm	No
113.808	4.15967 s	133.966 mm	No
131.366	***2.89912 s***	***5.20746 mm***	***Yes***

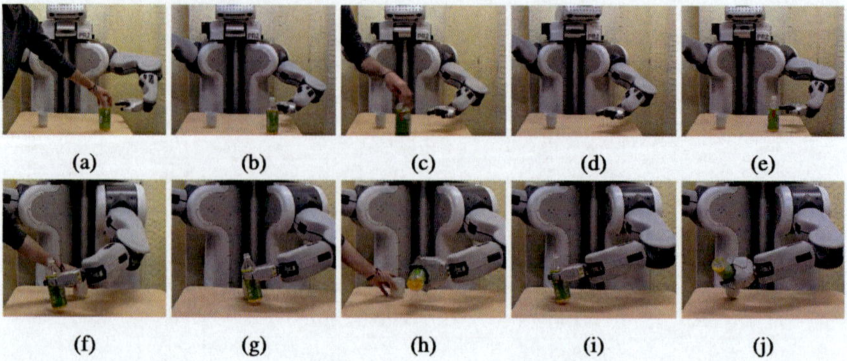

Fig. 12. Picking and pouring actions in the fourth experimental setup. The proposed system is able to differentiate between misrecognition results and actual environmental changes, ensuring stable execution and reactive handling of human interventions. (a) Human moves the bottle during the picking action. (b) Robot adapts to the new bottle position. (c) Human takes away the bottle during the picking action. (d) Robot searches for the bottle. (e) Robot adapts to the new bottle position. (f) Human moves the cup during the pouring action. (g) Robot adapts to the new cup position. (h) Human moves the cup during the pouring action. (i) Robot tilts the bottle up, and then adapts to the new cup position. (j) Robot finishes the pouring action.

5 Conclusions

In this paper we have proposed a framework that uses resumable interruptions to filter misrecognition results and ensure stability in reactive robotic systems acting on real world scenarios.

This technology is important because reactive systems are also over-sensitive to detection errors, potentially leading to chattering behavior and stagnation when misrecognition signals are received during task execution. In fact, experimentation has shown that such problems do occur when working with real world applications and standard robotic procedures. This difficulty has led to the increased usage of control substructures that inhibit reactivity, even among reactive frameworks – non-reactive structures are held as the standard control node in BehaviorTree.CPP [1], a popular behavior tree implementation library specialized for robotic usage.

When compared with the selective inhibition of environmental checks, the proposed method allows to attain enhanced reactivity and comparable stability under the same action module implementation. When compared with approaches that enforce zero resumption overhead on every module, the proposed system allows for simpler structures, without the added complexity of managing global variables, defining new recognition nodes, or administering extensive single-motion trees. When compared with standard filtering approaches, the "act first think later" nature of post-facto filtering allows to deliver signals faster and surely, while also providing prolonged coverage for misrecognition recovery.

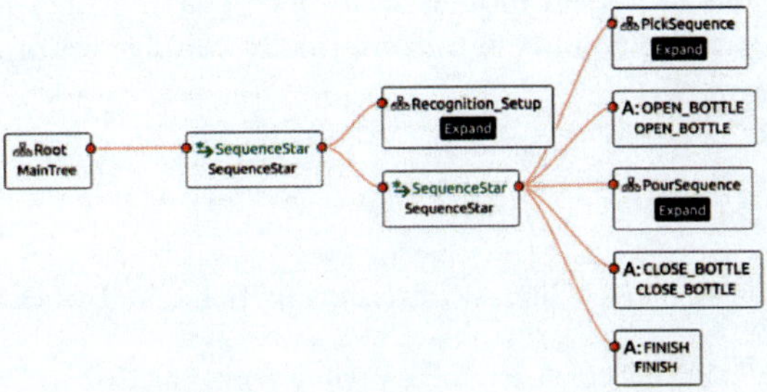

(a) Main task formulation using behavior trees.

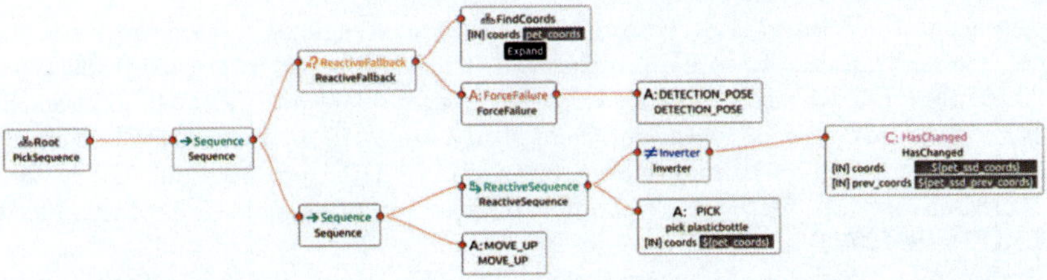

(b) The PickSequence subtree, describing a reactive formulation of the picking procedure.

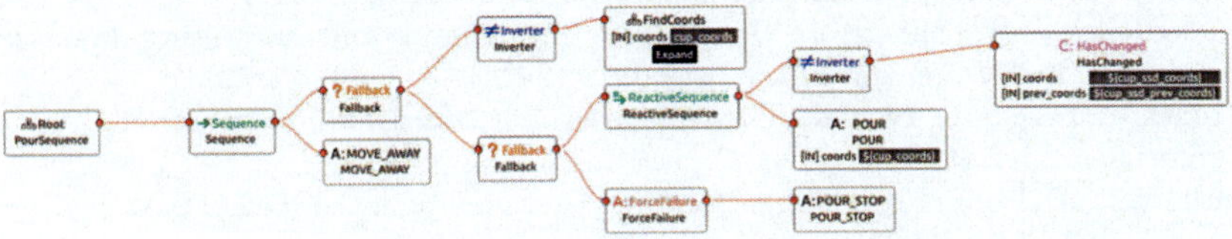

(c) The PourSequence subtree, describing a reactive formulation of the pouring procedure.

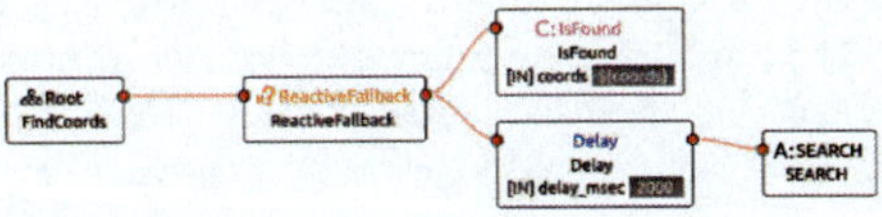

(d) The FindCoords subtree, which verifies if a RCNN coordinate is visible and start the searching action if it is not.

Fig. 13. Reactive task formulation using behavior trees, used in the third and fourth experiments. Actions are signalized by the prefix A, and conditions are signalized by a prefix C. The main task (a) includes three subtrees, describing the pick procedure (b), the pour procedure (c), and a detection procedure (d). Notice how reactive nodes (ReactiveFallback and ReactiveSequence) are used to intervene with the picking and pouring actions whenever the target object is moved during execution.

Experiments have shown that resuming actions up to two seconds later than their preemption was beneficial under the tested scenario, a time interval that would be typically unpractical when using standard filtering approaches.

The development of hybrid models with both pre and post-facto filtering, as well of adaptive systems capable of automatically tuning resumption parameters remains as future work.

References

1. BehaviorTree.CPP. https://github.com/BehaviorTree/BehaviorTree.CPP. Retrieved 2022
2. Bai, A., Russell, S.: Efficient reinforcement learning with hierarchies of machines by leveraging internal transitions. In: Twenty-Sixth International Joint Conference on Artificial Intelligence (2017)
3. Berenson, D., Srinivasa, S., Kuffner, J.: Task space regions: a framework for pose-constrained manipulation planning. Int. J. Robot. Res. **30**(12), 1435–1460 (2011)
4. Chakraborti, T., Sreedharan, S., Grover, S., Kambhampati, S.: Plan explanations as model reconciliation – an empirical study. In: 2019 14th ACM/IEEE International Conference on Human-Robot Interaction (HRI), pp. 258–266. IEEE (2019)
5. Colledanchise, M., Ögren, P.: Behavior Trees in Robotics and AI: An Introduction. CRC Press (2018)
6. Dudek, W., Węgierek, M., Szynkiewicz, W., Winiarski, T., et al.: Task harmonisation for a single-task robot controller. In: 2019 12th International Workshop on Robot Motion and Control (RoMoCo), pp. 86–91. IEEE (2019)
7. Goodfellow, I.J., Shlens, J., Szegedy, C.: Explaining and harnessing adversarial examples. arXiv preprint arXiv:1412.6572 (2014)
8. Holtz, J., Guha, A., Biswas, J.: Interactive robot transition repair with SMT. arXiv preprint arXiv:1802.01706 (2018)
9. Isla, D.: GDC 2005 Proceeding: Handling Complexity in the HALO 2 AI. Retrieved October, 21:2009 (2005)
10. Johnson, B., Kress-Gazit, H.: Analyzing and revising synthesized controllers for robots with sensing and actuation errors. Int. J. Robot. Res. **34**(6), 816–832 (2015)
11. Kim, K., Fainekos, G., Sankaranarayanan, S.: On the minimal revision problem of specification automata. Int. J. Robot. Res. **34**(12), 1515–1535 (2015)
12. Kurakin, A., Goodfellow, I., Bengio, S., et al.: Adversarial examples in the physical world (2016)
13. Li, T., Pan, J., Zhu, D., Meng, M.Q.-H.: Learning to interrupt: a hierarchical deep reinforcement learning framework for efficient exploration. In: IEEE International Conference on Robotics and Biomimetics (ROBIO), pp. 648–653. IEEE (2018)
14. Mateas, M., Stern, A.: A behavior language for story-based believable agents. IEEE Intell. Syst. **17**(4), 39–47 (2002)
15. Millington, I., Funge, J.: Artificial Intelligence for Games. CRC Press (2018)
16. Mohseni-Kabir, A., Veloso, M.M.: Robot task interruption by learning to switch among multiple models. In: IJCAI, pp. 4943–4949 (2018)
17. Murooka, M., Nozawa, S., Bando, M., Yanokura, I., Okada, K., Inaba, M.: Simultaneous planning and estimation based on physics reasoning in robot manipulation. In: 2018 IEEE International Conference on Robotics and Automation (ICRA), pp. 3137–3144. IEEE (2018)

18. Nedunuri, S., Prabhu, S., Moll, M., Chaudhuri, S., Kavraki, L.E.: SMT-based synthesis of integrated task and motion plans from plan outlines. In: 2014 IEEE International Conference on Robotics and Automation (ICRA), pp. 655–662. IEEE (2014)
19. Nilsson, N.J.: A mobile automaton: an application of artificial intelligence techniques. Technical report, SRI International, Menlo Park, CA, Artificial Intelligence Center, 1969
20. Nilsson, N.J.: Shakey the robot. Technical report, SRI International, Menlo Park, CA, 1984
21. Omar, S., Ngadi, A., Jebur, H.H.: Machine learning techniques for anomaly detection: an overview. Int. J. Comput. Appl. **79**(2) (2013)
22. Raptis, I.A., Sconyers, C., Martin, R., Mah, R., Oza, N., Mavris, D., Vachtsevanos, G.J.: A particle filtering-based framework for real-time fault diagnosis of autonomous vehicles. In: Annual Conference of the PHM Society, vol. 5 (2013)
23. Rigatos, G., Busawon, K.: Robotic Manipulators and Vehicles: Control, Estimation and Filtering, vol. 152. Springer (2018)
24. Rigatos, G.G.: Extended Kalman and particle filtering for sensor fusion in motion control of mobile robots. Math. Comput. Simul. **81**(3), 590–607 (2010)
25. Shon, T., Moon, J.: A hybrid machine learning approach to network anomaly detection. Inf. Sci. **177**(18), 3799–3821 (2007)
26. Sreedharan, S., Srivastava, S., Smith, D., Kambhampati, S.: Why can't you do that HAL? Explaining unsolvability of planning tasks. In: International Joint Conference on Artificial Intelligence (2019)
27. Wong, K.W., Ehlers, R., Kress-Gazit, H.: Correct high-level robot behavior in environments with unexpected events. In: Robotics: Science and Systems (2014)
28. Wong, K.W., Ehlers, R., Kress-Gazit, H.: Resilient, provably-correct, and high-level robot behaviors. IEEE Trans. Robot. **34**(4), 936–952 (2018)
29. Zachary Woodruff, J., Lynch, K.M.: Planning and control for dynamic, nonprehensile, and hybrid manipulation tasks. In: 2017 IEEE International Conference on Robotics and Automation (ICRA), pp. 4066–4073. IEEE (2017)

Gestural and Touchscreen Interaction for Human-Robot Collaboration: A Comparative Study

Antonino Bongiovanni, Alessio De Luca, Luna Gava, Lucrezia Grassi, Marta Lagomarsino, Marco Lapolla, Antonio Marino, Patrick Roncagliolo, Simone Macciò, Alessandro Carfì(✉), and Fulvio Mastrogiovanni

Department of Informatics, Bioengineering, Robotics, and Systems Engineering, University of Genoa, Via Opera Pia 13, 16145 Genoa, Italy
alessandro.carfi@dibris.unige.it

Abstract. Close human-robot interaction (HRI), especially in industrial scenarios, has been vastly investigated for the advantages of combining human and robot skills. For an effective HRI, the validity of currently available human-machine communication media or tools should be questioned, and new communication modalities should be explored. This article proposes a modular architecture allowing human operators to interact with robots through different modalities. In particular, we implemented the architecture to handle gestural and touchscreen input, respectively, using a smartwatch and a tablet. Finally, we performed a comparative user experience study between these two modalities.

1 Introduction

Research on human-robot collaboration (HRC) leapt forward in industrial scenarios with the introduction of collaborative robots (cobots) [1] such as the Kuka LBR iiwa, the Universal Robot UR5, or ABB YuMi (to mention just a few). Cobots are safe by design and make it possible for robots and humans to combine their skills, improving overall productivity, efficiency and flexibility while, possibly, reducing human stress and workload [2]. For a functional human-robot collaboration, safety is not the only requirement, and cobots should provide more intuitive interfaces moving away from classical ones such as teach pendants.

More prominent human-robot interfaces rely on touchscreen technology [3,4] that integrates into the same device the human to robot communication (touch) and the robot to human one (screen). Touch screen interfaces are classical graphical user interfaces (GUIs) where the presence of the physical screen acts as a gateway between the human and the robot. New technologies such as augmented and virtual reality provide alternatives for robots to communicate with humans.

Authors Antonino Bongiovanni, Alessio De Luca, Luna Gava, Lucrezia Grassi, Marta Lagomarsino, Marco Lapolla, Antonio Marino and Patrick Roncagliolo contributed equally.

I. Petrovic et al. (Eds.): IAS 2022, LNNS 577, pp. 122–138, 2023.
https://doi.org/10.1007/978-3-031-22216-0_9

These interfaces need novel approaches for human interaction, and researchers took inspiration from human-human communication exploring speech [5], gestures [6–8] and their combination [9–11].

Gestural- and speech-based interfaces are appealing since humans already use these modalities to interact with other humans. However, their usage in an intrinsically different scenario may lead to sub-optimal interaction results. For these reasons, researchers performed experiments to compare new interaction modalities with state-of-the-art ones, such as those based on touchscreens. Often, the number of participants of these studies is not significant, and they rely on Wizard of Oz (WoZ) experiments (i.e., experiments in which the subject interacts with an autonomous system, but in reality, an experimenter is operating it) [12].

(a) Touchscreen interface.

(b) Gesture-based interface.

Fig. 1. Two users using, respectively, a tablet and a smartwatch, highlighted by yellow circles, to interact with a robot.

Since previous studies highlighted them as one of the most promising approaches, we aim to contribute to this research field by evaluating the user experience and performance results of a gesture-based interface for HRC. We do this by comparing gestural (Fig. 1b) with touchscreen (Fig. 1a) interaction in a real-world HRC scenario. Touchscreen interaction has been picked as an optimal term of comparison because of its modern and reliable technology. For the experiment, although it could be suboptimal in the HRC scenario, we adopted a simple GUI to have optimal conditions for the touchscreen interface since a GUI is part of its usual setup. Part of the contribution of this work is the design and implementation of a software architecture capable of handling both communication modalities.

The paper is structured as follows. In Sect. 2 a brief review of HRC interfaces literature is presented. Section 3 describes the design principles adopted to design our software architecture. Section 4 describes, in detail, the architecture implementation for the gestural and touchscreen interaction and the logic managing the HRC. The setup and the experiment description are presented in Sect. 5, while in Sect. 6 the experimental results are presented and discussed. Conclusions follow.

2 Background

While humans and robots collaborate, they share information at different levels through explicit or implicit communication. Ideally, an interface for human-robot collaboration should mediate the two-way communication handling both explicit and implicit communication, but in this paper, we focus only on the explicit one. In this context, the interface allows the human to send direct commands to the robot and the robot answers providing appropriate feedback. Classical interfaces use screens to give the user visual feedback but, novel technologies, such as augmented reality (AR) and see-through displays [13–15], allow integrating the interface directly in the working environment.

At the same time, researchers explored different modalities to acquire human commands. Many studies focused on touchscreen interfaces for human-robot interaction, given their widespread adoption in the consumer market. Mateo et al. 2014 [3] proposed a tablet-based user interface for industrial robot programming, and a similar interface has been used to send high-level commands to robotic platforms [4]. However, touchscreen interfaces are suboptimal in scenarios where humans need free hands. Furthermore, the need for a screen to communicate limits the adoption of technologies such as AR that can only be used as a complement [9], and not fully substitute it. Therefore, developing new interfaces requires alternative communication modalities such as speech and gestures. Despite the extended research in speech analysis and recognition [16], the high noise level in industrial environments can jeopardize the correct function of speech-based interfaces. While gesture-based interfaces usually rely on upper-limb motions, possibly altering the human workflow. However, with gesture-based interfaces, users can send commands to the robot while handling tools; and without using a tablet or reaching a screen. Furthermore, interfaces can rely on gestures for scenarios where other communication modalities fail, e.g., underwater HRC [17].

Because of their advantages, researchers have explored the usage of gesture-based interfaces for industrial applications [6], search and rescue [11], and direct control of mobile robots [7]. Neto et al. 2019 [6] proposed a gesture-based interface using five inertial measurement units (IMUs), worn by the user, to send commands to an industrial manipulator. At the same time, other studies proposed interfaces using non-wearable devices, such as RGB [17] and RGB-D [7] cameras, or alternative wearable devices such as Electromyography (EMG) sensors

[11,18]. Although wearable devices can impede human motions and rely on batteries, their usage reduces the necessity for a structured environment, raises fewer privacy concerns, and does not suffer occlusions. IMUs are one of the most used sensing solutions for wearables because of their compact data stream, reduced price and small size. Independently by the sensor type, a gesture-based interface should process the collected data to identify motions with explicit communication intent and ignore normal user operations. This problem is known as gesture recognition, and for IMU data, it has been approached using Naive Bayes Classifiers, Logistic Regression and Decision Trees [19], Convolutional Neural Networks (CNNs) [20], Dynamic Time Warping (DTW) [21], Support Vector Machines (SVMs) [22], and Long Short-Term Memory Neural Networks (LSTM) [23].

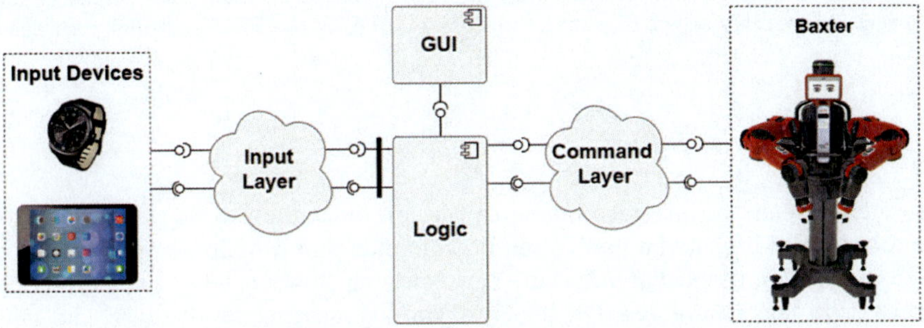

Fig. 2. A conceptual flow diagram of the proposed architecture.

The development of a new interface is not only a technological problem but, the human factor should be taken into consideration as well [24]. Therefore, an experimental validation should evaluate both the system performance and the user experience. For this reason, some studies proposed comparative analyses between different interaction modalities to determine which is the most suited given a target application. Taralle et al. 2015 [13] presented a comparative study between a gesture-based and touchscreen interface to control a small unmanned aerial vehicle (sUAV) while looking for a target in a video stream. This study relied on WoZ experiments for the gesture controls, and the results, over 20 volunteers, suggested a preference for the gestural interaction. At the same time, lower cognitive load and user preference for gestural and touch inputs have been found using WoZ experiments to compare touch, gesture, speech, and 3D pen in an industrial scenario [25]. Although these studies provide promising results, WoZ experiments overlook the effect of the system performances (e.g., accuracy, responsiveness) on the user experience. When compared in a real-world environment, reliable input modalities, such as touchscreens, outperform novel ones, such as gestures [12]. These results should not discourage the study of new interaction modalities but rather push to reach better system performances to (i) obtain interfaces for a more flexible human-robot collaboration and (ii) overcome the need for a screen in classical interfaces.

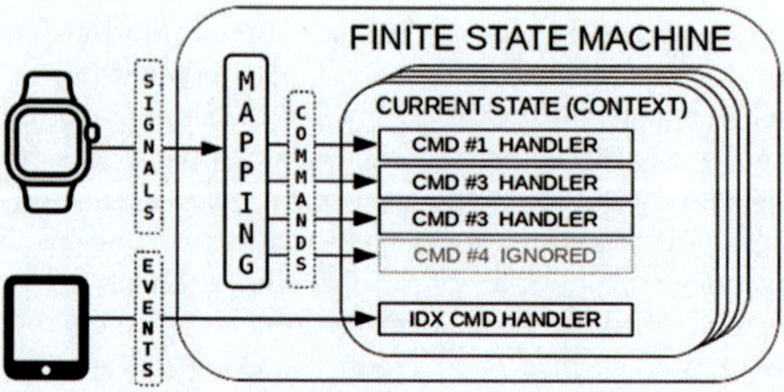

Fig. 3. Many heterogeneous discrete *signals* are mapped to one of the four FSM command handlers. Each FSM state can install the handlers, or ignore them (like command number 4 in this example). A secondary input scheme, i.e., *events*, uses valued signals (integers) to directly select one of the options offered by the FSM state for user selection.

3 Design Principles

As we have seen, an interface mediates the interaction between the human and the robot, allowing the human to send commands and provide appropriate feedback. In Fig. 2, a general architecture summarizing this concept is presented. The architecture consists of: a GUI displayed on a screen, an input layer, the main logic, and a command layer that communicates directly with the robot. Since our study focuses on the human to robot communication, we kept the robot to human one simple using only a screen. Since they are vastly adopted, we design a menu-based GUI listing all possible functionalities to minimize the novelty effect. Furthermore, menus allow an easy adaptation to different experimental scenarios.

The input layer collects the sensory information and processes them to recognize discrete human commands (i.e., arm gestures, keywords, or touchscreen pressures). Each communication modality has its dictionary D containing commands descriptions and associated identifiers. Once the sensory information matches the command description, the input layer returns as output the associated identifier. In some communication modalities, the input layer also generates additional information. For example, when the user presses a touchscreen, the input layer returns the identifier (i.e., screen pressure) and the related 2D position.

The logic layer is represented using a finite state machine (FSM) whose states describe different interaction stages. The set of states S ($S = \{s_1, \dots, s_i, \dots s_{|S|}\}$) composes the FSM, and the transitions from s_i ($s_i \in S$) to other states are described by the transitions set T_i. In our scenario, we recognize two states categories, namely *menu* and *action*. Menu states describe the GUI (Fig. 5 shows the GUI representation of some menu states), and when they are active, the interaction is limited to the menu navigation. Instead, action states implement the system functionalities, e.g., human teaching of a motion or robot execution of a task.

Transitions between FSM states are triggered either by the system (e.g., when the robot ends a task execution) or by human commands. As we have seen, different communication modalities can carry different information. Therefore, to make it possible to integrate various communication modalities, we designed two different schemas that a human command can follow to interact with the FSM: *signals* and *events* (see Fig. 3). With signals, for each state s_i, the system defines a one-to-one mapping between the dictionary commands and the transitions defined in T_i. Command handlers, defined for each FSM state (see Fig. 3), manage this mapping. With events, human commands are mapped directly to FSM states. Therefore, when the human performs the command associated with s_j, independently by the current FSM state, s_j is activated. This schema allows long jumps in the interface without a sequential transition through all the intermediate states. Usually, the number of states $|S|$ is higher than the number of transitions for a single state $|T_i|$, therefore using events needs bigger dictionaries or communication modalities that, carrying extra information, allow to associate more states to a single command. Signals are more appropriate for communication modalities where a big dictionary implies high user effort to recall commands (e.g., keyboard strokes or gestures). Instead, events can leverage communication modalities handling big dictionaries (e.g., vocal interaction) or providing extra information (e.g., touchscreen interaction).

4 Implementation

Following the design principles presented in the previous section, we have implemented our architecture (available open-source[1]) to interact with the dual-arm Baxter robot from Rethink Robotics [26] using gestures, sensed by a wrist-worn IMU and a touchscreen. The architecture uses the robot operating system (ROS) framework [27] to manage inter-module communication. Besides benefits in code reuse, this choice allows us to exploit the Baxter-related ROS APIs, which exposes services to acquire sensory data, control robot actuators, and record robot motions using Kinesthetic Teaching (KT) [28,29].

4.1 Graphical User Interface

Our Graphical User Interface arranges the menu's options vertically, and a red selector is used to highlight the selected option (see Fig. 5a). The FSM menu states describing the GUI are published on a ROS topic and converted, by a renderer, to a graphical representation. The GUI can be visualized on a screen either locally or remotely through a browser. Since the architecture decouples the GUI state representation (FSM) and its visual representation, it is possible to change the GUI appearance (e.g., textual user interfaces or GUI with different designs) or integrate other visualization devices (e.g., augmented reality headsets).

[1] Web: https://github.com/TheEngineRoom-UniGe/gesture_based_interface.

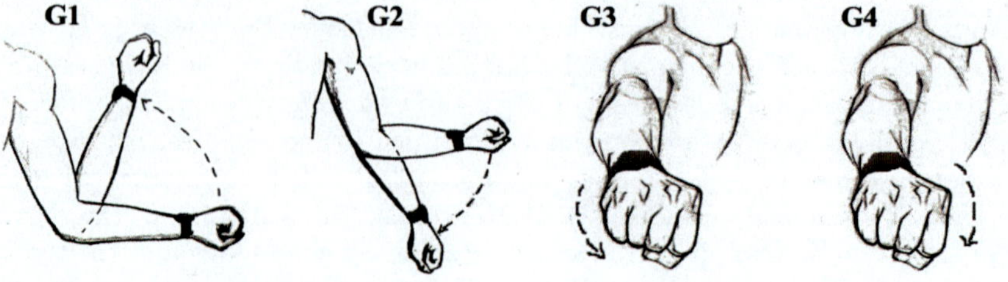

Fig. 4. The four gestures supported by our architecture: (G1) wrist up, (G2) wrist down, (G3) spike clockwise, and (G4) spike counter-clockwise.

4.2 Input

The two interaction modalities considered in this study are touchscreen and gestures. The two modalities are intrinsically different, and for the reasons previously presented, we use signals to handle gestures and events for the touchscreen. While using gestures, users navigate the menus sequentially, performing multiple gestures to reach and select an option. With the touchscreen instead, the user should press simply the corresponding virtual button in the GUI. In our experiment, the touchscreen interface is used on an iPad Air through Safari, and we developed a ROS node receiving the command *events* directly from the tablet.

For the gesture-based interface instead, we used SLOTH, a method taking advantage of LSTM neural networks to recognize gestures online. This method has been proposed for HRC scenarios and a ROS-compatible version, with a pre-trained neural network, is available on GitHub.[2] We developed an Android Wear 2 application to collect IMU data from an LG G Watch R smartwatch and publish them online to a ROS topic. Therefore, since the data acquisition pipeline differs from the one used in the original work, the SLOTH performances should be assessed again [30]. Every time a gesture is recognized, the SLOTH module sends a *signal* to the main logic activating an FSM transition. In our FSM implementation, the maximum number of possible transitions associated with a state is four ($\forall i \in \{1, \ldots, |S|\}, |T_i| \leq 4$), therefore the architecture presents only four command handlers (see Fig. 3). Since SLOTH recognizes between six gestures ($|D| = 6$), we ignore two of them. The command handlers map the others (see Fig. 4) as follows: command handler #1 maps wrist up (G1), command handler #2 maps wrist down (G2), command handler #3 maps spike clockwise (G3), and command handler #4 maps spike counter-clockwise (G4). Notice that the number of active command handlers in a state s_i corresponds to the number of possible transition T_i, and gestures mapped by inactive command handlers would be effectless.

[2] Web: https://github.com/ACarfi/SLOTH.

4.3 Logic

As previously described, we implemented the architecture logic using an FSM that the user can control using signals or events. When using the gesture interface, each gesture execution generates a signal triggering the corresponding state transition. In this case, whichever change in the GUI (e.g., selector up, selector down or option selection) corresponds to a state transition in the FSM. Users can navigate menu states using three gestures: G1 to select options, G2 to move down the selector, and G3 to move up the selector (in menu states G4 is effectless).

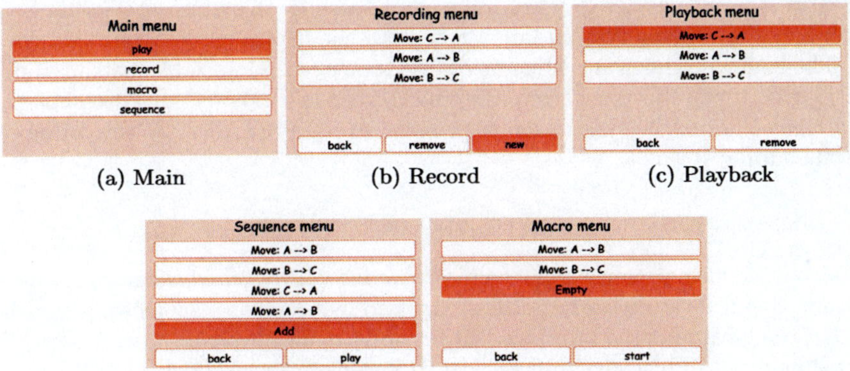

(a) Main (b) Record (c) Playback

(d) Sequential playback (e) Macro mode

Fig. 5. The menus for the four options provided by the interface.

Our architecture allows the user to control robot functionalities to record and playback end-effector trajectories. In particular, the main menu offers four options: *record*, *playback*, *sequential playback*, and *macro mode* (see Fig. 5a). Whenever one of these options is selected, the corresponding menu is opened.

The record menu (Fig. 5b) displays a list of recorded robot tasks (i.e., end-effector trajectories). The user can overwrite each task by selecting it or creating a new one using the corresponding option. By selecting a task from the list, the recording process, handled by the *record* action state, starts. In this state, a human operator can program a new task using KT (i.e., teaching a trajectory by physically guiding the robot) and use G4 to terminate the recording going back to the record menu.

The playback menu (Fig. 5c) lists all the saved tasks, which can be deleted by the human operator using the corresponding option. When a task is selected, the FSM transits to the *playback action* state, and the robot reproduces the associated trajectory. Here two commands are active: G2 pausing or resuming the task playback and G4 terminating the playback and moving back to the playback menu.

The sequential playback menu (Fig. 5d) allows the user to combine saved tasks in sequences to handle more complex behaviours. The user can remove or substitute a task from the sequence by selecting it. Furthermore, the *add* option brings to a selection sub-menu where the operator can append a task to the sequence, and the *run* option makes the FSM transit to the *playback action* state that reproduces the whole sequence.

Finally, the macro mode menu, see Fig. 5e, allows the user to associate one task to each of the three gestures G1, G2 and G3 (the correspondence task-gesture is ordered from top to bottom). The user can customize the mapping by selecting a specific slot and choosing a task from the list of available ones. When *run* is selected, the FSM transits to the *macro action* state, and the robot starts waiting for gestures. Every time the user performs one of the three mapped gestures, the robot executes the associated task. In this state, performing G4 terminates the interaction and moves back to the macro mode menu.

The logic does not vary when using the touchscreen interface. Instead of performing gestures, the user presses the touchscreen in a 2D position corresponding to a menu option, the system sends an event to the FSM, directly activating the corresponding state.

5 Experiment

Now that we have presented our architecture for multimodal human-robot interaction, we put forward our hypothesis inspired by previous researches. [Hypothesis H_1] the gesture-based interface will get positive user evaluation, and [$H_{1.1}$] the novelty effect will have a significant contribution. However, [H_2] the touchscreen interface will grant an overall better user experience [$H_{2.1}$] since it relies on a more stable technology and volunteers are accustomed to this kind of interface.

5.1 Experimental Setup

The experimental setup, whose conceptual sketch is represented in Fig. 6, includes: (i) a dual-arm Baxter manipulator, appropriately configured, (ii) a wooden, 3.7 cm edge cube, to be manipulated by the robot, (iii) the tablet or the smartwatch that human operators use, (iv) a table in front of the robot, on which two positions, namely *A* and *B*, are defined. Position *A* is where the cube is located at the beginning of the experiment. The distance between *A* and *B* is 60 cm. While performing the experiments, the human volunteer stands in front of the robot at a distance of approximately 1.5 m.

The architecture contains three pre-recorded tasks, namely MOVE: A → B, ACTION 1, reproducing a robot arm-waving to greet the user, and ACTION 2, performing a handover of the cube between the two Baxter's arms. These have been included to simulate the difficulty that using an already setup interface may imply.

5.2 Description of the Experiment

We performed a within-subject experiment divided into two phases, i.e., a trial using the tablet (touchscreen-based) and one wearing the smartwatch (gesture-based). The 25 participants, in a range between 18 and 40 years old, have been

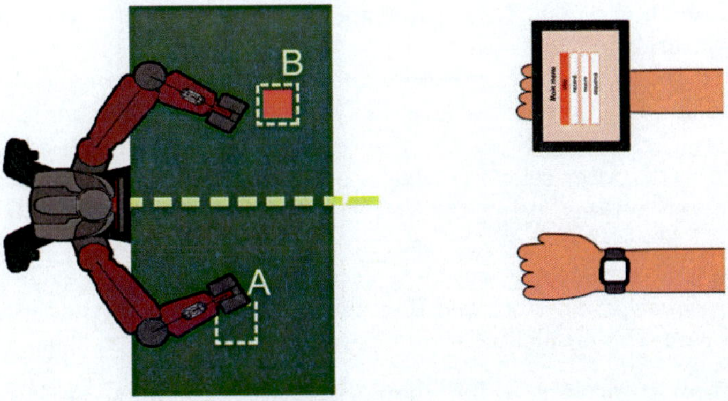

Fig. 6. Experimental scenario: a human holds the tablet or wears the smartwatch, while Baxter is behind a table where the pick-and-places of a wooden cube are performed.

divided into two groups. The two groups completed the experiment once with each device: the first group started using the smartwatch, and the second group with the tablet.

The experimental protocol consists of a series of transportation tasks of the wooden block from position A to position B and *vice versa*. The success in bringing back the cube from B to A and the overall quality of the robot motions are not evaluation criteria considered in this experiment. Instead, as previously anticipated, we keep track of the time each operator needs to complete the experiment. Before the experiment, all participants familiarized themselves with the gestural control and the KT for approximately 1.5–2 min. Whereas in the end, each participant has filled up the User Experience Questionnaire (UEQ) [31] and a custom made questionnaire asking familiarity with the devices, and whether they consider useful to keep on developing this project.

The experiment consists of four tasks, and it is imposed a *soft* limit of 5 min to complete them. The temporal constraint limits the experiment duration to prevent fatigue and distractions that may affect the final result. The experiment is composed of four ordered tasks. When a volunteer completes a task, an experimenter evaluates if the remaining time is sufficient for the next one. If the time is not enough, the experimenter stops the volunteer, and the experiment ends. The experimenter does not interrupt the volunteer task execution; therefore, some experiments can exceed the soft limit. The description of the ordered tasks follows:

1. Starting from the main menu, the participant must select the *playback* option, and reproduce the pre-recorded action MOVE: A $\rightarrow$ B, which commands the robot to move the cube from position A to position B.
2. Each participant must select the *record* option from the main menu to record a new task, i.e., MOVE: B $\rightarrow$ A, to bring the cube back from B to A. Using KT, enabled by cuffing Baxter's wrist, the volunteer physically guides the

robot through the task. Once the recording is finished, the wooden block is back to position A.

3. The third task consists in selecting the *macro* option from the main menu and associating G1 and G2 (see Fig. 4), or two buttons while using the touchscreen, respectively to the pre-recorded MOVE: A $\rightarrow$ B and the new MOVE: B $\rightarrow$ A. After selecting *play*, see Fig. 5e, the participant has to activate the two actions, using the corresponding inputs, to move the wooden block from A to B and *vice versa*.
4. The last task consists in selecting the *sequence* option from the main menu to create a sequence of actions and then selecting *play* to start the reproduction. The sequence should include the prerecorded tasks ACTION 1 and ACTION 2.

Notice that to preserve a fair comparison, the experiment design does not include factors that could disadvantage the touchscreen interface, e.g., the need to use working gloves or hold a tool.

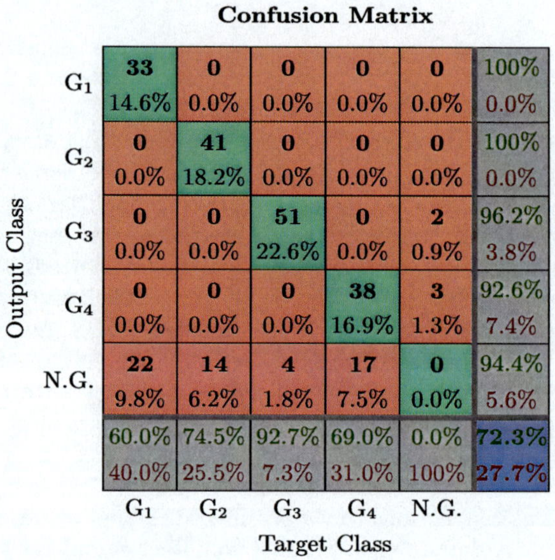

Fig. 7. Confusion matrix for online tests. The bottom row reports the *recall* measures, while the rightmost column reports the *precision* measures. The blue cell reports the overall accuracy.

6 Results

This section presents the SLOTH performances in our architecture and the results of the comparative study. We divide the study results between the evaluation of questionnaire answers and the analysis of the time needed to complete the experiment.

6.1 Assessment of SLOTH Performance

To assess the performance of SLOTH and decide if retraining the model, we asked eleven volunteers to perform five repetitions for all the four gestures considered in our experiment, see Fig. 4. We conducted the assessment using our Android infrastructure streaming IMU data at 10 Hz, as in the original work, directly to a workstation running SLOTH. The results are presented in the confusion matrix, Fig. 7. In general, the overall classification performance seems adequate for the study. In particular, the high level in precision indicates a reliable system that does not misinterpret gestures, preserving the correct association between gestures and commands, e.g., when the user performs G1, the system does not confuse it with G2, G3 or G4. However, the low recall levels suggest that user gestures may go undetected, e.g. when the user performs G1 has a 60% chance that is not detected. We expect this to negatively affect the user experience since the user will have to perform multiple gesture repetitions for a single command.

6.2 Analysis of Questionnaires

The UEQ [31], the questionnaire used during the experiments, uses 26 questions to assess different properties related to user experience. The questions are grouped into six user experience scales: (i) *attractiveness*, i.e., the overall impression about the interface; (ii) *perspicuity*, i.e., the easiness of learning how to use the interface; (iii) *efficiency*, i.e., the utility of the interface in completing the task; (iv) *dependability*, i.e., the perceived security and control over the interface; (v) *stimulation*, i.e., the engagement generated by the interface; and (vi) *novelty*, i.e., the perceived novelty and aroused interest. Five of these scales can then be grouped in *pragmatic qualities*, describing the task-related aspects of perspicuity, efficiency and dependability, and *hedonic qualities*, including non-task-related aspects, such as stimulation and novelty. Since all the volunteers were from Italy, we used the Italian version of the questionnaire. We analysed the questionnaire answers using the Microsoft Excel sheet provided together with the questionnaire. The tool gets the volunteers answers and calculates scaled values.

As a first step, we checked the value of the *Cronbach' s alpha coefficient*, which provides a measure of internal consistency, i.e., how closely related a set of items are as a group. The rule-of-thumb is that the alpha coefficient should be higher or equal than 0.7 to ensure scale reliability. Questionnaires related to the touchscreen interface have low alpha values only for dependency (0.35). Instead, for the gesture-based interface, we found low Cronbach's alpha for efficiency (0.55) and dependability (0.62). These findings can suggest a common misunderstanding of dependability items, but most probably are also influenced by the limited number of participants.

Figure 8 presents the comparative results between the touchscreen interface (tablet, in green) and the gesture-based one (smartwatch, in blue). The bar graph represents the mean for each considered user experience scale. According to the UEQ guidelines, mean values between −0.8 and 0.8 represent a neutral

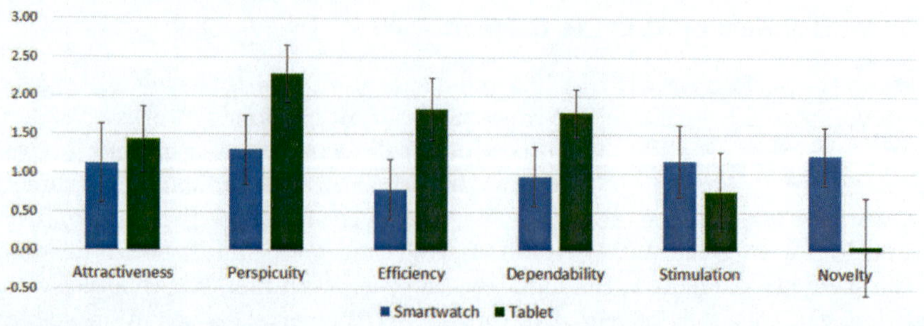

Fig. 8. Comparison between gesture-based (in blue) and touchscreen-based (in green) experiments, related to usability and user experience aspects.

evaluation of the corresponding scale, values greater than 0.8 represent a positive evaluation, and values less than −0.8 represent a negative evaluation. Instead, error bars represent the 95% confidence of scale means, i.e., the interval whereby 95% of the scale means are located if the experiment is repeated an infinite number of times. While comparing the values, if the two confidence intervals do not overlap, the difference is significant. Observing Fig. 8, a relevant difference is evident for pragmatic qualities (i.e., perspicuity, efficiency and dependability) where the tablet achieves higher evaluations. Instead, for hedonic qualities (i.e., stimulation and novelty), volunteers evaluated more positively the gesture-based interface.

Finally, from the custom made questionnaire, we found out that 64% of the participants were familiar with tablets, while only 12% with smartwatches. Moreover, while tablet usage assumes a touchscreen interaction, smartwatches are not typically used for gestural interaction. This difference, together with the non-optimal performances of the gesture recognition system, probably justifies the disadvantage of gestural interaction in the pragmatic scales and its advantage in the hedonic ones.

6.3 Time Analysis

Out of the 25 participants, only 5 completed the four tasks in the given time using the smartwatch, while 23 completed all the tasks using the tablet. This result is coherent with the participants' perception measured by the questionnaire about the overall interface efficiency. Figure 9 reports with box plots the statistics of the time spent to perform each of the four tasks with the tablet (in green) and the smartwatch (in blue). For both interfaces, we highlighted outliers in red. We point out that the box plots represent only the data of 23 volunteers since the subdivision, over the four tasks for the first two volunteers, have not been recorded. Furthermore, data for the sequence task using the gesture-based interface were available only for five participants since the others did not complete this task.

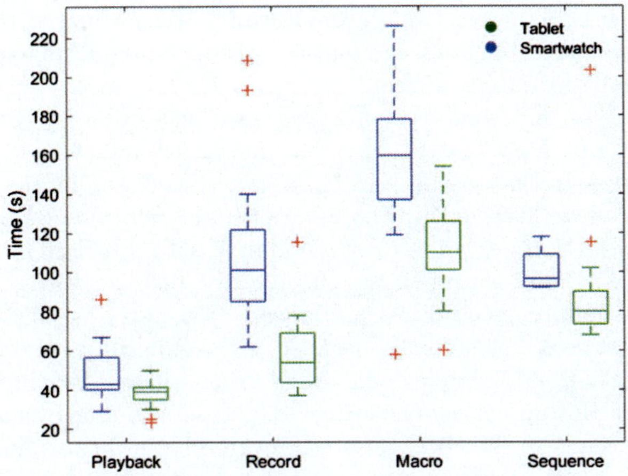

Fig. 9. Box plots of the time that each participant needed to complete the four experimental tasks. In green data about the touchscreen-based interaction, in blue data for the gesture-based one. We reported outliers in red.

The measured times do not measure only the user interaction but also include robot motions. This factor partially explains the higher variance for both touchscreen and smartwatch in the record and macro tasks. In fact, in these tasks, the time taken by the robot motion is different for each participant, since the MOVE: B → A task is recorded individually by each of them. Instead, in the playback and the sequence tasks, the used robot motions, pre-recorded by the experimenters, are the same for everyone, and therefore the variance is comparatively lower.

The playback task necessitates a few user inputs (i.e., selecting the playback option and then the motion to execute), and therefore the temporal difference between the two input modalities is limited. Instead, when the task becomes more complex (i.e., more user inputs are required to reach the end), the difference between the two input modalities increases due to the intrinsic difference between signals and events. This observation finds evidence in the timing results for the record and macro task but not for the equally complex sequence task. This result can be explained by observing that only five top-performing subjects reached this task using the smartwatch, and thus it suggests that the influence of volunteer skills is relevant. Therefore, more extensive user training could smooth the differences between gesture- and touchscreen-based interfaces.

7 Conclusions

In this work, we introduced a few designs principles for architectures supporting multiple human-robot interaction modalities. We presented the implementation of these design principles into an architecture mediating the human-robot interaction using both a smartwatch and a tablet. We evaluated the differences between the two communication modalities with a comparative experiment in a human-robot collaboration scenario.

The result of the experiments shows a positive evaluation for both interaction modalities [H_1 and H_2]. The smartwatch and tablet interface received better evaluations, respectively, on the hedonistic [$H_{1.1}$] and pragmatic scales [$H_{2.1}$]. These results can be justified by the non-optimal performances of the gesture recognition system and the volunteers' familiarity with touchscreen technologies. We also observed the user proficiency effect in the task completion time, where the most skilled volunteers obtained similar results with the two input modalities.

These are not conclusive results because of the experimental limitations: the small number of subjects, simple collaborative scenario and lack of extensive statistical analysis. However, they encourage more in-depth analyses and studies of alternative interfaces in the HRC context. Future studies in this field should aim to improve the technical performances of gesture-based technologies, consider more complex experimental setups and investigate new interaction modalities.

Acknowledgements. The authors would like to thank the teachers and students of the vocational education and training schools Centro Oratorio Votivo, Casa di Carità, Arti e Mestieri di Ovada for their contribution to the experiments. This work has been partially supported by the European Union Erasmus+ Programme via the European Master on Advanced Robotics Plus (EMARO+) program (grant agreement n. 2014-2616).

References

1. Colgate, J.E., Witaya, W., Peshkin, M.A.: Cobots: robots for collaboration with human operators. In: Proceeding of the ASME International Mechanical Engineering Congress and Exposition (IMECE), Atlanta, Georgia, USA, Nov 1996
2. Tsarouchi, P., Makris, S., Chryssolouris, G.: Human-robot interaction review and challenges on task planning and programming. Int. J. Comput. Integr. Manuf. **29**(8), 916–931 (2016)
3. Mateo, C., Brunete, A., Gambao, E., Hernando, M.: Hammer: an Android based application for end-user industrial robot programming. In: Proceedings of the 10th IEEE/ASME International Conference on Mechatronic and Embedded Systems and Applications (MESA), Senigallia, Italy, Sept 2014
4. Birkenkampf, P., Leidner, D., Borst, C.: A knowledge-driven shared autonomy human-robot interface for tablet computers. In: Proceedings of the 14th IEEE-RAS International Conference on Humanoid Robots (Humanoids), Madrid, Spain, Nov 2014
5. Poirier, S., Routhier, F., Campeau-Lecours, A.: Voice control interface prototype for assistive robots for people living with upper limb disabilities. In: Proceeding of the 16th IEEE International Conference on Rehabilitation Robotics (ICORR), Toronto, Canada, June 2019
6. Neto, P., Simão, M., Mendes, N., Safeea, M.: Gesture-based human-robot interaction for human assistance in manufacturing. Int. J. Adv. Manuf. Technol. **101**(1), 119–135 (2019)

7. Cicirelli, G., Attolico, C., Guaragnella, C., D'Orazio, T.: A kinect-based gesture recognition approach for a natural human robot interface. Int. J. Adv. Robot. Syst. **12**(3), 22 (2015)
8. Carfì, A., Mastrogiovanni, F.: Gesture-based human-machine interaction: taxonomy, problem definition, and analysis. IEEE Trans. Cybern. (2022)
9. Papanastasiou, S., et al.: Towards seamless human robot collaboration: integrating multimodal interaction. Int. J. Adv. Manuf. Technol. **105**(9), 3881–3897 (2019)
10. Wu, L., Alqasemi, R., Dubey, R.: Development of smartphone-based human-robot interfaces for individuals with disabilities. IEEE Robot. Autom. Lett. **5**(4), 5835–5841 (2020)
11. Gromov, B., Gambardella, L.M., Di Caro, G.A.: Wearable multi-modal interface for human multi-robot interaction. In: Proceedings of the 14th IEEE International Symposium on Safety, Security, and Rescue Robotics (SSRR), Lausanne, Switzerland, Oct 2016
12. Pohlt, C., Haubner, F., Lang, J., Rochholz, S., Schlegl, T., Wachsmuth, S.: Effects on user experience during human-robot collaboration in industrial scenarios. In: Proceedings of the IEEE International Conference on Systems, Man, and Cybernetics (SMC), Miyazaki, Japan, Feb 2018
13. Taralle, F., Paljic, A., Manitsaris, S., Grenier, J., Guettier, C.: Is symbolic gestural interaction better for the visual attention? Procedia Manuf. **3**, 1060–1065 (2015)
14. Ziegler, J., Heinze, S., Urbas, L.: The potential of smartwatches to support mobile industrial maintenance tasks. In: Proceedings of the 20th IEEE Conference on Emerging Technologies & Factory Automation (ETFA), Luxembourg, Sept 2015
15. Aromaa, S., Aaltonen, I., Kaasinen, E., Elo, J., Parkkinen, I.: Use of wearable and augmented reality technologies in industrial maintenance work. In: Proceedings of the 20th International Academic Mindtrek Conference, Tampere, Finland, Oct 2016
16. Novoa, J., Wuth, J., Escudero, J.P., Fredes, J., Mahu, R., Yoma, N.B.: DNN-HMM based automatic speech recognition for HRI scenarios. In: Proceedings of the ACM/IEEE International Conference on Human-Robot Interaction, Chicago, Illinois, USA, Feb 2018
17. Islam, M.J., Ho, M., Sattar, J.: Understanding human motion and gestures for underwater human-robot collaboration. J. Field Robot. **36**(5), 851–873 (2019)
18. Ahsan, M.R., Ibrahimy, M.I., Khalifa, O.O.: Electromyography (EMG) signal based hand gesture recognition using artificial neural network (ANN). In: Proceedings of the 4th IEEE International Conference on Mechatronics (ICOM), Kuala Lumpur, Malaysia, May 2011
19. Xu, C., Pathak, P.H., Mohapatra, P.: Finger-writing with smartwatch: a case for finger and hand gesture recognition using smartwatch. In: Proceedings of the 16th International Workshop on Mobile Computing Systems and Applications (HotMobile), Santa Fe, New Mexico, USA, Feb 2015
20. Kwon, M.C., Park, G., Choi, S.: Smartwatch user interface implementation using CNN-based gesture pattern recognition. Sensors **18**(9), 2997 (2018)
21. Hsu, Y.L., Chu, C.L., Tsai, Y.J., Wang, J.S.: An inertial pen with dynamic time warping recognizer for handwriting and gesture recognition. IEEE Sens. J. **15**(1), 154–163 (2015)
22. Wen, H., Ramos Rojas, J., Dey, A.K.: Serendipity: finger gesture recognition using an off-the-shelf smartwatch. In: Proceedings of the ACM Annual Conference on Human Factors in Computing Systems (CHI), San Jose, California, USA, May 2016

23. Carfì, A., Motolese, C., Bruno, B., Mastrogiovanni, F.: Online human gesture recognition using recurrent neural networks and wearable sensors. In: Proceedings of the 27th IEEE International Symposium on Robot and Human Interactive Communication (RO-MAN), Nanjing, China, Aug 2018
24. Wachs, J.P., Kölsch, M., Stern, H., Edan, Y.: Vision-based hand-gesture applications. Commun. ACM **54**(2), 60–71 (2011)
25. Profanter, S., Perzylo, A., Somani, N., Rickert, M., Knoll, A.: Analysis and semantic modeling of modality preferences in industrial human-robot interaction. In: Proceedings of the 28th IEEE/RSJ International Conference on Intelligent Robots and Systems (IROS), Hamburg, Germany, Sept 2015
26. Guizzo, E., Ackerman, E.: How rethink robotics built its new Baxter robot worker. IEEE Spectr. **7** (2012)
27. Quigley, M., et al.: ROS: an open-source robot operating system. In: Workshop on Open Source Software (ICRA), Kobe, Japan, May 2009
28. Kormushev, P., Calinon, S., Caldwell, D.G.: Imitation learning of positional and force skills demonstrated via kinesthetic teaching and haptic input. Adv. Robot. **25**(5), 581–603 (2011)
29. Carfì, A., Villalobos, J., Coronado, E., Bruno, B., Mastrogiovanni, F.: Can human-inspired learning behaviour facilitate human–robot interaction? Int. J. Soc. Robot. 1–14 (2019)
30. Stisen, A., et al.: Smart devices are different: assessing and mitigating mobile sensing heterogeneities for activity recognition. In: Proceedings of the 13th ACM Conference on Embedded Networked Sensor Systems (SenSys), Seoul, South Korea, Nov 2015
31. Hinderks, A., Schrepp, M., Mayo, F.J.D., Escalona, M.J., Thomaschewski, J.: Developing a UX KPI based on the user experience questionnaire. Comput. Stand. Interfaces **65**, 38–44 (2019)

Development and Evaluation of Fiber Reinforced Modular Soft Actuators and an Individualized Soft Rehabilitation Glove

Shota Kokubu[1], Pablo E. Tortós Vinocour[1], and Wenwei Yu[1,2](✉)

[1] Department of Medical Engineering, Graduate School of Engineering, Chiba University, Inage-ku, Yayoi-cho 1-33, Chiba 263-8522, Japan
s.kokubu@chiba-u.jp ccta0833@chiba-u.jp

[2] Center for Frontier Medical Engineering, Chiba University, Inage-ku, Yayoi-cho 1-33, Chiba 263-8522, Japan
yuwill@faculty.chiba-u.jp

Abstract. In soft rehabilitation gloves, Individualized support considering users' morphological and biomechanical characteristics is a next critical step towards use in rehabilitation practice. Compared with single-structure soft actuators for each finger, which require to be designed for each user, modular soft actuators enable the support for each joint by adjusting rigid connectors which are easier to fabricate. Several modular actuators have been developed, however, neither their support performance (bending angle, torque) nor individual adaptability has been appropriately evaluated. In this study, we proposed a new modular soft actuator that improved the support performance with enlarged chambers and fiber reinforcement. Moreover, we made a soft rehabilitation glove with the proposed actuators and conducted an objective evaluation for both actuators and the soft rehabilitation glove with a dummy finger and hand. As a result, the new modular soft actuators showed better support performance, and the soft rehabilitation glove can support multiple grasping profiles.

Keywords: Hand rehabilitation · Soft actuators · Fiber reinforcement

1 Introduction

The hand is one of the most important parts of the body that plays a role in many activities of daily living. Therefore, dysfunction of the hand can significantly negatively impact many activities of daily living and quality of life (QOL) of patients. The main causes of hand disability vary, including trauma from accidents and the aftereffects of strokes. In particular, it is reported that there are 80 million stroke patients globally [1]. In addition, the number of strokes is increasing due to the aging of the population in developed countries, and the demand

I. Petrovic et al. (Eds.): IAS 2022, LNNS 577, pp. 139–152, 2023.
https://doi.org/10.1007/978-3-031-22216-0_10

for post-stroke care, or functional recovery through hand rehabilitation, is also increasing.

In recent years, soft actuators and soft robots made of highly elastic materials have attracted attention to support rehabilitation. The reason is that soft actuators are flexible and lightweight due to their material properties. Moreover, their inherent viscoelasticity and high biocompatibility make them suitable for rehabilitating hands and fingers with elaborate structures. Several rehabilitation systems based on soft actuators have been developed so far. Most soft actuators used for hand rehabilitation are asymmetrically designed with air chambers inside. This asymmetry causes the actuator to bend in a specific direction when under air pressure. Polygerinos et al. developed a soft actuator with high-pressure resistance and bending performance by reinforcing a semicircular silicon cylinder with fibers around it, and applied it to a rehabilitation system [2]. Tarvainen et al. also improved the chamber structure of fiber-reinforced soft actuators and proposed a multi-pocket soft actuator with multiple bending sections. Other pioneering studies have also shown the potential of soft actuator applications for hand rehabilitation [3,4].

On the other hand, an important factor to consider in hand rehabilitation is individual adaptation. This is because there are individual differences in each joint's morphological and biomechanical properties in the hands of rehabilitation patients. It has been reported that even with the same cause of hand injury, there is variability in the joints and degree of functional impairment [5]. Therefore, rehabilitation systems need to support each joint individually and adapt to patient morphology [6]. However, most conventional soft actuators have a single-structure per finger, making it difficult to support each joint individually, and adaptation to individual differences is generally not considered [2–4]. To the best of our knowledge, the following are the studies on soft actuators that have the potential for individual adaptation. Yun et al. proposed a rehabilitation glove with an actuator module consisting of a small fiber-based chamber covered by a rigid shell [7]. Kokubu et al. also proposed a modular soft actuator for individual joints based on a fiber-reinforced soft actuator [8].

These studies verified the feasibility of individualized rehabilitation by modular soft actuators, however, neither their support performance (such as their bending angle and torque for joint support) nor their individual adaptability has been appropriately evaluated.

In this study, we proposed a new modular soft actuator that improved the bending support performance for individualized finger motion with enlarged chambers and matched fiber reinforcement. Moreover, we made a soft rehabilitation glove with the proposed new modular soft actuators and conducted an objective evaluation for both actuators and the soft rehabilitation glove with a dummy finger and a dummy hand for objective comparison and evaluation.

2 Methods

2.1 Soft Actuators

The previous modular soft actuator consists of an elastomeric body, an internal chamber, a silicone tube, a cotton thread wrapped around the outer circumference, and rigid parts at both ends (Fig. 1a) [8]. In addition, the connectors can be quickly made at any length by a 3D printer. This allows the modular soft actuator itself to remain unchanged, even if there are individual differences in the length of the user's fingers. However, since the bending performance of soft actuators is roughly proportional to the volume of their air chamber, likely, the previous modular soft actuators could not generate sufficient torque to bend the joints due to their small chamber volume. Besides, the existence of dead space in the actuator body, and instability of the connection between modules might affect the performance, too [8].

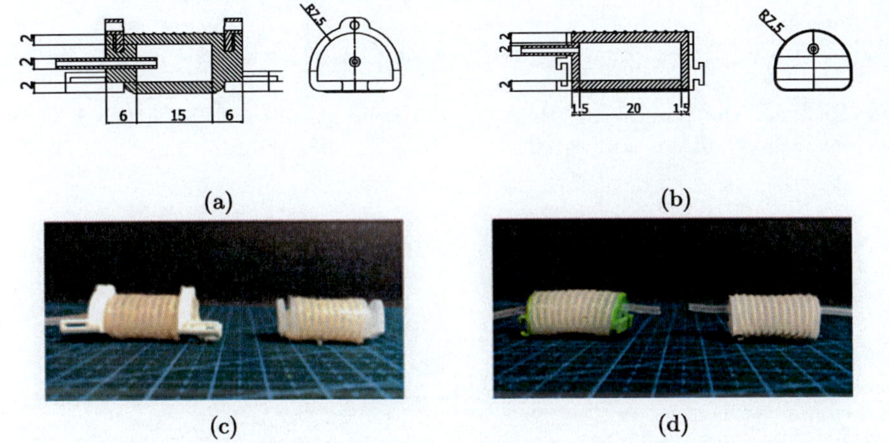

Fig. 1. Designs and prototypes of modular soft actuators. (a) CAD drawing of previous modular soft actuator [8] and (b) new modular soft actuator which has thinner silicone wall and larger chamber than the previous one. (c) Prototypes of previous modular soft actuator [8] and (d) new modular soft actuator. Left: Modular soft actuator with connector. Right: Modular soft actuator without connector.

Therefore, the new modular soft actuator was designed with the thinnest silicon walls at both ends, and no dead space in the air chamber, resulting in a larger chamber volume (Fig. 1b). Fiber reinforcement was employed to ensure the strength of the actuator to air inflation while improving bending performance. Reinforcing fibers prevent thin silicone walls from bursting due to internal air pressure. Hitching, which is specialized for bending motion, was adopted as the winding method of reinforcing fibers as same as the previous modular actuator [8,9]. Furthermore, instead of fitting the junction into the silicon groove as in the previous modular soft actuator, the new design has a thin junction that is glued to the silicon wall, so that the force generated during drive is transmitted to the junction without being dispersed, which is expected to improve force conductivity and save space.

Figure 1 shows the difference between this new design and one design of our previous work [8]. The length of the modular soft actuators was designed based on the average length of Japanese fingers, and the width was designed to fit the dorsal side of the fingers [10]. Both the new design and previous design were fabricated (Fig. 1c, d). The fabrication procedure is as follows (Fig. 2).

1. Molds were prepared for curing silicone with a 3D printer.
2. Silicone (Dragon Skin 10 NV (manufactured by SMOOTH-ON)) was poured into the mold, dried, and cured.
3. After curing, a tube was inserted and glued to a hole in the silicone body that had been made in advance.
4. 0.5 mm diameter cotton thread was wrapped along the grooves of the outer circumference of the silicone body using the hitching method as a fiber reinforcement.
5. 3D printed connectors were glued to both ends of the silicone body (only new modular soft actuator).
6. To fixate the reinforcing fibers, a thin layer of silicone was applied to the entire body, dried, and cured.

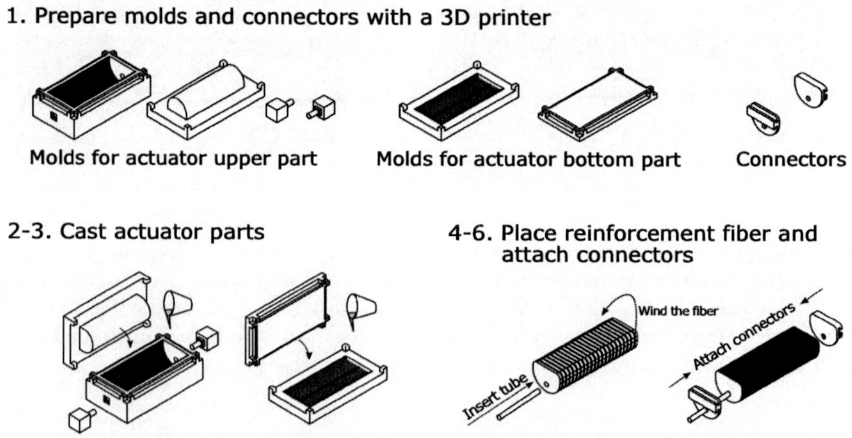

Fig. 2. Fabrication procedure of modular soft actuator

On the other hand, conventional single-structure soft actuators were designed and fabricated for comparison with the modular soft actuators. The single-structure soft actuators were redesigned with reference to the multi-pocket soft actuators introduced in Sect. 1 [4]. The redesigned single-structure soft actuator consists of an elastomeric body, an inner chamber, a silicone tube, and a cotton thread wrapped around the periphery, and no rigid parts are used. The reinforcement fibers were wrapped using the same method as for the modular soft actuator.

The single-structure soft actuator cannot accommodate different user hand dimensions, unlike the modular soft actuator. Therefore, in order to make a fair comparison with the modular soft actuators, the dimensions of the single-structure soft actuators were designed and fabricated based on the average hand length of Japanese people (Fig. 3a, b) [10]. The position of each chamber was designed to match the position of each joint.

(a) (b)

Fig. 3. Designs and prototypes of single-structure soft actuators. (a) CAD drawing. (b) Prototypes.

(a) (b)

Fig. 4. Designs and prototypes of dummy finger. (a) CAD drawing. (b) Prototype.

2.2 Dummy Finger and Hand

Dummy finger and hand were fabricated to evaluate the prototypes of the actuator. The dummy finger and hand reflect the morphological characteristics of the human hand fingers and can simulate the behavior of the actuator mounted to the hand finger. In consideration of safety during evaluation, measurements using a dummy finger and hand were conducted before conducting experiments on subjects.

The dummy finger was designed based on the average size of the Japanese index finger for evaluation in general size (Fig. 4) [10]. Each joint moves only in the direction of flexion and extension. The range of motion (ROM) is set at 80°, 100°, and 90° for the distal interphalangeal (DIP) joint, proximal interphalangeal (PIP) joint, and metacarpophalangeal (MCP) joint, respectively, based on a healthy person's data [11]. In addition, QR markers for measuring the ROM were provided at each joint and fingertip.

The dummy hand was made by assembling dummy fingers (Fig. 5). All finger and joint parts except the thumb's carpometacarpal (CM) joint have the same structure as the dummy finger. In the thumb's CM joint, a ball joint was added to the dummy hand's root to reproduce the movement of adduction and abduction. The dimensions of the dummy hand were determined with reference to the subject (a 24-year-old male) for comparison with actual hand measurements. All these parts were made with a 3D printer using polylactic acid (PLA) resin.

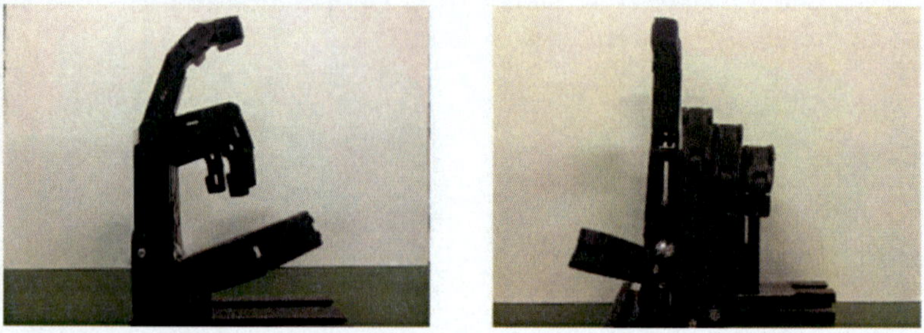

(a) (b)

Fig. 5. Prototype of dummy hand. (a) Side view. (b) Palm view.

2.3 Rehabilitation Glove

The configuration of the proposed rehabilitation glove and system for individualized hand joint support is shown in Fig. 6. This system was built with reference to several related studies dealing with soft actuators [8,12]. The whole system consisted of a control system, pneumatic, and rehabilitation glove, which consisted of the new modular soft actuators for each joint, fixed to the orthotic structure with 3D printed connectors. It can be mounted on the hand of patients using several bands. For each finger, three modular soft actuators are connected using 3D printed connectors, and each actuator can be driven independently, allowing individualized hand joint support. Also, for the thumb's CM joint, the new modular soft actuator has been modified to allow bending in both directions of adduction and abduction.

2.4 Characterization of Soft Actuators with the Dummy Finger

In order to confirm the performance of the modular soft actuator, characterization with the dummy finger was conducted. The two types of modular soft actuators, and the single-structure soft actuator were selected to compare. The dummy finger was used to simulate user's hand, and soft actuators' joint ROM (bending angle) and joint torque were measured as basic performance for rehabilitation support.

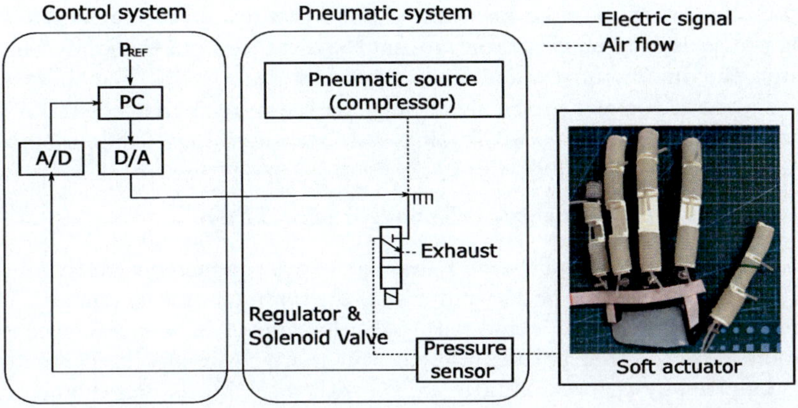

Fig. 6. System configuration. Left: Control system of air flow and pneumatic pressure. Center: Pneumatic system including a source, regulators, and sensors. Right: Rehabilitation glove including soft actuators.

ROM Joint ROMs were measured using a motion capture system and QR markers. The soft actuators were banded to the dummy finger, as shown in Fig. 7a. The QR markers were placed at a total of five locations: the base of the dummy finger, MCP joint, PIP joint, DIP joint, and fingertip. The air pressure in the soft actuator was from 0 kPa to 200 kPa in increments of 10 kPa and then reduced from 200 kPa to 0 kPa in increments of 10 kPa. The ROM of each joint was defined as the angle estimated by the QR marker on the joint and the three markers above and below it. Also, It was measured three times with 0 kPa as the reference angle.

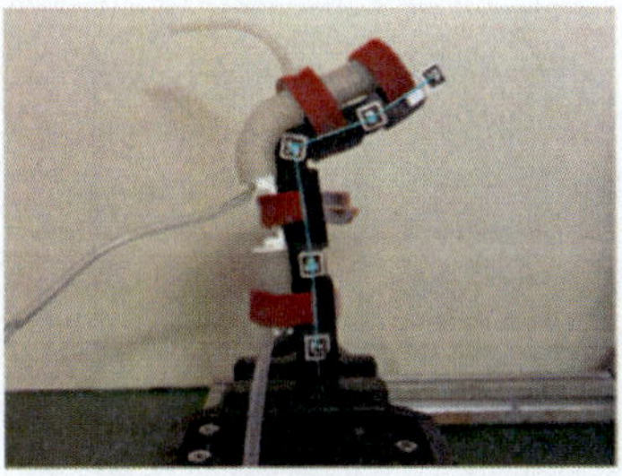

(a)

(b)

Fig. 7. Mesurement setups. (a) The dummy finger, which mounted the soft actuator, was fixed upright in the vise for ROM measurement. This shows the measurement of ROM at PIP joints. (b) The dummy finger, which mounted the soft actuator, was fixed to the load cell in a vise for torque measurement. This shows the measurement of torque at MCP joints.

Torque Joint torques were measured as the product of the force generated during pressurization of soft actuators and the distance from the joint. The soft actuator, the dummy finger, and load cell were set as Fig. 7b. The air pressure in the soft actuator was in the same range as the measurement of ROM. The torques of all DIP, PIP, and MCP joints were measured three times each.

2.5 Evaluation of Prototype Rehabilitation Glove

Measurements of index and thumb movements were conducted with the dummy hand and the subject (a 24-year-old male). Furthermore, measurements of the grasping motion were conducted with only the subject. The customized rehabilitation glove for that subject was used in the measurements. The subjects were supported by the rehabilitation glove with their fingers and thumb, and a camera captured their movements for evaluation.

Index and Thumb Movements The joint angles of the index finger and the thumb were measured while using the rehabilitation glove for motion support. The index finger's DIP, PIP, and MCP joints were independently supported for bending, and their joint angles were measured. For the thumb, the CM joint was supported in adduction and abduction, and its joint angle was measured. Changes in joint angles observed from the side view were used for evaluation. The air pressure in the soft actuator was set to 200 kPa when the soft actuator supported finger movement.

Grasping Motion The subject performed grasping motions for multiple objects, assuming similar motions to those used during daily activity. The objects to be grasped were a ball, a cup, a cylinder, and a stick, corresponding to the spherical grip, hook grip, cylindrical grip, and pulp pinch motions. The air pressure in the soft actuator was set to 200 kPa during the grasping, and it was checked whether the object could be grasped. The soft actuator that operates during the grasping motion was selected by trial and error, and the timing when actuators were pressurized was determined by the subject arbitrarily.

3 Results

3.1 Characterization of Actuators with the Dummy Fingers

This subsection shows the relationship between air pressure and ROM or torque for each joint and their maximal values.

ROM The measurement results of the ROMs of the modular soft actuators and the single-structure soft actuator are shown in Fig. 8. The hysteresis characteristic of non-linear control systems such as pneumatic circuits was observed in all results. The data of pressurization is lower than those of depressurization

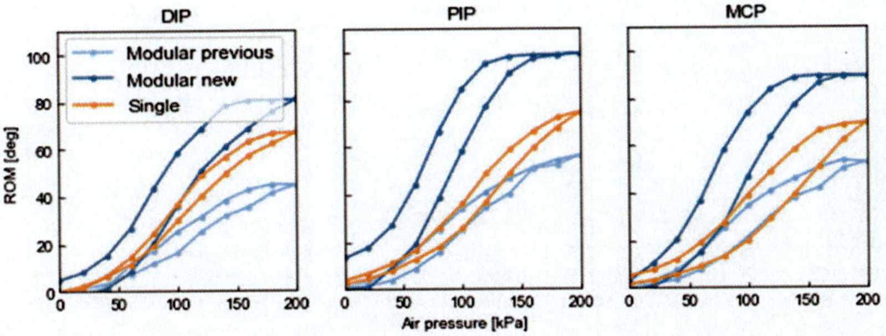

Fig. 8. The relationship between air pressure vs. ROM. Left: DIP joint. Center: PIP joint. Right: MCP joint.

Table 1. Maximal values of joint ROM

Joint	Actuator type		
	Modular previous	Modular new	Single-structure
DIP joint	45.0°	80.9°	66.8°
PIP joint	56.5°	99.6°	74.7°
MCP joint	69.4°	89.5°	69.4°

in all the curves in Fig. 8. Each curve tended to show larger values roughly in the order of new modular, single-structure, and previous modular soft actuator. The maximal values were 45.0°, 56.5°, and 69.4° for DIP, PIP, and MCP joints for previous modular soft actuator, 80.9°, 99.6°, and 89.5° for new modular soft actuator, and 66.8°, 74.7°, and 69.4° for single-structure soft actuator (Table 1).

Torque The results of the torques of the modular soft actuators and the single-structure soft actuator are shown in Fig. 9. The hysteresis is not as significant as in the results of ROM, though the same tendency was observed, that is, at the same pressure value, the measured torque of pressurization is lower than that of depressurization. Except for the MCP joint, the curves tended to show larger values in the order of new modular, previous modular, and single-structure soft actuator. Both types of modular soft actuators generated almost the same torque in the MCP joint. The maximal torque values for the DIP, PIP, and MCP joints, are 0.13 Nm, 0.13 Nm, and 0.16 Nm for the previous modular soft actuator, 0.17 Nm, 0.17 Nm, and 0.17 Nm for the new modular soft actuator, and 0.06 Nm, 0.09 Nm, and 0.09 Nm for the single-structure soft actuator, respectively (Table 2).

3.2 Evaluation of Prototype Rehabilitation Glove

This subsection shows the fingers' motion for different grasping profiles under the support of the rehabilitation glove.

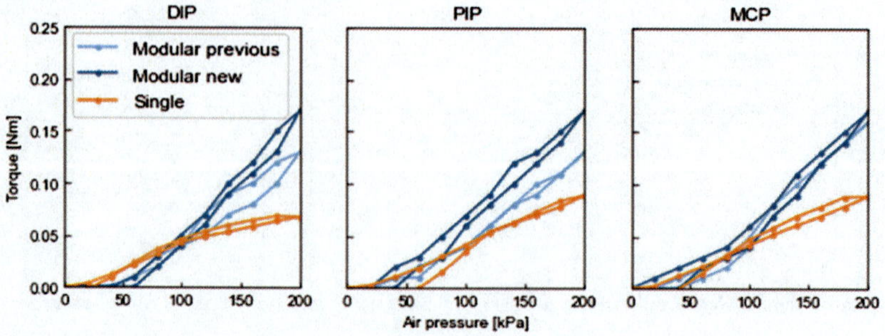

Fig. 9. The relationship between air pressure vs. torque. Left: DIP joint. Center: PIP joint. Right: MCP joint.

Table 2. Maximal values of joint torque

Joint	Actuator type			
	Modular previous	Modular new	Single-structure	
DIP joint	0.13	0.17	0.06	(Nmm)
PIP joint	0.13	0.17	0.09	(Nmm)
MCP joint	0.16	0.17	0.09	(Nmm)

Index and Thumb Movements Figures 10, 11 shows the bending of each joint of the index finger and thumb. It was confirmed that bending support was achieved for all joints of the dummy hand and the subject (Fig. 10). However, it was observed that during bending support of the DIP joint for the subject, flexion of the PIP joint was simultaneously induced.

From the observation of the thumb, it was found that adduction and abduction were possible from the initial state (Fig. 11). Although the initial position of the thumb was different for the dummy hand and the subject, they showed almost the same range of motion.

The maximal values of the index finger and thumb angle confirmed from the captures are shown in Table 3. Generally, the measured value with the dummy hand exceeded the measured value with the subject. It was also confirmed that the rehabilitation glove could support up to 60–90% of the ROM of a healthy person.

Grasping Motion The grasping of different objects supported by the rehabilitation glove are shown in Fig. 12. The experiment confirmed that the glove could grasp and lift each object by selecting an active new modular soft actuator. Also, the object did not fall even when the subject completely relaxed his hand.

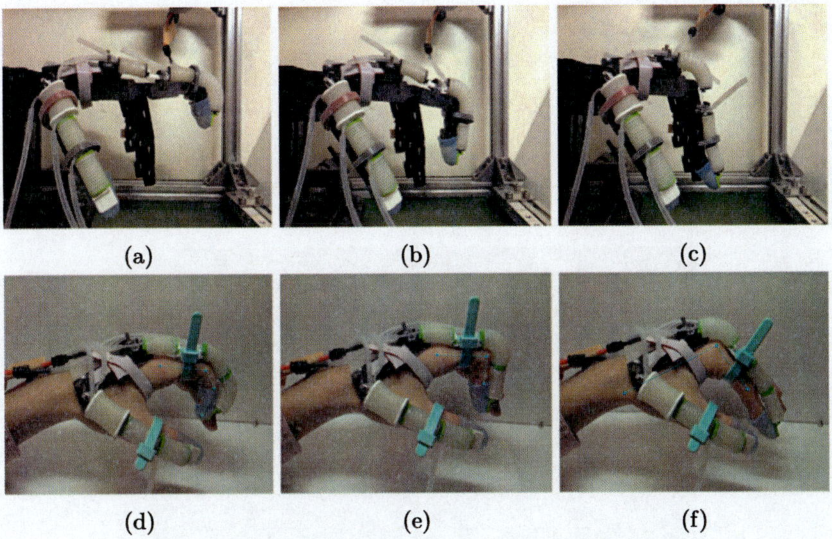

(a) (b) (c) (d) (e) (f)

Fig. 10. Index finger of the dummy hand and the subject under rehabilitation glove support. The dummy hand's (a) DIP joint, (b) PIP joint, and (c) MCP joint. The subject's (d) DIP joint, (e) PIP joint, and (f) MCP joint.

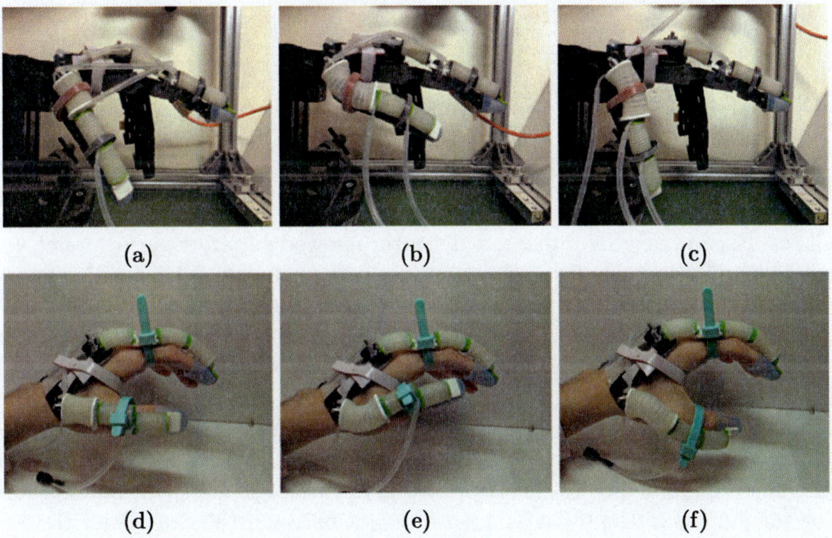

(a) (b) (c) (d) (e) (f)

Fig. 11. Thumb of the dummy hand and the subject under rehabilitation glove support. The dummy hand's CM joint at (a) initial position, (b) adduction and (c) abduction. The subject's CM joint at (d) initial position, (e) adduction and (f) abduction.

Table 3. Joint angles of index finger and thumb

	Joints of index			CM joint of thumb		
	DIP	PIP	MP	Initial	Adduction	Abduction
The dummy finger	63.8°	83.4°	83.0°	55.3°	22.2°	70.0°
The subject	51.5°	95.2°	80.0°	44.2°	10.2°	60.5°
ROM of healthy person [11]	80.0°	100.0°	90.0°	N/A	0°	90.0°

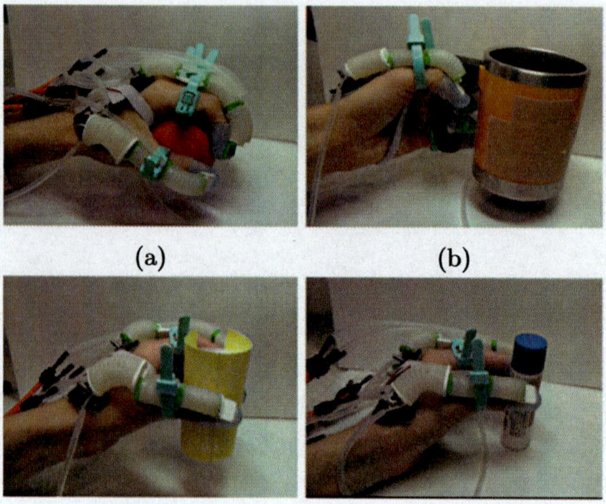

(a) (b) (c) (d)

Fig. 12. Grasping motions by rehabilitation glove. (a) Spherical grip, (b) hook grip, (c) cylindrical grip, and (d) pulp pinch

4 Discussion

In all results, the new modular soft actuator showed superior performance compared to the conventional single-structure soft actuator. Those differences in these results were particularly pronounced for torque, with a maximal difference of almost three times between the modular and single-structure.

The presence of the rigid material between chambers, which is one important design difference, may have influenced the torque generation of the actuators. Depending on the material of the components between the chambers, the following changes may have occurred during torque generation. The part between the chambers is connected by an elastic material in the single-structure soft actuator, so the torque generated by expansion diverges between the chambers. However, in the modular soft actuator, the torque generated by expansion is transmitted with little loss through rigid parts at both ends. Although these hypotheses need to be verified more rigorously by observing actuator deformation, replacing soft actuators with partially rigid parts may contribute to improved torque performance.

On the other hand, the ROM results showed a large difference in the comparison between the modular types. The internal design is the cause of that difference. The new modular soft actuator was designed with a larger chamber size than the previous one. Since the chamber is the part that expands under pressure, it is natural to assume that if this volume is large, the expansion rate will also be large. The ROM results showed that the larger the chamber, the easier it is for the actuator to bend. This is consistent with other studies that have already been reported [4]. Furthermore, with this improvement of the modular soft actuator, it obtained a higher bending torque than the existing modular soft actuator [8]. In particular, the maximal bending torque at 200 kPa was about 0.15 Nm for the modular soft actuator of Yun et al., and 0.17 Nm for the proposed new module soft actuator [7].

The results of the prototype rehabilitation glove showed that it provided motion support equivalent to 60% or more ROM of a healthy person and was capable of gripping multiple objects. The fact that we achieved multiple types of grasping indicates that the individual joint support is working well. This demonstrated the feasibility of the rehabilitation glove using modular soft actuators.

However, if the prototype rehabilitation glove is used in a clinical setting, it would require many pneumatic supply tubes to control each actuator. These may have the disadvantage of interfering with the human hand's movement and making each movement pattern more difficult to control. Our current solution is to have a simple control strategy for several major finger movement patterns. Future work requires more advanced control strategies and optimal placement of tubes on the glove.

In addition, the experiments were conducted with fixed-size dummy hands and a single subject, which naturally raises questions about whether the proposed glove is adaptable to a wide range of people. Therefore, we evaluated the modular soft actuator on variable-size dummy hands and showed that the individual joint supports worked well, as in this paper. These results were summarized and reported by Kokubu et al. [13].

5 Conclusion

In this study, we proposed a new modular soft actuator that can efficiently individually support each joint of the hand. We also prototyped a hand rehabilitation glove that supports the movement of all five fingers using the modular soft actuator.

The basic performance (bending angle, torque) of motion support of the modular soft actuator was evaluated and compared with one conventional single-structured soft actuator and one previous modular soft actuator, and showed superior performance in all aspects. It was also shown that the proposed modular soft actuator joint could contribute to individualized finger motion support by the test with a dummy finger.

Finally, the prototype rehabilitation glove was tested with one dummy hand and the hand of one subject, and each joint's ROM and gripping motion were confirmed. These results showed that the rehabilitation glove could support each joint for up to 60–90% of the ROM of a healthy person, making it possible to grasp multiple objects for supporting daily living activities.

References

1. World Stroke Organization: Global Stroke Fact Sheet 2016 (2016)
2. Polygerinos, P., Galloway, K.C., Savage, E., Herman, M., O'Donnell, K., Walsh, C.J.: Soft robotic glove for hand rehabilitation and task specific training. In: Proceedings - IEEE International Conference on Robotics and Automation, June 2015, pp. 2913–2919
3. Tarvainen, T., Yu, W.: Pneumatic multi-pocket elastomer actuators for metacarpophalangeal joint flexion and abduction-adduction. Actuators **6**(3), 27 (2017)
4. Shiota, K., et al.: Enhanced Kapandji test evaluation of a soft robotic thumb rehabilitation device by developing a fiber-reinforced elastomer-actuator based 5-digit assist system. Robot. Auton. Syst. **111**, 20–30 (2019)
5. Kamper, D.G., Rymer, W.Z.: Quantitative features of the stretch response of extrinsic finger muscles in hemiparetic stroke. Muscle Nerve **23**(6), 954–961 (2000)
6. Elaine, E.F., Cynthia, A.P., James, W.S.: Hand Splinting: Principles and Methods (Japanese Translation). The C. V. Mosby Company, Maryland Heights (1981)
7. Yun, S.S., Kang, B.B., Cho, K.J.: Exo-glove PM: an easily customizable modularized pneumatic assistive glove. IEEE Robot. Autom. Lett. **2**(3), 1725–1732 (2017)
8. Kokubu, S., Yu, W.: Developing a hybrid soft mechanism for assisting individualized flexion and extension of finger joints. In: Proceedings of the Annual International Conference of the IEEE Engineering in Medicine and Biology Society, EMBS, July 2020, pp. 4873–4877
9. Tarvainen, T., Fernandez-Vargas, J., Yu, W.: New layouts of fiber reinforcements to enable full finger motion assist with pneumatic multi-chamber elastomer actuators. Actuators **7**(2), 31 (2018)
10. Research Institute of Human Engineering for Quality Life: Human Hand Dimensions Data for Ergonomic Design 2010. Research Institute of Human Engineering for Quality Life, Osaka (2010)
11. Ueba, Y.: Hand Its Function and Dissection [Te sono kinou to kiabou] (in Japanese), 4th edn. KINPODO, INC. (2006)
12. Yap, H.K., Lim, J.H., Goh, J.C.H., Yeow, C.H.: Design of a soft robotic glove for hand rehabilitation of stroke patients with clenched fist deformity using inflatable plastic actuators. J. Med. Devices Trans. ASME **10**(4) (2016)
13. Kokubu, S., et al.: Evaluation of fiber-reinforced modular soft actuators for individualized soft rehabilitation gloves. Actuators **11**(3), 84 (2022)

Navigation

Minimum Displacement Motion Planning for Movable Obstacles

Antony Thomas(✉) and Fulvio Mastrogiovanni

Department of Informatics, Bioengineering, Robotics, and Systems Engineering, University of Genoa, Via All'Opera Pia 13, 16145 Genoa, Italy
antony.thomas@dibris.unige.it, fulvio.mastrogiovanni@unige.it

Abstract. This paper presents a minimum displacement motion planning problem wherein obstacles are displaced by a minimum amount to find a feasible path. We define a metric for robot-obstacle intersection that measures the extent of the intersection and use this to penalize robot-obstacle overlaps. Employing the actual robot dynamics, the planner first finds a path through the obstacles that minimizes the robot-obstacle intersections. The metric is then used to iteratively displace the obstacles to achieve a feasible path. Several examples are provided that successfully demonstrates the proposed problem.

1 Introduction

Classical robot motion planning problem finds a collision free path from a given initial pose to a desired goal pose of a robot amidst fixed obstacles [11]. This problem is well studied. However, in several situations a robot may need to move all (if no collision free path exists) or certain obstacles along its path. Mobile robots in a indoor environment may need to move obstacles to reach a goal. For example, open closed doors or push chairs around to reach a location. In the case of dynamic obstacles a robot may need to plan a path through these obstacles as they might clear a path while moving. Manipulator robots often rearrange or move obstacles aside to complete a task.

Such problems require the knowledge about the environment and the robot's motion. A collision between a robot and an obstacle occur when they intersect each other. If the robot has to clear its way to find a feasible path then the obstacle must be displaced such that there is no robot-obstacle intersection. In this paper, we investigate this problem, that is, finding a feasible path such that the obstacle displacements are minimized (see Fig. 1). Currently we do not focus on how the displacement happens and assume that the computed displacements can be achieved irrespective of the type of robot-obstacle interaction required to move the obstacles.

We present a new motion planning formulation in which the planner first finds a path through the obstacles such that the robot-obstacle intersections are minimized. To this end, the planner minimizes a weighted sum of the robot path length and a metric that measures the extent of robot-obstacle intersection.

I. Petrovic et al. (Eds.): IAS 2022, LNNS 577, pp. 155–166, 2023.
https://doi.org/10.1007/978-3-031-22216-0_11

This robot-obstacle intersection metric is then used to determine the minimum displacements for the obstacles so that the computed path is collision free. While computing the minimum displacements, we also take into account the fact that the displaced obstacles must themselves be collision free.

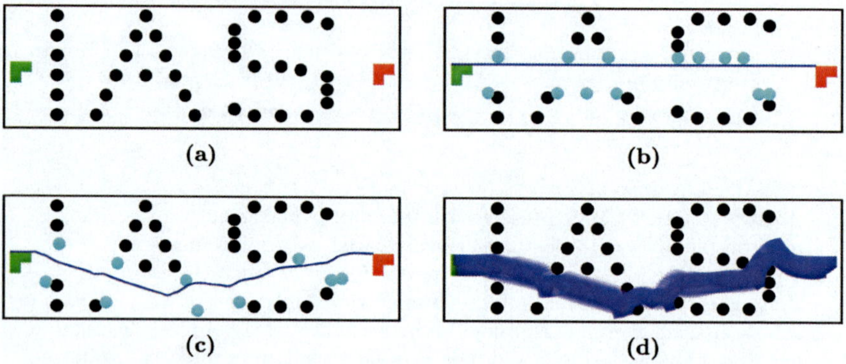

Fig. 1. (a) An L-shaped robot in the *IAS domain* has to move from left to right by displacing obstacles. (b) Robot takes the shortest path and the 13 displaced obstacles can be seen in cyan color. (c) Robot finds a path with smaller displacement, moving 10 obstacles. (d) An example of the planning phase of our approach where a path though the obstacles is obtained that minimizes the robot-obstacle overlaps.

2 Minimum Displacement Motion Planning

Let $\mathcal{C}$ denote the robot configuration space with $x^0, x^g \in \mathcal{C}$ denoting the start and goal states. Let $\mathcal{O} = \{o^i | \; 1 \leq i \leq n\}$ denote the set of movable obstacles in the environment with an associated displacement set $\mathcal{D} = \{d^i | \; 1 \leq i \leq n\}$ that represents the corresponding obstacle displacements. By abuse of notation we will use o^i to denote both the i-th obstacle as well as its state. Note that the displacements d^i in general may not always be translational alone and can include rotations or deformations. Thus d^i belongs to a displacement space of arbitrary dimension. In this paper, we will consider obstacles that can either perform translation alone or pure rotation alone about its center. Therefore d^i either represents pure translation or pure rotation. Initially, $d^i = 0$. We can now define the minimum displacement motion planning problem.

Definition 1. *The minimum displacement motion planning problem from* x^0 *to* x^g *finds a path* $x : [0,1] \rightarrow \mathcal{C}$ *with obstacle displacements* $d^1, \ldots, d^n$ *minimizing the cost* $C(x, \mathcal{D})$ *given by*

$$C = w^x c(x) + w^d \sum_{i=1}^{n} c^i(d^i) \tag{1}$$

such that

1. $x(0) = x^0$, $x(1) = x^g$
2. $\bigcup_{i \neq j} \left(o^i(d^i) \cap o^j(d^j) \right) = \{\emptyset\} \quad \forall i, j \in \{1, \ldots, n\}$
3. $x(t) \cap \bigcup_{i=1}^{n} o^i(d^i) = \{\emptyset\}$, $\forall t \in [0, 1]$

In (1), $c(x)$ is a function of the path length, $c^i(d^i)$ are positive displacement costs with w^x and w^d being the corresponding weights and $o^i(d^i)$ denotes the new obstacle locations. The first condition satisfies the endpoint constraints. The second condition ensures that the displaced obstacles are themselves collision free and the last condition guarantees the robot to not intersect with the displaced obstacles.

3 Algorithm for Minimum Displacement Motion Planning

In this section we present our planner for minimum displacement motion planning that employs a two stage optimization approach, namely, (a) the planning phase and (b) the refinement phase.

3.1 (a) Planning Phase

Let us consider a robot with a general dynamics model of the form

$$x_{k+1} = f(x_k, u_k) \tag{2}$$

where x_{k+1}, x_k are the robot states at time $k+1$, k respectively, and u_k is the applied control at k. We define a function $V(x', o^i)$ to measure the robot-obstacle intersection or overlap, where $x' = g(x)$ denote the relevant robot state (at any time instant) of interest and o^i is the obstacle state of interest. For example, if the robot and obstacle are spheres of radii r^r and r^o, respectively, then V is defined as

$$V(x', o) = \begin{cases} md & \text{if } \|x' - o\| \leq (r^r + r^o) \\ 0 & \text{otherwise.} \end{cases} \tag{3}$$

where

$$md = \|x' - o\| - (r^r + r^o) \tag{4}$$

with x', o denoting the centers of the robot and obstacle spheres, respectively and $\|\cdot\|$ denoting the L_2 norm. We note that if two spheres intersect/collide then $\|x' - o\| \leq (r^r + r^o)$. Thus $V(x', o)$ is essentially the collision condition and measures the extent of intersection or overlap when the robot and the obstacle collide. It is easily observed that the magnitude of md, that is, $|md|$ is actually the minimum distance by which the obstacle sphere is to be moved along the line joining the robot and obstacle sphere centers so as to make this overlap equal to zero. For non spherical objects, x' mostly comprises of the vertices and the center of the polyhedron and the $V(\cdot)$ is the collision condition between the two polyhedrons. However, in this paper we use minimum volume spheres as the underlying bounding volumes to tightly enclose a polyhedron as they provide an efficient approximation.

At each time instant k, the robot plans for L look-ahead steps minimizing an objective function

$$J_k = \sum_{l=0}^{L-1} c_l(x_{k+l}, \mathcal{O}) + c_L(x_{k+L}) \tag{5}$$

with c_l being the cost for each look-ahead step and c_L the terminal cost. As a result, we have the following immediate and terminal costs

$$c_l(x_{k+l}, \mathcal{O}) = \|x_{k+l}\|^2_{M_x} + \sum_{i=1}^{n} \|V^i\|^2_{M_i} \tag{6}$$

$$c_L(x_{k+L}) = \|x_{k+L} - x^g\|^2_{M_g} \tag{7}$$

where $\|x\|_S = \sqrt{x^T S x}$ is the Mahalanobis norm, M_u, M_g, M_i are weight matrices. The optimization problem can now be formally stated as

$$\min \sum_{l=0}^{L-1} \left[\|x_{k+l}\|^2_{M_x} + \sum_{i=1}^{n} \|V^i\|^2_{M_i} \right] + \|x_{k+L} - x^g\|^2_{M_g} \tag{8a}$$

s.t.

$$x_0 = x^0 \tag{8b}$$

$$x_{k+l} = f(x_{k+l-1}, u_{k+l-1}) \tag{8c}$$

$$x_T = x^g \tag{8d}$$

$$u_{k+l} \in U \tag{8e}$$

where (8e) constraints the control actions to lie within the feasible set of control inputs. The term $\|V^i\|^2_{M_i} = M_i(md^i)^2$ penalizes the robot-obstacle intersections and the optimization returns the robot trajectory $x_0, \ldots, x_T$ with minimum obstacle overlaps md^i at each time instant. As discerned before, the magnitude of md^i is actually the minimum displacement of the obstacle o^i required to make the robot-obstacle overlap equal to zero. However, each obstacle o^i is not overlapped by a single robot state x_k but l different discretized states since the robot trajectory is continuous. Moreover, a robot and the obstacle may be represented by m^1 and m^2 different bounding volume spheres, respectively. As a result, for each obstacle o^i, we have $md^{i^1}, \ldots, md^{i^p}$, $1 \leq p \leq l \times m^1 \times m^2$ different overlaps with the robot. The maximum among them is the required distance for zero overlap. The planning phase is summarized below:

Planning Phase:

1. Input: x^0, x^g, $\mathcal{O}$,$y^1 = 0, \ldots, y^n = 0$.
2. Compute $x_0, \ldots, x_T$ such that J_k is minimized at each $k \in [0, T]$.
3. Simulate $x_0, \ldots, x_T$ to compute $\{md^{ij}\}$, $1 \leq j \leq p$ for each o^i.
4. Compute $y^i = \max(\{md^{ij}\})$, to obtain the distance for each obstacle o^i.

Though we have computed the minimum distances to be moved, we have not yet ensured the condition 2., 3. in Definition 1, that is, the displaced obstacles being themselves collision free and the robot not colliding with the displaced obstacles while executing the computed trajectory. The refinement phase of the planner attends to these concerns.

3.2 (b) Refinement Phase

Displacing the obstacles by y^i does not ensure the obstacles being themselves collision free since it may introduce new overlaps with the obstacles. Similarly, in some cases moving the obstacle by y^i can lead to new intersections with the remaining robot trajectory. To tackle this, we iteratively increase the displacement magnitudes.

The refinement phase is essentially a displacement sampling process. We note here that the y^i computed is the minimum distance the obstacle sphere is to be moved along the line joining the robot and obstacle spheres. We note here that though a given obstacle may be represented using multiple spheres, y^i computed represents the maximum displacement from among all these spheres. The obstacle center is thus displaced by y^i along the maximum displacement direction. Each displaced obstacle is checked for collision with the robot and the other obstacles. If no displaced obstacles collide with the robot and other obstacles then the obtained y^i are the required minimum displacements. However, most often due to the nature of the environment and the continuous robot trajectory, some of the displaced obstacles introduce new overlaps, that is, collide either with the robot or other obstacles or both. A displacement $y^i = y^i + \Delta$ is computed and obstacle locations at a distance equal to the newly computed y^i from the initial obstacle location o^i are sampled and checked for collision. The displacements are incremented by Δ until no collisions are found. The refinement process is as follows:

Refinement Phase:
1. Input: $x_0, \ldots, x_T$, $\mathcal{O}$,$y^1, \ldots, y^n$, Δ, $d^1 = 0, \ldots, d^n = 0$.
2. While *true*
3. For each o^i such that $md^{ij} \neq 0$
4. Find y^i.
5. Run *Sample-Displacement.*
6. For *sample*
7. If $\bigcup_{i \neq j} \left(o^i(y^i) \cap o^j(y^j) \right) = \{\emptyset\}$ $\forall i, j \in \{1, \ldots, n\}$ and $x_t \cap \bigcup_{i=1}^{n} o^i(y^i) = \{\emptyset\}$ $\forall t \in \{1, \ldots, T\}$.

8. $d^i = y^i$.
9. Else
10. $y^i = y^i + \Delta$.

The *Sample-Displacement* subroutine samples obstacle locations at a distance of y^i from the initial obstacle location o^i. For a 2D scenario, the samples lie on a circle with center being the initial obstacle location before displacement, that is, o^i, and radius equal to the newly computed y^i. In a 3D scenario, the circle is replaced by a sphere and samples are points on the surface of the sphere. If the obstacle can only be rotated, then a discrete set of rotations are sampled. The subroutine runs as follows:

Sample-Displacement:
1. Input: $\mathcal{O}$, y^i.
2. For $N = 1, 2, \ldots$
3. If translation
4. Sample $o^i(y^i)$ such that $\|o^i(y^i) - o^i\| = y^i$.
5. If rotation
6. Sample $o^i(y^i)$, such that $0 \leq y^i < 2\pi$.
7. Return *sample* $o^i(y^i)$.

3.3 Optimality

The cost given in (1) is a combination of the sum of the costs in the planning phase and the incremental displacements computed in the refinement phase. Thus,

$$C = \sum_{k=0}^{T} \left(J_k + \sum_{i=1}^{n} (d^i - y^i) \right) \tag{9}$$

To prove any sort of optimality of our two stage planner, it should exhibit the optimal substructure property. A problem has an optimal substructure only if an optimal solution to the problem can be obtained from optimal solutions to its subproblems [1]. We assume that we have a mechanism by which we obtain the optimal values for the weights M_u, M_g, M_i. Let $M_u^\star, M_g^\star, M_i^\star$ give the optimal cost $J^\star = \sum_{k=0}^{T} J_k$. Let the distances returned by planning phase be $\{y^{i^\star}\}$. In the ideal scenario, that is, $\bigcap_{i=1}^{n} o^i(y^{i^\star}) = \{\emptyset\}$ and $\left(\bigcap_{t=1}^{T} x_t\right) \cap \left(\bigcap_{i=1}^{n} o^i(y^{i^\star})\right) = \{\emptyset\}$, and no refinement phase is required. Thus, $d^i = y^{i^\star}$. $J^\star$ is optimal because if there were another $J^+ \leq J^\star$, then we would use J^+ since the optimal weights M_u^+, M_g^+, M_i^+ are known to us.

If the refinement phase is to be run, that is, $d^i \neq y^{i^\star}$, then we find d^i using incremental sampling. This d^i can be made asymptotically closer to the actual value by choosing a sufficient increment $\Delta > 0$. Thus, for any $\epsilon > 0$ that forms an allowable limit on the suboptimality, $d^i \to d^{i^\star} + \epsilon$ with probability 1. Note that this is a direct consequence of our *Sample-Displacement* subroutine. The optimal cost is thus obtained as $C^\star = J^\star + \sum_{i=1}^{n} (d^i - y^{i^\star})$.

Let us consider any other solution J' for the planning phase with M_u', M_g', M_i' being the weights. Based on our assumption, any weights other than $M_u^\star, M_g^\star, M_i^\star$ gives $J' >= J^\star$ and $\sum y^{i'} >= \sum y^{i^\star}$. Now since the refinement phase proceeds in an incremental manner, we have $\sum(d^i - y^{i'}) >= \sum(d^i - y^{i^\star})$. Therefore we have, $C' = J' + \sum_{i=1}^{n}(d^i - y^{i'}) >= C^\star$, giving the desired result.

4 Examples and Applications

The nonlinear model predictive control (NMPC) based optimization in (8) is realized using the solver FORCESPRO [4,21] which generates optimized NMPC code. Note that the term V^i in the objective function (8a) is a conditional penalty term as defined in (3). Since the objective function has to be continuously differentiable (smooth), the switch between the penalty value and zero will have to be performed in a smooth manner. To this end, we employ a built-in smooth approximation function provided by FORCESPRO that computes the minimum between two quantities.[1] The optimal weights M_x, M_i, M_g vary with each problem and the tuning is currently achieved by trial and error. For all the experiments, we use a look-ahead $L = 21$ and an incremental displacement $\Delta = 0.01$. The offline code generation using FORCESPRO (planning phase) took around 16 s for the *IAS domain* with 35 obstacles. The performance is evaluated on an Intel® Core i7-10510U CPU@1.80GHz×8 with 16GB RAM under Ubuntu 18.04 LTS.

4.1 Theoretical Examples

The L-shaped robot example introduced in Fig. 1 has the following non-linear dynamics model

$$\begin{aligned} \dot{x}_x &= ucos(\theta) - vsin(\theta) \\ \dot{x}_y &= usin(\theta) + vcos(\theta) \\ \dot{\theta} &= \omega \end{aligned} \tag{10}$$

where $x = (x_x, x_y, \theta)$ is the robot state (location and orientation) with u, v being the linear velocities and w being the angular velocity. Figure 1b shows the shortest path trajectory with the planning phase displacement $\sum y^i = 13.08$. The refinement phase gave the final displacement of $\sum d^i = 18.52$ which is observed in Fig. 1b. On the other hand, Fig. 1c in which the displacements are heavily penalized gave $\sum y^i = 7.53$ with a final total displacement of $\sum d^i = 12.12$. Figure 1d shows an instance of the planning phase (line 3) where the computed trajectory is simulated to obtain the set of $\{md^{ij}\}$, that is, the set of robot-obstacle overlaps.

Figure 2 shows two instances of the IAS domain where 7 (Fig. 2a) and 10 (Fig. 2b) obstacles are displaced while the robot reaches its goal. Figure 2a gave

[1] https://forces.embotech.com/Documentation/modelling/index.html?#smooth-minimum.

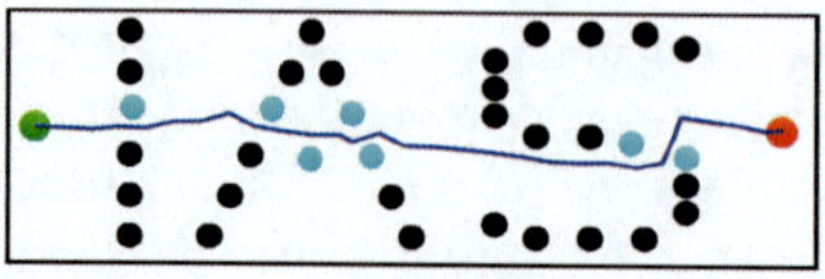
(a)

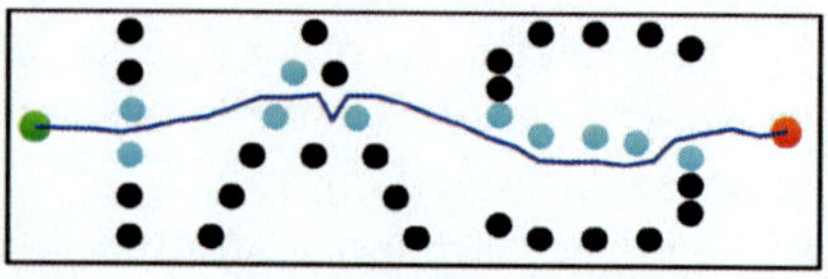
(b)

Fig. 2. A robot moving from left to right in the *IAS domain*. (a) A 7-obstacle solution. (b) A 10-obstacle solution with smaller displacement.

a planning phase displacement of $\sum y^i = 1.19$ and refinement phase displacement of $\sum d^i = 1.32$. However, Fig. 2b gave $\sum y^i = 1.01$ and a final total displacement of $d^i = 1.05$. This example demonstrates that minimum displacement of obstacles does not always imply minimum number of obstacles being displaced. We use the following non-linear dynamics for the robot

$$x_{k+1} = \begin{bmatrix} x_x \\ x_y \\ \theta \end{bmatrix}_{k+1} = \begin{bmatrix} x_x \\ x_y \\ \theta \end{bmatrix}_k + \begin{bmatrix} Dcos(\theta_k + T/2) + Ccos(\theta + (T+\pi)/2) \\ Dsin(\theta_k + T/2) + Csin(\theta + (T+\pi)/2) \\ T \end{bmatrix} \quad (11)$$

where $x = (x_x, x_y, \theta)$ is the robot state and D, C, T are the down-range, cross-range and turn control components, respectively.

Figure 3 exhibits an example of pure rotation. A point robot has to reach its goal by displacing two blocks that can only be rotated about their centers. The point robot follows the dynamics given in (11) and its $y-$axis component x_y is constrained to lie between the magenta lines. Figure 3b shows the bounding volume spheres used for both the blocks and Fig. 3c displays the corresponding solution. The total rotation computed is $\sum_{i=1}^{2} d^i = \sum_{i=1}^{2} y^i = 1.01$. We now relax the assumption of bounding volume spheres by using the collision condition between a rectangle and a point in (3) and compute the solution by penalizing the robot whenever it is inside the rectangle. The resulting trajectory is given in Fig. 3d with a total displacement of 0.98. Thus, our use of bounding volume spheres for enclosing obstacles is a reasonable assumption.

4.2 Applications

There exists several applications wherein a robot need to move through a cluttered space by rearranging or repositioning objects along its way– displacing chairs or other objects of interest. In Fig. 4 different instances of a robot navigation through the *IAS domain* among movable obstacles can be seen. The trajectory and the obstacle displacements are computed offline as described in Sect. 3. As the robot moves along the computed trajectory the obstacles are moved aside with the help of its arms. In the experiment reported in Fig. 4, the arm motions to displace the obstacles are teleoperated by a human.

Figure 5 shows the *sofa domain* wherein the robot has to reach its goal by pushing around the sofas. The resulting 4-sofa solution with minimum displacement is shown in Fig. 5c.

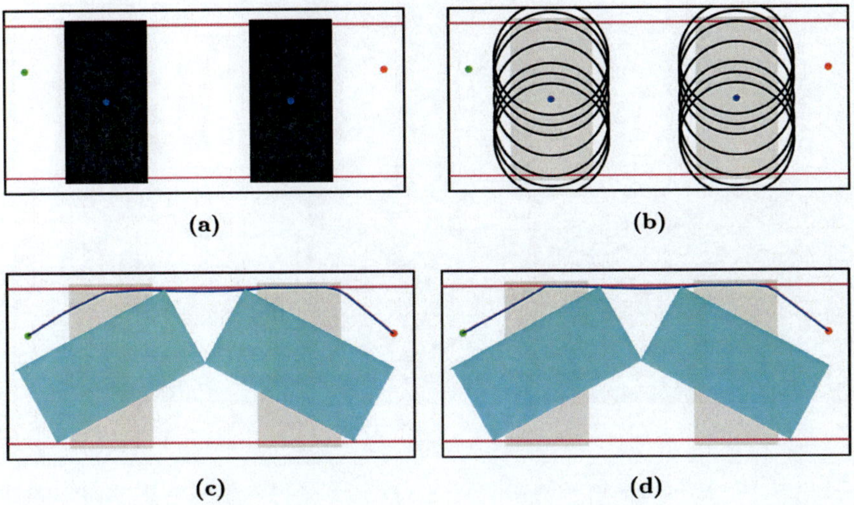

Fig. 3. (a) A point robot needs to move from left to right by moving the two blocks which can only be rotated about its centers (blue blobs). (b) Bounding volume spheres used to enclose the obstacles are shown. (c) Solution obtained using our approach. (d) Assumption of bounding volume spheres is relaxed and the solution obtained by minimizing robot-rectangle overlap is shown.

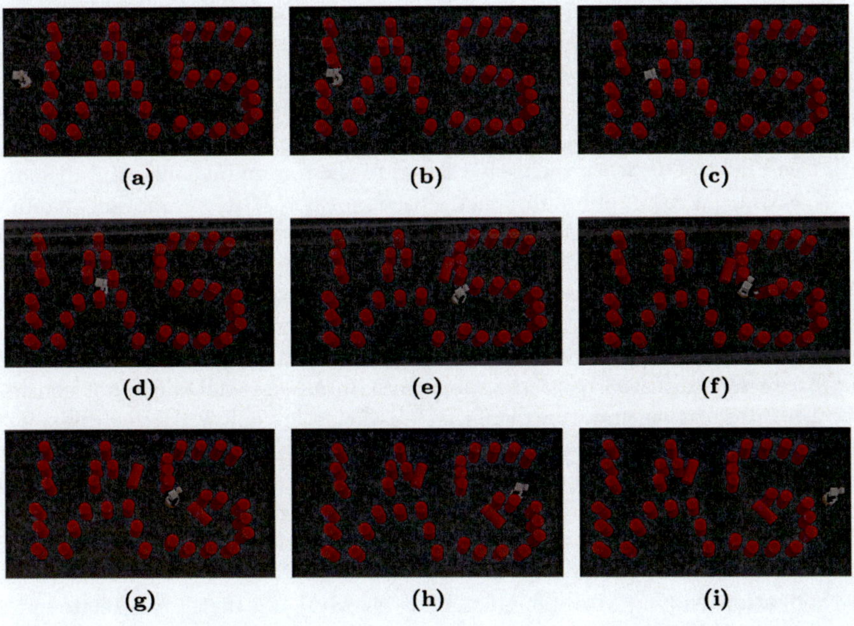

Fig. 4. Different instances of a robot navigating the *IAS domain* by moving obstacles to reach its goal.

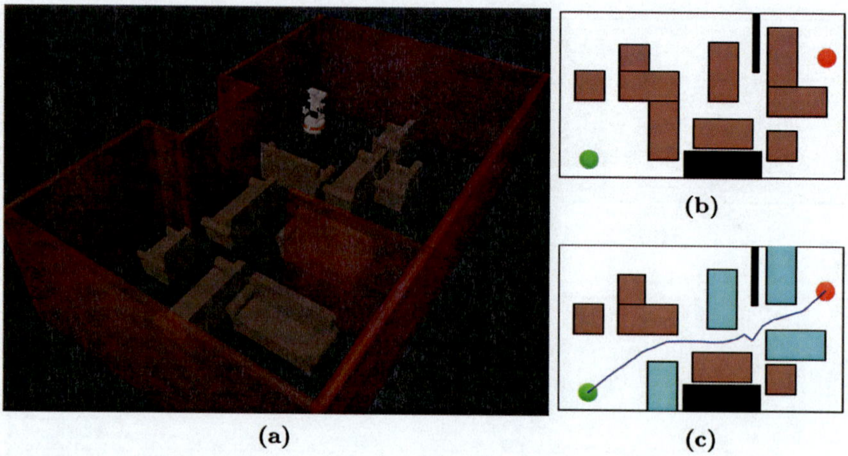

Fig. 5. (a) The *sofa domain* where the robot has to reach the goal by push around the sofas. (b) Map of the domain with green denoting the starting location of the robot and red being the goal location. (c) 4-obstacle solution with the displaced sofas shown in cyan.

5 Related Work

Minimum displacement motion planning problem is not an entirely new class of problems and bears resemblance several existing works in the literature. If all the displacements equal to zero then we have an instance of the classical motion planning or the Piano Mover's problem [13]. By tuning the weights w^x and w^d to heavily penalize longer paths, the problem can be reduced to finding the shortest path (see Fig. 1). Our work is closely related to the minimum constraint displacement (MCD) [6] and minimum constraint removal (MCR) [7] motion planning problems. Both MCD and MCR sample robot configurations and incrementally grows the roadmap (robot configurations connected by straight lines). We do not sample robot configurations and plan a path employing the actual robot dynamics to minimize the robot-obstacle overlaps. MCD also sample obstacle displacements and the iterations (extension of roadmap and displacement sampling) continue until the best cost solution is obtained. MCD does not consider obstacle interactions and the displaced obstacles are allowed to overlap with each other. MCR finds the path in the roadmap with the fewest geometric constraints or obstacles that need to be removed to connect the start state with the goal state. However, MCR does not direct where the obstacles should be moved. By encoding the obstacles with non-zero overlap to vanish our approach thus generalizes the MCR problem.

Navigation Among Movable Obstacles (NAMO) [12,15,19] is a related class of problems. NAMO is proved to be NP-hard [20] and most approaches solve a subclass of problems that selects a set of obstacles to be moved and displaces them to reconfigure the environment. Another related area is the field of task and motion planning [2,5,8,14,17,18]. Selecting the obstacles to be displace may be accomplished via task planning and the geometric constraints may be incorporated into the symbolic planner [14] or both the task and motion planning could be integrated by means of an efficient mapping between the two domains [2,18]. Manipulation among clutter or rearrangement planning in clutter [3,9,10,16] also bears resemblance to the problem proposed herein. [3,9] are often suboptimal in the sense that many objects may be moved than that is necessary due to their planner heuristic.

6 Conclusion

We have presented an approach for minimum displacement motion planning that minimizes obstacles displacements to find a feasible path. The proposed approach is a two stage process where we first find a path with minimum robot-obstacle overlaps. A metric is used to measure the extant of overlap and this in turn is used to incrementally displace the obstacles to achieve the required feasible path. By appropriately tuning the weights M_x, M_i, M_g, the planning phase can be made to return the optimal solution. However, the refinement phase may not always return the optimal solution. For example an ideal displacement of an obstacle could be a combination of translation and rotation. However, currently we assume that the obstacle can either only translate or only rotate and therefore may render the refinement phase suboptimal. We plan to relax this assumption in future work. We also plan to generalize this work by doing away with bounding volume spheres to use the actual polyhedral geometry while computing the robot-obstacle interactions.

References

1. Cormen, T.H., Leiserson, C.E., Rivest, R.L., Stein, C.: Introduction to algorithms. MIT Press (2009)
2. Dantam, N.T., Kingston, Z.K., Chaudhuri, S., Kavraki, L.E.: Incremental task and motion planning: a constraint-based approach. In: Robotics: Science and Systems (2016)
3. Dogar, M., Srinivasa, S.: A framework for push-grasping in clutter. In: Robotics: Science and Systems VII (2011)
4. Domahidi, A., Jerez, J.: Forces Professional. Embotech AG. https://embotech.com/FORCES-Pro (2014–2019)
5. Garrett, C.R., Lozano-Perez, T., Kaelbling, L.P.: FFRob: leveraging symbolic planning for efficient task and motion planning. Int. J. Rob. Res. **37**(1), 104–136 (2018)
6. Hauser, K.: Minimum constraint displacement motion planning. In: Robotics: science and Systems. vol. 6, p. 2 (2013)
7. Hauser, K.: The minimum constraint removal problem with three robotics applications. Int. J. Rob. Res. **33**(1), 5–17 (2014)
8. Kaelbling, L.P., Lozano-Pérez, T.: Integrated task and motion planning in belief space. Int. J. Rob. Res. **32**(9–10), 1194–1227 (2013)

9. Karami, H., Thomas, A., Mastrogiovanni, F.: A Task-Motion Planning Framework Using Iteratively Deepened AND/OR Graph Networks. arXiv preprint arXiv:2104.01549 (2021)
10. Krontiris, A., Bekris, K.E.: Dealing with difficult instances of object rearrangement. In: Robotics: Science and Systems. vol. 1123 (2015)
11. Latombe, J.C.: Robot Motion Planning. Kluwer Academic Publishers (1991)
12. Nieuwenhuisen, D., van der Stappen, A.F., Overmars, M.H.: An effective framework for path planning amidst movable obstacles. In: Algorithmic Foundation of Robotics VII, pp. 87–102. Springer (2008)
13. Reif, J.H.: Complexity of the mover's problem and generalizations. In: 20th Annual Symposium on Foundations of Computer Science (SFCS 1979). pp. 421–427. IEEE Computer Society (1979)
14. Srivastava, S., Fang, E., Riano, L., Chitnis, R., Russell, S., Abbeel, P.: Combined task and motion planning through an extensible planner-independent interface layer. In: Robotics and Automation (ICRA), IEEE International Conference on. pp. 639–646. IEEE (2014)
15. Stilman, M., Kuffner, J.J.: Navigation among movable obstacles: real-time reasoning in complex environments. Int. J. Humanoid Rob. **2**(04), 479–503 (2005)
16. Stilman, M., Schamburek, J.U., Kuffner, J., Asfour, T.: Manipulation planning among movable obstacles. In: Proceedings 2007 IEEE international Conference on Robotics and Automation. pp. 3327–3332. IEEE (2007)
17. Thomas, A., Mastrogiovanni, F., Baglietto, M.: Towards multi-robot task-motion planning for navigation in belief space. In: European Starting AI Researchers' Symposium. CEUR (2020)
18. Thomas, A., Mastrogiovanni, F., Baglietto, M.: MPTP: motion-planning-aware task planning for navigation in belief space. Rob. Auton. Syst. **141**, 103786 (2021). https://www.sciencedirect.com/science/article/pii/S0921889021000713
19. Van Den Berg, J., Stilman, M., Kuffner, J., Lin, M., Manocha, D.: Path planning among movable obstacles: a probabilistically complete approach. In: Workshop on the Algorithmic Foundations of Robotics VIII, WAFR, Guanajuato, Mexico. pp. 599–614. Springer (2009)
20. Wilfong, G.: Motion planning in the presence of movable obstacles. In: Proceedings of the 4th Annual Symposium on Computational Geometry, pp. 279-288. SCG '88, Association for Computing Machinery, New York, USA (1988). https://doi.org/10.1145/73393.73422
21. Zanelli, A., Domahidi, A., Jerez, J., Morari, M.: FORCES NLP: an efficient implementation of interior-point methods for multistage nonlinear nonconvex programs. Int. J. Control **93**(1), 13–29 (2020)

Mixed Use of Pontryagin's Principle and the Hamilton-Jacobi-Bellman Equation in Infinite- and Finite-Horizon Constrained Optimal Control

Jerome Weston[1(✉)], Domagoj Tolić[2], and Ivana Palunko[1]

[1] University of Dubrovnik, Ćira Carića 4, 20000 Dubrovnik, Croatia
jerome.weston@unidu.hr, ivana.palunko@unidu.hr
[2] RIT Croatia, Don Frana Bulića 6, 20000 Dubrovnik, Croatia
domagoj.tolic@croatia.rit.edu

Abstract. This paper proposes a framework for solving a class of nonlinear infinite- and finite-horizon optimal control problems with constraints. Establishment of existence and uniqueness of solutions to the Hamilton-Jacobi-Bellman (HJB) equation plays a crucial role in verifying well-posedness of a given problem and in streamlining numerical solutions. The proposed framework revolves around infinite-horizon Bolza-type cost functions with running costs exponentially decaying in time. We show Γ-convergence of solutions with such cost functions to the solutions of initial constrained (in)finite-horizon problems (that is, without running costs exponentially decaying in time). Basically, we demonstrate how to approximate solutions of (in)finite-horizon constrained optimal problems using our framework. Employing a solver based on the Pontryagin's Principle, we efficiently obtain optimal solutions for finite- and infinite-horizon problems. Efficiency of the proposed framework is demonstrated in simulation by solving a 3D path planing problem with obstacles for a full nonlinear model of an autonomous underwater vehicle (AUV).

1 Introduction

Building upon recent existence and uniqueness results regarding the Hamilton-Jacobi-Bellman (HJB) equation [2], this paper proposes a novel framework for solving a rather general class of infinite- and finite-horizon optimal control problems with constraints. The class of optimal problems considered herein includes nonlinear affine-in-control system dynamics and performance indices with locally bounded, Lipchitz and convex integral functions. Elliptic, and therefore convex, state constraints are considered and handled via the Valentine transformation. In addition, the admissible control signals belong to time-varying convex sets. We utilize the off-the-shelf software package DIDO [17], built upon the pseudo-spectral

Ivana Palunko: All authors are with LARIAT—Laboratory of intelligent and autonomous systems.

I. Petrovic et al. (Eds.): IAS 2022, LNNS 577, pp. 167–185, 2023.
https://doi.org/10.1007/978-3-031-22216-0_12

analysis and Pontryagin's Principle (PP), to numerically obtain the unique trajectories/controls starting from various initial positions. In a nutshell, given a constrained optimal control problem, we extend earlier HJB results devised for unconstrained problems and make a connection with PP to facilitate the application of PP-based solvers.

The proposed framework revolves around infinite-horizon Bolza-type cost functions with running costs exponentially decaying in time. This class of integral functions leads to existence and uniqueness of the HJB solutions [2], and mitigates the HJB solution-seeking process. We show Γ-convergence of solutions with these cost functions to the solutions of the original constrained problems whose running costs do not have to decay exponentially in time. Essentially, we demonstrate how to employ our infinite-horizon framework to approximate the original constrained optimal problems.

Prevalent approaches towards solving HJB are found in, among others, [1,2,8,12,13]. Owing to recent advances in computational power, a large portion of nowadays approaches is based on Dynamic Programming (DP) such as Approximate Dynamic Programming (ADP) and Reinforcement Learning (RL) [7]. ADP refers to model-based approaches whilst RL refers to model-free approaches. It is to be noted that the approach delineated herein is model based. However, ADP and RL do not deal well with constraints [12], but so-called soft constraints are typically imposed through penalty terms in cost functions, which is yet another incentive behind our framework. A somewhat more mature and more akin to our approach is Model Predictive Control (MPC) [8]. However, optimality of MPC solutions is typically traded for computational efficiency and stability guarantees. The present work aims for maintaining optimality when resorting to numeric solutions of HJB.

Examples of approaches concerning state-constrained optimal control are found in [3,11,19]. These approaches concern using PP and vary in terms of focus on measure spaces [3], state constraints at specific timestamps [11], and state-constrained infinite-horizon problems[19]. Compared to these works, our approach first approximates a finite-time, state constrained optimal control with an infinite-horizon version, converts this into an unconstrained problem through a formal transformation, and then applies HJB results to this new problem.

The principal contributions of this paper are fourfold: a) we extend previous results on HJB existence and uniqueness to constrained problems in order to b) prove optimality of solutions to HJB so that we can c) use a solver based on Pontryagin to efficiently compute this optimal control, and d) demonstrate this approach on finite-horizon problems of path planing in 3D with obstacles for the nonlinear AUV dynamics.

The remainder of this paper is organized as follows. Section 2 introduces the utilized notation and definitions. The infinite- and finite-horizon constrained optimal control problems considered herein are formulated in Sect. 3. Section 4 proposes methodology to successfully solve the problems of interest. Section 5 establishes relations between solutions obtained in Sect. 4 and the original finite-horizon problems from Sect. 3.

Section 6 provides numerical examples. The conclusions and future research avenues are in Sect. 7. Appendix brings several technical results aiding in paper self-containment.

2 Preliminaries

Going forward, we use the following conventions. We denote by $|\cdot|$, $|\cdot|_\infty$, $|\cdot|_m$, and $\langle\cdot,\cdot\rangle$ the Euclidean norm, the max norm, the matrix norm induced by the Euclidean norm, and the inner product in $\mathbb{R}^k$, respectively. For a nonempty set $C \subset \mathbb{R}^k$, we denote its closure by $\overline{C}$ and its boundary by ∂C. For matrix M, we denote its Moore-Penrose (left) inverse by $M^\dagger$.

For row vectors we use $r = [r_1 \ \dots \ r_{N_x}]$ and for column vectors we use $r = [r_1; \ \dots; \ r_{N_x}]$, where r_i's are scalars. When the argument of a function $f : \mathbb{R} \to \mathbb{R}$ is a row vector r, it means $f(r) = [f(r_1) \ \dots \ f(r_{N_x})]$. Likewise, for a column vector r, we have $f(r) = [f(r_1); \ \dots; \ f(r_{N_x})]$. For example, let row vector $r = [0 \ 1 \ 2 \ 3]$ and let $f : \mathbb{R} \to \mathbb{R}$ be defined as $f(x) = x^2$. Then $f(r) = [0 \ 1 \ 4 \ 9]$. Let $I(r)$ be a diagonal matrix where the diagonal entries are given by vector $r \in \mathbb{R}^{N_x}$. Unless stated otherwise, vectors are column vectors by default. Let $\{x^{(l)}\}$ denote a sequence of elements.

Define the matrix $\hat{x} \in SO(3)$ such that for $x \in \mathbb{R}^3$, we have

$$\hat{x} = \begin{bmatrix} 0 & x_3 & -x_2 \\ -x_3 & 0 & x_1 \\ x_2 & -x_1 & 0 \end{bmatrix}.$$

Let I and J be closed intervals in $\mathbb{R}$. Let μ be the Lebesgue measure. Denote by $L^1(I; \ J)$ the set of all J-valued Lebesgue integrable functions on I. We say that $f \in L^1_{loc}(I; \ J)$ if $f \in L^1(K; \ J)$ for any compact subset $K \subset I$. We denote by $\mathscr{L}_{loc}$ the set of all functions $f \in L^1_{loc}([0,\infty); \ \mathbb{R}^+)$ such that $\lim_{\sigma\to 0}\sup\{\int_J f(\tau)\,d\tau \ : J \subset [0,\infty), \ \mu(J) \le \sigma\} = 0$. We say that a function is C^1 if it is continuous and has continuous first derivatives in all variables. For a function f, denote $f^{(i)}$, $i \in \mathbb{N}$, as the ith time derivative.

A set-valued function $G : \mathbb{R}^k \rightsquigarrow \mathbb{R}^n$ is *lower semicontinuous* if, for all $x \in \mathbb{R}^k$, $\operatorname{Lim\,inf}_{y\to x} G(y) \subset G(x)$. Finally, given a family of functions $\{F_n\}$, we say that a function F is the Γ-limit of $\{F_n\}$ (denoted $F_n \xrightarrow{\Gamma} F$) if $F = F^{sup} = F^{inf}$, where $F^{inf}(x) = \inf\{\liminf_{\epsilon\to 0} F_\epsilon(x_\epsilon) \ : x_\epsilon \to x\}$, and $F^{sup}(x) = \inf\{\limsup_{\epsilon\to 0} F_\epsilon(x_\epsilon) \ : x_\epsilon \to x\}$. Note that $F^{inf} \le F^{sup}$.

2.1 Motivations for Using DIDO

In this paper, our main theorem involves establishing existence and uniqueness of controls and viscosity solutions for a HJB equation derived from a given optimal control problem. It is noted that DIDO is based on PP and thus it is beneficial to show a relation between PP and HJB.

Consider the method of characteristics for first order nonlinear PDEs. For general first order PDEs of the form

$$\begin{aligned} F(x, V, V_x) &= 0, x \in \Omega \subset \mathbb{R}^N \\ V(x) &= \overline{V}(x), x \in \partial\Omega \end{aligned} \tag{1}$$

with a continuous function $F : \mathbb{R}^N \times \mathbb{R} \times \mathbb{R}^N \to \mathbb{R}$ and endpoint data $\overline{V}(x)$, one can construct a solution V by solving the characteristic ODEs (substituting $p = V_x$)

$$\begin{aligned} \dot{x}_i &= F_{p_i}, i = 1, \cdots, n, \\ \dot{V} &= \sum_i p_i F_{p_i}, \\ \dot{p}_i &= -F_{x_i} - F_V p_i, \end{aligned} \tag{2}$$

with $x(0) = y$, $V(0) = V(y)$, $p(0) = V_x(y)$ for $y \in \partial\Omega$. It can be shown ([13, Sect. 7.2], [6, Chap. 8]) that the characteristic ODEs for HJB are the canonical ODEs for PP. It should be noted that the underlying derivation is based on sufficient smoothness of all functions involved – including V being C^1. Fortunately, characteristic ODEs can still be used with weaker assumptions on V ([13, Sect. 7.2], [6, Chap. 8]).

A side motivation to connect PP and HJB is the following. Controls generated by PP are, in general, *open-loop*, that is, $u = u(t)$. Controls given by HJB, in contrast, are *closed-loop*, that is, $u = u(t, x)$. Consequently, the concept of real-time optimal control (RTOC) can be used to transform an open-loop control, such as those constructed by `DIDO`, into a closed-loop control. The idea behind RTOC is to assume that an open-loop control exists beforehand and then to account for non-zero computational time of the optimal trajectory (i.e., updates based on given data). Methods that could be used to transform open-loop controls generated by `DIDO` into closed-loop controls are found in [16,18].

3 Problem Formulation

Before stating the problem, let us define our state space for ease of notation. Let $\mathbb{X}_s \subset \mathbb{R}^{N_s}$, with $N_s \in \{2, 3\}$, be the obstacle space and $\mathbb{X}_{ns} \subset \mathbb{R}^{k*N_s}$, with $k \in \{0, 1\}$, be the free space; for $k = 0$, one has $\mathbb{X}_{ns} = \emptyset$. Let $\mathbb{X} = \mathbb{X}_s \times \mathbb{X}_{ns}$ with both $\mathbb{X}_s$ and $\mathbb{X}_{ns}$ as open, connected, and bounded sets whose cluster points fall within $\mathbb{X}$. We take $\partial\mathbb{X}$ to be sufficiently smooth almost everywhere. The dimension of $\mathbb{X}$ is equal to $N_x = (k+1)N_s$. Here $\mathbb{X}_s$ represents the spatial region in which an autonomous system moves – typically in 2- or 3-dimensions. $\mathbb{X}_{ns}$ represents the space in which all other variables (i.e., velocities, rotations, etc.) are active.

We consider the following optimal control problem:

$$\begin{aligned} &x \in \mathbb{X} \subset \mathbb{R}^{N_x}, u(t) \in U(t) \subset \mathbb{R}^{N_u} \\ &\min_{(x,u)} \; J(t_0, t_f, x_f, x, u) = E(x(t_f), x_f) + \int_{t_0}^{t_f} l(s, x(s), u(s)) ds \end{aligned} \tag{3}$$

subject to

$$\dot{x} = a(t, x) + b(t, x)u,$$
$$h^L \leq h(x),$$
$$x(t_0) = x_0,$$

where $a : [t_0, t_f] \times \mathbb{X} \to \mathbb{R}^{(k+1)N_s}$ and $b : [t_0, t_f] \times \mathbb{X} \to \mathbb{R}^{(k+1)N_s \times N_u}$ are defined as follows for $x = [x_s; \ x_{ns}]$:

$$a(t, x) = \begin{bmatrix} a_{K_o}(t, x) \\ a_e(t, x) \end{bmatrix}, \ b(t, x) = \begin{bmatrix} b_{K_o}(t, x) \\ b_e(t, x) \end{bmatrix} \tag{4}$$

with functions $a_{K_o} : [t_0, t_f] \times \mathbb{X} \to \mathbb{R}^{N_s}$, $a_e : [t_0, t_f] \times \mathbb{X} \to \mathbb{R}^{k*N_s}$, $b_{K_o} : [t_0, t_f] \times \mathbb{X} \to \mathbb{R}^{N_s \times N_u}$, and $b_e : [t_0, t_f] \times \mathbb{X} \to \mathbb{R}^{k*N_s \times N_u}$ bounded, Lipschitz continuous, and convex (componentwise) in $x \in \mathbb{X}$ for almost all $t \geq t_0$. For $k = 0$, the dynamics reduce to $\dot{x} = a_{K_o}(t, x) + b_{K_o}(t, x)u$. It should be noted that many mobile agent models follow this structure. Furthermore, $U(t)$ is a compact and convex subset of $\mathbb{R}^{N_u}$ for each t. Also, we assume the dynamical system $\dot{x} = a(t, x) + b(t, x)u$ is controllable within $\mathbb{X}$ while satisfying all state constraints.

Define $h : \mathbb{X}_s \to \mathbb{R}^{n_o}$ componentwise as $h_i(x_s) = \sum_{j=1}^{N_s} \left(\frac{x_{s,j}(t) - \bar{x}_{i,\,j}}{r_{i,\,j}} \right)^2$ for $i = 1, \ldots, n_o$ where $\bar{x}_i = (\bar{x}_{i,\,1}, \ \bar{x}_{i,\,2}, \ \ldots, \ \bar{x}_{i,\,N_s})$, $r_i = (r_{i,1}, \ r_{i,2}, \ \ldots, \ r_{i,N_s})$ represents the center and principal axis lengths of an ellipsoid, respectively, and n_o represents the number of state constraints (that is, obstacles) in our model with $n_o \geq N_s$. (The case for $0 < n_o < N_s$ will appear in future works.) We define h^L componentwise as $h_i^L = (1 + d_i)^2$, where $d_i > 0$ represents a buffer for avoiding obstacles (for example, when autonomous agents are represented merely as points, d_i takes into account their actual dimensions so as to avoid collisions). Without loss of generality, it is also assumed that the obstacles are completely contained in $\mathbb{X}_s$ (just cut out the partially contained obstacles from $\mathbb{X}_s$). For further clarification, the inequality constraint $h^L \leq h(x)$ should be interpreted componentwise as $h_i^L \leq h_i(x)$ for $i = 1, \ldots, n_o$.

Let $l(t, x, u)$ be Lipschitz in x and u, lower semicontinuous in t, bounded for all $t \geq t_0$, $x \in \mathbb{X}$, $u(t) \in U(t)$, convex in x, and strictly convex in u. The terminal cost $E(\cdot)$ for finite-horizon problems is typically positive definite whereas it is absent in infinite-horizon problems. Clearly, we allow t_f to be finite or infinite. In what follows, we seek to approximate a solution to problem (3) in a reasonable way.

4 Methodology

In order to approximate (3), we exploit the following auxiliary optimal control problem:

$$\begin{aligned} & x \in \mathbb{X}\ ,\ u(t) \in U(t) \\ \min_{(x,u)} \ & J_\infty(t_0, x, u, \lambda) = \int_{t_0}^{\infty} l(s, x(s), u(s)) e^{-\lambda s} ds \\ & \text{subject to} \\ & \dot{x} = a(t,x) + b(t,x)u, \\ & h^L \leq h(x) \\ & x(t_0) = x_0, \end{aligned} \tag{5}$$

for $\lambda > 0$.

The approximation of (3) by (5) comes down to the following two factors:

1. How well $\int_{t_0}^{t_f} l(s, x(s), u(s)) e^{-\lambda s} ds$ approximates $\int_{t_0}^{t_f} l(s, x(s), u(s)) ds$ for some $\lambda > 0$ and interested t_f.
2. How well $\int_{t_f}^{\infty} l(s, x(s), u(s)) e^{-\lambda s} ds$ approximates $E(x(t_f), x_f)$ for some $\lambda > 0$ and interested final state x_f.

Further details are covered in Sect. 5.

4.1 Application to State- Constrained Problem

We can transform (5) into an unconstrained problem using the Valentine transformation [15,20]. Consider a slack function $\alpha : [t_0, \infty) \to \mathbb{R}^{n_o}$ such that

$$\tilde{h}(x_s(t)) + \frac{1}{2}(\alpha(t))^2 = 0, \tag{6}$$

where $\frac{1}{2}(\alpha(t))^2 = \frac{1}{2}[\alpha_1(t)^2, \ldots, \alpha_{n_o}(t)^2]^T$. We also consider the related differential equation

$$\frac{d}{dt}(\tilde{h}(x_s(t)) + \frac{1}{2}(\alpha(t))^2) = 0,$$

which can be reduced to

$$\nabla(\tilde{h}(x_s(t)))(a_{K_o}(t,x) + b_{K_o}(t,x)u) + I(\alpha(t))\dot{\alpha}(t) = 0, \tag{7}$$

where $I(\alpha)$ is an $n_o \times n_o$ diagonal matrix with the main diagonal equal to α. By the implicit function theorem, we can write the control u from (7) as $u = \Phi(t, x(t), \alpha(t), \dot{\alpha}(t))$ and define a new control $\tilde{u} = \dot{\alpha}$. Thus, we reach the following unconstrained problem

$$\begin{aligned} & (x,\alpha) \in \mathbb{X} \times \mathbb{A},\ \ \tilde{u} \in \tilde{U}(t,x,\alpha) \subset \mathbb{R}^{n_o} \\ \min_{(x,\tilde{u})} \ & \tilde{J}_\infty(t_0, x, \tilde{u}, \lambda) = \int_{t_0}^{\infty} \tilde{F}_\lambda(s, x(s), \alpha(s), \tilde{u}(s)) ds \end{aligned} \tag{8}$$

such that

$$\begin{aligned} \dot{x} &= a(t,x) + b(t,x)\Phi(t,x,\alpha,\tilde{u}), \\ \dot{\alpha} &= \tilde{u}, \\ x(t_0) &= x_0, \\ \alpha(t_0) &= [\sqrt{-2\tilde{h}_1(x_{s,0})}, \dots, \sqrt{-2\tilde{h}_{n_o}(x_{s,0})}]^\top, \end{aligned} \tag{9}$$

where $\mathbb{A} \subset \mathbb{R}^{n_o}$ and $\tilde{F}_\lambda(t,x,\alpha,\tilde{u}) := F_\lambda(t,x,\Phi(t,x,\alpha,\tilde{u}))$. We also define the corresponding Hamiltonian $\tilde{H}(t,x,\alpha,\tilde{u},p) := \tilde{F}_\lambda(t,x,\alpha,\tilde{u}) + \langle p, a(t,x) + b(t,x)\Phi(t,x,\alpha,\tilde{u})\rangle$.

Also, for our problem, we are able to give an explicit formula for Φ. From (7), we have

$$\begin{aligned} u &= -[\nabla(\tilde{h}(x_s(t)))b_{K_o}(t,x(t))]^\dagger * \\ &\quad *[\nabla(\tilde{h}(x_s(t)))a_{K_o}(t,x(t)) + I(\alpha(t))\tilde{u}(t)], \\ &= \Phi(t,x,\alpha,\tilde{u}). \end{aligned} \tag{10}$$

Guaranteeing existence and uniqueness of $\tilde{u}$ is a major step in proving our main result. To that end, we introduce the following lemma:

Lemma 1. *For a fixed* $(x,\alpha) \in \mathbb{X} \times \mathbb{A}$, *the control set* $\tilde{U}(t,x,\alpha) = \Phi^{-1}(t,x,\alpha,U(t))$ *is a compact and convex set for each* t.

Proof. We start by establishing the convexity of $\tilde{U}(t,x,\alpha)$. Note that $\tilde{U}(t,x,\alpha)$ is the preimage of $U(t)$ under Φ. Thus it suffices to show that, for any $\tilde{u}_1(t)$, $\tilde{u}_2(t) \in \tilde{U}(t,x,\alpha)$, the following holds for $\lambda \in [0,1]$ and each (x,α):

$$\tilde{u}_\lambda(t) := \lambda\tilde{u}_1(t) + (1-\lambda)\tilde{u}_2(t) \in \tilde{U}(t).$$

Call

$$A_0(t,x) = -[\nabla(\tilde{h}(x_s))b_{K_o}(t,x)]^\dagger * [\nabla(\tilde{h}(x_s))a_{K_o}(t,x)]$$

and

$$B_0(t,x,\alpha) = -[\nabla(\tilde{h}(x_s))b_{K_o}(t,x)]^\dagger * I(\alpha).$$

With this we have $\Phi(t,x,\alpha,\tilde{u}) = A_0(t,x) + B_0(t,x,\alpha)\tilde{u}$. Continuing from this, suppressing (x,α), we have

$$\begin{aligned} \Phi(t,\tilde{u}_\lambda(t)) &= A_0(t) + B_0(t)(\lambda\tilde{u}_1(t) + (1-\lambda)\tilde{u}_2(t)), \\ &= \lambda A_0(t) + (1-\lambda)A_0(t) \\ &\quad + B_0(t)(\lambda\tilde{u}_1(t) + (1-\lambda)\tilde{u}_2(t)), \\ &= \lambda\Phi(t,\tilde{u}_1(t)) + (1-\lambda)\Phi(t,\tilde{u}_2(t)), \\ &=: \lambda u_1(t) + (1-\lambda)u_2(t) =: u_\lambda. \end{aligned}$$

Since $U(t)$ is convex, it is clear that $u_\lambda \in U(t)$ and thus $\tilde{u}_\lambda \in \tilde{U}(t)$. Thus $\tilde{U}(t,x,\alpha)$ is convex. Compactness of $\tilde{U}(t,x,\alpha)$ follows from the fact that Φ is continuous in $\tilde{u}$.

Remark 1. It should be noted that equations (6) and (7) hold along the trajectories with further details held in [15,20]. Also, observe that for l, a and b satisfying the assumptions from Section III, the function $\tilde{l}(t,x,\alpha,\tilde{u}) = l(t,x,\Phi(t,x,\alpha,\tilde{u}))$ is convex and Lipschitz in (x,α) and strictly convex in $\tilde{u}$. Consequently, $\tilde{F}_\lambda$ inherits the favorable properties of F_λ.

Remark 2. It is possible that a symmetric obstacle configuration exists such that multiple optimal control/trajectory pairs could exist for a particular example of (5). We note that although the map Φ is affine in $\tilde{u}$, the Valentine transformation that changes (5) into (8)–(9) can allow for the multiple optimal pairs possible in a given example of (5) to be represented by the unique optimal pair for corresponding (8)–(9) via restrictions, projections, etc. (Further details will be included in the upcoming work)

4.2 Weak Solutions to HJB

We now present the main theorem of this paper, which establishes a unique weak solution of the HJB equation in the class of lower semicontinuous functions vanishing at infinity.

Theorem 1. *For the HJB equation derived from (8)–(9), there exists a unique viscosity solution $\hat{V}$ vanishing at infinity. In addition, there exists a unique control $\hat{u}$.*

Proof. In this proof, we establish existence and uniqueness of the control $\hat{u}$ based on the properties of the Hamiltonian $\tilde{H}$ and then establish uniqueness of $\hat{V}$ by showing that (8)–(9) satisfies a set of assumptions and by using a proof in the style of Proposition 2 from the Appendix.
Existence and Uniqueness of $\hat{u}$ Existence and uniqueness of $\hat{u}$ depends on the structure of $\tilde{H}$ and $\tilde{U}(t, x, \alpha)$. First we show continuity and strict convexity of $\tilde{H}$ in $\tilde{u}$. Both of these conditions are connected to the continuity and convexity of Φ in $\tilde{u}$. We note that function Φ is affine in $\tilde{u}$ for all $(t,\ x,\ \alpha)$ and thus continuous and convex in $\tilde{u}$. By definitions of $\tilde{F}_\lambda$ and $\tilde{f}$, we know that $\tilde{F}_\lambda$ is strictly convex and $\tilde{f}$ convex in $\tilde{u}$. Both functions are also continuous in $\tilde{u}$. As a result, $\tilde{H}$ is continuous and strictly convex in $\tilde{u}$. Together with Lemma 1, $\tilde{H}$ admits a unique minimizer $\hat{u}$.
Existence and Uniqueness of $\hat{V}$ For existence and uniqueness of $\hat{V}$, we check our problem against the assumptions required in Proposition 2.

A1 We note that both $\tilde{f}$ and $\tilde{F}_\lambda$ are continuous in $(t, \tilde{u})$ for all $x \in \mathbb{X}$, $\alpha \in \mathbb{A}$ and thus Lebesgue-Borel measurable in those arguments. For $\phi \in L^1([0,\infty);\mathbb{R})$, let us choose $\phi(t) = \tilde{l}_{-}e^{-\lambda t}$, where $\tilde{l}_{-} = \inf_{(t,x,\alpha,\tilde{u})} |\tilde{l}(t,x,\alpha,\tilde{u})|$.

A2 We need to establish a growth bound for a.e. $t > 0$ and for all $(x, \alpha) \in \mathbb{X}\times\mathbb{A}$, $\tilde{u}(t) \in \tilde{U}(t, x, \alpha)$. First, we rewrite the dynamics for easier computation of bounds.
From (10), we rewrite the dynamic function $\tilde{f}$ as

$$\tilde{f}(t, x, \overrightarrow{\alpha}, \tilde{u}) = A(t, x) + B(t, x, \alpha)\tilde{u}(t) \tag{11}$$

where

$$A(t, x) = a(t, x) - b(t, x)b_{K_o}(t, x)^\dagger \nabla \tilde{h(x_s)}^\dagger \nabla \tilde{h(x_s)} a_{K_o}(t, x)$$

and

$$B(t, x, \alpha) = -b(t, x)b_{K_o}(t, x)^\dagger \nabla \tilde{h}(x_s)^\dagger I(\alpha).$$

Going forward, we note that $\tilde{f}$ is bounded in all arguments (a.e. in t); thus, we have

$$|\tilde{f}(t, x, \alpha, \tilde{u})| \leq K_{\tilde{f}}(t)(1 + |(x, \alpha)|). \tag{12}$$

Next,

$$\begin{aligned} |\tilde{F}_\lambda(t, x, \alpha, \tilde{u})| &\leq \sup_{(x,\alpha,\tilde{u})} |\tilde{l}(t, x, \alpha, \tilde{u})| e^{-\lambda t}, \\ &\leq K_{\tilde{F}_\lambda}(t)(1 + |(x, \alpha)|), \end{aligned} \tag{13}$$

where $K_{\tilde{F}_\lambda}(t) := \sup_{(x,\alpha,\tilde{u})} |\tilde{l}(t, x, \alpha, \tilde{u})| e^{-\lambda t}$. Thus, we take $c \in L^1_{loc}([0, \infty); \mathbb{R}^+)$ as

$$c(t) = K_{\tilde{f}}(t) + K_{\tilde{F}_\lambda}(t).$$

A3 Define set-valued maps $\mathscr{F}(t, x, \alpha) := \{\tilde{f}(t, x, \alpha, \tilde{u}) : \tilde{u} \in \tilde{U}(t, x, \alpha)\}$ and $\mathscr{L}(t, x, \alpha) := \{\tilde{F}_\lambda(t, x, \alpha, \tilde{u}) : \tilde{u} \in \tilde{U}(t, x, \alpha)\}$. Since the functions $\tilde{f}$ and $\tilde{F}_\lambda$ are continuous for $(x, \alpha) \in \mathbb{X} \times \mathbb{A}$ and t a.e., one has that the maps $\mathscr{F}$ and $\mathscr{L}$ are continuous. To show the closedness of images, consider the sequence $\{(x_n, \alpha_n)\} \subset \mathbb{X} \times \mathbb{A}$ that converges to a point $(\bar{x}, \bar{\alpha}) \in \mathbb{X} \times \mathbb{A}$. It is clear that

$$\begin{aligned} &\mathscr{F}(t, x_n, \alpha_n) \to \mathscr{F}(t, \bar{x}, \bar{\alpha}), \\ &\mathscr{L}(t, x_n, \alpha_n) \to \mathscr{L}(t, \bar{x}, \bar{\alpha}). \end{aligned}$$

Convexity of the set

$$\{(\tilde{f}(t, x, \alpha, \tilde{u}), \tilde{F}(t, x, \alpha, \tilde{u}) + r) : \tilde{u} \in \tilde{U}(t, x, \alpha), r \geq 0\}$$

in (x, α) follows from the (componentwise) convexity of $\tilde{f}$ (which is derived from the componentwise convexity of f) and $\tilde{F}_\lambda$.

A4 To satisfy the assumption A4 of Proposition 2, we break this item into finding suitable bounds for $\tilde{F}_\lambda$ and $\tilde{f}$. First, fix t and $\tilde{u} \in \tilde{U}(t, x, \alpha)$. Next, for (x, α) and (y, β), we proceed with the bounds on $\tilde{F}_\lambda$. With the assumptions on $\tilde{l}$, we have $|\tilde{F}_\lambda(t, x, \alpha, \tilde{u}) - \tilde{F}_\lambda(t, y, \beta, \tilde{u})| = K_{\tilde{l}}(t)e^{-\lambda t}|(x, \alpha) - (y, \beta)|$.
In order to establish the appropriate bounds on $\tilde{f}$, we consider similar bounds on the matrix-valued functions A and B as given in $A2$. We note that since a and b are Lipschitz and bounded in x for a.e. $t \geq t_0$ and A is independent of α, we can establish Lipschitz bounds for A of the form

$$|A(t, x) - A(t, y)| \leq K_A(t)|(x, \alpha) - (y, \beta)|.$$

For the function B, we have

$$|(B(t, x, \alpha) - B(t, y, \beta))\tilde{u}(t)| \leq K_B(t)\tilde{u}_b|(x, \alpha) - (y, \beta)|,$$

where $\tilde{u}_b = \sup_{\tilde{u} \in \tilde{U}} |\tilde{u}|$. Thus, we conclude

$$|\tilde{f}(t, x, \alpha, \tilde{u}) - \tilde{f}(t, y, \beta, \tilde{u})| \leq C_{\tilde{f}}(t)|(x, \alpha) - (y, \beta)|, \tag{14}$$

where $C_{\tilde{f}}(t) = K_A(t) + K_B(t)\tilde{u}_b$. In other words, A4 is satisfied with

$$k(t) = K_{\tilde{l}}(t)e^{-\lambda t} + C_{\tilde{f}}(t).$$

A5 To show that $k \in \mathscr{L}_{loc}$, we note that the suprema of $K_{\tilde{l}}(t)$, $K_A(t)$, and $K_B(t)$ are finite. Thus we can say $k(t) \leq 3 * \max\{\sup_t K_{\tilde{l}}(t), \sup_t K_A(t), \sup_t K_B(t)\tilde{u}_b\} =: K$. For $J \subset [t_0, \infty)$ and $\sigma \in [t_0, \infty)$, we have

$$\int_J k(\tau)d\tau \leq K\mu(J),$$

$$\sup_{\{J:\mu(J)\leq\sigma\}} \int_J k(\tau)d\tau \leq \sup_{\{J:\mu(J)\leq\sigma\}} K\mu(J). \tag{15}$$

Letting σ approach zero shows that the RHS of (15) approaches zero and thus $k \in \mathscr{L}_{loc}$. Moreover, we get

$$\begin{aligned} \tfrac{1}{t}\int_{t_0}^{t}(c(\tau) + k(\tau))d\tau &\leq K\tfrac{t-t_0}{t} + \tfrac{1}{t}\int_{t_0}^{t} c(\tau)d\tau \\ &\leq (K + c_{max})(1 + \tfrac{t_0}{t}). \end{aligned} \tag{16}$$

As t approaches infinity, we note that the right-hand side of (16) is finite.

A6 We know from A2 that for $t \geq 0$, closed set $\mathbb{X}$, and all $x \in \partial\mathbb{X}$, it follows that $|\tilde{f}(t, x, \alpha, \tilde{u})| + |\tilde{F}_\lambda(t, x, \alpha, \tilde{u})| \leq c(t)(1 + \tilde{x}_m)$, where $\tilde{x}_m = \sup\{|(x, \alpha)| : (x, \alpha) \in \partial\mathbb{X} \times \partial\mathbb{A}\}$. Taking the supremum over $\tilde{u} \in \tilde{U}$ yields the assumption A6 with $q(t) = c(t)(1 + x_m)$.

A7 By the structure of $\tilde{F}_\lambda$, the assumption A7 readily follows.

A8 The idea of A8 is to show that for a point $(\bar{x}, \bar{\alpha}) \in \partial\mathbb{X} \times \partial\mathbb{A}$, there exists a trajectory that lies in the interior of $\mathbb{X} \times \mathbb{A}$ such that $(\bar{x}, \bar{\alpha})$ is reached. Prior to problem (8)–(9), we assumed that the dynamics $\dot{x} = a(t, x) + b(t, x)u$ are controllable (which satisfies A8 for (5)). Now we need to show that the controllability property holds also for $\dot{x} = \tilde{f}(t, x, \alpha, \tilde{u})$. Let us note that $\Phi(t, x, \alpha, \cdot)$ is a closed map (in the fourth variable). Since Φ is a closed map in $\tilde{u}$ and the original dynamics are controllable, there exists a control $\tilde{u}$ such that one can reach some final position $x_f \in \partial\mathbb{X}$ with a trajectory in the interior of $\mathbb{X}$ whose dynamics are governed by $\tilde{f}$.

Having established the assumptions A1–A8, it suffices to show that the value function $\hat{V}(t, x, \alpha) = \int_t^\infty \tilde{l}(s, x, \alpha, \hat{u})e^{-\lambda s}ds$ has the following properties: (*i*) $\hat{V}$ is lower semicontinuous, (*ii*) Dom $\hat{V}(t, \cdot, \cdot) \neq \emptyset$ for all large $t > t_0$, and (*iii*)

$$\lim_{t\to\infty} \sup_{(y,\beta)\in\mathrm{Dom}\,\hat{V}(t,\cdot,\cdot)} |\hat{V}(t, y, \beta)| = 0.$$

We know that $\tilde{l}$ is at least lower semicontinuous in t, x, and α and thus (*i*) is satisfied (e.g., see Remark 1). Similarly, since $\tilde{l}$ is bounded and defined in all necessary arguments, we have (*ii*). For (*iii*),

$$\lim_{t\to\infty} \sup_{(y,\beta)\in \text{Dom } \hat{V}(t,\cdot,\cdot)} |\hat{V}(t,y,\beta)| \leq \lim_{t\to\infty} \int_t^\infty \tilde{l}_+ e^{-\lambda s} ds,$$

where $\tilde{l}_+ = \sup_{(t,y,\beta)} |\tilde{l}(t,x,\alpha,\hat{u})|$. It is clear that the RHS converges to zero as $t \to \infty$ and thus we establish (*iii*). From here, we apply Proposition 2 with (1) $\Longrightarrow$ (2) and have the value function $\hat{V}$ as the unique viscosity solution. □

Remark 3. Note that the classical necessary and sufficient conditions for existence of solutions to (5) *without constraints* can be expressed through PP ([13, pp.102-3]) and HJB ([13, pp. 165]) respectively. We stress that existence and uniqueness results are applied to problem (8)–(9) rather than problem (5).

5 Solving the Original Problem

5.1 Convergence to Original Infinite-Horizon Problem

Up to this point, we have analyzed (5) by rewritting it as (8)–(9). Now we show how to exploit (5) to solve the infinite-horizon version of (3), that is, we set $t_f = \infty$ and $E(x_f) = 0$ in (3).

Recall that it is possible that non-unique trajectory/control pairs $\{(\tilde{x}, \tilde{u})\}$ exist that solve infinite-horizon (3). Given (x_λ, u_λ) as a solution of (5), we establish the conditions on (x_λ, u_λ) to be a sufficient approximation for a solution of infinite-horizon (3).

Proposition 1 *Consider the family of cost functions* $\{J_\infty : \mathbb{X} \times U(t) \times [0,\infty) \to \mathbb{R}\}$, *where* $J_\infty(x,u,\lambda) = \int_{t_0}^\infty l(s,x(s),u(s))e^{-\lambda s} ds$ *(λ acts as the indexing variable) with dynamics and constraints from (5) and* $J_\infty(x,u,0) = \int_{t_0}^\infty l(s,x(s),u(s))ds$ *(i.e., the cost function for the infinite-horizon variant of (3)). Let* (x_λ, u_λ) *be the associated trajectory/control pair that minimizes* $J_\infty(x,u,\lambda)$ *and* $l(t,x,u)$ *satisfy the conditions posed in Section III. Then* $J_\infty(x,u,\lambda) \xrightarrow{\Gamma} J_\infty(x,u,0)$ *and there exists a trajectory/control pair* (x^*,u^*) *such that* $(x_\lambda,u_\lambda) \to (x^*,u^*)$ *and* (x^*,u^*) *solves the infinite-horizon problem (3).*

Proof. To show Γ−convergence of $J(x,u,\lambda)$, we consider the standard arguments involving J^{sup} and J^{inf} (related to F^{sup} and F^{inf} from Section II). Since $e^{-\lambda t} \leq 1$ for $\lambda \in [0,\infty)$, we know $l(t,x(t),u(t))e^{-\lambda t} \leq l(t,x(t),u(t))$ and thus $\int_{t_0}^\infty l(s,x(s),u(s))e^{-\lambda s}ds \leq \int_{t_0}^\infty l(s,x(s),u(s))ds$. From here, it is clear that

$$J(x,u,0) \geq J^{sup}(x,u) := \limsup_{\epsilon\to 0} J(x_{1,\epsilon}, u_{1,\epsilon}, \epsilon) \tag{17}$$

for some sequence $(x_{1,\epsilon}(t), u_{1,\epsilon}(t)) \to (x(t),u(t))$. To get the necessary inequality

$$J(x,u,0) \leq J^{inf}(x,u) := \liminf_{\epsilon\to 0} J(x_{2,\epsilon}, u_{2,\epsilon}, \epsilon), \tag{18}$$

it suffices to define a sequence $(x_{2,\epsilon}(t), u_{2,\epsilon}(t))$ such that for a given $(x(t), u(t))$, $(x_{2,\epsilon}(t), u_{2,\epsilon}(t))$ converges to $(x(t), u(t))$ and (18) is satisfied.

To establish this sequence, we first consider target (x, u) such that $J(x, u, 0) < \infty$. We note that J is continuous in x, u, and ϵ. Thus, consider $(x_{2,\epsilon}(t), u_{2,\epsilon}(t)) := (x(t)+e_1(\epsilon), u(t)+e_2(\epsilon))$ such that $J(x_{2,\epsilon}, u_{2,\epsilon}, \epsilon) \geq J(x, u, 0)$ and $(e_1(\epsilon), e_2(\epsilon)) \xrightarrow{\epsilon \to 0} 0$ Here we have $\lim_{\epsilon \to 0} J(x_{2,\epsilon}, u_{2,\epsilon}, \epsilon) \geq J(x, u, 0)$. Hence, (18) is satisfied and we have shown that $J(x, u, \lambda) \xrightarrow{\Gamma} J(x, u, 0)$.

Next, we show that there exists a compact set C with $\inf_{(x,u)} J(x, u, \epsilon) = \inf_{(x,u) \in C} J(x, u, \epsilon)$ as $\epsilon > 0$. We note that, by the properties of the function $l(t, x(t), u(t))$, $J(x, u, \epsilon)$ is a bounded function. Thus for each $\epsilon > 0$, we can always find such a set C.

Owing to the previous part of the proof and since $\dot{x} = a(t, x) + b(t, x)u$ is controllable, there exists a pair (x^*, u^*) such that $\int_{t_0}^{\infty} l(s, x^*(s), u^*(s))ds$ is finite and minimal. By Theorem 1, for $\lambda > 0$, there exists a unique optimal trajectory/control pair (x_λ, u_λ) that solves the corresponding optimal control problem (5) with the corresponding λ value.

Since $J(x, u, \lambda) \xrightarrow{\Gamma} J(x, u, 0)$ and each pair (x_λ, u_λ) minimizes $J(x, u, \lambda)$ for a given λ, we reach $(x_\lambda, u_\lambda) \to (x^*, u^*)$ [10, Corollary 7.24]. □

Remark 4. One can interpret Proposition 1 as the result of when $\lambda \to 0$ in the sense that we are determining the efficacy of (x_λ, u_λ) when the cost function lacks the exponential decay in time (i.e., $\lambda = 0$).

Remark 5. We take the time to acknowledge the similarity of this approach to that of the "vanishing discount method" of ergodic control theory. In [4,5], one requires constructing a Lyapunov function to show precompactness of the family of cost functions $\mathbb{J} = \{J_\infty(\cdot, \cdot, \lambda)\}$. In short, one must rely on the dynamics of the problem (typically a stochastic differential equation, [9]) to obtain convergence for $\mathbb{J}$ and thus $\{(x_\lambda, u_\lambda)\}$. In our results, once Theorem 1 is applied, convergence only depends on the properties of $\mathbb{J}$. With respect to state constrained problems (see [14]), additional assumptions are required on the constraints. In this paper, state constraints are handled via the Valentine transformation and the results of Theorem 1 rather than with Γ-convergence. It should be noted that Γ-convergence can be used independently of any control framework, i.e. the behavior of a family functionals and their respective minimizers.

5.2 Approximation of Original Finite-Time Problem

For many practical problems, finite-horizon formulations are of interest, that is, t_f is finite and $E(\cdot) \not\equiv 0$ in (3). Essentially, we wish to form a relation between the infinite-time horizon problem (5) and the finite-time version of problem (3). Consider the new running cost functional $l'(t, x, u, \lambda)$ defined as

$$l'(t, x, u, \lambda) = \begin{cases} l(t, x, u) \text{ if } t \in [t_0, t_f], \\ \gamma(t, u, \lambda)E(x(t_f), x_f) \text{ if } t > t_f, \end{cases} \tag{19}$$

and thus

$$\int_{t_0}^{\infty} l'(s,x(s),u(s),\lambda)e^{-\lambda s}ds = \int_{t_0}^{t_f} l(s,x(s),u(s))e^{-\lambda s}ds + \int_{t_f}^{\infty} \gamma(s,u,\lambda)E(x(t_f),x_f)e^{-\lambda s}ds. \tag{20}$$

We choose $\gamma(t,u,\lambda)$ such that Theorem 1 holds and Proposition 1 is satisfied with the necessary modifications (e.g., restriction of one's interest to the interval $[t_0,t_f]$). In general the function α is not unique, so its choice is to personal preference. In the example section, we select $\gamma(t,u,\lambda)$ equal to $\lambda e^{\lambda t_f}(u^T R_s u+1)$ with a user-selected positive definite matrix R_s.

In order to assess the proposed approximation numerically, we consider the function

$$E_D(\lambda,t_f) := \frac{|J_\infty(t_0,\hat{x},\hat{u},\lambda) - J(t_0,t_f,\hat{x}_f,\hat{x},\hat{u})|}{J(t_0,t_f,\hat{x}_f,\hat{x},\hat{u})}, \tag{21}$$

with $(\hat{x},\hat{u})$ as the trajectory/control pair which minimizes J_∞. The function E_D acts as a (relative) error function between the given cost function J and the infinite-horizon approximation J_∞.

6 Example

This section presents the numerical example in which we are solving a path planing problem in 3D with obstacles for the nonlinear dynamics which represents the AUV. We present a test case for a finite-time application and continue with the calculation of E_D for multiple values of λ.

6.1 AUV

According to the more common notation of rigid-body models found in literature, this example uses ω for the state x in (3) and τ corresponds to u. Consider the problem

$$\min_{(\omega,\tau)} J(t_0,t_f,\omega,\tau) = \int_{t_0}^{t_f} \omega(s)^T Q\omega(s) + \tau(s)^T R\tau(s)ds \tag{22}$$

subject to

$$\begin{aligned} \dot{\omega_s} &= m^{-1}(\hat{\omega_s}(\omega_{ns}) - D_s\omega_s + \tau_s),\\ \dot{\omega_{ns}} &= I_{CG}^{-1}((\widehat{I_{CG}\omega_{ns}})(\omega_{ns}) - D_r\omega_{ns} + \tau_r),\\ (\omega(t_0),\omega(t_f)) &= (\omega_0,\omega_f), \end{aligned} \tag{23}$$

with $\omega = [\omega_s;\omega_{ns}]$ and $\tau = [\tau_s;\tau_r]$ where $\omega_s = [x;y;z] \in \mathbb{R}^3$ are the translation states, $\omega_{ns} = [\rho;\theta;\psi] \in [-\pi,\pi]^3$ are the rotational states and controls τ_s, $\tau_r \in \mathbb{R}^3$. We define m as the mass of the AUV robot, 3×3 matrices D_s, D_r are damping matrices, and I_{CG} is the inertia matrix which is symmetric and positive definite of the form

$$I_{CG} = \begin{bmatrix} I_x & -I_{xy} & -I_{xz} \\ -I_{yx} & I_y & -I_{yz} \\ -I_{zx} & -I_{zy} & I_z \end{bmatrix}.$$

We also include constraints

$$\begin{aligned} (\tfrac{x-8}{1.7})^2 + (\tfrac{y-8}{1.7})^2 + (\tfrac{z-8}{1.7})^2 \geq 1, \\ (\tfrac{x-1}{1.7})^2 + (\tfrac{y-1}{1.7})^2 + (\tfrac{z-2}{1.7})^2 \geq 1, \\ (\tfrac{x-4.5}{2.7})^2 + (\tfrac{y-5}{2.7})^2 + (\tfrac{z-8}{2.7})^2 \geq 1, \\ (\tfrac{x-5.5}{2.7})^2 + (\tfrac{y-5}{2.7})^2 + (\tfrac{z-2}{2.7})^2 \geq 1. \end{aligned} \tag{24}$$

We define constraints (24) in the form $I^{-1}(h_L)h(x_s) \geq I^{-1}(h_L)h_L$ for ease of numerical computation. Furthermore, $h_L = [1.7^2; 1.7^2; 2.7^2; 2.7^2]$ is understood with $d_i = 0.7$ (0.5 accounts for the radius of BlueROV, 0.2 gives additional buffing from obstacles; all values are given in meters). Define spaces $\mathbb{X} = [0, 10]^3 \times [-\pi, \pi]^3$ and $\mathbb{U}(t) = [-50, 50]^6$. We set values $Q = \text{diag}\{1; 1; 1; 50; 50; 50\}$, $R = \text{diag}\{1; 10; 10; 1; 5; 10\}$, $R_s = 0.02I_6$,
$I_x = I_y = 3.45$, $I_{xz} = I_{zx} = -1.28 \times 10^{-15}$, $I_z = 2.2$, $I_{xy} = I_{yx} = I_{yz} = I_{zy} = 0$, $D_s = \text{diag}\{0.1; 5; 5\}$, $D_r = \text{diag}\{5; 5; 0.1\}$, $\omega_0 = [10; 10; 10; 0; 0; 0]$, and $\omega_f = \mathbf{0_{6\times 1}}$.

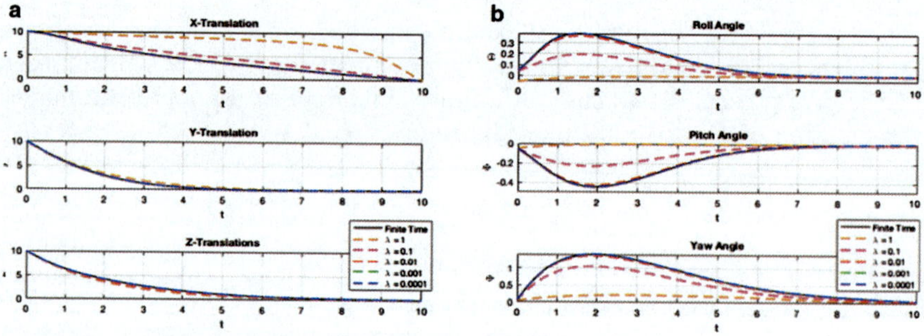

Fig. 1. a) Translations of AUV, b) Angles of AUV

For finite-time problem approximation, we choose $t_f = 10$ and also take into consideration the number of algorithm nodes (a_n) used for calculation with `DIDO`. In `DIDO`, a_n relates to the discretization of the time interval $[t_0, t_f]$ for computation of cost function $J(t_0, t_f, \omega, \tau)$.

Simulation results for the path planing problem in 3D with 3 obstacles for a AUV are shown in Figs. 1 and 2. As we observe in Fig. 1, as λ approaches 0, both translations and rotations converge to the resulting finite-time behavior. This behavior is also reflected in Table 1 with respect to E_D for both $a_n = 20$ and $a_n = 50$.

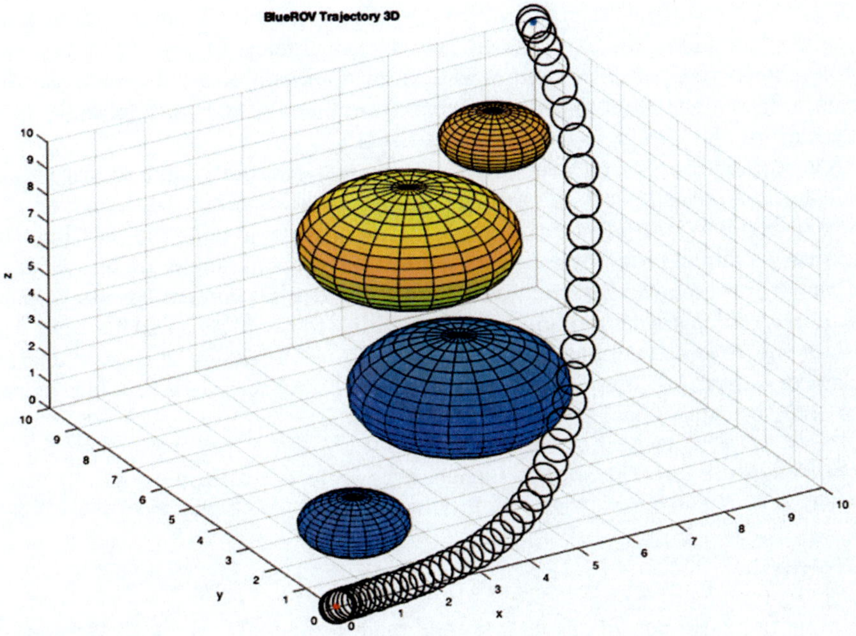

Fig. 2. Trajectory of AUV

Table 1. Relations between λ, $E_D(\lambda, t_f)$, and computational runtime with $a_n = 50$ and $a_n = 20$.

	$a_n = 50$		$a_n = 20$	
λ	$E_D(\lambda, t_f)$	runtime [s]	$E_D(\lambda, t_f)$	runtime [s]
1	0.84	185.44	0.83	58.34
0.1	0.26	149.72	0.25	37.84
0.01	0.03	148.23	2.73e-02	35.81
0.001	3e-03	141.58	2.8e-03	42.80
0.0001	2.56e-04	143.55	2.99e-04	45.67

7 Conclusions and Future Work

We have established uniqueness and existence results for a class of optimal control problem with constraints employing results for the HJB equation and Pontryagin's Principle. We have shown how Γ-convergence of cost functions can be used to solve constrained finite- and infinite-horizon optimal problems. Our framework is illustrated through solving a problem of 3D path planing with obstacles for the nonlinear dynamics of an AUV.

The principal research avenue is to investigate stability and robustness of the solutions obtained via the proposed framework. In other words, we do not want to *a priori* assume that the trajectories live in a bounded set, but the trajectory boundedness for initial conditions of interest ought to be obtained via the proposed optimization. Also, we plan to extend current results to path planning problems with dynamics corresponding to aerial and marine robots as well as the coverage problem for mobile agents with sensors such as cameras, sonars or lidars. We may also consider dynamic obstacles (e.g., mobile agents that intersect with floating debris or local wildlife).

Acknowledgments. This work is supported by project SeaClear, European Union's Horizon 2020 research and innovation programme under grant agreement No. 871295.

Appendix

In order to make this paper as self-contained as possible and for the reader's convenience, we establish some additional notation needed for this section. Let $B_r(x)$ be the closed ball in $\mathbb{R}^k$ of radius $r > 0$ centered at $x \in \mathbb{R}^k$. For a nonempty set $C \subset \mathbb{R}^k$, we denote its convex hull by $co(C)$.The negative polar cone of C is defined as $C^- := \{p \in \mathbb{R}^k : \langle p, c\rangle \le 0 \text{ for all } c \in C\}$. Denote the set distance from $x \in \mathbb{R}^k$ to C by $d_C(x) := \inf\{|x - y| : y \in C\}$.

For a nonempty set $C \subset \mathbb{R}^k$, collection of nonempty subsets $\{C_w\}_{w \in C_2}$, and $w_0 \in C_2$, let

$$\operatorname{Lim}\sup_{w \to w_0} C_w = \{v \in \mathbb{R}^k : \liminf_{w \to w_0} d_{C_w}(v) = 0\}$$

and

$$\operatorname{Lim}\inf_{w \to w_0} C_w = \{v \in \mathbb{R}^k : \limsup_{w \to w_0} d_{C_w}(v) = 0\}.$$

For $x \in \overline{C}$ define $T_C(x) := \{v \in \mathbb{R}^k : \liminf_{h \to 0+} d_A(x + hv)/h = 0\}$ and $N_C(x) := \operatorname{Lim}\sup_{y \to C^x} T_C(y)^-$. Finally, denote the epigraph of a function φ as epi φ.

Below we state the necessary assumptions and proposition required for the proof of Theorem 1.

A1 For all $x \in \mathbb{R}^n$, the mappings $\tilde{f}(\cdot, x, \cdot)$ and $\tilde{F}(\cdot, x, \cdot)$ are Lebesgue-Borel measurable and there exists $\phi \in L^1([0, \infty); \mathbb{R})$ such that $\tilde{F}(t, x, u) \ge \phi(t)$ for a.e. $t \ge 0$ and all $(x, u) \in \mathbb{R}^n \times \mathbb{R}^m$.

A2 There exists a function $c \in L^1_{loc}([0,\infty);\mathbb{R}^+)$ such that for a.e. $t \geq 0$ and for all $x \in \mathbb{R}^n$, $u \in U(t)$

$$|\tilde{f}(t,x,u)| + |\tilde{F}(t,x,u)| \leq c(t)(1+|x|).$$

A3 For a.e. $t \geq 0$ and all $x \in \mathbb{R}^n$, the set-valued map

$$y \to \{(\tilde{f}(t,y,u), \tilde{F}(t,y,u)) : u \in U(t)\}$$

is continuous with closed images, and the set

$$\{(\tilde{f}(t,x,u), \tilde{F}(t,x,u) + r) : u \in U(t),\ r \geq 0\}$$

is convex.

A4 There exists a function $k \in L^1_{loc}([0,\infty);\mathbb{R}^+)$ such that for a.e. $t \geq 0$ and for all $x, y \in \mathbb{R}^n$, $u \in U(t)$: $|\tilde{f}(t,x,u) - \tilde{f}(t,y,u)| + |\tilde{F}(t,x,u) - \tilde{F}(t,y,u)| \leq k(t)|x-y|$.

A5 $k \in \mathscr{L}_{loc}$ and $\limsup_{t\to\infty} \frac{1}{t}\int_0^t (c(s)+k(s))ds < \infty$.

A6 There exists a function $q \in \mathscr{L}_{loc}$ such that for a.e. $t \geq 0$

$$\sup_{u \in U(t)} (|\tilde{f}(t,x,u)| + |\tilde{F}(t,x,u)|) \leq q(t), \text{ for all } x \in \partial C$$

where $C \subset \mathbb{R}^n$ is closed.

A7 Let $V : [0,\infty) \times C \to \mathbb{R} \cup \{\pm\infty\}$ be the value function of (8)–(9). Then assume Dom $V \neq \varnothing$ and there exist $T > 0$ and $\psi \in L^1([T,\infty);\mathbb{R}^+)$ such that for all $(t_0, x_0) \in \text{Dom } V \cap ([T,\infty) \times \mathbb{R}^n)$ and any feasible trajectory-control pair $(x(\cdot), u(\cdot))$ on $[t_0,\infty)$, with $x(t_0) = x_0$,

$$|\tilde{F}(t, x(t), u(t))| \leq \psi(t) \text{ for a.e. } t \geq t_0.$$

A8 There exist η, $r > 0$ and $M \geq 0$ such that for a.e. $t > 0$ and any $y \in \partial C + \eta B_1(\mathbf{0})$, and any $v \in \tilde{f}(t,y,U(t))$, with $\inf_{n \in N^1_{y,\eta}} \langle n, v\rangle \leq 0$, we can find $w \in \tilde{f}(t,y,U(t)) \cap B_M(v)$ satisfying

$$\inf_{n \in N^1_{y,\eta}} \{\langle n, w\rangle,\ \langle n, w - v\rangle\} \geq r,$$

where $N^1_{y,\eta} := \{n \in \partial B_1(\mathbf{0}) : n \in \overline{co} N_C(x),\ x \in \partial C \cap B_\eta(y)\}$.

Proposition 2 (Theorem 3.3, [2]) *Let $W : [0,\infty) \times C \to \mathbb{R} \cup \{+\infty\}$ be a lower semicontinuous function such that Dom $V(t,\cdot) \subset$ Dom $W(t,\cdot) \neq \emptyset$ for all large $t > 0$ and*

$$\lim_{t\to\infty} \sup_{y \in Dom\ W(t,\cdot)} |W(t,y)| = 0. \tag{25}$$

Then the following statements are equivalent:

1. $W = V$;
2. W *is a weak (or viscosity) solution of HJB equation on* $(0, \infty) \times C$ *and* $t \to$ *epi* $W(t, \cdot)$ *is locally absolutely continuous.*

In addition, V *is the unique weak solution satisfying (25) with locally absolutely continuous* $t \to$ *epi* $V(t, \cdot)$.

References

1. Bardi M, Dolcetta, I. C.: Optimal Control and Viscosity Solutions of Hamilton-Jacobi-Bellman Equations. Birkhäuser Basel (2008). ISBN: 978-0-8176-4755-1
2. Basco, V., Frankowska, H.: Hamilton-Jacobi-Bellman equations with time-measurable data and infinite horizon. Nonlinear Diff. Equat. Appl. **26**(7) (2019). https://doi.org/10.1007/s00030-019-0553-y
3. Bonnet, B.: A Pontryagin maximum principle in Wasserstein spaces for constrained optimal control problems. ESAIM: COCV **25**, 52 (2019). https://doi.org/10.1051/cocv/2019044
4. Borkar, V., Budhiraja, A.: Ergodic control for constrained diffusions: characterization using HJB equations. SIAM J. Control Optim. **43**(4), 1467-92 (2004). https://doi.org/10.1137/S0363012902417619
5. Borkar, V., Budhiraja, A.: HJB equations for ergodic control problems for constrained diffusions in polyhedral domains. In: Proceedings of the 44th IEEE Conference on Decision and Control, 6668-72, Seville (2005). https://doi.org/10.1109/CDC.2005.1583233
6. Bressan, A., Piccoli, B.: Introduction to the Mathematical Theory of Control. American Institute of Mathematical Sciences (2007). ISBN: 9781601330024
7. Busoniu L., de Bruin T., Tolić D., Kober J., Palunko I.: Reinforcement learning for control: Performance, stability, and deep approximators. Annu. Rev. Control. **46**, 8–28 (2018). https://doi.org/10.1016/j.arcontrol.2018.09.005
8. Camacho E.F., Bordons, C.: Model Predictive Control. Springer, London (2007). ISBN: 9781852336943
9. Campillo, F.: Optimal ergodic control of nonlinear stochastic systems. In: Kree, P., Wedig, W. (eds.) Probabilistic Methods in Applied Physics, vol. 451, pp. 239–69. Springer, Berlin (1995). https://doi.org/10.1007/3-540-60214-3-59
10. Dal Maso, G.: An Introduction to Γ-Convergence, Birkhäuser Basel (1993). ISBN: 978-1-4612-0327-8
11. Dmitruk, A.V., Kaganovich, A.M.: Maximum principle for optimal control problems with intermediate constraints. Comput. Math. Model **22**, 180-215 (2011). https://doi.org/10.1007/s10598-011-9096-8
12. Gorges, D.: Relations between model predictive control and reinforcement learning. IFAC-PapersOnLine **50**(1), 4920–4928 (2017). https://doi.org/10.1016/j.ifacol.2017.08.747
13. Liberzon, D.; Calculus of Variations and Optimal Control Theory: A Concise Introduction. Princeton University (2012). ISBN: 9780691151878
14. Picarelli, A.: On Some Stochastic Control Problems with State Constraints. Mathematics [math]. ENSTA ParisTech (2015). English
15. Redfern, D. A., Goh, C. J.: Feedback control of state constrained optimal control problems. In: Doležal J., Fidler J. (eds.) System Modelling and Optimization. IFIP - The International Federation for Information Processing. Springer, Boston (1996). https://doi.org/10.1007/978-0-387-34897-1-53

16. Ross, I.M.: A Primer on Pontryagin's Principle in Optimal Control, Collegiate Publishers (2015). ISBN: 978-0984357116
17. Ross, I. M.: Enhancement to the DIDO optimal control toolbox, arXiv preprint (2020). https://doi.org/10.48550/arXiv.2004.13112
18. Ross, I.M., Gong, Q., Fahroo, F., Kang, W.: Practical Stabilization Through Real-Time Optimal Control. In: Proceedings of American Control Conference, pp.304–309, Minneapolis (2006). https://doi.org/10.1109/ACC.2006.1655372
19. Tauchnitz, N.: Pontryagin's Maximum Principle for Infinite Horizon Optimal Control Problems with Bounded Processes and with State Constraints (2020). https://doi.org/10.48550/arXiv.2007.09692
20. Valentine, F.A.: The Problem of Lagrange with Differential Inequalities as Added Side Conditions. Trace and Emergence of Nonlinear Programming, pp. 313–357. (2014). https://doi.org/10.1007/978-3-0348-0439-4-16

Comparing SONN Types for Efficient Robot Motion Planning in the Configuration Space

Lea Steffen(✉), Tobias Weyer, Katharina Glueck, Stefan Ulbrich, Arne Roennau, and Rüdiger Dillmann

FZI Research Center for Information Technology, 76131 Karlsruhe, Germany
steffen@fzi.de

Abstract. Motion planning in the configuration space (C-space) induces benefits, such as smooth trajectories. It becomes more complex as the degrees of freedom (DOF) increase. This is due to the direct relation between the dimensionality of the search space and the DOF. Self-organizing neural networks (SONN) with their famous candidate, the Self-Organizing Map, have been proven to be useful tools for C-space reduction while preserving its underlying topology. In this work, the approach is adapted from human motion data towards robots' kinematics. Furthermore, three additional models are implemented and compared to the existing ones with respect to their abilities to cover the used C-space of a robot and to preserve the topology. In total, the evaluation includes six SONN architectures, representing the consequent continuation of previous work. Generated Trajectories in the reduced SONN Output space were successfully tested in a robot simulation, providing a proof of concept for robot applications. The proposed method counteracts the Curse of Dimensionality for robots with many DOF, and thus a complete and optimal search algorithm can be used for path planning.

Keywords: Motion planning · Self-organization · C-space reduction · Wavefront algorithm

1 Introduction

While humans can easily perform target reaching without colliding with an obstacle, collision-free motion planning is a major challenge in robotics. A probate algorithm for trajectory planning is the Wavefront algorithm, a breadth-first search [23]. Its execution in the configuration space (C-space) is advantageous, however, it struggles due to the high dimensions. For a neural implementation within the C-space, neurons represent discrete configurations while synapses are applied for path planning. The impressive success of self-organizing neural networks (SONNs) is based on their simple definition, effective implementation and its features regarding clustering and visualization [5]. Hence, SONNs are a powerful and versatile tool capable to reduce the complexity of neural motion

I. Petrovic et al. (Eds.): IAS 2022, LNNS 577, pp. 186–203, 2023.
https://doi.org/10.1007/978-3-031-22216-0_13

control through pruning unnecessary neurons and synapses with simultaneous preservation of the topology. Several approaches of self-organization applied to motion control [2,9,20,25] have been introduced. Already at the beginning of the 90s, there were attempts to simplify robot movements by means of SONNs. In [20], a hierarchical architecture embodying many SONNs has been proposed to learn visuomotor coordination for a robot arm with 5 DOF. In [25] a representation of the joint and Cartesian positions of a robot arm is represented by one single network. A performance comparison of several learning algorithms being able to compute the inverse kinematics (IK) of a redundant robotic arm has been presented in [9]. A survey about self-organizing techniques for controlling robotic arms is given in [2]. The authors cover a broad spectrum from obstacle avoidance to hand-eye coordination and computation of IK. A general overview about neural networks for robot motion control, not limited to self-organization, is provided in [24].

In our previous paper [29], we presented a framework to reduce the dimension of a human arm's C-space with SONNs and a solid proof of concept that this reduced C-space is capable to support planning of trajectories. In this paper the consequent and targeted further development is reflected by analyzing the hypothesis obtained from the findings of [29]. Furthermore, an in depth evaluation about different SONN models is given, which allows to exclude some former models and introduce new ones. The proposed method counteracts the curse of dimensionality for robots with complex kinematics, allowing a complete and optimal search algorithm for path planning. Lastly, the shift from 7 DOF human motions to 6 DOF robot trajectories, not only shows the method's applicability for robot motion planning but also its easy transferability to other kinematics.

2 Self-Organizing Neural Networks

The self-organizing map (SOM) [14] represents an unsupervised learning model. Thus, a machine learning technique is applied to detect structural information and patterns in unlabeled data. SOMs create a discretized representation of the input space in a lower dimension, thereby input signals are mapped so that similar stimuli are close to each other, i.e. stimulate neurons that are neighbors [16]. The input layer is fully connected with the neurons in the competitive layer of the map, where learning takes place. The map's neurons are connected in an inhibitory mode, therefore new input hampers the neurons activity. For learning, an input vector $x = [x_1 \dots x_n]^T \in \mathbb{R}^n$ is presented in every step via the input layer to each neuron within the map. The neuron whose weight vector is most similar to the input signal is considered the winning best matching unit (BMU). It is common, but not mandatory, to use the Euclidean distance as a similarity metric [14]. For the BMU and in a less pronounced form, its neighboring neurons, the weights are shifted towards the input signal. With this strategy, the network slowly approximates the underlying topology of the input data. The SOM proved to be a sophisticated method for data exploration and has been gradually extended and refined in different directions. However, the original SOM

does not provide a complete topology preservation as it is folded in the multidimensional space due to the 'topological mismatch', as stated in [21]. Hence, all adjacent points in the reduced C-space should also be adjacent in the original C-space but not all adjacent points of the original Space are necessarily adjacent in the reduced C-space. Consequently, research has been focused on overcoming this issue over the last decades [36].
Besides the SOM, a fundamental representative of topology preserving self-organizing learning algorithms is the Neural Gas (NG) [18]. As NGs are unstructured and SOMs have a rigid structure their learning processes differ. While a winner-takes-all approach is applied for the SOM, for NGs rely in extended competitive learning. This means that the neurons, adapted to the input, besides the BMU, are the neurons whose weight vector is most similar to that of the BMU. A prerequisite for using SOMs and NGs is the required initial definition of the number of neurons. As it is not always predictable which grid size, and in the case of SOMs also its dimension, provides the optimal clustering, this is a challenging problem. Consequently, a method for detecting the required grid size during learning, named Growing Neural Gas (GNG) [11], has been proposed.

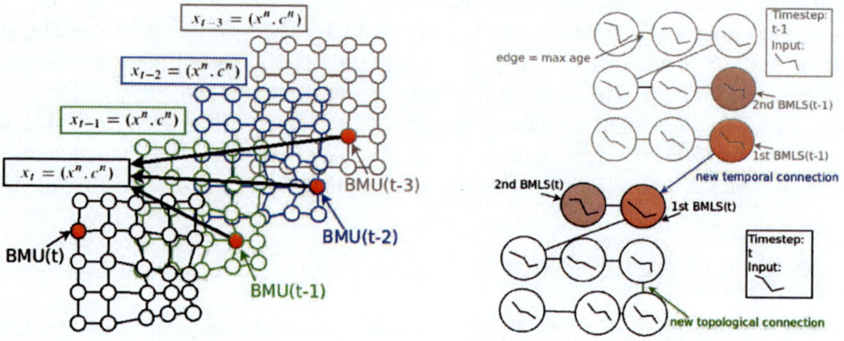

(a) γ-SOM (b) SGNG

Fig. 1. (a) The γ-SOM with $n = 4$, thus the γ-memory of depth 4 memorizes the three previous steps. (b) A SGNG's learning phase. After the BMLS for t is selected, a temporal connection is created between the current and previous BMLS. A topological connection, created between the 1st and 2nd BMLS at $t - 1$, is visible at t.

With regard to SOMs handling temporal characteristics, there are several extensions that are roughly divided into 5 classes in literature [13,33] *(1)* fixed-length windows [19,27], *(2)* specific sequence metrics [15,28], *(3)* statistical modeling incorporating appropriate generative models for sequences [3,32], *(4)* mapping of temporal dependencies to spatial correlation [8,26,38], *(5a)* SOMs with supplementary feedback connections to enable recurrent processing of time signals through integrating a neurons past activations like the Temporal Kohonen Map [4] and the recurrent SOM (RSOM) [17], *(5b)* SOMs allowing integration of

past states of other neurons as the Recursive SOM (RecSOM) [37], the SOM for structured data (SOMSD) [12], the MGNG [1] ,the Merge SOM (MSOM) [30] and its extensions the γ-SOM [6] (see Fig. 1a). A respective extension of the GNG was introduced in [7] as the γ-GNG.
A quite recent development, an extension of the GNG, taking a new approach, is the Segment GNG (SGNG) [34,35] (see Fig. 1b). This model differs strongly from other models because here, parts of the trajectory serve as input instead of joint angles. Overviews about SOM models are given in [13,22,33].

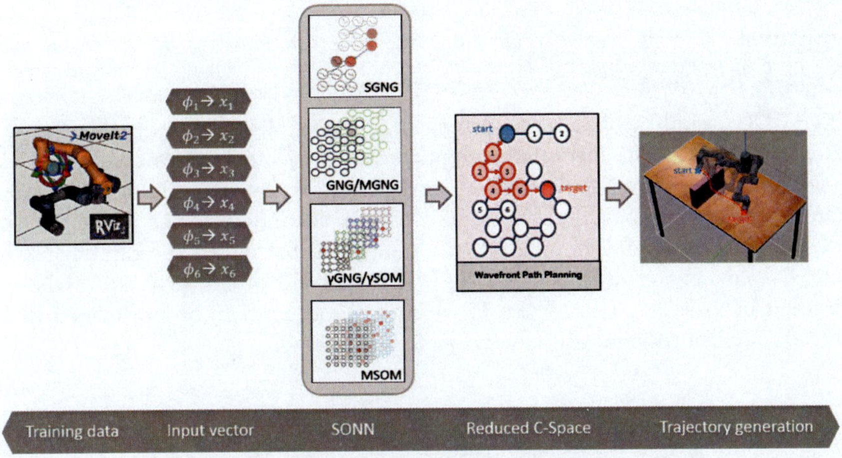

Fig. 2. A 6D input vector is derived from simulated UR3 trajectories. C-space dimensionality reduction is done via a selection of self-organized neural networks. Subsequently, motion planning is done in a reduced space by the Wavefront algorithm.

3 Methodology

Robotic motion planning is very robust in the high dimensional C-space but requires expensive computations and enormous amounts of memory. Because robotic arm movements are highly redundant, similar motions can be clustered to reduce the C-space [29]. In this work we build on these results and expand our previous study towards more potent models. Additionally, we transfer the approach from human motion data to a robot's kinematic. In Fig. 2, the conceptual sketch, including the new models and additional changes, are shown. The most important novelties are, first, the implementation of new SONN architectures as a consequence of the evaluation results in [29], yielding to a better coverage. Second, the training data, previously consisting of human trajectories from the Master Motor Map reference model [31], was replaced by real robot trajectories as visualized in Fig. 3. In addition, the different SONN types have been analyzed. Therefore, several hypotheses can be formulated, based on results of [29] and findings in literature [1,6,7,34,35]:

Hypothesis 1 *Densely sampled input sequences cause clustering as the same BMU is chosen for many iterations. Thus, sparser sampled data should be preferred.*

Hypothesis 2 *SONN models considering temporal context learn a better path preservation than versions which do not.*

Hypothesis 3 *The reason why GNG based models produce shorter paths in comparison to SOM based models is that GNGs establish many long connections which lead to shorter paths but also to large joint angle jumps.*

Hypothesis 4 *GNG-based models achieve a better coverage of the input space compared to SOM-based models.*

In Sect. 4.1 a qualitative and in Sects. 4.2, 4.3 and 4.4 a quantitative study is presented to verify the formulated hypotheses. Trajectories of a simulated UR3 are used to generate a 6D input vector, depicted at the left side of the graphic. Therefore, a sample based path planner, the RRTconnect of the Moveit library in ROS2, was applied. The main module, applied to reduce the dimensionality of the C-space is shown in the center. It consists of 1 of 6 alternative SONN, as described in Sect. 3.1. Lastly, the Wavefront algorithm can be performed in the reduced C-space to generate new trajectories.

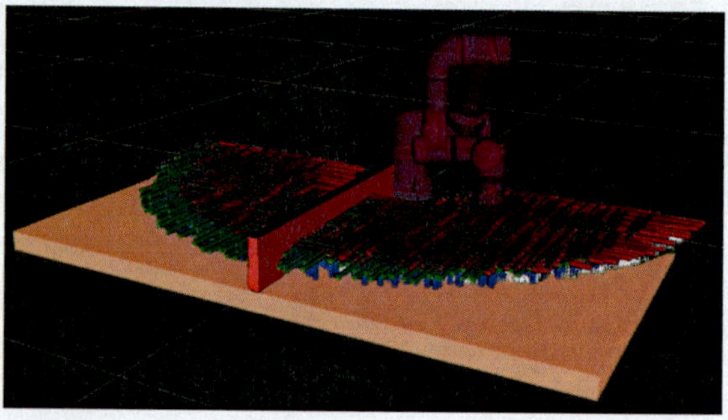

Fig. 3. Experimental setup, a UR3 mounted on a table with a barrier. Coordinate systems of start and target positions for every training trajectory are spread 5 cm over the table.

3.1 Model Selection

The reduced space must be a topology and path preserving map to be suited for path planning. A graph G forms a topology preserving map of a manifold M, if the mapping from M to G as well as the inverse mapping from G to M is neighborhood preserving. Only then, G forms a map on which adjacent locations, neurons in the context of SONNs, correspond to features neighboring on M and vice versa [21]. In other words, only with this definition of topology preservation it is possible to reduce the dimensionality of M. In this context, M is the C-space of the robot. And in addition, a topological path in G, the reduced space, has to be found and eventually re-transformed to a useful path in M. SOMs try to form a neighborhood preserving mapping from G to M, but not necessarily a neighborhood preserving mapping from M to G. Consequently, SOMs cannot theoretically provide a complete path preservation. However, it is very interesting from a scientific point of view to test these more rigid traditional SONN as well for the comparison of the models.

The network selection is based on the evaluation results from [29]. As listed in Table 1, it has been shown that path planning is not feasible in a reduced C-space generated with the NG and MNG. The reason is poor coverage of the input space and the lack of structure of these models. Thus, we did not continue both types in our study. The SOM and MSOM have deficiencies in regard to the coverage of the input space and they often generate paths with detours. Overall the best coverage was achieved with the growing versions, GNG and MGNG.

Table 1. The table depicts the evaluation results from [29]. As only small differences were found between the two basic networks, the SOM and the NG, and their respective merging versions, the MSOM and MNG, they are each listed in one category. The two growing version, GNG and MGNG, were also summarized due to great similarities, but there are certain differences in terms of their properties. The quantization error is abbreviated as QE.

	C-space coverage	Structure	QE
SOM/MSOM	Incomplete, but small joint angle distances well represented	Neighborhood of neurons can be used for path planning	High
NG/MNG	Incomplete	None	Very high
GNG/MGNG	Good (depending on the amount of input data)	Synapses can be used for path planning	Low

As the SOM and MSOM performed quite similarly, however, the MSOM was slightly superior, the MSOM was retained and the SOM was excluded. Due to their surprisingly well coverage of the input space and as they have shown some different characteristics, both the GNG and the MGNG were retained.

For an additional model, a solution that implements the memory structure very well is required. The γ-SOM model [6], as it belongs to class *(5a)* of Sect. 2, is designed for temporal sequence processing. It extends the original SOM algorithm by n context descriptors, referred to as the *γ-memory* with depth n, a short term memory structure. With the γ-SOM, it can be defined how many steps back are memorized by means of the memory depth as shown in Fig. 1a. If the memory depth is set to a single stage with $n = 1$, the γ-SOM is identical to the MSOM [30]. The authors of [6] claim that increasing depth does not cause a lost in resolution. A respective addition of a context vector including a memory function with adjustable depth also exists for the GNG. The γ-GNG [7] combines the network property of growing models to adapt the number of neurons with the advantages of a short term memory as context.

As the Segment GNG (SGNG) [35], an extension of the GNG, is said to have very good adaptation properties and a good temporary context representation, we included it into our study. The SGNG approximates small parts of the trajectories by segments, which sounds perfect for the purpose of trajectory learning in SONNs. The first aspect that distinguishes this model from the GNG, is that the network's basic unit is altered from a node to a linear segment between two nodes. Its second distinguishing feature is the addition of temporal connections between the winning neurons of two succeeding steps. This method is fundamentally different from all the previous ones as it does not consider the joint angles but instead partial trajectories, representing segments, as basic units of quantization. Hereby, the weight vector is irrelevant and the main unit is a segment. The segment quantizes portions of a trajectory, by increasing the number of included past samples and evaluate the obtained portions by determining the linearity. Consequently, there is no BMU in this approach, but instead a Best Matching Linear Segment (BMLS). A training sample consists of a trajectory portion $\varphi_i^{t-\tau}$. To determine its BMLS, a distance measure considering two characteristics of all segments S^i is applied. Firstly, d_{close}, the spatial closeness between each segment S^i and the training sample $\varphi_i^{t-\tau}$ is calculated, by means of their respective midpoints. Secondly, $d_{parallel}$ the degree of parallelism between S^i and the line joining the trajectory portion's extreme points $(\phi_t - \tau\phi_t)$ is determined through its cosine similarity. After selection, a temporal connection is established between the current and previous BMLS, as visualized in Fig. 1b. Additionally, a topological connection is established between the 1st and 2nd BMLS. In Fig. 1b, the topological connection created at time step $t-1$ is visible at time step t. A detailed description of the structure and learning properties of SGNG is given in [34,35].

3.2 Connection Reduction

In some rare cases connections cause jumps between successive joint angles due to their length. An intuitive solution to control the length of the connections and thereby avoid jumps, is to delete connections whose distance is above the threshold of 10°. In the course of this, the *Chebyshev distance*, the largest difference between the joint angles of two connected neurons, was used for distance calculation. Before the deletion of a synapse, it must be assured that the two respective neurons are still connected to each other indirectly. Otherwise, the reduced C-space would fragment. To check this condition, a copy of the connections is created, the connection is deleted and a wavefront is sent through the network. If a path is still found, the copied reduced connections are chosen.

4 Evaluation

The hypotheses postulated in Sect. 3, represent the foundation of the following qualitative (Sect. 4.1) and quantitative (Sects. 4.2, 4.3 and 4.4) analysis.

4.1 Qualitative Analysis

To evaluate the learning behavior of the SONN models, a qualitative visual analysis was conducted with a 3D visualization. Thus, only the two shoulder joints and the elbow joint were considered. Furthermore, only 15 random training trajectories were used.

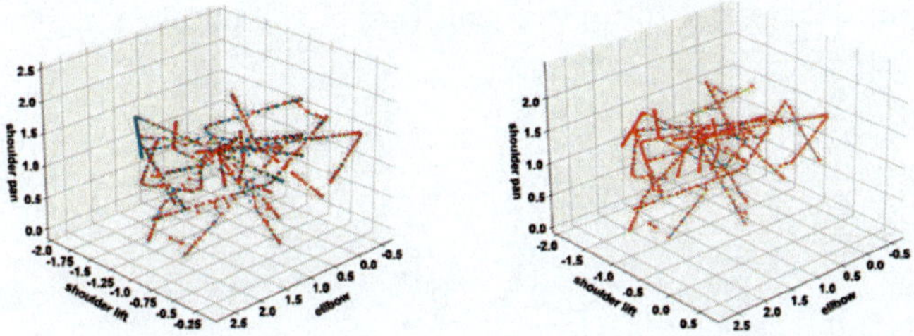

(a) sparse (b) dense

Fig. 4. GNG output spaces, visualized for three of the six joint angles, trained with original sparse vs. interpolated dense input data. Red scatter dots represent the learned weight vectors of the neurons. Lines between red dots indicate learned connections between neurons. The blue scatter dots represent input samples.

A comparison between sparsely sampled and densely sampled input data, to verify hypothesis 1, was carried out. Through repetitive interpolation (4x) the sample number was increased by factor 16. To represent the results of the comparison between sparse and dense input, the GNG model was chosen, but all other SONN types showed a very similar behavior. The output space of SONNs can be visualized by interpreting the neurons' weights as coordinates. The 3D input space results in weight vectors with three elements which can be displayed in a 3D graph. In Fig. 4a, the output space of a GNG with sparsely sampled input and in Fig. 4b, with densely sampled input is shown. Both variants achieve a good quantization of the input space, i.e. the red scatter dots representing the learned weight vectors approximate the blue scatter dots well. Additionally, the connections between the neurons reproduce a good topological estimation of the input trajectories. However, some topological connections between the neurons trained with sparsely sampled input trajectories appear to be broken at several positions. This leads to a splitting of the graph structure, causing problems for path planning. In contrast, the topological connectivity in the output space of densely sampled input trajectories corresponds almost exactly to the input space topology. A possible explanation is that splits occur with sparse input if neurons partially cover the input samples exactly. Hence, for two samples next to each other, these neurons will keep their positions in subsequent learning iterations while they also keep the same second BMU. Subsequently, only the connection to the constantly same second BMU is refreshed while the other connections to other neurons age and are eventually deleted. This effect cannot occur if the input space is densely sampled since there will always be enough samples around the neurons to generate changing BMU and second BMU pairs. Thus, a reasonable ratio between the number of input samples and the number of neurons is crucial. The graph splits if not enough neurons exist compared to the input samples, but trajectories cannot be properly approximated with too many neurons. However, any clustering of neurons for denser sampled input data, as postulated in hypothesis 1, could not be observed. On the contrary, denser sampled input data seems to prevent splitting of the topological structure.

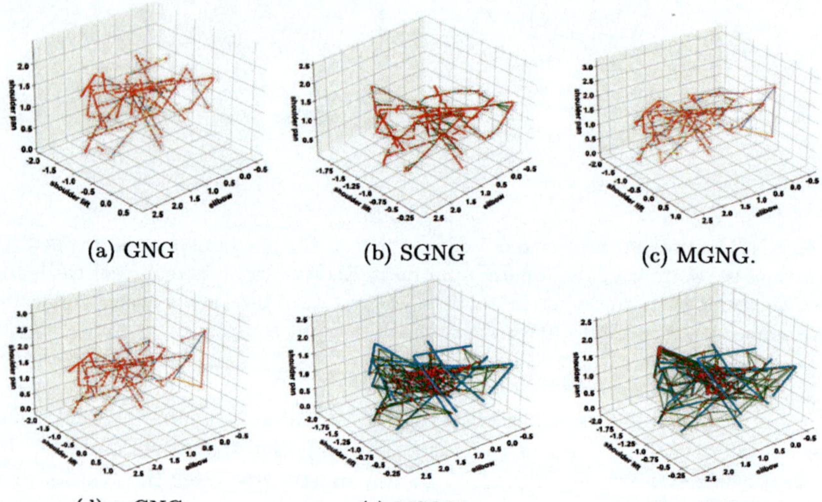

(a) GNG (b) SGNG (c) MGNG.

(d) γ-GNG (e) MSOM (f) γ-SOM

Fig. 5. SONN Comparison with 15 Trajectories. The red scatter dots represent the neurons' weight vectors and the lines indicate the learned connections. For the SGNG green lines indicate the topological connections between segments while segments are illustrated by red lines between two neurons.

As postulated in hypothesis 2 it is assumed, that SONN models which consider the temporal context should be able to learn a better path preservation in comparison to pure GNGs. In Fig. 5, it is depicted that all GNG based models produce a proper estimation of the input space. The basic GNG showed similar or even better path preservation than all models with temporal context. The SGNG shows several splits along individual trajectories. Furthermore, all models generated connections between close neurons of two closely passing trajectories. Technically, these are wrong spatio-temporal connections, which was expected for the pure GNG model, but not for the models considering the temporal context. However, these connections between different trajectories at close points allow a smooth switch between different taught trajectories during the path planning. In contrast to all GNG-based versions, the SOM based models perform significantly worse in terms of quantization and topological path preservation. Due to their topological mismatch the SOM-based models are somewhat folded in the higher dimensional space. This can be imagined like a carpet which tries to cover a 3D space and which is therefore folded and crumpled to generate some volume. Thereby, two carpet folds might be very close to each other in the 3D space but are far away from each other following the carpet course. This leads to long detours in the output space of the SOM models during the path finding (see Sect. 4.4). Furthermore, the lower dimensional structure of the SOMs is not well suited to map the topological structures of the higher dimensional input space. In Fig. 5e, f, the SOM-based models are displayed with the interpolated input data to illustrate their mismatch between input samples and learned neuron weights and the misleading connections between the neurons estimating different input trajectories, resulting from the topological mismatch. Thus, the SOM based models build longer connections than the GNG based models, contrary to hypothesis 3.

Table 2. Constant parameters for the experiments regarding the parameter setting. η_{BMU} is the factor for how strong the BMU is pulled towards the input vector. η_n refers to how strong its neighbors are pulled towards the BMU. λ defines the number of steps until a new neuron is inserted. ζ is the error counter and ϵ the max edge age.

Start size	η_{BMU}	η_n	λ	ζ	δ	ϵ
4	0.06	0.005	20	0.995	0.3	100

4.2 Parameter Setting

For space reasons, the parameter analysis has been carried out representatively with GNG. Constant parameters are listed in Table 2. A survey of SOM-specific quality measures is given in [10]. In the following, the QE is supplemented by the C-measure (CM), which provides information about neighborhood preservation between spaces. In Table 3, the results and the specific settings for individual experiments are listed. Obviously, the more neurons a network has, the higher the number of connections. If relating column *#N*, stating the final size of the network, with *QE* a correlation between an increased neuron number and an improved QE stands out. However, as this is not a uniform ratio, it can be assumed that other factors besides the network size exist. The results show that blocking neurons, as done for column *#B*, worsens the QE, which is somewhat surprising, since in [29] it was assumed that it is bad if single neurons are selected as the BMU more often in succession. Additionally, it was observed that blocked neurons increase the number of synapses and also worsen the CM. The larger the CM, the further away topologically distant neurons are from each other in terms of weight. This means that in networks where BMUs are blocked, more connections exist between neurons further apart in terms of weight in the output space, which can lead to joint angle jumps in the subsequent path planning.

Table 3. Parameter study for GNG. *#R* is the number of training runs, thus, how often a net sees each sample. *#B* gives the number of blocked neurons. 2x Int. means that the input data was interpolated two times, thus the amount of input data was increased by the factor 4. *#N* refers to the number of neurons of the trained network. The two last columns contain two values, firstly, the amount of connections and secondly, the respective CM. *# Red.* refers to the amount of connections after the reduction applied for the Chebyshev distance.

Test ID	#R	Norm	#B	Int.	#N	QE	# Con. / CM	# Red. / CM
1.1	1	euclid	0	No	2540	0.0852	7475/10.18	6262/12.80
1.4	1	euclid + cos	0	No	2540	0.0855	7515/9.60	6343/12.67
1.6	1	euclid + cos	2	No	2540	0.103	13967/7.40	10194/10.29
9.1	1	euclid + cos	2	2x	9965	0.0571	60990/9.42	55210/10.75
9.2	4	euclid + cos	2	No	10149	0.053	43902/10.13	39661/11.55
9.3	1	euclid + cos	0	2x	9964	0.0439	25479/14.40	24649/15.09
9.4	4	euclid + cos	0	No	10149	0.0436	24303/13.90	23310/15.39

The results of column *#R* (number of training runs) and *int. input* suggest that two-fold linear interpolation, quadrupling the samples per trajectory, improves the QE more than more frequent (4-fold) learning of the same samples, despite almost the same final network size. Comparing tests 1.* and 9.*, the effect of longer training while constantly inserting new neurons after a certain number of iterations becomes apparent. The QE decreases substantially while the CM increases with a larger network.

Regarding the CM, an increase of neurons and synapses does not automatically imply an improvement. This suggests that topological connections between distant neurons in the weight space still exist. One possible reason is that obsolete connections between neurons remain, which were connected before the network expanded. Another possible explanation is that the number of neurons is still too small to form a denser sampled paths and the neurons that are relatively far apart in the weight space are indeed still legitimately connected. However, reducing connections, as done in the last column of Table 3, leads to a significant increase of the CM. As only connections between neurons are deleted, which also have an alternative indirect connection, this suggests that the constant CM for an increasing network size, is caused by too many old connections. As some of the synapses above the threshold always remain, it can be assumed that the neurons are either not optimally connected to each other, or there are outliers

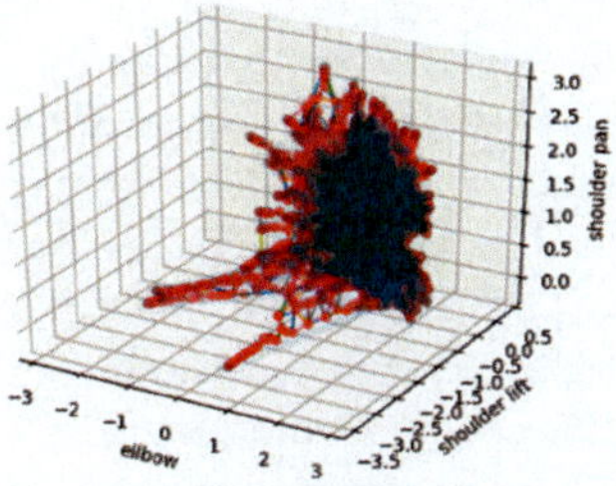

(a) GNG (Test 9.4)
no neuron blocking

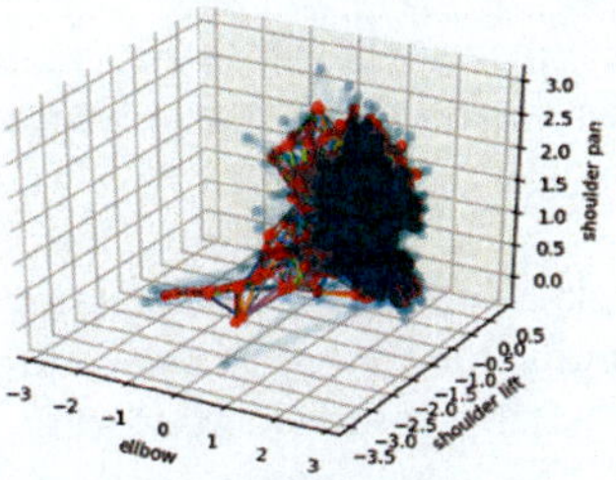

(b) GNG (Test 9.2)
blocked neurons

Fig. 6. Coverage plots of the GNG output spaces displayed for 3 joint angles but learned with 6D input vectors. The blue scatter dots are the input samples while the red once represent the learned weight vectors of the neurons. The lines between red dots indicate learned connections.

that are poorly connected to the other neurons.
Regarding the paths, generated with the different versions, the reduced connections lead to more neurons on the paths and thus, to higher path resolutions (see Fig. 8a). This correlates with the increasing CM. A similar correlation can be seen for bigger networks with more neurons. These larger networks imply higher path resolutions and a higher CM. Figure 6 shows plots regarding the coverage for experiments listed in Table 3. The GNG of test 9.4 in Fig. 6a shows a good coverage of the input space. In contrast, the neurons of test 9.2 in Fig. 6b are drawn towards the center, the input space is not covered as well and the QE is worse. This also discourages hypothesis 1, as blocking of BMUs for few iterations to enforce a more diverse BMU selection even worsened the results.

4.3 Comparison of the Different Models

Regarding all GNG-based models, γ-GNG, MGNG, SGNG, the parameters are from test 9.4 (see Table 3), as it generated the best trajectories. Hence, the GNG-based models of Table 5 had 4 training runs, no blocked neurons or interpolation.

Table 4. Parameter setting for the SOM-based models. The output dimension is defined by *dim*, σ is the neighborhood rate and η the learning rate.

#R	#N	Dim	σ	η	α	β
4	100×100 (10000)	2D	5.0	0.2	0.3	0.7

The parameters for the SOM-based models, stated in Table 4, were selected to maximize comparability. Respectively, the input data was not interpolated either. It is noticeable in Table 5 and Fig. 7 that coverage and QE are closely related. GNG, γ-GNG and SGNG have the best QE and also the best coverage of the input space. On the other hand the MGNG and especially the MSOM are clustered towards the center and obtain therefore a substantially worse QE. Therefore, hypothesis 4 is proven as correct. The biggest difference, however, seems to be in the connections. The MGNG, while having the worst CM, has established twice as many connections as any other GNG-based model. This indicates many long connections and therefore, large joint angle jumps and a rather bad path resolution. Referring to the coverage plots in Fig. 7, the GNG and the γ-GNG seem to have the best topology preserving connectivity between the neurons. Their connections are relatively short and mainly along the input trajectories. The other types show longer connections and outliers between different individual input trajectories. With the SOM variants no optimal topology preservation can take place, due to their 2D structure. Therefore, they are folded like a carpet in 3D space and do not run exclusively along the input trajectories. The SGNG contains substantially less neurons than other GNG based models. Apparently, many established connections and inserted neurons were deleted again during the training process.

Table 5. SONN comparison regarding their QE and CM. #N is the size of the trained networks, #N Con. the number of connections and #N Red. the number of reduced connections. As SOMs have a static number of synapses the reduction is only applied for GNG-based models.

type	#N	QE	# Con./CM	# Red./CM
GNG	10149	0.0436	24303/13.9	23310/15.38
MGNG	10149	0.0623	54571/7.25	46608/9.88
γ-GNG	10149	0.0458	29006/12.34	27902/13.75
SGNG	8968	0.0501	24589/8.973	22884/12.29
MSOM	10000	0.1149	19800/61.006	–
γ-SOM	10000	0.0575	19800/61.737	–

SOM based models show a substantially worse topology preservation compared to GNG based models. Firstly, many neurons lie between different input trajectories and the connections are linking wide distances between different input trajectories. Secondly, the mismatch between the topological dimensions becomes evidently and the "folding" of the 2D shaped SOMs in the 3D space can be clearly seen. Note, however, that the actual input space is 6D for which the topological mismatch becomes an even bigger problem. This also explains the MSOM's huge CM. Due to the folding effect, the paths in the output space can become ridiculously long for samples which are actually close in the input space.

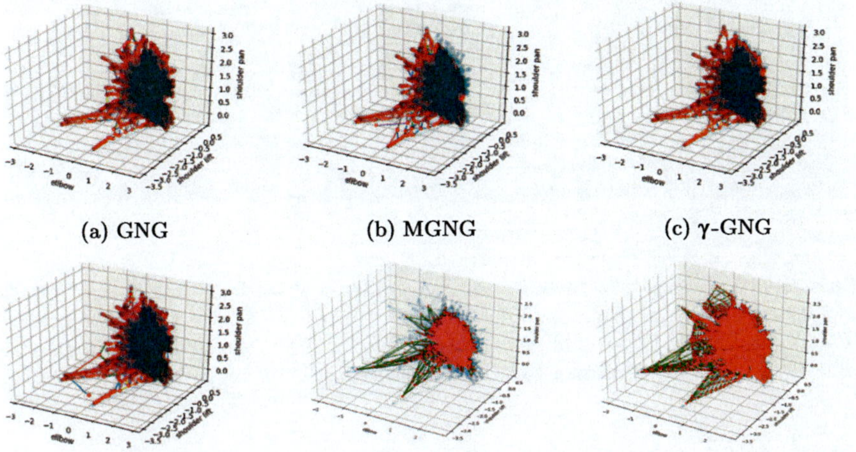

(a) GNG (b) MGNG (c) γ-GNG

(d) SGNG (e) MSOM (f) γ-SOM

Fig. 7. Plots of the learned output spaces for the different SONN models. The red scatter dots represent the learned weight vectors of the neurons. The lines between the red dots indicate the learned connections between the neurons, except for the SOM-based models which are connected to their topological neighbors. For parameters of GNG-based networks see Tables 2 and 3, for SOM-based models Table 4.

4.4 Path Analysis

To allow a fair analysis of the resulting path, exemplary generated motions in the obstacle-free task space with the reduced C-spaces were used, as depicted in Fig. 8. The graphic shows only the end effector's path for GNG, MSOM and γ-SOM, as all GNG-based networks produced similar results. From all GNG-based models, the MGNG achieved the worst path resolution. Which fits to the GNG's and γ-GNG's higher CM in comparison to the MGNG. Additionally, for the GNG a comparison between original and reduced connections, as introduced in Sect. 3.2, is displayed. Planning with the original longer connections results in a lower path resolution (27 vs. 32 neurons on the paths in Fig. 8a). Furthermore, the reduced connections seem to lead to more direct trajectories, while the original connections often produce wide-ranging detours.

The twists and loops in the paths of the SOM-based models in Fig. 8b, c are not surprising regarding the results from the previous SONN analysis in Sect. 4.3. Due to the topological mismatch between the 2D shape of the SOMs' output shape and the 6D input space, the SOMs are somewhat "folded". Hence, topologically neighbored neurons represent states which are also topologically neighbored in the input space. However, joint states which are close in the input space are not necessarily nearby located in the SOMs' output space. Thus, trajectories generated by SOM-based models contain many neurons and are relatively smooth with quite high path resolutions and few joint angle jumps but show great

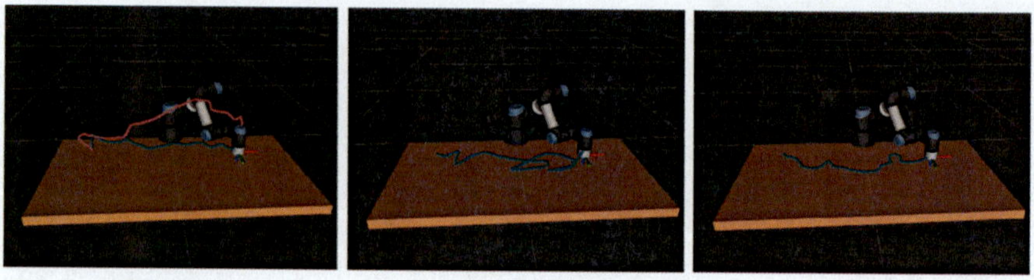

(a) GNG (b) MSOM (c) γ-SOM

Fig. 8. Comparison of generated motions with identical start and target configuration for different SONN models. The blue curve illustrates the Cartesian 3D trajectory of the end effector. In (a) an additional comparison between the path with original and reduced connections, regarding the GNG is provided.

detours and loops. Thereby, the γ-SOM outperforms the MSOM. Obviously, the many detours and loops of the trajectories generated by all SOM-based models are not practical. The reason for ridiculous long paths of the SOM based models lies especially in the topological mismatch between the 2D output space of the SOMs and the 6D input space.
The previous qualitative and quantitative analysis have shown that also the GNG-based models contain few long connections which can lead to joint angle jumps. However, they produce much less large joint angle jumps than SOM-based models, due to their flexible fitting to higher dimensional input spaces. Hence, the reason for shorter paths of the GNG-based in comparison to the SOM-based models, as it was observed in [29], is not that the GNGs produce many long connections with large joint angle jumps, as postulated in hypothesis 3. Actually, SOM-based models produce ridiculously long paths along their folded structures and GNG-based once show much more feasible and reasonable paths.

5 Conclusion

In this work, the approach presented in [29] is extended. As training data, 6 DOF robot trajectories are used instead of 7 DOF human motions. This is important for applying our approach to a real application and proves that a transfer of the method to different robot kinematics is possible. As the core part of this work, we conducted an evaluation about different SONN models for reducing the C-space. Based on their characteristics, we excluded three network models while maintaining the MSOM [30], GNG [11] and MGNG [1]. Respectively the SGNG [35], γ-SOM model [6] and γ-GNG [7] were introduced. Four hypotheses were formulated based on previous results as well as findings in literature. These were tested with a qualitative and quantitative analysis. Next to a deepened understanding of the learning behavior of the different SONN models with robot kinematics, the most important findings are that all SONN types reproduce the C-space properly but the best coverage was achieved by the GNG [11] and the γ-SOM model [6]. SOM- and GNG-based networks are equivalent in terms of coverage. However, SOM based models show a substantially worse topology preservation than GNG based models. Overall, the GNG and the γ-GNG showed the best results, while the CM values of the GNG seemed slightly superior. Furthermore, the deletion of extremely long connections could improve the trajectory quality and reduce long joint angle jumps on the path. It was shown in detail that SOMs lie folded in the multidimensional space, due to their 'topological mismatch' and therefore, not all adjacent points in the original C-space are also automatically adjacent in the reduced C-space. Based on our evaluation, GNG and γ–GNG are recommended for future projects with the presented approach. Additionally, it became apparent that the ratio between the number of neurons and samples plays an important role. While too few neurons lead to a poor resolution and coverage, too sparsely sampled data in combination with many neurons leads to splits in the graph structure of the output space of GNG based models.

The big advantage of this approach using SONNs in terms of motion planning is that due to the reduction of the search space an optimal and complete algorithm for finding the path can be used. This is due to the method's counteracting effect to the curse of dimensionality for robots with many DOF. With regard to obstacle integration, it should be emphasized that with this method, obstacles can be very easily taken into account in path planning. To do this, neurons whose configuration would lead to collisions must simply be blocked. The use of a complete and optimal search algorithm results in collision-free path planning with consideration of obstacles, without limiting the performance. How exactly the right neurons can be blocked, ideally in real time, is the subject of further research. Therefore, the system shall take static, and most importantly, also dynamic obstacles into account. A very fast and efficient transformation of obstacles from the Cartesian to the C-space is required. Finally, execution on a robotic demonstrator is planned, whereby integrating a vision component is necessary.

Acknowledgment. This research has been supported by the European Union's Horizon 2020 Framework Programme for Research and Innovation under the Specific Grant Agreement No. 945539 (Human Brain Project SGA3).

References

1. Andreakis, A., Hoyningen-Huene, N.V., Beetz, M.: Incremental unsupervised time series analysis using merge growing neural gas. Lecture Notes in Computer Science 5629 LNCS, pp. 10–18 (2009)
2. Barreto, G.D.A., Araújo, A.F.R., Ritter, H.J.: Self-organizing feature maps for modeling and control of robotic manipulators. J. Intell. Robot. Syst. **36**(4), 407–450 (2003)
3. Bishop, C.M., Hinton, G.E., Strachan, I.G.D.: GTM through time–Aston Research Explorer. ICANN, pp. 111–116 (1997)
4. Chappell, G.J., Taylor, J.G.: The temporal Kohønen map. Neural Netw. **6**(3), 441–445 (1993)

5. Cottrell, M., Olteanu, M., Rossi, F., Vialaneix, N.: Theoretical and applied aspects of the self-organizing maps. WSOM **11** (2016)
6. Estévez, P.A., Hernández, R.: Gamma SOM for temporal sequence processing. Lecture Notes in Computer Science 5629 LNCS, pp. 63–71 (2009)
7. Estévez, P.A., Vergara, J.R.: Nonlinear time series analysis by using gamma growing neural gas. Adv. Intell. Syst. Comput. **198**, 205–214 (2013)
8. Euliano, N.R., Principe, J.C.: A spatio-temporal memory based on SOMs with activity diffusion. In: Kohonen Maps, pp. 253–265. Elsevier (1999)
9. Fontinele, H.I., Melo, D.B., Barreto, G.A.: Local models for learning inverse kinematics of redundant robots: a performance comparison. Adv. Intell. Syst. Comput. **428**, 177–187 (2016)
10. Forest, F., Lebbah, M., Azzag, H., Lacaille, J.: A survey and implementation of performance metrics for self-organized maps. CoRR (2020)
11. Fritzke, B.: A growing neural gas network learns topologies. Adv. Neural Inf. Process. Syst. (1995)
12. Hagenbuchner, M., Sperduti, A., Tsoi, A.C.: A self-organizing map for adaptive processing of structured data. Trans. ANN **14**(3), 491–505 (2003)
13. Hammer, B., Micheli, A., Neubauer, N., Sperduti, A., Strickert, M.: Self organizing maps for time series. WSOM, 115–122. (2005)
14. Kohonen, T.: Self-organized formation of topologically correct feature maps. Biol. Cybern. **43**(1), 59–69 (1982)
15. Kohonen, T.: Self-Organizing Maps (1997)
16. Kohonen, T.: Essentials of the self-organizing map. Neural Netw. **37**, 52–65 (2013)
17. Koskela, T., Varsta, M., Heikkonen, J., Kaski, K.: Temporal sequence processing using recurrent SOM. KES **1** (1998)
18. Martinetz, T., Schulten, K.: A Neural-Gas Network Learns Topologies (1991)
19. Martinetz, T.: Competitive Hebbian learning rule forms perfectly topology preserving maps. In: ICANN. Springer, London (1993)
20. Martinetz, T., Ritter, H.J., Schulten, K.J.: Learning of visuomotor coordination of a robot arm with redundant degrees of freedom. Parallel Process. Neural Syst. Comput. (1990)
21. Martinetz, T., Schulten, K.: Topology representing networks. Neural Netw. **7**(3), 507–522 (1994)
22. Miljkovic, D.: Brief review of self-organizing maps. In: MIPRO. Institute of Electrical and Electronics Engineers Inc., pp. 1061–1066 (2017)
23. Naumov, V., Karova, M., Zhelyazkov, D., Todorova, M., Penev, I., Nikolov, V., Petkov, V.: Robot path planning algorithm. Int. J. Comput. Commun. **9** (2015)
24. Prabhu, S.M., Garg, D.P.: Artificial neural network based robot control: an overview. J. Intell. Robot. Syst **15**(4), 333–365 (1996)
25. Saxon, J.B., Mukerjee, A.: Learning the motion map of a robot arm with neural networks. IJCNN, 777–782 (1990)
26. Schulz, R., Reggia, J.A.: Temporally asymmetric learning supports sequence processing in multi-winner self-organizing maps. Neural Comput. **16**(3), 535–561 (2004)
27. Simon, G., Lendasse, A., Cottrell, M., Fort, J.C., Verleysen, M.: Double SOM for long-term time series prediction. WSOM, pp. 35–40 (2003)
28. Somervuo, P.J.: Online algorithm for the self-organizing map of symbol strings (2004)
29. Steffen, L., Glueck, K., Ulbrich, S., Roennau, A., Dillmann, D.: Reducing the Dimension of the Configuration Space with Self Organizing Neural Networks. ICARM (2021)

30. Strickert, M., Hammer, B.: Merge SOM for temporal data. Neurocomputing **64**, 39–71 (2005)
31. Terlemez, Ö., Ulbrich, S., Mandery, C., Do, M., Vahrenkamp, N., Asfour, T.: Master Motor Map (MMM)–framework and toolkit for capturing, representing, and reproducing human motion on humanoid robots. In: International Conference Humanoid Robots, pp. 894–901 (2015)
32. Tino, P., Kabán, A., Kabán, K., Sun, Y.: A Generative probabilistic approach to visualizing sets of symbolic sequences. Int. conf. Knowl. Discovery Data Min. **4**, 701–706 (2004)
33. Van Hulle, M.M.: Self-Organizing Maps. Tech. rep. (2012)
34. Vergara, J.R., Estévez, P.A.: A strategy for time series prediction using segment growing neural gas. WSOM (2017)
35. Vergara, J.R., Estévez, P.A., Serrano, Á.: Segment growing neural gas for nonlinear time series analysis. Adv. Intell. Syst. Comput. **428**, 107–117 (2016)
36. Villmann, T., Der, R., Herrmann, M., Martinetz, T.M.: Topology preservation in self-organizing feature maps: exact definition and measurement (1997)
37. Voegtlin, T., Dominey, P.F.: Recursive self-organizing maps. In: Advanced in SOM, pp. 210–215. Springer, London (2001)
38. Wiemer, J.C.: The time-organized map algorithm: extending the self-organizing map to spatiotemporal signals. Neural Comput. **15**, 1143–1171 (2003)

Physics-Based Motion Planning of a Fruit Harvesting Manipulator for Pushing Obstacles in a Cluttered Environment

Miyuki Nakamura[1](✉), Yosuke Kawasaki[1], and Masaki Takahashi[2]

[1] Graduate School of Science and Technology, Keio University, 3-14-1 Hiyoshi, Kohoku-Ku, Yokohama 223-8522, Japan
{nmiyuki73,y-kawasaki}@keio.jp

[2] Department of System Design Engineering, Faculty of Science and Technology, Keio University, 3-14-1 Hiyoshi, Kohoku-Ku, Yokohama 223-8522, Japan
takahashi@sd.keio.ac.jp

Abstract. Recently, the agricultural population is decreasing, and the demand for autonomous agricultural robots has increased. For a robot to successfully harvest fruits in diverse environments, various obstacles must be pushed away from its path. During the pushing motion, however, gravity acts on the fruits, causing them to move back and forth and return to their original state after being pushed away. This results in fruit-to-fruit contact, and fruit damage may occur. Therefore, a method to avoid obstacles while reducing the risk of fruit damage is required. However, the direct consideration of fruit damage is difficult because it requires many parameters that are difficult to observe. Herein, we propose a motion planning method by considering the potential energy of fruits as an evaluation index, which could be calculated based on observable parameters; this consideration could also suppress the reciprocating motion of fruits, thus preventing fruit damage. The proposed method was evaluated through simulations, and its effectiveness was verified.

Keywords: Physics-based motion planning · Kinodynamic motion planning · Manipulation · Harvesting robot · Cluttered environment

1 Introduction

In recent years, agricultural labor shortages, due to the decrease in agricultural population in developed countries and the increase in labor costs in developing countries have become a cause for concern. The introduction of agricultural robots would result in the automation and mechanization of farm work, which is expected to solve these problems. However, fruit harvesting is difficult to automate as selective harvesting is required, which relies heavily on human labor [1].

To address these issues, the research on and development of robots for fruit harvesting have been actively pursued [2, 3]. Particularly, in recent years, studies have been conducted on harvesting robots assuming an actual harvesting environment and considering the obstacles around the target fruit [4–6]. Luo et al. [7] proposed collision-free

I. Petrovic et al. (Eds.): IAS 2022, LNNS 577, pp. 204–215, 2023.
https://doi.org/10.1007/978-3-031-22216-0_14

autonomous motion planning for a grape harvesting robot to avoid obstacles. However, it cannot be applied to situations where the target fruit and obstacles are extremely near each other, such as in the case of strawberries or tomatoes, as avoiding the obstacles and harvesting the fruit becomes difficult. Zhang et al. [8] proposed a motion planning method to harvest the target fruit by cutting the leaves that acted as obstacles. However, when obstacles are immature, their cutting is considered inappropriate; thus, this method cannot be used to harvest the target fruit.

Xiong et al. [9] developed a strawberry harvesting robot that harvested the strawberries by pushing obstacles away. The approaching position was determined based on the locations of the target fruit and surrounding obstacles. The gripper was then raised vertically from below the target fruit. The opening and closing motions of the gripper tip were used to push away the obstacles. Sariah et al. [10] further studied this strawberry harvesting robot and proposed an efficient push-off motion using probabilistic movement primitives (ProMPs). However, these studies did not consider the possibility of the occurrence of fruit damage. While the target fruit can be harvested using pushing motion, a risk of fruit damage is involved in the process owing to the increased contact opportunities between the robot and the fruit and those between fruits. In particular, owing to gravity, fruits, such as strawberries, move back and forth to return to their original state after being pushed away. This further increases the chance of fruit-to-fruit contact and the risk of fruit damage. Since fruit quality is an important factor directly related to profitability, it is necessary to ensure that no fruit damage occurs while pushing away obstacles.

In this study, using the physics-based motion planning of a manipulator, pushing motion that ensures a reduction in the risk of fruit damage was generated by considering an appropriate evaluation index. With physics-based motion planning, the state of objects can be considered using a physics engine, which enables the generation of motions that consider the collisions between the manipulator and objects, such as the motion of pushing obstacles away. Moll et al. [11] and Akbari et al. [12] used physics-based motion planning to generate obstacle-pushing motions in a cluttered environment with many objects on a plane. However, for investigating the application of physics-based motion planning to a harvesting environment, reproducing the environment is a challenge. Additionally, to perform physical simulations for physics-based motion planning, it is necessary to create a simulation environment by observing the necessary object parameters in advance. However, in a dense harvesting environment, the fruit would be hidden by leaves, and fruit stalks would be intertwined, resulting in a complex structure. Therefore, there is a high possibility of the occurrence of observation and modeling errors, rendering the verification of whether the system can rectify these errors necessary.

The reproduction of a harvesting environment also becomes an issue when considering the evaluation indices for reducing the risk of fruit damage. It is known that the extent of fruit damage depends on the impact force between fruits, and especially large impact force needs to be avoided [13]. However, to calculate impact force, many parameters that are difficult to observe, such as fruit hardness and the vibration damping coefficient, are required for creating the simulation environment. Therefore, a different evaluation index for fruit damage based on observable parameters would be required.

To address this issue, we used the potential energy of fruits as an evaluation index, as it can be calculated based on observable parameters, such as their position and size. Furthermore, by reducing the potential energy of fruits, their reciprocating motion could be suppressed; consequently, fruit damage could be prevented. The proposed method was validated using simulations where the impact force between fruits could be calculated. It was confirmed that the proposed method could achieve the same level of performance as that achieved by the method that used the impact force between fruits as a direct evaluation index. In addition, we assumed an observation error in the verification environment and confirmed its effect.

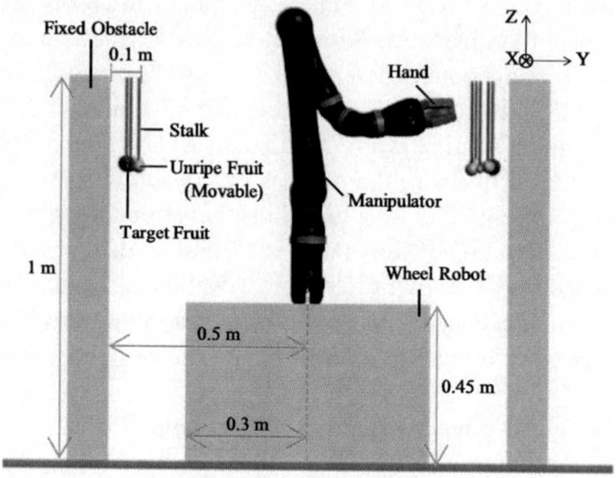

Fig. 1. Assumed environment

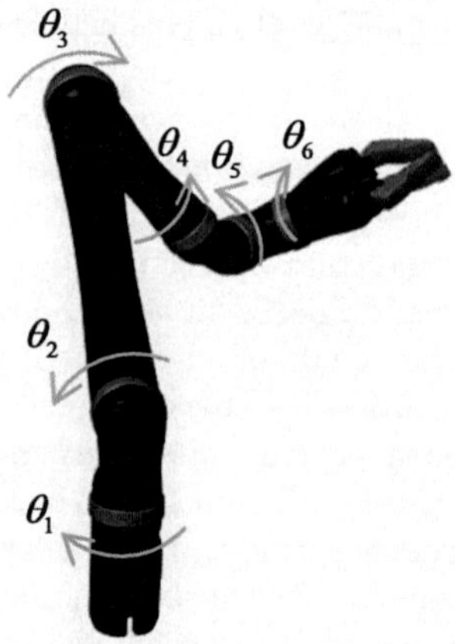

Fig. 2. Robot

2 Overview of the System

In this section, we describe the work to be assumed, the robots to be used, and the overall system flow.

2.1 Work

We assumed a harvesting scene for strawberries at a high altitude with unripe fruits acting as obstacles, as shown in Fig. 1. The robot approached the ground horizontally and generated motion to push the stalk of the unripe fruit away. In addition, three types of objects in the environment were considered: the target fruit, unripe fruit (movable obstacle), and fixed obstacle. A pendulum model was used for the target and unripe fruits to consider the effects of reciprocating motion.

2.2 Robot

We assumed a mobile manipulator, which is shown in Fig. 1. Jaco^2, manufactured by Kinova Robotics (Boisbriand, Quebec, Canada) [14], was used for this purpose. It is a 6-DOF manipulator, and its end effector is a hand with three fingers. In this study, only the joint angle $\boldsymbol{q} = [\theta_1, \theta_2, \theta_3]$ from the base to the third finger was moved to prevent combination explosions, and the hand was kept closed to prevent obstacles from entering into the hand (Fig. 2).

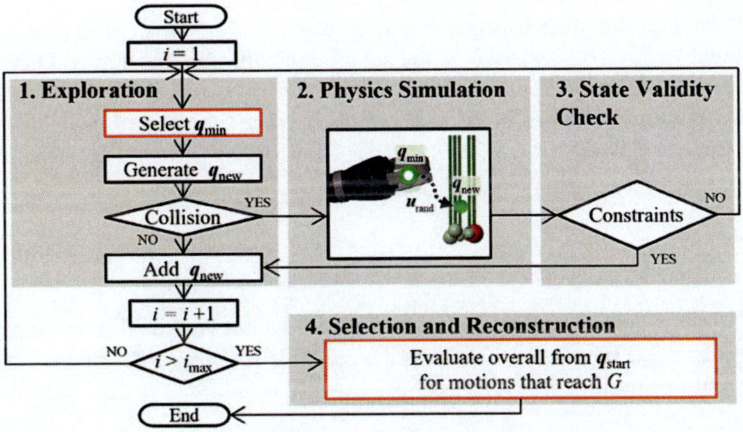

Fig. 3. System flow

2.3 System Flow

The flow of the proposed system comprised four stages, as illustrated in Fig. 3. First, the configuration space would be explored by expanding the tree structure, starting from the start node, $\boldsymbol{q}_{\text{start}}$. If collision occurred between the robot and the object, the robot would output a control input and move to the next stage of the physical simulation. In the third stage, the robot would determine whether the world state output from the physical simulation satisfied the validity constraints. This process would be repeated a certain number of times to complete the plan. In the final stage, an appropriate motion would be selected and reconstructed, and the result would be outputted. In this study, the selection of the node $\boldsymbol{q}_{\text{min}}$ at the first stage and that of the movement at the last stage were performed, considering the evaluation index to reduce the risk of fruit damage.

3 Physics-Based Motion Planning

In this section, we explain in detail each stage of the system shown in Fig. 3.

3.1 Exploration in the Configuration Space

We applied Kinodynamic Rapidly-Exploring Random Trees (RRT) [15] to expand the tree structure from $\boldsymbol{q}_{\text{start}}$ to the end point region G by random sampling.

First, a random sample $\boldsymbol{q}_{\text{rand}}$ was generated in the configuration space. For this $\boldsymbol{q}_{\text{rand}}$, n neighbor nodes that were small distance in the cartesian space were selected from the tree structure to generate a group of neighbor nodes $\boldsymbol{Q}_{\text{Near}} = \{\boldsymbol{q}_{\text{N}_1}, \boldsymbol{q}_{\text{N}_2}, \ldots, \boldsymbol{q}_{\text{N}_n}\}$, where n is a constant. Thereafter, the node with the smallest evaluation value was selected as $\boldsymbol{q}_{\text{min}}$ among $\boldsymbol{Q}_{\text{Near}}$. If the total number of nodes in the tree structure was less than n, all nodes would be selected for $\boldsymbol{Q}_{\text{Near}}$. If there were multiple nodes with $\boldsymbol{q}_{\text{min}}$, the node with the smallest from G in the cartesian space would be selected.

Next, random sampling was performed in the control input space to determine the control inputs $\boldsymbol{u}_{\text{rand}} \in U$, where U is the set of applicable control inputs. Thereafter, a new node $\boldsymbol{q}_{\text{new}}$ was generated based on Eq. (2), which was obtained by integrating the equation of motion described in Eq. (1), where Δt is the time step of the RRT, and $\boldsymbol{f}$ is the state transition function.

$$\dot{\boldsymbol{q}} = \boldsymbol{f}(\dot{\boldsymbol{q}}, \boldsymbol{u}) \tag{1}$$

$$\boldsymbol{q}_{\text{new}} = \boldsymbol{q}_{\text{min}} + \Delta t \boldsymbol{f}\left(\boldsymbol{q}_{\text{min}}, \boldsymbol{u}_{\text{rand}}\right) \tag{2}$$

In addition, a collision check between the manipulator and the fruit was performed from $\boldsymbol{q}_{\text{start}}$ to $\boldsymbol{q}_{\text{new}}$. If collision occurred, the physical simulation described in the next section would be performed. Otherwise, $\boldsymbol{q}_{\text{new}}$ would be added to the tree structure.

3.2 Physics Simulation

If a collision decision were made in the previous section, the control input $\boldsymbol{u}_{\text{rand}}$ from $\boldsymbol{q}_{\text{min}}$ to $\boldsymbol{q}_{\text{new}}$ would be outputted, and physics simulation would be performed based on $\boldsymbol{u}_{\text{rand}}$ using the physics engine. In this study, the physics engine Simscape Multibody [16] was used, which could simulate physical interactions, such as collision and momentum, by considering dynamic properties. The simulation environment could be generated by observing and inputting the necessary parameters of an object in advance.

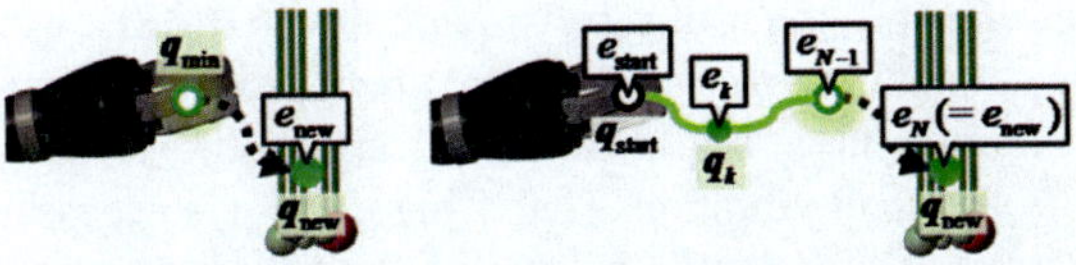

(a) Calculation of e_{new} (b) Calculation of E_{new}

Fig. 4. Calculation of an evaluation index

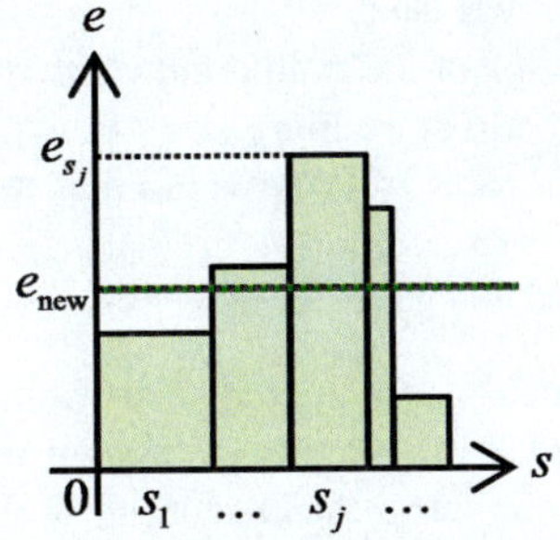

Fig. 5. Simulation Output

3.3 State Validity Check

As a result of the physical simulation in the previous section, the world state, including the manipulator and fruits, was outputted. For this state, judgment was made as to whether the validity constraints were satisfied; $\boldsymbol{q}_{\text{new}}$ would be added to the tree structure only if they were satisfied. In this study, the collisions between the manipulator and fixed obstacles and those between the manipulator and fruits were prohibited as validity constraints.

3.4 Motion Selection and Reconstruction

After the completion of the planning based on a predetermined number of samples, the motion with the smallest evaluation value was selected among those that reached G. Here, evaluation was performed for the entire motion using $\boldsymbol{q}_{\text{start}}$. However, since we used the Kinodynamic RRT to expand the tree structure randomly, it may have resulted in the addition of unnecessary detours between $\boldsymbol{q}_{\text{start}}$ and the start node of the pushing action, $\boldsymbol{q}_{\text{push}}$. Therefore, the result was outputted after the linear completion between $\boldsymbol{q}_{\text{start}}$ and $\boldsymbol{q}_{\text{push}}$.

4 Consideration of an Evaluation Index

In this section, we explain the definition and calculation method of each evaluation value when the potential energy of fruits was used as evaluation indices.

In the proposed method, the root-mean-square error (RMSE) value was defined as the evaluation value to minimize the contact force between fruits when considering their potential energy. A single large impact tends to cause more severe fruit damage than that caused by multiple small vibrations [13]. Therefore, it is necessary to prevent large impacts between fruits. The potential energy of the fruit increases when it is pushed to a state higher than its initial state. As the pushed fruit returns to its original state, its kinetic energy increases, resulting in the possibility of a large impact between the pushed fruit and the surrounding fruits. Therefore, it is necessary to suppress the large fluctuations in the position of the fruit; thus, the RMSE value that is sensitive to outliers and overestimates a large error was used.

The first step in the calculation of the evaluation value was to determine the potential energy e_{new} from $\boldsymbol{q}_{\text{min}}$ to $\boldsymbol{q}_{\text{new}}$ and store in $\boldsymbol{q}_{\text{new}}$, as shown in Fig. 4a. If there were no collisions from $\boldsymbol{q}_{\text{start}}$, the initial value of e_{start} would be substituted as follows:

$$e_{\text{new}} = e_{\text{start}}. \tag{3}$$

In contrast, if there were a collision, the physical simulation would run for Δt, and the result would be substituted. In this study, as shown in Fig. 5, using the potential energy e_j output at sampling interval s_j, the execution result was calculated as follows:

$$e_{\text{new}} = \sum s_j e_j / \sum s_j. \tag{4}$$

where $\sum s_j = \Delta t$. Thereafter, using the potential energy e_k stored in $\boldsymbol{q}_k$ from $\boldsymbol{q}_{\text{start}}$ to $\boldsymbol{q}_{\text{new}}$, as shown in Fig. 4b, the evaluation value was calculated as follows:

$$E_{\text{new}} = \sqrt{\frac{1}{N} \sum_{k=1}^{N} (e_k - e_{\text{start}})^2}, \tag{5}$$

where N represents the number of nodes from.

5 Evaluation

In this section, we confirm that the proposed method could reduce the risk of fruit damage in a simulation environment where the impact force between fruits can be calculated based on the parameters shown in Table 1.

5.1 Method

Evaluations were conducted with two comparison methods: Comparison 1, which minimized the change in the robot's joint angles and did not consider fruit damage; and Comparison 2, which minimized the sum of the impact force between fruits and directly considered fruit damage. The effectiveness of the proposed method was verified by evaluating the sum and maximum value of the impact force between fruits throughout the motion (Fig. 6).

5.2 Conditions

The initial state of the verification environment is illustrated in Fig. 7. We assumed a harvesting environment that each target fruit was surrounded by five immature fruits, as shown in Fig. 7a. In addition, we assumed that the initial state of the manipulator was changed, as shown in Fig. 7b, and that the harvesting environment was changed, as shown in Fig. 7c, to verify the robustness of the planned environment. Furthermore, we assumed an environment with the observation errors by shortening the length of the three fruit stalks (O_1, O_3, O_5) by 5 cm each, as shown in Fig. 7d, in contrast to the planned environment shown in Fig. 7a and checked its effect on the observation errors. The endpoint region G was located below the center of the petiole of the target fruit and within its radius. In addition, the randomness of the RRT was considered and each evaluation index was verified three times.

Table 1. Simulation parameters

			Evaluation 1, 2, 4	Evaluation 3
Motion Planning Condition	i_{max}		1000	
	U [deg/s]		[−5, 5]	
	Δt [s]		1	
Fruit	Dense [g/mm^3]		9×10^{-4}	
	Radius [mm]	Target fruit	20	
		Obstacle fruit 1	16	20
		Obstacle fruit 2	10	20
		Obstacle fruit 3	16	20
		Obstacle fruit 4	16	20
		Obstacle fruit 5	12	20

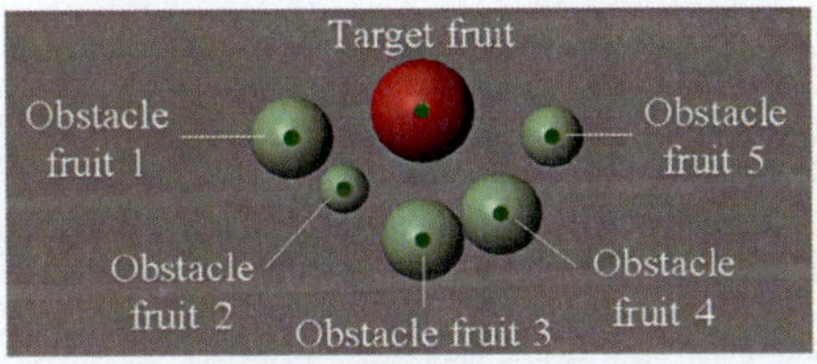

Fig. 6. Fruit location

5.3 Results

Table 2 lists the results of each evaluation item from evaluations 1–3, and the normalized value respected to that of Comparison 1 is shown in parentheses. Table 3 lists the success

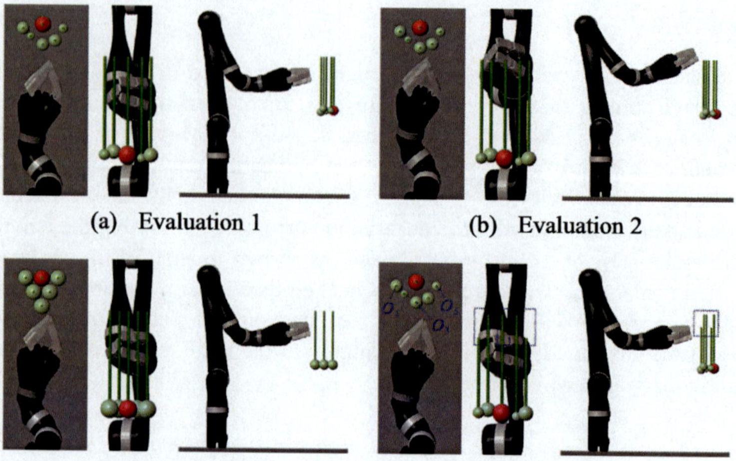

(a) Evaluation 1 (b) Evaluation 2

(c) Evaluation 3 (d) Evaluation 4

Fig. 7. Initial state of the evaluation environment

Table 2. Impact force between fruits [$\times 10^{-3}$ N]

		Comparison 1	Comparison 2	Proposed
Evaluation 1	Sum	44.03 (1.00)	4.10 (0.09)	5.53 (0.13)
	Max	1.01 (1.00)	0.42 (0.42)	0.38 (0.37)
Evaluation 2	Sum	73.44 (1.00)	6.41 (0.09)	6.55 (0.09)
	Max	0.88 (1.00)	0.73 (0.83)	0.59 (0.67)
Evaluation 3	Sum	698.14 (1.00)	40.29 (0.06)	68.00 (0.10)
	Max	2.86 (1.00)	1.59 (0.55)	0.80 (0.28)

Table 3. Success rate and impact force between fruits in evaluation 4

		Comparison 1	Comparison 2	Proposed
Success rate		1/3	3/3	3/3
Impact force between fruits [$\times 10^{-3}$ N] (The changed value respected to that of Eva.1)	Sum	55.09 (+11.06)	4.85 (+0.75)	6.10 (+0.58)
	Max	1.01 (+0.00)	0.65 (+0.23)	0.57 (+0.19)

rate and impact force between fruits in evaluation 4, and the changed value of the impact force respected to that of evaluation 1 is shown in parentheses. In addition, the motion trajectories of evaluations 1–4 are shown in Figs. 8, 9, 10 and 11, respectively; in these figures, the left side shows the state wherein the robot started to push away the obstacles, and the right side shows the state wherein the robot finished its operation.

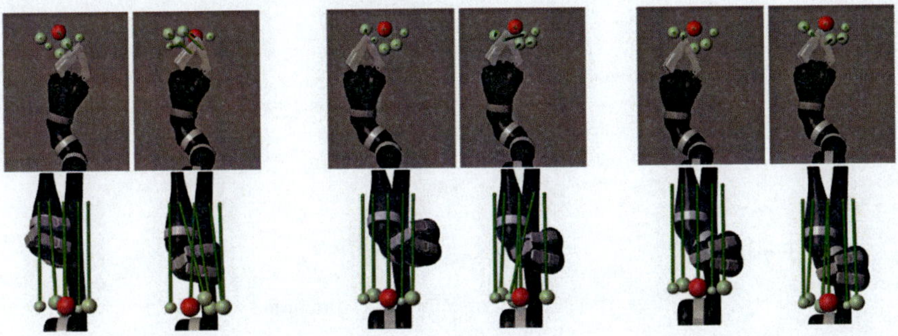

(a) Comparison 1 (b) Comparison 2 (c) Proposed

Fig. 8. Motion trajectory in evaluation 1

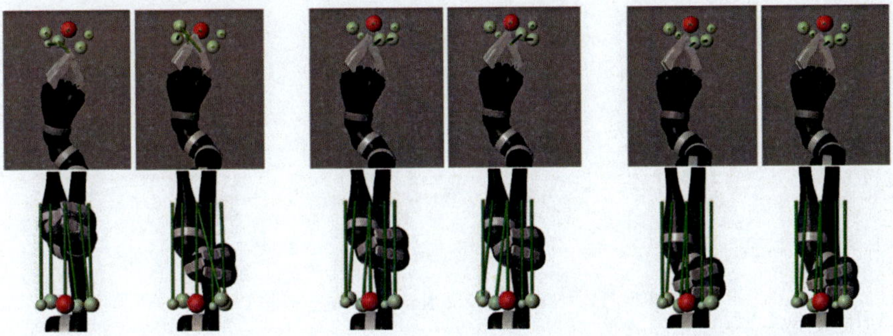

(a) Comparison 1 (b) Comparison 2 (c) Proposed

Fig. 9. Motion trajectory in evaluation 2

For evaluations 1–3, Table 2 shows that the sum and maximum value of the impact force of the proposed method were always less than those of Comparison 1, which did not consider fruit damage, and the normalized value was almost the same as those of Comparison 2, which directly considered fruit damage. Comparing the motion trajectories shown in Figs. 8, 9 and 10, it is evident that in the case of Comparison 1, the hand followed the shortest path and collided with the fruits at the top of the fruit stalk, resulting in significant fruit movement. In the proposed method, the hand was lowered to the bottom of the fruit stalk and then pushed away from the fruit. Consequently, the motion of the fruits was suppressed, and the total impact force between them was reduced. However, in Comparison 2, wherein the impact force of fruits was considered, several trends were observed, such as the pushing away of small fruits and the pushing away of a small number of fruits in a short duration. This suggests that when considering the impact force of the fruit directly, it was possible to generate motions with a reduced

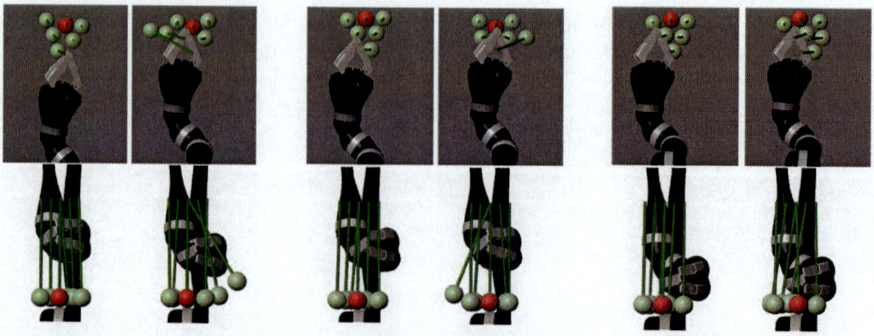

(a) Comparison 1 (b) Comparison 2 (c) Proposed

Fig. 10. Motion trajectory in evaluation 3

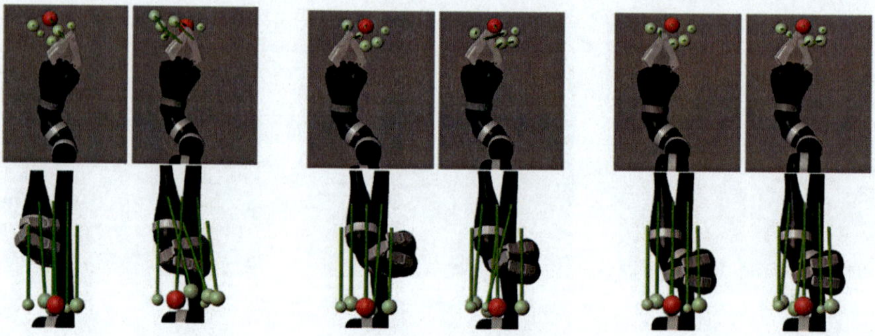

(a) Comparison 1 (b) Comparison 2 (c) Proposed

Fig. 11. Motion trajectory in evaluation 4

risk of fruit damage from various aspects, such as the mass of the fruit to be pushed, number of fruits, and time taken for these actions.

In evaluation 4, Comparison 1 could not reach the target fruit in two situations. However, when comparing the changed values between the proposed method and Comparison 2, the proposed method tended to exhibit a smaller changed value. In this study, the error was set by shortening the length of the fruit stalk, which tended to increase the amount of fruit variation. The action of pushing away the upper part of the obstacle was more likely to be affected by this error. Conversely, by considering potential energy, the lower part of the obstacle was pushed away, reducing the effect of this error.

6 Conclusion

In this study, we proposed a physics-based motion planning method using the potential energy of fruits as an evaluation index. Based on the observable parameters, the motion of pushing away obstacles with a reduced impact force between fruits was achieved. The reciprocating motion of the fruit was suppressed by considering its potential energy, and fruit damage was prevented. Actual verification using simulations indicated that the performance of the proposed method was equivalent to that of the method that directly considered fruit damage in terms of the total and maximum impact forces generated between fruits.

In the future, we would like to use the tendency of the pushing motion to lower the hand to the lower part of the fruit stalk and then start pushing it away as a heuristic for efficient search. In addition, we would like to confirm the effectiveness of the proposed method when modeling the fruit stalk as a continuum in order to approximate the actual harvesting environment.

References

1. Hayashi, S., et al.: Evaluation of a strawberry-harvesting robot in a field test. Biosys. Eng. **105**(2), 160–171 (2010)
2. Yaguchi, H., Nagahama, K., Hasegawa, T., Inaba, M.: Development of an autonomous tomato harvesting robot with rotational plucking gripper. In: IEEE/RSJ International Conference on Intelligent Robots and Systems (IROS), pp. 652–657 (2016)
3. Liu, Y., Yang, C., Ling, H., Mabu, S., Kuremoto, T.: A visual system of citrus picking robot using convolutional neural networks. In: IEEE International Conference on Systems and Informatics (ICSAI), pp. 344–349 (2018)
4. Van Henten, E.J., et al.: An autonomous robot for harvesting cucumbers in greenhouses. Auton. Robot. **13**(3), 241–258 (2002)
5. Botterill, T., et al.: A robot system for pruning grape vines. J. Field Rob. **34**(6), 1100–1122 (2017)
6. Cao, X., Zou, X., Jia, C., Chen, M., Zeng, Z.: RRT-based path planning for an intelligent litchi-picking manipulator. Comput. Electron. Agric. **156**, 105–118 (2019)
7. Luo, L., Wen, H., Lu, Q., Huang, H., Chen, W., Zou, X., Wang, C.: Collision-free path-planning for six-DOF serial harvesting robot based on energy optimal and artificial potential field. Complexity 2018 (2018)
8. Zhang, T., Huang, Z., You, W., Lin, J., Tang, X., Huang, H.: An autonomous fruit and vegetable harvester with a low-cost gripper using a 3D sensor. Sensors **20**(1), 93 (2020)
9. Xiong, Y., Yuanyue, G., From, P. J., "An obstacle separation method for robotic picking of fruits in clusters. Comput. Electron. Agric. **175** (2020)
10. Mghames, S., Hanheide, M., Ghalamzan, A.: Interactive movement primitives: planning to push occluding pieces for fruit picking. In: IEEE/RSJ International Conference on Intelligent Robots and Systems (IROS), pp. 2616–2623 (2020)
11. Moll, M., Kavraki, L., Rosell, J., Muhayyuddin.: Randomized physics-based motion planning for grasping in cluttered and uncertain environments. IEEE Rob. Autom. Lett. **3**(2), 712–719 (2017)
12. Akbari, A., Rosell, J., Muhayyuddin.: κ-PMP: enhancing physics-based motion planners with knowledge-based reasoning. J. Intell. Rob. Syst. **91**(3–4), 459–477 (2018)
13. Kitazawa, H., Ishikawa, Y., Lu, F., Hu, Y., Nakamura, N., Shiina, T.: Analysis of shock during strawberry transport and damage estimation. Hortic. Res. **9**(2), 221–227 (2010) (Japanese)
14. Kinova Robotics.: KINOVA™ Ultra lightweight robotic arm User Guide. https://www.kinovarobotics.com/sites/default/files/ULWSRA-JAC-UG-INT-EN%20201804-1.0%20%28KINOVA™%20Ultra%20lightweight%20robotic%20arm%20user%20guide%29_0.pdf. As of Feb. 2022
15. LaValle, S.M., Kuffner, J.J., Jr.: Randomized kinodynamic planning. Int. J. Rob. Res. **20**(5), 378–400 (2001)
16. MathWorks, "Simscape Multibody—MATLAB & Simulink. https://jp.mathworks.com/products/simscape-multibody.html. As of Feb. 2022

A Dynamics-Aware NMPC Method for Robot Navigation Among Moving Obstacles

Spyridon G. Tarantos(✉) and Giuseppe Oriolo

Dipartimento di Ingegneria Informatica, Automatica e Gestionale, Sapienza Università di Roma, via Ariosto 25, 00185 Roma, Italy
{tarantos,oriolo}@diag.uniroma1.it

Abstract. We present a novel method for mobile robot navigation among obstacles. Our approach is based on Nonlinear Model Predictive Control (NMPC) and uses a dynamics-aware collision avoidance constraint. The constraint, built upon the notion of *avoidable collision state*, considers not only the robot-obstacle distance but also their velocity as well as the robot actuation capabilities. To highlight the effectiveness of this constraint, we compare the proposed method with a version of the NMPC that uses a constraint purely based on distance information, showing that the first achieves better performance than the second, especially when the robot travels at higher speed among several moving obstacles. Results indicate that the method can work with relatively short prediction horizons and is therefore amenable to real-time implementation.

Keywords: Robot navigation · Mobile robots · Collision avoidance · NMPC

1 Introduction

In service applications, mobile robots must be able to navigate a variety of different environments. Apart from task fulfillment and kinodynamic feasibility, collision avoidance during motion is an essential requirement. This makes the use of real-time motion planning necessary.

In the literature, there is a huge number of motion planning methods. In [4] the authors categorize direct motion planning approaches as grid-based search methods [8,9], randomized probabilistic methods [15], artificial potential fields [14] and approaches based on the solution of an Optimal Control Problem (OCP) [1]. This last category includes NMPC, which is increasingly used as a motion generation technique. Using the robot dynamics as prediction model and through appropriate state and input constraints, NMPC solves an OCP at each time instant to compute an optimal motion which guarantees kinodynamic feasibility and collision avoidance. Moreover, it can be applied to both single

I. Petrovic et al. (Eds.): IAS 2022, LNNS 577, pp. 216–230, 2023.
https://doi.org/10.1007/978-3-031-22216-0_15

and multi-body robots with complex dynamics, a clear advantage over classical grid-based search approaches for navigation, such as the dynamic window approach [9]. Finally, thanks to the constant increase in computational capabilities and the use of solution approaches like real time iteration (RTI) [6] and appropriate software packages for the solution of the OCP like ACADO [13], it is possible to achieve real-time performance for NMPC.

Regarding the collision avoidance constraint, the structure of the environment affects its formulation. In some structured environments, collision avoidance can be often treated as a corridor navigation problem after proper reformulation of the equations of motion, see [11,16]. However, in the general case this is not possible. In [2], signed distance fields are used in order to generate collision-free trajectories. In [21], the authors proposed a reformulation of the generic collision avoidance constraint as a smooth nonlinear version more appropriate for numerical optimization. In [18], a soft collision avoidance constraint is introduced that considers obstacles described by general non-convex sets, while [7] emphasizes the importance of hard collision avoidance constraints in real-time NMPC.

It should be noted that the collision avoidance constraints in the aforementioned works are purely *distance-based*, without any consideration of the dynamic state of the robot. In an NMPC, the effectiveness of a distance-based collision avoidance method will critically depend on the length of the prediction horizon in relation to the robot actuation capabilities. The longer the prediction horizon, the earlier an imminent collision can be detected and averted without significant (sometimes prohibitive) actuation effort. On the other hand, real-time performance obviously does not allow arbitrarily long prediction horizons. In practice, since high-speed navigation requires the use of the robot dynamic model and high control frequency, the maximum achievable prediction horizon on typical robot processing platforms ends up being relatively short. Consequently, the use of a purely distance-based collision avoidance constraint may jeopardize the robot safety, since the danger of collision may be detected at a time when the robot does not have the necessary actuation power to prevent it.

To ensure an adequate look-ahead capability, one should consider both the robot state with respect to the environment and its actuation capabilities. In [3, 20], position, velocity and worst-case stopping time of the robot are considered in order to maintain a minimum clearance between the robot and the environment. In [10], the authors stress the importance of considering the whole robot state in order to guarantee safety and introduce the concept of *Inevitable Collision States* (ICS). In [17], the idea of ICS is exploited for anytime motion planning based on RRT. In [12], workspace obstacles are represented as velocity obstacles over a predefined time horizon, whose length depends on the robot state and actuation capabilities and is such that the robot can always avoid collision with the obstacle by a predefined maneuver. Finally, a forbidden velocity map is defined in [5] as the set of the robot prohibited velocities associated to each obstacle based on the state of the robot and its braking capabilities.

In this work, we take inspiration from the ICS concept to define the notion of *Avoidable Collision State* (ACS), i.e., a state from which the robot can avoid collision with a certain obstacle. Building on this, we propose an NMPC method for real-time motion generation which enforces collision avoidance through a constraint which essentially requires the robot to be in an ACS at all times. Extensive simulations on a differential-drive robot navigating in both static and dynamic environments clearly show the superiority of the proposed NMPC method with respect to distance-based formulations, especially when the robot must navigate at high speed in environments cluttered with moving obstacles. In particular, it is shown that, thanks to its dynamics-aware nature, the novel constraint works well with relatively short prediction horizons, making real-time performance achievable even on low-cost computational platforms.

The paper is organized as follows. Our navigation problem is formulated in Sect. 2. In Sect. 3 we outline the proposed NMPC approach, while in Sect. 4 we formally define the concept of ACS and derive the associated collision avoidance constraint. Simulation results for a differential-drive robot are presented in Sect. 5. Finally, some concluding remarks are offered in Sect. 6.

2 Problem Formulation

Consider a robotic system whose configuration $\boldsymbol{q}$ (the vector of generalized coordinates) takes values in an n-dimensional configuration space $\mathcal{C}$. The robot is assumed to be subject to k Pfaffian nonholonomic constraints, expressed as $\boldsymbol{A}^T(\boldsymbol{q})\dot{\boldsymbol{q}} = \boldsymbol{0}$, with $\boldsymbol{A}^T(\boldsymbol{q}) \in \mathbb{R}^{k\times n}$. After some manipulation [19], the dynamics of the robot can be written in state-space format as follows

$$\dot{\boldsymbol{x}} = \boldsymbol{f}(\boldsymbol{x}, \boldsymbol{u}) = \begin{pmatrix} \boldsymbol{G}(\boldsymbol{q})\boldsymbol{\nu} \\ \boldsymbol{M}^{-1}(\boldsymbol{q})(\boldsymbol{E}(\boldsymbol{q})\boldsymbol{u} - \boldsymbol{m}(\boldsymbol{q}, \boldsymbol{\nu})) \end{pmatrix}, \tag{1}$$

with the state defined as $\boldsymbol{x} = (\boldsymbol{q}, \boldsymbol{\nu})$, where $\boldsymbol{\nu}$ is the vector of the $m = n - k$ robot pseudovelocities, and the control input $\boldsymbol{u} \in \mathbb{R}^m$ is the vector of generalized forces applied by the m robot actuators. In the right hand side of (1), $\boldsymbol{G}(\boldsymbol{q})$ is a matrix whose columns span $\mathcal{N}(\boldsymbol{A}^T(\boldsymbol{q}))$, and we have set

$$\begin{aligned} \boldsymbol{M}(\boldsymbol{q}) &= \boldsymbol{G}^T(\boldsymbol{q})\boldsymbol{B}(\boldsymbol{q})\boldsymbol{G}(\boldsymbol{q}) \\ \boldsymbol{m}(\boldsymbol{q}, \boldsymbol{\nu}) &= \boldsymbol{G}^T(\boldsymbol{q})\boldsymbol{B}(\boldsymbol{q})\dot{\boldsymbol{G}}(\boldsymbol{q})\boldsymbol{\nu} + \boldsymbol{G}^T(\boldsymbol{q})\boldsymbol{n}(\boldsymbol{q}, \dot{\boldsymbol{q}}) \\ \boldsymbol{E}(\boldsymbol{q}) &= \boldsymbol{G}^T(\boldsymbol{q})\boldsymbol{S}(\boldsymbol{q}), \end{aligned}$$

being $\boldsymbol{B}(\boldsymbol{q}) \in \mathbb{R}^{n\times n}$ the inertia matrix, $\boldsymbol{n}(\boldsymbol{q}, \dot{\boldsymbol{q}}) \in \mathbb{R}^n$ the vector of velocity and gravitational terms, and $\boldsymbol{S}(\boldsymbol{q}) \in \mathbb{R}^{n\times m}$ the matrix that maps the actuator forces to forces performing work on the generalized coordinates.

We emphasize that the above dynamic model is general and appropriately describes mobile robots as well as robot manipulators, mobile manipulators and so on. In the particular case of an unconstrained (free-flying) robot, we simply have $\boldsymbol{G}(\boldsymbol{q}) = \boldsymbol{I}_n$ and $\boldsymbol{\nu} = \dot{\boldsymbol{q}}$.

The robot moves in an N-dimensional ($N = 2, 3$) workspace $\mathcal{W}$, populated by static and/or dynamic obstacles. We shall denote by $\mathcal{R}(\boldsymbol{q}) \subset \mathcal{W}$ the volume occupied by the robot at configuration $\boldsymbol{q}$, and by $\mathcal{O}_j(t) \subset \mathcal{W}$ the volume occupied by the j-th obstacle at time t ($j = 1, 2, \dots$).

A navigation task is assigned to the robot in terms of a vector $\boldsymbol{y} \in \mathcal{Y}$, which describes the position of a representative point on the robot and is related to the configuration via a forward kinematic map $\boldsymbol{y} = \boldsymbol{k}(\boldsymbol{q})$. The task is to drive $\boldsymbol{y}$ to a desired position $\boldsymbol{y}_d$ in the workspace.

Our problem is to generate in real-time a motion that:

1. drives the robot from any starting configuration $\boldsymbol{q}_s$ to any configuration realizing the task, i.e., a configuration belonging to $\mathcal{C}_g$, where $\mathcal{C}_g = \{\boldsymbol{q} \in \mathcal{C} : \boldsymbol{k}(\boldsymbol{q}) = \boldsymbol{y}_d\}$;
2. is kinodynamically feasible, in the sense that it is consistent with model (1) and respects existing constraints on both $\boldsymbol{x}$ (e.g., joint and velocity limits) and $\boldsymbol{u}$ (e.g., torque bounds);
3. avoids collisions between the robot and the obstacles, i.e., $\mathcal{R}(\boldsymbol{q}) \cap \mathcal{O}_j(t) = \emptyset$, for all t and $j = 1, 2, \dots$.

As for the information available for solving the problem, it is assumed that the robot is always aware of its own state and that an on-board sensor provides the position and velocity of all obstacles located inside the sensor field of view.

3 The Proposed NMPC Approach

Our approach relies on the use of an NMPC algorithm for real-time motion generation. NMPC solves an OCP at each discrete time instant; to allow efficient numerical solution, each OCP must be reduced to a Nonlinear Program (NLP).

Denote by H the prediction horizon, by δ the sampling interval and by $N = H/\delta$ the number of control intervals in the prediction horizon. We also denote by $\boldsymbol{x}_{k|i}$ and $\boldsymbol{u}_{k|i}$ the predicted robot state and control input at the discrete time instant t_{k+i} computed at t_k. The decision variables of the NLP to be solved at time t_k are the control inputs $\{\boldsymbol{u}_{k|0}, \dots, \boldsymbol{u}_{k|N-1}\}$ and the states $\{\boldsymbol{x}_{k|0}, \dots, \boldsymbol{x}_{k|N}\}$.

Our objective is to drive the task error to zero while using the least possible control effort. Denoting the predicted task error at t_{k+i} as $\boldsymbol{e}_{k|i} = \boldsymbol{y}_d - \boldsymbol{y}_{k|i}$, with $\boldsymbol{y}_{k|i}$ the position of the representative point at t_{k+i}, the running cost can then be expressed as:

$$V_{k|i}(\boldsymbol{x}_{k|i}, \boldsymbol{u}_{k|i}) = \boldsymbol{e}_{k|i}^T \boldsymbol{Q} \boldsymbol{e}_{k|i} + \dot{\boldsymbol{y}}_{k|i}^T \boldsymbol{P} \dot{\boldsymbol{y}}_{k|i} + \boldsymbol{u}_{k|i}^T \boldsymbol{R} \boldsymbol{u}_{k|i},$$

while the terminal cost will be

$$V_{k|N}(\boldsymbol{x}_{k|N}) = \boldsymbol{e}_{k|N}^T \boldsymbol{Q}_N \boldsymbol{e}_{k|N} + \dot{\boldsymbol{y}}_{k|N}^T \boldsymbol{P}_N \dot{\boldsymbol{y}}_{k|N}.$$

Here, $\boldsymbol{Q}$, $\boldsymbol{P}$ and $\boldsymbol{R}$ are positive definite weighting matrices for the task error, the velocity of the representative point and the control effort throughout the prediction horizon, while $\boldsymbol{Q}_N$ and $\boldsymbol{P}_N$ are positive definite weighting matrices for the task error and the velocity of the representative point at the final time instant.

The NLP to be solved at time instant t_k is then:

$$\min_{\substack{\boldsymbol{u}_{k|0},\ldots,\boldsymbol{u}_{k|N-1},\\ \boldsymbol{x}_{k|0},\ldots,\boldsymbol{x}_{k|N}}} \sum_{i=0}^{N-1} V_{k|i}(\boldsymbol{x}_{k|i}, \boldsymbol{u}_{k|i}) + V_{k|N}(\boldsymbol{x}_{k|N})$$

subject to:

$$\boldsymbol{x}_{k|0} - \boldsymbol{x}_k = 0$$

$$\boldsymbol{x}_{k|i+1} - \boldsymbol{F}(\boldsymbol{x}_{k|i}, \boldsymbol{u}_{k|i}) = 0, \quad i = 0, \ldots, N-1$$

$$\boldsymbol{x}_{\min} \leq \boldsymbol{x}_{k|i} \leq \boldsymbol{x}_{\max}, \quad i = 0, \ldots, N$$

$$\boldsymbol{u}_{\min} \leq \boldsymbol{u}_{k|i} \leq \boldsymbol{u}_{\max}, \quad i = 0, \ldots, N-1$$

collision avoidance constraints at $t_k, \ldots, t_{k+N}$

where $\boldsymbol{x}_k$ represents the current robot state, $\boldsymbol{F}(\cdot, \cdot)$ represents the right-hand-side of a discrete state-space model of the robot obtained via numerical integration (typically under the assumption of piecewise-constant control inputs), while $\boldsymbol{x}_{\min}, \boldsymbol{x}_{\max}$ and $\boldsymbol{u}_{\min}, \boldsymbol{u}_{\max}$ are respectively the lower/upper bounds on the state variables and on the control inputs.

As for the collision avoidance constraint, which is the main contribution of this work, it is discussed in full detail in the next section.

4 Collision Avoidance

The proposed collision avoidance constraint hinges upon the notion of Avoidable Collision State (ACS), i.e., a robot state from which it is possible to avoid collisions. In this section we will first give a formal definition of what an ACS is, and then derive the corresponding collision avoidance constraint. Although all computations are presented in a 2-dimensional workspace, the extension to the 3-dimensional case (e.g., for application to UAVs) is straightforward.

4.1 Preliminaries

The notion of ACS is obstacle-specific, in the sense that it characterizes the possibility for the robot to avoid a certain obstacle. In view of this, we first need to identify the obstacles for which there is an actual danger of collision given the current state of the robot.

We are going to use bounding spheres for both the robot and the obstacles. In particular, we take the smallest sphere that contains the robot volume $\mathcal{R}$, and denote its radius by ρ and its center by R. The position vector of R in the world frame is denoted by $\boldsymbol{r}$ and is related to the robot configuration via a forward kinematic map $\boldsymbol{r} = \boldsymbol{\sigma}(\boldsymbol{q})$. For its velocity we have: $\dot{\boldsymbol{r}} = \boldsymbol{J}(\boldsymbol{q})\boldsymbol{\nu}$ where $\boldsymbol{J}(\boldsymbol{q}) = \partial\boldsymbol{\sigma}(\boldsymbol{q})/\partial\boldsymbol{q}\boldsymbol{G}(\boldsymbol{q})$. Similarly, we use a sphere of radius ρ_j to envelop the generic obstacle $\mathcal{O}_j$, denoting its center by O_j, the corresponding position vector by $\boldsymbol{o}_j$ and its velocity by $\dot{\boldsymbol{o}}_j$. Finally, let $\boldsymbol{n}_j$ be the unit vector pointing from R to O_j, and $\boldsymbol{r}_j = \boldsymbol{r} - \boldsymbol{o}_j$ be the relative position of R with respect to O_j, so that the corresponding relative velocity is $\dot{\boldsymbol{r}}_j$. Refer to Fig. 1 for illustration.

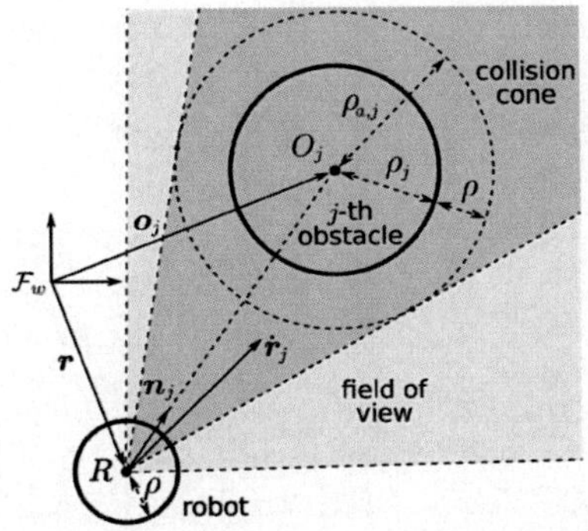

Fig. 1. Obstacle $\mathcal{O}_j$ is considered dangerous for the robot at a certain time if the relative velocity $\dot{\boldsymbol{r}}_j$ of the robot with respect to the obstacle lies inside the collision cone.

Now, augment the obstacle sphere by the radius of the robot sphere, denoting by $\rho_{a,j} = \rho_j + \rho$ the total radius. The *collision cone* is the cone defined by R and the tangents from R to the augmented obstacle. Obstacle $\mathcal{O}_j$ is *dangerous* at time t if $\dot{\boldsymbol{r}}_j(t)$ lies inside the collision cone (see Fig. 1). Simple geometrical arguments lead to the following condition for an obstacle to be dangerous:

$$h(\boldsymbol{x}, \boldsymbol{\xi}_j) = \boldsymbol{n}_j^T \frac{\dot{\boldsymbol{r}}_j}{\|\dot{\boldsymbol{r}}_j\|} - \frac{\sqrt{\|\boldsymbol{r}_j\|^2 - \rho_{a,j}^2}}{\|\boldsymbol{r}_j\|} \geq 0 \tag{2}$$

where $\boldsymbol{\xi}_j = (\boldsymbol{o}_j, \dot{\boldsymbol{o}}_j)$ is the j-th obstacle state.

4.2 Avoidable Collision States

By definition, to avoid an obstacle which is non-dangerous at time t the robot simply needs to keep its course. Therefore, we only need to characterize the possibility of avoiding obstacles that are dangerous at t. In particular, we will say that the robot is in an *Avoidable Collision State* (ACS) with respect to a dangerous obstacle if there exists at least one trajectory that originates from the current state, is kinodynamically feasible and avoids collision with the obstacle.

In principle, to conclude that a state is an ACS we should check all feasible trajectories emanating from it, until we find at least one that avoids collision. However, such a potentially exhaustive study is incompatible with a real-time application. We shall therefore look at one specific motion and ask ourselves if the robot can avoid collision during that motion. If the answer is positive, then it can be concluded that the current state is certainly an ACS.

In particular, in the presence of an obstacle $\mathcal{O}_j$ which moves along a given direction with constant speed and is dangerous at time t, we consider the robot moving in such a way that at each time instant $t' \geq t$ its relative velocity with respect to the obstacle, $\dot{\boldsymbol{r}}_j(t')$, projected on the original direction of collision, $\boldsymbol{n}_j(t)$, decreases with a constant rate α.

During this motion there will be eventually a time instant, denoted by t_v, in which the projection of the robot relative velocity on the original direction of collision will be zero, i.e. $\boldsymbol{n}_j^T(t)\dot{\boldsymbol{r}}_j(t_v) = 0$. A sufficient condition for this motion to be collision-free is:

$$\boldsymbol{n}_j^T(t)(\boldsymbol{o}_j(t') - \boldsymbol{r}(t')) \geq \rho_{a,j}, \quad \forall t' \in [t, t_v].$$

Denoting by $\gamma(t)$ the robot-obstacle clearance and considering that at time t the relation $\boldsymbol{n}_j^T(t)(\boldsymbol{o}_j(t) - \boldsymbol{r}(t)) = \gamma(t) + \rho_{a,j}$ holds, after simple substitution of $\rho_{a,j}$, the condition for collision-free motion becomes:

$$\boldsymbol{n}_j^T(t)(\boldsymbol{r}_j(t') - \boldsymbol{r}_j(t)) \leq \gamma(t), \quad \forall t' \in [t, t_v]. \tag{3}$$

Note that inequality (3) defines an admissible half-plane for the relative position of R with respect to O_j in which the robot has to remain for all $t' \in [t, t_v]$ (see Fig. 2).

In order to ensure that the considered motion can be implemented by the robot we will investigate whether the required deceleration of R, i.e. $\ddot{\boldsymbol{r}} = \boldsymbol{n}_j(t)\alpha$, projected on the robot actuation space lies within the actuation limits.

Throughout the considered motion, the position of R, projected on the original direction of collision $\boldsymbol{n}_j(t)$ is described at time $t' \geq t$ as

$$\boldsymbol{n}_j^T(t)\boldsymbol{r}(t') = \boldsymbol{n}_j^T(t)\boldsymbol{r}(t) + \boldsymbol{n}_j^T(t)\dot{\boldsymbol{r}}(t)(t' - t) + \frac{1}{2}\alpha(t' - t)^2 \tag{4}$$

and its projected velocity as

$$\boldsymbol{n}_j^T(t)\dot{\boldsymbol{r}}(t') = \boldsymbol{n}_j^T(t)\dot{\boldsymbol{r}}(t) + \alpha(t' - t). \tag{5}$$

As for the obstacle, the position of O_j at $t' \geq t$ projected on the original direction of collision is

$$\boldsymbol{n}_j^T(t)\boldsymbol{o}_j(t') = \boldsymbol{n}_j^T(t)\boldsymbol{o}_j(t) + \boldsymbol{n}_j^T(t)\dot{\boldsymbol{o}}_j(t' - t). \tag{6}$$

By solving the system of equations (3) for the equality, (4), (5) and (6), with $t' = t_v$, we obtain the minimum required deceleration of the robot in the original direction of collision:

$$\alpha = -\frac{1}{2}\frac{\left(\boldsymbol{n}_j^T(\dot{\boldsymbol{o}}_j - \dot{\boldsymbol{r}}(t))\right)^2}{\gamma(t)}.$$

Note that α depends on both the robot and the j-th obstacle state. Considering that the acceleration of R is:

$$\ddot{\boldsymbol{r}} = \boldsymbol{J}(\boldsymbol{q})\dot{\boldsymbol{\nu}} + \dot{\boldsymbol{J}}(\boldsymbol{q})\boldsymbol{\nu} \tag{7}$$

and that from (1) we can express $\dot{\boldsymbol{\nu}}$ as:

$$\dot{\boldsymbol{\nu}} = \boldsymbol{M}^{-1}(\boldsymbol{q})(\boldsymbol{E}(\boldsymbol{q})\boldsymbol{u} - \boldsymbol{m}(\boldsymbol{q}, \boldsymbol{\nu})), \tag{8}$$

we can obtain the required control inputs for this deceleration by substituting $\ddot{\boldsymbol{r}} = \boldsymbol{n}_j(t)\alpha$ and (8) in (7). Using the Moore-Penrose pseudoinverse, we get

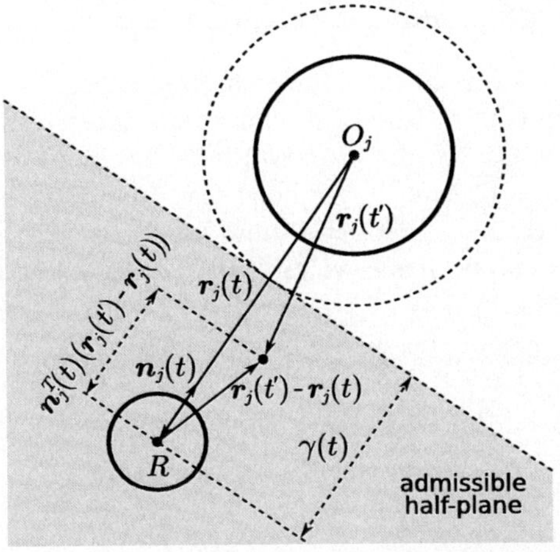

Fig. 2. The admissible half-plane for the relative position of robot R with respect to the obstacle O_j at time t', $t' \in [t, t_v]$.

the minimum norm control inputs needed in order to apply the deceleration $\ddot{\boldsymbol{r}} = \boldsymbol{n}_j(t)\alpha$ to R:

$$\boldsymbol{u}_\alpha(\boldsymbol{x}, \boldsymbol{\xi}_j) = \left(\boldsymbol{J}(\boldsymbol{q})\boldsymbol{M}^{-1}(\boldsymbol{q})\boldsymbol{E}(\boldsymbol{q})\right)^\dagger \boldsymbol{\beta}(\boldsymbol{x}, \boldsymbol{\xi}_j), \tag{9}$$

where

$$\boldsymbol{\beta}(\boldsymbol{x}, \boldsymbol{\xi}_j) = \boldsymbol{n}_j(t)\alpha - \dot{\boldsymbol{J}}(\boldsymbol{q})\boldsymbol{\nu} + \boldsymbol{J}(\boldsymbol{q})\boldsymbol{M}^{-1}(\boldsymbol{q})\boldsymbol{m}(\boldsymbol{q}, \boldsymbol{\nu}).$$

So the motion is kinodynamically feasible if the following condition is satisfied:

$$\boldsymbol{u}_{\min} \leq \boldsymbol{u}_\alpha(\boldsymbol{x}, \boldsymbol{\xi}_j) \leq \boldsymbol{u}_{\max}. \tag{10}$$

This condition can be used as a constraint in order to guarantee that the robot is always at an ACS.

Note that the ACS property for a state is strongly related to its being *safe* according to [10], i.e., not being an Inevitable Collision State (ICS). However, the two properties differ in two aspects:

- The number of obstacles considered by the property. An ICS is defined with respect to all the obstacles (or at least all the obstacles visible by the robot), a practice that obviously increases the computational time, while the ACS property is defined with respect to a specific (dangerous) obstacle;
- The way in which the property is established. To prove that a state is not ICS, in principle one has to search the whole control input set (or a finite subset [17]) to find a collision-free motion. On the other hand, to characterize a state as ACS we only look at the relative velocity of the robot with respect to the dangerous obstacle, and specifically investigate whether it is possible to stop motion along the robot-obstacle direction before collision.

4.3 Use of the ACS Condition in the NLP

Note that constraint (10) is suitable for enforcing collision avoidance since every robot motion that leads to collision will violate the constraint before the collision occurs. So we will use (10) in the proposed NLP applying it for each considered obstacle, for each time instant throughout the prediction horizon. In order to ensure that the constraint is inactive if non-dangerous obstacles are considered and to avoid using *if* statements, we multiply $\boldsymbol{u}_\alpha(\boldsymbol{x}, \boldsymbol{\xi}_j)$ as given by (9) by the sigmoid function $g(h(\boldsymbol{x}, \boldsymbol{\xi}_j)) = 1/(1 + e^{-\lambda h(\boldsymbol{x}, \boldsymbol{\xi}_j)})$, with λ being a constant value that tunes the steepness of the sigmoid function, getting:

$$\boldsymbol{u}_b(\boldsymbol{x}, \boldsymbol{\xi}_j) = g(h(\boldsymbol{x}, \boldsymbol{\xi}_j)) \left(\boldsymbol{J}(\boldsymbol{q})\boldsymbol{M}^{-1}(\boldsymbol{q})\boldsymbol{E}(\boldsymbol{q})\right)^{\dagger} \boldsymbol{\beta}(\boldsymbol{x}, \boldsymbol{\xi}_j).$$

So the constraint that will be applied in the NLP is:

$$\boldsymbol{u}_{\min} \leq \boldsymbol{u}_b(\boldsymbol{x}_{k|i}, \boldsymbol{\xi}_{j,k|i}) \leq \boldsymbol{u}_{\max}, \quad i = 0, \ldots, N, \quad j = 1, \ldots, n_o, \tag{11}$$

where $\boldsymbol{\xi}_{j,k|i}$ is the state of the j-th obstacle at t_{k+i} and n_o is the number of the considered obstacles in the NMPC.

5 Simulations

In order to show the effectiveness of the proposed method, we conducted a series of simulations in various static and dynamic environments. All the simulations were implemented in MATLAB on an Intel Core i9-9900K CPU at 3.60 GHz. For the numerical solution of the NLP required by our NMPC, we used the RTI scheme [6] implemented within the ACADO Toolkit [13].

The robotic platform used for our simulations is the differential-drive robot shown in Fig. 3, having length l = 0.60 m and width b = 0.30 m. The Center of Mass (CoM) is located at the geometric center of the vehicle, which is also the center R of the robot bounding sphere, whose radius is $\rho = \sqrt{l^2/4 + b^2/4}$. The rear wheels are located at a distance d = 0.25 m from the CoM, with a caster wheel in front for static balance. The vehicle mass and moment of inertia are respectively 50.0 kg and 1.41 kg·m^2, making this a rather heavy vehicle. The torque bounds for the wheel actuators are set to 2.5 Nm. The configuration vector of this robot is $\boldsymbol{q} = (x, y, \theta)$ and its dynamic model is that of a unicycle (e.g., see [19, Sect. 11.4]). The robot representative point for the navigation task is also placed at R. It is assumed that the robot is equipped with a line-of-sight sensor (such as a laser rangefinder) with an infinite field of view.

We wish to compare the proposed NMPC method, which relies upon a dynamics-aware (DA) collision avoidance constraint, with an NMPC that uses a purely distance-based (DB) collision avoidance constraint. In both methods, the NLP formulation is the same, apart from the collision avoidance constraint which in DB takes the form:

$$\left\|\boldsymbol{r}_{k|i} - \boldsymbol{o}_{j,k|i}\right\| \geq \rho_{a,j}, \quad i = 0, \ldots, N, \quad j = 1, \ldots, n_o,$$

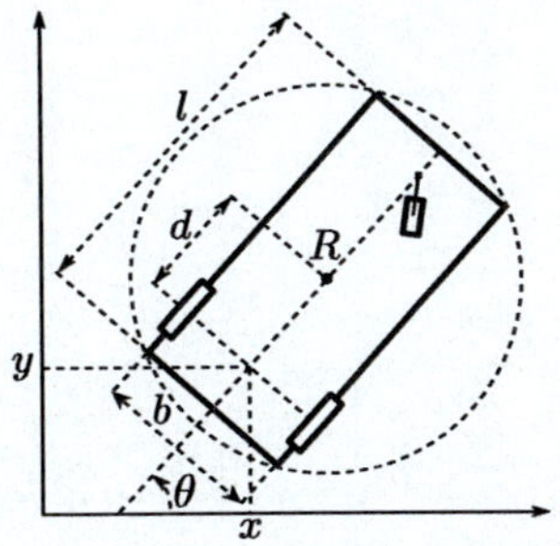

Fig. 3. The differential-drive robot used in simulation with its bounding sphere.

where $\boldsymbol{r}_{k|i}$ and $\boldsymbol{o}_{j,k|i}$ are the position of R and O_j at t_{k+i}, respectively. In both NMPCs we consider the $n_o = 5$ closest visible obstacles. The sampling interval for real-time control is $\delta = 31$ ms in both cases.

The two methods have been tested in 50 different environments (25 static and 25 dynamic), using 3 different values for the maximum driving velocity $v_{\max}$. The maximum steering velocity is always set at $\omega_{\max} = 20/3\, v_{\max}$ rad/s. The prediction horizons were chosen as large as possible to allow real-time performance for the two methods. In particular, for the DB method the prediction horizon is $H = 0.992$ s while for the DA method it is $H = 0.93$ s; in fact the DB method requires slightly less computations than the DA method.

We assess the performance of the two methods by taking into account the success rate as well as the quality of the motion according to the following criteria: (1) time t_g needed for the robot to reach the goal, (2) control effort $J_\tau = \int_0^{t_g} \|\boldsymbol{\tau}\|^2 dt$, (3) length l_p of the resulting path, (4) duration of the longest iteration $\delta_{\max}$ and (5) average iteration time $\bar{\delta}$.

Video clips of representative simulation results are shown in the accompanying video, also available at https://youtu.be/LS57E8jGoJk.

5.1 Static Environments

In the first group of simulations, the robot moves in 25 environments, each occupied by 10 static obstacles (e.g., see Fig. 4). The robot starts from $\boldsymbol{q}_s = (2, 2, \pi/3)$ and must reach $\boldsymbol{y}_g = (16, 15)$. Table 1 (top) summarizes the results of the DA and DB methods averaged over the 25 environments, for increasing values of the maximum driving velocity $v_{\max}$.

For lower $v_{\max}$ the two methods behave similarly having also the same success rate. An illustrative example of the robot motion with $v_{\max} = 0.9$ m/s is given in simulation 1 of the accompanying video. However, when $v_{\max} = 1.2$ m/s the success rate of the DB method drops to 88% while the DA method is hardly affected by the increase in the maximum velocity. Note that in this case, the

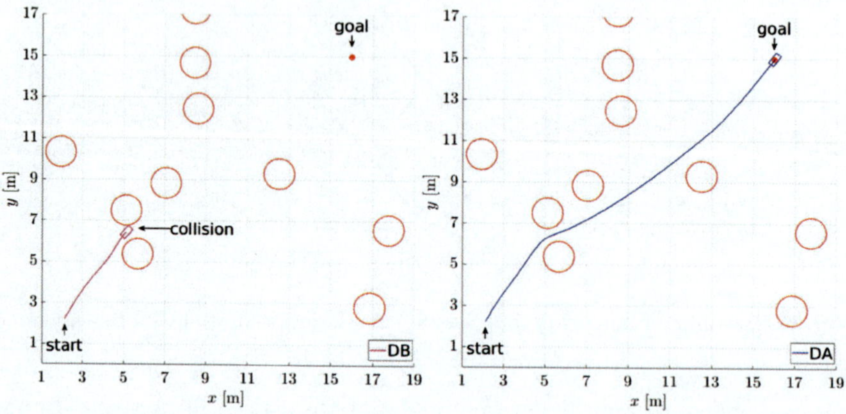

Fig. 4. Motions generated by the two methods in one of the static environments, with $v_{\max} = 1.2$ m/s. The DB method (left) cannot avoid collision with an obstacle, whereas DA (right) goes safely through the narrow passage and successfully reaches the goal. See also simulation 2 in the accompanying video.

Table 1. Averaged results over 25 environments for the proposed dynamics-aware (DA) method vs the distance-based (DB) method. Top: static environments, bottom: dynamic environments.

Static Environments						
	DB			DA		
$v_{\max}$ [m/s]; t_s [s]	0.9; 0.93	1.1; 1.116	1.2; 1.209	0.9; 0.93	1.1; 1.116	1.2; 1.209
success rate (%)	96	96	88	96	96	92
t_g [s]	27.16	24.29	25.07	26.63	24.76	23.98
J_τ [10^4 N^2m^2s]	1164.63	1495.56	1763.30	1046.58	1397.92	1601.14
l_p [m]	19.56	20.12	20.85	19.51	19.96	20.48
$\delta_{\max}$ [ms]	25.21	25.15	25.46	26.84	27.55	27.11
$\bar{\delta}$ [ms]	15.31	14.26	13.67	16.40	15.96	15.50

Dynamic Environments						
	DB			DA		
$v_{\max}$ [m/s]; t_s [s]	0.9; 0.93	1.1; 1.116	1.2; 1.209	0.9; 0.93	1.1; 1.116	1.2; 1.209
v_o [m/s]	0.45	0.55	0.6	0.45	0.55	0.6
success rate (%)	72	64	40	80	84	80
t_g [s]	29.65	25.52	24.38	30.61	29.78	27.19
J_τ [10^4 N^2m^2s]	1766.29	2001.42	2009.93	1625.94	2171.06	2510.07
l_p [m]	20.37	20.69	20.77	20.62	21.83	21.31
$\delta_{\max}$ [ms]	26.16	25.65	26.14	28.18	28.56	28.85
$\bar{\delta}$ [ms]	15.47	14.63	14.19	16.87	16.64	16.40

stopping time[1] t_s is significantly larger than the prediction horizon. This shows that under these conditions, unlike DA, a robot navigating with the DB method would require either a longer prediction horizon or greater braking capabilities in order to navigate safely. A representative example of the behaviour of the two methods is shown in Fig. 4 (also see simulation 2 in the accompanying video). In particular, it shows the trajectories generated by the two methods in one of the static environments with $v_{\max} = 1.2$ m/s. The DB method (left) runs into a collision while passing through a narrow passage, because the robot has accumulated too much speed in the initial part of the motion and avoiding the imminent collision would require more than the available torque (this corresponds to an unfeasibility for the NMPC). On the other hand, the DA method (right) starts the avoidance maneuver earlier thanks to the dynamics-aware constraint, which can detect a violation before the distance-based constraint does; therefore, DA is able to go through the narrow passage and reach the goal safely.

Looking at the other performance criteria of Table 1, one can observe that on the average the DA method produces trajectories that are slightly shorter and less energy-consuming than the DB method. As a counterpart, the DB method shows a slightly reduced duration of the longest iteration mainly due to the simplicity of the collision avoidance constraint.

5.2 Dynamic Environments

The second group of simulations take place in 25 dynamic environments, each occupied by 10 static and 10 moving obstacles. The moving obstacles are zigzagging at a speed $v_o = v_{\max}/2$, with direction changes of $\pi/3$ (toward the robot) after traveling a distance of 4.9 m. The results are presented in Table 1 (bottom).

The success rate of both methods is affected by the presence of the moving obstacles. However, the DA method is much more effective than the DB method, with success rates over 80% while DB goes as low as 40%.

Once again, for the lowest $v_{\max}$ the DB method has its higher success rate, yet lower than the DA method. In simulation 3 of the accompanying video we show the robot moving in a dynamic environment with $v_{\max} = 0.9$ m/s. Nevertheless, the success rate of the DB method decreases as $v_{\max}$ increases and plummets to 40% for $v_{\max} = 1.2$ m/s, revealing that the presence of moving obstacles in combination with fast robot motion affects significantly its performance. This is due to the fact that the maneuvers needed for collision avoidance become now more demanding, since a worst-case scenario may require the robot not only to stop, but also to accelerate in the opposite direction in order to avoid an imminent collision. Thus, by the time an imminent collision is detected in the DB case, there may not be enough actuation capability left to avoid the obstacle explaining why the DB method has such a low performance even in the case of the lowest maximum velocity. On the other hand, the DA method guarantees

[1] By stopping time t_s we denote the minimum time that a robot traveling on a straight line with speed $v_{\max}$ needs in order to stop.

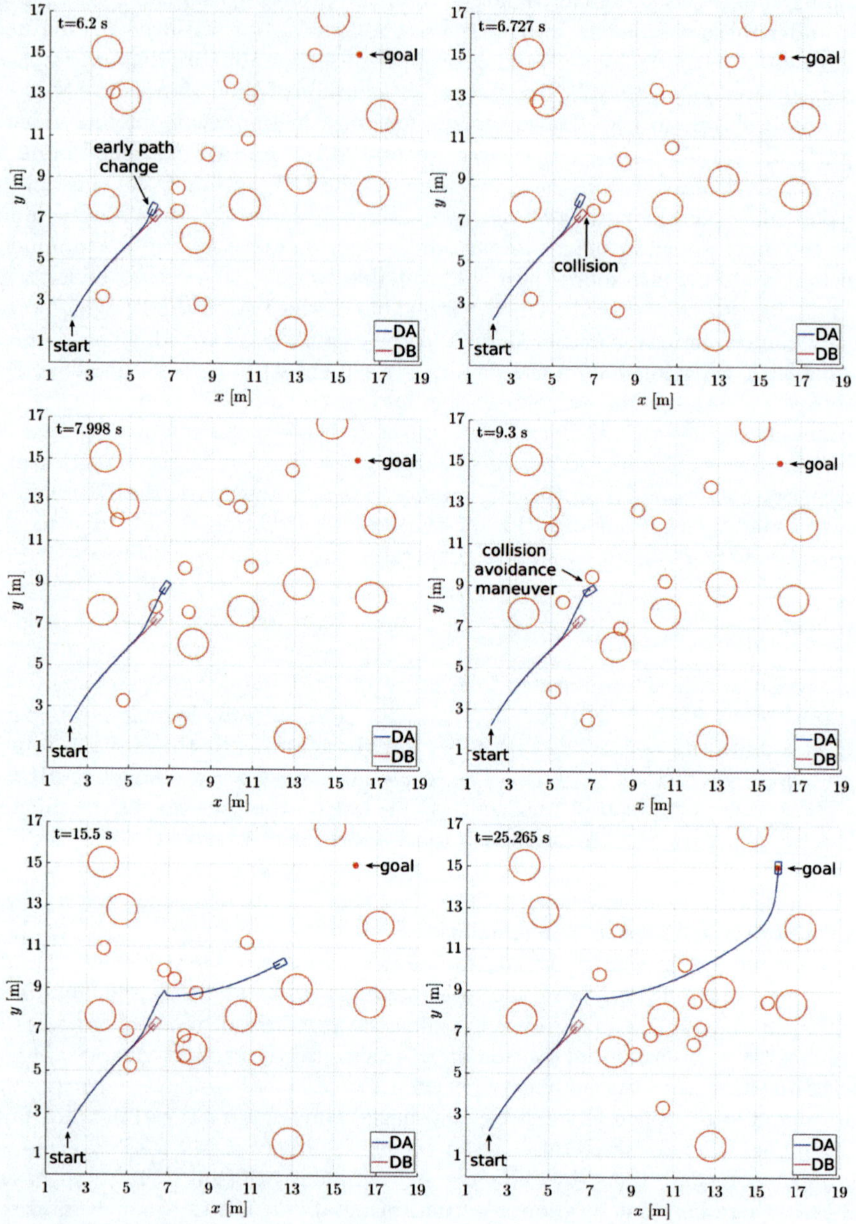

Fig. 5. Snapshots of the motion generated by the two methods in one of the dynamic environments, with $v_{\max} = 1.2$ m/s. As in the static environment, the DB method cannot avoid collision with an obstacle, whereas the DA method safely navigates to the goal. See also simulation 4 in the accompanying video.

that the robot is at all times in an ACS. As a result, the robot is able to veer off collision paths earlier.

For illustration, Fig. 5 (see also simulation 4 in the accompanying video) shows the behavior of the two methods in one of the dynamic environments through a series of snapshots. Again, we have here $v_{\max} = 1.2$ m/s. As in the static environment case, the DB method is not aware of the robot dynamic state, and therefore cannot prevent the collision with the moving obstacle. On the other hand, the DA method effectively reaches the goal avoiding nearby obstacles; note in particular how the early reaction in the presence of a moving obstacle averted the impending collision, and its elaborate avoidance maneuver in front of a moving obstacle combining a reverse motion and a quick reorientation.

In an attempt to further assess the effectiveness of the proposed method, we tested the performance of the two methods in a more cluttered version of the dynamic environment, increasing the number of moving obstacles to 15. The obstacles are zigzagging at a speed of 0.4 m/s, with direction changes of $\pi/3$ (toward the robot) after traveling a distance of 4.9 m. The robot maximum velocity is $v_{\max} = 1.2$ m/s. This simulation, whose results are only shown in simulation 5 of the accompanying video for compactness, confirms that the robot controlled by the dynamics-aware NMPC method safely reaches its goal even in this challenging environment, while the distance-based NMPC fails.

6 Conclusion

We have presented a novel real-time NMPC for robot navigation in environments populated by static and/or moving obstacles. Inspired by the concept of inevitable collision state, we defined the notion of ACS (avoidable collision state) and, based on this, formulated a hard constraint on the robot state guaranteeing that it can execute a collision avoidance trajectory in the presence of a dangerous obstacle. The method was compared with an NMPC that considers a purely distance-based collision avoidance constraint. The comparative simulations showed that the proposed method is generally more effective, especially when the robot navigates at high speed in cluttered dynamic environments.

Future work will be aimed at testing the proposed method on multi-body mobile robots (namely, mobile manipulators), which are already included in the class of robots considered in this paper. Another interesting possibility is to extend the proposed approach to flying robots.

References

1. Betts, J.: Survey of numerical methods for trajectory optimization. J. Guidance Control Dyn. **21**, 193–207 (1998)
2. Bonalli, R., Cauligi, A., Bylard, A., Pavone, M.: GuSTO: Guaranteed Sequential Trajectory optimization via Sequential Convex Programming. In: 2019 International Conference on Robotics and Automation, pp. 6741–6747 (2019)
3. Buizza Avanzini, G., Zanchettin, A.M., Rocco, P.: Constrained model predictive control for mobile robotic manipulators. Robotica, 1–20 (2017)

4. Choset, H., Lynch, K., Hutchinson, S., Kantor, G., Burgard, W., Kavraki, L., Thrun, S.: Principles of Robot Motion: Theory, Algorithms, and Implementations. MIT Press (2005)
5. Damas, B., Santos-Victor, J.: Avoiding moving obstacles: the forbidden velocity map. In: 2009 IEEE/RSJ International Conference on Intelligent Robots and Systems, pp. 4393–4398 (2009)
6. Diehl, M., Bock, H., Schlöder, J.P., Findeisen, R., Nagy, Z., Allgöwer, F.: Real-time optimization and nonlinear model predictive control of processes governed by differential-algebraic equations. J. Process Control **12**(4), 577–585 (2002)
7. Febbo, H., Jayakumar, P., Stein, J.L., Ersal, T.: Real-time trajectory planning for automated vehicle safety and performance in dynamic environments. J. Auton. Veh. Syst. 1(4) (2021)
8. Fiorini, P., Shiller, Z.: Motion planning in dynamic environments using velocity obstacles. Int. J. Rob. Res. **17**(7), 760–772 (1998)
9. Fox, D., Burgard, W., Thrun, S.: The dynamic window approach to collision avoidance. IEEE Rob. Autom. Mag. **4**(1), 23–33 (1997)
10. Fraichard, T., Asama, H.: Inevitable collision states. a step towards safer robots? In: 2003 IEEE/RSJ International Conference on Intelligent Robots and Systems, vol. 1, pp. 388–393 (2003)
11. Frasch, J.V., Gray, A., Zanon, M., Ferreau, H.J., Sager, S., Borrelli, F., Diehl, M.: An auto-generated nonlinear MPC algorithm for real-time obstacle avoidance of ground vehicles. In: 2013 European Control Conference, pp. 4136–4141 (2013)
12. Gal, O., Shiller, Z., Rimon, E.: Efficient and safe on-line motion planning in dynamic environments. In: 2009 IEEE International Conference on Robotics and Automation, pp. 88–93 (2009)
13. Houska, B., Ferreau, H.J., Diehl, M.: An auto-generated real-time iteration algorithm for nonlinear MPC in the microsecond range. Automatica **47**(10), 2279–2285 (2011)
14. Khatib, O.: Real-time obstacle avoidance for manipulators and mobile robots. In: 1985 IEEE International Conference on Robotics and Automation, vol. 2, pp. 500–505 (1985)
15. LaValle, S.M., James, J., Kuffner, J.: Randomized kinodynamic planning. Int. J. Rob. Res. **20**(5), 378–400 (2001)
16. Liniger, A., Domahidi, A., Morari, M.: Optimization-based autonomous racing of 1:43 scale RC cars. Optimal Control Appl. Methods **36**(5), 628–647 (2015)
17. Petti, S., Fraichard, T.: Safe motion planning in dynamic environments. In: 2005 IEEE/RSJ International Conference on Intelligent Robots and Systems, pp. 2210–2215 (2005)
18. Sathya, A., Sopasakis, P., Van Parys, R., Themelis, A., Pipeleers, G., Patrinos, P.: Embedded nonlinear model predictive control for obstacle avoidance using PANOC. In: 2018 European Control Conference, pp. 1523–1528 (2018)
19. Siciliano, B., Sciavicco, L., Villani, L., Oriolo, G.: Robotics: Modelling, Planning and Control. Springer (2009)
20. Zanchettin, A.M., Ceriani, N.M., Rocco, P., Ding, H., Matthias, B.: Safety in human-robot collaborative manufacturing environments: metrics and control. IEEE Trans. Autom. Sci. Eng. **13**(2), 882–893 (2016)
21. Zhang, X., Liniger, A., Borrelli, F.: Optimization-based collision avoidance. ArXiv abs/1711.03449 (2017)

Benchmark of Sampling-Based Optimizing Planners for Outdoor Robot Navigation

Fetullah Atas[1(✉)], Grzegorz Cielniak[2], and Lars Grimstad[1]

[1] Faculty of Science and Technology, Norwegian University of Life Sciences, Aas, Norway
fetullah.atas@nmbu.no

[2] Lincoln Centre for Autonomous Systems, University of Lincoln, Lincoln, UK
gcielniak@lincoln.ac.uk

Abstract. This paper evaluates Sampling-Based Optimizing (SBO) planners from the Open Motion Planning Library (OMPL) in the context of mobile robot navigation in outdoor environments. Many SBO planners have been proposed, and determining performance differences among these planners for path planning problems can be time-consuming and ambiguous. The probabilistic nature of SBO planners can also complicate this procedure, as different results for the same planning problem can be obtained even in consecutive queries from the same planner. We compare all available SBO planners in OMPL with an automated planning problem generation method designed specifically for outdoor robot navigation scenarios. Several evaluation metrics are chosen, such as the length, smoothness, and success rate of the resulting path, and probability distributions for metrics are presented. With the experimental results obtained, clear recommendations on high-performing planners for mobile robot path planning problems are made, which will be useful to researchers and practitioners in mobile robot planning and navigation.

Keywords: Sampling-based optimal planning · Path planning benchmark · Outdoor robot navigation

1 Introduction

Planning is a core component of an autonomous robot's software stack alongside perception and control. It is a process of estimating a series of consecutive, valid, collision-free states (or poses) that a robot can follow to safely get to the desired pose. Sampling-Based Optimal (SBO) planning has been a popular approach in recent years due to its effectiveness in planning in complex, high-dimensional state spaces typical in robotic manipulation scenarios. SBO planners aim to minimize or maximize a given objective function. The ultimate goal is to get a good plan with the lowest cost or highest reward depending on the selected objective

I. Petrovic et al. (Eds.): IAS 2022, LNNS 577, pp. 231–243, 2023.
https://doi.org/10.1007/978-3-031-22216-0_16

function. Unlike grid-cell-based methods, SBO planners efficiently explore high-dimensional state spaces using various *sampling* strategies. Most of the time, the objective function is a *cost* function, where the amount of cost depends on the path's different properties, such as its length, smoothness, or estimated energy consumption along the path.

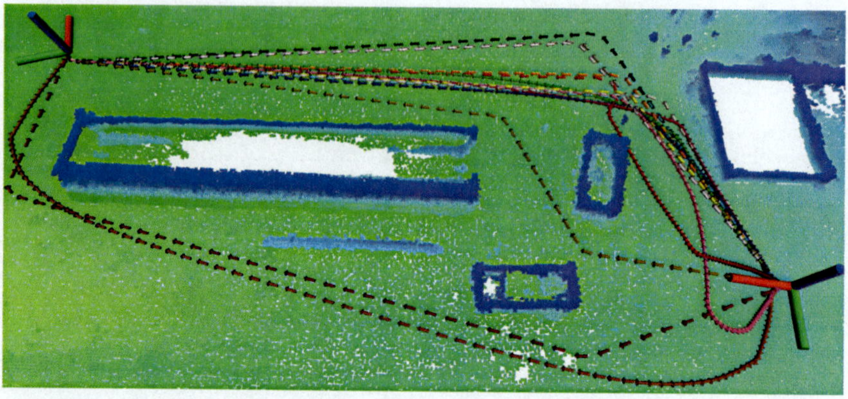

Fig. 1. Figure illustrates representative planning results from 16 distinct SBO planners in a real-world setting for a randomly generated planning problem. The dark blue points indicate various obstacles such as buildings, the colored paths generated by each planner are displayed with sequential arrows.

Since the breakthrough of SBO planners, initiated in [14], a large variety of SBO planners have been proposed. Most of these planners are built on top of RRT* and PRM* planners (e.g., RRTX, LazyPRM* InformedRRT*, etc.), more recently, another set of planners makes use of a structure that includes both graphs and trees (e.g., BIT*, ABIT*, AIT*). Some methods use parallelized approaches (e.g., CFOREST, AnytimePartShortening(APS)), where multiple instances of planners are run on different threads simultaneously while planners inform each other of the milestones they achieve. Out of all of these methods, it's critical to shed some light on the performance of various planner groups and their characteristics for outdoor robot navigation. SBO planner's efficient space exploration (by different sampling strategies) suits large-scale outdoor robot planning while also considering the kinematic constraints. There is a large variety of SBO planners that could be used for outdoor robot planning. However, there is little known about the performances of all these planners. Therefore it is vital to systematically benchmark and identify better performing planners specifically for outdoor robot planning tasks. The lack of work in literature that focuses on SBO path planning performance analysis, particularly for outdoor robot path planning, has been the motivation behind this work. The main contributions of this work are listed as follows:

- We propose a planner benchmarking methodology and experimental results. We have provided clear recommendations on the most suitable planners for outdoor robot motion planning scenarios.
- We use an automated planning problem generation method that can better capture the behavior of *probabilistic* SBO planners by observing results from many different planning problems.
- We present a performance analysis of state-of-the-art SBO planners with realistic planning times in real and simulated 3D environments.

2 Related Work

The number of SBO planners has increased steadily in the past decade. As this is the case, several benchmarking works have been proposed to investigate the performances of each planner in-depth for specific scenarios. Meijer et al. [19], investigated various OMPL planners for grasping execution tasks on different robotic arms. Based on critics such as; the number of exactly solved runs, path length, and computation times, recommendations were made for high-performing planners depending on the type of grasp execution, e.g., pick or place. Jedrzejczyk et al. [13] presented an empirical work focusing on benchmarking of various planners within MoveIt! framework [3]. The particular focus of the planning problem was related to a robotic harvester arm. The task was picking tomatoes in a simulated greenhouse environment. The authors used time of planning, time of generated plan simplification, and time plan execution as metrics for performance evaluation. The results for metrics are provided as *average* however it is not clear how many runs were performed to obtain the average values for metrics. Magalhaes et al. [18] benchmarked OMPL planners on a six-degree of freedom manipulator in a simulated vineyard for pruning tasks. The authors use the success rate, path length, planning time, and path clearance as their metrics. The authors conclude that most of the evaluated planners in OMPL perform poorly for demanding pruning tasks in agricultural setups.

Commonly, the above works focused on robotic *manipulation* scenarios; they used path length, success rate, and execution time as evaluation metrics. In our benchmark, we exclude the *execution time* from the metrics. We solely benchmark optimizing planners as we believe that the quality of generated path for an outdoor robot is crucial. The optimizing planners utilize all allowed time to get the most optimal path. Therefore time consumed by each planner is the same. We further discuss on this in Sect. 3.1.

More recent work by Heiden et al. [10] introduced a benchmarking tool for path planning of mobile robots. The authors present the Bench-MR platform for convenient performance evaluation of different sampling-based and grid-based path planning algorithms. To showcase the features of the proposed tool, the authors provide sample benchmarks from several planning algorithms. However, there is no large-scale comparison of different planners.

Moll et al. introduced a generic software infrastructure wrapping around the OMPL interface in [20], which allows straightforward comparison of OMPL planners considering different performance metrics. The tool has introduced many performance metrics making it convenient to evaluate performance for additional requirements. We have used this benchmarking tool to carry out experiments.

As can be seen from highlighted summaries above, the existing works on sampling-based planning benchmarking have focused on benchmarking planners for robotic manipulation. We also realize a lack of methodology that can reveal more profound performance differences of probabilistic planners that are being evaluated. We deal with these shortcomings by introducing a method (see Sect. 3.2). We also use an automated problem generation on the real and simulated environment that can generate a large number of valid planning problems. Both of these allow us to gain a clear picture of SBO planners' performance in the setting of unstructured outdoor path planning.

3 Problem Formulation

We highlight planners subject to evaluation in Table 1 and describe their unique characteristics shortly. All of the *optimizing* planners within OMPL are selected for evaluation. The reason for only choosing *optimizing* planners is that the quality of the acquired path is crucial for a mobile robot. An acquired path will not only be expected to be valid but also be shorter, smoother, and take less computation time to ensure minimal energy consumption and risk during the path following action.

The task of path planning is to find a set of consecutive states S_i that connects a start state S_s to a goal state S_g. All of the states belong to a physical space that the robot operates in; this will be called *Workspace* W. Within the Workspace W, a parameter space represents all possible state configurations of the robot, *State-space* S. However, states present in S must meet a condition to be valid. A state will be valid only if it does not collide with the physical space. Therefore final path must be in a subset state-space named *collision-free state-space* S_{free}. We can formally define a valid solution path, θ based on the above definitions.

$$\begin{gathered}\theta(0) = S_s, \theta(1) = S_g,\\ \theta(x) \in S_{free}, \forall x \in [0,1]\end{gathered} \tag{1}$$

To account for sampling-based optimal path planning, we must include an additional definition related to cost function c in Eq. 1

$$\begin{gathered}c : \sum \longrightarrow R_{\geq 0},\\ c(\theta*) = \min c(\theta), \theta(x) \in S_{free}, \forall x \in [0,1]\end{gathered} \tag{2}$$

where $c(\theta*)$ is an optimized path W.R.T given cost function c, which assigns a strictly positive cost to collision-free paths.

In our benchmark, planners will minimize a cost function c, where the cost is determined by the path's length and obstacle clearance properties. The obstacle clearance objective aims to steer the path away from obstacles. A weighted combination of these two objectives is used for determining cost c. The weights were set to 10 and 1 for path length and clearance objectives, respectively.

Table 1. List of evaluated SBO planners and their short descriptions.

Planner Name	Short Description
PRM* [14]	Number of nodes to be connected are not fixed but is determined automatically based on coverage of the state space.
LazyPRM* [2]	As in PRM*, the number of nodes to be connected is not fixed but is determined automatically based on coverage of the state space, differently, the edges of nodes are not immediately checked for the collision but are checked only when a better path to the goal is found.
RRT* [14]	An asymptotically-optimal version of RRT that is guaranteed to converge to an optimal solution if one exists. The optimality refers to a specified cost function, commonly a function that determines path length.
RRTsharp [1]	Bases on Rapidly-exploring Random Graphs(RRG), it improves the convergence rate by keeping stationary information of sampled vertices; by doing this, RRTsharp can achieve better estimations of potential optimal paths.
RRTX [22]	A variant of RRT*, the nodes in the tree are rewired only if the cost is being reduced by a specified margin(known as *epsilon*), this improves the convergence of solution.
InformedRRT* [8]	Narrow downs tree exploration space to boundaries of a prolate hyperspheroid, a prolate hyperspheroid is n-dimensional symmetric ellipse that is constructed around current optimal plan. As the search space is reduced to region around optimal plan, the converge rate improves significantly.
BIT* [7]	Uses a unified structure that is composed of both graph and trees, this leads to a few advantages such as being able to use heuristics like one in A* [9] to prioritize tree expansions towards the goal with high-quality paths.
ABIT* [27]	An extension of BIT*, Advanced BIT*(ABIT*), combines an advanced graph-based search with an RRT* like planner for faster convergence rate.
AIT* [26]	Another variant of BIT*, uses an asymmetric bidirectional search where both of the searches continuously inform each other, this leads to quicker initial solutions.
LBTRRT [24]	Allows for a continuous interpolation between RRT* and RRG algorithms for high-quality solutions (comparable to RRG and RRT*) with little running-time.
SST [16]	An enhancement of RRT* which tries to address optimal kinodynamic planning when there is no access to two-point boundary value problem(BVP) solver, BVP is required to get the desired property for systems with dynamics.
SPARS [5]	Built on top of PRM*, suggests finite-sized nodes within the roadmap which might provide near-optimality properties.
SPARStwo [4]	Extends on SPARS. Sparser graphs can provide solutions meeting near-optimality requirements. Reduced resources are used to solve the planning problem, SPARStwo also decreases the rate of node addition into the roadmap.
FMT* [12]	Utilizes recursive dynamic programming to form a tree from probabilistically sampled states.
CFOREST [21]	Wraps a parallelized framework around single-query SBO planners, e.g., RRT*. Separate trees are grown from different threads while these threads communicate with each other every time a thread finds a better path. Other threads are also informed, leading to a significant speedup of convergence. The result paths are shorter and smoother.
AnytimePart Shortening(APS) [17]	A generic method that wraps around one or more planner. The method applies an iterative path simplification, shortening process. It is capable of generating *a hybrid* path if they are to yield shorter solutions.

3.1 Evaluation Metrics

There are various metrics to assess the quality of resulting paths, such as path length, total execution time, availability of exact solution, etc. SBO planner's execution time is quite ambiguous and needs an explanation here. SBO planers use a *timeout* parameter; initially, some of the time is used to construct an initial valid plan (if one exists). After a valid plan is found, the remaining time is utilized for getting lower-cost paths, at the end, all given time is used. Since in this work we are interested in optimal paths, we cannot set execution time as a metric since SBO planners utilize all given time to construct the optimal plan. In experiments, all planner's timeout were set to **5 s**. Metrics of evaluation and their short descriptions are given as follows.

- *Average Solve Rate*: Number of exact solutions divided by the total number of problems.
- *Solution Length*: Real length of the resulting solution path.
- *Solution Smoothness*: The smoothness of the path is calculated by the sum of angles between consecutive path segments. Assume a path as sequence of segments where three states within the path form two segments. Using Pythagoras generalized theorem, one can calculate angles between consecutive segments for all intermediate states. Each angle is normalized by the summed lengths of two segments. Summation of all normalized angles gives final smoothness value for the path. The lower this value the smoother the path is. For instance, a perfectly straight line will result in a smoothness value of zero.

3.2 Evaluation Methodology

The planners subject to evaluation are not deterministic but probabilistic. Therefore from a single run, it is not possible to measure the performance of a planner. Hence we run planners on a high number of randomly generated planning problems, we then use normal distributions of results from each run to determine and rank the performances of planners.

Two-Sample Z-Test is a statistical method to evaluate the differences between two normal distributions. We use a *Two-Sample Z-Test* to find out whether the performance of a planner differs significantly compared to other planners, this helps us to obtain the proportion of difference from two normal distribution metrics. Additionally, we can visualize the performance of each planner against every other planner, as in Fig. 2. The formal equations of the Two-Sample Z-Test are written as follows;

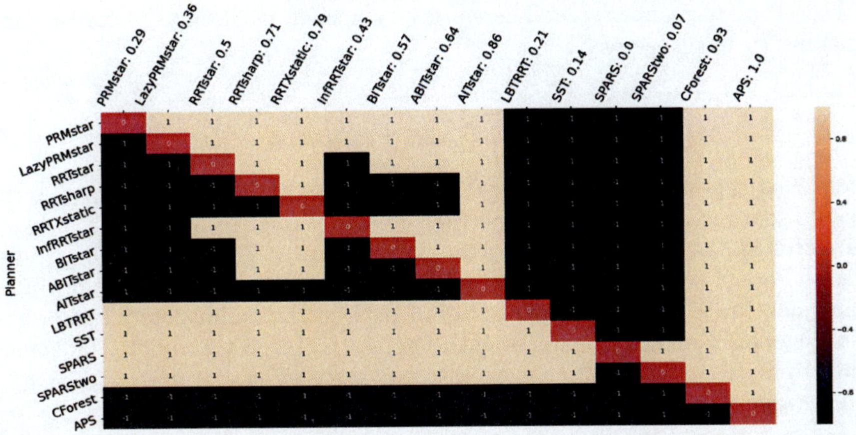

Fig. 2. A matrix depicting whether one can state that, a planner's performance in a column produced *significantly* shorter paths than a planner in a row with the confidence of $100 * (1 - a)\%$. The significance threshold a is set to 0.05. The normalized column-wise summation of matrix values is obtained on the vertical axis for each planner. In this case, the APS planner achieves the best score (1.0).

$$
\begin{aligned}
s &= \sqrt{\frac{\sigma_1^2}{N_1} + \frac{\sigma_2^2}{N_2}}, \\
z &= \frac{\mu_1 - \mu_2}{s}, \\
f(x) &= \frac{1}{\sigma_1\sqrt{2\pi}} e^{0.5(\frac{x-\mu_1}{\sigma_1})}, \\
P(X \leq z) &= \frac{1}{\sigma_1\sqrt{2\pi}} \int_{-\infty}^{z} f(x), dx
\end{aligned}
\qquad (3)
$$

where σ_1 and σ_2 are standard deviations, N_1 and N_2 are sample sizes and μ_1 and μ_2 are mean values of normal distributions for compared planners. Finally, $P(X \leq z)$, also named the p-value, determines the area under the normal distribution and assesses the significance of change between the two compared distributions. We perform the procedure of Z-Test for each planner against every other planner(including the planner itself); if the p-value for this comparison is below a threshold a, the result is 1 and -1 if otherwise. This results in a 2D skew-symmetric matrix with the size of the number of planners. The values are in the range of $[-1, 1]$, the column-wise summation of positive values in this matrix tells how many planners have been outperformed by the planner in each column(see Fig. 2). We acquire this matrix for both smoothness, and path lengths metrics since results from these two are presented with distributions. Considering multiple non-deterministic metric values, it is difficult to withdraw an overall performance score. Therefore we consider weighting the metrics and summing them respectively.

$$T = S_{rate}(w_{length}Z_{length} + w_{smooth}Z_{smooth}) \quad (4)$$

where T is the final score of the evaluated planner's performance, S_{rate} is the average solution rate of the planner(depicted in Table 3). Z_{length} and Z_{smooth} are scores of planners produced by the Two-Sample Z-Test (See the horizontal axis of Fig. 2). We assume that the length of a path has higher importance than smoothness on the overall quality of path in the Dubins state space. The resulting paths in Dubins are smooth enough to respect a robot's kinematic constraints. Therefore in Sect. 4 we assume values of w_{length} and w_{smooth} as 2 and 1, respectively.

4 Experiments

4.1 Scenarios and Evaluation Setup

The environment is represented with an OctoMap [11], a memory-efficient tree data structure that holds 3D occupancy information of the environment in a probabilistic manner. Collision checking is performed between a 3D bounding box that represents the robot body and a prebuilt OctoMap of the environment. For the collision check, we use the Flexible Collision Library(FCL) [23]. We evaluate planners on two different maps, one acquired from the Gazebo [15] simulator and one from a real environment, the map from the real environment was acquired with a 3D SLAM method named LIO-SAM in [25].

Both simulated and real environments consist of various objects such as buildings, lamp posts, plants, etc. The terrain is unstructured, though the simulated environment includes slightly more aggressive hills. We selected the maps to be of natural non-designated areas with 3D representations to account for the navigation of mobile robots in a cluttered outdoor terrains (see Fig. 1).

We use a randomized goal and start pose generation within defined state space boundaries. The validity of these start and goal poses was acquired by collision check status between a 3D box representing the robot's body and OctoMap at start and goal poses. To make sure that an actual valid plan exists between randomly generated start and goal poses, a planning request was made from start to goal pose with a sufficient amount of timeout (**10 s**). The actual benchmarking was initialized only if there existed a valid plan from random start to random goal pose, we repeat this process until we achieve **500** valid pairs of start and goal poses that define 500 unique planning problems.

The planners are evaluated in Dubin [6] state space to meet the kinematic structure of the Ackerman steering ground robot. Dubin's state-space particularly suits forward-only motions, adding more challenge to the planning problem. Dubin's state space ensures kinematic constraints are complied with by providing path segments that lay on curves in between start and goal poses. We must note that each planner's performance will be reasonable and similar if given enough time. Therefore we used a strict timeout of **5 s** to challenge and reveal performance differences among SBO planners. We generated **500** random planning problems with the minimum Euclidean distance between start and goal pose being **30 meters.** To reduce bias related to the probabilistic nature of planners, we requested each planner to solve each problem twice, making the total number of runs **1000** for each planner. Since we evaluated 16 SBO planners, 16000 runs were performed, each taking 5 s, resulting in a total computation time of 22.22 h on a laptop with Intel Core i7-6700HQ 2600 MHz MHz CPU with eight cores and 16GB of memory. All numbers reported in the following sections are result of 1000 runs.

In the following, we will interpret results metric-wise.

4.2 Interpreting Results Metric-Wise

The means and standard deviations of resulting path lengths are presented in Table 2. It is important to note that these values are acquired only from the exactly solved runs. For instance, the FMT* planner could not exactly solve any given problem. Therefore, it is not included in Table 2 and Table 3. Looking at Table 3, one can see that except for the FMT* planner, each planner's number of exact solutions is similar, meaning that the number of samples used to

Table 2. Means and deviations of acquired path length and smoothness metric from 1000 runs. A lower mean value indicates that planners achieve better performance for both metrics.

Planners	Length Mean	Length Dev.	Smoothness Mean	Smoothness Dev.
APS	51.8	15.3	0.001	0.004
CForest	54.7	16.3	0.005	0.008
RRTX	56.7	17.0	0.006	0.012
RRTsharp	56.5	17.1	0.006	0.012
RRT*	57.5	17.8	0.007	0.011
InfRRT*	57.6	18.1	0.008	0.011
AIT*	54.7	19.2	0.021	0.025
BIT*	56.6	17.2	0.016	0.021
ABIT*	56.6	17.5	0.017	0.020
LazyPRM*	60.0	19.4	0.013	0.021
PRM*	65.1	22.5	0.039	0.059
LBTRRT	67.2	21.2	0.048	0.053
SPARStwo	92.3	41.6	0.024	0.027
SPARS	98.1	44.6	0.035	0.040
SST	79.6	30.6	0.439	0.281
FMT*	–	–	–	–

Table 3. A table presenting the average solve rate and overall performance scores of planners calculated by the proposed method in Sect. 3.2. The planners are ranked with highest final score values.

Planner Name	Average Solve Rate	Overall Performance Score
APS	0.99	7.96
CForest	0.97	4.31
RRTX	0.94	3.14
RRTsharp	0.96	3.03
RRT*	0.96	2.94
InfRRT*	0.95	2.80
AIT*	0.93	2.06
BIT*	0.94	1.99
ABIT*	0.94	1.96
LazyPRM*	0.94	1.84
PRM*	0.99	1.05
LBTRRT	0.97	0.90
SPARStwo	0.97	0.78
SPARS	0.95	0.42
SST	0.92	0.37
FMT*	–	–

calculate values in Table 2 is similar for all planners. In Table 2, a lower mean value indicates a shorter path in the context of the path length metric. According to mean values, APS achieves 25.90% shorter paths than the rest of the planners, while SPARS achieves the longest paths. Both APS and CFOREST use a parallelization(see Table 1) *trick* to initiate multiple instances of a selected planner and run them separately on parallel threads. These different instances of planners continuously inform each other of achieved milestones, which seems to land them significant performance advantages towards more optimal paths within the given time. This can be seen from the horizontal axis of Fig. 2 and Table 2. Planners from RRT* and BIT* families perform similarly, except AIT* planner, which performs better than (6.59% shorter paths on average) the rest of the planners from RRT* and BIT* families, thanks to its asymmetric bidirectional search technique. Planners from the PRM* family, such as SPARS and SPARStwo, produced the longest paths. We argue that this is due to the planner's stricter attitude towards adding more nodes to the roadmap. We observed from a sample run that PRM*, a predecessor of SPARStwo, added 799 nodes to the roadmap while SPARStwo added only 30 nodes to the roadmap, leading to longer and less smooth paths overall.

The APS planner achieves the best performance in the path smoothness metric, while SST achieves the worst. We notice that, on average, the APS planner creates five times smoother paths compared to the next best performing planner CFOREST. Unlike in path length metric, planners of RRT* family outperform planners of BIT* family in path smoothness metric (see Table 2). Similar to their performance in path length metric, PRM* family planners do not particularly create smoother paths, making them average performers. From Table 2, we see that tree-based planners create smoother plans compared to graph-based planners. This makes sense since a tree-based planner adds the next sample state to a relevant pose within the tree structure whereas, in a graph-based planner, the next sample state could be placed to a less relevant pose, causing increased curvature of the final path.

4.3 Combining the Metrics for Absolute Ranking

We use the proposed methodology described in Sect. 3.2 to achieve an absolute performance score for each planner. It must be noted that we assume a higher weight for the path length metric. We assume that the path length importance is more significant for a mobile robot, still, it is possible to use different weights and recalculate final scores with given information in Table 2 and Table 3. As seen in the vertical label of Fig. 2, the score for each planner has been calculated by normalized column-wise summation, these accumulative scores are created for path length and smoothness metrics, which are denoted with Z_{length} and Z_{smooth} respectively.

With acquired Z_{length} and Z_{smooth}, Using Eq. 4, we achieve the final performance scores presented in Table 3. Note that the values present in Table 3 are achieved by comparing planners with each other, a low value does not necessarily mean a planner has failed to produce the desired result, but it instead means that, the planner has performed worse than other planners with greater final score values. Our primary focus in this paper was on the problem of mobile robot path planning with kinematic constraints. Our recommendation for users is to consult to the final scores of planners given in Table 3. Based on these results, one can judge the most suitable planners for mobile robot navigation on outdoor terrains. We were not able to acquire any exact solution from FMT* planner. It is unclear what might have caused this. More investigation for this particular planner is needed.

5 Conclusion

We evaluated all SBO planners in the OMPL with a systematic approach. We run each planner on 500 different planning problems twice, resulting in 1000 results that reveal each SBO planner's behavior. The evaluation setup was mainly aimed at the planning of outdoor mobile robots. The proposed evaluation methodology identifies the most suitable planners for planning in unstructured terrains. With provided probability distributions of metrics and their combinations, the method revealed performance differences among planners numerically. The APS planner was found to be the most consistent planner along with all metrics. Therefore we recommend the readers' APS planner, as it is the best performing SBO planner for mobile robot path planning. This paper does not propose a methodology to find optimal configuration parameters for each evaluated planner automatically. There are hundreds of parameters for all planners being assessed, making it challenging to find optimal values for each parameter. Therefore all planners were set for experiments with their default parameters. We plan to investigate parameter tuning for each planner methodologically in future works.

References

1. Arslan, O., Tsiotras, P.: Use of relaxation methods in sampling-based algorithms for optimal motion planning. In: 2013 IEEE International Conference on Robotics and Automation, pp. 2421–2428 (2013)
2. Bohlin, R., Kavraki, L.: Path planning using lazy PRM. **1**, 521–528 (2000)
3. Coleman, D., Sucan, I.A., Chitta, S., Correll, N.: Reducing the barrier to entry of complex robotic software: a moveit! case study. CoRR (2014). abs/1404.3785
4. Dobson, A., Bekris, K.E.: Improving sparse roadmap spanners. In: 2013 IEEE International Conference on Robotics and Automation, pp. 4106–4111 (2013)
5. Dobson, A., Krontiris, A., Bekris, K.E.: Sparse roadmap spanners. In: Frazzoli, E., Lozano-Perez, T., Roy, N., Rus, D. (eds.) Algorithmic Foundations of Robotics X, pp. 279–296. Springer, Berlin (2013)
6. Dubins, L.E.: On curves of minimal length with a constraint on average curvature, and with prescribed initial and terminal positions and tangents. Am. J. Math. **79**(3), 497–516 (1957)
7. Gammell, J.D., Barfoot, T.D., Srinivasa, S.S.: Batch informed trees (bit*): informed asymptotically optimal anytime search. Int. J. Rob. Res. **39**(5), 543–567 (2020)
8. Gammell, J.D., Srinivasa, S.S., Barfoot, T.D.: Informed rrt*: optimal incremental path planning focused through an admissible ellipsoidal heuristic. CoRR (2014). abs/1404.2334
9. Hart, P., Nilsson, N., Raphael, B.: A formal basis for the heuristic determination of minimum cost paths. IEEE Trans. Syst. Sci. Cybern. **4**(2), 100–107 (1968)
10. Heiden, E., Palmieri, L., Bruns, L., Arras, K.O., Sukhatme, G.S., Koenig, S.: Bench-MR: a motion planning benchmark for wheeled mobile robots. IEEE Rob. Autom. Lett. **6**(3), 4536–4543 (2021)
11. Hornung, A., Wurm, K.M., Bennewitz, M., Stachniss, C., Burgard, W.: Octomap: an efficient probabilistic 3d mapping framework based on octrees. Auton, Rob (2013)
12. Janson, L., Pavone, M.: Fast marching trees: a fast marching sampling-based method for optimal motion planning in many dimensions–extended version. CoRR (2013). abs/1306.3532
13. Jedrzejczyk, F., Bajer, J., Gawdzik, G., Główka, J., Sprońska, A.: Benchmark and analysis of path planning algorithms of "ros moveit!" for pick and place task in tomato harvesting. In: Szewczyk, R., Zieliński, C., Kaliczyńska, M. (eds.) Automation 2021: Recent Achievements in Automation. Robotics and Measurement Techniques, pp. 272–284. Springer International Publishing, Cham (2021)
14. Karaman, S., Frazzoli, E.: Sampling-based algorithms for optimal motion planning. CoRR (2011). abs/1105.1186

15. Koenig, N., Howard, A.: Design and use paradigms for gazebo, an open-source multi-robot simulator. In: 2004 IEEE/RSJ International Conference on Intelligent Robots and Systems (IROS) (IEEE Cat. No.04CH37566), vol. 3, pp. 2149–2154 (2004)
16. Li, Y., Littlefield, Z., Bekris, K.E.: Asymptotically optimal sampling-based kinodynamic planning (2016)
17. Luna, R., Şucan, I.A., Moll, M., Kavraki, L.E.: Anytime solution optimization for sampling-based motion planning. In: 2013 IEEE International Conference on Robotics and Automation, pp. 5068–5074 (2013)
18. Magalhães, S., Neves Dos Santos, F., Martins, R., Rocha, L., Brito, J.: Path planning algorithms benchmarking for grapevines pruning and monitoring, pp. 295–306 (2019)
19. Meijer, J., Lei, Q., Wisse, M.: Performance study of single-query motion planning for grasp execution using various manipulators. In: 2017 18th International Conference on Advanced Robotics (ICAR), pp. 450–457 (2017)
20. Moll, M., Şucan, I.A., Kavraki, L.E.: Benchmarking motion planning algorithms: an extensible infrastructure for analysis and visualization. IEEE Rob. Autom. Mag. **22**(3), 96–102 (2015)
21. Otte, M., Correll, N.: C-forest: Parallel shortest path planning with superlinear speedup. IEEE Trans. Rob. **29**(3), 798–806 (2013)
22. Otte, M., Frazzoli, E.: Rrtx: real-time motion planning/replanning for environments with unpredictable obstacles. In: WAFR (2014)
23. Pan, J., Chitta, S., Manocha, D.: Fcl: a general purpose library for collision and proximity queries. In: 2012 IEEE International Conference on Robotics and Automation, pp. 3859–3866 (2012)
24. Salzman, O., Halperin, D.: Asymptotically near-optimal RRT for fast, high-quality, motion planning. CoRR (2013). abs/1308.0189
25. Shan, T., Englot, B., Meyers, D., Wang, W., Ratti, C., Daniela, R.: Lio-sam: tightly-coupled lidar inertial odometry via smoothing and mapping. In: IEEE/RSJ International Conference on Intelligent Robots and Systems (IROS), pp. 5135–5142. IEEE (2020)
26. Strub, M.P., Gammell, J.D.: Adaptively informed trees (AIT*): fast asymptotically optimal path planning through adaptive heuristics. In: Proceedings of the IEEE International Conference on Robotics and Automation (ICRA), pp. 3191–3198, 31 May-31 Aug 2020
27. Strub, M.P., Gammell, J.D.: Advanced bit* (abit*): sampling-based planning with advanced graph-search techniques. In: 2020 IEEE International Conference on Robotics and Automation (ICRA), May 2020

3D Traversability Analysis in Forest Environments Based on Mechanical Effort

Afonso E. Carvalho[1], João Filipe Ferreira[1,2], and David Portugal[1](✉)

[1] Institute of Systems and Robotics, University of Coimbra, Coimbra, Portugal
afonso.carvalho@isr.uc.pt, jfilipe@isr.uc.pt, joao.ferreira@ntu.ac.uk, davidbsp@isr.uc.pt

[2] School of Science and Technology, Nottingham Trent University, Nottingham, UK

Abstract. Autonomous navigation in harsh and dynamic 3D environments poses a great challenge for modern Robotics. This work presents a novel traversability analysis and path-planning technique that processes 3D pointcloud maps to generate terrain gradient information. An analysis of terrain roughness and presence of obstacles is applied on the perceived environment in order to generate efficient paths. These avoid major hills when more conservative paths are available, potentially promoting fuel economy and reducing the wear of the equipment and the associated risks. The proposed approach outperforms existing techniques based on results in realistic 3D simulation scenarios, which are discussed in detail.

Keywords: Forestry robotics · Traversability analysis · 3d navigation · Path planning · UGV · Mechanical effort

1 Introduction

Forest maintenance is a labor-intensive, time-consuming, expensive, and ultimately hazardous activity, making it a prime candidate for automation. This requires reliable unsupervised navigation, which is a sizeable challenge considering that forests are highly unstructured, dynamic environments. As such, R&D efforts have been increasing, tackling key issues such as autonomous maintenance and management in fully [1] and semi-unstructured [2] forests, as well as more delicate environments including orchards [3] and plantation fields [4].

This work proposes an innovative path planning technique that uses the concept of mechanical effort [5] to generate efficient paths for a UGV, avoiding steep sections of terrain to potentially improve fuel economy and reduce mechanical wear by minimizing the mechanical effort that the robot is subject to. For this purpose, a 2D costmap is used to represent the perceived environment, on which traversability analysis and successive path planning is then performed. The proposed method was tested and compared against existing techniques on a realistic 3D Gazebo[1] simulation environment providing a solid basis for discussion.

[1] http://gazebosim.org/.

I. Petrovic et al. (Eds.): IAS 2022, LNNS 577, pp. 244–256, 2023.
https://doi.org/10.1007/978-3-031-22216-0_17

Existing literature on navigation and traversability analysis focuses on methods such as Elevation Maps [6], 2.5D-NDT [7], Octomaps [8] and fusion of complex physics-based planning with simplified approaches [9] to represent the foreseeable environment in a discrete way. Typically, traversability estimation is applied on these discrete representations by assigning costs to distinct portions of terrain. These include probabilistic [10], semantic segmentation [11] and neural network (NN) depth maps [12] approaches. Finally, classical planning algorithms, e.g. A* [13], recent methods, e.g. D* Lite [14], or those based on NNs [15], can generate a plan leading the agent to its target destination.

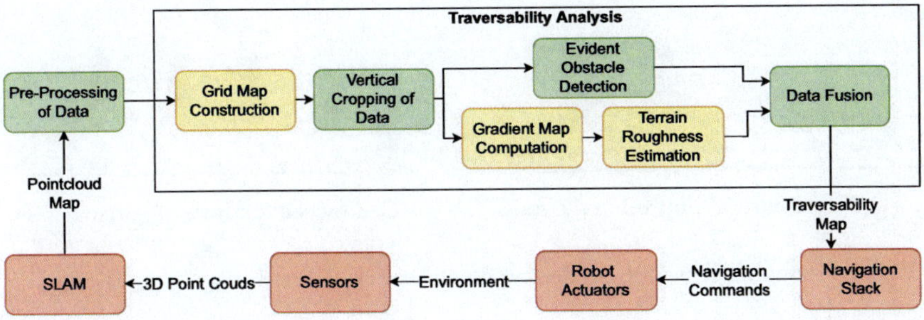

Fig. 1. Proposed System architecture. Green boxes: novel modules; Yellow boxes: modified modules; Red boxes: off-the-shelf modules used "as is".

2 System Architecture

Figure 1 presents an overview of the architecture developed, consisting of:

1. A pre-processing module responsible for down sampling the input data;
2. A traversability analysis block responsible for generating traversability maps;
3. A navigation technique that uses its contents to generate and execute paths.

All modules have been developed using ROS[2] and run concurrently.

Pre-Processing In this module, a 3D global pointcloud map of the environment available *a priori* (whose generation is out of the scope of this work) is filtered to remove points outside the robot's workspace. The raw pointcloud is decimated using a 3D voxel grid with a configurable downsample factor, which significantly reduces the number of points (up to 6.5×), depending on density.

Grid Map Construction This module is responsible for creating a 2D grid map. It takes as input the decimated pointcloud that has been processed before and creates a grid of configurable size and resolution where each 3D point is projected and assigned to a particular grid cell according to its location in space.

[2] Robot Operating System – http://ros.org.

Vertical Cropping of Data

Having then centered the decimated pointcloud in the robot's frame of reference, all points that are outside of a configured 3D box with no limit along the Z vertical axis are filtered. This is depicted in Fig. 2a, where we can observe that if we also crop along the Z-axis up to a certain height, it would result in the elimination of relevant information for path planning, such as trees on higher ground. Afterwards, the pointcloud is filtered vertically by performing a cell-wise analysis, where the lowest point (i.e. ground level) in each cell is found, and only points that stand up to a given amount above it are included. This heuristic is demonstrated in Fig. 2b.

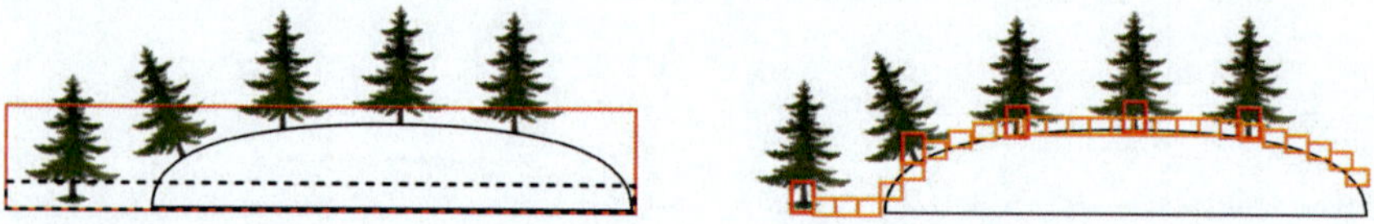

(a) Cropbox with a limit on the Z-axis. (b) Cell by cell analysis of vertical axis.

Fig. 2. Cropping heuristics. The one illustrated in (b) was ultimately selected.

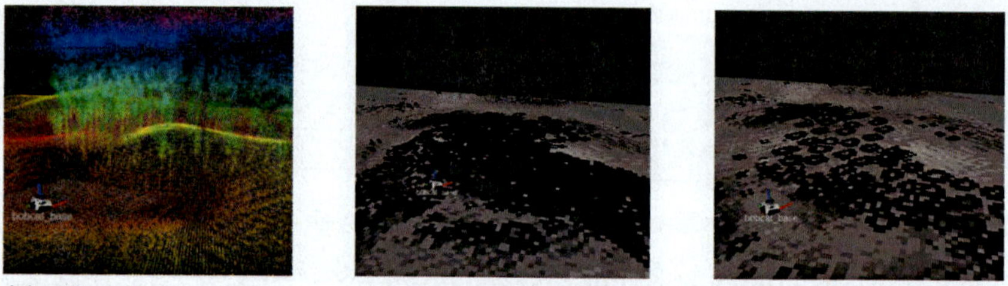

(a) Pointcloud to be analyzed. (b) Costmap before applying cropping heuristic. (c) Costmap after applying cropping heuristic.

Fig. 3. Impact of applying the 3D vertical cropping heuristic on the 2D costmap representation (X-Y plane), which outlines existing obstacles more clearly (see black cells).

In Fig. 3, we illustrate the impact of applying the heuristic on the generated costmap. The main goal is to reduce the number of points to be analyzed and consequently reduce the complexity and computational cost of the entire process.

Gradient Map Computation In this step, the terrain is analyzed using the gradient. For cells that contain more than one point, we apply the median on the set of points in that cell, and only then we apply the gradient to the resulting pointcloud, generating a map that estimates the overall roughness of the terrain.

Considering that our goal is to minimize traversing effort and fuel consumption, it is important to note that there is a strong correlation between the gradient of a terrain and the energy consumption that it requires to be traversed, as it provides information of the direction of greatest variation.

Since it is not possible to employ a function that describes the entire point distribution, we make use of an approximation of the gradient, computed using second order accurate central differences, and the resulting gradient values are clipped to a configurable threshold, i.e.:

$$\nabla f(x_m, y_n) = \begin{cases} \nabla f(x_m, y_n), & \|\nabla f(x_m, y_n)\| < \nabla_{th} \\ \nabla_{th}, & \textbf{otherwise} \end{cases}, \quad (1)$$

where $\nabla f(x_m, y_m)$ represents the gradient in a given cell and ∇_{th} represents the clipping threshold. This allows us to exclude both hills and depressions that exceed this configurable inclination limit. Figure 4 illustrates this step.

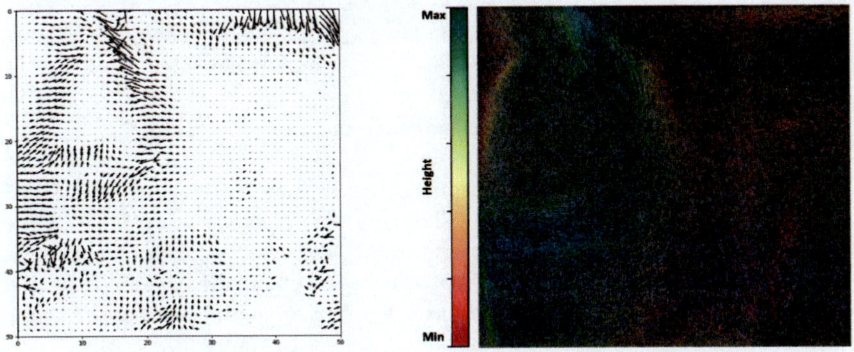

(a) Gradient Map. (b) Pointcloud map (top view).

Fig. 4. Output representation of the gradient analysis stage.

Evident Obstacle Detection In this stage, we analyze the filtered pointcloud generated from the vertical heuristic applied previously (see Figs. 1 and 2b). We compute the mean, variance and range of the heights of all points that form each cell, and apply the following heuristic that marks a cell as an evident obstacle:

$$obstacle = \begin{cases} \text{true}, & \begin{cases} \mu_{(x,y)} > \mu_{th} \\ \sigma^2_{(x,y)} > \sigma^2_{th} \\ (max_{(x,y)} - min_{(x,y)}) > \gamma \end{cases} \\ \text{false}, & \textbf{otherwise} \end{cases}, \quad (2)$$

where μ_{th}, σ^2_{th} and γ are the given thresholds for the mean, variance and range of the sample, respectively. By combining these, we are able to predict with a high degree of certainty the presence of obstacles.

Figure 5 illustrates the output of this step in a forest scenario. It becomes clear that the algorithm considers tree trunks as evident obstacles while discarding tree tops, thus generating a traversable 2D gridmap that would otherwise be unattainable.

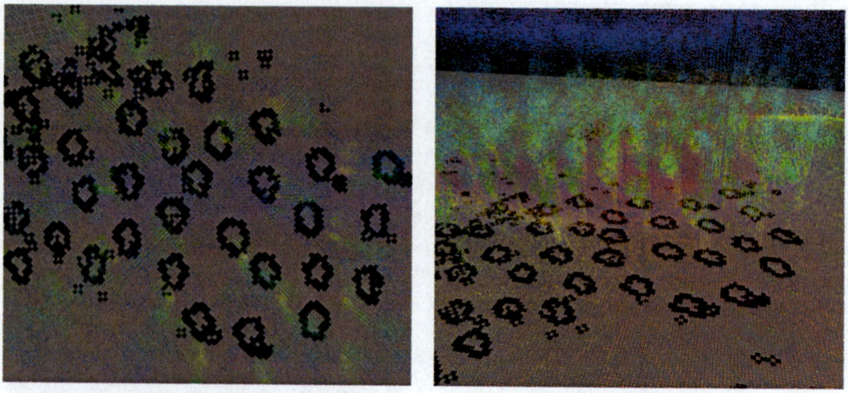

(a) Top view. (b) Side view.

Fig. 5. Evident obstacles detection on a forest scenario.

Terrain Roughness Estimation This module estimates terrain roughness using the previously generated gradient map. For this, we use the definition of mechanical effort $\gamma = \|\nabla f(x,y)\| \cdot \cos\theta$ from [5], where θ is the angle between the vector that connects the position of the robot with the position of the cell we are currently analyzing and the gradient vector in that cell.

The terrain roughness estimation module takes as input the gradient map and modifies it according to the mechanical effort equation, thus generating a 2D gridmap, in which the traversability cost increases as the robot's direction of movement aligns with the gradient and vice-versa.

Data Fusion Finally, this module outputs a single 2D global costmap that incorporates all the useful information contained on the *evident obstacles costmap* and the *terrain roughness costmap*. All cells identified as evident obstacles are marked as non-traversable ($cost = 100$) in the final costmap, and the remaining ones are marked with an integer ranging from 0 to 99, where 50 represents traversing even terrain, 0 represents the lowest possible traversing cost, and 99 the maximum possible cost of traversing a cell that is not an obstacle.

3 Experimental Evaluation

We have designed three realistic simulation scenarios in Gazebo to evaluate the performance of the proposed method. Figure 6 represents `Scenario 1`, which

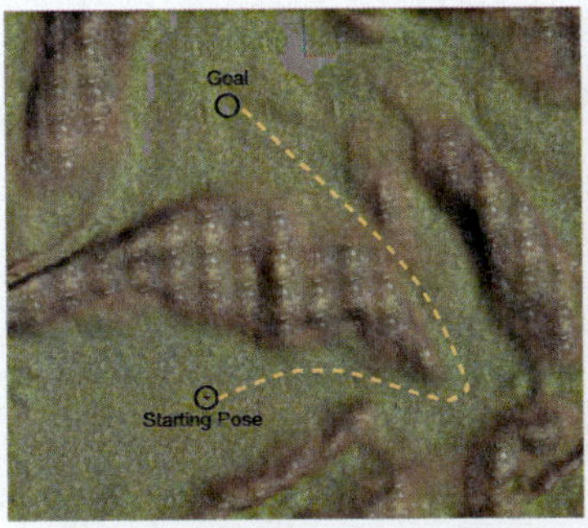

(a) Top view of Scenario 1.

(b) Side view of Scenario 1.

Fig. 6. First experimental scenario from two different perspectives, with starting pose, goal point and expected approximate path for our technique in yellow.

includes steep hills, making this an appropriate benchmarking scenario, rewarding path safety and mechanical effort over traveled distance and travel time, while still taking them into account. A second version of this scenario was designed, which we call `Scenario 1 remapped`, having the same spatial configuration, but with a 20% increase in height values, leading to a major difference: some slopes exceed the angular limits of the UGV, becoming untraversable.

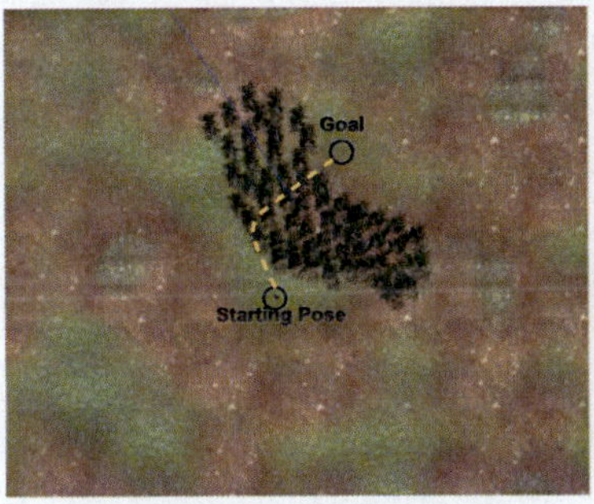

(a) Top view of Scenario 2.

(b) Front view of Scenario 2.

Fig. 7. Second experimental scenario from two different perspectives, with starting pose, goal point and expected approximate path for our technique in yellow.

Figure 7 represents `Scenario 2`, which besides hilly terrain, also includes a forest of fully grown pine trees. This poses an added challenge to the candidate techniques, as the UGV needs to be able to safely navigate within the trees without colliding with them, while choosing a traversable path to do so.

We have defined the following **experimental objectives**:

1 Demonstrate that our work manages to create and execute paths with a lower energetic cost when compared to others.
2 Verify that our method outperforms the method in which it is based.
3 Show that our method can run in real time.

In order to quantify performance and compare methods, we have defined the following metrics:

Elapsed Time The total time that it takes for the UGV to go from the starting pose to the end goal. A method should minimize the elapsed time without compromising other more important factors. This metric helps to sustain objective 2.

Travelled Distance The distance travelled by the UGV from the starting pose to the end goal. A technique should be rewarded for reducing the travelled distance, with some exceptions such as not compromising/worsening some other important metrics. This metric helps demonstrating objectives 1 and 2.

Mean Map Generation Time The time it takes to generate a new costmap. For the purpose of this work, we consider that any approach runs in real time if a new map is generated before the UGV reaches the end of the current one. This represents the difference between the UGV navigating known or unknown terrain. This metric helps demonstrating objectives 2 and 3.

Positive Height Variation The cumulative vertical distance travelled by the UGV. In general, a technique should attempt to minimize this metric, as the positive gradient of the terrain is the main factor affecting fuel economy. This metric helps to demonstrate objective 1.

Mean Effort The mean effort that a technique faces while executing the given path. This metric is measured by averaging all the absolute UGV's pitch orientation (θ) values along the executed trajectory. It gives us an indication of the mean vertical travelling angle, which has a direct correlation with the effort of the chosen path. The candidate techniques should aim at minimizing it. This metric helps to demonstrate objective 1.

Path Riskiness Index The path riskiness index metric indicates how risky the travelled path is. In this context, we define a "risky" situation as one where the UGV travels while exceeding any of its angular limits, and we quantify it as the percentage of time that the UGV exceeded said limits in the travelled path. This metric should be minimized and it helps demonstrating objective 2.

Roll Danger Index The mean percentage of roll-related risk that the UGV takes during the execution of its path. The absolute roll values are filtered with the following sigmoid function:

$$RDI = \frac{1}{1 + e^{-(0.5\phi - 13)}}, \tag{3}$$

which de-linearizes the penalty curve of the UGV's inclination. Considering the roll angle limit to be 35°, it is almost equally safe to travel with between 5o-10o of inclination. However, the same does not apply to the 5o difference between 25o-30o of inclination, because the stability of the vehicle rapidly decreases when approaching its angular limits. The sigmoid filter in (3) describes this behavior. **Pitch Danger Index** The mean percentage of pitch-related risk that the UGV takes during the execution of its path. Using the same rationale, the absolute pitch values are filtered through the following sigmoid function:

$$PDI = \frac{1}{1 + e^{-(0.25\theta - 6)}}. \tag{4}$$

Failure Rate We define the failure rate as the percentage of times that the path planner aborts the whole mission due to one of the following reasons:

- It has not been able to follow the selected path (e.g. an internal error in the navigation module).
- It has not been able to create a valid or feasible plan to the targeted goal.
- The specified timeout for the mission has been exceeded.
- The UGV rolled over.

This metric keeps track of failed runs, providing a solid measure of a system's reliability and robustness, helping to demonstrate objective 2.

4 Results and Discussion

Our system is compared against `Lourenço2020` [5], which introduced the original concept of mechanical effort, and `move_base_naive`, which applies the ROS navigation stack[3] with no observation sources and empty costmaps, following a straight path to the goal that, although expected to underperform other techniques, establishes a baseline for benchmarking. In this section, we discuss the findings of running the three methods on the scenarios described previously. To this end, 50 navigation trials were conducted in each scenario and a statistical analysis of the results is presented. In these experiments, our system uses a 200×200 m global costmap, while `Lourenço2020` uses a 15×15 m local costmap.

Table 1 and Fig. 8 present the results of the experiments in Scenario 1. Instead of traversing straight to the goal, which would represent a shorter elapsed time and travel distance at the cost of major risk (especially pitch related risk) and significantly higher mechanical effort, our system chooses the more conservative and levelled path available, which is around the hill. By doing so, it is able to significantly reduce the effort associated to that trajectory and potentially the fuel consumption, considering its relationship with terrain gradient (see [16]).

The pitch danger index is significantly lower than in other techniques, while the roll danger index is not that low, especially when compared to `move_base_naive`. This is justified since while the UGV travels around the hill, it tries to minimize the travelled distance without incurring in excessive risk,

[3] http://wiki.ros.org/move_base.

thus travelling on side slopes for the majority of the path, due to the morphology of the terrain in that area. By doing so, it only travels on average at 12.8% the maximum inclination allowed, while `Lourenço2020` travels at a considerably higher 26.6%, because it detects the presence of a significant slope ahead, but is unable to find a better path due to its short sensory horizon. Therefore, it travels sideways on the slope while attempting to find a better path, a behavior that is not ideal, considering that it tends to increase the danger and riskiness indexes and the elapsed time.

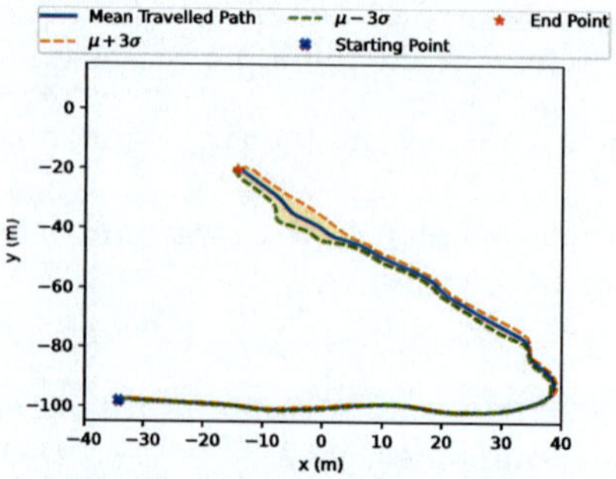

(a) 50 trajectories projected on the X-Y plane with our system.

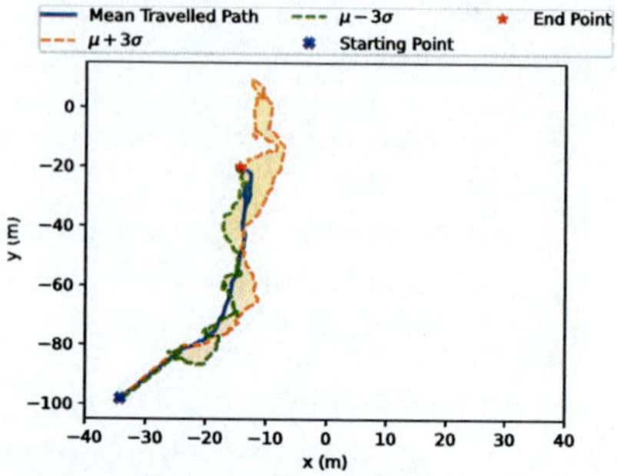

(b) 50 trajectories projected on the X-Y plane with `Lourenço2020` [5].

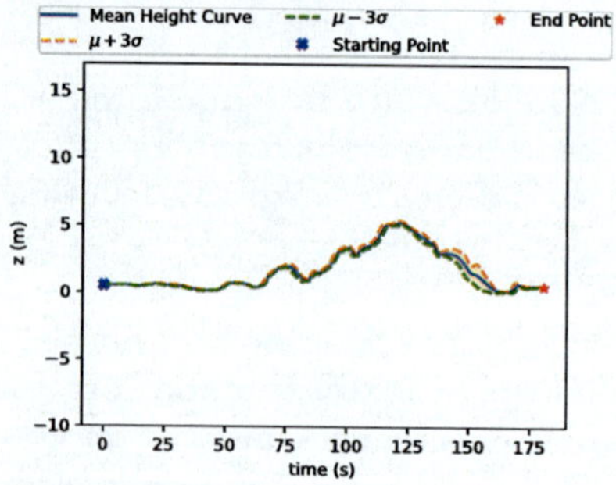

(c) 50 trajectories of height as a function of time with our system.

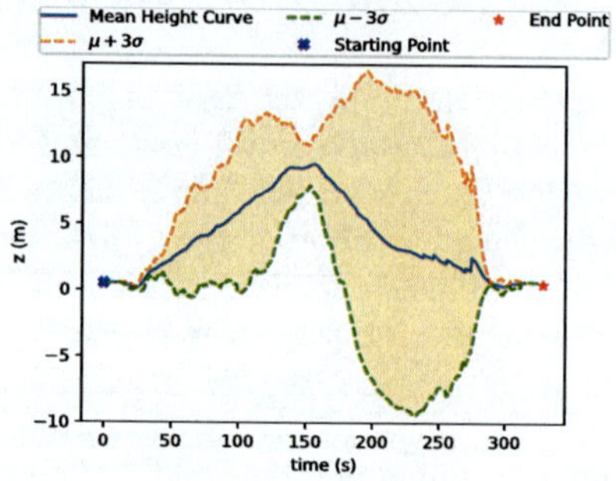

(d) 50 trajectories of height as a function of time with `Lourenço2020` [5].

Fig. 8. Resulting plots from 50 runs in Scenario 1 (Fig. 6) with our system in the left, and `Lourenço2020` [5] in the right.

Table 2 shows the results from running the same experiment in the remapped Scenario 1. We can see that by slightly increasing the slope angles, the other systems perform drastically worse, while ours presents similar results to Scenario 1. The large failure rates observed are no surprise, as this scenario penalizes any technique that chooses to ignore its large hills, reinforcing the importance of avoiding the steepest sections of terrain and ultimately showing the robustness and reliability of our system, while also supporting objectives 1 and 2.

Table 1. Results from 50 runs in Scenario 1 (Fig. 6). Best values highlighted in bold.

Metrics	Lourenço2020	move_base_naive	Our System
Failure Rate (%)	84.000	**2.000**	**2.000**
Mean Effort (°)	18.661 ± 1.828	12.805 ± 0.205	**5.601 ± 0.060**
Pitch Danger Index (%)	38.592 ± 4.892	25.406 ± 0.335	**2.814 ± 0.142**
Up Variation (m)	15.240 ± 1.886	8.877 ± 0.036	**8.183 ± 0.088**
Travelled Distance (m)	112.387 ± 8.066	**84.958 ± 0.068**	171.887 ± 0.343
Path Riskiness Index (%)	4.291 ± 4.094	**0.517 ± 0.650**	0.932 ± 0.212
Roll Danger Index (%)	26.580 ± 4.648	**0.042 ± 0.137**	12.843 ± 0.695
Mean Map Generation Time (seconds @ points)	8.2 @ 40.1k	N/A	18.3 @ 287k
Elapsed Time (s)	343.582 ± 14.886	**87.014 ± 12.689**	174.368 ± 4.184

Table 2. Results from 50 runs in Scenario 1 remapped. Best values highlighted in bold.

Metrics	Lourenço2020	move_base_naive	Our System
Failure Rate (%)	100.000	90.000	**10.000**
Mean Effort (°)	N/A	17.172 ± 0.902	**7.004 ± 0.301**
Pitch Danger Index (%)	N/A	32.131 ± 1.199	**5.807 ± 0.653**
Up Variation (m)	N/A	11.586 ± 0.050	**10.319 ± 0.476**
Travelled Distance (m)	N/A	**87.873 ± 0.399**	175.689 ± 1.894
Path Riskiness Index (%)	N/A	12.558 ± 0.853	**2.839 ± 1.134**
Roll Danger Index (%)	N/A	**0.807 ± 1.413**	12.788 ± 2.581
Mean Map Generation Time (seconds @ points)	8.9 @ 40.5k	N/A	18.3 @ 287k
Elapsed Time (s)	N/A	**25.299 ± 2.510**	173.037 ± 31.038

Lastly, Table 3 and Fig. 9 presents the results of the experiments in Scenario 2 (Fig. 7), arguably the most demanding scenario, considering not only that the UGV must avoid every tree in its path, but it should also choose the path with the lowest effort associated. The paths chosen by our technique do not lead to any situations where the inclination of the terrain is such that the UGV exceeds the roll and pitch limits, i.e. the path riskiness index is always zero.

Our system substantially outperforms the remaining ones in every single metric except in the travelled distance, which was slightly higher[4]. However, that small increase in travel distance resulted in a significant difference in the mean effort and vertical variation, the main variables of interest for our objective, therefore posing a very reasonable trade-off.

[4] A video of our system in Scenario 2 is available at https://youtu.be/2wSMTPBg0ZU.

Table 3. Results from 50 runs in Scenario 2 (Fig. 7). Best values highlighted in bold.

Metrics	Lourenço2020	move_base_naive	Our System
Failure Rate (%)	58.000	92.000	**6.000**
Mean Effort (°)	11.024 ± 3.500	15.188 ± 0.100	**3.595 ± 0.066**
Pitch Danger Index (%)	9.326 ± 8.016	23.300 ± 0.565	**0.794 ± 0.015**
Up Variation (m)	7.879 ± 0.496	7.035 ± 0.066	**3.714 ± 0.045**
Travelled Distance (m)	80.226 ± 9.643	**58.299 ± 0.497**	85.586 ± 0.493
Path Riskiness Index (%)	1.353 ± 5.998	0.767 ± 0.701	**0.000 ± 0.000**
Roll Danger Index (%)	0.297 ± 0.297	0.909 ± 0.469	**0.020 ± 0.001**
Mean Map Generation Time (seconds @ points)	12.1 @ 42.4k	N/A	20.2 @ 358k
Elapsed Time (s)	340.362 ± 123.149	182.514 ± 193.306	**88.266 ± 12.011**

As stated before, the consistency and robustness of our technique is patent in the results presented, especially in the failure rate and mean effort. Moreover, our system significantly outperforms Lourenço2020 [5] in Mean Map Generation Time. This metric is expressed in `seconds @ points`, providing the time it took to generate each map and the number of points that it contains, on average.

5 Conclusion

This work proposes a novel traversability analysis and path planning technique based on the mechanical effort concept introduced in [5]. It categorizes terrain according to the effort required to traverse it, while identifying key evident obstacles, thus generating efficient paths that avoid obstacles and major hills, potentially minimizing fuel consumption. A concurrent pipeline of obstacle detection and terrain roughness estimation is implemented and the system is optimized to execute in real time while performing a global analysis. This potentially allows a robot to plan in real-time far beyond its observable range if given an *a priori* map of the environment. Finally, the implemented system has been tested against other methods in a 3D realistic simulation engine, yielding very positive results and proving to be a strong competitor against other techniques.

We have accomplished the experimental objectives proposed, while also successfully addressing one of the weaknesses stated by Lourenço et al. [5], i.e. the map generation frequency was improved by 3 to 5 times for the scenarios tested.

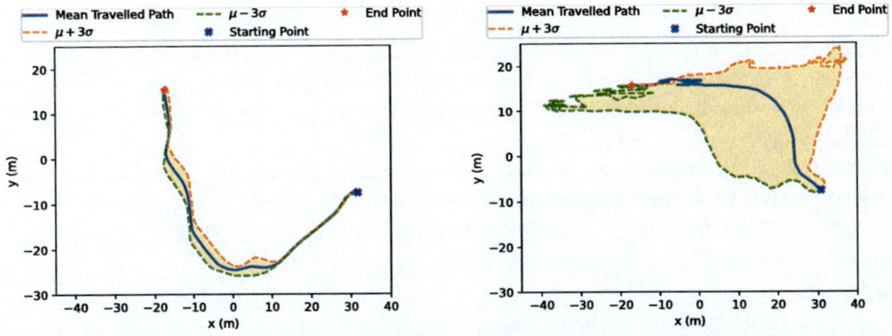

(a) 50 trajectories projected on the X-Y plane with our system.

(b) 50 trajectories projected on the X-Y plane with **Lourenço2020** [5].

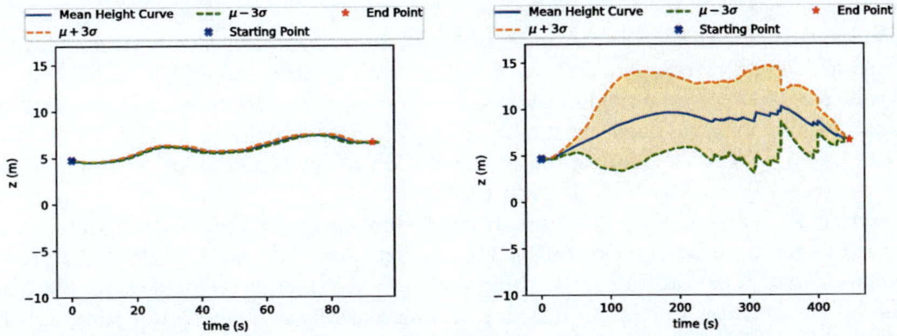

(c) 50 trajectories of height as a function of time with our system.

(d) 50 trajectories of height as a function of time with **Lourenço2020** [5].

Fig. 9. Resulting plots from 50 runs in Scenario 2 (Fig. 7) with our system in the left, and **Lourenço2020** [5] in the right.

Acknowledgements. We thank **Gonçalo Martins** for the invaluable insight and active support. This work was supported by the Safety, Exploration and Maintenance of Forests with Ecological Robotics (**SEMFIRE**, ref. CENTRO-01-0247-FEDER-03269) research project co-funded by the "Agência Nacional de Inovação" within the Portugal2020 programme.

References

1. Couceiro, M.S., Portugal, D., Ferreira, J.F., Rocha, R.P.: SEMFIRE: towards a new generation of forestry maintenance multi-robot systems. In: 2019 IEEE/SICE International Symposium on System Integration, SII 2019, pp. 270–276 (2019)
2. Mowshowitz, A., Tominaga, A., Hayashi, E.: Robot navigation in forest management. J. Rob. Mechatron. **30**(2), 223–230 (2018)
3. Shalal, N., Low, T., McCarthy, C., Hancock, N.: Orchard mapping and mobile robot localisation using on-board camera and laser scanner data fusion-part b: mapping and localisation. Comput. Electron, Agric (2015)
4. Choi, K.H., Han, S.K., Han, S.H., Park, K.-H., Kim, K.-S., Kim, S.: Morphology-based guidance line extraction for an autonomous weeding robot in paddy fields. Comput. Electron. Agric. 113 (2015)

5. Lourenço, D., Ferreira, J.F., Portugal, D.: 3D local planning for a forestry UGV based on Terrain gradient and mechanical effort. In: IEEE/RSJ International Conference on Intelligent Robots and Systems, Workshop on Perception, Planning and Mobility in Forestry Robotics (WPPMFR 2020), Las Vegas, NV, USA, Oct 29 (2020)
6. Fankhauser, P., Bloesch, M., Hutter, M.: Probabilistic terrain mapping for mobile robots with uncertain localization. IEEE RAL 3(4) (2018)
7. Yang, S., Yang, S., Yi, X.: An efficient spatial representation for path planning of ground robots in 3D environments. IEEE Access 6 (2018)
8. Hornung, A., et al.: OctoMap: an efficient probabilistic 3D mapping framework based on octrees. Auton. Rob. **34**(3), 189–206 (2013)
9. Krüsi, P., Furgale, P., Bosse, M., Siegwart, R.: Driving on point clouds: motion planning, trajectory optimization, and Terrain assessment in generic nonplanar environments. J. Field Rob. **34**(5), 940–984 (2017)
10. Fankhauser, P., Hutter, M.: A universal grid map library: implementation and use case for rough Terrain Navigation. In: ROS, vol. 1. Springer (2016)
11. Usui, K.: Data augmentation using image-to-image translation for detecting forest strip roads based on deep learning. Int. J. For. Eng. (2020)
12. Tai, L., Li, S., Liu, M.: A deep-network solution towards model-less obstacle avoidance. IEEE IROS **2016**, 2759–2764 (2016)
13. Hart, P.E., Nilsson, N.J., Raphael, B.: A formal basis for the heuristic determination of minimum cost paths. IEEE Trans. Syst. Sci. Cybern. **4**(2), 100–107 (1968)
14. Osmankovic, D., Tahirovic, A., Magnani, G.: All terrain vehicle path planning based on D* lite and MPC based planning paradigm in discrete space. IEEE/ASME International Conference on Advanced Intelligent Mechatronics, pp. 334–339, AIM (2017)
15. Ormahony, N., et al.: Deep Learning for Visual Navigation of Unmanned Ground Vehicles : A review. In: 29th Irish Signals and Systems Conference, ISSC 2018 (2018)
16. Svenson, G., Fjeld, D.: The influence of road characteristics on fuel consumption for logging trucks. In: HVTT12: 12th International Symposium on Heavy Vehicle Transport Technology, vol. 31, no. 5, pp. 526–536 (2012)

Two-step Planning of Dynamic UAV Trajectories Using Iterative δ-Spaces

Sebastian Schräder, Daniel Schleich(✉), and Sven Behnke

Institute for Computer Science VI, Autonomous Intelligent Systems, University of Bonn, Bonn, Germany
schleich@ais.uni-bonn.de

Abstract. UAV trajectory planning is often done in a two-step approach, where a low-dimensional path is refined to a dynamic trajectory. The resulting trajectories are only locally optimal, however. On the other hand, direct planning in higher-dimensional state spaces generates globally optimal solutions but is time-consuming and thus infeasible for time-constrained applications. To address this issue, we propose δ-Spaces, a pruned high-dimensional state space representation for trajectory refinement. It does not only contain the area around a single lower-dimensional path but consists of the union of multiple near-optimal paths. Thus, it is less prone to local minima. Furthermore, we propose an anytime algorithm using δ-Spaces of increasing sizes. We compare our method against state-of-the-art search-based trajectory planning methods and evaluate it in 2D and 3D environments to generate second-order and third-order UAV trajectories.

1 Introduction

In recent years, unmanned aerial vehicles (UAVs) have gained increasing interest for practical usage in many different fields like industrial inspection, agriculture, search & rescue, and delivery. Due to their ability for agile flight and high velocities, UAVs have huge potential for time-critical applications, e.g., in disaster-response scenarios. When flying close to obstacles, the systems' dynamics have to be considered during trajectory planning.

Liu et al. [1] directly incorporate UAV dynamics into a global planner. They use high-dimensional state lattices, from which they extract the optimal trajectory using A* and a heuristic based on the solution of a Linear Quadratic Minimum Time problem. Such approaches become infeasible for large environments, however, since searching high-dimensional state spaces is computationally expensive. Thus, a reduction of the state space size is necessary, either explicitly via pruning or implicitly using more informed search heuristics [2].

Many existing methods reduce the computational load by splitting the trajectory planning problem into two parts. A lower-dimensional geometric path can efficiently be found using search-based [3] or sampling-based [4] planners. After choosing an initial time allocation, this path can be refined to a high-dimensional

I. Petrovic et al. (Eds.): IAS 2022, LNNS 577, pp. 257–271, 2023.
https://doi.org/10.1007/978-3-031-22216-0_18

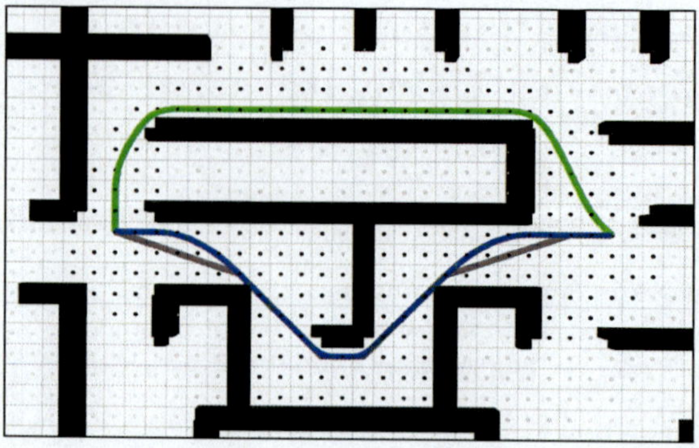

Fig. 1. Refining a 2D path (grey) to a dynamic trajectory. The blue trajectory is obtained by planning in a tunnel around the 2D path. The green trajectory is obtained by planning in the δ-Space (black dots). Although the green trajectory is more distant to the shortest 2D path, it can be tracked at higher velocities and has thus a shorter flight time.

dynamic trajectory [5]. To optimize the smoothness of high-dimensional trajectories, multiple methods haven been proposed, including gradient-based optimizers [6,7], B-splines [8], and neural networks [9]. These approaches only locally optimize the trajectory and are prone to local minima, however. The shortest geometric path towards the goal is not necessarily a good initialization for trajectory optimization, especially when it contains many turns or if we consider non-static initial states. A longer path might suit the UAV dynamics better, allowing higher velocities and thus resulting in shorter flight times.

In this work, we propose a method where the high-dimensional trajectory generation is not guided by a single shortest path but multiple near-optimal paths are considered. For this, we introduce the δ-Space, a set of paths whose length is within a certain sub-optimality bound δ. The motivation behind this definition is visualized in Fig. 1. By allowing larger deviations from the shortest low-dimensional path, we find trajectories that can be tracked at higher velocities and reduce the risk of local minima. Additionally, our method considers the structure of the environment and can efficiently prune dead ends, which is not possible if a tunnel with large radius around a single low-dimensional path is used.

In summary, the main contributions of this paper are:

- the introduction of δ-Spaces to prune high-dimensional state spaces while reducing the risk of local minima,
- an anytime algorithm allowing to iteratively increase the size of the δ-Space,
- and the application to second-order and third-order dynamic trajectory planning for UAVs, including an efficient search heuristic that reuses information from the δ-Space construction.

2 Related Work

Several approaches to reduce the size of high-dimensional state spaces have been proposed. A common method is to restrict the high-dimensional search to the vicinity of the solution of a lower-dimensional problem representation. For example, this can be done by only considering high-dimensional states, whose position components lie within a tunnel around a previously computed geometric path [10]. Such approaches are prone to local minima, however. The high-dimensional search might not even find a solution due to kinodynamic constraints if the tunnel size is too small.

Instead of pruning the state space, the solution of a lower-dimensional problem formulation can be used as a heuristic to guide the high-dimensional search. MacAllister et al. [11] solve a simplified spatial path-planning problem using breadth-first search to guide the planning, but they only consider 4D state lattices. Liu et al. [2] proposed to iteratively compute first-, second- and third-order UAV trajectories, where the lower-dimensional solution is used to guide the next-higher dimensional trajectory. This ensures that a solution can be found since large deviations from the low-dimensional trajectory are discouraged but possible. The resulting trajectories are no longer globally optimal, however. One approach to address the local minima problem of hierarchical methods was introduced by Ding et al. [12], who proposed a global B-spline-based kinodynamic search algorithm.

Another approach is to reduce the state space size globally using multiresolution. Du et al. [13] perform multiple A* searches in parallel, each at a different resolution of the state space. This approach has been extended to an anytime algorithm by Saxena et al. [14]. The idea of multiresolution has also be considered in combination with high-dimensional state lattices. González-Sieira et al. [15] adapt the resolution of a state lattice locally to the environment. This is done by grouping motion primitives that represent similar actions and considering only the longest collision-free primitive from each group. Schleich and Behnke [16] apply the idea of local multiresolution [17] to state lattices for faster replanning of dynamic UAV trajectories.

Besides the resolution, the state dimensionality can also be adapted to accelerate the planning. Gochev et al. [18] represent the state space locally at different dimensionalities. This allows them to only use high-dimensional state representations when necessary, i.e., in narrow areas of the environment, and to plan in a lower-dimensional space otherwise. Gochev et al. [19] extend this method with an incremental version of weighted A* to achieve a better performance. Vemula et al. [20] apply adaptive dimensionality to the problem setting of path planning with dynamic obstacles.

In this work, we propose a state space pruning technique similar to the tunnel method [10] but use multiple lower-dimensional paths to determine which high-dimensional states are pruned. This makes our method less prone to local minima. Additionally, we can iteratively increase the size of our state space efficiently in an anytime fashion to further improve the solution quality.

3 Method

Applying search-based planning methods to high-dimensional state spaces is computationally expensive. Thus, many methods reduce the complexity of such search problems by projecting the high-dimensional robot state onto a lower-dimensional space, where a solution can be found with less effort. This low-dimensional solution can then be used to accelerate high-dimensional planning, either by guiding the search or by pruning the state space. In the context of UAV trajectory planning, the first step usually consists of finding an optimal spatial path without dynamic consideration. Here, optimality is usually defined as minimizing the path length. However, if the shortest path contains many turns, it can only be tracked by the UAV at low velocities. A longer path might suit the UAV dynamics much better and can lead to shorter flight times. Thus, the high-dimensional trajectory generation should not be guided by a single shortest path. Instead, we propose to use the δ-Space, i.e., the union of all paths whose length is within a sub-optimality bound δ. The high-dimensional trajectory search can then be restricted to states whose position components are part of this space. In Sect. 3.1, we give a formal definition of the δ-Space and how it is used to solve high-dimensional search-problems.

The choice of δ significantly influences the performance of the proposed method. A small value results in fast planning times but yields only locally optimal solutions. When choosing high values, the search is less prone to local minima, but planning times are increased. In Sect. 3.2, we propose an efficient method how to iteratively increase δ while reusing the previous search results. Thus, an initial result can be found quickly using a small value of δ and the obtained solution can be subsequently improved in an anytime fashion.

Finally, we point out that δ-Spaces cannot only be used to prune high-dimensional state spaces. When generating the δ-Space, we already explore the structure of the environment. The obtained information can be integrated into a search heuristic to guide the higher-dimensional planning. Since this depends on the specific problem setting, we present an example for such a heuristic in Sect. 3.3, which describes the application of δ-Spaces to dynamic UAV trajectory planning.

3.1 δ-Space Definition

We consider a planning problem with a set of high-dimensional robot states $\mathcal{S}_h$, valid state transitions $\mathcal{E}_h \subset \mathcal{S}_h \times \mathcal{S}_h$, a cost function $c_h : \mathcal{E}_h \to \mathbb{R}_{\geq 0}$, an initial state s, and a goal state g. A trajectory $\mathcal{T}_{h,s\mapsto g}$ from s to g is a sequence $(s_0, \ldots, s_n)$ with $(s_i, s_{i+1}) \in \mathcal{E}_h$, $s_0 = s$, and $s_n = g$. The cost of a trajectory is defined as the sum of the individual state transition costs, i.e.,

$$\mathcal{C}_h(\mathcal{T}_{h,s\mapsto g}) := \sum_{i=0}^{n-1} c_h(s_i, s_{i+1}). \tag{1}$$

We are interested in finding a trajectory $\mathcal{T}^{opt}_{h,s\mapsto g}$ minimizing these costs. In the following, we use the shorthand notation $\mathcal{C}_h(s,g) := \mathcal{C}_h(\mathcal{T}^{opt}_{h,s\mapsto g})$ for the costs of an optimal trajectory from s to g.

To support the planning, we additionally define an abstract, lower-dimensional problem representation, consisting of states $\mathcal{S}_l$, transitions $\mathcal{E}_l \subset \mathcal{S}_l \times \mathcal{S}_l$ and costs $\mathcal{C}_l : \mathcal{E}_l \to \mathbb{R}$. A high-dimensional robot state can be transformed into an abstract one using the projection $\pi : \mathcal{S}_h \to \mathcal{S}_l$.

For a given instance of the planning task, we define the δ-Space $\mathcal{S}^\delta_l$ as

$$\mathcal{S}^\delta_l := \{s_l \in \mathcal{S}_l \mid \mathcal{C}_l(\pi(s), s_l) + \mathcal{C}_l(s_l, \pi(g)) \leq \mathcal{C}_l(\pi(s), \pi(g)) + \delta\} . \tag{2}$$

Note that this definition depends on the specific start and goal states of the current planning instance.

For solving the higher-dimensional planning problem, we restrict the state space to states for which the corresponding abstract state is part of the δ-Space. Thus, we consider the pruned state space

$$\mathcal{S}^\delta_h := \left\{s_h \in \mathcal{S}_h \mid \pi(s_h) \in \mathcal{S}^\delta_l\right\} . \tag{3}$$

Additionally, we define the closure $\overline{\mathcal{S}}^\delta_h$ of this set as the set of all states that can be reached from $\mathcal{S}^\delta_h$, i.e.,

$$\overline{\mathcal{S}}^\delta_h := \left\{s_1 \in \mathcal{S}_h \mid \exists s_0 \in \mathcal{S}^\delta_h : (s_0, s_1) \in \mathcal{E}_h\right\} . \tag{4}$$

Correspondingly, we define the set of possible state transitions

$$\mathcal{E}^\delta_h := \left\{e_h := (s_0, s_1) \in \mathcal{E}_h \mid s_0, s_1 \in \mathcal{S}^\delta_h\right\} \tag{5}$$

and

$$\overline{\mathcal{E}}^\delta_h := \left\{e_h := (s_0, s_1) \in \mathcal{E}_h \mid s_0, s_1 \in \overline{\mathcal{S}}^\delta_h\right\} . \tag{6}$$

To ensure that only states from $\mathcal{S}^\delta_h$ are considered during planning, we adjust the cost function to be

$$c^\delta_h(e_h) = \begin{cases} c_h(e_h), & \text{if } e_h \in \mathcal{E}^\delta_h \\ \infty, & \text{otherwise.} \end{cases} \tag{7}$$

States from the boundary of the state space, i.e., from $\overline{\mathcal{S}}^\delta_h \setminus \mathcal{S}^\delta_h$, are only needed for the anytime planning algorithm in Sect. 3.2 to determine possible states that will be added to the search space when increasing δ.

Although the definition of the δ-Space can be applied to continuous state spaces as well, we restrict ourselves in this work to discretized states. Thus, we can cast planning as graph search. After the δ-Space is constructed, the high-dimensional trajectory can be generated by using graph search algorithms like A* on the state lattice graph $\mathcal{G}^\delta_h(\overline{\mathcal{S}}^\delta_h, \overline{\mathcal{E}}^\delta_h)$ consisting of nodes $\overline{\mathcal{S}}^\delta_h$ and edges $\overline{\mathcal{E}}^\delta_h$.

To construct the δ-Space, we solve the lower-dimensional planning problem using A* search. According to (2), for each state $s_l \in \mathcal{S}_l$, we need to know both

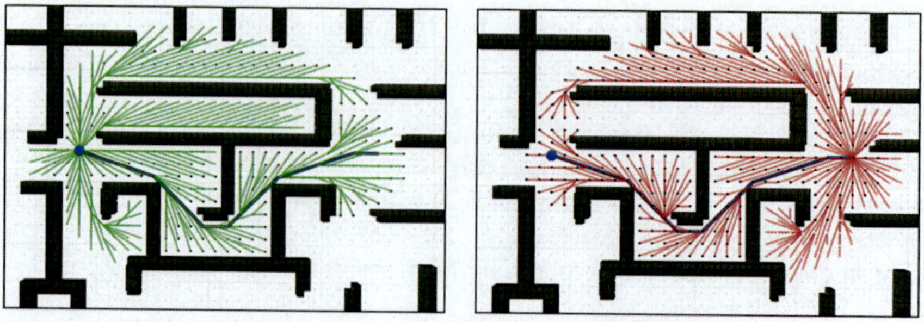

(a) Forward search (b) Backward search

Fig. 2. Lower-dimensional A* searches to calculate the δ-Space. The start state is marked with the blue circle. The nodes forming the δ-Space (black dots) are expanded by both forward and backward search. The blue line is thereby the shortest path.

the cost from the start state to s_l as well as the cost from s_l towards the goal. Thus, we use two searches, one forward search from the start and one backward search starting at the goal, as shown in Fig. 2. These searches are independent of each other and can be parallelized for a better performance. If a state s_l has been expanded by both searches, we can easily decide whether it belongs to the δ-Space by checking the criterion (2). To find all states of the δ-Space, we slightly adjust the A* algorithm: The f-value of a state s' is defined as the sum of the cost from start to s' and an heuristic value $h(s')$ estimating the remaining cost-to-go. If the heuristic function h used for the A* search is admissible, $f(s')$ represents a lower threshold on the costs of an optimal trajectory containing s'. Thus, after finding the optimal trajectory $\mathcal{T}^{opt}_{l,\pi(s)\mapsto\pi(g)}$, we extract its costs and continue to expand states until the next state to be expanded has an f-value exceeding $\mathcal{C}_l(\mathcal{T}^{opt}_{l,s\mapsto g}) + \delta$. For an admissible heuristic, we thus can ensure that all states of the δ-Space are expanded by the forward and backward searches.

3.2 Iterative δ-Spaces

In general, a small δ results in low planning times. However, larger values of δ are necessary to avoid local minima. In this section, we describe an anytime planning algorithm for the case where a planning time limit is set, e.g., when replanning at a fixed frequency. The anytime planner starts with a small δ to quickly generate a first solution and afterwards iteratively increases δ to further improve the solution until the planning time limit is reached.

As described above, the states within the δ-Space are found by executing forward and backward A* searches until f-values are reached that exceed $\mathcal{C}_l(\mathcal{T}^{opt}_{l,s\mapsto g}) + \delta$. For higher values of δ, the same searches are executed, but up to higher f-values. Since the f-values increase monotonically during A* searches, we can pause and later resume the searches for planning in δ-Spaces of increasing sizes.

Algorithm 1: Anytime Planning In Iterative δ-Spaces

Input: Start state s, goal state g, δ_0, step size $\Delta\delta$
Output: Trajectory $\mathcal{T}^{opt}_{\delta,s\mapsto g}$

1 $\mathcal{S}_l^{\delta_0} \leftarrow$ GenerateDeltaSpace($\pi(s), \pi(g), \delta_0$)
`/* get optimal Trajectory and final search state */`
2 $\mathcal{T}^{opt}_{\delta,s\mapsto g}, (\mathcal{O}_0, \mathcal{O}'_0, \mathcal{F}_0, \mathcal{G}_0) \leftarrow$ HighDimSearch($s, g, \mathcal{G}_h^{\delta_0}$)
3 $i \leftarrow 0$
4 **while** *There is time to improve the trajectory* **do**
5 $\quad \delta_{i+1} \leftarrow \delta_i + \Delta\delta$
6 $\quad \mathcal{S}_l^{\delta_{i+1}} \leftarrow$ UpdateDeltaSpace($\mathcal{S}_l^{\delta_i}, \delta_{i+1}$)
7 $\quad$ Update $\mathcal{F}_i, \mathcal{G}_i$ for all nodes in $\mathcal{O}'_i \cap \mathcal{S}_h^{\delta_{i+1}}$
8 $\quad \mathcal{O}_i \leftarrow \mathcal{O}_i \cup (\mathcal{O}'_i \cap \mathcal{S}_h^{\delta_{i+1}})$
9 $\quad \mathcal{T}^{opt}_{\delta,s\mapsto g}, (\mathcal{O}_{i+1}, \mathcal{O}'_{i+1}, \mathcal{F}_{i+1}, \mathcal{G}_{i+1}) \leftarrow$ HighDimSearch($s, g, \mathcal{G}_h^{\delta_{i+1}}, (\mathcal{O}_i, \mathcal{F}_i, \mathcal{G}_i)$)
10 $\quad i \leftarrow i + 1$
11 **return** $\mathcal{T}^{opt}_{\delta,s\mapsto g}$

The complete anytime planning algorithm is given in Algorithm 1. First, we calculate the δ-Space for the initial size δ_0 (Line 1) and plan a high-dimensional trajectory by searching the pruned state-lattice graph $\mathcal{G}_h^{\delta_0}$ using A* (Line 2). Here, we save the final state of the A* search, including the open list $\mathcal{O}_0$, as well as the set of f-values $\mathcal{F}_0$ and g-values $\mathcal{G}_0$. Additionally, we keep track of all states where the search hits the boundary of the δ-Space, i.e., all states within $\overline{\mathcal{S}}_h^{\delta_0} \setminus \mathcal{S}_h^{\delta_0}$ that are neighbours of expanded states. Since these states can only be reached via edges with infinite costs, they are not considered as valid states in the current search. However, they might become valid once δ is increased. Thus, instead of adding them to the open list, we add them to a separate list $\mathcal{O}'_0$ for later use.

If the planning time limit is not met yet, we can iteratively plan in larger δ-Spaces to further improve the trajectory (Line 4). First, we increase the size of the δ-Space by resuming the low-dimensional searches with increased δ (Line 5 and 6) as described above. Afterwards, we continue the high-dimensional search on the extended state space. Here, we first update the cost function for all states from $\mathcal{O}'_i$ that have been added to the larger δ-Space $\mathcal{S}_h^{\delta_{i+1}}$ (Line 7) and add them to the open list of the previous search (Line 8). Then, we execute the high-dimensional search on the extended search space (Line 9), starting with the already initialized open list, f-values and g-values from the previous search. Note, that we do not include the closed list. If the optimal trajectory contains states that were not part of the previous state space $\mathcal{S}_h^{\delta_i}$, the f-values of previously expanded states might be suboptimal. Thus, we need to allow one re-expansion per state in each iteration.

3.3 Application to UAV Trajectory Planning

We apply δ-Spaces to plan second-order and third-order UAV trajectories. Thus, we model the UAV state as 6-tuples $s = (\mathbf{p}, \mathbf{v}) \in \mathbb{R}^6$ or 9-tuples $s = (\mathbf{p}, \mathbf{v}, \mathbf{a}) \in \mathbb{R}^9$, consisting of a discretized 3D position $\mathbf{p}$, velocity $\mathbf{v}$, and acceleration $\mathbf{a}$. To compare against the State of the Art, we combine δ-Spaces with the search-based trajectory generation methods from [2,16]. In both works, the set of state transitions $\mathcal{E}_h$ consists of motion primitives $e_{\mathbf{u},\tau}$ which are generated by applying constant acceleration or jerk commands $\mathbf{u}$ over a short time interval τ. The corresponding costs are defined as the weighted sum of control effort and primitive duration:

$$c_h(e_{\mathbf{u},\tau}) = ||\mathbf{u}||_2^2\tau + \rho\tau. \tag{8}$$

For the low-dimensional problem representation, we project a high-dimensional state $s = (\mathbf{p}, \mathbf{v}, \mathbf{a})$ onto its position components $\mathbf{p}$ and use a simple 3D grid with a 26-connected neighbourhood as low-dimensional planning space.

When using search-based methods like A*, the choice of a good heuristic function is crucial for the planning performance. In this work, we propose a heuristic that efficiently uses information obtained by solving the lower-dimensional planning problem.

From the construction of the δ-Space, we directly obtain the distance from each state towards the goal along the shortest lower-dimensional path. We assume that the distance towards the goal is travelled on a straight line and the UAV reaches a zero velocity at the goal. Thus, we can assume that the UAV accelerates to the maximal reachable velocity, keeps this velocity for a certain time and then decelerates to stop at the goal. In the following, we describe how this can be incorporated into a search heuristic for second-order trajectories. The extension to third-order trajectories is left for future work.

For a pair of discretized 1D velocities v_1, v_2, let the minimum time needed to accelerate the UAV from v_1 to v_2 be t_{v_1,v_2}. The corresponding control effort is denoted by c_{v_1,v_2} and the distance the UAV moves during the acceleration by d_{v_1,v_2}. These values can efficiently be precomputed. During search, we calculate the maximal velocity that can be reached from an initial velocity v when travelling a distance of d as $v_{max} := \max\{v' | d_{v,v'} + d_{v',0} \leq d\}$. Then, we can estimate the flight time

$$T = \frac{d - d_{v,v_{max}} - d_{v_{max},0}}{v_{max}} + t_{v,v_{max}} + t_{v_{max},0} \tag{9}$$

and the corresponding control cost

$$c = c_{v,v_{max}} + c_{v_{max},0}. \tag{10}$$

For the initial velocity v, we choose the maximum of the absolute velocities along the independent axes. Note that this heuristic is not admissible since it overestimates the cost for diagonal movements. As our experiments show, this slightly increases the trajectory costs compared to admissible heuristics but it results in lower costs than the inadmissible heuristic used by Liu et al. [2].

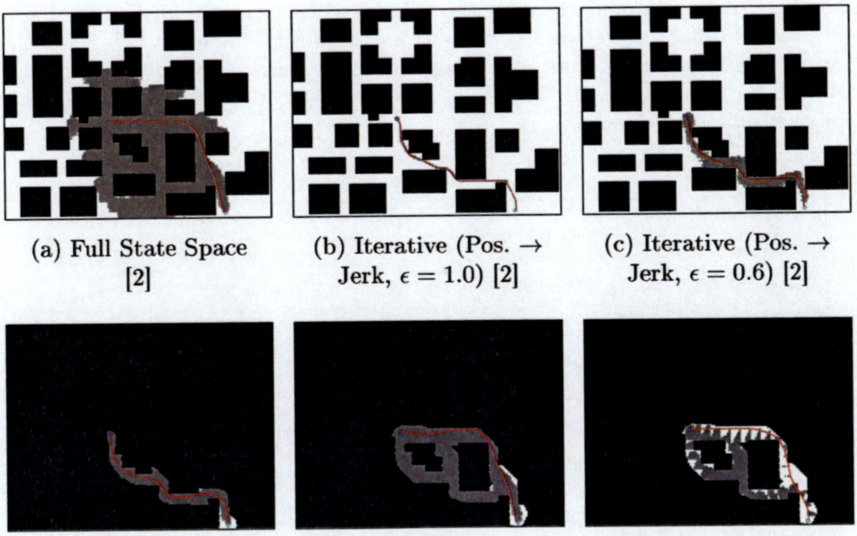

(a) Full State Space [2]

(b) Iterative (Pos. → Jerk, $\epsilon = 1.0$) [2]

(c) Iterative (Pos. → Jerk, $\epsilon = 0.6$) [2]

(d) Tunnel

(e) δ-Space

(f) δ-Space Heuristic

Fig. 3. Examples for generating third-order 2D trajectories with the method of Liu et al. [2]. The start is marked by the blue circle. Obstacles and pruned states are black. Expanded states are grey. (a) Searching the full state space finds the optimal trajectory but is computationally expensive. (b) The iterative heuristic with a high greediness parameter ϵ has low planning times but results in a slow trajectory which tightly follows the guiding path. (c) A lower greediness generates a faster trajectory at the cost of higher planning times, but the obtained solution is still close to the 2D path and thus only locally optimal. (d) The tunnel also results in a suboptimal trajectory. (e) δ-Spaces find the optimal trajectory while still having low planning times. (f) The additional δ-Space heuristic further improves the planning time.

4 Evaluation

In a first experiment, we use δ-Spaces on top of the method of Liu et al. [2] to plan third-order trajectories in a cluttered 2D environment of size 80×62 m. We use the source code[1] from Liu et al. [2] with the following parameters:

All other parameters were not changed and are the default ones from the implementation [2]. To reduce planning times, Liu et al. propose to generate a lower-order trajectory first, which can then be used as a heuristic to guide the higher-dimensional search. In this experiment, we choose a 2D path as guiding trajectory for a fair comparison against the δ-Space. The resulting trajectories for one example task are shown in Fig. 3 and the average planning performance over a set of 1000 tasks is summarized in Table 1. The iterative heuristic of Liu et al. uses a greediness parameter ϵ to control the deviation of the high-dimensional trajectory from the guiding 2D path. With a high parameter, the

[1] https://github.com/sikang/motion_primitive_library.

Table 1. Planning statistics for using the δ-Space with the method from Liu et al. [2] to generate third-order trajectories in a 2D environment. Planning time, number of expansions and trajectory costs are averaged over the tasks where all methods found a solution. Success denotes the fraction of tasks for which a solution was found within the allowed planning time.

	Success	Time	Expansions	Costs
Full State Space [2]	74.0%	214 ms	12 291	121.92
Iterative (Pos. $\rightarrow$ Jerk, $\epsilon = 0.6$) [2]	89.7%	334 ms	6 125	180.33
Iterative (Pos. $\rightarrow$ Jerk, $\epsilon = 0.7$) [2]	96.5%	233 ms	4 231	204.49
Iterative (Pos. $\rightarrow$ Jerk, $\epsilon = 0.8$) [2]	96.3%	97 ms	1 608	223.44
Iterative (Pos. $\rightarrow$ Jerk, $\epsilon = 0.9$) [2]	95.7%	76 ms	1 128	247.48
Iterative (Pos. $\rightarrow$ Jerk, $\epsilon = 1.0$) [2]	95.2%	**28 ms**	**260**	267.87
Tunnel ($r = 1.0$ m)	94.6%	50 ms	3 153	127.25
Tunnel ($r = 1.5$ m)	95.4%	58 ms	3 586	124.12
Tunnel ($r = 2.5$ m)	95.4%	67 ms	4 070	123.27
Tunnel ($r = 3.5$ m)	94.6%	72 ms	4 328	123.12
δ-Space ($\delta = 1.0$ m)	95.9%	115 ms	6 039	122.53
δ-Space ($\delta = 1.5$ m)	91.7%	139 ms	7 175	122.46
δ-Space ($\delta = 2.5$ m)	84.3%	182 ms	9 014	122.37
δ-Space Heuristic ($\delta = 1.0$ m)	**97.6%**	53 ms	1 980	117.79
δ-Space Heuristic ($\delta = 1.5$ m)	96.1%	62 ms	2 231	117.71
δ-Space Heuristic ($\delta = 2.5$ m)	**97.6%**	75 ms	2 484	**117.53**

ρ	τ	$v_{\max}$	$a_{\max}$	$u_{\max}$	du	Timeout
10	1.0 s	3 m/s	1 m/s^2	1 m/s^2	0.5 m/s^2	1 s

trajectory tightly follows the shortest path (Fig. 3b). This results in planning times that are significantly faster than for all other methods. However, these trajectories can only be traversed at low velocities and thus, their average costs are more than twice the global optimum. The costs can be reduced by lowering the greediness parameter, which results in higher planning times. Note that with a low greediniess of $\epsilon = 0.6$, the method of Liu et al. has significantly higher planning times than the δ-Space method, while still generating trajectories with higher costs. Figure 3c shows an example trajectory, which is only locally optimal since it still tightly follows the shortest 2D path. Similar trajectories but higher velocities are obtained when restricting the search to a tunnel around the 2D path (Fig. 3d). Our method (Fig. 3e) allows larger deviations from the shortest 2D path and generates a trajectory close to the optimal one. This result is obtained by using the same (inadmissible) heuristic that Liu et al. use when planning in the full state space (Fig. 3a). From the construction of the δ-Space we already know the 2D distances from all states to the goal. We can use these as a heuristic if we assume that the UAV flies at maximum velocity. This further reduces the planning times of our method and results in the lowest trajectory costs (Fig. 3f).

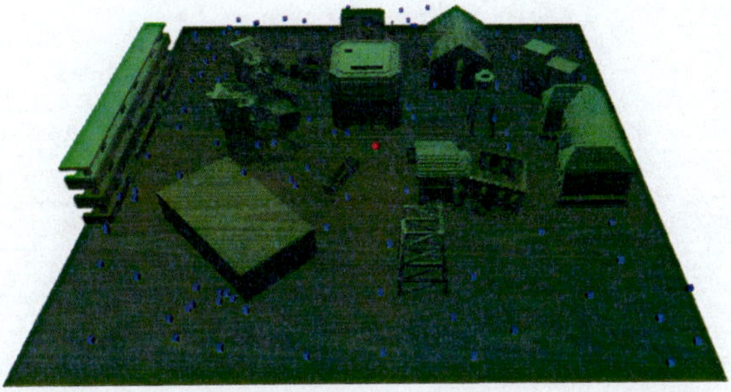

Fig. 4. Simulated 3D outdoor environment. The UAV starts at the map center (red circle) and plans trajectories to all goal positions (blue squares).

In a second experiment, we evaluate our method in a simulated outdoor environment (Fig. 4). Here, we use the method from [16] with the same parameters as in the original paper:

Note that we abort searches, if no solution within the first 1 000 000 expansions is found. Although we compare against multiresolutional state lattices, the δ-Space is applied on top of a uniform state lattice, since the δ-Space already reduces the size of the state space. The results are reported in Table 2. When using standard A* search, the δ-Space outperforms uniform planning in the full state space as well as the multiresolutional planning method from [16] with respect to planning time and success rates, while achieving trajectory costs that are close to the optimal ones. Using the δ-Space heuristic reduces the planning time even further. However, since the heuristic is not admissible, this comes at the cost of significantly increased path costs.

Note that, compared to uniform planning, multiresolution increases the success rates but has a slightly higher average planning time. This has already been reported in [16]. For some planning tasks, the suboptimal trajectory costs introduced by multiresolution require more expansions during A* search. To address this issue, a level-based expansion scheme was introduced in [16]. This reduces the planning times at the cost of more suboptimal trajectories. Here, multiresolution is significantly faster than δ-Space planning. This is due to the overhead of the low-dimensional search needed to generate the δ-Space, which took on average 0.266 s. The high-dimensional search itself is actually faster than multiresolutional planning. Overall, in open-spaced environments, multiresolution can be faster than δ-Space planning but results in higher trajectory costs. However, δ-Spaces can also be used in more complex environments (like the one in Fig. 3), where multiresolution is not applicable.

Table 2. Planning statistics for using the δ-Space with the method from [16] to generate second-order trajectories in a 3D environment. We report results for different expansion schemes during search: standard A* search and a level-based expansion scheme (Level-A*) proposed in [16], which prioritize expansions of states on higher resolution levels. Planning time, number of expansions and trajectory costs are averaged over the tasks where all methods found a solution. Success denotes the fraction of tasks for which a solution was found within the allowed planning time.

	Success	Time	Expansions	Costs
Uniform (A*)	91.8%	0.923 s	79 883	**251.95**
Multiresolution (A*) [16]	95.9%	1.096 s	81 700	259.55
Tunnel (A*)	**100.0%**	0.603 s	34 994	259.18
δ-Space (A*)	98.97%	0.834 s	55 138	253.05
δ-Space with Heuristic (A*)	**100.0%**	**0.324 s**	**4695**	264.64
Uniform (Level-A*)	97.9%	0.108 s	9099	262.82
Multiresolution (Level-A*) [16]	**100.0%**	**0.076 s**	4491	277.34
Tunnel (Level-A*)	**100.0%**	0.287 s	6450	268.91
δ-Space (Level-A*)	**100.0%**	0.324 s	6056	**262.14**
δ-Space with Heuristic (Level-A*)	**100.0%**	0.276 s	**479**	288.45

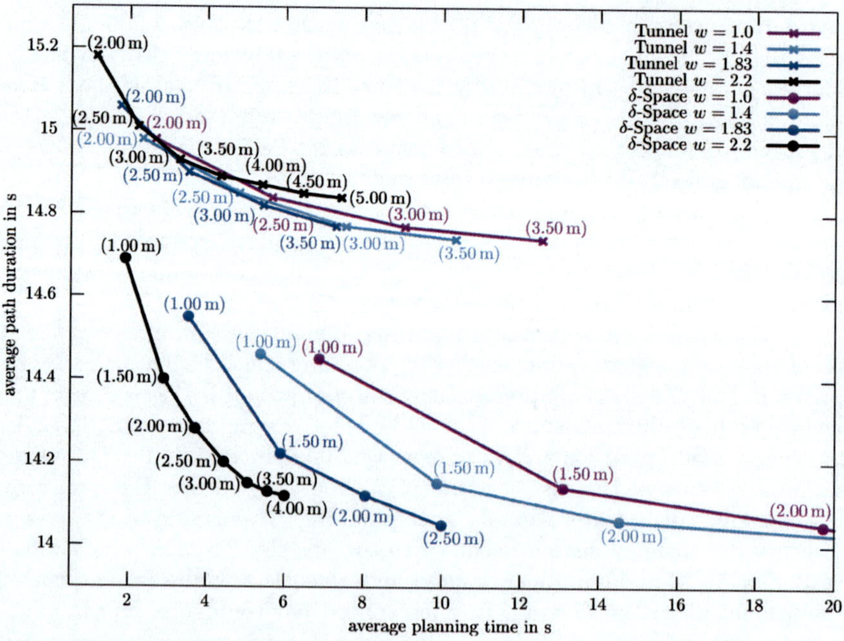

Fig. 5. The impact of weighted heuristics and anytime planning. A line represents an iterative search as described in Sect. 3.2.

ρ	τ	$v_{\max}$	$u_{\max}$	du	Timeout
16	0.5 s	4 m/s	2 m/s^2	2 m/s^2	1 000 000 expansions

Finally, we evaluate the anytime version of our planning algorithm. It iteratively increases the size of the δ-Space but can also be applied to plan in tunnels of increasing radius around the low-dimensional path. In this experiment, we investigate which state space representation is more suitable for the anytime planning algorithm. Furthermore, we analyse the influence of different heuristic weights that might speed up planning at the cost of sub-optimal solutions. Since the anytime algorithm requires significant adjustments of the source code of the high-dimensional planning algorithms, we implemented our own planning method for this experiment. Our planner generates third-order trajectories on a 3D map. For this, it uses motion primitives that were precomputed by the time-optimal trajectory generation method from [21]. Figure 5 shows the results for both the tunnel and the δ-Space with different heuristic weights. The colors of the graphs represents the used weight and a line represents an iterative search as described in Sect. 3.2. Here, it can be clearly seen that tunnel and δ-Space behave differently. The heuristic weight w does impact the tunnel only slightly. Increasing the tunnel radius does not improve the trajectory costs much and the costs start to stagnate after few iterations. On the other hand, the heuristic weight strongly influences the performance of the δ-Spaces. A larger weight significantly decreases the planning times. Although the trajectory costs are increased, this can be made up for by iteratively increasing the size of the δ-Space. Thus, using the δ-Space with a large heuristic weight generates trajectories with similar planning times as the tunnel method but significantly lower costs.

Table 3. Planning statistics for the anytime algorithm to generate third-order trajectories in a 3D environment.

		iterative planning time	accumulated iterative planning time	direct planning time	trajectory cost (iterative)	trajectory cost (direct)
δ-Space	$\delta = 1.0$ m	+3.5s	3.5s	3.5s	14.55s	14.55s
	$\delta = 1.5$ m	+2.4s	5.9s	4.7s	14.22s	14.25s
	$\delta = 2.0$ m	+2.2s	8.1s	6.3s	14.12s	14.17s
	$\delta = 2.5$ m	+1.9s	10.0s	7.7s	14.05s	14.09s
Tunnel	$r = 1.0$ m	+0.6s	0.6s	0.6s	16.01s	16.01s
	$r = 1.5$ m	+0.9s	1.5s	1.2s	15.27s	15.25s
	$r = 2.0$ m	+1.3s	2.8s	1.8s	15.03s	15.06s
	$r = 2.5$ m	+1.8s	4.6s	2.8s	14.90s	14.94s

Table 3 summarizes the planning performance for each iteration of the anytime algorithms using a heuristic weight $w = 1.83$. As expected, the accumulated planning time for iterative planning is higher than the time for directly planning in the corresponding state space. However, the relative increase is smaller for the δ-Space, compared to the Tunnel. This shows that δ-Spaces are more suitable for anytime planning. Interestingly, the trajectory costs are slightly lower for anytime planning. Since each iteration allows additional re-expansions, iteratively increasing δ can help to correct trajectories that are suboptimal due to the inadmissible heuristic.

5 Conclusion

In this paper, we propose a new method for pruning of high-dimensional state spaces. Instead of restricting the search to the vicinity of an optimal lower-dimensional path, we introduce the δ-Space, which represents the union of multiple paths within a suboptimality bound δ. Our experiments show that this significantly reduces planning times. In contrast to other state space reduction methods, δ-Spaces are less prone to local minima and can find trajectories close to the globally optimal solution. Furthermore, we introduced an anytime planning algorithm that efficiently plans in δ-Spaces of increasing sizes. Thus, a first initial solution can be found quickly, while subsequent iterations further improve the solution quality.

Acknowledgements. This work has been funded by the German Federal Ministry of Education and Research (BMBF) in the project "Kompetenzzentrum: Aufbau des Deutschen Rettungsrobotik-Zentrums (A-DRZ)".

References

1. Liu, S., Atanasov, N., Mohta, K., Kumar, V.: Search-based motion planning for quadrotors using linear quadratic minimum time control. In: Proceedings of IEEE/RSJ International Conference on Intelligent Robots and Systems (IROS) (2017)
2. Liu, S., Mohta, K., Atanasov, N., Kumar, V.: Search-based motion planning for aggressive flight in SE(3). In: Proceedings of IEEE International Conference on Robotics and Automation (ICRA) (2018)
3. E. Hart, P., J. Nilsson, N., Raphael, B.: A formal basis for the heuristic determination of minimum cost paths. Intell. Sigart Bull. SIGART **37**, 28–29 (1972)
4. LaValle, S.M.: Rapidly-exploring random trees: a new tool for path planning. Iowa State University, Computer Science Department (1998)
5. Nieuwenhuisen, M., Behnke, S.: Local multiresolution trajectory optimization for micro aerial vehicles employing continuous curvature transitions. In: Proceedings of IEEE/RSJ International Conference on Intelligent Robots and Systems (IROS) (2016)
6. Kalakrishnan, M., Chitta, S., Theodorou, E., Pastor, P., Schaal, S.: STOMP: Stochastic trajectory optimization for motion planning. In: Proceedings of IEEE International Conference on Robotics and Automation (ICRA) (2011)

7. Zucker, M., Ratliff, N., Dragan, A.D., Pivtoraiko, M., Klingensmith, M., Dellin, C.M., Bagnell, J.A., Srinivasa, S.S.: CHOMP: covariant Hamiltonian optimization for motion planning. Int. J. Rob. Res. **32**(9–10), 1164–1193 (2013)
8. Arney, T.: Dynamic path planning and execution using B-splines. In: 3rd International Conference on Information and Automation for Sustainability (2007)
9. Simon, D.: The application of neural networks to optimal robot trajectory planning. Rob. Auton. Syst. **11**(1), 23–34 (1993)
10. Stachniss, C., Burgard, W.: An integrated approach to goal-directed obstacle avoidance under dynamic constraints for dynamic environments. In: Proceedings of IEEE/RSJ International Conference on Intelligent Robots and Systems (IROS) (2002)
11. MacAllister, B., Butzke, J., Kushleyev, A., Pandey, H., Likhachev, M.: Path planning for non-circular micro aerial vehicles in constrained environments. In: Proceedings of IEEE International Conference on Robotics and Automation (ICRA) (2013)
12. Ding, W., Gao, W., Wang, K., Shen, S.: An efficient B-spline-based kinodynamic replanning framework for quadrotors. IEEE Trans Rob. **35**(6), 1287–1306 (2019)
13. Du, W., Islam, F., Likhachev, M.: Multi-resolution A*. In: 13th Annual Symposium on Combinatorial Search (2020)
14. Saxena, D.M., Kusnur, T., Likhachev, M.: AMRA*: Anytime multi-resolution multi-heuristic A*. arXiv preprint arXiv:2110.05328 (2021)
15. GonzÃlez-Sieira, A., Mucientes, M., BugarÃn, A.: Graduated fidelity lattices for motion planning under uncertainty. In: Proceedings of IEEE International Conference on Robotics and Automation (ICRA) (2019)
16. Schleich, D., Behnke, S.: Search-based planning of dynamic MAV trajectories using local multiresolution state lattices. In: Proceedings of IEEE International Conference on Robotics and Automation (ICRA) (2021)
17. Behnke, S.: Local multiresolution path planning. In: RoboCup 2003: Robot Soccer World Cup VII, pp. 332–343, Springer (2003)
18. Gochev, K., Cohen, B., Butzke, J., Safonova, A., Likhachev, M.: Path planning with adaptive dimensionality. In: 4th Annual Symposium on Combinatorial Search (2011)
19. Gochev, K., Safonova, A., Likhachev, M.: Incremental planning with adaptive dimensionality. In: Proceedings of the International Conference on Automated Planning and Scheduling (ICAPS) (2013)
20. Vemula, A., Muelling, K., Oh, J.: Path planning in dynamic environments with adaptive dimensionality. In: 9th Annual Symposium on Combinatorial Search (2016)
21. Beul, M., Behnke, S.: Fast full state trajectory generation for multirotors. In: Proceedings of International Conference on Unmanned Aircraft Systems (ICUAS), IEEE (2017)

End-to-End Path Estimation and Automatic Dataset Generation for Robot Navigation in Plant-Rich Environments

Yoshinobu Uzawa(✉), Shigemichi Matsuzaki, Hiroaki Masuzawa, and Jun Miura

Toyohashi University of Technology, Toyohashi, Aichi, Japan
y_uzawa@aisl.cs.tut.ac.jp

Abstract. This paper describes a method of estimating a path to follow from an image in plant-rich environments such as greenhouses and unstructured outdoor scenes. In such environments, there are several factors that make it difficult for robots to determine a path, such as plants covering the path and ambiguous path boundaries. Approaches based on segmentation of traversable regions cannot be applied to such environments because the regions may not be clearly defined or may be occluded. In this work, we propose a method of estimating a path from a single image in the end-to-end fashion. We also develop an automatic annotation method utilizing the robot's trajectory during the data acquisition phase. We conducted a real-world experiment of robot navigation and confirmed that the proposed method is capable of navigation in paths partially covered by plants. We also confirmed that proposed data annotation method can generate training data more efficiently than manual annotation.

Keywords: Mobile robot navigation · Convolutional neural network · End-to-end learning

1 Introduction

Autonomous robot navigation has been studied in various fields such as autonomous driving [1] and agriculture [2]. In autonomous navigation, a robot has to recognize a traversable region and determine a path to follow. Existing approaches include edge-based detection of lanes [3] and road marking [4,5], and segmentation of a traversable region [6,7].

In unstructured outdoor scenes with plants, however, such methods are not effective due to occlusion of traversable regions by plant parts. Figure 1 shows scenes where the traversable regions are covered by plant parts and not fully visible. Although it is difficult to estimate a traversable region in such scenes, it is relatively easy for humans to determine a path to follow. We suppose this is

I. Petrovic et al. (Eds.): IAS 2022, LNNS 577, pp. 272–284, 2023.
https://doi.org/10.1007/978-3-031-22216-0_19

(a) Forest pathway

(b) Greenhouse

Fig. 1. Scenes where traversable regions are not fully visible.

because we determine the direction to move based on appearance and location of plants and the ground without explicitly recognizing traversable regions. It is, therefore, reasonable for the robot to directly determine a path as humans would do in unstructured environments.

In this paper, we propose a method of estimating a path to follow directly from an image. We exploit a deep neural network (DNN) for estimating image coordinates of the start and the end point of the path in the image in the end-to-end fashion. In addition, to mitigate the burden of manual annotation on the training data, we also propose an automatic data annotation method that exploits the robot's trajectory during the data acquisition phase.

Our contributions are three-fold:

- An end-to-end path estimation method that is applicable to environments where traversable regions are not fully visible due to plants.
- An automatic data annotation method for efficient generation of training data.
- Experimental validation of the methods in a real-world robot navigation.

2 Related Work

2.1 Traversable Region and Path Detection

A major approach to path detection for autonomous robots and vehicles is to first identify a traversable region and then determine a path to follow. In urban scenes, lane detection based on edge features and road markings are actively studied. Matsushita et al. [8] proposed a method for detecting road boundary features via sensor fusion of a camera and a laser range finder (LRF). Chiku et al. [9] exploited a stereo camera for road boundary estimation. Wang et al. [4] applied an instance segmentation approach to segmenting the lines on the road. Lee et al. [5] proposed a method for segmentation of the lines and predicting a path. Another approach is semantic segmentation. Onozuka et al. [6] proposed a weakly-supervised segmentation of road regions recommended for traversal.

In unstructured scenes, Opiyo et al. [7] proposed a method to compute a path using texture information of the path region on an image in orchards. Phung et

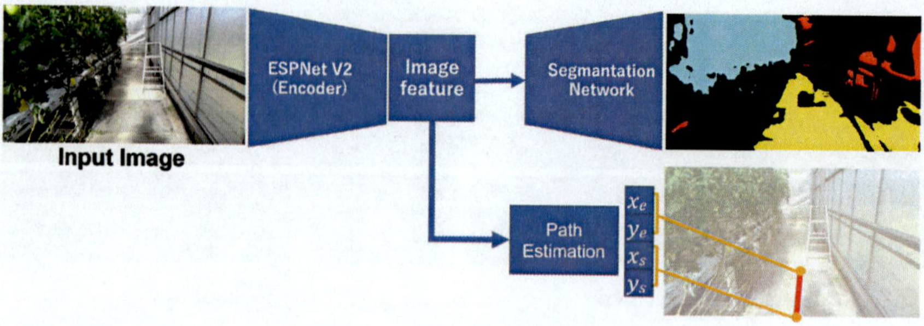

Fig. 2. Network architecture.

al. [10] proposed a method to detect a pedestrian lane automatically from the image vanishing point.

These methods assume that the traversable region is visible. Such an assumption does not always hold in unstructured scenes due to presence of occlusion by plants. In contrast, the proposed end-to-end path estimation is applicable to those scenes because we estimate the path direction directly from the image.

2.2 Automatic Data Annotation

A preferred approach to automatic acquisition of labeled data for model training is to utilize information of the robot's footprints. Meyer et al. [11] proposed a method of labeling image regions where a vehicle has traversed. Onozuka et al. [6] also adopted a similar approach to automatic labeling. Wellhausen et al. [12] proposed to utilize footprints of a four-legged robot to label traversable regions in images. Matsuzaki et al. [13] proposed an automatic labeling of image regions traversed by a robot. Giusti et al. [14] trained an image-based DNN classifier to decide the control of UAV. To this end, they proposed an automatic data annotation method utilizing three head-mounted cameras.

Since our task is estimation of a path in an image, the required labels are different from the aforementioned methods. This path estimation task allows a simple implementation of automatic annotation by directly utilizing a robot's trajectory generated by a Simultaneous Localization and Mapping (SLAM) method.

3 Overview of the Proposed Method

In this work, we adopt a multi-task deep neural network of semantic segmentation and path estimation (see Fig. 2). A base network is ESPNet v2 [15], an efficient semantic segmentation model, with some modifications described in [16]. A path estimation module is attached to the semantic segmentation network. Here, a path is defined as a straight line that a robot should follow. To estimate a path, the path estimation module takes a feature map from the encoder of the segmentation network and predicts four values, i.e., the x and the y coordinate value of the start and the end point of the path.

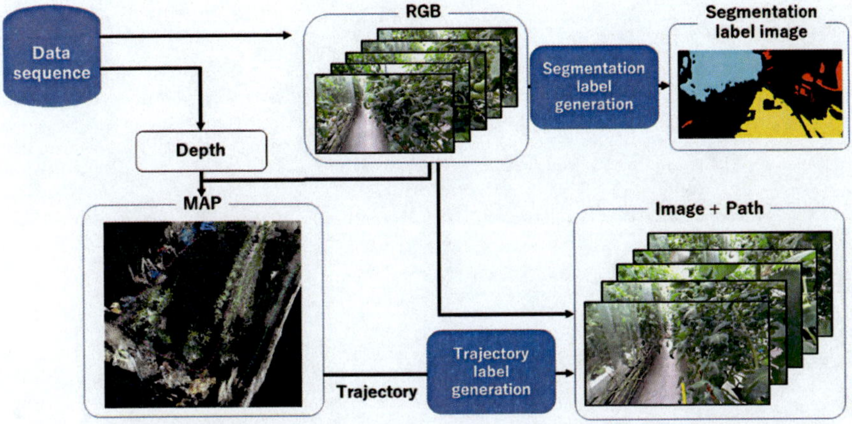

Fig. 3. Overview of the proposed method.

For training the path estimation module, we automatically generate training data from a sequence of RGB-D data acquired in the target environment by manually controlling the a robot in the target environment. We then reconstruct the robot's trajectory via SLAM and automatically label images by projecting the trajectory on the images. We describe the method in detail in the next sections.

4 Automatic Data Generation by Trajectory Estimation

To mitigate the burden of labeling on training images, we propose a method to automatically generate labeled images utilizing the robot's trajectory during data recording. The process of the data generation is shown in Fig. 3. First, we manually control the robot in the target environment while taking RGB-D data. Next, the robot's trajectory is estimated by RTAB-Map [17], an RGB-D SLAM method. The trajectory is represented by a sequence of the observation points where RGB-D data were acquired, which are hereafter referred to as "trajectory points." Labels of the path on the images are generated by projecting the robot trajectory to the image taken at each observation point.

The projection of a trajectory point in the world frame to the image is given by:

$$s \begin{pmatrix} x_i \\ y_i \\ 1 \end{pmatrix} = \mathbf{K} {}^c_r\mathbf{T}^{-1} {}^r_w\mathbf{T}^{-1} \begin{pmatrix} X_w \\ Y_w \\ Z_w \\ 1 \end{pmatrix}, \tag{1}$$

where (x_i, y_i) is a coordinate in the image frame and (X_w, Y_w, Z_w) is a coordinate of the trajectory point in the world frame. $\mathbf{K}$ is a camera matrix, ${}^c_r\mathbf{T}$ and ${}^r_w\mathbf{T} \in \mathbb{SE}(3)$ are transformation matrices representing the transformations from the

Fig. 4. A robot's trajectories projected on the images.

robot frame to the camera frame and from the world frame to the robot frame, respectively. We assume $\mathbf{K}$ and ${}^{c}_{r}\mathbf{T}$ to be calculated beforehand and fixed. ${}^{r}_{w}\mathbf{T}$ is given as the transformation from the world frame to the trajectory point estimated by SLAM.

Examples of the robot trajectories projected on images are shown in Fig. 4.

Using the projected trajectories, we compute the start and the end point of the path used as training labels. Figure 5 shows the definition of the points. Here we define the start point as a point at the bottom of an image closest to the robot on the path within the image. The end point is defined as a point at the 20% of the image height from the bottom, representing a point about 5–7 [m] ahead of the robot. The x coordinate of the start point is defined as the intersection of a horizontal line at the bottom of the image and a line drawn between the two consecutive projected points that are placed above and below the horizontal line. Note that the point below the horizontal line is out of the image field of view, but can be calculated by eq. (1). The x coordinate of the end point is computed in the same way as the start point at 20 % of the image height from the bottom. Examples of the computed paths are shown in Fig. 6. To limit the range of the values to be estimated by DNN independently from the image size, we normalize them by the width and the height of the image, respectively.

For data augmentation, we apply image cropping. Figure 7 shows the overview of the cropping. The images are cropped with a window with the size of 80% of the height and width of the original images. The window is shifted horizontally and vertically by a length of 10% of the width and height, respectively. If either of the start or the end point is out of the image region after cropping, the cropped image is not included in the dataset.

Fig. 5. The start and the end point.

Fig. 6. Computed paths.

5 Path Estimation

As described in Sect. 3, a path is estimated using a multi-task DNN. We assume a training set consisting of N images $\{I_i\}_{i=1}^{N}$ and corresponding ground truth normalized coordinate values of the start and the end point $\{\{p_i^j\}_{j\in\{x_s,y_s,x_e,y_e\}}\}_{i=1}^{N}$, where j denotes the ID of the predicted values with x_s, y_s, x_e, and y_e being the x and the y value of the start and the end point in the image frame, respectively. The path estimation module consists of a fully connected layer followed by a sigmoid layer. The module takes an intermediate feature map from the segmentation encoder and predicts x and y coordinate values of the start and the end point of the path.

The path estimation module takes a feature map of size $B \times C \times H \times W$ from the encoder of the segmentation network, where B, C, H, and W denote the batch size, the channel size, the height and the width of the feature map, respec-

Fig. 7. Data augmentation.

tively. The feature map is first reshaped to $B \times CHW$ to form one-dimensional feature vectors each of which corresponds to one of B input images. The reshaped feature map is then fed into the path estimation module. Since the ground truth coordinate values are normalized to [0, 1]., a sigmoid function is used as the final activation function.

The loss function for the path estimation module is defined as follows:

$$L_{path} = \sum_{i=1}^{N} \sum_{j \in \{x_s, y_s, x_e, y_e\}} p_i^j \log \hat{p}_i^j + (1 - p_i^j) \log (1 - \hat{p}_i^j), \tag{2}$$

where $\hat{p}_i^j$ denotes a predicted value for coordinate j of ith image. For the semantic segmentation module, a cross-entropy loss with uncertainty weighting, L_{rect}, is used as proposed in [16]. The complete loss function is given as follows:

$$L_{full} = L_{rect} + \lambda L_{path}, \tag{3}$$

where λ denotes a weight for the loss and set to 1.0 in the experiment.

In inference, an image is fed into the model and normalized coordinate values of the start and the end point are predicted. The pixel locations are then computed by multiplying the values by the width and the height of the image and used in the computation of the path in the robot frame for path following.

6 Robot Control Based on the Estimated Path

The process of the navigation is as follows;

1. Estimate the path from the image.
2. Project the estimated line to the robot frame.
3. Compute the linear and the rotational velocity to follow the line.
4. Send the control signal to the robot.

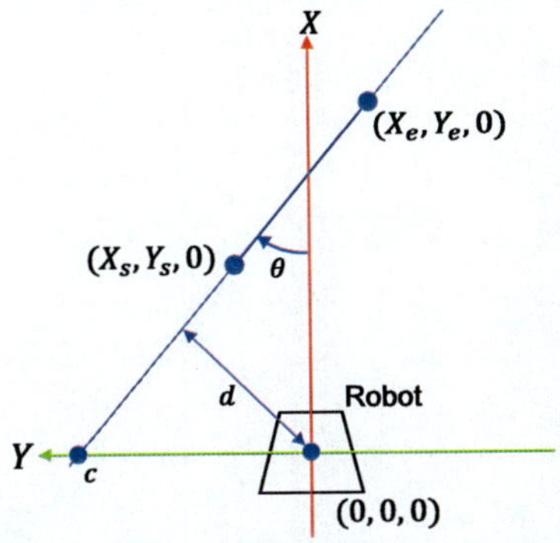

Fig. 8. Distance and angle between the robot and the path

We generate a path that the robot follows by transforming the start and the end point in the image.

Here, we define the X axis of the robot frame as the forward direction and the Y axis as the leftward direction. Figure 8 shows an example of a path in the robot frame. The path in the robot frame is given by:

$$(X_e - X_s)\, X + (Y_e - Y_s)\, Y + c = 0, \tag{4}$$

where c denotes the intercept of the line on the Y axis in the robot frame and X_s, Y_s, X_e, and Y_e denote X and Y coordinates of the start and the end point in the robot frame, respectively. The intercept c is then given by:

$$c = -\left(\left(X_e - X_s\right) X_s + \left(Y_e - Y_s\right) Y_s\right). \tag{5}$$

The distance a between the robot and the path is given by:

$$d = \frac{|c|}{\sqrt{(X_e - X_s)^2 + (Y_e - Y_s)^2}}. \tag{6}$$

The angle θ between the robot's heading and the direction of the path is given by:

$$\theta = \tan^{-1}\left(\frac{Y_e - Y_s}{X_e - X_s}\right). \tag{7}$$

The linear and the rotational velocity for the robot are then computed using d and θ. In this experiment, the linear velocity is fixed to 0.1 [m/s] for safety. The rotational velocity is given by:

$$\omega = \begin{cases} k_d d + k_\theta \theta & (c \geq 0) \\ -k_d d + k_\theta \theta & (c < 0) \end{cases}, \tag{8}$$

where k_d is a constant for the distance factor which is empirically set to 0.2, and k_θ is a coefficient for the angle set to 0.08.

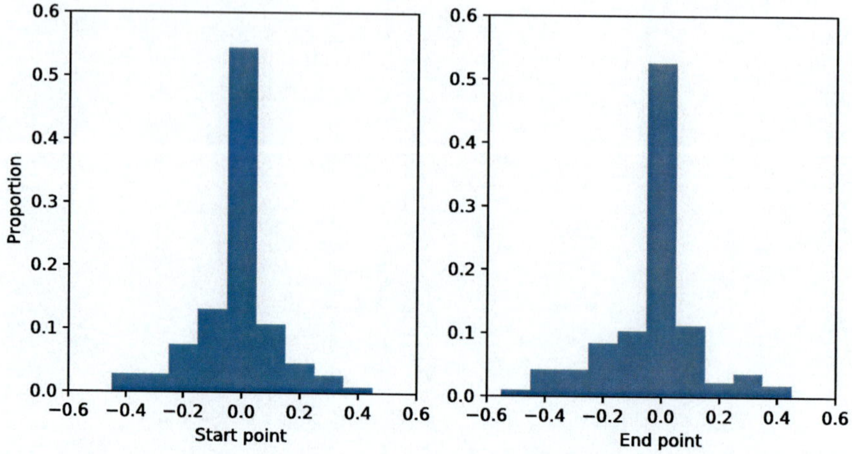

Fig. 9. Distributions of prediction errors on the start and end points. The horizontal axis indicates the degree of horizontal misalignment of the normalized start and end coordinates.

Table 1. The mean and the standard deviation of estimation errors

	Start mean (std)	End mean (std)
Manual	−0.0107 (0.1339)	−0.0363 (0.1591)
Auto (proposed)	0.0048 (0.1344)	−0.0270 (0.1573)

7 Experiments

7.1 Training with the Automatically Generated Dataset

We conduct the model training with the training dataset generated by the proposed method. RGB-D data are acquired through 35 min of manual operation of the robot inside a greenhouse. The automatic annotation produced 497 labeled images, and we adopted the data augmentation described in Sec. 4 to increase the training data to 4338. During the training, we also used random flipping, gaussian noise, gaussian blur, shifting RGB values, and modification of brightness and contrast using Albumentations library [18]. As for the test data, we manually annotated on 213 images in total taken on the same day as the training data as well as a different date and time. We set the batch size to 16, and the number of epochs to 100. The initial learning rate is 10^{-3} and decayed by the exponential learning rate scheduling [19]. Adam [20] is used as the optimization method.

Here, we evaluate the trained model based on the errors in the horizontal axis in the image, which is important for determining the direction of navigation. Figure 9 shows the distributions of errors between the x coordinates of the outputs for the test images and the corresponding ground truth labels. The

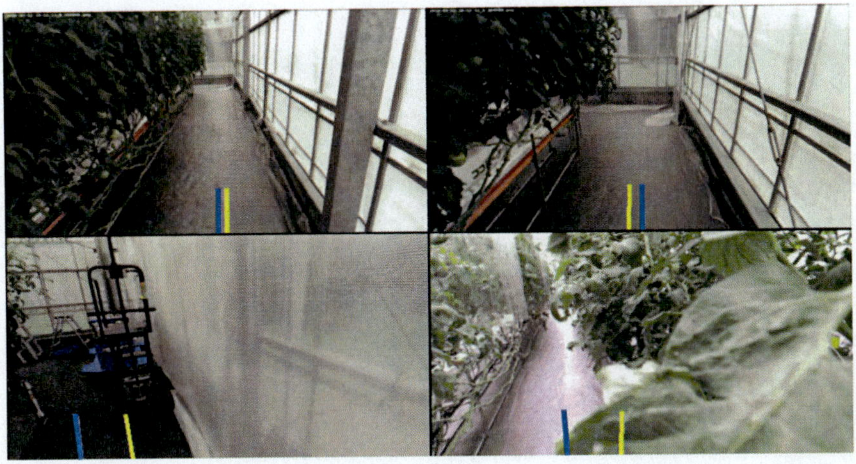

Fig. 10. Predictions by the path estimation module (yellow: prediction, blue: ground truth).

Fig. 11. Predictions during the experiment.

errors of prediction were within ±20% for approximately 80% of the test images. The examples of the prediction are shown in Fig. 10.

In addition, we compare the estimation accuracy with models trained on manually annotated labels. Table 1 shows the means and standard deviations of the prediction errors of the models trained on the labels produced by the proposed method as well as manual annotation. The prediction by the model trained on the automatically annotated data resulted in comparative performance. Manual annotation on the 497 training images took us approximately 2.5 h using a dedicated annotation tool. On the other hand, the proposed method only required 40 min in total for data acquisition, trajectory estimation, and data annotation to generate the same number of labeled images. The proposed method is, therefore, capable of generating training data more efficiently.

7.2 Navigation

We conducted a navigation experiment using the robot in the greenhouse that grows tomatoes in Toyohashi University of Technology. We used Revast Mercury as the mobile robot, and a laptop PC with Intel Core i7-6700HQ CPU, 32GB RAM, and NVIDIA GeForce GTX 970M with 6GB memory for inference.

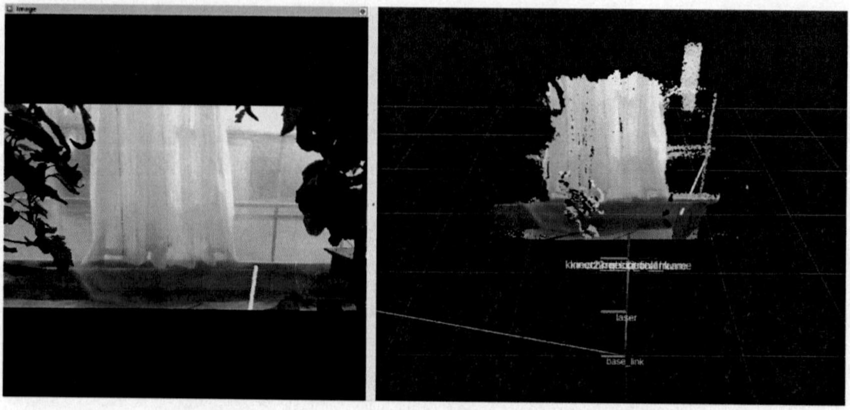

Fig. 12. An example of erroneous inference.

Figure 11 shows visualized estimated path during the experiment. We conducted five trials of navigation on different paths between crop rows with a length of approximately 10 [m]. In all the trials, the proposed system successfully navigated the robot through the path until near the end of the path. When approaching the end, however, the accuracy of the path estimation degraded as shown in Fig. 12 because the view is very different from the ones in the training set.

8 Conclusions and Future Work

In this paper, we proposed a method of image-based end-to-end path estimation and automatic data generation for robot navigation in plant-rich environments. Our system automatically labels images 10 times faster than manual annotation by exploiting the robot's trajectories. We also confirmed in the experiment that a real mobile robot with our proposed path estimation method consistently navigated the robot through the paths between plant lines.

A limitation of the current system is that the path estimation accuracy degrades near the end of the path possibly because of the limited FOV of the sensor. To mitigate the problem, possible options include recognizing the end of the path and switching to dead reckoning or another NN model when approaching the end point.

References

1. Ort, T., Paull, L., Rus, D.: Autonomous vehicle navigation in rural environments without detailed prior maps. In: IEEE International Conference on Robotics and Automation, pp. 2040–2047 (2018)

2. Ponnambalam, V.R., Bakken, M., Moore, R.J.D., Glenn Omhol Gjevestad, J., Johan From, P.: Autonomous crop row guidance using adaptive multi-roi in strawberry fields. Sensors **20**(18) (2020)
3. Zheng, J., Kargbo, A. H.: A robust lane detection using edge detection with symmetric molecules in visual perception for self-driving cars. Int. J. Eng. Res. Technol. **10**(07), 268–273 (2021)
4. Wang, W., Lin, H., Wang, J.: CNN based lane detection with instance segmentation in edge-cloud computing. J. Cloud Comput. **9**(1) (2020)
5. Lee, D.-H., Liu, J.-L.: End-to-end deep learning of lane detection and path prediction for real-time autonomous driving. Technical report (2021)
6. Onozuka, Y., Matsumi, R., Shino, M.: Weakly-supervised recommended traversable area segmentation using automatically labeled images for autonomous driving in pedestrian environment with no edges. Sensors **21**(2) (2021)
7. Opiyo, S., Okinda, C., Zhou, J., Mwangi, E., Makange, N.: Medial axis-based machine-vision system for orchard robot navigation. Comput. Electron. Agric. **185**, 106153 (2021)
8. Matsushita, Y., Miura, J.: On-line road boundary modeling with multiple sensory features, flexible road model, and particle filter. Rob. Auton. Syst. **59**(5), 274–284 (2011)
9. Chiku, T., Miura, J.: On-line road boundary estimation by switching multiple road models using visual features from a stereo camera. in: IEEE International Conference on Intelligent Robots and Systems, pp. 4939–4944 (2012)
10. Phung, S.L., Le, M.C., Bouzerdoum, A.: Pedestrian lane detection in unstructured scenes for assistive navigation. Comput. Vis .Image Underst. **149**, 186–196 (2016)
11. Meyer, A., Salscheider, N.O., Orzechowski, P.F., Stiller, C.: Deep semantic lane segmentation for mapless driving. In: IEEE International Conference on Intelligent Robots and Systems, pp. 869–875 (2018)
12. Wellhausen, L., Dosovitskiy, A., Ranftl, R., Walas, K., Cadena, C., Hutter, M.: Where should i walk? Predicting terrain properties from images via self-supervised learning. IEEE Rob. Autom. Lett. **4**(2), 1509–1516 (2019)
13. Matsuzaki, S., Masuzawa, H., Miura, J.: Image-based scene recognition for robot navigation considering traversable plants and its manual annotation-free training. IEEE Access **10**, 5115–5128 (2022)
14. Giusti, A., Guzzi, J., Ciresan, D.C., He, F.-L., Rodriguez, J.P., Fontana, Flavio, Faessler, M., Forster, C., Schmidhuber, Jurgen, Di Caro, Gianni, Scaramuzza, Davide, Gambardella, Luca M.: A machine learning approach to visual perception of forest trails for mobile robots. IEEE Rob. Autom. Lett. **1**(2), 661–667 (2016)
15. Mehta, S., Rastegari, M., Shapiro, L., Hajishirzi. H.: ESPNetv2: A light-weight, power efficient, and general purpose convolutional neural network. In: CVPR, pp. 9190–9200 (2019)
16. Matsuzaki, S., Miura, J., Masuzawa, H.:. Multi-source pseudo-label learning of semantic segmentation for the scene recognition of agricultural mobile robots. arXiv:2102.06386 (2021)
17. Labbe, M., Michaud, F.: RTAB-Map as an Open-Source Lidar and VisualSLAM Library for Large-Scale and Long-Term Online Operation. J. Field Rob. **36**(2), 416–446 (2019)
18. Buslaev, A., Iglovikov, V.I., Khvedchenya, E., Parinov, A., Druzhinin, M., Kalinin, A. A.: Albumentations: fast and flexible image augmentations. Information **11**(2) (2020)
19. Li , Z., Arora, S.:. An exponential learning rate schedule for deep learning. arXiv:1910.07454 (2019)

20. Kingma, D. P., Ba, J. L.: ADAM: a method for stochastic optimization. In: ICLR (2015)

Localization on Indoor Topological Maps—SCAM: Scale-Compatible Adaptive Monte-Carlo Localization

Zhikai Li(✉), Krittin Kawkeeree, and Marcelo H. Ang

National University of Singapore, Queenstown, Singapore
e0004076@u.nus.edu, krittin.kawkeeree@gmail.com, mpeangh@nus.edu.sg

Abstract. This paper introduces Scale Compatible Adaptive Monte-Carlo Localization (SCAM) to localize on topological maps, such as hand-drawn maps and floor plans. This enables fast modifications to maps of indoor spaces whereby the layout changes frequently via image editing instead of re-mapping the enviro-nment. SCAM uses x and y scale components in addition to a 2D pose to form a five-dimensional state, modifying Adaptive Monte-Carlo Localization (AMCL) significantly at the prediction and re-sampling step to account for the extra scale components. The scale components influence the projection of the LiDAR scan on each particle, improving the scan match from the LiDAR scan with the imperfect map. The performance of SCAM is tested with real-life data gathered on an empty hallway, a cluttered lab, and a populous lobby. SCAM is evaluated on hand-drawn maps, where the introduction of scale components in SCAM prevented major issues such as loss of localization and running over obstacles, but sees more minor issues such as gliding past obstacles. This result is further verified via repetitions of test cases and making modifications to the map while preserving the topology, and applied to a floor plan. On scaled point cloud maps, SCAM only introduces relatively small positional errors of 0.270m, 0.161m, 0.491m, and heading errors of 1.55°, 8.10°, 10.29° in the three test areas respectively.

Keywords: Localization · Simultaneous localization and mapping · Particle filter

1 Introduction

In autonomous vehicles (AV), Simultaneous Localization and Mapping (SLAM) maps the environment using the AV's sensors before an operation. A localization algorithm then tracks the state trajectory (pose) of the AV on the map.

In this paper, we explore the ability to localize without the prior mapping of the environment using SLAM. Instead, we utilize indoor topological maps normally unsuitable for autonomous navigation. This can include hand-drawn maps, shopping directories, or floor plans. Doing so also allows maps to be easily edited by modifying the map image directly when the layout changes, such

I. Petrovic et al. (Eds.): IAS 2022, LNNS 577, pp. 285–297, 2023.
https://doi.org/10.1007/978-3-031-22216-0_20

as when furniture is rearranged. These maps pose a challenge for localization algorithms as scale and geometry are only loosely represented and differ greatly from what the sensor observes of the environment.

In previous work, a gradient descent localization method was used that incorporated 'scale' as part of the AV's state during localization [1]. It was shown that the scale component of state was effective in managing affine distortions in the given hand-drawn map, but was sensitive to dynamic obstacles, which easily corrupts the scan matching and causes a loss in localization.

We propose SCAM, Scale-Compatible Adaptive Monte-Carlo Localization, a localization algorithm that reaps the effectiveness of the scale state and implements it over a popular localization algorithm, the Adaptive Monte-Carlo Localization (AMCL) [2], which retains some resilience in dynamic environments. This extends AMCL's capabilities beyond localization in metric maps as SCAM can be applied to floor plans, point clouds, or hand-drawn maps. Doing so, we make the following contributions:

- Demonstrate the effectiveness and advantages of incorporating the scale state into popular localization algorithms to handle imperfect maps.
- Show the viability of accommodating imperfect maps in dynamic environments, allowing localization in a new environment without prior mapping.

2 Background and Related Work

AV architectures commonly utilize maps generated by SLAM. LiDAR SLAM methods include FastSLAM [3], which proposes a probabilistic factorisation to separately localize self and environment object, [4], and LOAM [5], which uses LiDAR scan match for odometry and mapping. These methods generate point cloud maps representing the obstacles. Visual SLAM methods include MonoSLAM [6], which records SIFT features of the environment, and ORB-SLAM [7], which uses ORB features, a keyframe detection, and a bag of words mechanism. These generate feature maps or persistent obstacle maps of the environment. While LiDAR-Vision fusion SLAM, such as V-LOAM [8] are also gaining interest, a common characteristic of these methods is they require building a well-scaled and features rich map environment before they can localize into the built map. If a section of the map changes, that section must be re-mapped.

In SCAM, we explore the use of alternative maps, such as hand-drawn maps or floor plans that only roughly capture the layout of an indoor space. If localization is possible, the input map can be digitally edited if regions of the map change. Several works have explored the use of alternate maps.

Analytical approaches include [9], which uses a street directory on rural roads to elucidate untraversable vegetation via rough scan patches and repeatedly steer towards its goal. [10] uses a floor plan with hand-marked features, such as corners, to match during scan-matching. These two methods are constrained to specific environments and are unsuitable for the scope of generic indoor spaces. In [1], a gradient descent method was used to perform scan-match, allowing the freedom of scale state to stretch the LiDAR scan. The localization is jittery and easily corrupted by dynamic obstacles, but the scale state was effective in handling map inaccuracies and will be utilized in this paper.

Deep learning methods include Intention-Net [11] and Nav-Net [12], which takes a path or local intention on a floor plan and vision as inputs to a neural network to output the steering and acceleration of a mobile robot. In [13], vision from an AV camera suite and planned route on a street directory are used as inputs to a neural network to output localization and possible steering controls to navigate the road. While deep learning methods yield promising results, it is a black box, limiting flexibility and understanding. If the AV has to identify roadblocks or modify parts of the map, these methods will have difficulty adapting to such uses.

Approximate localization on an inaccurate map is the goal of this work. In robotics and AV, a Bayes Filter is ideally used to fuse odometry and LiDAR or vision but is often impractical due to unknown or overly complicated belief distribution and system model [14]. Kalman Filters overcome this by assuming a linear system model and Gaussian positional belief to solve the Bayes Filter analytically [15], while Extended Kalman Filters (EKF) linearize the system model about the current state to do the same [16]. VINS-Mono [17] is an example that uses EKF to fuse odometry and vision. However, a Gaussian position belief may not be suitable in this work, where the map can be heavily distorted.

Particle filters, such as Monte-Carlo Localization [18], project possible state particles forward in time by sampling the state transition probability, weighing them with observation probability and re-sample them. It does not assume a Gaussian belief, but only approximates the belief distribution. AMCL is a particle filter that dynamically adjusts the number of particles used during re-sampling [2]. A measure using Kullback-Leibler Divergence (KLD) is used to determine the sample count required for a probabilistic guarantee bounding the error in the reconstructed belief posterior [19]. AMCL is popular in mobile robotics, where in recent works like [20] and [21], AMCL has produced excellent results.

3 SCAM: Scale-Compatible Adaptive Monte-Carlo Localization

The goal of localization is to elucidate an agent's pose in a map using multiple sources of information. SCAM, built upon AMCL, follows its architecture closely as in Fig. 1. It is a particle filter that incorporates both the wheel encoder information and LiDAR scan to determine the state of the AV in a hand-drawn or point cloud map. In SCAM, each particle represents a five-dimensional state, $[x, y, \theta, s_x, s_y]$, representing its x, y position and θ heading with respect to the global frame, and scale states in the x and y direction. SCAM thus has an additional step of processing and propagating the scale state in each particle.

In the architecture shown in Fig. 1, the prediction and observation update steps are changed in SCAM and will be further described in the following subsections. The remainder of the architecture is the same as AMCL in [2].

Fig. 1. The architecture of SCAM Localization, with a visualization of how particles are treated at each stage, represented by red arrows. The architecture follows closely to AMCL [2]. The prediction stage propagates particles forward. The particles are then weighted by scan matching, where a thicker arrow represents larger weights assigned. Adaptive re-sampling is used to draw a new set of particles, with highly weighted particles being more likely to be sampled. Finally, the re-sampled particle set is clustered and a final hypothesis state, the black arrow, is chosen using the cluster.

3.1 Prediction

The prediction step uses odometry readings, such as wheel encoders, with information on the reliability of the source to form mean and covariance values for the state transition distribution. Each particle samples the state transition probability, $p(x_t \mid u_t, x_{t-1})$ to obtain a change in state and applies to itself. In SCAM, since the assumption of constant map scale is relaxed, the sampled change in the positional pose x and y must hence also be multiplied by a factor of the previous s_x and s_y state components respectively.

Odometry readings do not gather information about the change in scale, s_x and s_y, of the map. As such, a separate prediction scheme must be done for s_x and s_y. Poor scale sampling can result in the particles failing to converge. In SCAM, the topological map provided by the user is assumed to be created with a desired scale in mind, s^*. Due to inaccuracies of approximating the topological map, such as when drawing by hand, the map scale will often be close to, but not exactly s^*. This allows us to set s^* as a rough average of the map scale. During sampling, each particle samples $s_x, s_y \sim N(s^*, \sigma^2)$, with σ^2 as the variance of the scale prediction, interpreted as the how much the map scale deviates from s^*. A poorly created map with highly irregular scaling will have a larger σ^2.

3.2 Observation Update

In SCAM, each particle has a different s_x and s_y scale components. During the observation update, the LiDAR scan is recorded as a set of depth values, d, which must be resolved into d_x and d_y and scaled by s_x and s_y into c_x and c_y respectively before projection. This is shown in Fig. 2 and Eqns. 1 and 2.

$$c_x = s_x d_x = s_x d \sin\theta \tag{1}$$

$$c_y = s_y d_y = s_y d \cos\theta \tag{2}$$

The projected point is used in the likelihood computation of that particle by associating it with its nearest obstacle on the map. The distance between the nearest neighbor association must then be converted back from pixel distance to meters using s_x and s_y. Many implementations of particle filters then use the

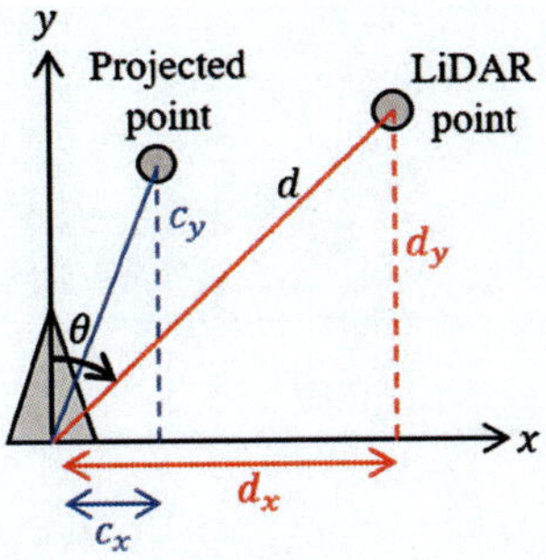

Fig. 2. References for projecting LiDAR scan a particle's local frame.

association distance to compute observation probability to weigh the particles by chaining the probability of observing each scan point. However, in SCAM, each scan point probability is scaled by a factor of s_x and s_y. Having k LiDAR scan points in a sweep causes all weights to be scaled by a factor of s_x^k and s_y^k respectively, resulting in numerical instability when s_x and s_y become small. As a precautionary measure, we limit $k = 100$ to not use the entire LiDAR scan.

In SCAM, the inverse of the sum of distances from each projected point to its associated obstacle is used to weigh the particle the scan is projected to. This gives a small weight to a particle if the LiDAR scan points projected to it are far from their associated obstacles, i.e. poor scan match, and vice versa. At the same time, this avoids numerical stability issues by avoiding the multiplication of small probability values, each influenced by s_x and s_y. Furthermore, using $s_x, s_y \sim N(s^*, \sigma^2)$ in the prediction step also ensures that s_x and s_y are unlikely to become small, as long as s^* is large enough.

4 Experimental Results and Discussion

4.1 Experimental Hardware and Set-up

Two SICK TIM551 2D LiDARs each covering a 270° field of view are mounted at the front of a wheelchair and used for experimentation. Experimental data is collected and played back over the Robot Operating System (ROS).

During experimentation, the wheelchair was driven around three environments - an empty corridor, a cluttered office environment, and a populous lobby area, shown in Fig. 3. In the empty corridor, the wheelchair observes mostly distinct walls and no dynamic obstacles. In the cluttered lab, the wheelchairs see narrow pathways, partially obstructed by dynamic obstacles like chairs, tables, and lab apparatus. The populous lobby features a large empty area with many dynamic obstacles like people walking around and some coffee tables. Multiple bags of data are collected for each environment and are used to evaluate SCAM.

Fig. 3. Test Environments.

Table 1. Error Count of Trajectories by AMCL and SCAM on Hand-Drawn Maps

Method	Area	Trials	(Minor) — Count of Problematic Trajectories → (Major)		
			Glide past Obst	Run over Obstacle	Lost Localization
AMCL	Corridor	10	2	3	3
	Office	5	3	2	1
	Lobby	5	2	3	1
SCAM	Corridor	10	5	1	0
	Office	5	4	1	0
	Lobby	5	3	1	0

4.2 Localization on Hand-Drawn Maps

SCAM is tested on poorly represented hand-drawn maps, along with AMCL to compare differences in localization trajectory introduced by the scale states. In such maps, pose error analysis is problematic as there may not a one-to-one correspondence between the environment and the map and generating a ground truth trajectory on the hand-drawn map is not possible. Instead, the trajectories produced by the algorithms are evaluated in the context of autonomous driving. Flaws in trajectories are identified, grouped by whether they have a major or minor impact on a full navigation stack.

Three common flaws are considered in increasing severity in Table 1, based on how the flaw can affect the overall navigation of an AV. The first flaw is gliding past the edges of an obstacle. This is a minor flaw as most AVs tend to have obstacle avoidance in their local planner to avoid collision when this flaw occurs. The AV's ability to navigate to a target location is hence likely not compromised. The second flaw is running through obstacles. This is a major flaw as it can cause problems for motion planners to navigate 'out' of the obstacle. The AV can sometimes recover from this via slowing down/stopping to take multiple readings of the surroundings and re-localize. In the event the AV is unable to, it runs into the last critical flaw in Table 1, loss of localization.

The results of SCAM applied on the empty corridor, cluttered office and populated lobby are shown in Figs. 4, 5, 6 respectively. The count of flaws observed in each environment is tabulated in Table 1.

From Table 1, SCAM avoids the loss of localization but often glides past corners of the walls. Critical loss of localization is prevented and major flaws of

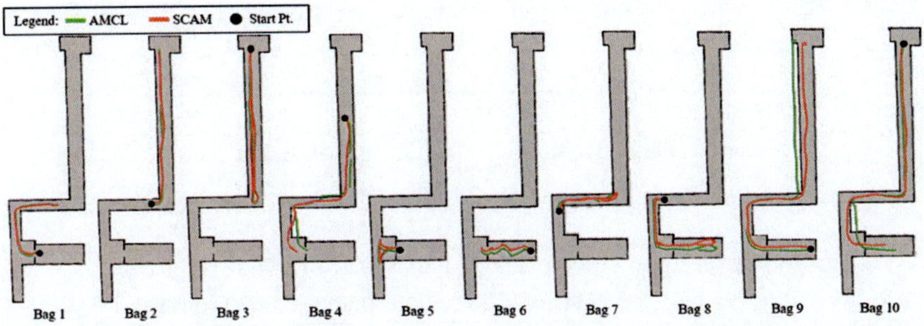

Fig. 4. State trajectories traced by AMCL (green) and SCAM (red) on a hand-drawn empty corridor map.

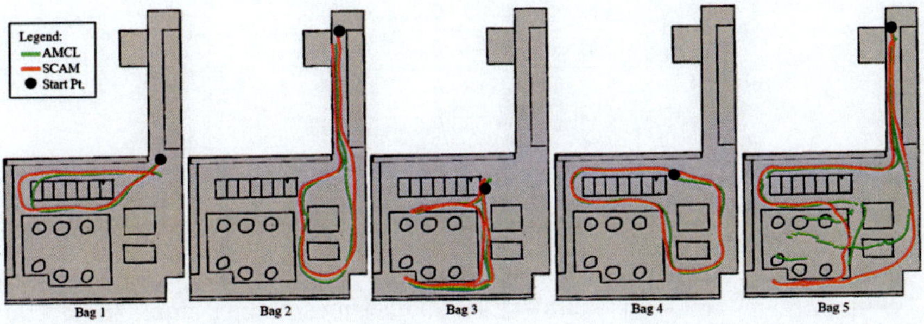

Fig. 5. State trajectories traced by AMCL (green) and SCAM (red) on a hand-drawn cluttered office map.

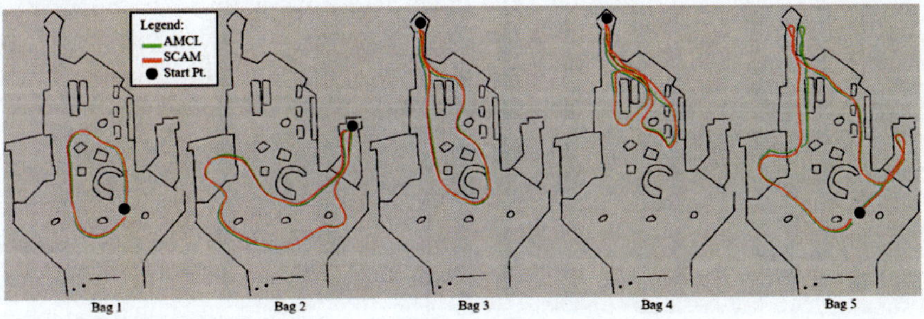

Fig. 6. State trajectories traced by AMCL (green) and SCAM (red) on a hand-drawn populous lobby map.

running over obstacles are reduced, mitigated into minor flaws of gliding along with the obstacles instead. Examples of this occurs in Bags 4, 9, 10 of Fig. 4, Bag 5 of Fig. 5 and Bag 5 of Fig. 6. In all mentioned test cases, AMCL does not consider dynamic changes in map scale and often produces a localized state trajectory moving out of the traversable areas. The introduction of the scale

Table 2. Standard Deviation, σ, of Localized Trajectory across 10 repetitions

Test case	Stat	$\sigma(x)$	$\sigma(y)$	$\sigma(\theta)$	$\sigma(s_x)$	$\sigma(s_y)$
SCAM						
Cluttered office	Mean	0.134	0.599	0.249	0.002	0.002
	Min.	0.006	0.007	0.003	0.0002	0.0003
	Max.	0.446	0.132	3.068	0.009	0.009
Lobby	Mean	1.272	0.687	0.137	0.003	0.003
	Min.	0.006	0.006	0.002	0.0003	0.0005
	Max.	4.592	6.279	2.183	0.017	0.013
AMCL						
Cluttered office	Mean	0.300	0.090	0.283	–	–
	Min.	0.002	0.003	0.002	–	–
	Max.	0.864	0.545	3.069	–	–
Lobby	Mean	0.218	0.131	0.076	–	–
	Min.	0.005	0.010	0.002	–	–
	Max.	1.047	0.518	2.441	–	–

state in SCAM allowed it to better respect the map topology, avoiding cases of running out of the map boundary but sometimes gliding along the obstacle boundaries instead.

When correcting localization, gliding often occurs as the scale components of the state are adjusting to the incorrect geometry or position of map objects. Over a few consecutive iterations, the hypothesis cluster thins out outside and grows in the areas inside the map boundary. During this transition period, the centroid of the hypothesis cluster slowly shifts back from outside the map bounds to within, sometimes causing the gliding along the wall to happen.

4.3 Consistency Checks of Localization Results

Understandably, particle filters are random and repeated runs of the same test case may not yield identical results. Hence, both SCAM and AMCL are each repeatedly run for 10 times on cases where the difference in introducing the scale state is most apparent - bag 5 of the cluttered office (shown in Fig. 5) and bag 5 of the populous lobby (shown in Fig. 6). The localized state outputs across different repetitions are consolidated and the standard deviations of the state outputs are collated at each timestamp are computed. The average, minimum and maximum standard deviations along the trajectory are reported in Table 2.

From Table 2, a small average standard deviation of less than 1m for both algorithms in most cases indicate the high consistency of state trajectories across different runs of the same bag. As the smallest object clusters in the environment are at least 2m in width, this means that no trajectory amongst the repetitions took a different detour from that in Fig. 6, thereby also implying that AMCL

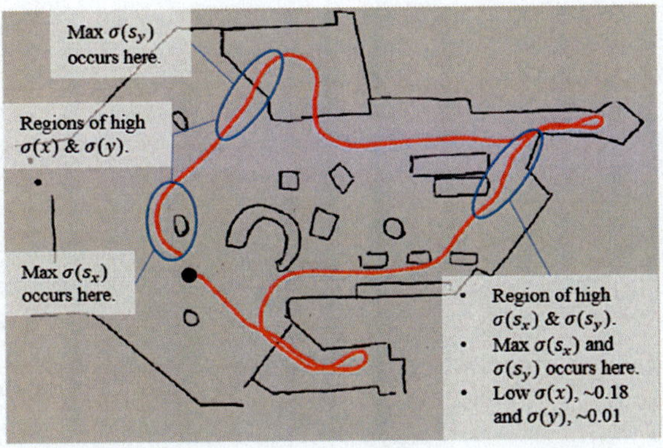

Fig. 7. Trajectory of SCAM on the Lobby environment, indication area with high standard deviation of particular state components.

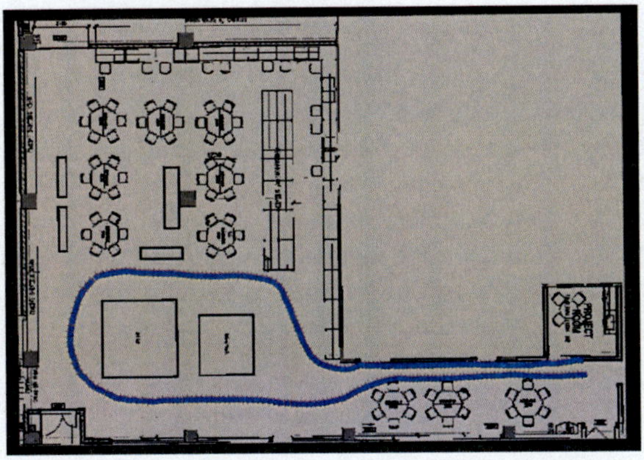

Fig. 8. Localized trajectory of SCAM on modified Floor Plan Map on the Cluttered Office. Reference lines were cleaned out and missing equipment are drawn as squares.

always localizes with the wrong trajectory. However, in SCAM, a large maximum standard deviation for x, y, s_x, and s_y was observed in the Lobby environment, prompting further investigation into where these maximum deviations occur.

From Fig. 7, regions of high $\sigma(x)$ and $\sigma(y)$ occur at parts of the lobby that are large, empty, have few features, and have most people walking around. Whilst $\sigma(x)$ and $\sigma(y)$ are large, this is not disruptive the topological path of the trajectory remains the same. Furthermore, the region of high $\sigma(s_x)$ and $\sigma(s_y)$ have low $\sigma(x)$ and $\sigma(y)$ and occurs at where AMCL loses localization, showing that having scale components of state reliably gives robustness to localization algorithms, enabling localization on distorted topological maps.

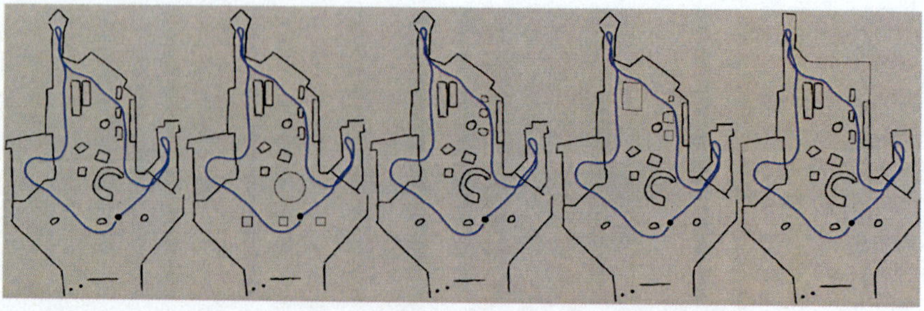

Fig. 9. Hand-Drawn Map of Lobby, with intentional distortions to the map, demonstrating that SCAM is robust to inaccurate representations

4.4 Applicability of SCAM

In SCAM, development work was done on hand-drawn maps as these often tend to more messy and distorted. In reality, floor plans or fire escape maps are often more accessible and SCAM can be applied on these with minimal edits, such as the floor plan map shown in Fig. 8.

SCAM also allows the map to be easily updated via image editing. If obstacles are moved, their new positions can be represented on the map with low precision. To demonstrate this resilience, Fig. 9 shows the same trajectory produced by SCAM across multiple maps with obstacles internationally drawn incorrectly. Hence, when updating a new layout on a map, one can edit the image free-hand, as SCAM is robust to low precision in obstacles' shape or position.

4.5 Increased Initial Pose Error Tolerance

When repeating the experiment multiple times with hand-drawn maps, it was observed that SCAM has some error tolerance to the initial pose estimate. Figure 10 compares multiple runs of SCAM and AMCL with different starting poses. SCAM is resilient and reliably converges to the correct trajectory, while AMCL is sensitive to and can lose localization when given a bad initial pose estimate.

This phenomenon can be explained by a more extensive scan match covered by particles varying their scales. For example, if the AV initializes too far from an observed feature, sampled particles with smaller scales will see a better scan-match. Subsequently, if the robot moves towards the feature, the scaled odometry will 'speed up' the AV towards the feature, recovering from the incorrect pose.

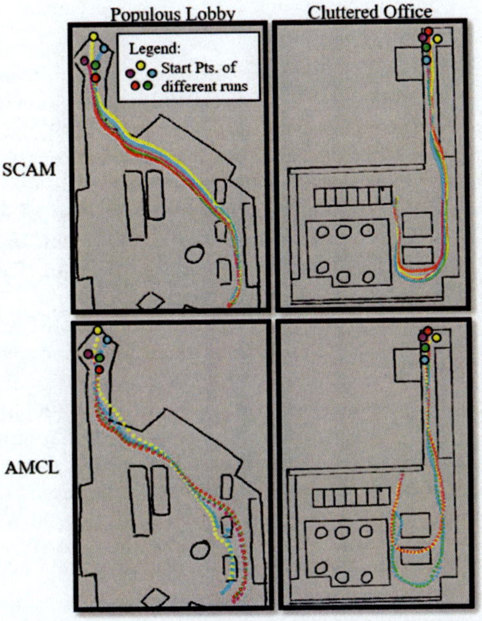

Fig. 10. State trajectories of AMCL and SCAM when with different initial pose estimates. Each colored trajectory has a different initial pose estimate.

5 Conclusion

Scale Compatible Adaptive Monte-Carlo Localization (SCAM) is proposed for 2D LiDAR localization in simplistic topological maps which are noisy or misrepresented, such as hand-drawn maps or floor plans. In hand-drawn maps, the scale components introduced in SCAM mitigate major flaws of localization loss or running over obstacles, but have more minor flaws of gliding past obstacles. This experiment is repeated to demonstrate its consistency and robustness to different distortions to the map. SCAM is also observed to have improved tolerance to poor initial pose estimates. Future work can use SCAM for map refinement or correlate image maps with a point cloud for AV user interfaces.

Acknowledgements. This research was supported by the National Research Foundation, Prime Minister's Office, Singapore, under its CREATE programme, Singapore-MIT Alliance for Research and Technology (SMART) Future Urban Mobility (FM) IRG. We also gratefully acknowledge the technical support of Nvidia Corporation through the Memorandum of Understanding with the Advanced Robotics Centre of the National University of Singapore on autonomous system technologies.

References

1. Li, Z., Ang, M.H., Rus, D.: Online localization with imprecise floor space maps using stochastic gradient descent. In: IEEE/RSJ International Conference on Intelligent Robots and Systems (IROS), Las Vegas, USA (2020)
2. Fox, D.: Adapting the sample size in particle filters through KLD-sampling. Int. J. Robot. Res. **22**(12), 985–1004 (2003)
3. Montemerlo, M., Thrun, S., Koller, D., Wegbreit, B.: FastSLAM: a factored solution to the simultaneous localization and mapping problem. In Eighteenth national conference on Artificial Intelligence, Edmonton, Alberta, Canada, pp. 593–598 (2002)
4. Montemerlo, M., Thrun, S., Koller, D., Wegbreit, B.: FastSLAM 2.0: An improved particle filtering algorithm for simultaneous localization and mapping that provably converges. In: IJCAI'03 Proceedings of the 18th international joint conference on Artificial intelligence, Acapulco, Mexico, pp. 1151–1156 (2003)
5. Zhang, J., Singh, S.: LOAM: lidar odometry and mapping in real-time. In: Robotics: Science and Systems Conference (RSS) (2014, July)
6. Davison, A.J., Reid, I.D., Molton, N.D., Stasse, O.: MonoSLAM: Real-time single camera SLAM. IEEE Trans. Pattern Anal. Mach. Intell. **29**(6), 1052–1067 (2007)
7. Mur-Artal, R., Montiel, J.M.M., Tardos, J.D.: ORB-SLAM: a versatile and accurate monocular SLAM system. IEEE Trans. Robot. **31**(5), 1147–1163 (2015)
8. Zhang, J., Singh, S.: Visual-lidar odometry and mapping: low-drift, robust, and fast. In: IEEE International Conference on Robotics and Automation (ICRA), Seattle, Washington, USA, pp. 2174–2181 (2015, May)
9. Ort, T., Paull, L., Rus, D.: Autonomous vehicle navigation in rural environments without detailed prior maps. In: IEEE International Conference on Robotics and Automation (ICRA), Brisbane, Australia, pp. 2040–2047 (2018)
10. Wang, X., Marcotte, R.J., Olson, E.: GLFP: Global localization from a floor plan. In: IEEE/RSJ International Conference on Intelligent Robots and Systems (IROS), Macau, China, pp. 1627–1632 (2019)
11. Gao, W., Hsu, D., Lee, W.S., Shen, S., Subramanian, K.: Intention-Net: integrating planning and deep learning for goal-directed autonomous navigation. In: 1st Annual Conference on Robot Learning, CoRL 2017, Mountain View, California, USA, pp. 185–194 (2017)
12. Karkus, P., Hsu, D., Lee, W.S.: Integrating algorithmic planning and deep learning for partially observable navigation. ArXiv, vol.abs/1807.06696 (2018, July)
13. Amini, A., Rosman, G., Karaman, S., Rus, D.: Variational end-to-end navigation and localization. In: IEEE International Conference on Robotics and Automation (ICRA), Montreal, Quebec, Canada, pp. 8958–8964 (2019)
14. Thrun, S., Burgard, W., Fox, D.: Probabilistic Robotics (Intelligent Robotics and Autonomous Agents). The MIT Press (2005)
15. Kalman, R.E.: A new approach to linear filtering and prediction problems. J. Basic Eng. **82**(1), 35–45 (1960)
16. Julier, S.J., Uhlmann, J.K.: A new extension of the Kalman filter to nonlinear systems. In: Signal Processing, Sensor Fusion, and Target Recognition VI, pp. 182–193 (1997)
17. Qin, T., Li, P., Shen, S.: VINS-Mono: a robust and versatile monocular visual-inertial state estimator. IEEE Trans. Robot. **34**, 1004–1020 (2018)
18. Fox, D., Burgard, W., Dellaert, F., Thrun, S.: Monte carlo localization: efficient position estimation for mobile robots. In: National Conference on Artificial Intelligence (AAAI), Orlando, Florida, USA, pp. 343–349 (1999)

19. Fox, D.: Kld-sampling: Adaptive particle filters. In: Advances in Neural Information Processing Systems, pp. 713–720. British Columbia, Canada, Jan, Vancouver (2001)
20. Guan, R.P., Ristic, B., Wang, L., Palmer, J.L.: Kld sampling with Gmapping proposal for Monte Carlo localization of mobile robots. Inf. Fusion **49**, 79–88 (2019)
21. Wasisto, I., Istiqomah, N., Trisnawan, I.K.N., Jati, A.N.: Implementation of mobile sensor navigation system based on adaptive Monte Carlo localization. In 2019 International Conference on Computer, Control, Informatics and its Applications: Emerging Trends in Big Data and Artificial Intelligence, IC3INA (2019)

Perception

Hyperspectral 3D Point Cloud Segmentation Using RandLA-Net

Isaak Mitschke[1(✉)], Thomas Wiemann[1,2], Felix Igelbrink[1], and Joachim Hertzberg[1,2]

[1] Osnabrück University, Institute of Computer Science, Berghoffstr. 11, 49090 Osnabrück, Germany
imitschke@uni-osnabrueck.de, twiemann@uni-osnabrueck.de, thomas.wiemann@dfki.de, figelbrink@uni-osnabrueck.de, joachim.hertzberg@dfki.de

[2] DFKI Niedersachsen Lab, Plan Based Robot Control, Berghoffstr. 11, 49090 Osnabrück, Germany

Abstract. Point clouds are commonly used in robotics to represent 3D maps. To gain further understanding of their content, it is useful to annotate such maps semantically. To segment 3D point clouds with RGB values, different solutions exist. In machine learning, pre-trained classifiers are used for this purpose. Since it is not always possible to differentiate between entities relying solely on RGB information, hyperspectral histograms can augment the 3D data. The aim of this work is to evaluate, if hyperspectral information can improve the segmentation results for ambiguous objects, e.g., streets, sidewalks, and cars using established deep learning methods. Given the reported performance on geometrical 3D data and the possibility to directly integrate point annotations, we extended the neural network RandLA-Net. In addition to the evaluation of RandLA-Net in this context, we also provide a reference dataset consisting of semantically annotated hyperspectral 3D point clouds.

Keywords: Point clouds · Semantic segmentation · Deep learning · Hyperspectral imaging

1 Introduction

Semantic maps are essential for autonomous robots to perform goal-directed and unsupervised tasks. While creating such maps, instances of known classes of objects have to be labeled in the recorded 3D environment model, which is typically a 3D point cloud. In analogy to classical image analysis, it is desirable to learn classifiers that automatically segment point clouds into previously defined classes. In recent years, advances in deep learning have led to the development of an increasing number of methods for segmenting point clouds with the help of neural networks [1]. Currently, one of the best performing neural networks for classifying 3D point clouds is RandLA-Net [2]. With RandLA-Net, it is possible

I. Petrovic et al. (Eds.): IAS 2022, LNNS 577, pp. 301–312, 2023.
https://doi.org/10.1007/978-3-031-22216-0_21

to segment large-scale high resolution point clouds, as shown in experiments on the Semantic3D datasets [3]. In addition to 3D coordinates, some scanning systems measure additional surface properties like color or reflectance [1]. In [4] a terrestrial 3D scanning system is described that captures hyperspectral images and maps the recorded spectra onto the recorded 3D points. Hyperspectral imaging has already been successfully applied the field of remote sensing to analyze aerial images. In this discipline, the hyperspectral information is used to accurately distinguish between different materials [5]. Deep learning using neuronal networks can further increase the segmentation precision [6]. With the development of hyperspectral laser scanners, it has become feasible to combine both modalities. However, the combination of hyperspectral data and 3D point clouds is rarely found in current research and the classification of such data is even less common. In this paper, we present a first step to apply deep learning to segment terrestrial hyperspectral point clouds. Given the promising results of RandLA-Net, our approach builds on top of that network structure. The novelty of this work is to combine the spatial information encoded in the point clouds with the additional information in the hyperspectral images. For that, we project the hyperspectral information directly on the point clouds rather than doing the classification separately on both domains. Because no benchmark datasets for classification on hyperspectral 3D point clouds are publicly available, we also provide a labeled reference dataset. Our experiments show that the classification quality on the combined data improves significantly compared to the existing RandLA-Net classifier.

2 Related Work

Classification and segmentation of 2D images has been successfully applied in many applications like driver assistance systems, medicine, and remote sensing [7]. Many of these image classification approaches rely on deep learning using neural networks. Similar to 2D images, the objective of classifying or segmenting point clouds is to assign a label to each 3D point in a point cloud. Labels of interest may be object classes like Cars and Trees, but can also coarser specifications in terms of object categories such as Vehicle or Vegetation. Using deep learning to analyze point clouds is a relatively new field of research. Over the last years, many promising approaches to solve this task have been proposed. A milestone in research on deep learning-based approaches was PointNet [8], presented in 2017, and its extension PointNet++ [9], which was published shortly afterwards. Since then, many other methods have been developed and an increasing number of benchmark datasets for the segmentation of point clouds have been made available to compare the results of different segmentation methods. Examples for such benchmark datasets are Semantic3D [3], S3DIS [10], Paris-Lille-3D [11] and SemanticKitti [12].

Guo et al. [1] evaluated and compared the performance of the most recent point cloud segmentation methods. They reported that the best results on almost all datasets were achieved by using RandLA-Net[2]. Similar to PointNet and

PointNet++, RandLA-Net also uses a point-based approach, where the raw, unorganized point cloud is fed into the network. RandLA-Net is based on an encoder/decoder architecture and takes n points with corresponding features as input. The encoder layer consists of dilated residual blocks followed by randomized downsampling. The dilated residual blocks are used to compute features taking the neighborhood of each point into account. Attentive pooling is then used to generate new aggregated features.

Hyperspectral imaging is the process of capturing digital images, where during measurement, the intensity of multiple wavelengths of light is sensed rather than three color channels (red, green, and blue) as in common digital imaging. Light visible to humans is physically electromagnetic radiation between 300 and 750 nm. By analyzing the spectral signature of targets, detailed assumptions about the chemical nature of the material may be inferred. The spectral signature, for example, can be utilized to differentiate between vegetation and non-vegetation. This is due to the fact that chlorophyll in vegetation absorbs light in the red and blue spectral bands in particular. In the near infrared spectrum, for example, vegetation largely reflects the light [13]. A simple way of classifying vegetation and non-vegetation based on this knowledge is to compute the Normalized Difference Vegetation Index (NDVI). Vegetation indices can also be used to determine the health of vegetation. This is used in earth observation to detect drought or disease [14]. Hyperspectral imaging as a tool for non-invasive analysis of objects is used in many domains. Examples are agriculture, food processing, astronomy, geology and environmental research [15–18]. One potential limitation of using hyperspectral data is the presence of noise in the data. The main sources for that are the image sensors themselves, which induce noise, as well as physical influences like illumination fluctuations and atmospheric effects [19]. The occurrence of noise negatively influences the performance when classifying hyperspectral images [20]. Therefore, smoothing techniques are commonly applied to enhance the recorded data. Popular examples are median, moving average or Savitzky-Golay [21] filters. Another commonly used noise reduction technique is the Minimum Noise Fraction Transform (MNF) , also sometimes referred to as Noise-Adjusted Principal Components Transform (NAPC). Unlike PCA, where the principal components maximize the variance, the MNF transform minimizes the noise content. The first component has the highest signal-to-noise ratio and the last component the lowest. The MNF transform was first introduced in 1988 by Green et al. [22] and in 1990 redesigned by Lee et al. [23] and renamed to NAPC.

Another problem when working with hyperspectral data are illumination differences, resulting in different spectral intensities within the same material class. This is caused by different physical properties of materials and the surface structure of the observed materials. Especially in the near infrared range [24], scattering effects occur, which lead to a high variance in different samples of one class. To minimize this variance, the Standard Normal Variant Transform (SNV) [25] or Multiplical Scatter Correction (MSC) [26] can be applied to equalize the data.

So far, there are only few approaches that use hyperspectral data combined with spatial data. Buckley et al. [27] use hyperspectral point clouds to classify geological materials in quarries. However, the classification is performed only on the hyperspectral images. The resulting segmentation is then used as texture for the 3D scene. Other approaches originate from the field of remote sensing, where aerial hyperspectral imagery is fused with airborne laser scans. In contrast to terrestrial laserscans, remote sensing data usually has a low spatial resolution. Some papers already use such sparse 3D point clouds to segment urban [28,29] or agricultural scenes [30]. But to our knowledge, no high resolution reference datasets exist, that allow to investigate the combination of large scale 3D point cloud segmentation and hyperspectral data in the context of deep learning.

3 The Hyperspectral Semantic Street Scene Dataset

In order to annotate point clouds with hyperspectral data, the camera and the laser scanner have to be co-calibrated. The extrinsic calibration, which describes the transformation from laser scanner to camera, is used to convert the 3D world coordinates into camera coordinates. Using the intrinsic calibration of the camera, the points are projected into pixel coordinates. Thus, a corresponding color value or, in the case of hyperspectral cameras, a corresponding spectrum can be assigned to each 3D point. Igelbrink et al. [4] developed a method for an ad-hoc calibration of a terrestrial 3D laser scanner and a hyperspectral camera. The calibration procedure is performed using an automated, marker-less method instead of using reference patterns. Igelbrink's method calculates the Normalized Mutual Information (NMI), which is a measure of the similarity between two images. A panorama is generated from the point intensity values of the 3D laser scan, using the cylindrical camera model, which is then compared with the hyperspectral panorama. With a perfect calibration, the NMI value is equal to 1. The Normalized Mutual Information metric is specifically well-suited for comparing multi-modal images, such as in this case an intensity image from the laser scanner and the hyperspectral image. An optimization procedure that maximizes the NMI is then utilized to find the best parameters for the calibration. The evaluation of the procedure shows that the resulting calibration is very accurate and is generated in a few seconds through GPU acceleration.

In order to train classifiers, a lot of training data is required. For all classes, numerous examples should be available, such that the classifiers have a high level of transferability to previously unseen data [31]. In this work, we chose an urban environment as a show-case to apply deep learning on the combined data for several reasons. First, urban environments are an obvious application scenario for mobile systems. Second, they are often challenging for classical image-based segmentation, as many elements like streets and sidewalks have similar colors. To overcome these ambiguities, the inclusion of additional information, e.g., spatial relations from the 3D point clouds is necessary. Third, we expect that the mixture of different materials like concrete, asphalt and vegetation will lead to significantly different spectral signatures, that will demonstrate the benefit of including the spectral domain into the classification process. Given these properties of the input data, a trained classifier should be able to differentiate between streets, sidewalks, cars, vegetation and buildings.

To verify these assumptions, we created a manually labeled reference dataset ("3D Hyperspectral Semantic Street Scene", 3DHSSS[1]). The dataset has been acquired with a Riegl VZ-400i terrestrial laser scanner with co-calibrated Resonon Pika L hyperspectral line camera. The measurement setup is shown in Fig. 1a. During scan acquisition, the laser scanner performs a 360°C rotation, which is used to create a hyperspectral panorama image from the single lines of the hyperspectral camera. The captured points are then projected into the panorama image, using the method described above [4]. Hence, each point that is visible from the hyperspectral camera, is associated with the corresponding spectral histogram consisting of 150 spectral values between 600 and 1000 nm nm. Since the horizontal field of view of the hyperspectral camera (48.5°) is less than that of the laser scanner (100°), not all 3D points are annotated with hyperspectral values. Areas without hyperspectral values were filtered out to make the evaluation of the results with and without hyperspectral data more comparable. As the reflected spectra are affected by external light conditions, the dataset includes scans taken at different periods in time and under varied weather conditions. This was done to enhance the stability and accuracy of the prediction models by providing training examples recorded under diverse illumination situations. The 3DHSSS dataset contains of two road sections in the city of Osnabrück , as shown in Fig. 1 (b). The first road section is located at the ICO Innovation Centre of the City of Osnabrück, Germany. The second is located near the Westerberg campus of Osnabrück University. The position difference between single scans varies from 15 to 50 m in each scan section. The complete dataset consists of 15 different scans with an average of 45 million points each, resulting in a total of approximately one billion points. In addition to the 3D coordinates, the laser scanner returns a distance-independent reflectance measure and pulse shape deviation values for each point, which are also included.

Within these hyperspectral 3D point clouds, we labeled instances of objects from the before mentioned categories using a customized 3D labeling tool. To obtain a ground truth suitable for training, the point clouds were manually segmented into 17 different classes: Street, Sidewalk, Building, Vehicle, Street Lamp, Fence, Sign, Sign Pole, Bollard, Driveway, Low Vegetation, Bush, Tree Crown, Tree Trunk, Barrier, Distribution Box and Scan Artifact. All classes except Fence and Distribution Box are present in all scans. These two exceptions are only visible in two of the 15 scans. Furthermore, the class Tree has been divided into the subclasses Tree Crown and Tree Trunk. The reason for this is that tree crowns are often undesirable in laser scans. Especially, when surface reconstruction is done, leaves and thin structures like trunks can lead to undesirable scattered artifacts. With this purpose in mind, we introduced these separate classes to support filtering.

[1] Available in the Robotic 3D Scan Repository: http://kos.informatik.uni-osnabrueck.de/3Dscans/.

(a) Sensor Setup (b) Dataset

Fig. 1. Visualization sensor setup (a) and the two scenes (b) in the 3HSSS data: the hyperspectral camera (Pika L) is mounted on the terrestrial laser scanner (Riegl VZ-400i). The acquisition of the laser scan and the hyperspectral image is synchronized. Figure (b) visualizes the *ICO* (top) and *Campus* (bottom) dataset. The left column of (b) shows aerial images of the respective environments. The right column of shows the combined point clouds combined from several scan positions, colorized with the point intensities provided by the laser scanner.

4 Semantic Classification with RandLA-Net

4.1 Pre-filtering

To prepare the hyperspectral data for classification, we perform several preprocessing steps. Since the laser scans contain many noisy points, we first apply the Statistical Outlier Removal algorithm (SOR)[32] to the recorded data to filter out artifacts. The algorithm was parameterized with the neighborhood size set to 16 and the standard deviation multiplier to 1. This setting provides a good performance in terms of outlier filtering while still preserving distant points. However, not all noise points can be eliminated with this heuristic approach. Another criterion to detect scan artifacts is to analyze the recorded intensity, as most recorded noise points have low intensity values. The problem is that some relevant materials like asphalt also absorb a lot of signal energy, hence also feature low intensity. Therefore, we decided not to apply a simple threshold filter. Instead, we examined whether these can also be reliably detected with the classification pipeline as separate class. This would allow to reliably remove them after classification without influencing the results of the relevant classes.

In a second step, the individual spectra are smoothed using the Savitzky-Golay filter. The window size was set to 5 and the degree of the polynomial to be fitted to 3. Next, the spectra are normalized using the SNV transformation, where each spectrum is centered around zero and divided by its standard deviation. Finally, the dimension of the hyperspectral data is reduced to 64 using Principal Component Analysis.

4.2 RandLA-Net Implementation and Parameterization

In our experiments, we used the RandLA-Net implementation of Open3D-ML [33]. To integrate the hyperspectral values into the existing software, we implemented a custom dataset class in the framework, which provides functionality for loading point clouds and hyperspectral images. The 3DHSSS dataset is stored in the well known HDF5 file format in a schema similar to [34]. In addition to the 3D point clouds, hyperspectral images and label sets, all required meta information, i.e., camera parameters, resolution and scanning position, are also stored in the HDF5 file. A custom dataset class loads the corresponding training and test data from the file. Before training, the point clouds are downsampled with a voxel grid filter to ensure uniform point density (voxel size: 50 mm). After downsampling, a kd-tree is built for each point cloud to support k-nearest neighbor search. This is required to speed up the computation, since RandLA-Net analyzes the neighborhood of a point when computing features. We configured RandLA-Net to use five encoder and decoder layers. After each encoder layer, the point cloud is reduced using random sampling. The first four layers reduce the points by a factor of 4 and the last layer by a factor of 2. The output feature dimension of each layer was set to $16 \rightarrow 64 \rightarrow 128 \rightarrow 256 \rightarrow 512$. RandLA-Net receives 2^{16} points with matching hyperspectral (reduced to a dimension of 64, using PCA) and reflectance values as input. The initial reduction of the features to 16 by the first layer bears the risk that potentially important information is lost. However, with higher dimensions, the complexity of the neural network increases enormously, which makes training on consumer graphics cards unfeasible. To obtain more training data and make the model more robust, we also apply a data augmentation step before training, which adds noise and rotates or mirrors the data randomly.

We trained to RandLA-Net models on the generated training datasets: one with and one without hyperspectral data to provide a baseline for comparison. The point cloud without hyperspectral values consists of the 3D coordinates and the corresponding distance-independent reflectance values. The point clouds of the validation set are then segmented with the trained classifier and the standard metrics Accuracy, Precision, Recall, (mean) Intersection over Union are computed to assess the classification result.

5 Evaluation

In this Section we discuss the results of the trained classifiers on point clouds containing both hyperspectral and regular point clouds. Figure 2 shows exemplary classification results. Both classifiers perform well, however the hyperspectral information can enhance the segmentation of geometrically similar objects. One example is the classification of tree trunks which are sometimes confused with lamps without hyperspectral information. Due to the fact that vegetation can be separated easily in the spectral domain, the additional information here helps the network to learn a classifier that exploits this information as well.

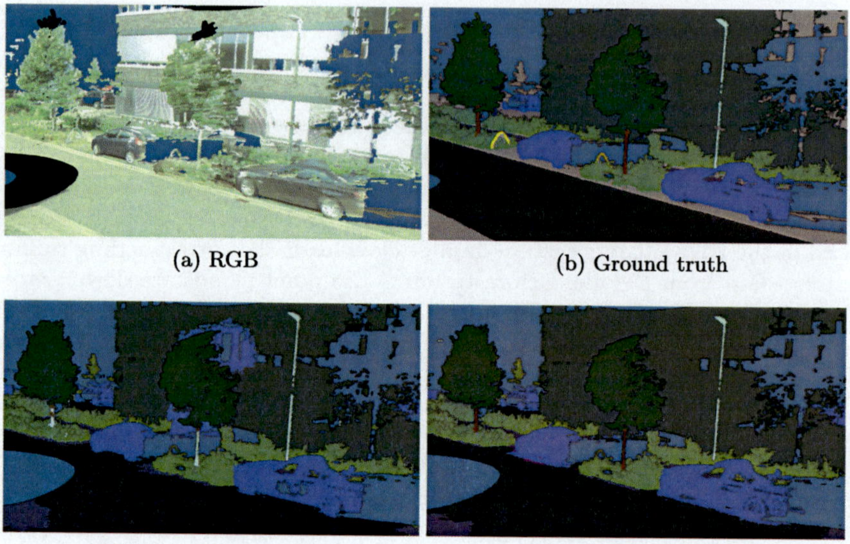

(a) RGB (b) Ground truth

(c) Without hyperspectral (d) with hyperspectral

Fig. 2. Segmentation results of one point cloud. The majority of the data is well classified both with and without hyperspectral data. However, it is noticeable that without hyperspectral data (c) errors occur when classifiers similar classes such as tree (brown) and streetlamp (light blue). These can be reduced by the additional hyperspectral data (d).

For quantitative evaluation, we first compare the average values of the evaluation metrics (cf. Table 1). Because the number of samples of the various classes in the dataset is partially imbalanced, the averages were calculated using the macro method as well as the micro method. The macro method calculates the metrics for each class individually and then averages them, with equal weighting for each class. In the micro method, the individual true positives (TP), true negatives (TN), false positives (FP), and false negatives (FN) of the classes are determined and then summed up to calculate the micro statistics.

Table 1 clearly shows that the model with additional hyperspectral data performed better on all metrics. In addition to these averages, the classification results for each separate class are shown in Table 2. Some classes listed at the bottom of the table, such as Bollard, Sign Pole, Barrier, and Fence, are classified very poorly in some cases because the number of samples in the dataset are too low. The proportion of samples of these classes in the dataset is less than 1%. When examining the individual class results, it is evident that the use

Table 1. Comparison of classification results for the model with and without hyperspectral data, using standard metrics: IoU (Intersection over Union),F1-Score, Accuracy, Precision and Recall. Due to imbalanced data, the averages of the metrics were calculated using both the micro and macro methods.

Spectral	Acc.	IoU (Macro)	IoU (Micro)	F1 (Macro)	F1 (Micro)	Precision (Macro)	Precision (Micro)	Recall (Macro)	Recall (Micro)
No	0.83	0.39	0.72	0.46	0.83	0.45	0.83	0.54	0.83
Yes	0.89	0.52	0.80	0.60	0.89	0.66	0.89	0.61	0.89

Table 2. Per class results of the classification models. The results are presented as Intersection over Union (IoU), Precision (Pre) and Recall (Rec)

	Without Hyperspectral			With Hyperspectral		
Class	IoU	Pre	Rec	IoU	Pre	Rec
Scan Artifact	0.15	0.16	**0.99**	**0.86**	**0.89**	0.97
Street	0.82	0.83	0.98	**0.89**	**0.90**	0.98
Vehicle	0.32	0.32	0.96	**0.85**	**0.86**	**0.98**
Sign	**0.82**	**0.94**	**0.86**	0.74	0.91	0.79
Building	0.63	0.98	0.64	**0.94**	**0.99**	**0.95**
Sidewalk	0.59	0.76	**0.72**	0.59	**0.87**	0.64
Driveway	0.03	0.08	0.04	**0.12**	**0.17**	**0.29**
Tree Trunk	0.17	0.25	0.34	**0.55**	**0.71**	**0.70**
Tree Crown	0.90	0.90	0.99	0.90	0.90	0.99
Bush	**0.85**	0.97	**0.87**	0.75	0.97	0.76
Low Vegetation	**0.79**	**0.84**	0.93	0.56	0.57	**0.97**
Street Lamp	0.43	0.49	0.79	**0.70**	**0.78**	**0.88**
Bollard	0.00	0.00	0.00	**0.01**	**0.02**	**0.01**
Sign Pole	0.00	0.00	0.00	**0.01**	**1.00**	**0.01**
Distribution Box	0.01	0.04	0.03	**0.13**	**0.34**	**0.17**
Barrier	0.00	0.00	0.00	0.00	0.00	0.00
Fence	0.01	0.02	0.04	**0.15**	**0.25**	**0.27**

of additional hyperspectral data can improve the quality of the classifier. For the classes Scan Artifact, Street and Vehicle the hyperspectral data results in less false positives (noticeable by the higher precision values).

Scan artifacts are detected successfully with and without hyperspectral data, though without hyperspectral data the false positive rate is very high. The models can therefore also serve as a filter to remove undesired scan artifacts. As discussed before, the metrics also reflect the observation that classification of tree trunks is improved. Without hyperspectral data, Tree Trunks are often misclassified as Street Lamps (see Fig. 2). Here again the hyperspectral data leads to a better distinction between similar objects with a different material composition.

6 Conclusion and Outlook

We have established an initial baseline to demonstrate that hyperspectral data as additional domain to spatial data can improve segmentation and classification with neural networks on 3D data. We showed that RandLA-Net can be extended with little effort to support the segmentation of hyperspectral 3D point clouds. The segmentation results on the provided manually labeled reference dataset prove that the inclusion of this additional data domain improves the segmentation results as expected. In future work, the data basis in terms of reference classes should be further improved to achieve a better generalization of the deep learning models. In particular, point clouds of regions containing more of the less common classes should be included. Additionally, similar classes should be combined, e.g., bush and tree crowns or barrier and bollard. Furthermore, it would also be beneficial to capture images at different seasons and more varying lighting conditions, since the spectral signature of materials is strongly dependent on the ambient illumination. Currently, we only use the standard version of RandLA-Net to include the hyperspectral data. Since hyperspectral data is high dimensional, the original network structure should be enhanced to support such data more efficiently, as it is currently designed to consider only few point attributes. Here, special encoder/decoder designs to pre-classify the hyperspectral data seems to be a promising approach for future developments.

References

1. Guo, Y., Wang, H., Hu, Q., Liu, H., Liu, L., Bennamoun, M.: Deep learning for 3d point clouds: a survey. IEEE Trans. Pattern Anal. Mach. Intell. **43**(12), 4338–4364 (2021)
2. Hu, Q., Yang, B., Xie, L., Rosa, S., Guo, Y., Wang, Z., Trigoni, N., Markham, A.: Randla-net: Efficient semantic segmentation of large-scale point clouds. In: Proceedings of the IEEE/CVF Conference on Computer Vision and Pattern Recognition (CVPR) (2020, June)
3. Hackel, T., Savinov, N., Ladicky, L., Wegner, J.D., Schindler, K., Pollefeys, M.: Semantic3d. net: a new large-scale point cloud classification benchmark. arXiv preprint arXiv:1704.03847 (2017)
4. Igelbrink, F., Wiemann, T., Püetz, S., Hertzberg, J.: Markerless Ad-hoc Calibration of a Hyperspectral Camera and a 3D Laser Scanner. Advances in Intelligent Systems and Computing. Springer International Publishing (2018)
5. Ghamisi, P., Plaza, J., Chen, Y., Li, J., Plaza, A.J.: Advanced spectral classifiers for hyperspectral images: a review. IEEE Geosci. Remote Sens. Mag. **5**(1), 8–32 (2017)
6. Audebert, N., Le Saux, B., Lefèvre, S.: Deep learning for classification of hyperspectral data: a comparative review. IEEE Geosci. Remote Sens. Mag. **7**(2), 159–173 (2019)
7. Dargan, S., Kumar, M., Ayyagari, M.R., Kumar, G.: A survey of deep learning and its applications: a new paradigm to machine learning. Arch. Comput. Methods Eng. **27**(4), 1071–1092 (2019)

8. Qi, C.R., Su, H., Mo, K., Guibas, L.J.: Pointnet: Deep learning on point sets for 3d classification and segmentation. In: Proceedings of the IEEE Conference on computer Vision and Pattern Recognition, pp. 652–660 (2017)
9. Qi, C.R., Yi, L., Su, H., Guibas, L.J.: Pointnet++: Deep hierarchical feature learning on point sets in a metric space. arXiv preprint arXiv:1706.02413 (2017)
10. Armeni, I., Sax, A., Zamir, A.R., Savarese, S.: Joint 2D-3D-Semantic Data for Indoor Scene Understanding. ArXiv e-prints (2017, February)
11. Roynard, X., Deschaud, J.E., Goulette, F.: Paris-Lille-3d: a large and high-quality ground-truth urban point cloud dataset for automatic segmentation and classification. Int. J. Robot. Res. **37**(6), 545–557 (2018)
12. Behley, J., Garbade, M., Milioto, A., Quenzel, J., Behnke, S., Stachniss, C., Gall, J.: Semantickitti: a dataset for semantic scene understanding of lidar sequences. In: Proceedings of the IEEE/CVF International Conference on Computer Vision (ICCV) (2019)
13. Manolakis, D., Lockwood, R., Cooley, T.: Hyperspectral Imaging Remote Sensing. Cambridge University Press (2016)
14. Xue, J., Su, B.: Significant remote sensing vegetation indices: a review of developments and applications. J. Sens. **2017** (2017)
15. Wu, D., Sun, D.W.: Advanced applications of hyperspectral imaging technology for food quality and safety analysis and assessment: a review-part i: fundamentals. Innovative Food Sci. Emerg. Technol. **19**, 1–14 (2013)
16. Van der Meer, F.D., Van der Werff, H.M., Van Ruitenbeek, F.J., Hecker, C.A., Bakker, W.H., Noomen, M.F., Van Der Meijde, M., Carranza, E.J.M., De Smeth, J.B., Woldai, T.: Multi-and hyperspectral geologic remote sensing: a review. Int. J. Appl. Earth Obs. Geoinf. **14**(1), 112–128 (2012)
17. Goetz, A.F.: Three decades of hyperspectral remote sensing of the earth: a personal view. Remote Sens. Environ. **113**, S5–S16 (2009)
18. Hege, E.K., O'Connell, D., Johnson, W., Basty, S., Dereniak, E.L.: Hyperspectral imaging for astronomy and space surveillance. In: Imaging Spectrometry IX, Vol. 5159. International Society for Optics and Photonics, pp. 380–391 (2004)
19. Vaiphasa, C.: Consideration of smoothing techniques for hyperspectral remote sensing. ISPRS J. Photogrammetry Remote Sens. **60**(2), 91–99 (2006)
20. Raczkowska, M.K., Koziol, P., Urbaniak-Wasik, S., Paluszkiewicz, C., Kwiatek, W.M., Wrobel, T.P.: Influence of denoising on classification results in the context of hyperspectral data: high definition FT-IR imaging. Anal. Chimica Acta **1085**, 39–47 (2019)
21. Savitzky, A., Golay, M.J.E.: Smoothing and differentiation of data by simplified least squares procedures. Anal. Chem. **36**(8), 1627–1639 (1964)
22. Green, A., Berman, M., Switzer, P., Craig, M.: A transformation for ordering multispectral data in terms of image quality with implications for noise removal. IEEE Trans. Geosci. Remote Sens. **26**(1), 65–74 (1988)
23. Lee, J., Woodyatt, A., Berman, M.: Enhancement of high spectral resolution remote-sensing data by a noise-adjusted principal components transform. IEEE Trans. Geosci. Remote Sens. **28**(3), 295–304 (1990)
24. Rinnan, A., van den Berg, F., Engelsen, S.B.: Review of the most common preprocessing techniques for near-infrared spectra. TrAC Trends Anal. Chem. **28**(10), 1201–1222 (2009)
25. Barnes, R.J., Dhanoa, M.S., Lister, S.J.: Standard normal variate transformation and de-trending of near-infrared diffuse reflectance spectra. Appl. Spectrosc. **43**(5), 772–777 (1989)

26. Geladi, P., MacDougall, D., Martens, H.: Linearization and scatter-correction for near-infrared reflectance spectra of meat. Appl. Spectrosc. **39**(3), 491–500 (1985)
27. Buckley, S.J., Kurz, T.H., Howell, J.A., Schneider, D.: Terrestrial lidar and hyperspectral data fusion products for geological outcrop analysis. Comput. Geosci. **54**, 249–258 (2013)
28. Weinmann, M.: Fusion of hyperspectral, multispectral, color and 3d point cloud information for the semantic interpretation of urban environments. Int. Arch. Photogrammetry Remote Sens. Spat. Inf. Sci. (2019)
29. Brell, M., Segl, K., Guanter, L., Bookhagen, B.: 3d hyperspectral point cloud generation: fusing airborne laser scanning and hyperspectral imaging sensors for improved object-based information extraction. ISPRS J. Photogrammetry Remote Sens. **149**, 200–214 (2019)
30. Roy Choudhury, M., Das, S., Christopher, J., Apan, A., Chapman, S., Menzies, N.W., Dang, Y.P.: Improving biomass and grain yield prediction of wheat genotypes on sodic soil using integrated high-resolution multispectral, hyperspectral, 3d point cloud, and machine learning techniques. Remote Sens. **13**(17) (2021)
31. Goodfellow, I., Bengio, Y., Courville, A.: Deep Learning. MIT Press (2016) http://www.deeplearningbook.org
32. Rusu, R.B., Marton, Z.C., Blodow, N., Dolha, M., Beetz, M.: Towards 3d point cloud based object maps for household environments. Robot. Auton. Syst. **56**(11), 927–941 (2008)
33. Zhou, Q., Park, J., Koltun, V.: Open3d: a modern library for 3d data processing. CoRR abs/1801.09847 (2018)
34. Wiemann, T., Igelbrink, F., Pütz, S., Hertzberg, J.: A file structure and reference data set for high resolution hyperspectral 3d point clouds. IFAC-PapersOnLine **52**(8), 403–408 (2019)

3D Semantic Scene Perception Using Distributed Smart Edge Sensors

Simon Bultmann(✉) and Sven Behnke

Institute for Computer Science VI, Autonomous Intelligent Systems, University of Bonn, Friedrich -Hirzebruch -Allee 8, 53115 Bonn, Germany
bultmann@ais.uni-bonn.de
https://www.ais.uni-bonn.de

Abstract. We present a system for 3D semantic scene perception consisting of a network of distributed smart edge sensors. The sensor nodes are based on an embedded CNN inference accelerator and RGB-D and thermal cameras. Efficient vision CNN models for object detection, semantic segmentation, and human pose estimation run on-device in real time. 2D human keypoint estimations, augmented with the RGB-D depth estimate, as well as semantically annotated point clouds are streamed from the sensors to a central backend, where multiple viewpoints are fused into an allocentric 3D semantic scene model. As the image interpretation is computed locally, only semantic information is sent over the network. The raw images remain on the sensor boards, significantly reducing the required bandwidth, and mitigating privacy risks for the observed persons. We evaluate the proposed system in challenging real-world multi-person scenes in our lab. The proposed perception system provides a complete scene view containing semantically annotated 3D geometry and estimates 3D poses of multiple persons in real time.

Keywords: Semantic scene understanding · Intelligent sensors and systems · Visual perception · Sensor fusion

1 Introduction

Accurate semantic perception of 3D scene geometry and persons is challenging and an important prerequisite for many robotic tasks, such as safe and anticipative robot movement in the vicinity of people as well as human-robot interaction. In this work, we propose a system for 3D semantic scene perception consisting of a network of distributed smart edge sensors. It provides a complete scene view containing semantically annotated 3D geometry and estimates 3D poses of multiple persons in real time.

We build upon our previous work on real-time 3D human pose estimation using semantic feedback to smart edge sensors [2]. While this existing pipeline is able to track poses of multiple persons in real time, it lacks modeling of other aspects of the scene, i.e. 3D geometry, object detections, and surface categorization. Semantically annotated 3D geometry, however, is required to explain and

I. Petrovic et al. (Eds.): IAS 2022, LNNS 577, pp. 313–329, 2023.
https://doi.org/10.1007/978-3-031-22216-0_22

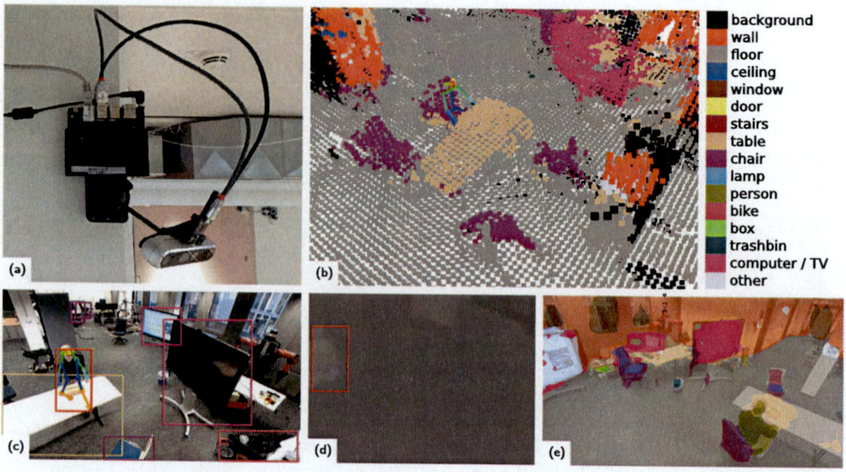

Fig. 1. Semantic perception with distributed smart edge sensors: (a) developed sensor node, (b) 3D semantic scene model with 3D human skeleton, (c) RGB and (d) thermal detections, (e) semantic segmentation. Person detections in red and skeleton keypoints colored by joint index. Occluded joints are marked in orange. CNN inference runs online on distributed sensors and semantic information is aggregated into an allocentric 3D scene model on the backend including 3D geometry (e.g., furniture, walls, floor) and 3D human pose.

predict interactions between persons and objects in the scene, and to handle occlusions. Temporal aggregation and fusion of semantic point clouds from multiple sensor perspectives further leads to a consistent and persistent 3D semantic scene model with the field of perception not being limited by the measurement range or occlusions of a single sensor.

To enable perception of these additional characteristics of the scene, the sensor network is extended with updated smart edge sensors with higher compute capabilities and greater flexibility w.r.t. the employed vision CNNs, as shown in Fig. 1. This enables to run object detection and semantic image segmentation together with human pose estimation on the sensors in real time. RGB-D cameras estimate 3D scene geometry and thermal cameras increase the person detection performance in low-light conditions. Semantic information from detections and image segmentation is fused into the point cloud computed from the depth image and 2D human joint detections are augmented with the depth measured at the keypoint location. Semantic point cloud and human poses are communicated to a central backend, where they are fused into an allocentric 3D metric-semantic scene model. Only the semantic information is sent over the network; the raw images remain on the sensor boards, significantly reducing the required bandwidth, and mitigating privacy issues for the observed persons.

The semantic point clouds from multiple viewpoints are aggregated into an allocentric map of 3D scene geometry and semantic classes on the backend. The map is further updated via ray-tracing to account for moving objects. 3D human poses are estimated in real time in the scene via multi-view triangulation. The allocentric 3D human poses are projected into the local camera views and sent back to the sensors as semantic feedback [2], where they are fused with the local detections. The 3D scene geometry enables to compute occlusion information for each joint in the respective camera view. This information is included into the semantic feedback from backend to sensors, improving the local scene model of each sensor by incorporating global context information. Unreliable, occluded joint detections can be discarded, and the local model is completed by the more reliable semantic feedback reprojected from the global, fused 3D model.

We evaluate the proposed system in experiments with challenging real-world multi-person scenes. In summary, our contributions are:

- The development of a smart edge sensor platform based on the Nvidia Jetson Xavier NX development kit and an RGB-D and thermal camera, running efficient vision CNN models for object detection and semantic segmentation together with human pose estimation on-device in real time;
- Temporal multi-view fusion of semantic point clouds from individual sensors into an allocentric semantic map of 3D scene geometry;
- The integration of multiple instances of the proposed novel sensor nodes into a network of distributed smart edge sensors for real-time multi-view 3D human pose estimation using semantic feedback [2], complementing the feedback from backend to sensor with occlusion information for human joints in the respective camera views, computed via ray-tracing through the estimated 3D scene geometry.

We make our implementation for both sensor boards[1] and backend[2] publicly available.

2 Related Work

Lightweight Vision CNNs for Embedded Hardware. Convolutional neural networks (CNNs) set the state-of-the-art for image processing and computer vision. On systems with restricted computational resources, like mobile embedded sensor platforms, however, lightweight, efficient models must be employed to achieve real-time performance. A popular approach is to replace classical backbone networks such as ResNets [10] with MobileNet [11,20] or EfficientNet [23] architectures, as the main computational load of CNN inference often lies in the backbone feature extractor. These architectures decrease the number of parameters and the computational cost significantly, e.g., by replacing standard convolutions with depthwise-separable convolutions.

For object detection on embedded devices, single-stage architectures such as SSD [13] or YOLO [17], which use predefined anchors instead of additional

[1] https://github.com/AIS-Bonn/JetsonTRTPerception.
[2] https://github.com/AIS-Bonn/SmartEdgeSensor3DScenePerception.

region proposal networks, were shown to be efficient. In our work, we employ the recently proposed MobileDets [26], that are optimized for embedded inference accelerators using the SSD architecture with MobileNet v3 backbone.

The DeepLab v3+ architecture [5] for semantic segmentation uses elements of MobileNets, such as depthwise-separable convolutions, for efficiency on embedded hardware and shows state-of-the-art performance on large, general datasets. We employ a DeepLab v3+ model with MobileNet v3 backbone in our work.

For human pose estimation, OpenPose [4] set a new standard by detecting body parts of multiple persons in an image and associating them to individuals via Part Affinity Fields (PAFs). This bottom-up approach scales well with the number of person detections. Top-down approaches, on the other hand, first detect individuals and then estimate body keypoints for each single-person crop. These approaches achieve higher accuracy and better scale-invariance, as the person detections are interpolated to a fixed input resolution before pose inference. Xiao et al. [25] propose an efficient CNN architecture consisting of a backbone feature extractor and deconvolutional layers. We adopt this architecture and replace the ResNet backbone with MobileNet v3 for better efficiency on embedded hardware.

Semantic Mapping. Semantic information about the environment is a prerequisite for many high-level robotic functions. For this, semantic mapping systems build an allocentric model of 3D scene geometry with semantic class information.

SemanticFusion [14] builds semantic maps from RGB-D camera input using surface elements (Surfels), where a Gaussian approximates the local point distribution. Pixel-wise class probabilities are obtained from the color image via semantic segmentation and fused into the map using a Bayesian approach assuming independence of individual measurements.

Dengler et al. [8] proposed an object-centric 2D/3D map representation for real-time service robotics applications, using RGB-D data as input. A geometric segmentation of small objects in the point cloud is obtained via Euclidean clustering. Stückler et al. [22] fuse probabilistic object segmentations from multiple RGB-D camera views into a voxel-based 3D map using a Bayesian framework.

Recently, Bultmann et al. [3] proposed a framework for online multi-modal semantic fusion onboard a UAV combining 3D LiDAR data with detections and semantic segmentation of 2D color and thermal images. The semantic point clouds are aggregated into an allocentric voxel-based map using poses from LiDAR odometry to transform multiple viewpoints into a common coordinate frame.

3D Human Pose Estimation. 2D human joint detection, inferred by image CNNs as introduced above, provides the input for 3D human pose estimation. 3D poses are recovered from 2D keypoint detections from multiple, calibrated camera views via variants of the Pictorial Structures Model (PSM) [9,15] or based on direct triangulation [6,18]. The PSM approaches are computationally expensive, due to a large discrete state space used in the optimization, restricting them

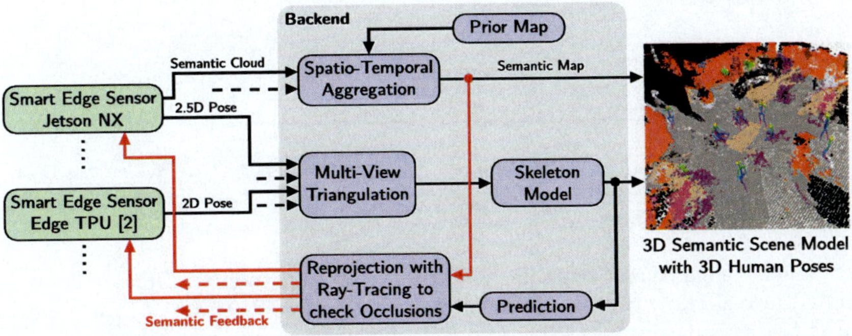

Fig. 2. Overview of the multi-sensor pipeline for 3D semantic mapping and human pose estimation: The Jetson NX smart edge sensors extend a sensor network from prior work [2] of nodes with lower compute capabilities. Semantic point clouds from multiple sensor views are aggregated into an allocentric 3D semantic map and 3D human poses are estimated in real time. The map is used to check reprojected joints for occlusion in the resp. sensor view via ray-tracing and this information is added to the semantic feedback sent to the smart edge sensors.

to offline processing. Triangulation-based approaches are more computationally efficient and enable 3D pose estimation for multiple persons in real time.

In previous work [2], we proposed a pipeline for real-time 3D human pose estimation using multiple calibrated smart edge sensors that perform 2D pose estimation on-device. Semantic pose information is transmitted to a central backend where multiple views are fused into a 3D skeleton via triangulation and an efficient, factor graph-based skeleton model. The fused allocentric 3D joint positions, after motion prediction to compensate for the pipeline delay, are reprojected into local views and sent back to the sensors as semantic feedback, where they are fused with the detected keypoint heatmaps. This enables the sensors to incorporate global context information into their local scene view interpretation. The pipeline delay is estimated as the difference of the timestamps of the current detection and the latest received feedback message on a sensor and updated using a moving average filter.

We build upon this work and extend the sensor network with new smart edge sensor nodes with significantly increased computational power and RGB-D cameras that enable the perception of 3D geometry. In addition to human pose estimation, object detection and semantic image segmentation are computed on the sensor boards and fused via 3D projection into a semantic point cloud. Semantic point clouds from multiple sensor views are fused into a sparse voxel hash-map with per-voxel full semantic class probabilities on the backend. The semantic map is used to obtain occlusion information for person keypoints in the local camera views, which is added to the semantic feedback to increase the robustness of the pose estimation pipeline.

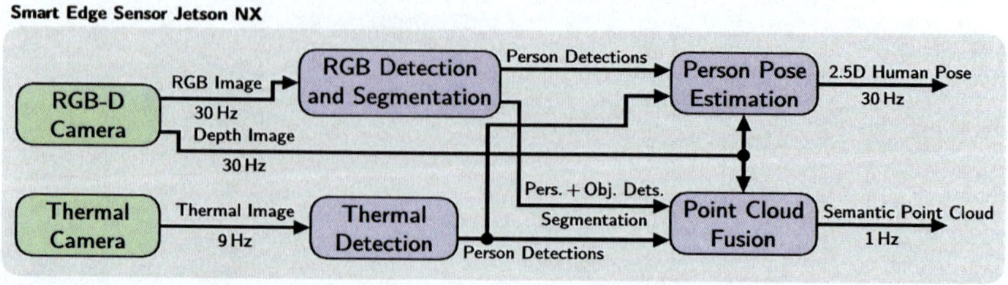

Fig. 3. Smart edge sensor semantic perception system overview. Human poses are estimated in real time, while the semantic point cloud of the static or slowly moving scene geometry is output at a lower frequency to save compute resources.

3 Method

Figure 2 illustrates the proposed multi-sensor pipeline for 3D semantic perception and human pose estimation combining two types of smart edge sensors. The proposed Jetson NX sensors are integrated into a sensor network from prior work [2], consisting of nodes based on the Google Edge TPU with lower compute capabilities and RGB image-only 2D human pose estimation, without local depth estimation. We consider a calibrated camera network, with known projection matrices from sensor to world coordinates, where the sensors are software-synchronized via NTP. Semantic mapping is only performed with the here proposed Jetson NX sensor nodes, while data from both sensor types is combined for 3D human pose estimation.

An overview of the proposed approach for semantic perception onboard each Jetson NX smart edge sensor is given in Fig. 3. We detail individual components of the data processing on each sensor board, as well as the fusion of multiple sensor views for 3D mapping and 3D human pose estimation in the following.

3.1 Smart Edge Sensor Hardware

We developed smart edge sensors based on the Nvidia Jetson Xavier NX developer kit[3] (cf. Fig. 1a). They are equipped with a 6-core ARM processor, 384 CUDA cores, and 8 GB of RAM. The Jetson NX embedded system achieves a CNN inference performance of 21 trillion operations per second (TOPS), a significant increase compared to the 4 TOPS of the sensor platform employed in our previous work [2]. For visual perception, we connect an Intel RealSense D455 RGB-D camera and a FLIR Lepton 3.5 thermal camera to the Jetson NX board.

3.2 Single-View Embedded Semantic Perception

Person and Object Detection. We employ the recent MobileDet architecture [26] for person and object detection. The RGB detector is trained on the COCO

[3] https://developer.nvidia.com/embedded/jetson-xavier-nx-devkit.

dataset [12] using *person* and 12 indoor object classes (e.g., *chair*, *table*, *computer/tv*), with an input resolution of 848×480 px. The same network architecture is used for the thermal detector, taking one-channel 8-bit gray-scale thermal images at the camera resolution of 160×120 px as input. The thermal detector is trained on the ChaLearn IPHD dataset [7], with annotations for the *person* class only.

Person Keypoint Estimation. We adopt a top-down approach for person pose estimation on the smart edge sensors, where crops of single persons are analyzed by the keypoint estimation CNN. The CNN architecture of Xiao et al. [25] is the basis of our person pose estimation, but we exchange the ResNet backbone with the significantly more lightweight MobileNet v3 feature extractor [11]. We train the pose estimation network on the COCO dataset [12] using person keypoint annotations.

Person detections from RGB and thermal images are forwarded to the keypoint estimation CNN. The RGB-D depth thereby is used to project detections from the thermal camera to the color image. Redundant detections of the same person in both modalities are filtered via non-maximum suppression (NMS). Each person crop is then resized to the fixed 192×256 input resolution of the keypoint estimation CNN and inference is run for all crops together in batched mode. Batch processing gives a significant improvement in the scaling of inference time with the number of persons compared to previous work [2], where the embedded hardware only supported processing a single crop at a time (cf. Sect. 4.4).

The pose estimation model outputs multi-channel images, called *heatmaps*, encoding the confidence of a joint being present at the pixel location. As single-person crops are processed, 2D joint locations are determined as global maxima of the respective heatmap channel. The RGB-D range image is used to augment the 2D keypoints to a 2.5D pose representation. For each joint, the median depth of a 5×5 px region around the joint location is obtained from the depth image. The local depth estimate enables the projection of keypoints into three-dimensional space but often suffers from noise and occlusions, as is further analyzed in Sect. 3.5. The 2.5D pose estimate for each detected person is sent to a central backend, where multiple sensor views are fused into a coherent 3D pose representation. The pose estimation pipeline runs with the highest real-time priority on the sensor boards, to enable tracking of dynamic human motions. To save computational resources, the person detector is run only once per second and the crops are updated based on the keypoint estimations between detector runs.

Semantic Segmentation. We adopt the DeepLab v3+ [5] architecture with MobileNet v3 [11] backbone for semantic segmentation. We train the model on the indoor scenes of the ADE20K dataset [27] and reduce the labels to 16 classes most relevant for the intended indoor application scenarios (cf. Fig. 1). The input image size is set to 849×481 px during inference, fitting to the 16:9 aspect ratio of our camera. The semantic segmentation is run only once per second, similar to the person and object detector, to save computational resources.

3.3 Multi-modal Semantic Point Cloud Fusion

We obtain a geometric point cloud by projecting the RGB-D range image into 3D. The point cloud is uniformly subsampled using a voxel-grid filter with 5 cm resolution to reduce the amount of data, economizing computational resources, and later network bandwidth for transmission to the backend. Sparse outlier measurements are further removed by a statistical outlier filter, as implemented in the PCL library [19]. A point is deleted when the distance to its neighbors is outside an interval defined by the mean and standard deviation of the entire point cloud.

Semantic information from RGB and thermal detections, as well as RGB semantic segmentation, is fused into the point cloud using a projection-based approach as proposed by Bultmann et al. [3]. For this, the points are projected into the segmentation mask inferred from the RGB image. The semantic class scores $\boldsymbol{c}_{\text{segm}} \in \mathbb{R}^C$ are obtained from semantic segmentation via bilinear interpolation at the projected point location. A normalized probability distribution over the employed $C = 16$ classes is then approximated by applying the soft-max operation:

$$p_i = \sigma(c_i) = \frac{\exp c_i}{\sum_{j=1}^{C} \exp c_j}, \quad (1)$$

obtaining $\boldsymbol{p}_{\text{segm}} \in \mathbb{R}^C$, with $p_i \in [0, 1]$ and $\sum_i p_i = 1$. If a projected point falls inside a detection bounding box in either thermal or color images, we further fuse the detector result with the semantic segmentation. We reconstruct the detection probability distribution $\boldsymbol{p}_{\text{det}}$ from the score for the detected class following the maximum entropy principle: The probability of the detected class p_{det} is given by the detector score and the remaining probability mass $1 - p_{\text{det}}$ is equally distributed over the remaining $C - 1$ classes. Both estimates are fused following the Bayesian update rule [14], assuming independence of segmentation and detection:

$$\boldsymbol{p}_{\text{fused}} = \frac{\boldsymbol{p}_{\text{segm}} \circ \boldsymbol{p}_{\text{det}}}{\sum_{i=1}^{C} p_{i,\text{segm}} \, p_{i,\text{det}}}, \quad (2)$$

with $\circ$ being the coefficient-wise product. For better numerical stability, we use the implementation of the Bayesian fusion in logarithmic form from [3].

As the detection bounding boxes are axis-aligned, border-effects have to be considered for non-rectangular or non-axis-aligned objects before detection fusion. Inclusion of all points projected into the bounding box in the fusion would falsely label points on the ground and in the background as the detected class. To alleviate this issue, the ground plane is removed and the remaining points are clustered in 3D Euclidean space by a distance threshold [3]. Only the clustered points are included into detection fusion.

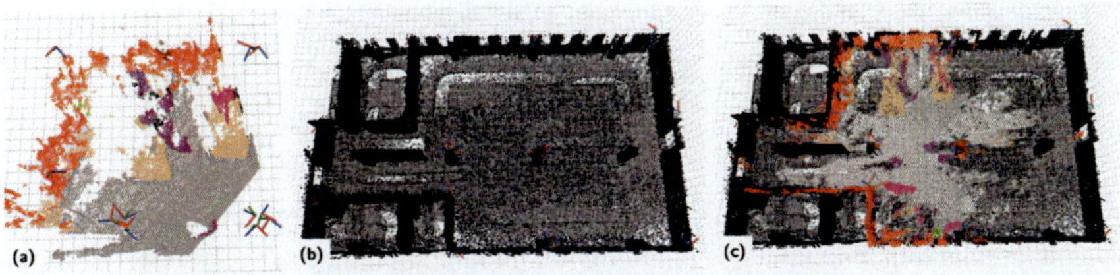

Fig. 4. 3D Semantic Mapping: (a) semantic point cloud of a single sensor, (b) prior map, (c) fused semantic map. The smart edge sensors send semantic point clouds of their resp. perspectives to the backend. Here, the fused map is initialized with a prior map and updated with the observations including semantic classes.

The output semantic point cloud includes the class probability vector and the argmax class color per point (cf. Fig. 4a) and is sent to the central backend over the network. It is computed at a reduced update frequency 1 Hz on the sensors, as it is targeted to observe the static or slowly moving scene geometry. Thus, computational resources are kept free for the real-time estimation of dynamic human motions in the pose estimation pipeline.

3.4 3D Semantic Mapping

Semantic point clouds from multiple calibrated camera perspectives are fused into an allocentric semantic map on the central backend, as illustrated in Fig. 4. For this, the 3D space is uniformly subdivided into cubic volume elements (voxels). We employ sparse voxel hashing [16] as a memory-efficient data structure.

For indoor environments, prior information on building structure is often easily available, e.g., via floor plans or 3D models. To incorporate this prior information, we initialize the scene model with a prior map of the empty building (Fig. 4b). Here, the prior map was obtained from aggregated laser scans of the empty rooms, but it could also be replaced, e.g., by a floor plan with a fixed wall height or an architectural CAD model of the building.

To include semantic information and current observations into the map, we transform the semantic point clouds from individual sensors (Fig. 4a) into global coordinates using the known camera calibration and bin the points into voxels of 10 cm side length. The semantic probabilities of all points falling into a voxel are fused probabilistically, using Bayes' rule [14], assuming independence between observations $P\left(l_i|X_k\right)$ for the semantic point cloud X_k with label l_i for class i:

$$P(l_i|X_{1:k}) = \frac{P\left(l_i|X_{1:k-1}\right) P\left(l_i|X_k\right)}{\sum_i P\left(l_i|X_{1:k-1}\right) P\left(l_i|X_k\right)}\,. \tag{3}$$

We again use the implementation of Bayesian fusion in logarithmic form from [3] for better numerical stability. Points labeled as *person* are not included in the semantic map, as dynamic human segments are tracked at a higher update rate via the 3D skeleton representation (cf. Sect. 3.5).

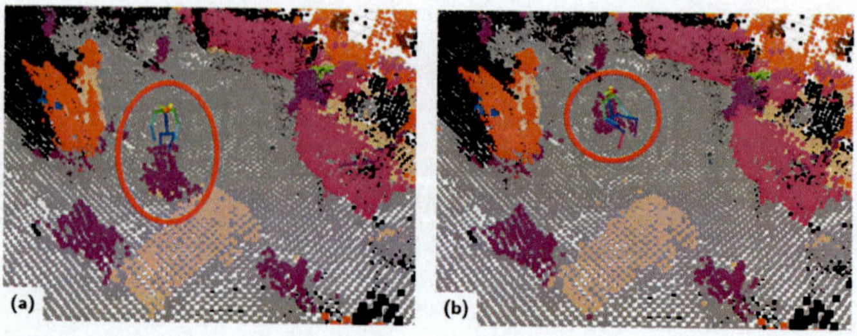

Fig. 5. Map update: 3D Semantic map and 3D person skeleton (a) before and (b) after moving a chair (highlighted with red circle). The semantic map is updated via ray-tracing to account for moving objects.

The fused semantic map (Fig. 4c) contains 3D geometry and semantic classes of the areas observed by the smart edge sensors and is completed by the prior information for currently unobserved areas.

To account for moving objects, we adopt a simple ray-tracing approach to update occupancy information of the voxels [21], as illustrated in Fig. 5. Starting from the sensor pose towards the measured voxels, we ray-trace using a 3D implementation of Bresenham's algorithm [1]. All voxels between the start and endpoint of the ray are updated as being free space, while the measured voxels are updated as being occupied. The semantic class probability is reset when a voxel state transitions from occupied to free.

3.5 3D Human Pose Estimation with Occlusion Feedback

The 3D joint positions of the detected persons are recovered from a set of 2D keypoint detections from multiple viewpoints via triangulation, and the result is refined using a factor graph skeleton model, as proposed in [2]. Furthermore, a semantic feedback channel from backend to sensors is implemented in our framework that enables the local semantic models of each sensor to incorporate globally-fused 3D pose information.

In this work, we add occlusion information for each joint to the semantic feedback, using the estimated 3D semantic map (cf. Fig. 2). We employ ray-tracing to check each joint for occlusion in the respective local sensor view. For this, we traverse the ray from the respective camera pose to the 3D joint through the 3D map using Bresenham's 3D line-search [1]. When the ray hits a minimum number of $k = 2$ occupied voxels, the joint is marked as occluded in the respective local view.

The benefits of the occlusion information for the local sensor model are illustrated in Fig. 6. Without occlusion feedback, heavy occlusion causes the pose estimation to collapse to the visible side only (Fig. 6b), which cannot be recovered by the feedback on the heatmap level [2]. With occlusion information (Fig. 6a), unreliable, occluded joint detections can be discarded, and the local model is completed by the more reliable semantic feedback. Completely occluded persons

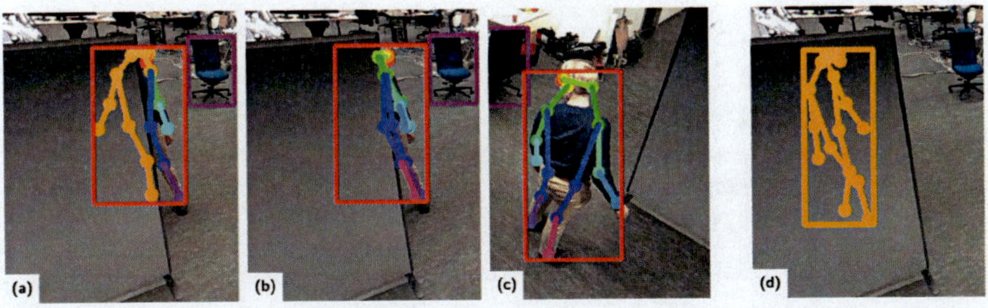

Fig. 6. Occlusion information in semantic feedback: Local 2D pose estimation (a) with and (b) without occlusion information via semantic feedback, (c) reference view without occlusion, (d) fully occluded person. Person detections in red and skeleton keypoints colored by joint index. Occluded joints are marked in orange. Heavy occlusion causes the pose estimation to collapse to the visible side only. With occlusion information, unreliable, occluded joint detections can be discarded, and the local model is completed by the more reliable semantic feedback.

can also be added back into the local model (Fig. 6d), making the sensor aware of persons that are going to re-appear in the future. Furthermore, the known occluded joints are excluded from multi-view triangulation in the next forward pass, as no new information can be gained from the respective sensor view. In Sect. 4.2, we show that the added occlusion information improves the overall consistency in terms of reprojection error.

We further investigate the reliability of the local depth estimate of skeleton joints from the Jetson NX smart edge sensors. The local depth enables estimating 3D joint positions from a single camera only, without dependence on other sensors. However, the RGB-D depth suffers from significant noise at larger distances and the depth measurement for a joint often is obstructed by occlusions or self-occlusions, as illustrated in Fig. 7. The local depth estimate results in a good approximation of the 3D skeleton for a front view, but is inaccurate in case of self-occlusion, e.g. from a side view. Multi-view triangulation is more robust to these issues but requires synchronization with other sensors.

The local depth estimate, however, can still be used as an indication to constrain the data association between cameras for multi-view triangulation. Person detections from different camera views are associated based on the epipolar distance of their joints using the efficient iterative greedy matching proposed by Tanke et al. [24]. Keypoint detections from one image are projected as epipolar lines into the other cameras, where the distance from corresponding joint detections to the epipolar line is used as data-association cost. When a depth estimate is available, including an uncertainty interval computed from the keypoint confidence and the distribution of local depth readings, the matching can be restricted to a line segment. This helps to resolve ambiguous situations, where keypoints from multiple persons have a low distance to the epipolar line but are located at different positions along the line. Keypoints located on the line segment close

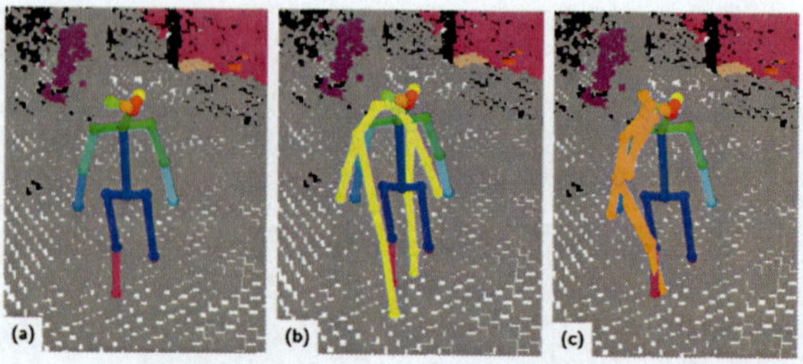

Fig. 7. Comparison of multi-view triangulation and local depth for estimating person keypoints in 3D: (a) multi-view triangulated 3D skeleton, (b) local depth from front view, and (c) local depth from side view. The local depth estimate results in a good approximation of the 3D skeleton for a front view, but is inaccurate in case of self-occlusion, e.g. from a side view. Multi-view triangulation is more robust but requires synchronization with other sensors.

to the projected depth estimate will receive lower data association cost while correspondences outside the projected depth interval will be discarded.

4 Evaluation

We evaluate the proposed system in challenging, cluttered real-world indoor scenes with multiple persons.

4.1 Implementation Details

Our sensor network consists of 20 smart edge sensors, thereof 4 based on the Jetson NX board, as introduced in this paper, and 16 based on the Google Edge TPU [2]. The boards are connected to mains power supply and the power consumption of an NX board is 20 – 25 W during inference, thereof ~ 5 W for powering the RGB-D camera, compared to 7 W for the Edge TPU board. The sensors cover an area of roughly 12×22 m. The cameras face downward towards the center and run 30 Hz. We conduct experiments with the proposed, extended sensor network, with 8 persons moving in the covered area, which are evaluated in the following using a sequence of 106 s containing ~3,000 frames per camera.

4.2 Quantitative Results

To analyze the consistency between local and globally-fused human pose estimation, we evaluate the error between 2D poses detected in the individual sensor views and reprojected fused 3D poses in Tab. 1. The reprojection error decreases for all joint classes when using the semantic feedback proposed in [2] over a

Table 1. Evaluation in real-world multi-person scenes with 20 cameras and 8 persons: Reprojection error (px) per joint class between detected 2D poses and fused 3D poses.

Feedback	Cams	Pers	Head	Hips	Knees	Ankls	Shlds	Elbs	Wrists	Avg
w/o fb	20	8	5.09	5.98	5.75	6.87	4.67	5.53	6.95	5.69
fb [2]	20	8	4.76	5.51	4.98	5.94	4.34	4.88	5.66	5.08
fb + occl	20	8	4.36	4.68	4.37	5.44	3.97	4.38	**5.04**	4.56
fb + occl. + local depth	20	8	**4.30**	**4.63**	**4.32**	**5.42**	**3.91**	**4.33**	**5.04**	**4.51**

Bold indicates lowest reprojection error (columnwise)

purely feed-forward pipeline. Adding the occlusion information, as proposed in this work, further decreases the reprojection error, as unreliable occluded keypoint detections can be discarded and excluded from multi-view triangulation. Constraining the data association using the local depth estimates of the RGB-D cameras gives a further small improvement. The proposed pipeline leads to the lowest reprojection error for all joint classes, amounting to 4.51 px on average, indicating that the consistency between local and globally-fused pose estimation increases through the semantic feedback with occlusion information and by using local depth estimates in the data association step for multi-view triangulation.

4.3 Qualitative Results

An exemplary scene of the real-world multi-person experiments is shown in Fig. 8. Local detections and pose estimation in two reference camera views are depicted together with the 3D semantic scene view. 3D poses of eight persons are estimated online, in real-time during the experiment. The semantic map represents the 3D geometry of the scene, fusing prior map and current sensor observations, including semantic class probabilities. Interactions between persons and objects in the scene, e.g., persons sitting on chairs, are explained in a physically plausible manner by the scene model.

4.4 Run-time Analysis

We analyze the run-time of CNN inference and the validation score on the respective training dataset (cf. Sect. 3.2) on different embedded hardware accelerators and for different numerical precision for the employed models in Tab. 2. Thermal detector and RGB segmentation are only executed on the Jetson NX, as the Edge TPU does not have enough computational power to run all models in parallel. The run-times on Jetson NX in 16-bit floating-point mode (fp16) are comparable to 8-bit quantized (int8) inference on the Edge TPU. The inference times roughly halve when using int8 precision on Jetson NX for the detectors and also decrease for the segmentation. For pose estimation, here stated for a single crop and batch size 1, the difference is less significant. The inference time is only about 4 ms, and the precision is less relevant compared to other overhead from

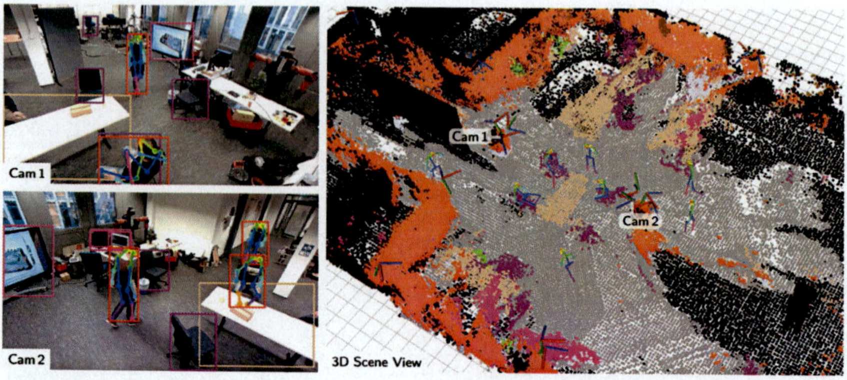

Fig. 8. Experiments in real-world multi-person scenes: Local detections of two reference sensor views and 3D semantic scene view with 3D poses of eight persons estimated in real-time. The geometry of the room is accurately represented by the map, fusing prior map (black) and current observations of smart edge sensors with their semantic classes (e.g., tables, chairs, computers). Interactions between persons and the scene, e.g., persons sitting on chairs (violet), are explained in a physically plausible way by the scene model.

the inference framework in this case. The validation score is given as bounding-box or keypoint mean average precision (mAP) for the *person* class as defined for the COCO dataset [12] for the detectors or pose estimation, respectively, and as mean intersection over union (mIOU) for the semantic segmentation. It decreases between 0.2 and 1.2% when using int8 precision instead of fp16. The slightly better performance of the RGB detector on the Edge TPU can be explained as it was trained for the *person* class only, while the detector used on Jetson NX was trained for *person* and 12 indoor object classes.

Table 3 shows the scaling of processing time for pose estimation, including CNN inference and post-processing, with an increasing number of person detections per image. During our experiments with 8 persons in the scene, a maximum of 6 persons were visible at a time in one camera. On the Edge TPU sensors [2], crops are processed one by one, as only a batch size of one is supported, and the runtime scales linearly. Up to three persons can be tracked at the full camera frame rate 30 Hz. On the Jetson NX platform, batch-processing is possible, and therefore the run-times scale sub-linearly. Up to five persons can be tracked at the full camera frame rate. For pose inference, there is only a small difference in run-time between fp16 and int8 mode on Jetson NX.

We run CNN inference in fp16 mode on the Jetson NX smart edge sensors during our online experiments to benefit from the higher numerical precision, as 8-bit quantization gives only little gains in run-time for the pose estimation with strong real-time constraints, and the fp16 inference time is sufficient for the other models that run with lower priority at 1 Hz update rate.

Table 2. Average inference time and validation score (given as mAP for detectors and pose estimation and mIOU for segmentation) of CNN models (batch size 1) on different embedded hardware and for different numerical precision.

Model	Input Res.	Edge TPU (int8)		Jetson NX (fp16)		Jetson NX (int8)	
		Time (ms)	Val. score (%)	Time (ms)	Val. score (%)	Time (ms)	Val. score (%)
RGB det	848 × 480	–	–	24.1	36.2	**11.8**	36.0
RGB det	640 × 480	21.5	**36.7**	–	–	–	–
Pose est	192 × 256	4.5	68.4	4.0	**69.3**	**3.5**	68.6
Thermal det	160 × 120	–	–	13.9	**25.4**	**6.0**	24.9
RGB segm	849 × 481	–	–	27.0	**50.0**	**20.0**	48.8

Bold indicates highest validation score resp. lowest inference time (rowwise)

Table 3. Average processing time for pose estimation (inference + post-processing) for increasing number of person detections per image. Batch processing can be used on Jetson NX while crops are processed one by one on the Edge TPU.

Sensor type	Precision	1 (ms)	2 (ms)	3 (ms)	4 (ms)	5 (ms)	6 (ms)
Edge TPU	int8	14.4	23.6	33.0	43.9	54.3	65.9
Jetson NX	int8	12.1	**15.0**	**18.0**	27.4	**31.0**	**38.8**
Jetson NX	fp16	**11.8**	17.4	21.9	**26.4**	31.7	41.5

Bold indicates lowest inference time (columnwise)

5 Conclusion

In this work, we presented a network of distributed smart edge sensors for 3D semantic scene perception, including static or slowly moving geometry as well as dynamic human motions. RGB-D and thermal camera images are processed locally on the sensor boards with vision CNNs for person and object detection, semantic segmentation, and human pose estimation. 2D human keypoint estimations, augmented with the RGB-D depth estimate, as well as semantically annotated point clouds are streamed from the sensors to a central backend, where multiple viewpoints are fused into an allocentric 3D semantic scene model. The individual sensors incorporate global context information into their local models via a semantic feedback channel. For this, the globally-fused 3D human poses are projected into the sensor views, where they are fused with the local detections. The estimated 3D geometry enables to add occlusion information for each joint to the semantic feedback, such that unreliable, occluded joint detections can be discarded on the sensors, and the local models can be complemented by the more reliable feedback joint positions. We built a sensor network of 20 smart edge sensors, thereof 4 based on the novel Jetson NX board, covering an area of about 12×22 m, and evaluated the proposed system in challenging, cluttered real-world scenes with up to 8 persons. Dynamic human motions are estimated in real time and the semantically annotated 3D geometry provides a complete scene view that also explains interactions between persons and objects in the scene.

Future work includes using the 3D semantic scene model and human poses estimated by the smart edge sensors to enable anticipative robot behavior and safe human-robot interaction in a shared workspace. Mobile sensor nodes could further be added to the sensor network for active exploration of areas not covered by the permanently installed sensors.

Acknowledgments. This work was funded by grant BE 2556/16-2 of the German Research Foundation (DFG) and Fraunhofer IAIS.

References

1. Amanatides, J., Woo, A.: A fast voxel traversal algorithm for ray tracing. In: 8th European Computer Graphics Conference and Exhibition (EuroGraphics) (1987)
2. Bultmann, S., Behnke, S.: Real-time multi-view 3D human pose estimation using semantic feedback to smart edge sensors. In: Robotics: Science and Systems (RSS) (2021)
3. Bultmann, S., Quenzel, J., Behnke, S.: Real-time multi-modal semantic fusion on unmanned aerial vehicles. In: European Conference on Mobile Robots (ECMR) (2021)
4. Cao, Z., Hidalgo, G., Simon, T., Wei, S.E., Sheikh, Y.: OpenPose: Realtime multi-person 2D pose estimation using part affinity fields. IEEE Trans. Pattern Anal. Mach. Intell. **43**(1), 172–186 (2021)
5. Chen, L.C., Zhu, Y., Papandreou, G., Schroff, F., Adam, H.: Encoder-decoder with atrous separable convolution for semantic image segmentation. In: European Conference on Computer Vision (ECCV), pp. 833–841 (2018)
6. Chen, L., Ai, H., Chen, R., Zhuang, Z., Liu, S.: Cross-view tracking for multi-human 3D pose estimation at over 100 FPS. In: IEEE Conference on Computer Vision and Pattern Recognition (CVPR), pp. 3276–3285 (2020)
7. Clapés, A., Jacques Junior, J.C.S., Morral, C., Escalera, S.: ChaLearn LAP 2020 challenge on identity-preserved human detection: Dataset and results. In: IEEE International Conference on Automatic Face and Gesture Recognition (FG), pp 801–808 (2020)
8. Dengler, N., Zaenker, T., Verdoja, F., Bennewitz, M.: Online object-oriented semantic mapping and map updating. In: European Conference on Mobile Robots (ECMR) (2021)
9. Dong, J., Jiang, W., Huang, Q., Bao, H., Zhou, X.: Fast and robust multi-person 3D pose estimation from multiple views. IEEE Conference on Computer Vision and Pattern Recognition (CVPR), pp. 7784–7793 (2019)
10. He, K., Zhang, X., Ren, S., Sun, J.: Deep residual learning for image recognition. In: IEEE Conference on Computer Vision and Pattern Recognition (CVPR), pp. 770–778 (2016)
11. Howard, A., Sandler, M., Chu, G., Chen, L.C., Chen, B., Tan, M., Wang, W., Zhu, Y., Pang, R., Vasudevan, V., Le, Q.V., Adam, H.: Searching for MobileNetV3. In: IEEE International Conference on Computer Vision (ICCV), pp. 1314–1324 (2019)
12. Lin, T.Y., Maire, M., Belongie, S., Hays, J., Perona, P., Ramanan, D., Dollár, P., Zitnick, C.L.: Microsoft COCO: Common objects in context. In: European Conference on Computer Vision (ECCV), pp. 740–755 (2014)

13. Liu, W., Anguelov, D., Erhan, D., Szegedy, C., Reed, S., Fu, C.Y., Berg, A.C.: SSD: Single shot multibox detector. In: European Conference on Computer Vision (ECCV), pp. 21–37 (2016)
14. McCormac, J., Handa, A., Davison, A., Leutenegger, S.: SemanticFusion: Dense 3D semantic mapping with convolutional neural networks. In: IEEE International Conference on Robotics and Automation (ICRA), pp. 4628–4635 (2017)
15. Qiu, H., Wang, C., Wang, J., Wang, N., Zeng, W.: Cross view fusion for 3D human pose estimation. In: IEEE International Conference on Computer Vision (ICCV), pp. 4341–4350 (2019)
16. Quenzel, J., Behnke, S.: Real-time multi-adaptive-resolution-surfel 6D LiDAR odometry using continuous-time trajectory optimization. In: IEEE/RSJ International Conference on Intelligent Robots and Systems (IROS), pp. 5499–5506 (2021)
17. Redmon, J., Divvala, S., Girshick, R., Farhadi, A.: You only look once: unified, real-time object detection. In: IEEE Conference on Computer Vision and Pattern Recognition (CVPR), pp. 779–788 (2016)
18. Remelli, E., Han, S., Honari, S., Fua, P., Wang, R.: Lightweight multi-view 3D pose estimation through camera-disentangled representation. In: IEEE Conference on Computer Vision and Pattern Recognition (CVPR), pp. 6039–6048 (2020)
19. Rusu, R.B., Cousins, S.: 3D is here: Point cloud library (PCL). In: IEEE International Conference on Robotics and Automation (ICRA) (2011)
20. Sandler, M., Howard, A., Zhu, M., Zhmoginov, A., Chen, L.C.: MobileNetV2: inverted residuals and linear bottlenecks. In: IEEE Conference on Computer Vision and Pattern Recognition (CVPR), pp. 4510–4520 (2018)
21. Schleich, D., Beul, M., Quenzel, J., Behnke, S.: Autonomous flight in unknown GNSS-denied environments for disaster examination. In: International Conference on Unmanned Aircraft Systems (ICUAS), pp. 950–957 (2021)
22. Stückler, J., Biresev, N., Behnke, S.: Semantic mapping using object-class segmentation of RGB-D images. In: IEEE/RSJ International Conference on Intelligent Robots and Systems (IROS), pp. 3005–3010 (2012)
23. Tan, M., Le, Q.: EfficientNet: Rethinking model scaling for convolutional neural metworks. In: International Conference on Machine Learning (ICML), pp. 6105–6114 (2019)
24. Tanke, J., Gall, J.: Iterative greedy matching for 3D human pose tracking from multiple views. In: German Conference on Pattern Recognition (GCPR), pp. 537–550 (2019)
25. Xiao, B., Wu, H., Wei, Y.: Simple baselines for human pose estimation and tracking. In: European Conference on Computer Vision (ECCV), pp. 466–481 (2018)
26. Xiong, Y., Liu, H., Gupta, S., Akin, B., Bender, G., Wang, Y., Kindermans, P.J., Tan, M., Singh, V., Chen, B.: MobileDets: Searching for object detection architectures for mobile accelerators. In: IEEE Conference on Computer Vision and Pattern Recognition (CVPR), pp. 3825–3834 (2021)
27. Zhou, B., Zhao, H., Puig, X., Xiao, T., Fidler, S., Barriuso, A., Torralba, A.: Semantic understanding of scenes through the ADE20K dataset. Int. J. Comput. Vis. **127**(3), 302–321 (2019)

Collision Warning by Rotating 2D LiDAR for Safe Crane Operation

Yusuke Kawasaki and Jun Miura(✉)

Department of Computer Science and Engineering,
Toyohashi University of Technology, Toyohashi, Japan
jun.miura@tut.jp

Abstract. Construction cranes are unanimously used for carrying very heavy items such as steel frames and concrete blocks at construction sites and harbors. One of the issues of using such large machines is how to prevent accidents, including collision of a person and the load. This paper proposes a collision detection and warning system composed of a long-range 2D LiDAR (light detection and ranging) and a rotary table. By rotating the LiDAR, the system covers a spherical field of view. Since the rotation speed is limited, however, we need to deal with the trade-off between the scanning cycle time and the area to be covered. We propose a method to detect collision with setting a warning margin volume around a danger volume. We present equations for determining the appropriate margin and the angular velocity of the table. We verified the system in simulation and in an actual scene.

Keywords: Crane safety · Collision detection · Rotating 2D LiDAR

1 Introduction

Construction cranes are unanimously used in many construction sites for carrying very heavy items such as steel frames and concrete blocks. They are also used at harbors for loading and unloading items with cargo ships. One of the issues of using such large machines is how to prevent accidents that are sometimes very severe. For example, it is reported that there were 94 fatalities related to construction machinery in 2019 in Japan.[1]

We are interested in improving safety in construction crane operation. In operating a large crane, the operator may miss dangerous situations like collision due to blind spots, occlusion, and lack of attention. Therefore, there is a strong demand for systems that can detect such situations and warn to the operator and people around the machine.

Neitzel et al. [1] analyze the characteristics of various construction cranes and survey safety devices and procedures. Kang et al. [2] developed methods for

[1] A report by Japan Construction Occupational Safety and Health Association, https://www.kensaibou.or.jp/public_relations/enforcement_plan/files/2021_enforcement_plan_07.pdf.

I. Petrovic et al. (Eds.): IAS 2022, LNNS 577, pp. 330–341, 2023.
https://doi.org/10.1007/978-3-031-22216-0_23

simulating and visualizing a construction process for the crane operator support. Guo et al. [3] discuss methods for identifying *Job Hazard Area* (JHA) using visualization techniques such as BIM (building information modeling) and 4D CAD. Gheisari and Esmaeili [4] discuss the use of UASs (unmanned aerial systems) for surveillance of construction sites. These work are mainly for preparation for increasing the safety.

Obstacle and person detection is one of the necessary functions of autonomous mobile robots. Cameras and 3D LiDARs are usually used for detecting obstacles [5–7]. Such sensors can achieve real-time detection but usually suffer from a limited field of view. Visual collision detection for manipulators is also popular, especially in a recent human-robot collaboration context [8,9]. Such approaches usually require extra sensors in the environment.

In our intended scenarios, a crane needs to detect possible collisions online. In addition, considering the size and the dynamics (e.g., a building grows day by day as construction steps proceed) of a construction site, it is not appropriate to install multiple sensors in the environment for covering the whole site. Therefore, we decided to put a wide-view and long-range sensing system on the boom of a crane. The system has a long-range 2D LiDAR put on a rotary table and gets a whole spherical scan by collecting 2D scans as the table is rotating. Since the rotation speed is limited, we need to deal with the trade-off between the scanning cycle time and the area to be covered. We propose a strategy for efficiently locating possible collisions by appropriately limiting the scanned area.

The contributions of the paper are twofold: (1) we present and formalize the collision warning method using a rotating 2D LiDAR for construction cranes; (2) we evaluate the method for various cases in simulation and real world experiments.

The rest of the paper is organized as follows. Section 2 describes the developed sensing system. Section 3 explains the idea and the details of the proposed collision warning approach. Section 4 describes our developed crane simulation environment and shows the results of applying the proposed method to collision warning. Section 5 shows preliminary experimental results at our campus. Section 6 concludes the paper and presents future work.

2 Sensor System Development and Preliminary Tests

Sensor systems for crane safety have to cover both a wide field of view and a long range as large construction cranes have booms over one hundred meters long. Omnidirectional cameras or a set of cameras may cover such a large field but range measurement may be error prone at a long distance. Some of existing 3D LiDARs can cover a long range but suffer from a limited field of view. Therefore, we developed a sensing system with a long-range 2D LiDAR (LD-LRS3611 by SICK) and a rotary table (by Orion Giken Co.) [10]. The measurable range of the LiDAR is 5 m–250 m, and the maximum rotating speed of the rotary table is 10°/s.

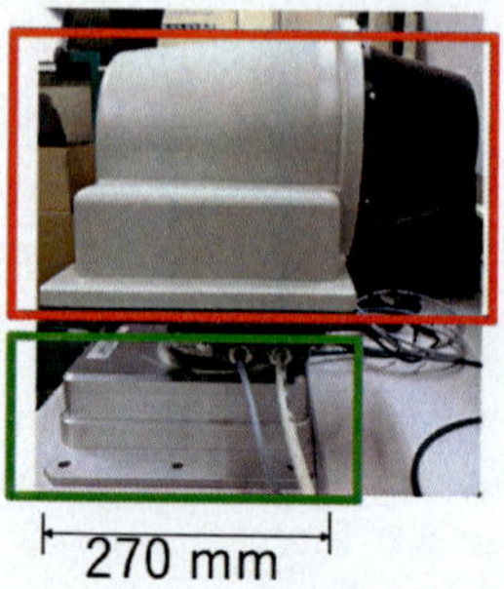

(a) LiDAR (red) is on a rotary table (green).

(b) The system hung by a crane.

Fig. 1. The developed sensing system.

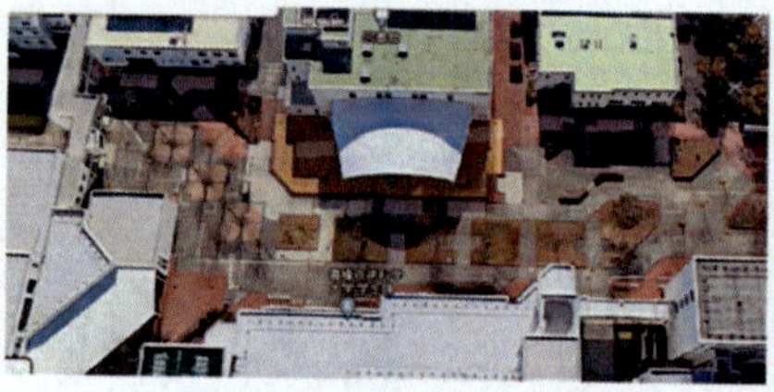

(a) A target scene. The sensing system is put at the center of the scene.

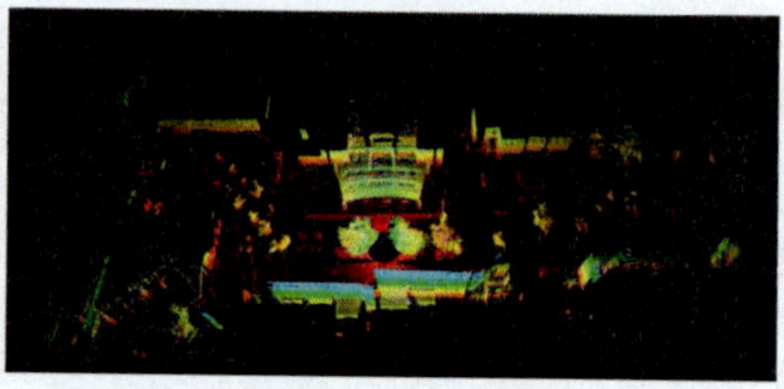

(b) Scanning result. Colors indicate the height from the ground plane.

Fig. 2. 3D scanning result with a fixed sensing system.

(a) An experimental site.

(b) Scanning result.

Fig. 3. 3D scanning result for a real site.

Figure 1 shows a view of the sensing system. Each 2D scan is recorded with the corresponding encoder reading of the rotary table so that a consistent set of scans is obtained. Figure 2 shows an example 3D scan of our campus.

As a preliminary test of the sensing system, we attached it to the load hung by a real crane (see Fig. 1(b)) and scanned a wide field. Figure 3 shows the target scene and a 3D scan result. The system can scan a very wide area; the

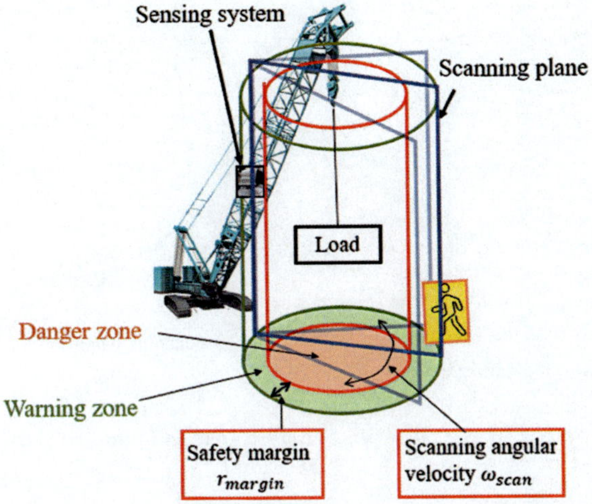

Fig. 4. Collision warning approach using a rotating 2D LiDAR.

maximum distance to 3D point reached 204.42 m. Such a system is indispensable for mapping a real construction scene. It is also noted that the scanned data are wavy due to a swinging motion of the hung load. This effect may exist even if we attached the system to the boom due to the vibration of the boom which is flexible to some extent. Compensation to the system's swinging motion is out of the scope of this paper.

3 Collision Warning

3.1 Approach

The purpose of proposed warning system is to detect and issue a warning to a person who (probably) unintentionally comes close to the load. Figure 4 shows the idea of proposed approach. We set a cylindrical danger volume where any person are requested not to enter. The radius of the volume will be given in advance from the size of the load and the predicted magnitude of swinging motion. The sensing system is set at some point of the crane boom and performs scanning continuously. Note that we assume to use some mechanism which always keeps the axis of rotary table to be vertical.

If we could rotate the LiDAR very fast, thereby obtaining a 3D scan instantaneously, it would be enough to watch the danger volume by appropriately rotating the LiDAR to cover it. In the case of slowly rotating LiDARs such as ours, since we can detect only one scanning plane at once, it is impossible to avoid a person to enter the danger zone. Therefore, we set a margin around the danger volume and try to detect a person in the margin zone.

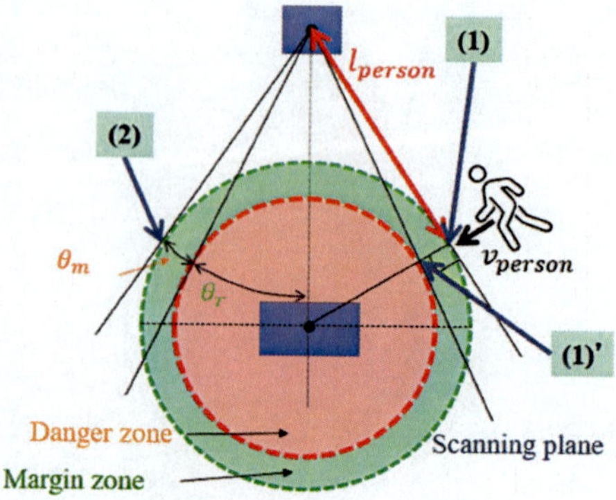

Fig. 5. Top view of the situation where a person is entering the margin.

We move the scanning plane (i.e., by rotating the rotary table) back and forth at a constant angular velocity ω_{scan} for covering the safety margin zone. There is a trade-off concerning the safety margin size r_{margin}. That is, a smaller margin makes the cycle time shorter but possible time duration for detecting a person also be come shorter, and vice versa for a larger margin. There is another constraint on the angular velocity. A high ω_{scan} may miss the person by passing the person volume without scan it. Since the r_{margin} and ω_{scan} are related, we first derive a condition on ω_{scan} and then that on r_{margin}.

3.2 Determining the Condition on Scanning Angular Velocity ω_{scan}

Figure 5 shows the top view of a situation where a person is entering the margin zone. Two angles, θ_r and θ_m, are halves of viewing angles of the danger zone and the margin zone, respectively, defined by the direction to the center of the load and two tangential lines. We consider the case where a person moves perpendicularly to the tangential line to the margin zone because the effective velocity becomes the highest.

Let w_{person} be the width of a person's body. Then, the viewing angle θ_{person} of the person is given by:

$$\theta_{person} = \tan^{-1}(w_{person}/l_{person}), \tag{1}$$

where l_{person} is the distance to the person's entry point to the margin zone. The instantaneous angular velocity of the person is similarly calculated as:

$$\omega_{person} = \tan^{-1}(v_{person}/l_{person}). \tag{2}$$

Next, let f_{scan} is the scanning frequency of the LiDAR. Since the relative angular velocity of the scanning with respect to the person is $\omega_{scan} + \omega_{person}$, the condition that at least one scan hits the person is given by:

$$\frac{\omega_{scan} + \omega_{person}}{f_{scan}} < \theta_{person}. \tag{3}$$

Then, we have:

$$\omega_{scan} < f_{scan} \cdot \theta_{person} - \omega_{person}. \tag{4}$$

Assuming that θ_m, θ_{person} and ω_{person} are small enough, we approximate the above inequality as:

$$\omega_{scan} < \frac{f_{scan} \cdot w_{person} - v_{person}}{l_{person}}. \tag{5}$$

3.3 Determining Safety Margin r_{margin}

We derive an equation which determines the safety margin r_{margin} for an angular velocity ω_{scan}. The worst case scenario is as follows (see Fig. 5). When a person enters the margin zone just after the scan reaches the rightmost point (point (1) in the figure), since the scan moves leftward, the scan will meet the person while the scan moves rightwards after making a turn at the leftmost point (point (2)). To detect the person in the margin zone, the scan needs to reach the rightmost point in the danger zone (point (1)') before the person reaches there. Let T_{scan} be the time for the scan to move from points (1) to (1)' via (2), and T_{person} be the one for the person to move from points (1) to (1)' directly. Then, the above condition is written as $T_{scan} \leq T_{person}$. Therefore, we get:

$$T_{scan} = \frac{4\theta_r + 3\theta_m}{\omega_{scan}} \leq \frac{\theta_m}{\omega_{person}} = T_{person}. \tag{6}$$

By solving the inequality, we get:

$$\theta_m \geq \frac{4\theta_r \omega_{person}}{\omega_{scan} - 3\omega_{person}}. \tag{7}$$

Assuming that θ_m, θ_{person} and ω_{person} are small enough, the minimum of the margin r^*_{margin} is given by:

$$r^*_{margin} = \frac{4\theta_r l_{person} v_{person}}{\omega_{scan} l_{person} - 3v_{person}}, \tag{8}$$

under the condition:

$$\omega_{scan} l_{person} > 3v_{person}. \tag{9}$$

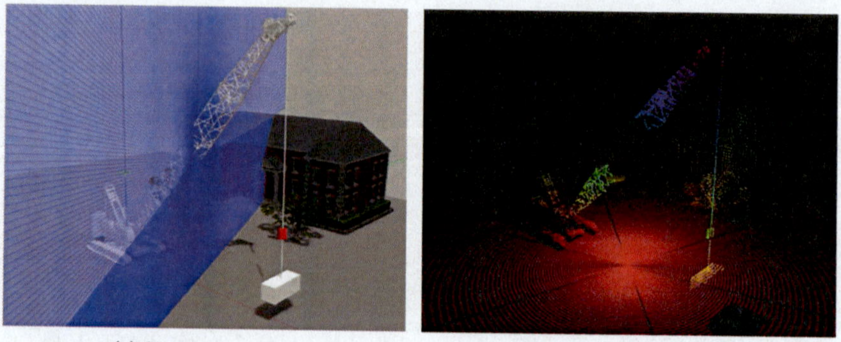

(a) Simulated scene. (b) Acquired point cloud.

Fig. 6. Crane and sensing system simulation.

3.4 Determining Safety Margin for Collision of Moving Load and Static Obstacles

Collision of the moving load and a static obstacle can also be detected by setting a margin zone. We can derive the minimum margin similarly to the moving person case by replacing the speed v_{person} of and the distance l_{person} to the person with those of the load, v_{load} and l_{load}, respectively. The minimum margin in this case is:

$$r^*_{margin} = \frac{4\theta_r l_{load} v_{load}}{\omega_{scan} l_{load} - 3v_{load}}. \tag{10}$$

The condition on the maximum angular velocity is obtained by:

$$\omega_{scan} < \frac{f_{scan} \cdot w_{obstacle} - v_{load}}{l_{load}}, \tag{11}$$

where $w_{obstacle}$ is the width of an obstacle under consideration. Note that we do not use the condition if the obstacle under consideration is large enough (e.g., walls); we use the maximum scanning speed of the system in such a case.

4 Experiments in Simulation

4.1 Crane and Sensor Simulator

We developed a crane simulation environment using Gazebo [11]. The model of the simulated crane is MasterTech-G series of crawler crane by Kobelco Construction Machinery Co., Ltd. Figure 6 shows a simulated scene of the crane and the sensing system with the corresponding point cloud.

Table 1. Calculation of ω_{scan} and r_{margin} for detecting a person approaching to the load.

l_{boom} (m)	v_{person} (m/s)	ω_{scan} (deg/s)	r_{margin} (m)
31.5	0.10	10	2.20
	0.30	10	31.03
96.0	0.10	5	1.13
	0.30	4	8.00
150.0	0.10	3	1.21
	0.30	2	5.70

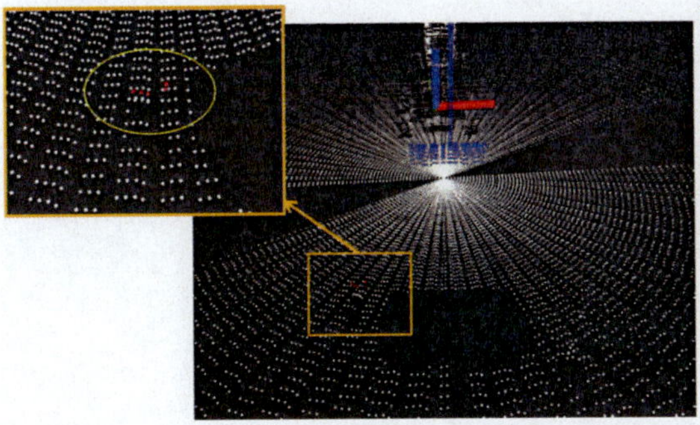

(a) Point cloud.

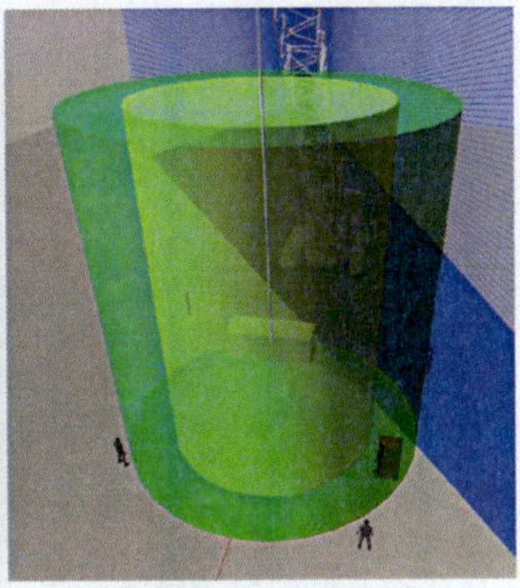

(b) Simulated scene.

Fig. 7. Person detection and collision warning example for static load and moving person. (a) The system extracts points from the load (blue) and those from the person (red). (b) Danger volume (yellow), warning volume (green), and detected person (red box).

4.2 Verifying Collision Warning in Simulation

We tested the proposed method using the simulator with the following parameters: the radius of the danger zone $r_d = 5$ m, the width of person $w_{person} = 28.0$ cm, and the scanning frequency $f_{scan} = 8.0$ Hz.

The first example is the detection of a person approaching to a static load. Table 1 shows pairs of ω_{scan} and r_{margin} for various combinations of boom length l_{boom} and velocity of person v_{person}. We set the upward angle of the boom to 60°. We suppose that the sensing system is attached at the right middle of the boom. Considering the upward angle, the distance to the load $l_{load} = l_{boom}/4$. Figure 7 shows that a person is correctly detected in the warning zone.

The second example is detection of static person when the load is moving. Table 2 shows pairs of ω_{scan} and r_{margin} for combinations of boom length l_{boom} and velocity of person v_{person}. The sensing system location and the upward angle of the boom are the same as before. Figure 8 shows that a person is correctly detected in the warning zone.

Table 2. Calculation of ω_{scan} and r_{margin} for detecting a person when the load is moving.

l_{boom} (m)	v_{person} (m/s)	ω_{scan} (deg/s)	r_{margin} (m)
31.5	0.10	10	1.03
	0.15	10	1.70
96.0	0.10	5	2.14
	0.15	4	5.60
150.0	0.10	3	4.48
	0.15	2	65.49

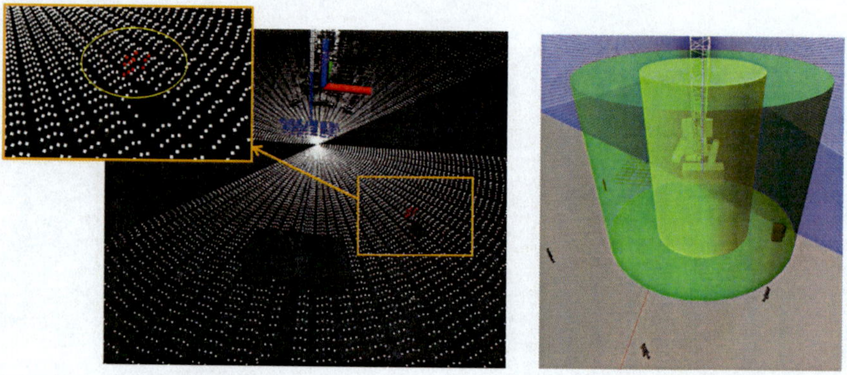

(a) Point cloud.

(b) Simulated scene.

Fig. 8. Person detection and collision warning example for static person and moving load. (a) The system extracts points from the load (blue) and those from the person (red). (b) Danger volume (yellow), warning volume (green), and detected person (red box).

5 Preliminary Real-World Experiments

5.1 Experimental Setup

We did a preliminary experiment of person detection using the proposed method and the sensing system. The experiment was carried out at our campus; we could not perform an on-site experiment with a real crane due to the current pandemic situation.[2]

We set the sensing system on the sixth floor of a building to observe persons on the ground. A virtual danger zone and margin zone were set as shown in Fig. 9(b). We used the following parameters: the radius of the danger zone $r_d = 3$ m, the width of person $w_{person} = 28$ cm, and the scanning frequency $f_{scan} = 8.0$ Hz. The distance to the load l_{load} was set based on the relative location between

[2] The data shown in Fig. 3 was obtained before the pandemic.

Table 3. Calculation of ω_{scan} and r_{margin} for real experimental setup.

v_{person} (m/s)	ω_{scan} (deg/s)	r_{margin} (m)
0.15	2	1.78
0.25	2	4.20
0.35	2	10.17
0.45	2	47.99

Table 4. Summary of detection results in a real scene.

v_{person} (m/s)	0.15	0.25
# of non-detected person.	0	0
# of persons detected in the margin zone.	6	4
# of Peron's detected in the danger zone.	3	4
Detection rate either in the margin or the danger zone	100.0%	100.0%
Detection rate in the margin zone	66.7%	50.0%

the sensor and the virtual danger zone, Table 3 shows the calculated angular velocity and margin for several person velocities.

5.2 Results

Figure 9 shows the case where the system successfully detected persons in the margin zone when $v_{person} = 0.15$ m/s. The yellow and the green circle in Fig. 9(b) show the danger and the margin zone, respectively. In Fig. 9(a), the green dotted line indicates the boundary of the margin zone, the red points correspond to the 3D points of the detected person, and the blue points are the ones from static objects eliminated from collision judgement in advance.

Figure 10 shows the case where the system successfully detected persons in the margin zone when $v_{person} = 0.25$ m/s. The blue points region in Fig. 10(a) and the green zone in Fig. 10(b) are larger than the one in the previous case.

Figure 11 shows the case the system failed to detect person in the warning zone but detected in the danger zone. The reason of this failure is that the actual person velocity was larger than the set value.

Table 4 summarizes the detection results in the real scene experiment. All of the persons were detected either in the margin or the danger zone. This shows that the scanning angular velocity ω_{scan} is set appropriately. Failure of detecting persons in the margin zone often occurred. This is mainly due to a higher velocity of walking persons than the set value (v_{person}) as a normal walking speed of a person is around 1.0 (m/s). It is, however, to set the v_{person} to such a normal walking speed is not reasonable because r_{margin} becomes very large.

In actual construction sites, we expect that a person in danger is rambling around the load, and the walking speed is much smaller than the normal walking

(a) 3D point cloud.

(b) Experimental scene.

Fig. 9. Person detection example 1 in the margin zone.

(a) 3D point cloud.

(b) Experimental scene.

Fig. 10. Person detection example 2 in the margin zone.

(a) 3D point cloud.

(b) Experimental scene.

Fig. 11. Person detection example in the danger zone.

speed. Therefore, the detection rate will be larger than the one in this experiment. If we need to deal with faster walking speed, it is necessary to use other sensors such as cameras and 3D LiDARs to limit the area to be scanned by this sensing system. However, an original problem of limited field of view of additional sensors may arise again in such as case.

6 Conclusions and Future Work

This paper has described a novel warning system for crane safety operations using a rotating long-range 2D LiDAR attached to the crane boom. We derived the equations for calculating both the appropriate angular velocity of the rotary table and the size of the warning margin. We verified the proposed method in simulated construction environment. We also did a preliminary experiment in a real environment. The experimental results show the feasibility of the proposed approach.

We are now planning to conduct real site experiments when every condition is satisfied. Fusion with other sensors such as cameras and effective presentation of detection results to the operator are among future work.

References

1. Neitzel, R.L., Seixas, N.S., Ren, K.K.: A review of crane safety in the construction industry. Appl. Occup. Environ. Hyg. **16**(12), 1106–1117 (2002)
2. Kang, S.-C., Chi, H.-L., Miranda, E.: Three-dimensional simulation and visualization of crane assisted construction erection process. J. Comput. Civ. Eng. **23**, 363–371 (2009)
3. Guo, H., Yu, Y., Skitmore, M.: Visualization technology-based construction safety management: a review. Autom. Constr. **73**, 135–144 (2017)
4. Gheisari, M., Esmaeili, B.: Applications and requirements of unmanned aerial systems (UASs) for construction safety. Saf. Sci. **118**, 230–240 (2019)
5. Montesano, L., Minguez, J., Montano, L.: Modeling the static and the dynamic parts of the environment to improve sensor-based Navigation. In: Proceedings of 2005 IEEE International Conference on Robotics and Automation (2005)
6. Arras, K.O., Mozos, O.M., Burgard, W.: Using boosted features for the detection of people in 2D range data. In: Proceedings of the 2007 IEEE International Conference on Robotics and Automation, pp. 3402–3407 (2007)
7. Asvadi, A., Premebida, C., Peixoto, P., Nunes, U.: 3D LiDAR-based static and moving obstacle detection in driving environments: an approach based on voxels and multi-region ground planes. Robot. Auton. Syst. **83**, 299–311 (2018)
8. Halme, R.-J., Lanz, M., Kämäräinen, J., Pieters, R., Latokartano, J., Hietanen, A.: Review of vision-based safety systems for human-robot collaboration. Proc. CIRP **72**, 111–116 (2018)
9. Maric, B., Jurican, F., Orsag, M., Kobacic, Z.: Vision based collision detection for a safe collaborative industrial manipulator. In: Proceedings of 2021 IEEE International Conference on Intelligence and Safety for Robotics, pp. 334–337 (2021)
10. Sakai, T., Koide, K., Miura, J., Oishi, S.: Large-scale 3D outdoor mapping and on-line localization using 3D-2D matching. In: Proceedings of 2017 IEEE/SICE International Symposium on System Integration (2017)
11. Koenig, N., Howard, A.: Design and use paradigms for Gazebo, an open-source multi-robot simulator. In: 2004 IEEE/RSJ International Conference on Intelligent Robots and Systems (IROS), vol. 3, pp. 2149–2154 (2004)

Semantic Classification in Uncolored 3D Point Clouds Using Multiscale Features

Michael Neumann, Dorit Borrmann, and Andreas Nüchter(✉)

Informatics VII—Robotics and Telematics, Julius Maximilian University of Würzburg, Würzburg, Germany
andreas.nuechter@uni-wuerzburg.de

Abstract. While the semantic segmentation of 2D images is already a well-researched field, the assignment of semantic labels to 3D data is lagging behind. This is partly due to the fact that prelabeled training data is only rarely available since not only the training and application of classification methods but also the manual labeling process are much more time-consuming in 3D. This paper focuses on the more classical approach of first calculating features and subsequently applying a classification algorithm. Existing handcrafted feature definitions are enhanced by using multiple selected reductions of the point cloud as approximations. This serves as input to train a well-studied random forest classifier. A comparison to a recently presented deep learning approach, i.e., the Kernel Point Convolution method, reveals that there are well-justified applications for both modern and classical machine learning methods. To enable the smooth conversion of existing 3D scenes to semantically labeled 3D point clouds the tool *Blender2Helios* is presented. We show that the therewith generated artificial data is a good choice for training real-world classifiers.

Keywords: Point cloud interpretation · 3D scanning · Semantics

1 Introduction

Semantic classification in 3D point clouds is the process of assigning semantic labels to each 3D point in a cloud. Some major differences to the same task for 2D images exist. Besides the considerably high volume of data, it lacks a fixed grid-structure. This makes it challenging to apply common algorithms for 2D images on point clouds. The variable density of the data is another difficulty. It highly depends on the distance to the laser scanner and is influenced by occlusions.

A few years ago a typical approach for classification in point clouds was a two-step procedure. First, hand-crafted local features of each point were calculated. For this, Principal Component Analysis (PCA) was applied on a number of neighbors. Originally, the size of this neighborhood, respectively the number of neighbors was fixed. In 2014 an improvement with varying neighborhood sizes, called *optimal neighborhood size selection* was proposed [1]. Up until 2018 various

I. Petrovic et al. (Eds.): IAS 2022, LNNS 577, pp. 342–359, 2023.
https://doi.org/10.1007/978-3-031-22216-0_24

other improvements were developed and published. For instance, [2,3] proposed the calculation of features on many different scales but with an adapted neighborhood definition. Second, common classification algorithms were applied on the features. While basically any general classifier is applicable, random forests showed high accuracies and became a de-facto standard.

With the availability of datasets with billions of labeled 3D points like *Semantic3D* [4] and capable hardware, deep learning emerged in this field of research. Thus, the manual definition of features became redundant. Some early approaches projected 3D points on 2D planes to create artificial images on which existing methods were applied. Meanwhile, a direct processing of 3D data is done most often. The published solutions range from *superpoint graphs* [5] to *RandLA-Net* [6] and *KPConv* [7]. The latter approach, which is very similar to well-known convolutional neural networks, is used in this paper for comparisons. Publicly available benchmark tests like the Semantic3D challenge [8] provide an overview of the methods that currently achieve the highest accuracies.

The current development towards deep learning for semantic labeling of 3D point cloud data raises the question, if and how much deep learning methods outperform classical approaches. Thus, this paper describes an improved approach using classical hand-crafted features. We propose the calculation of features on selected point cloud reductions as a high-performance approximation. A random forest is afterwards used for the classification. Finally, the results are compared to the output of KPConv, a proponent of the deep learning classifiers. For representative results, all parameters of the learners are tuned on only one dataset, a part of Semantic3D. Their performance is reviewed on a total of four different datasets. This resembles typical applications and differentiates this work from others. The used datasets consist of terrestrial (TLS) and mobile laser scans (MLS) of outdoor scenes.

One set of scans, the *Sim2Real* dataset, allows the examination of classifiers that were only trained with simulated data. For this purpose, we present the open-source software *Blender2Helios* [9] that allows to create semantically labeled training data of existing 3D scenes.

The paper is structured as follows. The next section summarizes random forests as the technical basis for the feature learning approach used in this paper. The following section introduces the tools and datasets used in the evaluation. Section 4 details the training of the classifier proposed in this paper while Sect. 4.3 introduces the related work used as comparison in the following evaluation in Sect. 5.

2 Random Forest Classification

Random forests (RFs) are a special form of bagging of decisions trees. We will use the definition of [10] for our work with RFs as it is very common. *Bagging*, short for *bootstrap aggregating*, is used to improve accuracy and stability. Multiple training sets, or bags, are sampled from the whole set of training data. They all have the size of the original training set. By sampling uniformly with replacement, $1 - \frac{1}{e} \approx 63.2\%$ of each subset are unique examples while the rest are duplicates. In each training iteration one set is used to train a decision tree. In the task of classification each tree then votes for its predicted class.

Compared to simply bagging of decision trees, only a subset of all attributes, typically of size $\lceil \sqrt{d} \rceil$ or $\lfloor \log_2 (d) + 1 \rfloor$ for d attributes, is taken into account for every split decision when training a random forest. This allows the algorithm to create more varying trees.

2.1 Parameters in Our RF Implementation

The number of trees in a forest, is a main parameter when constructing a RF. Probst and Boulesteix show in a benchmark study on over 300 datasets that the first 100 trees in a forest achieve the highest performance improvement [11]. In [12] different random forest implementations are tested on five datasets. The average classification accuracy in all their tests is within a one percent range when changing the parameter for the number of trees from 100 to 2,000. Meanwhile, using 100 trees has become a default setting for many random forest implementations that we also stick with.

Another important parameter is the tree depth d_t. It limits the length of a path from the root to all leaf nodes. While a high depth may result in overfitting, a too small depth leads to underfitting. Generally each tree should be able to predict all classes of a data set which gives a constraint of $d_t \geq \lceil \log_2 |C| \rceil - 1$ for a number of $|C|$ classes. Often this parameter is optimized together with the number of trees in a grid search.

2.2 Performance Evaluation Strategy

In our approach we calculate features for optimal local neighborhoods. In the preferred case one neighborhood will contain roughly the same points for many neighboring 3D points representing the same object. This results in nearly the same feature values for all of these points whereas other instances of similar objects will result in diverse values. For instance, the features of two different cars, both in one scan of a parking lot, might be very different due to various factors. They are influenced by occlusions and by the distance and angle between the laser scanner and the car.

For these reasons and because other strategies result in higher computation time, we use the classical *hold-out* strategy where the dataset is split into training, validation, and test data. Between these sets all scans shall be taken at different scenes. Where this is not feasible, at least very differing scanner positions are preferred.

2.3 Handling of Highly Imbalanced Data

The data that is used to train the random forests later is highly imbalanced. There are factors of 100 and 1,000 respectively between occurrences of different classes – even after deletion of very small classes. This is caused by the natural structure of the scenes. While ground and facades are very dominant in urban scenes, motorcycles, bushes, or trash cans are seen less often. Even if many instances of these infrequent objects are scanned, their small footprint results in a discrimination of their share in the 3D point cloud.

Learning a RF directly on this imbalanced data results in a classifier that has a bias towards these predominant classes. We choose to perform *down-sampling*, so each class in the training set is sampled only n_{min} times with n_{min} being the size of the smallest class. We prefer this over *up-sampling* as many data points of the dominant classes are already redundant and up-sampling of small classes is prone to over-fitting on the specific object instances in the data.

3 Tools and Datasets

The presented approach is evaluated using existing and newly created datasets that are introduced in this section. To help with the difficulty of recording and labeling real world data sets, we propose to enhance the training data by use of simulated scenes. Thus, the tools and methods used to generate this kind of data are first described here.

3DTK—The 3D Toolkit [13] is a toolkit for 3D point cloud processing. This paper is implemented as an extension of the open-source part of 3DTK and contributes the *scan2features* tool. Weka (*Waikato Environment for Knowledge Analysis*) is an open source machine learning software. We use its Attribute-Relation File Format for storing attributes and training of the random forests. Helios, the *Heidelberg LiDAR Operations Simulator*, is a software package for interactive real-time simulation and visualization of laser scanning surveys. Terrestrial, mobile and airborne laser scans are supported. Finally, we use Blender 2.81, an open-source 3D graphics software toolset.

Blender2Helios is a Blender 2.8 add-on that was developed for this work. It builds an interface between Blender and Helios. The add-on is released under GNU GPLv3. Code and documentation are available at [9]. Blender2Helios provides the possibility to build various scenes without much effort or even use existing ones. These scenes are then converted to Helios compatible XML files to be able to perform laser scan surveys subsequently. Also, semantic labels can be assigned easily by the collections feature in Blender.

Many benchmark datasets are available for 2D image classification. Often, modern machine learning classifiers achieve accuracies far over 95%. For 3D point clouds, however, less training data is available. Even well-known datasets often consist of only a few million points whereas a typical terrestrial laser scan already contains over ten million points. Most datasets like the *Oakland* [14] and the *IQmulus & TerraMobilita Contest* [15] datasets were created with mobile laser scanners. By design these are sparse compared to terrestrial laser scans. The datasets used in this paper are the following.

Oakland. The class distribution is given in Table 1. As the classifiers are tuned using another – the Semantic3D – dataset, we choose not to use the given validation data. We just use the training data to train the classifiers and the much bigger testing portion to test them.

Table 1. Class distribution in the Oakland dataset.

Subset	ID					
	Scatter/misc	Wire	Pole	Load/bearing	Facade	Total
Training	14,441	2,571	1,086	14,121	4,713	36,932
Validation	8,485	899	1,441	67,419	13,271	91,515
Testing	267,325	3,794	7,933	934,146	111,112	1,324,310

Table 2. Class distribution in the Semantic3D dataset (reduced-8).

ID	Class name	Number of points	
		Training	Testing
–	Unlabeled	156,046,453	11,607,777
1	Man-made Terrain	796,491,240	14,317,509
2	Natural Terrain	480,979,227	10,694,746
3	High Vegetation	135,634,808	4,675,871
4	Low Vegetation	99,971,504	3,021,894
5	Buildings	285,746,986	30,911,206
6	Hard scape	83,466,776	2,384,545
7	Artefacts	51,979,924	233,839
8	Cars	15,609,275	851,942
	Total	2,105,926,193	78,699,329

Fig. 1. Real parking lot scene in Würzburg.

Paris-rue-Madame. The dataset fully called *Paris-rue-Madame database: MINES ParisTech 3D mobile laser scanner dataset from Madame street in Paris* consists of two laser scans containing exactly 10 million points each. 624 objects are annotated and categorized in 26 classes. *Scan 1_2* are used for training while the other one (*scan 1_3*) is kept for testing. As many classes occur not at all or only rarely in both scans, we additionally eliminate very small classes.

Semantic3D. The highest quantity of training instances used in this research is provided by the *Semantic3D.net* [4] dataset. A static terrestrial laser scanner was used to create dense point clouds of outdoor scenes consisting of over four billion points. The scanned environments include churches, streets, railroad tracks, squares, villages, soccer fields, and castles in Central Europe.

Eight semantic classes were assigned by manual labeling. Table 2 pictures the distribution of class labels for the *reduced-8* data.

Sim2Real. We test the performance on our own data as well. In this paper we present one scene of a parking lot near the Department of Computer Science at the University of Würzburg. To determine the applicability of training classifiers on easy to generate, artificial scenes and using them for real world applications, we use two scenes. The first one was created in Blender using freely available object models. It was then sampled with a simulated laser scanner using Blender2Helios and Helios. Accuracy of the classifier is tested with

Fig. 2. Left: Blender scene created with freely available 3D models similar to Fig. 1. Right: Corresponding 3D point cloud generated with a simulated Riegl VZ-400 in Helios. Colors represent class labels.

Table 3. Semantic classes in our Sim2Real dataset. The dataset is split into training data created with *Helios* and real testing data.

ID	Class name	Examples	Number of points	
			Training	Testing
1	Man-made ground	Paved ground, asphalt	1,120,044	6,381,848
2	Natural ground	grass, grassland	128,813	209,347
3	Vegetation	Bushes, trees	204,011	1,823,701
4	Object	Lamps, poles, railings, stairs	16,900	64,575
5	Facade	Walls, buildings	280,063	754,867
6	Car	All vehicles	161,480	1,478,190
		Total	1,911,311	10,712,528

the manually labeled point cloud of the real parking lot. Figure 1 illustrates the on-site scenery. We use a division into six classes as shown in Table 3.

In Fig. 2 one can see the coarsely remodelled scene in Blender and its point cloud representation. For instance, paved/asphaltic ground is modeled by just a flat plane while a Blender *Displacement Modifier* in z direction combined with a *Voronoi texture* is used to generate small spikes that represent grass. As the classifiers show that they have learned the different ground types, no more effort was put into a finer remodeling. One should be able to reproduce our results with his or her own 3D scenes without trying to build an exact copy of the testing data.

4 Semantic Classification

In constrast to [1], we perform feature calculation on multiple differently reduced scans. The process of parameter selection is performed on the Semantic3D dataset. For this, the prelabeled training part is split into two parts. One part, containing twelve of the fifteen scans, is used for training while the remaining three scans (*untermaederbrunnen1*, *untermaederbrunnen3*, and *sg28_4*) are used for validation. The resulting parameters are then used to train and test the classifiers on all the introduced datasets including the Semantic3D reduced-8 challenge.

We first optimize the feature calculation on the basis of a standard random forest setup with 100 trees and a small tree depth of four. Similar configurations have already been used in literature [1,3] and proven to be sufficient to detect changes generated by different parameters in the feature calculation. Tuning of the RF is done afterwards.

To allow the testing of many configurations, features are calculated on only a reduced number of points. An octree-based reduction with a voxel size of 10 cm is done and features are only calculated for the resulting points. Afterwards, the data is subsampled to bring the classes to a uniform distribution with approximately 63,000 points per class. However, for the calculation of feature values all points are used.

4.1 Finding the Optimal Neighborhood Size

To calculate local features, a number of neighbors must be taken into account. For this set of points the principal components are distinguished using PCA. Typical approaches use a fixed number of k nearest neighbors [16], a sphere with a fixed radius [17], or a cylindrical neighborhood definition [18].

In [1] a state-of-the-art approach which is based on dimensionality features [19] is presented. Instead of defining a fixed neighborhood size, different ways to find an *optimal neighborhood size* k_{Opt} per point are compared. The best results are achieved by choosing a neighborhood size that minimizes the eigenentropy. Compared to the common default of using the nearest $k = 100$ neighbors for feature calculation, this method yields 2% to 14% higher overall accuracy scores.

The eigenentropy E_λ is given by the Shannon entropy of eigenvalues λ_i. Its calculation for the three-dimensional case is:

$$E_\lambda = -\Lambda_1 \ln(\Lambda_1) - \Lambda_2 \ln(\Lambda_2) - \Lambda_3 \ln(\Lambda_3). \tag{1}$$

using normalized eigenvalues to reduce the impact of the local point density:

$$\Lambda_i = \frac{\lambda_i}{\sum_{j=1}^{3} \lambda_j}. \tag{2}$$

When searching for the optimal neighborhood size, one minimizes the eigenentropy. Analyzing Equation (1) reveals that one-dimensional structures are favored, i.e., when one normalized eigenvalue is close to one and the others are close to zero. The maximum entropy of $-\ln(1/3) \approx 1.0986$ is reached by neighborhoods with equally distributed points along all three axes, resulting in $\lambda_1 = \lambda_2 = \lambda_3$ and, accordingly, $\Lambda_1 = \Lambda_2 = \Lambda_3 = 1/3$.

The optimal number of neighbors k_{Opt} is found by repeatedly calculating the eigenentropy of neighborhoods with a varying number of neighbors $k \in$

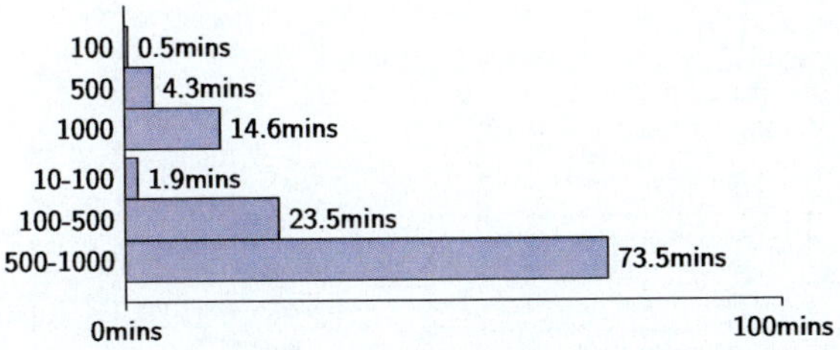

Fig. 3. Times for calculating features with different neighborhood sizes using one million 3D points in a typical outdoor scan. The first three measurements use a fixed neighborhood size (100, 500, 1000) whereas for the last three ones our optimal neighborhood definition is used.

Table 4. Impact of varying ranges for finding the optimal neighborhood size. The F_1 scores of our random forest classifier are shown.

	Terrain		Vegetation		Buildings	Hard scape	Artefacts	Cars	Accuracy (%)
N. size	Man-m.	Nat.	High	Low					
10–100	**0.916**	0.359	0.749	**0.307**	0.583	0.157	0.227	0.123	64.26
10–500	0.913	**0.368**	0.749	0.300	0.608	0.153	0.225	0.126	64.49
100–500	0.822	0.298	**0.789**	0.177	**0.729**	0.139	0.119	0.123	64.58
500–1000	0.904	0.249	0.783	0.229	0.715	**0.213**	0.355	0.122	**68.36**
1000–5000	0.880	0.107	0.752	0.289	0.688	0.024	**0.410**	**0.168**	66.46
10–5000	0.893	0.278	0.749	0.300	0.621	0.144	0.243	0.157	64.16%

$\{k_{Min}, k_{Min} + k_\Delta, \ldots, k_{Max}\}$. k_{Min}, k_{Max}, and k_Δ are parameters that need to be set by the user. Common values are $k_{Min} = 10$, $k_{Max} = 100$, and $k_\Delta = 1$. This includes the often fixed value $k = 100$ and limits the number of times the eigenvalues have to be calculated per point.

Figure 3 illustrates the time needed to query the neighborhoods of 1 million points. One can see that bigger neighborhoods with 500 or even 1,000 neighbors take longer to find and to perform the PCA on. Especially trying to find optimal neighborhood sizes within big ranges is very time consuming.

Table 4 shows the overall accuracy and class-wise F_1 scores when comparing different optimal neighborhood boundaries. Using between 500 and 1,000 neighbors for a bigger area of influence shows the highest accuracy. However, using that many neighbors for feature calculation is very time consuming as previously presented in Fig. 3.

We follow an approximation approach similar to [2]. Instead of calculating features on a large number of neighbors, a smaller number of neighbors is used within a reduced representation of the point cloud. This allows a fast calculation while still taking a wide spatial area into account. Octree-based reductions are used to create less dense point clouds.

$$\boldsymbol{features}(\boldsymbol{p_i}, \boldsymbol{P}) = \begin{bmatrix} \boldsymbol{features}_k^{3D}(\boldsymbol{p_i}, P) \\ \boldsymbol{features}_k^{2D}(\boldsymbol{p_i}, P) \\ \boldsymbol{features}_r^{Cyl}(\boldsymbol{p_i}, P) \end{bmatrix} \quad \boldsymbol{features}_k^{2D}(\boldsymbol{p_i}, \boldsymbol{P}) = \begin{bmatrix} Radius_k^{2D}(\boldsymbol{p_i}, P) \\ LocalDensity_k^{2D}(\boldsymbol{p_i}, P) \\ SumOfEigenvalues_k^{2D}(\boldsymbol{p_i}, P) \\ Ratio_k^{2D}(\boldsymbol{p_i}, P) \end{bmatrix} = \begin{bmatrix} ||\boldsymbol{q_k^{2D}} - \boldsymbol{p_i^{2D}}|| \\ \frac{k}{\pi \cdot Radius_k^{2D}(\boldsymbol{p_i}, P)^2} \\ \sum_{i=1}^{2} \lambda_i^{2D} \\ \frac{\lambda_1^{2D}}{\lambda_2^{2D}} \end{bmatrix}$$

$$\boldsymbol{features}_k^{3D}(\boldsymbol{p_i}, \boldsymbol{P}) = \begin{bmatrix} Linearity_k^{3D}(\boldsymbol{p_i}, P) \\ Planarity_k^{3D}(\boldsymbol{p_i}, P) \\ Scattering_k^{3D}(\boldsymbol{p_i}, P) \\ Omnivariance_k^{3D}(\boldsymbol{p_i}, P) \\ Anisotropy_k^{3D}(\boldsymbol{p_i}, P) \\ Eigenentropy_k^{3D}(\boldsymbol{p_i}, P) \\ SumOfEigenvalues_k^{3D}(\boldsymbol{p_i}, P) \\ ChangeOfCurvature_k^{3D}(\boldsymbol{p_i}, P) \\ Verticality_k^{3D}(\boldsymbol{p_i}, P) \\ LocalDensity_k^{3D}(\boldsymbol{p_i}, P) \\ Radius_k^{3D}(\boldsymbol{p_i}, P) \\ MaxHeightDifference_k^{3D}(\boldsymbol{p_i}, P) \\ HeightVariance_k^{3D}(\boldsymbol{p_i}, P) \end{bmatrix} = \begin{bmatrix} \frac{\Lambda_1 - \Lambda_2}{\Lambda_1} \\ \frac{\Lambda_2 - \Lambda_3}{\Lambda_1} \\ \frac{\Lambda_3}{\Lambda_1} \\ \sqrt[3]{\Lambda_1 \cdot \Lambda_2 \cdot \Lambda_3} \\ \frac{\Lambda_1 - \Lambda_3}{\Lambda_1} \\ -\sum_{i=1}^{3} \Lambda_i \cdot ln(\Lambda_i) \\ \sum_{i=1}^{3} \lambda_i \\ \Lambda_3 \\ 1 - |(\frac{\boldsymbol{e_3}}{||\boldsymbol{e_3}||})_{vert}| \\ \frac{k}{4/3 \cdot \pi \cdot Radius_k^{3D}(\boldsymbol{p_i}, P)^3} \\ ||\boldsymbol{q_k} - \boldsymbol{p_i}|| \\ \max_{1 \le i \le k}(q_{i,vert}) - \min_{1 \le i \le k}(q_{i,vert}) \\ \frac{1}{k} \sum_{i=1}^{k} (q_{i,vert} - \mu_{vert})^2 \end{bmatrix}$$

$$\boldsymbol{features}_r^{Cyl}(\boldsymbol{p_i}, \boldsymbol{P}) = \begin{bmatrix} NrPoints_r^{Cyl}(\boldsymbol{p_i}, P) \\ MaxHeightDifference_r^{Cyl}(\boldsymbol{p_i}, P) \\ HeightVariance_r^{Cyl}(\boldsymbol{p_i}, P) \\ HeightAboveMin_r^{Cyl}(\boldsymbol{p_i}, P) \end{bmatrix} = \begin{bmatrix} k^{Cyl} \\ \max_{1 \le i \le k^{Cyl}}(q_{i,vert}^{Cyl}) - \min_{1 \le i \le k^{Cyl}}(q_{i,vert}^{Cyl}) \\ \frac{1}{k^{Cyl}} \sum_{i=1}^{k^{Cyl}} (q_{i,vert}^{Cyl} - \mu_{vert}^{Cyl})^2 \\ p_{i,vert} - \min_{1 \le i \le k^{Cyl}}(q_{i,vert}^{Cyl}) \end{bmatrix}$$

Fig. 4. Features of point p_i in a 3D point cloud P.

4.2 Definition of Multiscale Features

Weinmann et al. also introduce a set of 21 features for every 3D point [1]. The definitions of the special geometric properties that are exploited are based on [20,21]. We slightly change the original feature definition of [1]. While the original features, for instance, include the absolute height of a 3D point, we calculate the height of a point with respect to the lowest point within a cylindrical neighborhood with a fixed radius.

The 21-dimensional feature vector for each point $\boldsymbol{p_i}$ in a point cloud P is given by the equations in Fig. 4. λ_i and λ_i^{2D} denote the i-th eigenvalue of the 3D, respectively 2D, covariance matrices of the k neighboring points $\{\boldsymbol{q_1}, \boldsymbol{q_2}, \ldots, \boldsymbol{q_k}\}$ with mean $\boldsymbol{\mu} = \frac{1}{k} \sum_{1 \le j \le k} \boldsymbol{q_j}$ around point $\boldsymbol{p_i}$. The neighbors are ordered by their distance to $\boldsymbol{p_i}$ so that $(||\boldsymbol{q_j} - \boldsymbol{p_i}|| \le ||\boldsymbol{q_l} - \boldsymbol{p_i}||$ for $j \le l)$ holds true. Two-dimensional data is created by projecting all the points $\boldsymbol{p_i}$ to two-dimensional points $\boldsymbol{p_i^{2D}}$ on the ground plane by ignoring the vertical axis. The neighbors $\{\boldsymbol{q_1^{2D}}, \boldsymbol{q_2^{2D}}, \ldots, \boldsymbol{q_k^{2D}}\}$ are again ordered by ascending distance to the reference point. We further order the eigenvalues and the corresponding eigenvectors $\boldsymbol{e_i}$ in a way that $\lambda_1 \ge \lambda_2 \ge \lambda_3 \ge 0$ and $\lambda_1^{2D} \ge \lambda_2^{2D} \ge 0$ hold true. Λ_i denotes the normalized equivalents of λ_i again. In all equations the subscript $_{vert}$ means the vertical dimension. For the cylindrical features we further let $\boldsymbol{q_j^{Cyl}}$ with $1 \le j \le k^{Cyl}$ be points within a cylindrical neighborhood with radius r around point $\boldsymbol{p_i}$. $\boldsymbol{\mu^{Cyl}} = \frac{1}{k^{Cyl}} \sum_{1 \le j \le k^{Cyl}} \boldsymbol{q_j^{Cyl}}$ denotes their mean. We also propose the feature calculation with different neighborhood sizes on differently reduced scans.

A prototype of the feature calculation tool *scan2features* is contributed to 3DTK. The four parameters k_{Reds}, k_{Mins}, k_{Maxs}, and k_{Deltas} impact the calculation of feature values. All four parameters hold lists of the same length, here denoted as k_{len}. For each $i \in \{1, \ldots, k_{len}\}$ an octree-reduced scan with a voxel size of $k_{Reds}[i]$ cm is generated. Only one randomly chosen point is kept per voxel. Within this cloud, denoted as $P_{k_{Reds}[i]}$, the 3D and 2D features ($\boldsymbol{features}_k^{3D}(\boldsymbol{p}_i, \boldsymbol{P})$ and $\boldsymbol{features}_k^{2D}(\boldsymbol{p}_i, \boldsymbol{P})$) of an optimal neighborhood between $k_{Mins}[i]$ and $k_{Maxs}[i]$ with a step size of $k_{Deltas}[i]$ are calculated. One defines equivalent scan reductions (cyl_{Reds}) and radii (cyl_{Radii}) of the same length used for the calculation of the cylindrical portion of the feature vector ($\boldsymbol{features}_r^{Cyl}(\boldsymbol{p}_i, \boldsymbol{P})$). All the feature values calculated in different reductions are concatenated to one feature vector at the end.

The best parameters for the approximate feature calculation are found as follows: A number of different reduction parameters are tested with $10 \le k_{Opt} \le 100$. Class-wise F_1 scores and accuracies are printed in Table 5. One can see that different reductions allow different classes to be recognized better. Voxel sizes over 80 cm seem to impact the accuracy negatively.

Using three different reductions shows a reasonable tradeoff between classification accuracy and computation time. Calculating features on point clouds reduced with voxel sizes of 2, 20, and 80 cm shows the best result with 72.42% accuracy. More reductions are imaginable but show only small improvements while leading to higher computation times.

Figure 5 illustrates the varying optimal neighborhood sizes in a whole laser scan for different reduction parameters. While in a detailed scan (2 cm voxel size) especially thin poles and sharp edges on cars show a high number of optimal neighbors, on coarser reduction scales other areas prefer big neighborhoods. Interestingly mainly medium-sized objects like bushes and cars favor a high number of neighbors in a scan reduced with a voxel size of 80 cm. Figure 6 (left and center) shows that the distribution of different neighborhood sizes is similar across all three reductions with a small bias towards big neighborhoods in highly reduced scans.

Table 5. Left: Impact of different and multiple reduction scales for feature calculation. The F_1 scores and accuracies of our random forest classifier are shown. Right: Impact of multiple cylindrical features and more trees in the RF.

Reduct-ion(s), k_{Reds}	Terrain Man-m.	Terrain Nat.	Vegetation High	Vegetation Low	Build-ings	Hard Scape	Arte-facts	Cars	Accu-racy
0	0.946	0.356	0.758	0.335	0.578	0.011	**0.394**	0.132	**66.47%**
1	**0.947**	0.349	0.739	**0.337**	0.547	0.046	0.385	0.129	65.24%
2	0.939	**0.394**	0.744	**0.337**	0.571	0.047	0.363	0.141	65.68%
5	0.932	0.390	0.743	0.330	0.580	0.052	0.265	0.134	65.21%
10	0.916	0.359	0.749	0.307	0.583	**0.157**	0.227	0.123	64.26%
20	0.873	0.374	0.755	0.260	**0.708**	0.139	0.189	0.176	66.41%
40	0.848	0.315	**0.765**	0.187	**0.708**	0.135	0.166	0.164	65.20%
80	0.895	0.277	0.741	0.224	0.679	0.071	0.177	0.191	66.32%
160	0.767	0.048	0.688	0.258	0.618	0.028	0.220	**0.205**	58.31%
2, 10	0.923	0.431	0.764	**0.329**	0.603	0.064	0.350	0.135	65.84%
2, 20	0.933	**0.468**	0.777	0.318	0.627	0.070	0.324	0.130	67.34%
2, 80	**0.938**	0.371	**0.789**	0.325	**0.697**	**0.081**	**0.362**	**0.225**	**70.76%**
2, 10, 80	0.926	0.377	0.805	0.343	0.735	0.152	**0.382**	**0.212**	71.99%
2, 20, 80	**0.931**	**0.429**	**0.812**	0.321	0.740	0.122	0.379	0.196	**72.42%**
2, 8, 32	0.920	0.411	0.774	**0.385**	0.729	**0.158**	0.359	0.186	69.51%
5, 20, 80	0.926	0.402	0.804	0.340	**0.744**	0.125	0.317	0.202	72.09%
10, 20, 40	0.917	0.424	0.791	0.306	0.733	0.154	0.247	0.197	70.46%

	# Trees	Cylindrical Features: cyl_{Reds} (cyl_{Radii}) in cm
A	100	10 (25)
B	100	2, 20, 80 (25, 150, 500)
C	500	2, 20, 80 (25, 150, 500)

	Weighted Avg. F_1	Accuracy
A	0.752	72.42%
B	**0.786**	**76.81%**
C	0.784	76.61%

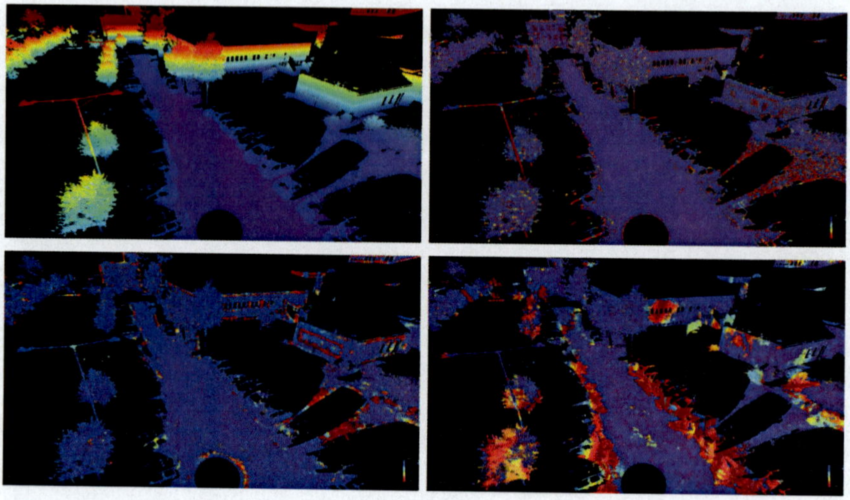

Fig. 5. Illustration of optimal neighborhood sizes k_{Opt} on differently reduced scans. The color gradient indicates the found optimal neighborhood size from $k_{Min} = 10$ (blue) to red $k_{Max} = 100$ (red). From top left to bottom right: Reference image colored by point height. Reduction with 2 cm voxel size. Reduction with 20 cm voxel size. Reduction with 80 cm voxel size.

For the previously shown results a standard RF setup with 100 trees and a tree depth of four are used. After determining a good set of features, the classifier itself needs to be tuned. The selected evaluations in Table 5 prove that our expectations are correct and more than 100 trees do not significantly impact the accuracy of the classifiers. The table also shows an example of the impact of calculating multiple cylindrical features on point cloud reductions. Calculating those features on three different scan reductions (with voxel-sizes of 2, 20, and 80 cm and cylinder radii of 25, 150, and 500 cm) increases the classification accuracy by over 4% compared to only one cylindrical feature set. A more detailed analysis can be found in [22]. Therefore, the only additionally tuned parameter is the tree depth. Figure 6 (right) shows its influence. The overall accuracy increases from 75.96% (depth of 3) to 81.95% (depth of 63). The maximum of 63 is given by the number of 21 attributes we calculate on three different reductions. To avoid overfitting we keep the trees small and pick a tree depth of 20 for the later comparison. It corresponds to 81.59% overall accuracy on our validation data.

4.3 Related Work: Deep Learning for 3D Point Cloud Classification

There are different approaches of using deep learning algorithms for classifying 3D point clouds. Thomas et al. assign these into four different categories [7].

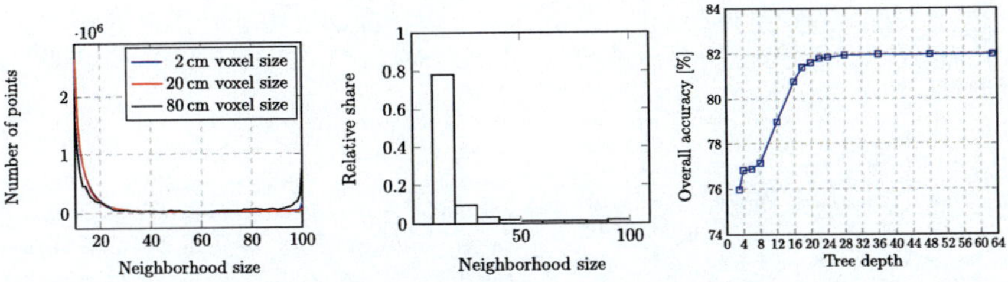

Fig. 6. Distribution of optimal neighborhood sizes k_{Opt} on different reduction scales. Input was the outdoor scan already shown in Fig. 5. Left: Absolute distribution of k_{Opt} on different reduction scales. Middle: Histogramm showing the relative distribution of k_{Opt} sizes in the scan reduced with voxel size 20 cm. Right: Influence of the tree depth on the random forest's accuracy.

For comparison we chose Kernel Point Convolution (*KPConv*) which belongs to the category of *point convolution networks*. Other well-known representatives are *Pointwise CNN* [23] and *SpiderCNN* [24].

KPConv operates directly on point clouds without any intermediate representation. Due to the lack of a fixed grid structure, the convolution weights of KPConv are located in Euclidean space by so-called *kernel points*. These are applied on the input points close to them.

The authors describe their *KP-FCNN* as the applicable method for our point-wise classification task in 3D point clouds. It implements an encoder-decoder CNN consisting of convolutional blocks. First, higher-order features are calculated step-by-step on reductions of the point cloud. Then, the original detailedness is achieved by nearest upsampling in the decoder part. Skip links are used to transfer information between layers, skipping the network parts in between.

5 Results and Comparison

In this section we show a comparison of the different classifiers applied on the four datasets. The settings of the feature calculation procedure are as previously shown: $k_{Reds} = cyl_{Reds} = (2, 20, 80)$ and $cyl_{Radii} = (25, 150, 500)$ with between 10 and 100 optimal neighbors. Random forests with 100 trees and a tree depth of 20 are trained.

5.1 Classification Results

Oakland. Applied on the Oakland dataset our RF classifier shows an accuracy of 95.27% which is around 3% higher than the results using the original single-scale approach described in [1]. Our approach also minimizes the assumptions made, like a consistent height of the laser scanner. The detailed result of our random forest as well as class-wise F_1 scores and the overall accuracies of the different classifiers are printed in Table 6. KPConv performs slightly better than

Table 6. Left: Performance of our RF on the Oakland dataset. Overall accuracy: 95.27%. Right: Comparison of F_1 scores and accuracies of the classifiers on the Oakland dataset.

Our ID	Orig. ID	Class Name	Prec.	Recall	F-Meas.
1	1004	Scatter/Misc	0.953	0.922	0.937
2	1100	Wire	0.101	0.894	0.181
3	1103	Pole	0.376	0.751	0.501
4	1200	Load/Bearing	0.999	0.994	0.996
5	1400	Facade	0.893	0.696	0.782
	Weighted Avg.		0.974	0.953	0.961

	Class F_1 Score					Over.
Classifier	**1**	**2**	**3**	**4**	**5**	**Accur.**
Original RF	0.878	0.165	0.364	0.978	0.749	92.2%
Our RF	0.937	0.181	**0.501**	**0.996**	0.782	95.3%
KPConv	**0.966**	**0.186**	0.429	0.992	**0.813**	**95.7%**

Table 7. Comparison of F_1 scores and accuracies of the classifiers on the Paris-rue-Madame dataset.

	Class $\mathbf{F_1}$ Score						Overall
Cline2-8 Classifier	1	2	3	4	5	6	Accuracy (%)
Original RF	0.960	0.932	0.672	**0.036**	0.206	0.105	90.1
Our RF	0.983	0.965	0.897	0.000	0.202	0.304	96.4
KPConv	**0.989**	**0.988**	**0.972**	0.000	**0.733**	**0.767**	**98.6**

our RF approach. It has to be noted that this dataset only contains 1,086 points belonging to class *pole*. Due to sub-sampling, our RF classifier is therefore only trained on 5,430 equally distributed points.

Paris-rue-Madame. Due to the small number of pedestrians in the training data, the data is subsampled to 3,656 examples per class for our RF classifier. Compared to the original paper we achieve an over 6% higher accuracy of 96.44% as presented in Table 7. Another 2% are achieved by KPConv.

Figure 7 shows the outputs of our random forest and KPConv after applying them to the depicted test data. Region A in the figure shows an area where the high recall value belonging to the class *cars* of our random forest classifier can be seen. However, also the points belonging to the ground around the cars are assigned the same label. In constrast, KPConv labels cars more conservatively leading to a higher precision score but introducing some errors where *facades* are predicted. The traffic sign in region B is detected with a high accuracy by both methods. The motorcycle in area C is not detected by either classifier. KPConv, however, detects other motorcycles that are not presented in the illustration. The pedestrian in sector D is ignored by both approaches.

Especially the bad matching of pedestrians has to be noted. Precision and recall values are zero for this class as shown in Table 8. We interpret this as an over-fitting on the seen object instances due to the fact that only three pedestrians are part of the training data.

Fig. 7. Paris-rue-Madame labeled test data for reference. Colors indicate class affiliations. *Blue*: facade, *yellow*: ground, *red*: cars, *dark green*: pedestrians, *rose*: motorcycles, *light blue*: traffic signs. Left: Ground truth. White circles indicate regions that are later discussed in the text. Middle: Output of our random forest classifier. Right: Output of KPConv.

Fig. 8. Sim2Real labeled test data for reference. Colors indicate class affiliations. *Yellow*: man-made ground, *orange*: natural ground, *green*: vegetation, *light blue*: object, *dark blue*: facade, *red*: car. Left: Semantic labels in the Sim2Real test data. Annotation was done manually. Middle: Output of our random forest classifier. Right: Output of KPConv.

Sim2Real. The simulated laser scan created with Blender, Blender2Helios, and Helios are subsampled to 16,900 examples per class for training our RF. Table 9 shows the detailed performance of our RF classifier. Evaluation is done with the manually labeled real scan from the parking lot in Würzburg.

Results of the different classifiers are shown in Table 9. Our RF and KPConv reach the same overall accuracy of 94.21%.

Figure 8 shows an excerpt of the test data. Region A in the figure shows a car that is correctly classified by us, but is treated like vegetation by KPConv. The recognition of area B, the upper part of a street light, is incorrectly classified by both methods. We have no explanation why KPConv detects a facade in sector C which is paved ground. However, in part D, KPConv detects the building with a much higher accuracy. Area E is probably the most difficult section. The grassland has an irregular incline and is peppered with smaller plants. Big parts of it are correctly classified by the random forest. KPConv, in constrast, heavily labels the points as cars or vegetation.

Semantic3D Challenge. Semantic3D is the dataset containing the most points for training and testing.

To tune the parameters for our RF, an additional split of the supplied training data is used. Now for testing all the prelabeled data is used to create an equally distributed set of features to train a classifier with best performance.

Table 8. Performance of our RF on the Paris-rue-Madame dataset. Overall accuracy: 96.44%.

Our ID	Orig. ID	Class name	Precision	Recall	F-measure
1	1	Facade	0.984	0.982	0.983
2	2	Ground	0.994	0.937	0.965
3	4	Cars	0.816	0.995	0.897
4	9	Pedestrians	0.000	0.000	0.000
5	10	Motorcycles	0.291	0.154	0.202
6	14	Traffic signs	0.200	0.635	0.304
	Weighted Avg.		0.968	0.964	0.965

Table 9. Left: Performance of our RF classifier on the Sim2Real dataset. Overall accuracy: 94.21%. Right: Comparison of F_1 scores and accuracies of the classifiers on Sim2Real.

ID	Class Name	Prec.	Recall	F-Meas.
1	Manmade Ground	1.000	0.973	0.986
2	Natural Ground	0.784	0.762	0.773
3	Vegetation	0.876	0.961	0.916
4	Lamps/Railings	0.893	0.304	0.454
5	Buildings/Walls	0.959	0.650	0.775
6	Cars	0.827	0.986	0.900
	Weighted Avg.	0.947	0.942	0.940

	Class F_1 Score						Over.
Classifier	**1**	**2**	**3**	**4**	**5**	**6**	**Accur.**
Our RF	0.986	**0.773**	**0.916**	**0.454**	0.775	**0.900**	**94.2%**
KPConv	**0.990**	0.364	0.888	0.373	**0.946**	0.876	**94.2%**

KPConv is already provided with a training procedure for the Semantic3D challenges. As the authors of [7] already put much effort into tuning, this is just adapted to our needs. The only change is the removal of the color information to allow a fair comparison of the classifiers.

The results of our submissions to the Semantic3D reduced-8 challenge are given in Table 10. For comparison the original KPConv results that use color information are also depicted. F_1 scores and accuracies are shown in Table 11. One can see that KPConv provides better results than our RF on this large dataset.

5.2 Run Times

For the comparison of run times we use a *Lenovo Yoga 720-15IKB* that is equipped with an *Intel Core i5-7300HQ* CPU (4x 2.50 GHz) and 24 GB of RAM. For KPConv an *Asus GeForce RTX 2080 Ti ROG STRIX* with 11 GB of video memory is plugged into a *Razer Core X* external GPU graphics card enclosure. The Thunderbolt connection is a small bottleneck and allows only for utilizing the GPU up to approximately 70–80%. Of course, modern workstations will perform better than our hardware. However, the times shown in Table 12 already provide a coarse reference.

Table 10. Results of the Semantic3D reduced-8 submissions: IoU.

Classifier	Colors used?	Intersection over union for class 1	2	3	4	5	6	7	8	Average IoU
Our RF	No	0.886	0.839	0.595	0.371	0.861	0.187	0.203	0.398	0.542
KPConv	No	**0.982**	**0.905**	0.814	0.373	0.943	0.319	0.609	**0.824**	0.721
KPConv	Yes	0.909	0.822	**0.842**	**0.479**	**0.949**	**0.400**	**0.773**	0.797	**0.746**

Table 11. Results of the semantic3D reduced-8 submissions: F_1 scores and accuracies.

Classifier	Colors used?	Class $\mathbf{F_1}$ score 1	2	3	4	5	6	7	8	Overall accuracy
Our RF	No	0.940	0.912	0.746	0.541	0.926	0.315	0.337	0.569	86.49%
KPConv	No	**0.991**	**0.950**	0.897	0.543	0.971	0.484	0.757	**0.903**	**93.51%**
KPConv	Yes	0.952	0.902	**0.914**	**0.648**	**0.974**	**0.571**	**0.872**	0.887	92.86%

Table 12. Training times of our RF and the KPConv classifier.

Number of points	Step	Time taken (approx.) Our RF[a]	KPConv
2 million	Feature calculation	40 s (100,000)	7 h
	Classifier training	40 s (100,000)	
10 million	Feature calculation	3 min (400,000)	13.5 h
	Classifier training	4 min (400,000)	

[a]The numbers in brackets show the number of points that we are able to choose equally distributed from the whole scan for training of the RF. However, all points have impact on the calculation of feature values

The numbers indicate that calculating handcrafted features and training a random forest is much faster than the neural network approach. It has to be mentioned that the parameters for KPConv are optimized for the huge Semantic3D dataset. By changing the learning rate and adapting the stop criterion a speedup of the KPConv learning process seems likely. However, it can be seen that the RF approach adapts better to varying dataset sizes. Application of our classifier on a point cloud with 10 million points takes around 1 h for feature calculation and 9 min to apply the RF.

6 Conclusions and Future Work

The main objective of this research is the comparison of different approaches for semantic classification in 3D point clouds. One can see that classical methods with handcrafted features still have a right to exist. With our optimization of using multiple reduced scans to approximate different neighborhood sizes they show very similar results compared to the deep learning approach on small- to medium-sized datasets. Their short runtime for training makes them an excellent choice for creating a good baseline or for choosing a reasonable set of training data. Due to their low hardware requirements, they can be used on cheap and even mobile hardware.

On the big *Semantic3D* dataset the deep learning approach performs better. We justify this by its capability of learning a more complex model. This comes with the downside of long training times and increased hardware prerequisites.

With the development of *Blender2Helios*, an interface between an easily usable 3D suite and the LiDAR simulation software *Helios* is created. Due to the good compatibility of *Blender*, many existing scenes can now be converted to 3D point clouds. As labeling in *Blender* is much easier than in the point cloud, we hope to promote the creation of highly detailed and labeled 3D point clouds. In this paper we show with the *Sim2Real* dataset that artificial data can be used to train classifiers that are later used in real-world applications. We will dive deeper into this in future work and hope to see some applications of our tool in future publications.

Needless to say, a lot of work remains to be done. Instead of down-sampling the training data, other preprocessing strategies might positively impact the accuracy of our approach. Also many different neighborhood definitions and sets of handcrafted features exist. It is conceivable that other combinations achieve higher accuracies than the ones we have tested.

Acknowledgments. The autors would like to acknowledge Prof. Bernhard Höfle and the 3DGeo Research Group of Heidelberg University for the software HELIOS and Prof. Thomas P. Kersten from Hafencity University of Hamburg for supporting our work.

References

1. Weinmann, M., Jutzi, B., Mallet, C.: Semantic 3d scene interpretation: a framework combining optimal neighborhood size selection with relevant features. ISPRS Ann. Photogrammetry Remote Sens. Spat. Inf. Sci. **2**(3), 181 (2014)
2. Hackel, T., Wegner, J., Schindler, K.: Fast semantic segmentation of 3d point clouds with strongly varying density. ISPRS Ann. Photogram. Remote Sens. Spat. Inf. Sci. (2016)
3. Thomas, H., Goulette, F., Deschaud, J.-E., Marcotegui, B.: Semantic classification of 3d point clouds with multiscale spherical neighborhoods. In: Proceedings of the IEEE International Conference on 3D Vision (3DV'8)
4. Hackel, T., Savinov, N., Ladicky, L. Wegner, J.D., Schindler, K., Pollefeys, M.: Semantic3d. net: a new large-scale point cloud classification benchmark. arXiv preprint arXiv:1704.03847 (2017)
5. Landrieu, L., Simonovsky, M.: Large-scale point cloud semantic segmentation with superpoint graphs. In: Proceedings of the IEEE Conference on Computer Vision and Pattern Recognition, pp. 4558–4567 (2018)
6. Hu, Q., Yang, B., Xie, L., Rosa, S., Guo, Y., Wang, Z., Trigoni, N., Markham, A.: RandLA-Net: efficient semantic segmentation of large-scale point clouds. arXiv preprint:1911.11236 (2019)

7. Thomas, H., Qi, C.R., Deschaud, J.-E., Marcotegui, B., Goulette, F., Guibas, L.J.: KPConv: flexible and deformable convolution for point clouds. In: Proceedings of the IEEE International Conference on Computer Vision (ICCV'19)(2019)
8. Hackel, T., Savinov, N., Ladicky, L., Wegner, J.D., Schindler, K., Pollefeys, M
9. Neumann, M.: Blender2Helios—github repository. https://github.com/neumicha/Blender2Helios (2020)
10. Breiman, L.: Random forests. Mac. Learn. **45**(1), 5–32 (2001)
11. Probst, P., Boulesteix, A.-L.: To tune or not to tune the number of trees in random forest. J. Mach. Learn. Res. **18**(1), 6673–6690 (2017)
12. Amrehn, M., Mualla, F., Angelopoulou, E., Steidl, S., Maier, A.: The random forest classifier in Weka: discussion and new developments for imbalanced data. arXiv preprint arXiv:1812.08102 (2018)
13. Nüchter, A., Borrmann, D., Schauer, J.: 3DTK—the 3d toolkit. http://threedtk.de (2019). Accessed 1 Feb 2020
14. Munoz, D., Bagnell, J.A., Vandapel, N., Hebert, M.: Contextual classification with functional max-margin markov networks. In: 2009 IEEE Conference on Computer Vision and Pattern Recognition. IEEE, pp. 975–982 (2009)
15. Vallet, B., Brédif, M., Serna, A., Marcotegui, B., Paparoditis, N.: TerraMobilita/iQmulus urban point cloud analysis benchmark. Comput. Graph. **49**, 126–133 (2015)
16. Lalonde, J.-F., Vandapel, N., Huber, D.F., Hebert, M.: Natural terrain classification using three-dimensional ladar data for ground robot mobility. J. Field Rob. **23**(10), 839–861 (2006)
17. Lee, I., Schenk, T.: Perceptual organization of 3d surface points. Int. Arch. Photogrammetry Remote Sens. Spat. Inf. Sci. **34**(3/A):193–198 (2002)
18. Filin, S., Pfeifer, N.: Neighborhood systems for airborne laser data. Photogram. Eng. Remote Sens. **71**(6), 743–755 (2005)
19. Demantké, J., Mallet, C., David, N., Vallet, B.: Dimensionality Based Scale Selection in 3d Lidar Point Clouds. The International Archives of the Photogrammetry, Remote Sensing and Spatial Information Sciences (2011)
20. Toshev, A., Mordohai, P., Taskar, B.: Detecting and parsing architecture at city scale from range data. In: 2010 IEEE Computer Society Conference on Computer Vision and Pattern Recognition. IEEE, pp. 398–405 (2010)
21. Mallet, C., Bretar, F., Roux, M., Soergel, U., Heipke, C.: Relevance assessment of full-waveform lidar data for urban area classification. ISPRS J. Photogram. Remote Sens. **66**(6), S71–S84 (2011)
22. Neumann, M.: Semantic classification in uncolored 3D point clouds using multiscale features. Master's thesis, Julius Maximilian University of Würzburg (2020)
23. Wu, Z., Song, S., Khosla, A., Yu, F., Zhang, L., Tang, X., Xiao, J.: 3D ShapeNets: a deep representation for volumetric shapes. In Proceedings of the IEEE Conference on Computer Vision and Pattern Recognition (CVPR '15) (2015)
24. Xu, Y., Fan, T., Xu, M., Zeng, L., Qiao, Y.: SpiderCNN: deep learning on point sets with parameterized convolutional filters. In: Proceedings of the European Conference on Computer Vision (ECCV), pp. 87–102 (2018)

On the Evaluation of RGB-D-Based Categorical Pose and Shape Estimation

Leonard Bruns(✉) and Patric Jensfelt

KTH Royal Institute of Technology, Division of Robotics, Perception and Learning, Stockholm, Sweden
{leonardb,patric}@kth.se

Abstract. Recently, various methods for 6D pose and shape estimation of objects have been proposed. Typically, these methods evaluate their pose estimation in terms of average precision and reconstruction quality in terms of chamfer distance. In this work, we take a critical look at this predominant evaluation protocol, including metrics and datasets. We propose a new set of metrics, contribute new annotations for the Redwood dataset, and evaluate state-of-the-art methods in a fair comparison. We find that existing methods do not generalize well to unconstrained orientations and are actually heavily biased towards objects being upright. We provide an easy-to-use evaluation toolbox with well-defined metrics, method, and dataset interfaces, which allows evaluation and comparison with various state-of-the-art approaches (https://github.com/roym899/pose_and_shape_evaluation).

Keywords: Pose estimation · Shape reconstruction · RGB-D-based perception

1 Introduction

We consider the problem of pose and shape estimation on a per-category level. Classic grasp and motion planning methods often assume that full knowledge of pose and shape is available, making them difficult to apply with partial sensor information. Estimating the full shape and pose promises to bridge this gap from partial sensor information to an actionable representation. Although shape reconstruction itself is sufficient for some tasks, categorical pose estimation additionally provides a reference frame. This categorical reference frame could, for example, further enable alignment of objects and pose-dependent grasp computation (e.g., an upside-down mug has to be grasped differently from an upright mug).

Over the last two years, various learning-based categorical pose and shape estimation methods have been introduced. Most methods are built and evaluated on the two datasets proposed by [26]. The CAMERA dataset is a large dataset of real RGB-D tabletop scenes with synthetic objects generated on top of the table. The REAL dataset is a smaller real-world dataset of tabletop sequences

I. Petrovic et al. (Eds.): IAS 2022, LNNS 577, pp. 360–377, 2023.
https://doi.org/10.1007/978-3-031-22216-0_25

with objects that have been scanned and tracked for the purpose of evaluating categorical pose estimation. Notably, both datasets only contain upright objects, which opens the question of how well existing methods generalize to less constrained settings.

To answer this question, we contribute a set of annotations to evaluate unconstrained 6D pose and shape estimation. Our annotations consist of meshes and poses for handheld objects of three categories in the Redwood dataset [8]. In that dataset, the objects are freely rotated in front of the camera, and the orientations vary significantly more than in the datasets by [26].

Most methods evaluate pose estimation by following the same evaluation protocol as initially proposed by [26]. However, the method in [26] combines mask detection and pose estimation into a single network and therefore evaluates pose estimation with *average precision* (AP), which is a common detection metric. However, many of the subsequent methods assume the mask to be an input to their method, making AP an unnatural evaluation metric, as it makes the results unnecessarily difficult to interpret. Therefore, we propose a set of simpler metrics that use ground-truth masks and categories to evaluate pose estimation.

Pose estimation with shape reconstruction was first demonstrated by [4,24]. Both methods independently used the chamfer distance as their reconstruction metric, which is now commonly used by methods performing shape reconstruction. However, [23] showed that the chamfer distance is *not* a good measure of reconstruction quality and other metrics better correlate with perceived reconstruction quality. Therefore, in this work, we advocate for a new set of reconstruction metrics and propose a new evaluation protocol for both shape reconstruction and pose estimation.

To summarize, our contributions are:

- a well-defined evaluation protocol,
- a challenging set of novel annotations to evaluate unconstrained pose and shape estimation,
- a fair evaluation of various state-of-the-art methods, and
- an open source evaluation toolbox for the task of categorical pose and shape estimation.

2 Related Work

Since the introduction of the BOP benchmark suite [15], much progress has been made in the task of instance-level pose estimation, where a mesh of the target object is available. While such pose estimation with known meshes has made remarkable progress, the more general task of category-level pose estimation has only recently received more attention. Compared to instance-level pose estimation categorical pose and shape estimation is more difficult due to the large possible variations in shape and appearance even for a single category.

Wang et al. [26] introduced the first deep learning-based method to address the 6D pose estimation problem at the per-category level. They introduced two datasets: a synthetic dataset that combines real scenes with meshes from the

ShapeNet dataset [3] and a smaller real-world dataset that is mainly used for fine-tuning and evaluation. Their method is based on the *normalized object coordinate space* (NOCS), in which objects of one category have a common alignment. The projection of the NOCS coordinates in the image plane (also called NOCS map) is predicted by extending Mask R-CNN [14] with an additional head. From this prediction, the 6D pose and scale can be estimated by employing the Umeyama algorithm [25] with RANSAC [13] for outlier removal.

The NOCS map was predicted using only RGB information. Since geometry typically varies less than appearance for a fixed category, several methods were proposed to more directly incorporate the observed point set into the prediction. Chen et al. [4] introduced *canonical shape space* (CASS), which regresses the orientation and position directly from the cropped image and the set of observed points. As a byproduct of their method, they also reconstruct the full canonical point set. Tian et al. [24] introduced *shape prior deformation* (SPD), which uses a canonical point set and predicts a deformation based on the observed RGB-D information. CR-Net [27] and SGPA [5] are two extensions of the original idea of SPD. CR-Net uses a recurrent architecture to iteratively deform the canonical point set, and SGPA uses a transformer architecture to more effectively adjust the canonical point set. Recently, good results were demonstrated by only training on synthetically generated views of ShapeNet meshes [2].

Other methods such as [6,18] predict pose and bounding box without reconstructing the full shape of the object. For the evaluation presented in this work, we limit ourselves to methods that perform both reconstruction and pose estimation, although our evaluation protocol could in principle be used for pure pose estimation methods as well.

Aside from these RGB-D-based methods, more RGB-based methods have been proposed. Chen et al. [7] proposed an analysis-by-synthesis framework in which the latent representation of a generative model is iteratively optimized to fit the observed color image. The generative model allows to generate novel views of the object, but a full reconstruction is not readily available. Lee et al. [17] introduced a framework for estimating a mesh from an RGB image, while Engelmann et al. [10] propose reconstructing shapes in a representation-agnostic way by classifying the closest matching object from a database.

Evaluation The most established benchmark dataset for categorical pose estimation is the REAL275 dataset proposed by [26]. We will take a critical look at that dataset in Sect. 3.3 and show that it only evaluates a constrained set of orientations, hiding inherent difficulties of the task, such as multimodal orientation distributions. [26] also proposed *average precision* as a metric to evaluate pose estimation.

To evaluate shape reconstruction most papers currently use chamfer distance (CD) [2,4,24], which was introduced to measure the difference of point sets by [12]. However, [23] noted that CD is not robust to outliers, that is, the distance of outliers affects the metric. Therefore, the authors advocate using a robust thresholded metric such as F-score [16] to measure the quality of reconstruction.

3 Evaluation Protocol

In this section, we will discuss the existing and proposed evaluation protocol. We will start by formally defining the problem of categorical pose and shape estimation. We will then discuss metrics to evaluate and compare different solutions to this problem. Finally, we discuss the evaluation datasets.

3.1 Problem Definition

Let $\mathbf{I} \in \mathbb{R}^{H \times W \times 3}$ be an RGB image, $\mathbf{D} \in \mathbb{R}^{H \times W}$ be a depth map, and $\mathbf{P} \in \mathbb{R}^{3 \times 4}$ be the projection matrix of the associated camera. Further, let ${}^i\mathbf{T}_j$ be the homogeneous transformation matrix, that transforms a point ${}^j\mathbf{p}$ from frame j to frame i, that is, ${}^i\mathbf{p} = {}^i\mathbf{T}_j {}^j\mathbf{p}$. Note that depending on the context, ${}^i\mathbf{T}_j$ can also be interpreted as the 6D pose of frame j in frame i. Let ${}^i\mathbf{R}_j$ and ${}^i\mathbf{t}_j$ further denote the rotation matrix and translation vector of which ${}^i\mathbf{T}_j$ is composed.

We will use o to denote the object's coordinate frame and c to denote the camera's coordinate frame. We will use $\mathcal{O}$ to denote a 3D object and $\mathcal{B}(\mathcal{O})$ to denote the axis-aligned bounding box of $\mathcal{O}$ in $\mathcal{O}$'s frame o. We will assume that the center of $\mathcal{B}(\mathcal{O})$ is at the origin of frame o. We further assume that transforms can be applied to 3D objects and bounding boxes, for example, ${}^\mathrm{c}\mathcal{O} = {}^\mathrm{c}\mathbf{T}_\mathrm{o}\mathcal{O}$. Following this notation, note that there is a difference between $\mathcal{B}({}^\mathrm{c}\mathcal{O})$ and ${}^\mathrm{c}\mathbf{T}_\mathrm{o}\mathcal{B}(\mathcal{O})$. The first one is an *axis-aligned bounding box* (AABB), the second is an *oriented bounding box* (OBB).

Problem 1. (Categorical Pose and Shape Estimation) Given $(\mathbf{I}, \mathbf{D}, \mathbf{P})$ imaging an object $\mathcal{O}$ of known category c at pose ${}^\mathrm{c}\mathbf{T}_\mathrm{o}$, and given the mask $\mathbf{M}$ of visible points of the object in the image, find estimates $\widetilde{\mathcal{O}}$ and ${}^\mathrm{c}\widetilde{\mathbf{T}}_\mathrm{o}$ of $\mathcal{O}$ and ${}^\mathrm{c}\mathbf{T}_\mathrm{o}$, respectively.

Similarly, one could define the problems of categorical pose estimation (estimate ${}^\mathrm{c}\mathbf{T}_\mathrm{o}$ only) and categorical pose and size estimation (estimate ${}^\mathrm{c}\mathbf{T}_\mathrm{o}$ and $\mathcal{B}(\mathcal{O})$). Extensions to multiple images are possible by introducing a world coordinate system, but are not further considered in this work.

3.2 Metrics

Various metrics exist to assess how well a method solves Problem 1. Currently, the predominant evaluation metric is *average precision* [2,4–6,17,18,26,27] for pose estimation and *chamfer distance* [2,4,5,12,17,27] for shape reconstruction. In this section, we will introduce these metrics and discuss several issues with them. Subsequently, we will advocate for *precision* (contrary to average precision) and *F-score* to evaluate pose estimation and shape reconstruction, respectively.

We first define similarity measures for transforms and objects. These are later used to define the evaluation metrics for Problem 1.

Definition 1. *Let* $d({}^{c}\mathbf{T}_{o}, {}^{c}\widetilde{\mathbf{T}}_{o})$ *denote the translation error between the ground-truth transform and the estimated transform, that is,*

$$d({}^{c}\mathbf{T}_{o}, {}^{c}\widetilde{\mathbf{T}}_{o}) = \|{}^{c}\mathbf{t}_{o} - {}^{c}\widetilde{\mathbf{t}}_{o}\|_2. \tag{1}$$

Definition 2. *Let* $\delta({}^{c}\mathbf{T}_{o}, {}^{c}\widetilde{\mathbf{T}}_{o})$ *denote the rotation error between the ground-truth transform and the estimated transform, that is,*

$$\delta({}^{c}\mathbf{T}_{o}, {}^{c}\widetilde{\mathbf{T}}_{o}) = \left| \frac{\mathrm{trace}({}^{c}\mathbf{R}_{o}\, {}^{c}\widetilde{\mathbf{R}}_{o}^{-1}) - 1}{2} \right|. \tag{2}$$

Definition 3. *Let* $\mathrm{IoU}(\mathcal{B}_1, \mathcal{B}_2)$ *denote the true intersection over union (IoU) of two tight, oriented bounding boxes [1]. Further, let the axis-aligned IoU between two objects be defined by*

$$\mathrm{IoU}^{+}({}^{c}\mathcal{O}, {}^{c}\widetilde{\mathcal{O}}) = \mathrm{IoU}\left(\mathcal{B}({}^{c}\mathcal{O}), \mathcal{B}({}^{c}\widetilde{\mathcal{O}})\right), \tag{3}$$

and the true IoU using tight, oriented bounding boxes by

$$\mathrm{IoU}({}^{c}\mathcal{O}, {}^{c}\widetilde{\mathcal{O}}) = \mathrm{IoU}\left({}^{c}\mathbf{T}_{o}\mathcal{B}(\mathcal{O}), {}^{c}\widetilde{\mathbf{T}}_{o}\mathcal{B}(\widetilde{\mathcal{O}})\right). \tag{4}$$

The current evaluation protocol [26] uses axis-aligned IoU^{+} instead of oriented IoU, although the former is less accurate. Our implementation follows [1] and computes oriented IoU.

Chamfer Distance In the context of shape reconstruction *chamfer distance* (CD) was introduced by [12] to differentiably measure the difference of point sets.

Definition 4. *Let* $\mathcal{S} \subset \mathbb{R}^3$ *and* $\widetilde{\mathcal{S}} \subset \mathbb{R}^3$ *denote point sets sampled from the surfaces of* $\mathcal{O}$ *and* $\widetilde{\mathcal{O}}$*, respectively. We define CD as*

$$\mathrm{CD}(\mathcal{S}, \widetilde{\mathcal{S}}) = \frac{1}{2|\mathcal{S}|} \sum_{\mathbf{x} \in \mathcal{S}} \min_{\mathbf{y} \in \widetilde{\mathcal{S}}} \|\mathbf{x} - \mathbf{y}\|_2 + \frac{1}{2|\widetilde{\mathcal{S}}|} \sum_{\mathbf{y} \in \widetilde{\mathcal{S}}} \min_{\mathbf{x} \in \mathcal{S}} \|\mathbf{x} - \mathbf{y}\|_2. \tag{5}$$

It is easiest to interpret as the mean Euclidean distance from a point in one point set to the closest point in the other set. Note that slightly different CD versions exist, such as squared versions and ones using the sum instead of arithmetic mean.

In Fig. 1 we visualize potential issues with CD as an evaluation metric. Consider the two mugs with different handles (denoted 1 and 2) as reconstructions of the mug without handle (denoted GT). The relative quality difference of these reconstructions measured by CD varies significantly depending on the number of samples. Furthermore, a large number of samples is required for CD to converge.

Notably, the number of ground-truth samples is left unspecified by most methods. This problem becomes further amplified because methods perform reconstruction by predicting a varying number of samples as noted by [2]. Therefore, we discourage further use of CD for evaluation purposes.

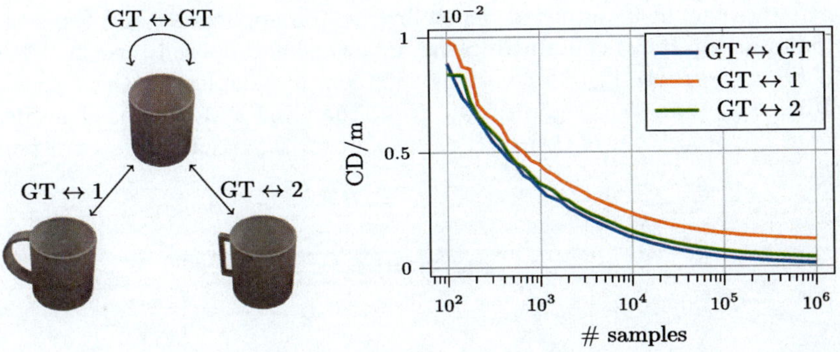

(a) Meshes (b) Chamfer distance for varying number of samples

Fig. 1. Visualization of the effect of varying number of samples on the chamfer distance. We consider two reconstructions (denoted by 1 and 2) of the ground truth (denoted by GT). Note that particularly the relative difference between $\mathrm{CD}(\mathcal{S}_{\mathrm{GT}}, \mathcal{S}_1)$ and $\mathrm{CD}(\mathcal{S}_{\mathrm{GT}}, \mathcal{S}_2)$ varies significantly. This is because the majority of the error stems from sparse sampling, not from actual differences in geometry. All mugs have been scaled to be 10 cm tall.

Reconstruction F-score Following [23], we advocate using the F-score instead of CD to evaluate shape reconstruction. Furthermore, note that we can evaluate shape reconstruction in the object frame or in the camera frame, taking into account ${}^{\mathrm{c}}\mathbf{T}_{\mathrm{o}}$ and ${}^{\mathrm{c}}\widetilde{\mathbf{T}}_{\mathrm{o}}$. Although previous methods evaluated in the object frame (i.e., assuming perfect alignment based on the canonical reference frame), we believe it is better to evaluate posed reconstruction in the camera frame, as it correlates more directly with downstream usability of the full estimate.

Definition 5. *Let $\mathcal{S} \subset \mathbb{R}^3$ and $\widetilde{\mathcal{S}} \subset \mathbb{R}^3$ denote point sets sampled from the surfaces of $\mathcal{O}$ and $\widetilde{\mathcal{O}}$, respectively. Given an application-specific threshold Δ, we define reconstruction recall as*

$$r_\Delta = \frac{1}{|\mathcal{S}|} \sum_{\mathbf{x} \in \mathcal{S}} \left[\min_{\mathbf{y} \in \widetilde{\mathcal{S}}} \|\mathbf{x} - \mathbf{y}\|_2 < \Delta \right] \tag{6}$$

and reconstruction precision as

$$p_\Delta = \frac{1}{|\widetilde{\mathcal{S}}|} \sum_{\mathbf{y} \in \widetilde{\mathcal{S}}} \left[\min_{\mathbf{x} \in \mathcal{S}} \|\mathbf{x} - \mathbf{y}\|_2 < \Delta \right], \tag{7}$$

where $[\cdot]$ denotes the Iverson bracket. Finally, we define F-score as the harmonic mean of precision and recall

$$F_\Delta = \frac{2}{p_\Delta^{-1} + r_\Delta^{-1}}. \tag{8}$$

Note that Δ should be adjusted depending on the application and the sensor. For the table-top items contained in the datasets, we propose to use $\Delta = 1\,\text{cm}$.

In Fig. 2 we show $F_{1\text{cm}}$ for varying numbers of samples for the meshes in Fig. 1(a). $F_{1\text{cm}}$ converges significantly faster than CD and can easily be interpreted as the percentage of correct (i.e., error below $\Delta = 1\,\text{cm}$) surfaces or points [23].

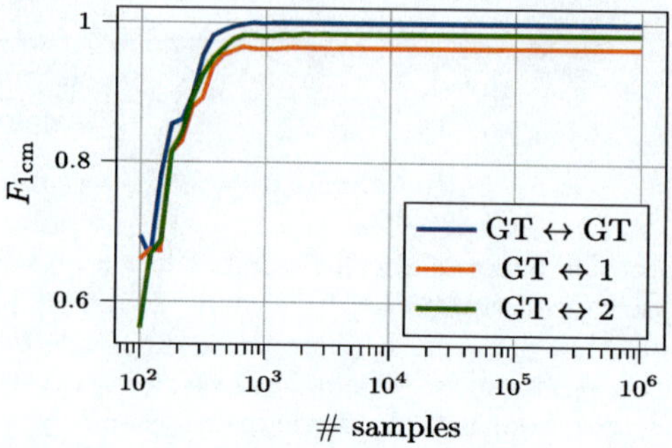

Fig. 2. Visualization of the effect of varying number of samples on the $F_{1\text{cm}}$ metric. Note that compared to CD (see Fig. 1) $F_{1\text{cm}}$ converges significantly faster.

All metrics so far (i.e., $d, \delta, \text{IoU}, \text{CD}$, and F_Δ) assess the quality of a single estimate. Next, we discuss average precision and precision, which attempt to summarize a method's performance on a dataset. In principle, one could also compute different averages of the aforementioned metrics, but those are typically affected by outliers and hard to interpret in comparison to thresholded evaluation metrics that classify estimates as true or false.

Average Precision *Average precision* (AP) summarizes precision-recall curves in a single value [22] and has been the standard evaluation metric for object detection on the PASCAL VOC [11] and COCO datasets [19]. In general, average precision is calculated based on the interpolated precision-recall curve, which is constructed by varying a confidence threshold.

Wang et al. [26] proposed to use AP with different thresholds on IoU^+, d, and δ to evaluate their pose estimation (all specified thresholds must hold for a prediction to count as true positive). Their method includes a Mask R-CNN architecture to detect objects and therefore had a confidence threshold to compute AP. However, *none* of the following pose and shape estimation methods include such a confidence threshold. Instead they all assume $\mathbf{M}$ to be given as stated in Problem 1.

To still follow the same evaluation protocol as [26], all other methods rely on the same, suboptimal Mask R-CNN predictions that [26] provided. This protocol effectively limits the achievable AP due to wrong classifications, missing detections and poor masks. Furthermore, AP is inherently difficult to interpret compared to simpler metrics.

Therefore, we believe that AP is unnecessary to compare pose and shape estimation methods. Rather, simpler metrics, such as precision (see below), should be used, assuming that mask $\mathbf{M}$ and category c are provided.

Precision We propose to use precision, contrary to average precision, to assess categorical pose and shape estimation.

Definition 6. *Given inputs* $(\mathbf{I}^i, \mathbf{D}^i, \mathbf{P}^i, \mathbf{M}^i, c^i)$*, ground truths* $(\mathcal{O}^i, {}^{\mathrm{c}}\mathbf{T}_{\mathrm{o}}^i)$*, and associated predictions* $(\widetilde{\mathcal{O}}^i, {}^{\mathrm{c}}\widetilde{\mathbf{T}}_{\mathrm{o}}^i)$ *with* $i = 1, ..., N$*, let precision be defined as*

$$P = \frac{\sum_{i=1}^{N} \left[\mathrm{c}(\mathcal{O}^i, {}^{\mathrm{c}}\mathbf{T}_{\mathrm{o}}^i, \widetilde{\mathcal{O}}^i, {}^{\mathrm{c}}\widetilde{\mathbf{T}}_{\mathrm{o}}^i) \right]}{N}, \tag{9}$$

where $[\cdot]$ *denotes the Iverson bracket and* c *determines whether a prediction is correct or not based on a single or multiple thresholds on* IoU, δ, d *or* F_Δ.

That is, precision measures the percentage of correct estimates based on thresholds on translation error, rotation error, IoU and F-score. Note that we can use this simpler metric instead of average precision because we decouple pose estimation from detection and classification.

Summary We propose to evaluate categorical pose and shape estimation methods by calculating precision at varying thresholds of d, δ and F_Δ. Furthermore, this evaluation procedure can be adjusted for categorical pose estimation by using d and δ only and for categorical pose and size estimation by using IoU instead of F_Δ. The thresholds for these metrics must be adjusted based on application requirements, sensor accuracy, and annotation quality. Furthermore, when combining multiple thresholds, care should be taken that they are roughly equally strict.

Note that when evaluating δ and IoU, extra care must be taken with respect to the categories containing symmetric objects. We follow [26] and ignore rotations around the up-axis for the bottle, bowl, and can categories. Further issues with ambiguities are discussed in Sect. 5.

3.3 Datasets

So far, most methods have been evaluated on the synthetic CAMERA25 dataset and on the smaller real-world dataset REAL275 [26]. Since we are more interested in real-world performance, we only include the REAL275 dataset in our evaluation protocol. Next, we will give an overview over the REAL275 dataset, and our new annotations for the Redwood dataset [8].

REAL275 The REAL dataset was proposed by Wang et al. [26] and consists of 4300 training images (7 video sequences) and 2750 test images (6 video sequences). The dataset contains 6 categories (bottle, bowl, camera, can, laptop, and mug) and contains 4 to 7 objects per scene. Meshes for each object are provided, obtained using an RGB-D reconstruction algorithm. Since we are primarily interested in evaluation, we will focus on the evaluation split, called REAL275, from here.

Fig. 3. Examples of REAL275 samples for the 6 object categories. Note that all objects are positioned upright on a table.

Figure 3 shows point sets with their corresponding ground-truth annotations. Note that all objects are upright on planar surfaces. Similar constrained orientations can be found in the training splits of the CAMERA and REAL datasets. Figure 6(a) visualizes the distribution of orientations contained in the REAL275 dataset. Note that such constraints, present in training and test data, can significantly simplify the learning problem as pose and shape ambiguities disappear (e.g., upright or upside-down can).

Redwood To evaluate methods on less constrained orientations, we contribute annotations for a set of images in the Redwood dataset [8]. The Redwood dataset contains sequences of handheld objects being freely rotated in front of the camera. No ground-truth reconstructions are provided for these sequences.

We annotated pose and shape for 3 categories (bottle, bowl, mug) for 5 sequences each. These annotations were created by manually creating OBBs in multiple frames and exploiting potential symmetries of the object. Alignment of OBBs with previous annotations was sped up and refined by using the iterative closest point (ICP) algorithm. For each of the annotated sequences we took a subset of 5 frames covering various orientations. We will refer to this set of annotations as REDWOOD75.

Fig. 4. Manual annotations for Redwood dataset. The left column shows the cropped, accumulated point sets (including symmetries) extracted from annotated bounding boxes. The middle columns shows the voxel grid after carving. The right column shows the extracted mesh, overlaid with the point set.

To reconstruct the shape from the partially occluded and noisy depth data, we start from a dense voxel grid inside the bounding box and apply voxel carving using the annotated frames to remove hands and other temporary occlusions. The remaining voxel grid contains only voxels that are not observed to be free in any of the annotated frames. From this voxel grid, we extract a mesh and apply Laplacian smoothing. Figure 4 visualizes the annotation process.

Note that this method only approximates the real shape and is sensitive to misaligned bounding boxes and missing depth data. Especially thin surfaces and details, such as mug handles, are difficult to extract accurately due to sensor noise. Additionally, alignment errors can easily accumulate, resulting in too large or too small objects. However, the annotations are accurate enough to evaluate the performance of current methods on unconstrained orientations.

To produce the final ground truth, we compute tight bounding boxes based on the extracted meshes. Figure 5 shows examples of the final annotations. In Fig. 6 we compare the orientation distribution of REDWOOD75 and REAL275. Note that the orientations in REDWOOD75 are significantly less constrained than in REAL275.

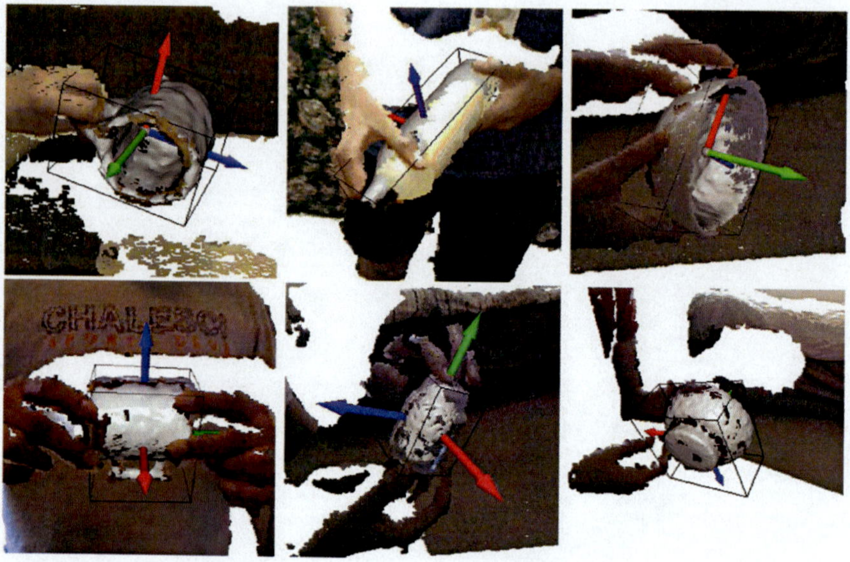

Fig. 5. Examples of REDWOOD75 samples.

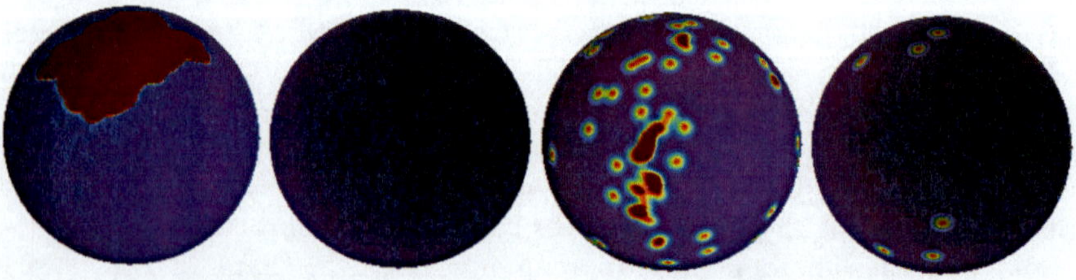

(a) REAL275 (b) REDWOOD75

Fig. 6. Distribution of the up-axis in REAL275 and REDWOOD75 datasets. REDWOOD75 covers a larger variety of orientations.

4 Experiments

We follow our proposed evaluation protocol and compare three methods for shape and pose estimation: CASS [4], SPD [24], and ASM-Net [2]. CASS and SPD are well-established baselines for this task and were both trained on the CAMERA and REAL datasets. Unlike the other methods, ASM-Net [2] was trained on synthetic ShapeNet [3] renderings only. All methods estimate 6D pose and reconstruct a point set of varying density.

For all methods, we closely followed the published inference code and verified that our method interface produced results similar to their evaluation code. We found that only SPD's published model achieved the same qualitative results as shown in their publication. CASS' reconstructed point sets were significantly worse except for the laptop category and ASM-Net often predicted negative scales, which causes some reconstructions to be upside down, while the object frame o is predicted in the correct orientation.

We have implemented the metrics, and interfaces to the datasets and methods described in the previous sections using Open3D [28] and PyTorch [21]. We open-source our code as a benchmarking toolbox, with the goal of simplifying fair comparison with state-of-the-art methods. We plan to extend the toolbox as new methods are released.

Qualitative Results Figure 7 shows randomly selected results on the REAL275 and REDWOOD75 datasets. On REAL275, all methods perform pose estimation with a similar quality as shown in the respective publications. On REDWOOD75, on the other hand, only ASM-Net shows limited generalization capability. CASS and SPD predict upright objects consistent with the orientation distribution of REAL275 independent of the input.

The shape reconstructions of SPD are qualitatively the best. ASM-Net's reconstructions are often flipped, but surfaces typically align well. As noted above, CASS completely fails to reconstruct any object except laptops. For both ASM-Net and CASS it is unclear whether this performance difference stems from errors in the code or if the published model weights are suboptimal.

Quantitative Results We now present the results using the metrics introduced in Sect. 3.2. In Table 1 we report precision with thresholds of varying strictness. To assess pose estimation independent of shape estimation, we use $5°, 1\,\text{cm}$ and $10°, 2\,\text{cm}$. To further include shape reconstruction, we use $5°, 1\,\text{cm}, 0.8$ and $10°, 2\,\text{cm}, 0.6$ for δ, d, and $F_{1\text{cm}}$, respectively. We picked these tuples of thresholds such that all thresholds in a tuple are roughly equally strict. In the past, some methods used pairs such as $10°, 5\text{cm}$, or $10°, 10\text{cm}$, where only the $10°$ threshold practically mattered.

The results from Table 1 confirm the qualitative observations from before. CASS in particular performs poorly on the shape reconstruction metrics. Note that for both datasets there is still a lot of room for improvement. Typically,

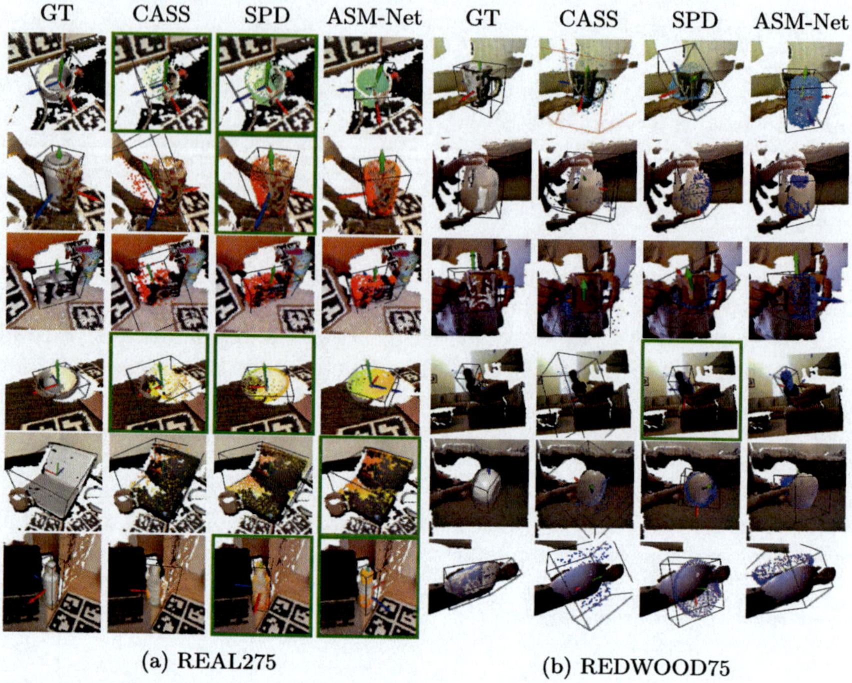

Fig. 7. Randomly selected results on REAL275 and REDWOOD75 datasets. Results that are considered correct under $\delta = 10°, d = 2\,\text{cm}, F_{1\text{cm}} = 0.6$ thresholds are highlighted.

Table 1. Precision at varying position, orientation and F-score thresholds.

	REAL275			REDWOOD75		
	CASS	SPD	ASM-Net	CASS	SPD	ASM-Net
10°,2 cm	0.331	**0.535**	0.331	0.013	0.2	**0.307**
5°,1 cm	0.073	**0.205**	0.069	0.000	0.013	**0.080**
10°,2 cm,0.6	0.031	**0.471**	0.215	0.000	**0.173**	**0.173**
5°,1 cm,0.8	0.000	**0.170**	0.050	0.000	0.013	**0.053**

The best results are highlighted in **bold**

significantly fewer than 50% of the estimates are of sufficient quality to be considered correct in pose and shape. This shows that categorical pose and shape estimation is still an open problem, especially for unconstrained orientations. The performance gap between the two datasets is especially noticeable for SPD, which performs better than ASM-Net on REAL275 but worse on REDWOOD75.

To gain further insight into the estimation quality of the methods, we show detailed results for varying thresholds in Fig. 8. It can be seen that the difference between the two datasets is more pronounced for orientation-based thresholding.

This confirms the issue of constrained orientations discussed in Sect. 3.3 (see Fig. 6). ASM-Net, which was trained on synthetic data (the exact distribution was not specified but is likely less constrained than the CAMERA and REAL datasets), performs best in this metric.

In Table 2 we further report the precision per category with the more lenient pose and shape estimation thresholds $\delta = 10°, d = 2\,\text{cm}, F_{1\text{cm}} = 0.6$. Note that CASS' reconstructions are very sparse and noisy, and therefore rarely reach $F_{1\text{cm}} > 0.6$ (see also Fig. 8). On REAL275, all methods fail at the camera category which contains significantly more shape variation than the other categories. Note that, despite these relatively lenient thresholds, all methods fail to sufficiently recover pose and shape for most of the REDWOOD75 samples.

Run Time We also compute the mean run time of each method per prediction. We include the transfer time from CPU to GPU, but exclude the computation of the metrics. In our experiments[1], CASS required 22.3 ms, SPD 199 ms, and ASM-Net 58.6 ms.

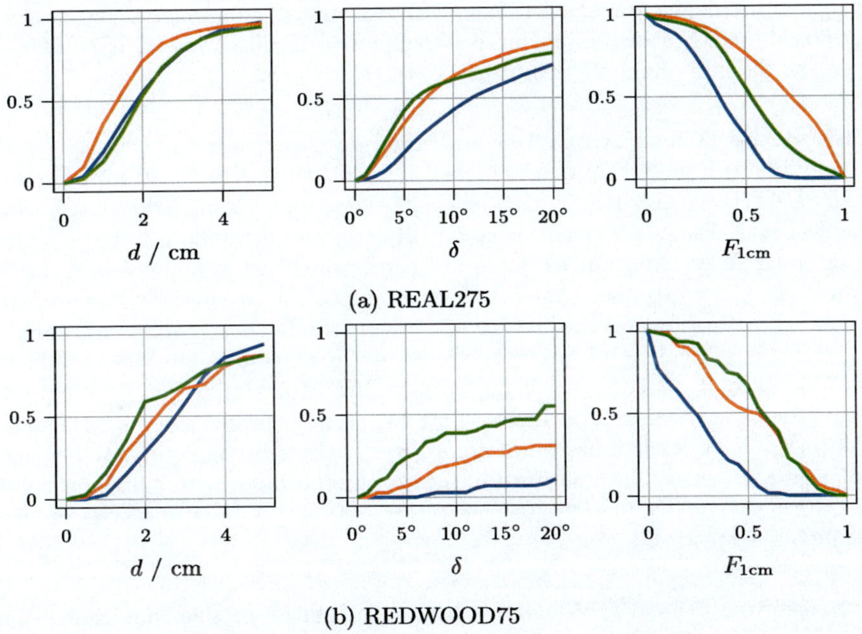

Fig. 8. Detailed precision results for CASS (—), SPD (—) and ASM-Net (—) for varying thresholds of position, orientation and F-score thresholds.

[1] Intel Core i7-6850K CPU, NVIDIA TITAN X (Pascal) GPU.

Table 2. Precision for different categories with $\delta = 10°, d = 2\,\text{cm}, F_{1\text{cm}} = 0.6$.

	REAL275						REDWOOD75		
	Bottle	Bowl	Camera	Can	Laptop	Mug	Bottle	Bowl	Mug
CASS	0.002	0.093	0.001	0.030	0.0	0.068	0.000	0.000	0.000
SPD	**0.610**	**0.892**	**0.052**	**0.863**	**0.218**	**0.246**	**0.320**	0.160	**0.040**
ASM-Net	0.167	0.137	0.023	0.587	0.133	0.229	0.160	**0.360**	0.000

The best results are highlighted in **bold**

5 Limitations

Comparability The results from Sect. 4 suggest that training on synthetic data currently generalizes better to unconstrained orientations. This is expected, since synthetic data allows accurate, unconstrained generation of training data. This opens the question of whether CASS or SPD would generalize better or worse when trained on the same synthetic data as ASM-Net. Since the training code for CASS and ASM-Net has not been published, it is difficult to perform further comparisons. In general, since methods currently vary significant parts of training datasets, architecture, pose parameterization, and losses, it is difficult to assess the individual contributions of a single change.

Multimodal Distributions Unconstrained pose estimation introduces significant difficulties in the task, which were hidden due to the constraints present in the CAMERA and REAL datasets. Consider, for example, the bottom left mug in Fig. 5. From the given view, it is difficult to tell which way the opening of the mug faces. Another example are cans, which are geometry-wise nearly symmetric. Currently there are only few works [9,20] that consider this problem of ambiguous poses and evaluation of such methods that predict multimodal posteriors is difficult. One possible way of evaluating such methods would be to allow methods to generate N hypotheses. Precision could then be computed for the best and worst hypothesis. A strong method would generate the same hypothesis N times if there is no ambiguity. If there is ambiguity, the correct hypothesis would still be contained in the set of hypotheses with a high probability. Alternatively, likelihood-based metrics might be suitable if methods provide likelihoods instead of hypotheses.

Dataset Size The REDWOOD75 dataset is limited in size, but the results suggest a clear lack of generalization capability of current approaches. This shows the need for larger datasets for unconstrained pose and shape estimation. It is an open question how such a dataset could be collected in the most efficient way.

6 Conclusion and Outlook

In this work, we have discussed the limitations of the current evaluation protocol prevalent in the field of categorical pose and shape estimation. In particular, existing datasets contain only a heavily constrained set of orientations, which simplifies the problem by removing pose and shape ambiguities. Furthermore, existing evaluation metrics are suboptimal and unnecessarily difficult to interpret. To alleviate these problems, we propose a new set of metrics applicable to both the established REAL275 and our proposed REDWOOD75 dataset, which contains a large variety of orientations. We apply our evaluation protocol to three methods and confirm limited generalization capability as suggested by the constrained orientations in their training data.

Our experiments suggest that there is a need for larger, high-quality datasets for unconstrained pose and shape estimation as well as for methods that can handle unconstrained orientations and the resulting pose ambiguities in a principled way.

Acknowledgements. This work was partially supported by the Wallenberg AI, Autonomous Systems and Software Program (WASP) funded by the Knut and Alice Wallenberg Foundation.

References

1. Ahmadyan, A., Zhang, L., Ablavatski, A., Wei, J., Grundmann, M.: Objectron: A large scale dataset of object-centric videos in the wild with pose annotations. In: Proceedings of the IEEE/CVF Conference on Computer Vision and Pattern Recognition. pp. 7822–7831 (2021)
2. Akizuki, S., Hashimoto, M.: ASM-Net: Category-level pose and shape estimation using parametric deformation. In: Proceedings of the British Machine Vision Conference (2021)
3. Chang, A.X., Funkhouser, T., Guibas, L., Hanrahan, P., Huang, Q., Li, Z., Savarese, S., Savva, M., Song, S., Su, H., Xiao, J., Yi, L., Yu, F.: ShapeNet: An information-rich 3D model repository. Tech. Rep. 1512.03012, arXiv preprint (Dec 2015)
4. Chen, D., Li, J., Wang, Z., Xu, K.: Learning canonical shape space for category-level 6D object pose and size estimation. In: Proceedings of the IEEE/CVF Conference on Computer Vision and Pattern Recognition, pp. 11973–11982 (2020)
5. Chen, K., Dou, Q.: SGPA: Structure-guided prior adaptation for category-level 6D object pose estimation. In: Proceedings of the IEEE/CVF International Conference on Computer Vision, pp. 2773–2782 (2021)
6. Chen, W., Jia, X., Chang, H.J., Duan, J., Shen, L., Leonardis, A.: FS-Net: Fast shape-based network for category-level 6D object pose estimation with decoupled rotation mechanism. In: Proceedings of the IEEE/CVF Conference on Computer Vision and Pattern Recognition, pp. 1581–1590 (2021)
7. Chen, X., Dong, Z., Song, J., Geiger, A., Hilliges, O.: Category level object pose estimation via neural analysis-by-synthesis. In: European Conference on Computer Vision, pp. 139–156. Springer, Berlin (2020)
8. Choi, S., Zhou, Q.Y., Miller, S., Koltun, V.: A large dataset of object scans. arXiv preprint arXiv:1602.02481 (2016)
9. Deng, H., Bui, M., Navab, N., Guibas, L., Ilic, S., Birdal, T.: Deep Bingham networks: Dealing with uncertainty and ambiguity in pose estimation. arXiv preprint arXiv:2012.11002 (2020)

10. Engelmann, F., Rematas, K., Leibe, B., Ferrari, V.: From points to multi-object 3D reconstruction. In: Proceedings of the IEEE/CVF Conference on Computer Vision and Pattern Recognition, pp. 4588–4597 (2021)
11. Everingham, M., Van Gool, L., Williams, C.K., Winn, J., Zisserman, A.: The PASCAL visual object classes (VOC) challenge. Int. J. Comput. Vis. **88**(2), 303–338 (2010)
12. Fan, H., Su, H., Guibas, L.J.: A point set generation network for 3D object reconstruction from a single image. In: Proceedings of the IEEE Conference on Computer Vision and Pattern Recognition, pp. 605–613 (2017)
13. Fischler, M.A., Bolles, R.C.: Random sample consensus: a paradigm for model fitting with applications to image analysis and automated cartography. Commun. ACM **24**(6), 381–395 (1981)
14. He, K., Gkioxari, G., Dollár, P., Girshick, R.: Mask R-CNN. In: Proceedings of the IEEE International Conference on Computer Vision, pp. 2961–2969 (2017)
15. Hodaň, T., Sundermeyer, M., Drost, B., Labbé, Y., Brachmann, E., Michel, F., Rother, C., Matas, J.: BOP challenge 2020 on 6D object localization. In: European Conference on Computer Vision, pp. 577–594. Springer (2020)
16. Knapitsch, A., Park, J., Zhou, Q.Y., Koltun, V.: Tanks and temples: benchmarking large-scale scene reconstruction. ACM Trans. Graph. **36**(4) (2017)
17. Lee, T., Lee, B.U., Kim, M., Kweon, I.S.: Category-level metric scale object shape and pose estimation. IEEE Robot. Autom. Lett. **6**(4), 8575–8582 (2021)
18. Lin, J., Wei, Z., Li, Z., Xu, S., Jia, K., Li, Y.: DualPoseNet: Category-level 6D object pose and size estimation using dual pose network with refined learning of pose consistency. In: Proceedings of the IEEE/CVF International Conference on Computer Vision, pp. 3560–3569 (2021)
19. Lin, T.Y., Maire, M., Belongie, S., Hays, J., Perona, P., Ramanan, D., Dollár, P., Zitnick, C.L.: Microsoft COCO: Common objects in context. In: Proceedings of the European Conference on Computer Vision, pp. 740–755 (2014)
20. Manhardt, F., Arroyo, D.M., Rupprecht, C., Busam, B., Birdal, T., Navab, N., Tombari, F.: Explaining the ambiguity of object detection and 6D pose from visual data. In: Proceedings of the IEEE/CVF International Conference on Computer Vision, pp. 6841–6850 (2019)
21. Paszke, A., Gross, S., Massa, F., Lerer, A., Bradbury, J., Chanan, G., Killeen, T., Lin, Z., Gimelshein, N., Antiga, L., et al.: PyTorch: an imperative style, high-performance deep learning library. Adv. Neural Inf. Process. Syst. **32**, 8026–8037 (2019)
22. Salton, G., McGill, M.J.: Introduction to modern information retrieval. McGraw Hill (1983)
23. Tatarchenko, M., Richter, S.R., Ranftl, R., Li, Z., Koltun, V., Brox, T.: What do single-view 3D reconstruction networks learn? In: Proceedings of the IEEE/CVF Conference on Computer Vision and Pattern Recognition, pp. 3405–3414 (2019)
24. Tian, M., Ang, M.H., Lee, G.H.: Shape prior deformation for categorical 6D object pose and size estimation. In: European Conference on Computer Vision, pp. 530–546. Springer, Berlin (2020)
25. Umeyama, S.: Least-squares estimation of transformation parameters between two point patterns. IEEE Trans. Pattern Anal. Mach. Intell. **13**(04), 376–380 (1991)
26. Wang, H., Sridhar, S., Huang, J., Valentin, J., Song, S., Guibas, L.J.: Normalized object coordinate space for category-level 6D object pose and size estimation. In: Proceedings of the IEEE/CVF Conference on Computer Vision and Pattern Recognition, pp. 2642–2651 (2019)

27. Wang, J., Chen, K., Dou, Q.: Category-level 6D object pose estimation via cascaded relation and recurrent reconstruction networks. In: Proceedings of the IEEE/RSJ International Conference on Intelligent Robots and Systems, pp. 4807–4814 (2021)
28. Zhou, Q.Y., Park, J., Koltun, V.: Open3D: A modern library for 3D data processing. arXiv preprint arXiv:1801.09847 (2018)

On Hand-Eye Calibration via On-Manifold Gauss-Newton Optimization

Marta Čolaković-Benčerić(✉), Juraj Peršić, Ivan Marković, and Ivan Petrović

University of Zagreb Faculty of Electrical Engineering and Computing Laboratory for Autonomous Systems and Mobile Robotics (LAMOR), Zagreb, Croatia
{marta.colakovic,juraj.persic,ivan.markovic,ivan.petrovic}@fer.hr

Abstract. Perception of autonomous robotic systems is highly dependent on accurate data fusion of multiple heterogeneous sensors. However, to maximally exploit the advantages of such setups, sensor data fusion necessitates accurate extrinsic calibration. In this paper, we propose a novel derivation of the Gauss-Newton based iterative on-manifold batch solution to the hand-eye calibration problem. By adopting a special Euclidean group formulation of the objective function, we derive exact and approximate solutions and validate them via synthetic and real-world experiments. The results show that the accuracy of the proposed approximate solutions is on par with the exact solution and alternative on-manifold iterative solutions. Moreover, due to the near commutativity of the hand-eye problem in low noise scenarios, the proposed 0th order approximation achieves up to 4 times faster execution time, thus opening up practical possibilities of utilization in more complex optimization techniques.

Keywords: Hand-eye calibration · Lie groups · Odometry-based calibration · On-manifold optimization

1 Introduction

In order to successfully navigate the environment and complete varied and complex tasks, autonomous robotic systems are often equipped with multiple heterogeneous sensors. Fusing information gathered by such sensors relies on accurate intrinsic, extrinsic, and temporal calibrations. Intrinsic calibration involves building a sensor model and determining the internal parameters, while temporal calibration estimates the time delays between measurements made by different sensors of a robotic system. Extrinsic calibration, on the other hand, is the process of determining the rigid transformation between different sensors, or a sensor and a part of the robotic system to which it is rigidly attached. This is usually achieved by finding correspondences in data coming from multiple sensors. The source of data correspondences can be a manufactured target or the environment itself, resulting in the division of calibration approaches into target-based and

I. Petrovic et al. (Eds.): IAS 2022, LNNS 577, pp. 378–391, 2023.
https://doi.org/10.1007/978-3-031-22216-0_26

targetless. Target-based extrinsic calibration methods use the manufactured target's features to find correspondences in sensor data [15,24,25], which are then used to evaluate the closed-form solution or to iteratively optimize the calibration parameters. Conversely, targetless methods rely on information gathered from the environment.

A special class of targetless methods relies on ego-motion for correspondence registration. These methods can be (i) SLAM-based [5,18,22], with correspondence registration relying on features, or (ii) odometry-based, leveraging incremental motion of each sensor [6,17]. In this paper, we focus on the odometry-based methods, for which a common mathematical formulation is called the hand-eye problem. One highly studied case of the hand-eye problem can be modeled by the equation $\mathbf{AX} = \mathbf{XB}$, and we can group related methods into decoupled and simultaneous. Decoupled methods first solve for the rotational component of the rigid transform, making the translational part of the hand-eye problem linear. To obtain closed-form solutions, in [26,29], the authors used the angle-axis representation and in [8,13] rotation was represented as unit quaternions, while in [23] the closed-form solution was found through $\mathfrak{so}(3)$ Lie algebra representation of rotation. In [19], the hand-eye problem was formulated using dual-quaternions reaching a least squares problem subject to a linear constraint. However, unlike simultaneous approaches, these solutions ignore the non-linear coupling of the translational and rotational components, thus propagating rotational error to the translational component of the calibration. Additionally, quaternion and dual-quaternion representation of rotation necessitates renormalization of the state vector. Such attributes of the rigid transform motivated further research into various possible parameterizations of the transformation components.

The first formulation of a simultaneous solution to the hand-eye problem was presented in [7] by using screw-motion representation of the rigid transformation. Subsequently, in [13], the authors proposed a quaternion-based simultaneous iterative solution optimized with the Levenberg-Marquardt algorithm. Other simultaneous solutions used the dual-quaternion representation of rigid motion [4,9,11]. In [9], calibration parameters are solved for using the singular value decomposition (SVD) method. Similarly, in [11], the authors offered a closed-form linear solution along with an iterative solution based on the Levenberg-Marquardt algorithm. Finally, in [4], the authors have formulated the dual-quaternion-based hand-eye problem as a maximum likelihood optimization problem, again, solving it with the Levenberg-Marquardt algorithm. An alternative approach was proposed in [1], where the Kronecker matrix product was used to reach a linear solution to the hand-eye problem. The proposed solution does not guarantee orthogonality of the estimated rotation matrix when simultaneously estimating rotational and translational components of the rigid transform. Recently, a multi-step iterative quaternion and Lie algebra based solution was proposed in [21] using the Levenberg-Marquardt algorithm for finding the optimal parameters. In [12], the authors formulated the hand-eye problem as a bound-and-branch algorithm using epipolar constraints. Optimization was carried out by linear programming reaching a globally optimal solution with respect to the L_∞ norm. Approaching the hand-eye problem with the errors-in-variables paradigm, in [16], the hand-eye problem was formulated on the $SO(3)\times\mathbb{R}^3$ group and based the optimization algorithm on the Gauss-Helmert model. The authors of [20] framed the hand-eye problem as a non-linear least squares problem on the $SO(3)\times\mathbb{R}^3$ group. Similarly, a novel error model for non-linear optimization on the $SO(3)\times\mathbb{R}^3$ group was proposed in [28]. Finally, the objective function of the hand-eye problem was formulated directly on the *SE(3)* group in [14], where an approximate, yet analytical Jacobian of the objective function was introduced, allowing for on-manifold optimization by the Gauss-Newton algorithm. Rigid transformations were represented as members of the $\mathbb{R}^6$ vector space isomorphic to the $\mathfrak{se}(3)$ Lie algebra.

In this paper, we derive an iterative on-manifold batch optimization algorithm for hand-eye calibration based on the Gauss-Newton method. The hand-eye problem is formulated directly on the *SE(3)* group. That guarantees group closure and removes the need for normalization or orthogonalization during the optimization procedure caused by the presence of noise, while retaining a minimal number of degrees of freedom of the state vector. We derive an exact analytic Jacobian matrix of the objective function and a novel approximate Jacobian matrix necessary for evaluating the Gauss-Newton iterative solution. The proposed solutions are validated and compared against alternative approaches on synthetic and real-world experiments.

2 Mathematical Background

In this section, we introduce the necessary mathematical background of the *SE(3)* group and the associated $\mathfrak{se}(3)$ Lie algebra. Following the notation used in [2], motions, i.e., rigid transforms, are represented by $\mathbf{T} \in SE(3)$ – an object composed of a rotation matrix $\mathbf{R} \in SO(3)$ and a translation vector $\mathbf{t} \in \mathbb{R}^3$

$$SE(3) : \mathbf{T} = \begin{bmatrix} \mathbf{R} & \mathbf{t} \\ \mathbf{0}^T & 1 \end{bmatrix} \in SE(3) \subset \mathbb{R}^{4\times 4} \mid \mathbf{R} \in SO(3), \mathbf{t} \in \mathbb{R}^3. \quad (1)$$

The $SO(3)$ special orthogonal Lie group is the group of all rotations about the origin of 3D Euclidean space, $\mathbb{R}^3$. It is a non-abelian group closed under matrix multiplication. Specifically, that means two members of the group are generally non-commutative. The exceptions being the identity element and the inverse of a particular element

$$\mathbf{R}_1, \mathbf{R}_2 \in SO(3): \quad \mathbf{R}_2\mathbf{R}_1 \neq \mathbf{R}_2\mathbf{R}_1, \quad \mathbf{R}_1\mathbf{R}_1^{-1} = \mathbf{R}_1^{-1}\mathbf{R}_1, \quad \mathbf{R}_1\mathbf{I} = \mathbf{I}\mathbf{R}_1. \quad (2)$$

Transitively, the $SE(3)$ group is non-abelian with equivalent exceptions. Generally, a Lie algebra of a Lie group is a tangent vector space at the identity paired with its associated Lie bracket, with the $SO(3)$ Lie algebra being denoted by $\mathfrak{so}(3)$. Similarly, an $\mathfrak{se}(3)$ Lie algebra element $\boldsymbol{\xi}^\wedge$ is composed of an $\mathfrak{so}(3)$ Lie algebra element $\boldsymbol{\phi}^\wedge$ and a vector $\boldsymbol{\rho} \in \mathbb{R}^3$,

$$\boldsymbol{\xi}^\wedge = \begin{bmatrix} \boldsymbol{\phi}^\wedge & \boldsymbol{\rho} \\ \mathbf{0}^T & 0 \end{bmatrix}, \quad \boldsymbol{\phi}^\wedge = \begin{bmatrix} \phi_x \\ \phi_y \\ \phi_z \end{bmatrix}^\wedge = \begin{bmatrix} 0 & -\phi_z & \phi_y \\ \phi_z & 0 & -\phi_x \\ -\phi_y & \phi_x & 0 \end{bmatrix}. \tag{3}$$

Finally, the elements of the vector space isomorphic to the tangent Lie algebra are equal to $\boldsymbol{\xi} = \left[\boldsymbol{\rho}\ \boldsymbol{\phi}\right]^T$.

The exponential mapping connecting the elements of the tangent vector space isomorphic to $\mathfrak{se}(3)$ with the *SE(3)* Lie group is defined as

$$\text{Exp} : \mathbb{R}^6 \rightarrow SE(3)\ ;\ \boldsymbol{\xi} \mapsto \mathbf{T} = \text{Exp}(\boldsymbol{\xi}) = \exp(\boldsymbol{\xi}^\wedge) = \begin{bmatrix} \exp(\boldsymbol{\phi}^\wedge) & \mathbf{V}(\boldsymbol{\phi})\boldsymbol{\rho} \\ \mathbf{0}^T & 1 \end{bmatrix}, \tag{4}$$

with $\mathbf{V}(\boldsymbol{\phi})$ being the left *SO(3)* group Jacobian. Inversely, the logarithmic mapping transforms the group element into the tangent vector and is defined as

$$\text{Log} : SE(3) \rightarrow \mathbb{R}^6\ ;\ \mathbf{T} \mapsto \boldsymbol{\xi} = \text{Log}(\mathbf{T}) = \log(\mathbf{T})^\vee = \begin{bmatrix} \mathbf{V}^{-1}(\boldsymbol{\phi})\mathbf{t} \\ \text{Log}(\mathbf{R}) \end{bmatrix}, \tag{5}$$

with $\boldsymbol{\phi} = \text{Log}(\mathbf{R})$.

Defining the Lie bracket as $[\mathbf{A}, \mathbf{B}] = \mathbf{AB} - \mathbf{BA}$, allows us to write the Lie brackets of $\mathfrak{se}(3)$ and $\mathfrak{so}(3)$

$$[\boldsymbol{\xi}_1^\wedge, \boldsymbol{\xi}_2^\wedge] = \boldsymbol{\xi}_1^\wedge \boldsymbol{\xi}_2^\wedge - \boldsymbol{\xi}_2^\wedge \boldsymbol{\xi}_1^\wedge = \left(\boldsymbol{\xi}_1^\curlywedge \boldsymbol{\xi}_2\right)^\wedge, \tag{6}$$

$$[\boldsymbol{\phi}_1^\wedge, \boldsymbol{\phi}_2^\wedge] = \boldsymbol{\phi}_1^\wedge \boldsymbol{\phi}_2^\wedge - \boldsymbol{\phi}_2^\wedge \boldsymbol{\phi}_1^\wedge = \left(\boldsymbol{\phi}_1^\wedge \boldsymbol{\phi}_2\right)^\wedge. \tag{7}$$

where the '$\curlywedge$' operator is defined as

$$\boldsymbol{\xi}^\curlywedge = \begin{bmatrix} \boldsymbol{\rho} \\ \boldsymbol{\phi} \end{bmatrix}^\curlywedge = \begin{bmatrix} \boldsymbol{\phi}^\wedge & \boldsymbol{\rho}^\wedge \\ \mathbf{0}_{3\times3} & \boldsymbol{\phi}^\wedge \end{bmatrix}. \tag{8}$$

The solution for $\mathbf{C}$ to $\exp(\mathbf{A})\exp(\mathbf{B}) = \exp(\mathbf{C})$, for non-commutative $\mathbf{A}$ and $\mathbf{B}$, is given by the *Baker-Campbell-Hausdorff* (BCH) formula,

$$\mathbf{C} = \log(\exp(\mathbf{A})\exp(\mathbf{B})) = \mathbf{A} + \mathbf{B} + \frac{1}{2}[\mathbf{A}, \mathbf{B}] + \text{ h.o.t.} \tag{9}$$

Due to linearity of the '$\vee$' operator, the BCH formula can be expressed in terms of members of vector spaces isomorphic to the associated Lie algebra, $\mathfrak{se}(3)$,

$$\log(\exp(\boldsymbol{\xi}_1^\wedge)\exp(\boldsymbol{\xi}_2^\wedge))^\vee = \boldsymbol{\xi}_1 + \boldsymbol{\xi}_2 + \frac{1}{2}\boldsymbol{\xi}_1^\curlywedge \boldsymbol{\xi}_2 + \text{h.o.t.} \tag{10}$$

Another important concept is the adjoint. The adjoint is a linear transform that maps the Lie algebra elements at $\mathbf{X}$ to Lie algebra elements at the identity, thus connecting the local and global elements. The adjoint of a Lie group at an element $\mathbf{X}$ is defined as

$$\boldsymbol{\tau}^\wedge \mapsto \text{Ad}_{\mathbf{X}}\left(\boldsymbol{\tau}^\wedge\right) \triangleq \mathbf{X}\boldsymbol{\tau}^\wedge\mathbf{X}^{-1}. \tag{11}$$

Linearity of the adjoint action allows us to define the equivalent matrix operator that maps local tangent vectors to their global counterparts, ${}^{\mathbf{X}}\mathcal{T} \mapsto {}^{\mathcal{E}}\mathcal{T} = \mathbf{Ad_X}{}^{\mathbf{X}}\mathcal{T}$. It can be shown, [2,27], that the adjoint matrix of a rigid transformation $\mathbf{X} = \exp(\boldsymbol{\xi}_{\mathbf{X}}{}^\wedge) \in SE(3)$ evaluates to

$$\mathbf{Ad_X} = \mathcal{T}_{\mathbf{X}} = \sum_{n=0}^{\infty} \frac{1}{n!} \left(\boldsymbol{\xi}_{\mathbf{X}}{}^{\curlywedge}\right)^n = \begin{bmatrix} \mathbf{R_X} & \mathbf{t_X}^\wedge \mathbf{R_X} \\ \mathbf{0}_{3\times 3} & \mathbf{R_X} \end{bmatrix}, \tag{12}$$

while the adjoint matrix of a rotation matrix $\mathbf{R} \in SO(3)$ is simply $\mathbf{R}$.

Finally, given that the special Euclidean group is closed under matrix multiplication we define left (global) and right (local) multiplicative increment as

$$\delta\boldsymbol{\xi} \oplus \mathbf{X} = \exp((\delta\boldsymbol{\xi})^\wedge)\exp(\boldsymbol{\xi}^\wedge) \approx \exp((\boldsymbol{\xi} + \mathcal{J}_l^{-1}(\boldsymbol{\xi})\delta\boldsymbol{\xi})^\wedge), \tag{13}$$

$$\mathbf{X} \oplus \delta\boldsymbol{\xi} = \exp(\boldsymbol{\xi}^\wedge)\exp((\delta\boldsymbol{\xi})^\wedge) \approx \exp((\boldsymbol{\xi} + \mathcal{J}_r^{-1}(\boldsymbol{\xi})\delta\boldsymbol{\xi})^\wedge), \tag{14}$$

where $\mathcal{J}_l$ and $\mathcal{J}_r$ denote the left and right Jacobian of the $SE(3)$ group with the inverses being equal to

$$\mathcal{J}_l^{-1}(\boldsymbol{\xi}) = \sum_{n=0}^{\infty} \frac{B_n}{n!} \left(\boldsymbol{\xi}^{\curlywedge}\right)^n, \quad \mathcal{J}_r^{-1}(\boldsymbol{\phi}) = \mathcal{J}_l^{-1}(-\boldsymbol{\phi}). \tag{15}$$

B_n denote the Bernoulli numbers and the analytic closed-form expressions for $\mathcal{J}_l$, $\mathcal{J}_r$ and their inverses can be found in [3].

3 Proposed Method

In this section, we formulate the batch calibration problem on the $SE(3)$ Lie group and derive the exact and approximate Gauss-Newton solutions. The hand-eye calibration problem is modeled by the equation

$$\mathbf{AX} = \mathbf{XB}. \tag{16}$$

The unknown fixed rigid transformation between two sensors is represented by $\mathbf{X}$ and motion segments of a pair of sensors by $\mathbf{A}$ and $\mathbf{B}$.

To determine the optimal $\mathbf{X}$ from a set of measured motion segments $\mathbf{A}_i$, $\mathbf{B}_i$, $i = 1, \ldots, N$, it is necessary to minimize the objective function defined as the sum of $SE(3)$ distances

$$\mathbf{X}^* = \underset{\mathbf{X} \in SE(3)}{\arg\min} \frac{1}{2} \sum_{i=1}^{N} \|\mathbf{f}_i\|_2^2, \tag{17}$$

with $\|\cdot\|_2^2$ denoting the Euclidean norm of a vector. Expressing the hand-eye problem via (16) leads us to define $\mathbf{f}_i$ as follows

$$\mathbf{f}_i : SE(3) \rightarrow \mathbb{R}^6 \; ; \; \mathbf{f}(\mathbf{X}) = \log(\mathbf{A}_i{}^{-1}\mathbf{X}\mathbf{B}_i\mathbf{X}^{-1})^\vee. \tag{18}$$

To find the optimal $\mathbf{X}^*$ and determine the Gauss-Newton step we expand the distance function around the current calibration estimate, $\mathbf{X}$,

$$\mathbf{f}_i(\delta\boldsymbol{\xi}_\mathbf{X} \oplus \mathbf{X}) \approx \mathbf{e}_i(\mathbf{X}) + \mathbf{G}_i(\mathbf{X})\delta\boldsymbol{\xi}_\mathbf{X}, \tag{19}$$

where $\mathbf{G}_i(\mathbf{X})$ denotes the Jacobian matrix comprised of Lie derivatives of the distance vector function $\mathbf{f}_i$ at the current estimate $\mathbf{X}$, and $\mathbf{e}_i(\mathbf{X})$ denotes the error. That allows us to define the Gauss-Newton update as

$$\delta\boldsymbol{\xi}_\mathbf{X} = -\left(\sum_{i=1}^{N} \mathbf{G}_i^T \cdot \boldsymbol{\Sigma}_i^{-1} \cdot \mathbf{G}_i\right)^{-1} \sum_{i=1}^{N} \mathbf{G}_i^T \cdot \boldsymbol{\Sigma}_i^{-1} \cdot \mathbf{e}_i, \tag{20}$$

with $\boldsymbol{\Sigma}_i$ being the covariance matrix of each measurement. The Gauss-Newton update is applied multiplicatively to the left in accordance to the relation (13). For brevity the subscript i is omitted from the remainder of the derivation.

The exact formulation of the error function equals the distance function at the current estimate, $\mathbf{e}_{\mathbf{Exact}}(\mathbf{X}) = \mathbf{f}(\mathbf{X}) = \log(\mathbf{A}^{-1}\mathbf{X}\mathbf{B}\mathbf{X}^{-1})^\vee$, while the Jacobian matrix can be found through its definition,

$$\mathbf{G}_{\mathbf{Exact}} = (\mathcal{L}_\varepsilon \mathbf{f}(\mathbf{X})) \triangleq \lim_{\delta\boldsymbol{\xi}_\mathbf{X} \to \mathbf{0}} \frac{\mathbf{f}(\delta\boldsymbol{\xi}_\mathbf{X} \oplus \mathbf{X}) - \mathbf{f}(\mathbf{X})}{\delta\boldsymbol{\xi}_\mathbf{X}} : SE(3) \to \mathbb{R}^{6\times 6}. \tag{21}$$

The numerator in (21) is equal to

$$\begin{aligned}\mathbf{f}(\delta\boldsymbol{\xi}_\mathbf{X} \oplus \mathbf{X}) - \mathbf{f}(\mathbf{X}) &= \log(\mathbf{A}^{-1}\delta\mathbf{X}\mathbf{X}\mathbf{B}\mathbf{X}^{-1}\delta\mathbf{X}^{-1})^\vee - \log(\mathbf{A}^{-1}\mathbf{X}\mathbf{B}\mathbf{X}^{-1})^\vee \\ &= \left(\log(\mathbf{A}^{-1}\delta\mathbf{X}\mathbf{X}\mathbf{B}\mathbf{X}^{-1}\delta\mathbf{X}^{-1}) - \log(\mathbf{A}^{-1}\mathbf{X}\mathbf{B}\mathbf{X}^{-1})\right)^\vee \\ &= \log\left(\mathbf{A}^{-1}\delta\mathbf{X}\mathbf{X}\mathbf{B}\mathbf{X}^{-1}\delta\mathbf{X}^{-1}\mathbf{X}\mathbf{B}^{-1}\mathbf{X}^{-1}\mathbf{A}\right)^\vee.\end{aligned} \tag{22}$$

First we define the tangent vector of $\delta\mathbf{X}$ as $\delta\boldsymbol{\xi}_\mathbf{X} = \log(\delta\mathbf{X})^\vee$. Then by applying the identity $\mathbf{X}\exp(\boldsymbol{\xi}^\wedge)\mathbf{X}^{-1} = \exp(\mathbf{X}\boldsymbol{\xi}^\wedge\mathbf{X}^{-1})$, and the definition of the adjoint (11) to (22), we reach

$$\mathbf{f}(\delta\boldsymbol{\xi}_\mathbf{X} \oplus \mathbf{X}) - \mathbf{f}(\mathbf{X}) = \log\left(\mathbf{A}^{-1}\exp(\delta\boldsymbol{\xi}_\mathbf{X}{}^\wedge)\exp((-\mathcal{T}_\mathbf{X}\mathcal{T}_\mathbf{B}\mathcal{T}_X{}^{-1}\delta\boldsymbol{\xi}_\mathbf{X})^\wedge)\mathbf{A}\right)^\vee. \tag{23}$$

We simplify the expression further by using the definition the left increment (13) and the inverse of the left *SE(3)* group Jacobian (15),

$$\begin{aligned}\mathbf{f}(\delta\boldsymbol{\xi}_\mathbf{X} \oplus \mathbf{X}) - \mathbf{f}(\mathbf{X}) &= \log\left(\mathbf{A}^{-1}\exp((\delta\boldsymbol{\xi}_\mathbf{X} - \mathcal{T}_\mathbf{X}\mathcal{T}_\mathbf{B}\mathcal{T}_X{}^{-1}\delta\boldsymbol{\xi}_\mathbf{X})^\wedge)\mathbf{A}\right)^\vee \\ &= \log\left(\exp((\mathcal{T}_\mathbf{A}{}^{-1}(\mathbf{I}_6 - \mathcal{T}_\mathbf{X}\mathcal{T}_\mathbf{B}\mathcal{T}_X{}^{-1})\delta\boldsymbol{\xi}_\mathbf{X})^\wedge)\right)^\vee.\end{aligned} \tag{24}$$

From (24) we deduce the exact Jacobian of the distance function

$$\mathbf{G}_{\mathbf{Exact}}(\mathbf{X}) = \mathcal{T}_\mathbf{A}{}^{-1}(\mathbf{I}_6 - \mathcal{T}_\mathbf{X}\mathcal{T}_\mathbf{B}\mathcal{T}_X{}^{-1}). \tag{25}$$

3D rotations are generally non-commutative, transitively making rigid transformations generally non-commutative. Due to the nature of the hand-eye problem, in zero noise environments $\mathbf{A}^{-1}$ and $\mathbf{X}\mathbf{B}\mathbf{X}^{-1}$ are identical and, thus, commutative. If the motion segment measurements are only slightly noisy, we view

$\mathbf{A}^{-1}$ and $\mathbf{XBX}^{-1}$ as nearly commutative, making $\|[\mathbf{A}^{-1}, \mathbf{XBX}^{-1}]^\vee\|_2^2$ small. In order to reduce computational complexity in calculating the optimization step, we develop an approximate solution justified by the previous argument.

Following a similar procedure to [14], we apply the increment to the left to the current estimate of the calibration parameters $\mathbf{X}$ in the distance function,

$$\mathbf{f}(\delta\boldsymbol{\xi}_\mathbf{X} \oplus \mathbf{X}) = \log\left(\exp(-\boldsymbol{\xi}_\mathbf{A}{}^\wedge)\exp(\delta\boldsymbol{\xi}_\mathbf{X}{}^\wedge)\mathbf{X}\exp(\boldsymbol{\xi}_\mathbf{B}{}^\wedge)\mathbf{X}^{-1}\exp(-\delta\boldsymbol{\xi}_\mathbf{X}{}^\wedge)\right)^\vee. \tag{26}$$

Again, using the identity $\mathbf{X}\exp(\boldsymbol{\xi}^\wedge)\mathbf{X}^{-1} = \exp(\mathbf{X}\boldsymbol{\xi}^\wedge\mathbf{X}^{-1})$, and the definition of the adjoint (11) we reach

$$\mathbf{f}(\delta\boldsymbol{\xi}_\mathbf{X} \oplus \mathbf{X}) = \log\left(\exp(-\boldsymbol{\xi}_A{}^\wedge)\exp(\delta\boldsymbol{\xi}_\mathbf{X}{}^\wedge)\exp((\underbrace{\mathcal{T}_\mathbf{X}\boldsymbol{\xi}_B}_{\boldsymbol{\xi}_{XB}})^\wedge)\exp(-\delta\boldsymbol{\xi}_\mathbf{X}{}^\wedge)\right)^\vee. \tag{27}$$

Repeating the previous step gives the following formulation of the perturbed distance function

$$\mathbf{f}(\delta\boldsymbol{\xi}_\mathbf{X} \oplus \mathbf{X}) = \log\left(\exp(-\boldsymbol{\xi}_\mathbf{A}{}^\wedge)\exp((\mathcal{T}_{\delta\mathbf{X}}\boldsymbol{\xi}_\mathbf{XB})^\wedge)\right)^\vee. \tag{28}$$

After applying the BCH formula on $\mathfrak{se}(3)$, as in (10), to the first order of approximation, we are left with

$$\mathbf{f}(\delta\boldsymbol{\xi}_\mathbf{X} \oplus \mathbf{X}) \approx -\boldsymbol{\xi}_\mathbf{A} + \mathcal{T}_{\delta\mathbf{X}}\boldsymbol{\xi}_\mathbf{XB} - \frac{1}{2}\boldsymbol{\xi}_\mathbf{A}{}^\curlywedge\mathcal{T}_{\delta\mathbf{X}}\boldsymbol{\xi}_\mathbf{XB}. \tag{29}$$

For small $\|\delta\mathbf{X}\|$ the adjoint can be approximated with $\mathcal{T}_{\delta\mathbf{X}} \approx \mathbf{I}_6 + \delta\boldsymbol{\xi}_\mathbf{X}{}^\curlywedge$. Along with the equality $\mathbf{x}^\curlywedge\mathbf{y} = -\mathbf{y}^\curlywedge\mathbf{x}$, that leads us to

$$\mathbf{f}(\delta\boldsymbol{\xi}_\mathbf{X} \oplus \mathbf{X}) \approx \underbrace{-\boldsymbol{\xi}_\mathbf{A} + \boldsymbol{\xi}_\mathbf{XB} - \frac{1}{2}\boldsymbol{\xi}_\mathbf{A}{}^\curlywedge\boldsymbol{\xi}_\mathbf{XB}}_{\mathbf{e}_1} + \underbrace{\left(-\boldsymbol{\xi}_\mathbf{XB}{}^\curlywedge + \frac{1}{2}\boldsymbol{\xi}_\mathbf{A}{}^\curlywedge\boldsymbol{\xi}_\mathbf{XB}{}^\curlywedge\right)}_{\mathbf{G}_1}\delta\boldsymbol{\xi}_\mathbf{X}. \tag{30}$$

The 0th order error and Jacobian functions can be extracted from the 1st order formulation

$$\mathbf{e}_0(\mathbf{X}) = -\boldsymbol{\xi}_\mathbf{A} + \boldsymbol{\xi}_\mathbf{XB}, \quad \mathbf{G}_0(\mathbf{X}) = -\boldsymbol{\xi}_\mathbf{XB}{}^\curlywedge. \tag{31}$$

To determine the influence of the approximation order in the BCH formula on the calibration accuracy, we compare results reached by using the exact formulation, the 1st and the 0th order approximate formulations. Approximate Gauss-Newton solutions were derived by leveraging the near-commutative nature of $\mathbf{A}^{-1}$ and $\mathbf{XBX}^{-1}$ matrices making up the distance function on the $SE(3)$ group and applying the BCH formula. Additionally, the novel formulation of the perturbed distance function through the adjoint transformation on $SE(3)$ made the proposed approximate solutions computationally less costly than the approximate solution previously suggested in [14].

4 Experimental Results

Here we present results of the performed experiments on synthetic and real world datasets where we investigated the accuracy of calibration with varying orders of approximation, rigid motion representations and noise levels. In previous work, [14,16,20], regarding on-manifold optimization and the hand-eye problem, it was assumed that two sides, or at least the rotational parts, of (16) are commutative without it being explicitly stated. In noisy environments that is not strictly true; however, with the following experiments we will demonstrate that in low noise data the Gauss-Newton solution derived with the assumption of commutativity produces sufficiently accurate results.

We compare the following approaches. First, the optimization algorithms based on the proposed approximate methods derived from the 0th order approximation, Eq. (31), and the 1st order approximation, Eq. (30), of the objective function using the BCH formula, are referred to as **SE3_0** and **SE3_1**, respectively. A similar method proposed in [14] is referred to as **SE3_JL**. To determine the influence of rigid motion representation on the calibration outcome, we have re-derived a solution on the commonly used $SO(3) \times \mathbb{R}^3$ group [20,29] following a similar procedure to the one used in Sect. 3

$$\mathbf{e}_{SO3R3}(\mathbf{X}) = \begin{bmatrix} (\mathbf{I}_3 - \mathbf{R_A})\mathbf{t_X} + \mathbf{R_X}\mathbf{t_B} - \mathbf{t_A} \\ -\boldsymbol{\phi}_\mathbf{A} + \mathbf{R_X}\boldsymbol{\phi}_\mathbf{B} \end{bmatrix}, \tag{32}$$

$$\mathbf{G}_{SO3R3}(\mathbf{X}) = \begin{bmatrix} \mathbf{I}_3 - \mathbf{R_A} & -(\mathbf{R_X}\mathbf{t_B})^\wedge \\ \mathbf{0} & -(\mathbf{R_X}\boldsymbol{\phi}_\mathbf{B})^\wedge \end{bmatrix}. \tag{33}$$

We denote the algorithm, based on the error function and Jacobian on $SO(3) \times \mathbb{R}(3)$, by **SO3R3**. Finally, the algorithm based on (25), the exact calculation of the error and Jacobian functions, is denoted by **Exact**.

All of the methods were implemented in *Python3* and ran on a machine with i7-6700HQ CPU at 2.6 GHz × 8 and 16 GB of 2133 MHz MHz DDR4 RAM. Furthermore, all of the methods were initialized with $\mathbf{X}_0 = \mathbf{I}_4$, with the ground truth calibration being $\mathbf{t}_{\mathbf{GT}}^T = [0.0 \ -0.20 \ 0.0]$ m, $\boldsymbol{\phi}_{\mathbf{GT}}^T = [-1.21 \ -1.21 \ -1.21]$ rad.

The following error metrics are used to evaluate the results

$$E_\mathbf{R} = \| \log\left(\mathbf{R}_{\mathrm{GT}}\mathbf{R}_{\mathrm{est}}^T\right)^\vee \|_2^2, \tag{34}$$

$$E_\mathbf{t} = \|\mathbf{t}_{\mathrm{GT}} - \mathbf{t}_{\mathrm{est}}\|_2^2. \tag{35}$$

In (34) and (35), $\mathbf{R}_{\mathrm{GT}}$ and $\mathbf{t}_{\mathrm{GT}}$ denote components of the ground truth rigid transform between sensors, and $\mathbf{R}_{\mathrm{est}}$ and $\mathbf{t}_{\mathrm{est}}$ components of the estimated rigid transformation.

4.1 Simulation Results

In the first synthetic experiment, we have generated a reasonably informative, yet achievable, 3D synthetic trajectory shown in Fig. 1. It was inspired by the

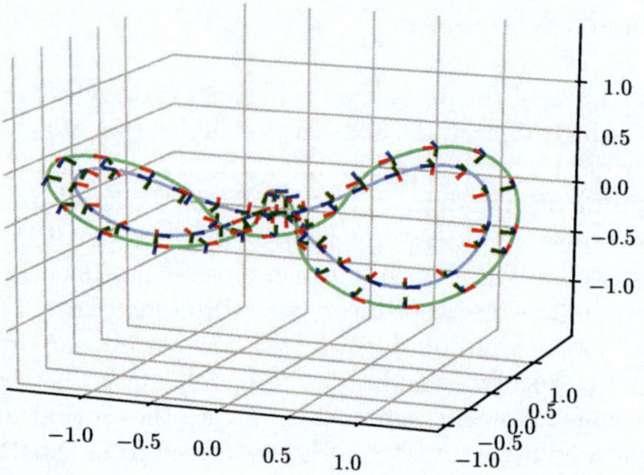

Fig. 1. The synthetic 3D Bernoulli lemniscate trajectory. Green and blue parametric curves represent trajectories of two sensors with associated local coordinate systems.

Bernoulli lemniscate plane curve and the 3D trajectory simulator *Bernoulli-Lemniscate* from the *SRRG* toolbox [10]. Our 3D trajectory is described by the following set of parametric equations, with $a = 1.5\,\mathrm{m}$ denoting the half-width in the x axis direction

$$x(t) = a\frac{\cos(t)}{1+\sin^2(t)}, \tag{36}$$

$$y(t) = \sin(t)x(t), \tag{37}$$

$$z(t) = \cos(t)y(t). \tag{38}$$

The trajectory of the second sensor was generated by applying the $\mathbf{X}_{\mathrm{GT}}$ transformation. After extracting equally sampled ($\Delta t = 0.02\,\mathrm{s}$) motion segments $\mathbf{A}_i$, $\mathbf{B}_i$, $i = 1, \ldots, 315$, from the synthetic trajectory, Gaussian noise was added in the local frame. Each instance of the Gaussian noise was extracted from a distribution $\mathcal{N}(0, \boldsymbol{\Sigma})$, $\boldsymbol{\Sigma} = \sigma^2\mathbf{I}_6$, with units of σ being in meters for the translational parameters $\boldsymbol{\rho}$ and radians for the rotational parameters $\boldsymbol{\phi}$. The standard deviation of noise σ in our experiments varied from 0.001 to 0.009 for both the translational and rotational part.

To analyze the performance of the benchmarked methods we have generated 100 trajectories per noise level, with noise drawn from the aforementioned distribution. The results are shown in Fig. 2. It is evident that, within the applied noise range, methods based on *SE(3)* and *SO(3)* $\times\ \mathbb{R}^3$ representation of rigid motions, produce almost identical results. Comparing the performance of the exact solution with the 0th and 1st order solutions, we can see that in this noise and motion range the 1st order approximate methods **SE3_1** and **SE3_JL** are effectively equivalent to the exact solution. Equally, the 0th order approximate method yields similarly accurate solutions within the aforementioned noise range.

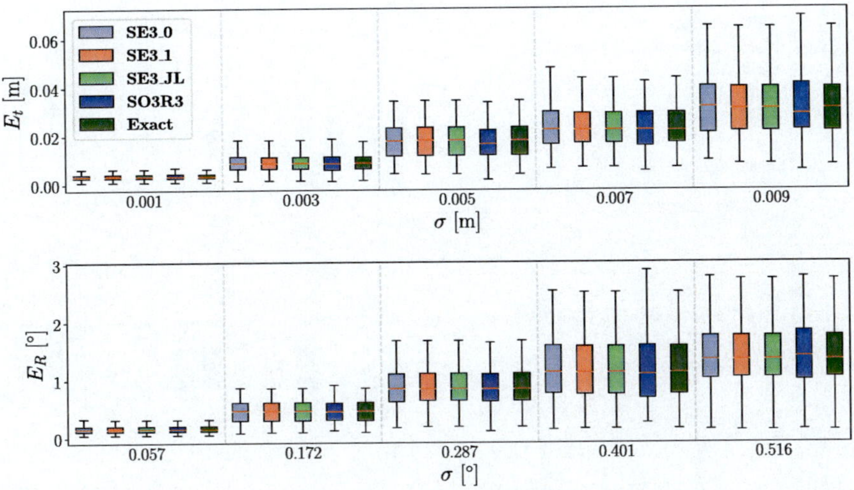

Fig. 2. Translational (top) and rotational (bottom) error of the benchmarked approaches with increasing noise for the Bernoulli lemniscate trajectory. The translational (top) and rotational (bottom) error are shown on the y axis in units of [m] and [°]. The standard deviation of noise σ is shown on the x axis.

To eliminate the influence of trajectory informativeness on the calibration accuracy, in the second synthetic experiment we have analyzed the methods' performance on 100 randomly generated trajectories drawn form a Gaussian distribution. Randomly generated trajectories are generally the most informative possible trajectories but are difficult to achieve in real life situations. The distribution was adjusted to have a similar range of translational ($0.01 - 0.05$ m) and rotational displacements ($0 - 7$ °) as a distribution of motion segments of the 3D Bernoulli lemniscate trajectory. Again, noise drawn from a Gaussian distribution ($\mathcal{N}(0, \mathbf{\Sigma})$, $\mathbf{\Sigma} = \sigma^2\mathbf{I}_6$) was added in the local frame. Randomly generated trajectories consist of $N = 315$ motion segments, lasting $\Delta t = 0.02$ s. The results are shown in Fig. 3. Testing the algorithms on randomly generated trajectories has confirmed that all of the on-manifold optimization schemes achieve similar accuracy in the studied noise range, with the calibration results being slightly improved by the more informative random trajectories.

In order to examine the runtime of the five benchmarked methods, we have averaged runtime over 100 randomly generated trajectories over the full noise range. The importance of our approximate *SE(3)* based solutions is evident when comparing the execution times of the benchmarked methods as shown in Fig. 4. The **Exact** and **SO3R3** methods have a longer time of execution due to multiple matrix multiplication operations needed to evaluate the Jacobian and error matrices. Additionally, the **SE3_JL** method proposed in [14] has the longest runtime. That is caused by the complex structure of the analytical expression for $\mathcal{J}_l^{-1}$, used to calculate the approximate Jacobian matrix. On average, all of

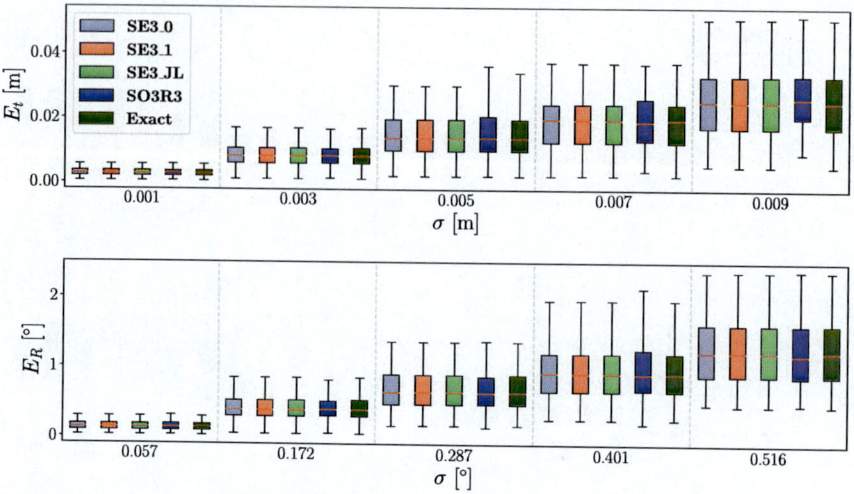

Fig. 3. Translational (top) and rotational (bottom) error of benchmarked approaches with increasing noise for randomly generated trajectories. The translational (top) and rotational (bottom) errors are shown on the y axis in units of [m] and [°]. The standard deviation of noise σ is shown on the x axis.

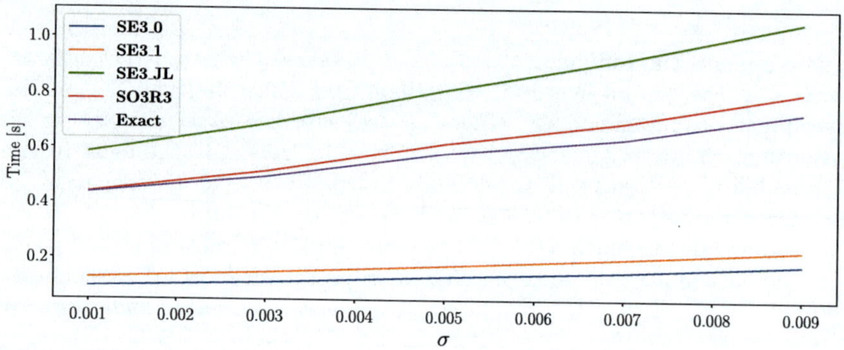

Fig. 4. Evaluation time of the benchmarked methods in seconds averaged over 100 noise instances.

the these methods achieved convergence in an equal number of iterations. The comparatively low execution time of the proposed approximate methods, **SE3_0** and **SE3_1**, opens up the practical possibility of using them in more complex optimization techniques. Although the optimization using regression models that account for the measurement noise, e.g., [16], are generally more accurate, they are time consuming, making the fast analytic Jacobians key in formulating an on-line solution.

Table 1. Calibration results on the RGB-D dataset.

Method	E_t (m)	E_R (°)	Runtime (s)
SE3_0	0.01614	1.206	**0.1400**
SE3_1	0.01553	1.195	0.1735
SE3_JL	0.01553	1.191	0.8283
SO3R3	0.02088	**1.182**	0.6295
Exact	**0.01541**	1.195	0.5526

Bold font highlights the results of the best performing algorithms in each category

4.2 Real World Data

For the real-world data experiment, we used the dataset presented in [4]. It includes 200 estimated motion segments of two rigidly mounted Xtion RGB-D cameras and the ground truth calibration between them, with motion estimation performed by the KinectFusion algorithm. The ground truth calibration was provided with the dataset and the covariance matrices of each measurement are assumed to be constant. From Table 1 we can conclude that all of the methods based on the *SE(3)* representation of rigid motion, **SE3_0**, **SE3_1**, **SE3_JL**, **SO3R3**, and **Exact**, performed similarly well at determining the rotational calibration parameters. However, the **SO3R3** method yielded somewhat less accurate translational calibration parameters than the *SE(3)* based methods. Again, the benefits of using proposed approximate methods, **SE3_0** and **SE3_1**, becomes evident when comparing the runtime. The number of iterations across the benchmarked methods was similar, with **SE3_0** and **SE3_1** having significantly smaller runtime. Again, the conclusions drawn from the synthetic experiments have been confirmed with real-world data. The accuracy of the *SE(3)* based methods is largely the same, while the $SO(3) \times \mathbb{R}^3$ based method has slightly higher rotational accuracy but lower translational accuracy. The **SE3_0** and **SE3_1** proposed methods vastly outperform other on-manifold optimization methods in regard to execution time.

5 Conclusion

In this paper, we have proposed an on-manifold solution to the batch hand-eye calibration problem. Framing the hand-eye problem on the *SE(3)* manifold guarantees the group closure of the solution and removes the need for renormalization or reorthogonalization during the optimization process. That in turn lowers the computational cost. Additionally, formulating the calibration problem directly on the *SE(3)* manifold allowed us to derive analytic exact and approximate Jacobians needed to calculate the iterative Gauss-Newton solution. The derived analytic expressions make for an efficient optimization scheme, all the while retaining the accuracy of conventional optimization approaches. We validated the proposed solutions and compared them to alternative approaches on synthetic and real-world data. The results explicitly demonstrate that in low noise situations the hand-eye problem can be considered near commutative, and that the novel approximate on-manifold iterative solution is the most efficient out of the benchmarked solutions, while retaining on par performance.

For future work, the superior efficiency of the proposed method opens up the practical possibility of extending the Gauss-Newton based solution to more complex regression models.

Acknowledgment. This research has been supported by the European Regional Development Fund under the grant KK.01.2.1.02.0119 (A-UNIT).

References

1. Andreff, N., Horaud, R., Espiau, B.: On-line hand-eye calibration. In: Second International Conference on 3-D Digital Imaging and Modeling (3DIM '99), pp. 430–436 (1999)
2. Barfoot, T.: State estimation for robotics. Cambridge University Press, Cambridge, United Kingdom New York, NY, USA (2017)
3. Barfoot, T.D., Furgale, P.T.: Associating uncertainty with three-dimensional poses for use in estimation problems. IEEE Trans. Rob. **30**(3), 679–693 (2014)
4. Brookshire, J., Teller, S.: Extrinsic calibration from per-sensor egomotion. In: Proceedings of Robotics: Science and Systems, Sydney, Australia (2012, July)
5. Carrera, G., Angeli, A., Davison, A.J.: SLAM-based automatic extrinsic calibration of a multi-camera rig. In: 2011 IEEE International Conference on Robotics and Automation (ICRA), pp. 2652–2659 (2011)
6. Censi, A., Franchi, A., Marchionni, L., Oriolo, G.: Simultaneous calibration of odometry and sensor parameters for mobile robots. IEEE Trans. Rob. **29**(2), 475–492 (2013)
7. Chen, H.: A screw motion approach to uniqueness analysis of head-eye geometry. In: Proceedings of 1991 IEEE Computer Society Conference on Computer Vision and Pattern Recognition, pp. 145–151 (1991)
8. Chou, J.C.K., Kamel, M.: Finding the position and orientation of a sensor on a robot manipulator using quaternions. Int. J. Rob. Res. **10**(3), 240–254 (1991)
9. Daniilidis, K., Bayro-Corrochano, E.: The dual quaternion approach to hand-eye calibration. In: Proceedings of 13th International Conference on Pattern Recognition, IEEE (1996)
10. Della Corte, B., Andreasson, H., Stoyanov, T., Grisetti, G.: Unified motion-based calibration of mobile multi-sensor platforms with time delay estimation. IEEE Rob. Autom. Lett. **4**(2), 902–909 (2019)
11. Dornaika, F., Horaud, R.: Simultaneous robot-world and hand-eye calibration. IEEE Trans. Rob. Autom. **14**(4), 617–622 (1998)
12. Heller, J., Havlena, M., Pajdla, T.: Globally optimal hand-eye calibration using branch-and-bound. IEEE Trans. Pattern Anal. Mach. Intell. **38**(5), 1027–1033 (2016)
13. Horaud, R., Dornaika, F.: Hand-eye calibration. Int. J. Rob. Res. **14**(3), 195–210 (1995)
14. Hu, X., Olesen, D., Per, K.: Calibration of extrinsic transformation using manifold optimization. IFAC-PapersOnLine **52**(8), 124–129 (2019)

15. Huang, J.K., Wang, S., Ghaffari, M., Grizzle, J.W.: Lidartag: a real-time fiducial tag system for point clouds. IEEE Rob. Autom. Lett. **6**(3), 4875–4882 (2021)
16. Huang, K., Stachniss, C.: Extrinsic multi-sensor calibration for mobile robots using the Gauss-Helmert model. In: 2017 IEEE/RSJ International Conference on Intelligent Robots and Systems (IROS), pp. 1490–1496 (2017)
17. Ishikawa, R., Oishi, T., Ikeuchi, K.: Lidar and camera calibration using motions estimated by sensor fusion odometry. In: 2018 IEEE/RSJ International Conference on Intelligent Robots and Systems (IROS), pp. 7342–7349 (2018)
18. Levinson, J., Thrun, S.: Automatic online calibration of cameras and lasers. In: Robotics: Science and Systems (RSS) (2013)
19. Malti, A., Barreto, J.P.: Robust hand-eye calibration for computer aided medical endoscopy. In: 2010 IEEE International Conference on Robotics and Automation (ICRA), pp. 5543–5549 (2010)
20. Nguyen, H., Pham, Q.C.: On the covariance of $\boldsymbol{X}$ in $\boldsymbol{AX} = \boldsymbol{XB}$. IEEE Trans. Rob. **34**(6), 1651–1658 (2018)
21. Pachtrachai, K., Vasconcelos, F., Chadebecq, F., Allan, M., Hailes, S., Pawar, V., Stoyanov, D.: Adjoint transformation algorithm for hand–eye calibration with applications in robotic assisted surgery. Ann. Biomed. Eng. **46**(10), 1606–1620 (2018)
22. Pandey, G., McBride, J.R., Savarese, S., Eustice, R.M.: Automatic extrinsic calibration of vision and lidar by maximizing mutual information. J. Field Rob. **32**(5), 696–722 (2015)
23. Park, F.C., Martin, B.J.: Robot sensor calibration: solving AX=XB on the Euclidean group. IEEE Trans. Rob. Autom. **10**(5), 717–721 (1994)
24. Peršić, J., Marković, I., Petrović, I.: Extrinsic 6DOF calibration of a radar–LiDAR–camera system enhanced by radar cross section estimates evaluation. Rob. Auton. Syst. **114**, 217–230 (2019)
25. Richardson, A., Strom, J., Olson, E.: Aprilcal: Assisted and repeatable camera calibration. In: 2013 IEEE/RSJ International Conference on Intelligent Robots and Systems, pp. 1814–1821 (2013)
26. Shiu, Y.C., Ahmad, S.: Calibration of wrist-mounted robotic sensors by solving homogeneous transform equations of the form AX=XB. IEEE Trans. Rob. Autom. **5**(1), 16–29 (1989)
27. Solà, J., Deray, J., Atchuthan, D.: A micro Lie theory for state estimation in robotics (2018, December). arXiv e-prints. arXiv:1812.01537
28. Strobl, K.H., Hirzinger, G.: Optimal hand-eye calibration. In: 2006 IEEE/RSJ International Conference on Intelligent Robots and Systems (IROS), pp. 4647–4653 (2006)
29. Tsai, R.Y., Lenz, R.K.: A new technique for fully autonomous and efficient 3D robotics hand/eye calibration. IEEE Trans. Rob. Autom. **5**(3), 345–358 (1989)

YOLOPose: Transformer-Based Multi-object 6D Pose Estimation Using Keypoint Regression

Arash Amini, Arul Selvam Periyasamy(✉), and Sven Behnke

Autonomous Intelligent Systems, University of Bonn, Bonn, Germany
periyasa@ais.uni-bonn.de

Abstract. 6D object pose estimation is a crucial prerequisite for autonomous robot manipulation applications. The state-of-the-art models for pose estimation are convolutional neural network (CNN)-based. Lately, Transformers, an architecture originally proposed for natural language processing, is achieving state-of-the-art results in many computer vision tasks as well. Equipped with the multi-head self-attention mechanism, Transformers enable simple single-stage end-to-end architectures for learning object detection and 6D object pose estimation jointly. In this work, we propose YOLOPose (short form for You Only Look Once Pose estimation), a Transformer-based multi-object 6D pose estimation method based on keypoint regression. In contrast to the standard heatmaps for predicting keypoints in an image, we directly regress the keypoints. Additionally, we employ a learnable orientation estimation module to predict the orientation from the keypoints. Along with a separate translation estimation module, our model is end-to-end differentiable. Our method is suitable for real-time applications and achieves results comparable to state-of-the-art methods.

Keywords: Object pose estimation · Scene understanding · Vision transformer · Object detection

Autonomous robotic object manipulation in real-world scenarios depends on high-quality 6D object pose estimation. Such object poses are also crucial in many other applications like augmented reality, autonomous navigation, and industrial bin picking. In recent years, with the advent of convolutional neural networks (CNNs), significant progress has been made to boost the performance of object pose estimation methods. Due to the complex nature of the problem, the standard methods favor multi-stage approaches, i.e., feature extraction followed by object detection and/or instance segmentation, target object crop extraction, and, finally, 6D object pose estimation. In contrast, Carion et al. [2] introduced DETR, a Transformer-based single-stage architecture for object detection. In

Arash Amini, Arul Selvam Periyasamy are equal contribution.

I. Petrovic et al. (Eds.): IAS 2022, LNNS 577, pp. 392–406, 2023.
https://doi.org/10.1007/978-3-031-22216-0_27

our previous work [1], we extended the DETR model with the T6D-Direct architecture to perform multi-object 6D pose direct regression. Compared to multi-stage CNN-based methods that employ components like bounding box proposals, region of interest (RoI) pooling, non-maximum suppression (NMS) to construct end-to-end differentiable pipelines, the T6D-Direct model learns object detection and 6D pose estimation jointly. Taking advantage of the *pleasingly parallel* nature of the Transformer architecture, the T6D-Direct model predicts 6D poses for all the objects in an image in one forward-pass. Despite the advantages of the architecture and its impressive performance, the overall 6D pose estimation accuracy of T6D-Direct, which directly regresses translation and orientation components of the 6D object poses, is inferior to state-of-the-art CNN-based methods, especially in rotation estimation. Instead of directly regressing the translation and orientation components, the keypoint-based methods predict the 2D pixel projection of 3D keypoints and use the PnP algorithm to recover the 6D pose. We extend our T6D-Direct model to learn sparse 2D-3D correspondences. Our proposed model performs keypoint direct regression instead of the standard heatmaps for predicting the spatial position of the keypoints in a given RGB image and uses a multi layer perceptron (MLP) to learn the orientation component of 6D object pose from the keypoints. Another independent MLP serves as the translation estimator. In short, our contributions include:

1. A Transformer-based real-time single-stage model for multi-object monocular 6D pose estimation using keypoint regression,
2. a learnable rotation estimation module to estimate object orientation from a set of keypoints to develop an end-to-end differentiable architecture for pose estimation, and
3. results comparable to state-of-the-art pose estimation methods on the YCB-Video dataset while being capable of real-time inference (Fig. 1).

1 Related Work

1.1 RGB Object Pose Estimation

The recent significant progress in the task of 6D object pose estimation from RGB images is driven—like in many computer vision tasks—by deep learning methods. The methods for the object pose estimation from RGB images can be broadly classified into three major categories, namely direct regression methods, keypoint-based methods, and refinement-based methods. Direct regression methods formulate the problem of pose estimation as a regression of continuous translation and rotation components, whereas keypoint-based methods predict the location of projection of some of the specific keypoints or the 3D coordinates of the visible pixels of an object in an image and use the PnP algorithm to recover the 6D pose from the estimated 2D-3D correspondences. The PnP algorithm is used in conjunction with RANSAC to improve the robustness of the pose estimation.

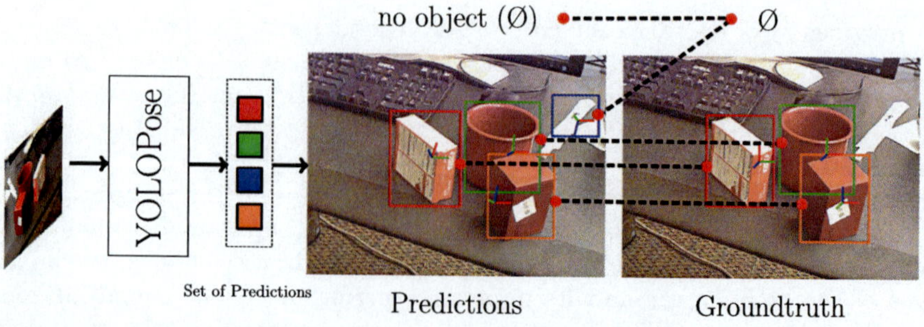

Fig. 1. Proposed YOLOPose approach. Our model predicts a set with a fixed cardinality N. Each element in the set corresponds to an object prediction and after predicting all the objects in the given input image, the rest of the elements are padded with Ø as no object predictions. The predicted and the groundtruth sets are matched using bipartite matching and the model is trained to minimize the Hungarian loss between the matched pairs. Our model is end-to-end differentiable.

Some examples for direct regression methods include [1,22,29,30] and examples for keypoint-based methods include [8,9,21,24,28]. One important detail to note is that, except for [1] all the other methods use multi-staged CNNs. In the first stage, the model performs object detection and/or semantic or instance segmentation to detect the objects in the given RGB image. Using the object detections from the first stage, a crop containing the target object is extracted. In the second stage, the model predicts the 6D pose of the target object from the image crop. In terms of the 6D pose prediction accuracy, keypoint-based methods perform considerably better than the direct regression methods [7], though this performance gap is shrinking [1].

The third category of the pose estimation methods is refinement-based. These methods formulate the task of pose estimation as iterative pose refinement, i.e., the target object is rendered according to the current pose estimate, and a model is trained to estimate a pose update that minimizes the pose error between the groundtruth and the current pose prediction. Refinement-based methods [11,15, 19,23,26] achieve the highest pose prediction accuracy among all categories [7].

1.2 Learned PnP

Given a set of 3D keypoints and their corresponding 2D projections and the camera intrinsics, the PnP algorithm is used to recover the 6D object pose. The standard PnP algorithm [5] and its variant EPnP [13] are used in combination with RANSAC to improve the robustness against outliers. Both PnP and RANSAC are not trivially differentiable. In order to realize an end-to-end differentiable pipeline for the 6D object pose estimation, Wang et al. [29], and [8] proposed a learning-based PnP module. Similarly, Li et al. [14] introduced a learnable 3D Lifter module to estimate vehicle orientation. Lately, Chen et al. [3] proposed to

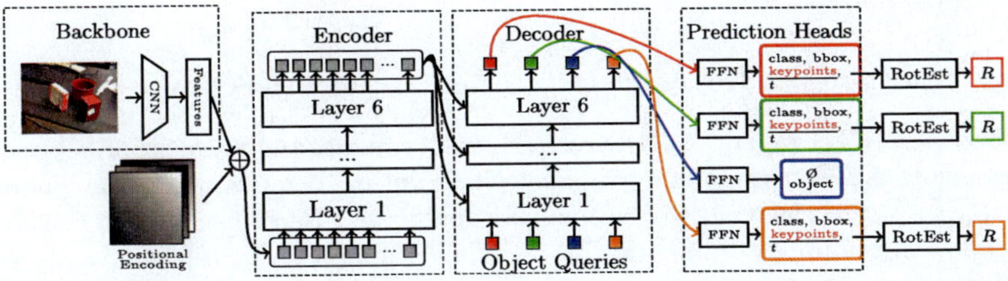

Fig. 2. YOLOPose architecture in detail. Given an RGB input image, we extract features using the standard ResNet model. The extracted features are supplemented with positional encoding and provided as input to the Transformer encoder. The encoder module consists of 6 standard encoder layers with skip connections. The output of the encoder module is provided to the decoder module along with N object queries and the decoder module also consists of 6 standard decoder layers with skip connections generating N output embeddings. The output embeddings are processed with FFNs to generate a set of N elements in parallel. Each element in the set is a tuple consisting of bounding box, class probability, translation and interpolated bounding box keypoints. A learnable rotation estimation module is employed to estimate object orientation R from the predicted keypoints.

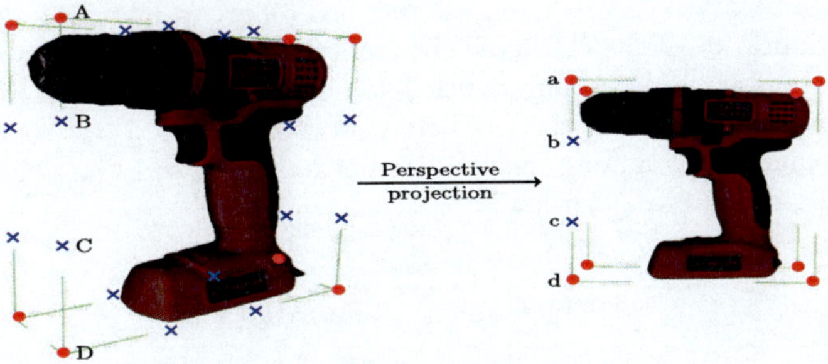

Fig. 3. Interpolated bounding box points. Bounding box points are indicated with red dots, and the interpolated points are indicated with blue crosses. The cross-ratio of every four collinear points is preserved during perspective projection,e.g., the cross-ratio the points A, B, C, and D remains the same in 3D and, after perspective projection, in 2D.

differentiate PnP using the implicit function theorem. Although a generic differentiable PnP has many potentials, due to the overhead incurred during training, we opt for a simple MLP that estimates the orientation component given the 2D keypoints (Fig. 2).

2 Method

2.1 Multi-object Keypoint Regression as Set Prediction

Following DETR [2] and T6D-Direct [1], we formulate the problem of keypoint regression as a set prediction problem. Given an RGB input image, our model outputs a set of elements with a fixed cardinality N. Each element in the set is a tuple containing 2D bounding boxes, class probability, translation, and keypoints. The 2D bounding boxes are represented with the center coordinates, height, and width proportional to the image size. The class probability is predicted using a softmax function. We regress the translation component directly, where our translation representation follows Xiang et al. [30]. The exact choice of the keypoints is discussed in Sect. 2.2. The number of objects present in an image varies. Therefore, to enable output sets with fixed cardinality, we choose N to be larger than the expected maximum number of objects in an image in the dataset and introduce a null object class $\emptyset$. The $\emptyset$ class is analogous to the background class used in semantic segmentation models. In addition to predicting the corresponding classes for objects present in the image, our model is trained to predict $\emptyset$ class for the rest of the elements in the set.

A set is an unordered collection of elements. To facilitate comparing the groundtruth set and the predicted set, we use bipartite matching [2,10,27] to find the permutation of the predicted elements that minimizes the matching cost. Given the $\emptyset$ class padded groundtruth set y of cardinality N containing labels $y_1, y_2, ..., y_n$ for n elements, the predicted set denoted by $\hat{y}$, we search for the optimal permutation $\hat{\sigma}$ among the possible permutations $\sigma \in \mathfrak{S}_N$ that minimizes the matching cost $\mathcal{L}_{match}$. Formally,

$$\hat{\sigma} = \arg\min_{\sigma \in \mathfrak{S}_N} \sum_{i}^{N} \mathcal{L}_{match}(y_i, \hat{y}_{\sigma(i)}). \tag{1}$$

Although each element of the set is a tuple containing four components, bounding box, class probability, translation, and keypoints, we use only the bounding box and class probability components to define the cost function. In practice, omitting the other components in the cost function definition does not hinder the model's ability in learning to predict the keypoints.

2.2 Keypoints Representation

An obvious choice for 3D keypoints is the eight corners of the 3D bounding box [20]. Peng et al. [21] argued that predicting the projection of 3D bounding boxes of an object might be difficult for a CNN-based model since the projection might lie outside of the object boundary in the RGB image. To alleviate this issue, they proposed to use the Farthest Point Sampling algorithm (FPS) to sample eight keypoints on the surface of the object meshes. Li et al. [14] defined the 3D representation of an object as sparse interpolated bounding boxes (IBBs), depicted in Fig. 3, and exploited the property of perspective projection that the cross-ratio of every four collinear points in 3D (A, B, C, and D as illustrated in Fig. 3) is preserved under perspective projection in 2D [6]. The cross-ratio consistency is enforced by an additional component in the loss function that the model learns to minimize during training. We further investigate these keypoints representations in Sect. 4 and present our results in Table 3.

2.3 RotEst

For each object, from the estimated pixel coordinates of the 32 keypoints (the eight corners of the 3D bounding box and the 24 intermediate bounding box keypoints), the RotEst module predicts the object orientation represented as the 6D continuous representation in SO(3) [31]. Formally, the RotEst module takes a 64-dimensional vector (32 pixel coordinates) and generates a 6D object orientation estimate. We implement the RotEst module using six fully connected layers with hidden dimension 1024 and dropout probability of 0.5.

2.4 Loss Function

The Hungarian loss consists of four components: class probability loss, bounding box loss, keypoint loss, and pose loss. The class probability loss and the bounding box loss follow the DETR model [2].

2.4.1 Class Probability Loss The class probability loss function is the standard negative log-likelihood. Since we choose the cardinality of the set to be higher than the expected maximum number of objects in an image, the Ø class appears disproportionately often. Thus, we weigh the loss for the Ø class with a factor of 0.4.

2.4.2 Bounding Box Loss We use a weighted combination of the Generalized IoU (GIoU) [25] and ℓ_1-loss with 2 and 5 factors, respectively, for the bounding box loss.

2.4.3 Keypoint Loss Having the groundtruth K_i and the model output $\hat{K}_{\hat{\sigma}(i)}$ our keypoints loss can be represented as:

$$\mathcal{L}_{keypoints}(K_i, \hat{K}_{\hat{\sigma}(i)}) = \gamma||K_i - \hat{K}_{\hat{\sigma}(i)}||_1 + \delta\mathcal{L}_{\mathcal{CR}}, \tag{2}$$

where γ and δ are hyperparameters. The first part of the keypoints loss is the ℓ_1 loss, and for the second part, we employ the cross-ratio loss $\mathcal{L}_{\mathcal{CR}}$ provided in Eq. 3 to enforce the cross-ratio consistency in the keypoints loss as proposed by Li et al. [14]. This loss is self-supervised by preserving the cross-ratio of each line to be 4/3. The reason is that after camera projection of the 3D bounding box on the image plane, the cross-ratio of every four collinear points remains the same.

$$\mathcal{L}_{\mathcal{CR}} = Smooth\ell_1(\mathrm{CR}^2 - \frac{||c-a||^2||d-b||^2}{||c-b||^2||d-a||^2}), \tag{3}$$

where CR^2 is chosen since $||.||^2$ can be easily computed using vector inner product. Given four collinear points A, B, C and D and their predicted 2D projections a, b, c, and d, the groundtruth cross-ratio CR is defined as:

$$\mathrm{CR} = \frac{||C-A||\;||D-B||}{||C-B||\;||D-A||} = \frac{4}{3}. \tag{4}$$

2.4.4 Pose Loss We supervise the rotation R and the translation t individually via employing PLoss and SLoss from [30] for rotation and ℓ_1 loss for translation:

$$\mathcal{L}_{pose}(R_i, t_i, \hat{R}_{\sigma(i)}, \hat{t}_{\sigma(i)}) = \mathcal{L}_{rot}(R_i, \hat{R}_{\sigma(i)}) + ||t_i - \hat{t}_{\sigma(i)}||_1, \tag{5}$$

$$\mathcal{L}_{rot} = \begin{cases} \frac{1}{|\mathcal{M}_i|} \sum\limits_{\mathrm{x}_1 \in \mathcal{M}_i} \min\limits_{\mathrm{x}_2 \in \mathcal{M}_i} ||R_i \mathrm{x}_1 - \hat{R}_{\sigma(i)} \mathrm{x}_2||_1 & \text{if sym,} \\ \frac{1}{|\mathcal{M}_i|} \sum\limits_{\mathrm{x} \in \mathcal{M}_i} ||R_i \mathrm{x} - \hat{R}_{\sigma(i)} \mathrm{x}||_1 & \text{otherwise,} \end{cases} \tag{6}$$

where $\mathcal{M}_i$ indicates the set of 3D model points. Here, we subsample 1.5K points from meshes provided with the dataset. R_i is the groundtruth rotation, and t_i is the groundtruth translation. $\hat{R}_{\sigma(i)}$ and $\hat{t}_{\sigma(i)}$ are the predicted rotation and translation, respectively.

2.5 Model Architecture

The proposed YOLOPose architecture is inspired by T6D-Direct [1]. The model consists of a ResNet backbone followed by an encoder-decoder based Transformer and MLP prediction heads to predict a set of tuples described in Sect. 2.1. CNN architectures have several inductive biases designed into them [4,12]. These strong biases enable CNNs to learn efficient local spatial features in a fixed neighborhood defined by the receptive field to perform well on many computer vision tasks. In contrast, Transformers, aided by self-attention, are suitable for learning spatial features over the entire image. This makes the Transformer architecture ideal for multi-object pose estimation. In this section, we describe the individual components of the YOLOPose architecture.

2.5.1 Backbone Network We use a ResNet50 backbone for extracting features from the given RGB image. For an image size of height H and width W, the backbone network extracts 2048 low-resolution feature maps of size H/32×W/32. We then use 1×1 convolution to reduce the 2048 feature dimensions to a lower number of d dimensions. The standard Transformer models are designed to process vectors. Therefore, to enable processing the d×H/32×W/32 features, we vectorize them to $d \times \frac{H}{32}\frac{W}{32}$.

2.5.2 Positional Encodings Multi-head self-attention, the core component of the Transformer model, is permutation-invariant. Thus, the Transformer architecture ignores the order of the input vectors. We employ the standard solution of supplementing the input vectors with absolute positional encoding following [2] to provide the Transformer model with spatial information of the pixels. The positional embeddings are added elementwise to the backbone feature vectors before feeding them to the Transformer encoder as input.

2.5.3 Encoder The Transformer encoder module consists of six encoder layers with skip connections. Each layer performs multi-head self-attention of the input vectors. The embeddings used in our model are 256-dimensional vectors.

2.5.4 Decoder From the encoder output embedding and N positional embedding inputs, the decoder generates N output embeddings using the multi-head self-attention and cross-attention mechanisms, where N is the cardinality of the predicted set. Unlike the fixed positional encoding used in the encoder, we use learnable positional encoding in the decoder, called *object queries*. From the N decoder output embeddings, we use feed-forward prediction heads to generate a set of N output tuples independently.

2.5.5 FFN For each object query decoder output, we use four feed-forward prediction heads shared across object queries to predict the class probability, bounding box, translation, and keypoints independently. Prediction heads are straightforward three-layer MLPs with hidden dimension 256 in each layer and ReLU activation.

3 Evaluation

In this section, we evaluate the performance of our proposed YOLOPose model and compare it with other state-of-the-art 6D pose estimation methods.

Fig. 4. Qualitative results on YCB-V test set. Top row: The predicted IBB keypoints overlaid on the input images. Bottom row: Groundtruth and predicted object poses visualized as object contours in green and blue colors, respectively.

3.1 Dataset

We use the challenging YCB-Video (YCB-V) [30] dataset to evaluate the performance of our model. YCB-V provides bounding box, segmentation, and 6D pose annotation for 133,936 RGB-D images. Since our model is RGB-based, we do not use the provided depth information. The dataset is generated by capturing video sequences of a random subset of objects from a total of 21 objects placed in tabletop configuration. There are 92 video sequences, out of which twelve are used for testing and the rest are used for training. The objects used exhibit varying geometric shapes, reflectance properties, and symmetry. Thus, YCB-V is a challenging dataset for benchmarking 6D object pose estimation methods. YCB-V also provides high-quality meshes for all 21 objects. Mesh points from these objects are used in computing the evaluation metrics discussed in Sect 3.2. Hodaň et al. [7] provided a variant of YCB-V[1] as a part of the BOP challenge in which the centers of the 3D bounding boxes are aligned with the origin of the model coordinate system, and the groundtruth annotations are converted correspondingly. We use the BOP variant of the YCB-V dataset. Additionally, we use the COCO dataset [17] for pretraining our model on the task of object detection.

3.2 Metrics

Xiang et al. [30] proposed area under the curve (AUC) of the ADD and ADD-S metrics for evaluating the accuracy of non-symmetric and symmetric objects, respectively. Given the groundtruth 6D pose annotation with rotation and translation components R and t, and the predicted rotation and translation components $\hat{R}$ and $\hat{t}$, the ADD metric is the average ℓ_2 distance between the subsampled mesh points $\mathcal{M}$ in the groundtruth and the predicted pose. In contrast, the symmetry aware ADD-S metric is the average distance between the closest subsampled mesh points $\mathcal{M}$ in the groundtruth and predicted pose. A predicted pose is considered correct if the ADD metric is below 0.1 m. The ADD(-S) metric is the combination of ADD-S for symmetric objects and ADD for the non-symmetric objects. Formally,

$$\text{ADD} = \frac{1}{|\mathcal{M}|} \sum_{x \in \mathcal{M}} \|(Rx + t) - (\hat{R}x + \hat{t})\|, \tag{7}$$

$$\text{ADD-S} = \frac{1}{|\mathcal{M}|} \sum_{x_1 \in \mathcal{M}} \min_{x_2 \in \mathcal{M}} \|(Rx_1 + t) - (\hat{R}x_2 + \hat{t})\|. \tag{8}$$

3.3 Hyperparameters

The γ and δ hyperparameters in $\mathcal{L}_{keypoints}$ (Eq. (2)) are set to 10 and 1, respectively. While computing the Hungarian loss, the pose loss component is weighted

[1] https://bop.felk.cvut.cz/datasets/.

down by a factor 0.02. The cardinality of the predicted set N is set to 20. The model takes images of the size 640 × 480 as input and is trained using the AdamW optimizer [18] with an initial learning rate of $2e^{-4}$ for 335K iterations. The learning rate is decayed to $2e^{-5}$ after 271K iterations, and the batch size is 32. Moreover, gradient clipping with a maximal gradient norm of 0.1 is applied. In addition to YCB-V dataset images, we use the synthetic dataset provided by PoseCNN for training our model.

3.4 Results

Table 1. Comparison of keypoints-based method with state-of-the-art methods on YCB-V. P.E=1 denotes one model for all objects, whereas P.E.=(N) denotes the usage of object specific models. The symmetric objects are denoted by *, and the best results are shown in bold.

Method	PoseCNN [30]		PVNet [21]	GDR-Net [29]		T6D-Direct [1]		YOLOPose (Ours)		DeepIM [15]	
P.E.	1		N	1		1		1		1	
Metric	AUC of ADD-S	AUC of ADD(-S)	AUC of ADD(-S)	AUC of ADD-S	AUC of ADD(-S)	AUC of ADD-S	AUC of ADD(-S)	AUC of ADD-S	AUC of ADD(-S)	AUC of ADD-S	AUC of ADD(-S)
master_chef_can	84.0	50.9	81.6	96.6	71.1	91.9	61.5	91.3	64.0	93.1	71.2
cracker_box	76.9	51.7	80.5	84.9	63.5	86.6	76.3	86.8	77.9	91.0	83.6
sugar_box	84.3	68.6	84.9	98.3	93.2	90.3	81.8	92.6	87.3	96.2	94.1
tomato_soup_can	80.9	66.0	78.2	96.1	88.9	88.9	72.0	90.5	77.8	92.4	86.1
mustard_bottle	90.2	79.9	88.3	99.5	93.8	94.7	85.7	93.6	87.9	95.1	91.5
tuna_fish_can	87.9	70.4	62.2	95.1	85.1	92.2	59.0	94.3	74.4	96.1	87.7
pudding_box	79.0	62.9	85.2	94.8	86.5	85.1	72.7	92.3	87.9	90.7	82.7
gelatin_box	87.1	75.2	88.7	95.3	88.5	86.9	74.4	90.1	83.4	94.3	91.9
potted_meat_can	78.5	59.6	65.1	82.9	72.9	83.5	67.8	85.8	76.7	86.4	76.2
banana	85.9	72.3	51.8	96.0	85.2	93.8	87.4	95.0	88.2	91.3	81.2
pitcher_base	76.8	52.5	91.2	98.8	94.3	92.3	84.5	93.6	88.5	94.6	90.1
bleach_cleanser	71.9	50.5	74.8	94.4	80.5	83.0	65.0	85.3	73.0	90.3	81.2
bowl*	69.7	69.7	89.0	84.0	84.0	91.6	91.6	92.3	92.3	81.4	81.4
mug	78.0	57.7	81.5	96.9	87.6	89.8	72.1	84.9	69.6	91.3	81.4
power_drill	72.8	55.1	83.4	91.9	78.7	88.8	77.7	92.6	86.1	92.3	85.5
wood_block*	65.8	65.8	71.5	77.3	77.3	90.7	90.7	84.3	84.3	81.9	81.9
scissors	56.2	35.8	54.8	68.4	43.7	83.0	59.7	93.3	87.0	75.4	60.9
large_marker	71.4	58.0	35.8	87.4	76.2	74.9	63.9	84.9	76.6	86.2	75.6
large_clamp*	49.9	49.9	66.3	69.3	69.3	78.3	78.3	92.0	92.0	74.3	74.3
extra_large_clamp*	47.0	47.0	53.9	73.6	73.6	54.7	54.7	88.9	88.9	73.3	73.3
foam_brick*	87.8	87.8	80.6	90.4	90.4	89.9	89.9	90.7	90.7	81.9	81.9
MEAN	75.9	61.3	73.4	89.1	80.2	86.2	74.6	**90.1**	**82.6**	88.1	81.9

Table 2. Results on YCB-V.

Method	ADD(-S)	AUC of ADD-S	AUC of ADD(-S)	Inference Time [s]
CosyPose† [11]	–	89.8	**84.5**	0.395
PoseCNN [30]	21.3	75.9	61.3	–
SegDriven [9]	39.0	–	–	–
Single-Stage [8]	53.9	–	–	–
GDR-Net [29]	49.1	89.1	80.2	0.065
T6D-Direct [1]	48.7	86.2	74.6	**0.017**
YOLOPose (Ours)	**65.0**	**90.1**	82.6	**0.017**

† Refinement-based method.

In this section, we present the quantitative and qualitative results of our method. In Table 1, we provide the quantitative per class area under the accuracy curve (AUC) of the ADD-S and ADD(-S) metrics. Except for DeepIM, a

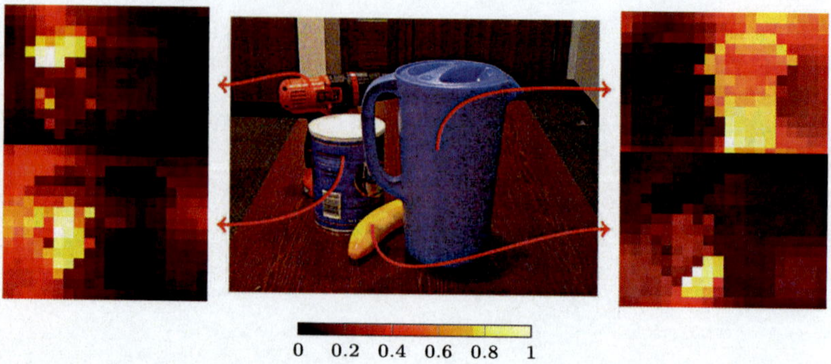

Fig. 5. Encoder self-attention. We visualize the self-attention maps for four pixels belonging to four objects in the image. Note that for each object, roughly all the corresponding pixels are attended.

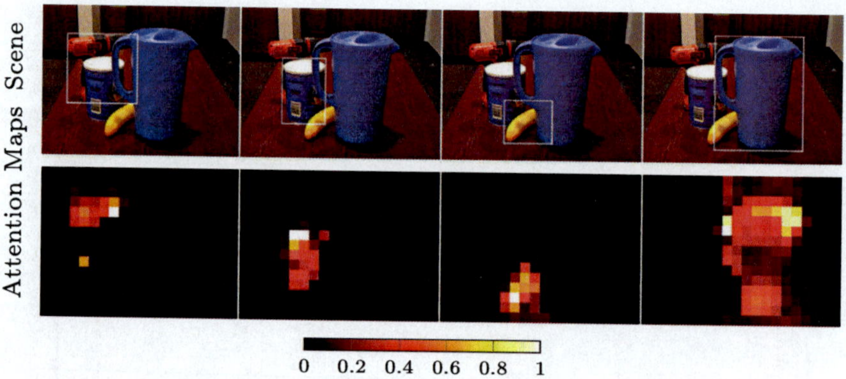

Fig. 6. Top: Object detections predicted by bounding boxes in a given image. Bottom: Decoder cross-attention maps for the object queries corresponding to the predictions in the first row.

refinement-based method and PVNet, an indirect method, all other methods estimate the 6D pose directly. Our method outperforms all of the competing approaches. Additionally in Table 2, we present Average Recall (AR) of ADD(-S) 0.1d, and AUC of ADD-S and ADD(-S) of the state-of-the-art methods. In terms of the AR of ADD(-S) and AUC of ADD-S metrics, our method achieves state-of-the-art results among the pose estimators. Note that the pose refinement approach, CosyPose, achieves the best result in terms of the AUC of ADD(-S) metric. However, in terms of the approach pose refinement methods are orthogonal to the pose estimation methods and can benefit from the improved pose estimation accuracy. Furthermore, pose estimation models with admissible accuracy avoid the need for training an additional pose refinement model and enable faster inference time. We present exemplar qualitative results in Fig. 4. In Fig. 5, we visualize encoder self-attention of four different pixels belonging to four different objects and in Fig. 6, we visualize the decoder cross-attention corresponding to four different object detections. In both visualizations, the attended regions correspond to the spatial position of the object in the image very well.

3.5 Inference Time Analysis

In terms of the inference speed, one of the major advantages of our architecture is that the feed-forward prediction networks (FFN) can be executed in parallel for each object. Thus, irrespective of the number of objects in an image, our model generates pose predictions in parallel. In Table 2, we present the inference time results for 6D pose estimation. Our method operates at ~59 FPS on an NVIDIA 3090 GPU and Intel 3.70 GHz CPU and is, hence, suitable for real-time applications.

4 Ablation Study

In contrast to the standard approach of estimating the 2D keypoints and using PnP solver—which is not trivially differentiable—to estimate the 6D object pose, we use the learnable RotEst module to estimate the object orientation from a set of predicted interpolated keypoints. In this section, we analyze the effectiveness of our RotEst module and the choice of the keypoint representation.

4.1 Effectiveness of Keypoints Representations

We compare various keypoint representations, namely 3D bounding box keypoints (BB), random keypoints sampled using the FPS algorithm and our representation of choice, the interpolated bounding box keypoints (IBB). We use the OpenCV implementation of the RANSAC-based EPnP algorithm with the same parameters to recover 6D object pose from the predicted keypoints. Since EPnP does not contain any learnable components, this experiment serves the goal of evaluating the ability of the YOLOPose model to estimate different keypoint representations in isolation. YOLOPose is trained using only the ℓ_1 loss in the case of BB and FPS representations, whereas in the case of IBB representation, ℓ_1 is combined with the cross-ratio loss described in Sect. 2.4.3. In our experiments presented in Table 3, when used in conjunction with the EPnP solver, the FPS keypoints performed worse than all other representations. In contrast, the IBB keypoints representation yields the best performance. We conjecture that the additional cross-ratio loss employed helps our model in learning the IBB keypoint projections better.

4.2 Effectiveness of RotEst

After deciding on the keypoint representation, we compare the performance of the learnable feed-forward rotation and translation estimators against the analytical EPnP algorithm. Based on the observation that the rotation and translation components impacted by different factors [16], we decide to estimate rotation and translation separately. As shown in Table 3, using only the rotation from the EPnP result and directly regressing the translation improved the accuracy significantly. In general, RotEst performs slightly better than using EPnP orientation and direct translation estimation. Furthermore, the RotEst module and the translation estimators are straightforward MLPs and, thus, do not add much execution time overhead. This enables YOLOPose to perform inference in real-time.

Table 3. Ablation study on YCB-V. We present the comparison results of different keypoint representations and the effectiveness of RotEst. The top section of the table corresponds to different keypoint representations in combination with the standard EPnP algorithm, and the bottom section corresponds to the effectiveness of the learnable RotEst module using IBB keypoints.

Method	ADD(-S)	AUC of ADD(-S)
FPS + EPnP	31.4	56.9
handpicked + EPnP	31.5	55.7
IBB + EPnP	**56.0**	**74.7**
IBB + EPnP for R; head for t	63.9	82.3
IBB + RotEst for R and head for t	**65.0**	**82.6**

5 Discussion & Conclusion

We presented YOLOPose, a Transformer-based single-stage multi-object pose estimation method using keypoint regression. Our model jointly estimates bounding boxes, class labels, translation vectors, and pixel coordinates of 3D keypoints for all objects in the given input image. Employing the learnable RotEst module to estimate the object orientation from the predicted keypoints coordinate enables the model to be end-to-end differentiable. We evaluated our model on the widely-used YCB-Video dataset and reported results comparable to the state-of-the-art approaches while being real-time capable. In the future, we plan to extend our model to video sequences and exploit temporal consistency to improve the pose estimation accuracy further.

Acknowledgment. This work has been funded by the German Ministry of Education and Research (BMBF), grant no. 01IS21080, project "Learn2Grasp: Learning Human-like Interactive Grasping based on Visual and Haptic Feedback".

References

1. Amini, A., Periyasamy, A.S., Behnke, S.: T6D-Direct: Transformers for multi-object 6D object pose estimation. In: DAGM German Conference on Pattern Recognition (GCPR) (2021)
2. Carion, N., Massa, F., Synnaeve, G., Usunier, N., Kirillov, A., Zagoruyko, S.: End-to-end object detection with transformers. In: European Conference on Computer Vision (ECCV), pp. 213–229 (2020)
3. Chen, B., Parra, A., Cao, J., Li, N., Chin, T.J.: End-to-end learnable geometric vision by backpropagating pnp optimization. In: IEEE/CVF Conference on Computer Vision and Pattern Recognition (CVPR), pp. 8100–8109 (2020)
4. Cohen, N., Shashua, A.: Inductive bias of deep convolutional networks through pooling geometry. In: International Conference on Learning Representations, ICLR 2017, Toulon, France, OpenReview.net (2017)
5. Gao, X., Hou, X., Tang, J., Cheng, H.: Complete solution classification for the perspective-three-point problem. IEEE Trans. Pattern Anal. Mach. Intell. (PAMI) **25**, 930–943 (2003)
6. Hartley, R., Zisserman, A.: Multiple View Geometry in Computer Vision, 2 edn. Cambridge University Press (2004). https://doi.org/10.1017/CBO9780511811685
7. Hodaň, T., Sundermeyer, M., Drost, B., Labbé, Y., Brachmann, E., Michel, F., Rother, C., Matas, J.: BOP challenge 2020 on 6D object localization. In: European Conference on Computer Vision (ECCV), pp. 577–594 (2020)
8. Hu, Y., Fua, P., Wang, W., Salzmann, M.: Single-stage 6D object pose estimation. In: IEEE/CVF Conference on Computer Vision and Pattern Recognition (CVPR), pp. 2930–2939 (2020)
9. Hu, Y., Hugonot, J., Fua, P., Salzmann, M.: Segmentation-driven 6D object pose estimation. In: IEEE Conference on Computer Vision and Pattern Recognition (CVPR), pp. 3385–3394 (2019)
10. Kuhn, H.W.: The Hungarian method for the assignment problem. Naval Res. Logistics Quart. **2**(1–2), 83–97 (1955)
11. Labbe, Y., Carpentier, J., Aubry, M., Sivic, J.: CosyPose: Consistent multi-view multi-object 6D pose estimation. In: European Conference on Computer Vision (ECCV) (2020)
12. LeCun, Y., Bengio, Y., et al.: Convolutional networks for images, speech, and time series. Handb. Brain Theor. Neural Netw. **3361**(10), 1995 (1995)
13. Lepetit, V., Moreno-Noguer, F., Fua, P.: EPnP: An accurate o(n) solution to the PnP problem. Int. J. Comput. Vis. (IJCV) **81**(2), 155 (2009)
14. Li, S., Yan, Z., Li, H., Cheng, K.T.: Exploring intermediate representation for monocular vehicle pose estimation. In: IEEE/CVF Conference on Computer Vision and Pattern Recognition (CVPR), pp. 1873–1883 (2021)
15. Li, Y., Wang, G., Ji, X., Xiang, Y., Fox, D.: DeepIM: Deep iterative matching for 6D pose estimation. In: European Conference on Computer Vision (ECCV), pp. 683–698 (2018)
16. Li, Z., Wang, G., Ji, X.: CDPN: Coordinates-based disentangled pose network for real-time RGB-based 6-DoF object pose estimation. In: Proceedings of the IEEE International Conference on Computer Vision (ICCV), pp. 7678–7687 (2019)
17. Lin, T.Y., Maire, M., Belongie, S., Hays, J., Perona, P., Ramanan, D., Dollár, P., Zitnick, C.L.: Microsoft COCO: common objects in context. In: European Conference on Computer Vision (ECCV), pp. 740–755 (2014)

18. Loshchilov, I., Hutter, F.: Decoupled weight decay regularization. In: International Conference on Learning Representations (ICLR) (2017)
19. Manhardt, F., Kehl, W., Navab, N., Tombari, F.: Deep model-based 6D pose refinement in RGB. In: European Conference on Computer Vision (ECCV), pp. 800–815 (2018)
20. Oberweger, M., Rad, M., Lepetit, V.: Making deep heatmaps robust to partial occlusions for 3D object pose estimation. In: European Conference on Computer Vision (ECCV) (2018)
21. Peng, S., Liu, Y., Huang, Q., Zhou, X., Bao, H.: PVNet: Pixel-wise voting network for 6DOF pose estimation. In: IEEE Conference on Computer Vision and Pattern Recognition (CVPR), pp. 4561–4570 (2019)
22. Periyasamy, A.S., Schwarz, M., Behnke, S.: Robust 6D object pose estimation in cluttered scenes using semantic segmentation and pose regression networks. In: International Conference on Intelligent Robots and Systems (IROS) (2018), 10.1109/IROS.2018.8594406
23. Periyasamy, A.S., Schwarz, M., Behnke, S.: Refining 6D object pose predictions using abstract render-and-compare. In: IEEE-RAS International Conference on Humanoid Robots (Humanoids), pp. 739–746 (2019)
24. Rad, M., Lepetit, V.: BB8: A scalable, accurate, robust to partial occlusion method for predicting the 3D poses of challenging objects without using depth. In: IEEE International Conference on Computer Vision (ICCV), pp. 3828–3836 (2017)
25. Rezatofighi, H., Tsoi, N., Gwak, J., Sadeghian, A., Reid, I., Savarese, S.: Generalized intersection over union: a metric and a loss for bounding box regression. In: IEEE Conference on Computer Vision and Pattern Recognition (CVPR), pp. 658–666 (2019)
26. Shao, J., Jiang, Y., Wang, G., Li, Z., Ji, X.: PFRL: Pose-free reinforcement learning for 6D pose estimation. In: IEEE Conference on Computer Vision and Pattern Recognition (CVPR) (2020)
27. Stewart, R., Andriluka, M., Ng, A.Y.: End-to-end people detection in crowded scenes. In: IEEE Conference on Computer Vision and Pattern Recognition (CVPR), pp. 2325–2333 (2016)
28. Tekin, B., Sinha, S.N., Fua, P.: Real-time seamless single shot 6D object pose prediction. In: IEEE Conference on Computer Vision and Pattern Recognition (CVPR) (2018)
29. Wang, G., Manhardt, F., Tombari, F., Ji, X.: GDR-Net: Geometry-guided direct regression network for monocular 6D object pose estimation. In: IEEE Conference on Computer Vision and Pattern Recognition (CVPR) (2021)
30. Xiang, Y., Schmidt, T., Narayanan, V., Fox, D.: PoseCNN: A convolutional neural network for 6D object pose estimation in cluttered scenes. arXiv:1711.00199 (2017)
31. Zhou, Y., Barnes, C., Lu, J., Yang, J., Li, H.: On the continuity of rotation representations in neural networks. In: IEEE Conference on Computer Vision and Pattern Recognition (CVPR), pp. 5745–5753 (2019)

People Tracking in Panoramic Video for Guiding Robots

Alberto Bacchin(✉), Filippo Berno, Emanuele Menegatti, and Alberto Pretto

Intelligent Autonomous Systems Laboratory (IAS—Lab), Department of Information Engineering, University of Padova, Via Gradenigo 6/B, 35131 Padova, Italy
{bacchinalb,emg,alberto.pretto}@dei.unipd.it

Abstract. A guiding robot aims to effectively bring people to and from specific places within environments that are possibly unknown to them. During this operation the robot should be able to detect and track the accompanied person, trying never to lose sight of her/him. A solution to minimize this event is to use an omnidirectional camera: its 360° Field of View (FoV) guarantees that any framed object cannot leave the FoV if not occluded or very far from the sensor. However, the acquired panoramic videos introduce new challenges in perception tasks such as people detection and tracking, including the large size of the images to be processed, the distortion effects introduced by the cylindrical projection and the periodic nature of panoramic images. In this paper, we propose a set of targeted methods that allow to effectively adapt to panoramic videos a standard people detection and tracking pipeline originally designed for perspective cameras. Our methods have been implemented and tested inside a deep learning-based people detection and tracking framework with a commercial 360° camera. Experiments performed on datasets specifically acquired for guiding robot applications and on a real service robot show the effectiveness of the proposed approach over other state-of-the-art systems. We release with this paper the acquired and annotated datasets and the open-source implementation of our method.

1 Introduction

Guiding robots were one of the first examples of the successful application of autonomous mobile robots outside the domain of academic research [1]. In large public spaces like museums, hospitals, hotels, and many others, a guiding robot can help people to reach the desired destination by accompanying them. To achieve this behavior, it is important to be able to detect and track people in the robot space, even if they are temporarily occluded. The robot's on-board cameras usually have a reduced Field of View (FoV) which limits the tracking capabilities since the target person can easily get out of the FoV, for example when the robot is turning. Omnidirectional cameras can solve this problem

Alberto Bacchin and Filippo Berno: The authors equally contribute to this paper.

I. Petrovic et al. (Eds.): IAS 2022, LNNS 577, pp. 407–424, 2023.
https://doi.org/10.1007/978-3-031-22216-0_28

Fig. 1. (Left): The dual-fisheye omnidirectional camera mounted on the robot used in the experiments; (Right): an example of equirectangular panoramic image acquired during the experiments. This image is obtained by applying a cylindrical projection to the two images acquired by the fisheye cameras and by a subsequent image stitching phase.

thanks to the 360° FoV which always frames the whole surrounding environment. Nowadays several low-cost, lightweight, and high-quality omnidirectional cameras (also called 360° cameras) are available in the market (e.g., left part of Fig. 1), enabling plenty of new applications and challenges for mobile robots. In this work, we target the use of an omnidirectional camera for people detection and tracking, applied to the specific domain of guiding robots. We build upon the monocular people following framework proposed by Koide *et al.* [2], we call here *MonoPTrack*, adapting it to equirectangular panoramic videos (e.g., right part of Fig. 1). *MonoPTrack* is based on *OpenPose* [3], a popular deep neural network for human skeleton detection. The panoramic images provided by an omnidirectional camera could be directly supplied as input to this system. Unfortunately, due to the high computing resources required, *OpenPose* is unable to perform inference in real-time on full resolution panoramic images on standard GPUs (Graphics Processing Units). On the other hand, performing inference on a downscaled image can lead to false-negative detections or wrong person re-identification. Moreover, the 360° FoV comes at the cost of severe image distortions. In our case, the dimensions of a person in the image decrease very rapidly with the distance (e.g., right part of Fig. 1) so that in a downscaled image *OpenPose* cannot detect anyone more than 3–4 meters away from the camera. Furthermore, when mapping the entire surrounding scene to an equirectangular image, a *horizontal wrap-around connection* is introduced. The wrap-around structure of panoramic images is made clear when a target subject (e.g., a person) exits from one side of the image and reappears at the same time on the opposite side. The phenomenon easily leads traditional tracking strategies to failure.

In this paper we propose specific solutions to address the aforementioned problems. In particular, we faced the challenges deriving from the high resolution of panoramic images and from the distortion introduced by cylindrical projection by introducing two simple but effective strategies, called *ROI* and

TILES, that boost the detection and re-identification performance of existing people pose estimation systems. We also introduced a tracking strategy that allows to explicitly address the wrap-around structure of panoramic images. We evaluated the proposed techniques on a real world dataset, showing the benefits in terms of detection robustness, efficiency, and tracking accuracy.

The main contributions of this paper are:

- Two new strategies for addressing the challenges associated with detecting people in panoramic video such as high resolutions and projective distortion, enabling the use of omnidirectional cameras on real-time robotic applications;
- An effective solution for tracking people in 3D space starting from the equirectangular representation provided by an omnidirectional camera;
- A comparison of the proposed method against a current state-of-the-art method;
- An open-source implementation of our system and the acquired datasets are made publicly available with this paper:

https://github.com/bach05/PanoramicVideoPeopleTracking.git

The rest of the paper is organized as follows. Sec. 2 contains a review of related works. In Sec. 3, we introduce the proposed *ROI* and *TILES* algorithms and the modification to a conventional tracking system to cope with panoramic videos. Sec. 4 describes the evaluation procedure, the experiments performed, and a discussion of the results. Finally, conclusions and future research directions are provided in Sec. 5.

2 Related Work

2.1 People Detection

Traditional approaches for detecting people rely on searching for a candidate region using a sliding window approach. For each region, a set of features are extracted and used to classify it. In [4], authors used HAAR-LBP and HOG cascade classifier to detect people in surveillance videos. Since scanning the whole image multiple times with a sliding window approach is not so efficient, Geronimo *et al.* [5] proposed to use a calibrated camera to determine which regions are placed on the ground and so suitable for containing a person.

However, nowadays deep learning-based methods have overcome the traditional approaches [6]. Zhao *et al.* [7] proposed a Faster R-CNN architecture optimized for people detections. The Faster R-CNN [8] architecture is inspired by the traditional object detection pipeline. In fact, it uses a Region Proposal Network to identify the regions of interest in the image and a classification network to assign the correct label to each proposed region. Other approaches are based on YOLO [9], another very popular object detector based on deep learning. In [10], authors used a YOLO person detector to locate people in the scene, relying on the centroid of the bounding boxes. Once the person is detected, through a calibrated camera, it is also possible to reconstruct her/his 3D position [11,12].

Anyway, for human detection, bounding boxes do not always provide enough information. For these reason, human pose or skeleton estimation have been proposed [3,13]. In this case, the aim is to predict the position of N human body joints. This richer representation allows more precise localization in 3D space [2,14] without need of additional hardware such as RGB-D cameras [15,16] and laser range-finders [17].

2.2 People Tracking

The tracking task consists in associating consecutive detections of the same object. Optical flow can be used to estimate the motion of the objects for tracking purposes [18]. However, being a pixel-level computation, optical flow is not enough reliable and it works only in the image space. Instead, it is common to use an efficient state estimation algorithm, such as a Kalman Filter, to track detections. Ardiyanto et al. [19] exploited an Unscented Kalman Filter (UKF) to track people pose in the 3D world space. Using only the pose may not be robust enough, especially in crowded environments where trajectories are close and often intercept. Adding visual features for data association have been demonstrated to be more robust [20]. Also, Deep Learning has been explored for tracking [21].

Despite over the years there was a huge interest in people detection and tracking, only a few works focused on the possibility of exploiting panoramic videos. In [22], authors propose a people tracking system on panoramic videos obtained from a system of multiple cameras. They focused on real-time processing, relying on a HOG detector and an optical flow detector tracker. A more advanced approach, based on Siamese Networks, is proposed by Tai et al. [23]. Li et al. [24] instead used YOLOv3 and ResNet50 to detect and track people in panoramic videos. However, all of them are limited to the image space and so the applicability in robotics applications remains unexplored. Recent studies used a couple of calibrated omnidirectional cameras to achieve accurate 3D people tracking in panoramic videos [25]. The proposal of Yang et al. [26], called *Multi-person Panoramic Localization and Tracking (MPLT)*, instead relies on four calibrated perspective cameras, positioned such that they cover a field of view of 360° to simulate an omnidirectional camera. The OpenPifPaf framework [13] is used for skeleton detection and a pinhole camera model is used to map detections in the 3D space, under the assumption that the height of a person is almost constant. The tracking system models each couple of tracks with two costs, one associated with the appearance of the person and one associated with the distance of the trajectories estimated with a Kalman Filter. The tracking is performed by finding the associations that minimize the costs. However, compared to our approach, multi-camera configurations require more complex calibration processes and higher computational costs.

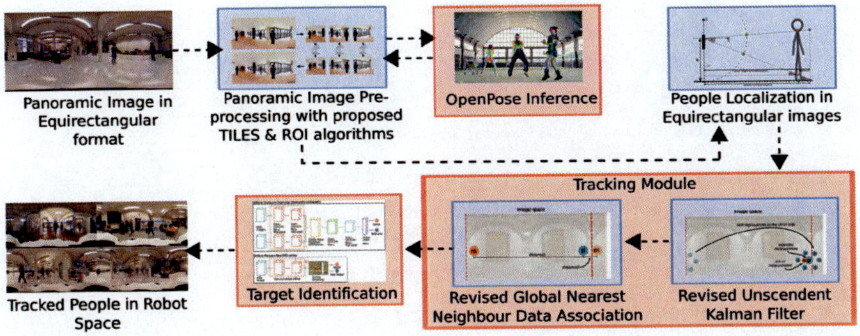

Fig. 2. Block diagram of the proposed pipeline. Blue blocks highlight the aspects we contribute on, red blocks instead are taken from previous methods.

3 Methodology

3.1 Baseline

Our system is based on the *MonoPTrack* framework [2], with new specific strategies aimed at improving the robustness of the person detection front-end and the target tracking sub-system. Thanks to parallel pipelines and small-sized processed images, the proposed algorithms also made it possible to improve runtime performance even on low-end computers, such as those that typically equip service robots. The front-end of *MonoPTrack* is represented by the OpenPose [3] skeleton detection. The neck and ankles joints are used to estimate the position of a person in the 3D space, given a calibrated camera. An UKF is used in the tracking process. UKF [27] is a sampling based-method designed to overcome the limitation of a traditional Kalman Filter that works only for linear models. The idea is to select a set of representative points from the starting distribution, called *Sigma Points*, apply the non-linear transformation to each point, and finally estimate a Gaussian distribution from the sampled points. Tracking is completed with a Global Nearest Neighbour (GNN) data association. The last module of the original pipeline is the re-identification module which learns appearance features of the target person. Our work focused on three points of the original pipeline. (i) Before skeleton detection, we introduced the *ROI* and *TILES* pre-processing algorithms to ensure better runtime performance while boosting the detection rate also in case of people far from the camera, hence projected in a small region of the panoramic image. (ii) We modified the camera model to deal with equirectangular panoramic images in order to correctly map from image space to robot space; here we also introduced ankle height correction, to improve the accuracy of person localization. (iii) Finally, we modified the tracker to cope with the *wrap-around* connections. A scheme of the proposed pipeline can be found in Fig. 2.

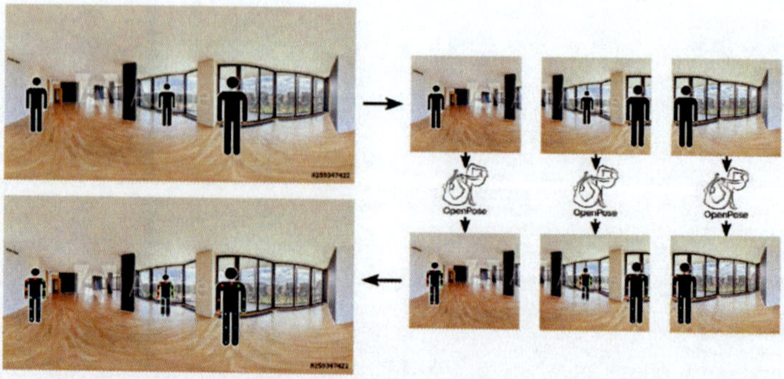

(a) Graphical representation of the *TILES* approach. Each tile is processed by a dedicate instance of *OpenPose* and then all the detections are fused.

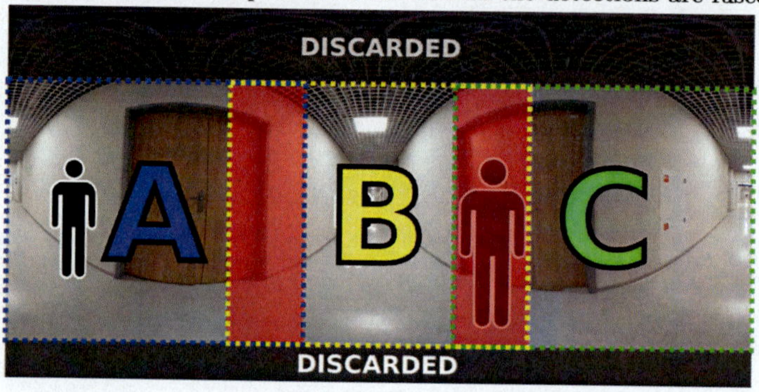

(b) The subdivision of the image into tiles; the upper and lower zones are discarded because no useful information is contained there. The three tiles are marked with A, B and C, while overlapping zones are in red.

Fig. 3. The working principle of the *TILES* method.

3.2 The TILES Approach

The *TILES* algorithm focuses on the problem of efficiently inferring skeleton detection from high-resolution images. The aim is to reduce the resolution while parallelizing the inference. We divide the image into three equal parts. Each portion, namely tile, will be analyzed by a dedicated instance of *OpenPose* and results will be merged. The division can split a person over two consecutive tiles, preventing the detection. To solve this issue, tiles are designed to be slightly overlapping, as shown in Fig. 3. The overlapping area is 150-pixel wide, which is the width of a person 1 m away from the robot. We take 1 m as worst case since is rare that a person stays closer to a moving robot.

A person in the overlapping area can be detected twice. To avoid double detection, before delivering the set of detections, we compute a bounding box containing shoulder, neck, and hip joints around each person. For each pair

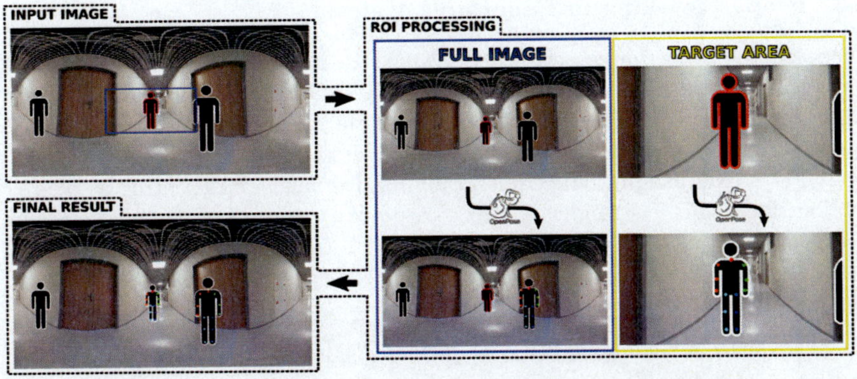

Fig. 4. The image shows the working principle of the *ROI* method. One instance of *OpenPose* processes a low-resolution version of the full image. Another processes a portion of the full image, containing the target person, outlined in red.

of adjacent tiles and each pair of bounding boxes (B_1, B_2), we compute the following score:

$$S(B_1, B_2) = \frac{|B_1 \cap B_2|}{|\min(B_1, B_2)|} \tag{1}$$

If $S(B_1, B_2) \geq \sigma_1$, we merge B_1 and B_2 in a single detection. People at different distances from the camera and with different physical appearances (i.e. different joint positions) will be bounded with different sized boxes. This strategy allows fusing double detection with a low overhead since it does not need additional information like extracting visual features from people.

3.3 The ROI Approach

The *TILES* algorithm, thanks to the parallelization, provides a good processing speed while improving detection accuracy of further people. The *ROI* (i.e., *Region Of Interest*) approach explores the possibility of further improvements, under the specific case of a guidance robot that usually focuses on a specific target person. The first frame is re-scaled and processed only with a single open *OpenPose* inference to obtain the target. Identifying a meaningful target person is beyond the scope of this work, so we promote as target the first detected person. Then, *ROI* performs two *OpenPose* inferences for each frame, see Fig. 4. The first on the whole panoramic image, but re-scaled to lower resolution. The second is performed on a specif region of interest, from the full-resolution image, which contains the target person indicated by the re-identification module on the previous frame. We use the same method illustrated in Sect. 3.2 to merge double detections. Focusing on a specific region, the network can detect a person very far from the camera, at the cost of worse performances in multi-person detection because of the re-scaling.

3.4 People Tracking in Panoramic Videos

Once skeleton detection is performed, we use the information of neck and ankles joints to estimate the location of each perceived person. This usually involves a calibrated camera, but we instead exploited the peculiar characteristics of the equirectangular image, assuming that the camera is positioned orthogonal to the ground at a fixed height. The equirectangular format is representing a sphere on a flat surface, so it is straightforward to use polar coordinates (θ, φ, ρ), as shown in Fig. 5.

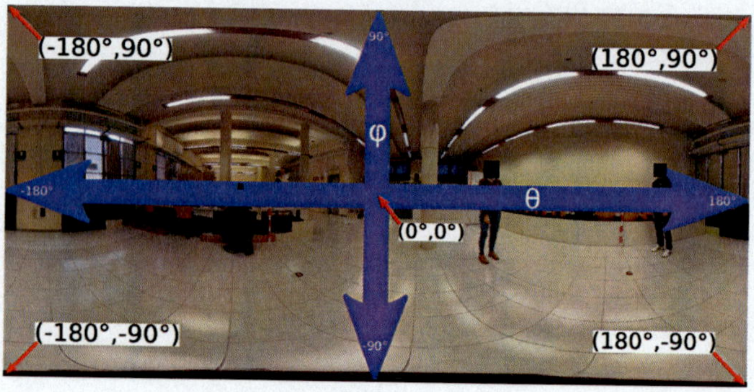

Fig. 5. In this pictures, we can see how the equi-rectangular format is mapped into polar coordinates. The column determines the angle θ, the row determines the angle φ.

Given the middle point between the two ankles joints (α) in the 3D space, its polar coordinates can be computed from its image coordinates (x_α, y_α):

$$\theta_\alpha = 180 - \frac{FoV_H}{I_W} x_\alpha \quad \varphi_\alpha = 90 - \frac{FoV_V}{I_H} y_\alpha \quad \rho_\alpha = \frac{h-k}{tan(\varphi_\alpha \cdot \frac{\pi}{180})} \tag{2}$$

where FoV_H and FoV_V are horizontal and vertical fields of view in degrees, I_W x I_H is the panoramic image resolution in pixels, h is the height of the camera from the ground and k is the average ankle height from the ground for a human, both in meters. Thanks to the neck joint (N), it is possible to estimate the person height:

$$h_N = h + \rho_N \cdot tan(\varphi_N \cdot \frac{\pi}{180}) \tag{3}$$

The position of a person P in Euclidean coordinates is:

$$\begin{pmatrix} X_P \\ Y_P \\ Z_P \end{pmatrix} = \begin{pmatrix} \rho_\alpha \cdot cos(\theta_\alpha) \\ \rho_\alpha \cdot sin(\theta_\alpha) \\ h_N \end{pmatrix} \tag{4}$$

The neck joint and the height estimation in the state allow updating the filter even if the ankles are occluded for more stable tracking. Fig. 6 shows visually the aforementioned geometric transformations.

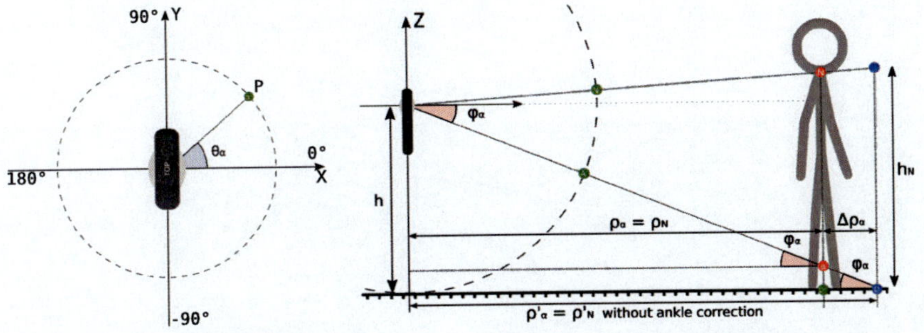

(a) The system from above. (b) The system from the side.

Fig. 6. The geometry behind the mapping from image space to 3D space. The dashed circumference represents the image plane of the omni-directional camera. α' and N' are the points in the image that corresponds to the ankle and neck joints. α and N are the corresponding points in the robot space. h is the height of the camera above the ground. h_N is the neck height which approximates person's height. On the right, it is also shown how the ankle height correction improves the final estimation of the person distance from the camera.

The *wrap-around* problem, introduced in Sect. 1, forces to modify the UKF and the data association strategy in the tracking module. As explained in Sec. 3.1, UKF uses the *Sigma points* to deal with non-linear functions. When a person is closed to the boundaries of the equirectangular image, it may happen that *Sigma points* from the starting distribution in the image would be mapped on two different sides. This fact has disruptive effects on the estimation of the predicted distribution, see Fig. 7a. To solve, we shift the *Sigma points* fallen on the opposite side of the image by increasing their x-coordinate by an amount equal to the width of the image. This results in a correct average calculation, as shown in Fig. 7b. If the resulting average is outside the image space, it is sufficient to subtract the image width from the x coordinate.

Since the data association for the tracking takes place in the image space, the same problem related to the *wrap-around* could arise. To tackle the issue, we reformulated the equation to compute the distance of two points, P_1 and P_2, in the image space:

$$d(P_1, P_2) = \sqrt{(\min(x_{P_1} - x_{P_2}, image_W - (x_{P_1} - x_{P_2})))^2 + (y_{P_1} - y_{P_2})^2} \quad (5)$$

where I_W is the width of the image in pixels, (x_{P_1}, y_{P_1}) and (x_{P_2}, y_{P_2}) are the Cartesian coordinates of the points in the image space.

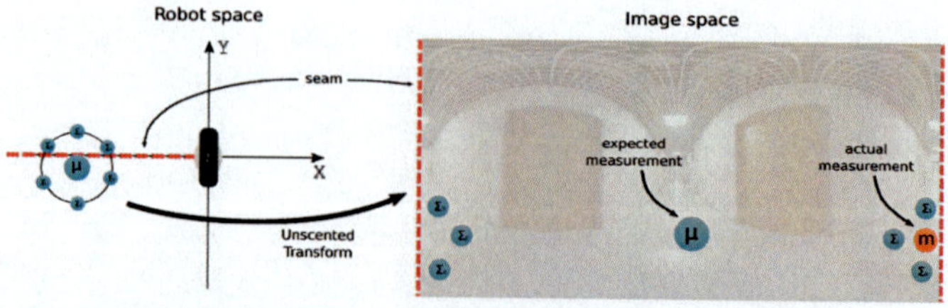

(a) Wrong mean estimation due to wrap-around.

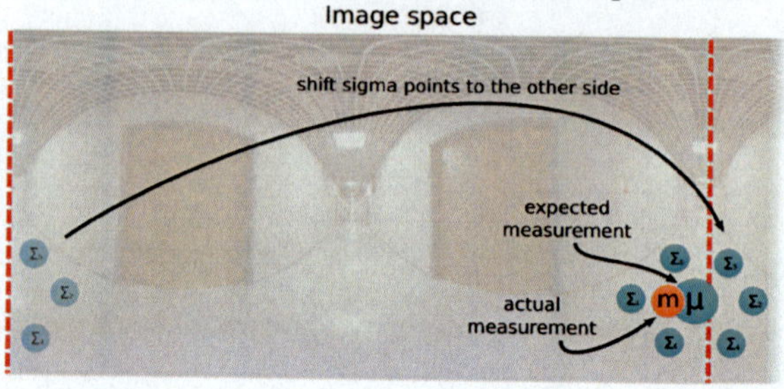

(b) Corrected estimation.

Fig. 7. The figures shows the correction applied to the UKF. Σ_i are the sigma points, μ is the mean of the sigma points. (a) Disruptive impact of Sigma points mapped on the other side of the image. (b) Correction applied to obtain the correct mean estimation.

4 Experimental Evaluation

4.1 Experimental Setup

We collected a dataset of 14 videos of variable lengths, between 6 s and 134 s. Each frame can contain up to 5 people. The dataset was acquired with a Ricoh Theta Z1,[1] a dual-fisheye omnidirectional camera that allows a 360° horizontal, 180° vertical FoV and real-time streaming at 30 fps via USB connection. The rendered image is an equirectangular projection, generated by the camera itself, with a resolution of 1920×960 pixels. The recordings include both indoor and outdoor environments, with one or more people moving around the camera. In some videos the camera is static, in some others the camera was mounted on a mobile robot to simulate a guiding task. The distance from the ground and the camera sensor was 1.2 m.

The dataset images have been manually annotated with the neck and ankles joints of the target person, since we aim a guiding application where the robot focuses on a single person. The videos were not densely annotated, namely, we provided annotations every 10 or 20 frames which are enough for the evaluation of a tracking system. To avoid biases due to inaccurate assumptions, for example if the camera is not perfectly perpendicular, the ground truth of the target 3D position is computed using the model presented in Sec. 3.4 from the annotation

[1] https://theta360.com/it/about/theta/z1.html.

of neck and ankles joints on the images. We ended up with a total of 3200 annotated frames. A subset of the videos containing 1108 annotated frames was used for parameter tuning, the rest for the test.

We set a threshold $\theta_1 = 0.9$ for Eq. 1 after experimental tuning. For *ROI* algorithm, we set the size of the region of interest to 576×192 pixel, while the full image was re-scaled to 640×320 pixels. We used default parameters for the UKF and the tracker.

The evaluation was performed offline on the pre-recorded test set with a PC equipped with an NVIDIA GTX 1650 GPU, 16 GB of RAM, and an Intel i7-4790 CPU. We also compared the performance of our proposal with the *Multi-person Panoramic Localization and Tracking (MPLT)* [26] introduced in Sect. 2.2. Since *MPLT* works with four calibrated cameras to provide a panoramic view, not directly with omnidirectional cameras, we split the equirectangular image into four portions of size 480×960 pixels. Each portion is meant to cover the area in front, behind, at the left, and the right of the camera, without overlapping. This approximates quite well the original configuration. Moreover, the code provided by the authors tries to load in memory the whole video before processing it, with a consequent saturation of the RAM with our dataset. We have reorganized the processing pipeline to elaborate the video frame by frame, as in real-time systems like our proposal. We have used the proposed camera model, adjusting the parameters for our camera.

4.2 Evaluation Metrics

To design the evaluation metrics, we started from ETISEO metrics [28]. Given the practical application of a guiding robot, we propose three metrics that mainly focus on the performance in target tracking.

The Normalized Detection Time, called $M1$, measures the percentage of time during which the target is detected by the system and is given by the following equation:

$$M1(v) = \frac{1}{N_v} \sum_{i=1}^{N_v} \frac{T_i}{F_i} \tag{6}$$

where N_v is the number of people annotated in the video v ($N_v = 1$ in our case), T_i is the number of frames in which person i is tracked, F_i is the number of frames for which person i is annotated. $M1$ takes values between 0 and 1, high values are better.

$M2$, the Tracking Turnover, evaluates the persistence of the ID assigned to the target. This allows testing the continuity of the target tracking, since every time the tracking is lost and then restarted a new ID is assigned. $M2$ is defined by the following formula:

$$M2(v) = \frac{1}{N_v} \sum_{i=1}^{N_v} \frac{1}{Frag_i} \tag{7}$$

where N_v is the number of people annotated ($N_v = 1$ in our case), $Frag_i$ is the number of times a new ID is assigned to the target. Larger is better.

$M3$, the Localization Error, evaluates the ability to precisely localize the target in the robot space. It consists of the average Euclidean distance between the target location estimated by the tracker and the ground truth location:

$$M3(v) = \frac{1}{N_v} \sum_{i=1}^{N_v} \sqrt{(X_i - X_{GT,i})^2 + (Y_i - Y_{GT,i})^2} \tag{8}$$

where $(X_{GT,i}, Y_{GT,i})$ is the ground truth location of the target and (X_i, Y_i) is the predicted location on frame i. N_v is the number of annotated frames for video v. Smaller values are better.

4.3 Results and Discussions

Performance Since our target is an application in guiding robotics, our goal is a system with good accuracy in tracking a target subject, while being able to run in real-time on low-end GPUs like the ones can find in a laptop. To assess the potentiality of our method, we also evaluated the three metrics against *MPLT*, a state-of-the-art approach in panoramic video tracking introduced in Sect. 2.2. The results are shown in Table 1.

From the Normalized Detection Time (M1) results in Table 1a, it is possible to evaluate the strength of the detection capabilities. It is worth noting that *MPLT* performs poorly in the sequences *vid*2, 3 and 4. These sequences were recorded outdoor, letting the target go very far from the camera, up to 7.5 m. The higher distance, the smaller the person size with respect to the rest of the image. *ROI* and *TILES* detectors, on the other hand, remain reliable even when the target is very far from the camera, with slightly better results for ROI. The advantage is probably due to the fact in ROI, the detector can focus in a smaller region which means less noise around the person. The Tracking Turnover (M2) in Table 1b instead shows the capability of the system to keep track of the target. The difference between the three approaches, in this case, is less noticeable. *MPLT* uses a traditional Kalman Filter and visual features to track, while our proposals use the UKF which is often more robust for modeling non-linear functions [29]. *MPLT* is also penalized by the poorer results in M1. For a better understanding of the performance in tracking, we also provide a visual representation in Fig. 8 that confirms the previous considerations. In outdoor, where people can move further from the camera, *MPLT* is more inaccurate and unstable. In indoor, the differences are less marked. The Localization Error (M3) in Table 1c shows the accuracy of the systems in estimating the position of the target in the robot space. The results are clearly in favor of our systems. While *MPLT* just uses the center of the detection as a reference point to estimate the distances assuming the height of a person is almost constant, our approach is using three joints and continuously updating the person height estimation. *ROI* and *TILES* contribute in providing an higher accuracy in human pose detection which is critical for an accurate localization in the robot space. Especially *ROI*

Table 1. The following tables collect the results from each tracking metric. We show the performance in each video and the average over the whole dataset. The processing speed on a GTX 1650 is shown too. Best results are in bold.

(a) M1, Normalized Tracking Time

Video ID	TILES	ROI	MPLT
vid1	0.97	0.97	**0.98**
vid2	0.84	**0.99**	0.32
vid3	**0.99**	**0.99**	0.63
vid4	**0.98**	**0.98**	0.53
vid5	**0.99**	**0.99**	0.91
vid6	**1.00**	0.96	0.90
vid7	0.93	0.94	**0.99**
vid8	**0.96**	0.92	0.90
vid9	**0.97**	0.96	0.95
vid10	**0.98**	0.96	0.96
Average	0.96	**0.97**	0.81

(b) M2, Tracking Turnover

Video ID	TILES	ROI	MPLT
vid1	**1.00**	**1.00**	**1.00**
vid2	0.50	**1.00**	0.50
vid3	**1.00**	**1.00**	**1.00**
vid4	0.50	**1.00**	0.50
vid5	**1.00**	**1.00**	**1.00**
vid6	**1.00**	**1.00**	**1.00**
vid7	**0.50**	0.33	**0.50**
vid8	**0.50**	**0.50**	**0.50**
vid9	**0.50**	**0.50**	**0.50**
vid10	**1.00**	**1.00**	**1.00**
Average	0.75	**0.83**	0.75

(c) M3, Localization Error

Video ID	TILES	ROI	MPLT
vid1	**0.077**	0.087	0.178
vid2	**0.394**	0.408	0.496
vid3	0.341	**0.326**	0.344
vid4	0.465	**0.406**	0.312
vid5	**0.121**	0.13	0.373
vid6	0.305	**0.248**	0.402
vid7	0.261	**0.224**	0.149
vid8	0.271	**0.238**	0.653
vid9	**0.241**	0.244	0.362
vid10	0.296	**0.254**	0.652
Average	0.277	**0.256**	0.392

(d) Processing speed

System	FPS
TILES	13.7
ROI	**18.7**
MPLT	1.12

provides an average accuracy of around 0.25 m, which is 65% more precise than *MPLT*.

The Table 1d shows the results for the processing speed. Despite we ran the tests on a low-end GPU (an NVIDIA GTX 1650), the results show that our methods overcome the compared method in processing speed, running between 10 and 20 times faster and reaching frame rates suitable for a robotic system with real-time constraints. The parallel inferences for the human pose detection - which is the most demanding task in the pipeline in terms of processing time - enables to speed up the elaboration of high-resolution images. *ROI* is 5 fps faster than *TILES* since it performs only two inferences in parallel, on smaller

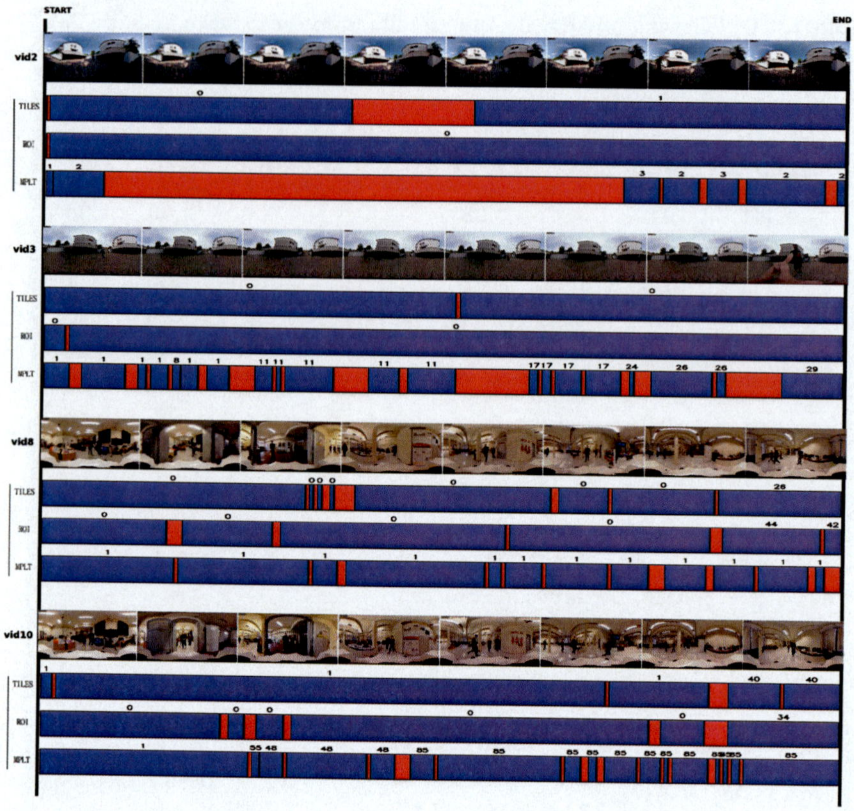

Fig. 8. A visual representation of the performance of the systems in 4 samples from our dataset. Blue means the target is correctly tracked, red instead the target is lost. Above each bar, we indicate the current ID assigned by the tracker.

images. On the other hand, *TILES* treats equally every person in the scene and it is more suitable for multi-person tracking. It is worth noticing that *MPLT* is using a ResNet-50 [30] as backbone, while our approaches are based on a lighter MobileNetV2 [31]. Also the tracking step in *MPLT* takes more time, around 12 ms per frame, while our method takes around 1 ms.

The Impact of Fisheye Distortion The major problem in low-cost omnidirectional cameras is related to the fisheye lens. This type of lens allows to obtain a wider field of view while reducing the number of sensors in the camera but, at the same time, it introduces strong distortions. It is worth studying the impact of these distortions. The main problem is a rapid decrease in the size of objects with distance from the camera. In addition to making detection more hard, the accuracy of the localization is also reduced. As reported in Fig. 9, small errors in joint detection lead to large localization errors as the distance from the camera increases.

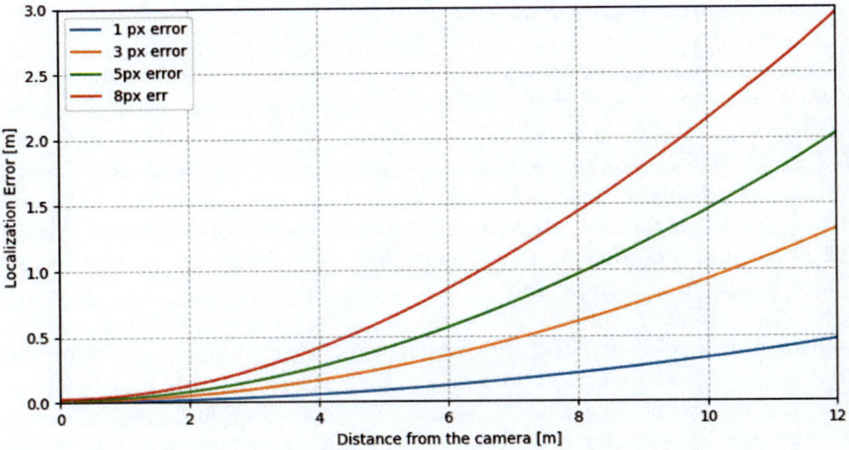

Fig. 9. Localization Error vs Distance from the camera, with different localization error in the image space.

The higher detection accuracy provided by our methods has a twofold benefit, as shown in Fig. 10. While *MPLT* fails to detect subjects further than 4 m, we can detect people 7–8 m away from the camera. More interesting, the more accurate joint estimation in the image provided by *ROI* and *TILES* can mitigate the growth of the localization error. With *MPLT*, the error rapidly increases and overcomes 1 m when the person is 2.5 m away from the camera. Our methods are able to maintain the error under 1 m until 6 m for *TILES* and 7 m for *ROI*. This difference is also due to the camera model used in *MPLT*, which probably cannot handle the strong distortion of fisheye lenses. The advantage of *ROI* on *TILES* is explainable with the focus in the target region, as discussed above.

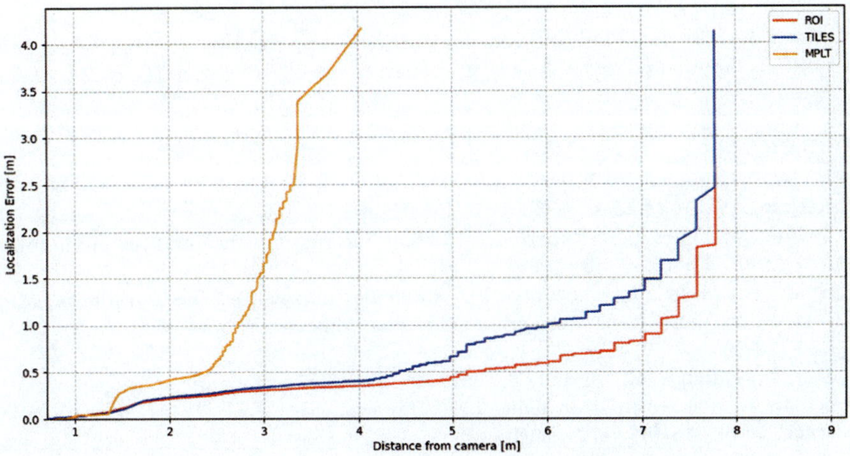

Fig. 10. Localization Error vs Distance from the camera in different systems.

5 Conclusions and Future Work

We have proposed an extension of a previous work on people detection and tracking, suitable to work with low-cost commercial omnidirectional cameras and targeting the application of guiding robotics. We introduced two algorithms, *TILES* and *ROI*, to tackle the challenges in panoramic videos such as high resolutions and fisheye distortions. Both the proposed algorithms overcome the current state-of-the-art in terms of detection accuracy, localization accuracy, and real-time processing according to our evaluations. *ROI* shows higher performances both in accuracy and speed, but it is optimized for a single target person. In fact, it is specifically designed to work on guiding robots, which usually take one person as a reference. *TILES* instead, with a very limited deterioration in performance, is a more generic approach, able to track multiple people with the same accuracy as the target. Moreover, the modified UKF and GNN allow us to solve the problems related with the *wrap-around* to achieve an higher tracking stability. Since there is no large availability of panoramic video datasets, we also collected and labeled a dataset to evaluate our proposal against a state of art framework for people detection ad tracking in panoramic videos.

This work has demonstrated that is possible to easily adapt a traditional framework for people detection and tracking to work with panoramic videos. The use of low-cost omnidirectional cameras can be very effective on robots, where the small FoV often limits the system's capabilities. A 360° FoV is particularly suitable when dealing with dynamic obstacles and targets like people, who can easily exit the FoV of traditional cameras. Given that, as future work we would like to do more extensive tests with a larger dataset, containing multi-person annotations. Relaxing the assumption that the camera is placed at a fixed height and orthogonal to the ground is desirable to enable larger and quicker adoption of our system. Further studies will involve application on a real guiding robotic system, aiming to use both the information of the target and the other people to manage the navigation of the robot.

Acknowledgements. This research was partially supported by MIUR (Italian Minister for Education) under the initiative "Departments of Excellence" (Law 232/2016).

References

1. Burgard, W., Cremers, A.B., Fox, D., Hähnel, D., Lakemeyer, G., Schulz, D., Steiner, W., Thrun, S.: Experiences with an interactive museum tour-guide robot. Art. Intell. **114**(1), 3–55 (1999)
2. Koide, K., Miura, J., Menegatti, E.: Monocular person tracking and identification with on-line deep feature selection for person following robots. Rob. Auton. Syst. **124** (2020)
3. Cao, Z., Hidalgo, G., Simon, T., Wei, S., Sheikh, Y.: Openpose: realtime multi-person 2d pose estimation using part affinity fields. IEEE Trans. Pattern Anal. Mach. Intell. **43**(01), 172–186 (2021)

4. Aguilar, W.G., Luna, M.A., Moya, J.F., Abad, V., Ruiz, H., Parra, H., Lopez, W.: Cascade classifiers and saliency maps based people detection. In: De Paolis, L.T., Bourdot, P., Mongelli, A. (eds.) Augmented Reality, Virtual Reality, and Computer Graphics, pp. 501–510. Springer International Publishing, Cham (2017)
5. Geronimo, D., Sappa, A., López, A., Ponsa, D.: Pedestrian detection using adaboost learning of features and vehicle pitch estimation. In Proceedings of the IASTED Int. Conf. on Visualization, Imaging and Image Processing, pp. 400–4005 (2006)
6. Ji, X., Fang, Q., Dong, J., Shuai, Q., Jiang, W., Zhou, X.: A survey on monocular 3d human pose estimation. Virtual Reality Intell. Hardware **2**(6), 471–500 (2020)
7. Zhao, X., Li, W., Zhang, Y., Gulliver, T.A., Chang, S., Feng, Z.: A faster RCNN-based pedestrian detection system. In: 2016 IEEE 84th Vehicular Technology Conference (VTC-Fall), pp. 1–5 (2016)
8. Ren, S., He, K., Girshick, R., Sun, J.: Faster r-cnn: Towards real-time object detection with region proposal networks. In: Cortes, C., Lawrence, N., Lee, D., Sugiyama, M., Garnett, R. (eds.) Advances in Neural Information Processing Systems, vol. 28 (2015)
9. Redmon, J., Divvala, S., Girshick, R., Farhadi, A.: You only look once: unified, real-time object detection. In: 2016 IEEE Conference on Computer Vision and Pattern Recognition (CVPR), pp. 779–788 (2016)
10. Ahmad, M., Ahmed, I., Adnan, A.: Overhead view person detection using yolo. In: IEEE 10th Annual Ubiquitous Computing. Electronics Mobile Communication Conference (UEMCON), vol. 2019, pp. 0627–0633 (2019)
11. Sharma, S., Ansari, J.A., Krishna Murthy, J., Madhava Krishna, K.: Beyond pixels: Leveraging geometry and shape cues for online multi-object tracking. In: 2018 IEEE International Conference on Robotics and Automation (ICRA), pp. 3508–3515 (2018)
12. Niu, Y., Xu, Z., Xu, E., Li, G., Huo, Y., Sun, W.: Monocular pedestrian 3d localization for social distance monitoring. Sensors **21** (2021)
13. Kreiss, S., Bertoni, L., Alahi, A.: OpenPifPaf: Composite Fields for Semantic Keypoint Detection and Spatio-Temporal Association. IEEE Trans. Intell. Transp. Syst. 1–14 (2021, March)
14. Bertoni, L., Kreiss, S., Alahi, A.: Monoloco: Monocular 3d pedestrian localization and uncertainty estimation. In: 2019 IEEE/CVF International Conference on Computer Vision (ICCV), pp. 6860–6870 (2019)
15. Kollmitz, M., Eitel, A., Vasquez, A., Burgard, W.: Deep 3d perception of people and their mobility aids. Rob. Auton. Syst. **114**, 29–40 (2019)
16. Munaro, M., Menegatti, E.: Fast RGB-D people tracking for service robots. Auton. Robots **37** (2014)
17. Zheng, K., Wu, F., Chen, X.: Laser-based people detection and obstacle avoidance for a hospital transport robot. Sensors **21**(3) (2021)
18. Choi, W., Savarese, S.: Multiple target tracking in world coordinate with single, minimally calibrated camera. In: Daniilidis, K., Maragos, P., Paragios, N. (eds.) Computer Vision—ECCV 2010, pp. 553–567. Heidelberg, Springer, Berlin Heidelberg, Berlin (2010)
19. Ardiyanto, I., Miura, J.: Partial least squares-based human upper body orientation estimation with combined detection and tracking. Image Vis. Comput. **32**(11), 904–915 (2014)
20. Liu, Y., Yin, J., Yu, D., Zhao, S., Shen, J.: Multiple people tracking with articulation detection and stitching strategy. Neurocomputing **386**, 18–29 (2020)

21. Ondrúška, P., Posner, I.: Deep tracking: seeing beyond seeing using recurrent neural networks. In: Proceedings of the Thirtieth AAAI Conference on Artificial Intelligence. AAAI'16, AAAI Press, pp. 3361–3367 (2016)
22. Thaler, M., Bailer, W.: Real-time person detection and tracking in panoramic video. In: 2013 IEEE Conference on Computer Vision and Pattern Recognition Workshops, pp. 1027–1032 (2013)
23. Tai, K.C., Tang, C.W.: Siamese networks based people tracking for 360-degree videos with equi-angular cubemap format. In: 2020 IEEE International Conference on Consumer Electronics—Taiwan (ICCE-Taiwan), pp. 1–2 (2020)
24. Li, J., Zang, Z., Xie, H., Wang, G.: Implementation of person tracking system in panorama based on personalized distribution. In: 2020 39th Chinese Control Conference (CCC), pp. 7123–7128 (2020)
25. Shere, M., Kim, H., Hilton, A.: 3d multi person tracking with dual 360° cameras. In: 2020 IEEE International Conference on Image Processing (ICIP), pp. 2765–2769 (2020)
26. Yang, F., Li, F., Wu, Y., Sakti, S., Nakamura, S.: Using panoramic videos for multi-person localization and tracking in a 3d panoramic coordinate. In: ICASSP 2020—2020 IEEE International Conference on Acoustics, Speech and Signal Processing (ICASSP), pp. 1863–1867 (2020)
27. Wan, E., Van Der Merwe, R.: The unscented kalman filter for nonlinear estimation. In: Proceedings of the IEEE 2000 Adaptive Systems for Signal Processing, Communications, and Control Symposium (Cat. No.00EX373), pp. 153–158 (2000)
28. Nghiem, A.T., Bremond, F., Thonnat, M., Valentin, V.: Etiseo, performance evaluation for video surveillance systems. In: 2007 IEEE Conference on Advanced Video and Signal Based Surveillance, pp. 476–481 (2007)
29. Vandyke, M., Schwartz, J., Hall, C.: Unscented kalman filtering for spacecraft attitude state and parameter estimation. Adv. Astronaut. Sci. **119** (2004)
30. He, K., Zhang, X., Ren, S., Sun, J.: Deep residual learning for image recognition. In: 2016 IEEE Conference on Computer Vision and Pattern Recognition (CVPR), pp. 770–778 (2016
31. Sandler, M., Howard, A., Zhu, M., Zhmoginov, A., Chen, L.C.: Mobilenetv2: Inverted residuals and linear bottlenecks. In: 2018 IEEE/CVF Conference on Computer Vision and Pattern Recognition, pp. 4510–4520 (2018)

Clustering-Based Refinement for 3D Human Body Parts Segmentation

Leonardo Barcellona, Matteo Terreran(✉), Daniele Evangelista, and Stefano Ghidoni

Department of Information Engineering DEI, University of Padova, Padova, Italy
{barcellona,terreran,evangelista,ghidoni}@dei.unipd.it

Abstract. A common approach to address human body parts segmentation on 3D data involves the use of a 2D segmentation network and 3D projection. Following this approach, several errors could be introduced in the final 3D segmentation output, such as segmentation errors and reprojection errors. Such errors are even more significant when considering very small body parts such as hands. In this paper, we propose a new algorithm that aims to reduce such errors and improve 3D segmentation of human body parts. The algorithm detects noise points and wrong clusters using DBSCAN algorithm, and changes the labels of the points exploiting the shape and position of the clusters. We evaluated the proposed algorithm on the 3DPeople synthetic dataset and on a real dataset, highlighting how it can greatly improve the 3D segmentation of small body parts like hands. With our algorithm we achieved an improvement up to 4.68% of IoU on the synthetic dataset and up to 2.30% of IoU in the real scenario.

Keywords: Human parsing · 3D semantic segmentation · 3D clustering

1 Introduction

Human perception is a key element in a lot of new challenges, for example, in human-robot collaboration, where any physical barrier between the human operator and the robot is removed. In this scenario, the robot needs to precisely detect the human body to avoid dangerous situations (e.g., collisions), and improve the overall task planning [1]. Another example where human perception is fundamental is motion prediction, where the objective is to estimate future movement trajectories of human body [2].

Such applications need high precision and reliability when they estimate people and their pose. Common methods to estimate people in a scene are based on people detectors and skeletal tracking algorithms [3,4], which provide just a schematic representation of the human body made of joints and links. Considering instead a volumetric approach (e.g., segmenting human body parts in the 3D space) could provide a more informative representation of the person combining

I. Petrovic et al. (Eds.): IAS 2022, LNNS 577, pp. 425–440, 2023.
https://doi.org/10.1007/978-3-031-22216-0_29

both semantics and volume information. For example, authors in [5] propose the use of semantic segmentation techniques to guarantee human safety during a human-robot collaborative task: by estimating which pixels in the image correspond to body parts (e.g., hands) it is possible to control the robot to take objects passed by the human while avoiding to collide with his/her hands.

Semantic segmentation of human body parts is also known as human parsing (HP) in the literature. State-of-the-art solutions for human parsing are mainly based on deep neural networks [6–8] trained on popular datasets such as Pascal-Person-Part [9] or LIP [10]. The majority of these methods work only with 2D images, due to the availability of many human parsing dataset with annotations for RGB images. Only recently, 3D data started to be directly exploited for training deep models that accurately detect human body parts [7,11,12]. However, the number of public 3D human parsing datasets of 3D is still limited, making it difficult to develop models that work directly on 3D data. A common strategy to tackle 3D human parsing is, hence, to reuse models trained on RGB images exploiting additional depth information to project the 2D body parts segmentation onto the 3D space. For example, we used such an approach in a previous work [13] where we developed a camera network system for 3D body parts segmentation based on SCHP network [6]. In [13] we focused on human-robot collaborative scenarios and we also proposed a manually labelled dataset acquired in a real scenario to evaluate the performance of the system.

When projecting the 2D segmentation results onto the 3D space two main types of errors should be taken into account: a 2D segmentation error of the human parsing model and a depth projection error; we named the latter Depth-Segmentation Association (DSA), which represents the wrong association between color and depth pixels. Even in calibrated cameras, the depth value of a pixel is not perfectly accurate (e.g., error in the sensor measurements), especially near the edges of small objects, so the color information may not be projected correctly. This problem is particularly significant in human body parts segmentation because, for example, the hands and fingers have a very small shape and the depth may be wrongly estimated. This may negatively affect the applicability of projecting the 2D segmentation results to 3D point clouds, especially for applications which require high accuracy and precision in segmentation.

The main effect of the DSA error is the creation of small isolated groups of 3D points, named clusters, with a different label compared to the neighboring ones. The number of isolated clusters increases in the presence of multiple people in the scene, because some wrongly projected points of a person may be attached to other people. Moreover, some clusters may also be created by segmentation errors that overestimate the dimension of the body parts. A visual examples of such errors is shown in Fig. 1, where the circles highlight the small clusters created during the association between depth and segmentation mask.

In this paper, we propose a new algorithm that reduces both DSA and small segmentation errors. Such errors corresponds to clusters of 3D points in the point cloud with a wrong associated label. Our proposed algorithm aims to find such

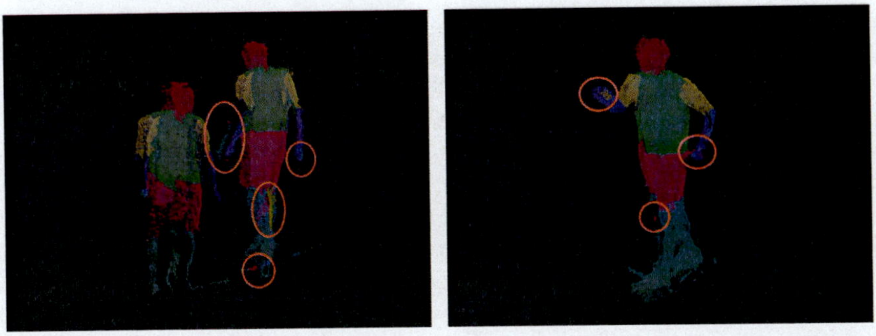

(a) (b)

Fig. 1. Two examples of projection errors. The orange circles highlight regions with wrong labels. The figure on the left (a) is extracted in a scene with two people and the label of the yellow points on the legs belongs to the second person but these points are wrongly associated with the depth. In the other figure (b) there is only one person, but there are still some error on the hands.

clusters by means of the DBSCAN algorithm [14], and to correct their labels exploiting the relationships between the body parts.

Experiments have been performed considering both synthetic data and a real scenario, highlighting a general improvement of the segmentation thanks to our algorithm. In particular, they show a marked benefit of the algorithm in improving the segmentation of smaller body parts (e.g., arms and hands), which are the parts most prone to errors described.

Another main contribution of the paper is the adaptation of the multi-view synthetic 3DPeople dataset [12] for the human parsing task. To our knowledge, there are no works that exploit such dataset to address human parsing on 3D data. Finally, it is presented a comparison of the state of the art human parsing networks for the 3D segmentation through projection using depth images.

The reminder of the paper is organized as follows. Section 2 is dedicated to describe the works related to human parsing, 3D projection and clustering algorithms. In Sect. 3, the proposed algorithm is described in detail. Section 4 reports the experiments and Sect. 5 is reserved for the conclusions.

2 Related Works

2.1 Human Parsing

Detecting body parts is essential. For example, it guarantees both safety measures and efficient interaction in human-robot collaboration. Some systems rely their human perception on a pose estimator [15], that only detects human keypoints around the joints. Other systems use semantic segmentation techniques to make a robot perceive the environment, labelling each pixel of an image according to the class it represents. When the task addresses human body parts, it is called human parsing [10].

The set of body parts targeted by human parsing depends on the representation used as reference. The ATR dataset [16] considers clothes worn alongside body parts (e.g., trousers, shoes or arms) and contains mainly people from the front that are completely visible. The LIP dataset [10] also addresses the task distinguishing between body limbs and clothes, but contains images of people varying positions and actions. The extension of LIP containing instance information is called CHIP [10]. Finally, Pascal-Person-Part dataset [9] distinguishes between body limbs (e.g., head, torso, arms, legs) without considering the clothes. This last representation is strictly related to pose estimation [17] and may be useful in a lot of applications. In fact, the authors of [5] exploit the Pascal-Person-Part representation to pass objects from human hands to robot grippers.

State of the art human parsing solutions are based on deep learning. Several models achieve good performance in the Pascal-Person-Part test dataset. The standard metric to compare the networks is the mean intersection over union (IoU), a ratio that measures how the mask detected deviates from the ground truth mask. SCHP [6] is one of the first convolutional neural networks able to reach a high IoU in Pascal-Person-Part. It leverages the context information and the edges information to increase the performance of the segmentation. CDCL [7] is a network with performance similar to SCHP, but obtained in a completely different way. It merges the pose information to train the network in a synthetic dataset keeping good performance also in real datasets. Another network worth mentioning is Grapy [8], which inherited the idea of Graphonomy [18] and trains the network in the Pascal-Person-Part dataset boosting the performance using heterogeneous dataset such as the ATR dataset and CHIP dataset.

2.2 3D Human Parsing

Semantic segmentation is frequently addressed in 3D data processing. There are a lot of examples in literature that directly segment point clouds in 3D, but the majority do not divide people into body parts. Some of them, such as [19,20], target object part segmentation. However, there are no point clouds of human bodies among the datasets they target.

The authors of [11] propose a multi-view dataset that also contains human point clouds. The labels of the points are divided among twenty classes focusing on the clothes worn instead of body parts.

Synthetic datasets are a possible solution to fill the scarcity of 3D human parsing datasets that target body parts. Moreover, their ground-truth does not present human errors, because are not annotated by hand. They retrieve the labels from body models rendered on a background image [21]. 3DPeople [12] is an example of a multi-view synthetic dataset, rendered from four points of view.

Directly projecting the segmentation mask extracted from the RGB image using the camera parameters is another solution to avoid the lack of 3D annotated data. This approach is exploited by [22]. In [13] we use this approach in a human-robot collaboration scenario, exploiting the good performance of human parsing networks in 2D and re-proposing them in 3D.

2.3 Clustering Algorithms and Applications

When the point cloud labels are generated from the projection of the 2D data, the 3D points form groups that share the labels. These are called clusters. Clustering detection is an unsupervised learning task aiming at finding the best set of clusters given a set of data and a distance function. The distance function is used for computing how close two points are and the most common function is the euclidean distance.

In the literature, there are a lot of clustering algorithms. One of the most used is K-means [23] that tries to divide the points based on a given number of clusters and the mean position of the centers. Another important algorithm is DBSCAN [14] that divides the points into clusters based on the density. Clustering is one of the most investigated topics in computer vision and all these algorithms are currently applied [24–27]. In [24] the authors use DBSCAN as a superpixel extractor. DBSCAN is used to extract clusters with similar colors and positions. Then, the clusters are merged using a distance function to form the superpixels. In [25] the authors detect humans from lidar point clouds using an online learning technique based on clustering, while in [26] the authors divide human body parts using clustering. In [27] the human body is divided into limbs using k-means.

3 Clustering-Based Algorithm for 3D Segmentation Refinement

Human body parts segmentation is mainly addressed considering 2D images as input. A common strategy is to project such 2D segmentation masks by means of depth information. This procedure allows to reuse existing 2D human parsing models, but introduces errors (i.e., 2D segmentation and DSA errors) when the labels are wrongly associated with the 3D points. The consequence is the creation of small groups of points with incorrect labels. In this work, we aim to solve the aforementioned errors by proposing the procedure described in Algorithm 1. The algorithm takes as input a segmented point cloud, obtained by projecting in 3D the predicted segmentation masks of a 2D human parsing model, and a set of people instances in the point cloud. The final output is a point cloud with the same number of points and refined labels.

In our algorithm, people instances are described in terms of 3D bounding boxes, generated as in [13] using a people detection network and the 2D segmentation mask from each single camera. Such boxes are also used to remove outliers, namely points too far from the body person. For each person in the scene, a point cloud is obtained considering all the points inside the corresponding 3D bounding box. Then the algorithm process each person point cloud with a clustering step based on DBSCAN, which detects all the clusters of body parts and provides also groups of points marked as noise (i.e., not belonging to any of the detected clusters). For each detected cluster a set of features is extracted, considering also as a feature the temporary label assigned to the cluster itself.

Algorithm 1: Clustering Correction

```
Input  : start_point cloud, boxes
Output: corrected_point cloud
people_point clouds ← null
for i=0 i < boxes.size do
    person ← filter_using_box(start_point cloud, boxes[i])
    clusters, noise ← DBSCAN(person)
    clusters_features ← features_extractor(clusters, person)
    graph ← graph_extractor(clusters_features)
    cluster_features ← merge_nodes(graph, cluster_features)
    cluster_features ← remove_nodes(graph, cluster_features)
    corrected_person ← correct_clusters(clusters, cluster_features)
    corrected_person ← reassign_noise(corrected_person, noise)
    people_point clouds.push_back(corrected_person)
end
corrected_point cloud ← fuse_poitclouds(start_point cloud, people_point
clouds)
```

Moreover, a graph representing all the relations between clusters is computed. The temporary label is then updated if the cluster is contained in another one, or if it is not connected to the graph. After these, all the points previously classified as noise are updated according to the label of the nearest cluster. Finally, all the corrected person's point clouds are merged. In this last step, if the same point is wrongly associated with two or more persons and there are discordant labels, the original label is kept.

The main steps are described in detail in the following subsections. Section 3.1 is dedicated to DBSCAN and the procedure to detect the clusters. Section 3.2 explains the features and the creation of the graph to store the position of the clusters. Finally, Sect. 3.3 highlights the correction performed.

3.1 Human Body Clusters Detection

In the first step of the algorithm, we aim to group 3D points based on common features, in order to highlight all the points which do not actually belong to human limbs or have been assigned to a wrong labels. In general, the number of clusters is not known a priori and should be estimated from each person point cloud. For example, the hands are fused together when they are touching and divided when they are not. Furthermore, the DSA error increases the number of clusters. The distance metric chosen to distinguish the clusters is the Euclidean distance, which is increased when two points do not share a label so as to separate not only by point density, but also by the initially assigned label. From the description of the constraints, the chosen clustering method is DBSCAN [14], that does not need the number of clusters as input and is also capable of detecting noise points, that are points which do not belong to any clusters.

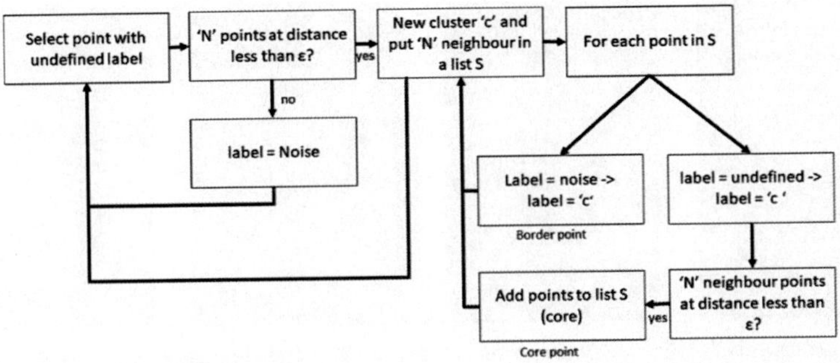

Fig. 2. A schematic overview of the DBSCAN algorithm [14]. The whole pipeline is iterated until there are unlabelled points.

In particular, DBSCAN separates points based on density considering two main parameters: the distance ϵ used to consider two points adjacent, and the number of adjacent points N needed for a point to be considered as core points. These parameters could be considered as the scale of near points and the minimum density of a cluster. A schematic overview of the DBSCAN clustering algorithm is shown in Fig. 2, highlighting the main steps performed. The algorithm starts from a random point. If such point has N or more points around with a distance $d \leq \epsilon$, it is then considered a *core point* of the cluster. Otherwise, if the random point is close to at least one core point, it is associated to the cluster and considered a border point. Finally, if a point is not a core point and is not close to any core points, it is considered as noise. The final output of the DBSCAN algorithm is a set of clusters and a set of "noise" points not associated to any of the cluster found. An example of the clusters found by DBSCAN is provided in Fig. 3b, showing the result on a person point cloud. Note that in the input point cloud shown in Fig. 3a, some points of the hands have been assigned to a wrong label (either due to a segmentation or a projection error). All these mislabeled points are detected by the DBSCAN algorithm and assigned to different clusters, highlighting the set of points which needs a label refinement in the next steps of the algorithm.

3.2 Feature Extraction and Graph Construction

After the clustering step, we investigate a suitable set of features in order to understand which cluster is correctly labelled and which is not. We extract from each cluster three main features: the actual label (shared by the whole cluster), the cluster centroid and the cluster radius. The radius of the cluster is computed as the maximum distance between the centroid and the points belonging to the cluster. The centroid and the radius do not change value during the execution of the algorithm, while the label is updated as exploited in Sect. 3.3.

To keep information about the relation between clusters, we use a graph. The graph is a set of nodes and edges. In our case, the nodes are the clusters, while

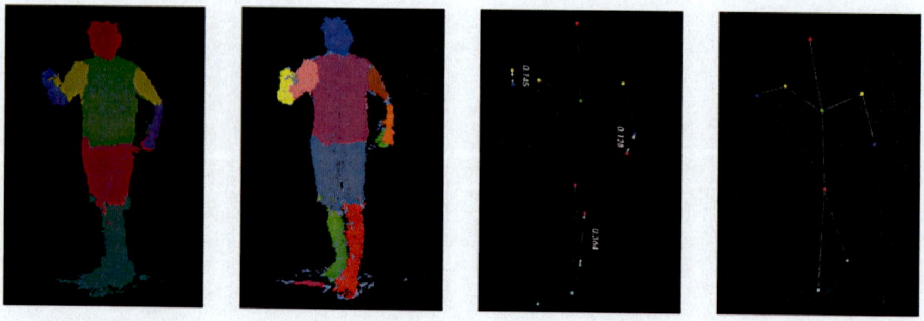

(a) Point Cloud (b) Clusters (c) Graph (d) Tree

Fig. 3. The cluster detected by DBSCAN and the graph constructed from them. Figure (a) is the starting point cloud. In figure (b) every color correspond to a cluster, while turquoise points the noise. Figure (c) shows the graph constructed from the clusters and figure (d) the best tree matching the person.

the edges keep information about their relative position. A graph representing an human body has a particular structure: a “torso” node connected with a “head” node, two “upper arm” nodes and two “upper legs” nodes. Each upper arm is connected to a lower arm and each upper leg to a lower leg. This particular graph is a tree and we construct it starting from the biggest cluster of label torso as root and, then, connecting the arms, the hands, the legs and the head.

In our graph structure there are three types of edges distinguished by these three cases: a cluster is inside another, a cluster intersects another or a cluster belongs to the best tree that represents the human body. The last edges is computed using the torso as root as previously reported while the radius and the centroid are used to check the first two type of edges. Fig. 3 shows a visualization of the constructed graph and the best tree representing the person.

3.3 Segmentation Label Refinement

The final step is the refinement of the segmented point cloud by correcting the labels of the noise points. The nodes of the body tree are unchangeable clusters because they represent the best body structure. The nodes of the graph that are contained in another cluster are updated, inheriting the label of the biggest cluster containing them with a different label. This is the “merge_nodes” step (M) of the pseudocode reported in Algorithm 1.

The clusters not connected to the best matching tree are labelled as background during the “remove_nodes” step (R). If a valid tree is not found, this phase, is not executed. Finally, the point cloud is refined assigning the new labels of the clusters to the points. The final step is re-labelling the points that were classified as noise. They inherit the label of the nearest point of the corrected point cloud. The search is performed using the kdtree search.

4 Experiments

The following section presents a set of experiments to evaluate the effectiveness of the proposed algorithm. In the first subsection, our segmentation refinement algorithm is tested using the synthetic multi-view dataset [12] presented in the related works, while the second subsection is focused on the evaluation of the algorithm on data acquired in a real scenario (i.e., the annotated segmentation masks from [13]).

In all the experiments, performances are evaluated in terms of intersection over union (IoU), which is computed as the ratio between the number of points with a correct label and the total number of points in the point cloud. We also compute the mean IoU of the body parts (BIoU) because the focus of the work is on correcting them. Other two reported metrics are the precision and the F1 score, which represent the number of true points of a class divided by the number of points of that class and the harmonic mean between precision and recall.

We analyze the improvement of each step of the algorithm. We indicate with "M" the merge phase, where the clusters are fused in one bigger group. We name "N" the label reassignment of the noise points. With "M - N", we refer to the algorithm with both the previous steps. Finally, "M - R - N" refers to the whole algorithm, whit an outliers-removal, as described in Algorithm 1.

4.1 Performance on Synthetic Data

Synthetic datasets offer a large amount of labelled data, which can be acquired and annotated in an easier and quicker way than real datasets. For example, the 3DPeople synthetic dataset [12] contains 2.5 Million frames of 80 subjects performing different actions, 40 female and 40 male models. The dataset contains high variability, with a a large range of distinct body shapes, skin tones and clothing outfits, and provides RGB-D data under different viewpoints. Regarding annotations, the dataset offers ground truth 2D segmentation masks for clothes and human body, the latter divided into fourteen classes distinguishing among the rest, left and right limbs.

(a) View 1 (b) View 2 (c) View 3 (d) Point cloud

Fig. 4. Example of the dataset created by [12]. Images (a)–(c) show the three of the four views and (d) shows the reconstructed point cloud generated by SCHP predictions [6].

For our experiments, we grouped the body parts annotation of the 3DPeople dataset in a set of 6 classes, namely *Head*, *Torso*, *Upper arms*, *Lower arms*,

Upper legs and *Lower legs*. We then computed point clouds from the RGB-D data provided in the 3DPeople dataset for each camera; for each pair of RGB-D data the point cloud with body parts segmentation is obtained by projecting the labels predicted by the SCHP [6] network, while ground truth point cloud are obtained by projecting the dataset annotations. Figure 4 shows an example of a model rendered on a background image from multiple points of view and the point cloud generated by using SCHP predictions. In our experiments we used the last ten male and female models to test our segmentation refinement algorithm, following the suggestion of 3DPeople dataset's authors but increasing the number of models used as test set. In particular, 1440 frames were extracted with different models, clothes, light, background and pose.

The human parsing network used for evaluating the refinement algorithm is SCHP [6]. However, one problem is the difficulty to reuse the solutions obtained in real datasets in synthetic ones or vice-versa. For example, a human parsing network trained on Pascal-Person-Part [9] may not be suitable for segmenting the 3DPeople synthetic dataset and vice-versa. For this reason, the available pretained model is not suited and a new training of the SCHP network is required before testing the proposed algorithm on the synthetic dataset. The 30 models of males and females not used for testing the algorithm were used for this purpose. 1716 frames were extracted from these to match the number of images used in Pascal-Person-Part [9] and the labels were reduced to match the set of six classes considered in the Pascal-Person-Part dataset and listed before.

The point clouds to be refined to test the proposed algorithm were computed using 1440 test frames of 3DPeople RGB-D data and the segmentation masks predicted by SCHP after the new training. Table 1 shows the results of the proposed algorithm applied to these point clouds. The parameters of DBSCAN, used in the first clustering phase of the algorithm, are 1.0 as minimum distance and 40 as the number of neighbours points. The columns show the intersection over union of the body classes, the average IoU, the average IoU of the body classes, the average precision and the F1 score.

(a) Start (b) Clusters (c) M (d) N (e) M-N (f) M-R-N

Fig. 5. From the left to the right: the starting cloud, the clusters, the results of the merge step, the results of the noise relabeling, the results of the merge and noise relabeling, the results of the whole algorithm. The point clouds are extracted from 3DPeople [12].

Table 1. The performance on the 3DPeople synthetic dataset. First columns show IoU per class. Last columns show the global performance in terms of mean IoU, mean IoU of body classes, average precision and F1 score. The second line of each row shows the change between the reference and the output of the algorithm as reported in Sect. 3. M is the merge phase of the algorithm, N is the noise relabelling phase, M-N is the merge phase with the noise relabelling phase and M-R-N is the whole algorithm.

Type of scene	Head	Torso	Upper arms	Lower arms	Upper legs	Lower legs	Background	mIoU	BIoU	AP	F1
Reference	73.68	76.84	66.26	68.96	69.98	75.86	99.05	75.8	71.93	91.85	86.00
M	73.59	76.96	66.57	72.97	70.22	77.1	99.09	76.51	72.90	93.09	86.49
	-0.09	0.12	0.31	4.01	0.24	1.24	0.04	0.71	0.97	1.24	0.49
N	74.21	77.07	66.94	71.78	70.30	76.97	99.09	76.62	72.88	93.00	86.56
	0.53	0.23	0.68	2.82	0.32	1.11	0.04	0.82	0.95	1.15	0.56
M - N	73.66	77.06	66.62	72.27	70.27	77.11	99.09	76.58	72.83	93.03	86.54
	-0.02	0.22	0.36	3.31	0.29	1.25	0.04	0.78	0.90	1.18	0.54
M - R - N	73.75	77.13	66.76	73.64	70.41	77.83	99.12	76.95	73.25	93.6	86.79
	0.07	0.29	0.50	4.68	0.43	1.97	0.07	1.15	1.32	1.75	0.79

The results reported in Table 1 are divided in rows to highlight the contribution of each step of the proposed algorithm. The letter "M" denotes the merge function that associates the labels to the cluster wrapping the initial cluster entirely. The letter "N" denotes the noise reassignment contribution and "M - N" represents the contribution of both merge and noise reassignment. The final row, denoted as "M-R-N", shows the results of the whole algorithm. For each main row, we also reported the performance increase or decrease (in green or red respectively) compared to the baseline, shown in the first row. Some output examples of the various steps of the algorithm are also depicted in Fig. 5, together with the input point cloud and the clusters detected by DBSCAN.

From the results obtained on the synthetic dataset, it can be shown that the proposed algorithm leads to a visible improvement on some classes, especially the "Lower arms" class. Indeed, this is the class that is most frequently misclassified by the SCHP network (i.e., segmentation error). Moreover, since the lower arms class includes also the hands, for this class we have in general also a high DSA error, due to the difficulty to correctly estimate the depth of the points around the hands. The proposed algorithm is therefore able to reduce the effect of both these errors, allowing to improve the segmentation of difficult classes such as hands. The background class has a good IoU affecting also the mean, but we are more interested in the body classes because the algorithm mainly focuses on improving them. The mean Body IoU (BIoU) is the average IoU of the body classes and may better highlight the improvement of the algorithm, with an increase of up to 1.32% of IoU. Moreover, the lower arms class reaches an improvement of 4.68%. From the table, each part of the algorithm contributes an increase in the results.

4.2 Performance on Real Data

The proposed algorithm proves to be effective in the experiments on synthetic data, showing an increase in performance on challenging classes. On the

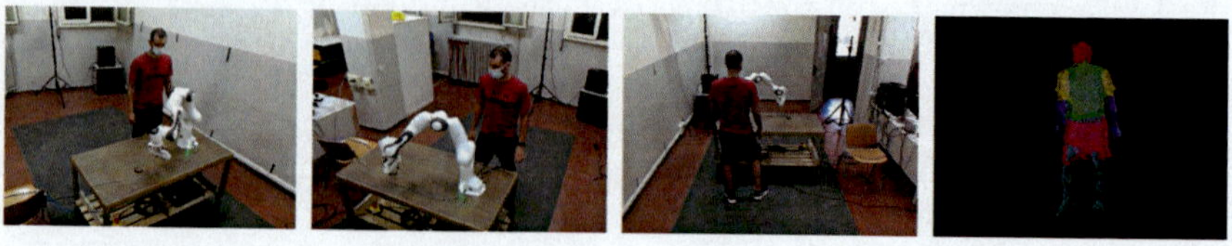

(a) View 1 (b) View 2 (c) View 3 (d) Point cloud

Fig. 6. Example of the dataset created by [13]. Images (a)–(c) show the three view and (d) shows the ground truth point cloud.

Table 2. Performance on the real dataset. First columns show IoU per class. Last columns show the global performance in terms of mean IoU, mean IoU of body classes, average precision and F1 score. The second line of each row show the change between the reference and the output of the algorithm as reported in section 3. M is the merge phase of the algorithm, N is the noise relabelling phase, M-N is the merge phase with the noise relabelling phase and M-R-N is the whole algorithm.

SCHP [6]

Type of scene	Head	Torso	Upper arms	Lower arms	Upper legs	Lower legs	Background	mIoU	BIoU	AP	F1
Reference	80.47	76.69	57.19	50.52	71.22	59.07	99.15	70.61	65.86	81.92	82.07
M	79.77	76.56	57.48	50.75	71.91	61.32	99.09	70.98	66.30	84.74	82.35
	-0.70	-0.13	0.29	0.23	0.69	2.25	-0.06	0.37	0.44	2.82	0.28
N	80.63	77.07	57.81	51.21	72.66	60.3	99.2	71.3	66.61	82.86	82.56
	0.16	0.38	0.62	0.69	1.44	1.23	0.05	0.69	0.75	0.94	0.49
M - N	81.11	77.18	57.59	52.03	72.91	60.92	99.2	71.56	66.96	83.14	82.76
	0.64	0.49	0.40	1.51	1.69	1.85	0.05	0.95	1.10	1.22	0.69
M - R - N	81.11	77.24	57.61	51.97	72.95	61.37	99.21	71.64	67.04	83.38	82.81
	0.64	0.55	0.42	1.45	1.73	2.30	0.06	1.03	1.18	1.46	0.74

CDCL [7]

Type of scene	Head	Torso	Upper arms	Lower arms	Upper legs	Lower legs	Background	mIoU	BIoU	AP	F1
Reference	86.02	81.56	71.83	70.46	81.05	72.89	99.70	80.46	77.30	88.41	88.96
M	85.47	83.41	71.94	71.98	79.87	71.73	99.32	80.53	77.40	91.09	89.00
	-0.55	1.85	0.11	1.52	-1.18	-1.16	-0.38	0.07	0.10	2.68	0.04
N	86.36	84.43	72.40	72.08	81.58	73.54	99.47	81.41	78.40	89.61	89.53
	0.34	2.87	0.57	1.62	0.53	0.65	-0.23	0.95	1.10	1.20	0.57
M - N	86.33	84.50	72.38	72.43	81.63	73.54	99.48	81.47	78.47	89.71	89.56
	0.31	2.94	0.55	1.97	0.58	0.65	-0.22	1.01	1.17	1.30	0.60
M - R - N	86.33	84.52	72.38	72.51	81.63	73.38	99.48	81.46	78.46	89.76	89.56
	0.31	2.96	0.55	2.05	0.58	0.49	-0.22	1.00	1.16	1.35	0.60

Grapy [8]

Type of scene	Head	Torso	Upper arms	Lower arms	Upper legs	Lower legs	Background	mIoU	BIoU	AP	F1
Reference	86.13	84.84	67.82	64.79	79.13	71.05	99.48	79.04	75.63	87.46	87.87
M	84.92	84.09	67.18	64.2	79.09	70.63	99.37	78.45	75.02	89.31	87.52
	-1.21	-0.75	-0.64	-0.59	-0.04	-0.42	-0.11	-0.59	-0.61	1.85	-0.35
N	86.16	84.73	67.71	64.66	80.3	71.69	99.36	79.25	75.88	87.97	88
	0.03	-0.11	-0.11	-0.13	1.17	0.64	-0.12	0.21	0.25	0.51	0.13
M - N	86.24	84.73	67.69	64.77	80.44	71.77	99.5	79.31	75.94	88.00	88.03
	0.11	-0.11	-0.13	-0.02	1.31	0.72	0.02	0.27	0.31	0.54	0.16
M - R - N	86.24	84.81	67.69	64.82	80.41	71.22	99.5	79.25	75.87	88.04	87.99
	0.11	-0.03	-0.13	0.03	1.28	0.17	0.02	0.21	0.24	0.58	0.12

other hand, we also tested the proposed method on a more challenging dataset composed by real data. We used the dataset from [13] which includes manually annotated RGB-D data of a real setup. Such dataset shows up to three people moving in a scene with obstacles or a robot occluding the views. The scenes are recorded from three points of view using Microsoft Kinect One RGB-D cameras. Figure 6 shows a scene from such a dataset, where a person is standing near a robotic arm, and the corresponding ground truth point cloud.

Similarly to the previous experiments, we extracted the point cloud projecting the segmented labels from the human parsing networks. In addition to the SCHP architecture [6], we considered also two other human parsing networks, namely CDCL [7] and Grapy [8]. We did not use these networks in the previous section because there are no open source implementations suitable for training such models on custom datasets. For all the three networks, we used the pre-trained weights provided by the authors, obtained after training on the Pascal-Person-Part [9] dataset. DBSCAN parameters were set to 0.05 as the maximum distance between near points and 40 as the number of neighbour points of a core point. The change is due to the different distances of the points generated by the Kinect sensors with respect to the case of synthetic depth.

(a) Start (b) Clusters (c) M (d) N (e) M-N (f) M-R-N

Fig. 7. From the left to the right: the input point cloud, the clusters, the results of the merge step, the results of the noise relabeling, the results of the merge and noise relabeling, the results of the whole algorithm. The input point cloud is taken from a real scenario [13].

Some output examples of each step in the refinement algorithm are depicted in Figure 7, showing also the input point cloud, the clusters detected by DBSCAN. Note as the noise relabeling step helps correcting some labels but the hands are not affected by it, while the merge step is able to correct it. A quantitative evaluation of the algorithm performance is reported in Table 2. For all the networks considered, our segmentation refinement approach leads to an improvement of the BIoU metric, but there are some differences in the results. In particular, the improvement is highly marked for the first two human parsing networks using all the steps of the proposed algorithm, while in case of the Grapy network the results are mostly unchanged. In particular, the merge phase for this last network does not seem effective. This means that the network is less affected by segmentation errors that go outside the edges of the objects. Secondly, the most affected body part changes based on the network used, meaning that they tend to overestimate the segmentation of some specific classes more frequently. SCHP finds more problems of over-segmentation of lower arms and legs, as found on the synthetic data, while CDCL has more segmentation problems with lower arms and torso classes. Grapy only tends to over-segment the upper legs. In this set of experiments it is worth noticing that the real dataset is more challenging than the synthetic data, because people are recorded in occluded scenarios that may interfere with the perception, dividing people's shape and decreasing the effect of the merge procedure.

5 Conclusions

When addressing human body parts segmentation on 3D data, a common approach to deal with the lack of large 3D human parsing dataset consists in the use of 2D human parsing networks: given a input image, such networks predict a segmentation mask that is then projected in 3D by using the corresponding depth information. The resulting 3D segmentation output is however affected by several errors, namely a segmentation error and a projection error due to inaccuracies of the depth information. In this paper, we proposed a new algorithm to tackle such errors and improve the final 3D segmentation results. Our algorithm is based on a clustering technique to identify and localize 3D points which have been assigned to a wrong label or that represent noise. All the clusters are then used to define a graph structure, which allows to refine the overall segmentation based on relationships between the clusters. The proposed algorithm has been evaluated on both synthetic and real data, showing the effectiveness of our approach to refine 3D body parts segmentation outputs. In particular, the algorithm shows major improvements when considering small body parts such as lower arms and hands, which are often misclassified by 2D human parsing networks. Especially the hands are also difficult to be projected in 3D, due to the inaccuracies of the depth information around the borders of small body parts (e.g., fingers). As demonstrated experimentally, the proposed algorithm is able to reduce the effect of both segmentation and projection errors, allowing to improve the 3D segmentation of body parts, even for complex shapes like hands. This is an interesting result, especially for human-robot collaboration applications, where our segmentation refinement approach can be very useful to the robot to get a more accurate representation of the people in the workcell, enabling a close and direct collaboration with the human. As a future research direction we will further investigate how to improve 3D body parts segmentation accuracy when several points of view are available, combining all the segmentation information by means of Bayesian fusion techniques. Moreover, we will focus on the integration of the segmentation refinement algorithm in a real robotic workcell to test our algorithm during a real human-robot collaboration task.

Acknowledgment. This research has received funding from the European Union's Horizon 2020 research and innovation program under grant agreement No. 101006732.

References

1. Merckaert, K., Convens, B., Wu, C.J., Roncone, A., Nicotra, M.M., Vanderborght, B.: Real-time motion control of robotic manipulators for safe human–robot coexistence. Robot. Comput. Integr. Manuf. **73**, 02223 (2022)
2. Zanchettin, A.M., Casalino, A., Piroddi, L., Rocco, P.: Prediction of human activity patterns for human–robot collaborative assembly tasks. IEEE Trans. Ind. Inform. **15**(7), 3934–3942 (2018)
3. Cao, Z., Hidalgo Martinez, G., Simon, T., Wei, S., Sheikh, Y.A.: Openpose: real-time multi-person 2d pose estimation using part affinity fields. IEEE Trans. Pattern Anal. Mach. Intell. (2019)
4. Carraro, M., Munaro, M., Burke, J., Menegatti, E.: Real-time marker-less multi-person 3D pose estimation in RGB-depth camera networks (2017)
5. Rosenberger, P., Cosgun, A., Newbury, R., Kwan, J., Ortenzi, V., Corke, P., Grafinger, M.: Object-independent human-to-robot handovers using real time robotic vision. IEEE Robot. Autom. Lett. **6**(1), 17–23 (2020)
6. Li, P., Xu, Y., Wei, Y., Yang, Y.: Self-correction for human parsing. IEEE Trans. Pattern Anal. Mach. Intell. (2020)
7. Lin, K., Wang, L., Luo, K., Chen, Y., Liu, Z., Sun, M.T.: Cross-domain complementary learning using pose for multi-person part segmentation. IEEE Trans. Circuits Syst. Video Technol. **31**(3), 1066–1078 (2020)
8. He, H., Zhang, J., Zhang, Q., Tao, D.: Grapy-ML: graph pyramid mutual learning for cross-dataset human parsing. Proc. AAAI Conf. Artif. Intell. **34**, 10949–10956 (2020)
9. Chen, X., Mottaghi, R., Liu, X., Fidler, S., Urtasun, R., Yuille, A.: Detect what you can: detecting and representing objects using holistic models and body parts. In: Proceedings of the IEEE Conference on Computer Vision and Pattern Recognition, pp. 1971–1978 (2014)
10. Gong, K., Liang, X., Zhang, D., Shen, X., Lin, L.: Look into person: self-supervised structure-sensitive learning and a new benchmark for human parsing. In: Proceedings of the IEEE Conference on Computer Vision and Pattern Recognition (2017)
11. Yu, Z., Yoon, J.S., Lee, I.K., Venkatesh, P., Park, J., Yu, J., Park, H.S.: Humbi: a large multiview dataset of human body expressions. In: Proceedings of the IEEE/CVF Conference on Computer Vision and Pattern Recognition, pp. 2990–3000 (2020)
12. Pumarola, A., Sanchez, J., Choi, G., Sanfeliu, A., Moreno-Noguer, F.: 3DPeople: modeling the geometry of dressed humans. In: International Conference in Computer Vision (ICCV) (2019)
13. Terreran, M., Barcellona, L., Evangelista, D., Ghidoni, S.: Multi-view human parsing for human-robot collaboration. In: 2021 20th International Conference on Advanced Robotics (ICAR), pp. 905–912. IEEE (2021)
14. Schubert, E., Sander, J., Ester, M., Kriegel, H.P., Xu, X.: DBSCAN revisited, revisited: why and how you should (still) use DBSCAN. ACM Trans. Database Syst. (TODS) **42**(3), 1–21 (2017)
15. Wang, Y., Ye, X., Yang, Y., Zhang, W.: Collision-free trajectory planning in human–robot interaction through hand movement prediction from vision. In: 2017 IEEE-RAS 17th International Conference on Humanoid Robotics (Humanoids), pp. 305–310. IEEE (2017)
16. Liang, X., Liu, S., Shen, X., Yang, J., Liu, L., Dong, J., Lin, L., Yan, S.: Deep human parsing with active template regression. IEEE Trans. Pattern Anal. Mach. Intell. **37**(12), 2402–2414 (2015)

17. Xia, F., Wang, P., Chen, X., Yuille, A.L.: Joint multi-person pose estimation and semantic part segmentation. In: Proceedings of the IEEE Conference on Computer Vision and Pattern Recognition, pp. 6769–6778 (2017)
18. Gong, K., Gao, Y., Liang, X., Shen, X., Wang, M., Lin, L.: Graphonomy: universal human parsing via graph transfer learning. In: Proceedings of the IEEE/CVF Conference on Computer Vision and Pattern Recognition (2019)
19. Qi, C.R., Su, H., Mo, K., Guibas, L.J.: Pointnet: deep learning on point sets for 3D classification and segmentation. In: Proceedings of the IEEE Conference on Computer Vision and Pattern Recognition, pp. 652–660 (2017)
20. Qi, C.R., Yi, L., Su, H., Guibas, L.J.: Pointnet++: deep hierarchical feature learning on point sets in a metric space. arXiv preprint arXiv:1706.02413 (2017)
21. Varol, G., Romero, J., Martin, X., Mahmood, N., Black, M.J., Laptev, I., Schmid, C.: Learning from synthetic humans. In: Proceedings of the IEEE Conference on Computer Vision and Pattern Recognition, pp. 109–117 (2017)
22. Navaneet, K., Mandikal, P., Agarwal, M., Babu, R.V.: CAPNet: continuous approximation projection for 3D point cloud reconstruction using 2D supervision. Proc. AAAI Conf. Artif. Intell. **33**, 8819–8826 (2019)
23. Hartigan, J.A., Wong, M.A.: Algorithm as 136: a k-means clustering algorithm. J. Roy. Stat. Soc. Ser. C (Appl. Stat.) **28**(1), 100–108 (1979)
24. Shen, J., Hao, X., Liang, Z., Liu, Y., Wang, W., Shao, L.: Real-time superpixel segmentation by DBSCAN clustering algorithm. IEEE Trans. Image Process. **25**(12), 5933–5942 (2016)
25. Yan, Z., Duckett, T., Bellotto, N.: Online learning for human classification in 3D lidar-based tracking. In: 2017 IEEE/RSJ International Conference on Intelligent Robots and Systems (IROS), pp. 864–871. IEEE (2017)
26. Buys, K., Cagniart, C., Baksheev, A., De Laet, T., De Schutter, J., Pantofaru, C.: An adaptable system for RGB-D based human body detection and pose estimation. J. Visual Commun. Image Representation **25**(1), 39–52 (2014)
27. Haggag, H., Hossny, M., Haggag, S., Nahavandi, S., Creighton, D.: Efficacy comparison of clustering systems for limb detection. In: 2014 9th International Conference on System of Systems Engineering (SOSE), pp. 148–153. IEEE (2014)

Autonomous Exploration for 3D Mapping Using a Mobile Manipulator Robot with an RGB-D Camera

Ryota Uomi[1], Ayanori Yorozu[2](✉), and Akihisa Ohya[2]

[1] Graduate School of Science and Technology, University of Tsukuba, 1-1-1 Tennodai, Tsukuba 305-8573, Japan
uomi-r@roboken.cs.tsukuba.ac.jp

[2] Faculty of Engineering, Information and Systems, University of Tsukuba, 1-1-1 Tennodai, Tsukuba 305-8573, Japan
{yorozu,ohya}@cs.tsukuba.ac.jp

Abstract. This study focuses on an unknown environment exploration method for generating a 3D occupancy grid map using a mobile manipulator robot with an RGB-D camera. To determine the moving destination and camera pose, a boundary between known and unknown regions is detected as a frontier, and a frontier observation map is created to record the number of frontiers that can be observed at a certain position and pose. Subsequently, based on the frontier observation map, an approach to efficiently generate a map in a short travel distance and time by actively changing the camera posture while moving is explored. However, if the robot simply sets the position by using the largest number of frontiers as the destination, it may have to move back and forth to the same location many times when missing some frontiers, resulting in more time to generate the map. To solve this problem, a method of modifying the frontier observation map based on the distance and angle to preferentially observe the frontier near the robot and in the direction where the robot is moving, respectively, is proposed. The effectiveness of the modification process in determining the destination in terms of the length of the path traveled and map generation time for several environments is verified by simulations. In addition, the effect of actively changing the camera posture while the robot is moving is evaluated.

Keywords: Mobile manipulator robot · 3D mapping · Autonomous exploration

1 Introduction

In recent years, robots that operate in human living environments, such as cleaning robots, have been put to practical use [1]. In addition, mobile manipulator robots have been proposed to perform tasks such as cleaning [2]. To achieve localization and understand the space in which a robot can move and plan its movements, it is necessary to generate a three-dimensional (3D) map of the environment. As shown in Fig. 1, this study focuses on a method for automatically generating a 3D grid map with detailed object locations

I. Petrovic et al. (Eds.): IAS 2022, LNNS 577, pp. 441–454, 2023.
https://doi.org/10.1007/978-3-031-22216-0_30

and shapes for robots to operate in a human living space where various objects are located. Exploration methods for automatically generating a map while moving through the environment have been proposed for mobile robots equipped with 2D-LiDAR or cameras, such as cleaning robots [3–10]. However, in an environment with objects of various heights, such as tables and chairs, there are areas that cannot be observed if the sensor is fixed to the robot. Therefore, this study aims to investigate an exploration method for generating a 3D map using a mobile manipulator robot equipped with an arm having an RGB-D camera at its tip, as shown illustrated in Fig. 2. The robot has its arm mounted on a mobile cart with a small height. By raising the arm posture, the robot can observe the top of a desk or table. By lowering the arm posture, the robot can not only observe the underside of a desk or table but also move around the underside. In this study, we propose a method for efficiently searching the environment in a short time and distance by actively changing the posture of the arm carrying the camera, considering the observation range and mobility.

Fig. 1. 3D occupancy grid map for service robot operation in a human living space

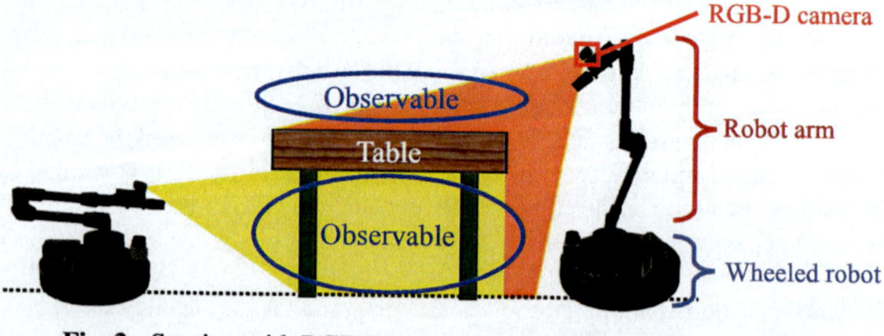

Fig. 2. Sensing with RGB-D camera on a mobile manipulator robot

For map generation, it has been suggested to divide the environment into occupied, free, and unknown areas and to continue searching until no more unknown areas exist in the environment [4]. Yamauchi defined an unknown region adjacent to the free area as a frontier and proposed a searching method by repeatedly moving to a large frontier close to the robot as the destination until there are no more frontiers in the environment [4]. In this study, we also use a frontier-based exploration method. However, the RGB-D camera used in this study has a narrow angle of view, and the map is updated during movement. Several methods for exploring a 3D environment with a camera have also been recommended [11–16]. Similar to our study, exploration methods using an RGB-D camera mounted on a mobile manipulator robot have been put forward [15, 16]. However, these methods do not take advantage of the ability of changing the camera pose while the robot is moving.

In this study, we propose an exploration method for generating a frontier observation map, which records the number of frontiers that can be observed at a certain position and posture so that the camera can observe many frontiers each time the map is updated, and then sequentially determining the moving destination. However, if the robot simply sets the position by using the largest number of frontiers as the destination, it may have to move back and forth to the same location many times when some frontiers are missed, causing more time to generate the map. To solve this problem, we propose a method of modifying the frontier observation map based on the distance and angle to preferentially observe the frontier near the robot and in the direction that the robot is moving, respectively. The effectiveness of the modification process in determining the destination in terms of the length of path traveled and map generation time for several environments is evaluated by simulations. In addition, the effect of actively changing the camera posture while the robot is moving is verified.

2 Related Works

In automatic map generation, the boundary between known and unknown areas of the map being created is defined as a frontier. A frontier-based method in which observations are made by moving toward the frontier has been proposed [4] and is often used. Yamauchi suggested a method of searching for the nearest frontier as a destination for a robot equipped with a two-dimensional omnidirectional ranging sensor [4]. However, if the frontier closest to the robot is considered as the destination, the final path will overlap and the movement will be inefficient. To solve this problem, it was recommended to create a histogram of the directions in which the frontiers can be seen from the current position of the robot and to filter the histogram to prioritize the frontiers in the directions where many frontiers can be observed [5]. When the environment is large, it requires a long time to search the entire area. Therefore, a method of searching the frontier as a candidate destination by rapidly-exploring random trees (RRT) and determining the next destination based on the clustering of the frontier and the unknown number of grids around the frontier has been suggested [6].

Another method that integrates the simultaneous localization and mapping (SLAM) approach and considers the uncertainty of the map to determine the next movement based on the expected value of entropy has also been proposed [7–10]. In contrast to the frontier-based approach, this method can allow the exploration of both unknown environments and places that have already been passed but where the map information is uncertain. However, LiDAR has been mostly applied to mobile robots, and there have been few reports of its application to mobile manipulators with cameras that can change their posture, which is the subject of this study.

Several methods have been suggested for constructing 3D maps using cameras. Using a stereo camera, to observe the frontier, a method of camera trajectory planning based on a perpendicular view of the frontier plane has been recommended [11]. Using an RGB-D camera mounted on a drone, a method of clustering the locations where frontiers can be observed and observing objects on the ground from a high position using the location where most frontiers can be observed has been proposed [12].

An approach to automatically generate 3D maps from RGB-D cameras fixed to a mobile robot has also been put forward. Ding and Fang recommended to sequentially determine the trajectory of a robot based on the expected value of entropy from the candidates of the trajectory of the robot. When there is no frontier around the robot, the robot switches its behavior so that the destination is the location where the frontier can be observed in the entire map [13].

Exploration methods using a mobile manipulator robot equipped with an RGB-D camera, similar to that used in this study, have been proposed. A frontier-based method for determining the viewpoint of an autonomous rescue robot was suggested in [15]. However, only the camera viewpoint was controlled by the arm, and the crawler robot was not moving. A method for creating a 3D map of the surroundings of a building using a mobile manipulator robot equipped with an RGB-D camera has been put forward in [16]. In this method, the robot moves around the building while directing the camera to follow the unknown area and switches between two strategies: rough and detailed observation. However, the purpose is only to explore the area around the building and not to move through objects of various heights or underneath them as in the human living environment, which is the target of this study.

Here, we use a frontier-based approach and propose an exploration method to generate a frontier observation map that records the number of frontiers that can be observed at a certain position and posture. Whenever the map is updated, the frontier is moved to the most observable position as the destination. Taking advantage of the ability to change the camera pose, we propose a method of searching while changing the camera pose in the direction where the frontier is most observable at that location while moving.

3 Overview of the Proposed Exploration

This study focuses on the mobile manipulator robot shown in Fig. 2. The robotic arm is mounted on a two-wheeled mobile cart. An RGB-D camera is attached to the end of the robot arm to obtain the 3D distance data. Figure 3 displays an overview of the proposed exploration method in an unknown environment for mapping. We propose a method of searching while sequentially changing the destination each time the map is updated, taking into consideration that the map is sequentially updated by movements, instead of determining the next destination after moving to a previously determined destination, as in previous studies [4]. In addition, considering the viewing angle of the camera, the destination of the movement and the position and orientation angle of the camera are determined such that many unobserved areas can be viewed. As exhibited in Fig. 3, the system is divided into two processes: map generation and robot movement.

In map generation, the target 3D occupancy grid map is updated based on the 3D point cloud data acquired from the RGB-D camera. Each grid of the map contains attribute information of free, occupied, and unknown areas. After the map is updated, unknown grids adjacent to the free areas are detected as frontiers. If no more observable frontiers in the environment can be detected, the mapping is completed. If it exists, a frontier observation map is generated to observe the unknown area. A frontier observation map

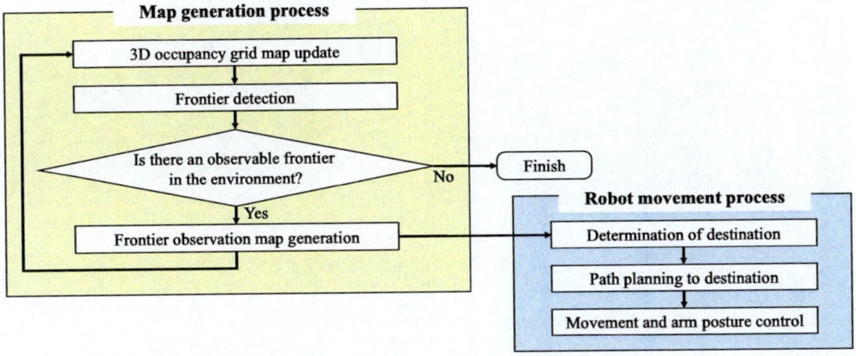

Fig. 3. Flow of proposed exploration

is a four-dimensional grid map that records the number of frontiers that can be observed at the camera position and yaw angle.

During the robot movement, whenever the frontier survey map is updated, the robot determines the next destination and moves to the destination by changing the posture of the camera arm. The processes of updating the map and moving the robot are executed asynchronously until there are no observable frontiers in the environment. To determine the destination of the robot's movement, several methods are considered, and their characteristics and other factors are experimentally verified and evaluated by simulations in several environments.

4 Frontier Observation Map Generation

4.1 Frontier Detection

In this study, the existing 3D occupancy grid mapping framework: Octomap [17] based on the 3D point cloud of the surrounding environment acquired from the camera is used, as displayed in Fig. 4(a). Figure 4(a) depicts the occupancy grid map when the map is constructed by scanning the table and wall with an RGB-D camera. Grids with obstacles, without obstacles, and not observed by sensors are assigned the statuses of occupied, free, and unknown, respectively. An unknown grid adjacent to a free grid is detected as a frontier grid. Figure 4(b) illustrates the detected frontier grids in Fig. 4(a) as blue grids.

4.2 Frontier Observation Map Generation

The frontier observation map is a four-dimensional grid map that records the number of frontiers that can be observed at the camera position coordinates (x, y, z) and yaw angle θ, as illustrated in Fig. 5. In this study, roll and pitch angles are not considered. As shown in Fig. 5, the number of frontier grids that can be observed in the field of view of the camera with observation distance V_d [m] and viewing angle V_θ [deg] is recorded in each grid of the frontier observation map. If there is an obstacle grid on the line connecting the camera position to the frontier grid, then the frontier grid is not counted in the frontier observation map.

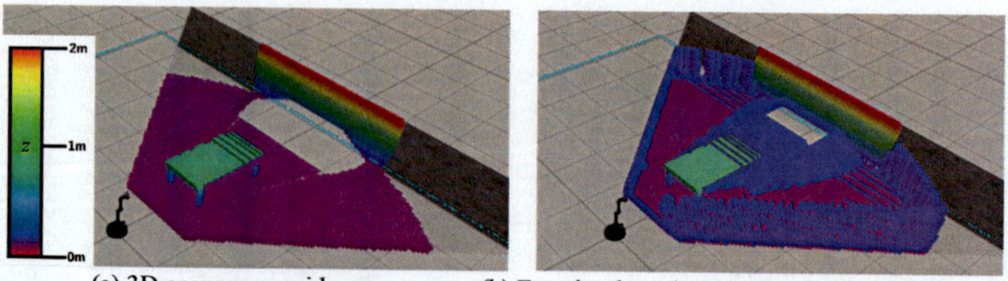

(a) 3D occupancy grid map (b) Frontier detection: the blue grids are frontiers

Fig. 4. 3D occupancy grid map built using Octomap and frontier detection

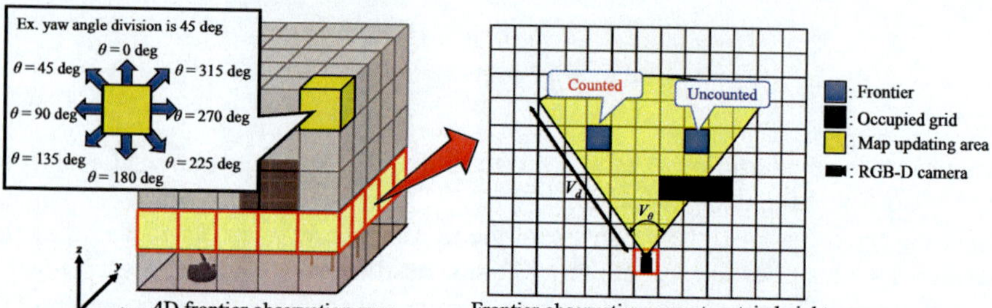

Fig. 5. Frontier observation map

5 Determination of Travel Destination and Arm Posture Control

5.1 Determination of Travel Destination

During the robot movement process, when the frontier observation map is updated, the destination is determined based on the number of observable frontier grids recorded in the frontier observation map. We follow the previous studies [4, 12] and consider the position with the largest number of observable frontier grids as the next destination. However, because the map is updated as the user moves, if the destination is based on the position and angle with the largest number of observable frontier grids,

The destination will significantly change every time the map is updated, resulting in a long travel route and time. For example, if the destination is set to the position and angle where the robot can observe a distant frontier, even though there is still a frontier near the robot, the robot will eventually have to travel back to the position and angle of the frontier that it fails to observe.

Therefore, we propose a method of determining the destination based on the current position and angle of the robot after modifying the value of the frontier observation map. Two modification methods are considered. First, to preferentially observe the remaining frontiers near the robot, the value of the observation map is modified with respect to the distance d from the robot, as follows:

$$m^{d}_{x,y,z,\theta} = \frac{m_{x,y,z,\theta}}{d_{x,y}} \tag{1}$$

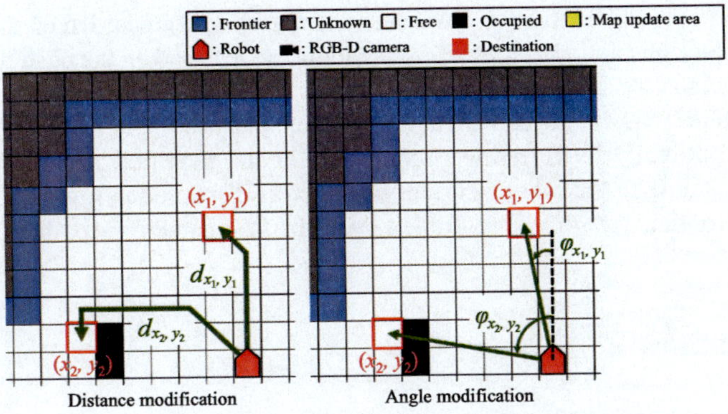

Fig. 6. Determination of moving destination

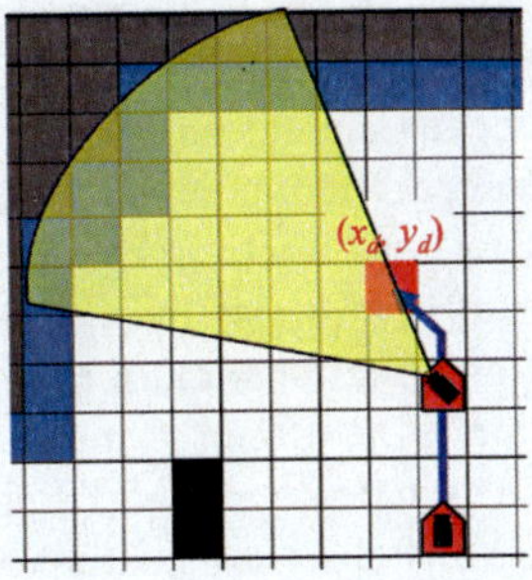

Fig. 7. Path planning and arm posture control

Where $m_{x,y,z,\theta}$ is the value of the frontier observation map at position (x, y, z) and camera yaw angle θ. As depicted in Fig. 6, the distance $d_{x,y}$ is calculated as the distance traveled to avoid obstacles. Second, to preferentially observe the direction in which the robot is moving, the value of the observation map is modified with respect to the yaw angle direction of the robot's mobile cart, as expressed in the following equation:

$$m^{\varphi}_{x,y,z,\theta} = m_{x,y,z,\theta}\left(\cos\varphi_{x,y} + 2\right) \tag{2}$$

where $\varphi_{x,y}$ represents the direction of the grid at position (x, y) from the front of the robot. For the modified frontier observation map, the location with the maximum frontier observation is determined as the destination of the movement.

However, the effect of this modification may vary depending on the environment in which the search is conducted. In this study, we experimentally verify the effects of four cases, namely, no modification, modification of distance, modification of angle, and modification of both distance and angle, in several environments by simulations.

5.2 Moving to the Destination and Controlling the Arm Posture

After determining the destination, the path from the current position to the destination is calculated using the A* algorithm [18], as illustrated in Fig. 7, and a dynamic window approach [19] is used to avoid the obstacles (occupancy grid) recorded on the map. The robot also avoids obstacles in areas where it cannot move based on the posture of the arm (minimum height constraint). During movement, if there are observable frontiers, the arm is controlled to point the camera at the position and yaw angle where it can observe the frontier optimally by referring to the value of the modified frontier observation map corresponding to the robot's current position, as displayed in Fig. 7. If there is no observable frontier, the arm is controlled such that the camera faces the front of the robot.

6 Verifications

6.1 Verification Scenarios

Three verifications of the proposed exploration method were carried out. The first was verifying the effect of actively changing the camera posture (yaw angle) while moving according to the frontier observation. We investigated the effect of pointing the camera in the direction that observes the largest number of frontiers during the proposed movement on the movement path and map generation time. The second was to verify the effect of modifying the distance and angle from the robot in determining the destination. We evaluated the effects of four cases, that is, no modification, modification of distance, modification of angle, and modification of both distance and angle, in several environments. Finally, we verified if the proposed method could generate a 3D map in environments where there are tables and other objects that make it impossible to observe the entire area without changing the height and posture of the camera. The verification was performed using ROS and the Gazebo [20] simulator. The robot used in this study is illustrated in Fig. 2. The two-wheeled mobile robot is Kobuki®[21], which is manufactured by YUJIN ROBOT and is equipped with a 5-DoF robot arm WidowX 250 Robot Arm® [22] manufactured by InterbotiX. An Intel ® RealSense™ D435 [23] is mounted as an RGB-D camera at the end of the arm. No noise is added to the depth information in the simulation. Table 1 presents the parameters of the map generation and robot motion. The z-direction of the frontier observation map is set to two levels, namely, low ($z = 0.3$) and high ($z = 0.7$), and the yaw direction is set to calculate the number of frontier observations every 45°.

6.2 Results and Discussion

A. Effect of actively changing the camera posture

The test environment illustrated in Fig. 8(a) was used. It is surrounded by walls 4 m high, 10 m wide, and 10 m long. Ten red square pillars 1 m high, 0.5 m long, and 0.5 m wide are randomly placed. We compared the method in which the yaw angle of the camera is always fixed to the direction in front of the robot with the proposed method of pointing the arm in the direction of the maximum observable frontier.

Figure 8(b) and (c) show the results of the path traveled by the robot (red line) for each method. As indicated in Table 2, the path length and map generation time are

Table. 1 Parameters of map generation and robot motion

Parameter	Value
Grid size of occupancy grid map	0.1 m
Grid size of frontier observation map (*x*- and *y*-direction)	0.2 m
Steps of *z*-direction of frontier observation map	2 (0.3 m and 0.7 m)
Decomposition angle in the θ direction of frontier observation map	8 (45° each)
Distance range of RBG-D camera used for map update: V_r	5.0 m
Horizontal viewing angle of RGB-D camera used for map update: V_θ	85.2°
Maximum speed of robot	0.1 m/s
Maximum angular velocity of robot	0.8 rad/s

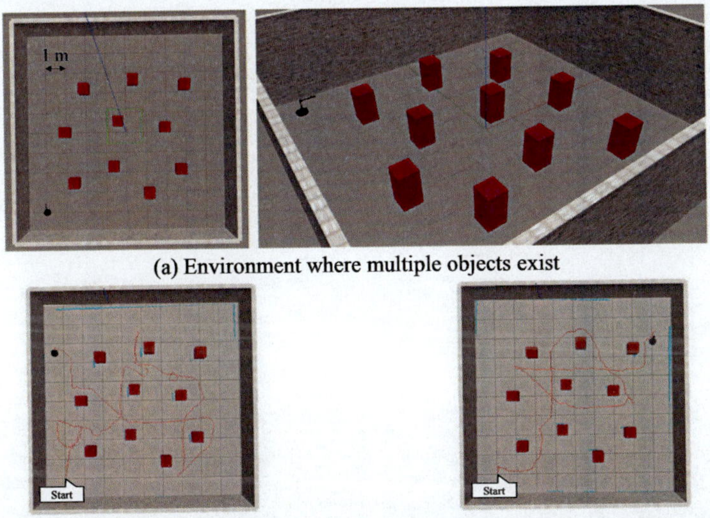

(a) Environment where multiple objects exist

(b) Result of travel path with method where the camera yaw angle is fixed

(c) Result of travel path with method where the camera yaw angle is variable (proposed)

Fig. 8. Traveled path results of verification A

shorter when moving with the camera pointed in the direction where the frontier can be observed optimally than when the camera is fixed. As exhibited in Fig. 8(b), there are many back-and-forth movements in the environment when the camera direction is fixed. It is believed that fixing the camera causes the robot to overlook the nearby frontier, and consequently, the robot often returns to the place where it previously passed. In contrast, actively changing the camera pose in response to the observed number of frontiers

Table. 2 Results of verification A

Method	Travel path length [m]	Travel time [s]
Yaw angle of camera is fixed	42.59	527.84
Yaw angle of camera is variable (proposed)	31.52	455.95

reduces the number of missed frontiers and unnecessary round-trip movements, thereby allowing for more efficient exploration motion.

B. Effect of modifying the value of frontier observation for determining the destination
The effects of distance and angle modifications were examined in the environments shown in Fig. 8(a), where multiple objects are randomly located, and in the corridor like environment shown in Fig. 10(a). Due to the noise given to the simulator, the robot does not move in the same way. The results of traveled path are different for each execution. In this verification, three trials were conducted for each modification method and mean and SD of the travel distance and time were obtained.

Table. 3 Results of verification B in multiple objects environment

Method	Distance modification	Angle modification	Travel path length mean ± SD [m]	Travel time mean ± SD [s]
1	-	-	39.51 ± 1.36	453.68 ± 34.43
2	Yes	-	24.74 ± 2.51	345.58 ± 71.45
3	-	Yes	23.08 ± 3.87	265.71 ± 33.04
4	Yes	Yes	32.10 ± 2.62	427.23 ± 43.75

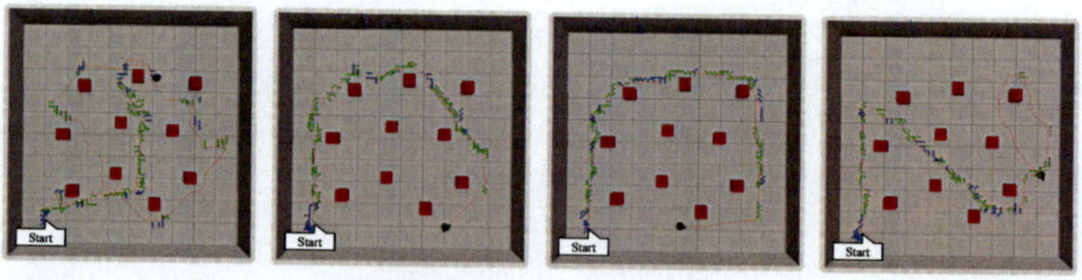

(a) No modification (b) Distance modification (c) Angle modification (d) Distance and angle modification

Fig. 9. Traveled path results of verification B in multiple objects environment

Table 3 shows the results of the length and time of the path traveled until the map is generated in Fig. 8(a) environment. Figure 9 shows the one example of results of traveled path of the robot (red line) and target camera pose (blue arrows: $z = 0.7$, green arrows: $z = 0.3$) in each method. As shown in Table 3, two methods with distance or angle modifications can significantly reduce the travel path length and time compared to the method without modifications that the maximum frontier observation position is set to the destination in the environment where multiple objects were randomly located. As shown in Fig. 9(b) and (c), it was confirmed that each method searched with a short path without overlap of paths. On the other hand, the method with both modifications also reduced the path length and time compared to the method without modifications, but the results were not as good as the method with only distance or angle modifications.

Table 4 shows the results of the length and time of the path traveled until the map is generated in Fig. 10(a) environment. Figure 10(b) to (e) show the one example of results of traveled path of the robot (red line) target camera pose (blue arrows: $z = 0.7$, green arrows: $z = 0.3$) in each method. As shown in Fig. 10 and Table 4, it was confirmed that the method of angle modification that prioritizes the frontier of the moving direction had the shortest path length and map generation time. On the other hand, the method with distance modification resulted in a longer travel path and time than the method without modifications. As shown in Fig. 10(b), in environments where there are fork in the corridor and large sets of frontiers occur in different directions, the destination switches while moving to a certain set of frontiers. This is expected to result in a round trip along the same path, increasing the path length and time. The method with both distance and angle modifications resulted in a slightly shorter travel path and time than the method without modifications, but not a significant improvement.

Table. 4 Results of verification B in corridor-like environment

Method	Distance modification	Angle modification	Travel path length mean ± SD [m]	Travel time mean ± SD [s]
1	-	-	27.66 ± 2.14	350.23 ± 43.64
2	Yes	-	33.34 ± 1.15	424.62 ± 68.24
3	-	Yes	22.47 ± 1.97	281.25 ± 46.50
4	Yes	Yes	25.97 ± 1.37	337.34 ± 29.16

These results indicated that the method of angle modification, which prioritizes the frontier in the moving direction of the robot to determine the destination, resulted in the shortest path and time for exploration in both multiple objects located environment and corridor-like environment. If the destination is determined by giving priority to the frontier in the direction the robot is moving with angle modification, the possibility of overlooking the frontier around the robot increases. However, by actively changing the camera pose to frontier observable direction, the possibility of missing the frontier around the robot can be reduced. That was the reason why the robot was able to explore efficiently without back-and-forth movements. On the other hand, when both distance and angle modifications were applied at the same time, it was confirmed that the features of both methods canceled out and effective search could not be achieved.

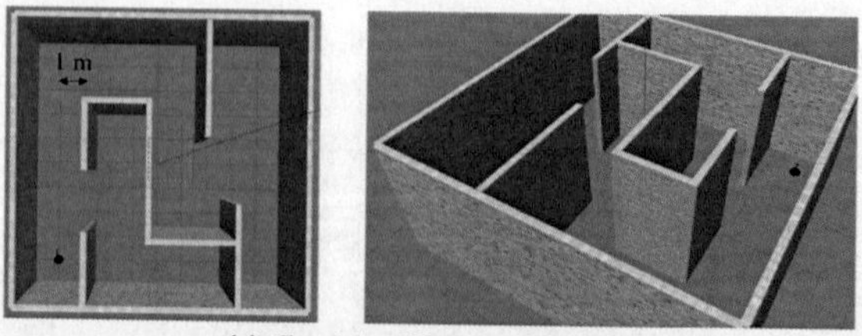

(a) Corridor-like environment

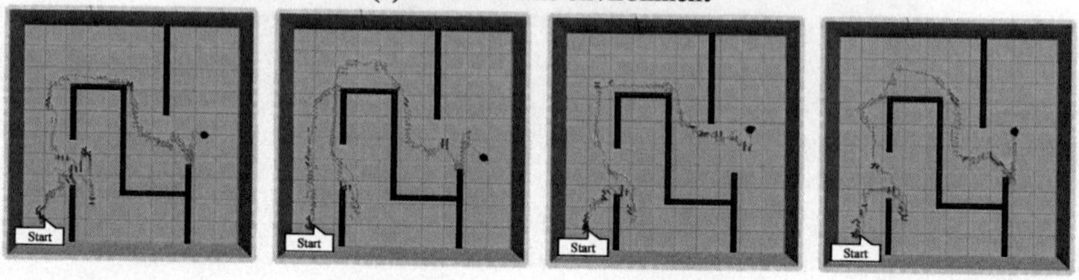

(b) No modification (c) Distance modification (d) Angle modification (e) Distance and angle modification

Fig. 10. Traveled path results of verification B in corridor-like environment

C. 3D map generation using the proposed exploration

The effect of proposed exploration method was examined in the environment shown in Fig. 11 (a), where four tables were located, and it was impossible to observe the top and bottom of the tables without changing the camera height. The angle modification method to determine the destination was used.

Figure 11(b) shows a scene where the robot was moving near the table. On the left side of Fig. 11(b), the camera posture was raised to observe the upper side of the table. Then, as shown in the right side of Fig. 11(b), the motion was planned to move under the table by lowering the posture of the arm to observe the frontiers on the wall. It was confirmed that the proposed method can generate a 3D environment map as shown in Fig. 11(c) by searching while changing the height of the camera viewpoint.

7 Conclusion

In this study, we proposed an unknown environment exploration method for 3D occupancy grid map generation using a mobile manipulator robot with an RGB-D camera. In the proposed method, the boundary between known and unknown regions is detected as a frontier, and a frontier observation map is created to record the number of frontiers that can be observed at a certain position and posture. When the frontier observation map is updated, the moving destination is determined and then the robot moves to the destination while changing the position and posture of the camera based on the observable frontier. However, when the destination is simply the position where most frontiers can be observed, the robot may move back and forth to the same location when some frontiers are missed. Therefore, we proposed a method of modifying the frontier observation in terms of the distance and angle to preferentially observe frontiers near the robot and in the direction of the robot movement, respectively. Through simulations, it was confirmed that by actively changing the camera pose according to the frontier observation map, the traveled path length and mapping time can be reduced. Furthermore, the method of angle modification, which prioritizes the frontier in the moving direction of the robot to determine the destination, resulted in the shortest path for exploration. In addition, by changing the position and posture of the camera based on the frontier observation map, the robot can perform exploration motions, such as observing above and below the table, which cannot be accomplished when the camera height is fixed, or passing under the table by lowering the posture of the arm.

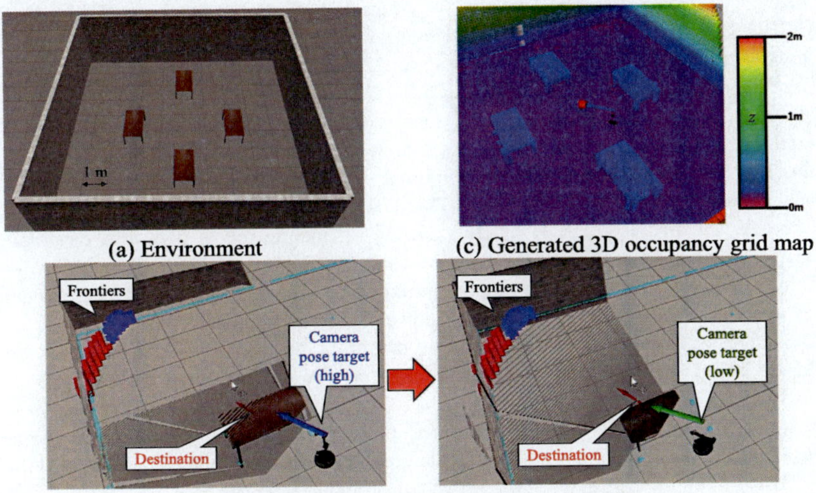

(a) Environment (c) Generated 3D occupancy grid map

(b) A scene when the robot moved near the table

Fig. 11. Verification C: 3D map generation by changing camera pose

Acknowledgement. This work was supported by JSPS KAKENHI Grant Number 21K14102.

References

1. Zachiotis, G.A., Andrikopoulos, G., Gornez, R., Nakamura, K., Nikolakopoulos, G.: A survey on the application trends of home service robotics. In: 2018 IEEE International Conference Robotics Biomimetics, pp. 1999–2006 (2018)
2. Yamamoto, T., Terada, K., Ochiai, A., Saito, F., Asahara, Y., Murase, K.: Development of Human Support Robot as the research platform of a domestic mobile manipulator. Robomech J. **6**(1), 1–15 (2019). https://doi.org/10.1186/s40648-019-0132-3
3. Stachniss, C.: Robotic mapping and exploration, Springer (2009)
4. Yamauchi, B.: A frontier-based approach for autonomous exploration, in: Proc. 1997 IEEE International Symposium on Computational Intelligence in Robotics and Automation CIRA'97. 'Towards New Computational Principle for Robotics Automation', pp. 146–151 (1997)

5. Mobarhani, A., Nazari, S., Tamjidi, A.H., Taghirad, H.D.: Histogram based frontier exploration. In: 2011 IEEE/RSJ International Conference on Intelligent Robots Systems, pp. 1128–1133 (2011)
6. Umari, H., Mukhopadhyay, S.: Autonomous robotic exploration based on multiple rapidly-exploring randomized trees. In: 2017 IEEE/RSJ International Conference Intelligent Robots Systems, pp. 1396–1402. IEEE (2017)
7. Stachniss, C., Grisetti, G., Burgard, W.: Information gain-based exploration using rao-blackwellized particle filters. In: Robotics Science Systems, pp. 65–72 (2005)
8. Ila, V., Porta, J.M., Andrade-Cetto, J.: Information-based compact pose SLAM. IEEE Trans. Robot. **26**, 78–93 (2010)
9. Valencia, R., Miró, J.V., Dissanayake, G., Andrade-Cetto, J.: Active pose SLAM. In: 2012 IEEE/RSJ International Conference Intelligent Robotics Systems, pp. 1885–1891 (2012)
10. Vallvé, J., Andrade-Cetto, J.: Active pose SLAM with RRT*. In: 2015 IEEE International Conference Robotics Automation, pp. 2167–2173 (2015)
11. Shade, R., Newman, P.: Choosing where to go: complete 3D exploration with stereo. In: 2011 IEEE International Conference Robotics Automation, pp. 2806–2811 (2011)
12. Lu, L., Redondo, C., Campoy, P.: Optimal frontier-based autonomous exploration in unconstructed environment using RGB-D sensor. Sensors (Switzerland). **20**, 1–16 (2020)
13. Chen, Y., Huang, S., Fitch, R.: Active SLAM for mobile robots with area coverage and obstacle avoidance. IEEE/ASME Trans. Mechatron. **25**, 1182–1192 (2020)
14. Ding, J., Fang, Y.: Multi-strategy based exploration for 3D mapping in unknown environments using a mobile robot. In: 2019 Chinese Control Conference, pp. 4732–4738 (2019)
15. Dornhege, C., Kleiner, A.: A frontier-void-based approach for autonomous exploration in 3d. In: 2011 IEEE International Symposium Safety Security Rescue Robotics, pp. 351–356 (2011)
16. Ramanagopal, M.S., Nguyen, A.P., Le Ny, J.: A motion planning strategy for the active vision-based mapping of ground-level structures. IEEE Trans. Autom. Sci. Eng. **15**, 356–368 (2018)
17. Hornung, A., Wurm, K.M., Bennewitz, M., Stachniss, C., Burgard, W.: OctoMap: an efficient probabilistic 3D mapping framework based on octrees. Auton. Robots. **34**, 189–206 (2013)
18. Hart, P.E., Nilsson, N.J., Raphael, B.: A formal basis for the heuristic determination of minimum cost paths. IEEE Trans. Syst. Sci. Cybern. **4**, 100–107 (1968)
19. Fox, D., Burgard, W., Thrun, S.: The dynamic window approach to collision avoidance. IEEE Robot. Autom. Mag. **4**, 23–33 (1997)
20. Gazebo. http://gazebosim.org/. Access 2021
21. YUJIN ROBOT. http://kobuki.yujinrobot.com/. Access 2021
22. Interbotix. https://www.trossenrobotics.com/m/interbotix. Access 2021
23. Intel RealSense Depth Camera D435. https://www.intel.com/content/www/us/en/products/sku/128255/intel-realsense-depth-camera-d435/specifications.html. Access 2021

Reaching Motion Planning with Vision-Based Deep Neural Networks for Dual Arm Robots

Satoshi Hoshino(✉) and Ryota Oikawa

Department of Mechanical and Intelligent Engineering, Utsunomiya University, 7-1-2 Yoto, Utsunomiya, Tochigi 321-8585, Japan
hoshino@cc.utsunomiya-u.ac.jp

Abstract. Dual arm robots have been attracting attention from the view point of factory automation. These robots are basically required to reach their hands toward the respective target objects, simultaneously. Therefore, we focus on motion planning with vision-based deep neural networks. Given an RGB-D camera mounted on a robot, object images are fed as the inputs to the reaching motion planner based on convolutional neural network, CNN. For multiple objects, the depth of each object in the image is useful information to determine a reaching target. If the objects are close to each other, however, the depth becomes similar. For this challenge, we propose to generate the target object image through instance segmentation and order classifier. In the experiment with multiple objects, we show that the robot is able to reach both the hands toward the target objects by using the target object images.

Keywords: Dual arm robot · Reaching · Motion planning · Deep neural networks

1 Introduction

Dual arm robots have been attracting attention from the view point of factory automation. These robots are basically required to reach their hands toward the respective target objects, simultaneously. Reaching motion planning is to find a series of collision-free configurations to grasp an object. The final configuration is calculated from the object position measured using sensors. In this paper, deep neural networks are used to solve these planning issues comprehensively. Given an RGB-D camera sensor mounted on top of a robot as shown in Fig. 1, object images are used as the inputs to the networks.

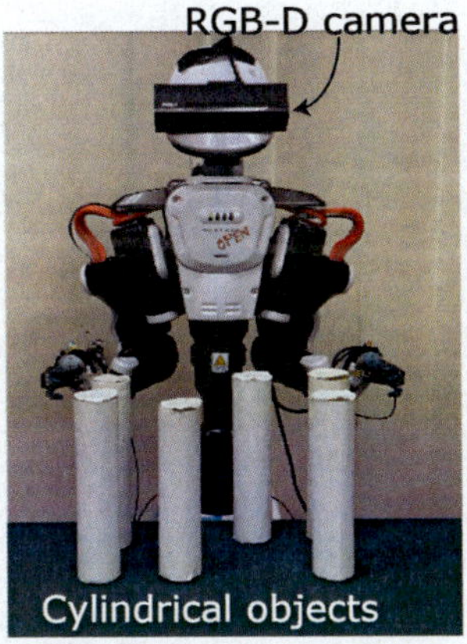

Fig. 1. Dual Arm Robot Equipped with RGB-D Camera

In advance, deep neural networks are trained to be motion planners. View-based teaching and playback between a supervisor and robot is an

I. Petrovic et al. (Eds.): IAS 2022, LNNS 577, pp. 455–469, 2023.
https://doi.org/10.1007/978-3-031-22216-0_31

effective way of network training [1,2]. This is called imitation learning [3]. We have introduced an object recognition process into the simple motion teaching and playback framework without network training [4]. However, the robots used only one hand. For dual arm robots, motion planners with multimodal inputs, such as raw RGB images, sound spectrum, and joint angles, have been proposed [5,6]. However, multimodal inputs increase the training data set and therefore the difficulty of network training.

On the other hand, our motion planner is based on convolutional neural network, CNN [7]. Instead of the multimodal inputs acquired using various sensors, only RGB images are the inputs. For network training, first off, a reaching motion toward a target object is instructed by a supervisor. During the motion instruction, the robot registers the images and motions as the inputs and outputs. After the instruction phase, network parameters of the CNN are optimized in the learning phase so as to predict suitable motion outputs for the image inputs. The detailed descriptions of imitation learning are given in [8].

Imitation learning is highly dependent on the training data set. In addition, the instruction cost increases in response to the number of objects. Therefore, we use a single object in the motion instruction phase. However, since the robot is instructed to reach each of the hands toward the single object, it is difficult to predict a suitable motion for an image with multiple objects. For this image, the robot is required to determine a sole reaching target and ignore the other objects. The image with the target object is called a target object image.

For various objects in clutter, it is possible to determine a target through object recognition [9]. For the same objects as shown in Fig. 1, the depth of each object is useful information. 3D point clouds obtained with an RGB-D camera have been used for detecting object grasping points [10]. If the objects are close to each other, however, the depth becomes similar. In this case, it is difficult to extract the point cloud of each object through distance-based clustering [11]. In [12], two RGB-D cameras were used to have point clouds from different perspectives. For this challenge inherent in using a single camera, we propose to generate the target object image through instance segmentation [13] and order classifier. Once the target(s) is/are determined, the robot plans the reaching motion.

In Sect. 2, we describe the approach to generate the target object image without using depth information. In this approach, Mask R-CNN is used for instance segmentation (Sect. 3), and CNN is used for the reaching order classifier (Sect. 4). In Sect. 5, the target object images are used as the inputs to the grasping classifier and motion planner based on CNNs. Through the experiment in Sect. 6, we show that the robot is able to reach both the hands toward the respective targets by using the target object images. In Sect. 7, we finally conclude this paper.

2 Approach to Generate Target Object Images

The robot, as shown in Fig. 1, is allowed to have RGB-D images of objects. In addition to RGB information about the objects, the depth is also useful information to determine a reaching target. Fig. 2 illustrates a target object image generated from RGB and depth images.

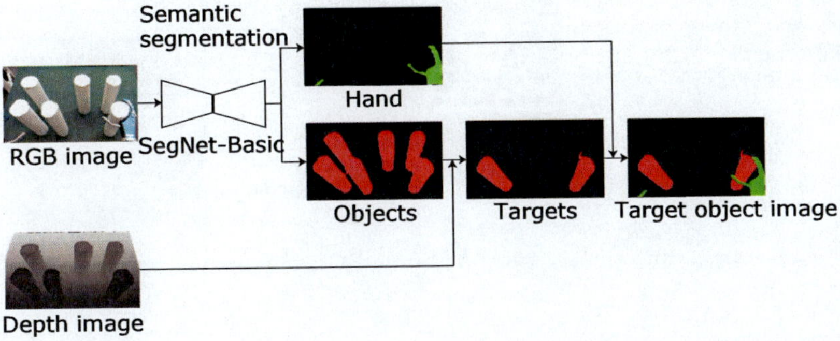

Fig. 2. Target object image generated from RGB-D information

From the RGB image, the robot hands (green) and objects (red) are segmented through semantic segmentation [14]. In this paper, SegNet-Basic [15] is applied for semantic segmentation. The segmented objects are separated from each other through connected-component labeling [16], and merged with the depth information. As a result, the objects closest to the right and left hands are determined as the targets, while the other objects are colored by black. Finally, the target objects and robot hands are merged in the target object image.

If the objects are close to each other, however, the depth of the cluster[1] becomes similar. In this case, it is difficult to determine a sole target from the cluster with the similar depth. For this challenge, we propose to generate the target object image through instance segmentation and order classifier, as illustrated in Fig. 3.

From the RGB image, each of the objects is separately segmented through instance segmentation based on Mask R-CNN. Compared to the object segmentation image in Fig. 2, a single object is segmented in each image. The segmentation image is further classified into the reaching order with the right or left hand label, R_i or L_j. According to the index, i and j, the object with the lowest number is determined as the target. Finally, the objects in the corresponding segmentation images are merged together. This output is the target object image generated without using depth information.

3 Instance Segmentation Based on Mask R-CNN [13]

Unlike semantic segmentation, even the same cylindrical objects are separately segmented through instance segmentation. Figure 4 illustrates the architecture

[1] Since these objects are given the same label and not separated through connected-component labeling, a cluster is formed by the objects in the image.

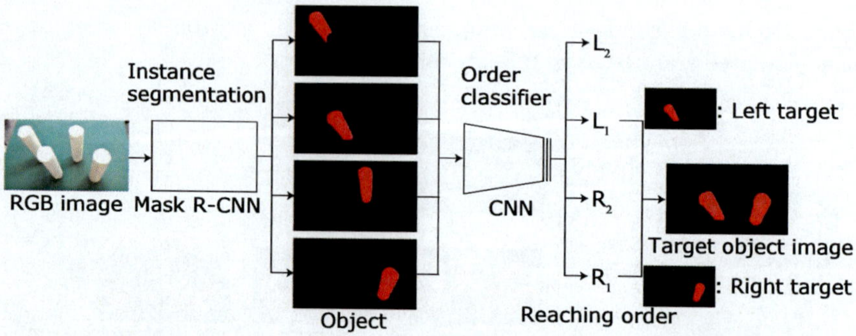

Fig. 3. Target object image generated from RGB information without depth

of Mask R-CNN used for instance segmentation. As can be seen from the output images, the architecture has two branches for object detection and segmentation.

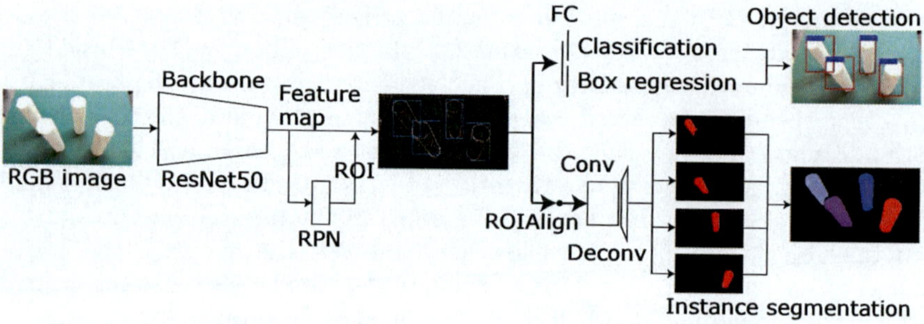

Fig. 4. Architecture of mask R-CNN for instance segmentation

In the convolutional backbone, image features are extracted. The backbone is based on ResNet50 [17]. The extracted feature map is further fed as the input to the region proposal network, RPN, composed of convolution layers. From the RPN, the region of interest, ROI, for each object is detected. Hence, the image with the feature map and ROI with a rectangular shape is obtained. In the upper branch composed of the fully-connected layers, the object class in each ROI and the bounding box are displayed in the output image. In the lower branch, the feature map is partially clipped by each ROI and changed to a fixed size. The image feature is then decoded through the convolution and deconvolution layers. Finally, a single object is segmented as a result of instance segmentation. The objects separately segmented from the others are indicated by different colors.

4 Reaching Order Classifier Based on CNN

4.1 Architecture of CNN

For the segmentation images in the lower branch of Fig. 4, the objects closest to the right and left hands are determined as the targets. In this regard, however, RGB is the only information available to generate the target object image. Therefore, we utilize the appearance feature based on location and size of objects in the segmentation image. For each object, the segmentation image is classified into the reaching order including the right or left hand label. This is a function of the order classifier. Figure 5 illustrates the architecture of CNN used as the order classifier.

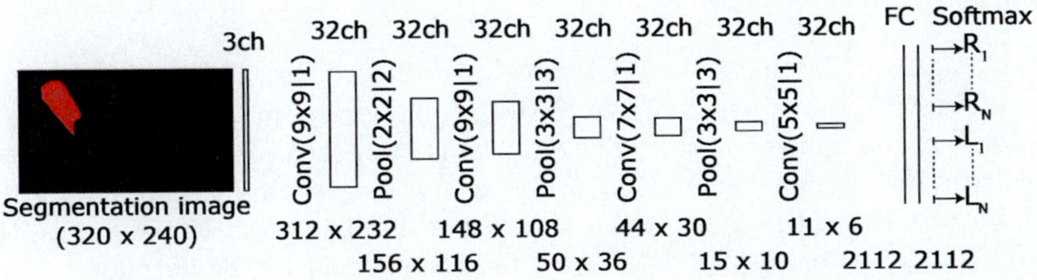

Fig. 5. Architecture of reaching order classifier based on CNN

For a segmentation image, the appearance feature is extracted through the convolution and pooling layers. In each layer, the kernel size, $n \times m$, and stride, l, are represented as $(n \times m \mid l)$. From the extracted image feature, normalized probabilities for the order outputs are calculated through fully-connected layers and softmax function. Finally, R_i or L_j with the highest probability is the classified reaching order for the object in the segmentation image. Excepting the output layer, ReLU is used in each of the convolution and fully-connected layers.

All the segmentation images are classified into the reaching order with the right or left hand label, R_i or L_j. As for an object classified into R_i, the robot uses the right hand. According to the index, i, the object with the lowest number is determined as the target for the right hand. The target object for the left hand, L_j, is also determined in the same manner.

4.2 Network Training

For network training of the order classifier as illustrated in Fig. 5, a data set is acquired from motion instructions provided by a supervisor. In the motion instruction phase, a single object is respectively placed on a workbench. During the instruction for each hand, the robot registers the RGB images. Given a trained Mask R-CNN in Fig. 3, both the hand and object in the RGB image are segmented through instance segmentation. This segmentation image is used as the input in the training data set. For each hand, the object is placed in N positions. Hence, the segmentation image input with the object is annotated with the reaching order class, 1 to N, as the output depending on the distance from the hand. Figure 6 illustrates an example of the training data set for the reaching order classifier. The object is placed in the closest positions.

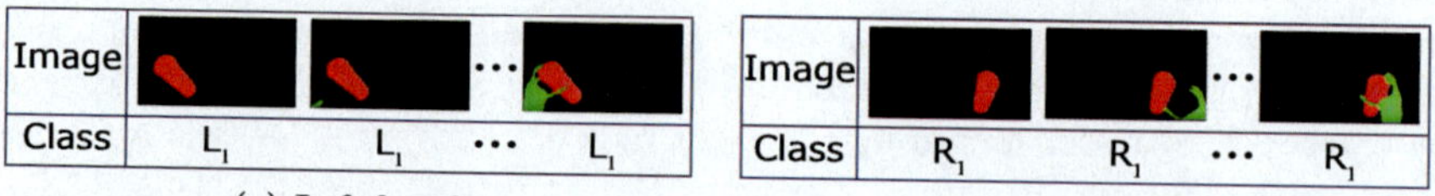

(a) Left hand (b) Right hand

Fig. 6. Data set with objects closest to right and left hands for reaching order classifier based on CNN

In Fig. 6(a), the object is placed in the left-hand side. In Fig. 6(b), the object is placed in the right-hand side. The data set is composed of inputs and outputs, which are the segmentation images and reaching order including the right or left hand label. Since the object is placed in the positions closest to the hands, the image inputs are annotated with R_1 or L_1, respectively, as the outputs.

In the learning phase, network parameters of the CNN are optimized so as to classify the image inputs into the correct reaching order including the hand label. For optimization, the loss function is defined by the cross-entropy, and Adam [18] is used to update the network parameters.

5 Reaching Motion Planning for Dual Arm Robot

5.1 Overview of Reaching Motion Control

The robot based on the motion planner is enabled to determine the amount of movement of both the hands for the image input in each step. Furthermore, the robot grasps the target objects after a series of reaching motions. For this purpose, a grasping classifier is developed in addition to the motion planner. Figure 7 illustrates the overview of the control procedure.

From the RGB image with multiple objects, the target object image is generated as described in the previous sections. Before motion planning, the target object image is fed as the input to the grasping classifier. The robot grasps the objects when the classified hand status is close. On the other hand, the target object image is further fed as the input to the reaching motion planner as long as the classified hand status is open. From the motion planner, the amount of movement of both the hands, $(\Delta x_R, \Delta y_R, \Delta z_R)$ and $(\Delta x_L, \Delta y_L, \Delta z_L)$, toward the target objects are derived as the outputs. According to Δs, the hands' positions in the next step are given. The robot then moves the hands to the positions.

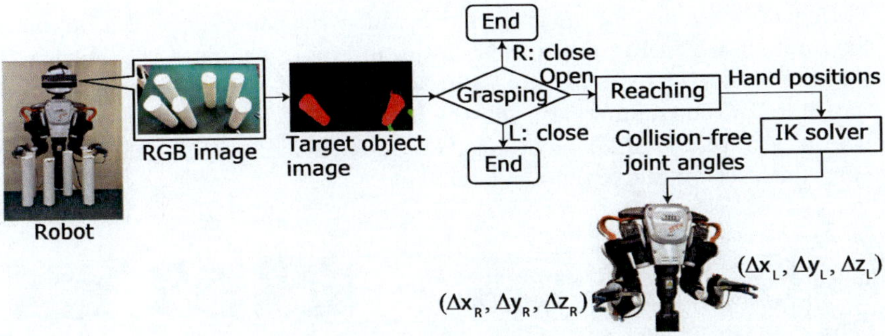

Fig. 7. Overview of control procedure for reaching motion

For this purpose, joint angles are calculated by solving the inverse kinematics, IK, problem including self-collision avoidance. This control ends when both the hands close.

5.2 Grasping Classifier Based on CNN

In each control step, the hand status, open or close, for the image input acts as a trigger of the following motion planning. Once the hand closes and grasps the target, the robot is no longer allowed to control the hand. Figure 8 illustrates the architecture of CNN used as the grasping classifier.

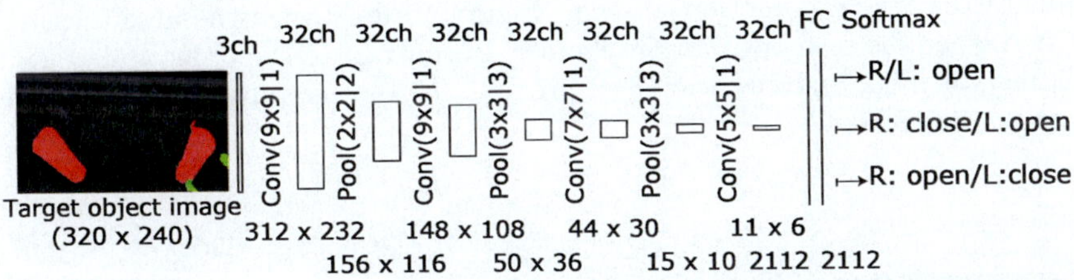

Fig. 8. Architecture of grasping classifier based on CNN

The target object image is used as the input to the grasping classifier. The image feature is extracted through the following convolution and pooling layers. After that, the flatten feature is fed as the input to the fully-connected layers. The output layer is composed of three status. Note that the robot is not allowed to open the right (left) hand again after grasping the target object even if the hand status is R: open/L: close (R: close/L: open) in the next step. For normalized probabilities through the softmax function, the hand status with the highest probability is the classification result.

For network training of the grasping classifier, the same segmentation images as the ones shown in Fig. 6 are used as the inputs in the data set. During the instruction for each hand, the robot registers the hand status, i.e. open or close, in each step. In the training data set, the hand status class is used as the output for the object placed in N positions. Figure 9 illustrates an example of the training data set for the grasping classifier.

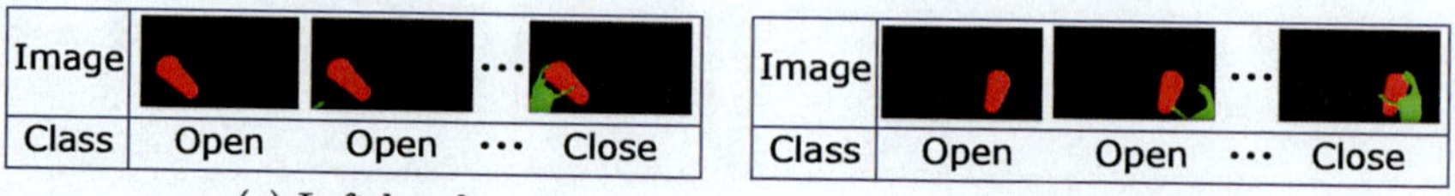

(a) Left hand (b) Right hand

Fig. 9. Data set with objects closest to right and left hands for grasping classifier based on CNN

For each object, the image inputs are annotated with the left hand status in Fig. 9(a) and right hand status in Fig. 9(b), as the outputs. In the learning phase, network parameters of the CNN are optimized so as to classify the image inputs into the correct hand status. For optimization, the loss function is defined by the cross-entropy, and Adam is used to update the network parameters.

5.3 Reaching Motion Planner Based on CNN

If the hand status is open in Fig. 8, the target object image is further fed as the input to the reaching motion planner. Figure 10 illustrates the architecture of CNN used as the reaching motion planner. In contrast to the order and grasping classifiers in Figs. 5 and 8, the CNN predicts the motion outputs for the image inputs.

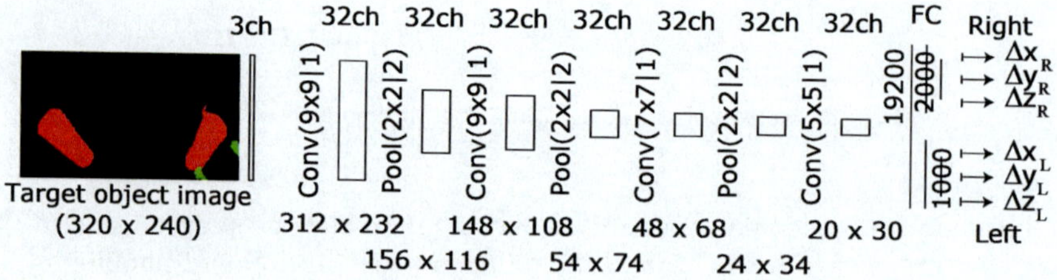

Fig. 10. Architecture of reaching motion planner based on CNN

The architecture has two branches in the fully-connected layers to the right and left hands independently. In doing so, the robot is enabled to control the right and left hands simultaneously for the same image input with the two target objects. Each of the outputs from the two branches indicates the amount of movement of the hand, Δ, in the x, y, and z directions. For your information, the robot failed in planning the reaching motions when the CNN without the branches in the fully-connected layers[2] was used.

For network training, a supervisor instructs the robot to reach the right or left hand toward a single object placed in N positions, respectively. During the motion instruction, the robot registers the RGB images and hand positions. Figure 11 illustrates an example of the training data set for the motion planner. The object is placed in the closest positions.

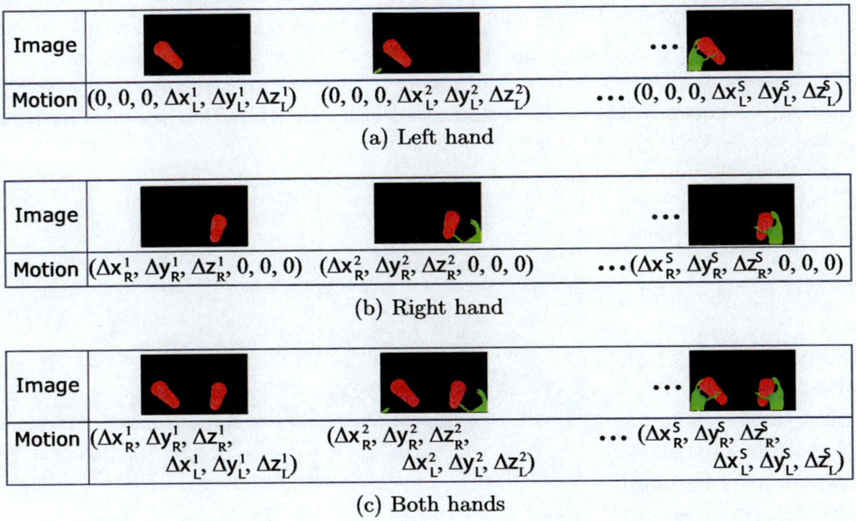

Fig. 11. Data set with objects closest to right and left hands for motion planner

As well as the data sets in Figs. 6 and 9, the segmentation images are used as the inputs in the training data sets of the left hand (Fig. 11(a)) and right hand (Fig. 11(b)). From a series of the hand positions, the amount of changes between the two consecutive positions are divided into X, Y, and Z axes, and are used as the outputs, $(\Delta x_R, \Delta y_R, \Delta z_R, \Delta x_L, \Delta y_L, \Delta z_L)$. Furthermore, the data sets of the right and left hands are merged together as illustrated in Fig. 11(c). This is the training data set in consideration of both the hands.

[2] In the CNN architecture without the branches, the amount of movement of both the hands, Δs, were derived from the six units in the output through the same fully-connected layers composed of 19200, 4000, and 1000 units.

In the learning phase, network parameters of the CNN are optimized so as to predict the correct motion outputs for the image inputs. For optimization, Adam is used to update the network parameters. The loss function is defined by the mean squared error, MSE, comparing the motion outputs from the CNN and the instructed motion outputs in the data set.

6 Reaching Experiments for Multiple Objects

6.1 Settings

In this experiment, reaching tasks are imposed on a dual arm robot. For multiple objects placed on a workbench, the robot reaches the right and left hands toward the respective targets, simultaneously. In Fig. 7, MoveIt [19] is used as the IK solver. Figure 12 shows the environment. The dual arm robot, Nextage [20], produced by Kawada Robotics is equipped with an RGB-D camera. The robot has the hands with three fingers.

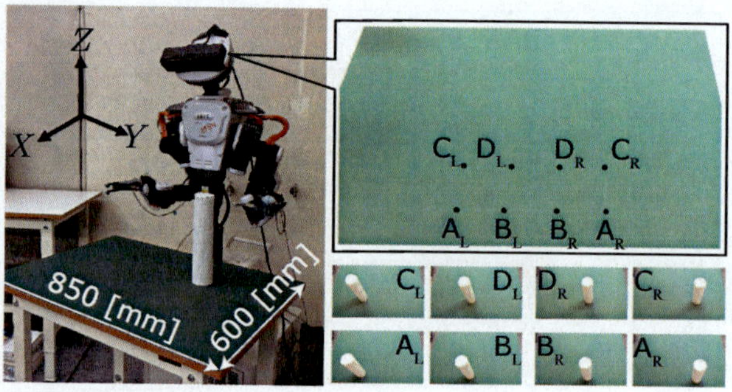

Fig. 12. Experimental Environment

On the workbench, a cylinder with a diameter of 70 mm is placed as an object. In consideration of the motion range, the object is placed in four positions marked •, i.e. $N = 4$, for each hand in the motion instruction phase. $\mathrm{A_R}$ to $\mathrm{D_R}$ indicate the positions for the right hand, and $\mathrm{A_L}$ to $\mathrm{D_L}$ indicate the positions for the left hand. The robot was instructed to reach each of the hands toward the object one by one. These motion instructions prevent the robot from crossing the hands. For each object, the robot registered 69 images; thus $S = 69$ in Fig. 11.

The training data sets for the order and grasping classifiers are composed of $69 \times 4 \times 2 = 552$ segmentation image inputs and annotated outputs. In the data set for the order classifier, the reaching order including the right or left hand label was annotated depending the distance. The hand status was opened during the former steps, whereas it was closed during the latter steps. In the data set for the grasping classifier, therefore, the hand status was annotated according to the step. Table 1 shows these annotated outputs. The training data set for the motion planner is composed of 1656 image inputs and motion outputs in consideration of the additional combination of both the hands, $69 \times 4 \times 4 = 1104$, as illustrated in Fig. 11(c).

For comparison, two target object images are generated from RGB-D and RGB information as illustrated in Figs. 2 and 3. All the deep neural networks, i.e. SegNet-Basic, Mask R-CNN, and three CNNs, are trained in the learning

Table 1. Annotated outputs in data sets for order and grasping classifiers

	Object	Left-hand side				Right-hand side			
	Class	A_L	B_L	C_L	D_L	A_R	B_R	C_R	D_R
Order w/ hand label: R or L		L_1	L_2	L_3	L_4	R_1	R_2	R_3	R_4
Hand status	Open (steps)	26	44	29	45	26	44	29	45
	Close (steps)	43	25	40	24	43	25	40	24

phase using the data sets. As the reaching task, cylindrical objects are placed as shown in Fig. 13. All the objects are close to each other. The objects, represented as R and L, are placed in the right-hand side and left-hand side, respectively.

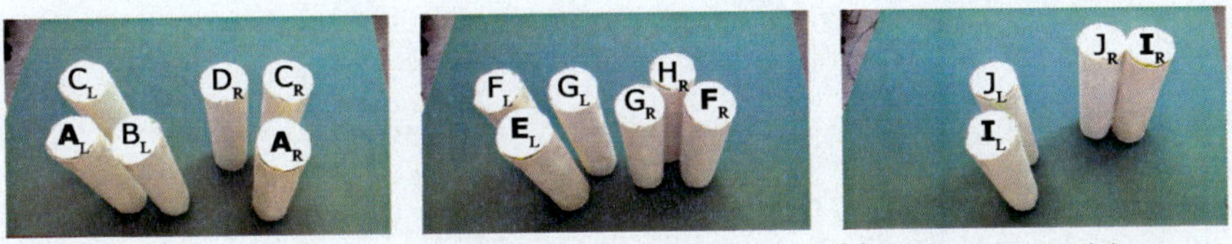

(a) Known positions (b) Unknown positions inside instruction area (c) Unknown positions outside instruction area

Fig. 13. Reaching tasks for multiple objects close to each other

In Fig. 13(a), six objects are placed in the positions same as the object in the motion instruction phase, as shown in Fig. 12. Therefore, these positions are known to the robot. In Fig. 13(b) and (c), on the other hand, six and four objects are placed in the positions inside and outside the instruction area unknown to the robot.

6.2 Results

In Fig. 13, the objects with the bold letters are the first targets. For the different target object image inputs, the robot based on the same grasping classifier and motion planner reached the hands toward the respective targets, simultaneously. Figure 14 shows the reaching trajectories of the robot hands when the two target object images, TOIs, generated from RGB-D (TOIs-RGBD) and RGB (TOIs-RGB) were used as the inputs. S represents the start position.

In Fig. 14(a), while the right hand reached the target, A_R, the left hand stopped short of the target, A_L. The robot stopped reaching the left hand because the hand status for the image input was classified into close. In Fig. 14(b), the right hand also stopped in the middle of reaching for the same reason. Moreover, the hand did not move toward the target, F_R. On the other hand, the left hand collided with the target, E_L. In contrast to these results of TOIs-RGBD, the

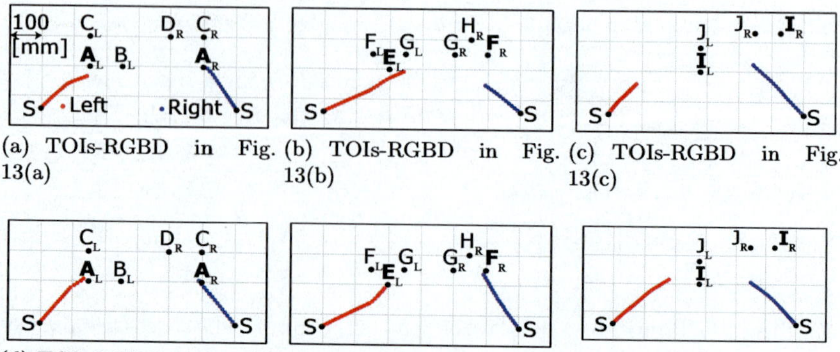

(a) TOIs-RGBD in Fig. 13(a) (b) TOIs-RGBD in Fig. 13(b) (c) TOIs-RGBD in Fig. 13(c)

(d) TOIs-RGB in Fig. 13(a) (e) TOIs-RGB in Fig. 13(b) (f) TOIs-RGB in Fig. 13(c)

Fig. 14. Comparison of reaching trajectories depending on target object images, TOIs-RGBD and TOIs-RGB, in X-Y plane

robot successfully reached the hands toward the targets by using the TOIs-RGB as can be seen in Fig. 14(d) and (e). The result of Fig. 14(e) indicates the generalization capabilities of the grasping classifier and motion planner enough for the task. In Fig. 14(c) and (f), however, the robot stopped both the hands in the middle of reaching regardless of the target object images.

In order to discuss the reaching results, we focus on the two target object images shown in Fig. 15. As for the TOIs-RGBD, the depth was converted into a grayscale image. Each pixel in the grayscale image has a depth value of $[0, 255]$. Thus in the target object images, the pixels corresponding to the segmented objects with the minimum depth value, $d_{\min}$, to $d_{\min} + 2$ are displayed.

From the results in Fig. 15(a) to (c), we can see that the sole target was not determined for the objects close to each other. As for the objects, $\mathrm{A_L}$ in Fig. 15(a), $\mathrm{F_R}$ in Fig. 15(b), and $\mathrm{I_R}$ in Fig. 15(c), these were wrongly determined as the targets together with $\mathrm{B_L}$, $\mathrm{G_R}$ and $\mathrm{H_R}$, and $\mathrm{J_R}$. In consequence, the left hand in Fig. 14(a) and right hand in Fig. 14(b) failed in the reaching tasks. In another experiment, the same target object images were generated even if the Mask R-CNN was applied for instance segmentation instead of SegNet-Basic and connected-component labeling in Fig. 2, as long as the depth information was used. On the other hand, the sole target was determined as can be seen in Fig. 15(d) to (e). It is noticeable that the targets were correctly determined even for the objects placed in the unknown positions, as shown in Fig. 15(e) and (e). Table 2 shows the classified reaching order for each object.

The objects in Fig. 13(a) were classified into the correct reaching order as annotated in the data set. As for the objects placed in the unknown positions in Fig. 13(b) and (c), the segmentation images were classified into the reaching order with the correct hand label. From the classification results, it was shown that the object with the same hand label, R or L, and the lowest number was determined as the target. If two or more objects are classified into the same

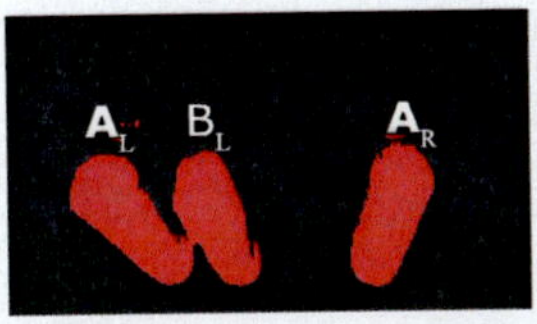
(a) TOI-RGBD in Fig. 13(a)

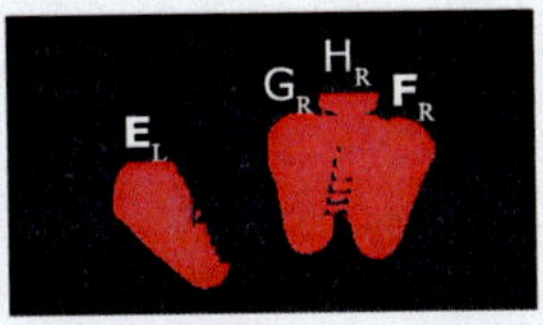
(b) TOI-RGBD in Fig. 13(b)

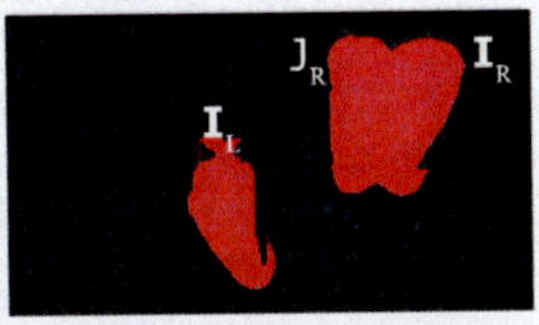
(c) TOI-RGBD in Fig. 13(c)

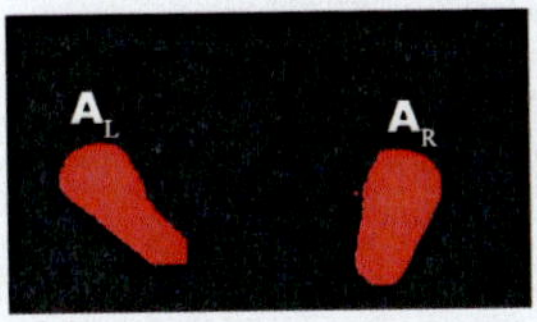
(d) TOI-RGB in Fig. 13(a)

(e) TOI-RGB in Fig. 13(b)

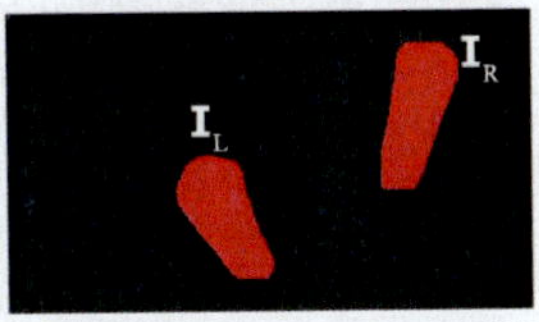
(f) TOI-RGB in Fig. 13(c)

Fig. 15. Comparison of generated target object images, TOIs-RGBD and TOIs-RGB

Table 2. Classified reaching order for objects in segmentation images

Objects	Fig. 13(a)						Fig. 13(b)						Fig. 13(c)			
	$\mathbf{A}_L$	B_L	C_L	$\mathbf{A}_R$	C_R	D_R	$\mathbf{E}_L$	F_L	G_L	$\mathbf{F}_R$	G_R	H_R	$\mathbf{I}_L$	J_L	$\mathbf{I}_R$	J_R
Order w/ hand label	L_1	L_2	L_3	R_1	R_3	R_4	L_1	L_3	L_4	R_1	R_2	R_4	L_2	L_4	R_3	R_4

reaching order, the normalized probabilities in the output layer in Fig. 8 are compared, and the one with the highest probability is determined as the target.

From the results described above, it was shown that the target object images sufficient for reaching motion planning were generated without using depth information of objects. As shown in Fig. 14(f), however, the robot closed the hands and stopped reaching even though the correct target object image was fed as the input to the grasping classifier. This result indicates that the generalization capability was not applicable to the objects outside the instruction area.

The robot continued the reaching tasks for all the objects. Since the robot failed in the tasks for the objects shown in Fig. 13(c), we focus on the results for the objects shown in Fig. 13(a) and (b). After each trial, the current targets are removed from the workbench. For instance, the next targets in the second trial in Fig. 13(a) are C_R and B_L. The task is completed when the robot grasps both the target objects. In Table 3, the result is represented as S or F, success or failure, for each target.

From the results listed in the first row, we can see that the robot with the TOIs-RGBD sometimes failed in the reaching tasks. These failures were concentrated in the first and second trials. This is because that the objects were close to each other, and the target object images were not generated correctly as shown in Fig. 15(a) to (c). On the other hand, the robot with the TOIs-RGB succeeded in all the reaching tasks as can be seen from the results listed in the second row.

Table 3. Reaching results: S (Success) and F (Failure)

	Fig. 13(a)						Fig. 13(b)					
	1st trial		2nd trial		3rd trial		1st trial		2nd trial		3rd trial	
Input	A_R	A_L	C_R	B_L	D_R	C_L	F_R	E_L	G_R	F_L	H_R	G_L
TOIs-RGBD	S	F	S	S	S	S	F	S[2]	S	F	S	S
TOIs-RGB	S	S	S	S	S	S	S	S	S	S	S	S

Although the left hand collided with the target as shown in Fig. 14(b), the robot then grasped the object.

Therefore, the effectiveness of the proposed approach to generate the target object images was shown for the multiple objects close to each other. For the objects outside the instruction area, the generalization capability of the grasping classifier is required to be improved. We believe that the robot will be able to classify the image inputs into the correct hand status by expanding the instruction area and placing the object in more positions.

7 Conclusions

In this paper, we focused on reaching motions of dual arm robots. For motion planning, the robot is required to determine a target from multiple objects. For this challenge, we proposed to generate target object images without using depth information. The target object images were generated through the instance segmentation and reaching order classifier. Moreover, the presented motion planner based on CNN with two branches enabled the robot to control the right and left hands simultaneously toward the respective targets. All the deep neural networks were trained through supervised imitation learning. In the reaching experiment, the robot successfully determined the targets from the objects close to each other. Using the target object images as the inputs, the robot succeeded in all the reaching tasks for the objects placed in the known and unknown positions within the instruction area. Therefore, the effectiveness of the proposed approach to generate the target object images was finally shown.

References

1. Maeda, Y., Moriyama, Y.: View-based teaching/playback for industrial manipulators. In: IEEE International Conference on Robotics and Automation, pp. 4306–4311 (2011)
2. Maeda, Y., Nakamura, T.: View-based teaching/playback for robotic manipulation. ROBOMECH J. **2**(2) (2015)
3. Argall, B.D., et al.: A survey of robot learning from demonstration. Robot. Autonom. Syst. **57**, 469–483 (2009)
4. Hoshino, S., Urayama, K.: Teach and playback for robotic handling through object recognition. In: IEEE/SICE International Symposium on System Integration, pp. 582–583 (2021)

5. Noda, K., et al.: Multimodal integration learning of robot behavior using deep neural networks. Robot. Auton. Syst. **62**(6), 721–736 (2014)
6. Yang, P., et al.: Repeatable folding task by humanoid robot worker using deep learning. IEEE Robot. Autom. Lett. **2**(2), 397–403 (2017)
7. Lecun, Y., et al.: Gradient-based learning applied to document recognition. Proc. IEEE **86**(11), 2278–2324 (1998)
8. Hoshino, S., et al.: Imitation learning based on data augmentation for robotic reaching. In: Annual Conference of the Society of Instrument and Control Engineers of Japan, pp. 417–424 (2021)
9. Zeng, A., et al.: Robotic pick-and-place of novel objects in clutter with multi-affordance grasping and cross-domain image matching. Int. J. Robot. Res. 1–14 (2019)
10. Zapata-Impata, B., et al.: Using geometry to detect grasping points on 3D unknown point cloud. In: International Conference on Informatics in Control, Automation and Robotics, pp. 292–299 (2017)
11. Rusu, R.B., Cousins, S.: 3D is here: point cloud library (PCL). In: IEEE International Conference on Robotics and Automation, pp. 1–4 (2011)
12. Ten Pas, A., Platt, R.: Using geometry to detect grasp poses in 3D point clouds. Robot. Res. **1**, 307–324 (2018)
13. Kaiming, H., et al.: Mask R-CNN. IEEE International Conference on Computer Vision, pp. 2980–2988 (2017)
14. Badrinarayanan, V., et al.: SegNet: a deep convolutional encoder-decoder architecture for image segmentation. IEEE Trans. Pattern Anal. Mach. Intel. **39**(12), 2481–2495 (2017)
15. Badrinarayanan, V., et al.: SegNet: A Deep Convolutional Encoder-Decoder Architecture for Robust Semantic Pixel-Wise Labelling (2015). arXiv, 1505.07293
16. Dillencourt, M.B., et al.: A general approach to connected-component labeling for arbitrary image representations. J. ACM **39**(2), 253–280 (1992)
17. He, K., et al.: Deep residual learning for image recognition. IEEE Conference on Computer Vision and Pattern Recognition, pp. 770–778 (2016)
18. Kingma, D., Ba, J.: Adam: a method for stochastic optimization. In: International Conference for Learning Representations, pp. 1–15 (2014)
19. Chitta, S., et al.: Moveit! [ROS topics]. IEEE Robot. Autom. Mag. **19**(1), 18–19 (2012)
20. Nextage Open. www.kawadarobot.co.jp/en/nextage/

Kinematic Calibration of a Collaborative Robot by a Marker Based Optical Measurement Procedure

Oliver Rettig(✉), Silvan Müller, and Marcus Strand

Baden-Wuerttemberg Cooperative State University, Karlsruhe, Germany
Oliver.Rettig@dhbw-karlsruhe.de
https://www.karlsruhe.dhbw.de/rahmlab

Abstract. Static robot calibration determines the parameters of a mathematical model that approximates as closely as possible the relationship between the end-effector pose of a robot and its corresponding actuated joint variables. The manufacturer of a robot delivers such a set of parameters but in principle more accurate own measurements can be used to determine optimized parameters. The purpose of this paper is 1. to show how much the position accuracy of robots of type UR5e from Universal Robots can be increased by application of a marker based optical measurement procedure based on circle fits and 2. what are the limitations of this procedure. Furthermore it is described how to use so called modified Denavit-Hartenberg parameters instead of the typically used distal/classical definition to achieve comparability between different measurements. Overall it is shown that with the calibration procedure the errors can be reduced in practice but only less for a a brand-new robot and with usage of the manufacturer calibration data set. The remaining errors show a specific structure, which suggests that the mechanical stability of the robot limits the positioning accuracy and not the accuracy of the measurement procedure to determine the DH-parameters.

Keywords: Marker based optical measurement · Robot kinematics precision and accuracy · Denavit-Hartenberg

1 Introduction

Static robot calibration determines the parameters of a mathematical model that approximates as closely as possible the relationship between the end-effector pose of a robot and its corresponding actuated joint variables. Universal Robots, the manufacturer of the collaborative robots of type UR5e, delivers such a set of individual measured parameters for each robot. Different to more expensive industrial robots these measurements are done by so called Dual-Robot-Calibration, described in the US patent US20130079928A1 and are determined as (distal) Denavit-Hartenberg (DH) parameters. The described procedure allows to determine these parameters with the help of a second robot and avoids the need of the very expensive laser-tracking equipment [1,3,4] typically used for industrial robots.

I. Petrovic et al. (Eds.): IAS 2022, LNNS 577, pp. 470–482, 2023.
https://doi.org/10.1007/978-3-031-22216-0_32

For the UR5e the manufacturer specifies the precision (as defined in ISO 3534:2014) with 0.03 mm. In the documentation the word repeatability is used to focus on differences between repetitions of the same task. The position accuracy (as defined in ISO 3534:2014) of this type of robot is much less and not specified in the manufacturers documentation. We can expect position accuracy values in the order of centimeters based on nominal DH-parameters [2] and in the order millimeters based on manufacturer calibration (personal communication). Some values can be found also in the literature [14].

As a low-price alternative to laser-trackers in [5] the application of a marker based optical system is described to determine DH-parameters. This method determines position and orientation of the joint axes, independent from each other. The results can be used to determine DH-parameters. In the literature this is called "screw-axis measurement method" [8]. In the typical case of hinge joints one segment moves approximately in a circle around the adjoining segment. Mathematically a so called circle-fit method can be applied to estimate the joint axis position and orientation from the tracked segment poses.

In [5] it was shown that the measured DH parameters are similar to the nominal parameters of an robot of UR5e. But the comparison was difficult because of the DH characteristics, which yields to big variations in the d and r parameters in the case of almost parallel consecutive joint axes.

It was assumed that sufficient measurement accuracy should be reachable by a marker based optical measurement procedure to determine DH-parameters for UR5e.

In [7] mean absolute positioning errors of 0.15 mm with a variability of 0.015 mm for static measurements are reached with spherical markers diameter of 12.7 mm and already 2012 Yang et al. [6] show that $20\,\mu$m movements can be measured with acceptable accuracy and precision within a volume of $400 \times 300 \times 300\,\text{mm}^3$.

In consideration of this result reaching even higher accuracy should be possible than by the Dual-Robot-Calibration, the manufacturer calibration is based on. This study is a follow-up study to improve the already described measurement procedure, to analyse and discuss its limitations and especially to test, if an improvement of the position accuracy can be reached in practice.

2 Hardware, Data Aquisition and Algorithms

This section describes the important technical details, features and improvements in comparison to the previous publication [5]. It helps to understand the overall procedure and shares experiences which are useful especially for readers interested in starting similar experiments.

2.1 Experimental Setting and Preparation

The measurements are done in the Robot- and Human Motion Lab (RaHM-Lab[1]), where a Vicon system for optical markerbased 3D motion capture is fix

[1] https://www.karlsruhe.dhbw.de/rahmlab.

mounted. Three-dimensional positions of spherical markers, in the sight of more than one camera simultaneously, can be determined via triangulation and the markers are tracked with a sample rate 120 Hz. Around each camera LEDs emitting infrared light are placed which is reflected by the markers which boost the contrast very much. A robot arm of type UR5e (Universal Robots) and the appropriate Controller CB 5.6 are mounted on a heavy weight table (Vention, Canada, based on aluminium profiles), Fig. 1.

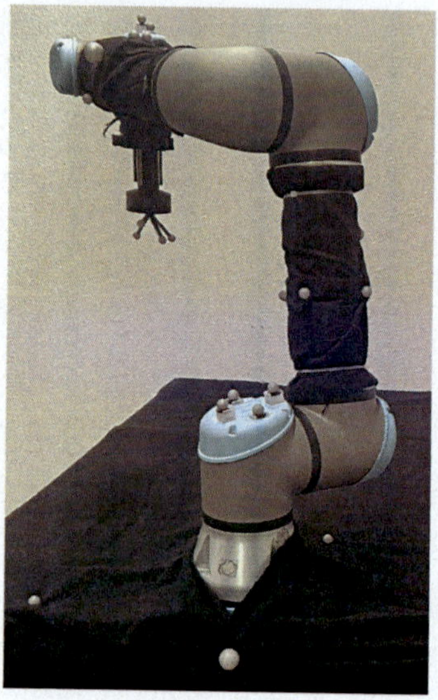

Fig. 1. A robot arm of type UR5e (Universal Robotics) mounted on a heavy weight table. IR reflecting components are masked with black tissue

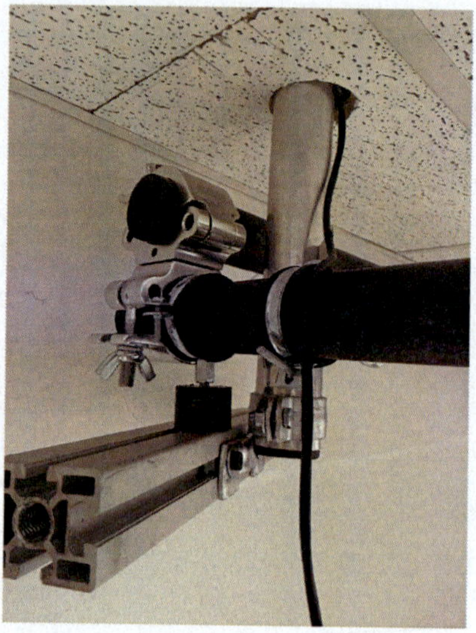

Fig. 2. Carbon pipes mounted at the ceiling via shock absorber. The carbon pipes are mechanically isolated from the aluminium pipe.

To avoid light reflections the table and the robot segments are masked with a black tissue (LightTeknik: Backdrop) as used as event technology. The windows of the lab are masked with a specific plastic film (Velken) which blocks the infrared light of the sun. Eleven Vicon-Vero-v2.2 Megapixel cameras (Vicon Motion Systems Ltd UK) are mounted at the ceiling to avoid vibrancy (Fig. 2) with a distance of 140 cm to the table. The cameras are clamped on custom-build pipes based on carbon fibres, which builds a quadratic structure with the measures $3 \times 3\,\text{m}^2$. The previously used aluminium pipes show too much lengthening with changes of the room temperature. The new pipes are build on carbon fibres twisted with an angle of 20° to the longitudinal direction, crossed in two bundles. This specific manufacturing results in no more measurable changes in the lengthening of the pipes with changes of the room temperature. The pipes are screwed together and the emerged form is mounted via shock absorber (Fig. 2) to the supporting tubes to the ceiling. The pipes are filled with quartz sand to increase the mass of the form and therefore increases the inertness against vibrancy, which can be coupled in from the ceiling.

2.2 Camera Calibration

The metric of the Vicon system is strongly defined by an electronic calibration device called "Vicon Active Wand V2"[2] (Fig. 3). This is build up on plastic material and instead of precise spherical markers LEDs are used. The manufacturer delivers this tool with a file, which includes the distances between the LEDs but these distances are not individually measured. This tool is typically not used for applications, where submillimeter accuracy is needed. Calibration of the tool is difficult because LEDs are used and the light emitting points can not be determined easy with submillimeter accuracy. That's why we have implemented a better defined calibration artifact. In the literature such artifacts are called metrology bar since their stiffness and thermal expansion characteristics allow deflection of less than 0.01 mm. With ASTM E3064 [10] a standard exists, which makes it possible to compare the accuracy of different systems. The tool (Fig. 4) is build on precision steel and shows less temperature dependence than the "Vicon Active wand". The position of the mounted spheres and it distances are determined in a coordinate measurement system with accuracy in the order of micrometer.

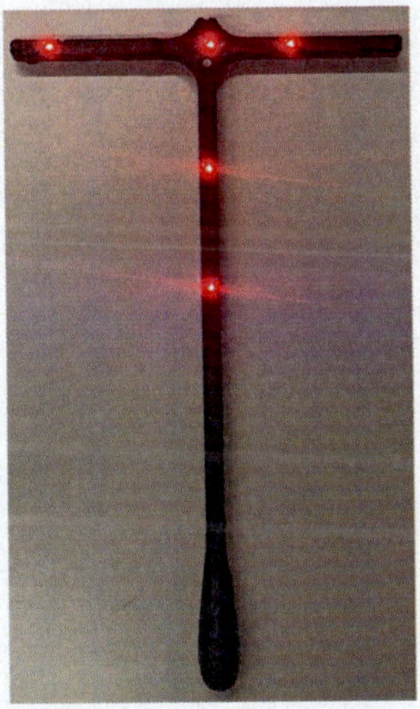

Fig. 3. Calibration camera calibration tool Active-Wand-V2 from Vicon.

Fig. 4. Custum-build camera calibration tool "Passive-Wand-V1" build from precession steel.

[2] https://www.vicon.com/hardware/devices/calibration/.

The markers are organised in clusters of three till five markers and the software Tracker[3] (Vicon) is used. Further details are described in [5].

Marker-clusters are automatically labeled and streamed via Vicons DataStream API. The software JDataAquisition[4] is used for triggering of the robot control and collecting the data. To do this, ROS is integrated, by implementation of the "Measurement System Service Provider Interface", of this software and a Java-Wrapper[5] for the DataStream API. Robot pathplaning is done with the software RPS[6] (ArtiMinds) which generates UR script code based on a set of waypoints or by given joint angles.

2.3 Referencing of the Coordinate System of the Robot

For referencing of the coordinate systems of the robot and the Vicon system several methods are tried. Simply placing markers oriented at the housing of the robot resulted in differences up to half a degree between the y-axes of both systems. The method used in [5] was based on the robots end-effector position, measured by the robot itself in 400 static positions. For these positions a singular value decomposition method was implemented to determine the best fit with the tracked positions by the Vicon system. The disadvantage of this method was that it is based on the forward kinematics of the robot and its limitations. That's why a new method was invented based on the robots base- and shoulder joint axes, which are determined by circle fits.

The base joint axis ($\vec{j}_{base}$) determined from data of the shoulder segment moving around the base with an outstretched robot is defined as the z-axis the UR5e. The base joint is than moved in its zero position (Fig. 7). The shoulder joint axis ($\vec{j}_{shoulder}$, referenced in the base marker cluster) is orthogonalized to the z-axis by application of a cross product. With a further cross product to the base joint axis the y-axis is determined. And from this a further cross-product defines the x-axis.

$$\vec{\sigma}(t) = \begin{pmatrix} x_0^{base} \\ y_0^{base} \\ z_0^{measured} \end{pmatrix} \tag{1}$$

$$\vec{z} = \vec{j}_{base} \tag{2}$$

$$\vec{y} = (\vec{j}_{base} \times \vec{j}_{shoulder}) \times \vec{j}_{base} \tag{3}$$

$$\vec{x} = \vec{z} \times \vec{y} \tag{4}$$

$$\Sigma = \{\vec{\sigma}, \vec{x}, \vec{y}, \vec{z}\} \tag{5}$$

---

[3] https://www.vicon.com/software/tracker/.

[4] http://jdataaquisition.orat.de.

[5] https://github.com/MobMonRob/JViconDataStream2.

[6] https://www.artiminds.com/de/produkte-und-leistungen/robot-programming-suite-basic/.

Three markers placed on the table (near the robots base flange) defines the plane of the mounting flange. The point, where the z-axis of the robot intersects with this plane defines the origin of the robots coordinate system. In other words the origin is defined as a point on the base joint axis in the height of the flange (mounted on the table), which means that mainly the z-coordinate of this point ($z_0^{measured}$) is measured.

2.4 Measurement of Kinematics and Detection of Static Phases

To estimate joint axes, the circle-fit method SARA [9] is used. This method is well-known in biomechanics but new in the field of robotics. Instead of methods fitting points on a circle it is based on complete frames based on a cluster of markers which moves around an axis defined also by a marker cluster. A discussion about advantages of this method is available in [5]. The joint axes are determined separately. That's why data with moving of a single joint only are used (Fig. 5). The phases are defined by the corresponding joint angle trajectories based on streamed values from the UR5e robot itself. The derivatives are determined and the phases are defined by including frames for which the derivative is bigger than 50°/s. This allows to restrict the phases to movements with constant velocity.

Furthermore we drive the robot to a set of waypoints with 10 cm distance in a raster, organized in 6 planes (5 rows and 10 columns, Fig. 6), which results in 300 static poses inside a working volume of a cuboid with 0.5 m × 1 m × 0.8 m. At each point in this cuboid the robot stays for one second and the pose of all segments are tracked synchronously by the Vicon system and by the robot built in sensors.

For higher robustness of the detection of the static phases (time intervals during the robot stays in a defined waypoint), a new procedure different to [5] (which is based on marker based movement measurements only) is implemented. The IO-Ports of the UR-controller are used and in the robot program a signal is set, if the robot stops in a static pose and deleted before the robot starts moving to the next pose. The IO-signal is read from the Vicon Lab-Lock hardware and therefore can be read synchronously with the tracked and streamed marker position data. As in the old procedure from these raw-phases only "sub"-phases are used starting in the mid of the raw-phases and ending at 80% of the width of the complete raw phase.

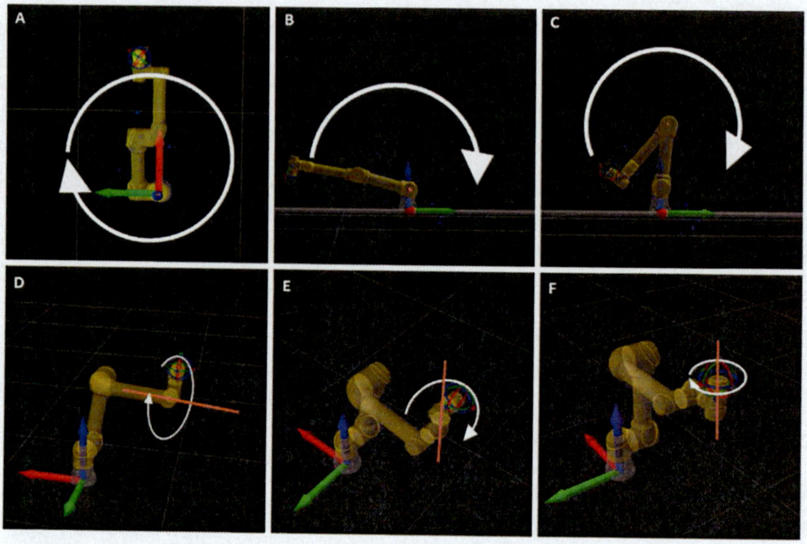

Fig. 5. Rotating each of the 6 joints separately (a–f) in a fix pose of the respectively other joints. a: Base joint, b: Shoulder joint in two different configurations of the base segment, rotated 90° to each other, c: Elbow, d: Wrist-1, e: Wrist-2 and f: TCP.

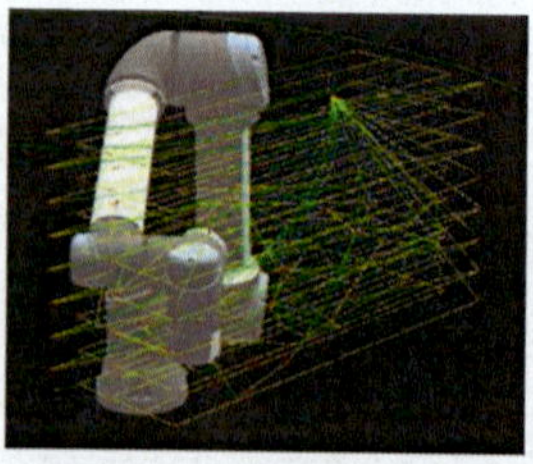

Fig. 6. Organized in 8 planes with 5 rows and 10 columns, a sequence of all together 400 points are reached. In this study only 6 of the planes with 300 points altogether are used.

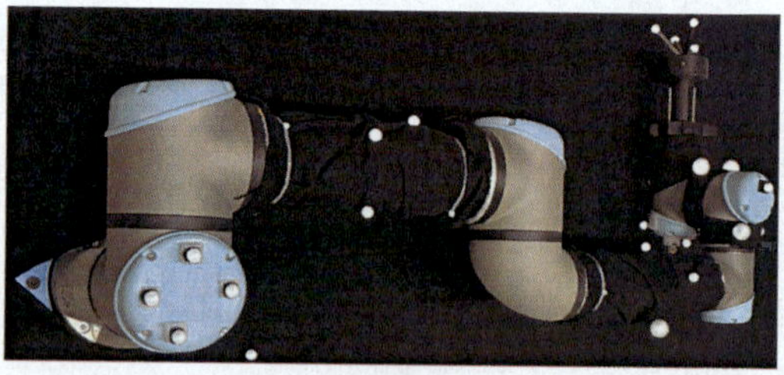

Fig. 7. Static outstretched pose of the robot as described and also used by Universal robots as reference pose. Attached Spherical markers with 14 mm diameter. IR reflecting components are masked with black tissue.

3 Experiments

3.1 Determination of Joint Axes and DH-parameters

To test day-to-day precision the measurement procedure to determine the joint axes are repeated three times at different days in a week. All of the joint axes are referenced in its adjacent base cluster markers coordinate system. After the circle fits are done the robot is moved into its zero-position (Fig. 7). This pose is defined by moving all joints to $\theta_i = 0°$ (measured by the robot itself). In this position the segment axes (between adjacent joint axes) should be all be orthogonal to one of the axes of the coordinate system of the UR. The angle between the segments (between adjacent joint axes) and its corresponding axis of the robots coordinate system define the $\Delta\theta_i$ values. In this pose also the orientation of the joint axes themselves are determined as two (projection) angles each, which describe the angles between the joint axis and two orthogonal axes of the robots coordinate system.

To get comparable d- and r-parameters of the DH-convention a so called "modified" version [11,12] is used. In this convention the segment between two adjacent joint axes is orthogonal only to one of the two joint axes. This allows to define freely the d-parameters of almost parallel adjacent joint axes. We have set $d_2 = d_3 = 0$. The implementation is based on Plücker-coordinates for describing the joint axes as lines [13].

3.2 Moving the Robot Based on Optimized Modified DH-parameters

As an overall test, the determined DH-parameters are used to move the robot to the 300 poses in the described grid (simulating a typical working volume). For each grid position (Fig. 6) the difference between the target position and the position tracked by the Vicon system is determined. For all of the points in a plane of the grid mean and standard deviation of the differences to the target values are calculated to get an impression about the structure of the errors. The differences are calculated for each of the three coordinate axes of the Robot coordinate system separately. In the frontal view there are six planes with 50 points and constant z-component each. In the sagital view there are ten planes with 30 points and constant y-component each. And in the frontal view there are five planes with 60 point and constant x-component each.

4 Results

The calibration accuracy reached for the Vicon cameras is typically better than 0.1 mm (Table 1). Spherical markers are tracked by the Vicon system in static cases with signal noise of maximal amplitude (precision) of ± 0.02 mm in 300 grid points. The usage of the new calibration tool (Fig. 4) changes the metric of the measurement system in the same order of 0.02 mm in the measurement volume

Table 1. Example camera calibration data set. Maximum error of the eleven Vicon Vero 2.2 cameras in [mm].

Camera	1	2	3	4	5	6	7	8	9	10	11
World	0.0697182	0.0890782	0.0668268	0.071927	0.126107	0.0990812	0.0877552	0.0871235	0.0219467	0.0601198	0.0847558

used for this study and therefore has a meaningful effect on the accuracy. But for the overall accuracy of the system we expect values only in the range of 0.1 mm. Table 2 shows the orientation of three joint axes against its corresponding two orthogonal axes of the robots coordinate system for the static zero pose (Fig. 7) and a pose with 90°-rotated shoulder joint.

The differences between the two poses are much higher than the differences between days measurements. Repeatability (precision) for the last is better than 0.1°. Comparable big differences are found, if the circle-fit to determine the shoulder joint axis is done in a 90° rotated base segment. Table 3 shows the (modified) DH-parameters, determined by the marker-based measurement procedure, for one trial in comparison to the values based on the nominal DH-parameters given by the manufacturer.

Table 2. Orientation of the joint axes in the robots static zero pose and a pose with 90° rotated shoulder segment.

Joint	δ_1[°]	δ_2[°]	δ_1^{90}[°]	δ_2^{90}[°]
Shoulder	−0.16	00.00	−0.22	0.04
Elbow	−0.31	−0.13	−0.05	0.09
Wrist 1	−0.47	−0.05	−0.12	0.58

Table 3. Nominal modified-DH-parameters from the manufacturer compared with measured ones.

mDH	r [mm]		d [mm]		α[°]	
Joint	**nominal**	**calibrated**	**nominal**	**calibrated**	**nominal**	**calibrated**
Base	00.0	00.000	162.5	162.424	00.0	00.000
Shoulder	00.0	−0.157	00.0	00.000	90.0	89.950
Elbow	−425.0	−425.146	00.0	00.000	00.0	0.120
Wrist1	−392.0	392.135	133.3	134.007	00.0	−0.255
Wrist2	00.0	00.063	99.7	99.639	−90.0	−89.978
TCP	00.0	00.090	99.6	99.382	90.0	89.973

The α values which describe the angles between adjacent joint axes are identical to the corresponding DH-parameters. The modified-DH-parameters d_i and r_i in principle correspond to the DH-parameters d_{i+1} and r_i but for DH-parameters of a calibrated robot these parameters can have very big values for the almost parallel joint axes of the shoulder, elbow and wrist 1. However in the modified Denavit Hartenberg model the parameters d_i which correspond to the almost parallel joint axes can be set to arbitrary fix values. Furthermore the first and the last d-parameter d_0 and d_5 can not be determined from robots joint motions.

Table 4. Differences between measured and target TCP-positions for 300 poses in the working volume grid. The data is averaged in planes in the three orthogonal axes of the grid. Data is shown for the frontal (x), sagital (y) and transversal (z) planes. Also the nominal values without calibration is given.

	$\Delta x \pm$ std [mm]		$\Delta y \pm$ std [mm]		$\Delta z \pm$ std [mm]	
Plane	nom.	cal.	nom.	cal.	nom.	cal.
1	-1.60 ± 0.78	-1.37 ± 0.34	-1.71 ± 3.84	1.62 ± 0.36	-2.71 ± 0.39	-1.94 ± 0.41
2	-1.17 ± 0.76	-0.98 ± 0.31	0.19 ± 0.62	1.61 ± 1.90	-2.86 ± 0.42	-2.06 ± 0.44
3	-1.23 ± 0.74	-1.02 ± 0.31	3.84 ± 0.60	0.87 ± 1.10	-2.95 ± 0.40	-2.13 ± 0.41
4	-1.03 ± 0.68	-0.84 ± 0.28	3.62 ± 0.54	0.32 ± 0.45	-2.91 ± 0.45	-2.09 ± 0.47
5	-0.89 ± 0.52	-0.72 ± 0.27	3.85 ± 0.50	0.22 ± 0.34	-3.17 ± 0.44	-2.34 ± 0.45
6			1.17 ± 0.62	0.22 ± 0.34	-3.22 ± 0.42	-2.39 ± 0.43
7			4.29 ± 0.58	-0.03 ± 0.22		
8			4.67 ± 0.40	-0.03 ± 0.25		
9			2.29 ± 0.43	-0.45 ± 0.26		
10			4.14 ± 0.39	-0.63 ± 0.26		

The described segment lengths are defined by an adjacent segment only from one side (distal for the base segment, proximal for the TCP wrist 2 segment). The other sides are defined by a geometrical fix points (the flange of the base segment to mount the robot on the table and the flange at the TCP to mount the tool) only. The first one can be calipered with an accuracy of 0.1 mm (based on three markers, as explained in Sect. 2.3) but for the last one we have no measurement method available because the geometrical point is optical not available without unmounting the TCP marker cluster.

Table 4 shows the overall averages and standard deviations of the differences between marker based measurements and achieved positions measured with the robot sensors, in the frontal, sagital and transveral planes of the measured grid. The values are components of the three axes of the robots coordinate system for its corresponding planes. With calibration in all cases the differences are smaller than for an uncalibrated robot as expected. The achieved positions measured with the robot sensors are almost identical to the target positions because only static positions are compared.

5 Discussion

The between-days reproducibility/precision was high enough to be able to use the measurement procedure as a robust method for robot recalibration, e.g. of older robots for which the original manufacturer calibration does not fit well any more. The joint axes determined by circle-fit methods are very reproducible and it is surprising, that for the two different joint poses (zero pose and 90° rotation of the shoulder segment) bigger differences (Table 2) are found than between-days measurements. For more distal joints the differences between the joint axes and the corresponding axes of the robots coordinate system increases. That is expected because the differences in the joint sequence have to be summed up. The projection into two orthogonal planes can partialy obscures this effect. $\theta_2 = 0°$ because the shoulders joint axis is used to define the robots coordinate system.

The differences between nominal and measured modified DH-parameters are less than 0.4 mm respectively 0.6°. Between-days precision is much better but the differences between determination of the parameters in different static poses are much higher corresponding to the axes orientations shown in Table 2. It looks that circle-fit estimation is very accurate and joint laxity is the main limitation. Joint axes are determined well in the pose in which they are determined but in other poses the alignment changes. Maybe to get the optimized set of DH-parameters it is better to use only poses of the working volume for axis circle fitting, instead of moving a complete circle at once in a specific joint configuration (even outside of the working volume, Table 4) as done in this study.

Differences between the measured and the target TCP positions in order of centimeters as expected for an uncalibrated robot (usage of nominal DH parameters) could not be found in the working volume of the defined raster. The maximum value found was 9 mm. With individual calibration, differences are found in the order of millimeters as expected.

The overall test of the robot calibration is the comparison of the measured and the target values in a typical working volume. Table 4 compares the differences splited into components for three orthogonal planes for an uncalibrated (nominal values) and the calibrated robot. The structure of the errors are different for three planes. The uncalibrated robot shows maximum errors in y-direction, the calibrated robot however shows maximum errors in the z-direction. Standard deviations are very similar for all planes in the case of the calibrated robot and they are reduced much for x- and y-direction but not for the z-direction (in comparison to the uncalibrated robot). In z-direction only a small increase of the error for planes with higher z-component is shown. For the working volume it makes sense to reduce the errors in z-direction by subtracting the average difference from the target-position for further optimization. For the sagital planes an interesting error pattern is observable. Starting at a mid-position (plane 5) for planes with higher and also with lower y-values the error increases. For frontal planes the error increases with planes further away from the origin of the robots coordinate systems origin. This implies that the error increases, if the robot arm is more stretched. The structural dependency of the errors from the planes cauld not be explained in this study but it is found also in the literature, e.g. in the context of the usage of machine learning to compensate for the errors. E.g. in [15] for errors in each of the three planes a separate neuronal network is trained. That is why we think the structural differences have not its origin in our specific robot or the specific measurement procedure.

6 Conclusion and Future Work

Major improvements of the measurement procedure cauld be reached by the custom-build camera calibration tool, detection of static phases and the new method for referencing the coordinate systems of the robot and the Vicon measurement system. The usage of modified DH-parameters has allowed to make parameter-sets comparable. The study shows that the marker based measurement technique is able to uncover the structure of inaccuracies in the robot positioning which assumed have their origin in mechanical limitation of the robot and its mounting.

On the other side, the results also show that despite of in principle sufficient accuracy of measurement technique, it is difficult to achive a higher positioning accuracy for a brand-new robot of type UR5e in practice. The manufactory calibration data, delivered with a brand-new UR5e, looks sufficient accurate reflecting the mechanical limitations of the robot. The positioning accuracy of a brand-new UR5e looks mainly limited by nonlinear effects produced by mechanical instabilities and not by a limitation of the accuracy of the manufacturer calibration. But custom made kinematic calibration of DH-parameters by marker-based measurement procedures can be very useful to improve the positioning accuracy of a rough-handled robot.

Under the perspective of this study increasing the positioning accuracy of a collaborative robot should focus to compensate for nonlinear effects. There is potential also of a brand-new robot and machine learning algorithms look very promising to do this.

References

1. Nubiola, A., Bonev, I.A.: Absolute calibration of an ABB IRB 1600 robot using a laser tracker. Robot. Comput.-Integr. Manuf. **29**, 236–245 (2013)
2. www.universal-robots.com/articles/ur/application-installation/dh-parameters-for-calculations-of-kinematics-and-dynamics/
3. Corbel, D., Company, O., Nabat, V., et al.: Geometrical calibration of the high speed robot Part 4 using a laser tracker. In: MMAR 2006–12th IEEE International Conference on Methods and Models in Automation and Robotics 28–31 Aug (2006)
4. Alici, G., Shirinzadeh, B.: A systematic technique to estimate positioning errors for robot accuracy improvement using laser interferometry based sensing. Mech. Machine. **40**(8), 879–906 (2005)
5. Rettig, O., Müller, S., Strand, M.: A marker based optical measurement procedure to analyse robot arm movements and its application to improve accuracy of industrial robots. In: Intelligent Autonomous Systems 16, Volume 412 of the Lecture Notes in Networks and Systems series (2021)
6. Yang, P., Sanno, M., Brüggemann, G., Rittweger, J.: Evaluation of the performance of a motion capture system for small displacement recording and a discussion for its application potential in bone deformation in vivo measurements. Proc. Inst. Mech. Eng. H. **226**(11), 838–47 (2012)
7. Merriaux, P., Dupuis, Y., Boutteau, R., Vasseur, P., Savatier, X.: A study of Vicon system positioning performance. Sensors **17**(7), 1591 (2017)

8. Santolaria, Gines, M.: Uncertainty estimation in robot kinematic calibration. Robot. Comput.-Integr. Manuf. **29**, 370–384 (2013)
9. Ehrig, R.M., Taylor, W.R., Duda, G.N., Heller, M.O.: A survey of formal methods for determining functional joint axes. J. Biomech. **40**(10), 2150–2157 (2007)
10. ASTM International: 100 Barr Harbor Drive, PO Box C700, West Conshohocken: Standard Test Method for Evaluation the Performance of Optical Tracking Systems that Measure Six Degrees of Freedom (6DOF) Pose. PA 19428–2959, United States
11. Granja, M., Chang, N., Granja, V., Duque, M., Llulluna, F.: Comparison between standard and modified Denavit-Hartenberg methods in robotics modelling. In: Proceedings of the 2nd World Congress on Mechanical, Chemical, and Material Engineering (MCM'16) Budapest, Hungary, 22–23 August 2016. Paper No. ICMIE 118
12. Li, N., Ping, X.: Research on D-H parameter modeling methods. Int. J. Comput. Sci. Control Eng. **7**(1), 8–16 (2019)
13. Chittawadigi, R.G., Hayat, A.A., Saha, S.K.: Geometric model identification of a serial robot. In: 3rd IFToMM International Symposium on Robotics and Mechatronics, 2–4 Oct (2013)
14. Jiang, Y., Yu, L., Jia, H., Zhao, H., Xia, H.: Absolute positioning accuracy improvement in an industrial robot. Sensors **20**(16), 4354 (2020)
15. Hsiao, J.-C., Shivam, K., Lu, I.-F., Kam, T.-Y.: Positioning Accuracy improvement of industrial robots considering configuration and payload effects via a hybrid calibration approach. IEEE Access (2020)

Randomized Robotic Visual Quality Inspection with In-hand Camera

Zvezdan Lončarević[1,2], Simon Reberšek [1], Aleš Ude[1], and Andrej Gams[1](✉)

[1] Department of Automatics, Biocybernetics and Robotics, Jožef Stefan Institute, Jamova cesta 39, 1000 Ljubljana, Slovenia
{zvezdan.loncarevic,simon.rebersek,ales.ude,andrej.gams}@ijs.si
[2] Jožef Stefan International Postgraduate School, Ljubljana, Slovenia

Abstract. Robotic visual inspection is based on manually pre-defined postures (relation) between the camera and the object, and the robot either moves the object in front of the camera, or the in-hand camera around of the object. Either the complete object, or—in order to save time—some critical parts are checked. The path of the robot is typically always the same, determined in advance. But, in order to check only a random subset out of all possible aspects of the product, we need to generate all possible transitions between all possible aspects, and then choose the corresponding subset of motions. In this paper, we experimentally evaluate various motion planning algorithms to autonomously generate transitions between different, random inspection postures for guiding an in-hand camera with a robot in a confined space. This allows for completely random sequences of motion through predefined postures for visual quality inspection. Results show that for our use-case Transition-based RRT outperformed other available motion planners.

Keywords: Visual quality inspection · Robot motion planning · Autonomous robot · Industrial robotics

1 Introduction

Zero defect manufacturing (ZDM), as a concept of production strategy, is being implemented more and more in production lines and factories [1]. One of the reasons is the availability of data, hardware [2] and techniques stemming from Industry 4.0 [1]. Anomalies that appear in any phase of the production process may degrade the quality of produced parts/products. Thus, as a part of ZDM, inspecting the quality of actions is used also in the beginning and middle of the manufacturing cycle and not only at the very end in order to use/reject the final product [1].

One type of inspection is visual quality inspection, and a plethora of visual sensors have been employed for visual inspection and quality control processes in industrial settings [3,4]. These include time of flight, structured light, stereo vision, line scan cameras, and also the monocular vision with a standard area

I. Petrovic et al. (Eds.): IAS 2022, LNNS 577, pp. 483–494, 2023.
https://doi.org/10.1007/978-3-031-22216-0_33

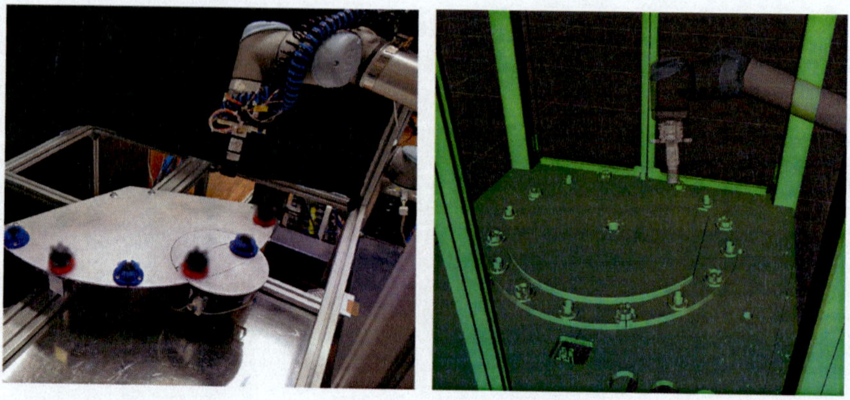

Fig. 1. Physical implementation of the visual quality inspection cell for the use-case (left), and RViz visualization (right).

scan camera or an RGB camera. Robots are often employed in order to achieve the desired relation between the visual sensor (camera) and the object of inspection, specifically when the object needs to be inspected from different viewpoints [5].

Application of robot-supported visual quality inspection can follow different concepts, depending on the requirements. For example, the product or semi-product can be checked from a few discrete viewpoints or a continuous visual stream can be acquired [5]. Another division is in the type of motion—the robot can either move an in-hand camera around the object of inspection, or it can move the object around the fixed camera. When holding the camera, the robot can place it so that there are no obstacles in the line of sight [6]. A robot with an in-hand camera can adapt its position and the camera viewpoint in order to be able to see the required area of the inspected object. Appropriate positioning of the camera is crucial for proper visual inspection. However, with a relatively confined space, which is often the case due to other production and inspection elements, the robot by itself may not be able to achieve optimal viewpoints for all the required surfaces of the object. Thus, robot in-hand camera motion is often combined with a rotation of the object of inspection. Camera location is often the subject of optimization for quality inspection [7,8]. An algorithm for robot-supported autofocus for a fixed-focus camera was described in [9].

Once the visual data is acquired, it can be processed by different methods, including advanced deep learning methods [10,11]. Visual quality inspection in any stage of the production should accurately detect and appropriately classify the errors on the final (semi) product, at the pace of the production. The pace of the production is often relatively high and leaves little time for the robot to move the camera around the object. Therefore, fast and precise motion of the robot is required. Even so, some production processes, such as injection moulding, cannot wait, and therefore unless specific extensions to the production line are implemented, there is simply not enough time to check all. Examples of extensions include buffers, where the products wait for inspection, or employing parallel checking with several cameras on several robots. Introducing all of these might increase production costs and therefore a trade-off is struck between the possible number of defects not being detected and the effort put into quality inspection.

In this paper, we experimentally evaluate a use-case in which there is not enough time and resources to fully inspect every aspect of the product with an in-hand camera. The use-case is shown in Fig. 1. But, in order to reduce the number of defects as much as possible, we still want to check as many parts of the product as possible. To avoid always checking the same aspects of the product, we propose to randomly change between the required inspection viewpoints. This has the added problem that the motion of the robot, and by extension the in-hand camera, does not follow pre-defined trajectories. While all possible motions could be generated and then called from memory, there might be too many of them for practical use. We therefore propose to use motion-planners to generate trajectories as required.

Different motion planners exist, based on different approaches and with different results of the application [12]. In this paper, we show how we can experimentally determine the correct motion planner for the given use-case through an in-depth statistical analysis of a set of readily available motion planners and tools [13]. The results show that choosing the correct motion planner can lead to fast, practically instantaneous motion generation in confined spaces. By checking only some aspects of the object we save time and reduce the number of faulty items that are shipped. However, not all defects can be detected if all possible aspects are not checked.

2 Use-Case

As briefly outlined in the Introduction, the presented use-case deals with visual quality inspection, where a robot guides the camera to a set of pre-defined postures. The postures are determined so that the camera can see certain aspects of a product/workpiece. Because there are too many postures to reach them in the allocated time, the robot guides the camera to only a small, random sub-set of the given postures. Thus, for every production cycle, we need to generate trajectories between the selected postures. If there are k postures, there are

$$s = k \times (k - 1) \tag{1}$$

trajectories between them—every posture is connected with every other. One could argue that the connection between postures $(k, k-1)$ is the same as between $(k-1, k)$, and there are actually only half as many. This is spatially true, but not temporally, and the motion needs to be recomputed. This can be avoided if we consider that the sequences can only be increasing, i.e., one can only go from posture 4 to postures 5 and more, and not to postures with a lower number. However, this makes it less general and has other implications, for example, if we would actually want motion $(k-1,k)$. This is further elaborated in the Discussion.

Specifically, in this use-case there are 10 postures. They are depicted in Fig. 2. Having 10 poses means that according to Eq. (1), one would have to pre-compute and store 90 different motions. This is manageable, but again reduces the generality of the approach and quickly becomes unsustainable with a larger number of postures.

Fig. 2. Ten postures for visual quality inspection from the given use-case, depicted in simulation.

The question we are answering in this paper is how to select a proper motion planner to accurately and quickly generate trajectories between the given postures, and how much this can save.

3 Motion Planning

The goal of autonomous planners is to enable robots to automatically compute their motions from descriptors of tasks and models acquired from sensors [14]. In recent years considerable progress was made in making faster and more robust algorithms.

By the definition, the task of path-planner is to "find a collision-free motion between an initial (start) and a final configuration (goal) within a specified environment" [15], see Fig. 3.

Characterization of motion planning problem can be done according to the task it addresses, properties of the robot and properties of the algorithm [14]. Motion planning heavily depends on the degrees of freedom (DoFs) of the task space and configuration space of the robot. When configuration space is defined, we have to be aware of constraints (joint limits) of the robot and if the robot is

Fig. 3. Path planner inputs and outputs

modelled using kinematic or dynamic equations in order to know if the output of the planner needs to be torques (forces) or angles (offsets) of joints.

The robot path planning problem in literature is usually divided into classical methods and heuristic methods. In the following, we introduce to the main ideas to the most widely used planning algorithms.

Classical approaches were very popular in solving robot motion planning tasks because it is observed that either a result would be obtained, or it would be confirmed that a result does not exist [16]. The major drawback of this approach is high computational cost and failure to respond to the uncertainties of the environment so it is less preferred for real-time implementation. This is why for motions in dynamic environments, where objects are not stationary, heuristic methods based on artificial intelligence (AI) are becoming more popular. Classical approaches include Cell Decomposition (CD), Potential Field, Sampling based methods (Probabilistic Road-Map and Rapidly-Exploring Random Trees) and Subgoal Networks. In CD method, robot's workspace is divided into regions called cells and based on this division, adjacency graph where each node corresponds to one cell and each edge corresponds to the connection between adjacent cells is formed. At first, we determine in which cell (node) are starting and goal positions and then using algorithms such as Dijkstra's or A-star we find the shortest connection. The main idea of Potential field method is to present obstacles with repulsive and the goal with attractive forces. With doing so, the robot goes in the direction of the negative of the gradient of the potential field until it reaches its goal.

Sampling-based methods for planning (SBP) have been widely used recently because of their capability to solve time critical real-world planning problems. They are unique by the fact that planning is done in configuration space of the robot. SBP is trying to capture the connectivity between randomly sampled points. The most widely used algorithms of this subgroup are Probabilistic Road-Map (PRM) [17] and Rapidly-Exploring Random Trees (RRT) [18,19]. Both algorithms are based on the same idea of connecting randomly sampled points, but different in the manner they construct a graph connecting the points [20]. The main advantage of these two methods is that they are capable of finding fast solutions for difficult problems, and disadvantage is that the solutions could be suboptimal because of random sampling.

Heuristic algorithms have become more popular because of the increased requirement of computational power with classical approaches when dimensionality of the problem is increased. Some of the most widely used heuristic methods are Artificial Neural Networks, Fuzzy logic and nature inspired algorithms such as Genetic Algorithms (GA), Particle Swarm Optimization (PSO)[21] and Ant Colony Optimization (ACO).

Many planning algorithms are readily available in different software packages, for example in MoveIt! [22]. Planning libraries, such as the Open Motion planning library (OMPL) [23], provide geometric and control-based planners. Other options exist, for example the Stochastic Trajectory Optimization for Motion Planning (STOMP) [24], Gradient Optimization Techniques for Efficient Motion Planning (CHOMP) [25], or Search-based planning library (SBPL) [26].

4 Experimental Evaluation

The use-case, presented in Section 2, calls for inspection of 3 random positions on the product, of 10 possible positions. The visualization and the physical implementation are depicted in Fig. 1.

For the evaluation of different available motion planners, we used tools available within the Robot Operating System (ROS) framework. This includes MoveIt! [22], RViz for visualization, and Planner arena [13] for benchmarking of different planners. We tested different motion planners available within the Open Motion planning library (OMPL) [23], which offers also constrained planning [27]. The goal of the evaluation was to find planners which can generate collision-free motion trajectories in a fast and reliable manner. We considered 21 different planners, given in Table 1, which generated trajectories between 10 random pairs of poses, 5 times for each pair.

We checked various parameters to determine which planner might be most suitable for the given use-case. As explained in [13], there is no general solution. First, we checked if the planners reach the desired pose and all did. Next, we checked the average time required for planning. We can see in Fig. 4 that some of the planners took much longer (durations in Fig. 4 are cut at 2 s to reserve data details of shorter planning times) and in fact timed out at the predefined time of 10 s. Those were immediately ruled out because the aim is to have as fast as possible trajectory generations. The planners in Figures are marked with numbers and their corresponding names are given in Table 1.

Next, we checked trajectory smoothness, where results closer to zero is better, shown in Fig. 5. Smoothness S takes three consecutive way-points and the angle formed between them. The value is calculated as a square of sum of all the angles calculated

$$S = \sum_{i=2}^{n-1} \left(2\left(\pi - \arccos \frac{d_{i\text{-}2,i\text{-}1}^2 + d_{i\text{-}1,i}^2 - d_{i\text{-}2,i}^2}{2d_{i\text{-}2,i\text{-}1}d_{i\text{-}1,i}}\right)\right)^2. \tag{2}$$

Here n is the number of way-points, $d_{i,i-1}$ is the distance between way-points with indices i and $i-1$ and $S = 0$ if two way-points are aligned.

Figure 6 further shows the length L of the calculated trajectory. The length is given as the sum of angles traveled by each of the joints

$$L = \sum_{i=0}^{l-1} |q_i - q_{i0}|, \tag{3}$$

Table 1. List of tested planning algorithms

Planner #	Planner name
1	BiEST (bidirectional version of EST)
2	ProjEST (projection-based version of EST)
3	LazyPRM
4	LazyPRM*
5	SPARS (SPArse roadmap spanner algorithm)
6	SPARS2
7	SBL (single-query, bi-directional and lazy collision checking)
8	EST (expansive space trees)
9	LBKPIECE (lazy BKPIECE)
10	BKPIECE (bidirectionnal KPIECE)
11	KPIECE (kinematic planning by interior-exterior cell exploration)
12	RRT (rapidly-exploring random trees)
13	RRT connect
14	RRT*
15	T-RRT (transition-based RRT)
16	PRM (probabilistic roadmap)
17	PRM*
18	FMT (fast marching tree algorithm)
19	PDST (path-directed subdivision trees)
20	STRIDE (search tree with resolution independent density estimation)
21	BiTRRT

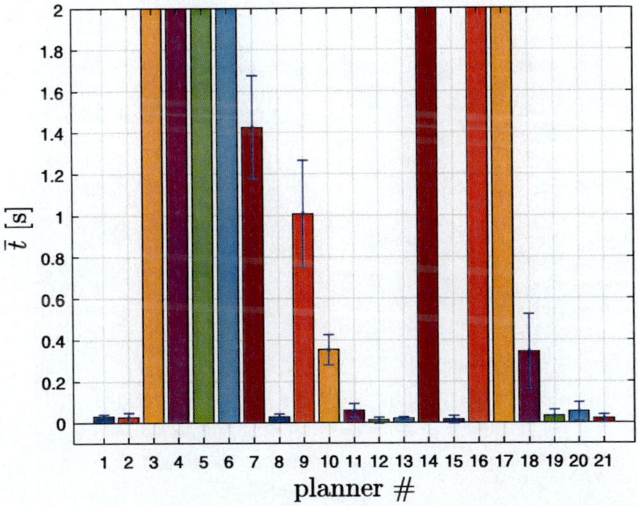

Fig. 4. Average duration for the calculation of transition trajectory between 10 random pairs of poses, 5 times per pair, for 21 motion planning algorithms listed in Table 1. The times are cut at 2 s, while the calculations for some planners took longer, up to pre-defined 10s calculation cut-of time.

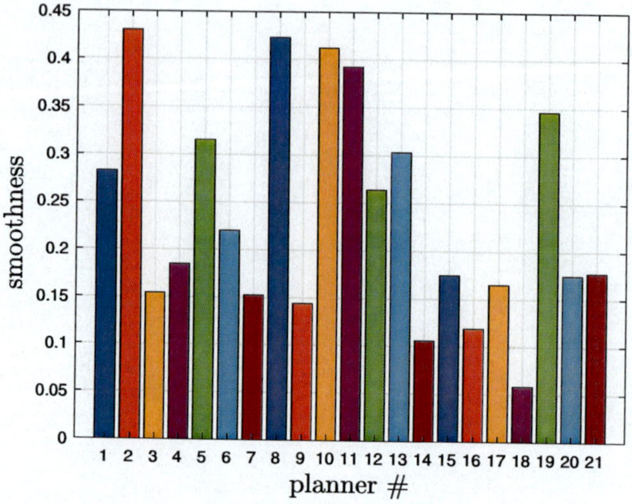

Fig. 5. Average smoothness of the calculated trajectory, calculated by Eq. (2) between 10 random pairs of poses, 5 times per pair, for 21 motion planning algorithms listed in Table 1.

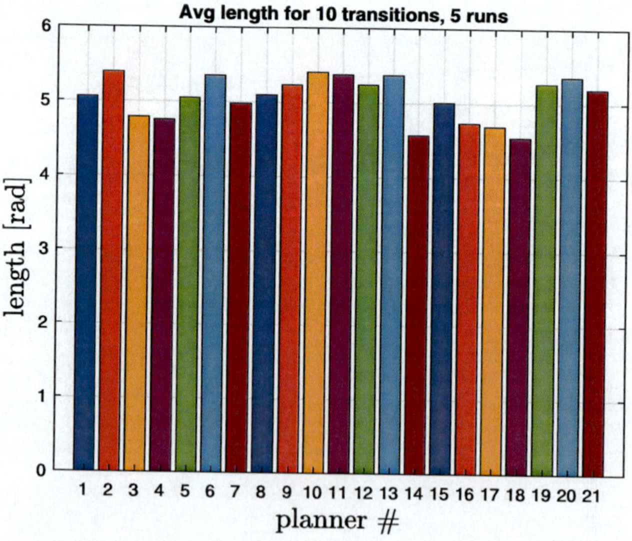

Fig. 6. Average length, defined by Eq. (3), of the calculated trajectory between 10 random pairs of poses, 5 times per pair, for 21 motion planning algorithms listed in Table 1.

with l the number of robot joints, q_i the i-th joint goal position and q_{i0} the i-th joint start position.

Table 2. Time of travel between example sets of vantage points.

Vantage point #	times (s)
1,2,3,4,5,6,7,8,9,10	33.27
10,6,2	12.25
3,5,9	13.69
1,2,3	10.21
7,10,3	11.97
4,9,1	11.35
8,3,5	12.07
5,6,10	12.88
2,4,8	10.98
6,7,8	11.53
9,1,7	13.5
mean	12.04 ± 1.04

Looking at the smoothness and length of these planners, we can see that planner 15, i.e., T-RRT (Transition-based RRT [28]) is the best in both of these categories amongst the remaining 7 planners that were not excluded based on the calculation time.

T-RRT combines the exploration strength of the RRT algorithm, which rapidly grows random trees toward unexplored regions of the space, with the efficiency of stochastic optimization methods that use transition tests to accept or reject any new potential state. It does not provide any hard optimality guarantees, but usually results in a short, low-cost path.

Given the selected motion planner, we evaluated how much time can be saved by the proposed system if the robot visited three out of all 10 possible vantage points. The robot postures at these 10 vantage points are shown in Fig. 2. The results for the complete series of vantage points and for 10 transitions between three random vantage points are shown in Table 2.

The minimum time needed to pass through the shortest paths between all 10 of them, with sufficient time given to capture good quality images, is 33.27 s. On the other hand, visiting only 3 vantage points takes on the average 12.04 s ± 1.04 s. Still images of real-world experiments for two random sub-sets of postures are given in Fig. 7.

5 Discussion

As stated in [13], there is no general solution for which motion planner to use and when. In this paper, we have shown how we can evaluate and choose the

Fig. 7. Sequences of still-images depicting the real-world execution of the generated trajectories for motions through two different sets of postures, with a UR-10 robot with an in-hand camera.

most appropriate planning algorithm for a given use-case. In our example, it was Transition-based RRT.

It should be noted that in the presented use-case, the selection of camera postures and vantage points over the object of inspection is performed randomly from a set of predefined postures. If we assume an even distribution of defects over time and over the possible defect locations, then randomly checking 30% of the locations will result in detecting 30% of the errors. While this way we miss most of the defects, we still reduce the number of delivered defective products by these 30%. In some cases, this might be enough. However, such checking is more useful to detect an increase in the number of defects. If such an increase is detected, we need to stop production and eliminate the cause of the defects.

Assuming an even distribution of defects is highly unlikely. It is far more likely that defects will appear more often at some location than the other, for example, because of machine wear. Thus, randomized checking will not be effective anymore, and directed inspection should be performed. ZDM approaches try to predict and eliminate the defects. This paper should be considered in the scope of possible prediction of defects, for example, from a set of production parameters. Given enough data and a correlation between the production parameters and defects, a deep neural network can be trained to predict the chances of defects appearing at locations. Then, the proposed algorithm can be used to generate the motion between not-anymore-random inspection postures, but postures ordered by the predicted probability. Similar, targeted coverage path planning was shown, for example, in [29]. For our use-case, which is based on a real-world injection-moulding production line, prediction of defects requires really a lot of data and remains open for future work.

6 Conclusion

In the paper, we have shown how we can generate motion trajectories for visual quality control for a random sub-set of given robot-guided camera positions. Because moving between postures of a random subs-set of even a small set requires new trajectories every time, these have to be planned every time. We have shown how different motion planning algorithms can be evaluated and we have shown which was best for our use-case. This is useful if we can only check a sub-set of inspection points, for example, because of time constraints. However, the main usefulness of this paper is as preparation for directed motion between vantage points. For example, a prediction of possible defects can be used to determine the sub-set of vantage points that need to be visited. While the prediction of possible defects remains a work for the future, the given paper gives the required robotic ground-work for the implementation.

References

1. Psarommatis, F., May, G., Dreyfus, P.-A., Kiritsis, D.: Zero defect manufacturing: state-of-the-art review, shortcomings and future directions in research. Int. J. Prod. Res. **58**(1), 1–17 (2020)
2. Ahuett-Garza, H., Kurfess, T.: A brief discussion on the trends of habilitating technologies for industry 4.0 and smart manufacturing. Manufact. Lett. **15**, 60–63 (2018)
3. Herakovič, N.: Robot vision in industrial assembly and quality control processes. In: Robot Vision, IntechOpen, Rijeka, Ch. 26, pp. 501–534 (2010)
4. Cho, C.-S., Chung, B.-M., Park, M.-J.: Development of real-time vision-based fabric inspection system. IEEE Trans. Indus. Electron. **52**(4), 1073–1079 (2005)
5. Lončarević, Z., Gams, A., Reberšek, S., Nemec, B., Škrabar, J., Skvarč, J., Ude, A.: Specifying and optimizing robotic motion for visual quality inspection. Robot. Comput.-Integr. Manuf. **72**, 102200 (2021)
6. Montironi, M., Castellini, P., Stroppa, L., Paone, N.: Adaptive autonomous positioning of a robot vision system: application to quality control on production lines. Robot. Comput.-Integr. Manuf. **30**(5), 489–498 (2014)
7. Triggs, B., Laugier, C.: Automatic camera placement for robot vision tasks. In: IEEE International Conference on Robotics and Automation (ICRA), Nagoya, Japan, pp. 1732–1737 (1995)
8. Zang, C., Hashimoto, K.: A flexible camera positioning strategy for robot-based visual inspection applications. In: Conference on Mechatronics and Automation (ICMA), pp. 527–532. IEEE, China, Chengdu (2012)
9. Bevec, R., Gašpar, T., Ude, A.: Robot-driven autofocus control mechanism for an in-hand fixed focus camera. In: Aspragathos, N.A., Koustoumpardis, P.N., Moulianitis, V.C. (eds.) Advances in Service and Industrial Robotics, pp. 551–559. Springer International Publishing, Cham (2019)
10. Villalba-Diez, J., Schmidt, D., Gevers, R., Ordieres-Meré, J., Buchwitz, M., Wellbrock, W.: Deep learning for industrial computer vision quality control in the printing industry 4.0. Sensors (Basel, Switzerland) 19(18)
11. Zavrtanik, V., Kristan, M., Skočaj, D.: Reconstruction by inpainting for visual anomaly detection. Pattern Recogn. **112**, 107706 (2021)

12. Mac, T.T., Copot, C., Tran, D.T., De Keyser, R.: Heuristic approaches in robot path planning: a survey. Robot. Autonom. Syst. **86**, 13–28 (2016)
13. Moll, M., Sucan, I.A., Kavraki, L.E.: Benchmarking motion planning algorithms: an extensible infrastructure for analysis and visualization. IEEE Robot. Autom. Mag. **22**(3), 96–102 (2015)
14. Choset, H., Lynch, K., Hutchinson, S., Kantor, G., Burgard, W., Kavraki, L., Thrun, S.: Principles of Robot Motion: Theory, Algorithms, and Implementations. MIT Press (2005)
15. Gasparetto, A., Boscariol, P., Lanzutti, A., Vidoni, R.: Path Planning and Trajectory Planning Algorithms: A General Overview. Mechanisms and Machine Science
16. Patle, B., Pandey, G.L.D.A., Parhi, D., Anne, J.: A Review: On Path Planning Strategies for Navigation of Mobile Robot. Defence Technology
17. Kavraki, L., Latombe, J.-C.: Randomized preprocessing of configuration space for path planning: articulated robots. In: Proceedings of IEEE/RSJ International Conference on Intelligent Robots and Systems, vol. 3, pp. 1764–1771 (1994)
18. Karaman, S., Frazzoli, E.: Sampling-based algorithms for optimal motion planning. Int. J. Robot. Res. **30**(7), 846–894 (2011)
19. Lee, J., Kwon, O., Zhang, L., Yoon, S.: A selective retraction-based rrt planner for various environments. IEEE Trans. Robot. **30**(4), 1002–1011 (2014)
20. Karaman, S., Frazzoli, E.: Sampling-based algorithms for optimal motion planning. Int. J. Robot. Res. **30**(7), 846–894 (2011)
21. Mac, T., Copot, C., Tran, D., Keyser, R.: Heuristic approaches in robot path planning: a survey. Robotics and Autonomous Systems
22. Coleman, D., Sucan, I., Chitta, S., Correll, N.: Reducing the Barrier to Entry of Complex Robotic Software: A Move It! Case Study (2014). arXiv:1404.3785
23. Şucan, I.A., Moll, M., Kavraki, L.E.: The open motion planning library. IEEE Robot. Autom. Mag. **19**(4), 72–82 (2012)
24. Kalakrishnan, M., Chitta, S., Theodorou, E., Pastor, P., Schaal, S.: Stomp: stochastic trajectory optimization for motion planning. In: IEEE International Conference on Robotics and Automation, pp. 4569–4574 (2011)
25. Ratliff, N., Zucker, M., Bagnell, J.A., Srinivasa, S.: Chomp: gradient optimization techniques for efficient motion planning. In: IEEE International Conference on Robotics and Automation, pp. 489–494 (2009)
26. Likhachev, M.: Search-Based Planning Library (sbpl) (2016)
27. Kingston, Z., Moll, M., Kavraki, L.E.: Exploring implicit spaces for constrained sampling-based planning. Int. J. Robot. Res. **38**(10–11), 1151–1178 (2019)
28. Jaillet, L., Cortes, J., Simeon, T.: Transition-based rrt for path planning in continuous cost spaces. In: IEEE/RSJ International Conference on Intelligent Robots and Systems, pp. 2145–2150 (2008)
29. Glorieux, E., Franciosa, P., Ceglarek, D.: Coverage path planning with targetted viewpoint sampling for robotic free-form surface inspection. Robot. Comput.-Integr. Manuf. **61**, 101843 (2020)

Articulated Objects: From Detection to Manipulation—Survey

Petra Pejić(✉), Valentin Šimundić, Matej Džijan, and Robert Cupec

Faculty of Electrical Engineering, Computer Science and Information Technology, Osijek, Croatia
petra.pejic@ferit.hr

Abstract. Robotic manipulation of articulated objects, such as opening and closing doors, drawers, and cabinets, appears to be an emerging research topic in robotics. This is evidenced by a large number of recent research papers on this topic, as well as annotated datasets of articulated objects and their parts. This article reviews the state of the art and compares image-processing-based methods and benchmark datasets for detecting and segmenting articulated objects, determining their joint parameters, and finally manipulating articulated objects by a robot.

1 Introduction

Objects consisting of parts with one or more translational or revolute joints are called articulated objects. In order to interact with articulated objects, e.g., open doors, drawers, or cabinets, unscrew bottle caps, open laptops, or handle tools such as scissors or pliers, the robot must recognize their moving parts. An example of an articulated object is shown in Fig. 1. In recent research, this topic is being explored more and more. As far as we know, it has not been systematized yet. Therefore this article provides an overview of this research area.

When it comes to vision-based handling of articulated objects and their moving parts, there are various approaches in the relevant research. Usually, some kind of machine learning is applied. In both learning and testing, most algorithms take point clouds (PC) as input, generated either from CAD 3D models or from RGB-D images. Methods that work in 2D are rarely found. The developed algorithms are usually used in robotics for various autonomous manipulation tasks.

Existing research in this area addresses several subproblems related to articulated objects: recognition and segmentation of articulated objects and their moving parts, estimation of their joint configurations, and performing some kind of manipulation with the articulated objects. Thus this article is roughly divided into three main sections, each dealing with one of the above problems, although they sometimes overlap and a clear distinction between them is not always easy. In Sect. 2, methods for detecting and segmenting parts of articulated objects are

This work has been fully supported by the Croatian Science Foundation under the project number IP-2019-04-6819.

I. Petrovic et al. (Eds.): IAS 2022, LNNS 577, pp. 495–508, 2023.
https://doi.org/10.1007/978-3-031-22216-0_34

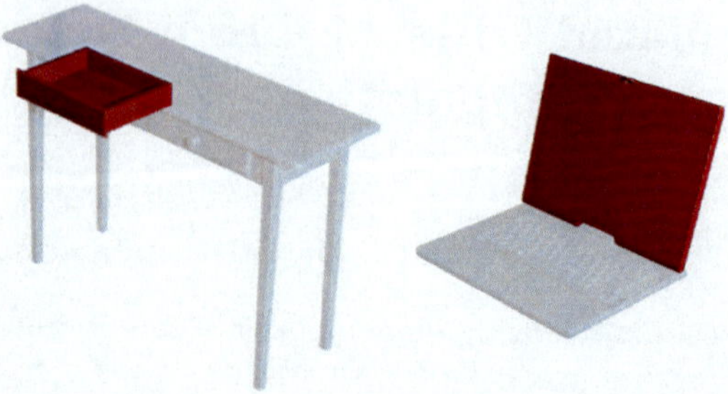

Fig. 1. Articulated objects with moving parts (colored red): a drawer that moves along a translation axis and a laptop display that rotates around a rotation axis. The models are taken the PartNet dataset (Mo et al. 2019))

analyzed. Methods for determining joint parameters of articulated objects and their moving parts are reviewed in Sect. 3. Section 4 presents methods that focus on manipulation with articulated objects to determine or evaluate their joint parameters. The paper is concluded in Sect. 5.

2 Detection and Segmentation of Articulated Objects

While there are many recent methods for semantic segmentation of 2D and 3D data and methods for part segmentation, for this survey, we consider methods that can be used for detecting moving parts of articulated objects. Several such methods are summarized in Table 2, some of which are described below.

A very common problem is the detection of doors and drawers and their handling. Llopart et al. (2017) propose a method which uses both RGB and depth data to detect handles on doors and cabinets. They first use RGB data to detect a region of interest (ROI) that contains a single door or drawer. Then they detect a single handle in that ROI in two ways, by color and by geometry. They first segment the ROI based on color using the k-means algorithm to get the handle segmentation. They then project the ROI onto the PC, detect the dominant plane, and, assuming that it is door or a drawer plane, extract the points that are not on that plane and consider them as points of a handle. Once they have extracted the handle both ways, they simply fuse the extractions using the logical operator AND extracting only the points extracted by both methods. Finally, they extract the key data for handle manipulation. This method achieves an average error rate of 0.96% in handle detection for a distance of 0.5 m, 3.11% for 1 m and 10.55% for 1.5 m. This method works best under certain conditions with single color, planar doors and drawers. The authors also show that the method does not work as well at distances greater than 1 m and from certain angles. The proposed approach, which combines both visual segmentation and planar model extraction, outperforms either method on its own.

The detection of other articulated objects has also been reported in the literature. Huang et al. (2021) propose a method that takes a set of PCs sampled

Table 1. Datasets of Articulated Objects

Name	Properties	Joint parameters
ImageNet (Deng et al. 2009)	3.2 million cleanly annotated images spread over 5247 categories in 12 subtrees	×
REAL275 (Wang et al. 2019a)	3 unique instances (3D mesh) in 6 object categories, 2750 RGB-D frames of 6 real scenes, each with at least 5 object instances	×
ShapeNet (Chang et al. 2015)	51,300 unique PC models in 55 classes	×
Shape2Motion (Wang et al. 2019b)	2440 models in 45 shape categories (from ShapeNet and 3D Warehouse)	Manually labeled motion parts and the corresponding motion attributes
PartNet-Mobility (Mo et al. 2019)	A continuation of ShapeNet and PartNet (Mo et al. 2019), 2346 models in 46 shape categories	Tree-structured joint system (hinge, slider and screw motions)
RBO (Martín-Martín et al. 2018)	358 RGB-D video sequences (67') of humans interacting with 14 articulated objects	Annotations of poses of the rigid parts and kinematic state of the articulated objects

from the same articulated object as input, segments the PCs into rigid parts and infers the motion for each part. First, the point correspondences between all PC pairs must be determined. To do this, a 3D scene flow is predicted for each PC pair using a deep neural network. Then, relative motion segmentation matrices are predicted for each pair of PCs. A relative motion segmentation matrix shows which points of a pair of PCs belong to the same rigid part. These matrices are used to predict the number of parts and the part segmentation. Finally, the motion of each part is estimated followed by joint pose estimation. During the test time, this procedure is iterated several times. In terms of segmentation accuracy, they achieved 66.7% multi-scan mIoU on the SAPIEN dataset (Xiang et al. 2020) and 90.7% multi-scan mIoU on their DynLab dataset. DynLab consists of 8 different scenes composed of moving solid objects.

While the aforementioned method (Huang et al. 2021) requires multiple PCs as input, Kawana et al. (2021) propose a method that uses only one PC as input. Moreover, their method is unsupervised. Their generative unsupervised method segments PCs of articulated objects into parts and predicts the poses of each of these parts. First they encode the PC using a PointNet (Qi et al. 2016)-based architecture (Michel et al. 2019) to obtain latent vectors for shape and pose. The pose vectors are then decoded to obtain part poses, and these poses are used together with the latent vector for shape to obtain a composite implicit field. This implicit field divided into parts models the shape of the object and its

Table 2. Methods for part detection and pose estimation of articulated objects and their parts

Method	Methodology	Inference input	Training dataset	Output
Llopart et al. (2017)	CNN & k-means	RGB-D image	ImageNet	Segmentation of handle points
Weng et al. (2021)	NN	Number of parts, PC sequence, initial part pose	NOCV-REAL275 & Sapien	9 DoF pose for the current frame for each part
Huang et al. (2021)	NN	Multi-view PC	(Yi et al. 2016) & ShapeNet	Part segmentation and their SE(3) transformations
Michel et al. (2015)	Random forest	RGB-D scan, number of parts and types of joints	Generated dataset using 3D models	6D pose of each part
Kawana et al. (2021)	NN	Point cloud	Shape2Motion and SAPIEN	Implicit field and SE(3) pose for each part
Pavlasek et al. (2020)	NN and Markov random field	Complex RGB-D image, articulation constraints, 3D geometry (URDF), 3D mesh models	Custom real dataset and ImageNet	6D pose of each part of each object
Yi et al. (2018)	NN	Pair of point clouds	(Yi et al. 2016)	Part segmentation
Zeng et al. (2021)	NN	RGB-D image and segmentation images of two object parts	PartNet-Mobility	Motion residual flow and part connectedness
Gadre et al. (2021)	NN	RGB images of an object with different joint configurations	PartNet-Mobility	Motion mask for a part that moved

moving parts. For both pose and shape, they utilize category-common decoders, whose outputs are combined with instance-specific decoders for prediction. They evaluate their method using five categories from two datasets: Motion (Wang et al. 2019a) (oven, eyeglasses, laptop, and washing machine) and SAPIEN (Xiang et al. 2020) (drawer). They achieve a mIoU of 81.0%. They also verify their approach on real data for one category, laptop, from the RBO dataset (Martin-Martin et al. 2019) and the Articulated Object Dataset (Michel et al. 2015). Although their method is unsupervised, it shows comparable performance in estimating the pose parts as the supervised baseline. Despite their model being trained on synthetic data, the authors claim it can be applied to real-world data.

Unlike other methods that combine RGB and depth data, Gadre et al. (2021) propose an approach for detecting and segmenting object parts using only RGB data. The method uses two types of networks. The Interaction Network computes the “push” and “hold” actions from an RGB image and the current memory of the parts. The next observation of the object, along with the first, is fed into the Part Network, which uses the information about actions and both observations to derive motion masks for the moving part. Finally, the History Aggregation module uses newly discovered and segmented parts to update the existing parts in memory. The evaluation included numerous object categories, both in simulation and with real data. Objects used for the evaluation included pliers, microwaves, eyeglasses, lighters, and multilinks (three-part generated objects) from the PartNet mobility dataset. The method averaged 60.16% mIoU across all simulated test categories. Although the method does not need to know the object category, it only assumes structures with no more than three links and restricts the direction of motion to be parallel to the ground.

3 Estimation of Joint Parameters and Configuration

Joint parameters specify the direction of translation or axis of rotation of a moving part with respect to the rest of the object. The joint configuration contains information about the current state of the joint variables, i.e., the angle of each revolute joint and the translation along each prismatic joint. The comparison with the ground truth of the joint variables and the orientation and position of the joint axes is usually given as the mean and standard deviation of the error. Estimation of such parameters is essential for articulated objects and distinguishes recognition of articulated from rigid objects. In this section, several methods for estimating the joint parameters and determination of joints' configuration of articulated objects are presented. The methods are summarized in Table 3.

Typical results of such methods are shown in Fig. 2. These results were obtained using the NOCS-based (Wang et al. 2019a) method described in Li et al. (2020) with data from the PartNet-Mobility dataset (Mo et al. 2019). For a new object, from a known category, represented by a single depth image, this network estimates the part segmentation, pose prediction and joint parameters estimation. Figure 2a shows a rendered drawer object with parts highlighted. The parts of the same object are segmented in Fig. 2b and the directions of the joint axes for each drawer are shown in Fig. 2c.

Liu et al. (2020) proposed an unsupervised, model-free articulated object pose estimation system which estimates the 6D pose of each rigid body in an articulated object with their joint parameters using only video sequences as input, the geometric constraints and 2D dense correspondence. The method was developed for finding the joint parameters of robotic arms, but it can be applied to several other categories without additional training. It is not possible to determine the pose or articulation form a single view, though.

The idea of estimating the pose of an articulated object based on an image sequence, without having a CAD model of the object was also explored by Jain et al. (2021a) (DUST-net) and Jain et al. (2021b) (ScrewNet). Their approaches differ mainly in architecture, but have similar properties. Both methods use sequences of depth images in which object parts move relative to each other to compute a category-independent articulation model and a distribution over joint parameters and joint configurations. They estimate the articulation properties of objects, such as cabinets, drawers, microwaves, toasters and ovens, with uncertainty estimates, which is useful in robotic manipulation. In addition to revolute,

Table 3. Methods for estimation of joint parameters of articulated objects

Method	Methodology	Inference input	Training dataset	Output
Liu et al. (2020)	Geometric formulation and deep learning (GC-Pose)	Video sequence	PartNet-Mobility, Craves, and their own dataset	6D pose of object joints and joint parameters
Jain et al. (2021a) and 2021b	NN (DUST-net and ScrewNet)	Sequences of depth images of object parts moving relative to each other	PartNet-Mobility and Abbatematteo et al. (2020)	Joint parameters, and model uncertainty estimates
Wang et al. (2019b)	NN (Shape2Motion)	3D shape PC	Shape2Motion	Mobility-oriented segmentation and motion attributes
Desingh et al. (2019b)	Markov random field	3D models with articulation constraints and observed 3D sensor data	Real world data	Object-part pose beliefs
Li et al. (2020)	NN (ANCSH)	3D PC from a single depth image	Shape2Motion and SAPIEN	Part pose, segmentation, joint parameters and joint state
Zhang et al. (2021)	NN (Strobe-Net)	Sparse RGB images with different articulation	Their dataset, SAPIEN, and Shape2Motion	Joint parameters of object parts, 3D reconstruction
Noguchi et al. (2021)	NN	Video sequence of moving articulated objects with foreground mask	ZJU-MoCap	Parts and joints discovery; novel views rendering

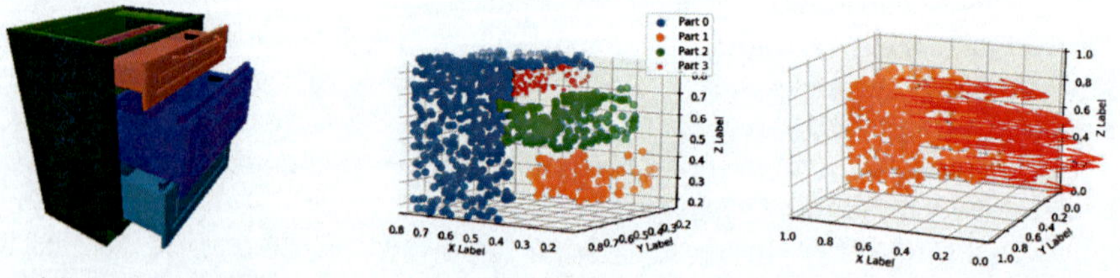

(a) Synthetic render (b) Part segmentation (c) Direction of joint axis

Fig. 2. Typical result of part detection and estimation of joint parameters

prismatic and rigid motion, ScrewNet can estimate the helical model (motion of a screw). Moreover, these networks do not require the category of the joint model in advance. Because they use a single network for all common joint models, they are more computationally and data efficient. Besides on the PartNet-Mobility dataset (Mo et al. 2019), methods are evaluated on three real-world objects. However, DUST-Net and ScrewNet can only directly predict parameters for 1-DOF joint models, otherwise additional image segmentation is required to remove all irrelevant object parts.

Wang et al. (2019b) proposed Shape2Motion, an annotated dataset containing various objects, from toys to household items, and method consisting of two deep neural networks that compute mobility-oriented segmentation and associated motion attributes based on the single 3D point cloud model. They keep on mind that semantic parts often do not match motion parts, and that a motion part can possess multiple motions. By working with point clouds representations, this method may not be able to represent highly detailed structures. Also, it is not analyzed on scanned real-world objects.

Desingh et al. (2019b) propose an approach to estimate the poses of articulated objects, using an efficient nonparametric belifes propagation algorithm. The system takes a 3D point cloud from a sensor and an object geometry model as input and provides belief samples in the continuous pose domain, that are used to estimate the pose of an object part, that will eventually be used in robotic manipulation (Desingh et al. 2019a). The required geometry models consist of a 6D pose and manually defined joint parameters. The method is evaluated on two real world objects, a cabinet with three drawers and Fetch robot. It is shown that the method works even in the presence of occlusion.

Zhang et al. (2021) proposed StrobeNet, a neural network that reconstructs an animatable 3D shape from a set of sparse RGB images by predicting the canonical point cloud, part segmentation and joint position and orientation in NOCS, as well. It does not require category-level templates. This method has several limitations. Although theoretically possible, it does not translate adequately to real-world scenes and assumes a fixed number of joints per category.

Table 1 provides an overview of the datasets created to evaluate methods for detecting articulated object and estimating their joint parameters. These benchmark datasets typically include annotations of parts and joint parameters.

Methods for computing joint parameters of articulated objects using robot manipulation are presented in Sect. 4.

4 Robot Manipulation with Articulated Objects

An important aspect of this research is the study of autonomous interaction of robots with articulated objects. In general, two cases have been studied in which a robot manipulates with an articulated object: (i) manipulation with the goal of estimating the joint parameters of the object, and (ii) manipulation with the goal of evaluating the joint parameters of the object. While in the first case a kinematic model is created by the manipulation, in the second case the kinematic model obtained by other means is evaluated by the manipulation. The manipulation is evaluated by the robot's attempts to successfully interact with an object based on its kinematic model.

All methods described in the following subsections are summarized in Table 4. The table shows which methods were used to estimate or evaluate the joint parameters of objects, followed by the inputs and outputs of each method and whether they use a simulation in any step of the process.

4.1 Manipulation for Estimation of Kinematic Parameters

Sturm et al. (2010) address the problem of estimating joint parameters from the end-effector trajectory of the robot. The robot records the trajectory of its end effector over time to estimate a kinematic model. A type of impedance control, an equilibrium point control, is used in conjunction with the estimated kinematic model to generate smooth Cartesian equilibrium point trajectories for opening doors. Compared to their previous work, the authors concluded that the use of RANSAC-based model estimators for revolute models outperforms the gradient descent minimizer in terms of average trajectory prediction error. The results show that for 40 attempts to open an object (cabinet doors, a dishwasher, a drawer, and a sliding door were used), the method was successful 92.5% of the time. It is also shown that remembering previous models further lowers the prediction errors. Ruhr et al. (2012) limited the previous method to opening doors and drawers in kitchen environments. Their approach consists of several components, with handle detection added alongside the manipulation methods of Sturm et al. (2010). Although the robot successfully opened a cabinet in 56.2% of the trials, the achieved probability of correct model selection is over 95%. The main cause of failure in opening doors and drawers is that the robot is too weak to apply the force to open them, followed by slipping of the gripper.

Katz et al. (2014) presented an approach that uses the scripted motion of the manipulator to interact with objects. The robot's motion makes the objects physically move, which can reveal their kinematic properties. Unlike Sturm et al. (2010), who assume that only one object part can move at a time, this method identifies multiple rigid bodies in the scene. The identification of rigid bodies makes the method more general than Ruhr et al. (2012), as the latter focuses only on doors and drawers. The method is decomposed into three subtasks: (i) segmentation based on color, texture and spatial features, (ii) depth reconstruction from an RGB image, and (iii) joint detection based on a comparison between the trajectories of two parts. The method yielded the kinematic structure in 100% of the experiments. The position and orientation of the joints were detected in 30 out of 31 experiments. The evaluation was performed on: a prismatic cabinet door, a train toy, a spinning wheel, an elevator door, a shelf on wheels, a door and a box. A similar approach was used in Katz et al. (2013), using an RGB-D sensor instead of the RGB sensor, which allowed the use of SIFT features and KLT feature tracking while eliminating bundle adjustment.

Hausman et al. (2015) proposed a human demonstration based approach that combines visual observations and manipulation feedback. Estimation is performed by selecting actions to derive an articulation model using a Bayesian update formula implemented as a particle filter. Each particle in the particle filter represents an articulation model type and its associated parameters. No force or torque sensor is used to measure forces in the robot's wrist. Instead, an effort controller is implemented to detect whether the robot motion is expected for a particular type of articulation model or whether a resistive motion is present. Experimental scenarios include: a cabinet door, two free-body erasers, a locked drawer to represent a rigid joint, and a stapler as a revolute joint. The results show that an efficient way to reduce the entropy of the particle distribution is to use the KL-divergence to select the best short motion of the robot that most reduces the uncertainty of the model.

Mo et al. (2021) use neural network output in the form of "actionable" pixel probabilities to select an area on the object that is likely to move when pushed or pulled on. Unlike other methods, this method is not limited to handles when pushing on an object as it allows pushing on any movable part of the object. The authors used only a two-fingered flying gripper that can be initialized in any position and orientation to manipulate the object in the scene. The synthetic dataset used to train and evaluate the network was created in the SAPIEN simulator, using buckets, kitchen pots, safes tables, and washing machines. The highest F-score obtained by the method on the test data is 26.61% (the push action). However, the method is limited to only six types of preprogrammed gripper motions and does not explicitly model an object as articulated.

Arduengo et al. (2021) approached door opening with a trial-and-error strategy. After detecting the door and the centroid of the door handle with the YOLO detector, the robot aligns its hand normal with the door normal at the point where it grasps the handle. An appropriate motion to open the door is derived by a framework that uses the Bayesian model. The opening of the door is performed by a framework called Task Space Region (TSR), which models specific path planning constraints based on the kinematic model of the object and the pose constraints of the end effector. The robot detects if the handle is unlocked by rotating its hand in both directions and opens the door. The results show that the method has a success rate of 87% in learning the kinematic model, achieving better results than the directly comparable method of Ruhr et al. (2012), with no problems in applying force to open the doors. The failures occur when the gripper slips off the handle.

4.2 Manipulation for Evaluation of Joint Parameters

Abbatematteo et al. (2020) used robotic manipulation as a way to evaluate the previously estimated kinematic model. The robot records observations that are used to estimate the position and orientation of the object. A fixed grasp is provided by a human on the object's handle and a series of waypoints are computed as a function of the estimated object kinematics, end-effector pose and desired object configuration. Based on these estimates, the mobile robot is able to generate kinematic motion plans that achieve the desired manipulations in a pre-computed map of a room. The experiments were performed on two objects: a microwave and a drawer.

Mittal et al. (2021) considered the planning and control of a mobile manipulator as a holistic optimization problem. The control of the base and arm was combined with representation of the environment as a Euclidean signed distance field derived from an occupancy map to avoid obstacles while performing

Table 4. Manipulation methods for estimating or evaluating joint parameters

Method	NN	Input	Output	Simulation
		Estimation methods		
Sturm et al. (2010)	×	Robot trajectory	Joint parameters	×
Ruhr et al. (2012)	×	Point cloud, robot trajectory	Joint parameters	×
Katz et al. (2014)	×	RGB	Joint parameters	×
Katz et al. (2013)	×	RGB-D	Joint parameters	×
Hausman et al. (2015)	×	RGB-D, AR fiducial tags	Joint parameters	×
Mo et al. (2021)	✓	RGB-D	Interaction hotspots	✓
Arduengo et al. (2021)	✓	RGB-D	Joint parameters	✓
		Evaluation methods		
Abbatematteo et al. (2020)	✓	RGB-D, grasp on handle	Motion plan	×
Mittal et al. (2021)	✓	RGB-D, grasp on handle	Motion plan	✓
Xu et al. (2021)	✓	RGB-D	Interaction position, push/pull action	✓

an articulated object manipulation task. The kitchen environments, in which the system was evaluated, were created in NVIDIA Isaac Sim, using the PartNet dataset to design the layouts. The evaluation showed a 93.8% success rate (including washing machine, drawer and oven) on 60 attempts to open objects.

Xu et al. (2021) used a different approach to pull or push the articulated object with a suction-based gripper, using a single-policy network that discovers possible manipulation strategies from RGB-D images (UMPNet). The network is trained and tested with data from interaction trials simulated in Pybullet. The network outputs an Array-of-Time (AoT), a label indicating whether a robot action returns the object to its previous state or to a new state. This is used to infer the direction in which the suction cup should push or pull to achieve the specified configuration. While the previous methods only include drawers and objects with some kind of doors, this method is more general in that it includes other types of objects with moving parts. The authors evaluated their approach on four different objects: a box, a laptop, a microwave and a stapler. However, this approach does not explicitly model objects as articulated and is most similar to Mo et al. (2021) for selecting the interaction area rather than estimating the pose of each part.

5 Conclusion

This paper reviews recent methods that can be used to detect articulated objects and determine their joint parameters using computer vision and robot manipulation. It can be noted that the mentioned methods are usually based on neural networks (NN) or classical image processing methods. The computations are performed in 3D, mostly using point clouds as input. The result is a segmentation of the articulated objects into parts, a 6D pose of the parts or joint parameters. The methods are evaluated on benchmark datasets. Note that all methods, except for Wang et al. (2019b), in Tables 2 and 3 provide some kind of evaluation on real data. Some methods are evaluated in SAPIEN (Xiang et al. 2020), a realistic and physics-rich simulated environment that includes a large-scale set for articulated objects from PartNet-Mobility (Mo et al. 2019). It seems that there are many application and research paths dealing with articulated objects, and there is a lack of consistency in comparing the results obtained.

An interesting subfield of the discussed research area is the recognition of the human body, which can also be considered as an articulated object. Due to the large practical application of this problem, there is a considerable amount of research in this area, e.g. Deng et al. (2015), Fouhey et al. (2014), Güler et al. (2018), Wu et al. (2019), Zimmermann et al. (2018). However, this research is not the focus of this survey.

References

Abbatematteo, B., Tellex, S., Konidaris, G.: Learning to generalize kinematic models to novel objects. In: Proceedings of the Conference on Robot Learning, pp. 1289–1299. PMLR (2020)

Arduengo, M., Torras, C., Sentis, L.: Robust and Adaptive Door Operation with a Mobile Robot **14**(3), 409–425 (2021)

Chang, A.X. et al.: Shapenet: An Information-Rich 3d Model Repository (2015). arXiv:1512.03012

Deng, B., Lewis, J.P., Jeruzalski, T., Pons-Moll, G., Hinton, G., Norouzi, M., Tagliasacchi, A.: NASA: Neural Articulated Shape Approximation (2020)

Deng, J., Dong, W., Socher, R., Li, L.-J., Li, K., Fei-Fei, L.: Imagenet: A large-scale hierarchical image database. In: 2009 IEEE Conference on Computer Vision and Pattern Recognition, pp. 248–255. IEEE (2009)

Desingh, K., Lu, S., Opipari, A., Jenkins, O.C.: Efficient nonparametric belief propagation for pose estimation and manipulation of articulated objects. In: Science Robotics (2019a)

Desingh, K., Lu, S., Opipari, A., Jenkins, O.C.: Factored pose estimation of articulated objects using efficient nonparametric belief propagation. In: 2019 International Conference on Robotics and Automation (ICRA), pp. 7221–7227. IEEE (2019b)

Fouhey, D.F., Delaitre, V., Gupta, A., Efros, A.A., Laptev, I., Sivic, J.: People watching: human actions as a cue for single view geometry. In: People Watching (2014)

Gadre, S.Y., Ehsani, K., Song, S.: Act the part: learning interaction strategies for articulated object part discovery. In: ICCV (2021)

Güler, R.A., Neverova, N., Kokkinos, I.: Dense-Pose: Dense Human Pose Estimation in the Wild. (2018). arXiv:1802.00434

Hausman, K., Niekum, S., Osentoski, S., Sukhatme, G.S.: Active articulation model estimation through interactive perception. In: 2015 IEEE International Conference on Robotics and Automation (ICRA), pp. 3305–3312. Seattle, WA, USA. IEEE (2015)

Huang, J., Wang, H., Birdal, T., Sung, M., Arrigoni, F., Hu, S.-M., Guibas, L.: Multi-BodySync: Multi-Body Segmentation and Motion Estimation via 3D Scan Synchronization, pp. 7104–7114 (2021)

Jain, A., Giguere, S., Lioutikov, R., Niekum, S.: Distributional Depth-Based Estimation of Object Articulation Models (2021a). arXiv:2108.05875

Jain, A., Lioutikov, R., Chuck, C., Niekum, S.: ScrewNet: category-independent articulation model estimation from depth images using screw theory. In: 2021 IEEE International Conference on Robotics and Automation (ICRA), pp. 13670–13677. IEEE (2021b)

Katz, D., Kazemi, M., Andrew Bagnell, J., Stentz, A.: Interactive segmentation, tracking, and kinematic modeling of unknown 3D articulated objects. In: 2013 IEEE International Conference on Robotics and Automation, pp. 5003–5010. IEEE, Karlsruhe, Germany (May 2013)

Katz, D., Orthey, A., Brock, O.: Interactive perception of articulated objects. In: Khatib, O., Kumar, V., Sukhatme, G. (eds.), Experimental Roboticspp, pp. 301–315, vol. 79. Springer, Berlin (2014)

Kawana, Y., Mukuta, Y., Harada, T.: Unsupervised Pose-Aware Part Decomposition for 3D Articulated Objects. arXiv: 2110. 04411 (2021)

Li, X., Wang, H., Yi, L., Guibas, L.J., Lynn Abbott, A., Song, S.: Category-level articulated object pose estimation. In: 2020 IEEE/CVF Conference on Computer Vision and Pattern Recognition (CVPR), pp. 3703–3712. IEEE (2020)

Liu, Q., Qiu, W., Wang, W., Hager, G.D., Yuille, A.L.: Nothing But Geometric Constraints: A Model-Free Method for Articulated Object Pose Estimation (2020). arXiv:2012.00088

Llopart, A., Ravn, O., Andersen, N.A.: Door and cabinet recognition using convolutional neural nets and real-time method fusion for handle detection and grasping. In: 2017 3rd International Conference on Control, Automation and Robotics, ICCAR 2017, pp. 144–149 (2017)

Martin-Martin, R., Eppner, C., Brock, O.: The RBO dataset of articulated objects and interactions. Int. J. Robot. Res. **38**, 027836491984431 (2019)

Martín-Martín, Roberto, C.E., Brock, O.: The RBO Dataset of Articulated Objects and Interactions (2018). arXiv:1806.06465

Mescheder, L., Oechsle, M., Niemeyer, M., Nowozin, S., Geiger, A.: Occupancy networks: learning 3D reconstruction in function space. In: Proceedings of the IEEE/CVF Conference on Computer Vision and Pattern Recognition (CVPR) (2019)

Michel, F., Krull, A., Brachmann, E., Yang, M.Y., Gumhold, S., Rother, C.: Pose estimation of kinematic chain instances via object coordinate regression. In: British Machine Vision Association and Society for Pattern Recognition (2015), pp. 181.1-181.11

Mittal, M., Hoeller, D., Farshidian, F., Hutter, M., Garg, A.: Articulated Object Interaction in Unknown Scenes with Whole-Body Mobile Manipulation (2021). arXiv:2103.10534

Mo, K., Guibas, L., Mukadam, M., Gupta, A., Tulsiani, S.: Where2Act: From Pixels to Actions for Articulated 3D Objects (2021). arXiv:2101.02692

Mo, K., Zhu, S., Chang, A.X., Yi, L., Tripathi, S., Guibas, L.J., Su, H.: PartNet: a large-scale benchmark for fine-grained and hierarchical part-level 3D object understanding. In: The IEEE Conference on Computer Vision and Pattern Recognition (CVPR) (2019)

Noguchi, A., Iqbal, U., Tremblay, J., Harada, T., Gallo, O.: Watch It Move: Unsupervised Discovery of 3D Joints for Re-Posing of Articulated Objects (2021). arXiv preprint arXiv:2112.11347

Pavlasek, J., Lewis, S., Desingh, K., Jenkins, O.C.: Parts-based articulated object localization in clutter using belief propagation. In: IEEE International Conference on Intelligent Robots and Systems, pp. 10595–10602 (2020)

Qi, C.R., Su, H., Mo, K., Guibas, L.J.: PointNet: Deep Learning on Point Sets for 3D Classification and Segmentation (2016)

Ruhr, T., Sturm, J., Pangercic, D., Beetz, M., Cremers, D.: A generalized framework for opening doors and drawers in kitchen environments. In: 2012 IEEE International Conference on Robotics and Automation. IEEE, St Paul, MN, USA, pp. 3852–3858 (May 2012)

Sturm, J., Jain, A., Stachniss, C., Kemp, C.C., Burgard, W.: Operating articulated objects based on experience. In: 2010 IEEE/RSJ International Conference on Intelligent Robots and Systems (2010). IEEE, Taipei, pp. 2739–2744

Wang, H., Sridhar, S., Huang, J., Valentin, J., Song, S., Guibas, L.: Normalized Object Coordinate Space for Category-Level 6D Object Pose and Size Estimation, pp. 2637–2646 (2019a)

Wang, X., Zhou, B., Shi, Y., Chen, X., Zhao, Q., Xu, K.: Shape2Motion: joint analysis of motion parts and attributes from 3D shapes. In: Proceedings of the IEEE/CVF Conference on Computer Vision and Pattern Recognition (CVPR) (2019b)

Weng, Y., Wang, H., Zhou, Q., Qin, Y., Duan, Y., Fan, Q., Chen, B., Su, H., Guibas, L.J.: CAPTRA: Category Level Pose Tracking for Rigid and Articulated Objects from Point Clouds (2021). arXiv: 2104.03437

Wu, Y., Kirillov, A., Massa, F., Lo, W.-Y., Girshick, R.: Detectron2 (2019). www.github.com/facebookresearch/detectron2

Xiang, F. et al.: SAPIEN: a simulated part-based interactive environment. In: The IEEE Conference on Computer Vision and Pattern Recognition (CVPR) (2020)

Xu, Z., He, Z., Song, S.: UMPNet: Universal Manipulation Policy Network for Articulated Objects (2021) arXiv:2109.05668

Yi, L., Huang, H., Liu, D., Kalogerakis, E., Su, H., Guibas, L.: Deep part induction from articulated object pairs. In: SIGGRAPH Asia 2018 Technical Papers, SIGGRAPH Asia 2018. Association for Computing Machinery, Inc (2018)

Yi, L. et al.: A scalable active framework for region annotation in 3D shape collections. In: SIGGRAPH Asia (2016)

Zeng, V., Lee, T.E., Liang, J., Kroemer, O.: Visual identification of articulated object parts. In: IEEE International Conference on Intelligent Robots and Systems, pp. 2443–2450 (2021)

Zhang, G., Litany, O., Sridhar, S., Guibas, L.: StrobeNet: Category-Level Multiview Reconstruction of Articulated Objects (2021). arXiv: 2105.08016

Zimmermann, C., Welschehold, T., Dornhege, C., Burgard, W., Brox, T.: 3D human pose estimation in RGBD images for robotic task learning. In: 2018 IEEE International Conference on Robotics and Automation (ICRA) (2018). Brisbane, QLD

Learning

Learning Sequential Latent Variable Models from Multimodal Time Series Data

Oliver Limoyo(✉), Trevor Ablett, and Jonathan Kelly

University of Toronto Institute for Aerospace Studies (UTIAS), Toronto, Canada
{oliver.limoyo,trevor.ablett,jonathan.kelly}@robotics.utias.utoronto.ca

Abstract. Sequential modelling of high-dimensional data is an important problem that appears in many domains including model-based reinforcement learning and dynamics identification for control. Latent variable models applied to sequential data (i.e., latent dynamics models) have been shown to be a particularly effective probabilistic approach to solve this problem, especially when dealing with images. However, in many application areas (e.g., robotics), information from multiple sensing modalities is available—existing latent dynamics methods have not yet been extended to effectively make use of such multimodal sequential data. Multimodal sensor streams can be correlated in a useful manner and often contain complementary information across modalities. In this work, we present a self-supervised generative modelling framework to jointly learn a probabilistic latent state representation of multimodal data and the respective dynamics. Using synthetic and real-world datasets from a multimodal robotic planar pushing task, we demonstrate that our approach leads to significant improvements in prediction and representation quality.Furthermore, we compare to the common learning baseline of concatenating each modality in the latent space and show that our principled probabilistic formulation performs better. Finally, despite being fully self-supervised, we demonstrate that our method is nearly as effective as an existing supervised approach that relies on ground truth labels.

Keywords: Data fusion and machine learning · Robot vision · Intelligent perceptions

1 Introduction

Sequential modelling of high-dimensional data is a challenging problem that is encountered in many domains such as visual model-based reinforcement learning [1] and image-based dynamics identification for control [2]. Recently, there has been broad interest in studying the problem through the lens of generative latent variable models. These methods embed high-dimensional observations into a lower-dimensional representation space, or *latent space*, often by means of a variational autoencoder (VAE) [3,4], where the (latent) dynamics can be identified in a self-supervised manner. Importantly, latent dynamics models form the

I. Petrovic et al. (Eds.): IAS 2022, LNNS 577, pp. 511–528, 2023.
https://doi.org/10.1007/978-3-031-22216-0_35

backbone of many recent methods for image-based control, reinforcement learning [5,6], and motion planning with complex robotic systems [7,8]. When dealing with images as a single modality, this approach has been shown to be particularly effective in various existing works [2,9–12]. However, in many robotic systems, images are accompanied by data from multiple additional sensing modalities with varying characteristics, that is, as *multimodal* data. These multimodal data may contain complementary information, patterns, and useful statistical correlations across modalities. Prior research has demonstrated the advantages of capturing multimodal information in the contexts of classification and regression of multimedia data [13,14], learning from demonstration [15], localization [16], and control policy learning [17]. Conversely, the use of multimodal data has not yet been studied in the context of learned latent dynamics models.

In this paper, we present a novel probabilistic framework for learning latent dynamics from multimodal time series data. Inspired by the multimodal variational autoencoder (MVAE) architecture [18], we employ a product of experts [19] to encode all data modalities into a shared probabilistic latent representation while jointly learning the dynamics in a self-supervised manner with a recurrent neural network (RNN) [20].

For validation and demonstration purposes, we use simulated and real-world datasets collected from a robotic manipulator pushing task involving three heterogenous data modalities: images from a camera, force and torque readings from a force-torque sensor, and proprioceptive readings from the manipulator encoders. Figure 1 provides a visual summary of our approach.

Our main contributions are as follows:

1. a formulation of a variational inference model for self-supervised training of latent dynamics models specifically with multimodal time series data;
2. experimental results demonstrating that our method of incorporating multimodal data in the latent dynamics framework has superior representative and predictive capability when compared to the baseline of simply concatenating each modality [21–24];
3. experimental results demonstrating that our self-supervised method achieves results comparable to an existing supervised method [25], which requires ground-truth labels, when used to capture task-relevant state information and dynamics; and
4. an open-source implementation of our method and experiments in PyTorch [26].[1]

2 Related Work

In this section, we survey papers related to the modelling of high-dimensional sequential data with latent variable models and learning with multimodal data in machine learning and robotics. We pay particular attention to vision as it is one of the most commonly-encountered modalities across many domains.

[1] https://github.com/utiasSTARS/visual-haptic-dynamics.

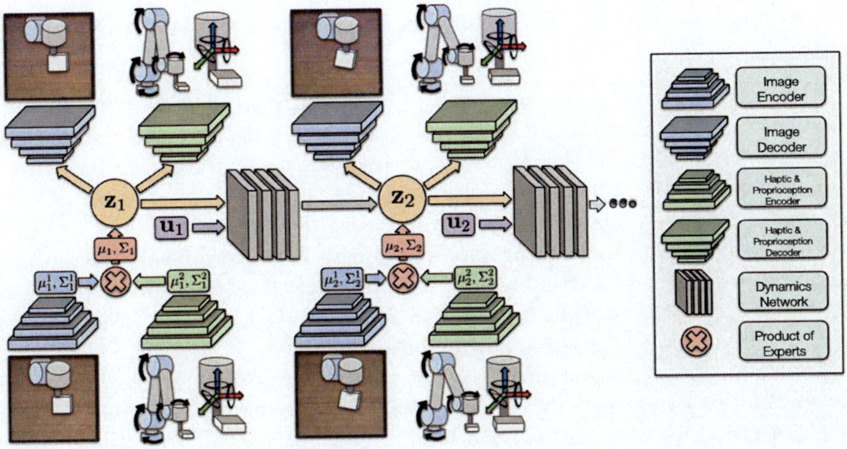

Fig. 1. We learn a sequential latent variable model for multimodal time series data (e.g., image, proprioceptive, and haptic data). Each modality is separately encoded into a probabilistic latent distribution, in this case a Gaussian distribution parametrized by $\boldsymbol{\mu}$ and $\boldsymbol{\Sigma}$, and then combined into a joint representation using a product of experts. Simultaneously, we learn the respective (latent) dynamics based on the control inputs $\mathbf{u}$.

Latent Dynamics Early deep dynamical models [10] used the bottleneck of a standard autoencoder as the compressed state from which the dynamics were learned in a tractable manner. Probabilistic extensions using the VAE were studied later [2,27]. A significant amount of work has been carried out with sequential image data in the context of learned probabilistic state space models [11,28] and differentiable filtering [29]. These approaches attempt to combine the structure and interpretability of probabilistic graphical models with the flexibility and representational capacity of neural networks. Probabilistic graphical models have been used as a way to impose structure for fast and exact inference [12,30] and to filter out novel or out-of-distribution images using a notion of uncertainty that comes from generative models [31]. The closely-related topic of image-based transition models, or *world models*, has been studied in the reinforcement learning literature [1,32]. Our work is an extension to these latent dynamics models that makes them amenable to multimodal data.

Multimodal Machine Learning Machine learning has been applied to the problem of learning representations and patterns of multimodal data for various downstream tasks. A good summary of the existing literature focused on applications involving multimedia data (e.g., video, text, and audio) is provided in [14]. Probabilistic methods have also been applied to model the joint and conditional distributions of non-sequential multimodal data. Examples of these include the Restricted Boltzman Machine (RBM) [13,33] and the VAE [18,34,35]. Our work is most similar to the latter of these two approaches. We build upon the MVAE [18], but, critically, we apply and extend the framework to the sequential setting so that it is amenable to the application of capturing the latent dynamics of multimodal data.

Multimodal Learning for Robotics Our work is most similar in nature to [24], where the authors use audio data to augment a deterministic state-based forward (or dynamics) model. However, in [24], the audio data are taken from a previous random interaction and do not provide causally-related information to the forward model—the audio data is simply used to augment the representation. In contrast, we directly model the dynamics of the observed multimodal data. We also do not assume a relaxed deterministic state-based setting and instead learn a probabilistic representation from raw multimodal data directly (as opposed to dealing with difficult-to-acquire state labels).

Similar to the approach presented in this paper, other groups have investigated the use of learned differentiable filters with multimodal measurement models of vision, proprioceptive, and haptic data, relying on ground-truth annotations [25]. However, in many cases ground-truth labels are expensive or impossible to acquire, which may hinder the scalability of such methods. We leverage recent work in variational inference and devise a self-supervised generative approach to bypass this limitation. We do so by maximizing a proper lower bound of the marginal likelihood of the data itself.

Other works have also leveraged the MVAE architecture for learned localization with multimodal data [16]. Closer to our work, the authors of [36] demonstrate a technique to learn a notion of 'intuitive physics' in a self-supervised manner by applying the MVAE architecture as a generative model of multimodal sensor measurements. Specifically, future sensor measurements resulting from interaction with objects are decoded based on an encoding of the current sensor measurements. In our work, as opposed to directly decoding future transitions, we learn a dynamics model based on the compressed latent space; we therefore have the choice to predict while remaining in a low-dimensional space and without having to decode, which saves a significant amount of computation and memory.

3 Methodology

We begin by presenting a baseline sequential latent variable model in Sect. 3.1 and then demonstrate how to extend this framework to multimodal data in Sect. 3.2.

3.1 Sequential Latent Variable Models

We consider a sequence of observations $\mathbf{x} = \{\mathbf{x}_t\}_{t=1}^T$ of a single modality with respective control inputs $\mathbf{u} = \{\mathbf{u}_t\}_{t=1}^{T-1}$. We then introduce latent variables $\mathbf{z} = \{\mathbf{z}_t\}_{t=1}^T$ to create a joint distribution $p_\theta(\mathbf{x}, \mathbf{z}|\mathbf{u})$, where θ are the learnable parameters of our distributions that are parametrized by neural networks. We factorize the joint distribution of the generative process as $p_\theta(\mathbf{x}, \mathbf{z}|\mathbf{u}) = p_\theta(\mathbf{x}|\mathbf{z}, \mathbf{u})\, p_\theta(\mathbf{z}|\mathbf{u})$, where:

$$p_\theta(\mathbf{z}|\mathbf{u}) = p(\mathbf{z}_1) \prod_{t=2}^{T} p_\theta(\mathbf{z}_t|\mathbf{z}_{t-1}, \mathbf{u}_{t-1}), \tag{1}$$

and

$$p_\theta(\mathbf{x}|\mathbf{z}, \mathbf{u}) = \prod_{t=1}^{T} p_\theta(\mathbf{x}_t|\mathbf{z}_t). \tag{2}$$

We model the latent dynamics with the distribution $p_\theta(\mathbf{z}_t|\mathbf{z}_{t-1}, \mathbf{u}_{t-1})$ and the observation or measurement model with $p_\theta(\mathbf{x}_t|\mathbf{z}_t)$. The distribution $p(\mathbf{z}_1)$ is an arbitrary initial distribution with high uncertainty. The goal of learning is to maximize the marginal likelihood of the data or the evidence, given for a single sequence as

$$p_\theta(\mathbf{x}|\mathbf{u}) = \int p_\theta(\mathbf{x}|\mathbf{z}, \mathbf{u})\, p_\theta(\mathbf{z}|\mathbf{u}) d\mathbf{z}, \tag{3}$$

with respect to the parameters θ. Unfortunately, in the general case with this model, the posterior distribution used for inference, $p(\mathbf{z}|\mathbf{x}, \mathbf{u})$, is intractable. A common solution from recent work in variational inference [3,4] is to introduce a recognition or inference model $q_\phi(\mathbf{z}|\mathbf{x}, \mathbf{u})$ with parameters ϕ to approximate the intractable posterior. This leads to the following lower bound on the marginal log-likelihood or the evidence lower bound (ELBO):

$$\ln p_\theta(\mathbf{x}|\mathbf{u}) \geq \mathbb{E}_{q_\phi(\mathbf{z}|\mathbf{x},\mathbf{u})}[\ln p_\theta(\mathbf{x}|\mathbf{z}, \mathbf{u})] - \mathrm{KL}(q_\phi(\mathbf{z}|\mathbf{x}, \mathbf{u}) \| p_\theta(\mathbf{z}|\mathbf{u})). \tag{4}$$

Maximizing this lower bound can be shown to be equivalent to minimizing the KL-divergence between the true posterior $p_\theta(\mathbf{z}|\mathbf{x}, \mathbf{u})$ and the recognition model $q_\phi(\mathbf{z}|\mathbf{x}, \mathbf{u})$. The resulting optimization objective, as denoted by Eq. 4, is based on an expectation with respect to the distribution $q_\phi(\mathbf{z}|\mathbf{x}, \mathbf{u})$, which itself is based on the parameters ϕ. As is typically done, we restrict $q_\phi(\mathbf{z}|\mathbf{x}, \mathbf{u})$ to be a Gaussian variational approximation. This enables us to use stochastic gradient descent (i.e., using Monte Carlo estimates of the gradient) via the reparameterization trick [3] to optimize the lower bound. The specific choice for the factorization of $q_\phi(\mathbf{z}|\mathbf{x}, \mathbf{u})$ varies depending on the application (i.e., prediction, smoothing, or filtering). Given our intended applications of prediction we choose to only use causal (i.e., current and past) information for inference,

$$q_\phi(\mathbf{z}|\mathbf{x}, \mathbf{u}) = \prod_{t=1}^{T} q_\phi(\mathbf{z}_t|\mathbf{x}_{\leq t}, \mathbf{u}_{<t}). \tag{5}$$

3.2 Multimodal Sequential Latent Variable Models

We now extend and generalize the sequential latent variable model defined above to the multimodal case. We consider N sequences of separate observations or modalities, $\mathbf{X} = \{\mathbf{x}^n\}_{n=1}^N$, where we assume that each sequence is of equivalent length T: $\mathbf{x}^n = \{\mathbf{x}_t^n\}_{t=1}^T$. As done in the previous section, we include the respective control inputs $\mathbf{u} = \{\mathbf{u}_t\}_{t=1}^{T-1}$ and again introduce a set of latent variables $\mathbf{z} = \{\mathbf{z}_t\}_{t=1}^{T}$ as a joint lower-dimensional latent space containing some underlying dynamics of interest. The final joint distribution factorizes as $p_\theta(\mathbf{X}, \mathbf{z}|\mathbf{u}) = p_\theta(\mathbf{X}|\mathbf{z}, \mathbf{u})\, p_\theta(\mathbf{z}|\mathbf{u})$. We choose to define the generative process as follows:

$$\begin{aligned} p_\theta(\mathbf{X}|\mathbf{z}, \mathbf{u}) &= \prod_{n=1}^{N} p_\theta(\mathbf{x}^n|\mathbf{z}, \mathbf{u}) \\ &= \prod_{n=1}^{N}\prod_{t=1}^{T} p_\theta(\mathbf{x}_t^n|\mathbf{z}_t), \end{aligned} \tag{6}$$

and with $p_\theta(\mathbf{z}|\mathbf{u})$ remaining the same as shown in Eq. 1. The marginal likelihood is then

$$p_\theta(\mathbf{X}|\mathbf{u}) = \int p_\theta(\mathbf{X}|\mathbf{z}, \mathbf{u})\, p_\theta(\mathbf{z}|\mathbf{u}) \mathrm{d}\mathbf{z}, \tag{7}$$

and the respective ELBO, for a single sequence, is then:

$$\ln p_\theta(\mathbf{X}|\mathbf{u}) \geq \mathbb{E}_{q_\phi(\mathbf{z}|\mathbf{X},\mathbf{u})}[\sum_{\mathbf{x}^n \in \mathbf{X}} \ln p_\theta(\mathbf{x}^n|\mathbf{z}, \mathbf{u})] - \mathrm{KL}(q_\phi(\mathbf{z}|\mathbf{X}, \mathbf{u}) \| p_\theta(\mathbf{z}|\mathbf{u})). \tag{8}$$

In order to decide on the factorization of $q_\phi(\mathbf{z}|\mathbf{X}, \mathbf{u})$, we draw inspiration from the MVAE architecture [18], and base our inference network on the structure of the true multimodal posterior $p(\mathbf{z}|\mathbf{X}, \mathbf{u})$:

$$\begin{aligned} p(\mathbf{z}|\mathbf{X}, \mathbf{u}) &= \frac{p(\mathbf{z}|\mathbf{u})\, p(\mathbf{X}|\mathbf{z}, \mathbf{u})}{p(\mathbf{X}|\mathbf{u})} \\ &= \frac{p(\mathbf{z}|\mathbf{u})}{p(\mathbf{x}^1, ..., \mathbf{x}^N|\mathbf{u})} \prod_{n=1}^{N} p(\mathbf{x}^n|\mathbf{z}, \mathbf{u}) \\ &= \frac{p(\mathbf{z}|\mathbf{u})}{p(\mathbf{x}^1, ..., \mathbf{x}^N|\mathbf{u})} \prod_{n=1}^{N} \frac{p(\mathbf{z}|\mathbf{x}^n, \mathbf{u}) p(\mathbf{x}^n|\mathbf{u})}{p(\mathbf{z}|\mathbf{u})} \\ &= \frac{\prod_{n=1}^{N} p(\mathbf{x}^n|\mathbf{u})}{p(\mathbf{x}^1, ..., \mathbf{x}^N|\mathbf{u})} \cdot \frac{\prod_{n=1}^{N} p(\mathbf{z}|\mathbf{x}^n, \mathbf{u})}{\prod_{n=1}^{N-1} p(\mathbf{z}|\mathbf{u})} \\ &\propto \frac{\prod_{n=1}^{N} p(\mathbf{z}|\mathbf{x}^n, \mathbf{u})}{\prod_{n=1}^{N-1} p(\mathbf{z}|\mathbf{u})}. \end{aligned} \tag{9}$$

Based on the last term in Eq. 9, the final factorization of the joint posterior is then a quotient between a product of the individual modality-specific posteriors and the prior. Accordingly, we choose our inference model to be $q(\mathbf{z}|\mathbf{X}, \mathbf{u}) = \frac{\prod_{n=1}^{N} q(\mathbf{z}|\mathbf{x}^n,\mathbf{u})}{\prod_{n=1}^{N-1} p(\mathbf{z}|\mathbf{u})}$. We also use the same representation reformulation trick from the MVAE [18] and set $q(\mathbf{z}|\mathbf{x}^n, \mathbf{u}) = \tilde{q}(\mathbf{z}|\mathbf{x}^n, \mathbf{u})\, p(\mathbf{z}|\mathbf{u})$ in order to produce a simpler and numerically more stable product of experts,

$$q_\phi(\mathbf{z}|\mathbf{X}, \mathbf{u}) = p_\theta(\mathbf{z}|\mathbf{u}) \prod_{n=1}^{N} \tilde{q}_\phi(\mathbf{z}|\mathbf{x}^n, \mathbf{u}), \tag{10}$$

where $\tilde{q}_\phi(\mathbf{z}|\mathbf{x}^n, \mathbf{u})$ would be equivalent to the standard single modality posterior shown in Eq. 5. Finally, we can further factorize the posterior in Eq. 10 into a more intuitive form for our sequential setting:

$$q_\phi(\mathbf{z}|\mathbf{X}, \mathbf{u}) = \underbrace{p(\mathbf{z}_1) \prod_{n=1}^{N} \tilde{q}_\phi(\mathbf{z}_1|\mathbf{x}_1^n)}_{q_\phi(\mathbf{z}_1|\mathbf{x}_1^1,\ldots,\mathbf{x}_1^N)} \prod_{t=2}^{T} \underbrace{\left(p_\theta(\mathbf{z}_t|\mathbf{z}_{t-1}, \mathbf{u}_{t-1}) \prod_{n=1}^{N} \tilde{q}_\phi(\mathbf{z}_t|\mathbf{x}_{\leq t}^n, \mathbf{u}_{<t}) \right)}_{q_\phi(\mathbf{z}_t|\mathbf{x}_{\leq t}^1,\ldots,\mathbf{x}_{\leq t}^N,\mathbf{u}_{<t})}$$

$$= \prod_{t=1}^{T} q_\phi(\mathbf{z}_t|\mathbf{x}_{\leq t}^1, \ldots, \mathbf{x}_{\leq t}^N, \mathbf{u}_{<t}). \tag{11}$$

Our factorization reveals that, at every time step of Eq. 11, each data modality is first separately encoded into a Gaussian distribution by its own inference model. A product is then taken of each modality-specific distributions and the prior, which is also the transition distribution (i.e., the dynamics) of our latent space. Interestingly, we recover a similar form to the commonly-used recurrent state space model [1], where the transition distribution is included in the posterior. The product of distributions at each time step are not generally solvable in closed form. However, by assuming Gaussian distributions for the prior dynamics and each modality-specific inference distribution, we end up with a final product of Gaussians for which a closed-form analytical solution does exist. Conveniently, a product of Gaussians is also itself a Gaussian [37].

4 Experiments

In this section, we present implementation details and an empirical evaluation of our method. For our experiments, we chose to study a common multimodal task: planar pushing with a robotic manipulator using vision, haptic, and proprioceptive data from a camera, a force-torque sensor, and joint encoders, respectively.

Planar pushing involves complex contact dynamics that are difficult to model with vision alone, while the multimodal sensor data produced are highly heterogeneous in dimension and quality.

We compared five different models to study the effects of multimodality on this type of task. The models were, in order: (1) vision-only (denoted by V), (2) visual-proprioceptive (denoted by VP), (3) visual-haptic (denoted by VH), (4) visual-haptic-proprioceptive (denoted by VHP), and (5) a commonly-used baseline where the latent embedding of each data modality is simply concatenated (denoted by VHP-C).

4.1 Datasets

We use both a synthetic and a real-world dataset as shown in Fig. 2. We provide more details on the data and the collection procedure below.

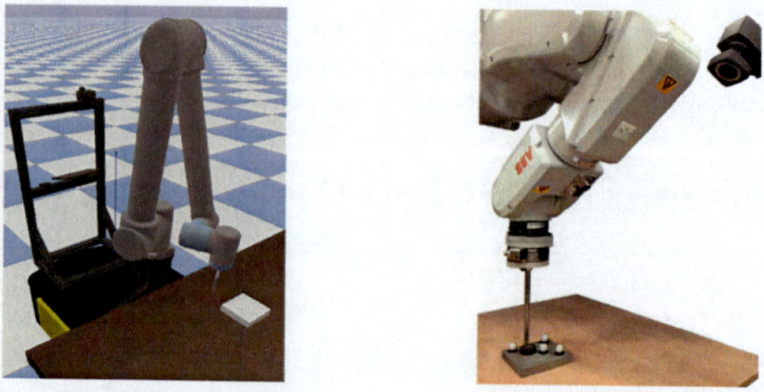

(a) PyBullet pushing simulation. (b) Real-world pushing task.

Fig. 2. Environments from which the datasets were collected. Real-world setup image from existing MIT pushing dataset [38].

Simulated Manipulator Pushing We used PyBullet [39] to generate data from a simulated manipulator pushing task. We collected grayscale images of size 64 × 64 pixels, $\mathbf{x}_t^1 \in \mathbb{R}^{64\times 64}$, with pixel intensities rescaled to be in the range of zero to one. The proprioceptive data consisted of the Cartesian position and velocity of the end-effector, while the haptic data included force and torque measurements along three axes. We combined the haptic and proprioceptive data into a single second modality when both were used, $\mathbf{x}_t^2 \in \mathbb{R}^{12}$, and otherwise used one of the two, $\mathbf{x}_t^2 \in \mathbb{R}^6$. We argue that this is a reasonable decision since both proprioceptive and haptic data have similar characteristics, as opposed to, for example, image data. The control inputs were the end-effector velocity commands along the planar x and y directions, $\mathbf{u}_t \in \mathbb{R}^2$.

A total of 4,800 trajectories were collected. The image data was recorded at a frequency of 4 Hz and force-torque and proprioceptive data at a frequency 120 Hz. We concatenated sequences of measurements in order to keep a consistent number of time steps between modalities. We used 4,320 trajectories for training and held out 480, or 10%, for evaluation. Each trajectory was of length $T =$ 16 (i.e., 16 images and 16 × 32 (concatenated) force-torque and proprioception measurements in total). The number of measurements per discrete time step was based on the respective frequency of each data source. The object being pushed

(a) Ellipse-1 on Delrin. (b) Ellipse-2 on plywood. (c) Ellipse-3 on ABS. (d) Ellipse-3 on polyurethane.

Fig. 3. Downsampled images generated from the real-world MIT pushing dataset using existing tools [40]. Realistic lighting and material textures are also rendered.

was a single square plate. In order to collect training data, we used a policy with actions drawn from a fixed Gaussian distribution. We used the same initial object position for each trajectory.

MIT Pushing The MIT pushing dataset [38] consists of high-fidelity real robot pushes carried out on various material surfaces and with various object shapes. Our dataset was a subset of trajectories with three different ellipse-shaped plates and four different surface materials (Delrin, plywood, ABS, and polyurethane). We followed the experimental protocol of the authors in [25].

We used the code provided by previous work [40] to preprocess the data and to artificially render the respective images of the trajectories (since no image data was collected as part of the original dataset). In Fig. 3, we show four examples of rendered images based on the real-world pushing data (i.e., object pose, end-effector pose, and force-torque data). We note that the rendered images are not completely realistic (e.g., the images are not occluded and only the manipulator's tip is rendered). However, for our purposes, the dataset provided an adequate starting point for preliminary experiments and comparisons.

We downsampled the images to 64 × 64 pixels and transformed them to grayscale, $\mathbf{x}_t^1 \in \mathbb{R}^{64\times64}$, with pixel intensities rescaled to be in the range of zero to one. The proprioceptive data consisted of the x and y Cartesian coordinates of the end-effector. The haptic data consisted of force measurements along the x and y (in-plane) axes and the torque measurements about the z axis. We combined the haptic and proprioceptive data to form a single second modality, $\mathbf{x}_t^2 \in \mathbb{R}^5$, if both were used, and otherwise used one or the other, $\mathbf{x}_t^2 \in \mathbb{R}^3$ or $\mathbf{x}_t^2 \in \mathbb{R}^2$. The control inputs were position commands along the x and y axes, $\mathbf{u}_t \in \mathbb{R}^2$. The images were recorded at a frequency of 18 Hz and the force and proprioceptive data were recorded at a frequency of 180 Hz. We used a subset of 2,332 trajectories out of the total dataset (2,099 trajectories for training and 233, or approximately 10%, held out for evaluation). Each trajectory was of length $T = 48$ (i.e., 48 images and 48 × 10 (concatenated) force-torque and proprioception measurements in total). The data were collected using several pre-planned pushes with varying velocities, accelerations, and contact points and angles.

4.2 Network Architecture and Training

We parametrized our models with neural networks. For the image encoder, we used a fully convolutional neural network based on the architecture of [32]. The respective decoder was a matching deconvolutional network. For the proprioceptive and force-torque data, we used a simple 1D convolutional architecture for the encoder and a 1D deconvolutional network for the decoder.

Our transition function or dynamics model, $p_\theta(\mathbf{z}_t|\mathbf{z}_{t-1}, \mathbf{u}_{t-1})$ was parameterized as a single-layer GRU network [20] with 256 units that produced linear transition matrices (i.e., $\mathbf{z}_t = \mathbf{A}_t\mathbf{z}_{t-1} + \mathbf{B}_t\mathbf{u}_{t-1}$, where $\mathbf{A}_t$ and $\mathbf{B}_t$ are outputs of the network), as done in previous work [11,12,31]. During training, we sampled mini-batches of 32 trajectories. We applied weight normalization [41] to all of the network layers except for the GRU. We used ReLU activation functions for all of the networks. Our latent space for all experiments consisted of 16-dimensional Gaussians with diagonal covariance matrices. We applied the `Adam` optimizer [42] with a learning rate of 0.0003 and a gradient clipping norm of 0.5 for the GRU network.

4.3 Image Prediction Experiments

In order to compare the models, we first evaluated the quality of the generated image sequences relative to the known ground truth images as a proxy for prediction quality. We began by encoding an initial amount of data into the latent space and then predicted future latent states with our learned dynamics models and known control inputs. Given these predicted states, we then decoded and generated future predicted images. We computed the squared error (SE) in terms of pixels, structural similarity index measure (SSIM), and peak signal-to-noise ratio (PSNR) between the ground truth images and predicted images. Accurate predictions translate into a lower SE, higher SSIM, and higher PSNR.

Table 1. Quantitative prediction quality for the synthetic dataset. The best results are highlighted in bold.

Model	RMSE (↓)	SSIM (↑)	PSNR (↑)
V	3.243	0.955 ± 0.04	28.639 ± 5.84
VP	1.758	0.982 ± 0.02	33.935 ± 5.93
VH	2.956	0.961 ± 0.04	29.684 ± 5.70
VHP-C	2.911	0.962 ± 0.04	29.700 ± 5.39
VHP	**1.749**	$\mathbf{0.983 \pm 0.02}$	$\mathbf{34.202 \pm 6.19}$

We compared the predicted images' quality with their respective ground truth for four models trained with various subsets of modalities. We show one standard deviation for the SSIM and PSNR values and calculate the average score across the entire predicted horizon

Table 2. Quantitative prediction quality with the MIT dataset as described in Table 1. The best results are highlighted in bold.

Model	RMSE (↓)	SSIM (↑)	PSNR (↑)
V	1.917	0.907 ± 0.02	30.806 ± 1.66
VP	1.850	0.923 ± 0.03	31.395 ± 2.28
VH	1.839	0.922 ± 0.03	31.298 ± 2.03
VHP-C	1.846	0.923 ± 0.03	31.333 ± 2.12
VHP	**1.734**	$\mathbf{0.927 \pm 0.02}$	$\mathbf{31.741 \pm 1.84}$

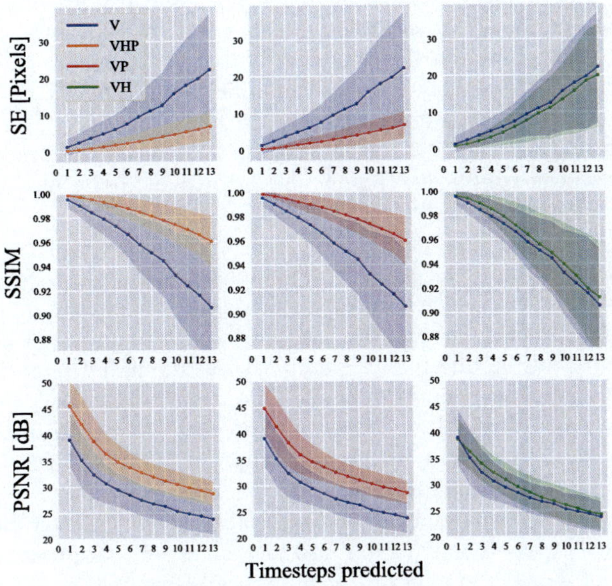

(a) Results from the synthetic dataset.

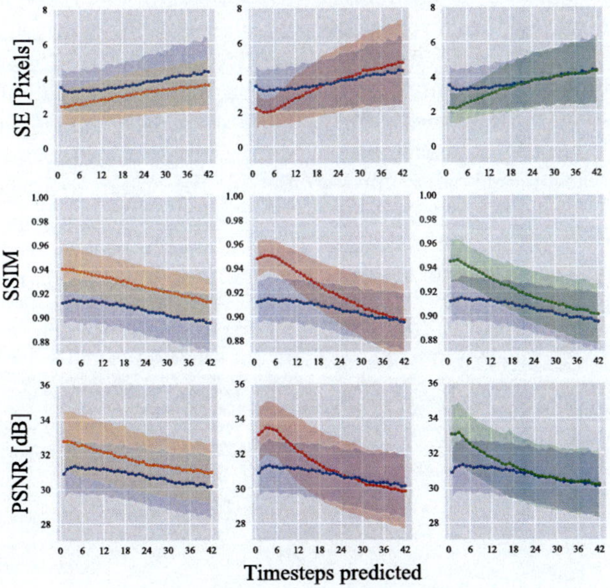

(b) Results from the MIT dataset.

Fig. 4. Graph of prediction quality over multiple prediction horizons from both datasets. We compare each model to the baseline vision-only (V) model. We plot the mean values of the SE, SSIM and PSNR with one standard deviation shaded.

For the synthetic data, we first conditioned on four frames and predicted the next 11 frames in the sequence. Figure 4a is a visualization of the mean score, with one standard deviation shaded, based on all of the held out test data. For clear comparisons, we overlay each multimodal model (i.e., VHP, VP, and VH) on top of the baseline vision-only model (i.e., V).

Additionally, we compile the average score over the entire predicted horizon and summarize the results in Table 1. We also include the baseline results of simply concatenating all modalities, labelled as VHP-C.

Overall, the VHP model performed the best for all three metrics when averaged over all prediction lengths, but it was closely followed by the VP model. All multimodal models outperformed the baseline, vision-only model, although the improvement in the case of the VH model was marginal. We hypothesize that this result may be due to the quality of the simulated force torque sensor data and the simulated contact dynamics. In this case, proprioception was the most effective second modality, as demonstrated by the performance of the VP model. However, using vision, haptic and proprioceptive data together with the VHP model led to the best results.

Notably, simply concatenating each modality (i.e., VHP-C) yielded poorer performance when compared to our product of experts approach (i.e., VHP). The product of experts appears to have an appealing 'filtering' inductive bias for prediction problems. The product of experts formulation enables decisions to be made about when to use each modality at each time step based on the respective uncertainties. Further, each expert can selectively focus on a few dimensions without having to cover the full dimensionality of the state. Finally, a product of experts produces a sharper final distribution than the individual expert models [19]. This idea is well known in the context of digit image generation: one low-resolution model can generate the approximate overall shape of the digit while other local models can refine segments of the stroke with the correct fine structure [19]. Figure 5 is a visualization of a selected roll-out or prediction that demonstrates a typical failure mode of the vision-only model. As shown by the later frames generated by the vision-only model in the red box outline, the over-

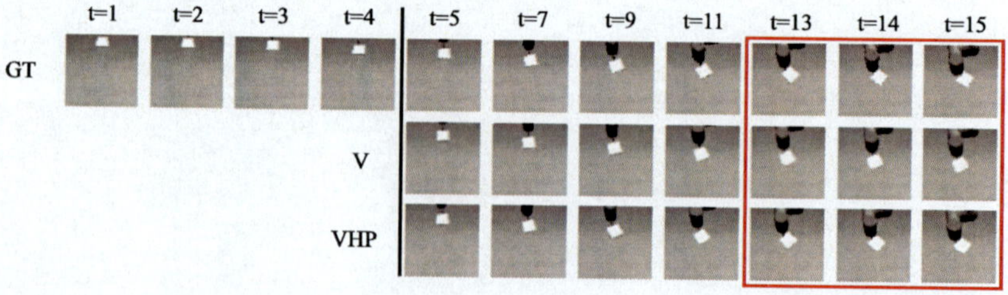

Fig. 5. A prediction example that demonstrates a failure mode of the vision-only (V) model trained on the synthetic dataset; the model fails to fully predict the object's true motion (GT) and instead predicts less counterclockwise rotation in the later frames (denoted by red box). In contrast, the multimodal model (VHP) correctly predicts the object's motion.

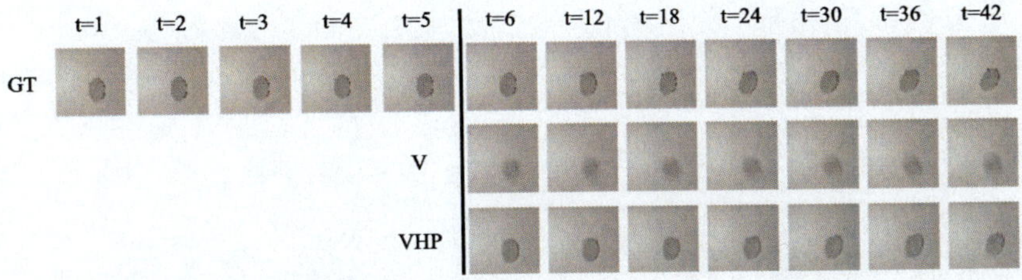

Fig. 6. A prediction example that demonstrates a different failure mode of the vision-only (V) model trained on the MIT dataset; the model is unable to produce 'crisp' predictions of the object form and instead outputs a blurry average at all time steps. The multimodal model (VHP) is able to predict the slight angle of the ellipse (clockwise), matching the ground truth (GT).

all motion of the object is almost correct, but the finer-scale changes (e.g., the exact amount of rotation) are not well captured. Our VHP model, however, was able to capture these smaller changes.

We ran similar image prediction experiments for the MIT pushing dataset. In this case, we first conditioned on five frames and predicted the next 37 frames in the sequence. As shown by Fig. 4b, both the VP and VH models performed slightly better than the full VHP during the early time steps according to all three metrics (SE, SSIM and PSNR). However, over a longer prediction horizon the VHP model was clearly superior. For the real-world data, unlike in the synthetic case, the VH model significantly outperformed the baseline V model. Table 2 lists the average scores over the entire prediction horizon. Consistent with the previous synthetic experiment, simply concatenating the modalities (VHP-C) led to poorer performance when compared to the product of experts model (VHP). Each multimodal model performed better than the baseline vision-only model. We visualize another failure mode of the vision-only model in Fig. 6. In this case, we observe that the vision-only model failed to capture the more subtle dynamics of the object, which ended up angled and tilted clockwise; the model defaulted to outputting a blurry average at the approximate location of the object. On the other hand, our VHP model was able to produce a relatively crisp and accurate prediction of the object pose.

4.4 Regression Experiments

We further evaluated the predictive capability of the models with a downstream regression experiment that measured how well the model was able to capture and predict the underlying position of the object being pushed. As is commonly done in the self-supervised representation learning literature [43,44], we trained separate regressors on top of the frozen representations and dynamics models to do this.

We first encoded an initial latent state $\mathbf{z}_i$ (i.e., a filtered state) from a set of observations $\mathbf{x}_{1:i}$. Then, using our learned dynamics and h known future

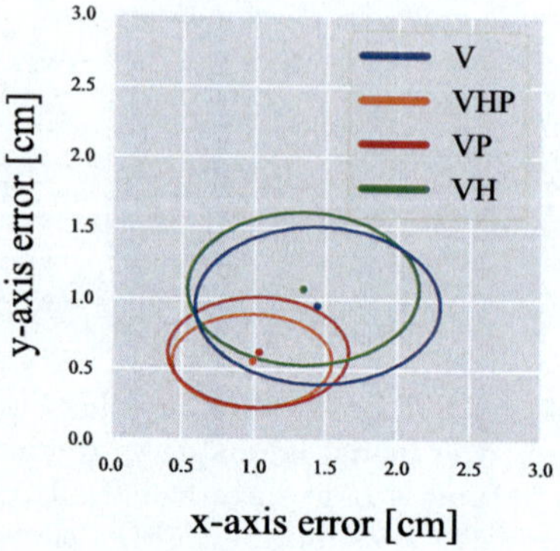

(a) Results from the synthetic dataset.

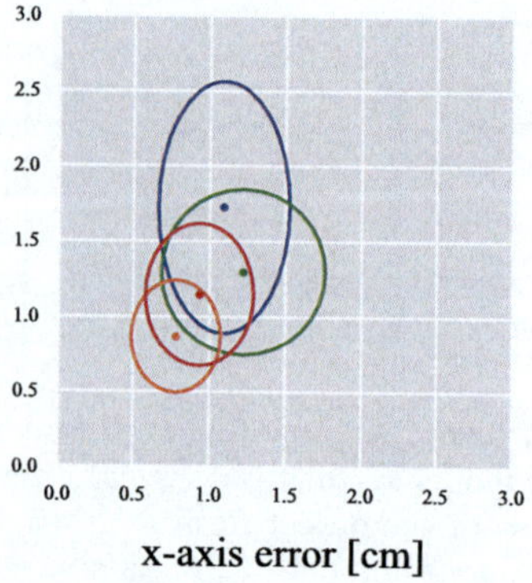

(b) Results from the MIT dataset.

Fig. 7. Mean translation errors for regressing the object's x and y coordinates from the predicted latent states. The ellipses represent one standard deviation.

controls $\mathbf{u}_{i:i+h}$, we predicted h future latent states $\hat{\mathbf{z}}_{i+1:i+1+h}$ (i.e., predicted states). Using the predicted states as inputs, we regressed the ground truth positions of the object, while keeping the weights of all of our previously-learned networks frozen. Poor regression results would indicate lower correlation between the predicted latent states and the object's ground truth position. This in turn would imply that the learned latent representation did not encode all of the necessary information to represent the state of the object, or that the learned dynamics of the object were inaccurate, or both. We kept the same prediction horizons from the image prediction experiments in Sect. 4.3.

For our results with synthetic data, we found ordinary least squares to be adequate for regression. Figure 7a shows that the mean and variance of the absolute translation errors from the regression are lowest for the full VHP model. We (again) observe that the inclusion of haptic data alone (VH) as an extra modality in the model did not lead to huge improvements. On the other hand, using proprioception (VP) did lead to significant improvements. The inclusion of both proprioceptive and haptic data (VHP) outperformed using proprioceptive data alone (VP).

For the MIT pushing dataset, we trained a separate simple neural network with a single hidden layer of size 50 to regress the position of the object. The real-world results generally match our synthetic results, as shown by Fig. 7b. Including any sort of multimodal data reduced the mean and variance of the absolute translation errors when compared to the results from the vision-only model.

As an additional benchmark with the MIT dataset, we compared the accuracy of our regressed values of the position of the object to those computed by a related approach on differentiable filtering [25]. Using the same dataset, the authors trained a variety of differentiable filters to directly regress the positions of the objects from the multimodal sensor data in a supervised manner using ground truth labels. Because the multimodal differentiable filters were trained with ground truth annotations of the objects' locations, while our model is completely self-supervised, the results provide a reasonable estimate of an upper bound on the expected performance. Table 3 demonstrates that our regression results, from both the predicted and filtered states, are comparable.

Table 3. Root-mean-square error (RMSE) of the objects' regressed positions with the MIT pushing dataset.

Model	RMSE [cm] (↓)
EKF (Sup. w/ GT, filt.)	1.33
PF (Sup. w/ GT, filt.)	1.14
LSTM (Sup. w/ GT, filt.)	2.32
VHP (Self-sup., filt.) (Ours)	1.80
VHP (Self-sup., pred.) (Ours)	1.96

We compared both the filtering (filt.) and prediction (pred.) results from our self-supervised multimodal model (VHP) to various types of differentiable filters from [25]

5 Conclusions and Future Work

We have presented a self-supervised sequential latent variable model for multimodal time series data. Our probabilistic formulation extends existing latent dynamics models, a key backbone of many methods in vision-based control and model-based reinforcement learning, to multimodal sensor data. We provided a case study of a manipulator pushing task where visual, haptic, and proprioceptive data streams were available. For latent dynamics in particular, we demonstrated that a principled probabilistic formulation for fusing sequential multimodal data performed significantly better than the common baseline of directly concatenating each modality in the latent space. Additionally, our learned self-supervised approach was shown to be competitive with an existing multimodal differentiable filtering method that relies on supervised ground truth labels. As interesting avenues for future work, we plan to further investigate the properties of the learned uncertainties used for multimodal fusion and to use our model for control policy learning.

References

1. Hafner, D., Lillicrap, T., Fischer, I., Villegas, R., Ha, D., Lee, H., Davidson, J.: Learning latent dynamics for planning from pixels. Int. Conf. Machine Learn. (ICML) 2555–2565 (2019)
2. Watter, M., Springenberg, J.T., Boedecker, J., Riedmiller, M.: Embed to control: a locally linear latent dynamics model for control from raw images. In: Advances in Neural Information Processing Systems (NeurIPS), pp. 2746–2754 (2015)
3. Kingma, D.P., Welling, M.: Auto-encoding variational Bayes. In: Bengio, Y., LeCun, Y. (eds.) International Conference on Learning Representations (ICLR) (2014)
4. Rezende, D.J., Mohamed, S., Wierstra, D.: Stochastic backpropagation and approximate inference in deep generative models. In: International Conference on Machine Learning (ICML), pp. 1278–1286 (2014)
5. Hafner, D., Lillicrap, T.P., Norouzi, M., Ba, J.: Mastering atari with discrete world models. In: International Conference on Learning Representations (ICLR) (2021)
6. Rafailov, R., Yu, T., Rajeswaran, A., Finn, C.: Offline reinforcement learning from images with latent space models. In: Learning for Dynamics and Control (L4DC), pp. 1154–1168. PMLR (2021)
7. Ichter, B., Pavone, M.: Robot motion planning in learned latent spaces. IEEE Robot. Autom. Lett. **4**(3), 2407–2414 (2019)
8. Lippi, M., Poklukar, P., Welle, M.C., Varava, A., Yin, H., Marino, A., Kragic, D.: Latent space roadmap for visual action planning of deformable and rigid object manipulation. In: IEEE/RSJ International Conference on Intelligent Robots and Systems (IROS) (2020)
9. Zhang, M., Vikram, S., Smith, L., Abbeel, P., Johnson, M., Levine, S.: Solar: deep structured representations for model-based reinforcement learning. In: International Conference on Machine Learning (ICML), PMLR, pp. 7444–7453 (2019)
10. Wahlström, N., Schön, T.B., Desienroth, M.P.: From pixels to torques: policy learning with deep dynamical models. In: ICML'15 Deep Learning Workshop (2015)
11. Karl, M., Sölch, M., Bayer, J., van der Smagt, P.: Deep variational Bayes filters: unsupervised learning of state space models from raw data. In: International Conference on Learning Representations (ICLR) (2017)
12. Fraccaro, M., Kamronn, S., Paquet, U., Winther, O.: A disentangled recognition and nonlinear dynamics model for unsupervised learning. In Guyon, I., von Luxburg, U., Bengio, S., Wallach, H.M., Fergus, R., Vishwanathan, S.V.N., Garnett, R. (eds.) Advances in Neural Information Processing Systems (NeurIPS), pp. 3601–3610 (2017)
13. Ngiam, J., Khosla, A., Kim, M., Nam, J., Lee, H., Ng, A.Y.: Multimodal deep learning. In: International Conference on Machine Learning (ICML) (2011)
14. Baltrušaitis, T., Ahuja, C., Morency, L.: Multimodal machine learning: a survey and taxonomy. IEEE Trans. Pattern Anal. Machine Intell. **41**(2), 423–443 (2019)
15. Hristov, Y., Ramamoorthy, S.: Learning from demonstration with weakly supervised disentanglement. In: International Conference on Learning Representations (ICLR). (2021)
16. Zhou, K., Chen, C., Wang, B., Saputra, M.R.U., Trigoni, N., Markham, A.: VMLoc: Variational fusion for learning-based multimodal camera localization. In: Association for the Advancement of Artificial Intelligence Conference (AAAI) (2021), vol. 35, pp. 6165–6173

17. Sergeant, J., Suenderhauf, N., Milford, M., Upcroft, B.: Multimodal deep autoencoders for control of a mobile robot. In: Li, H., Kim, J. (eds.) Proceedings of Australasian Conference on Robotics and Automation, Australia, pp. 1–10 (2015)
18. Wu, M., Goodman, N.: Multimodal generative models for scalable weakly-supervised learning. In: Advances in Neural Information Processing Systems (NeurIPS), vol. 31 (2018)
19. Hinton, G.E.: Training products of experts by minimizing contrastive divergence. Neural Comput. **14**(8), 1771–1800 (2002)
20. Cho, K., van Merriënboer, B., Bahdanau, D., Bengio, Y.: On the properties of neural machine translation: encoder-decoder approaches. In: SSST-8 Eighth Workshop on Syntax, Semantics and Structure in Statistical Translation, pp. 103–111 (2014)
21. Kiela, D., Bottou, L.: Learning image embeddings using convolutional neural networks for improved multi-modal semantics. In: Proceedings of the 2014 Conference on Empirical Methods in Natural Language Processing (EMNLP), pp. 36–45 (2014)
22. D'mello, S.K., Kory, J.: A review and meta-analysis of multimodal affect detection systems. ACM Comput. Surv. (CSUR) 47(3) (2015) 1–36
23. Regneri, M., Rohrbach, M., Wetzel, D., Thater, S., Schiele, B., Pinkal, M.: Grounding action descriptions in videos. Trans. Assoc. Comput. Linguist. **1**, 25–36 (2013)
24. Gandhi, D., Gupta, A., Pinto, L.: Swoosh! Rattle! Thump!-actions that sound. In: Robotics: Science and Systems (RSS) (2020)
25. Lee, M., Yi, B., Martin-Martin, R., Savarese, S., Bohg, J.: Multimodal sensor fusion with differentiable filters. In: IEEE/RSJ International Conference on Intelligent Robots and Systems (IROS) (2020)
26. Paszke, A., Gross, S., Massa, F., Lerer, A., Bradbury, J., Chanan, G., Killeen, T., Lin, Z., Gimelshein, N., Antiga, L., Desmaison, A., Kopf, A., Yang, E., DeVito, Z., Raison, M., Tejani, A., Chilamkurthy, S., Steiner, B., Fang, L., Bai, J., Chintala, S.: PyTorch: an imperative style, high-performance deep learning library. In: Wallach, H., Larochelle, H., Beygelzimer, A., d'Alché Buc, F., Fox, E., Garnett, R. (eds.) Advances in Neural Information Processing Systems (NeurIPS), pp. 8024–8035. Curran Associates, Inc. (2019)
27. Banijamali, E., Shu, R., Ghavamzadeh, M., Bui, H.H., Ghodsi, A.: Robust locally-linear controllable embedding. In: International Conference on Artificial Intelligence and Statistics (AISTATS), vol. 84, pp. 1751–1759 (2018)
28. Krishnan, R.G., Shalit, U., Sontag, D.: Structured inference networks for nonlinear state space models. In: Association for the Advancement of Artificial Intelligence Conference (AAAI), pp. 2101–2109 (2017)
29. Haarnoja, T., Ajay, A., Levine, S., Abbeel, P.: Backprop KF: Learning discriminative deterministic state estimators. In: Lee, D., Sugiyama, M., Luxburg, U., Guyon, I., Garnett, R. (eds.) Advances in Neural Information Processing Systems (NeurIPS), vol. 29, pp. 4376–4384. Curran Associates, Inc. (2016)
30. Johnson, M.J., Duvenaud, D., Wiltschko, A.B., Adams, R.P., Datta, S.R.: Composing graphical models with neural networks for structured representations and fast inference. Adv. Neural Inform. Proces. Syst. (NeurIPS) **29**, 2946–2954 (2016)
31. Limoyo, O., Chan, B., Maric, F., Wagstaff, B., Mahmood, R., Kelly, J.: Heteroscedastic uncertainty for robust generative latent dynamics. IEEE Robot. Autom. Lett. **5**(4), 6654–6661 (2020)
32. Ha, D., Schmidhuber, J.: Recurrent world models facilitate policy evolution. Adv. Neural Inform. Proces. Syst. (NeurIPS) 2450–2462 (2018)
33. Srivastava, N., Salakhutdinov, R.: Multimodal learning with deep boltzmann machines. J. Machine Learn. Res. **15**(1), 2949–2980 (2014)

34. Suzuki, M., Nakayama, K., Matsuo, Y.: Joint multimodal learning with deep generative models. arXiv preprint arXiv:1611.01891 (2016)
35. Vedantam, R., Fischer, I., Huang, J., Murphy, K.: Generative models of visually grounded imagination. In: International Conference on Learning Representations (ICLR) (2018)
36. Rezaei-Shoshtari, S., Hogan, F.R., Jenkin, M., Meger, D., Dudek, G.: Learning intuitive physics with multimodal generative models. In: Association for the Advancement of Artificial Intelligence Conference (AAAI) (2021)
37. Cao, Y., Fleet., D.J.: Generalized product of experts for automatic and principled fusion of Gaussian process predictions. In: NIPS'14 Modern Nonparametrics 3: Automating the Learning Pipeline Workshop (2014)
38. Yu, K., Bauzá, M., Fazeli, N., Rodriguez, A.: More than a million ways to be pushed. A high-fidelity experimental dataset of planar pushing. In: IEEE/RSJ International Conference on Intelligent Robots and Systems (IROS), pp. 30–37. IEEE (2016)
39. Coumans, E., Bai, Y.: Pybullet, A Python Module for Physics Simulation for Games, Robotics and Machine Learning (2016–2019). www.pybullet.org
40. Kloss, A., Schaal, S., Bohg, J.: Combining learned and analytical models for predicting action effects from sensory data. Int. J. Robot. Res. (2020)
41. Salimans, T., Kingma, D.P.: Weight normalization: a simple reparameterization to accelerate training of deep neural networks. In: Advances in Neural Information Processing Systems (NeurIPS) (2016)
42. Kingma, D.P., Ba, J.: Adam: a method for stochastic optimization. In Bengio, Y., LeCun, Y. (eds.) International Conference on Learning Representations (ICLR) (2015)
43. Grill, J.B., Strub, F., Altché, F., Tallec, C., Richemond, P., Buchatskaya, E., Doersch, C., Avila Pires, B., Guo, Z., Gheshlaghi Azar, M., Piot, B., kavukcuoglu, k., Munos, R., Valko, M.: Bootstrap your own latent–a new approach to self-supervised learning. In: Advances in Neural Information Processing Systems (NeurIPS), vol. 33, pp. 21271–21284 (2020)
44. Kornblith, S., Shlens, J., Le, Q.V.: Do better imagenet models transfer better? In: IEEE Conference on Computer Vision and Pattern Recognition (CVPR), pp. 2661–2671 (2019)

Pushing the Limits of Learning-Based Traversability Analysis for Autonomous Driving on CPU

Daniel Fusaro[1(✉)], Emilio Olivastri[1], Daniele Evangelista[1], Marco Imperoli[2], Emanuele Menegatti[1], and Alberto Pretto[1]

[1] Department of Information Engineering, University of Padova, Padova, Italy
{fusarodani,olivastrie,evangelista}@dei.unipd.it
[2] FlexSight Srl, Padova, Italy
marco.imperoli@flexsight.eu

Abstract. Self-driving vehicles and autonomous ground robots require a reliable and accurate method to analyze the traversability of the surrounding environment for safe navigation. This paper proposes and evaluates a real-time machine learning-based Traversability Analysis method that combines geometric features with appearance-based features in a hybrid approach based on a SVM classifier. In particular, we show that integrating a new set of geometric and visual features and focusing on important implementation details enables a noticeable boost in performance and reliability. The proposed approach has been compared with state-of-the-art Deep Learning approaches on a public dataset of outdoor driving scenarios. It reaches an accuracy of 89.2% in scenarios of varying complexity, demonstrating its effectiveness and robustness. The method runs fully on CPU and reaches comparable results with respect to the other methods, operates faster, and requires fewer hardware resources.

1 Introduction

Traversability Analysis is a fundamental task in the field of robotics and autonomous driving, in particular, this is one of the key activities that an autonomous vehicle has to perform to achieve an effective navigation in every type of scenario.

The ability to correctly detect non-traversable regions of the terrain is closely related to the type of vehicle on which the analysis is performed, and, according to [1], its computational complexity increases with the diversity of the surrounding terrain. As in [2], a grid cell is said to be *Non Traversable* if its features exceed certain thresholds for the vehicle specification: sizes, workloads, risk awareness and so on.[1]

A binary terrain classification that discriminates between *Traversable* and *Non Traversable* areas needs to be reliable and efficient. The reliability requirement is essential, it affects how much the vehicle is able to move in the environment, and it is strictly connected to the robot safety. Accordingly, the real-time

[1] Note that this definition differs from the *driveability* stated in [3].

I. Petrovic et al. (Eds.): IAS 2022, LNNS 577, pp. 529–545, 2023.
https://doi.org/10.1007/978-3-031-22216-0_36

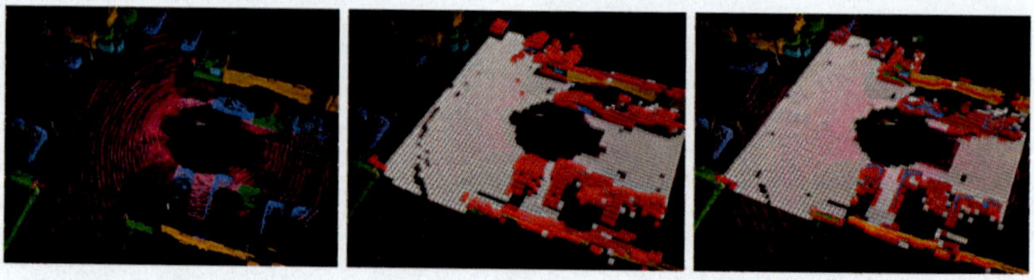

(a) (b) (c)

Fig. 1. From left to right: the original SemanticKITTI Labeled cloud in a frame from Scenario 00, and the corresponding Ground Truth and predicted traversability grids respectively (white points are traversable, red ones are not)

requirement needs special attention as well, since this may affect the reaction speed of the vehicle.

In this paper, we propose a real-time machine-learning-based grid segmentation method, in which 3D LiDAR data and 2D camera images are used for the traversability analysis. Our method is designed to work either with a single point cloud or with sequences of registered point clouds. In the simplest version, a point cloud is arranged in buckets of 2D points, thus creating a 2D grid such that its reference frame coincides with the LiDAR center. This setting is suitable for getting immediate traversability feedback, at the price of a low cloud density. Considering a sequence of point clouds and estimating the relative motions between them by using point cloud registration algorithms such as ICP (Iterative Closest Point) or a LiDAR odometry system such as [4], it is possible to obtain a denser local map, thus a more informative 2D grid. If a calibrated camera is available, it is possible to re-project the point cloud into the camera image plane, so collecting also the intensity and/or color information of the points. Our goal is to classify each cell of the collected grid: we compute a set of features representing the group of points in each cell, then by using such features we evaluate the level of traversability of the cells using a binary classifier.

The challenge of this work is to use geometric and appearance-based features in a mixed urban environment: roads, sidewalks, mixed vegetation (e.g., trees, near-road vegetation, urban parks, forests), dynamic agents and many other urban elements that make the classification of the 3D cloud more difficult and ambiguous.

The contributions of this work can be summarized as follows: (i) the development of a hybrid approach for a SVM-based traversability analysis that includes four new geometric features along with a set of appearance-based features; (ii) the comparison of the proposed approach with state-of-the-art semantic segmentation based methods on a large public dataset; (iii) the implementation code is being released for public usage.[2]

[2] Code available at: https://bitbucket.org/flexsight/traversabilityanalysis.

2 Related Work

The concept of traversability has been proposed in the literature for measuring the general characteristics of roads [5], but it has also been used for the analysis of the human driving style [3]. In the context of autonomous driving, there is no explicit formulation of traversability, some usage of this concept can be found instead in the motion planning context where "traversability maps" are used for the classification of a cell-divided map into "traversable" or "non-traversable" categories [6–9].

Different grid-based approaches for 3D terrain traversability analysis have been proposed in the literature. A complete and comprehensive review of those methods is presented in [1]. This survey gives a detailed overview of traversability analysis methods for autonomous vehicles within environments with highly variable complexity. In particular, the survey puts emphasis on exteroceptive sensory data processing methodologies that are presented in two main categories, namely *geometry-* and *appearance-based*.

In [10], a method for traversability analysis in urban and extra-urban scenarios is proposed. In the paper, authors evaluate the performances of various approaches based on point clouds and color data in various scenarios. They also train a SVM classifier that uses both geometric- and appearance-based local descriptors (features). Experimental results show that purely geometric-based features already provide enough accuracy, but the integration of different types of features, i.e., descriptors taken from 2D camera image processing, increases the robustness of the classifier and improves its classification accuracy.

Another purely geometric-based approach is proposed in [11]. The LiDAR point cloud is transformed into a new 2D representation called *LiDAR Histogram*. In the new domain, a road line estimation can be applied to get the actual road projected into the 3D data, and the same can be done for obstacles and water hazard segmentation.

A semi-supervised learning approach is presented in [12]. This approach uses a more compact feature vector and a grid representation to compute the traversability score. The classifier is trained only on positive samples, assumed to be truly traversable.

A traversability mapping method that uses Bayesian Generalized Kernel inference is found within the LeGO-LOAM framework [13]. This work proposes an approach for solving the traversability analysis problem producing 2.5D elevation grid maps. It exploits Bayesian Generalized Kernel (BGK) inference to solve the sparse data problem encountered during terrain mapping (it uses only single LiDAR scans). Then, the traversability analysis is performed on those locations intersected by the current LiDAR scan points, and elsewhere it is estimated by BGK traversability inference.
Deep learning frameworks are one of the most used instruments to solve problems that we are not able to model. There are many Deep-Learning-based solutions for traversability analysis.

In [14] images are used for Off-road traversability analysis. Authors used a double segmentation using CNNs. The first one divides the pixels in the image between trasversable and non trasversable, and then the trasversable pixels are re-segments into subclasses in order to asses the quality of the traversability based on the type of the terrain.

In [15] RGB images are used for traversability map prediction. In particular the input are the past frame, the current frame and the velocity inputs that pass through 2 CNNs, one predicts how the static elements of the image will change in the next frame (SNet), while the other predicts how the dynamic elements of the scene will change (Dnet). The predicted image is fed then to a third net that will produce the traversability map (GONet).

In [16] instead of determining the traversability for specific types of robot, it labels which terrain is trasversable and also for which type of robot (e.g. legged, wheeled, flying). In [17,18] both images and point clouds have been fed to deep networks in order to extract as much as information from the environment to obtain more robust traversability analysis. In fact, in [17] has been shown that this approach is effective even in unstructured environments.

In [19] a different learning approach has been exploited. A behaviour-based method to solve traversability analysis in an offroad scenario has been proposed. This method falls in the category of Deep Inverse Reinforcement Learning because the cost function is extracted from experts demonstrations, and then that cost function is used in order to learn a optimal policy. In [20], Deep Inverse Reinforcement Learning has been used in order to predict the trajectories of the other agents in the scene. Using this information to produce a traversability map that also takes into account future events. The predicted map will allow safer navigation.

When developing robust autonomous driving methods, two additional elements need to be taken into account, in particular, the availability of large datasets and the usage of good metrics for testing their performance. The former is essential especially if learning-based approaches are used. An exhaustive survey of available datasets for autonomous driving benchmarking is given in [21], including the SemanticKITTI dataset [22] that is used in this paper.

3 Method

The proposed approach leverages handcrafted geometrical and appearance-based features and Support Vector Machine (SVM) models. In machine learning and pattern recognition, a feature is an individual measurable property or characteristic of a phenomenon [23]. A geometric feature is calculated based on geometric properties, while an appearance feature is calculated starting from color or intensity properties. Handcrafted features are not new in the literature, and they may be overcome by self-learned features in the hidden layers of Deep Convolutional Neural Networks. However, considering comparable hardware, approaches that use these self-learned features are still slower than simple handcrafted features, and certainly not so accessible to a deep understanding of the whole process. Often the working mechanisms of the hidden layers' kernels of a Deep Neural Network are not so easily interpretable [24]. Since they need large amount of

data to properly generalize the self-learned decision rule, another drawback is made by the lack of rich and public labeled dataset.
In the following, a *feature vector* of a set of points $S = \{p_1, p_2, ..., p_n\}$ with $p_i \in \mathrm{R}^3$ is expressed in the form of a vector of features

$$\mathcal{F}(S) = [f_1, f_2, ..., f_m]^T$$

where m is the number of features. This vector of fixed dimension is used to train a SVM model. SVMs are popular decision models used in classification and regression problems. As stated in [23], an important property of SVMs is that the determination of the model parameters corresponds to a convex optimization problem, and so any local solution is also a global optimum. For the sake of brevity, the details on SVM models can be found in [25]. It is sufficient to say that training a SVM model means to find an optimal decision rule to assess the label of a feature vector. The so-called *kernel trick* helps in finding such decision rule by transforming the original data into a higher dimensional space. By experimental results, we have found that RBF kernel was the best performing kernel with respect to Linear and Polynomial kernels. It performed good both in training accuracy and testing accuracy. For this reason, we used that one in the proposed approach.

3.1 Grid-Based Space Representation

The traversability of the environment is evaluated using Digital Elevation Maps (DEMs), which are a set of uniformly discretized planar grid cells on the ground: at each cell is assigned a height value, as described in [13]. Here, the grid is squared and it maintains a fixed size during a simulation. Given a grid resolution r and a sensor maximum range of assessment *maxRange*, the grid will have $l = \frac{maxRange}{r}$ cells in each side, leading to l^2 cells in total as can be seen in Fig. 1b, c.

3.2 Geometric-Based Feature Extraction

The geometric-based features $\mathcal{F}_g(S) = [f_1, f_2, ..., f_r]$ are computed based on the geometrical properties of a set of 3D points S. Most of these features are calculated using the eigenvalues and the eigenvectors of the correlation matrix of the points in S. This matrix is computed based on the relative space coordinates of the points in S, and expresses the dependency between such points. It is a 3×3 symmetric positive semi-definite matrix with all real elements, so all of its eigenvalues are real and non-negative. Listing the eigenvalues in descending order $\lambda_1 \geq \lambda_2 \geq \lambda_3$, the corresponding eigenvectors v_1, v_2 and v_3 assume a spatial significance. As explained in [10], the eigenvector v_1 represents the direction of maximum variance of the points in S and v_2 represents the direction of the second maximum variance of the points. On the contrary, v_3 represents the smallest direction of variance but, at the same time, assuming that the points are arranged in a smooth plane in the space, v_3 is also normal to that plane. In the context of terrain traversability analysis, where roads are generally planar, this assumes a certain relevance. Thus, assuming that $\lambda_1 \neq 0$, among others we use all the features proposed in [10]: In our approach, four new features are

$$\text{linearity} = \frac{\lambda_1 - \lambda_2}{\lambda_1} \qquad \text{sphericity} = \frac{\lambda_3}{\lambda_1}$$

$$\text{planarity} = \frac{\lambda_2 - \lambda_3}{\lambda_1} \qquad \text{omnivariance} = \sqrt[3]{\lambda_1 \lambda_2 \lambda_2}$$

$$\text{anisotropy} = \frac{\lambda_1 - \lambda_3}{\lambda_1} \qquad \text{eigenentropy} = \sum_{i=1}^{3} \lambda_i \log \lambda_i$$

$$\text{sum of eigenvalues} = \sum_{i=1}^{3} \lambda_i \qquad \text{curvature} = \frac{\lambda_3}{\sum \lambda}$$

$$\text{angle} = \arccos(\boldsymbol{n} \cdot \hat{z}) \qquad \text{goodness of fit} = \min(\sigma_i(C_q))$$

$$\text{roughness} = \sigma_z^2 = \frac{1}{k} \sum_{i=1} k(z_i - \bar{z}_q)^2 \qquad \text{normal vector} = \boldsymbol{n}$$

$$\text{unevenness} = \frac{|r_q|}{k} \qquad \text{surface density} = \frac{k}{d_m^2}$$

introduced: Zeta difference, Internal density, Curvity and Volume.
The *Zeta Difference* feature represents the maximum difference along the elevation direction, expressed by the normal to the plane fitted to the whole point cloud. This is particularly relevant, because non traversable regions (e.g., obstacles) tend to have a wide variance along the normal direction, while points in traversable regions (e.g., roads) tend to have a very small height difference. In order to obtain a consistent metric, each point inside a cell is projected to a line, passing by the origin (the LiDAR frame origin) and having direction equal to the normal vector v. The projection p_p of a point p is obtained as follows:

$$p_p = \frac{vv^T}{v^T v} p.$$

Let M and m be the points of a cell whose projections have higher and lower values of z. Then

$$z_{diff} = |M_z - m_z|.$$

The *Internal density* feature expresses the discretized distribution of points internally to a cell. Let's define r_i as the *internal resolution* of a cell. Thus, we can subdivide the cell internally in $\frac{r}{r_i}$ cells. Then, the feature is obtained by counting how many internal cells are occupied by at least one point and dividing this by the total number of internal cells. This may be discriminant of some partially filled grid cells, or some obstacles, in which the points are very compacted.

The *Curvity* feature expresses how much the LiDAR scans are arranged as smooth circular curves inside the cells. This is motivated by the observation that in traversable regions (e.g. roads) the LiDAR scan is circularly smooth, since the road is a plain surface and there are no obstacles corrupting its smoothness. In non traversable regions (e.g. grass), the scans are very chaotic, sparse, not smoothly circular curved. To express this property quantitatively, we compute a discrete histogram of the Euclidean distance of each point from the LiDAR sensor. The number of bins of the histogram is a finite number n_b. Then, the curvity feature is the number of empty bars. The reason behind this is that all the points in a smooth circular curve have pretty much the same Euclidean distance, and when the distances are discretized, they get counted in a single bin, leading to many empty bins. Generalizing to multiple smooth circular curves, only some bins are occupied, while many are still empty. In a chaotic situation, bins tend to be rarely left empty.

The *Volume* feature expresses the volume of the smallest prism which contain all the points inside each cell. By definition, it's height is equal to the Z_{diff} feature. Empirically points inside obstacle regions tend to occupy more volume than points in traversable ones.

3.3 Appearance-Based Feature Extraction

Purely geometric features are often not sufficient to accurately discriminate cell classes. As shown in [10], on average hybrid features, based on both geometry and appearance, perform better than geometric features. This fact is also confirmed by our ablation study (see Sect. 4.6).
We compute the appearance-based features $\mathcal{F}_a(S) = [f_1, f_2, ..., f_r]$ based on the color properties of a set of points S. Unfortunately, due to the temporal desynchronization between the LiDAR and the camera and the small inaccuracies in the calibration, the 3D points are not always projected in the correct image position. To mitigate this inaccuracy, we obtain the appearance-based features of each cell by projecting a 3D prism built over the cell into the RGB image and retrieving the color information from the set of pixels within the polygonal projection of such prism (see Fig. 2).

Specifically, we detect for each cell the eight vertices of the prism which encloses all the points belonging to such cell. The height of the prism is by definition the value of the Z_{diff} feature, and the base is obtained by looking at the distribution along x and y directions of the points. Then we project the eight vertices and check that, for each prism, at least one fits inside the image (outliers are corrected according to the image size). Finally we convert the color of each pixel within the set S from the RGB color space to the Hue-Saturation-Value (HSV) color space. The appearance-based feature used in our approach is a HSV histogram of the pixels inside the polygonal projected prism. The histogram is defined by a number of buckets for each color channel: let $\#H$, $\#S$, $\#V$ be such quantities. The buckets $[H_b, S_b, V_b]$ corresponding to a pixel $[p_H, p_S, p_V]$ are obtained by quantizing the HSV values according to the number of buckets:

$$[H_b, S_b, V_b] = \left[\lfloor \frac{p_H}{\#H} \rfloor, \lfloor \frac{p_S}{\#S} \rfloor, \lfloor \frac{p_V}{\#V} \rfloor \right]$$

Fig. 2. Color propagation at the beginning of *Scenario 00*. The top image is the 28th Left Camera image of the Scenario 00 with the 3D prisms projected (colors were chosen randomly for visualization purposes only) and the three bottom images are the colored grids at frames 03, 08 and 28, respectively (the color of each cell is obtained setting for each H, S and V channel the value of the fullest buckets). It can be seen that at the beginning few cells are colored by the intersection of grid and camera image, but as the car proceeds new cells get colored.

When all pixels have voted to the histogram, all the buckets are normalized by the number of pixels that have voted, so values are in $[0, 1]$ range.

Color Propagation A single image doesn't provide enough color information: the amount of grid cells that can be correctly projected to the image frame is usually around 15% of the total cells. For this reason a color propagation method which exploits the odometry of the vehicle helps in providing more information. Once the appearance-based features of the cells of a scan have been computed, in the following scan these features are translated to the cells which occupies their position in the current situation, and those cells are updated using the new color information (see Fig. 2).

3.4 Point Clouds Integration

To augment the prediction input information, n subsequent point clouds are integrated forming a denser point cloud. The *LiDAR* reference frame, rigidly linked to the vehicle, is in motion with respect to a fixed reference frame we call *world*. By using point set registration or LiDAR odometry algorithms, it is possible to estimate the rigid body transformations that relate a sequence of LiDAR readings. When a point cloud P is obtained from the sensors, expressed in the *LiDAR* coordinate frame, it is transformed to the *world* frame, as it is considered a fixed reference frame. The traversability grid is also expressed in the frame *world*. In a stationary situation (once all n point clouds have been integrated), the SVM model is able to assess the traversability of the environment. In our method we exploit the ground truth ego-motion estimation, but other methods can be used, for example LeGO-LOAM framework [4] as a LiDAR odometry front-end for point cloud integration.

3.5 Point Sorting into the Grid Cells

Each point of the integrated point cloud is assigned to the cell it belongs to by looking at the point's 2D coordinates. If it lies outside the grid, it is ignored. The row and the column of the cell to which a point belongs are calculated as follows:

$$\begin{pmatrix} col \\ row \end{pmatrix} \approx \frac{1}{r} \begin{pmatrix} x - grid_x \\ y - grid_y \end{pmatrix}$$

where r is the map resolution, $grid_x$ and $grid_y$ the coordinates of the bottom-right cell of the grid expressed in the *world* frame. A cell is then considered *unpredictable* if it contains fewer points than a threshold value, otherwise it is considered *predictable*.

At this stage, a feature vector is computed for each *predictable* cell c, let's call it $\mathcal{F}_c$. A trained SVM model can predict the traversability value of the cell c using $\mathcal{F}_c$ as input.

3.6 Outliers Filter

Given the difficulty of the task, the SVM model can sometimes produce some outliers. For example, in some isolated cells of a plain traversable road region, it can predict a non traversable value. For this reason, a simple yet really effective signaling-based post processing filter is applied. Here we make the assumption of ignoring very small obstacles (size $\leq$ resolution r). Once all the cells have been assigned a traversability value by the SVM model, let us call it t_{ij}^0 with $i, j \in [0, ..., l]$, each cell is assigned a new value, let it be t_{ij}^1. This is computed considering the predicted label of the cell t_{ij}^0 weighted with a factor w and the labels of the 8-neighbors of the cell c_{ij} weighted with a factor 1. t_{ij}^1 is then the most frequent label.

Consider a cell t_{ij}^0 labeled as *non traversable* ($\bar{T}$) surrounded by all *traversable* (T) cells. Then the label-distribution for such cell would be, using $w = 3$, $\{\bar{T}, \bar{T}, \bar{T}, T, T, T, T, T, T, T, T\}$. Then, having 3 $\bar{T}$ and 8 T, we will assign $t_{ij}^1 = T$, that is, the cell will be labeled as *traversable*, even if the SVM-model labeled it as *non traversable*.

4 Experiments

To evaluate the performance of the proposed approach, the SemanticKITTI dataset is used [22]. All the experiments of this method are tested on an AMD

Ryzen™ 75,800 H, 8 ∗ 3.2 GHz, 16 GB RAM platform. All the Deep Learning based approaches are tested on an Intel® Core™) i9-10920X CPU (3.50 GHz) with Nvidia Titan RTX. All algorithms are implemented in ROS Noetic, in C++, in a Linux environment, with no GPU support and no CPU parallelization.

4.1 SemanticKITTI Dataset

The dataset used to evaluate the approach proposed is SemanticKITTI, a public dataset for Semantic Scene Understanding using LiDAR Sequences. This dataset is based on the KITTI Odometry Benchmark [26]. It comes up with 11 different scenarios, named *scenario00-scenario10*, in which a vehicle with the sensor setup installed on board is driven within an urban context, in low traffic conditions. There are sometimes dynamic agents (other vehicles moving around, people, etc.) and natural elements (grass, parks, trees, etc.). Among the info provided by the dataset, the method makes use of:

- LiDAR clouds, where each point carries the 3D Euclidean coordinates expressed in the LiDAR frame along with the RGB value of the class it belongs to;
- Left RGB Camera Images;
- Calibration Matrix, expressing the relative pose between the left RGB camera frame and the LiDAR frame;
- Ground Truth ego-motion estimation.

4.2 Traversability Ground Truth Extraction

The points in the SemanticKITTI dataset are not labeled with the *traversable* and *not traversable* labels. Indeed, the definition of traversability is in itself quite ambiguous. In order to obtain a ground truth dataset to evaluate the method, a pre-processing of the labels was needed. We considered as *traversable* the points having the label corresponding to one of the following classes of the SemanticKITTI dataset: *road, sidewalk, parking, lane marking, other ground.* Since our method refers to grid-based traversability analysis, we need to generalize the concept at a grid cell level.

The label of a cell is calculated by looking at the LiDAR points that fits inside each squared cell. Let m be the minimum number of points for a cell to be considered predictable. In our approach we set $m = 2$. If a cell contains less than m points, it is labeled as *unpredictable.* Otherwise, if the cell contains at least 2 *non traversable* points, the cell is labeled as *non traversable.* Otherwise, the cell is labeled *traversable.* This is done because some outliers are collected in the SemanticKitti Dataset: sometimes a fully *traversable* cell (e.g., road cells) contains an outlier marked as *non traversable.* By filtering out based on at least 2 *non traversable* points, we get rid of such outliers. The Ground Truth collected in this way comes with drawbacks and benefits. Consequences include:

- There may be points labeled as *traversable* that are actually under a car.

- A number of $m = 2$ points is sometimes not sufficient for some geometric properties to completely discriminate cells. This can be solved by using appearance-based features and a maxRange relatively close to the vehicle. If the car proceeds at high speeds, the sparsity of the cloud may be insufficient. In our approach this is solved by clouds integration and training using a bigger maxRange than the one used during test evaluation.
- Semantic-based methods produce a segmented cloud as output. Such methods produce a label for each point, so in order to compare our approach with their results we need to transfer the labels in the same way as we did for the SemanticKitti Ground Truth.
- Ground Truth for integrated clouds involve the formation of phantoms of points (e.g., when a car moves fast, the resulting integrated cloud maintains the points in all the subsequent frames). In order to be consistent with the different configurations, we decided to build the Ground Truth upon integrated or single cloud based on the need of each system. It is advisable that in the future a phantom-recognizer system should get rid of such outliers.

At this stage, each feature vector used in the previous sections can now be assigned a ground truth traversability label.

4.3 Training and Parameters

The experiments have been carried out using scripts that (a) automatically extract features for training, (b) train the model and (c) test the performances on unknown data. Each scenario of the dataset is used in the first 5 s to produce some amount of features, to obtain a dataset to train a SVM model. The training is carried out in a grid search approach. The fixed parameters are listed in Table 1.

Table 1. List of the used parameters

Grid length	$maxRange = 12\,\text{m}$
Map resolution	$r = 0.4\,\text{m}$
Internal cell resolution	$r_i = 0.2\,\text{m}$
Min points in a predictable cell	$min = 2$
Point clouds integrated	$n = 3$
Curvity histogram bars	$n_b = 160$
HSV histogram buckets	#H=32, #S=8, #V=48

Using a grid based approach in a 5-folds validation setup, the best parameters for each model were selected. The best model was then retrained using the whole training set. Since the first 5 s of each scenario are used for the training phase, the performance evaluation is done using the following 50 s of the sequences.

4.4 Metrics

To compute the metrics we used the number of True Positives (TP), True Negatives (TN), False Positives (FP), False Negatives (FN) and Unknowns (UNK). The latest are those having a label in the ground truth but unpredictable by the method. Let $TOT = TP + TN + FP + FN + UNK$, the performance is measured using the following metrics:

- Accuracy (Acc):

$$Acc = \frac{TP + TN}{TOT}$$

- Intersection over Union (IoU):

$$IoU = \frac{TP}{FP + FN + TP}$$

- F1 score: (harmonic mean of precision and recall)

$$F1 = \frac{2TP}{2TP + FP + FN}$$

- Rates over the total number of samples:

$$(TPR, TNR, FPR, FNR) = \frac{(TP, TN, FP, FN)}{TOT}$$

- Latency of the system inferencing the labels: it refers to the time taken to process one unit of data provided that only one unit of data is processed at a time. The metric is calculated by averaging the inference times of each label. Recall that the LiDAR data rate is 10 Hz, thus the maximum time allowed for inferencing a cloud is 100ms.

4.5 Comparison

The experiments carried out aim to emphasize the performances brought by the contributions of this work, in particular the use of the 4 new geometric-based features and the appearance-based features, the post-processing filter and the low latency due to the SVM models.

We tested our framework against different recent methods: RangeNet++ [27] (comparing many backbones) and PointNet [28].

RangeNet++ is a segmentation-based network: it outputs the label of each point of the cloud that was given as input. Every RangeNet++ model was trained from scratch using as learning labels only two classes (traversable and not traversable) by remapping the original labels as explained in section [27]. RangeNet++ has been trained using different backbones (DarkNet and Squeezeseg, in different versions). To compare them with the proposed approach we treat the RangeNet++ predicted labels as if they were the ground truth, assigning labels to cells as described in Sect. 4.2. PointNet, instead, is a classification network. When given the set of points belonging to each cell the network predicts

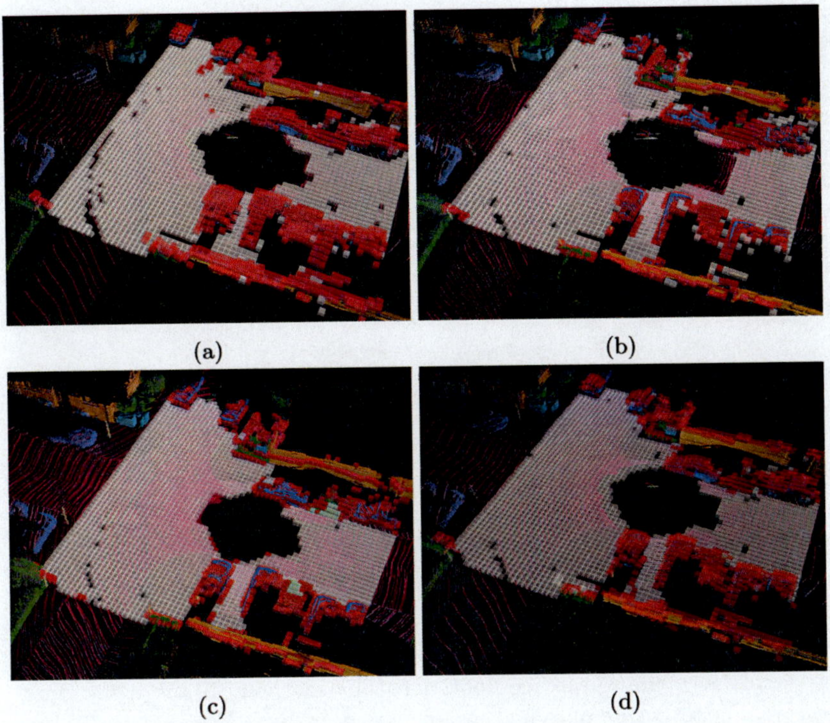

Fig. 3. Figure 3a depicts an example of Ground Truth Traversability (white points are traversable, red ones are not), Fig. 3b is the result of the method proposed, Fig. 3c the result of the PointNet, Fig. 3d the result of RangeNet++ (with DarkNet53-1024px backbone).

the corresponding label. In order to do so, we need to create a new dataset from SemanticKitti where points were packed up to the cell they belong to, together with their ground truth label. A fixed-length vector of points is fed to the network (by trial and errors, we found that the best accuracy is reached with a fixed length of 256) and the result of the network is the label of each cell (Fig. 3).

Table 2 shows that the results of the proposed method is comparable with the other methods results. Deep Learning based method have slightly better accuracy, but it comes at the cost of a bigger latency and expensive required hardware. SVM latency is around 19 ms, while the other methods' latency vary across 20–105 ms for RangeNet++ and 2000ms for PointNet.

The overhead of the system is very important in this context. If the LiDAR and the camera provide new information with a rate 10 Hz, only 100ms are available for the traversability computation. But 100ms is even too much time: very often an Autonomous Driving system has to compute many other things, and so the available time is a smaller portion of the total available. Not to mention the time needed for the system to contextualize the results: for path

Table 2. Average results on SemanticKITTI among scenarios 00-10 (all values are percentages)

Experiment	Acc	mIoU	F1	FPR	TPR	FNR	TNR	Latency (ms)
Trav_analysis	89.2	84.9	91.4	10.2	93.6	6.4	89.8	**19**
Darknet21	92.7	86.0	92.2	**3.1**	88.3	11.7	**96.9**	40
Darknet53	**93.4**	**87.4**	**93.0**	3.7	90.2	9.8	96.3	105
Darknet53-512px	91.6	83.8	90.9	3.8	86.9	13.1	96.2	25
Darknet53-1024px	93.1	86.9	92.7	3.4	89.3	10.7	96.6	55
Squeezeseg	90.1	81.4	89.4	5.2	85.3	14.7	94.8	20
SqueezesegV2	92.3	85.5	91.9	4.3	88.1	11.9	95.7	25
PointNet	90.0	87.4	93.0	7.3	93.8	6.3	92.7	2000

planning and other tasks occupancy grids like the one proposed are used, and so the system needs also to compute such information from the pure segmentation cloud.

It is clear how the best accuracy is reached by RangeNet++ using the Darknet53-1024px backbone. It's beaten only by the DarkNet53 in TNR and FPR rates. Both are slower compared to our method. If the same duration metric was meant to be obtained, the accuracy of RangeNet++ would be very similar to our method, while the mIoU drops lower our mIoU. By comparing [27]'s results and ours, we identify the same so-called *label re-projection problem.* Thus the Ground Truth extraction may penalize this drawback of the method. The same extraction affects out method: when planar *non traversable* regions are met illumination drastically changes.

PointNet is rather slow and not easily exploitable in a real context, although it performs very well from an accuracy point of view.

The proposed approach is feasible: the latency is very low and the accuracy is competitive. The whole process of clouds integration, feature extraction, SVM prediction and grid development takes around 60ms, perfectly suitable for real-time real-world application. The hardware needed is a CPU-only PC, a 3D LiDAR scan and an RGB Camera.

4.6 Ablation Study

In order to demonstrate the effectiveness of the appearance-based features in a hybrid approach in contrast to a purely geometric-based method, we have conducted an ablation study. As shown in Table 3, the purely geometric method performs worse than the hybrid one. This happens because several *Not Traversable* flat, equally distributed cells, e.g. containing pretty flat grass points, cannot be distinguished from *Traversable* road cells using only geometric information.

Table 3. Average results on SemanticKITTI among scenarios 00-10 (all values are percentages)—ablation study to demonstrate the effectiveness of the appearance-based features

Experiment	Acc	mIoU	F1	FPR	TPR	FNR	TNR	Latency (ms)
Trav_analysis	89.2	84.9	91.4	10.2	93.6	6.4	89.8	19
Only geom	85.8	79.5	87.9	11.8	89.4	10.6	88.2	13

5 Conclusions

In this paper, a real-time Machine Learning Traversability Analysis method has been proposed, which combines geometric and appearance based features. The method has been evaluated on a public dataset and the performances have been compared with state-of-the-art methods. The proposed method is comparable with the others in terms of accuracy metrics, while it is faster and cheaper in terms of hardware resources since it runs completely on CPU without the need for high-end GPU like other methods. The tests have shown the effectiveness of the newly introduced features, the exploited grid-based data integration method and the simplified framework make the overall approach easy to integrate within real autonomous driving pipelines with minimal effort but with competitive results in terms of classification of traversable areas in the map. Possible improvements can be obtained with the integration of self-learning subsystems, in particular for image-based scene understanding, to better exploit the color information when available.

Acknowledgment. This project has received support from the FlexSight Srl company. Part of this work has been supported by MIUR (Italian Minister for Education) under the initiative "Departments of Excellence" (Law 232/2016).

References

1. Papadakis, P.: Terrain traversability analysis methods for unmanned ground vehicles: a survey. Eng. Appl. Artif. Intel. **26**(4), 1373–1385 (2013)
2. Lee, H., Chung, W.: A self-training approach-based traversability analysis for mobile robots in urban environments. IEEE Int. Conf. Robot. Autom. (ICRA) **2021**, 3389–3394 (2021)
3. Bekiaris, E., Amditis, A.J., Panou, M.C.: Drivability: a new concept for modelling driving performance. Cogn. Technol. Work **5**, 152–161 (2003)
4. Shan, T., Englot, B.: Lego-loam: lightweight and ground-optimized lidar odometry and mapping on variable terrain. In: 2018 IEEE/RSJ International Conference on Intelligent Robots and Systems (IROS), pp. 4758–4765 (2018)
5. Tsukada, Y., Okutani, T., Itsubo, S., Tanabe, J.: Evaluation of roads network in japan from viewpoint of driveability. Proc. Eastern Asia Soc. Transp. Stud. **2007**, 336–336 (2007)

6. Leonard, J.J., How, J.P., Teller, S.J., Berger, M., Campbell, S., Fiore, G.A., Fletcher, L., Frazzoli, E., Huang, A.S., Karaman, S., Koch, O., Kuwata, Y., Moore, D.C., Olson, E., Peters, S.C., Teo, J., Truax, R., Walter, M.R., Barrett, D., Epstein, A., Maheloni, K., Moyer, K., Jones, T., Buckley, R., Antone, M.E., Galejs, R., Krishnamurthy, S., Williams, J.: A perception-driven autonomous urban vehicle. In: The DARPA Urban Challenge (2009)
7. Sivaraman, S., Trivedi, M.M.: Dynamic probabilistic drivability maps for lane change and merge driver assistance. IEEE Trans. Intel. Transp. Syst. **15**, 2063–2073 (2014)
8. Kim, K., Kim, B., Lee, K.-J., Ko, B., Yi, K.: Design of integrated risk management-based dynamic driving control of automated vehicles. IEEE Intel. Transp. Syst. Mag. **9**, 57–73 (2017)
9. Thrun, S., Montemerlo, M., Dahlkamp, H., Stavens, D., Aron, A., Diebel, J., Fong, P., Gale, J., Halpenny, M., Hoffmann, G., Lau, K., Oakley, C., Palatucci, M., Pratt, V., Stang, P., Strohband, S., Dupont, C., Jendrossek, L.-E., Koelen, C., Mahoney, P.: Stanley: the robot that won the darpa grand challenge. J. Field Robot. **23**(01), 661–692 (2006)
10. Bellone, M., Reina, G., Caltagirone, L., Wahde, M.: Learning traversability from point clouds in challenging scenarios. IEEE Trans. Intel. Transp. Syst. **19**, 296–305 (2018)
11. Chen, L., Yang, J., Kong, H.: Lidar-histogram for fast road and obstacle detection. In: 2017 IEEE International Conference on Robotics and Automation (ICRA), pp. 1343–1348 (2017)
12. Suger, B., Steder, B., Burgard, W.: Traversability analysis for mobile robots in outdoor environments: a semi-supervised learning approach based on 3d-lidar data. In: IEEE International Conference on Robotics and Automation (ICRA), vol. 2015, pp. 3941–3946 (2015)
13. Shan, T., Wang, J., Englot, B., Doherty, K.A.J.: Bayesian generalized kernel inference for terrain traversability mapping. In: CoRL (2018)
14. Viswanath, K.S., Singh, K., Jiang, P., Sujit, P.B., Saripalli, S.: Offseg: a semantic segmentation framework for off-road driving. In: 2021 IEEE 17th International Conference on Automation Science and Engineering (CASE), pp. 354–359 (2021)
15. Hirose, N., Sadeghian, A., Xia, F., Martin-Martin, R., Savarese, S.: Vunet: dynamic scene view synthesis for traversability estimation using an rgb camera. IEEE Robot. Autom. Lett. 1(1) 2019
16. Hosseinpoor, S., Torresen, J., Mantelli, M., Pitto, D., Kolberg, M., Maffei, R., Prestes, E.: Traversability analysis by semantic terrain segmentation for mobile robots. In: 2021 IEEE 17th International Conference on Automation Science and Engineering (CASE), pp. 1407–1413 (2021)
17. Paz, D., Zhang, H., Li, Q., Xiang, H., Christensen, H.I.: Probabilistic semantic mapping for urban autonomous driving applications. In: 2020 IEEE/RSJ International Conference on Intelligent Robots and Systems (IROS), pp. 2059–2064 (2020)
18. Guan, T., He, Z., Manocha, D., Zhang, L.: TTM: Terrain Traversability Mapping for Autonomous Excavator Navigation in Unstructured Environments (2021). ArXiv, vol. abs/2109.06250
19. Zhu, Z., Li, N., Sun, R., Zhao, H., Xu, D.: Off-road autonomous vehicles traversability analysis and trajectory planning based on deep inverse reinforcement learning. In: 2020 IEEE Intelligent Vehicles Symposium (IV), pp. 971–977 (2020)

20. Jung, C., Shim, D.H.: Incorporating multi-context into the traversability map for urban autonomous driving using deep inverse reinforcement learning. IEEE Robot. Autom. Lett. **6**(2), 1662–1669 (2021)
21. Yin, H., Berger, C.: When to use what data set for your self-driving car algorithm: An overview of publicly available driving datasets. In: 2017 IEEE 20th International Conference on Intelligent Transportation Systems (ITSC), pp. 1–8 (2017)
22. Behley, J., Garbade, M., Milioto, A., Quenzel, J., Behnke, S., Stachniss, C., Gall, J.: Semantickitti: a dataset for semantic scene understanding of lidar sequences. In: 2019 IEEE/CVF International Conference on Computer Vision (ICCV), pp. 9296–9306 (2019)
23. Bishop, C.M.: Pattern Recognition and Machine Learning (Information Science and Statistics). Springer, Berlin (2006)
24. Fan, F., Xiong, J., Li, M., Wang, G.: On interpretability of artificial neural networks: a survey. IEEE Trans. Radiat. Plasma Med. Sci. **5**, 741–760 (2021)
25. Cristianini, N., Shawe-Taylor, J.: An Introduction to Support Vector Machines and Other Kernel-based Learning Methods. Cambridge University Press, Cambridge (2000)
26. Geiger, A., Lenz, P., Urtasun, R.: Are we ready for autonomous driving? the kitti vision benchmark suite. In: 2012 IEEE Conference on Computer Vision and Pattern Recognition, pp. 3354–3361 (2012)
27. Milioto, A., Vizzo, I., Behley, J., Stachniss, C.: Rangenet++: fast and accurate lidar semantic segmentation. 2019 IEEE/RSJ International Conference on Intelligent Robots and Systems (IROS), pp. 4213–4220 (2019)
28. Qi, C., Su, H., Mo, K., Guibas, L.J.: Pointnet: deep learning on point sets for 3d classification and segmentation. In: 2017 IEEE Conference on Computer Vision and Pattern Recognition (CVPR), pp. 77–85 (2017)

Sensor-Based Navigation Using Hierarchical Reinforcement Learning

Christopher Gebauer, Nils Dengler, and Maren Bennewitz(✉)

Humanoid Robots Lab, University of Bonn, Bonn, Germany
maren@cs.uni-bonn.de

Abstract. Robotic systems are nowadays capable of solving complex navigation tasks. However, their capabilities are limited to the knowledge of the designer and consequently lack generalizability to initially unconsidered situations. This makes deep reinforcement learning (DRL) especially interesting, as these algorithms promise a self-learning system only relying on feedback from the environment. In this paper, we consider the problem of lidar-based robot navigation in continuous action space using DRL without providing any goal-oriented or global information. By relying solely on local sensor data to solve navigation tasks, we design an agent that assigns its own waypoints based on intrinsic motivation. Our agent is able to learn goal-directed navigation behavior even when facing only sparse feedback, i.e., delayed rewards when reaching the target. To address this challenge and the complexity of the continuous action space, we deploy a hierarchical agent structure in which the exploration is distributed across multiple layers. Within the hierarchical structure, our agent self-assigns internal goals and learns to extract reasonable waypoints to reach the desired target position only based on local sensor data. In our experiments, we demonstrate the navigation capabilities of our agent in two environments and show that the hierarchical structure seriously improves the performance in terms of success rate and success weighted by path length in comparison to a flat structure. Furthermore, we provide a real-robot experiment to illustrate that the trained agent can be easily transferred to a real-world scenario.

1 Introduction

In the last decade, the consumer market reveals an increasing demand for mobile robots to assist a human user in a diverse set of housekeeping tasks. While the general benefit for the consumer is out of question, most systems have a major limitation: They do not improve their behavior over time and are poor at generalization, in particular when facing situations that were not considered beforehand. The idea of machine learning is very inspiring in that case, because

This work has partially been funded by the European Commission under grant agreement number 964854–RePAIR–H2020-FETOPEN-2018–2020 and by the Deutsche Forschungsgemeinschaft (DFG, German Research Foundation) under Germany's Excellence Strategy, EXC-2070—390732324—Phenorob.

I. Petrovic et al. (Eds.): IAS 2022, LNNS 577, pp. 546–560, 2023.
https://doi.org/10.1007/978-3-031-22216-0_37

it allows to search for the optimal solution and improve it over time in case of unconsidered situations or changes in expectation. Especially, deep reinforcement learning (DRL) is appealing as it does not require the optimal solution to learn from but only feedback from the environment (Fig. 1).

Fig. 1. We train a hierarchical reinforcement learning agent to identify and reach its target only based on local sensor information, without providing any global or goal-oriented information. At the beginning of the training, the agent is not aware how the target looks like and where it is placed. This figure shows how the agent successfully navigates to and identifies the hidden target (green pole) in a randomly generated maze, as indicated by the black trajectory. The small green markers represent the internal goals generated by our agent.

For robot navigation, the current state of the art uses global information about the environment in form of a map and is aware of the position of the target [18]. For reinforcement learning approaches, this information is used during runtime to provide insight to the surrounding obstacles [20,21] or the heading direction of the surrounding humans [17]. In our work, we aim at designing a navigational reinforcement learning agent that is not provided with any goal-oriented or global information and that assigns its own waypoints based on intrinsic motivation, without relying on any expert knowledge. Recently, Mirowski et al. [19] trained a navigation agent without providing target information to the agent with very promising results. However, the authors select the control signals from a discrete action space, which generally limits the fineness of the robot's motion capabilities. In this paper, we present an approach to learning sensor-based navigation in *continuous* action space. To address the complexity of the continuous action space, we deploy a hierarchical agent structure [14,15], which distributes the exploration across multiple layers. The task of the agent is to navigate to a hidden target in an unknown environment, where the target is uniquely identifiable, but unknown at the beginning of the training. We hereby consider a wheeled robot base equipped with a 2D lidar. To efficiently incorporate the high-dimensional information of the 2D lidar, we pretrain an autoencoder [10], which provides a dense state representation to the hierarchical agent.

Our main contribution is a DRL agent capable of solving navigation tasks in continuous action space without relying on any global or target information. As we show in the experimental evaluation, our agent learns to reach the hidden target in unknown environments solely based on local sensor data and while receiving only sparse feedback in form of delayed rewards when reaching the target. We evaluate the performance in two different environments and show the seriously increased performance of our hierarchical structure in comparison to a flat agent regarding the success rate and success weighted by path length. Furthermore, we transfer the trained agent into a real-world scenario and demonstrate the applicability of the learned policy on a real robot.

2 Related Work

While classical approaches for robot navigation, e.g., the dynamic window approach [8], work very well in many situations [18], it is much more desirable to have algorithms that search for the optimal solution by itself [25]. To address this idea, Pfeiffer et al. [20] used a target-driven approach that learns to steer a mobile robot collision-free to reach a relative goal pose. Furthermore, Tai et al. [26] proposed to combine reinforcement learning for local navigation with a classical global path planning algorithm to reach a target position. Stein et al. [24] on the other side, improved the global path planning by incorporating machine learning in the process of evaluating possible future subgoals. The authors proposed to predict the expected cost for each of the possible subgoals and decrease the overall path length by choosing the most promising ones.

In contrast to the approaches above, Jaderberg et al. [11] introduced an agent that solves a robot navigation task end-to-end by only providing sensor information. The agent has to explore an unknown maze to find hidden targets. Mirowski et al. [19] extended this approach by improving the architecture of the agent and adding a stacked long short-term memory structure. Even though learning on raw sensor data works, Sax et al. [23] showed that a well-defined state space is crucial when addressing complex navigation tasks and proposed to use the latent space of different pre-trained autoencoders. Chaplot et al. [5] designed an object-oriented navigation agent based on semantic exploration. This helps to understand where certain objects are more likely to be found and therefore improves the navigation capabilities. While all these methods seem promising in general, they are limited to a discrete action space and Dhiman et al. [7] questioned if such agents are capable of solving the navigation problem with comparable performance to classical hand-crafted approaches. Whaid et al. [28] trained in continuous action space and compare different observations for a flat agent structure. The authors showed that lidar information have significant impact on the performance in learning-based navigation tasks.

To increase the capabilities of an agent in complex tasks, Vezhnevets et al. [27] introduced a hierarchical agent based on deep reinforcement learning, where a high-level manager sets subgoals on low frequency and a worker executes low-level actions at each time step following these subgoals. Levy et al. [15] introduced off-policy corrections based on hindsight experience replay [2] and reformulated the hierarchical approach to an arbitrary number of hierarchical layers. The authors demonstrated the effectiveness in the context of robot navigation for up to three hierarchical layers. However, in their approach the agent received the position of the goal as part of the state representation. Li et al. [16] introduced a hierarchical approach to simultaneously control a robot arm and the base it is attached to. While this again demonstrates the incredible capabilities of hierarchical structures, also here the state contains the location of the goal. Kulkarni et al. [14] proposed a hierarchical agent where a top-level value function learns a policy over a given set of intrinsic goals, and a lower-level function learns a policy over discrete actions to satisfy the goals.

In this paper, we introduce a hierarchical reinforcement learning approach to solve the problem of sensor-based robot navigation in continuous action space in unknown environments. In contrast to the related work, we do not provide any target information within the state space and our agent identifies its own internal goals based on observations and experience from the past training.

3 Our Approach

In this section, we first introduce the navigation task to be solved. Then, we give the preliminaries regarding reinforcement learning and the general extension to a hierarchical structure. Next, we describe our agent and its structure in the context of the navigation task, as well as the reward function. Finally, we discuss the countermeasures for the instabilities due to concurrent training of all layers.

3.1 Navigation Task

As we explicitly avoid any global information, our agent is designed to solve general navigation problems with a designated target, which is not known to the agent at the beginning of the training. Target information is only obtained during the training using the provided reward. Thereby, the target needs to be uniquely recognizable by the agent from sensor data and the agent must know when the target has been reached.

In our application, a wheeled robot observes the external environment using a 2D lidar sensor and measures its own state based on odometry. We refer to these observations as o_{ext} and o_{int}, respectively. The robot state corresponds to a pose measurement in Cartesian space, i.e., we only consider the relative offset to the most recent measurement, thereby assuming neglectable noise. The output of the hierarchical agent is a pair of translational and rotational velocities, which are sent to the robot base.

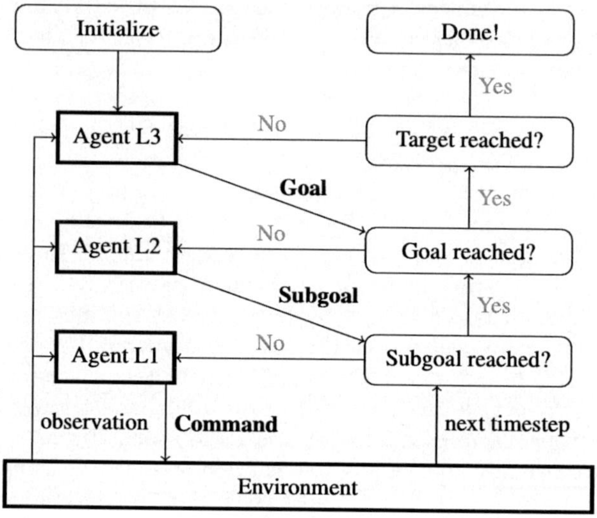

Fig. 2. The information flow of our hierarchical reinforcement learning structure using one mid-level agent, i.e., I = 3. The algorithm is initialized on the top left. The top-level agent (L3) provides a relative goal position based on the provided observation, with the purpose to reach the hidden target. Based on the provided goal and a current observation, the mid-level agent (L2) provides a subgoal. The low-level agent (L1) sends commands, i.e., translational and rotational velocities, to the robot base, based on the provided subgoal and a current observation.

3.2 Preliminaries

We define our reinforcement learning setup as a partially observable Markov Decision Process (POMDP) [12] in a hierarchical structure of I layers. The hierarchical POMPD is a tuple of ($\mathcal{S}$, $\mathcal{O}$, $\mathcal{A}$, $\mathcal{R}$, $\mathcal{T}$, γ, I) components described below. The state $s_t \in \mathcal{S}$ describes the distinct and explicit configuration of the environment in timestep t and is not directly accessible for the agent. We simplify the full hierarchical POMDP into a layer-wise representation ($\mathcal{S}$, $\mathcal{O}_i$, $\mathcal{A}_i$,

$\mathcal{R}_i$, $\mathcal{T}_i$, γ, Δt_i), where i represents the current layer and denotes from now on layer-dependent quantities. With this step we assume that the structure of the underlying process in each layer is identical. Furthermore, it outlines that each layer operates on a different time scale Δt_i, i.e., a complete episode for layer i equals one time step for layer $i + 1 \leq \mathrm{I}$.

Proceeding with layer i, the agent receives in timestep t_i an observation $o_{t,i} \in \mathcal{O}_i$ and selects an action $a_{t,i} \in \mathcal{A}_i$ according to its policy $\pi(\theta_i) : \hat{\mathcal{O}}_i \rightarrow \mathcal{A}_i$, where θ_i parametrizes the policy. The extended observation space $\hat{\mathcal{O}}_i$ for $i < \mathrm{I}$ is computed by combining the observation space with the current action from the layer above

$$f_i : \mathcal{O}_i \times \mathcal{A}_{i+1} \rightarrow \hat{\mathcal{O}}_i.$$

For $i = \mathrm{I}$ it is equal to the observation space itself, i.e., $\hat{\mathcal{O}}_\mathrm{I} = \mathcal{O}_\mathrm{I}$. After action $a_{t,i}$ has been executed, the agent receives a reward $r_{t,i} \in \mathcal{R}_i$ and the next observation $o_{t+1,i}$. The transition is selected according to the state-transition function $\mathcal{T}_i : \mathcal{S} \times \mathcal{A}_i \times \mathcal{S}$. The optimal policy $\pi(\theta_i)$ maximizes the discounted reward $R_{t_i} = \sum_{j=0}^{T_i} \gamma^j r_{t_i + j\Delta t_i}$, where T_i are the number of steps to reach the final timestep of the current episode in layer i and $\gamma \in [0, 1]$ is the discount factor.

3.3 Agent Structure

In our application, we use a hierarchical agent with three layers as visualized in Fig. 2. The *highest layer* addresses the navigation task with high-level abstraction, i.e., finding the hidden target within the unknown environment by proposing internal goals. All other layers have the purpose to reach the generated internal goals. The action for the highest layer is a relative position in Cartesian space, which we refer to as the *goal.* Since our aim is to avoid any global information, we use only the external observation as input to generate an action for Layer 3:

$$a_{t,3} = \pi(o_{ext,t,3}, \theta_3) \quad \text{and} \quad d_3 = p_3(o_{ext,t,3})$$

where the quantity d_3 is an indicator that signalizes whether the target has been reached, based on a predictor p_3 that is trained on past experience using the external observation.

The *middle layer* breaks down reaching the distant goal from the highest layer to approaching closer subgoals. In addition to the relative position in Cartesian space, this layer provides preferred velocities, i.e., to represent slowing down in narrow areas. In the following, we will refer to the combination of both as the *subgoal* provided by Layer 2. As input we use the external observation but reduce the line of sight to increase the attention for nearby obstacles. Additionally, based on the internal state and the action from the layer above, we compute a progress using the function f_2, which is used as further input for the policy to generate an action. In more detail, we determine the progress by transforming the position of the goal obtained from the layer above into the current base frame of the mobile robot, i.e., providing its current relative position. The quantity d_2 signalizes if the relative position of the goal given from the layer above has been reached and is computed based on the progress:

$$a_{t,2} = \pi\left(o_{ext,t,2}, prog_2, \theta_2\right) \quad \text{and} \quad d_2 = p_2(prog_2)$$
$$\text{with} \quad prog_2 = f_2(o_{int,t}, a_{t,3})$$

The *lowest layer* directly communicates with the mobile base and sends control signals in the form of translational and rotational velocities. It receives only the progress computed using the function f_1 based on the subgoal from the layer above and the internal observation as input to generate the action:

$$a_{t,1} = \pi\left(prog_1, \theta_1\right) \quad \text{and} \quad d_1 = p_1(prog_1)$$
$$\text{with} \quad prog_1 = f_1(o_{int,t}, a_{t,2})$$

The external observation is not needed as input here, because the subgoals are learned to be reachable collision-free, due to the received penalties when collisions occur, i.e., the agent automatically learns to set subgoals such that the lowest layer is capable of reaching those collision-free. The quantity d_1 signalizes if the position of the relative position of the subgoal $a_{t,2}$ has been reached and is computed based on the progress.

3.4 Reward Function

We divide the reward into two categories: An external reward provided from the environment r_{env} and an intrinsic reward with two different origins, based on which layer is receiving it. The *external* reward from the environment r_{env} is only sparse: The robot receives a greater positive reward when it has reached the target and a smaller penalty when a collision occurs. In all other timesteps this reward is zero. We only provide r_{env} to the layers that use the external observations from the robot's sensor, i.e., Layer 3 and Layer 2. From r_{env}, Layer 2 learns to generate subgoals that are reachable collision-free by the lowest layer. For Layer 2 and Layer 1 that receive the goal and subgoal, the *intrinsic* reward r_{prog} is computed based on the progress towards the goal or subgoal. It consists of a smaller dense reward, i.e., the distance change normalized by the initial distance to the goal or subgoal, and a greater reward when it is reached by the robot. To have an intrinsic reward for the top layer that does not receive any goal, we grant a small reward r_{act} correlating with the magnitude of the action, i.e., the distance to the chosen goal, to emphasize doing anything. In our experiments, we choose the environment reward r_{env} to be a magnitude above the intrinsic motivation, as it increases the convergence speed. The reward function is summarized as follows:

$$r_{t,i} = \begin{cases} r_{env} + r_{act} & , i = 3 \\ r_{env} + r_{prog} & , i = 2 \\ r_{prog} & , i = 1 \end{cases}$$

3.5 Reinforcement Learning

In this section, we first present the preprocessing of the sensor data and then describe the learning strategy for the agent.

Feature Extraction We use a 2D-lidar sensor as external sensor and build our preprocessing pipeline upon our previous work [10], i.e., we convert raw lidar data into a local occupancy grid, which is encoded using a variational autoencoder (VAE) [13]. This has two benefits in contrast to autoencoding raw lidar scans. First, the reconstruction capabilities are significantly increased, which allows to reconstruct more accurately complex geometric shapes. Furthermore, the grid map can be cropped without any need of replacing the new out-of-range measurements. This reduces the information content provided to the agent when only information of nearby obstacles is required as for the middle layer of the agent (see Sect. 3.3) and helps decreasing the size of the latent space.

Learning Algorithm and Training Strategy All three agents of our hierarchical learning structure are optimized with the twin delayed deep deterministic policy gradient (TD3) algorithm [9]. We parametrize all policies with small dense neural networks, consisting of two layers.

For training, we apply curriculum learning [3] and increase the difficulty step by step, i.e., the target is placed closer and easier to reach by the robot in early stages of the training. An episode ends, when the target has been reached, a collision occurs, or the maximal number of steps for the top layer is reached. If one of the two other layers reaches its maximal number of steps, a new goal is provided to that layer.

Note that training the full hierarchical agent at once can lead to instabilities since even if the state-transition probability $\mathcal{T}_i$ for $i > 1$ is defined stationary, it will change when the policy of the layer below is adopted. Therefore, we use the hindsight actions introduced by Levy et al. [15] and assume wherever the agent ended up was the intended goal position so that the reward for reaching it is experienced.

In our experiments, we noticed that these additional interactions are essential at the early stage of the training, but not anymore as soon as the lower layers start to converge. Therefore, we stop collecting these interactions when a certain reward is reached during training.

4 Experimental Evaluation

In this section, we evaluate our approach in the context of navigation capabilities within simulation and transfer it into a real-world scenario afterwards. At first, we specify our setup including the agent and the simulation environments. Next, we show the performance when only training the top agent. Finally, we compare the performance of our hierarchical agent to a flat agent and transfer the trained agent into a real-world scenario.

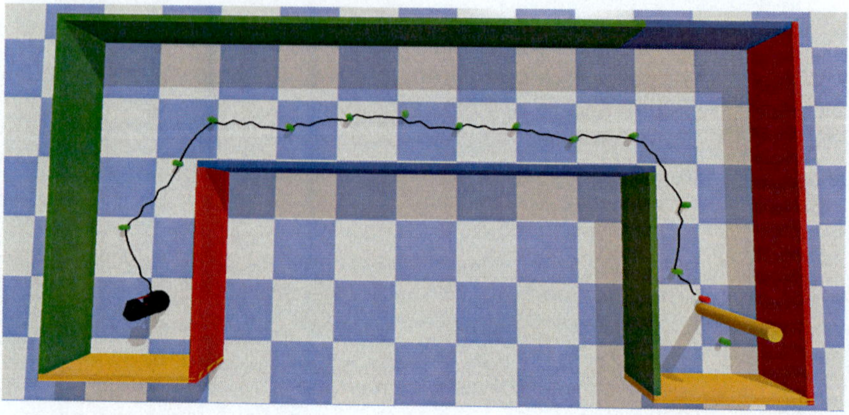

Fig. 3. Visualization of the *rooms* environment and trajectory in simulation, the yellow pole indicates the hidden target and the green markers represent the internal goals generated by the highest layer of our agent.

4.1 Setup

Environment We test our agent in two environments with different degrees of complexity. As external sensor we use a 2D lidar scanner with a maximal range of 10 m. Both environments consist of interconnected corridors or rooms with one pole representing the hidden target. The positions of the robot and of the target are drawn randomly at the beginning of each episode. The first environment consists of up to three interconnected rectangular rooms or corridors, see Fig. 3. The width and length of the rooms as well as the composition are randomly drawn at the beginning of each episode. We will refer to this environment as *rooms*. The second environment consists of a random maze, which we refer to as *mazes* and which is visualized in Fig. 1. The composition is drawn randomly, but the size is always fixed. For the simulation we used pybullet [6] and wrapped it with Gym [4].

Agent In our implementation, the generated goal from the highest layer and the subgoal from the middle layer are represented by the relative offset in x- and y-coordinates. For the highest layer, we use a maximal range of 1 m and for the middle layer a range of 0.3 m. The middle layer provides a target velocity in translational direction in addition to the relative waypoint to reach in order to learn to slow down in narrow passages. For this layer, we reduce the line of sight of the lidar sensor to 2 m, which still includes the position of the generated goal from the top layer but already reduces the dimension of the latent space by a factor of 4. Thus, we train an own VAE for the middle layer. The lowest layer sends velocity commands in translational and rotational direction to the mobile base. As time scale for the lowest layer, we use 5 Hz. We set the discount factor for the reward to 0.95.

4.2 Top Layer Training

We first focus on testing different network configurations for the top-level agent. In particular, we compare the effect of concurrently continuing the training of the encoder weights with the agent and incorporating the reinforcement learning loss. Furthermore, we investigate the influence of adding recurrent layers to the structure or the agent.

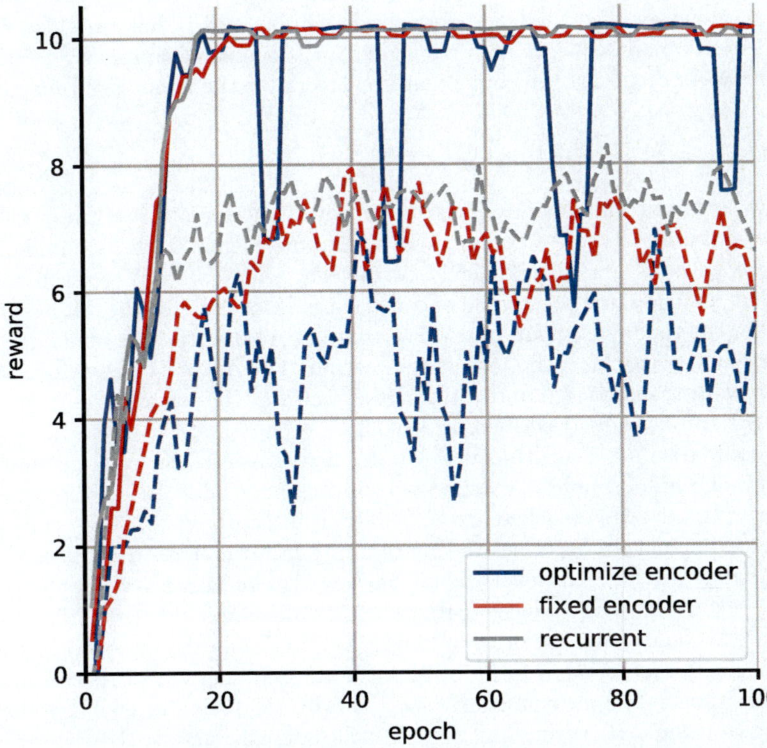

Fig. 4. Comparison of different setups for the network structure of the top-level agent in the *rooms* (solid) and *mazes* (dashed) environments in terms of collected reward over the number of epochs. Training the encoder of the lidar data concurrently (blue) leads to instabilities, while the recurrent layer (gray) has minimal effect. See text for a more detailed description. All curves are averaged over three runs and smoothed with a savgol filter [22].

As can be seen in Fig. 4, all configurations lead to similar results for the *rooms* environment. Only when training the encoder concurrently slight performance drops occur, which is originated in the fact that the agent constantly has to adapt to the changing state representation of the lidar scan. While it is neglectable in the case of the *rooms* environment, it is crucial for the *mazes* environment. Concurrently adopting to the changing state representation leads to a decreased overall performance in terms of collected reward. However, for lidar-based agents that is no problem, because data can easily be generated from any environment upfront, i.e., an encoder can usually be pretrained and kept fixed later on. In our previous work [10], we even noticed great generalization capabilities, which means that retraining is usually not required. Therefore, we will keep the encoder weights fixed. In contrast, the recurrent agent slightly improves the performance in the *mazes* environment, especially in the early stages of the training. However, the final performance is similar and it has no effect in the *rooms* environment. Therefore, we decided for the sake of simplicity to use the dense network structure without recurrent layers for the agent from now on.

4.3 Comparative Evaluation

Next, we evaluate the performance of our full hierarchical structure in the two different environments. For comparison, we train an agent with a dense network structure without any hierarchical abstraction, which we will refer to as *flat agent*. We use the same pretrained encoder of the laser data for the flat agent that we use for the top layer of our approach. As input, we pass the encoded lidar scan, the current velocity, and the last chosen action. The output consists of velocity commands in translational and rotational direction. The reward identical to the one of the top layer as described in Sect. 3.4.

As comparison, we use the success rate of the agents. An episode counts as success if the hidden pole is reached without collisions and before the maximal number of timesteps is reached. From Table 1 it is clearly visible that our hierarchical agent structure increases the success rate in both environments compared to the flat structure. However, when looking at the success weighted by the normalized inverse path length (SPL) [1] it becomes visible that the increased success rate comes with the price of longer, i.e., less smooth trajectories, see also the resulting trajectories in Fig. 1 or in our accompanying video. As can be seen, our SPL drops in comparison to the success rate relatively more as it is the case for the flat agent. However, our SPL is still seriously higher than the SPL of the flat agent. To achieve smoother trajectories, in the future we plan to include reward terms that penalize larger orientation changes resulting from the chosen actions.

With these experiments, we have demonstrated that when using our hierarchical agent for navigation providing any global or target information can be avoided. Furthermore, in terms of success rate we achieve an improved performance in comparison to the flat agent. However, from the success rates in Table 1 it can be seen that navigation in complex environments such as *mazes* is still a challenging task.

Table 1. Success rate and success weighted by the normalized inverse path length (SPL) [1] for the *rooms* and *mazes* environment.

Environment	Metric	Flat	Our	Random Walker
Rooms	Success	0.664	**0.969**	0.016
	SPL [1]	0.573	**0.782**	0.015
Mazes	Success	0.342	**0.581**	0.003
	SPL [1]	0.306	**0.452**	0.003

As can be seen, our agent seriously increases the success rate and the SPL in comparison to the flat agent. The random walker, executing randomly drawn actions, is more than one magnitude below the flat and our agent. All values are obtained by averaging over 1000 trials.

Fig. 5. Real-world test environment in our lab where the yellow pole represents the target to reach.

4.4 Transfer to Real-World

Finally, we evaluate the transferability of our hierarchical agent to a real-world scenario after training in simulation. Note that neither the encoder nor the agent itself have been trained on real-world data. For evaluation we built a setup similar to the *rooms* environment, see Fig. 5. As mobile base we used a TurtleBot2i that was also used in simulation. We placed the target and the robot on 15 different positions and started our hierarchical algorithm. To unambiguously determine the end of an episode, we let the algorithm run until the robot touched the target.

In each episode, our agent successfully reached the target, i.e., we achieved a success rate of 1.0 and an SPL of 0.742. In simulation, our agent reached in a room with the shape of the real environment in Fig. 5, also an success rate of 1.0 and an SPL of 0.738 averaged over 1000 trials. Thus, the results obtained in the real-world scenario are comparable to the ones from simulation.

5 Conclusion

In this paper, we presented a hierarchical structure for learning navigation tasks using deep reinforcement learning. Our agent learns to reach a hidden target in unknown environments by self-assigning internal goals and choosing waypoints to reach the target position only based on local sensor data. In this way, we avoid providing any global or goal-directed information and generate an intrinsically motivated agent that discovers its own targets while receiving only a delayed reward. To exploit the full capabilities of a mobile robot, we use the continuous action space of target velocities, which also simplifies the transfer to real-world scenarios.

In experiments with a wheeled base equipped with a 2D lidar, we showed that our agent is able to solve navigation tasks using only local sensor data and demonstrated that the hierarchical structure clearly outperforms a flat agent in terms of success rate and success weighted by path length in two different simulated environments. Furthermore, we successfully transferred the trained agent to a real-world scenario.

References

1. Anderson, P., Chang, A.X., Chaplot, D.S., Dosovitskiy, A., Gupta, S., Koltun, V., Kosecka, J., Malik, J., Mottaghi, R., Savva, M., Zamir, A.R.: On Evaluation of Embodied Navigation Agents (2018)
2. Andrychowicz, M., Wolski, F., Ray, A., Schneider, J., Fong, R., Welinder, P., McGrew, B., Tobin, J., Pieter Abbeel, O., Zaremba, W.: Hindsight experience replay. In: Advances in Neural Information Processing Systems (2017)
3. Bengio, Y., Louradour, J., Collobert, R., Weston, J.: Curriculum learning. In: Proceedings of the International Conference on Machine Learning (ICML) (2009)
4. Brockman, G., Cheung, V., Pettersson, L., Schneider, J., Schulman, J., Tang, J., Zaremba, W.: OpenAI Gym (2016)
5. Chaplot, D.S., Gandhi, D., Gupta, A., Salakhutdinov, R.: Object goal navigation using goal-oriented semantic exploration. In: Proceedings of the Conference on Neural Information Processing Systems (NIPS) (2020)
6. Coumans, E., Bai, Y.: PyBullet, a Python module for physics simulation for games, robotics and machine learning (2016–2019). www.pybullet.org
7. Dhiman, V., Banerjee, S., Griffin, B., Siskind, J.M., Corso, J.J.: A Critical Investigation of Deep Reinforcement Learning for Navigation (2018)
8. Fox, D., Burgard, W., Thrun, S.: The Dynamic Window Approach to Collision Avoidance. Robot. Autom. Mag, IEEE (1997)

9. Fujimoto, S., van Hoof, H., Meger, D.: Addressing function approximation error in actor-critic methods. In: Proceedings of the International Conference on Machine Learning (ICML) (2018)
10. Gebauer, C., Bennewitz, M.: The Pitfall of More Powerful Autoencoders in Lidar-Based Navigation (2021)
11. Jaderberg, M., Mnih, V., Czarnecki, W.M., Schaul, T., Leibo, J.Z., Silver, D., Kavukcuoglu, K.: Reinforcement Learning with Unsupervised Auxiliary Tasks. CoRR (2016)
12. Kaelbling, L.P., Littman, M.L., Cassandra, A.R.: Planning and acting in partially observable stochastic domains. In: Artificial Intelligence (1998)
13. Kingma, D.P., Welling, M.: Auto-encoding variational bayes. In: International Conference on Learning Representations (ICLR) (2013)
14. Kulkarni, T., Narasimhan, K., Saeedi, A., Tenenbaum, J.: Hierarchical deep reinforcement learning: integrating temporal abstraction and intrinsic motivation. In: Proceedings of the Conference on Neural Information Processing Systems (NIPS) (2017)
15. Levy, A., Konidaris, G., Platt, R., Saenko, K.: Learning multi-level hierarchies with hindsight. In: International Conference on Learning Representations (ICLR) (2019)
16. Li, C., Xia, F., Martín-Martín, R., Savarese, S.: Hrl4in: hierarchical reinforcement learning for interactive navigation with mobile manipulators. In: Proceedings of the Conference on Robot Learning (CoRL) (2020)
17. Liu, L., Dugas, D., Cesari, G., Siegwart, R., Dubé, R.: Robot navigation in crowded environments using deep reinforcement learning. In: Proceedings of the IEEE/RSJ International Conference on Intelligent Robots and Systems (IROS) (2020)
18. Macenski, S., Martin, F., White, R., Ginés Clavero, J.: The marathon 2: a navigation system. In: Proceedings of the IEEE/RSJ International Conference on Intelligent Robots and Systems (IROS) (2020)
19. Mirowski, P., Pascanu, R., Viola, F., Soyer, H., Ballard, A.J., Banino, A., Denil, M., Goroshin, R., Sifre, L., Kavukcuoglu, K., Kumaran, D., Hadsell, R.: Learning to navigate in complex environments. In: International Conference on Learning Representations (ICLR) (2017)
20. Pfeiffer, M., Schaeuble, M., Nieto, J.I., Siegwart, R., Cadena, C.: From Perception to Decision: A Data-driven Approach to End-to-end Motion Planning for Autonomous Ground Robots. CoRR (2016)
21. Regier, P., Gesing, L., Bennewitz, M.: Deep reinforcement learning for navigation in cluttered environments. Proceedings of the International Conference on Machine Learning and Applications (CMLA) (2020)
22. Savitzky, A., Golay, M.J.E.: Smoothing and differentiation of data by simplified least squares procedures. Analytical Chemistry (1964)
23. Sax, A., Zhang, J.O., Emi, B., Zamir, A.R., Guibas, L.J., Savarese, S., Malik, J.: Learning to navigate using mid-level visual priors. In: Proceedings of the Conference on Robot Learning (CoRL) (2019)
24. Stein, G.J., Bradley, C., Roy, N.: Learning over subgoals for efficient navigation of structured, unknown environments. In: Proceedings of the Conference on Robot Learning (CoRL) (2018)
25. Sutton, R.: The Bitter Lesson (2019). www.incompleteideas.net/IncIdeas/BitterLesson.html
26. Tai, L., Paolo, G., Liu, M.: Virtual-to-real deep reinforcement learning: continuous control of mobile robots for mapless navigation. In: 2017 IEEE/RSJ International Conference on Intelligent Robots and Systems (IROS) (2017)

27. Vezhnevets, A.S., Osindero, S., Schaul, T., Heess, N., Jaderberg, M., Silver, D., Kavukcuoglu, K.: FeUdal networks for hierarchical reinforcement learning. In: Proceedings of the International Conference on Machine Learning (ICML) (2017)
28. Wahid, A., Stone, A., Chen, K., Ichter, B., Toshev, A.: Learning Object-Conditioned Exploration Using Distributed Soft Actor Critic (2020)

Traction Optimization for Robust Navigation in Unstructured Environments Using Deep Neural Networks on the Example of the Off-Road Truck Unimog

Patrick Wolf, Pankaj Deoli(✉), Satish Kumar Thangellapally, and Karsten Berns

Robotics Research Laboratory, Department of Computer Science, TU Kaiserslautern, Kaiserslautern, Germany
{patrick.wolf,deoli,berns}@cs.uni-kl.de, thangell@rhrk.uni-kl.de

Abstract. Traction control is fundamental for self-driving vehicles' safe and robust off-road control. Aspects such as surface material heavily affect slippage and driving behavior and requires consideration for a forward-thinking driving method. This paper presents an approach for optimized traction control based on deep convolutional neural networks, combined with OpenStreetMap data to determine ground surface properties. The semantic surface segmentation provides information about estimated surface friction coefficients. The driving behavior adjusts and maximizes traction by selecting the best-suited gear selection, differential lock control, and tire pressure. Additionally, steering and breaking adjust to the ground friction. The demonstration uses the autonomous off-road truck U5023 in a forest-like environment close to a landfill.

Keywords: Off-road robotics · Navigation · Deep learning · Behavior-based control · Traction control

1 Introduction

Off-road and Commercial vehicles navigate and work in mining, construction, defense, and forestry domains, where the working conditions are often harsh, surprising, and often dangerous. Terms such as safety, cost, productivity, vehicle maintenance, and environment encourage the deployment of autonomous systems. Simultaneously, there is a dire need for labor in these work areas. Accordingly, autonomous commercial off-road vehicles aims to significantly impact future civil construction, agriculture, and the search and rescue domain. They

This work was founded by the ministry of economy, transport, agriculture and viticulture Rhineland-Palatinate, Germany (MWVLW RLP).

I. Petrovic et al. (Eds.): IAS 2022, LNNS 577, pp. 561–579, 2023.
https://doi.org/10.1007/978-3-031-22216-0_38

enable the accomplishment of complex tasks and the fulfillment of versatile roles currently limited to human operators.

A key challenge remains the safe, reliable, and robust off-road navigation in unstructured environments like construction sites and forestry areas. Typically, there exists none to only limited infrastructure like paved roads, lane indicators, or likewise. Unfortunately, local object and surface aspects as varying traction substantially impacts the navigation outcome. Unexpected tire slip and traction loss are a severe danger to the robot and the environment. Therefore, there is a need to identify the ground conditions early and adjust the driving behavior accordingly. Approaches in literature, as proposed by Wolf et al. [1], consider the geometrical surface shape for estimating and predicting vehicle properties for control adjustment. However, there is no consideration of high-level semantic information during driving. In contrast, human (expert) drivers can do so by experience and visual appearance of the surface. Additionally, they consider aspects like temperature, weather, humidity for driving adequately.

The perception of relevant environmental properties and scene evaluation skills are not easily transferable to robot control due to the high complexity of environments. A feasible approach to reduce complexity is the consideration of surface types to estimate and map the ground friction coefficient. Therefore, semantic segmentation estimates the ground surface material. A friction estimation uses the segmentation information to derive particular surface properties during navigation. A global map representation (here OpenStreetMap) complements the segmentation system to store surface properties recognized before and to supplement control if the detection quality is below a threshold. Additionally, mapping allows exchanging the driving experience between robots since different vehicles can access previous surface annotations. The approach's demonstration uses the autonomous off-road truck U5023 in a forest-like environment.

This paper introduces a traction estimation concept during navigation. Therefore, Sect. 2 presents related work to semantic segmentation of off-road scenes. Details on the demonstrator robot and its degrees of freedom are available in Sect. 3. Section 4 shows the traction optimization concept and shows how a surface class and its associated traction value acts on the robot control systems. Section 5 presents details on the segmentation and training process. Finally, Sect. 6 provides experiments and Sect. 7 summarizes and concludes.

2 Related Work

Although the recent trends in Computer vision/Deep learning have contributed a lot towards environment perception in urban scenarios, the same cannot be said about off-road environments. Off-road pathways categorize with varying surface types (having different surface coefficients), and therefore, the estimation of correct surface type aids in safe and reliable traversability. The approaches developed by researchers (for off-road surface type detection) are either relatively minor, are based on multi-sensors usage, or are limited to just detecting a driveable pathway. Mei et al. [2] presented an approach towards scene adaptive vision-based road detection for a robot's navigation in off-road environments. The authors formulated the problem as detecting the region in front of the robot that is a) *mechanically traversable*, b) *apt to be chosen by humans* and c) *able to see far-field road conditions.* However, the authors of [3] proposed a contrastive-based learning method to discriminate off-road regions with different traversability, thereby aiding in fine-grained semantic segmentation and mapping for enhanced scene perception. Utilizing only 3D point cloud, the authors of [4,5] presented approaches towards terrain mapping and off-road driveable area extraction, thereby depicting the traversable regions in off-road environments. Not only detecting the surface type but roughness score of a path can also aid in providing an additional factor for surface type estimation as seen by Stavens and Thrun [6], where the authors presented a self-supervised terrain roughness estimator from laser-range data for high-speed navigation. Another viewpoint presented by the authors of [7] uses Deep Maximum Entropy Inverse Reinforcement learning to analyze the traversability and trajectory planning for off-road navigation. The authors developed two CNNs, namely *RL ConvNet* and *Svf ConvNet* to achieve enhanced forward reinforcement learning to incorporate the vehicle's kinematics. Working towards the problem of scene segmentation, the authors of [8–10] presented datasets, namely RUGD, RELLIS-3D and FREIBURG FOREST DATASET respectively, with classes such as a creek, trail, construction site, grass, mulch. Nevertheless, the true richness of different surface labels is still limited. Therefore, to tackle this issue of unbalanced quality and quantity, Seff and Xiao [11] proposed a two-stage network *OFFSEG* wherein the results (only the traversable region) of stage 1 (i.e., polling of all classes into four main classes) forward to a stage 2, which further categorizes the region as one of grass, puddle, dirt, gravel through k-means clustering.

While the classical-camera (or online) based datasets provide a benchmark for researchers, another database, i.e., Open Street Maps (OSMs), provides an offline pool of ground areas. OSM is a standardized geodata database containing roads, railways, buildings, or other landmarks. Also, the geographical nodes contain additional information with tags such as speed limit, surface (asphalt, dirt, grass, gravel, sand), land usage, which is accessible by passing the geo-coordinates [12]. The authors from [11] used features from street view images to present a visual scene understanding, whereas [13] uses OSM data, vehicle pose sensors, and camera to produce road annotation for training which can be later used for network training.

In summary, most of the work done in the field of off-road surface type estimation has been limited to detecting traversable or non-traversable paths, and even though these maps above contain information about road surface's type, not every road in these databases are tagged (around 40% is tagged). Therefore this paper advances towards optimizing the traction of an autonomous vehicle with appropriate surface type estimation using semantic segmentation and offline maps.

3 Off-Road Robot Unimog

The demonstrator vehicle for illustrating the traction optimization approach is an autonomous Unimog U5023. The vehicle is highly off-road-capable and therefore suited for methodology transfer to vehicles operating in more moderate terrains. The U5023 is a versatile, multipurpose vehicle with additional degrees of freedom, which are highly relevant for traction control:

controlable differential locks there exist different configurations as rear-wheel drive (2 WD), all-wheel drive (4 WD), rear-axis locked (RAL), and all-axis locked (AAL) to control the slip in a specific terrain
gearbox with off-road gears the U5023 has 8 forward and 6 reverse gears. In addition there is a off-road gear group available with 8 additional forward and backward gears.
adjustable tire-pressure the robot's tire pressure is adjustable to avoid stuck situations on surfaces such as mud or sand. There exist predefined tire pressure modes as highway, cross country, sand-mud-snow, and emergency, which can engage.
anti-lock braking system depending on the terrain, the U5023s ABS should be activated or deactivated.

The frame bends diagonally up to 60 cm, there is an axle articulation up to 30°, plus a high ground clearance of 46 cm. A low center of gravity, large embankment angles, plus a static tilt angle of 38° enables climbing steep slopes. The U5023 uses a mix of various depth sensors as stereo cameras, 2D- and 3D-laser scanners, and localization devices (see Fig. 1 and Table 1). The described poses are relative to the kinematic center, which is on the ground below the rear axis center. A unique feature concerning the sensor poses is the twisting of the axles, frame, and cabin. Therefore, the relative positions and orientations to the kinematic center cannot be assumed to be constant but depend on the vehicle's position in the terrain, making a later data fusion and consistency preservation more challenging.

Fig. 1. Autonomous off-road vehicle Unimog U5023. *Photo* MWVLW RLP/Brenner.

Table 1. Selection of U5023 depth sensor setup.

Sensor	x [m]	y [m]	z [m]	Φ [°]	ψ [°]	φ [°]
ZED (f)	4.91	0.00	1.55	– 2.00°	19.00°	0.00°
ZED (b)	– 0.95	0.00	1.25	– 0.10°	27.00°	180.00°
Realsense D435 (l)	4.90	1.40	1.35	0.00°	35.00°	45.10°
Realsense D435 (r)	4.91	– 1.4	1.34	0.00°	30.00°	– 45.00°
RS-LiDAR-16	4.90	– 0.40	1.46	– 0.36°	– 0.31°	0.00°
RS-BPearl (l)	2.40	1.32	1.89	– 89.40°	2.00°	0.00°
RS-BPearl (r)	2.40	– 1.32	1.89	91.43°	2.00°	0.00°

There exist multiple stereo cameras and 3D laser scanners for a full surround perception. Sensor poses are relative to the kinematic center of the vehicle. (f/ l/ r/ b) indicates a front, left, right, or back mounting.

4 Traction Opimization Concept Based on Surface Data

Traction depends on various conditions as the wheel's dynamic load the tractive and frictional forces. The friction directly affects the navigation behavior of a robot. The breaking distance d, and therefore the safe navigation velocity v concerning obstacles, depends on the ground friction μ:

$$F = \mu m g \tag{1}$$

$$d = \frac{mv^2}{2F} = \frac{v^2}{2\mu g} \tag{2}$$

$$v = \sqrt{2\mu g d} \tag{3}$$

where F notes the breaking force, g standard gravity, and m the mass of the robot. Accordingly, the safe velocity reduces and breaking distance increases for smaller values of μ. Literature correlate surface types and friction coefficients as shown for selected examples in Table 2.

Table 2. Exemplary coefficient of friction for road surfaces [14].

Surface type	Coefficient of friction	Surface type	Coefficient of friction
Gravel and dirt road	0.35	Wet concrete	0.60
Wet, grassy field	0.20	Snow	0.20–0.25
Dry asphaltic concrete	0.65	Ice	0.10–0.15
Wet asphaltic concrete	0.50	Loose moist dirt	0.60–0.65
Dry concrete	0.75		

Therefore, it is valuable knowledge to know the surface type for determining the robot's traction. Besides the surface material (e.g., asphalt, gravel, dirt), additional properties such as humidity or temperature affect the friction.

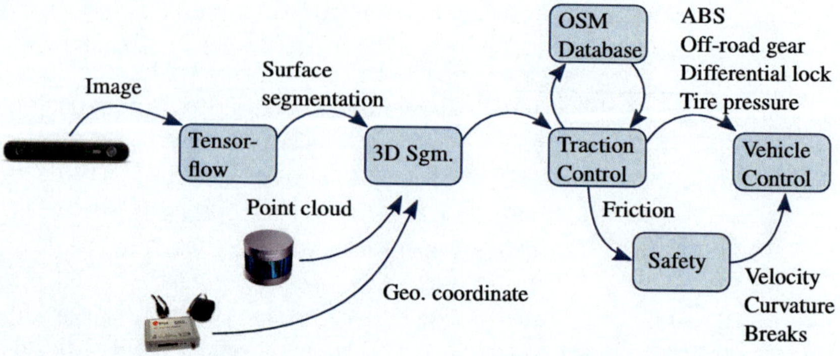

Fig. 2. Concept for traction optimization. Image segmentation extracts ground classes. Distance and geographical information convert the segmentation results to 3D segmentations, which the traction control uses to set the U5023s degrees of freedom. OSM data supports the process if there is insufficient segmentation data available.

A human-inspired approach is visually observing the environment and categorizing the ground. Dependent on the existing surface conditions, a human driver adjusts driving. Accordingly, a related concept for traction-optimization on a robot shows Fig. 2. The fundamental idea is combining visual and spatial surface information with database knowledge to control the robot's degrees of freedom and adjust driving. The robot perceives the surroundings through a camera, which provides an image stream. These data segment surface types using DNNs. Supplementary distance data allows for 3D segmentation relative to the current robot pose. The localization enables projecting this data into a global, geographical reference coordinate system. The traction control considers the segmentation information and cross-validates using OSM surface data according to the detection quality and confidence. If there exists no surface information, the robot annotates the existing surface in the database.

4.1 Off-Road Surface Data

Surface segmentation relies on two main components: semantic segmentation and database information. Accordingly, the initial task is to define the relevant ground classes that the control system can consider. In literature, there are different off-road datasets available (see Table 3), with varying amounts of off-road classes.

RUGD and RELLIS-3D datasets are most promising regarding the finely divided distinction of categories. By far, the RUGD dataset is the largest, with a total of 7453 images with an image resolution of 688×555 (training-4779, validation-733, and test images-1924). It has 24 classes and covers the considerable variation in the off-road areas representing four main categories: creek, park, trail, and village. Example images cover distinct rough terrains like dirt, grass, sand, asphalt, water, rock bed, concrete, gravel, bush, and mulch. RUGD

Table 3. Off-road datasets overview.

Name	Dataset images	Resolution	Total available Classes
RUGD [8]	7,546	688×555	24
RELLIS-3D [9]	6,235	1920×1200	20
DeepScene [10]	366	880×487	6

All these datasets have pixel-wise annotated RGB images.

dataset has a large quantity of data with variation, but this is not enough to solve the task of off-road perception. The first mentioned classes contribute more than 90% to the entire dataset and are imbalanced in data distribution. Since the focus is segmenting the ground surface, this dataset is converted, comprising all the ten traversable classes and all non-traversable regions combined into one (see Table 4).

RELLIS-3D (similar to the RUGD dataset) is a multimodal dataset for off-road scenes containing 6235 images (training-3302, validation-983, and test-1672) and 20 classes (including void). Supplementary to RUGD, it has additional mud, rubble piles, puddles, and deep water classes. RELLIS-3D also has inherited class imbalance, with 4 out of 20 classes (grass, sky, tree, and bushes) contributing to 94% of the pixel data. In [16], state-of-the-art models developed for the cityscapes dataset [17] showed low performance on the RUGD dataset. Likewise, the ResNet-based architecture [18] tested on this dataset showed a mere 32.5% mIoU compared to the performance of 80% mIoU in the cityscape dataset. It highlights that even the best-performing model on cityscapes cannot be directly used for off-road scene interpretation due to the inherent class imbalance of the dataset.

Similarly, the authors of [19] show that the best performing models on the cityscapes test set are significantly underperforming due to the significant class imbalance and the unstructured data with unclear boundaries. Also, compared to RUGD, the whole RELLIS-3D has no to significantly less variation in the dataset concerning the different scenes in the off-road environment, and almost every image has one class (grass), leading to poor performance of the model. RUGD and RELLIS-3D merge a new dataset with 12 classes, optimized for surface type segmentation, to overcome the problem of class imbalance and coverage (see Table 4). It overcomes certain limitations that few classes are only in one of the datasets (e.g., sand, gravel, rock). Nonetheless, both the datasets have low pixel presentation for almost all classes. Therefore, similar classes combine rock, rock-bed, and rubble.

There exist map-based ground materials like the OpenStreetMap[1] surface tags (see Table 5), which are supplementary to image segmentation data for ground surface determination. OSM surface data supports perception if the detection quality of the image segmentation is below a threshold.

[1] https://www.taginfo.openstreetmap.org/tags.

Table 4. Overview of merged (RUGD + RELLIS-3D) dataset consisting of 12 classes.

RUGD class	RELLIS-3D class	Final class	Class number
Non-traversable (void)	Non-traversable (void)	Non-traversable (void)	0
Dirt	Dirt	Dirt	1
Sand	n/a	Sand	2
Grass	Grass	Grass	3
Water	Water/puddle	Water	4
Asphalt	Asphalt	Asphalt	5
Gravel	n/a	Gravel	6
Mulch	n/a	Mulch	7
Rock-bed/rock	Rubble	Rock/rock-bed/rubble	8
Bush	Bush	Bush	9
Concrete	Concrete	Concrete	10
n/a	Mud	Mud	11

Table 5. Examples of OSM tags[1] with the keyword `surface`.

OSM class	Count	OSM class	Count	OSM class	Count
Asphalt	18,505,945	Unpaved	8,673,174	Paved	3,819,266
Ground	2,548,189	Concrete	1,960,246	Paving stones	1,834,928
Gravel	1,724,686	Dirt	1,071,037	Grass	865,272
Compacted	654,912	Sand	383,875	Sett	243,401
Fine gravel	233,116	Cobblestone	170,360	Wood	136,952
Earth	129,741	Concrete plates	86,687	Pebblestone	78,633
Clay	22,288	Mud	21,466		

Accordingly, OSM contains a large number of surface tags. However, a class balance is again observable, as the example of asphalt or unpaved surfaces shows. Additional challenges are the incompleteness of OSM and the imprecisions of annotations. OMS has (initially) a comparatively low accuracy and is less frequently updated in off-road environments. A solution to increase map quality and robustness simultaneously is the active node/edge annotation of OSM through segmentation data during runtime. Therefore, OSM data provides memory to segmentation to cross-validate the current results and automatically update non-annotated parts of the map. An odds control the label selection approach [20] to avoid wrong segmentations and short-term errors to disturb the maps.

4.2 Traction-Dependent Vehicle Configuration

The robot performs spatial queries on the outlines of the planned tractory outlines to receive semantic segmentation and OSM surface information. Accordingly, the robot's degrees of freedom adjust as illustrated in Table 2. Dependent on the particular degree of freedom, different update times exist since the physical counterparts on the robot must adjust. For example, the differential locks require a few seconds to engage. In contrast, a tire pressure adaptation demands several minutes. Therefore, frequent switching between lowest and full pressure is not possible. If an actual critical situation occurs (e.g., extreme mud), the vehicle must stop before proceeding.

Additionally, the control software adjusts. Slow down systems as proposed by Ropertz et al. [21] correct the maximum allowed velocity according to the friction-adjusted time to collision (Table 6).

Table 6. Vehicle configuration change according to ground friction.

OSM surface	Segment. surface	Differential lock	Tire pressure	Off-road gears	ABS type	Friction
Asphalt	Asphalt	2 WD	Highway	Off	Off	0.65
Concrete, concrete plates	Concrete	2 WD	Highway	Off	Off	0.62
Dirt	Dirt	4 WD	Cross country	On	Off	0.35
Gravel	Gravel	RAL	Cross country	On	Off	0.55
n/a	Mulch	RAL	Cross country	On	Off	0.6
Grass	Grass, bush	RAL	Cross country	On	Off	0.35
Sand	Sand	AAL	Mud, sand	On	Off	0.2
n/a	Rock/rock bed/rubble	AAL	Mud, sand	On	On	0.65

5 Semantic Segmentation

Low computation expense, better accuracy, and low memory consumption are relevant parameters to train an autonomous system successfully. The research goal was to achieve real-time performance (> 16 FPS), having fewer parameters with a good segmentation accuracy.

SwiftnetRN-18 [22] accomplishes the task of semantic segmentation with the help of encoder-decoder arrangement wherein a Spatial Pyramid pooling (SPP) is present between the encoder-decoder. This SPP module has the characteristics of a large receptive field along with a lightweight decoder (which helps in up-sampling) to impart semantic information. The *Encoder* has the task of transforming the input image into visual features which are semantically dense. The architecture uses ResNet-18, one of the lightest models depicting at-par performance in image recognition. Skip connections in ResNet counteract vanishing gradient issues. The *Decoder* up-samples dense features to match with the image's original resolution. The low-resolution features are (at first) up-sampled to the identical resolution as the higher resolution features using bi-linear interpolation. The element-wise summation of up-sampled input features and lateral encoder features is then combined with a 3×3 convolution. The softmax layer at the end of the decoder outputs the pixel-wise class probability for the input image.

The *Spatial Pyramid Pooling (SPP)* module is a simplified version of PSPNet [18] wherein the stacked dense feature map from each layer of the encoder is fed into the SPP module, thereby combining these multi-scale feature maps from the encoder to enlarge the receptive field and minimize loss due to spatial imprecision. Owing to all the above characteristics as well as depicting state-of-the-art performance on Cityscapes dataset [17] (mIoU: 75% and 39.9 FPS), SwiftNetRN-18 was chosen as the baseline architecture for our problem at hand.

5.1 Network Training

Initially, the experiments were performed with the help of RUGD and RELLIS-3D datasets. The performance compared with the cityscapes dataset. Data augmentations based on geometry and color transformations are used to tackle the issue of class imbalance, partially improving the learning behavior of the model. Augmentations techniques such as *Random Crop, Random brightness contrast + Hue saturation value, Horizontal flip, Horizontal flip + random flipping + ISO noise, ISO Noise + Random brightness contrast + Random sum flare, random crop + Random tone curve* and *Center crop* are used. Many loss functions are intended to handle the issue of class disparity explicitly. However, with SwiftnetRN-18, a loss function based on Cross-entropy is used.

For achieving real-time performance and good accuracy, a batch size of 12 was used for training purposes. The original image resolution is reduced to 550×550 and 768×768 for RUGD and RELLIS datasets. The models have trained with adam optimizer with the decay of $1e^{-4}$ and a learning rate of $4e^{-4}$ at the initial stage. Ultimately, it dropped to $1e^{-6}$ for the final epoch with cosine delay. The choice of loss functions is significant incorrectly evaluating the model's performance. A pixel-wise cross-entropy loss is a generic loss used for semantic segmentation, which checks the pixel-wise loss for the predicted output with the ground truth and averages the loss. Nevertheless, since most of the datasets are skewed, this loss function becomes biased. Therefore, to tackle this problem (considering an unstructured and unbalanced environment), a specialized loss function apply (see Table 7).

The base loss function used in SwiftNetRN-18 is Cross-entropy, an ordinary loss for many machine learning models. As depicted from Table 7, it is visible that CE-based loss functions such as CE, WCE, and combo losses have performed well. It is mainly due to off-road data, where the environment consists of patches of multiple regions. Among these, *Focal loss* with the mIoU of 45% and a loss

Table 7. Comparison of class balancing loss functions on reduced RUGD dataset based on mean IoU, mean Precision and mean Recall.

Loss fucntion	mIoU	mPrecision	mRecall	Total Loss
CE loss	0.42	0.54	0.47	0.0038
WCE loss	0.41	0.53	0.47	0.09
Dice loss	0.34	0.39	0.38	0.04
Combo (dice+BCE) Loss	0.44	0.55	0.49	0.289
Focal loss	**0.45**	**0.56**	**0.50**	**0.07**
Tversky loss	0.3468	0.38	0.38	0.03
Focal Tversky loss	0.35	0.38	0.38	0.06

of 0.07 showcased the best results. Therefore for further experiments, focal loss with gamma = 4 and alpha = 2 is used.

With augmentations performed, the final dataset comprised 38,232 images. The model was initially trained for 200 epochs with 673,200 training steps with a cosine decay (initially $4e^{-4}$ and reducing to $1e^{-6}$) and a weight decay set to $1e^{-4}$. Both training and validation set now have a broad distribution of classes and hence was chosen as a base dataset. With the significant learning rate, the model has learned quickly to reduce loss and improve on mIoU and other metrics. With the focal loss at the beginning of the training, all classes are equally treated, and easy classes have helped minimize the total loss. However, as the training progress, there is a vast difference between the total loss of easy and challenging examples. The learning is reduced, and we can see that the model remained almost constant with minimal loss reduction. Total loss has reduced from around 0.2–0.07 at step 319,500 (Epoch 100), and the learning rate has been reduced to half $2e^{-4}$. From here on the model has improved only a little until step 492,540 (approximately till Epoch-150) and in the last 50 epochs, the model has not shown any noticeable improvement. However, due to the focal loss, the prediction on the complex examples with less pixel representation has improved.

The loss fluctuations are due to the ambiguous complex scenes in off-road environments and similar-looking classes (grass-bushes, dirt-gravel-mulch in some scenarios). Technically these classes are very different, but the inherent natural similarities in the classes like grass-bushes have the same color and similar texture. Furthermore, the tall grass is confused for bushes. Six out of 12 classes have 0.5% pixel data for training, and because of low pixel representation, the model has shown minimal learning during the later stages. The model treated this data as noise and could not pick the pattern. The classes with 1–10% of pixel representation have performed moderately better than minor classes with 20–50% of mIoU. As a result of all these factors, the model has performed well on major classes with higher pixel representation in the train set, and also these classes have strong, easy-to-identify features. Therefore, the loss has initially decreased drastically until these classes are predicted with a higher probability. For the later stages, the model struggled to learn.

6 Experiments

The approach's performance tests used the robot Unimog U5023 (Sect. 3). The vehicle traveled through a forest-like environment containing asphalt roads, gravel tracks with varying surface conditions (grass, fallen branches), and an off-road section with little trees and mud. A primary requirement of the traction control, and therefore segmentation systems, is real-time capabilities. On an NVIDIA Geforce RTX 2080 Ti GPU, the model achieves a mIoU of 29% (with 52.5% mIoU without minor classes) and a frame rate of 55 frames-per-second. The real-time capabilities of the network are feasible through the upgraded SwiftNet architecture.

6.1 Performance on Dataset

This section shows the qualitative performance of the network on the test dataset. It applied color mapping of the RUGD datasets shown in Fig. 3. The RUGD test set consists of 1924 images in which the testing has been performed.

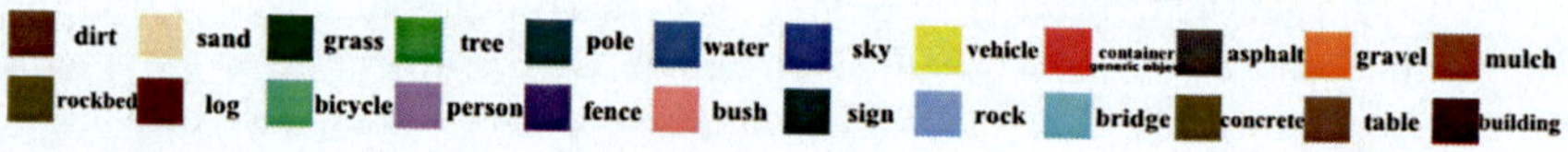

Fig. 3. RUGD color mappings. All the results from the experiments section are color coded according to this RUGD color mappings [16].

A good network's prediction capability could be achieved on the data set, where in clear and distinct areas exist (Fig. 4, Table 8).

However, the actual performance of any off-road network can be evaluated on low-illumination, occlusion, and ambiguous areas. Usually, real-world versatility cannot be captured adequately by datasets. Therefore, real-world performance is usually significantly worse than expected by the training and validation stages. In our case, the network struggles to perform at par due to the low differentiation probability in the regions such as gravel, mulch, rock bed and grass, and bushes. Due to these poor distinction characteristics, pixels accumulate and show as a single class (see Fig. 5). Natural conditions like leaf-fall, low illumination, algae-covered rocks, and other ambiguous characteristics introduce similar objects during the training data and let the network learn these characteristics.

Nevertheless, since it is sometimes tricky for human beings to accurately judge the surface type, teaching a network to do the same would take some serious work under the cost of abandoning real-time performance and lightweight DNNs.

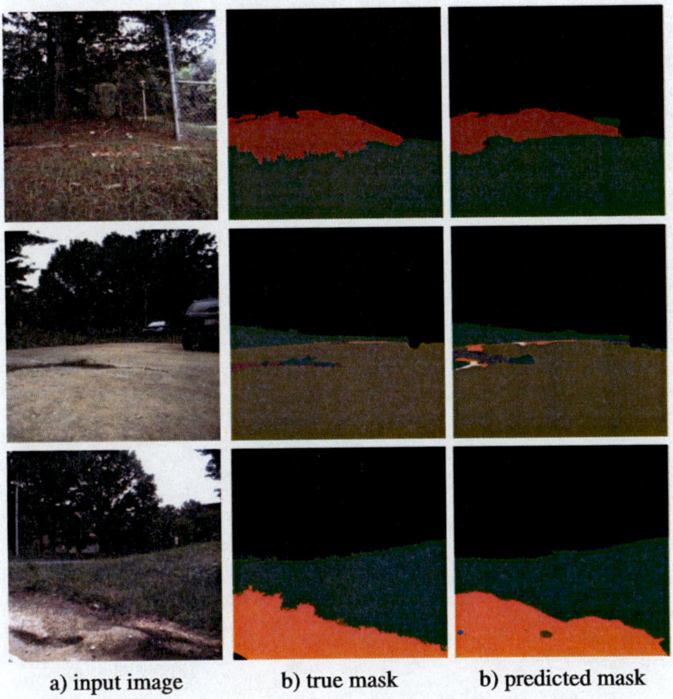

Fig. 4. Results on RUGD test set [8] showing good prediction results.

Table 8. Results on RUGD test set.

Class	Pixel (%)	Mean IoU (%)	Mean precision (%)	Mean recall (%)
Void	51.07	95.18	98	97
Dirt	0.13	0.4	3.4	0.6
Sand	0.16	4	6	4
Grass	23.52	80	82	97
Water	0.10	5	7	14
Asphalt	1.88	4.7	9	8.8
Gravel	8.51	45.7	51	81.5
Mulch	11.89	62.5	90	66.8
Rock-bed	0.06	0	0	0
Bush	2.20	27.2	68	31.2
Rock	0.12	0.6	1.2	0.1
Concrete	0.36	23.7	96	23.9
Mean	–	29	47.3	35.4
***Mean	–	52.5	66.63	63.71

***Mean indicates mean values calculated without considering minor classes (< 1% pixel representation)

Fig. 5. Results on RUGD test set [8] showing false prediction in some patches.

6.2 Performance on Vehicle Data

Additional tests used the autonomous Unimog U5023 and its sensor setup. Results show Figs. 7 and 8. The training dataset is affected by poor class imbalance and limited quality considering the segmentation model. Despite this, the model performs decently and can segment main surfaces such as grass, gravel, rock-bed, mulch, bushes, and dirt very well. Three observations exist:

Class variability and versatility—Even a tiny variation in the environment can negatively affect the final class prediction for the same surface. An example shows Figs. 6 and 7, row 5, wherein the prediction is varying between the classes gravel, dirt and rock bed. Understanding the textural differences between these surfaces can understand the results from segmentation. A loose accumulation of rock particles is called gravel, where small granular gravel is about 2–4 mm in size instead of pebble gravel, which is about 4–64 mm in size. Rock bed, however, is the solid rock that lies underneath the loose surface layer. Bedrock may be overlapped or covered with various types of fractured and worn-out rock materials such as soil, dirt, and gravel. These characteristics make it challenging for a network to extract meaningful features from this complex block and learn, directing towards slight variations in the environment leading to a whole different class of surface.

Static ROI—In most test runs, the model has performed decently to predict surfaces. However, in some cases, even when the prediction was correct, the wrong dominant class is selected (as visualized from last of Fig. 8). These variations can be minimized if we consider the surface class only in the tire track, thereby avoiding unwanted areas. Concentrating on the arc that is in sync with the vehicle's turning radius aids in a dynamic region of interest.
Poor lighting—A rich representation of the surfaces affected by poor lighting should be covered in a dataset to perform at-par in real-world scenarios. Offline maps can attend this issue but only to some extent.

Fig. 6. Sample images from RUGD dataset [8] depicting similar looking classes: gravel (orange) versus rock bed (green).

Since an image contains many surface classes, the most dominant class within a region of interest (ROI) supports robot control if a degree of freedom requires adjustment. This procedure minimizes control switches for changing off-road conditions. Therefore, an ROI (here fixed) indicates the most relevant region for driving, and the most frequent class label within this ROI defines the ground class and, therefore, the friction.

7 Conclusion

The work introduced a framework for traction optimization using the heavy off-road robot U5023. The U5023s degrees of freedom and safety margins were adapted according to the estimated ground friction during testing. Therefore, deep neural networks and OpenStreetMap data are applied to determine the surface class. A combined test and training dataset was derived using RUGD and RELLIS-3D datasets. Accordingly, the existing class imbalance, which results from the limited amount of training data in a versatile off-road environment, reduced using a combined dataset of ground-specific off-road classes. Additional surface information of OSM is associated with the corresponding image segmentation data. Further, a modified version of SwiftNet was used to train the network and accomplish real-time performance.

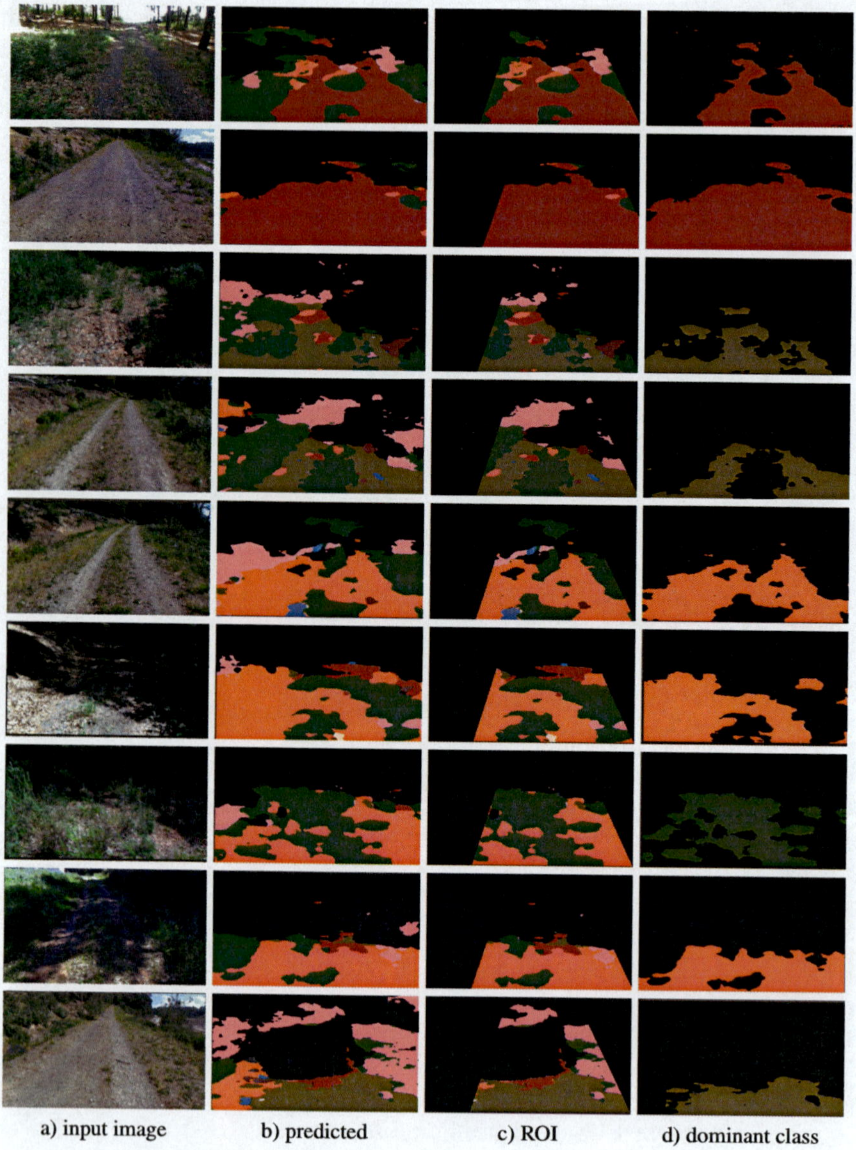

Fig. 7. Sample results on U5023 test data during forest trials.

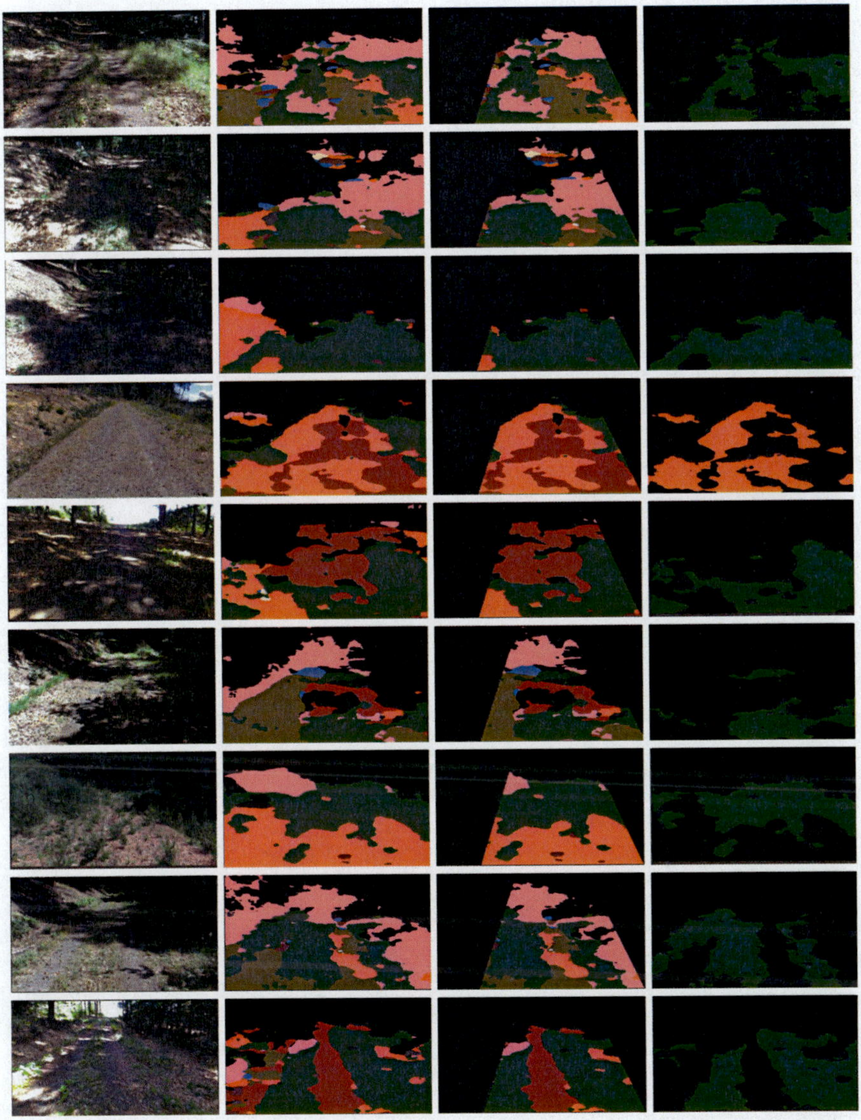

Fig. 8. Sample images showing detection failures on U5023 test data.

The experimentation showed satisfactory image segmentation results. There are cases of non-distinct or undistinguishable ground features. However, these unclear detections had no negative impact on control since they had similar properties like a common friction coefficient. OSM data supported the perception system for low-quality segmentation results and used the traction control system to cross-validate the DNN results. However, it showed that off-road map data was initially somewhat incomplete, and only a few regions were tagged with surface information so that the initial support was limited. Nonetheless, the update and annotation system reduces this limitation while driving a scene. The traction optimization procedure enables a more nuanced differentiation of environmental conditions and, therefore, an optimization of the vehicle to this environment, which was not done before.

Future work addresses the OSM database. Here, trials over different seasons and different weather conditions should prove the long-term performance of this system. According to the temporal conditions, OSM and image segmentation showed similar results for new annotated scenes. It can be expected that the ground surface type differs during season changes. E.g., snow on a track covers the visual features. Therefore, such a remembrance ability should be valuable. Additionally, weather conditions such as humidity and temperature affect traction. Currently, only the surface type defines the expected traction. Therefore, additional environmental effects should complete the surface type for friction determination.

References

1. Wolf, P., Ropertz, T., Berns, K., Thul, M., Wetzel, P., Vogt, A.: Behavior-based control for safe and robust navigation of an unimog in off-road environments. In: Commercial Vehicle Technology 2018. Proceedings of the 5th Commercial Vehicle Technology Symposium–CVT 2018, Springer Vieweg, Kaiserslautern, Germany, pp. 63–76, 13–15 Mar 2018
2. Mei, J., Yu, Y., Zhao, H., Zha, H.: Scene-adaptive off-road detection using a monocular camera. IEEE Trans. Intel. Transp. Syst. **19**(1), 242–253 (2018)
3. Gao, B., Hu, S., Zhao, X., Zhao, H.: Fine-grained off-road semantic segmentation and mapping via contrastive learning. In: IEEE/RSJ International Conference on Intelligent Robots and Systems (IROS), vol. 2021, pp. 5950–5957 (2021)
4. Broggi, A., Cardarelli, E., Cattani, S., Sabbatelli, M.: Terrain mapping for off-road autonomous ground vehicles using rational b-spline surfaces and stereo vision. In: IEEE Intelligent Vehicles Symposium (IV), vol. 2013, pp. 648–653 (2013)
5. Gao, B., Xu, A., Pan, Y., Zhao, X., Yao, W., Zhao, H.: Off-road drivable area extraction using 3d lidar data. In: IEEE Intelligent Vehicles Symposium (IV), vol. 2019, pp. 1505–1511 (2019)
6. Stavens, D., Thrun, S.: A self-supervised terrain roughness estimator for off-road autonomous driving. In: CoRR, vol. abs/1206.6872 (2012). Available: arxiv.org/abs/1206.6872
7. Zhu, Z., Li, N., Sun, R., Xu, D., Zhao, H.: Off-road autonomous vehicles traversability analysis and trajectory planning based on deep inverse reinforcement learning. In: IEEE Intelligent Vehicles Symposium (IV) vol. 2020, pp. 971–977 (2020)

8. Wigness, M., Eum, S., Rogers, J.G., Han, D., Kwon, H.: A rugd dataset for autonomous navigation and visual perception in unstructured outdoor environments. In: IEEE/RSJ International Conference on Intelligent Robots and Systems (IROS), vol. 2019, 5000–5007 (2019)
9. Jiang, P., Osteen, P.R., Wigness, M.B., Saripalli, S.: RELLIS-3D dataset: data, benchmarks and analysis. In: CoRR (2020), vol. abs/2011.12954. [Online]. Available: arxiv.org/abs/2011.12954
10. Valada, A., Oliveira, G., Brox, T., Burgard, W.: Deep multispectral semantic scene understanding of forested environments using multimodal fusion. In: International Symposium on Experimental Robotics (ISER) (2016)
11. Seff, A., Xiao, J.: Learning from maps: Visual common sense for autonomous driving. In: ArXiv (2016), vol. abs/1611.08583
12. Fleischmann, P., Pfister, T., Oswald, M., Berns, K.: Using openstreetmap for autonomous mobile robot navigation. In: Proceedings of the 14th International Conference on Intelligent Autonomous Systems (IAS-14), Shanghai, China, 3–7 July 2016
13. Laddha, A.G., Kocamaz, M.K., Navarro-Serment, L.E., Hebert, M.: Map-supervised road detection. In: 2016 IEEE Intelligent Vehicles Symposium (IV), pp. 118–123 (2016)
14. Noon, R.K.: Engineering Analysis of Vehicular Accidents. CRC Press (1994)
15. Maturana, D., Chou, P.-W., Uenoyama, M., Scherer, S.: Real-time semantic mapping for autonomous off-road navigation. In: Field and Service Robotics. Springer, pp. 335–350 (2018)
16. Wigness, M., Eum, S., Rogers, J.G., Han, D., Kwon, H.: A rugd dataset for autonomous navigation and visual perception in unstructured outdoor environments. In: International Conference on Intelligent Robots and Systems (IROS) (2019)
17. Cordts, M., Omran, M., Ramos, S., Rehfeld, T., Enzweiler, M., Benenson, R., Franke, U., Roth, S., Schiele, B.: The cityscapes dataset for semantic urban scene understanding. In: 2016 IEEE Conference on Computer Vision and Pattern Recognition (CVPR), pp. 3213–3223 (2016)
18. Zhao, H., Shi, J., Qi, X., Wang, X., Jia, J.: Pyramid scene parsing network. In: CVPR (2017)
19. Jiang, P., Osteen, P., Wigness, M., Saripalli, S.: Rellis-3d Dataset: Data, Benchmarks and Analysis (2020)
20. Thrun, S.: Learning occupancy grid maps with forward sensor models. Auton. Robots **15**, 111–127 (2003)
21. Ropertz, T., Wolf, P., Berns, K.: Behavior-based low-level control for (semi-) autonomous vehicles in rough terrain. In: Proceedings of ISR 2018: 50th International Symposium on Robotics. VDE VERLAG GMBH, Munich, Germany, pp. 386–393, 20–21 June 2018
22. Orsic, M., Kreso, I., Bevandic, P., Segvic, S.: In defense of pre-trained imagenet architectures for real-time semantic segmentation of road-driving images. In: 2019 IEEE/CVF Conference on Computer Vision and Pattern Recognition (CVPR), pp. 12599–12608 (2019)

Flattening Clothes with a Single-Arm Robot Based on Reinforcement Learning

Hassan Shehawy(✉), Daniele Pareyson, Virginia Caruso, Andrea Maria Zanchettin, and Paolo Rocco

Department of Electronics, Informatics and Bio-engineering, Polytechnic University of Milan, Milan 20133, Italy
{hassanmahmoud.elhanash,daniele.pareyson,virginia.caruso, andreamaria.zanchettin,paolo.rocco}@polimi.it

Abstract. Using reinforcement learning has enabled robots to learn how to accomplish a wide range of tasks without explicit instructions. In this paper, we use a single-arm robot for the flattening of a piece of cloth which is crumpled and placed on a table. We create a simulation environment with a gripper and a piece of cloth to learn a policy for the robot to choose the best action based on the observation of the environment. The policy is then transferred to a real robot and successfully tested. We also introduce our method on the recognition of the corners of the cloth using computer vision which includes comparing classic computer vision approach to a deep learning one. We use an ABB robot and a 2D camera for the experiments and PyBullet software for the simulation.

Keywords: Reinforcement learning · Robotics · Deformable objects

1 Introduction

Reinforcement learning is very useful for robots to learn how to accomplish tasks and has found many applications in various domains [1]. While deep learning has been used with robotics in clothes manipulation in several works [2–4], the challenging nature of simulating the interaction with clothes limits the number of works in reinforcement learning. Furthermore, the perception and manipulation of clothes poses many challenges for robots because of the infinite degrees of freedom a cloth has.

Previous works that used robots with clothes had different kind of tasks like folding clothes [5], dressing assistance [6] and tidying a room [7]. Various designs of the robotic systems were used in such works. The authors in [8] used a dual-arm robot while in [9] they used a single-arm one. Different sensors and approaches were used for identifying clothes and their state which included cameras [10,11], haptics [12] and hybrid techniques [13]. In [14], the authors present an extensive and thorough framework for the grasping of garments and discuss in depth the grippers used in this context. In [15] the authors presented a method for creating a motion plan that consists of a series of primitives for cloth folding.

I. Petrovic et al. (Eds.): IAS 2022, LNNS 577, pp. 580–595, 2023.
https://doi.org/10.1007/978-3-031-22216-0_39

The task of flattening a cloth with a robot has appeared in different works. In [16], the authors developed an algorithm for flattening a piece of cloth. They used both a 2D image for the first stage to detect corners and then the depth information to detect wrinkles. In [17], a heuristic-based approach was developed to do the flattening which was based on the removal of wrinkles that were detected using range maps. They extended their work in [18] where they developed a surface analysis method for describing wrinkles which is based on computing the curvature of the surface. They used RGB-D images using stereo cameras and they showed the significance of the quality of the depth image on the results. They used a dual-arm robot for validating their approach. They developed their shape analysis method further in [19] and used a clustering-based approach for detecting and quantifying wrinkles.

Teaching robots to accomplish clothes-related tasks also appeared in multiple works. In [20] random-forest algorithm was used to map extracted features of a cloth to a robot's manipulator optimal control action. They identified 3 tasks to test their approach and they included cloth flattening using 3D camera and dual-arm robot. In [21], the authors used reinforcement learning for teaching a robot how to assist in wearing a T-shirt. In [22], they extended their work to teach a robot the motor skills to wear a T-shirt by using motion capture system. In [23], they used depth camera instead and developed a framework for observing Human-Cloth topological relationship. In [24] the authors identified 3 tasks for a robot that included manipulating a towel and used a simulation software to let the robot learn through reinforcement learning how to accomplish the tasks successfully.

When describing actions in the context of a robotics application, the values of the robot's joints are usually used although grasping and releasing points were used in some works [25]. In this paper we describe the actions in the same way a human would think of when flattening a towel. Actions are described as series of grasping of the towel from one point and extending it in some direction till the towel is flattened. Although the task of flattening was considered in previous works, the majority of them relied heavily on the wrinkles of the cloth and on depth images. The contribution of the paper is the development of the reinforcement learning model with our setup of a single-arm robot flattening a towel while moving in a planar space and using solely color images. Furthermore, the description of the towel's state in terms of visibility of its corners is a novel approach in this context.

The rest of the paper is organized as follows. The problem is described as a reinforcement learning problem with its elements in the next Section. Section 3 describes our approach that includes the identification of the towel corners, the algorithm we used and the structure of the neural network used to implement it, and the simulation setup. Finally the experiments section describes the setup of the experiment and the results.

2 Problem Description

Given a crumpled cloth placed on a flat surface, the goal is to flatten it as a prerequisite for other operations such as folding or ironing. One edge of the towel is constrained, and the free corners of the towels are considered the optimum grasping points. Computer vision is used for perception and determining the cloth state while we use reinforcement learning to teach the robot which actions to take based on the state of the cloth. Figure 1 shows the environment setup in simulation and in real world scenario, where a robot arm is interacting with a piece of cloth.

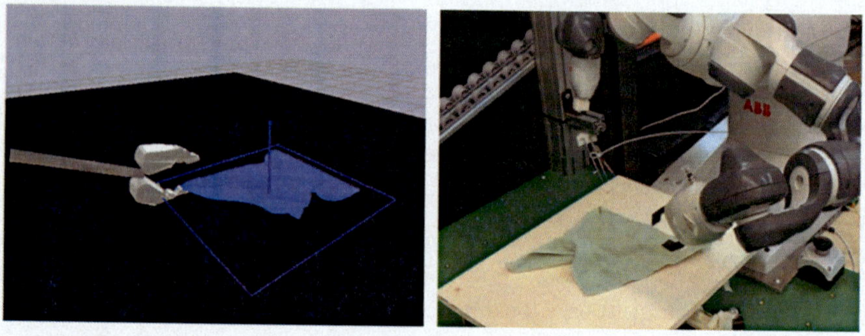

Fig. 1. Environment setup in simulation (left) and in the real test-bed (right) The setup of the environment in the simulation software (PyBullet) with a gripper and piece of cloth on a table (left) and the robot used in real world experiment ready to grasp the piece of cloth (right).

In reinforcement learning, we have an agent (the robot here) in an environment taking an action, $a \in \mathcal{A}$ where $\mathcal{A}$ is the action space. The action leads to a change in the environment state, $s \in \mathcal{S}$, where $\mathcal{S}$ is the state space. Based on the state, the agent receives a reward that comes from the reward function r and the goal is to learn a policy π that maximizes the accumulative reward. The action-value function estimates the likely return of an action a_t in a state s_t following a policy π and denoted as $\mathcal{Q}^{\pi}(s_t, a_t)$. Learning that function is one of the commonly used algorithms in reinforcement learning and known as Q-Learning. We use the Deep Deterministic Policy Gradient (DDPG) algorithm [26] which is built on the Deterministic Policy Gradient algorithm [27]. DDPG is model-free and allows for the insertion of human demonstration that can significantly improve the performance.

2.1 State

The state of the cloth is described using a vector comprised of 5 values that represent the coordinates of 2 corners and the area of the cloth. The restrained edge is used to create a virtual rectangle corresponding to the cloth and a reference point of XY axes is placed in the center of this rectangle.

2.2 Actions

The action that could be taken is grasping a point of the cloth and moving it in some direction. Therefore, the action can be described as a trajectory between a grasping point and another one for dictating a direction.

2.3 Rewards

Rewards can be given after each iteration or at the end of the episode. We chose to set the reward after the completion of the episode, and it's based on the area of the cloth. If the detected surface of the cloth has an area $> 93\%$ of the cloth area, the reward is set to 1 and the episode is successfully terminated. Otherwise, the reward is 0 and it is considered a failure.

3 Methodology

In this section we describe the visual perception pipeline for determining the state, the structure of the neural network used to implement the DDPG and how we built the simulation.

3.1 Visual Perception

Using a 2D camera, an RGB image can be obtained for the towel and the goal is to determine if corners are visible or not and their location, if applicable. We begin with computer vision approach where the largest contour is extracted from the image (that would correspond to the towel) and then corners are identified using Harris corner detector [28]. Figure 2 shows an example of a correct identification and another one where an extra corner was detected. Wrinkles can add more challenges in this approach of detecting corners.

We used deep learning as an alternative approach for finding the corners if they are present in the image. We created a dataset of 15,000 images generated from Blender software to train the model and an extension of this was made for real towel pictures with 2,400 images. The dataset included images for portion of the towel where a corner exists and other images where there are no corners. Figure 3 shows examples of images from synthetic (left) and real (right) images, and images with visible corners (left) and images where there are no corners (right). Images were made at different resolutions, rotated, and towel/background color was varied to allow for successful classification with new data. Images were split as 70 % for training and 30% for testing.

Figure 4 shows the structure of the convolutional neural network we used which has as input 28×28 image. It is composed of the following layers:

- Five layers each composed of:
  - Convolution with a 3×3 filter. The number of filters is chosen as 32, 64, 128, 256 and 512 for the 5 layers.
  - Leaky ReLU with $\alpha = 0.1$

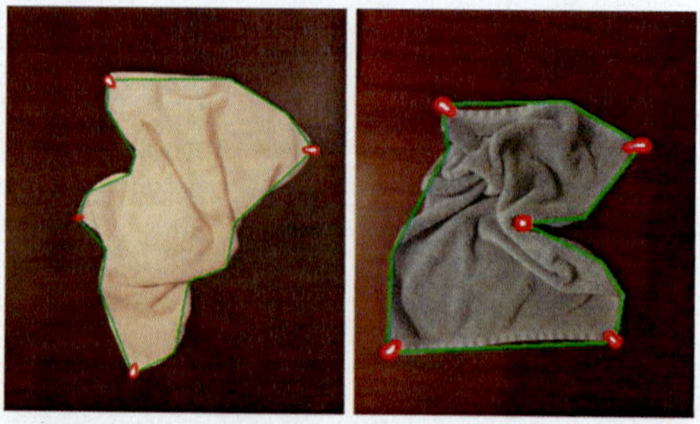

Fig. 2. Corners detection An example of a successful identification of corners (left) and an excessive number of corners identification (right).

Fig. 3. Part of the images dataset used for corners identification. Images used with the neural network for corner identification included images from simulation (left) and from real world (right). Dataset had images with visible corners (left) and others without corners (right).

  - Max Pooling
- Flattening layer with output size of 512 to allow for the handling in the following dense layers.
- Four dense layers with linear activation and output sizes of 256, 128, 64 and 28.
- Leaky ReLU layer with $\alpha = 0.1$.
- Output dense layer with 2 outputs neurons and softmax activation function.

Figure 5 shows the training of our model with the simulated (left) and real images (right). The accuracy vs epoch graph for simulated images (left) shows that we could get an accuracy of 98% while with the real ones (right) we got 93%. Even though the model provided promising results, the resulted identified corners were sometimes incorrect. Factors contributing to this problem included challenging lighting condition, shadows, non-smooth background, crumples and wrinkles. To solve this problem, we added an extra step of post-processing by setting a certain threshold of distance from the identified contour that bounds the cloth. Any point that is at a larger distance than this threshold is ignored. Figure 6 shows an example of this case with an input image (left) that has some points wrongly identified as corners and an output image (right) where points identified at a certain distance from the contour are ignored. Figure 7 shows some images with challenging conditions that have been post-processed with correct identification of corners and shows the robustness of our method in tackling the issues with the images.

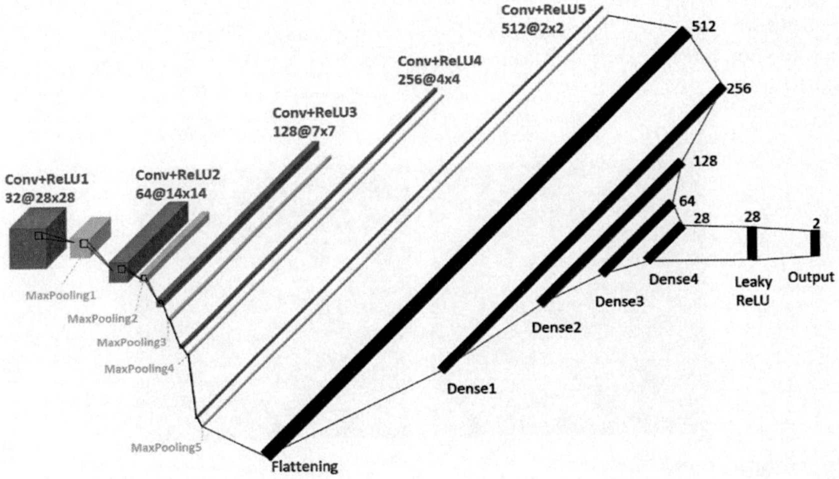

Fig. 4. CNN Model for Corners Identification The convolutional neural network model used for identifying corners in an image. Convolution layers included 3 × 3 filters with increasing number of them from 32 to 512, ReLU and MaxPooling followed by dense layers and leaky ReLU to the final softmax output layer.

To compare our CNN-based method against the computer vision approach that is based on Harris corner detector, we created a dataset of 100 synthetic images and 100 real ones. Table 1 shows a comparison between the two methods using the definitions of precision, recall and F1-score below:

$$Precision = \frac{TP}{TP + FP}, Recall = \frac{TP}{TP + FN},$$
$$F1 - Score = \frac{2 * Precision * Recall}{Precision + Recall}$$

where TP is True Positive, FP is False Positive and FN is False Negative.

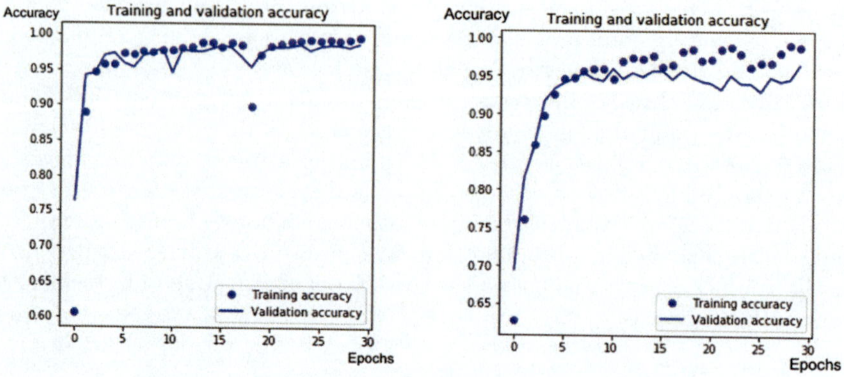

Fig. 5. Model accuracy for simulated (left) and real images (right). Accuracy vs epoch graph for our model using synthetic images from Blender (left) and real world images (right).

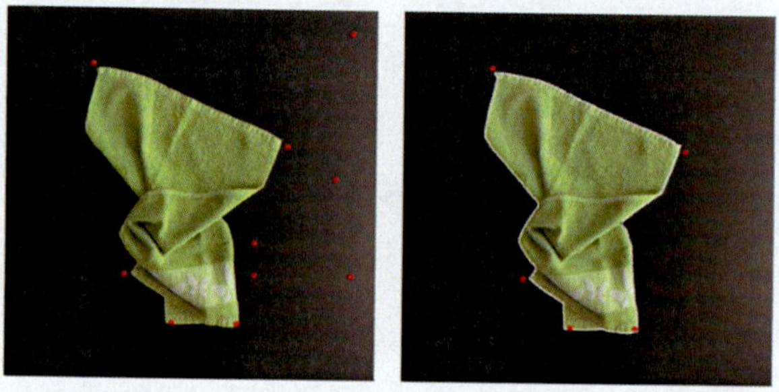

Fig. 6. Post-processing of identified corners. An example of post-processing. Identified corners (left) and after removing those at a certain distance from the contour (right).

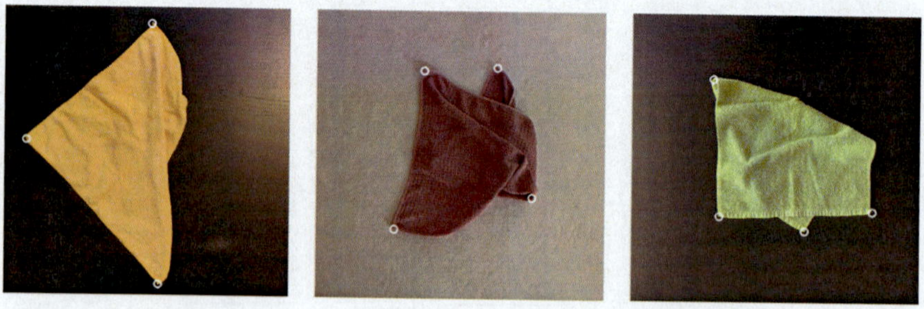

Fig. 7. Post-processed images for corners identification. Examples of images that have been post-processed after using the proposed CNN model. Lighting conditions, shadows, wrinkles and crumples add to the challenges of successfully identifying corners and our proposed post-processing method shows robustness against these conditions.

Table 1. Comparing CV approach with CNN

Sim/real	Harris corner detector-based approach	Deep learning using CNN
Synthetic images	Precision: 0.78	Precision: 0.81
	Recall: 0.85	Recall: 0.95
	F1-score: 0.81	F1-Score: 0.87
Real images	Precision: 0.59	Precision: 0.61
	Recall: 0.78	Recall:0.79
	F1-score: 0.66	F1-Score: 0.69

Comparing our CNN-based method with another approach that is based on Harris corner detector. Comparison includes the synthetic and real images using 100 images per each category.

3.2 Neural Network Structure for DDPG

The DDPG algorithm has 2 elements: the actor and the critic networks. The actor is the one that takes observations as input and outputs the actions. The critic is parameterized by θ^Q and it tries to learn $\mathcal{Q}$ by minimizing the Bellman loss. The structure of the network is shown in Fig. 8. The main hyper-parameters to consider are batch-size, sampling ratio, discount rate and learning rate. The mini-batch size is set to 1024. The sampling ratio between demonstrations and transitions ρ, is set to 0.125 and kept fixed along with the training: at every step, a number $N_D = 128$ of demonstrations are collected from the demo-buffer, while 896 experienced transitions are sampled from the main replay buffer. While the discount rate γ is generally set close to 1, setting it to 0.8 in our work gave better results. The learning rate τ, which determines how fast the target networks are updated was selected as 0.005.

To collect the demonstration data, actions are performed and all transition tuples (s_i, r_i, s_{i+1}, d_i) are recorded and stored. In these demonstrations, the towel lies slightly crumpled, with a single corner out of position. Therefore, only one action needs to be accomplished in order to complete the flattening: grasping the corner which is out of position and pulling it towards the right zone. At the end of these transitions, when the manoeuvre is performed, the garment is flattened, the act is positively rewarded and the episode is brought to an end. Sometimes the corner is not pulled exactly to the target position but very close, in order to provide some noise and keep some wrinkles. 1024 of these demonstration transitions are directly stored in the replay buffer, so that they provide examples of the correct behaviour.

3.3 Simulation

We use the PyBullet software for the simulation where a modeled cloth is placed on a table in an environment with a gripper. An initial pre-training phase is

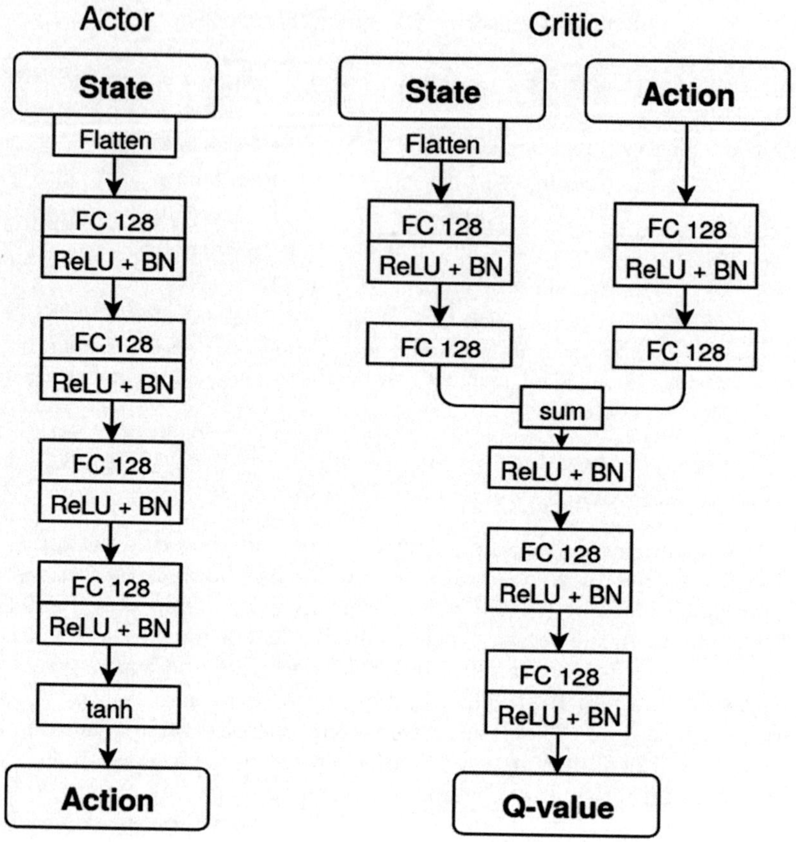

Fig. 8. NN Structure for implementing DDPG. Neural Network structure for the DDPG. Actor and critic have 4 layers each has 128 neurons. ReLU and Batch Normalization are added after each layer.

developed to prepare some basic guidelines that could help the agent in the main task.

The algorithm operates with a rudimentary environment which has an identical setup to the principal version, but is made up of a 2D geometrical representation of the towel. A bi-dimensional polygon is reproduced in a planar space and simulates a very primitive cloth-like figure, where the vertices represent the corners of the towel. The environment is shown in Fig. 9. When the grasping position, determined by the first 2 parameters of the action, falls close to a free corner, this one is moved towards a target position, defined by the second 2 parameters. When the manoeuvre does not allow to perform the manipulation, it is ignored and not executed. Despite not being executed, these moves can still contribute to the agent's learning, because the transition is stored and the model gets ordinarily updated.

Fig. 9. Environment for the pre-training phase. The virtual environment that is used as a pre-training phase. 2D polygon is created to represented the cloth and the arrows represent the direction in which the cloth is grasped to. The numbers correspond to the ratio of the visible area of the cloth to the total area of the cloth.

This pre-training phase is very quick and allows the agent to be trained with a preliminary behaviour policy; it lasts about 2800 episodes, each made of 100 steps. An exploration noise is added with a Gaussian function $N(\theta,\sigma)$ to the chosen action, where σ is parametrised and varies from $\sigma_{max} = 0.4$ to $\sigma_{min} = 0.05$, along with the episodes, according to an exponential function whose slope is determined by a decay-parameter: decay $= 400$:

$$\sigma = \sigma_{min} + (\sigma_{max} - \sigma_{min})e^{-\frac{n_{ep}}{decay}}$$

This noise allows to better explore the environment in the initial phases, and to increase the probability to grasp a corner in case the grasping point is too far from the potential pinning point. The learning starts after the first 10 episodes of 100 steps, since at least 896 transitions are required to fill the memory buffer and boot the algorithm. The graph in Fig. 10 (top) shows how the agent quickly learns to reach the goal, reinforces its behaviour and improves from the first episodes, constantly reaching the reward.

The policy reaches a level of accuracy of 80% only after two hundred episodes, and is kept running so that it can consolidate its behaviour. The average reward immediately gets close to 1 and remains over 0.9 for the entire duration of the training, with some slight oscillations. The reason for such a fast learning is that the environment is very simple, the parameters are carefully tuned, and the target positions of the free corners are the same in every episode, since an edge is anchored and the final targeted state has always the same values.

After the pre-training phase, the agent is connected to the main environment with the 3D physics engine PyBullet. The setup here is similar to the previous one, but a physical interaction occurs and the towel model is a 3D-simulation. The filter on the grasping point in the action slightly changes: the pinning point is not required to be only within a certain range from the corner, but it also needs to be located inside the towel contour itself, so that a solid grasp can be granted. This condition now is relevant. In the absence of it, the pinning point could fall outside the surface of the towel, leading to a grasping failure. However, due to some instabilities in the towel-gripper interaction inside PyBullet, it sometimes

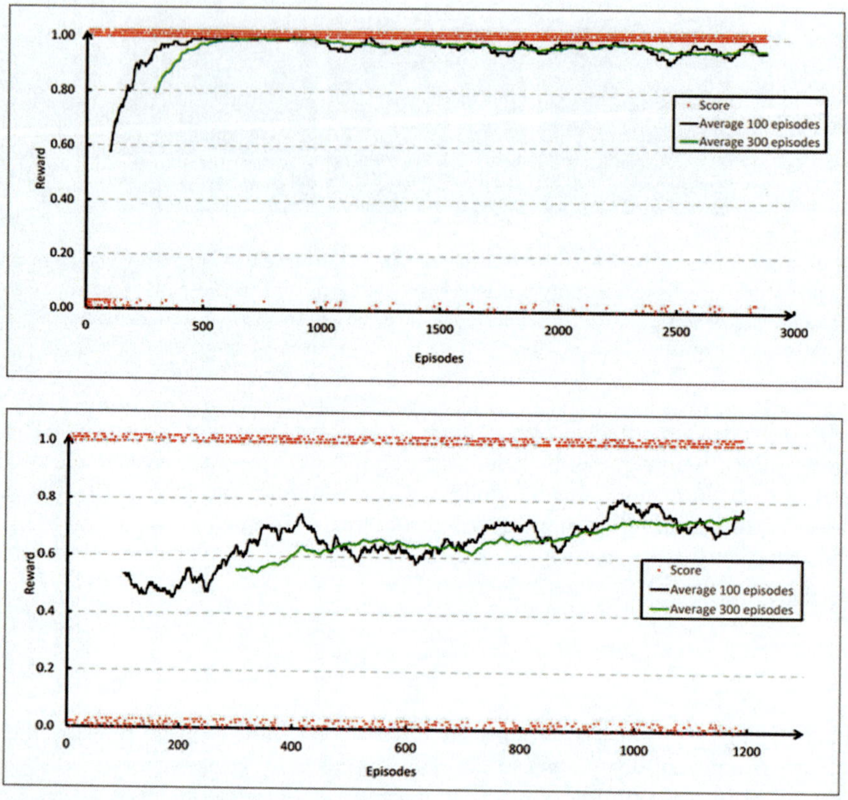

Fig. 10. Learning curve for geometric (top) and physical (bottom) model. The reward vs epoch graph shows how the agent learns to reach its goal. The red dots are the reward per each episode (either positively (+1) or negatively (+0)) while the black and green curves are the average rewards per 100 and 300 epochs respectively. The curve for the pre-training phase (top) shows quick learning as it's a very simple environment while for the physical model (bottom) it has less values for the rewards as it is more complex model.

occurs that the towel slips from the gripper and is not pulled through the full path of the gripper, as planned. 1200 executions are carried out, after which learning starts to stabilise and does not improve any further. Every episode is made of 300 steps, but generally lasts much less because the task is accomplished before the end of the process. Each feasible manoeuvre is fulfilled in about 20 seconds, while for non-executed ones it takes only 2 seconds for the learning update. The average number of actions per episode performed is 20, but decreases as the learning improves. Each task lasts around 3 minutes and the whole training goes on for 60 hours.

Both the new dynamics of the cloth and the interaction between the soft and the rigid body, constitute a further challenge for learning and the current policy is no longer optimal. The pre-training phase, however, allows the agent to avoid actions which may be totally incorrect, such as pushing the towel towards the anchored edge or miscalculating at all the directions of pulling. The training phase shows significant improvements and the behaviour of the agent becomes more efficient in achieving the final goal, both in terms of speed and robustness. Figure 10 (bottom) shows the reward vs epoch curve which increases for the whole learning up to a limit of 0.75. At the end of the learning, the policy flattens the towel and takes only a few steps for it. It is also robust enough to correct possible misalignment in the orientation of the towel. Figure 11 shows an example of an episode with the sequence of actions which flatten the towel.

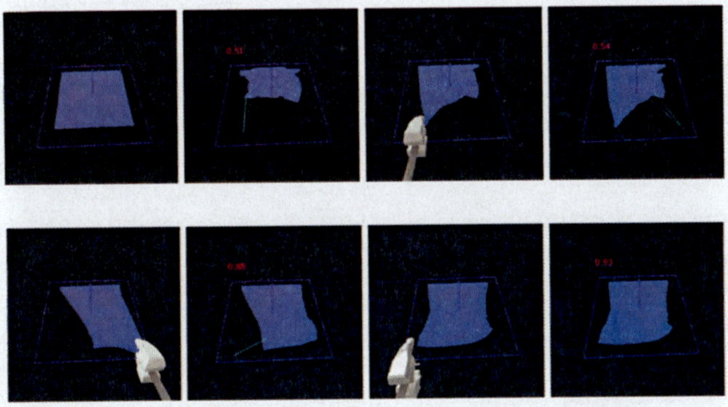

Fig. 11. Successful sequence of actions in simulation. An example of an epoch inside the simulation software showing a sequence of actions that led to the flattening the towel where the episode is rewarded.

4 Experiments

We use an ABB robot (IRB-14000 YuMi) for the experiments which is a dual arm one, yet we only use a single arm. We also do not use the robot's off-the-shelf fingers, instead we design and 3D print new ones to suit the needs of grasping clothes. To achieve an action, the gripper reaches a grasping point, fingers are closed and then the gripper moves towards a certain release point where fingers are opened. The output of the model is the set of grasping and releasing locations for the gripper which together with auxiliary points are used to generate a trajectory for the gripper. The points and the trajectory are illustrated in Fig. 12 where the robot begins from a predefined point so that it will not create any occlusion for the camera.

To evaluate the trained model, 50 episodes are carried out. As for the training phase, each episode is positively rewarded only when the visible area of the towel exceeds 93%. At the beginning of each episode the towel is creased and arranged in different ways. The percentage of visible area at the beginning of the task

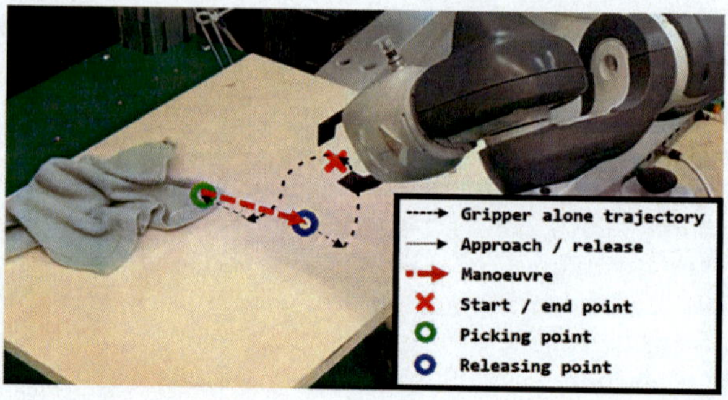

Fig. 12. Robot with a trajectory. The trajectory for the robot to accomplish the task. Robot always starts from a fixed position marked with a red x. Black curves are the robot's trajectory to reach the picking point (marked with a green circle) and move back to the initial point after reaching the release point (marked with blue circle). Robot grasps the cloth at the picking point and moves it to the releasing point.

is always kept to a value lower than 50%, so that the agent actually starts its work from a challenging configuration. In 30 episodes (60% of the cases), the two free corners are kept visible, either lying outwards or being flipped inwards; in 10 episodes (20% of the total) only one corner is hidden; in the remaining 20% neither can be seen. Table 2 summarizes the results obtained in the experiments, which shows the rate of successfully accomplishing the task in the different scenarios. The successful flattening is achieved in 26 out of 50 episodes, with an overall success rate of 52%.

Table 2. Summary of results

Visible corners	Episodes	Successes	Success ratio (%)
2	30	21	70
1	10	3	30
0	10	2	20
Total	50	26	52

The success ratio of accomplishing the task in 3 different cases (2, 1 and 0 visible corners). Clearly as more corners are visible, the success ratio is higher.

5 Discussion

The overall accomplishment rate is lower than the one obtained with the simulations. When the cloth did not have a complex configuration, the agent easily managed to complete the task. In most of the encountered situations, the arm was able to flatten the towel using very few manoeuvres (an average of 4 steps), knowing what corner to grasp and where to pull it. In these episodes, the success ratio results were consistent with the expectations, with a rate of 70%. The performance gets worse if the initial layout of the towel is more entangled and the success rate drops to 20–30% as most of the actions do not produce an efficient untwisting of the folded edge.

There are two main types of failures that occur during the validation test. The first and more common situation happens when the agent is not able to generate a feasible action. Since the model is deterministic, the action is identically repeated at the successive step. Some Gaussian noise generally allows to overcome this issue, and helps the agent to select a feasible action. However, when the grasping point is completely incorrect, after a number of attempts without any effect, the episode ends and the task is considered as un-accomplished. The second type of error is a grasping failure, which is caused either by a weak pinch, or by an infeasible grasp. In the first case, the gripper cannot robustly hold the cloth while pinching it, and the action needs to be repeated. Even if a second instance is always allowed, and typically fixes the event, sometimes, this happens multiple times and becomes critical for the accomplishment of the goal. In the second case, rather similarly to the previous one, it happens that a large wrinkle forms, and the arm cannot perform the grasp. Even if the action is repeated, it is very unlikely that the error will be recovered. The episode therefore ends with a failure, even in those cases where the state is extremely close to the reward. Some further interruptions rarely occur when two or more layers of the towel are pinned together and the unexpected thickness perceived by the gripper gives origin to error alerts in the servo-motors of the hand. A trade-off is thus required to have a tight but soft pinch between the robot fingers.

6 Conclusions

Using the reinforcement learning framework that is based on a Markov Decision Process with a single-arm robot and camera it was possible to manipulate a towel. Despite the limitations in physics engines simulation of cloth-like objects, we managed to build an environment for the robot to interact with the towel and determine its state. The DDPG algorithm was used, with tuned parameters and supported by human demonstrations. Consequently the robot could successfully exploit the collected experiences and learned how to flatten a towel with constrained edge. The results show how the training stage has efficiently reinforced the correct behaviour, both in the simulations and in the real-world application, and frequently leads to the accomplishment of the task in most episodes. Some challenges in the validation phase emerge and highlight the sensitivity of the model to the initial configuration of the towel. The main restraint is probably caused by the physics engine, which did not allow to accurately randomise the initial configuration of the towel. Another challenge that comes from the DDPG algorithm is that it is very sensitive to the variation of the parameters, and requires a large number of attempts.

References

1. Polydoros, A.S., Nalpantidis, L.: Survey of model-based reinforcement learning: applications on robotics. J. Intel. Robot. Syst. **86**(2), 153173 (2017)
2. Mariolis, I., Peleka, G., Kargakos, A., Malassiotis, S.: Pose and category recognition of highly deformable objects using deep learning. In: IEEE International Conference on Advanced Robotics (ICAR), pp. 655–662 (2015)
3. Yuan, W., Mo, Y., Wang, S., Adelson, E.H.: Active clothing material perception using tactile sensing and deep learning. In: IEEE International Conference on Robotics and Automation (ICRA), pp. 4842–4849 (2018)
4. Saxena, K., Shibata, T.: Garment recognition and grasping point detection for clothing assistance task using deep learning. IEEE/SICE International Symposium on System Integration (SII), pp. 632–637 (2019)
5. Doumanoglou, A., Stria, J., Peleka, G., Mariolis, I., Petrik, V., Kargakos, A., Wagner, L., Hlavac, V., Kim, T.K., Malassiotis, S.: Folding clothes autonomously: a complete pipeline. IEEE Trans. Robot. **32**(6), 1461–1478 (2016)
6. Yamazaki, K., Oya, R., Nagahama, K., Okada, K., Inaba, M.: Bottom dressing by a life-sized humanoid robot provided failure detection and recovery functions. In: 2014 IEEE/SICE International Symposium on System Integration. SII, vol. 2014, pp. 564–570 (2014)
7. Yamazaki, K., Ueda, R., Nozawa, S., Mori, Y., Maki, T., Hatao, N., Okada, K., Inaba, M.: System integration of a daily assistive robot and its application to tidying and cleaning rooms. In: IEEE/RSJ 2010 International Conference on Intelligent Robots and Systems, IROS 2010–Conference Proceedings, pp. 1365–1371 (2010)
8. Stria, J., Prusa, D., Hlavac, V., Wagner, L., Petrik, V., Krsek, P., Smutny, V.: Garment perception and its folding using a dual-arm robot. In: IEEE International Conference on Intelligent Robots and Systems, pp. 61–67 (2014)
9. Petrík, V., Smutný, V., Krsek, P., Hlaváč, V.: Single arm robotic garment folding path generation. Adv. Robot. **31**(23–24), 1325–1337 (2017)
10. Yamazaki, K., Nagahama, K., Inaba, M.: Daily clothes observation from visible surfaces based on wrinkle and cloth-overlap detection. In: Proceedings of the 12th IAPR Conference on Machine Vision Applications, MVA 2011, pp. 275–278 (2011)
11. Sun, L., Aragon-camarasa, G., Rogers, S., Siebert, J.P.: Autonomous clothes manipulation using a hierarchical vision architecture. IEEE Access **6**, 76646–76662 (2018)
12. Clegg, A., Yu, W., Erickson, Z., Tan, J., Liu, C., Turk, G.: Learning to navigate cloth using haptics. In: IEEE International Conference on Intelligent Robots and Systems, pp. 2799–2805. Institute of Electrical and Electronics Engineers Inc. (2017)
13. Yuan, W., Mo, Y., Wang, S., Adelson, E.H.: Active clothing material perception using tactile sensing and deep learning. In: IEEE International Conference on Robotics and Automation (ICRA), pp. 4842–4849. IEEE (2018)
14. Borrás, J., Alenyá, G., Torras, C.: A grasping-centered analysis for cloth manipulation. IEEE Trans. Robot. **36**(3), 924–936 (2020)
15. Lakshmanan, K., Sachdev, A., Xie, Z., Berenson, D., Goldberg, K., Abbeel, P.: Constraint-Aware Motion Planning Algorithm A, for Robotic Folding of Clothes, Experimental Robotics. Springer Tracts in Advanced Robotics, vol. 88. Springer, Heidelberg (2013)
16. Willimon, B., Birchfield, S., Walker, I.: Model for Unfolding Laundry Using Interactive Perception pp. 4871–4876. IROS (2011)

17. Sun, L., Aragon-Camarasa, G., Cocshott, P., Rogers, S., Siebert, J.P.: A Heuristic-Based Approach for Flattening Wrinkled Clothes. TAROS (2013)
18. Sun, L., Aragon-Camarasa, G., Rogers, S., Siebert, J.P.: Accurate garment surface analysis using an active stereo robot head with application to dual- arm flattening. Robot. Autom. (ICRA) 185–192 (2015)
19. Sun, L., Camarasa, G.A., Khan, A., Rogers, S., Siebert, P.: A precise method for cloth configuration parsing applied to single-arm flattening. Int. J. Adv. Robot. Syst. 13(2) (2016)
20. Jia, B., Pan, Z., Hu, Z., Pan, J., Manocha, D.: Cloth manipulation using random-forest-based imitation learning. IEEE Robot. Autom. Lett. **4**(2), 2086–93 (2019)
21. Tamei, T., Matsubara, T., Rai, A., Shibata, T.: Reinforcement learning of clothing assistance with a dual-arm robot. In: IEEE-RAS International Conference on Humanoid Robots, pp. 733–738 (2011)
22. Matsubara, T.: Reinforcement learning of motor skills with non-rigid materials using topology coordinates. Adv. Robot. **27**(7), 513–524 (2013)
23. Koganti, N., Tamei, T., Matsubara, T., Shibata, T.: Real-time estimation of human-cloth topological relationship using depth sensor for robotic clothing assistance. In: IEEE International Workshop on Robot and Human Interactive Communication, pp. 124–129 (2014)
24. Matas, J., James, S., Davison, A.J.: Sim-to-real reinforcement learning for deformable object manipulation. In: 2nd Conference on Robot Learning (CoRL 2018) (2018)
25. Wu, Y., Yan, W., Kurutach, T., Pinto, L., Abbeel, P.: Learning to Manipulate Deformable Objects without Demon-Strations (2019). ArXiv:1910.13439
26. Lillicrap, T.P., Hunt, J.J., Pritzel, A., Heess, N., Erez, T., Tassa, Y., Silver, D., Wierstra, D.: Continuous control with deep reinforcement learning. In: CoRR (2015)
27. Silver, D. et al.: Deterministic policy gradient algorithms. In: International Conference on Machine Learning (2014)
28. Harris, C., Stephens, M.: A combined corner and edge detector. In: Alvey Vision Conference (1988)
29. Večerík, M., Hester, T., Scholz, J., Wang, F., Pietquin, O., Piot, B., Heess, N., Rothörl, T., Lampe, T., Riedmiller, M.: Leveraging demonstrations for deep reinforcement learning on robotics problems with sparse rewards. In: CoRR (2017)

Robot Simulations

A Monte Carlo Framework for Incremental Improvement of Simulation Fidelity

D. M. Lyons[1(✉)], J. Finocchiaro[2], M. Novitzky[2], and C. Korpela[2]

[1] Fordham University,Bronx, New York, USA
dlyons@fordham.edu
[2] United States Military Academy,West Point, New York, USA
{james.finocchiaro,michael.novitzky,christopher.korpela}@westpoint.edu

Abstract. Robot software developed in simulation often does not behave as expected when deployed because the simulation does not sufficiently represent reality—this is sometimes called the 'reality gap' problem. We propose a novel algorithm to address the reality gap by injecting real-world experience into the simulation. It is assumed that the robot program (control policy) is developed using simulation, but subsequently deployed on a real system, and that the program includes *a performance objective* monitor procedure with scalar output. The proposed approach collects simulation and real world observations and builds conditional probability functions. These are used to generate paired roll-outs to identify points of divergence in simulation and real behavior. From these, *state-space kernels* are generated that, when integrated with the original simulation, coerce the simulation into behaving more like observed reality. Performance results are presented for a long-term deployment of an autonomous delivery vehicle example.

1 Introduction

Simulation tools are widely used in robot program development, whether the program/controller is built by hand or using machine learning. At the very least, simulation allows a robot programmer to eliminate obvious program flaws. The availability of physics engines [1] has produced simulations that can more accurately model physical behavior, making it more attractive to use simulation in conjunction with machine learning techniques [2,3] to develop robot programs. However, a robot program validated with simulation, when operating in a real, unstructured environment may come across phenomena that its designers just did not know to include in the simulation, even though the phenomenon could in fact be simulated if it were known a-priori to be relevant. Examples of this

The views and conclusions contained herein are those of the authors and should not be interpreted as necessarily representing the official policies or endorsements, either expressed or implied, of the U.S. Department of Defense or the U.S. Government. Lyons was partially supported by DL-47359-15016 from Bloomberg LP.

I. Petrovic et al. (Eds.): IAS 2022, LNNS 577, pp. 599–616, 2023.
https://doi.org/10.1007/978-3-031-22216-0_40

kind of simulation *'reality gap'* include inaccurate robot joint parameters, surface friction, object masses, sizes and locations.

This paper addresses closing the *reality gap* for simulations used to develop robot programs by combining *sim-to-real* [4] and *real-to-sim* [5] approaches. We will assume that the simulation is a black-box and we present a framework to coerce simulation behavior to more closely resemble real experience.

The simulation has configuration parameters ϕ. If deployment of the robot program (π) results in failure, as determined by a performance monitor, then our overall objective is for ϕ to be updated by the deployment experience and π redeveloped to handle that experience. We expect this virtuous cycle of simulation, deployment and improvement to iterate. The simulation will be modeled as a transition function $T_{\phi}(s'|s,a)$ where s is the current sensor data from the simulation, a is the action to be carried out, s' is the resulting sensor data after the action is taken, and ϕ is a setting of the configuration parameters for the simulation. Real world experience will also be modeled as a transition function $T_r(s'|s,a)$. Two important novel aspects of our work are (1) a domain independent proposal for ϕ (as opposed to the more domain specific examples of domain randomization, e.g., [6,7]) and (2) a Monte Carlo method to modify ϕ based on a direct comparison of estimated transitions functions (as opposed to the comparison of observations).

Fig. 1. Modified Traaxis platform used in field trials (left); simulation model (right).

The remainder of the paper is laid out as follows. Section 2 is a review of related work. Section 3 presents our method and architecture. Section 4 details an introductory, simulation-only example. Section 5 describes the field trial, using simulation for design and a customized Traaxis platform for deployment, Fig. 1. Section 6 presents our results in two parts. The first part addresses the unexpected environment change observed in deployment. The second part addresses unexpected sensor performance. We show that data collected during deployment can be used to make the simulation behave more realistically, iteratively narrowing the reality gap. Conclusions and next steps are presented in Sect. 6.

2 Related Work

Researchers using learning techniques to train robot programs on simulation have developed several approaches to the simulation reality gap. The *sim-to-real* approach addresses the issue of moving a policy trained in simulation to real hardware; Hofer [4] discusses this new direction and its open challenges, one key one of which is improving the accuracy of the simulation tools used—where we aim to contribute.

Estimating a model of an observed system is typically a system identification [8] problem. The problem addressed here differs in two ways: Firstly, the system addressed is not the robot to be controlled, but rather the environment in which the robot will be embedded. Secondly, the model to be constructed is an 'add-on' model for an existing 3D simulation that employs its own blackbox model of the environment to overrides the simulation in certain parts of the state space and make its performance more accurate. In this sense, the work is similar to the *real-to-sim* approach, a complementary approach [5] in which real-world information is transferred to a simulation. We employ a composite local model approach, e.g., [9], to address the integration of reality and simulation.

Producing more realistic behavior does not just concern an improved physics engine and rendering technology, but also concerns handling uncertainty about which specific environment the robot will encounter. Benjamin [10] and Lyons [11] proposed an approach for cognitive robotics where a simulation is used to predict the environment state and to visually compare video with simulation predicted imagery. Differences are used to modify the simulation, changing object appearance, positions and velocities to match the observed video. While this approach has similarities to mapping (e.g., SLAM), a key difference is its attempt to insert object descriptions from camera information into a 3D simulation so as to be able to leverage the predictive properties of the simulation. The approach relied on a graybox view of the simulation and its modeling of objects.

Christian et al. [12] use a simulation to learn policies for a number of tasks and consider transferring the policy from simulation to real robot. They note that the simulation policy is generally correct in high-level gist but fails on some lower-level details. Yu [13] proposes a two stage approach to policy training for bipedal locomotion: a presimulation step to ballpark the system identification for simulation, followed by a more accurate tuning at deployment. Permana [14] trains a CNN on synthetic imagery for visual detection of ground casualties, and handles the sim-to-real issue by injecting noise, downsampling, segment removal and other changes to the simulation data.

Adopting the domain randomization approach, Peng et al. [2] use random modification of 95 simulation parameters to show that even low fidelity simulation can be used to train a robot to push a puck successfully over a wide range of real-world situations. Cheboter et al. [15] interleave many simulation runs with much fewer real-world runs, and modify the simulation configuration so that it simulates a range of situations more similar to observed experience. Our proposed approach is most like that of Chebotar. However, instead of using the domain randomization—establishing a range of initial parameter values—we follow an approach more similar to Benjamin and Lyons [11] who modify simulation state during a simulation to close the reality gap. Our approach differs from sim-to-real approaches in putting the emphasis on improving the simulation via real-to-sim so that the sim-to-real step becomes less onerous. It differs from [11] in adopting a blackbox view of the simulation.

3 Approach

The simulated and physical environments are modeled as Markov Decision Processes (MDP) $M_{sim} = (S, A, T_{sim}, R)$ and $M_{phy} = (S, A, T_{phy}, R)$ that differ only in their transition function. The robot program implements a policy π that can be applied to either MDP. The sensors available to the robot program are $SN = sn_0, sn_1, \ldots, Sn_n$ with value sets $SV = Sv_0, Sv_1, \ldots, Sv_n$. The control outputs available are $AN = an_0, an_1, \ldots, an_m$ with value sets $AV = Av_0, Av_1, \ldots, Av_m$. The actions available to the robot are any setting of the control outputs

$$A = Av_0 \times Av_1 \times \ldots Av_m \tag{1}$$

and the set of states S is

$$S = Sv_0 \times Sv_1 \times \ldots Sv_n \tag{2}$$

A reward function, R, will be used as a measure of when the robot program is achieving its performance objectives on M_{sim} or M_{phy}. The software to measure R can either be hand-written by a software engineer, generated automatically from task specifications [16], or, if the program was generated by learning, copied from the learning reward module.

The MDP transition functions are defined

$$T_{sim}, T_{phy} : S \times S \times A \rightarrow [0, 1] \tag{3}$$

and are interpreted in their usual way as the conditional probability $T(s'|s, a)$ of transitioning from state s to state s' when action a is carried out. These functions encapsulate any difference between what occurs when an action a is carried out in the simulation in a state s and when the same action a is carried out in state s in the real environment.

3.1 Kernel Generation

Once the program is ready to be deployed, we collect data from multiple runs of the final version of the robot program through the simulation:

$$H_{sim} = \{(s_i, a_i, r_i, s_i') : 0 \leq i \leq I_{max}\} \tag{4}$$

for all action a taken in state s then resulting in state s' and reward r in a run of the simulation. Similarly, H_{phy} is collected for the final program in deployment – representing real experience. Using a frequentist approach, T_{sim} is estimated from H_{sim}, and T_{phy} from H_{phy}. The state to state transition function P that is the combination of the T (for each function) with π, the control software is:

$$\begin{array}{ll} B(s'|s,a) = T(s'|s,a)\pi(a|s), & a \in A, s, s' \in S \\ P(s'|s) \quad = \sum_{a \in A} B(s'|s,a) & \end{array} \tag{5}$$

Let $SR(s)$ be the set of state, action, probability tuples that directly transition from state s to state s' in P:

$$SR(s) = \{(s', a, B(s'|s,a)) : B(s'|s,a) > 0\} \tag{6}$$

Since a state only includes what is observable to the robot, there may be hidden dynamics in the real world that result in $SR_{phy}(s)$ differing from $SR_{sim}(s)$. Our first objective is to identify this difference. We define the policy roll-out from a state s as a sequence:

$$Roll(s) = \begin{cases} SR(s) \neq \emptyset & (s', a, p).Roll(s') \; (s', a, p) \in SR(s) \\ else & \perp \end{cases} \tag{7}$$

where '.' is sequence concatenation and $\perp$ indicates end of sequence. A number of *paired roll-outs* are calculated $Roll_{sim}(s)$ and $Roll_{phy}(s)$ and compared up to the point at which they are considered to diverge. The distance between two states is defined by the weighted L1 norm $c(s, s')$:

$$c(s, s') = \sum_i w_i |s_i - s'_i|, \; \sum_i w_i = 1 \; w_i \in \Re, \tag{8}$$

States differ if their distance is greater than a threshold ϵ_c. If states differ, and they have sufficiently different probabilities p under each transition function, this is called *a roll-out divergence*. For each divergence detected, we construct *a state-space kernel* that we will use to modify the behavior of the simulation in that region of its state space so that it is more similar to observed behavior of the physical environment.

Next we will describe how the kernel is built, then how the simulation is modified by a kernel. The key pieces of information in a kernel k are:

- The *region of state space* in which it is active: $e_k = N(0, \sigma^2)$ is a univariate normal distribution of distance from s, the state preceding the divergence, where $c_k(.) = c(s, .)$ is used to calculate the scalar distance between s and any other state. For any state s', $e_k \circ c_k(s')$ is measure of how active kernel k is at s'.
- The *divergence probability distribution*: $\wp_k = (p_s, p_p)$, where $\delta = (s, a, p)$ is the common roll-out point just before divergence and $\delta_s = (s_s, p_s, a)$ in $Roll_{sim}(s_0)$ and $\delta_p = (s_p, p_p, a)$ in $Roll_{phy}(s_0)$ are the points after divergence.
- The *kernel transfer function*: $f_k : S \times A \rightarrow S$. This mapping is a composition of linear functions f_{k,sn_i} for each sensor $sn_i \in SN$ constructed by least-squares fit to the state and action data in the sequence starting with the

state preceding divergence and including a small number of successor states. We will restrict the kernel to be active in a small region of the state space, and since the transfer function estimate is in a small region of the state space, we argue that this can serve as a piecewise linear approximation of a more complex function. For now, we omit functions that transform a state back to itself. We discuss this useful special case later in the paper.

Algorithm 1 shows the algorithm for kernel construction.

Algorithm 1 Generate Kernels

procedure GENKER(P_{sim}, P_{phy})
  $S_0 = States(P_{sim}) \cap States(P_{phy})$
  while $i < N_{rollouts}$ **do**
    $s_0 \sim U(S_0)$ // sample uniform distrib
    $r_s, r_p = Roll_{sim}(s_0), Roll_{phy}(s_0)$
    $\delta, \delta_s, \delta_p = divergence(r_s, r_p)$ // state δ followed by δ_s in r_s & δ_p in r_p.
    if $\delta \neq null$ **then** // construct kernel at δ
      $e = N(s, \sigma^2)$ // $\delta = (s, a, p)$ Eq. (6)
      $d = normalized(p_s, p_p)$ // where $\delta_s = (s_s, a_s, p_s)$, sim. for δ_p
      $f = linearfit(\delta, \delta_p, \ldots)$ // arg. is a short subseq of r_p
      $AddKernel(e, d, f)$
    end if
  end while
end procedure

3.2 Kernel Manager

We consider the simulation as a black box process, but one over which we have some control. For example, in [15] a black-box simulation is configured by selecting initial parameters ϕ from a distribution P_ϕ. They attempt to learn the distribution P_ϕ that best matches traces of real stored experience. We will also assume that we have access to a set of parameters ϕ for the black box simulation but more similar to those of [11] which effect the simulation *on simulation each time-step* and not just initialization (Fig. 2).

At each simulation time step, the robot program sends its selected action a to the simulation. While the internal state of the simulation is unknown, it generates the sensor output values, the M_{sim} state s, and this is communicated to the *Kernel Manager*. The kernel manager checks first to see whether the state s is within the spatial scope for any kernel k by evaluating $e_k \circ c_k(s) > \epsilon_c$ (for small ϵ_c) to see if that kernel is active. The kernel manager selects the first active kernel, if any are active, and uses it to modify the state of the simulation, using the simulation parameters ϕ so that the simulation returns $f_k(s, a)$. The kernel manager itself has no knowledge of the internal simulation state and dynamics. Knowledge of the simulation is encapsulated in the design of the simulation

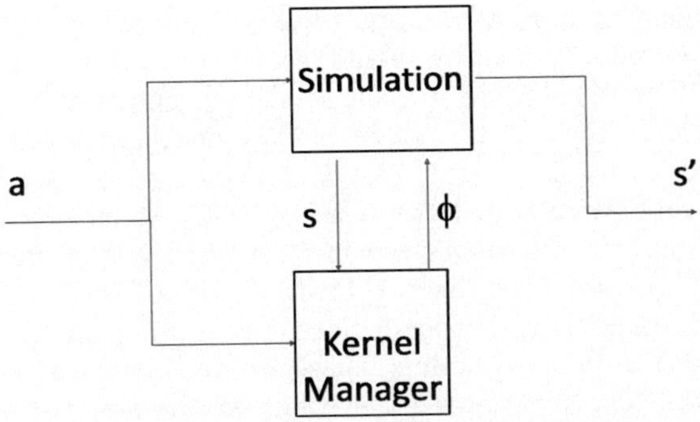

Fig. 2. Architecture for kernel modified simulation

configuration parameters ϕ so that values of ϕ will modify simulation internal state to ensure that $f_k(s, a)$ is generated.

Although a kernel is only active in a small region of the state space, the effect of any change to the simulation state is potentially long lasting. This results in both a spatial and temporal generalization effect. A modification of an object location for example, will result in sensors picking up the revised location no matter the pose of the robot—a spatial generalization. Other simulation entities that subsequently interact with the object (e.g., collide) will do so at the new location—a temporal generalization. Both of these generalizations leverage the simulation dynamics applied to new experience gained from execution of the robot program in a real environment.

4 Delivery Vehicle

We will consider an autonomous delivery vehicle as a long-term service deployment example. The vehicle will acquire goods from a storage location and deliver to two outlying sites, returning again to the start location. For this paper, we will simplify the task to just the transit portion of the problem. The vehicle repeatedly makes the trip from stores to each delivery location and then back to stores in a long-term outdoor deployment; each trip is one mission. The performance measure for the vehicle is that it reach each goal g_i within a specified position accuracy ϵ_i and time limit d_i:

$$perf(t) = \begin{cases} 1 & |p(t) - g_i| < \epsilon_i \ \& \ d(t) \leq d_i \\ 0 & |p(t) - g_i| > \epsilon_i \ \& \ d(t) \leq d_i \\ -1 & d(t) > d_i \end{cases}$$

where $p(t), d(t)$ are the position of the robot and time since last goal, respectively.

During the design phase, the robot program is tested in simulation using what is known about the problem a-priori. The deployment vehicle is a modified Traaxis platform with Ackermann steering (described in more detail later) operating in large open field area. Figure 3(a, b) show the intended deployment platform and a Gazebo simulation of the platform (using the open source Ackermann_vehicle model[3]). Figure 3(c) shows the Gazebo simulation of the open field area with the three locations (store and two delivery sites) shown by lighter colored patches.

The task is programmed in Python using ROS (Robot Operating System). The main loop sends the the robot to each of the waypoints in turn and initializes the performance monitor with the goal location, time deadline and accuracy. A saturated spring model is used to calculate speed and a bicycle pursuit model used to generate Ackermann steering. These are linearly transformed to Traaxis motor commands. The position of the robot at the start of each mission is selected from a uniform distribution and a small amount of zero mean noise is injected into the speed and steering signals. GPS data was simulated by providing Gazebo model position information at a rate 20 Hz and vehicle orientation was calculated from GPS velocity. Vehicle pitch sensing was also simulated from model information.

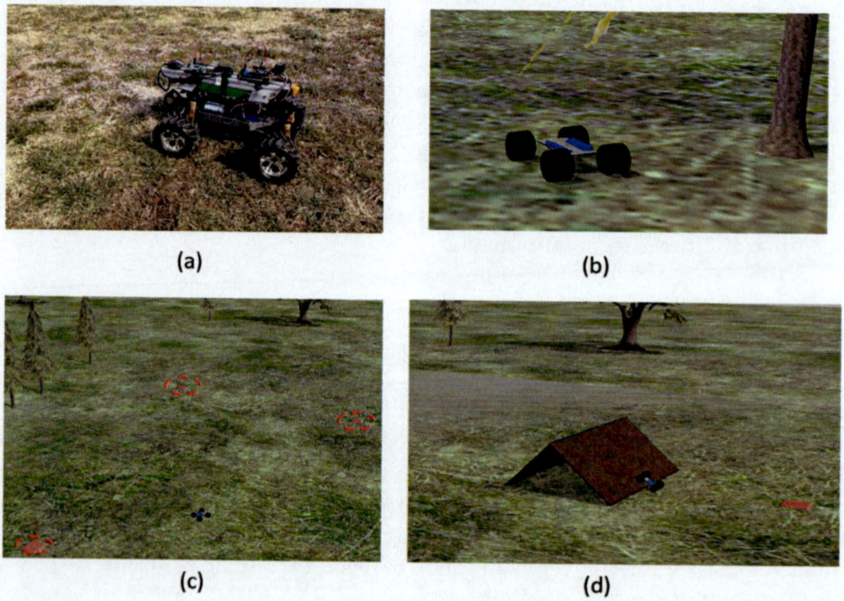

Fig. 3. Real robot (a), simulated robot (b); simulated delivery terrain with waypoints highlighted (c), unexpected terrain (d)

As discussed, the initial robot program could be handwritten or developed using a method such as reinforcement learning or any combination of these. All we insist is that the performance measure is in place and the robot program

[3] www.github.com/trainman419/ackermann_vehicle-1.

(once ready for deployment) is validated in the simulation. The performance measure is used to assign a reward and for the purposes of this paper the two phrases 'reward' and 'performance measure' can be considered synonymous.

The Average Total Reward (ATR), the sum of all rewards divided by number of missions, is used as a measure of success. Figure 4 shows the ATR graphs for the design and deployment of this example. The graph labeled "Design" is the ATR from validating the robot program in the simulation. At this point in the design process, the software designers have all the information they consider necessary to have the program fulfill their understanding of the specification. Thus, we consider the ATR graph to be an objective characterization of success (Fig. 5).

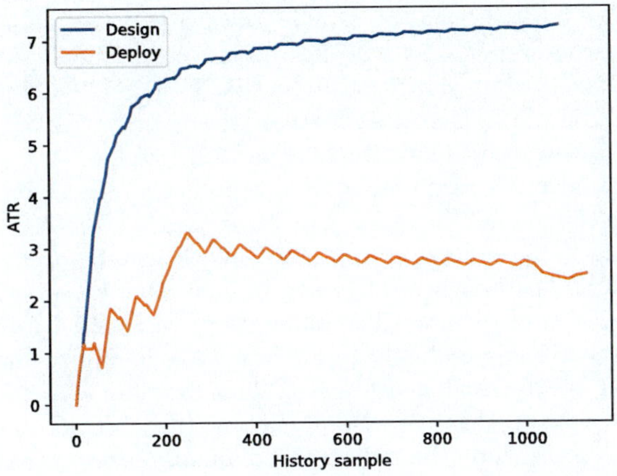

Fig. 4. ATR graphs for robot program at design and deployment for example

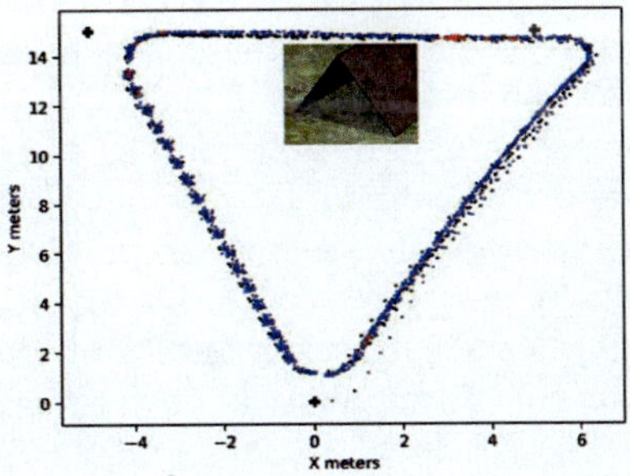

Fig. 5. Position scatter graph, H_{sim} (black), H_{phy} (blue); calculated kernel posn (red). '+' marks goal locations. 'Unexpected' ramp location shown.

During a long-term deployment, the environment may exhibit phenomena not anticipated by software designers. For this experiment, we added a ramp between two of the three waypoints (Fig. 3(d)). The vehicle can navigate the ramp, but it does slow it down and will almost consistently cause it to fail to meet one of its deadlines. The ATR graph in Fig. 4, labeled "Deploy" shows the performance of the control software in this modified simulation (which is taking the place of a real deployment for our example purposes). It is lower than the design ATR. We consider this as an empirical indication that the robot program is not meeting its design specification; however, we can't say yet what is different about the environment at deployment time that caused this to happen.

The robot program is instructed to record H (Eq. 4). The state in this example includes the robot odometry and pitch. It is important that all sensing information is recorded, not just that (currently) used in the task. The action value is the robot speed and steering topics. The reward is the value from the performance metric. H_{sim} and H_{phy} were collected for the missions graphed in Fig. 4. The kernel generation algorithm was run on this data and 38 unique kernels were constructed (omitting duplicates and candidate divergences that did not have sufficient information to estimate the transfer function).

The robot program is rerun through the simulation receiving configuration commands from the Kernel Manager (Fig. 2), generating the ATR graph labeled "Kernels" in Fig. 6. It is similar to the "Deploy" graph for an initial period but overall it is not as poorly performing as real deployment while still much worse than the "Design" curve. The simulation, modified by the deployment experience, now much more closely represents what happens when the delivery robot operates in its physical environment. This provides a tool for designers to redesign the program so that it will work better in the physical environment.

The robot program could be redesigned manually, using the kernel-modified simulation for testing. For this example, we chose to use a SARSA reinforcement learning algorithm in conjunction with the kernel-modified simulation to select modified control values that maximized the reward. The resulting robot program issued faster velocity values when non-zero pitch values were encountered. The redesigned program, the "Redesign/K" graph in Fig. 6, shows substantially improved performance with respect to the "Kernels" graph.

Finally, the redesigned robot program was run again in the "Deploy" situation—that is, in the simulation that included the unexpected ramp, and without any Kernel manager input. The graph in Fig. 6 labeled "Redeploy" shows the ATR for this case. It is higher than the original "Deploy" graph and close, but not identical to, to the original "Design" graph. The simulation, enhanced by the kernels automatically extracted from prior physical experience, is now a better reflection of the physical environment and can be used to produce a robot program that will work better in the physical environment.

5 Field Test

To evaluate the proposed approach, we prototyped a robot program for a three GPS waypoint mission using Gazebo and ROS. The robot program was then

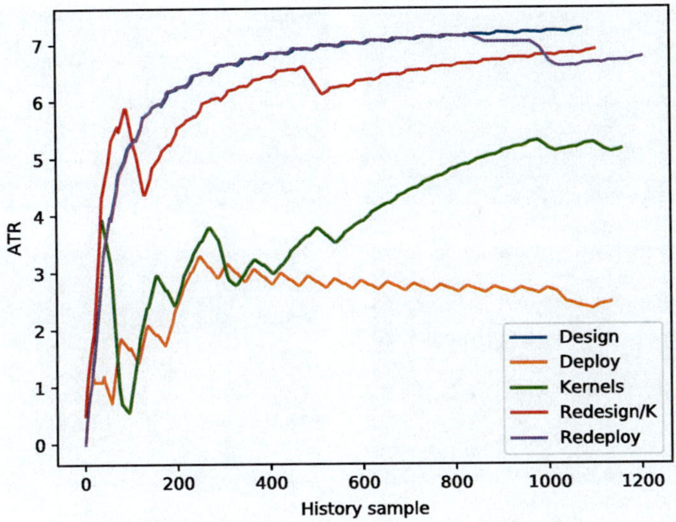

Fig. 6. ATR graphs for robot program redesign

deployed to a physical robot for traversing the GPS waypoints in an open field, with the unexpected situation that the robot sometimes encountered a steep ramp between two of the waypoints. The mission differed from that of the previous section in that the deadline and accuracy for performance related to the entire mission, from first to last waypoint, instead of per waypoint, and of course in that the deployment was not a simulation. The short deployment trial experience was used to generate state-space kernels, which were used to reduce the reality gap for the simulation. The experimental procedure and mechanism is described in more detail below, and the results are presented in the next section.

5.1 Design and Simulation

Design and simulation work was completed using ROS Indigo on a Dell Latitude 3460 laptop. The mission program was written in Python 2.7 and simulated using Gazebo with the open source Ackermann_vehicle model as before, Fig. 7(a), and with the same waypoint coordinates, Fig. 7(b). The robot program used a saturated spring model to generate velocities towards waypoints, and a bicycle pursuit model to generate the steering and speed for the vehicle, running 10 Hz. However, preliminary experimentation with the real platform showed that the GPS data could certainly not be provided 20 Hz, as it was in the previous section. Instead, GPS data was simulated by providing Gazebo model position information at a rate 1 Hz, and vehicle orientation was calculated from GPS velocity. Vehicle pitch sensing was calculated as before.

To compensate for the slow GPS, Ackermann kinematics were used to conservatively interpolate position between GPS samples. A small amount of uniformly

Fig. 7. Simulation robot (a) and course (b), and real robot (c) and course (d)

distributed noise was added to the steering and speed signals to simulate moving on bumpy grass. The performance requirement was that the vehicle complete the course within a time deadline of 39 s (empirically established for the experiment) and with an accuracy of 2m of the final waypoint. The robot program was instrumented to collect all sensor, control and reward signals at 0.5 Hz.

5.2 Description of the Field Trials

The field trials were performed at the River Courts located at the United States Military Academy (USMA) in West Point, NY. The Robotics Research Center (RRC) at USMA uses the River Courts to perform robotics testing for aerial and ground robots. To facilitate testing, they have a trailer designed to be used/moved to remote locations for field trials and provide researchers with electricity, wifi, and climate control.

A significantly modified Traaxas Stampede 4×4 VXL, a four-wheel drive remote controlled vehicle, was used as the base of our platform for conducting experiments. Attached to the constructed frame was a Raspberry Pi 3 B+ and an Arduino Mega. The Arduino Mega served as a ROS node for the Raspberry Pi and returned IMU data from an Adafruit LSM9DS1 sensor [17].

The ROS mission program developed in simulation was also used for the field trials, with the exception that GPS location and vehicle pitch were read from topics published by the Arduino node. The relationship between the PWM control signals and the mission program steering and speed signals was established empirically and very coarsely. The lack of careful calibration between simulation and physical platform was purposeful, since closing that gap is part of the objective of the research reported here.

6 Results

The data collected from the design and deployment phase is shown in Fig. 8. The scatter plot in Fig. 8(a) shows all the logged position samples of the simulated robot during the design phase of the work. The track differs from that in Fig. 5 because of the slower GPS rate: While the straight line segments are similar, the turns at each waypoint show some overshoot.The scatter graph of robot locations of the physical robot during deployment is shown in Fig. 8(b), and it differs from both of the prior simulation graphs.

The Average Total Reward (ATR) graph shows that the simulation performs well according to its performance monitor. The deployed robot program fails to meet the performance requirement however, as evidenced by its decreasing ATR graph. Of course, we engineered the field trial to force this effect by introducing unexpected terrain, a physical ramp between waypoints (5,15) and (−5,15), Fig. 3(d).

As described in Algorithm 1, the data collected from design and deployment (Eq. 4) was used to generate the transition functions P_{sim} and P_{phy} (Eq. 6). P_{sim} had 951 state transitions, and P_{phy} had 942 transitions. The vast majority of both had a branching factor of 1. There were 882 states in common between the two, where states are compared as described in Eq. (8) and its accompanying text. One hundred paired roll-outs were conducted and divergent states identified. From this, 15 unique divergences were identified and kernels developed for them. Each of the kernels included a linear transfer function mapping velocity and sensor state to a new sensor state—position, orientation and pitch. Any kernel whose transfer function mapped each state to itself was omitted—for now. These will be discussed in a subsequent section.

6.1 Kernels Non-identity Transfer Functions

Fig. 9(a) shows the kernel mean location overlayed on the deployment tracks. Recall that this location is the one immediately preceding the divergence—and on the track overlay, many kernels are clustered around and just after the waypoint preceding the ramp.

The robot program was then rerun in the simulation but with the Kernel Manager active. Figure 9(b) shows a number of ATR graphs including the original design and deployment graphs. The graph labeled "Kernels" shows the result of executing the original robot program but in the kernel modified simulation. The performance is much worse than the original performance, but not as severe as the actual deployment.

There are many ways that a simulation could be coerced into failing the performance measure. But unless the mechanism failure allows a designer or learning algorithm to redesign and test in simulation and have this generalizes to improved real performance, the method would have limited use. In Sect. 4, the modified simulation was used in conjunction with a SARSA reinforcement learning algorithm to generate an improved robot program that generated faster velocities when non zero pitch was detected. That same same solution was applied

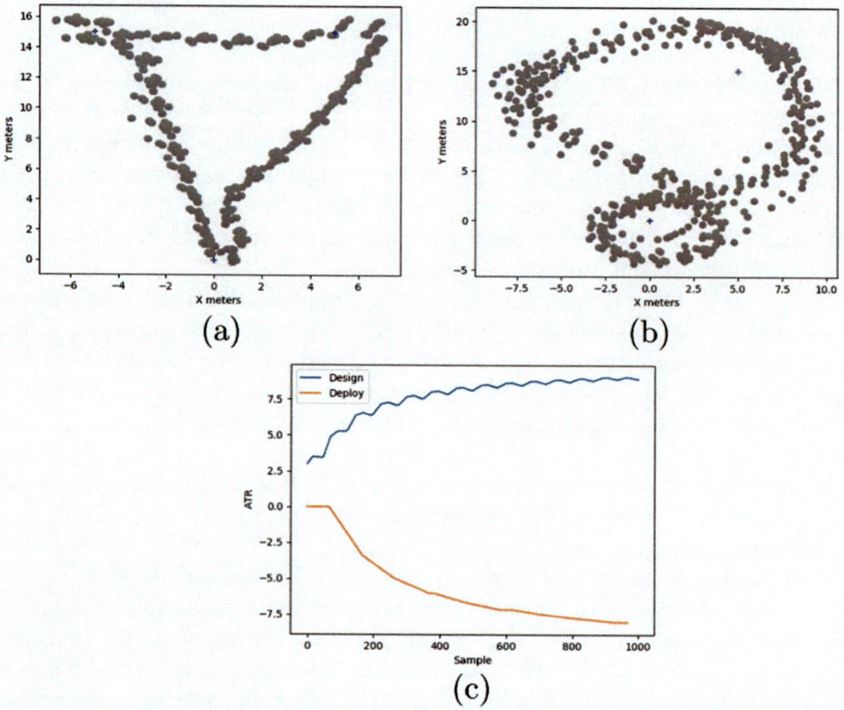

Fig. 8. Data from design and deployment trials (a) scatter plot of 2D position in design, (b) in deployment, and (c) combined ATR graph. Waypoint locations are marked as '+'.

here, but the velocity was manually adjusted so that the performance measure was achieved in the modified simulation. Our approach is agnostic as to the specific method of robot program improvement (machine learning in Sect. 4 and manual modification here).

This improved robot program was tested on the kernel modified simulation, producing the ATR graph labelled "Redes/K" in Fig. 9(b). The modified robot program shows improved performance over the original program, approaching within ~25% of the original performance. The modified program was redeployed to the robot, and that performance is shown in the ATR graph labeled "Redeploy" in Fig. 9(b). The improvement in behavior transfers from simulation to reality, supporting the argument that robot programs (implementations of control policies) developed with the kernel mechanism transfer well to reality.

6.2 Kernels with Identity Transfer Functions

Our inital analysis rejected any kernels with identity transfer functions. This allowed us to identify the behavior divergences in roll-outs related to the introduction of the ramp. However, these divergences don't address the reason that

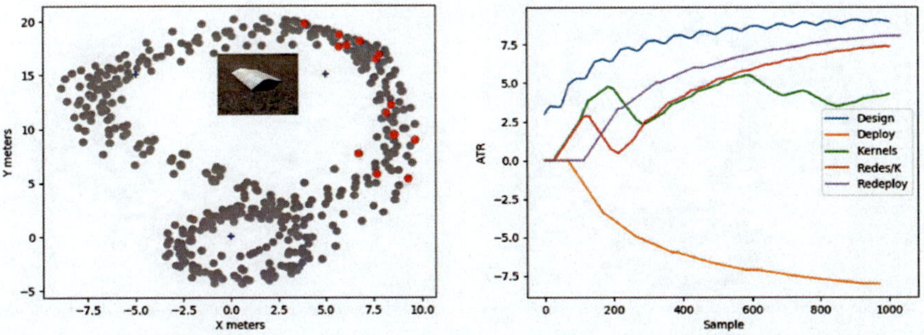

(a) (b)

Fig. 9. Deployment track data (a) overlay showing calculated kernel locations in red, where ramp is shown at its approx. deployed location, (b) ATR graph for kernels, redesign and redeployment.

Fig. 8(a) is quite different in appearance from Fig. 8(b). To capture this, we need to discuss kernels that have an identity transfer function—these represent divergences in which the position sensor did not report any update of the position in the deployment roll-out.

This phenomenon is observed because the deployment GPS rate is lower than was expected at design time. However, the predictive model used at design time is unable to compensate, so that the position invariably lags, resulting in extensive overshoots. Unfortunately, our existing approach will not work here. If a kernel predicts no motion of the platform, then in addition to the sensors being caused to report no motion, the simulation is also modified to force no motion. The only way to distinguish a kernel from a sensor divergence, which we are discussing now, from that from a motor divergence, which is what we have addressed up until now, is to utilize a sensor timestamp, extending state as

$$S = Stv_0, \times Stv_1 \times \ldots Stv_n \qquad Stv_i = Sv_i \times T \tag{9}$$

For a set of times $T = \{t_0, t_1, \ldots\}$. Each roll-out will thus include timestamped sensor data. If f_{k,sn_i} is the transfer function for active kernel k sensor sn_i, and the kernel manager applies this to the current state $s = (stv_0, stv_1, \ldots, stv_n)$, then the modified state $s' = f_k(s, a)$ has $stv'_i = f_{k,sn_i}(stv_i, a)$. Finally, since $stv_i = (sv_i, t_i)$, if f_{k,sn_i} is an identity transfer function, then $t_i = t'_i$ and this is unambiguously from a sensor divergence. Any other transfer function is from a motor divergence. For convenience, these will be referred to as sensor kernels and motor kernels respectively.

The kernel manager treats these cases slightly differently: an active sensor kernel modifies the sensor data only, whereas an active motor kernel modifies sensor data and simulation state. When sensory divergences are *now not removed* from the one hundred paired roll-outs, 54 sensory kernels are identified in addition to the motor kernels already identified. Where the 15 motor kernels relate principally to the unexpected ramp, the 54 sensory kernels relate principally to

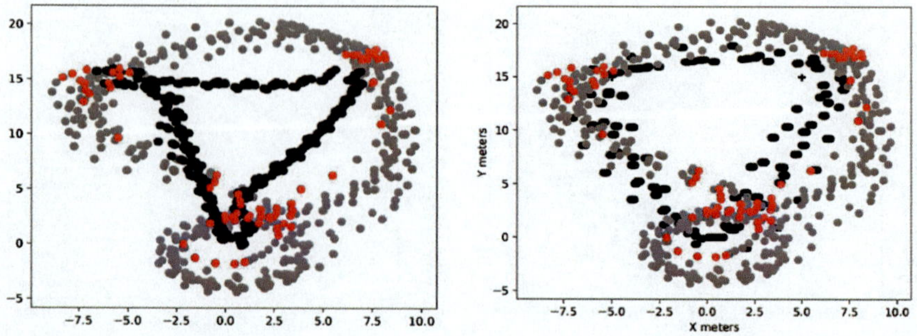

(a) (b)

Fig. 10. Sensory Kernels: (a) overlay showing design track (black), deploy track (gray) and calculated kernel locations (red), (b) overlay of kernel modified design track (black) and original deploy track (gray).

the slower than expected GPS. Figure 10(a) shows the location of the sensory kernels. They are focused around the turns at the waypoints. Figure 10(b) shows an overlay of the original deployment track (gray) with the new, kernel modified simulation tracks. The mission program was not changed to produce this track, which is now more similar to the original deployment; the change results from the effect of the sensory kernels in Fig. 10(a) extracted using our approach.

7 Conclusions

This paper has proposed a novel Monte-Carlo approach to the reality gap problem. The approach collects simulation and real world observations and builds conditional probability functions for them. These are used to generate paired roll-outs and to look for points of divergence in behavior. The divergences are employed to generate state-space kernels coercing the simulation into behaving more like observed reality within a region of the state space. Our results support not just that the kernel approach can force the simulation to behave more like reality, but that the modification is such that a robot program with an improved control policy tested in the modified simulation also performs better in the real world.

The kernel managed simulation in our field trials did not produce as poor an ATR graph as the actual deployment Fig. 9. This is principally due to restricting the scope of each state-space kernel to a small region of state space with $e_k \circ c_k(s') < \epsilon_c$. The advantage is that it allows us to leverage a fast function approximation method. The disadvantage is that it could require making a lot of kernels to capture the real environment with high fidelity. We argue however that there is an advantage in understating the effect of experience gained in each iterative deployment since it produces a sequence of smaller learning obstacles rather than a single large obstacle. This is a useful learning strategy [18].

The work most similar to ours is that of [15]. A crucial point of difference is that [15] (and some others) address the reality gap problem with a principled domain randomization approach, providing a range of environment for policy development, and generating a policy that is robust along the right dimensions of variability. We address the same problem but from the perspective of making each simulation run more closely resemble reality. This is reflected in how the simulation is "wrapped" by each approach: our approach requires a more invasive configuration—access to the Gazebo model information—but we argue that our configuration is less application specific and more easily generalized. We are especially interested in generalization to higher-dimensional state-spaces.

The quality of the improvement that our method makes is proportional to the divergence information that it has to work with. An avenue of future study is development of control strategies that improve divergence information by exploring more of the state space during deployment.

References

1. Reckhous, M., Hochgeschwender, N., Paulus, J., Shakhimardanov, S., Kraetzschmar, G.: An overview about simulation and emulation in robotics. In: International Conference on Simulation, Modeling & Programming for Autonomous Robots, pp. 365–374 (2010)
2. Peng, X.B., Andrychowicz, M., Zaremba, W., Abbeel, P.: Sim-to-real transfer of robotic control with dynamics randomization. In: IEEE International Conference on Robotics and Automation (2018)
3. Moeckel, R., Perov, Y., Nguyen, A., Vespignani, M., Bonardi, S., Pouya, S., Sproewitz, A., van den Kieboom, J., Wilhelm, F., Ijspeert, A.: Gait optimization for roombots modular robots–matching simulation and reality. In: IEEE/RSJ International Conference on Intelligent Robots and System (2013)
4. Hofer, S.E.A.: Perspectives on Sim2Real Transfer for Robotics: A Summary of the RSS 2020 Workshop (2020). arXiv:2012.03806 [cs.RO]
5. Takayama, Y., Ratsamee, P., Mashita, T.: Reduced simulation: real-to-sim approach toward collision detection in narrowly confined environments. Robotics 10(131) (2021)
6. Tobin, J.M., Zaremba, W., Abbeel, P.: Domain randomization and generative models for robotic grasping. In: IEEE/RSJ International Conference on Intelligent Robots and System (2017)
7. Dehban, A., Borrego, J., Figueiredo, R., Moreno, P., Bernardino, A., Santos-Victor, J.: The impact of domain randomization on object detection: a case study on parametric shapes and synthetic textures. In: IEEE/RSJ International Conference on Intelligent Robots and System, pp. 2593–2600 (2019)
8. Ljung, L.: Perspectives on system identification. In: 17th IFAC World Congress, vol. 41, no. 2 (2008)
9. Verdult, V., Ljung, L., Verhaegen, M.: Identification of composite local linear state-space models using a projected gradient search. Int. J. Cont. **75**, 16–17 (2001)
10. Benjamin, P., Lyons, D., Funk, C.: A cognitive approach to vision for a mobile robot. In: SPIE Multisensor. Multisource Inf, Fusion (2013)
11. Lyons, D., Nirmal, P.: Navigation of uncertain terrain by fusion of information from real and synthetic imagery. In: SPIE Multisensor. Multisource Inf, Fusion (2012)

12. Christiano, P., Shah, Z., Mordatch, I., Schneider, J., Blackwell, T., Tobin, J., Abbeel, P., Zaremba, W.: Transfer from Simulation to Real World Through Learning Deep Inverse Dynamics Model. arXiv 1610.03518 (2016)
13. Yu, W., Kumar, V.C., Turk, G., Liu, C.K.: Sim-to-real transfer for biped locomotion. In: IEEE/RSJ International Conference on Intelligent Robots and System (2019)
14. Saputra, R.P., Rakicevic, N., Kormushev, P.: Sim-to-real learning for casualty detection from ground projected point cloud data. In: IEEE/RSJ International Conference on Robotics and System (2019)
15. Chebotar, Y., Handa, A., Makoviychuk, V., Macklin, M., Issac, J., Ratliff, N., Fox, D.: Closing the sim-to-real loop: adapting simulation randomization with real world experience. In: IEEE International Conference on Robotics and Automation (2019)
16. Bartocci, E., Falcone, Y.: Lecture Notes in Computer Science. In: Lectures on Runtime Verification. LNCS, vol. 10457. Springer, Cham. https://doi.org/10.1007/978-3-319-75632-5
17. Adafruit lsm9ds1 imu (2020). www.adafruit.com/product/3387
18. Wang, R., Lehman, J., Clune, J., Stanley, K.O.: Paired Open-Ended Trailblazer (Poet): Endlessly Generating Increasingly Complex and Diverse Learning Environments and Their Solutions (2019). ArXiv, abs/1901.01753

Real2Sim or Sim2Real: Robotics Visual Insertion Using Deep Reinforcement Learning and Real2Sim Policy Adaptation

Yiwen Chen, Xue Li, Sheng Guo(✉), Xian Yao Ng, and Marcelo H. Ang

Advanced Robotics Center, National University of Singapore, Singapore, Singapore
{yiwen.chen,E0576004}@u.nus.edu

Abstract. Reinforcement learning has shown a wide usage in robotics tasks, such as insertion and grasping. However, without a practical sim2real strategy, the policy trained in simulation could fail on the real task. There are also wide researches in the sim2real strategies, but most of those methods rely on heavy image rendering, domain randomization training, or tuning. In this work, we solve the insertion task using a pure visual reinforcement learning solution with minimum infrastructure requirement. We also propose a novel sim2real strategy, Real2Sim, which provides a novel and easier solution in policy adaptation. We discuss the advantage of Real2Sim compared with Sim2Real.

Keywords: Deep reinforcement learning · GAN · Transfer learning · Sim2Real

1 Introduction

Insertion is an essential step in many assembly processes in manufacturing lines. Insertion task automatically executed by industrial robots is a frontier topic that has been studied for a long time. In this work, we are motivated to achieve comparably good performance but minimize the infrastructure cost (from more than \$500 to \$10), prerequisites, human labor intensity, and maximize the noise resistance.

1.1 Minimum Infrastructure and Prerequisites

Traditional control methods have high requirements for the precision of sensors and robots. The deployment of such methods usually depends on the complex and tuned setup. Some strict preconditions are assumed, such as the position of the hole being fixed and known [1], target pose of the gripper is assumed given [8], target image [15] with the peg already in the hole, or the background color is simple [7] and flatten [16], or light quality is fixed [15]. As for additional sensors,

X. Li and S. Guo are equally contributed.

I. Petrovic et al. (Eds.): IAS 2022, LNNS 577, pp. 617–629, 2023.
https://doi.org/10.1007/978-3-031-22216-0_41

some methods use high-precision depth camera [16], torque-force sensors [13] Those complex infrastructure and prerequisites severely limit practical applications. Noise and disturbance in our daily environments may easily defeat the fragile pre-assumptions. Therefore, there is an urgent need to design a method with minimum infrastructure and prerequisites. In our work, only one RGB low-cost camera is used to realize robotics insertion tasks instead, quite simple and adaptive. Due to the expensive human-labor cost, we hope there is a solution that involves as little as possible human labor intensity. Generally, data labeling such as trajectory generation supervised training [12] and detection training [7], or human demonstrations [7] requires human intervention. Therefore, in this work, we also aim to minimize human intervention with unsupervised method.

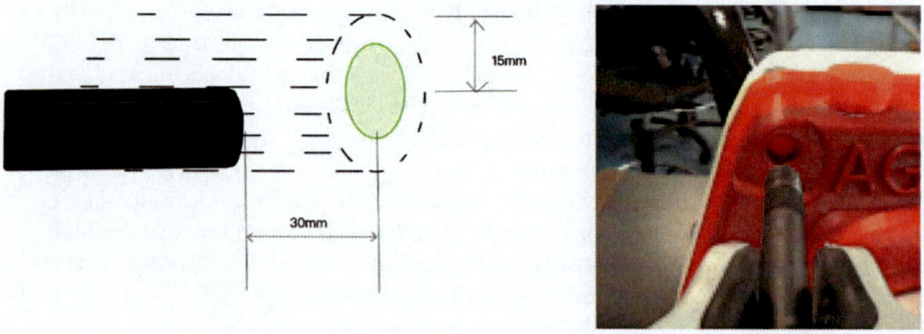

(a) Insertion Start

(b) Insert with RGB

Fig. 1. Task Description

1.2 Real2Sim or Sim2Real

There has been increasing research work on robotic insertion focusing on learning-based methods [2,15]. However, simulation learning-based methods may result in a serious decrease in the success rates when deployed to real situations without an effective sim2real strategy [2]. Therefore sim2real adaptation efforts have been made to narrow the gap between the simulation and the reality [14,19].

However, as for the sim2real strategy, the challenge here is that it is hard to exactly rendering the real-world industrial scene in the simulator. Previous sim2real work tries to enrich the high-quality simulation to match the real world in the training phase, which requires a huge effort in data preparation and costs significantly more. Domain randomization has also been discussed, but it may still fail when there is a huge gap between simulation and the real environment. With a generative model, it is also hard to control the quality of generated image [14].

Our work is unique by applying an inverse idea, real2sim in the test phase. The real scene will be transformed into an image akin to the simulator before it is used as the input for insertion, as shown in Fig. 2.

1.3 Contributions

Summarily, our main contributions are as follows:

- With minimum infrastructure and prerequisites, we solve the robotics insertion problem with raw images from a monocular RGB camera. Our approach significantly reduced the infrastructure cost from more than $500 to only $10, and minimize the human intervention from heavy labeling to unsupervised training to achieve a comparable result with the benchmarks.
- A unique real2sim approach is proposed to solve the last mile of the algorithm implementation and cancels the background and environmental noise. It provides a brand novel view in policy adaptation.

2 Task Description

The objective of this work is to develop a novel and practical approach applicable to the insertion task with uncertainty of hole pose, with only given a real-time camera video stream (RGB Images). The algorithm should be trained in virtual environment, the simulation, and tested in reality.

A manipulator mounted with a real-time camera on the hand, capturing images in Fig. 1b, is used for this insertion task. Camera is the only observation sensor. The relative position and orientation between the camera and gripper is fixed and known. The diameter of the hole is 11mm, with 1mm clearance to the bolt. The bolt is assumed to be firmly held by two fingers of the robot all the time. The initial position of the bolt in the gripper falls randomly in the range of a cylinder concentric with the hole, with a diameter of 20mm and a height of 3cm. During the whole process, the insertion speed in z direction (hole axis) keeps constant while the speed of x and y directions are provided by a policy network (deep neuron network). If the insertion depth reaches 5mm, the task is considered as successfully completed.

3 Related Works

3.1 Robotics Insertion

Traditional methods of insertion tasks considerably rely on human ingenuity. The standard control method, such as the PID controller, heavily depends on engineers' manual modeling and tuning, which lacks both accuracy and adaptability while huge efforts and time are consumed [15]. Active compliance control, though greatly contributes to the accuracy of the insertion task, remains unpractical to apply in most industrial scenes [2]. By contrast, deep reinforcement learning enables robotics to learn insertion skills, solving insertion tasks efficiently and robustly [5]. And [8] proposed DDPGfd (Deep Deterministic Policy Gradient from Demonstration). However, strong prerequisites given in [8] limit the applicability in noisy reality. Different from specializing on one or a

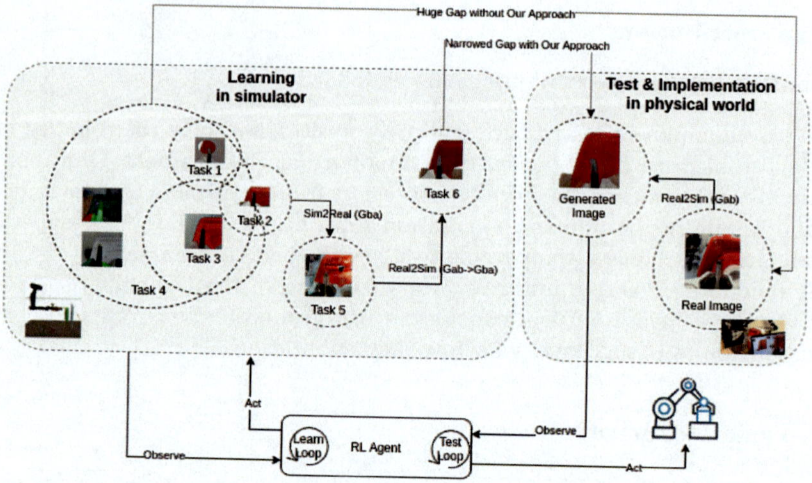

Fig. 2. Insertion task domain adaptation and mixing task learning

few specific insertion tasks, [16] propose an off-policy meta reinforcement learning method named probabilistic embeddings for actor-critic RL (PEARL), which enable robotics to learn from the latent context variables encoding salient information from different kinds of insertion, resulting in a rapid adaptation when doing a new task. In spite of the success of multi-stage insertion tasks, significant challenges are remained for further applying to a more complex environment.

3.2 Reinforcement Learning and Sim2real Adaptation

Model-based Reinforcement Learning and Model-free RL have drawn wide attention. Trust-region methods in reinforcement learning have been broadly used in solving robotics control and complex strategy learning [17,18]. Domain randomization [19] shows that it is possible to transfer a simulated algorithm to a real-world scenario, which is also extensively discussed in the transfer learning research field. As for the combination of Sim2Real and deep RL, [14] considered improving Cycle GAN with Q loss. Another notable sim2real domain adaptation research given in recent years is the cubic solving with RGB image and robotics hands [4]. It leverages the advantages of deep reinforcement learning and domain adaptation methods called automatic domain randomization (ADR). GAN methods and domain randomization are also discussed in our approaches.

4 Our Approach

4.1 Overview

The main idea of our approach is to successfully learn the insertion action, called policy, in the simulation, and then to transfer this strategy, by building the sim2real and real2sim transferring pipeline with generative model and domain randomization tricks, to implement it into a real robotics insertion scenario. As for the learning part, we adopt mixing learning to increase its stability and robustness for various tasks and improve training efficiency. The algorithm is described in the algorithm 1. In the inference part, the real physical robot would take it is as working under simulation, which is generated by the GAN model, even though it is working in a real physical world.

Algorithm 1: Task Mixing Learning and Real2Sim

Initialize:;
Trained GAN model $G_{\phi ab}$ and $G_{\phi ba}$;
Defined Tasks $T_1, T_2, T_3, T_4, T_5(G_{\phi ba}), T_6(G_{\phi ab}, G_{\phi ba})$;
Vectored Mixing Task $VT = [T_1, T_2, T_i ...]$;
Policy Model $a \sim \pi_\theta(a|o)$;
Simulating observation o_B;
Real observation o_A;
while *Training Loop* **do**
    Sample action $a \sim \pi_\theta(a|o_B)$;
    $o, r = VT(a)$;
    if *reach training steps* **then**
        θ_{new} = Optimize PPO objective wrt θ;
        $\theta = \theta_{new}$;
    end
end
while *Inference Loop* **do**
    get o_A from RGB camera;
    Generate image $o_B = G_{\phi ab}(o_A)$;
    Sample action $a \sim \pi_\theta(a|o_B)$;
    Input a to robot action server;
end

4.2 Reinforcement Learning for Insertion Action Learning

Proximal Policy Optimization (PPO) In this approach, we adapt the advantage of PPO algorithm [18]. The main idea of PPO is to restrict the policy update stepsize, by using a clipping factor ϵ. TRPO approach [17] gives objective function as following

$$r_t = \left[\frac{\pi_\theta\left(a_t|s_t\right)}{\pi_{\theta old}\left(a_t|s_t\right)}\right]$$

$$L_{(\theta)} = \hat{E}_t\left[r_t(\theta)\hat{A}_t\right]$$

And PPO modified this objective function with clipping factor ϵ, making is much more easier to implement[18]. The new objective function is defined as following.

$$L_{(\theta)} = \hat{E}_t \left[min(r_t(\theta)\hat{A}_t, clip(r_t(\theta), 1-\epsilon, 1+\epsilon)\hat{A}_t) \right]$$

The policy output of the reinforcement learning has the dimension of 2, which ranges from -1 to 1. It gives the x, y displacement in every time step. Observation dimension is $(128, 128, 3)$ what to describe the RGB data subscribed from the hand camera.

4.3 Real2Sim and Sim2Real Domain Adaptation

A generative adversarial network (GAN) [6,10] has drawn wide attention in recent years, and there has shown wide applications of the GAN model in the industry, like image domain transfer or sim2real learning [14]. In Fig. 3, the original image from the real domain can be transferred into the simulation domain with the model G_{ab}, and with G_{ba} it will be transferred back to the real domain. Training objective cycle GAN is to reduce the discrimination loss by updating the Discriminator while increase the discrimination loss with updating the Generator.

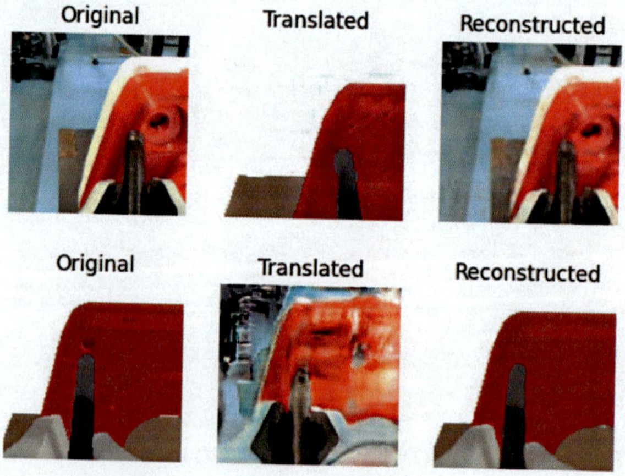

Fig. 3. Sim and real transfer

And in our solution, we do not only implement the sim2real in our training for better robustness, but also adapt the advantage of real2sim as well. Our robot aims to have the capability to transfer the real scenario to the simulation scenario, which narrows the gap between the realistic and simulation (Fig. 2).

4.4 Task Mixing, Randomization, and Vectorization

In the simulation, multi-tasks have been designed for better and more robust learning performance. Paralleled training with multi-process is adopted in this approach, which significantly speeds up the training efficiency, from 15 fps to 200 fps (depending on the number of CPU cores). Details of each task can be found in Sect. 5.2. Task randomization and mixing adopts the advantage of meta-learning [9,16], which helps the algorithm to learn more important and transferable features.

Domain Randomization Domain randomization [19] has been widely used in the real2sim transfer, seeing related work in Sect. 3 .

Vectorization and Mixing- Efficient Paralleled Learning Vectored multiple environments and various tasks aim to increase the robustness of our algorithm. Multiple environments are running in every step of action prediction. Actions are vectored and passed through all n environments, and it will return n observations [11]. Multiple CPU will be used for environments for image rendering, which significantly increase the training efficiency.

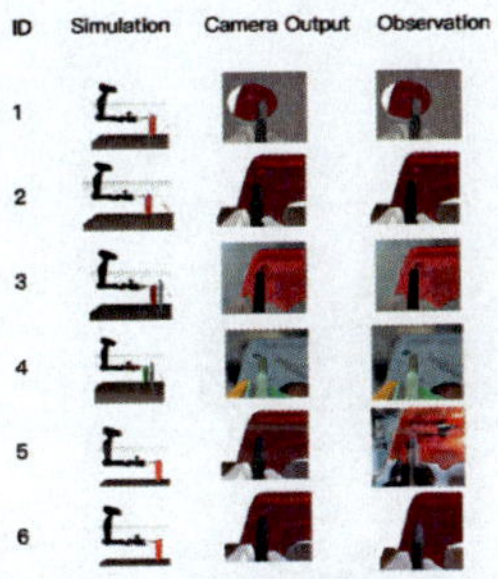

Fig. 4. Simulation task designs

5 Experiment

5.1 Experiment Overview

Our experiments are mainly given out in following areas to show the result: test in the simulation, the performance improvement with vectored environments, mixing learning method, and the performance under real scenario.

5.2 Task Design

We have designed 6 tasks for training, as discribed in following list and Fig. 4. Task 1: It narrows the path search field using a searching allowed area. The initial position of the gripper is fixed. Generally, task 1 is a simplified task for the insertion. The disadvantage of this task is that CNN may not observe enough information, and it doesn't have a good capability to defeat noise, such as light and background variation. Task 2: In task 2,3,4 the front protect shell is invisible but gives a collision. The camera is able to capture the image of the whole scenario. In task 2, the insert behavior starts with a random position. Task 3: In this task, the gear box will change its color, following the idea of domain randomization [19] Task 4: It is a more difficult task than Task 3. In this task, the case color, bolt color, light, background color, texture will change randomly in a larger range. Task 5: Generative Model is used in this task. All the image input will be transferred to the domain of Realistic Images. The RL agent will access the "fake real" image in this task to learn actions. An important feature such as the hole might disappear in the generated model. This will be discussed in the Sect. 5.7. Task 6: Generative model is used to transfer the image from b to a, and successively from a to b. This task aims to help the agent to learn the a domain features given by the generative model, which is slightly different from the real simulation environment image (such as in Fig. 3). Task 7: It generates the noise by GAN network, aiming to help RL learning under a uncertain environment (Fig. 5).

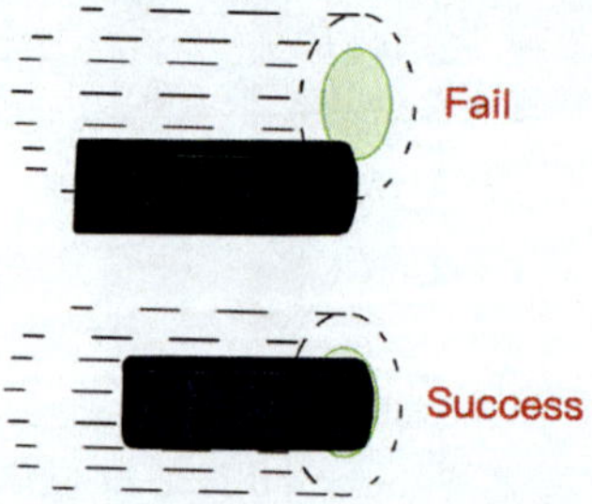

Fig. 5. Terminal condition

Terminal Condition Terminal signal will be triggered. If the bolt touches the insertion detector placed at the 0.5 cm depth of the hole, it will send the success signal to the simulation that the insertion is done.

5.3 Convergence of Reinforcement Learning on Tasks

We test PPO algorithm under our different tasks, and the results are shown in the Fig. 6. It shows a good stability to solve all of these 6 task. And the task 7 is

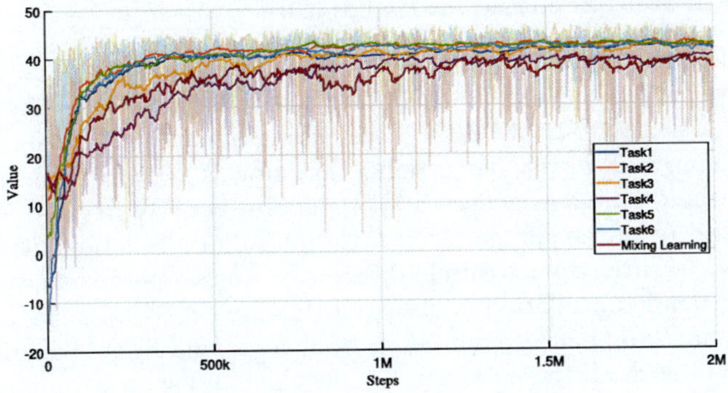

Fig. 6. Success on different Tasks

only for noise adding, and it will not be tested separatelyy. Mixing learning task introduced in Sect. 5.4 is also tested. As the curve shows, task 4 shows a higher difficulty level, which costs more time to converge to the optimal policy.

5.4 Mixing Learning

For testing the mixing learning for the performance improvement, we first give metrics to evaluate the robustness, noted as σ. These metrics are relative to the robustness to solve all the tasks. A higher robustness index shows a better probability to solve wider tasks. It also illustrates the average return per episode per task. Seeing in the Fig. 7, with mixing learning methods, the RL agent shows the highest robustness dealing with various tasks. Therefore, we adopt this model in real scenario test.

$$\sigma = \frac{\sum_{T=0}^{T_{max}} \sum_{e=0}^{e_{max}} \sum_{t=0}^{t_{max}} r_i}{T_{max} e_{max}}$$

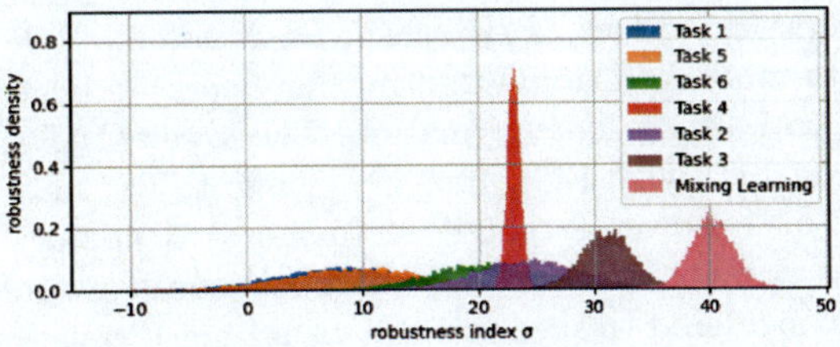

Fig. 7. Robustness test with different learning methods

5.5 Test in Real Scenario and Real2Sim

The work pipeline is introduced above in Fig. 2. With this pipeline, we tested 13 times of insertion under a real scenario, and it shows a successful insertion with 12 times, with a success rate of 92.3%. Fine-tuning of camera position is necessary for reproducing this result, since policy learning is a camera position-specific policy (see the discussion in Sect. 5.7). Real2Sim is necessary for this pipeline. Policy learned from the simulation can totally lose its functionality in the real, since the real feature appears totally differently. A good performance of real-time noisy background cancellation is observed. In this experiment, we intentionally give camera noise by background noise, front noise and light changing noise. As shown in the Fig. 8, all the noise can be cancelled and show a good performance of Real2Sim transferring.

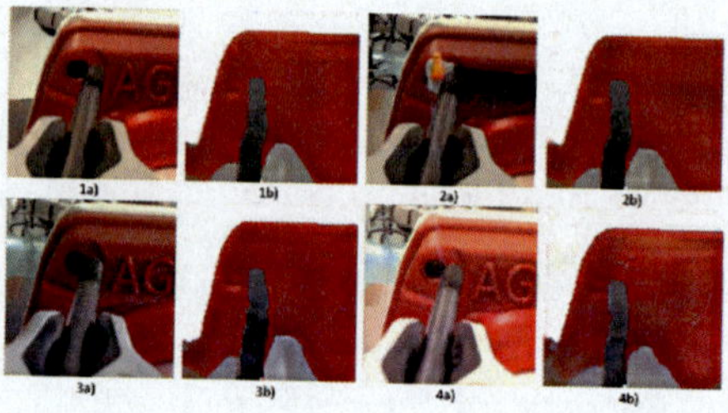

Fig. 8. Real2Sim adaptation and noise cancellation

5.6 Benchmark Comparison

The comparison of our method with others is shown in Table 1. HLI (Human Labor Intensity) includes data labeling (scores 2 points) and human demonstration (scores 1 point). EC (Estimated Cost) gives this approach's cost (RGB Camera \$10, RGB-D Camera \$350, Torque/Force Sensors \$500). The lower MIS (Minimum Infrastructure Score) score requires less cost and setup complexity (with RGB camera = 1 point, T/F sensor = 2 points, RGB-D camera = 2 points). MAS (Minimum Additional-information Score) indicates to what degree the methods rely on hard to get information. The lower MAS score requires less hard to get information (with RGB image = 1 point, Torque/Force = 2 points, Target-End Effector Pose or Target-Pose Image = 3 points). The final score of MIS or MAS is divided by assuming all subelements are included. In VNR, three factors are considered to measure the visual noise resistance of a method. The resistance to various background changes, environmental light changes, and camera color bias all score 1 point. Benchmark scores are given with our best knowledge of their methods.

According to the table result, our method's advantages lie in these aspects: First, our method performs well without any human intervention, while it is required of other methods, such as data labeling or human demonstrations, which are done manually and time-consuming. Second, We have successfully applied the sim2real in simulation learning-based method. In contrast, some of the others directly train robots in reality that security risks remain when training on real robots and the sampling efficiency is poor. Third, our method realizes insertion tasks with minimum infrastructure and additional information, only one RGB camera. In contrast, others are much more complex and expensive and bring huge financial burdens and hinder the industrial application. Forth, the VNR of our method scores full marks, which is much better than most other methods in resisting visual noise. In summary, our method achieves a relatively high success rate of insertion task while maintaining competitive position variability in a visually noisy environment and avoid the safety problems of real-machine training through the sim2real-based method with reduction of system complexity and cost.

Table 1. Benchmark Comparison. HLI represents human labor intensity. S2R represents whether the sim2real-based method is applied. EC represents an estimated cost. MIS and MAS represent minimal infrastructure scores and minimal additional information scores respectively. Acc means accuracy, PV denotes position variability and Prc. represents the precision of a method. VNR means visual noise resistance. The arrow up indicates that the score of the value of the indicator is positively correlated with the performance, representing the higher the better, vice versa.

Methods	HLI ↓	S2R	EC ↓	MIS ↓	MAS ↓	Acc ↑	PV(mm) ↑	Prc.(mm) ↓	VNR ↑	Sensors or Additional Info.
G. S. et al. [15], 2019	2	No	510$	0.6	0.67	64%~84%	2	1	0/3	Goal Pose image, RGB Cam., T/F
B. C. et al. [2], 2020	0	**Yes**	500$	0.4	0.56	60%~100%	5	1	–	T/F, Goal Pose
M. V. et al. [7], 2019	3	No	510$	0.6	0.33	77%~97%	**40**	0.5	2/3	RGB Cam, T/F
J. X. et al. [5], 2019	0	No	500$	0.4	0.22	**100%**	0	0.02	–	T/F
G. S. et al. [16], 2020	2	**Yes**	850$	0.8	0.67	**100%**	6	0.1	–	RGB-D Cam., T/F, Goal Pose
T. I. et al. [13], 2017	0	No	500$	0.4	0.22	**100%**	3	**0.01**	-	T/F
X. C. et al. [3], 2020	2	**Yes**	–	–	–	82%	–	–	2/3	Pick and Pouring tasks
A. H. et al. [12], 2020	2	**Yes**	**10$**	**0.2**	**0.11**	73%~100%	–	30	**3/3**	**RGB Cam.**
Ours	**0**	**Yes**	**10$**	**0.2**	**0.11**	**92.3%**	**30**	1	**3/3**	**RGB Cam.**

5.7 Discussion and Future Work

We discuss following issues and future works, Camera Position Fixed Policy , Important Feature Vanishing , Force Detection and Fusion, Dedicated Observation Network Design, Higher Precise and Stereo Camera, Robustness. 1. The current observation network is based on vanilla CNN, it does not have the capability to deal with time-based problems. It is also not deep enough and the architecture is not well-tuned. It is also important to note that even though we adapt domain randomization in our tasks design, the camera position is set almost fixed (variance is quite small). A strong precondition is made that the relative position of the camera and gripper is assumed given. This precondition may cause strict effort given upon the tuning position and orientation of the camera to match the setup in the simulation. The policy will show decrease performance with the changing of the camera position or orientation. And this problem will also be addressed in our future work. 2. During the Cycle GAN training and inference, we found the important features can be lost during the domain transferring, which was also indicated by [14]. Even though our approach shows it is capable to finish the insertion without those features, we believe in future work it is better to improve the visual network and GAN learning techniques. 3. Force detection and learning are not integrated into our algorithm, but which is important during the insertion and protects the hole quality. In future work, we are aiming to design a force fusing network to fuse the force information into reinforcement learning. 4. Currently, the version of the observation network inherited from the advantage of nature CNN, a 3 layers vanilla network. The performance can be potentially improved if a deeper network is implemented. 5. Currently insertion precision in the real scenario is about 1mm clearance. We are also considering using a stereo camera to provide richer image information, aiming to mimic human insertion behavior. An intelligent agent with a stereo camera is able to learn the representation of distance, and that could potentially improve the insertion precision and robustness. 6. The robustness of this approach can be potentially further improved in future work.

6 Conclusion

We have proposed a practical approach to solve the real insertion task under a noisy environment, with only RGB camera information. We provide the novel transfer learning method Real2Sim, for fast and easier policy adaptation. It shows a decent performance with 1mm clearance with fine-tuning of camera position. In future work, we will work on using stereo cameras and give a higher camera invariable robustness and precision.

Acknowledgement. This research is supported by the Agency for Science, Technology and Research (A*STAR), Singapore, under its AME Programmatic Funding Scheme (Project #A18A2 b0046). The computational work for this article was partially performed on resources of the National Supercomputing Centre, Singapore (https://www.nscc.sg). Thanks for the technical help from Jing Wei, Puang En Yen, Shanxiang Fang.

References

1. A Learning-Based Framework for Robot Peg-Hole-Insertion, volume Volume 2 of Dynamic Systems and Control Conference. 10 2015.V002T27A002.
2. Camilo, B.-H.C., Petit, D., Ramirez-Alpizar, I.G., Harada, K.: Variable compliance control for robotic peg-in-hole assembly: a deep reinforcement learning approach (2020)
3. Chen, X., Ghadirzadeh, A., Bjorkman, M., Jensfelt, P.: Adversarial feature training for generalizable robotic visuomotor control. In: 2020 IEEE International Conference on Robotics and Automation (ICRA), pp. 1142–1148 (2020)
4. Akkaya, I., et al.: Solving Rubik's cube with a robot hand (2019)

5. Xu, J., et al.: Feedback deep deterministic policy gradient with fuzzy reward for robotic multiple peg-in-hole assembly tasks. IEEE Trans. Ind. Inf. **15**(3), 1658–1667 (2019)
6. Zhu, J.-Y., et al.: Unpaired image-to-image translation using cycle-consistent adversarial networks (2020)
7. Vecerik, M., et al.: A practical approach to insertion with variable socket position using deep reinforcement learning. In: 2019 International Conference on Robotics and Automation (ICRA), pp. 754–760 (2019)
8. Vecerik, M., et al.: Leveraging demonstrations for deep reinforcement learning on robotics problems with sparse rewards (2018)
9. Finn, C., Abbeel, P., Levine, S.: Model-agnostic meta-learning for fast adaptation of deep networks (2017)
10. Goodfellow, I., Pouget-Abadie, J., Mirza, M., Xu, B., Warde-Farley, D., Courville, A., Bengio, Y.: Generative Adversarial Networks. Sherjil Ozair (2014)
11. Ashley, H., et al.: Stable baselines. https://github.com/hill-a/stable-baselines (2018)
12. Hamalainen, A., Arndt, K., Ghadirzadeh, A., Kyrki, V.: Affordance learning for end-to-end visuomotor robot control (2019)
13. Inoue, T., De Magistris, G., Munawar, A., Yokoya, T., Tachibana, R.: Deep reinforcement learning for high precision assembly tasks. In: 2017 IEEE/RSJ International Conference on Intelligent Robots and Systems (IROS), pp. 819–825 (2017)
14. Rao, K., Harris, C., Irpan, A., Levine, S., Ibarz, J., Khansari, M.: Reinforcement learning awaresimulation-to-real, Rl-cyclegan (2020)
15. Schoettler, G., Nair, A., Luo, J., Bahl, S., Apari-cio Ojea, J., Solowjow, E., Levine, S.: Deep reinforcement learning for industrial insertion tasks with visual inputs and natural rewards (2019)
16. Schoettler, G.: Nair, A., Levine, S., Solowjow, E.: Meta-reinforcement learning for robotic industrial insertion tasks. Juan Aparicio Ojea (2020)
17. Schulman, J., Levine, S., Jordan, M.I., Abbeel, P.: Trust region policy optimization. Philipp Moritz (2017)
18. Schulman, J., Wolski, F., Radford, A., Klimov, O.: Proximal policy optimization algorithms. Prafulla Dhariwal (2017)
19. Tobin, J., Fong, R., Ray, A., Schneider, J.: Zaremba, W., Abbeel, P.: Domain randomization for transferring deep neural networks from simulation to the real world (2017)

Towards Synthetic Data: Dealing with the Texture-Bias in Sim2real Learning

Jelena Tabak(✉), Marsela Polić, and Matko Orsag

Faculty of Electrical Engineering and Computing, University of Zagreb, 10000 Zagreb, Croatia
jelena.tabak@fer.hr, marsela.polic@fer.hr, matko.orsag@fer.hr

Abstract. In this paper we test and ultimately confirm the texture-bias hypothesis of the state of the art method for semantic segmentation, DeepLabv3+. However, our results show that even though texture bias is obvious, shape still plays an important role in the overall system performance. To perform the analysis, we propose using a synthetic dataset with highly realistic textures for sim2real learning in a leaf segmentation task. The labeled synthetic plant datasets used in the evaluation are created using a 3D modeling software, Blender, utilizing a custom designed procedural generation pipeline. We validate the accuracy of the models on real images of sweet peppers and we introduce an improved method for 6DOF pepper pose estimation which utilizes the presented leaf segmentation model. Accuracy of the pepper pose estimation method is experimentally validated.

Keywords: Texture-bias · Sim2real learning · Synthetic data

1 Introduction

Though consumer preference for organic food is growing, labor intensive production is the key factor limiting the supply. Robotising the organic agriculture, which is one of the goals of SpECULARIA project [26], would not only enable large scale organic food production but also alleviate the cost of food overall. One of the main challenges in replacing human labour input with robots is the development of a valid perception method. Namely, robots have to be capable of understanding the cluttered greenhouse environment, which can be achieved using state-of-the-art convolutional neural networks (CNNs).

CNNs achieve outstanding results in both object detection and segmentation tasks, but deployment of such deep learning perception methods is challenging mainly due to the labor intensive data acquisition and labeling procedure. This has motivated researchers to turn to synthetic dataset generation. It has been shown that augmenting real datasets with synthetic ones improves the accuracy of deep learning models in case of insufficient amount of annotated real data [4,9].

I. Petrovic et al. (Eds.): IAS 2022, LNNS 577, pp. 630–642, 2023.
https://doi.org/10.1007/978-3-031-22216-0_42

However, models trained only on synthetic datasets often poorly generalise to real data and bridging the gap between synthetic and real data remains an open research problem [10,25].

Some recent studies [12,20] indicate that, contrary to previous findings [29], CNNs strongly rely on texture cues when performing classification tasks on well known large scale datasets, such as ImageNet. It has been speculated that texture-bias is the main cause of domain gap between synthetic and real datasets [17]. If that is the case, using realistic textures in data synthesis could lead to improvement in generalisation of models trained exclusively on synthetic data.

1.1 Related Work

Since the advent of CNNs [18], deep convolutional models have shown an outstanding performance in various perception tasks, such as image classification, object detection and segmentation. Still, it is not yet quite clear how CNNs reach their decisions. As neural networks are biologically inspired, the difference between their architecture and human visual system has been broadly researched [11,16] and some recent studies have shown that CNNs differ from humans in many crucial aspects [3]. Namely, neural networks tend to rely more on textural cues and local features than on global features, such as object shape [2,5].

Since authors in [12] have demonstrated the texture-bias of state-of-the-art CNNs in image classification task, many researches have proposed different methods for building classification models which would be more shape-biased and consequently more robust to different perturbations, including adversarial attacks. Most of the existing work focuses on training the model with the combination of the original data and data generated using different style-transfer methods [22,34]. However, it is also possible to overcome the texture-bias by choosing more natural augmentation techniques [13]. Namely, authors in [13] suggest that CNNs are not inherently texture-biased and that the existing bias is the consequence of the unnatural augmentation techniques, such as random crop. Not only the augmentation of the training data but also changes in the model architecture can lead to the dominance of the shape-bias in the models. Some authors propose mimicking human cognitive models by either using Gabor filters in the first convolutional layer [21] or using multiple streams as an input to the network [23].

While all of the aforementioned methods increase shape-bias of the models, such models are not necessarily more accurate than their texture-biased counterparts [6], except in the case of a few-shot classification, where the model has seen only a small limited number of labeled examples [28]. However, though the models trained on ImageNet tend to have lower accuracy score if they are trained to rely more on shape than on texture, it has been reported that such models are more robust and better at generalisation [6,12]. Researches have addressed that issue by combining the texture-biased and shape-biased models in both training and prediction stage [8,19,24].

Most of the work in the field focuses on the object classification models, but some researches have tackled the problem of texture-bias in the semantic segmentation task as well. The authors in [15] have shown that the accuracy of deep semantic segmentation models significantly deteriorates when they are trained with the data generated using style-transfer method proposed in [12]. Training with data generated using alternative style transfer methods can result in higher shape-bias in state-of-the-art semantic segmentation models while retaining the accuracy [1,17].

It is assumed that the texture-bias of the neural networks and insufficiently realistic textures in synthetic datasets are the main causes of domain gap in sim2real learning [17,33]. This hypothesis is in line with findings of authors in [32] who reported state-of-the-art segmentation score on the leaf segmentation task using combination of real and synthetic data. Authors in [32] claim that the real data contributed to the texture diversity, while the synthetic data increased geometry variance. As annotating dataset on the pixel level is highly time consuming task we tested weather it is possible to bridge the gap in sim2real learning by using highly realistic textures in leaf segmentation task.

1.2 Contributions

We propose bridging the sim2real gap by using highly realistic textures in procedural synthetic dataset generation pipeline without trying to overcome the existing texture-bias. Instead we aim to utilize the bias to increase the performance of our system. At the same time, we provide evidence of texture-bias in the model using synthetic datasets described in the next section. We claim the following contributions: (i) demonstration of severe texture-bias in state-of-the-art semantic segmentation architecture, (ii) experimental validation of sim2real learning without fine-tuning the model on real dataset, and (iii) improvement of deep learning-based method for sweet pepper pose estimation using leaf segmentation model trained on publicly released synthetic dataset. The claimed contributions are discussed in detail in the remainder of this paper.

2 Method

In this work, synthetic plant datasets are generated using 3D modeling software, Blender [14]. Using synthetic data alleviates dataset generation and labeling procedure, but models trained on purely synthetic data often perform poor on real data, which is known as sim2real gap. One possible cause of sim2real gap is the texture-bias. We confirm the presence of the texture-bias in the state-of-the-art semantic segmentation model using synthetic pepper and strawberry plant datasets, chosen based on the significant difference between the texture and the shape of pepper and strawberry leaves. Working with the synthetic datasets generated with the same parameters enabled us to have a fair comparison of the performance of different models when testing the texture-bias hypothesis. This approach ultimately enables us to explore the effect the leaf shape has on training the detection model.

2.1 Datasets

The following datasets are rendered combining the shape and the texture of pepper and strawberry leaves: *PP* (pepper shape and pepper texture), *PS* (pepper shape and strawberry texture), *SS* (strawberry shape and strawberry texture) and *SP* (strawberry shape and pepper texture). An example image from each of the four datasets is shown in Fig. 1. The datasets are publicly released and can be downloaded from https://www.kaggle.com/datasets/jele38/synthetic-plant-datasets [31].

Fig. 1. Examples from the rendered datasets: *PP* (top-left, pepper shape and pepper texture), *PS* (top-right, pepper shape and strawberry texture), *SS* (bottom-left, strawberry shape and strawberry texture) and *SP* (bottom-right, strawberry shape and pepper texture).

Stem, pot and leaves were modeled from scratch. Photographs of pepper and strawberry leaves were used to generate textures for the leaf models. This approach resulted in leaf models with highly realistic textures and natural-looking shape. The custom dataset generation pipeline is made publicly available and can be downloaded from https://github.com/larics/blender_synthetic_data_generator [30]. The pipeline is implemented in Blender through its embedded Python interpreter as a Python script, automatically generated both images and segmentation masks. There was always one plant per image. In each pass, 20–40 leaves were randomly sampled and spawned along the stem curve. When spawning leaves, their size was randomized by setting Scale Randomness factor to 0.18, meaning that the size of the leaf instances varies from 88% of the maximum leaf size to the maximum leaf size. A single pass consisted of 20 frames. Camera position and light conditions differed across the frames. For each dataset, 50 passes were rendered, resulting in 1000 images per dataset. Images were rendered at 640×640 pixels. Background was uniform in 10% of the images, while the background for the remaining 90% of the images was randomly chosen among 100 images of different textures and environments. This prevented the model from overfitting and learning that all that is green in the image belongs to the leaf class.

Fig. 2. Examples of images from the real dataset (shown on the left) along with the manually annotated masks (shown on the right). The real dataset, used as test dataset, consisted of images of a single plant in the experimental environment (top-left) and images of plants in the cluttered greenhouse environment (bottom-left).

Manually annotated dataset of real images of sweet pepper plants was used as a test dataset. The dataset consisted of 41 images, out of which 10 images represent single plants in the experimental environment, with minimum environmental distractors, and 31 images represent plants in the cluttered greenhouse environment. Leaves from the plants in the real dataset were not used for generating textures for the synthetic datasets. Samples from the real dataset, along with the manually annotated masks, are shown in Fig. 2.

2.2 Semantic Segmentation Neural Network

State-of-the-art semantic segmentation architecture, DeepLabv3+ [7], pretrained on ImageNet was trained via transfer learning in all of the reported experiments. The network was trained in TensorFlow 2 environment. Size of input images was 640×640. While the training datasets were rendered at 640×640, the real images downscaled from 2268 × 2268. The weights of all layers were tuned during the training process. The batch size of 4 was used across the experiments with the maximum number of epochs being set to 200. Early stopping was employed in order to prevent the network from overfitting. Adam optimizer was used with the learning rate of 2e-5. Keras implementation of categorical crossentropy was used as a loss function. As we have been dealing with the unbalanced datasets in which the percentage of pixels belonging to one class (background) was significantly higher than the percentage of pixels belonging to the other class (leaf), the weights for the classes have been calculated based on the ratio of pixels belonging to a single class and the total number of pixels, and used in the training process.

3 Experimental Validation

Two sets of experiments were conducted to investigate the texture-bias in sim2real semantic segmentation learning. The experiments focus on detection of sweet pepper plant leaves in real images, using deep learning models trained on synthetic data. The synthetic data is generated using two plant species that significantly differ in shape and in texture, namely pepper and strawberry. Then, generalization ability of the segmentation models trained on these datasets is investigated with tests on a dataset of real pepper plants.

3.1 Texture-Bias in Semantic Segmentation Model

Semantic segmentation model, DeepLabv3+, was trained on each of the rendered synthetic datasets and the resulting models were tested on manually labeled real images of pepper plants. Training and evaluation procedure is conducted for leaf class only. The same network architecture and training parameters, described in the previous section, were employed when training each of the models. The test results, obtained on the test dataset consisting of manually labeled real pepper plants, are presented in Table 1. The accuracy is reported as Intersection over Union (IoU) for the class leaf.

Table 1. Pepper leaf segmentation results

Train set	*PP*	*PS*	*SS*	*SP*
IoU	0.63	0.5	0.45	0.61

IoU score on the real pepper images dataset for the model trained with the *PP* dataset suggests that sim2real transfer learning is possible even in case of a complete lack of real data if the synthetic data closely matches the expected distribution of the real data, including realistic textures, lightening conditions and camera positions. Predictions of the model with the *PP* dataset are shown in Fig. 3. A high IoU score was expected when training the model with the *SP* dataset as well, thanks to the texture-bias and the realistic pepper leaf texture applied to strawebery leaves models. However, it was not expected that it will differ from the result obtained with the *PP* dataset in only 2 percent points.

IoU score on the pepper test dataset was predictably significantly lower when training with the datasets with the strawberry texture. As to the best of our knowledge, a similar experiment has not yet been conducted, making this a strong confirmation of texture-bias in semantic segmentation CNN based neural networks. However, the consistent differences in IoU scores between the models trained on the datasets which differ in shape, but share the same texture suggests that the texture is not the only cue for the prediction. As factors other than the shape of the leaves were kept constant when generating the datasets, it is safe to conclude that the deployed network architecture does take the global features in consideration, but not as much as the local ones.

Fig. 3. Results of leaf segmentation in 2D. State-of-the-art semantic segmentation architecture, DeepLabv3+, trained on synthetic dataset, is used for segmentation. Input images are shown on the left and images with predicted masks are shown on the right.

3.2 Sim2real Approach for Leaf Segmentation in Robotic Harvesting

In order to validate the efficiency of sim2real learning for the practical application in the field of robotics, leaf segmentation model trained with the *PP* dataset was used to improve the sweet pepper picking method presented in [27]. The presented method relies on a CNN for pepper detection and geometric model fitting for 6DOF pepper pose estimation. Experimental robotic harvesting setup is shown in Fig. 4. Robot manipulator, equipped with an RGB-D camera, records the plant from multiple predefined search poses. In each search pose, both image and point cloud are recorded. For each detected pepper, its pose is estimated by fitting the cylinder model on the subset of the original point cloud, extracted using detection output from the CNN. Estimated pepper centroid, height and radius are used to plan the grasp. The point cloud of the pepper usually contains other plant parts as well, which can deteriorate the cylinder fitting procedure, especially in the case of occluding leaves. Here, we examine the possibility of improving the accuracy and the robustness of pepper pose estimation by introducing leaf segmentation step in the detection pipeline.

Fig. 4. Robotic harvesting setup from [27]. Collaborative robot Franka Emika, equipped with an RGB-D camera, records the plant from multiple search poses. Convolutional neural network is employed for pepper detection while centroid c, height h and radius of the pepper are estimated by fitting the cylinder model on the recorded point cloud.

Pepper pose estimation was improved by removing the points which belong to the leaf class from the point cloud before estimating the pose of the pepper. For each detected pepper, organised point cloud containing the pepper is filtered using the segmentation mask obtained with the leaf segmentation model. The results of filtering 3D point cloud by using 2D mask obtained from the corresponding image are shown in Fig. 5. The filtered point cloud is used as an input for the geometric model fitting. By removing leaf voxels from the original point cloud before the geometric model fitting, fitted cylinder model estimates the exact pepper pose and size more closely, which consequently improves the grasping precision.

In order to compare the precision of the pose estimation for the highly occluded peppers with and without the leaf segmentation, we conducted 10

Fig. 5. Results of leaf segmentation in 3D. Original image (top-left) is passed through 2D semantic segmentation network resulting in leaf mask (top-right). Corresponding point cloud (bottom-left) is filtered using the obtained leaf mask (bottom-right).

experiment repetitions for a single pepper both with and without the leaf segmentation. The pepper was detected from the same initial search positions of the robot. The results presented in Table 2 show the mean and the standard deviation of the estimated variables for a single pepper in both cases. As the accurate position of the pepper is unknown, the standard deviation is taken as a precision measure.

Table 2. Pepper detection precision results

	Without leaf segmentation		With leaf segmentation	
	Mean	Std	Mean	Std
x [mm]		5.2		3.4
y [mm]		7.1		7.6
z [mm]		12.6		7.2
r [mm]	25.5	5.6	23.0	6.2
h [mm]	109.6	17.4	107.9	8.9

The difference in the deviation of the pepper position in three axis is at a millimeter level, with the biggest improvement in the z axis. Regarding the geometric model, the difference of the dissipation of both estimated radius and height of the pepper is again at a millimeter level, with significant improvement for the height estimation. Adding leaf segmentation step slightly increased the deviation of the radius and the deviation in the y axis. In our setup, as seen in Fig. 4, global y axis is parallel with the vector pointing from the robot base frame to the pepper plant. Without the leaf segmentation step, leaves which occlude the plant are always taken in consideration when fitting the geometric model, which results in bigger estimated radius when compared to the radius estimated with the leaf segmentation step. However, as not all leaf voxels are filtered in all views, as seen in Fig. 5, leaf voxels which remain unfiltered in some views slightly increase the deviation in the y axis and the deviation of the radius. Although the deviation of two estimated variables did increase, the detected pepper is smaller, better localised and thus easier to grasp.

4 Conclusion

In this work, texture-bias of the state-of-the-art semantic segmentation model, DeepLabv3+, was shown by training leaf segmentation models on four different synthetic plant dataset. The evaluation was conducted using images generated by combining textures and shapes of pepper and strawberry leaves, which significantly differ both in shape and in texture. Synthetic images were rendered in Blender, using custom procedural dataset generation pipeline. All models were tested segmenting pepper leaves on real images of sweet pepper plants. As expected, models trained on images of leaves with strawberry texture performed significantly worse than the ones trained on images of leaves with pepper texture. Relatively small difference in IoU score between the model trained on the dataset with pepper shape and texture and the model trained on the dataset with strawberry shape and pepper texture implies that the model is highly texture-biased. At the same time, the difference in IoU score between the models trained on the datasets generated using the same texture, but of different shape, indicates that semantic segmentation models do not rely solely on local, textural cues.

As the texture-bias of the semantic segmentation model is so prominent, using realistic textures in synthetic datasets might be a better approach for sim2real learning than training the model to become shape-biased. In this work, sim2real learning was validated on the leaf segmentation task using generated synthetic dataset with highly realistic textures. The obtained model was deployed in robotic harvesting procedure resulting in improved pepper pose estimation, which was experimentally verified.

Though the accuracy of the trained model was satisfying and it was shown that it can be successfully applied in robotic harvesting scenario, the IoU score of 63% suggests that the lack of realistic textures in synthetic datasets might not be the only cause of domain gap in sim2real learning. Improvement could be introduced by creating more realistic models for stem and pot, using backgrounds which correspond to the backgrounds expected in the experimental scenario and diversifying shape of the leaf models with additional morphological transformations.

Acknowledgments. This work has been supported by Croatian Science Foundation under the project Specularia UIP-2017-05-4042 [26]. The work of doctoral student Jelena Tabak has been supported in part by the "Young researchers' career development project- training of doctoral students" of the Croatian Science Foundation funded by the European Union from the European Social Fund.

References

1. Azad, R., Fayjie, A.R., Kauffmann, C., Ben Ayed, I., Pedersoli, M., Dolz, J.: On the texture bias for few-shot cnn segmentation. In: Proceedings of the IEEE/CVF Winter Conference on Applications of Computer Vision (WACV), pp. 2674–2683 (January 2021)
2. Baker, N., Lu, H., Erlikhman, G., Kellman, P.: Deep convolutional networks do not classify based on global object shape. PLOS Comput. Biol. **14**, e1006613 (12 2018)
3. Baker, N., Lu, H., Erlikhman, G., Kellman, P.J.: Local features and global shape information in object classification by deep convolutional neural networks. Vision Res. **172**, 46–61 (2020)
4. Barth, R., IJsselmuiden, J., Hemming, J., Van Henten, E.J.: Data synthesis methods for semantic segmentation in agriculture: a capsicum annuum dataset. Comput. Electronics Agric. **144**, 284–296 (2018)
5. Brendel, W., Bethge, M.: Approximating CNNs with bag-of-local-features models works surprisingly well on imagenet. In: International Conference on Learning Representations (2019)
6. Brochu, F.: Increasing shape bias in imagenet-trained networks using transfer learning and domain-adversarial methods. CoRR abs/1907.12892 (2019). http://arxiv.org/abs/1907.12892
7. Chen, L.C., Zhu, Y., Papandreou, G., Schroff, F., Adam, H.: Encoder-decoder with atrous separable convolution for semantic image segmentation. In: Proceedings of the European Conference on Computer Vision (ECCV) (September 2018)
8. Co, K.T., Muñoz-González, L., Kanthan, L., Glocker, B., Lupu, E.C.: Universal Adversarial Robustness of Texture and Shape-Biased Models. arXiv e-prints arXiv:1911.10364 (2019)
9. Di Cicco, M., Potena, C., Grisetti, G., Pretto, A.: Automatic model based dataset generation for fast and accurate crop and weeds detection. In: 2017 IEEE/RSJ International Conference on Intelligent Robots and Systems (IROS), pp. 5188–5195. IEEE (2017)
10. Doersch, C., Zisserman, A.: Sim2real transfer learning for 3d human pose estimation: motion to the rescue. In: Wallach, H., Larochelle, H., Beygelzimer, A., d'Alché-Buc, F., Fox, E., Garnett, R. (eds.) Advances in Neural Information Processing Systems, vol. 32. Curran Associates, Inc (2019)
11. Geirhos, R., Jacobsen, J.H., Michaelis, C., Zemel, R., Brendel, W., Bethge, M., Wichmann, F.: Shortcut learning in deep neural networks. Nat. Mach. Intell. **2**, 665–673 (11 2020)
12. Geirhos, R., Rubisch, P., Michaelis, C., Bethge, M., Wichmann, F., Brendel, W.: Imagenet-Trained CNNs are Biased Towards Texture; Increasing Shape Bias Improves Accuracy and Robustness. ArXiv abs/1811.12231 (2019)
13. Hermann, K., Chen, T., Kornblith, S.: The origins and prevalence of texture bias in convolutional neural networks. In: Larochelle, H., Ranzato, M., Hadsell, R.,

Balcan, M.F., Lin, H. (eds.) Advances in Neural Information Processing Systems, vol. 33, pp. 19000–19015. Curran Associates, Inc (2020)
14. Hess, R.: Blender Foundations: The Essential Guide to Learning Blender 2.6. Focal Press (2010)
15. Islam, M.A., Kowal, M., Esser, P., Jia, S., Ommer, B., Derpanis, K.G., Bruce, N.D.B.: Shape or Texture: Understanding Discriminative Features in CNNs. CoRR abs/2101.11604 (2021). https://arxiv.org/abs/2101.11604
16. Kheradpisheh, S.R., Ghodrati, M., Ganjtabesh, M., Masquelier, T.: Deep networks can resemble human feed-forward vision in invariant object recognition. Sci. Rep. **6**, 32672 (2016)
17. Kim, M., Byun, H.: Learning Texture Invariant Representation for Domain Adaptation of Semantic Segmentation, pp. 12972–12981 (2020)
18. Krizhevsky, A., Sutskever, I., Hinton, G.: Imagenet classification with deep convolutional neural networks. Neural Inf. Process. Syst. **25** (2012)
19. Li, Y., Yu, Q., Tan, M., Mei, J., Tang, P., Shen, W., Yuille, A., Xie, C.: Shape-texture debiased neural network training. In: International Conference on Learning Representations (2021)
20. Malhotra, G., Bowers, J.: The contrasting roles of shape in human vision and convolutional neural networks. In: Goel, A., Seifert, C., Freksa, C. (eds.) Proceedings of the 41st Annual Conference of the Cognitive Science Society (2019)
21. Malhotra, G., Evans, B.D., Bowers, J.S.: Hiding a plane with a pixel: examining shape-bias in CNNs and the benefit of building in biological constraints. Vision Res. **174**, 57–68 (2020)
22. Mishra, S.K., Shah, A., Bansal, A., Choi, J., Shrivastava, A., Sharma, A., Jacobs, D.W.: Learning Visual Representations for Transfer Learning by Suppressing Texture. CoRR abs/2011.01901 (2020). https://arxiv.org/abs/2011.01901
23. Mohla, S., Nasery, A., Banerjee, B., Chaudhuri, S.: Cognitivecnn: Mimicking Human Cognitive Models to Resolve Texture-Shape Bias. CoRR abs/2006.14722 (2020). https://arxiv.org/abs/2006.14722
24. Nam, H., Lee, H., Park, J., Yoon, W., Yoo, D.: Reducing Domain Gap by Reducing Style Bias. arXiv e-prints arXiv:1910.11645 (2019)
25. Pashevich, A., Strudel, R., Kalevatykh, I., Laptev, I., Schmid, C.: Learning to augment synthetic images for sim2real policy transfer. In: 2019 IEEE/RSJ International Conference on Intelligent Robots and Systems (IROS), pp. 2651–2657 (2019)
26. Polic, M., Ivanovic, A., Maric, B., Arbanas, B., Tabak, J., Orsag, M.: Structured ecological cultivation with autonomous robots in indoor agriculture. In: 2021 16th International Conference on Telecommunications (ConTEL), pp. 189–195 (2021)
27. Polic, M., Tabak, J., Orsag, M.: Pepper to fall: a perception method for sweet pepper robotic harvesting. In: Intelligent Service Robotics (2021)
28. Ringer, S., Williams, W., Ash, T., Francis, R., MacLeod, D.: Texture Bias of CNNs Limits Few-Shot Classification Performance. CoRR abs/1910.08519 (2019). http://arxiv.org/abs/1910.08519
29. Ritter, S., Barrett, D., Santoro, A., Botvinick, M.: Cognitive Psychology for Deep Neural Networks: A Shape Bias Case Study (2017)
30. Tabak, J., Polic, M., Orsag, M.: Synthetic Dataset Generation Pipeline. https://github.com/larics/blender_synthetic_data_generator (2021). Accessed 22 Apr 2022
31. Tabak, J., Polic, M., Orsag, M.: Synthetic Plant Datasets. https://www.kaggle.com/datasets/jele38/synthetic-plant-datasets (2022). Accessed 22 Apr 2022
32. Ward, D., Moghadam, P., Hudson, N.: Deep Leaf Segmentation Using Synthetic Data. CoRR abs/1807.10931 (2018). http://arxiv.org/abs/1807.10931

33. Zaech, J.N., Dai, D., Hahner, M., Gool, L.V.: Texture underfitting for domain adaptation. In: 2019 IEEE Intelligent Transportation Systems Conference (ITSC), pp. 547–552 (2019)
34. Zhang, Y., Zhanga, Y., Xu, Q., Zhang, R.: Learning robust shape-based features for domain generalization. In: IEEE Access, pp. 1–1 (2020)

On Scene Engineering and Domain Randomization: Synthetic Data for Industrial Item Picking

Moritz Weisenböhler[1,2(✉)], Björn Hein[1,2], and Christian Wurll[1]

[1] Karlsruhe University of Applied Sciences, Karlsruhe, Germany
hebj0001@h-ka.de, wuch0001@h-ka.de
[2] Karlsruhe Institute of Technology, Karlsruhe, Germany
wemo0001@h-ka.de

Abstract. Synthetic data for training deep neural networks is increasingly used in computer vision. Different strategies, such as domain randomization or domain adaptation, exist to bridge the gap between synthetic training data and the real application. Despite recent progress and gain in knowledge in this area, the following question remains: How much adjustment to reality is required and which degree of randomization is useful for transferring precise object detectors to real use cases? In this paper, we present a detailed study with more than 100 datasets and 2,700 trained convolutional neural networks (CNNs), comparing the influence of different degrees of manual optimization (scene engineering) and domain randomization techniques. To distinguish precision and robustness, the trained object detectors are evaluated on different domain shifts with respect to scene environment and object appearance. Using the example of robot-based industrial item picking, we show that the scene context and structure as well as realistic textures are crucial for the simulation to reality transfer. The combination with well-chosen randomization parameters, especially lighting and distractor objects, improves the robustness of the CNNs at higher domain shifts.

Keywords: Synthetic image data · Object detection · Scene engineering · Domain randomization · Item picking · Industrial application

1 Introduction

Deep neural networks have become a key technology in modern computer vision. However, training convolutional neural networks (CNNs) for object detection or instance segmentation requires a lot of labeled data. Although there are free data collections, such as COCO [1] or ImageNet [2], these can not be used for applications where specific custom objects have to be detected. Therefore, in most industrial use cases, real-world training data must be generated manually with great effort. Synthetically generated data can be used to simplify the data

I. Petrovic et al. (Eds.): IAS 2022, LNNS 577, pp. 643–660, 2023.
https://doi.org/10.1007/978-3-031-22216-0_43

acquisition process. Simulation based image generation is comparatively inexpensive, easily scalable and produces very precise annotations. However, domain differences between synthetic and real data, the so called domain gap, lead to challenges when transferring the synthetically trained CNNs to real-world applications.

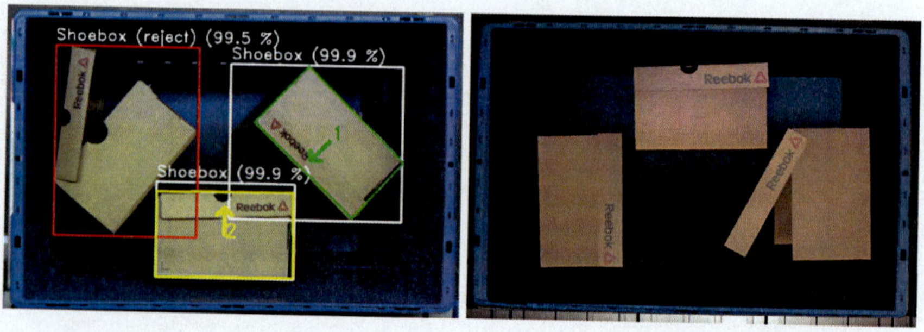

Fig. 1. In the examined use case, shoeboxes have to be localized and classified regarding opened lids (left). The corresponding object detectors are trained with synthetically generated data (right) and evaluated on real-world images.

Different methods to overcome this domain gap have been developed and examined. A typical approach is the fine-tuning of models that have been pretrained on a real-world domain (transfer learning). Domain adaptation by adjusting synthetic data to the real domain is also a common method. This can be achieved through manual optimization of the simulation, which is referred to as *scene engineering* (SE) in the following, or through post-processing, for example with generative adversarial networks (GANs) [3] or autoencoders. A more recent approach with increasing popularity is the use of domain randomization, where the simulation scenes are heavily randomized regarding appearance and structure. The idea is to train the CNNs on such a broad domain that the properties of the real data are included in it [4].

While the domain adaptation with GANs requires a wide selection of real-world images, manual scene optimization requires only a few samples. However, the manual adjustment of the simulation scene to the reference data is time-consuming and demands expert knowledge. Randomization, on the other hand, is very easy to use and requires neither expert knowledge nor sample data from the real application. However, it is still uncertain to what extent and how reliably the transfer from simulation to reality can be improved with these techniques. It is questionable whether randomization can completely replace scene engineering in real-world tasks, especially since a broad randomization potentially increases the complexity of the task the neural network has to learn.

Since it is difficult to answer these questions with general validity, in the following we focus on robot-based industrial item picking use cases in production and logistics automation. Many industrial automation tasks are well suited for the use of synthetic training data: the tasks are mostly related to very specific items, not present in publicly available datasets. While real-world training data is only rarely available, Computer-aided design (CAD) models and even realistic object textures, on the other hand, often exist. Since industrial plants are planned in detail and are easier to control than the free environment (e.g. compared to autonomous driving), much knowledge and certainty about the scene structure exists. Therefore, the comparatively static and homogeneous scene structure and background, lighting conditions and camera properties simplify the creation of simulation scenes.

In this work, synthetic data generation is examined using the object detection task of a real item picking solution for the fashion industry [5] shown in Fig. 1. In this use case tightly packed shoeboxes have to be precisely localized and categorized into pickable and non-pickable items. The latter are accidentally opened boxes that lead to handling problems and must be reliably detected and excluded. Compared to other computer vision tasks, this use case is less complex in terms of number and variety of classes and objects. However, it is very representative of many industrial automation applications, such as automated item picking or computer vision-based quality assurance and inventory. These industrial use cases often share specific challenges such as:

- **Tightly packed scenes**: The items are typically presented in a tightly packed load carrier (LC). Nonetheless, a secure detection and exact separation of the individual items is required.
- **Non-rigid or connected objects**: The objects to be detected are often completely (e.g. clothing) or partially (e.g. shoebox and lid) non-rigid, making the use of model-based computer vision difficult.
- **Distinguish similar objects**: Regularly, multiple similar articles with only few feature differences have to be distinguished. For example, when classifying different states of an object (e.g. opened vs. closed shoeboxes).

In this paper, we present a detailed study on scene engineering and domain randomization for synthetic image data. Our main contributions are as follows:

- We illustrate the capabilities of scene engineering and domain randomization using four degrees of domain shift in a real industrial item picking task.
- We systematically examine the relevance of single randomization approaches and parameters with respect to precision and robustness..
- We present our data generation approach for generating highly randomized yet highly comparable data set variations.

In the remainder of the paper, we first discuss related work and then describe the methods we used to generate synthetic data, as well as the training and evaluation of the CNNs. We then present and discuss the quantitative results of our base and ablation experiments. Finally, we present our results and conclude clear recommendations for domain matching and randomization in comparable industrial use cases.

2 Related work

Training deep neural networks for computer vision with synthetic training data has been carried out for applications in the area of vision based gripping pose determination [4,6–8] and robot control [9–13], localizing and classifying toy and household objects [14–17] or industrial parts [15,18–20] as well as object detection and instance segmentation in autonomous driving [21–26]. Since the transfer of neural networks is a central challenge when training with synthetic data, different approaches have been taken to reduce the domain gap to the real world.

2.1 Cut and Paste

Cut and paste methods are comparatively easy to use. Here single or multiple objects were rendered in front of a random background image, without considering the spatial scene structure. Missing shadows and physically incorrect positioned items potentially degrade the domain transfer [27–29]. Following works encountered this problem by using complex background textures, intelligent object positioning or blurring at the object outline [14,30].

2.2 Domain Adaptation

Using full-stack simulations requires an elaborate process of model generation and scene construction. The advantage of this approach is that the objects are correctly positioned and seamlessly inserted into the scene. In combination with high-quality rendering engines, the generated images achieve almost photorealistic appearance [21–24]. Others applied meta algorithms to optimize the relevant simulation parameters [31–33]. Here, the parameters are either adjusted to generate images closer to a reference domain or optimized with respect to the loss of trained and evaluated CNNs.

An alternative to computationally expensive rendering is the usage of post-processing methods for domain adaptation through image transformation. Since the rise of generative adversarial networks, these models are increasingly used to reduce the domain gap between synthetic and real images. Either synthetic images are transferred directly to the real domain, making them more realistic [11,18,34–39], or both synthetic and real images are transferred to a third domain called canonical, which is a segmentation map of objects [10].

2.3 Domain Randomization

Tobin et al. introduced the term *domain randomization* (DR) [4] as a use case abstraction technique used to bridge the gap between real and synthetic data. They assumed that synthetic data does not need much scene context or photo-realistic rendering if enough data variety is produced through extensive randomization. They showed that the transfer from simulation to reality can be solved using pre-trained models and simple randomization strategies. In the vision-based grasping task they examined, the most relevant randomization parameters were texture colors and patterns, camera positions and distractor objects. However, it is questionable whether their findings are applicable in real industrial applications, since they only used simple textures and geometric shapes (e.g. pyramids, cubes or cylinders).

Hinterstroisser et al. used DR for object detection in cluttered scenes [15]. Instead of simulating the complete scene with all surroundings, an object was rendered in front of a background image. Blurred object boundaries allowed a softer blending into the background, which was very important in their experiments. Their central finding was a huge performance boost by freezing the feature extractor while training, but the success of this approach could not be confirmed by [25]. In a subsequent publication they used a purely synthetic background layer consisting of several tightly spaced CAD models [16]. They achieved best results with image blurring and by randomizing the light color, whereas variations of the camera parameters were less relevant. However, their use case was very unstructured and did not include any scene context.

James et al. evaluated different randomization strategies in the context of vision based robot grasping [9]. They examined synthetically trained CNNs for object location without using any real-world data or pre-trained weights. Their ablation study proved that camera randomization and rendering shadows are important aspects for the simulation to reality transfer. Using complex textures instead of single colors was little less relevant and distractor objects were only crucial when also occurring in the real-world data.

Tremblay et al. used DR in the more complex task of car detection for autonomous driving [25]. The performance was evaluated on the real-world dataset KITTI [40]. They combined several randomization strategies with high quality models and rendering. Their results show that well chosen randomization parameters, especially lighting and textures, are transferable to complex real-world scenarios and can even beat synthetic close domain replicas such as VKITTI [21]. However, they trained only one-class detectors and discovered shortcomings for tightly positioned cars due to the missing scene context. For most picking tasks, the ability to precisely separate tightly arranged objects is crucial.

For the detection of cars, Prakesh et al. combined DR with adjusted parameter distributions depending on the scene context [26]. Focusing on the placement of objects and distractors, they achieved better results with this trade-off than with context-independent DR.

Instead of targeting general training and network parameters, such as dataset size or augmentation strategies, we will focus on direct simulation and image generation parameters. Other researchers have done comparative and ablation studies explicitly evaluating the influence of randomization properties in their use cases:

- **Scene context**: Using object embedding and positioning strategies [16,24,26,32]
- **Textures:** Using real-world textures applied to objects and serving as background images [4,13,16,24,25,28,29,41–43]
- **Colors and patterns**: Using random single color, gradient, checkerboard or Perlin noise textures for objects and background [4,9,17,25,42,44]
- **Camera**: Variation of the view by changing the camera pose or focal length [4,9,13,16,17,44]
- **Lighting**: Randomization of the number and position of lights in the scene as well as their intensity and color [4,15,16,25,26,43,44]
- **Distractors**: Placing random distractor objects in the scene foreground or background [4,9,25,43,44]
- **Rendering quality**: Rendering with different quality, e.g. with and without shadows [9,24]

In the present paper, we systematically evaluated and compared all these randomization parameters in a single comprehensive study, taking into account different domain shifts.

3 Method

The target of our work is to compare the performance scores s of different synthetic datasets D, to provide insights into the influence of scene design approaches and randomization parameters. The complete process, consisting of data generation, model training and evaluation is visualized in Fig. 2. In total,

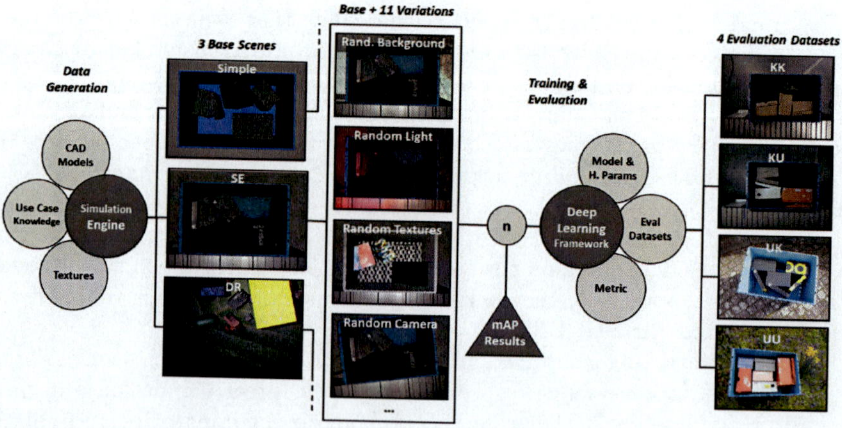

Fig. 2. The overall data generation, model training and evaluation process. Twelve datasets (base and variations) are generated for each of the three base scenes. For each dataset a CNN training is repeated n times and tested with four evaluation sets.

more than 100 data generation runs, 2,700 trainings and 11,300 evaluations were conducted.

For each base scene B_i where $i \in \{1, 2, 3\}$ and parameter variation p_j where $j \in \{1, \dots, 12\}$ a dataset $D_{i,j}$ is generated using the generation function g:

$$D_{i,j} = g(B_i, p_j) \tag{1}$$

The training of a CNN is repeated N times for each dataset $D_{i,j}$, with hyper-parameters h and a random seed r using a training function t. Each CNN is evaluated using metric m and evaluation dataset E_k where $k \in \{1, \dots, 4\}$. A performance score $s_{i,j,l}$ is calculated by averaging the evaluation results:

$$s_{i,j,k} = \frac{1}{N} \sum_{n=1}^{N} m\Big(t(D_{i,j}, h, r_n), E_k\Big) \tag{2}$$

3.1 Data Generation

The synthetic data generation is implemented using the Unity game engine which is expanded by custom components for scene randomization and data generation. Randomizations are defined by *manipulators*, determining the degrees of freedom of the scene. In each generation step, the manipulators randomly exchange parameters within a specified range.

The three evaluated base scenes (B) represent different abstraction levels of the real use case:

- A manually **engineered scene (SE)**, which was optimized with great effort: original textures for shoeboxes and the load carrier have been used, the camera pose and lighting have been adjusted to the real use case. The lids of the boxes were modeled with a hinge joint, allowing realistic positioning through physics simulation.
- A highly **simplified scene (Simple)**, created with less effort: simple single-color textures and no environment models were used. While the camera and lights were not randomized, the physics-based positioning process is the same as in the SE scene.
- A strongly **domain randomized scene (DR)**, where the real scene context and textures are completely neglected. Instead of realistic object positing, the shoeboxes were randomly scaled and placed in the viewport with a random lid angle. The camera pose as well as the light intensities and colors were strongly randomized.

The data generation system (g) produces scene instances, image renderings and annotations for multiple scene variations in a double-looped process, as described in Algorithm 1. Using this approach, the corresponding images of different dataset variations only differ in a single degree of freedom or randomization parameter. Nevertheless, the individual image instances within the same dataset variation have well-distributed features with multiple degrees of freedom (randomizations). Therefore, our method enables the generation of highly

Algorithm 1 Data Generation Algorithm

Input: Base scenes B, Scene variations v, Randomization parameters p_B and p_v
Output: datasets D

for all base scenes B_i **do**
    Apply base randomizations p_{B_i}
    Run the physics engine until stagnation
    Store scene parameters c_{B_i}
    for all scene variations v_j of B_i **do**
        Apply variation randomizations p_{v_j}
        Generate images and annotations d_{v_j}
        Append d_{v_j} to D_{B_i,v_j}
        Reset parameters to c_{B_i}
    end for
end for

comparable datasets, which is an essential feature to achieve conclusive findings when comparing individual parameter variations.

To investigate the influence of the individual randomization parameters (p), 11 additional variations were created for each base scene, as listed in Table 1. Datasets of 14 k, 28 k, and 56 k annotated images were generated for each of the resulting 36 simulation scenes. Additional datasets were used to examine specific scene properties, e.g. scenes without a load carrier or with an unrealistic camera viewport.

Table 1. Examined randomization parameters with two or three possible manifestations (A-C), resulting in 11 variations (scene structure is invariable for a base scene). "Realistic" means a slight randomization according to the real use case.

Parameter	Manifestation			Var.
	A	B	C	
Light intensity & color	Static	Realistic	Random	2
Camera pose	Static	Realistic	Random	2
Object textures	Colored	Realistic	Random	2
Background texture	Colored	Realistic	Random	2
Scaling LC or objects	No	Yes	–	1
Distractor objects	No	Yes	–	1
Rendering quality	Low	High	–	1
Scene structure	Realistic	Random	–	0

3.2 Training and Evaluation

For the training of the neural networks (t) the TensorFlow Object Detection API [45] was used. A custom wrapper module simplifies the repetition and evaluation of trainings with different datasets, models and hyperparameters. Two well-studied models were used for this study: Faster-RCNN [46] Inception-ResNet-v2 [47] (in the following FRCNN) and SSD [48] MobileNetV2 [49] with a Feature Pyramid Network [50] (in the following SSDMN). For both models pre-trained weights (COCO), a cosine decay learning rate schedule and several augmentation strategies were used, e.g. random crop, horizontal flip or hue adjustments.

In preparation of the study, an hyperparameter (h) search was conducted with regard to learning rates and training steps. The best leaning rates were 0.002 (FRCNN) and 0.005 (SSDMN). We found that saturation and overfitting occurs earlier with FRCNN than with SSDMN. While the real-world performance of FRCNN did not improve after 14 k images, SSDMN reached the peak after 28 k images. Using bigger dataset sizes or passing images multiple times regularly decreased the performance of FRCNN.

Metric (m): To measure the performance of the resulting CNNs the bounding box based mean average precision (mAP) over both classes is used. The mAPs are averaged over the training repetitions and reported in the range from 0.0 to 100.0. When comparing two performance scores, the signed difference of the averaged mAPs is used.

Evaluation Datasets (E): For the examination of the generated synthetic datasets (performance of the resulting CNN) a set of 960 real images split over four evaluation datasets was used, as illustrated in Fig. 2. These datasets deviate in different degrees from the real use case in terms of scene (camera, light and environment) and object appearance (texture, shape, size and lid type):

- Known scene and objects (**KK**): The images belong to the real use case.
- Known scene and unknown objects (**KU**): This dataset represents changes in the range of articles (e.g. emergence of articles of different appearance).
- Unknown scene and known objects (**UK**): The images represent environment changes (e.g. unexpected lighting conditions).
- Unknown scene and objects (**UU**): By combining article and environment changes, this dataset is the greatest challenge for the robustness of the CNNs.

Real Comparison Dataset: The maximum performance reached by synthetically trained CNNs depends on the task complexity. To allow an independent assessment, an additional training dataset with 6,454 real images was generated, manually annotated and tested with all evaluation datasets.

4 Experiments

We compared the two network models FRCNN and SSDMN for a selection of datasets and hyperparameters. As expected the smaller SSDMN reaches a lower performance for every training and evaluation dataset, e.g. a mAP of 56.2 compared to 75.3 for FRCNN (trained 26 k steps on SE and evaluated on KK). The absolute performance difference between the models differs depending on the datasets and the hyperparameters. However, we found similar relative differences when comparing the base scenes and the corresponding variations. The differences were slightly more pronounced for SSDMN, while the performance of FRCNN was more stable. We can conclude that the findings, presented in the following, are largely independent of the specific model.

When training models with exactly the same parameters and dataset we noticed high performance fluctuations up to a mAP difference of 5.0. In order to increase the evidence of the results, each training and evaluation was repeated 11 times on average. Instead of reporting the best results, the mAPs were averaged. This reduced the fluctuations to less than 0.5. The exact number of repetitions (5 to 100) was dynamically adapted to the significance of the results. Performance fluctuations between the compared dataset variations can largely be excluded due to the data generation process.

Using pre-trained network weights was very important for reality transfer in our use case. Training FRCNN with random weight initialization decreased the mAP from 75.3 to 19.8. This finding is consistent with other studies [4], although it is particularly pronounced in our examination.

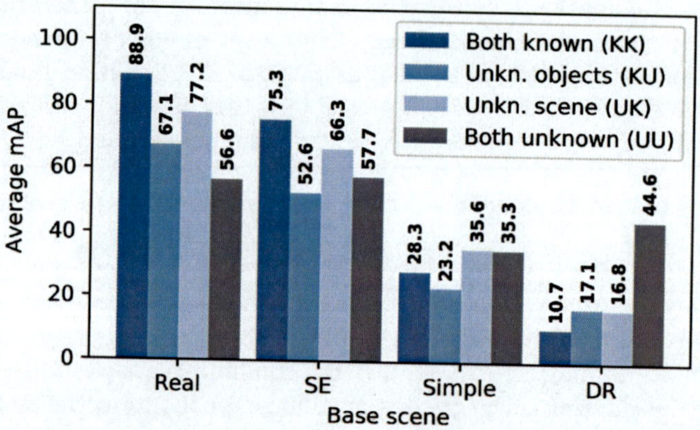

Fig. 3. Comparison of CNNs trained with base scene datasets (including real comparison dataset), tested on four evaluation datasets with different degrees of domain shift.

4.1 Base Scenes

With a mAP of 75.3, the **SE** scene reached the best results of the synthetic datasets for the known domain (KK), as seen in Fig. 3. Here, the **real comparison** dataset achieved a mAP of 88.9. Both, the SE as well as the real dataset, better generalize to unknown scenes (UK) than to unknown objects (KU). Due to the higher robustness through slight randomization, SE (mAP 57.7) beats the real dataset (mAP 56.6) for the fully unknown domain (UU).

Trained on the **simple** dataset the detectors perform worse on all evaluation datasets (mAP between 23.2 and 35.6), but better on unknown (UK and UU) than on known scenes. This is mainly due to better illumination in these datasets. These findings prove the importance of high quality CAD models, textures and rendering. Moreover, the simple scene misses the complexity and distraction of the real use case.

The **DR** dataset achieves the lowest precision on the known domain (KK), which proves the relevance of the scene context in tasks with tightly packed objects. However, it performs slightly better evaluated on unknown scenes (UK, mAP 16.8) and unknown objects (KU, mAP 17.1). This was to be expected since both are closer to the DR training data. When evaluated on the combination unknown scenes and objects (UU), DR (mAP 44.6) outperformed the simple dataset, which demonstrates the increased robustness due to the strong randomization.

4.2 Ablation Study: Scene Variations

In the following the single randomization parameters, listed in Table 1, are examined using the different variations of the base scenes. Additionally, special evaluations regarding important performance factors are presented. The focus is on the ablation of the SE approach tested on known data (KK), as shown in Fig. 4. The corresponding ablation results of the DR and simple scenes are shown in Fig. 5. The relevant results of the other three evaluation datasets (KU, UK, UU) are reported only in the text.

Textures and Colors: Using realistic textures clearly outperformed the other two strategies in our experiments. Using single color (−8.4) or random textures (−22.2) reduced the mAP of the SE base scene on known objects. The degradation due to random textures was less pronounced in the evaluation with unknown shoeboxes (KU), but was still present (−8.7). Adding realistic textures also had largest impact on the DR and simple scenes when detecting the known objects.

To investigate the effects of the tightly packed arrangements, a simplified evaluation dataset with only single unknown items was used (not indicated in a figure). Here, the CNNs trained on the real dataset (mAP 84.2) are outperformed by those trained on the SE dataset with random textures (mAP 85.2). This shows that the randomization of the texture improves the robustness of the resulting detectors, but leads to difficulties regarding the separation of single instances.

Camera and Viewport: For the variations of the camera no evident changes were observed when training with the SE dataset: mAP ±0.4 for static and strongly randomized camera. The randomization of the camera improved the mAP (+1.7) only for completely unknown data (UU). By choosing a unrealistic

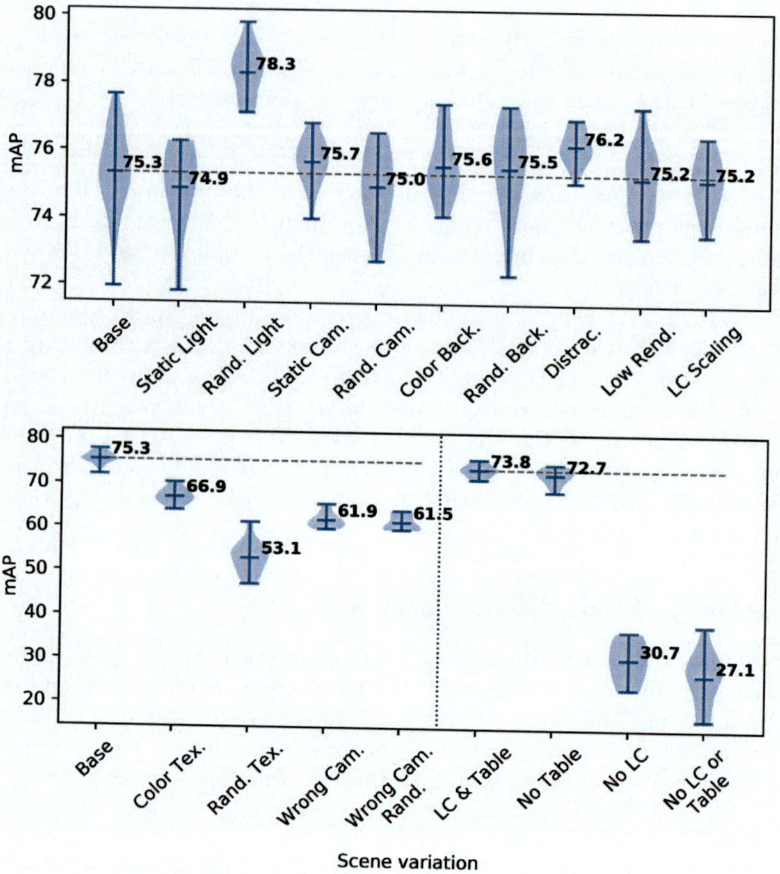

Fig. 4. Comparison of CNNs trained on variations of the SE base scene, including special datasets. The detectors are evaluated on known scene and objects (KK).

(wrong) camera origin, but where the objects are still visible, we were able to show that the viewport is of great importance: mAP -13.4 with realistic and -13,8 with strong randomization.

Lighting: A strong randomization of the lights regarding intensity and color achieved the best results for all datasets, whereas static lights always decreased the performance. The greatest impact was measured for the SE dataset (mAP +3.0). A central reason for the continuous improvement through light randomization is the difficulty of reconstructing realistic lighting conditions in simulation.

Scene Environment: In our study the scene environment consists of the load carrier (LC), a metallic table, the floor texture and randomly textured geometric distractor objects. Most relevant for a high performance of the trained object

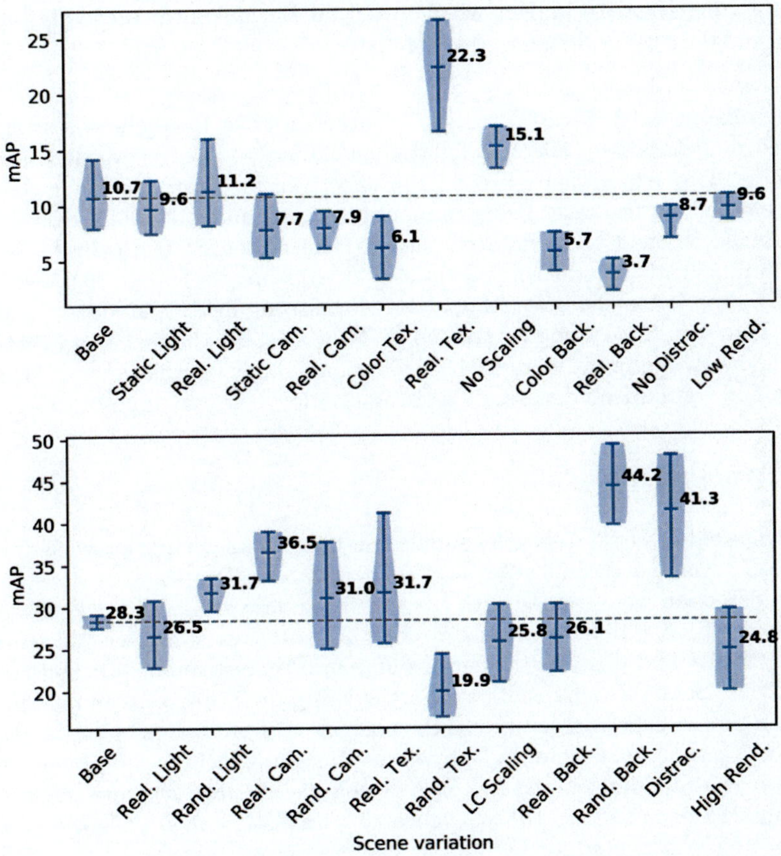

Fig. 5. Comparison of CNNs trained on different variations of the DR (top) and Simple (bottom) scene, evaluated on known scene and objects (KK).

detector was the inclusion of the LC (mAP −43.1 without). In contrast, random scaling of the LC had no measurable effect. Other scene surroundings such as the table were much less important (mAP −1.1 without table).

Distractor objects in the scene background improved the resulting CNNs for the SE scene: mAP +0.9 for known (KK) and +1.6 for unknown scene (UK). Random background textures, on the other hand, did not result in any significant improvement. The simple scene in contrast clearly benefited from both, randomized background (mAP +15.9) and distractors (mAP +13.0), proving the lack of complexity in the data.

Rendering Quality: Drastically reducing the rendering quality in Unity removed shadows and downgraded edge quality. While not noticeably affecting the performance on the known evaluation dataset, the lower rendering quality had a strong negative impact when tested on the unknown scene and objects (UU): mAP −3.2 for SE and −14.0 for DR.

Pre-training and Fine-Tuning: As other studies have shown, using pre-trained models is very effective for the simulation to reality transfer of neural networks. Our experiments proved the effectiveness of pre-trained models for our item picking use case. Using random initialization instead of model weights pre-trained on the COCO dataset, the mAP decreased dramatically (−55.6).

Training with a reduced real dataset of 100 images (mAP 79.6) was only slightly superior to the best SE data set with strong light randomization (mAP 78.3). However, fine-tuning SE trained CNNs with these 100 real images achieved a significantly higher mAP of 84.4, which is only 4.5 lower than the performance of the real comparison dataset of 6,454 images.

5 Conclusions

We compared three strategies for simulation-based data generation with different degrees of manual optimization and randomization. Deep neural networks for object detection were trained with the generated data and evaluated on multiple real datasets with different levels of domain shift. Our study was designed for comparability and provides evident insights into the precision and robustness of the CNNs. Some of our results support existing or intuitive assumptions with quantitative measurements, such as the usefulness of pre-trained models. Beyond that, we were able to gain and prove new insights, such as the superiority of realistic over randomized textures even on completely unknown objects. We were also able to demonstrate the relevance of a realistic camera viewport, which cannot be compensated for by random camera poses.

Our investigations prove that strong domain randomization as discussed in [4,9,15,25] cannot replace scene engineering in real industrial use cases with well known and comparatively static properties. Especially the scene context, in our use case the physically plausible item placement inside a load carrier, is key for a successful simulation to reality transfer in our and comparable use cases. This is consistent with the findings of [26]. In addition, the comparison with the simplified scene showed that beyond the scene context, a high degree of realism and complexity is necessary to train precise CNNs. In particular, realistic object models and textures as well as camera viewports are of crucial importance. Especially for larger domain gaps, the use of distractor objects and random backgrounds brought verifiable improvements.

Our ablation studies allows the classification individual randomized properties in relation to different domain gaps. We could show that the randomization of appropriate features with well-chosen parameters improves the robustness of the trained neural networks significantly. Since the reproduction of the lighting conditions is complex, we recommend a strong randomization of the lights. While a positive influence could also be determined for distractor objects, all other randomization properties did not bring any significant improvement for the manually constructed and optimized scene (SE).

Since scene engineering still requires a lot of manual effort, future work should focus on the automation of scene adaptation. This can target the automated parametrization of the simulation, e.g. lighting and scene structure, as well as the automated search for well suited randomization properties. In addition, the relevance of the scene context should be set in relation to photo-realistic rendering and post-processing with GANs. Finally, we found surprisingly large variations in performance when training with synthetic data compared to real data. Although we encountered this situation with a high number of training repetitions, additional investigations into the reliability of synthetically trained CNNs are necessary.

References

1. Lin, T.-Y., Maire, M., Belongie, S., Hays, J., Perona, P., Ramanan, D., Dollár, P., Zitnick, C.L.: Microsoft COCO: Common Objects in Context. In: Fleet, D., Pajdla, T., Schiele, B., Tuytelaars, T. (eds.) ECCV 2014. LNCS, vol. 8693, pp. 740–755. Springer, Cham (2014). https://doi.org/10.1007/978-3-319-10602-1_48
2. Deng, J., Dong, W., Socher, R., Li, L.J., Li, K., Fei-Fei, L.: Imagenet: a large-scale hierarchical image database. In: 2009 IEEE Conference on Computer Vision and Pattern Recognition. pp. 248–255. IEEE, Piscataway, NJ (2009)
3. Goodfellow, I.J., Pouget-Abadie, J., Mirza, M., Xu, B., Warde-Farley, D., Ozair, S., Courville, A., Bengio, Y.: Generative adversarial nets. In: Proceedings of the 27th International Conference on Neural Information Processing Systems - Volume 2. pp. 2672–2680. NIPS'14, MIT Press, Cambridge, MA, USA (2014)
4. Tobin, J., Fong, R., Ray, A., Schneider, J., Zaremba, W., Abbeel, P.: Domain randomization for transferring deep neural networks from simulation to the real world. In: 2017 IEEE/RSJ International Conference on Intelligent Robots and Systems (IROS), pp. 23–30 (2017)
5. Weisenboehler, M., Wurll, C.: Automated item picking for fashion articles using deep learning. In: ISR 2020; 52th International Symposium on Robotics, pp. 1–8 (2020)
6. Tobin, J., Biewald, L., Duan, R., Andrychowicz, M., Handa, A., Kumar, V., McGrew, B., Ray, A., Schneider, J., Welinder, P., Zaremba, W., Abbeel, P.: Domain randomization and generative models for robotic grasping. In: 2018 IEEE/RSJ International Conference on Intelligent Robots and Systems (IROS), pp. 3482–3489 (2018)
7. Fang, K., Bai, Y., Hinterstoisser, S., Savarese, S., Kalakrishnan, M.: Multi-task domain adaptation for deep learning of instance grasping from simulation. In: 2018 IEEE International Conference on Robotics and Automation (ICRA), pp. 3516–3523 (2018)
8. Zuo, G., Zhang, C., Liu, H., Gong, D.: Low-quality rendering-driven 6d object pose estimation from single RGB image. In: 2020 International Joint Conference on Neural Networks (IJCNN), pp. 1–8 (2020)
9. James, S., Davison, A.J., Johns, E.: Transferring end-to-end visuomotor control from simulation to real world for a multi-stage task. arXiv:1707.02267 (2017)

10. James, S., Wohlhart, P., Kalakrishnan, M., Kalashnikov, D., Irpan, A., Ibarz, J., Levine, S., Hadsell, R., Bousmalis, K.: Sim-to-real via sim-to-sim: Data-efficient robotic grasping via randomized-to-canonical adaptation networks. In: 2019 IEEE/CVF Conference on Computer Vision and Pattern Recognition (CVPR), pp. 12619–12629 (2019)
11. Ho, D., Rao, K., Xu, Z., Jang, E., Khansari, M., Bai, Y.: Retinagan: an object-aware approach to sim-to-real transfer. arXiv:2011.03148 (2020)
12. Sadeghi, F., Levine, S.: Cad2rl: Real single-image flight without a single real image. In: Amato, N., Systems, R.S.A. (eds.) Robotics: Science and System XIII. Robotics: Science and Systems Online Proceedings (2017)
13. Sadeghi, F., Toshev, A., Jang, E., Levine, S.: Sim2real viewpoint invariant visual servoing by recurrent control. In: 2018 IEEE/CVF Conference on Computer Vision and Pattern Recognition (CVPR 2018), pp. 4691–4699. IEEE, Piscataway, NJ (2018)
14. Georgakis, G., Mousavian, A., Berg, A., Kosecka, J.: Synthesizing training data for object detection in indoor scenes. In: Amato, N., Systems, R.S.a. (eds.) Robotics: Science and System XIII. Robotics: Science and Systems Online Proceedings (2017)
15. Hinterstoisser, S., Lepetit, V., Wohlhart, P., Konolige, K.: On Pre-trained Image Features and Synthetic Images for Deep Learning. In: Leal-Taixé, L., Roth, S. (eds.) ECCV 2018. LNCS, vol. 11129, pp. 682–697. Springer, Cham (2019). https://doi.org/10.1007/978-3-030-11009-3_42
16. Hinterstoisser, S., Pauly, O., Heibel, H., Martina, M., Bokeloh, M.: An annotation saved is an annotation earned: Using fully synthetic training for object detection. In: 2019 IEEE/CVF International Conference on Computer Vision Workshop (ICCVW), pp. 2787–2796 (2019)
17. Dehban, A., Borrego, J., Figueiredo, R., Moreno, P., Bernardino, A., Santos-Victor, J.: The impact of domain randomization on object detection: a case study on parametric shapes and synthetic textures. In: 2019 IEEE/RSJ International Conference on Intelligent Robots and Systems (IROS), pp. 2593–2600. IEEE, [Piscataway, NJ] (2019)
18. Lee, Y., Chuang, C., Lai, S., Jhang, Z.: Automatic generation of photorealistic training data for detection of industrial components. In: 2019 IEEE International Conference on Image Processing (ICIP), pp. 2751–2755 (2019)
19. Noh, S., Back, S., Kang, R., Shin, S., Lee, K.: Automatic detection and identification of fasteners with simple visual calibration using synthetic data. In: 2020 25th IEEE International Conference on Emerging Technologies and Factory Automation (ETFA), vol. 1, pp. 38–44 (2020)
20. Magaña, A., Wu, H., Bauer, P., Reinhart, G.: Posenetwork: pipeline for the automated generation of synthetic training data and cnn for object detection, segmentation, and orientation estimation. In: 2020 25th IEEE International Conference on Emerging Technologies and Factory Automation (ETFA), vol. 1, pp. 587–594 (2020)
21. Gaidon, A., Wang, Q., Cabon, Y., Vig, E.: Virtualworlds as proxy for multi-object tracking analysis. In: 2016 IEEE Conference on Computer Vision and Pattern Recognition (CVPR 2016), pp. 4340–4349. IEEE, Piscataway, NJ (2016)
22. Johnson-Roberson, M., Barto, C., Mehta, R., Sridhar, S.N., Rosaen, K., Vasudevan, R.: Driving in the matrix: Can virtual worlds replace human-generated annotations for real world tasks? In: 2017 IEEE International Conference on Robotics and Automation (ICRA), pp. 746–753. IEEE Computer Society, Danvers (MA) (2017)

23. Tsirikoglou, A., Kronander, J., Wrenninge, M., Unger, J.: Procedural modeling and physically based rendering for synthetic data generation in automotive applications. arXiv:1710.06270 (2017)
24. Abu Alhaija, H., Karthik Mustikovela, S., Mescheder, L., Geiger, A., Rother, C.: Augmented reality meets computer vision: efficient data generation for urban driving scenes. Int. J. Comput. Vis. **126**(9), 961–972 (2018)
25. Tremblay, J., Prakash, A., Acuna, D., Brophy, M., Jampani, V., Anil, C., To, T., Cameracci, E., Boochoon, S., Birchfield, S.: Training deep networks with synthetic data: Bridging the reality gap by domain randomization. In: 2018 IEEE/CVF Conference on Computer Vision and Pattern Recognition Workshops (CVPRW 2018). pp. 1082–10828. IEEE, Piscataway, NJ (2018)
26. Prakash, A., Boochoon, S., Brophy, M., Acuna, D., Cameracci, E., State, G., Shapira, O., Birchfield, S.: Structured domain randomization: bridging the reality gap by context-aware synthetic data. In: 2019 IEEE International Conference on Robotics and Automation (ICRA), pp. 7249–7255 (2019)
27. Peng, X., Sun, B., Ali, K., Saenko, K.: Learning deep object detectors from 3d models. In: 2015 IEEE International Conference on Computer Vision (ICCV), pp. 1278–1286. IEEE (2015)
28. Su, H., Qi, C.R., Li, Y., Guibas, L.J.: Render for cnn: Viewpoint estimation in images using cnns trained with rendered 3d model views. In: 2015 IEEE International Conference on Computer Vision (ICCV), pp. 2686–2694. IEEE (2015)
29. Rudorfer, M., Neumann, L., Krüger, J.: Towards learning 3d object detection and 6d pose estimation from synthetic data. In: 2019 24th IEEE International Conference on Emerging Technologies and Factory Automation (ETFA), pp. 1540–1543 (2019)
30. Dwibedi, D., Misra, I., Hebert, M.: Cut, paste and learn: Surprisingly easy synthesis for instance detection. In: 2017 IEEE International Conference on Computer Vision (ICCV), pp. 1310–1319 (2017)
31. Ruiz, N., Schulter, S., Chandraker, M.: Learning to simulate. In: International Conference on Learning Representations (2019). https://openreview.net/forum?id=HJgkx2Aqt7
32. Khirodkar, R., Yoo, D., Kitani, K.M.: Vadra: visual adversarial domain randomization and augmentation. arXiv:1812.00491 (2018)
33. Kar, A., Prakash, A., Liu, M.Y., Cameracci, E., Yuan, J., Rusiniak, M., Acuna, D., Torralba, A., Fidler, S.: Meta-sim: learning to generate synthetic datasets. In: 2019 International Conference on Computer Vision. pp. 4550–4559. IEEE, Piscataway, NJ (2019)
34. Bousmalis, K., Irpan, A., Wohlhart, P., Bai, Y., Kelcey, M., Kalakrishnan, M., Downs, L., Ibarz, J., Pastor, P., Konolige, K., Levine, S., Vanhoucke, V.: Using simulation and domain adaptation to improve efficiency of deep robotic grasping. In: 2018 IEEE International Conference on Robotics and Automation (ICRA), pp. 4243–4250. IEEE (2018)
35. Hoffman, J., Tzeng, E., Park, T., Zhu, J.Y., Isola, P., Saenko, K., Efros, A., Darrell, T.: Cycada: cycle-consistent adversarial domain adaptation. In: International Conference on Machine Learning, pp. 1989–1998 (2018). https://proceedings.mlr.press/v80/hoffman18a.html
36. Nogues, F.C., Huie, A., Dasgupta, S.: Object detection using domain randomization and generative adversarial refinement of synthetic images. arXiv:1805.11778 (2018)

37. Park, N., Mohammadi, M., Gorde, K., Jajodia, S., Park, H., Kim, Y.: Data synthesis based on generative adversarial networks. Proc. VLDB Endow. **11**(10), 1071–1083 (2018). https://doi.org/10.14778/3231751.3231757
38. Lin, Y., Suzuki, K., Takeda, H., Nakamura, K.: Generating synthetic training data for object detection using multi-task generative adversarial networks. ISPRS Ann. Photogrammetry, Rem. Sens. Spat. Inf. Sci. **2**, 443–449 (2020)
39. Rao, K., Harris, C., Irpan, A., Levine, S., Ibarz, J., Khansari, M.: Rl-cyclegan: reinforcement learning aware simulation-to-real. In: Proceedings, 2020 IEEE/CVF Conference on Computer Vision and Pattern Recognition, pp. 11154–11163. IEEE Computer Society, Conference Publishing Services, Los Alamitos, California (2020)
40. Geiger, A., Lenz, P., Stiller, C., Urtasun, R.: Vision meets robotics: the kitti dataset. Int. J. Robot. Res. **32**(11), 1231–1237 (2013)
41. Dörr, L., Brandt, F., Meyer, A., Pouls, M.: Lean training data generation for planar object detection models in unsteady logistics contexts. In: 2019 18th IEEE International Conference On Machine Learning and Applications (ICMLA), pp. 329–334 (2019)
42. Sarkar, K., Varanasi, K., Stricker, D.: Trained 3d models for CNN based object recognition. In: Proceedings of the 12th International Joint Conference on Computer Vision, Imaging and Computer Graphics Theory and Applications, pp. 130–137. SCITEPRESS - Science and Technology Publications (2017)
43. Khirodkar, R., Yoo, D., Kitani, K.: Domain randomization for scene-specific car detection and pose estimation. In: 2019 IEEE Winter Conference on Applications of Computer Vision (WACV), pp. 1932–1940, IEEE (2019)
44. Grün, S., Höninger, S., Scheikl, P.M., Hein, B., Kröger, T.: Evaluation of domain randomization techniques for transfer learning. In: 2019 19th International Conference on Advanced Robotics (ICAR), pp. 481–486 (2019)
45. Huang, J., Rathod, V., Sun, C., Zhu, M., Korattikara, A., Fathi, A., Fischer, I., Wojna, Z., Song, Y., Guadarrama, S., Murphy, K.: Speed/accuracy trade-offs for modern convolutional object detectors. In: Staff, I. (ed.) 2017 IEEE Conference on Computer Vision and Pattern Recognition (CVPR), pp. 3296–3297. IEEE, Piscataway, NJ (July (2017)
46. Ren, S., He, K., Girshick, R., Sun, J.: Faster R-CNN: towards real-time object detection with region proposal networks. IEEE Trans. Pattern Anal. Mach. Intell. **39**(6), 1137–1149 (2017)
47. Szegedy, C., Ioffe, S., Vanhoucke, V., Alemi, A.A.: Inception-v4, inception-resnet and the impact of residual connections on learning. In: Proceedings of the Thirty-First AAAI Conference on Artificial Intelligence. pp. 4278–4284. AAAI'17, AAAI Press (2017)
48. Liu, W., Anguelov, D., Erhan, D., Szegedy, C., Reed, S., Fu, C.-Y., Berg, A.C.: SSD: Single Shot MultiBox Detector. In: Leibe, B., Matas, J., Sebe, N., Welling, M. (eds.) ECCV 2016. LNCS, vol. 9905, pp. 21–37. Springer, Cham (2016). https://doi.org/10.1007/978-3-319-46448-0_2
49. Sandler, M., Howard, A., Zhu, M., Zhmoginov, A., Chen, L.C.: Mobilenetv2: inverted residuals and linear bottlenecks. In: 2018 IEEE/CVF Conference on Computer Vision and Pattern Recognition (CVPR 2018). pp. 4510–4520. IEEE, Piscataway, NJ (2018)
50. Lin, T.Y., Dollar, P., Girshick, R., He, K., Hariharan, B., Belongie, S.: Feature pyramid networks for object detection. In: Staff, I. (ed.) 2017 IEEE Conference on Computer Vision and Pattern Recognition (CVPR), pp. 936–944. IEEE, Piscataway, NJ (2017)

Hybrid Intelligence for Visual Identity Design: A Case Study

Bojan Jerbić[1](✉), Marko Švaco[1], Filip Šuligoj[1], Bojan Šekoranja[1], Josip Vidaković[1], Marija Turković[1], Mihaela Lekić[1], Borjan Pavlek[2], Bruno Bolfan[2], Davor Bruketa[2], Dina Borošić[2], and Barbara Bušić[2]

[1] Faculty of Mechanical Engineering and Naval Architecture, University of Zagreb, Zagreb, Croatia
bojan.jerbic@fsb.hr
[2] Bruketa&Žinić&Grey, Zagreb, Croatia
https://crta.fsb.hr/

Abstract. Design usually relies on human ingenuity, but the past decade has seen the field's toolbox expanding to Artificial Intelligence (AI) and its adjacent methods, making room for hybrid, algorithmic creations. This article aims to substantiate the concept of hybrid intelligence– that of natural and artificial intelligence—in the active co-creation of a visual identity, describing a case study of the Regional Center of Excellence for Robotic Technology (CRTA) which opened on 750 m^2 in June 2021 within the University of Zagreb. The visual identity of the Center comprises three separately devised elements, each representative of the human-AI relationship and embedded in the institution's logo. Firstly, the letter "C" (from the CRTA acronym) was created using a Gaussian Mixture Model (GMM) applied to (x, y) coordinates that the neurosurgical robot RONNA, CRTA's flagship innovation, generated when hand-guided by a human operator. The second shape of the letter "C" was created by using the same (x, y) coordinates as inputs fed to a neural network whose goal was to output letters in a novel, AI-generated typography. A basic feedforward neural network with two hidden layers was chosen for the task. The final and third design element was a trajectory the robot RONNA makes when performing a brain biopsy. As CRTA embodies a state-of-the-art venue for robotics research, the hybrid intelligence approach was used to accentuate the importance of human-robot collaboration which is at the core of the newly opened Center, illustrating the potential of reciprocal and amicable relationship that humans could have with technology.

Keywords: Hybrid intelligence · Neural networks · Computer art · Visual identity · Robotics

1 Introduction

The main goal of the University of Zagreb's CRTA, the Regional Center of Excellence for Robotic Technology (Fig. 1.), is to establish a reference center for research, development, and educational activities in the field of robotics and artificial intelligence. Activities of

I. Petrovic et al. (Eds.): IAS 2022, LNNS 577, pp. 661–670, 2023.
https://doi.org/10.1007/978-3-031-22216-0_44

the Center are focused on research and development of advanced robotic applications, especially in industry and medicine, but also in other areas of human activity where traditional automation and human labor should be replaced by autonomous and intelligent systems. An important step in such a venture is to enable all stakeholders of the robotics revolution, i.e., the general public and not exclusively scientists, to participate in the shaping of a new society that is to stand in close interaction with technology.

Fig. 1. A view of collaborative robotic systems next to assembly lines in the Laboratory for Autonomous Systems, one of the three research laboratories at CRTA

RONNA (Fig. 2.), a robot designed to perform brain surgery, was the 'co-creator' of the new visual identity for the Center. RONNA is a project successfully realized within the Faculty of Mechanical Engineering and Naval Architecture, University of Zagreb, which became a part of the Faculty and University identity. The clinical application of RONNA is intended for minimally invasive stereotactic procedures such as biopsies, stereoelectroencephalography, epilepsy surgeries, deep brain stimulation, and tumor resections [1, 2]. RONNA has performed numerous medical procedures since its first application in March 2016 in University Hospital Dubrava, one of Croatia's largest hospitals, where it has been undergoing clinical trials ever since. Although proven in neurosurgery, RONNA had to 'learn' how to write and interpret letters, a building block of the Center's visual identity.

2 Related Work

Since the 2010s, hundreds of art collectives had sprung out of the idea of AI-generated art, a notable commercial example being the Obvious collective from Paris [3]. Obvious relied on Robbie Barrat's code for Generative Adversarial Networks (GAN), a type of neural network architecture, to 'paint' portraits of the fictional Belamy family after being trained on more than 15,000 18th-century-style portraits and later on to sell one

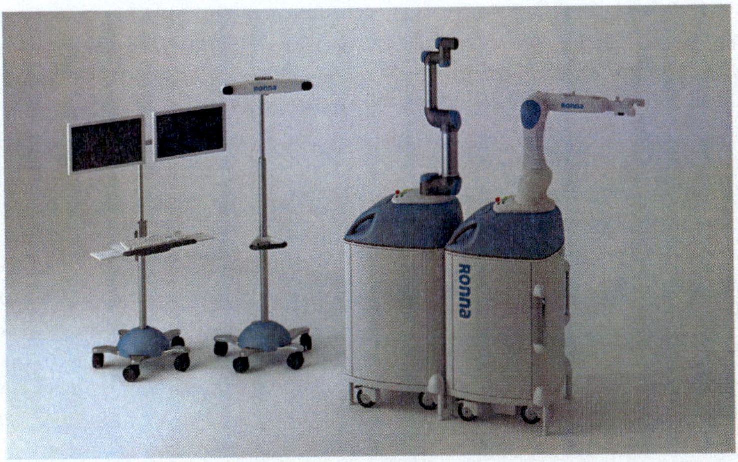

Fig. 2. The 3rd generation of the robotic system RONNA with an assistant (left) and master robot (right)

at Christie's, an NYC auction house, for $432,500. Questions of who gets credit for AI-based art have already been raised [4, 5]—man or machine, the engineer or the artist—hinting at the substantial human effort needed to structure the underlying databases and instructions for such seemingly autonomous systems to work, and to the interdependence of that AI-art communities have with the pioneering work of computer scientists. Though GANs are the current 'go-to' choice for most visual artists, some initiatives explored the idea of physical, robotic agents participating in the creative process. Roomba Paintings by Bobby Zokaites (2006) [6] showed how household robots could be transformed into painters by 'replacing their dust collectors with paint reservoirs'.

Apart from art, the design and communication realms have experienced a steady flow of AI innovations as well. In 2017, seven million jars of Nutella were designed by an algorithm which pulled from a database of dozens of patterns and colors to create unique packaging designs. Nutella's manufacturer Ferrero titled the project Nutella Unica [7] and worked on it with the Ogilvy & Mather Italia advertising agency. The limited edition of Nutella's label came in thousands of pattern and color combinations, each stamped with its unique code so it could be authenticated by collectors. Airbnb, the home-sharing company, developed a machine learning system [8] that can classify hand-drawn design components and turn sketches into prototypes with engineering specifications—almost instantaneously. In 2018, Lexus, the luxury division of Toyota, released an advert that had been scripted entirely by an artificial intelligence algorithm and shot by an Oscar-winning director Kevin Macdonald [9]. The algorithm was trained on 15 years of Cannes Lions International award-winning automotive ads, coupled with other data points on consumer behavior. A minute-long film is telling a story about a 'Takumi', an honorary title for a Lexus master craftsman, who completes the new Lexus ES car, which, as soon as it is released into the world, is facing an imminent crash. At a deciding moment, the car's automatic emergency braking system is activated, demonstrating the utility of intelligent features built into the car and leading the inhouse PR experts to conclude, *'This world-first collaboration between AI and a renowned craftsman (is) aimed to test the boundaries of how humans and machines can work together in perfect harmony, exploring the importance of intuition in the relationship between the two.'* The logo design process has also been impacted, though in limited capacity, because of the particular, aggravating properties of logos themselves. Unlike images, logos are hard to label since they contain a very low number of properties that are at the same time rich in meaning and suggestion [10], as emphasized by Sage et al. who built a logo synthesis GAN algorithm trained on more than 600,000 web-scraped logos.

Most of the aforementioned initiatives were carried out *in silico*, or in the virtual space, without interaction with elements in the natural environment such as humans. These projects framed the importance of AI-based algorithms in art and design as (quasi)independent and isolated creators, though some of them pointed to the symbiotic potential that AI could realize only when actively collaborating with humans—in hybrid intelligence i.e. 'interspecies' collaboration—'on equal terms', as working colleagues taking into consideration the other side's feedback. The term 'interspecies collaboration' was previously used almost entirely in the context of collaboration of humans and animals or plants [11, 12] while this article aims to stretch that definition to cover AI-based algorithms which are proving to have ever-increasing agency, forming a sort of 'species' with an established relationship to their evolutionary predecessors, i.e., humans.

3 Method

The approach to devising a visual identity for CRTA was to form a graphic solution consisting of three elements—each of which, in its own way, connects man and artificial intelligence. The components were chosen in collaboration with the Bruketa&Žinić&Grey agency and were as follows: 1. a hand-guided letter "C" (the first letter of the CRTA acronym) realized through the movement of RONNA's robotic arm; 2. an artificially generated letter "C" realized as a neural network output; 3. a brain surgery trajectory of the robot RONNA (Fig. 3.).

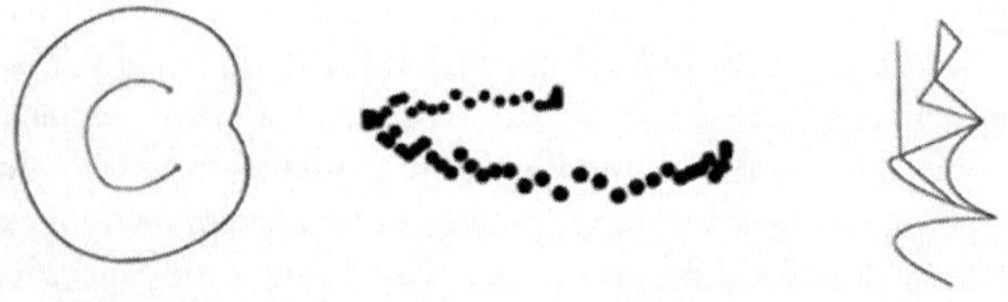

Fig. 3. A tripartite strategy for the visual identity of CRTA—from left to right: a hand-guided letter "C", an artificially-generated letter "C", and a brain surgery trajectory by RONNA

3.1 A Hand-Guided Letter "C"

The staff at CRTA, professors and students, each hand-guided the RONNA assistant robot to write the letter "C" in their own gestures and on a whiteboard (Fig. 4.), resulting in nine depictions of the letter, each unique to the person hand-guiding the robot and thus making the CRTA staff a part of the visual identity of the institution they would soon be actively involved with. The robot recorded (x, y) coordinates generated during its movement through space [Table 1.] with a fixed step of 10 Hz.

Fig. 4. A member of CRTA staff hand-guiding the robot

Table 1. Example of robot coordinates for a single letter "C"

−0.253, 0.341	−0.255, 0.345	−0.307, 0.424	−0.515, 0.282
−0.255, 0.345	−0.255, 0.346	−0.326, 0.426	−0.527, 0.256
−0.255, 0.345	−0.257, 0.351	−0.348, 0.426	−0.537, 0.231
−0.255, 0.345	−0.260, 0.357	−0.372, 0.422	−0.545, 0.205
−0.255, 0.345	−0.263, 0.367	−0.394, 0.414	−0.550, 0.183
−0.255, 0.345	−0.268, 0.378	−0.419, 0.403	−0.548, 0.161
−0.255, 0.345	−0.271, 0.387	−0.443, 0.385	−0.540, 0.143
−0.255, 0.345	−0.276, 0.398	−0.462, 0.366	−0.527, 0.130
−0.255, 0.345	−0.281, 0.409	−0.481, 0.340	−0.515, 0.122
−0.255, 0.345	−0.291, 0.418	−0.498, 0.312	−0.505, 0.119
…	…	…	

A Gaussian Mixture Model (GMM) was applied to the hand-guided (x, y) coordinates of the robot trajectory, as described and implemented by Vidaković et al. [13]. The resulting effect was a smooth and vectorized curve of the letters written through human-robot collaboration. The generalized letter "C" was used as a baseline for further graphic interventions by the human designer and ultimately formed the dominant part of the visual identity of the whole Center. The figure below (Fig. 5.) shows a GMM used for describing two-dimensional spatial curves with 10 normal distributions. Each normal distribution represents a field that 'attracts' the trajectories. A generalized curve was therefore given by approximation, using the coordinates of centers of normal distributions, and finally vectorized (Fig. 6.).

3.2 An Artificially-Generated Letter "C"

All hand-guided motifs were analyzed through an artificial neural network [14, 15] that gave its interpretation of the motifs by creating new representations of the input

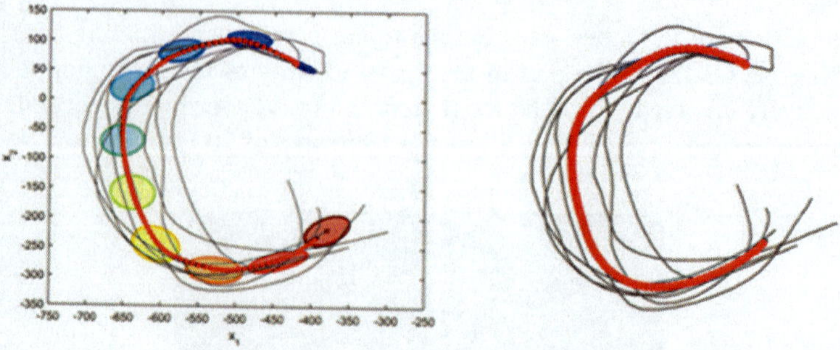

Fig. 5. Curve generalization using GMM

Fig. 6. Example of a vectorized curve of a single hand-guided letter derived from nine other letters

letters. By overlapping the human and artificially created motifs, a unique graphic art was generated—specific to the time, space, humans, and robots that participated in the creative process. The goal of the neural network was to learn to create different kinds of letters. However, since there were only a few input values (the dataset consisted of only nine letters generated during nine human-robot collaboration instances), the algorithm was constructed in a way that makes combinations of input and output pairs and in that way it was artificially enlarged to 56 input-output pairs in total [Table 2]. The input-output pairs of images with the same indices (n-n) were excluded from the dataset as the network would not be able to learn from these.

Table 2. Artificial enlargement of the input dataset using nine letters to form 56 input-output pairs

(Input Image – Output Image) pairs							
1–2	2–1	3–1	4–1	5–1	6–1	7–1	8–1
1–3	2–3	3–2	4–2	5–2	6–2	7–2	8–2
1–4	2–4	3–4	4–3	5–3	6–3	7–3	8–3
1–5	2–5	3–5	4–5	5–4	6–4	7–4	8–4
1–6	2–6	3–6	4–6	5–6	6–5	7–5	8–5
1–7	2–7	3–7	4–7	5–7	6–7	7–6	8–6
1–8	2–8	3–8	4–8	5–8	6–8	7–8	8–7

The dataset was divided into two groups—the training set which had different combinations of input and output pairs, and the test set. The input of a letter "C" written in one way would result in a different font of the same letter. The test set consisted of only one image, because the training set needed to be as large as possible and, in theory, only one output image needed to be forwarded to the graphic designer for final adjustments. The neural network was a back-propagating network with two hidden layers. The number of neurons in each layer was 10. The network had a learning rate of 0.0001 and the number of iterations was 10,000. The hyperbolic tangent activation function was used, and the optimization function was stochastic gradient descent. The cost function used for training was the following:

$$\text{Train Cost} = \frac{\sum (y_{\text{predicted}} - y_{\text{actual}})^2}{2}$$

Weights (w) and biases (b) were initialized as random values from a uniform distribution in the range of [0, 1] for weights and [0, 0.1] for biases. The formulas for each layer were as follows:

x == input [first layer]
a1 = tanh (input x w1 + b1 [1st hidden layer]
a2 = tanh (a1 x w2 + b2 [2nd hidden layer]
output == y = a2 x w3 + b3 [last layer]

The input and output values were the robot coordinates in 2D space (x, y), i.e., the horizontal whiteboard. Since all the inputs and outputs (x, y) were in the range of [-1, 1], there was no need to normalize them. The artificially generated letter "C" representing a dotted trajectory (Fig. 7.) was then handed to the designers at the Bruketa&Žinić&Grey agency for final adjustments. The "C" was extruded randomly to form the z-axis and eventually turned into a new 3D object which could be viewed in 360 °. The neural network output was then rearranged on the canvas of the logos in various configurations, therefore embedding the AI-generated typography into the core design of the Center.

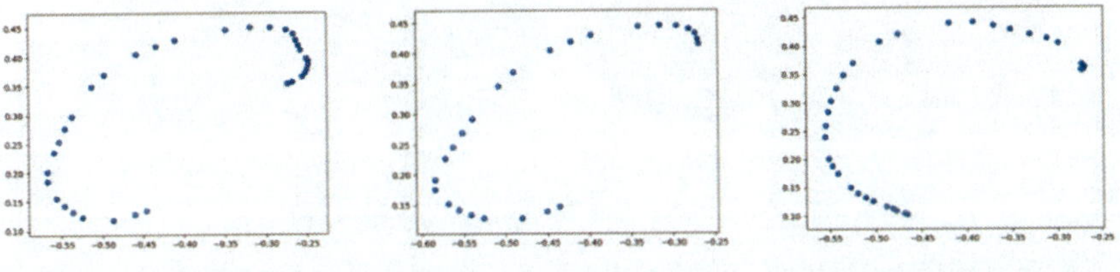

Fig. 7. Examples of neural network outputs of the letter "C" in a novel, AI typography

3.3 A Brain Surgery Trajectory by RONNA, CRTA'S Flagship Robotic Innovation

A neurosurgery trajectory (Fig. 8.) was generated by recording points of RONNA performing a patient localization and brain biopsy procedure with a fixed time step of 10 Hz. The points were interpolated and finally formed a continuous trajectory.

Fig. 8. Neurosurgery trajectory of the robot RONNA – from left to right: a 3D trajectory of brain surgery by RONNA, a 2D vector representation of the trajectory, and a simplified vector representation of the trajectory

The final design elements shown in Fig. 6., Fig. 7., and Fig. 8. Were then assembled to form the definite visual identity of CRTA shown in Fig. 10. And Fig. 11., an endeavor which required various post-processing steps executed in parallel (Fig. 9.).

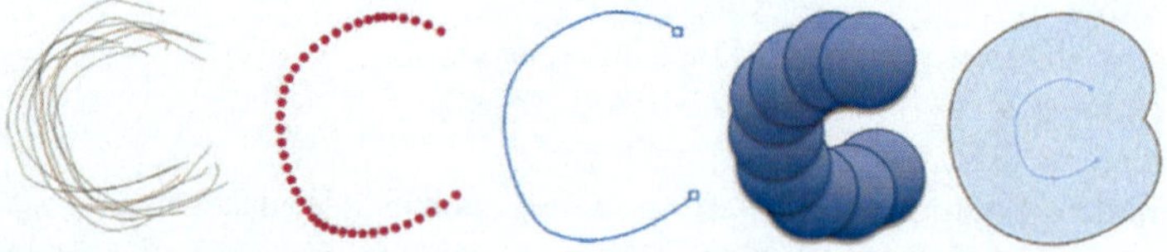

Fig. 9. Different shapes of letters specific to the way they were created—from left to right: hand-guided letters, neural network input, vectorized generalization of hand-guided robot trajectories, interventions by a human designer

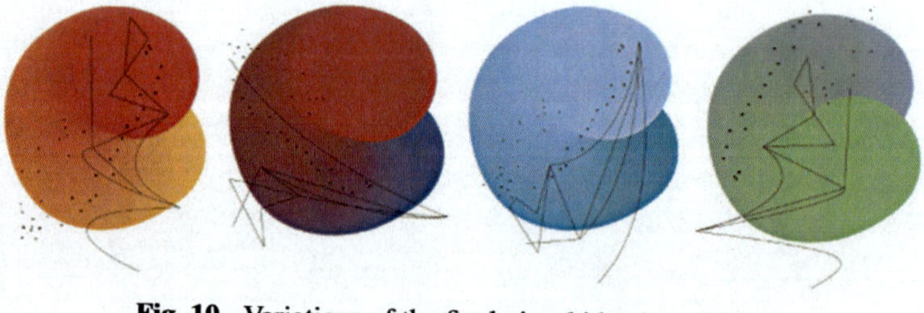

Fig. 10. Variations of the final visual identity of CRTA

4 Conclusion

We demonstrated how neural networks and statistical methods could be applied to produce a unique, authentic and highly representative visual identity of an institution in close collaboration with humans. Three design elements were created: a vectorized curve based on hand-guided letters "C", a reinterpretation of the letter "C" in a novel, AI typography, and a neurosurgery trajectory of the robot RONNA. The incorporation

Fig. 11. Applications of CRTA's symbolic logo in office stationery and the physical space of the Center

of various artificially generated design components into the final design was crucial in communicating all important aspects of the newly opened Regional Center of Excellence for Robotic Technology (CRTA). We believe this case study could serve as a valuable precedent for branding initiatives aimed at leveraging the immense expressive reserves of statistical and AI tools for designers.

Acknowledgements. The authors would like to acknowledge the support of the European Union's European Regional Development Fund which financed the project CRTA.

References

1. Jerbić, B., Švaco, M., Chudy, D., Šekoranja, B., Šuligoj, F., Vidaković, J., Dlaka, D., Vitez, N., Župančić, I., Drobilo, L., Turković, M., Žgaljić, A., Kajtazi, M., Stiperski, I.: RONNA G4—Robotic Neuronavigation: A novel robotic navigation device for stereotactic neurosurgery. In: Handbook of Robotic and Image-Guided Surgery, 599–625 (2020). https://doi.org/10.1016/B978-0-12-814245-5.00035-9
2. Dlaka, D., et al.: Frameless stereotactic brain biopsy: a prospective study on robot-assisted brain biopsies performed on 32 patients by using the RONNA G4 system. Int. J. Medical Rob. Comput. Assist. Surg. **17**(3), e2245 (2021). https://doi.org/10.1002/RCS.2245
3. Christie, S.: Is artificial intelligence set to become art's next medium?. https://www.christies.com/features/a-collaboration-between-two-artists-one-human-one-a-machine-9332-1.aspx (2018). Accessed 20 Jan 2022
4. Epstein, Z., Levine, S., Rand, D.G., Rahwan, I.: Who gets credit for AI-generated art?. IScience **23**(9), 101515 (2020). https://doi.org/10.1016/J.ISCI.2020.101515
5. McCormack, J., Gifford, T., Hutchings, P.: Autonomy, Authenticity, Authorship and Intention in computer generated art. In: Lecture Notes in Computer Science (Including Subseries Lecture Notes in Artificial Intelligence and Lecture Notes in Bioinformatics), 11453 LNCS, 35–50 (2019). https://doi.org/10.1007/978-3-030-16667-0_3
6. Zokaites, B.: Roomba paintings: 2006. http://www.bobbyzokaites.com/roomba-paintings (2006). Accessed 20 Jan 2022
7. Aouf, R. S. for Dezeen.: Algorithm designs seven million different jars of Nutella. https://www.dezeen.com/2017/06/01/algorithm-seven-million-different-jars-nutella-packaging-design/ (2017). Accessed 20 Jan 2022
8. Benjamin Wilkins for Airbnb.: Sketching interfaces—generating code from low fidelity wireframes (2017). https://airbnb.design/sketching-interfaces/. Accessed 20 Jan 2022

9. Lexus Europe Newsroom (2018). https://newsroom.lexus.eu/driven-by-intuition-car-by-lexus-story-by-artificial-intelligence-camera-by-oscar-winning-director/. Accessed 20 Jan 2022
10. Sage, A., Agustsson, E., Timofte, R., Van Gool, L.: Logo synthesis and manipulation with clustered generative adversarial networks. In: Proceedings of the IEEE Computer Society Conference on Computer Vision and Pattern Recognition, 5879–5888 (2017). https://doi.org/10.1109/CVPR.2018.00616
11. Jevbratt, L.: Interspecies collaboration—making art together with nonhuman animals. Presented at the Minding Animals Conference in Newcastle, Australia in July (2009). http://jevbratt.com/home_texts.html
12. Fischer, D.: Art Between Species: Two Case Studies of Animal's Agency in Interspecies Art, pp. 67–92 (2020). Leiden University Centre for the Arts in Society. https://hdl.handle.net/1887/123086
13. Vidaković, J., Jerbić, B., Šekoranja, B., Švaco, M., Šuligoj, F.: Learning from demonstration based on a classification of task parameters and trajectory optimization. J. Intell. Rob. Syst. **99**(2), 261–275 (2019). https://doi.org/10.1007/s10846-019-01101-2
14. Svozil, D., Kvasnička, V., Pospíchal, J.: Introduction to multi-layer feed-forward neural networks. Chemom. Intell. Lab. Syst. **39**(1), 43–62 (1997). https://doi.org/10.1016/S0169-7439(97)00061-0
15. Hayou, S., Doucet, A., Rousseau, J.: On the impact of the activation function on deep neural networks training, pp. 2672–2680. PMLR (2019). https://proceedings.mlr.press/v97/hayou19a.html

Multi-robot Systems

State-Aware Layered BTs—Behavior Tree Extensions for Post-Actions, Preferences and Local Priorities in Robotic Applications

Guilherme de Campos Affonso(✉), Kei Okada, and Masayuki Inaba

The University of Tokyo, 7-3-1 Hongo,, Bunkyo-ku Tokyo, Japan
affonso@jsk.imi.i.u-tokyo.ac.jp

Abstract. In this paper we propose State-Aware Layered BTs, a behavior tree extension that facilitates the implementation of post-actions, preferences, and local priorities for usage in robotics applications. These procedures are hard to be defined on standard behavior tree formulations because behavior trees lack the ability to structurally retain any information on previous states. Therefore, the execution of localized heuristics can only be attained through the inference or emulation of previous states, which can be imprecise, cause reactivity losses, or introduce added complexity to the structure. In this work we cope with this problem by (i) adding a native operator which accesses the previous execution states of individual nodes; (ii) adding separate layers of interaction which expand the operator functionality and are used to describe multiple goals within the same task. The validity of the proposed system is verified through extensive analysis of a series of annotated examples.

Keywords: Behavior trees · Robotics · Reactive robotics · Task formulation · Control structures

1 Introduction

Advances in machine vision and control structure technologies have guided robots from the execution of iterated procedures in production lines to the adaptive manipulation in unstructured warehouse environments and now towards the reactive handling of events in dynamic social environments.

Dynamic environments are characterized by their volatility, being able to be changed at any time due to the action of humans-in-the-loop or other exogenous agents. Under such conditions, correct task execution relies on (i) the continuous monitoring of the environment through robust detection algorithms with real-time capabilities; and (ii) reactive control structures determining how to handle possible environmental changes.

Tick-driven switching structures give a common formulation approach for such reactive control structures. Switching structures are formulations where the

I. Petrovic et al. (Eds.): IAS 2022, LNNS 577, pp. 673–689, 2023.
https://doi.org/10.1007/978-3-031-22216-0_45

action to be executed at a certain time is a direct function of the environment condition at that time. Tick-driven means that the decision process is triggered at a fixed (sufficiently small) time interval, being opposed to event-driven methods where a new decision process is requested upon the occurrence of certain events, such as action completion. Considering the set of environmental conditions at a given time t as C_t and the action to be executed at that time as A_t, switching structures can be described as $A_t = f(C_t)$. Because this formulation does not depend on any information on time $t-1$, we say that such switching structures are *stateless*.

Although the iterated environmental assertions in tick-driven switching structures allow for the early detection of environmental changes and provide maximized reactivity, its stateless nature poses challenges when dealing with robotic applications. This is because a generalized robotic procedure has its own heuristics determining how it can be safely interrupted, and how it can potentially revert any environmental side effects. For example, interrupting a walking motion requires the robot to regain its balance, interrupting a manipulation procedure might require the robot to place the object down and free its hands before engaging in the next action, and it may be desirable for a robot cleaning the dishes to close the tap when switching to another action. Within a purely stateless framework, such heuristics cannot be linked to the interrupted action, and instead must be inferred by every other action in an overly complicated composition. It is also important to notice that such interruption signals are not limited to unexpected events. Since shifts are triggered by any environmental change, even the accomplishment of a desired condition will interrupt the action in a greedy fashion. In general, the lack of states poses challenges when:

1. defining post-actions which dictate how an action should be safely interrupted;
2. stating preferences which are not directly related to task completion; and
3. imposing local priorities which locally subvert a more general rule.

Here, the term "post-action" is used as an equivalent to the aforementioned interruption heuristics. We opt for this nomenclature to better represent interruptions caused by both expected and unexpected environmental changes.

In order to cope with this problem, in this paper we propose State-Aware Layered Behavior Trees, an extension of behavior trees aimed at robotic applications which include previous state information and updated syntax for better representation of post-actions, preferences, and local priorities. Previous state information is provided to the user as a native operator which accesses the previous execution state of a specific or generalized node. Such state operators function as condition nodes, which can be actively used in the tree formulation to define state-dependent behavior. Better isolation and enhanced expressiveness are provided to this state-dependent behavior by introducing a new syntax where different tree formulations are organized into separate layers. This new syntax not only facilitates the usage of the state operator, but also provides a more fundamental method for defining preferences and local priorities.

In Sect. 2 we provide a brief overview of behavior trees and their applications in robotics. We also discuss some approaches used to dodge their stateless nature and define post-actions. In Sect. 3 we formally present State-Aware Layered Behavior Trees, and explain how it relates to other alternatives when defining state-dependent behavior. In Sect. 4 we thoroughly examine an example task scenario, showing how the proposed system is able to define post-actions, global and local rules, and generalized preferences. Finally, in the Conclusion section we highlight our method's achievements and future prospects.

2 Past Work

2.1 Basic Syntax of Behavior Trees

Behavior trees are a control structure articulated as a rooted tree. They work by propagating a tick signal from root to leaves and executing any nodes to which the tick signal has reached. Within the tree formulation, inner nodes are called control nodes and organize how the tree should be transversed, while leaf nodes are called execution nodes and hold the actual description of relevant procedures. In classical formulation, control nodes are usually limited to either sequence nodes or fallback nodes. Sequence nodes work as an *and* operator, propagating ticks to its child from left to right until a failure is received or until all of them return SUCCESS. Fallback nodes work as an *or* operator, propagating ticks to its child from left to right until a success is received or until all of them return FAILURE. Execution nodes can be either condition nodes, which assert a certain predicate's validity in a synchronous fashion, or action nodes, which implement a potentially asynchronous behavior. Action nodes return RUNNING to signalize an ongoing process, a non-blocking operation which allows for the behavior tree engine to process other nodes at the following tick.

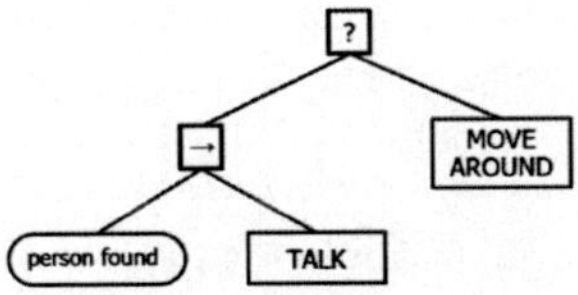

(a)

```
t1. ¬(person found) : MOVE AROUND
t2. ¬(person found) : MOVE AROUND
t3.   (person found) : TALK
t4. ¬(person found) : MOVE AROUND
```

(b)

Fig. 1. Example of formulation and execution of a simple conversational agent using behavior trees. (a) Tree formulation using Sequence (→) and Fallback (?) nodes. Conditions nodes (e.g. *person found*) are represented by curved edge blocks and lowercase, and action nodes by sharp edge blocks and capital letters (e.g. TALK, MOVE AROUND). In this formulation the agent verifies if a person is found at every tick. In case a person is found, the agent executes the TALK behavior. Otherwise, it executes the MOVE AROUND behavior. (b) Runtime example of this formulation, across four ticks. This example illustrates how the agent can promptly react to environmental changes and select the appropriate behavior at each time frame.

A simple annotated example is given in Fig. 1, where an autonomous agent is programmed to move around until a human is found, and then engage in a conversation. Here, right arrows (→) represent sequence nodes, and question marks (?) represent fallback nodes. Figure 1(b) illustrates a hypothetical runtime scenario across four ticks, showing how the agent is capable of dynamically adapting to changes in the environment.

In order to improve usability, parallel control nodes, decorators, and a common variable pool known as the Blackboard [10,11,17] have also been introduced to the standard behavior tree formulation. Parallel nodes work by simultaneously propagating ticks to all of its child and calculating a return value based on user-defined thresholds. Decorator is the generalized term for control nodes that accept a single child node and perform some kind of manipulation to its ticking pattern or return value. Some examples of decorator nodes include delaying execution, inverting the return value, re-ticking the node a given number of times, and other dedicated procedures. Finally, the blackboard functions as a global variable pool, which is useful for sharing information between nodes. Further documentation on behavior tree formulation and design is provided in [1,2,9].

2.2 Benefits of Behavior Trees Over Other Control Architectures

Behavior trees were first developed by the gaming industries [15,19] to provide a more modular alternative to state machines when designing the behavior of non-playable characters. Reactivity was not originally a core feature of behavior trees, but it was included in later formalization [6,11,18].

Behavior trees have been shown to be generalizations of the subsumption architecture and decision trees [8], and the introduction of global variables through the blackboard also allowed to prove mathematical equivalence to Finite State Machines [4,18,22]. When comparing behavior trees and finite state machines, there is a tradeoff between the implicit representation of transitions against the implicit representation of states as inherent features of the structure formulation. In behavior trees, abbreviating explicit transitions have allowed for enhanced modularity, as well as overall simplicity when dealing with reactive tasks.

Overall, Merrill [20] describes behavior trees as having a strong balance between implementation ease, visualization, and adaptability. Such benefits have been highly attractive to developers in many fields, making behavior trees a popular implementation approach for gaming, AI, and robotics research [12,14].

2.3 Behavior Tree Extensions for Robotics

Rovida et al. [24,25] have proposed *extended behavior trees* (eBTs), an extension that encapsulates the prerequisites and success conditions of each action into a single module. One of the core goals of this approach is to allow for integration with high-level semantic planners, which can then be used to dynamically generate eBTs as an output.

The idea of joining prerequisites and success conditions into the action module is also present in the proposition of *conditional behavior trees* (CBTs) [13] and in work by Colledanchise et al. [7]. Colledanchise et al. also similarly point to the need for integration with a high-level planner and for the dynamic modification of tree structures at runtime.

The above work indicates a common need in robotics for a more goal-oriented and adaptive design. However, the incorporation of high-level planners does not directly address the problem of statelessness and post-action definition, which is the main objective of this paper.

2.4 Approaches for Handling Statelessness

When considering robotics applications, the stateless nature of behavior trees has posed several challenges, to which many extensions and ad hoc solutions were proposed.

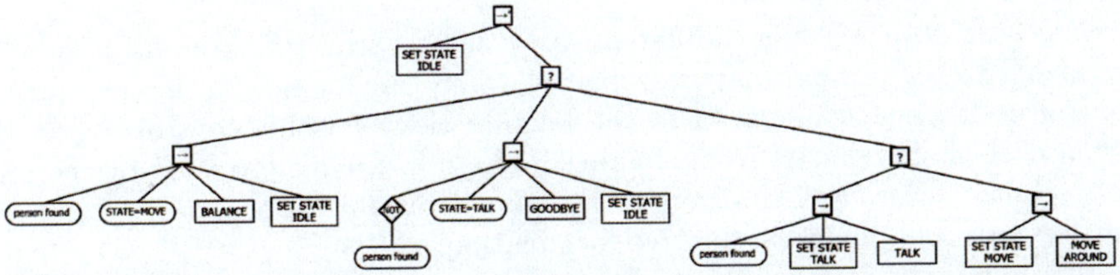

(a) Behavior tree formulation of a conversational agent including post-actions. Here, state-dependent behavior is achieved by storing state information in a global variable (SET STATE and STATE= nodes). Although functional, this method can result in overly complex structures, like the one above.

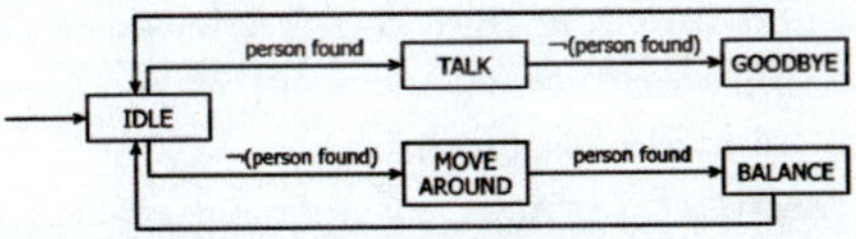

(b) State machine formulation of the same conversational agent. In this case, inherently managing the state information as a part of the control structure allows for a much more concise formulation.

Fig. 2. Comparison between state machine (a) and behavior tree (b) formulations of a simple conversational agent including post-actions. Here, the agent implemented in Fig. 1a is extended to ensure that the robot maintains proper balance when the moving procedure is interrupted (BALANCE action), and that it greets the human when leaving a conversation (GOODBYE action).

Arguably the most simple handling strategy is to manually store state information as a global variable on the blackboard. Although functional, this method results in added complexity to the structure, due to the necessity of including nodes to manage and access the blackboard variable. This is illustrated in Fig. 2a, where the addition of SET STATE and STATE= nodes have resulted in a rather cumbersome formulation of this otherwise trivial conversational agent. When considering this strategy, it is not hard to imagine cases where state machines would provide a much clearer description, therefore partially defeating the purpose of adopting behavior trees as a more concise structural approach. Figure 2b exemplifies this by showing a state machine formulation of the previous conversational agent. Here, the state machine formulation is able to achieve equivalent functionality in a much clearer design.

Another approach is given by the introduction of stateful nodes [16,18], which are capable of retaining their previous execution state in a non-reactive fashion. This is achieved by remembering the current progress of the ticking signal across different time frames, resuming from the last preemption point instead of re-propagating the signal across all of its child. As a result, nodes that have already returned SUCCESS or FAILURE are not executed again, ensuring state progression within a non-reactive framework. Compared with the global variable approach, being able to assign and manage state information autonomously within the node formulation itself allows to adopt simplified structures. However, there is a clear trade-off for reactivity, since the structure does not verify prior conditions during execution.

Klöckner [16] presents a different approach to achieving stateful nodes, by managing state information with separate decorators for each action and for the control node itself. Compared to the introduction of whole control nodes, the use of decorators benefits from enhanced design flexibility provided by the finer grain control. However, similarly to the global variable approach, this can add unnecessary complexity to the tree formulation.

The requirement for state progression in robotics is so pronounced that in BehaviorTree.CPP [3], a major implementation library focused on robotics applications, stateful nodes are enforced as the standard transition behavior. BehaviorTree.CPP also differentiates between two types of stateful node formulations: one that never resets and another that resets whenever the control node itself returns FAILURE. In this work we use $\rightarrow^*$ and $?^*$ referring to the latter implementation of sequence and fallback nodes.

Another common approach for defining post-actions is to directly assign interruption callbacks on the action module implementation code. Although being a simple and functional solution, such interruption callbacks are completely external to the tree structure, which leads to a loss in representation and reactivity. The loss in representation is due to the fact that such callbacks cannot to be visualized from the tree structure. The loss in reactivity is due to the fact that the tree remains unresponsive to any external changes during the execution of such callbacks.

Outside the robotics field, the inability of behavior trees to define local rules and shifting priorities have also been pointed out by several authors [9,15,20]. Such inability is due to both the stateless nature of behavior trees and their fixed priorities in the fallback node formulation. Among the many proposed extensions for addressing this problem, we highlight Switching Styles [15] and Hinted Behavior Trees [21] as methods that also allow for defining state-dependent behavior in a more general manner. Both of these approaches are based on having multiple behavior tree formulations for the same task and switching them according to a given state. In Switching Styles the whole tree formulation is replaced, while in Hinted Behavior Trees only a few nodes are swapped.

3 State-Aware Layered Behavior Trees

3.1 Introducing State Operators

Our first proposition is to introduce a dedicated operator for accessing previous state information. This dedicated state operator, which we represent with the letter δ, takes a single argument and is defined as follows.

$$\delta(NodeId)_t = \begin{cases} \text{SUCCESS}, & \text{if } S(NodeId)_{t-1} = \text{RUNNING} \\ \text{FAILURE}, & \text{otherwise} \end{cases}$$

$$\delta(NodeType)_t = \begin{cases} \text{SUCCESS}, & \text{if } \exists NodeId \in NodeType, S(NodeId)_{t-1} = \text{RUNNING} \\ \text{FAILURE}, & \text{otherwise} \end{cases}$$

Here, $NodeId$ signalizes a unique identifier for an action node or substructure in the tree, and $NodeType$ signalizes any action node or subtree belonging to a certain type. $S(NodeId)_{t-1}$ signalizes the execution state of a certain node in time $t-1$. In the examples explored in this paper we refer to identifier arguments unless we explicitly use a TYPE prefix. That is, $\delta(TALK)$ refers to the unique action node TALK, rather than any node which executes the TALK action. The latter is represented by $\delta(TYPE(TALK))$ instead.

The implementation of this operator requires a computational engine which:

1. attributes identifiers to each action and control node;
2. stores node states, which can be either SUCCESS, FAILURE, RUNNING, or IDLE, which is the default state for unticked nodes and is restored after execution.

Both of these requirements are already satisfied by the BehaviorTree.CPP engine. In this case, the state operator simply provides an interface that allows the user to access those values.

It is important to note that the state operator focuses on the RUNNING statement, since the original behavior tree formulation already provides sufficient support for modeling transitions related to the SUCCESS and the FAILURE return values. The handling of the RUNNING value, on the other hand, is largely ignored by classic control nodes, yet essential to assert the interruption of a certain action.

Compared to the manual management of state information through global variables, the proposed operator offers an automated interface for storing and updating values. This can lead to a simplified formulation, as illustrated in Fig. 3. Compared to the purely autonomous tracking of state through stateful nodes, the proposed system has the advantage of maintaining high reactivity.

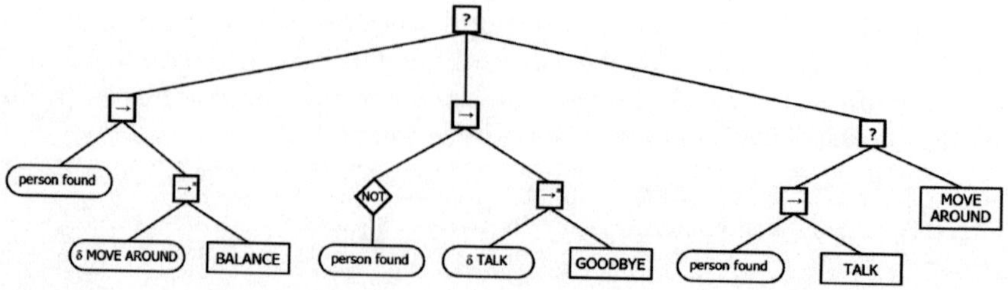

Fig. 3. Behavior tree formulation equivalent to the one in Fig. 2a using the proposed state operator, represented by the letter δ. Automatically tracking the previous execution state of each node and providing this information to the user allows to abbreviate the manual assignment of global variables. This results in an improvement in simplicity when defining post-actions and other state-dependent behavior.

However, the introduction of this operator alone still has some major limitations. First, it is highly unstable. Since each node's execution state and, therefore, the operator's return value changes with each tick, it becomes almost mandatory to couple the state operator with a stateful node. Second, the proposed operator still adds a higher than desirable complexity to the overall framework, generating structures that can be hard to visualize. Third, it compromises the overall modularity of the structure by adding separated inter-related nodes. To address these limitations, we introduce priority layers and expand the state operator as explained in the following subsection.

3.2 Introducing Priority Layers and Expanding the State Operator

Our second proposition is to organize multiple subtree formulations into different layers, which are ordered and executed according to a fixed priority. Each layer stack contains one main layer, which describes the main task objective, and zero or more subordinate layers, which describe other procedures that are not directly related to task completion. The idea of priority layers is inspired by the Subsumption Architecture [5] and Free-Flow Architectures [23].

Interactions between multiple layers are determined by propagating the tick from the bottom-most (right-most) layer until the main tree at each tick, finishing when one of the layers returns RUNNING or when the main layer is ticked. Executing layers from the bottom up (from right to left) is equivalent to adding sequence or fallback nodes at the left-most side of the root node (Fig. 4). This is expected, as modularity and hierarchical interactions between subtrees were a base concern of the original behavior tree design. However, when considering the proposed state operator, the introduction of separate layers have major benefits over the simple adding of nodes in the standard formulation, as described below.

First, separate layers provide isolation and stability to the state operator. This is because it becomes possible to administer the states of each layer individually, meaning that the propagation of a tick will not necessarily alter the state of all nodes. More specifically, this means that nodes in higher levels of

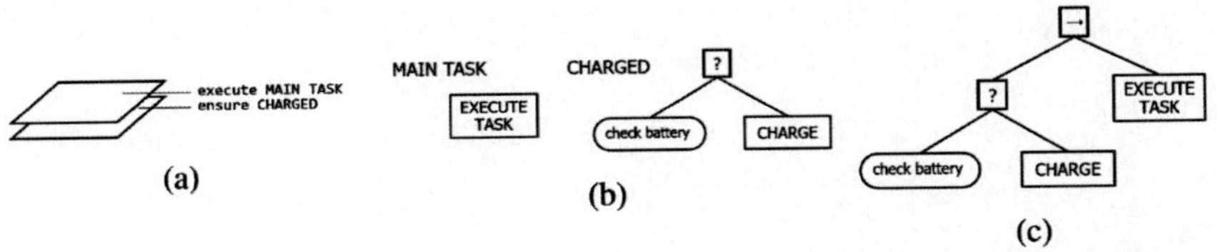

Fig. 4. Example of a simple layered behavior tree and its standard formulation equivalent. (a) The behavior tree is organized into two layers. The first layer is the main layer, which is responsible for ensuring task completion. The second layer is a subordinate layer, which describes another objective, in this case charging the robot. (b) Example behavior tree formulation of the main and subordinate layers. Within a single tick, layers are evaluated from bottom up (right to left) until a layer returns RUNNING or the main layer is ticked. (c) Standard behavior tree formulation of the same task. Stacking layers in layered behavior trees is equivalent to adding nodes to the left of the root node in a standard behavior tree formulation.

the tree can retain the RUNNING state, and, therefore, ensure consistent return values for the state operator throughout the execution of lower level layers.

Second, separate layers improve readability by dividing and labeling semantically related groups. This could also be attained by using subtrees, although enforcing this kind of division as a system requirement can be regarded as a good practice.

Third, separate layers improve modularity by making it easier to associate inter-related nodes. Since the typical implementation of a subordinate layer will have a clear goal that is closely connected with a certain type of action, it becomes possible to trace dependency links between actions and layers. Such dependencies could then be verified manually through human assertion or autonomously by adding dependency tags to each layer and compile-time checks. Without separate layers, it would be required to verify dependencies on a node-to-node basis, which is not always trivial. For example, in the hypothetical case in which we want to remove the CHARGE action from the definition provided in Fig. 4, it is easy to acknowledge that the CHARGED subordinate layer depends on the CHARGE action and should therefore be removed as well. However, it is less intuitive to evaluate if the *check battery* node should also be removed under this specific context.

Since it is not possible to directly position procedures described in subordinate layers within a specific context of the main task, we expand the state operator definition as follows. These changes enable the state operator to accept condition nodes present in the main tree as inputs, and are used to better describe unique transitions and to further enhance interactions between layers.

$$\delta(CondId)_t = \begin{cases} \text{SUCCESS}, & \text{if } S(CondId)_t = \text{SUCCESS} \\ \text{FAILURE}, & \text{otherwise} \end{cases}$$

$$\delta(\neg CondId)_t = \begin{cases} \text{SUCCESS}, & \text{if } S(CondId)_t = \text{FAILURE} \\ \text{FAILURE}, & \text{otherwise} \end{cases}$$

$$\delta(CondType)_t = \begin{cases} \text{SUCCESS}, & \text{if } \exists CondId \in CondType, S(CondId)_t = \text{SUCCESS} \\ \text{FAILURE}, & \text{otherwise} \end{cases}$$

$$\delta(\neg CondType)_t = \begin{cases} \text{SUCCESS}, & \text{if } \exists CondId \in CondType, S(CondId)_t = \text{FAILURE} \\ \text{FAILURE}, & \text{otherwise} \end{cases}$$

Here, *CondId* signalizes a unique identifier for a condition node present in the main tree, and *CondType* signalizes any condition node present in the main tree which is of that type. Note that for condition nodes the operator is defined for the actual execution state in time t, rather than the previous state in $t-1$. Since standard behavior trees provide no alternatives to molding interactions between different layers, it is required to define operators for asserting both SUCCESS and FAILURE outcomes. This is attained by also defining operations on the negation of the given argument.

Verifying the state of a certain type of condition in the main tree instead of simply asserting its value allows to check if the given condition is a current requirement for the execution of the main task. This kind of precondition analysis is comparable to the methods presented in Sect. 2.3, with the differential that we opt for an imperative approach instead of a declarative one. That is, in our framework preconditions are not strictly included as part of the task definition and used as input to a goal-oriented planner, but are rather loosely inferred according to the tree transversal and utilized by the user conform necessity.

Considering the extended state operator, interactions between layers are formalized by the following algorithm.

```
Transverse the tick through the main layer, without starting any action node.
if layer returns SUCCESS or FAILURE then
    return layer return value
end if
for all subordinate layers, from bottom up (right to left) do
    Tick the layer.
    if layer returns RUNNING then
        return RUNNING
    end if
end for
Tick the main layer.
Update the previous state values for each node.
return main layer return value
```

The proposed implementation requires a computation engine which:

1. attribute identifiers to each node;
2. store both current and previous states of all nodes in the main layer, which can be either SUCCESS, FAILURE, RUNNING, or IDLE;
3. take into account decorators when updating the current state of condition nodes in the main tree.

The BehaviorTree.CPP implementation engine currently only supports the first requirement. The second requirement could be easily fulfilled by adding a

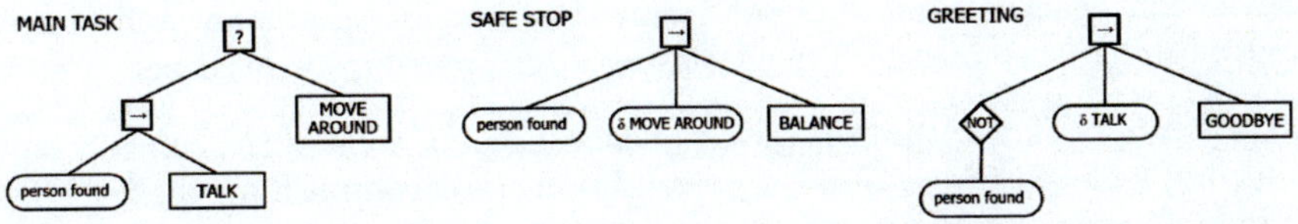

Fig. 5. Behavior tree formulation equivalent to Figs. 2a and 3 using State-Aware Layered Behavior Trees. The separation provided by different layers yields a structure that is easier to visualize and understand. The ability to administer the state of each layer individually also provides stability to the state operator, avoiding the necessity for stateful nodes.

slot variable to hold previous state values, which are then updated at each tick. Handling decorators when updating the state of condition nodes requires a more dedicated approach, but is also easily achievable without structural changes.

Figure 5 shows how State-Aware Layered Behavior Trees can be used to define structures equivalent to the ones in Figs. 2a and 3. Here, the formulation is divided into three parts:

i the main tree, which is the unaltered definition provided in Fig. 1(a);
ii a subordinate layer which ensures that the moving behavior is safely interrupted;
iii a subordinate layer which ensures that the robot greets the human when leaving a conversation.

When compared with the sole use of the state operator in Fig. 3, separating the tree structure into different layers yields a structure that is simpler and easier to visualize. Furthermore, using different layers also allows stabilizing the state operator output, which avoids the inclusion of stateful nodes.

More elaborate usage of the proposed formulation during the definition of post-actions, preferences, and local priorities is explored in the next section.

4 Case Analysis

In this section we consider the example task of taking a beverage from the fridge and delivering it to a human. This task is formulated as the behavior tree displayed in Fig. 6. The task definition is designed to ensure reactivity during the attainment of each sub-goal of the main task. Here, the robot will ensure that it is holding a can and then deliver it. If it is not holding any can, it will go to the fridge, open it, and then pick a can. For this analysis we assume condition nodes that are based solely on environment observation (no mediation through direct variable assignment) and temporally extended actions. Under such assumptions, the transversal of the tree is mainly driven by environmental changes rather than by the return value of action nodes.

Our main objective is to include commands for the robot to close the fridge door after picking the can, in a manner that is consistent with the behavior tree formulation, reactive, and does not require the manual assignment and management of global variables. Adding a closing door procedure is hard because:

1. state progression within the pick sequence is lost as soon as the can is grasped;
2. having a closed fridge is not a precondition for accomplishing the fetch task; and
3. inferring the previous robot state based on current environment condition is inaccurate.

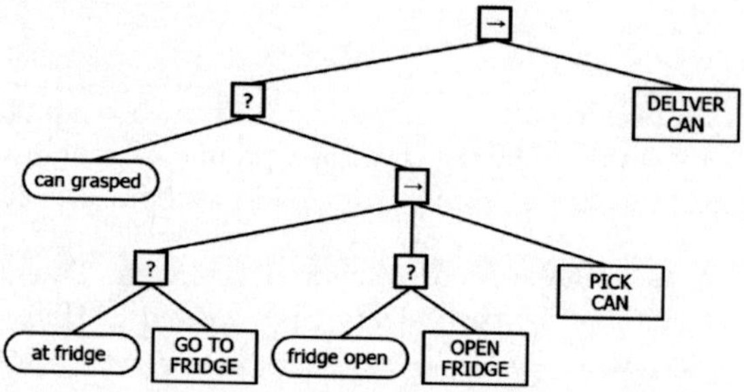

Fig. 6. Behavior tree formulation for a task where the robot fetches beverages from the fridge and delivers them to a human. The implementation adopts a reactive approach that constantly checks for the attainment of sub-goals and engages in actions that can satisfy them if they do not hold. Namely, if the robot is not holding a can it will try to pick it, if it is not near the fridge it will move towards it, and if the fridge is closed the robot will open it.

Three naïve implementations are shown in Fig. 7, in order to better highlight each of these limitations. In the first implementation approach, the closing action is introduced after the pick sequence (Fig. 7a). Since this block is interrupted as soon as the robot successfully grasps the can, the closing action is never reached. In the second implementation approach, the closing action is introduced below the root sequence, before the deliver procedure (Fig. 7b). When considering the situation where the robot obtains the can from a place other than the fridge— e.g. is given by a human or bought in a store –, this implementation would require the robot to go to the fridge and close it before engaging in the delivering procedure. Finally, in the third implementation approach, the closing action is also introduced below the root sequence and before the deliver procedure, but this time with a check to ensure that the robot is at the fridge and that the fridge is open before engaging in closing it (Fig. 7c). Here we will consider the situation where, after moving to the fridge to pick up the can, the robot is interrupted by a human. The human, also wanting to use the fridge, opens it and gives the robot the can, hoping that it will move away. Under the given formulation, however, the robot would then deliberately close the fridge door which was opened by the human before leaving.

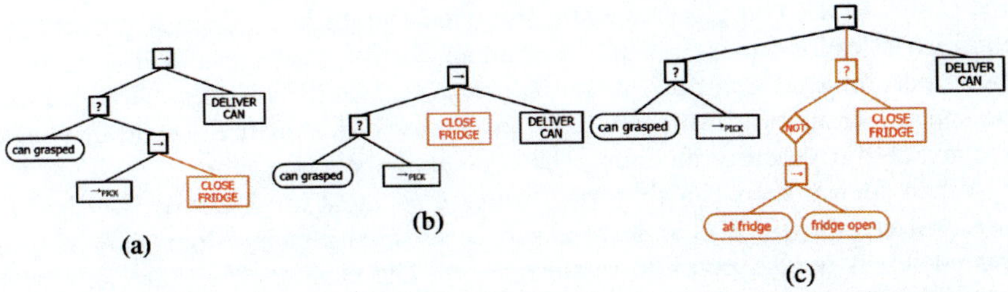

Fig. 7. Naïve approaches for including a CLOSE FRIDGE procedure (red) into the task definition. Here, $\rightarrow_{PICK}$ represents the subtree delimited by the sequence node right to the *can grasped* condition in the original formulation. (a) Including the closing action immediately after the pick sequence. The robot fails to start the action since the block immediately exits when the can is grasped. (b) Including the closing action below the root sequence, before the deliver action. This implies that closing the fridge door is a precondition for delivering the can, which is not true. (c) Including the closing action before the deliver action, but also including condition nodes which inference the previous state based on the current environment condition. This yields the most robust implementation approach, but is still unable to infer the correct state under certain scenarios.

The above limitations reflect a more subtle request: close the fridge door, *when you have opened it*. This assertion is fundamentally state-dependent, and only attainable by introducing some sort of state mechanism into the otherwise stateless behavior tree structure.

Figure 8 illustrates how the proposed state operator and layered behavior trees can be used to define such state-dependent behavior. Here, the closing procedure is defined as a post-action, which is executed after the transition from the PICK action through the *can grasped* condition. This implementation behaves similarly to state transitions in state machines. When compared to single tree formulations (Fig. 8b), the layered approach (Fig. 8a) provides a clear separation between subtrees which is easy to visualize and interpret. Here, the main

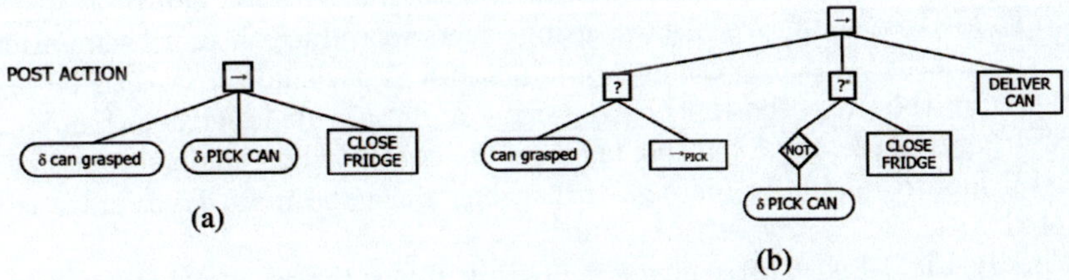

Fig. 8. Introducing the CLOSE FRIDGE procedure as a post-action to the pick node, both as a separate layer (a) or as a single tree structure (b). The inclusion of state operators allows to define the closing procedure as a progression from the PICK node, which yields correct execution even under highly dynamic and social scenarios.

tree ensures task completion, while the subordinate layer defines the interruption heuristics for when the PICK action succeeds—closing the fridge. Here, the state operator is also applied to the *can grasped* condition in Fig. 8a to pinpoint the specific transition from PICKing to *grasped.* This ensures scalability as the tree grows and possibly includes more nodes of the same type.

When focusing on the first two limitations mentioned above (*i.e.* ignoring cases where the robot is strictly asked to close the fridge door *only* if it has opened it), it is also possible to incorporate the closing procedure as a global rule which is locally subverted, or as a generalized preference.

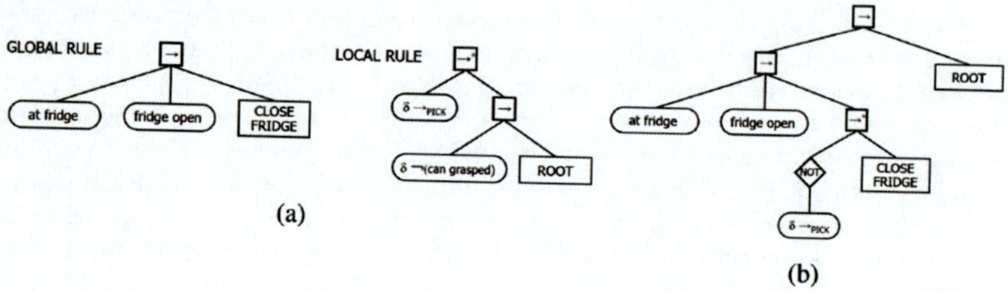

Fig. 9. Introducing the CLOSE FRIDGE procedure as a global rule, which is then locally subverted during the pick sequence. (a) Introducing both the global and the local rules as subordinate layers. Here, ROOT represents the main tree structure defined in Fig. 6. (b) Equivalent single-tree formulation of the layered approach. Both the globalized and the localized rules are merged into the same substructure.

Figure 9 illustrates an implementation approach using global and local rules. Here, closing the fridge door when the robot is near the fridge and the fridge is opened is defined as a global rule. This rule is then locally ignored during the pick sequence, allowing for the robot to actively open the fridge and pick the can. When the robot successfully grasps the can and finishes the pick sequence, the global rule reemerges, causing the robot to close the fridge door. Figure 9a shows an implementation adding subordinate layers for both the global and local rule definition, while Fig. 9b shows how the formulation can be translated into a single tree structure. Again, using layers introduces a more semantically divided structure which is easier to understand and visualize. It is to be noted that, within the proposed method, ignoring a global rule is harder than simply overwriting it. Here, the local rule layer references the whole main tree structure, and still has to manually manage transitions since the main layer state is not updated.

Finally, Fig. 10 illustrates how it is possible to use the extended state operator to implement the closing procedure as a generalized preference. Here, the robot checks if conditions of the type *fridge open* are transversed in the main tree, and closes the fridge if they are not. Analyzing the condition nodes activated in the main tree allows to infer the required conditions for the current execution step in the main task. This information can then be used to verify if a certain

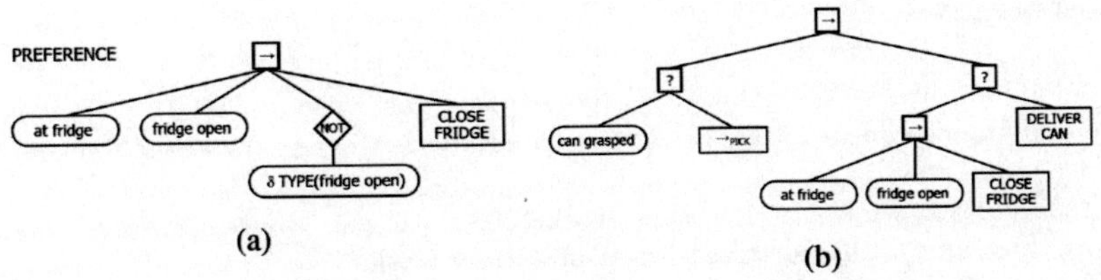

Fig. 10. Introducing the close fridge procedure as a generalized preference. Verifying if a certain condition type has been transversed by the main tree allows inferring the requirements of the current execution step. The user can then check these requirements against the side effects of a certain procedure, allowing to define generalized preferences that are met whenever they do not conflict with the main task. (a) Introducing the closing procedure as a subordinate layer. Using the extended operator to check the transversed conditions of the main tree allows for a formulation that emulates the deliberate planning and integration of sub-goals. (b) Equivalent single-tree formulation of the layered approach. A preference can be expressed by including the callback procedure in every action node whose transversal path does not include a certain condition. For the *fridge open* condition in this specific case, this means adding the callback to the GO TO FRIDGE and DELIVER CAN actions. However, since the transversal path of the GO TO FRIDGE action includes the $\neg$ *at fridge* condition, which is conflicting with the callback procedure, it can be abbreviated.

procedure is consistent with the main task at a certain point. In this case, since the closing action results in the negation of the *fridge open* state, the action is not executed when the main task explicitly requires for the fridge door to be opened. Therefore, in this formulation the robot *prefers* to close the fridge door whenever possible, but does not hesitate to open it when necessary.

5 Conclusions

In this paper we proposed State-Aware Layered Behavior Trees, a behavior tree extension that allows for the intrinsic representation of state-dependent behavior. The extension is based on the introduction of a state operator which can access previous execution states of specific or generalized nodes, and of layered interactions that provide better isolation and stability to the proposed operator.

Compared with alternative approaches, the proposed method is beneficial because it poses no compromises in reactivity, and because it attains simpler formulation by automatically updating the previous states and organizing substructures according to their objectives. These advantages are illustrated in two use case scenarios—a conversational agent and a fetching task –, demonstrating how the proposed system facilitates the definition of post-actions, local priorities, and general preferences.

However, the proposed system does reduce the modularity of the original behavior tree formulation by introducing inter-related nodes. The development of other reactive control structures that can fundamentally address to this limitation while also retaining a stateful nature is left for future work.

References

1. Behavior Trees — Unreal Engine Documentation (2022). https://docs.unrealengine.com/4.27/en-US/InteractiveExperiences/ArtificialIntelligence/BehaviorTrees/
2. Behavior trees for AI: How they work (2022). https://www.gamedeveloper.com/programming/behavior-trees-for-ai-how-they-work
3. BehaviorTree.CPP (2022). https://github.com/BehaviorTree/BehaviorTree.CPP. Retrieved 2022
4. Andrew Bagnell, J., Cavalcanti, F., Cui, L., Galluzzo, T., Hebert, M., Kazemi, M., Klingensmith, M., Libby, J., Liu, T.Y., Pollard, N., et al.: An integrated system for autonomous robotics manipulation. In: 2012 IEEE/RSJ International Conference on Intelligent Robots and Systems, pp. 2955–2962, IEEE (2012)
5. Brooks, R.A.: Achieving Artificial Intelligence Through Building Robots. Technical report, Massachusetts Inst of Tech Cambridge Artificial Intelligence Lab (1986)
6. Champandard, A.J., Dunstan, P.: The behavior tree starter kit. In Game AI Pro, pp. 27–46. CRC Press (2013)
7. Colledanchise, M., Almeida, D., Ögren, P.: Towards blended reactive planning and acting using behavior trees. In: 2019 International Conference on Robotics and Automation (ICRA), pp. 8839–8845, IEEE (2019)
8. Colledanchise, M., Ögren, P.: How behavior trees modularize hybrid control systems and generalize sequential behavior compositions, the subsumption architecture, and decision trees. IEEE Trans. Robot. **33**(2), 372–389 (2016)
9. Colledanchise, M., Ögren, P.: Behavior Trees in Robotics and AI: An Introduction. CRC Press (2018)
10. Colvin, R., Grunske, L., Winter, K.: Probabilistic timed behavior trees. In: International Conference on Integrated Formal Methods, pp 156–175. Springer, Heidelberg (2007)
11. Flórez-Puga, G., Gomez-Martin, M.A., Gomez-Martin, P.P., Díaz-Agudo, B., González-Calero, P.A.: Query-enabled behavior trees. IEEE Trans. Comput. Intell. AI Games **1**(4), 298–308 (2009)
12. Ghzouli, R., Berger, T., Johnsen, E.B., Dragule, S., Wasowski, A.: Behavior trees in action: a study of robotics applications. In: Proceedings of the 13th ACM SIGPLAN International Conference on Software Language Engineering, pp. 196–209 (2020)
13. Giunchiglia, E., Colledanchise, M., Natale, L., Tacchella, A.: Conditional behavior trees: definition, executability, and applications. In: 2019 IEEE International Conference on Systems, Man and Cybernetics (SMC), pp. 1899–1906, IEEE (2019)
14. Iovino, M., Scukins, E., Styrud, J., Ögren, P., Smith, C.: A survey of behavior trees in robotics and AI. arXiv preprint arXiv:2005.05842 (2020)
15. Isla, D.: GDC 2005 Proceeding: Handling Complexity in the HALO 2 AI. Retrieved October, 21:2009 (2005)
16. Klöckner, A.: Behavior trees with stateful tasks. In: Advances in Aerospace Guidance, Navigation and Control, pp. 509–519. Springer, Heidelberg (2015)
17. Lim, C.-U., Baumgarten, R., Colton, S.: Evolving behaviour trees for the commercial game defcon. In: European Conference on the Applications of Evolutionary Computation, pp. 100–110. Springer, Heidelberg (2010)
18. Marzinotto, A., Colledanchise, M., Smith, C., Ögren, P.: Towards a unified behavior trees framework for robot control. In: 2014 IEEE International Conference on Robotics and Automation (ICRA), pp. 5420–5427, IEEE (2014)

19. Mateas, M., Stern, A.: A behavior language for story-based believable agents. IEEE Intell. Syst. **17**(4), 39–47 (2002)
20. Merrill, B.: Building utility decisions into your existing behavior tree. In: Game AI Pro, 127 (2013)
21. Ocio, S.: Adapting AI behaviors to players in driver San Francisco: hinted-execution behavior trees. In: Eighth Artificial Intelligence and Interactive Digital Entertainment Conference (2012)
22. Ogren, P.: Increasing modularity of UAV control systems using computer game behavior trees. In: AIAA Guidance, Navigation, and Control Conference, p. 4458 (2012)
23. Kenneth Rosenblatt, J., Payton, D.: A fine-grained alternative to the subsumption architecture for mobile robot control. In: Proceedings of the IEEE/INNS International Joint Conference on Neural Networks, vol. 2, pp. 317–324 (1989)
24. Rovida, F., Grossmann, B., Krüger, V.: Extended behavior trees for quick definition of flexible robotic tasks. In: 2017 IEEE/RSJ International Conference on Intelligent Robots and Systems (IROS), pp. 6793–6800, IEEE (2017)
25. Rovida, F., Wuthier, D., Grossmann, B., Fumagalli, M., Krüger, V.: Motion generators combined with behavior trees: a novel approach to skill modelling. In: 2018 IEEE/RSJ International Conference on Intelligent Robots and Systems (IROS), pp. 5964–5971, IEEE (2018)

Multi-agent Coordination Based on POMDPs and Consensus for Active Perception

Frano Petric(✉), Marijana Peti, and Stjepan Bogdan

Faculty of Electrical Engineering and Computing, University of Zagreb, Unska 3, Zagreb, Croatia
frano.petric@fer.hr

Abstract. This paper deals with coordination in multi-agent systems based on Partially Observable Markov Decision Processes (POMDP). The well-known POMDP framework has been extended with consensus protocol, allowing for decentralized sharing of beliefs improving the local decision making of individual agents and eliminating the need to track the entire history of actions in the system to achieve coordination. To emphasize the importance of belief update after an observation is made in the system, we propose a novelty-biased consensus protocol. The efficiency of the proposed framework is demonstrated on the multi-agent system, which consists of three agents carrying out search missions in a simulation environment. The results obtained show that the mission completion time is significantly reduced without degradation in the success rate, while using simple POMDPs for individual agents.

1 Introduction

Multi-agent systems are becoming a frequent subject of research, especially when it comes to mission planning. One of the main reasons for this is that, in general, a complex mission, which could be a burden for a single agent, could be partitioned into several simple tasks, and each one of them distributed among several agents, thus reducing the workload of an agent and in the case of some missions, achieving parallelism in execution, which in turn could significantly reduce the time required to complete the mission.

During the autonomous execution of a mission, an agent often needs to make sequential decisions based on a state of the environment, its own state and observations. An appropriate mathematical framework for modeling sequential decisions is Markov Decision Processes (MDP) [1]. However, the observations taken by an agent are often based on data from sensors that cannot be fully trusted, meaning that an agent must act under some level of uncertainty. To incorporate these phenomena in the model, Markov Decision Processes are extended to the Partially Observable Markov Decision Processes (POMDP) [2]. This mathematical framework is used to model sequential decision-making processes that take into account the partial observability of the environment.

I. Petrovic et al. (Eds.): IAS 2022, LNNS 577, pp. 690–705, 2023.
https://doi.org/10.1007/978-3-031-22216-0_46

Although it is very powerful, the basic POMDP framework suffers from the fact that the policy for each agent in the group is determined and executed without taking into account actions of others. However, POMDPs have been extended to multi-agent systems, with the Decentralized POMDP (Dec-POMDP) being a prominent example [3–5]. However, solving Dec-POMDPs has been shown that to belong to the class of Non-deterministic Exponential-time (NEXP) hard problems [6]. Therefore, currently there is no formal algorithm that provides the optimal solution to a general Dec-POMDP, only approximate solutions for small-scale problems exist, generally focusing on techniques encompassing memory-bounded history-dependent policies [7] or optimizing policy graphs [8] [9]. Another extension of POMDPs for multi-agent systems are Interactive POMDPs (I-POMDPs) [10], which, similar to Dec-POMDPs, can only be solved approximately and for finite nesting of other agent models.

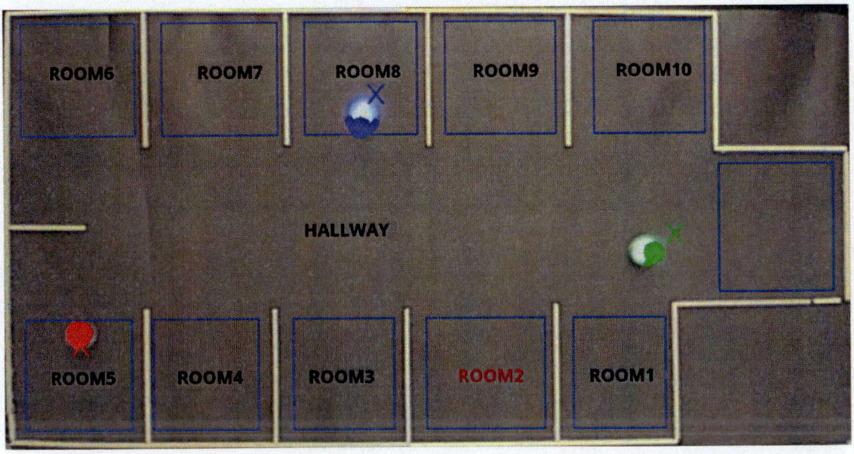

Fig. 1. A team of robots is searching for a fire in a building. The location of a fire is depicted by red text. In the envisioned scenario, the layout of the building is known in advance and only one room is on fire.

There have been attempts to find a solution to the Dec-POMDP problem through introduction of consensus protocol. In [11] authors demonstrate that by using consensus within Dec-POMDP framework, agents are able to reach a common policy independently, while history dependence and computational time are significantly reduced. Compared to [11], our approach uses significantly simpler belief space (standard POMDP belief in our approach compared to belief space including possible actions and history in [11]). Additionally, we only use an offline solution of a single POMDP for each agent, greatly reducing computation complexity. Authors in [12] use the consensus algorithm to fuse the Bayesian updates from neighbouring agents into an updated local belief. Although the approach is focused solely on monitoring, without decision making, the authors show convergence guarantees, proving that sharing beliefs via consensus protocol converges under assumptions of strictly non-negative initial values in belief vectors, and strongly connected graph of the agent network.

Building on these findings, we present and evaluate a decentralized approach for the policy determination in multi-agent systems that, instead of Dec-POMDP, exploits POMDP formalism in such a way that the POMDP model is defined independently for each agent in the group. Formulating and solving a POMDP for each of the agents establishes a new paradigm of decentralized planning and decentralized execution, compared to Dec-POMDPs which rely on centralized planning with decentralized execution. The proposed approach requires only sharing of information about the belief of the environmental states so that agents can agree on a common value using the consensus protocol [13]. Furthermore, we introduce variations in weights of edges of an underlying consensus graph so that more recent data exchanged between the agents has a greater impact on decision making than older information.

We perform extensive simulation experiments with three agents in an information gathering scenario, and implement several interesting scenarios with Sphero robots. Compared to models used in benchmarking Dec-POMDP solvers, the number of states in the proposed multi-agent system is comparable to largest Dec-POMDPs used in literature[1], but since most readily available solvers rely on policy iteration, direct comparison to Dec-POMDP solution of the proposed scenario was not performed since the proposed algorithm requires the solution in form of action-value functions. Following this introductory section, we introduce POMDPs and consensus in Sect. 2 and our methodology in Sect. 3. Section 4 presents agent POMDP models and the proposed communication strategy. We discuss the simulation results in Sect. 5, which is followed by a brief conclusion and comments on future work.

2 Preliminaries

2.1 Partially Observable Markov Decision Processes

A POMDP models an agent decision process where the agent cannot directly observe the system state. A POMDP is defined as a tuple (S, A, O, T, Ω, R), where S is a set of hidden states of the system, A is a set of actions that can be performed, O is a set of observations generated by the system, T denotes the conditional transition probability between states depending on action $p(s^k|s^{k-1}, a^{k-1})$, Ω defines the conditional observation probability $p(o^k|s^k, a^{k-1})$, while R is a reward function. Since the state of the system is hidden to the decision maker, the decision maker maintains the belief state $\mathbf{b}$ as a probability distribution over S. Thus, $b(s)$ is the probability that a system is in state $s \in S$ and is updated as follows:

[1] http://masplan.org/problem_domains.

$$b^k(s^k) = p(s^k|b^{k-1}, a^{k-1}, o^k) \tag{1}$$
$$= \frac{\Omega(o^k, s^k, a^{k-1}) \sum_{s^{k-1} \in S} b^{k-1}(s^{k-1}) T(s^k, a^{k-1}, s^{k-1})}{\eta}$$

where η is a normalizing constant.

Based on the current value of belief b, the decision maker selects the *best* action with respect to the expected reward. A mixed observability Markov decision process is a special case of POMDP [14]. Fully observable state components are represented as a single state variable X, while the partially observable components of the system state are represented as a different state variable Y, factoring the state space of the model, $S = X \times Y$, and hence $T = T_x \times T_y$. The full observability of some state components can then be exploited for more efficient solving of the POMDP, as implemented in the SARSOP [15] solver, which is used in this work.

2.2 Consensus Protocol

One of the ways to formally represent interactions between agents is by using graph theory. In such a formalism, each node in a graph stands for a state of an agent—a particular property of an agent (position, speed, etc.) or its perception of the environment (measurements of temperature, pressure, etc.). On the other hand, edges, connecting nodes, describe relations between the underlying states. Such a framework provides a powerful tool to describe various phenomena emerging in networked systems. A particularly interesting setup is related to the problem of consensus between agents in multi-agent systems. Let us consider a set of N agents with the state x_i of an agent i described as

$$\dot{x}_i(t) = u_i(t), \tag{2}$$

where u_i is a control input of agent i. Let us assume that the agents exchange information about their states through the underlying graph with the adjacency matrix $\mathbf{A}$ having element $a_{ij} = 1$ if agent j is in interaction with agent i, and $a_{ij} = 0$ otherwise. In a more general case, elements a_{ij} are time dependent, $a_{ij}(t)$, since they can represent some property of the interactions between agents (a quality of the communication channel [16], a quality of measurements, a trust between agents [17], etc.). In some applications they can attain negative value [18] (in case of a non-cooperative agent) or complex value [19] (for modeling motion on a flat surface). For the system of N agents, described by (2), we would like to define control input $u_i(t)$ and determine properties of underlying graph, so that condition

$$x_i(t) \rightarrow x_s, \forall i \;\; as \;\; t \rightarrow \infty \tag{3}$$

is satisfied. Consensus protocol [13], that defines control input $u_i(t)$ as

$$u_i(t) = \sum a_{ij}(t)[x_j(t) - x_i(t)] \tag{4}$$

fulfils (3) if underlying graph is connected $\forall t$. By including (4) in (2) a discrete form of the consensus protocol can be written as

$$x_i(k+1) = x_i(k) + \varepsilon \sum a_{ij}(k)[x_j(k) - x_i(k)] \tag{5}$$

where ε is discretization parameter that depends on the sampling interval T_s and the discretization method (for a simple Euler method $\varepsilon = T_s$).

3 Methodology

The aim of this work is to investigate whether we can leverage consensus-based communication to improve the efficacy of a multi-agent system in which each agent is driven by a POMDP policy. The scenario we use to showcase the improved performance of a POMDP-based multi-agent system with belief sharing is an exploration scenario in which a team of robots has to find a fire in one of the rooms of a building. The scenario is shown in Fig. 1.

Each agent is driven by a policy obtained from a POMDP that is solved offline. The POMDP for the agent considers that each agent is equipped with imperfect sensors that may not detect the fire when entering the room which is on fire, but we do not consider localization and navigation errors. We assume obstacle and collision avoidance are handled by low-level control.

3.1 Measured Variables

To measure the efficacy of the proposed system, we focus on two key variables: the time taken to finish the search and whether the search was successful. Since each agent decides on its own, based on the belief of a POMDP, when to end the search (see Sect. 4.1), under the assumption of constant communication among agents we consider the mission to be over as soon as one of the agents concludes that it has collected enough information to end the search.

The success of the search is measured by the belief of the agent over the location of the fire at the end of a mission. Let there be N rooms with the fire being in $room_i$. Then, the belief of an agent over locations of fire is $b_j, j = 1 \ldots N$, and the success of a mission is defined as follows:

$$success = \begin{cases} True & , \text{ if } \frac{b_i}{max_{j \neq i}(b_j)} > K \\ False & , \text{ if } \frac{b_i}{max_{j \neq i}(b_j)} \leq K \end{cases}, \tag{6}$$

where K can be set according to a desired level.

3.2 Simulating Multiple Missions with Sensing Failures

To investigate the behavior of agents belief, we simulate sensing inaccuracies and failures. We focus on failures to detect the fire and do not consider false positives in rooms where there is no fire. Each time an agent enters the room that is on fire (denoted RF), there are three possible outcomes: room observed and fire detected RF_{fire}, room observed and fire not detected RF_{clear}, room not observed $RF_{unknown}$. When an agent enters a room that is clear of fire (denoted RC), two scenarios are possible: room observed and fire is not detected RC_{clear} and room not observed $RC_{unknown}$ (scenario in which an agent observes fire in a clear room RC_{fire} is not possible in the proposed simulation setup). To simulate sensing uncertainty and failures, we define two probabilities: $p_{detection}$ which corresponds to expected sensor recall and $p_{failure}$ which denotes sensing failure rate. Then, each time an agent enters the room on fire, the probabilities of simulating a scenario are defined as:

$$p(RF_{fire}) = (1 - p_{failure}) * p_{detection} \tag{7}$$

$$p(RF_{clear}) = (1 - p_{failure}) * (1 - p_{detection}) \tag{8}$$

$$p(RF_{unknown}) = p_{failure} \tag{9}$$

Each time an agent enters the room clear of fire, the probability of simulating one of the scenarios is as follows:

$$p(RC_{fire}) = 0 \tag{10}$$

$$p(RC_{clear}) = 1 - p_{failure} \tag{11}$$

$$p(RC_{unknown}) = p_{failure} \tag{12}$$

By changing $p_{detection}$ and $p_{failure}$ we evaluate the performance of the multi-agent system under different conditions.

4 Multi-agent Framework Based on the POMDP and Consensus

4.1 Single-agent POMDP Model

We formulate the POMDP for an agent with two goals in mind: (a) the agent should be able to decide which room to visit next based on the current location and the potential benefits and costs of visiting another room; (b) the agent should be able to decide to end the search once sufficient information about the fire location has been collected. To track the location of the agent, we formulate a set of states S_L that correspond to key locations in the building shown in Fig. 1. The agent always starts the search from the state that corresponds to the location in the hallway.

To track the location of the fire, we use a set of states S_F. Since the location of the agent and that of the fire are not coupled, the agent's POMDP has the states $S = S_L \times S_F$. As already mentioned, we do not consider navigation and localization errors, which translates to S_L being fully observable. The states S_F are structured to correspond to each of the rooms in the building, extended by an additional state used to track the progress of information gathering. We call this state *unknown* and the agent starts the search with a large belief that the

Table 1. States, actions and observations of a single-agent POMDP

Sets of an agent POMDP	Cardinality
$S_L = \{hallway,\ room_i,\ i = 1 \ldots 10\}$	11
$S_F = \{unknown,\ fire_i,\ i = 1 \ldots 10\}$	11
$S = S_L \times S_F$	121
$A = \{scout_i,\ i = 1 \ldots 10,\ end\}$	11
$O = \{fire\ detected,\ no\ fire,\ unknown\}$	3

location of the fire is unknown. As the search progresses, the belief over this state should decrease.

The actions that the agent may take are only entering and scouting of the rooms, and there is exactly one scouting action for each of the rooms in the building. Additionally, the agent can perform a *return home* action, signaling the end of the search (denoted *end* for brevity). The observations that the agent can observe while scouting a room correspond to the scenarios described in Sect. 3.2: *fire detected, no fire, unknown.* Table 1 summarizes the POMDP for a single agent.

Transition matrices for the states S_L are formulated under the assumption of perfect localization and navigation:

$$p\left(s_L^{k+1} | a^k = scout_i, s_L^k\right) = \begin{cases} 1 & , s_L^{k+1} = room_i \\ 0 & , s_L^{k+1} \neq room_i \end{cases}. \tag{13}$$

The transition matrix for the action *end* is set to always return the agent to the starting position. Transition matrices for states S_F must be set up to encapsulate how much information is the agent expected to gather when scouting a room. We model this by a probability that the agent will stay in the state *unknown*, p_u:

$$\begin{aligned} & p\left(s_F^{k+1} | a^k = scout_i, s_F^k = unknown\right) \\ & = \begin{cases} p_u & , s_F^{k+1} = unknown \\ \frac{1-p_u}{N_{rooms}} & , s_F^{k+1} = fire_i \end{cases}. \end{aligned} \tag{14}$$

Scouting cannot change the fire location which is modeled by:

$$\begin{aligned} & p\left(s_F^{k+1} | a^k = scout_i, s_F^k \neq unknown\right) \\ & = \begin{cases} 1 & , s_F^{k+1} = s_F^k \\ 0 & , s_F^{k+1} \neq s_F^k \end{cases}. \end{aligned} \tag{15}$$

For all actions $scout_i$, the observation probabilities are defined in Table 2. The observation probabilities in Table 2 are carefully tuned to result in the agent increasing belief over the $s_F = unknown$ state if such an observation

Table 2. Observation probabilities for action $scout_i$, $p(o^{k+1}|a^k = scout_i, s_F^{k+1})$, for different states s_F^{k+1}.

	$s_F^{k+1} = fire_{j=i}$	$s_F^{k+1} = fire_{j\neq i}$	$s_F^{k+1} = unknown$
$p(o^{k+1} = fire\ detected)$	0.90	0.09	0.088
$p(o^{k+1} = no\ fire)$	0.09	0.90	0.895
$p(o^{k+1} = funknown)$	0.01	0.01	0.015

is obtained (compare probabilities of $p(o^{k+1} = unknown)$ for different states s_F^{k+1}), while decreasing the belief over $s_F = unknown$ when one of the other observations is received. The transition and observation values for the action *end* are formulated to reset the agent's belief to the initial belief over states S_F.

The final step in designing the single-agent POMDP is to define a reward model. We factor the reward in two sets: R_L rewarding the agent based on the states S_L and R_F rewarding the agent based on the states S_F. The reward parameters in R_F are set to penalize the action *end* when $b(unknown)$ is large, while scouting actions have the same cost regardless of the location of the fire. The parameters of R_L are designed to encourage sequential scouting of rooms (i.e. going from $room_i$ to $room_{i+1}$, then to $room_{i+2}$ and so on) and penalize staying in the same room (i.e. performing an action $scout_i$ while in $room_i$).

We use the same POMDP for each of the agents in the team, with the slight difference of rewarding the different action in the starting location (*hallway*) in order to avoid sending all agents to the same room at the start, but multiple agents going to the same room is theoretically possible since the only negotiation is performed on the level of belief, although this was not observed in the experiments.

4.2 Novelty-Biased Consensus

The main motivation for the use of consensus in decision making based on the POMDP formalism is to make a multi-agent system more efficient in carrying out a mission. Sharing information among the agents is a way in that direction, but it is fundamental to know what information is to be exchanged and how.

The approach presented here is based on the average consensus in which information is equally important and is equally treated no matter when and from which agent it originates. In a scenario where certain information is acquired (observed) only by a single agent and only for a short period of time, that can be a problem from the point of view of an agent introducing new information into the system. Due to the averaging in the consensus, this new information is considered the same way as existing (old) information, which in turn will cause degradation in the overall belief update of the system. To prevent this we introduce dynamic changes of adjacency matrix elements, an approach that could be considered as a variation of our earlier work on the trust-based consensus [17].

To introduce the bias towards novel information in the system, we change the value of $a_{ij}(k)$ each time the agent i gets new observation. The logic behind the change is to reduce the influence of neighbouring agents (holding old observations) on the agent i (who has new information based on the most recent observation).

Formally, all elements of the row i of the matrix $\mathbf{A}$, corresponding to the agent that obtained new observation at $k = \tau$, attain values

$$a_{ij}(k+1) = \gamma_i a_{ij}(k),\ 0 < \gamma_i < 1. \tag{16}$$

where γ_i is the novelty factor for agent i.

The updated value a_{ij}^{k+1} value remains constant until the agents reach consensus, and then returns to the value it had before the observation was acquired, a_{ij}^{k}. The process is repeated each time some of the agents get new observation. If a new observation occurs before consensus is reached for the previous observation, elements of the corresponding row in $\mathbf{A}$ will be changed according to (16), meaning that more than one row in $\mathbf{A}$ may have values between 0 and 1. The value of novelty factor γ_i, related to agent i, depends on the quality of the agent's observations—lower values of γ_i correspond to higher quality of observations, but it is important to note that scaling a row of the matrix $\mathbf{A}$ by $\gamma_i \neq 0$ does not affect the stability of the consensus algorithm, even in case of two agents simultaneously changing their adjacency matrix. Note here, that the initial values in $\mathbf{A}$ can encode different agent abilities (e.g. an agent is more/less *trustworthy* than others).

4.3 Integration of Novelty-Biased Consensus and POMDPs

First, for each of the agents the POMDP is solved offline and a policy, which is a mapping from belief to actions, is calculated and implemented in a POMDP controller. Each agent uses the same POMDP with slight differences for achieving different starting action. The agents communicate via a discrete form of consensus (5), implemented on each component of the belief vector $b(S_F)$. Upon acquisition of a new observation, the agent that observed the environment updates belief according to Eq. (1) and changes the matrix $\mathbf{A}$ according to Eq. (16). Once consensus over b is reached with the neighbouring agents, matrix $\mathbf{A}$ is reset to the default values and the negotiated belief b^* is sent to the POMDP controller and b^* is used to calculate the next action for the agent. In this way, the agent is able to integrate information acquired by other agents (obtained through b^*) into its own decision making process. The behavior of a single agent is shown in Fig. 2. The procedure is repeated until the POMDP controller of one of the agents selects the action *end*.

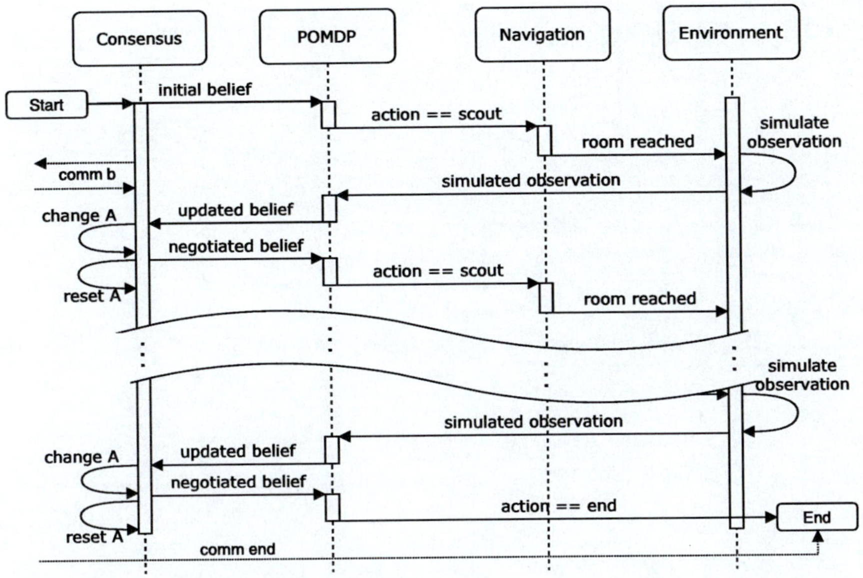

Fig. 2. Sequence diagram of the behavior of an agent in the system.

5 Results

As already mentioned, the analysis of the efficacy of the proposed multi-agent system is carried out by simulations. We use 3 agents, and search for a fire in a building with 10 rooms (see Fig. 1). Duration of each navigating action are simulated based on average times observed in physics-based simulations of UAVs in ROS and Gazebo. For the default values of the observation simulator we set $p_{detection} = 0.9$ (denoted as p_d on-wards), $p_{failure} = 0.05$ (denoted as p_f on-wards), and $p_u = 0.7$. The consensus, and thus the communication between agents, runs at a frequency 50 Hz, requiring around 50 KB/s of bandwidth in a fully connected network topology (measured up to the accuracy of the ROS bandwidth measuring tools). We set the novelty factor $\gamma_i = 0.2$ for each agent.

Figure 3 shows the evolution of belief in the proposed multi-robot framework when fire is in a room $s_F = room_i$. We track $b(s_F = room_i)$, $b(s_F = unknown)$ for each of the agents in the team. Additionally we show the maximum value for beliefs that the fire is one of the other rooms $max\,\{b(s_F = room_{j \neq i})\}$, since it is used to validate the success of the search mission.

It can be observed that the system starts with large belief that the location of fire is *unknown*. As the search progresses, the belief that the fire is located in $room_i$ overtakes the belief that the fire location is *unknown*. The belief that the fire is in $room_i$ at the end of search is significantly larger than the maximum value of the other belief components, indicating a successful search.

Next, we analyze the performance of the system in terms of the way agents communicate: (a) belief sharing via novelty-biased consensus among the agents

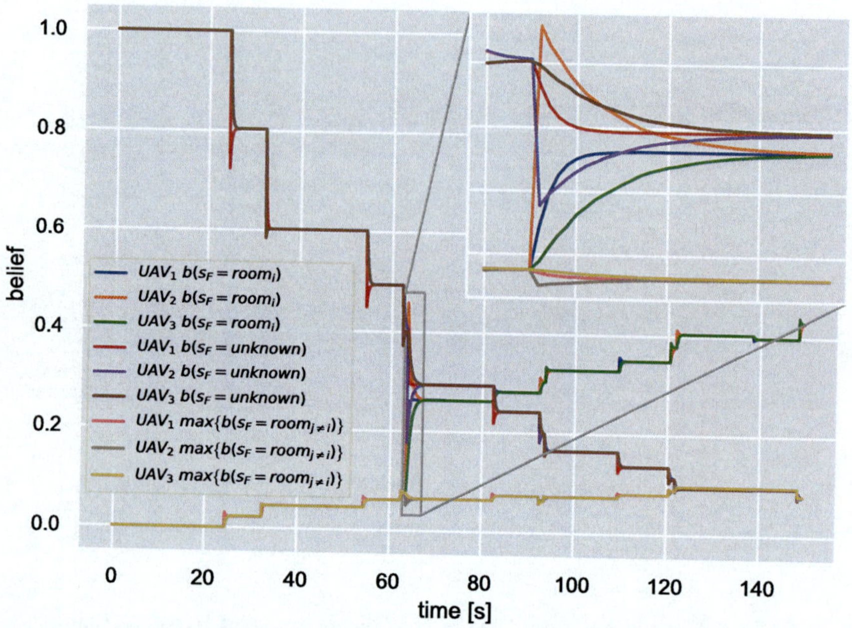

Fig. 3. The evolution of the belief for a search mission with 3 agents when fire is in $room_i$. Sharp changes in beliefs indicate times when the observation was made in the team. The zoomed-in portion of the graph shows the convergence of the belief components to the final value.

proposed in this paper (denoted *N-B cons.*); (b) belief sharing via a standard average consensus (denoted *Avg cons.*); (c) agents do not communicate (denoted *No comm*). The results for 100 iterations for each scenario are shown in Fig. 4, focusing on the time required to finish the mission.

The evolution of belief over the location of fire in Fig. 4 shows that there is no significant difference in the final belief when using different communication setups, but the time needed to complete the search is drastically reduced when using novelty-biased consensus, especially compared to a standard average consensus protocol where the old information is treated as equally informative as the newest observation, which worsens performance.

Next, we evaluate the performance of the system with less precise sensors by changing the values of p_d and p_f. We evaluate both the success of the search and the time needed to complete the search, as shown in Fig. 5. As expected, with lower sensing reliability we observe an increase in the time needed to complete the mission, with unsuccessful missions taking longer than successful ones. Also according to expectation, the success rate of the search decreases with lower sensing reliability, as shown in Table 3.

Table 3 summarizes the results of 1000 iterations with 10 different simulation setups, including different topology, communication type and detection capabil-

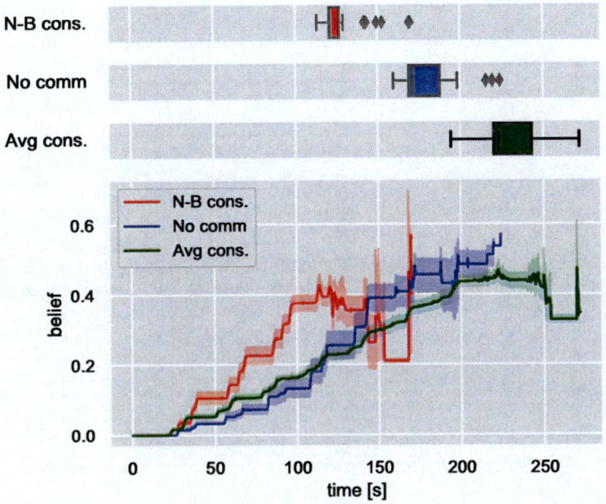

Fig. 4. Bottom part of the graph shows average belief with a 1σ confidence interval for a room under fire with different types of communication. The upper part of the graph shows box-plots of the recorded mission duration, showing significant improvements when the novelty-biased consensus is used.

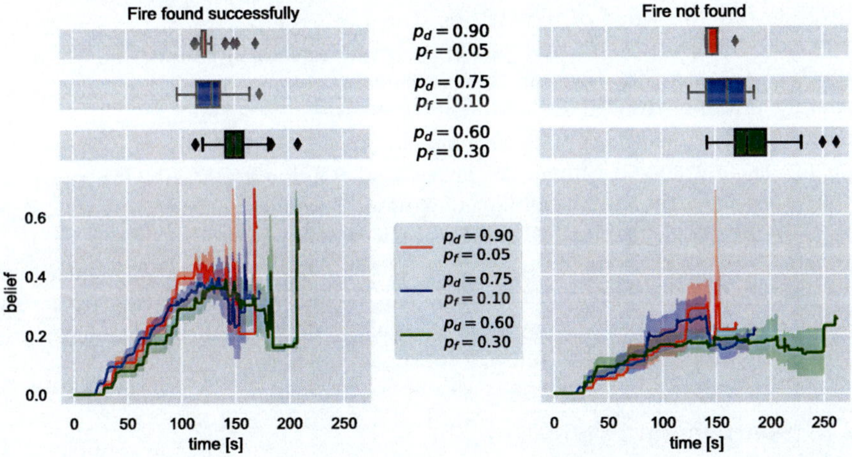

Fig. 5. Belief of agents when searching for a fire with different sensor precision parameters. The left side of the graph shows the average beliefs with 1σ confidence interval, combined with the distribution of mission duration for missions considered successful with $K = 3$, while the right side shows the same outcomes for unsuccessful missions.

Table 3. Comparison of the times needed to finish the mission and success rates with various setups. Each setup was run 100 times for a total of 1000 simulation runs.

Topology	Communication type	Detection parameters	Success rate (%)	Time to finish (s)
Fully connected	Average consensus	$p_d = 0.90 p_f = 0.05$	100	229.55 ± 14.87
		$p_d = 0.60 p_f = 0.30$	75	316.94 ± 44.05
	Novelty-biased consensus	$p_d = 0.90 p_f = 0.05$	94	126.39 ± 11.96
		$p_d = 0.75 p_f = 0.10$	84	132.49 ± 19.53
		$p_d = 0.60 p_f = 0.30$	69	160.96 ± 28.60
Star topology	Novelty-biased consensus	$p_d = 0.90 p_f = 0.05$	95	128.34 ± 11.83
Ring topology	Novelty-biased consensus	$p_d = 0.90 p_f = 0.05$	95	126.19 ± 15.53
No communication		$p_d = 0.90 p_f = 0.05$	94	177.78 ± 15.48
		$p_d = 0.75 p_f = 0.10$	89	190.27 ± 26.65
		$p_d = 0.60 p_f = 0.30$	64	223.28 ± 42.68

ity. Along with the aforementioned decrease in success rate with lower sensor accuracy, it can also be observed that there is no significant difference between different network topologies for the proposed framework, due to the fact that the agents share belief at the rate 50 Hz, which means that the negotiation time is negligible compared to the navigation time of agents. Another notable observation from Table 3 is that using average consensus results in better success rate at the lowest sensor accuracy settings, compared to the proposed framework. This indicates that the value of $\gamma = 0.2$ used in the experiments is too low for the quality of the information that the error-prone agents introduce, but it is also important to note that this slight improvement in success rate comes at a cost of nearly doubled time of search. Scaling up requires formulating and solving an additional POMDP for each new agent. Since POMDPs are of same size for each agent, and solved offline, we argue that solving POMDPs for any given scenario scales linearly with number of agents. Computational complexity of the consensus protocol also scales linearly with number of nodes in the network. The bandwidth requirements at best (ring topology) scale linearly with number of agents. In the worst case scenario, each agent communicating with every other agent, the bandwidth requirements scale with the square of the number of agents, indicating that the proposed approach scales well.

5.1 Experimental Evaluation

Based on the simulation experiments, we have identified 4 different scenarios to implement with Sphero robots, to evaluate performance of the system under stochastic distribution of observation events caused by imperfect navigation of the robots. The robots are tracked and controlled via a simple image processing algorithm, while observations were simulated according to procedure described in Sect. 3.2 since Sphero robots do not have sensors. The results are summarized in Table 4. Detailed description of the experiments, accompanied by videos of each experiment and belief graphs is available at https://sites.google.com/view/pomdp-consensus-experiments.

Table 4. Results of Sphero experiments. Four types of scenarios are evaluated. Each scenario is evaluated for all three considered communication types, accompanied with information about the duration of the search and whether the search was successful, with unsuccessful searches denoted with *.

Scenario	Fire location	Error type	Communication type	Time to finish (s)
1	r_2	None	No communication	46.80
			Average consensus	68.70
			Novelty-biased consensus	37.70
2	r_3	False positive $agent_2$ in r_4	No communication	53.70
			Average consensus	64.10
			Novelty-biased consensus	33.40
3	r_2	False negative $agent_1$ in r_2	No communication	44.80
			Average consensus	74.10
			Novelty-biased consensus *	51.60
4	r_3	None	No communication	54.30
			Average consensus	67.70
			Novelty-biased consensus	35.50

Since in three of the four scenarios the search results with success regardless of communication type, with times needed for search inline with the simulation results, we briefly discuss the scenario in which one of the agents makes an error in observing fire by not detecting the fire (false negative), while the other agents make correct observations. In this scenario, the search without communication and with sharing beliefs via the average consensus results in a successful search, while novelty-biased consensus communication propagates the incorrect observation throughout the system in such manner that correct observations from other agents cannot compensate for the error. This confirms that the proposed approach may not be suitable for situations in which some or all agents are significantly more error-prone than modeled in their corresponding POMDPs. The video accompanying this paper can be found on YouTube [20].

6 Conclusion and Future Work

We have shown that the introduction of communication between agents in the form of sharing beliefs improves the performance of the multi-robot system compared to a team driven by the same POMDP policy but without communication. We have also shown that the multi-robot team performs better when using the proposed novelty-biased consensus than compared to average consensus algorithm, when dealing with higher quality data. The research presented in this paper can be considered as a stepping stone towards the decentralization and cooperation in multi-robot systems driven by simple POMDPs. In our future work, we plan to experiment with a similar setup using I-POMDPs, and will investigate whether the assumption of a common belief among agents can resolve infinite nesting of interactive beliefs in such models.

Acknowledgment. This work is funded by European Commission Horizon 2020 Programme through project under G. A. number 810321, named Twinning coordination action for spreading excellence in Aerial Robotics - AeRoTwin[21]. Work of Frano Petric was supported by the European Regional Development Fund under the grant KK.01.1.1.01.0009 (DATACROSS).

References

1. Puterman, M.L. (ed.): Markov Decision Processes. Wiley, Hoboken (1994)
2. Monahan, G.E.: State of the art—a survey of partially observable Markov decision processes: theory, models, and algorithms. Manage. Sci. **28**(1), 1–16 (1982)
3. Frans, A.: Oliehoek and Christopher Amato, 1st edn. A Concise Introduction to Decentralized POMDPs. Springer Publishing Company, Incorporated (2016)
4. Omidshafiei, S., Agha-mohammadi, A., Amato, C., How, H.P.: Decentralized Control of Partially Observable Markov Decision Processes Using Belief Space Macro-Actions. CoRR, abs/1502.06030 (2015)
5. Amato, C.: Decision-making under uncertainty in multi-agent and multi-robot systems: planning and learning. In: Proceedings of the Twenty-Seventh International Joint Conference on Artificial Intelligence. International Joint Conferences on Artificial Intelligence Organization (2018)
6. Bernstein, D.S., Zilberstein, S., Immerman, N.: The Complexity of Decentralized Control of Markov Decision Processes. CoRR, abs/1301.3836 (2013)
7. Amato, C., Chowdhary, G., Geramifard, A., Kemal Ure, N., Kochenderfer, M.J.: Decentralized control of partially observable markov decision processes. In: 52nd IEEE Conference on Decision and Control, IEEE (2013)
8. Lauri, M., Pajarinen, J., Peters, J.: Information gathering in decentralized pomdps by policy graph improvement. In: Autonomous Agents and Multiagent Systems (AAMAS) (2019)
9. Amato,C., Konidaris, G., Anders, A., Cruz, G., How, J.P., Kaelbling, L.P.: Policy search for multi-robot coordination under uncertainty. Int. J. Robot. Rese. **35**(14), 1760–1778 (2016)
10. Doshi, P., Gmytrasiewicz, P.J.: A Framework for Sequential Planning in Multiagent Settings. CoRR, abs/1109.2135 (2011)
11. Zhang, W., Chen, X., Ma, L.: Online planning for multi-agent systems with consensus protocol. In: Proceedings of the 33rd Chinese Control Conference, pp. 1126–1131 (2014)
12. Lalitha, A., Javidi, T., Sarwate, A.D.: Social learning and distributed hypothesis testing. IEEE Trans. Inf. Theor. **64**(9), 6161–6179 (2018)

13. Olfati-Saber, R., Murray, R.M.: Consensus problems in networks of agents with switching topology and time-delays. IEEE Trans. Autom. Control **49**(9), 1520–1533 (2004)
14. Ong, S.C.W., Png, S.W., Hsu, D., Lee, W.S.: Planning under uncertainty for robotic tasks with mixed observability. Int. J. Robot. Res. **29**(8), 1053–1068 (2010)
15. Brock, O., Trinkle, J., Ramos, F.: SARSOP: Efficient Point-Based POMDP Planning by Approximating Optimally Reachable Belief Spaces, pp. 65–72. MIT Press (2009)
16. Lončar, I., Babić, A., Arbanas, B., Vasiljević, G., Petrović, T., Bogdan, S., Mišković, N.: A heterogeneous robotic swarm for long-term monitoring of marine environments. Appl. Sci. **9**(7), 1388 (2019)
17. Haus, T., Tolić, D., Mikulski, D.G., Lewis, F.L., Bogdan, S., Palunko, I.: Trust-based self-organising network control. IET Control Theor. Appl. **8**(18), 2126–2135 (2014)
18. Altafini, C.: Consensus problems on networks with antagonistic interactions. IEEE Trans. Autom. Control **58**(4), 935–946 (2013)
19. Dong, J.-G., Qiu, L.: Complex Laplacians and Applications in Multi-agent Systems (2015)
20. Multi-agent Coordination Based on POMDPS and Consensus. https://youtu.be/CRDzkafnGwU
21. Aerotwin Project. https://aerotwin.fer.hr/. Accessed 25 Mar 2021

KI5GRob: Fusing Cloud Computing and AI for Scalable Robotic System in Production and Logistics

Yongzhou Zhang[1,2](✉), Gergely Sóti[1,2](✉), Björn Hein[1,2], and Christian Wurll[1]

[1] Hochschule Karlsruhe—University of Applied Sciences, 76133 Karlsruhe, Germany
yongzhou.zhang@h-ka.de

[2] Karlsruhe Institute of Technology, 76131 Karlsruhe, Germany

Abstract. Robotics and AI are essential components in current and future production scenarios. For example, in object handling, breakthroughs in AI have achieved revolutionary progress. As systems scale and the complexity of tasks increase, the use of deep learning in industrial robotics applications leads to specific challenges in terms of scalability, reliability, and safety. Some of these challenges can be overcome by utilizing cutting-edge technologies in cloud computing such as on-demand computing, massive parallelization, microservices, and the DevOps pipeline. With KI5GRob, we propose a novel approach to facilitate the development of robotic applications fusing cloud computing and AI. As example application for the proposed cloud approach, KI5GRob investigates machine learning methods for multi-modal sensor data processing and transfer learning to solve industrial robotic manipulation tasks. The goal of the project is to develop a microservice based software architecture that enables on-demand deployment of such learning-based methods as well as cloud-based motion planning and robot control. This paper gives a high-level overview of the goals and the ongoing research activities in KI5GRob. The first results about the overall hardware and software architecture are also presented.

Keywords: Robotic manipulation · Machine learning · Multi-modal sensor fusion · Transfer learning · Cloud robotics · Microservice · Motion planning · Robot control

1 Introduction

The goal of traditional automation in production is to increase the throughput and accuracy for repetitive tasks, involving customized hard- and software solutions. As a result, the application scope of such robotic systems is rather narrow and is also restricted to specific tasks in static environments. Recent advancements in robotics technology lead to significantly broader application possibilities in production and logistics to meet the increasing demand in commercial

Y. Zhang and G. Sóti—equally contributed.

I. Petrovic et al. (Eds.): IAS 2022, LNNS 577, pp. 706–723, 2023.
https://doi.org/10.1007/978-3-031-22216-0_47

and consumer segments in the era of online shopping and customized products. The objective of automation has shifted from solving simple, repetitive tasks to dealing with complex manipulation tasks in more and more unstructured and dynamic environments, leading to further challenges regarding safety, reliability robustness, and scalability [1].

A substantial progress was made possible in diverse robotic tasks, in particular autonomous object manipulation, through recent breakthroughs in artificial intelligence. Deep learning techniques have dramatically improved robotic perception such as localization [2], generic object detection, and instance segmentation [3]. Data-driven grasping approaches enhance the ability of robots to deal with unknown, deformable objects, even in cluttered environments [4]. Reinforcement learning enables robots to learn new skills in simulation, which can then be transferred to the real world [5,6]. Utilizing multiple modalities like vision and touch with deep learning approaches has shown potential of using robots to perform difficult and complex manipulation tasks [7]. However, there are still many open questions and challenges to applying deep learning techniques in the context of robotics [8].

The performance of learning-based algorithms in robot systems relies heavily on a massive amount of data and extensive computational resources. As the diversity of applications and the complexity of algorithms increase rapidly, robots demand more on-board computing capacity, which is yet limited and hard to upgrade. In this case, cloud computing can be used for powerful data processing, large-scale storage, system monitoring, and robot fleet management. In the context of robot learning tasks, inference and especially training can be accelerated by massive computational resources from the cloud and the possibility of massive parallelization. VertexAI from Google Cloud is an MLOps platform for training ML models, and for sharing and storing data [9]. However, deploying all the services in a distant cloud is also restricted due to the high network latency, data security, and privacy. To complement cloud computing, edge computing [10] and fog computing [11] are being introduced to bring cloud computing services closer to users. In the context of manufacturing and industrial robotics, this means closer to the shop floor.

Regarding robotic applications, pure cloud-based solutions cannot fulfill all requirements such as low-latency and the real-time condition for low-level control processes, high bandwidth to stream camera and LIDAR data, or security of sensitive data. Therefore, a hybrid (cloud, edge, embedded devices) computing paradigm seems to be the key for future robotic systems, especially for scalable multi-robot systems. Furthermore, as the size and complexity of software grows rapidly, deployment takes more time and effort, especially, for large heterogeneous robot system in different environments. ROS2, which was built on the success of ROS 1, has adopted DDS (Data Distributed Service) in its communication middleware, which simplifies the deployment for multi-robot systems. However, an individual bash script is still needed to be developed for each application, to download code, resources, and to coordinate a multitude of launch files. In cloud native computing, microservice based architectures and DevOps

[12] are widely and successfully used to build a modern, scalable cloud applications. Transferring those technologies into robotic software development is a key enabler for scalable AI-powered robotic system in the production.

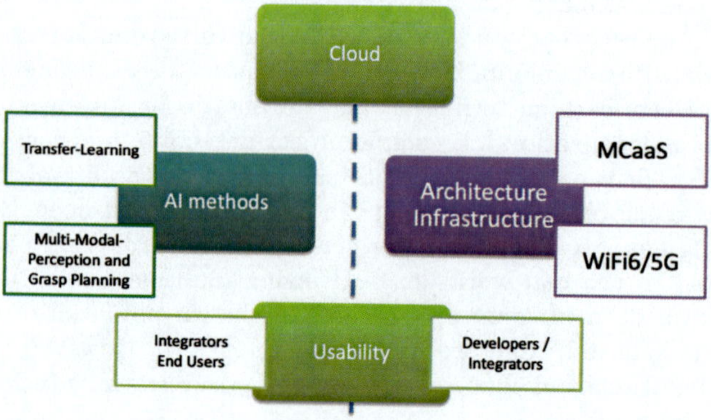

Fig. 1. The core idea of KI5GRob is to bring cloud concepts to robotics, in particular to investigate the interplay between cloud, AI methods and robotics, the required technical infrastructure and the impact on developers, integrators and end users.

In the research project KI5GRob we investigate how cloud and AI technologies can be combined with robotics in a beneficial way (s. Fig. 1). We therefore propose a component-oriented hardware architecture coupled with a microservice-oriented software architecture. In this framework, the DevOps paradigm is applied with the goal of improving the usability, accessibility, and maintainability of AI technologies in robotic applications. We aim to facilitate the development of robotic applications in production and logistics. On one hand, we investigate the use of robots in various manipulation tasks by using suitable AI methods for multi-modal sensor signal processing and interpretation. In particular, autonomous pick-and-place tasks with emphasis on accurate placing and task-level transfer learning are in our focus. On the other hand, we explore the software, hardware architecture, and communication infrastructure requirements to accelerate the development and simplify the deployment of AI-enabled robotic applications using cloud-based motion planning and robot control.

The paper is organized as follows: Sect. 2 discusses the state of the art of some of the key technical topics addressed in KI5GRob. Section 3 gives an overview of key elements of the KI5GRob project. On the cloud side, these are the hardware concept, the microservice architecture, the software deployment strategies or the cloud-based motion control architecture. On the AI side, these are object manipulation, multimodal perception, and transfer learning. The paper concludes with a brief conclusion in Sect. 4.

2 State of the Art

In order to achieve the goals of KI5GRob, several interdisciplinary areas are investigated. This section provides insights into the current state of the art on relevant research topics.

2.1 Cloud Robotics

No robot is an island. Almost all robots are connected to networks that provide extensive computing resources and exchange data with other systems or users [13]. The concept of cloud robotics was proposed firstly by J. Kuffner in 2010 [14]. It can be broadly defined as follows: any robot or automation system that relies on either data or code from a network to support its operation, i.e., where not all sensing, computation, and memory are integrated into a single standalone system [15]. This definition involves robot systems in various areas that can benefit from cloud computing: self-driving cars, networked mobile robots, AGVs in a warehouse in logistics, assembly robots in production, and service robots in public or household environments. Robots can gain potential benefits from the following fields by using cloud computing [15–18]:

- Big data: robotics related data including images, videos, maps, audio, that can facilitate machine learning and perform various tasks, e.g., grasping, motion planning.
- Cloud computing resources: learning-based methods require computing power for training and inference, that is available in the cloud and can be used on-demand.
- Collective learning: collected and shared data from multiple physical instances can facilitate robot learning.
- Human computation: using crowd sourcing access to remote human expertise for analyzing images, remote control, and imitation learning.

Regarding the research work about system architectures, Arumugam et al. proposed a software framework called DAvinCi [19] to build a map by using the FastSLAM algorithm in Map/Reduce based on the Hadoop cluster and gained a significant performance improvement. In the RoboEarth [20] project, a platform called Rapyuta [21] has been developed, which enables creating the robotic application in the cloud platform and provides access to the knowledge database from RoboEarth. Moreover, Xia et al. proposed a microservice-based software architecture for robots applied in an intelligent environment and demonstrated a prototype with the visual SLAM process [22]. In the context of deployment, a software framework called FogROS is presented by Chen et al. to allow rapid deployment of ROS system into cloud environments [23].

2.2 Cloud Based Motion Planning and Control

Motion planning is an essential task in robotics and can enormously benefit from using cloud computing with appropriate massive parallelization. Ichnowski et al. analyzed the economic benefits of cloud-based motion planning in terms of service life, operational efficiency, and battery-based operation time [24]. In their following paper, they introduced a cloud-based motion planning algorithm with parallelized construction of roadmaps for power-constrained robots [25]. Bekris et al. also proposed to use a pre-computed roadmap which is computed in the cloud, as well as locally, to perform flexible manipulation tasks [26]. The research challenges regarding networking and resource allocation have also been addressed in the papers above.

Additionally, the concept "RobotControl as a Service" proposed in [27] has indicated the feasibility of control of industrial robots on a mixed infrastructure that consists of local power-constrained devices and on-demand cloud computing. For multi-robot systems in unknown and complex scenarios, Zagradjanin et al. have presented an approach to improve global path planning in terms of multi-criteria by shifting heavy computation to the cloud [28].

2.3 Robotic Manipulation and Transfer Learning

Beside the aforementioned cloud aspects, AI aspects are also investigated in KI5GRob focusing especially on pick-and-place applications. Classical and most current industrial solutions for pick-and-place tasks consist of three main steps: object pose estimation, grasp synthesis, and pose displacement. Solutions for pose estimation and grasp synthesis can be divided into two main approaches: analytical and data-driven. Analytical solutions typically require 3D models of objects and data-driven methods need a large amount of labeled training data for achieving a robust performance [4]. Scaling these methods for many different or even unknown objects is not feasible, as creating 3D models and labeling training data requires significant amount of time and effort, resulting in ever increasing costs. A substantial part in pick-and-place tasks is determining the displacement of objects for a successful execution. It is a challenging problem itself, as it heavily depends on the prior grasp: different grasps on the same object result in different displacements. Additionally, in the case of novel or unknown objects, the robot would not know what to do with something it has never seen before [29].

A promising approach is sim-to-real transfer learning, where labeled training data is generated in a simulation with subsequent training with only a few real-world examples. However, this approach still needs 3D object models. Additionally, in many cases, models trained on simulated data do not generalize well to real-world applications [6].

Regardless of the data acquisition method, two fundamental approaches have emerged in learning robot manipulation without explicit object pose estimation. An end-to-end learning approach, using a velocity-like control of the end-effector [30,31], and the usage of predefined manipulation primitives where the task is to select and parametrize the appropriate primitives [29,32–34].

In the context of bin-picking, where the goal is to empty a cluttered bin, without any constraints on where or how to place the grasped objects, self-supervised reinforcement learning using real-world objects has been the focus of many publications. These methods overcome the need for manually created 3D models or training data while also showing promising results regarding generalization to unknown objects. Levine et al. trained 14 robots simultaneously for over two months to achieve 82.5% grasp success rate on novel objects [30]. Kalashnikov et al. even managed to achieve a 96% grasp rate with a similar setup while also implicitly learning pre-grasp manipulation, like pushing objects to re-position them for better grasps [31]. A well-known framework for estimating grasp poses is DexNet 4.0 [32]. It learns ambidextrous grasp policies for a parallel-jaw and a suction cup gripper on synthetic datasets using domain randomization and analytic physics and geometry. In real-world experiments it clears bins of up to 25 novel objects with greater than 95% reliability at a rate of more than 300 mean picks per hour. These methods however, do not deal with the problem of displacement.

2.4 Sensor Data Processing

The recent development of high-performance deep learning methods for computer vision enabled a surge of improvements in robot learning, hence most methods for robot manipulation mainly rely on visual sensor data. However, several approaches leverage additional sensor modalities, such as tactile and force-torque in object handling tasks. Lee et al. extended visual data with force feedback to improve robot control for the peg-insertion task using a self-supervised deep reinforcement learning method [35]. GelSight sensors, which are vision-based tactile sensors, were used by Calandra et al. in combination with visual data to improve grasp robustness by learning to adjust grasps if necessary [36]. Similarly, Hogan et al. proposed a deep learning based method to estimate grasp quality and apply a re-grasp policy based on tactile sensors being able to estimate a correct grasp outcome with an average accuracy of 75% for novel objects and improving grasp success rates by an average relative increase of 70% [37].

Navarro et al. showed that it is possible to grasp objects with virtually no displacement using a closed-loop servoing approach based on tactile-proximity-sensors to pre-align the gripper [38]. Time-of-Flight sensors on the fingers of a gripper were used by Sasaki et al. to adjust the pose of the fingers relative to objects and support surface to improve the robustness of robotic grasping [39].

Vision, force, and tactility are vastly different sensor modalities, yet humans still have the remarkable ability to interpret and fuse them. This includes having expectations like how heavy or soft an object is, or even what an object would look like from another perspective. Being able to do this could prove extremely helpful for motion and manipulation planning. Generative Query Networks tackle the vision side of this problem with a self-supervised learning framework for perceiving and predicting a scene [40]. From a few images (and their corresponding perspectives) of a scene, the model is able predict the same scene from previously unobserved viewpoints. A different kind of perception was achieved by Alagi et al. by using capacitive proximity sensors – a modality not available to humans – to recognize and differentiate between multiple materials using deep neural networks [41].

3 The KI5GRob Approach

KI5GRob focuses on fusing cloud computing and AI for scalable robot systems in production and logistics, to facilitate the development process of new robotic applications. On one hand, we investigate machine learning algorithms and methods, including transfer learning, for multi-modal sensor data processing in the context of industrial manipulation tasks, in particular pick-and-place tasks. On the other hand, we propose a new hardware architecture coupled with a microservice-oriented software architecture to make use of cloud technologies, to clarify the benefits of learning-based approaches, and to identify suitable ways of using them in industrial settings. For this purpose, we equip our lab with high-speed wireless communication infrastructure, design a novel software architecture, and apply the DevOps principles to develop intelligent robot system.

Research topics in KI5GRob include flexibly configurable and decoupled robot systems, microservice-based software architecture, deployment in hybrid environments, cloud-based motion planning and robot control, reliable robotic object handling based on multi-modal sensor data as well as task-level and sim-to-real transfer learning with high-fidelity simulation.

3.1 Decoupled Robot Hardware

In classical industrial robotic automation, each robot and its accessories, such as gripper, camera, or safety components, are controlled either by the robot controller or a separate computer. Each robot system is usually developed for one specific application as a closed-loop system. Once the robot is deployed in production, it is technically challenging to upgrade or adapt it for other use-cases. Currently, caused by the trend of mass customization of products and rapid product change-over, the tasks for robots have become more diverse. Thus, a flexible and straightforward configurable plug-and-play system would be favorable.

In KI5GRob, our lab is designed in a completely decoupled way. As depicted in Fig. 2, the gripper is not controlled as usual by the robot controller or onboard computer but works as a stand-alone component and can be controlled over local network. Each component is individually controllable and configurable. All hardware components communicate via wired or wireless networks instead of the internal communication process inside one computer. The system's configuration for one application will be realized by cloud-based software.

Using the loosely coupled hardware pattern, the monolithic hardware system can be decomposed into lightweight, function-focused independent small systems that can be individually developed and maintained. Each component provides its independent services for the whole system. One of the purposes of this concept is that the hardware for new robotic applications can be assembled fast and reliably depending on the requirements. Additionally, a physical decoupled

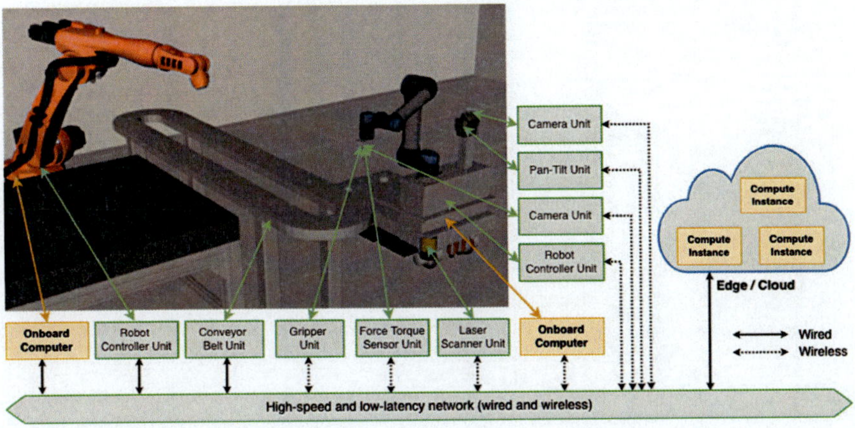

Fig. 2. The stationary robot is connected directly via cable to the the local lab network. The mobile robots and accessories such as grippers and cameras are equipped with wireless communication components (WiFi, 5G). This decoupled robot hardware is compatible with the microservice-based software and enables flexible deployment and easy configuration.

hardware system is essential for the microservice-based software architecture. By doing this, hardware-specific software can be applied as a service locally, or be deployed flexibly in the cloud. One possible drawback of this design is that the system's performance may be constrained by the bandwidth and latency of wireless network infrastructure. However, this bottleneck should be well compensated with the development of 5G technology.

3.2 Microservice-Based Software Architecture

The software of classical industrial robotic applications is usually designed following the "top-down design with stepwise refinement" concept and deployed as a monolithic system. Such as the widely used framework ROS, each functional module, e.g., motion controller, camera, driver, can be developed as an individual ROS package. ROS packages can be easily deployed on multiple connected machines (see example in Fig. 3a). The reusability and reliability of software are significantly increased by using the ROS framework and its design pattern. However, as the number of robots in an application grows and the required computing resources increase, two problems need to be considered. One is that each software update must be installed on every robot manually or automatically with specific tools. Secondly, additional computers have to be equipped for using intelligent and sophisticated algorithms, especially machine learning-based approaches.

To solve these problems, we propose a microservice-based software architecture for building scalable robotic applications that can run in a modern, hybrid computing infrastructure consisting of onboard devices, edge, and cloud. The

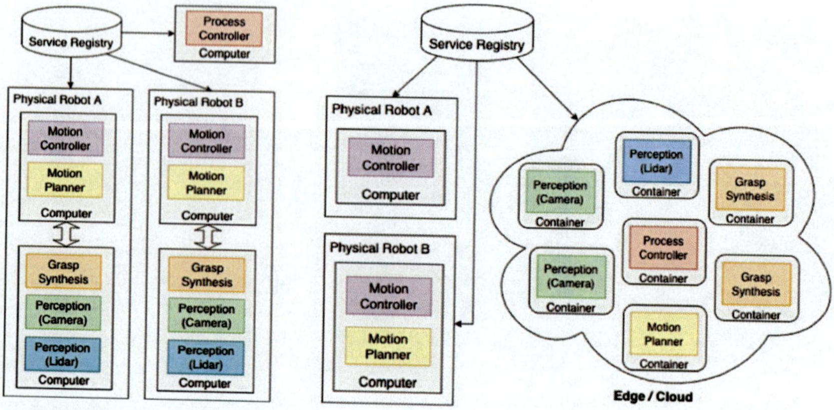

(a) Distributed monomer architecture

(b) Microservice-based architecture

Fig. 3. Classical distributed monomer architecture and modern microservice-based architecture which enables a flexible deployment across onboard computer and edge/cloud infrastructure

microservice architecture is widely and successfully used in cloud-native applications and has shown extensive advantages regarding productivity, development costs, and flexibility. As depicted in Fig. 3b, the entire software is divided into small services that can be developed individually and deployed in an isolated environment. The significant difference compared with distributed monomer architecture is the scalability and the possibility of flexible deployment of each service. The communication between services can be realized by REST [42], DDS [43], or gRPC [44], depending on application requirements. Currently, we are rebuilding a first set of software modules such as robot driver (e.g., KUKA EKI), object detection, grasping synthesis from ROS1 into ROS2 framework. Each service is packaged into a Docker container to be deployed and orchestrated by Kubernetes. Additionally, we also investigate the service granularity in the context of concrete tasks.

3.3 Software Deployment in Hybrid Environment

The containerized software packages can now be flexibly deployed and orchestrated in a cluster using the microservice-based architectural pattern. In our prototype, we use a self-managed Kubernetes [45] cluster for this purpose. Our cluster infrastructure contains several onboard computers installed near robots, VMs in edge (on-prem) servers, and VMs from a public cloud provider. Each computer unit can be flexible and dynamically added as a node or removed from the cluster. Calico [46] is used as the networking solution to enable communication between all pods across multiple machines without NAT (Network Address Translation). Besides using available open-source frameworks, several valuable tools are in development to lower the barrier to adopting our proposed deployment pipeline. The high-level overview of software stacks is depicted in Fig. 4.

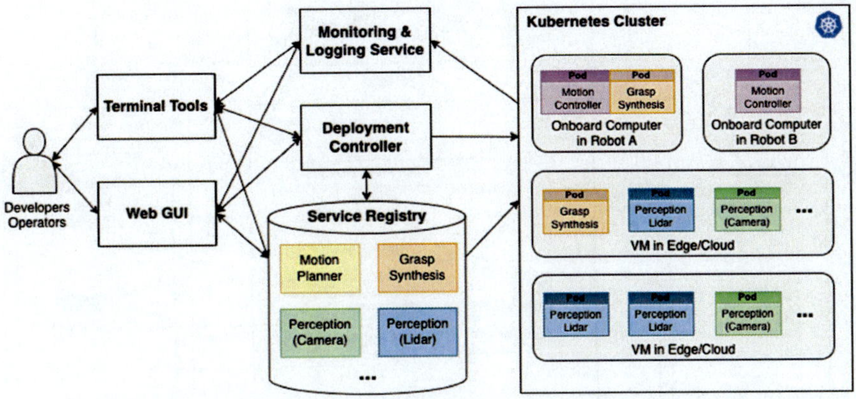

Fig. 4. Deployment and orchestration via Kubernetes

In KI5GRob, we adopt the up-to-date ROS2 [47] as a basic framework to develop the services, such as motion planning and grasp planning. ROS-based libraries and tools are widely used in robotic applications, and the on-developing ROS2 is expected to be a de facto standard in the future. We propose a service registry to store the metadata for deployment, including container image address, node preferences, scheduling policies, and resource allocation strategies. Furthermore, a terminal tool for Linux and a Web-based GUI, which provides easy access to the service registry, deployment controller, and cluster monitoring, will be developed.

Another crucial ongoing research topic is the optimization of pod scheduling and auto-scaling across the cluster based on system status for efficient computing resource utilization and high service availability. Robotic applications are typically more complicated than a deployment of pure cloud-native applications in the cloud, which is usually based on computing resource metrics such as the CPUs and memories. For example, the latency and traffic congestion caused by wireless communications and the limited bandwidth must be taken into account. In addition, the real-time requirement of low-level motion control is also a significant challenge of robotic applications.

3.4 Motion Control as a Service

Considering the cloud-based approach, it is only a natural research goal in KI5GRob to explore the possibilities for teams of robots. There are many very interesting tasks, that can be solved by using several robots cooperatively in a

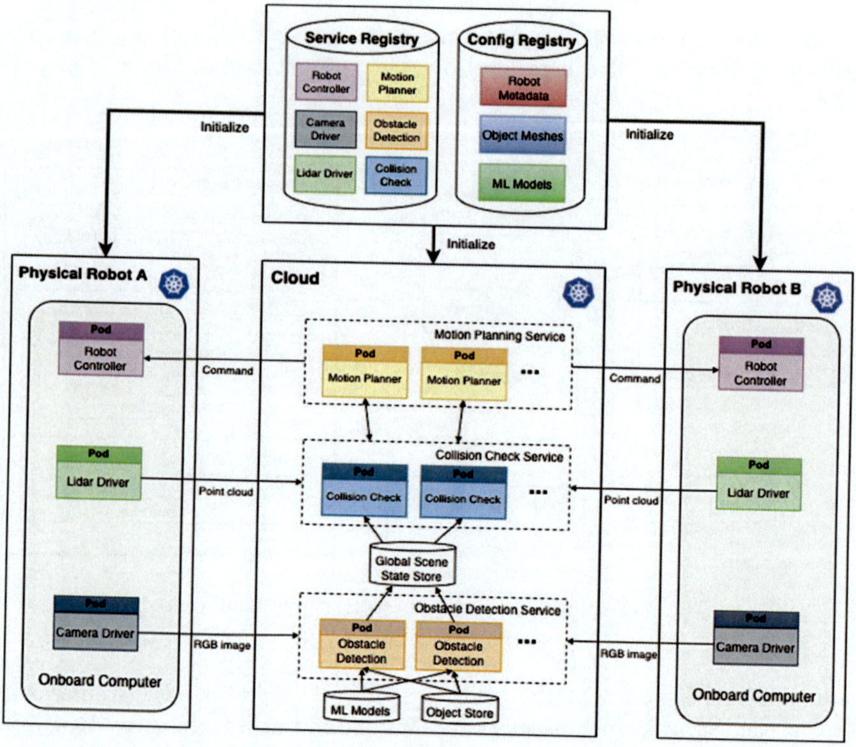

Fig. 5. Motion Control as a Service (MCaaS)

shared workspace, for example, object handover between two robots or carrying heavy load by multiple robots. In contrast to humans, who are excellent at solving cooperative tasks in a decentralized manner, KI5GRob investigates how quasi-centralized control can solve such tasks in the cloud. Coordinated multi-agent motion planning and control in a dynamic environments is a very interesting challenge: the motion trajectory has to be generated online, a state shared between the robots can be used very advantageously here, also to make changes in the environment known to all participants in order to react appropriately. An optimal solution here probably does not lie in a strictly centralized or completely decentralized structure. In order to achieve the highest possible flexibility, we are investigating the benefits of microservice-based software architecture. Thus, based on our hardware and software architecture, we propose a cloud-based approach for coordinated motion control of multi-robot systems and call it in the following Motion Control as a Service (MCaaS, s. Fig. 5). For demonstration purposes, we are currently developing MCaaS to control several stationary and mobile robots simultaneously. The MCaaS consists of multiple fully meshed services which can be scaled on demand in the hybrid cloud infrastructure. The overview of the services and the rough workflow are depicted in Fig. 5. In the initialization phase, all software packages are automatically deployed on the robot's on-board computer and VMs in cloud depending on application configuration. In order to increase the code reuseability, the configuration is abstracted from code and stored in a separate registry. Core services for motion planning such as obstacle detection that uses ML-algorithms, collision check, and motion planning are preferentially deployed in the cloud. NoSQL like MongoDB and in memory database like Redis can be used for information exchange between different stateless services. Moreover, the communication between ROS-based services such as robot controller, camera driver is implemented with DDS.

Being at the beginning of KI5GRob we expect Motion Control as a Service to scale from single robots to a fleet of robots. Therefore we use motion planning and control as one of the central tasks in KI5GRob to evaluate the new architecture. Future research focuses on the safety, efficiency, scalability, and service availability of MCaaS.

3.5 Robotic Manipulation

From people's perspective, object manipulation tasks are often seen as mundane, but assembling kits or even packing items into shipping boxes still remain rather challenging problems in robotics. One can consider a reason for this discrepancy to be the difference in quality, nature and composition of the manipulators, the human hand and different robotic grippers. It was shown however, that multiple domestic and non-domestic tasks, that do not seem feasible while using autonomously acting robots, can be easily solved via teleoperation [48,49].

In general, object handling can be divided into four main components: actuators, sensors, sensor signal interpretation and actuation. There can be significant differences in terms of mechanics between a human hand and different grippers, still, people are rather skillful when using gripper-like tools (e.g., toy gripper, grill tong). While human motion is somewhat inaccurate and has a low repetition accuracy i.e., signatures are similar but not the same, its adaptation based on sensory information and its use of environmental features makes it very agile and robust. In contrast, robots are built to achieve a very high repetition accuracy. This suggests, that solving manipulation tasks in robotics is mainly dependent on sensors, their interpretation and their coupling with actuation. When objects are handled by people, vision is a very important source of information. This is especially true for some manipulation states (e.g., pre-grasp). However, during manipulation, tactile and force feedback are just as, or even more important [50–53], both of which can be available for robotic manipulation. Of course, the nature of sensors available in robotics is somewhat different to those in a biological body, they can be augmented with additional modalities, like depth vision, proximity perception and exact joint position measurements.

An other plausible assumption is that people have a reasonable understanding of the environment and the manipulation objects. Fragile or liquid-filled objects are handled with more care. Even collision avoidance is context dependent: when packing items into a shipping box, one would initially avoid collisions with other items already in the box however, during placement contact between the packaged items might be desirable in order to make use of the available space efficiently. This is often done by exploiting collisions and contacts, allowing them to guide the movements while placing or aligning items. This requires a considerable understanding of the extent of the manipulated item, the box, the other items in the box and their dynamics. This can be called experience, and is most likely gathered in our whole lifetime by executing vastly different tasks successfully or less successfully.

Effective and robust robotic object manipulation is an incredibly challenging problem in robotics research. As mentioned before, in KI5GRob we tackle a specific type of object manipulation: pick-and-place, with particular focus on accurate positioning while placing. Our goal is to achieve a reliable and accurate task execution. Aligned with this, other related but more constrained tasks e.g., pick-and-pack and peg-in-hole are investigated, too. We mainly focus on multi-modal sensor data processing in combination with task-level and sim-to-real transfer learning. Currently, we investigate the Transporter Network architecture [33], that can learn to solve pick-and-place tasks from only a few demonstrations, however with constrained accuracy. In particular, we work on extending the model to achieve a higher precision for object-placements. A mandatory byproduct of this project is the evaluation of classical and learning-based approaches in terms of effectiveness and usability. An essential part of the research is to clarify the benefits of learning-based approaches and to identify suitable ways of using them in industrial settings.

3.6 Multi-modal Perception

The first step towards successful robotic object handling is perception, in other words, sensor data interpretation. In the scope of KI5GRob we use color and depth vision, tactile-proximity-sensors and force-torque sensors. While the main contribution towards a successful scene understanding is expected to come from color and depth vision, there are multiple modalities that have a clear correlation. Starting with multiple cameras and point clouds, the inclusion of near-field proximity sensor data is a rather obvious choice. This provides not only an estimation to one modality given an other, but even with noisy sensor-data, a proper combination of multiple inaccurate measurements leads to a more accurate estimation of the reality. Including temporal dependencies using e.g., filtering methods even enables a combination with the mechanic and dynamic nature of tactile and force-torque sensors. Finally, the described correlation model can tune itself in a self-supervised manner, since data is streamed whenever any robot system is used, regardless of whether it is a classical method (e.g., unscented Kalman filter) or a learning-based method (e.g., Generative Query Network).

Sensor data can be processed for different purposes: a unified scene perception model, high-level task planning or feedback for low-level closed loop control. Perception methods used for high-level planning in KI5GRob can be divided into two categories: object-centric and task-centric. The former includes classical and deep learning approaches, requires however objects to be known in advance. It includes object detection and pose estimation, which is then processed further for e.g., grasp pose estimation. The latter is mainly data-driven as it maps the perceived sensor data or scene to e.g., task specific gripper poses. This is expected to be generalizable to unseen objects. The currently investigated method for object manipulation, the Transporter Network architecture, falls into the task-centric category. A utilization of the different sensor modalities include their usage in closed loop control during manipulation: proximity sensor data for pre-grasp alignment, force-torque sensor data to exploit contact and collision while placing. In this context, we plan to investigate classical control methods, and reinforcement learning based methods. Tactile and force-torque sensor data can further be used for grasp-quality estimation to initialize regrasping if necessary.

By using three different sensor modalities, color, distance, and force-torque, in KI5GRob we aim to improve the robustness and accuracy of robotic manipulation tasks.

3.7 Transfer Learning

A goal of KI5GRob is to use cloud computing as enabler for robotic applications with focus on flexible and robust object handling tasks. This includes the implementation and deployment of solutions for new tasks or objects with as little effort as possible from users' side. For this, we implement a learning from demonstration robot programming approach extended with a repository, containing a cascade of trained machine learning models that are able to solve specific tasks. Whenever a new task is defined, first the models in the repository are tested and the best fitting models are adapted if required. In the context of pick-and-place, the set of observed sensor data and the corresponding pick and place poses offer a suitable task description for training, testing and comparing both object-centric and task-centric models.

This task description is essentially a solution demonstration. It can either be acquired from a simulation environment where everything is known, or from recording data from user guided demonstrations (e.g., hand guiding or teleoperation on real hardware). Depending on the task complexity and whether there are already well fitting models in the knowledge repository, additional demonstrations may be required for a successful model adaptation. This can be regarded as a sort of curriculum learning or rather task-level transfer learning. In case of simulated environments, a final sim-to-real adaptation with data from a real robot setup might be required. After adaptation (both task-level and sim-to-real) the models are stored in the repository.

This knowledge repository based approach is only possible using cloud computing, as both deep learning models and high-fidelity simulations require vast amounts of computational and storage resources. As an additional benefit, cloud technologies enable sharing and building up experience between different robots (even in different locations) for object manipulation. With task-level and sim-to-real transfer learning KI5GRob takes a step towards using learning-based approaches for robotic manipulation tasks that meet industry standards.

4 Conclusion

This paper provides a comprehensive insight into the ongoing research activities of the project KI5GRob. The novel decoupled robot hardware architecture, the microservice-based software architecture, and the deployment pipeline based on Kubernetes are presented. On one hand, we currently concentrate on the development of cloud-based software framework, tools for deployment, and rebuilding available software packages for MCaaS. On the other hand, we use multi-modal sensor data in combination with task-level and sim-to-real transfer learning to perform typical industrial pick-and-place tasks with focus on accurate placement. Future work will increasingly target the extensive scalability of robotic systems in unstructured industrial environments with a high degree of autonomy. Our goal is to facilitate the development of robotic applications by using the advantages of cloud technologies and AI methods.

Acknowledgement. The project "KI5GRob - Cloud computing and 5G as enablers for new sensor technologies in robot-based production and logistics" is funded by the Federal Ministry of Education and Research (BMBF) within the scope of the German Federal Government's "FH Kooperativ" program. KUKA, Schmalz, Roboception, and AAT provide support with their practice-oriented knowledge and experience as our industrial project partners.

References

1. Christensen, H.I., Amato, N.M., Yanco, H.A., Matarić, M.J., Choset, H., Drobnis, A.W., Goldberg, K., Grizzle, J.W., Hager, G., Hollerbach, J.M., Hutchinson, S., Krovi, V.N., Lee, D., Smart, B., Trinkle, J.C., Sukhatme, G.S.: A roadmap for us robotics—from internet to robotics 2020 edition. Found. Trends Robot. **8**(2021), 307–424 (2020)
2. Yu, C., Liu, Z., Liu, X.J., Xie, F., Yang, Y., Wei, Q., Fei, Q.: Ds-slam: a semantic visual slam towards dynamic environments. In: 2018 IEEE/RSJ International Conference on Intelligent Robots and Systems (IROS), pp. 1168–1174. IEEE (2018)
3. Liu, L., Ouyang, W., Wang, X., Fieguth, P., Chen, J., Liu, X., Pietikäinen, M.: Deep learning for generic object detection: a survey. Int. J. Comput. Vis. **128**(2), 261–318 (2020)
4. Bohg, J., Morales, A., Asfour, T., Kragic, D.: Data-driven grasp synthesis—a survey. IEEE Trans. Robot. **30**(2), 289–309 (2013)
5. Andrychowicz, O.M., Baker, B., Chociej, M., Jozefowicz, R., McGrew, B., Pachocki, J., Petron, A., Plappert, M., Powell, G., Ray, A., et al.: Learning dexterous in-hand manipulation. Int. J. Robot. Res. **39**(1), 3–20 (2020)
6. Bousmalis, K., Irpan, A., Wohlhart, P., Bai, Y., Kelcey, M., Kalakrishnan, M., Downs, L., Ibarz, J., Pastor, P., Konolige, K., et al.: Using simulation and domain adaptation to improve efficiency of deep robotic grasping. In: IEEE International Conference on Robotics and Automation (ICRA), pp. 4243–4250. IEEE (2018)
7. Billard, A., Kragic, D.: Trends and challenges in robot manipulation. Science **364**(6446) (2019)
8. Sünderhauf, N., Brock, O., Scheirer, W., Hadsell, R., Fox, D., Leitner, J., Upcroft, B., Abbeel, P., Burgard, W., Milford, M., Corke, P.: The limits and potentials of deep learning for robotics. IJRR **37**(4–5), 405–420 (2018)

9. Cloud, G.: VertexAI. https://cloud.google.com/vertex-ai. Accessed: 2022-02-03
10. Khan, W.Z., Ahmed, E., Hakak, S., Yaqoob, I., Ahmed, A.: Edge computing: a survey. Future Gener. Comput. Syst. **97**, 219–235 (2019)
11. Mouradian, C., Naboulsi, D., Yangui, S., Glitho, R.H., Morrow, M.J., Polakos, P.A.: A comprehensive survey on fog computing: state-of-the-art and research challenges. IEEE Commun. Surv. Tutorials **20**(1), 416–464 (2018)
12. Ebert, C., Gallardo, G., Hernantes, J., Serrano, N.: Devops. IEEE Softw. **33**(3), 94–100 (2016)
13. Society, I.R.&.A.: Technical Committees: Networked Robots. Accessed: 2022-02-03
14. Kuffner, J.: Cloud-enabled humanoid robotics. In: IEEE-RAS International Conference on Humanoid Robotics (2010)
15. Kehoe, B., Patil, S., Abbeel, P., Goldberg, K.: A survey of research on cloud robotics and automation. IEEE Trans. Autom. Sci. Eng. **12**(2), 398–409 (2015)
16. Goldberg, K., Kehoe, B.: Cloud robotics and automation: a survey of related work. IEEE Trans. Autom. Sci. Eng. **12** (2013)
17. Saha, O., Dasgupta, P.: A comprehensive survey of recent trends in cloud robotics architectures and applications. Robotics **7**(3) (2018)
18. Song, D., Tanwani, A.K., Goldberg, K.: Chapter 24: Networked-, cloud-and fog-robotics (2019)
19. Arumugam, R., Enti, V.R., Bingbing, L., Xiaojun, W., Baskaran, K., Kong, F.F., Kumar, A.S., Meng, K.D., Kit, G.W.: DAvinCi: A Cloud Computing Framework For Service Robots, pp. 3084–3089. ICRA (2010)
20. Waibel, B.M., Beetz, M., Civera, J., Andrea, R.D.: RoboEarth—a world wide web for robots. IEEE Robot. Autom. Mag. 69–82 (2011)
21. Hunziker, D., Gajamohan, M., Waibel, M., D'Andrea, R.: Rapyuta: the RoboEarth cloud engine. In: ICRA, pp. 438–444. IEEE (2013)
22. Xia, C., Zhang, Y., Wang, L., Coleman, S., Liu, Y.: Microservice-based cloud robotics system for intelligent space. Robot. Auton. Syst. **110**, 139–150 (2018)
23. Chen, K.E., Liang, Y., Jha, N., Ichnowski, J., Danielczuk, M., Gonzalez, J., Kubiatowicz, J., Goldberg, K.: FogROS: an adaptive framework for automating fog robotics deployment. In: IEEE International Conference on Automation Science and Engineering, 2021-Aug, pp. 2035–2042 (2021)
24. Ichnowski, J., Prins, J., Alterovitz, R.: The economic case for cloud-based computation for robot motion planning. In: ISRR, pp. 59–65. Springer International Publishing, Berlin (2017)
25. Ichnowski, J., Prins, J., Alterovitz, R.: Cloud-based motion plan computation for power-constrained robots. Algorithmic Found. Robot. 96–111 (2020)
26. Bekris, K., Shome, R., Krontiris, A., Dobson, A.: Cloud automation: precomputing roadmaps for flexible manipulation. IEEE Robot. Autom. Mag. **22**(2), 41–50 (2015)
27. Vick, A., Vonásek, V., Pěnička, R., Krüger, J.: [RCaaS] Robot control as a service—towards cloud-based motion planning and control for industrial robots. In: 2015 10th International Workshop on Robot Motion and Control, RoMoCo 2015, pp. 33–39 (2015)
28. Making, D., Full, U., Method, C.: Cloud-based multi-robot path planning in complex and crowded environment with multi-criteria. Symmetry (2019)
29. Berscheid, L., Meißner, P., Kröger, T.: Self-supervised learning for precise pick-and-place without object model. IEEE Robot. Autom. Lett. **5**(3), 4828–4835 (2020)
30. Levine, S., Pastor, P., Krizhevsky, A., Ibarz, J., Quillen, D.: Learning hand-eye coordination for robotic grasping with deep learning and large-scale data collection. Int. J. Robot. Res. **37**(4–5), 421–436 (2018)

31. Kalashnikov, D., Irpan, A., Pastor, P., Ibarz, J., Herzog, A., Jang, E., Quillen, D., Holly, E., Kalakrishnan, M., Vanhoucke, V., et al.: Qt-opt: scalable deep reinforcement learning for vision-based robotic manipulation. arXiv preprint arXiv:1806.10293 (2018)
32. Mahler, J., Matl, M., Satish, V., Danielczuk, M., DeRose, B., McKinley, S., Goldberg, K.: Learning ambidextrous robot grasping policies. Sci. Robot. **4**(26) (2019)
33. Zeng, A., Florence, P., Tompson, J., Welker, S., Chien, J., Attarian, M., Armstrong, T., Krasin, I., Duong, D., Sindhwani, V., et al.: Transporter networks: rearranging the visual world for robotic manipulation. arXiv preprint arXiv:2010.14406 (2020)
34. Seita, D., Florence, P., Tompson, J., Coumans, E., Sindhwani, V., Goldberg, K., Zeng, A.: Learning to rearrange deformable cables, fabrics, and bags with goal-conditioned transporter networks. In: 2021 IEEE International Conference on Robotics and Automation (ICRA), pp. 4568–4575. IEEE (2021)
35. Lee, M.A., Zhu, Y., Zachares, P., Tan, M., Srinivasan, K., Savarese, S., Fei-Fei, L., Garg, A., Bohg, J.: Making sense of vision and touch: learning multimodal representations for contact-rich tasks. IEEE Trans. Robot. **36**(3), 582–596 (2020)
36. Calandra, R., Owens, A., Jayaraman, D., Lin, J., Yuan, W., Malik, J., Adelson, E.H., Levine, S.: More than a feeling: learning to grasp and regrasp using vision and touch. IEEE Robot. Autom. Lett. **3**(4), 3300–3307 (2018)
37. Hogan, F.R., Bauza, M., Canal, O., Donlon, E., Rodriguez, A.: Tactile regrasp: grasp adjustments via simulated tactile transformations. In: 2018 IEEE/RSJ International Conference on Intelligent Robots and Systems (IROS), pp. 2963–2970. IEEE (2018)
38. Navarro, S.E., Schonert, M., Hein, B., Wörn, H.: 6d proximity servoing for preshaping and haptic exploration using capacitive tactile proximity sensors. In: 2014 IEEE/RSJ International Conference on Intelligent Robots and Systems, pp. 7–14. IEEE (2014)
39. Sasaki, K., Koyama, K., Ming, A., Shimojo, M., Plateaux, R., Choley, J.Y.: Robotic grasping using proximity sensors for detecting both target object and support surface. In: 2018 IEEE/RSJ International Conference on Intelligent Robots and Systems (IROS), pp. 2925–2932. IEEE (2018)
40. Eslami, S.A., Rezende, D.J., Besse, F., Viola, F., Morcos, A.S., Garnelo, M., Ruderman, A., Rusu, A.A., Danihelka, I., Gregor, K., et al.: Neural scene representation and rendering. Science **360**(6394), 1204–1210 (2018)
41. Alagi, H., Heilig, A., Navarro, S.E., Kroeger, T., Hein, B.: Material recognition using a capacitive proximity sensor with flexible spatial resolution. In: 2018 IEEE/RSJ International Conference on Intelligent Robots and Systems (IROS), pp. 6284–6290. IEEE (2018)
42. Fielding, R.T.: Architectural styles and the design of network-based software architectures (2000)
43. Group, O.M.: Data Distribution Service (DDS), version 1.4. https://www.omg.org/spec/DDS/1.4. Accessed: 2022-02-03
44. grpc.io: gRPC. https://grpc.io. Accessed: 2022-02-03
45. Luksa, M.: Kubernetes in Action. Simon and Schuster (2017)
46. Tigera, Inc.: Calico. https://www.tigera.io/project-calico/. Accessed: 2022-02-03
47. Maruyama, Y., Kato, S., Azumi, T.: Exploring the performance of ros2. In: 2016 International Conference on Embedded Software (EMSOFT), pp. 1–10 (2016)
48. Wyrobek, K.A., Berger, E.H., Van der Loos, H.M., Salisbury, J.K.: Towards a personal robotics development platform: rationale and design of an intrinsically safe personal robot. In: 2008 IEEE International Conference on Robotics and Automation, pp. 2165–2170. IEEE (2008)

49. Brüggemann, B., Röhling, T., Welle, J.: Coupled human-machine telemanipulation. Procedia Manuf. **3**, 998–1005 (2015)
50. Postma, A., Zuidhoek, S., Noordzij, M.L., Kappers, A.M.: Differences between early-blind, late-blind, and blindfolded-sighted people in haptic spatial-configuration learning and resulting memory traces. Perception **36**(8), 1253–1265 (2007)
51. Heller, M.A.: Picture and pattern perception in the sighted and the blind: the advantage of the late blind. Perception **18**(3), 379–389 (1989)
52. Kormi-Nouri, R.: The role of movement and object in action memory: a comparative study between blind, blindfolded and sighted subjects. Scand. J. Psychol. **41**(1), 71–76 (2000)
53. Jenmalm, P., Johansson, R.S.: Visual and somatosensory information about object shape control manipulative fingertip forces. J. Neurosci. **17**(11), 4486–4499 (1997)

A Correlated Random Walk Model to Rapidly Approximate Hitting Time Distributions in Multi-robot Systems

Yi Zhang, Daniel Boley(✉), John Harwell, and Maria Gini

Computer Science and Engineering, University of Minnesota, Minneapolis, MN 55455, USA
{zhan2854,harwe006,gini}@umn.edu

Abstract. Multi-robot systems are frequently used for tasks involving searching, so it is important to be able to estimate the searching time. Yet, simulation approaches and real-world experiments to determine searching time can be cumbersome and even impractical. In this work, we propose a correlated-random-walk based model to efficiently approximate hitting time distributions of multi-robot systems in large arenas. We verified the computational results by using ARGoS, a physics-based simulator. We found that the Gamma distribution can provide a good fit to the hitting time distributions of random walkers.

Keywords: Multi-robot systems · Correlated random walk · Hitting time distribution

1 Introduction

Operators can release multiple robots to search for a target within a predefined area. We denote this process as "searching" because robots don't know the target location. Robot searching has many real-world applications (e.g. foraging, rescuing, providing supplies, etc.) [14,19]. In this paper, we define the searching time as the time required by at least one of the robots to first reach the target. For many tasks, especially the time-sensitive ones, an estimate of searching time is often necessary. Furthermore, having an estimated searching time can help in the design of multi-robot systems that are more efficient at searching. Experimental investigation involving real robots (see [10] for an example) is often costly and can be even impractical; simulation (e.g. ARGoS [26]) can also be time-consuming, especially when the searched region is large. In this work, we aim to develop a computation model that can rapidly approximate the searching time.

Hitting time (HT) or first-passage time is the average time to first visit a node while doing a random walk (RW) from another node in a given network [21,22]. Many results for HT exist. In particular, the average time to reach a given node can be computed based on the fundamental matrix associated with the probability transition matrix of the network [13], by treating the target node as an

I. Petrovic et al. (Eds.): IAS 2022, LNNS 577, pp. 724–736, 2023.
https://doi.org/10.1007/978-3-031-22216-0_48

absorbing state. Boley et al. [5] showed how to obtain average HTs from any node to any node in a directed graph at once using the assymetric graph Laplacian, and later showed one way to obtain this rapidly using sparse matrix methods [4]. In the HT literature, there are also papers about HT higher moments [9,17], upper bounds [6], and distributions [20]. In addition, there are studies on the speed up of random walker search when there are multiple searchers [1,11,25] and random walker collisions [2,12].

To the best of our knowledge, none of the methods cited above has been applied directly to multi-robot systems or swarm robots. Thus, we aim to leverage some of these results to develop a computation model that can rapidly and accurately approximate the HT in the multi-robot searching process. Due to the forward persistence of robot motion, Jeong et al. [16] used correlated random walk (CRW, i.e. RW where the state transition depends on both the location and orientation of the walker [24]) to approximate robot movements in their network-based model. This is the approach used here.

Our work makes three main contributions to the field of multi-robot systems and stochastic process. First, our work is a pioneer attempt to apply HT results to analyze multi-robot searching processes. Second, we developed a computation model that can efficiently approximate the HT distributions for multi-robot searching processes, and we verified the computation results with the ARGoS simulator. Third, we found that the Gamma distribution can provide a good fit to CRW HT distributions.

The rest of the paper is organized as the follows. Section 2 describes the problem setting. Section 3 provides the Correlated Random Walk (CRW) computation results. Section 4 compares the computation results with the ARGoS robot simulation results. Section 5 provides a discussion and concludes the paper.

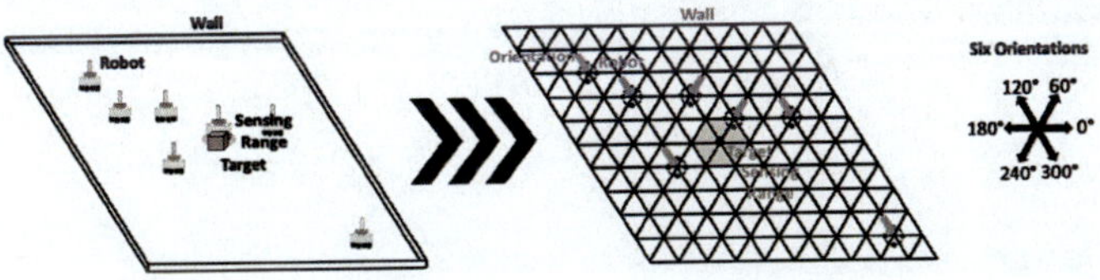

Fig. 1. Problem settings. Arena shape is a parallelogram with sides of equal length. Target location is the center. Left: a multi-robot searching example. Middle: the corresponding representation in our computation model (discretization by equilateral triangles). Right: six orientations of the robot.

2 Problem Settings

The problem settings are shown in Fig. 1 with details in the caption. We used parallelogram for arena shape because it simplifies the construction of the probability transition matrix. We discretized the arena using equilateral triangles instead of squares because the former approach leads to six possible orientations for robots and thus increases flexibility, while keeping all link lengths the same. When a robot is within a certain distance from the target (sensing range), it can sense the target and pick it up. We model the robot motion as a correlated random walk (CRW), implemented as an ordinary random walk with states encoding both robot position and orientation from the previous state. The nodes in the network are the states (position, orientation). The edges are state transitions, each weighted by the transition probability. At each position of the discretized arena, a robot has six possible orientations (Fig. 1 right).

Figure 2 (right side) provides an example of the state transitions in CRW associated with the robot R1 on the left. Currently, R1 is in state (H, 0°) (circled in red). R1 can move forward to reach position I with orientation 0°. Hence the new state of R1 would be (I, 0°). R1 can also make a small turn (60°) to reach D (if turning left) or M (if turning right). The new states would be (D, 60°) or (M, 300°) respectively. The robot can also make a big turn (120°, right or left) or go backward. If near the wall, the robot might also reflect. Since the robot tends to move forward and resists making large turns, the state transition probabilities are set to heavily favor forward motion, as listed in Table 1. Note that each robot could encounter six situations: interior within the arena (R1), move along an obtuse corner (R2), move along a sharp corner (R3), collide into an obtuse corner (R4), move parallel to a wall (R5), or collide into a wall (R6).

Transition probabilities associated with these situations are shown in Table 1, which contains examples for each situation. In the column "Example," "R1 (0°) H → I" denotes the state transition of R1 from position H with orientation 0° to location I (Fig. 2). The remaining examples follow the same convention. Note that we used a transition probability setting known to reduce HT, but other probability settings can also be used. After specifying the probabilities, we constructed a probability transition matrix **P** that can be easily scaled for arenas of any size. Note that we use bold symbols to denote vectors or matrices in this paper. An $N \times N$ arena leads to $6(N+1)^2$ states, and the associated adjacency matrix will be $6(N+1)^2 \times 6(N+1)^2$.

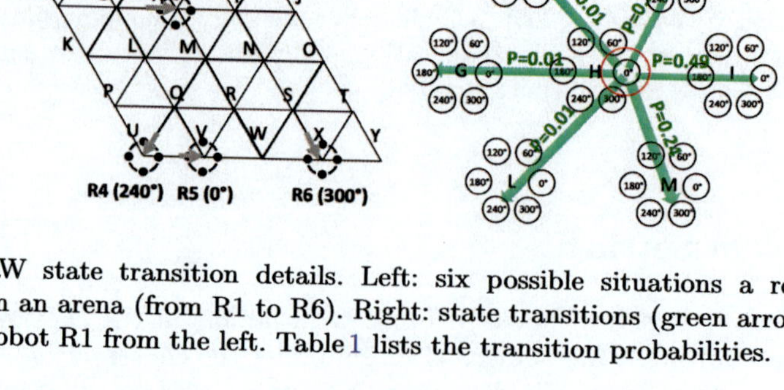

Fig. 2. CRW state transition details. Left: six possible situations a robot could encounter in an arena (from R1 to R6). Right: state transitions (green arrows) associated with robot R1 from the left. Table 1 lists the transition probabilities.

Table 1. Transition probabilities for different situations encountered by robots.

Situation	Example	Probability	Situation	Example	Probability
I, forward	R1 (0°) H → I	0.49	PW, small turn	R5 (0°) V → R	0.45
I, backward	R1 (0°) H → G	0.01	PW, big turn	R5 (0°) V → Q	0.06
I, small turn	R1 (0°) H → D	0.24	ASC, big turn	R3 (120°) A → B	0.5
I, big turn	R1 (0°) H → C	0.01	ASC, backward	R3 (120°) A → F	0.5
CW, reflect	R6 (300°) X → T	0.6	AOC, backward	R2 (120°) E → J	0.25
CW, backward	R6 (300°) X → S	0.15	AOC, small turn	R2 (120°) E → D	0.25
CW, small turn	R6 (300°) X → Y	0.15	AOC, big turn	R2 (120°) E → I	0.5
CW, big turn	R6 (300°) X → W	0.1	COC, backward	R4 (240°) U → Q	0.34
PW, forward	R5 (0°) V → W	0.45	COC, big turn	R4 (240°) U → P	0.33
PW, backward	R5 (0°) V → U	0.04			

I: inside, CW: collide toward the wall, PW: parallel with the wall, ASC: along the sharp corner, AOC: along the obtuse corner, COC: collide toward the obtuse corner.

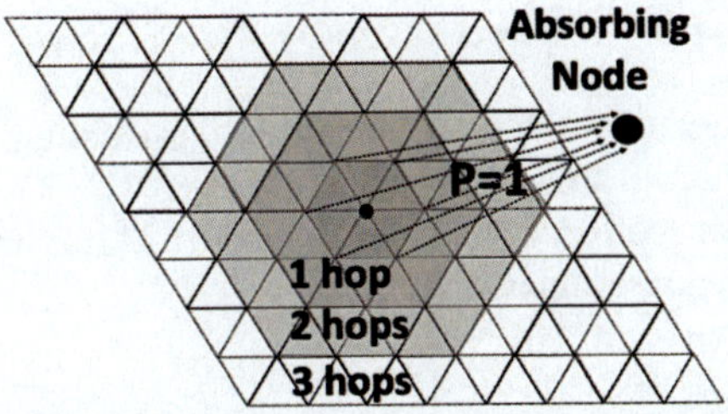

Fig. 3. Robot sensing ranges (e.g. 1-hop, 2-hop and 3-hop).

Our modelling approach of robot sensing ranges is shown in Fig. 3. When robots sense the target, they will reach the target following the shortest path. To model this behavior, we add an extra absorbing node, and we adjust the network so that whenever the robot/random walker hits the sensing range boundary (shaded region), it will be absorbed by the extra absorbing node with a probability of 1. The connectivity shown in Fig. 3 corresponds to the case with 1-hop sensing range. This method will always be valid if the robot starts outside the sensing range.

3 Correlated Random Walk Computations

Computation of HT mean and variance. Using the transition matrix $\mathbf{P}$ associated with this network, the HT mean $\mathbf{HT}_\mu$ and variance $\mathbf{HT}_{\sigma^2}$ from states outside the sensing range to the absorbing state can be determined. Following [13], we partition $\mathbf{P}$ to get $\mathbf{Q}$, the probability transition matrix corresponding to the non-absorbing states, via:

$$\mathbf{P} = \begin{bmatrix} \mathbf{Q} & \mathbf{R} \\ \mathbf{0} & \mathbf{I} \end{bmatrix} \tag{1}$$

$\mathbf{R}$ is associated with the absorbing state(s) and $\mathbf{I}$ is the identity matrix. The fundamental matrix $\mathbf{N}$ of $\mathbf{P}$ is [13]:

$$\mathbf{N} = (\mathbf{I} - \mathbf{Q})^{-1} \tag{2}$$

$\mathbf{HT}_\mu$ can be determined by:

$$\mathbf{HT}_\mu = \mathbf{Nc} \tag{3}$$

where $\mathbf{c}$ is a vector of all 1s. After rearranging, we get:

$$(\mathbf{I} - \mathbf{Q})\mathbf{HT}_\mu = \mathbf{c} \tag{4}$$

$\mathbf{HT}_\mu$ can be efficiently solved for by iterative methods such as Restarted GMRES [4,23], even if $\mathbf{I} - \mathbf{Q}$ is extremely large (e.g. 100,000 × 100,000). $\mathbf{HT}_{\sigma^2}$ can be calculated by [17]:

$$\mathbf{HT}_{\sigma^2} = (2\mathbf{N} - \mathbf{I})\mathbf{HT}_\mu - \mathbf{HT}_\mu^2, \tag{5}$$

where $\mathbf{HT}_\mu^2$ means elementwise squaring. After rearranging, we get:

$$(\mathbf{I} - \mathbf{Q})\mathbf{HT}_{\sigma^2} = 2\mathbf{HT}_\mu - (\mathbf{I} - \mathbf{Q})\mathbf{HT}_\mu - (\mathbf{I} - \mathbf{Q})\mathbf{HT}_\mu^2 \tag{6}$$

$\mathbf{HT}_{\sigma^2}$ can be computed similarly as before, and subsequently the HT standard deviation $\mathbf{HT}_\sigma$ can be obtained.

Some computation results are shown in Fig. 4. All computations were done in Python using the sparse matrix format.

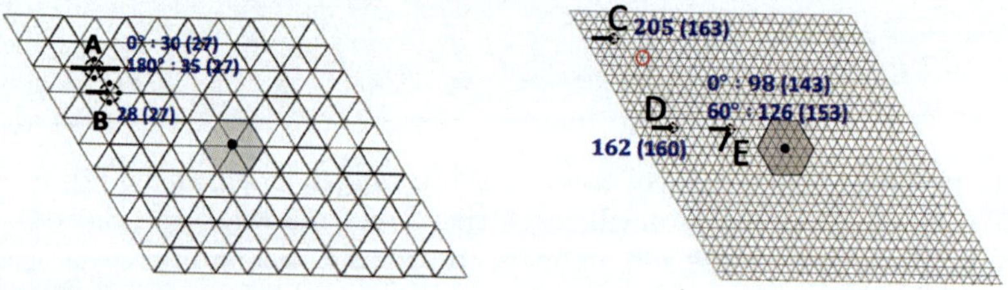

Fig. 4. Computed HT mean & standard deviation for a single robot starting from various initial positions. Blue numbers are $\mu(\sigma)$, on a 10×10 arena (left) and a 30×30 arena (right).

Single robot HT distribution. We used CRW simulations to determine HT distributions. Specifically, we started a random walker on the constructed network and let it move (i.e. state transition) with the probabilities specified in $\mathbf{P}$. We recorded HTs in 300 random trials to get a distribution.

The CRW simulation allows us to obtain the HT distribution. However, it is still time-consuming. We observed that all the obtained CRW HT distributions resemble the shape of a Gamma probability density function (PDF) $F(x)$:

$$F(x) = \frac{\beta^\alpha}{\Gamma(\alpha)} x^{\alpha-1} e^{-\beta x} \text{ with } \alpha = \frac{(\text{mean}(x))^2}{\text{var}(x)}, \beta = \frac{\text{var}(x)}{\text{mean}(x)}, \tag{7}$$

where x is the random variable ($HT - HT_{\min}$ in this case), and α, β are the shape and rate parameters written in terms of the mean and variance of x. The hitting time cannot be less than the shortest path length $HT_{\min}$, hence we shift the Gamma distribution by this amount. Thus, based on Eq. 7, we can get single robot HT distributions using HT_{σ^2} (or HT_{σ}) and HT_{μ} by leveraging the Gamma distribution. Note that the Gamma PDF assigns positive probabilities to HTs smaller than the minimum HT ($HT_{\min}$) required for the robot to reach the target. To address this issue, we subtract $HT_{\min}$ from HT_{μ} to get HT_{μ^*}, and we fit the gamma PDF by HT_{μ^*} and HT_{σ}. Last, we shift the entire Gamma distribution to the right by $HT_{\min}$.

In Fig. 5, we show the single robot HT distributions obtained by CRW simulation and Gamma distribution corresponding to 7 initial robot states in Fig. 4. All settings (e.g. arena size and robot locations) are shown in Fig. 4. The Gamma distribution parameters were determined from the mean and variance using Eq. 7 obtained using Eqs. 4 and 6. Note that the HT distributions from CRW simulation are discrete, yet the Gamma distribution is continuous. To enable the comparison of distributions, the discrete distribution is converted to its continuous counterpart by summing discrete probabilities in small intervals, and then dividing the sums by the interval size. The same change is applied in the sequel to discrete distributions from CRW or ARGoS simulations. Clearly, a Gamma distribution can closely approximate HT distributions for a single robot/random walker in the computation model.

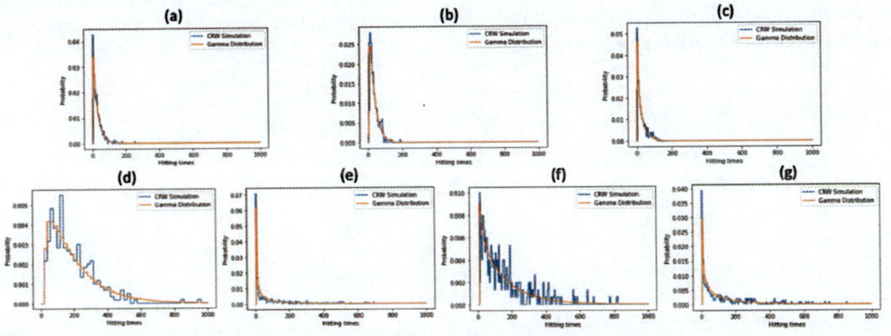

Fig. 5. Single robot HT distributions obtained by CRW simulation and Gamma distribution. (a) to (g) correspond to the initial robot states of (A, 0°), (A, 180°), (B, 0°), (C, 0°), (E, 0°), (D, 0°), and (E, 60°) shown in Fig. 4, respectively.

Multi-robot HT distribution. Next, we determined the multi-robot HT distributions using the computation model. First, we obtained the cumulative probability density function (CDF) $C\left(HT_{1\,robot}\right)$ corresponding to the PDF $F\left(HT_{1\,robot}\right)$ obtained with the Gamma distribution. Then, based on each value of $C\left(HT_{1\,robot}\right)$ (we denote a single value as $\dot{C}\left(HT_{1\,robot}\right)$), we can determine the cumulative probability for at least one robot out of n robots to reach the target within certain HT $\dot{C}\left(HT_{n\,robots}\right)$ by:

$$\dot{C}\left(HT_{n\,robots}\right) = 1 - \left(1 - \dot{C}\left(HT_{1\,robot}\right)\right)^{n} \tag{8}$$

Note that this equation relies on the assumption that the robots/random walkers are independent. This holds in our computation model. Strictly speaking, Eq. 8 also requires that all robots/random walkers have the same initial state. However, Eq. 8 is still a good approximation when different initial robot states are similar and lead to similar HT distributions. Note that Eq. 8 can be easily extended to the case when different robots start with different states. At this point, the computation model is able to approximate HT distributions for multiple robots using the computed HT_{μ} and HT_{σ} for a single robot. We verified this approach using the CRW simulation of multiple robots (Fig. 6). The agreement between the results corroborates our computation approach.

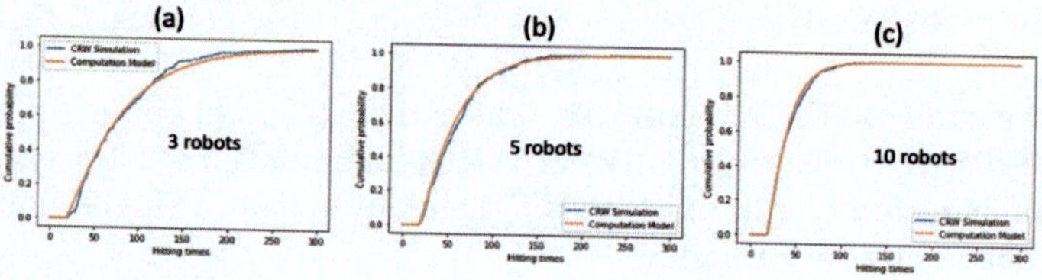

Fig. 6. Multi-robot HT CDF distributions determined by CRW simulation and computation. Arena 30 × 30, sensing range 3-hop, starting location of the robots circled in red in Fig. 4 with initial orientation 0°.

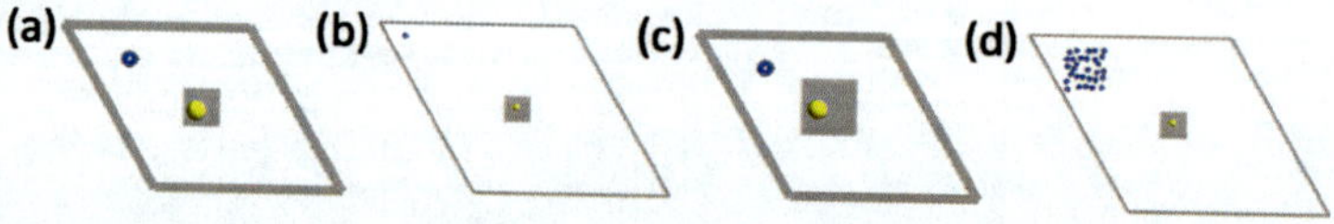

Fig. 7. Examples of ARGoS settings. (a) Small arena, sensing range 1, robot position point A (Fig. 4) (b) Big arena, sensing range 3, robot position point C (Fig. 4) (c) Sensing range 3. (d) 30 robots in the upper left corner.

4 Verification by ARGoS

Next, we use the ARGoS simulator to verify the computation model. Example screenshots of ARGoS simulation are in Fig. 7. In ARGoS, we set the robot motion parameters so that it takes approximately 10 and 30 time steps respectively for the robot to travel the distances corresponding to the wall lengths of small and big arenas. The robot has a probability of 0.5 to move forward or making 60° turns (0.25 for turning upward or downward). Larger turns are impractical and not allowed. Note that we assigned very small probabilities (e.g. 0.01) for the random walker to make larger turns in the computation model mainly to ensure the network connectivity. To mimic the robot sensing range, we place the nest in the center of the arena. After the simulation starts, the time step when at least one robot first enters the nest is recorded as the HT. To mimic larger sensing ranges, we simply increase the nest size. The starting locations and orientations of the robots are set to match the computation settings as well. Note that multiple robots always start within a confined region located on the upper-left side of the arena with orientation 0°. These initial states are similar, and we verified that they lead to similar single-robot HT distributions. Note that the selection of the confined region and initial state is arbitrary as long as it is kept similar in computation and simulation.

Comparisons between ARGoS simulation and computation results for HT distributions of a single robot with different initial states are shown in Fig. 8. HT distributions with different sensing ranges from two methods are in Fig. 9. HT CDFs corresponding to different numbers of robots are in Fig. 10.

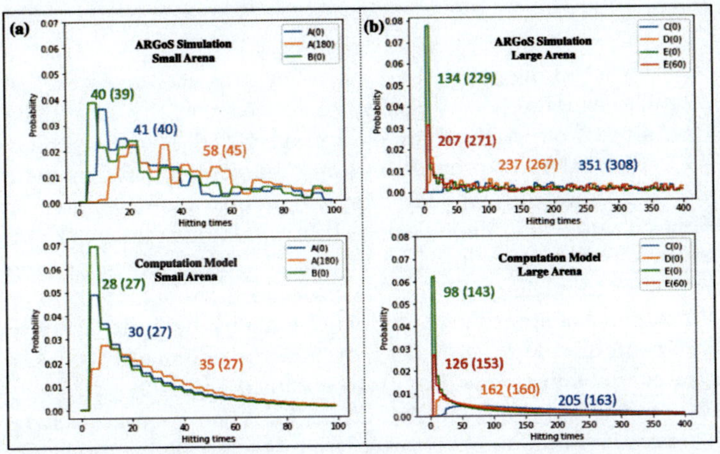

Fig. 8. Comparisons between CRW computation and ARGoS simulation results for single-robot HT distributions. (a) 10 × 10 arena (b) 30 × 30 arena. Robot starting states in the legend. For example, A(0) means the initial state of position A (Fig. 4) and 0° orientation. Value pairs "mean (standard deviation)" from the computation model and ARGoS are next to the curves with the same colors.

Improved computation model. As shown in Figs. 8, 9 and 10, the computation model and ARGoS simulation yield similar HT distributions under all circumstances we investigated. The computation model correctly predicts the ranking of HTs of a single robot with different initial states (Fig. 8) and the

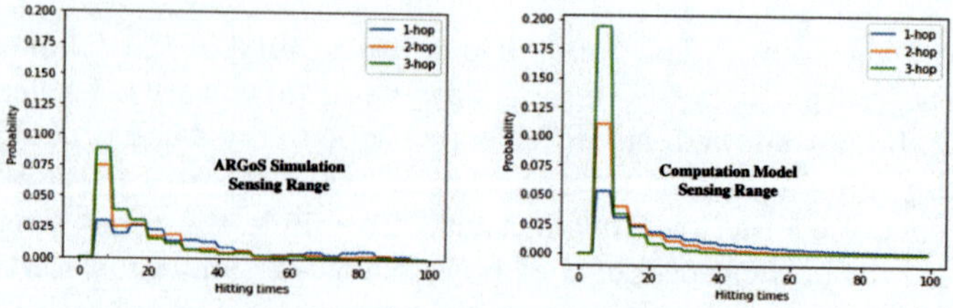

Fig. 9. ARGoS vs CRW: HT distribution for a single robot on a 10 × 10 arena with various sensing ranges, starting from point A.

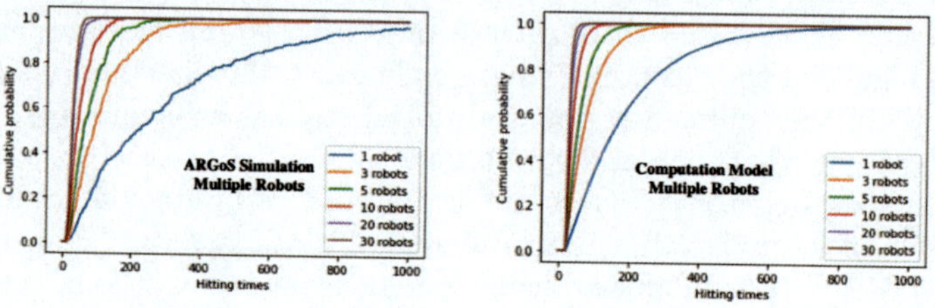

Fig. 10. ARGoS vs CRW: cumulative HT distributions for 30 × 30 arena, sensing range 3, with various number of robots.

speed up of the searching process as the robot sensing range (Fig. 9) and the number of robots increase (Fig. 10). Yet, we noticed that the value pairs of HT_μ and HT_σ obtained from ARGoS are on average 50% larger than the computed ones (Fig. 8). This led us to propose an improved computation model.

The aforementioned discrepancy is mostly due to the extra step associated with turning 60° in ARGoS simulation. In a random walk model it takes one step to advance in the any direction, but in ARGoS it takes one step to turn 60° and one step to advance. If the 60° turn has a probability of 0.5, then on average the number of steps in ARGoS will be higher by 50%. Hence we simply scale up the computed HT_μ and HT_σ by 50%. We applied this change to the computation model and redrew Fig. 10 to obtained Fig. 11.

As shown in Fig. 11, the HT distributions from two approaches agree better, especially when the robot starts far away from the target (e.g. point C and D in Fig. 4). Since our computation model works well on arenas of two different sizes, we tested its applicability to larger arenas. In the application of our improved computation model we aim to answer the following question: *In a* 150 × 150 *arena, how many robots (sensing range 1, initial position near arena upper left corner, initial orientation* 0°*) are required to have a probability of 0.8 that at least one robot will reach the target in the arena center within 500 time steps.*

This problem is related to many real-world tasks (e.g. foraging, rescuing, explosive removal). To solve it, first, we constructed the network associated with the robot movement in a 150 × 150 arena. This resulted in a 136, 806 × 136, 806 sparse matrix. We solved Eqs. 4 and 6 with restarted GMRES (20 restart itera-

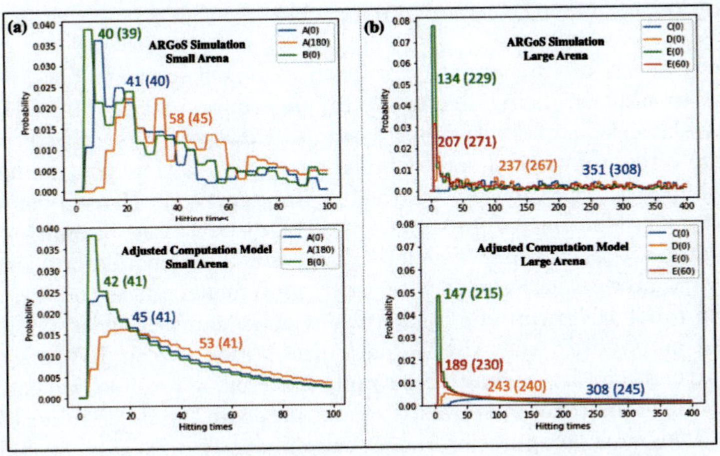

Fig. 11. Comparisons between ARGoS and the improved computation model.

tions) [4,23] within 3 min using the CPU from Google Colab. Then, we can get HT_μ and HT_σ of a single robot with almost every initial state. We are only interested in one initial state (upper left corner, 0°), and its associated HT_μ and HT_σ are 13,356 and 12,021 respectively. We scaled up these values by 50% and determined that the robot takes a minimum of 140 steps the target. Based on these values, we fitted the Gamma CDF for a single robot, and then applied Eq. 8 to determine the multi-robot HT CDFs. We varied the number of robots (Fig. 12), and found that 200 robots are needed to have a probability of 0.8 that at least one robot will reach the target within 500 steps. ARGoS may take hours or days to answer the same question, and testing with real robots will be even more challenging.

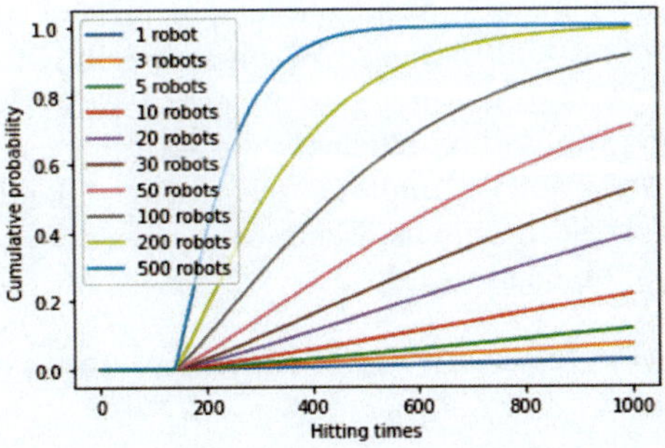

Fig. 12. Computed HT CDFs of multi-robot systems in a 150 × 150 arena.

5 Discussion and Conclusions

Searching time is often a quantity of interest, especially when tasks are time-sensitive. In addition, being able to rapidly determine Hitting Times (HTs) can facilitate the design of multi-robot systems. However, experiments with real robots are often impractical, and detailed simulation is time-consuming.

In this paper, we propose a Correlated Random Walk (CRW) computation model that can efficiently approximate the HT distributions of multi-robot systems in large arenas. We first tessellate a large arena with equilateral triangles to form a large network associated with robot movements. Based on the network, the single robot HT mean and standard deviation can be efficiently computed. We found and verified that the Gamma distribution can fit RW HT distributions. Furthermore, by leveraging a simple probability formula, we enabled the computation of multi-robot HT distributions based on the single robot HT distributions. Thus, our computation model can efficiently approximate multi-robot HT distributions. We verified the computation results by ARGoS simulation, which resembles real-world robot searching. We further improved the computation model by scaling up the computed mean and standard deviation of HT.

We found that the Gamma distribution can accurately approximate RW HT distributions when given only mean and standard deviation. This result itself is useful for stochastic processes, and can be used in fields involving stochastic processes (e.g. transportation, web, and animal modelling) [3,7,8,15,22].

Our model can approximate the ARGoS simulation results. Thus, we could approximate the optimal searching HT and associated parameter settings (e.g. transition probabilities) by performing a grid-search over the entire parameter space to determine the minimal HT. There are some sources of inaccuracy in our model. For example, our model ignores interactions between robots (e.g., avoidance, collisions), assuming robots are pair-wise independent. The inter-robot communication and issues with search area overlapping, collision, and under-utilization of robots will be investigated in the future. In the future, we will also consider the situation when all robots are required to reach the target. Other future directions include adding multiple targets and obstacles to the model, considering robot malfunctions [18], and modelling flying robots in a 3-D arena.

In conclusion, we proposed an efficient computation model that can rapidly approximate HT distributions of multi-robot systems. The model can be readily applied to solve real-world problems. Meanwhile, this study opens a wide range of research directions for future work.

Acknowledgements. This research was partially supported by the University of Minnesota Robotics Institute.

References

1. Alon, N., Avin, C., Koucký, M., Kozma, G., Lotker, Z., Tuttle, M.R.: Many random walks are faster than one. Comb. Probab. Comput. **20**(4), 481–502 (2011)

2. Barlow, M.T., Peres, Y., Sousi, P.: Collisions of random walks. Ann. l'IHP Probab. Stat. **48**(4), 922–946 (2012)
3. Besenczi, R., Bátfai, N., Jeszenszky, P., Major, R., Monori, F., Ispány, M.: Large-scale simulation of traffic flow using Markov model. PLOS One **16**(2) (2021)
4. Boley, D.: On fast computation of directed graph Laplacian pseudo-inverse. Linear Algebra Appl. **623**, 128–148 (2021)
5. Boley, D., Ranjan, G., Zhang, Z.L.: Commute times for a directed graph using an asymmetric Laplacian. Linear Algebra Appl. **435**(2), 224–242 (2011)
6. Brightwell, G., Winkler, P.: Maximum hitting time for random walks on graphs. Random Struct. Algorithms **1**(3), 263–276 (1990)
7. Codling, E.A., Bearon, R.N., Thorn, G.J.: Diffusion about the mean drift location in a biased random walk. Ecology **91**(10), 3106–3113 (2010)
8. Codling, E.A., Plank, M.J., Benhamou, S.: Random walk models in biology. J. R. Soc. Interface **5**(25), 813–834 (2008)
9. Dayar, T., Akar, N.: Computing moments of first passage times to a subset of states in Markov chains. SIAM J. Matrix Anal. **27**(2), 396–412 (2005)
10. Dimidov, C., Oriolo, G., Trianni, V.: Random walks in swarm robotics: an experiment with kilobots. In: International Conference on Swarm Intelligence, pp. 185–196. Springer, Berlin (2016)
11. Efremenko, K., Reingold, O.: How well do random walks parallelize? In: Dinur, I., Jansen, K., Naor, S., Rolim, J. (eds.) Approximation, Randomization, and Combinatorial Optimization. Algorithms and Techniques, vol. 5687. Springer, Berlin (2009)
12. Gaudillière, A.: Collision probability for random trajectories in two dimensions. Stoch. Process. Appl. **119**(3), 775–810 (2009)
13. Grinstead, C.M., Snell, J.L.: Introduction to Probability. American Mathematical Society (1997)
14. Hamann, H.: Modeling swarm systems and formal design methods. In: Swarm Robotics: A Formal Approach, pp. 95–127. Springer, Berlin (2018)
15. Hill, N., Häder, D.P.: A biased random walk model for the trajectories of swimming micro-organisms. J. Theor. Biol. **186**(4), 503–526 (1997)
16. Jeong, M., Harwell, J., Gini, M.: Analysis of exploration in swarm robotic systems. IAS-16 (2021)
17. Kemeny, J.G., Snell, J.L.: Finite Markov Chains. Springer, Berlin (1976)
18. Khalastchi, E., Kalech, M.: Fault detection and diagnosis in multi-robot systems: a survey. Sensors **19**(18) (2019)
19. Lancaster, J.P., Gustafson, D.A.: Predicting the behavior of robotic swarms in search and tag tasks. Procedia Comput. Sci. **20**, 77–82 (2013)
20. Lau, H.W., Szeto, K.Y.: Asymptotic analysis of first passage time in complex networks. EPL (Europhys. Lett.) **90**(4), 40005 (2010)
21. Lovász, L.: Random walks on graphs. Combinatorics **2**(1–46), 4 (1993)
22. Masuda, N., Porter, M.A., Lambiotte, R.: Random walks and diffusion on networks. Phys. Rep. **716**, 1–58 (2017)
23. Morgan, R.B.: GMRES with deflated restarting. SIAM J. Sci. Comput. **24**(1), 20–37 (2002)
24. Nain, R., Sen, K.: Transition probability matrices for correlated random walks. J. Appl. Probab. **17**(1), 253–258 (1980)
25. Patel, R., Carron, A., Bullo, F.: The hitting time of multiple random walks. SIAM J. Matrix Anal. **37**(3), 933–954 (2016)

26. Pinciroli, C., Trianni, V., O'Grady, R., Pini, G., Brutschy, A., Brambilla, M., Mathews, N., Ferrante, E., Di Caro, G., Ducatelle, F., et al.: ARGoS: a modular, multi-engine simulator for heterogeneous swarm robotics. In: IEEE/RSJ International Conference on Intelligent Robots and Systems, pp. 5027–5034. IEEE (2011)

Synthesis and Monitoring of Complex Tasks for Heterogeneous Robots in an Industry 4.0 Scenario

Francesco Trotti(✉), Eros Ghignoni, Riccardo Muradore, and Alessandro Farinelli

University of Verona, Verona 37134, Italy
francesco.trotti@univr.it, eros.ghignoni@univr.it, riccardo.muradore@univr.it, alessandro.farinelli@univr.it

Abstract. Synthesis and tasks execution is a well known problem pertaining to different fields. With the advent of Industry 4.0, the need to schedule and execute very complex tasks that involve different machinery or robots has become a key topic. In this article we propose a methodology that allows to define and monitor the execution of high-level tasks for a heterogeneous multi-robot system involved in assembly tasks. With the proposed approach, a state-chart is automatically synthesized starting from the definition of high-level tasks. The state-chart is then used to monitor online the operations that the robots must execute. The proposed methodology has been validated and tested in a research facility that reproduces a complete production line involving collaborative robotic arms, mobile robots and manufacturing machines.

Keywords: Industry 4.0 · Robotics · Task synthesis · State-chart

1 Introduction

The Industry 4.0 revolution is based on the use of advanced digital technologies to improve the production system. A key goal for Industry 4.0 is to increase the flexibility of the production process by collecting data from the plant and by enhancing the autonomy level of the machines (e.g., mobile robots and manipulators).

A standard framework for Industry 4.0 involves the use of several sensors installed in the machines (such as cameras to recognize objects and/or production defects), that transmit data to a management software (Enterprise Resource Planning, ERP) for processing. This software communicates directly with the Manufacturing Execution System (MES) for scheduling the operations to be performed on the production line by heterogeneous robots and devices [1,2].

In this paper we address the *task scheduling and planning problem* by proposing a methodology tailored on a realistic Industry 4.0 scenario. In particular, we consider the ICE (Industrial Computer Engineering) laboratory [3], which is

I. Petrovic et al. (Eds.): IAS 2022, LNNS 577, pp. 737–749, 2023.
https://doi.org/10.1007/978-3-031-22216-0_49

a research infrastructure that faithfully replicates a manufacturing production line equipped with the most modern machinery and automation methods. The ICE lab shown in Fig. 1 is composed of a conveyor belt, several robotic bays, an automatic vertical warehouse and two mobile collaborative robots.

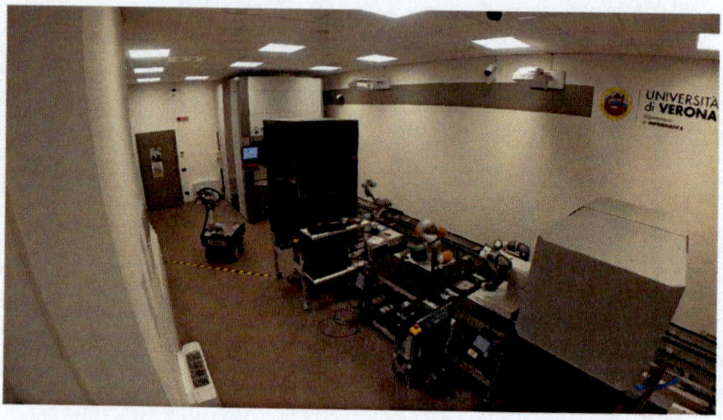

Fig. 1. ICE laboratory at University of Verona.

We address the problem to control multiple machines/robots from different manufacturers while maintaining a high reconfigurability of the task in execution. The task we consider is the assembly of an electronic gadget that requires several operations, e.g., pick-up and delivery of various components, alignments of small parts, screwing operations, and so forth.

Scheduling and task planning are well known topics and many solutions are available in the literature such as the Petri Net Plan framework [4,5], ROS-Plan [6,7], ML formalism [8], Semantic map [9]. In this paper we propose a methodology that allows us to solve this problem in a realistic industrial scenario using a simple and direct approach. This framework makes possible to integrate state-of-the-art approaches for specific problems (such as the ones cited above) and to investigate novel solutions increasing reliability and effectiveness of the plant.

The proposed methodology is based on different levels of abstraction, starting from the high level description of the tasks, up to the execution of each action by the robots or the machines. Initially, we define a pool of generic *primitives* that the machines/robots are able to execute. These primitives are then invoked by a dynamically generated *state-chart* devised by a high-level task definition, according to [10]. The procedure guarantees the reconfigurability of the operations of the machines and, at the same time, the possibility to control different machines and robots, without having to rely on proprietary implementations. With this methodology it is possible to concentrate all efforts on finding more efficient scheduling systems without having to spend most of the time and resources in programming the low level behaviors of the machines/robots which requires specialized personnel.

The paper is organized as follows: in Sect. 2 we show how the proposed methodology works by analyzing all the layers of abstraction. In Sect. 3 we apply this methodology to a recipe (i.e., group of tasks) that the production line has to perform, while in Sect. 4 we describe the experimental results showing the execution of the different tasks. Conclusions are drawn in Sect. 5.

2 Synthesis of State-Charts for Complex Tasks

The main idea of the presented approach is to be able to control and coordinate a group of industrial robots by describing a set of high-level tasks. Once these tasks are defined, the system automatically generates state-charts containing the sequence of operations that the robots have to execute.

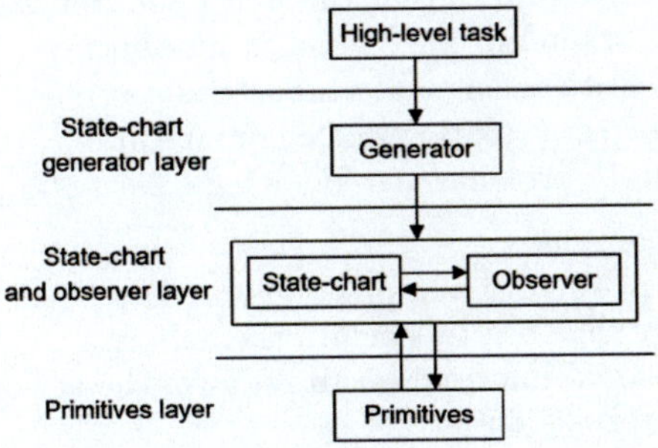

Fig. 2. Abstraction layers schema.

Our approach is based on the definition of three abstraction layers:

- *State-chart generator layer*: its purpose is to generate a state chart from a high-level task in a dynamic way.
- *State-chart and observer layer*: the generated state chart defines the different operations to be performed by the robots. The Observer is aware of the operations that the robots have to execute, and depending on their state, it makes the corresponding state chart evolve.
- *Primitives layer*: the layer containing all the primitive operations that each robot can execute.

In Fig. 2 the three layers are represented with their interactions.

2.1 State-Chart Generator Layer

The state-chart synthesizer receives as input the high-level task and generates a state-chart. Consequently, it is necessary to define a syntax for the task that allows the state-chart synthesizer to parse the instructions and create states that will compose the final finite states machine.

The grammar is based on some keywords that allow to synthesize a graph of instructions while maintaining the temporal relationships between one operation and the following one. To ensure time relationships, the order of understanding the task must be known by the layer of the state-chart synthesizer.

The task's syntax must consist of the definition of an action, a subject (the entity that has to perform the action), some parameters, and elements of conjunction for the concatenation of following actions. An example of task is:

"Move base from X to Y and move arm from Z to W"

In this case the keywords obtained by interpreting the task are: "move" that identifies the action, "base" and "arm" identify the subject that must perform the action, after "from/to" are expressed all the parameters of the action and the keyword "and" allows the temporal conjunction of two, or more, actions. The temporal relationship of the task is maintained by simply interpreting the task from left to right, which implies that the first operation concerns the motion of the mobile base and, when finished, the arm movement.

Each state-chart state consists of a structure containing the necessary information to ensure a correct evolution and communication with the primitive layer. In particular, each state provides the following information:

- id_state: status identifier;
- action: the name of the instruction to execute;
- subject: the entity that has to execute the action;
- parameters: additional information to perform the action;
- edges: list of the subsequent states;
- trigger: describes when it is possible to make a transition from this state to another.

2.2 State-chart and Observer Layer

To ensure the task is executed correctly, this layer is in continuous communication with the primitives layer of the robots, as shown in Fig. 2. The state-chart is a finite state machine where each state represents the primitive operation to perform and the edges are the triggers that evolve the current state to the next one. The Observer is the entity that is placed between the state-chart and the primitives layer that continuously checks the evolution of the operations performed by the robots and generates the triggers that make the state-chart evolve over time.

2.3 Primitives Layer

The primitives layer is where all the operations robots and devices can execute are defined; such operations can be atomic or not. When the state chart is in a state where an action is defined, a mapping between the information containing the state and the pool of primitives available is made. Referring to the example above, the first part of the sentence consists of an action, "move", a device, "base", and parameters, "from X to Y". These fields are mapped as

Move base from X to Y → state-chart → **function move_base**(X,Y)

The types and quantity of primitives that this layer provides depend on the application and the desired generalization level of the system. Many atomic primitives lead to a high generalization and a wider reconfigurability of the tasks, whereas high level primitives improve abstraction and make the implementation of complex tasks easier. It is then necessary to find out the proper trade-off according to the specific industrial scenario.

3 Industry 4.0 Application

The task we want to automate consists in assembling an electronic gadget composed of a bottom cover, a top cover (both made using the 3D printing process, see Fig. 3a) and an electronic board with battery. The two covers need to be screwed.

(a) 3D printed prototype (b) Robotic bay

Fig. 3. Images representing the gadget and the assembly process.

The mobile robot RB-Kairos, equipped with an UR5, takes the components from the vertical warehouse and places them on the conveyor that will bring the pieces to the robotic bay for the assembly process. At the robotic bay there are two cobots as shown in Fig. 3b: the ABB YuMi puts the two covers one over the other, while the KUKA LBR iiwa14 screws them.

Unfortunately, the setup is not fully ready: the industrial screwdriver on the KUKA robot with the automatic screw dispenser is not working yet, and we are still waiting for the electronic board. Nevertheless, the corresponding tasks have being designed but they are not shown in the companion video.

The high-level tasks that the robots have to perform to complete the recipe are listed below.

RB-KAIROS tasks

RB1. *move base to warehouse and detect UR5 top_cover and pick UR5 top_cover from warehouse if detected*

RB2. *move base to conveyor and place UR5 to top_cover pallet*

RB3. *move base to warehouse and detect bottom_cover and pick UR5 bottom_cover from warehouse if detected*

RB4. *move base to conveyor_bay5 and place UR5 bottom_cover to pallet* and move base to home

KUKA LBR iiwa 14 tasks

K1. *move KUKA to idle_position and wait KUKA gadget_assembled*

K2. *move KUKA to gadget_detection_position and detect KUKA gadget_holes*

K3. *move KUKA to pick_screw_pose and screw KUKA screw1 if detected and move KUKA to pick_screw_pose and screw KUKA screw2 if detected*

ABB YuMi tasks

Y1. *move YuMi to idle_position and wait YuMi top_cover and pick YuMi top_cover from pallet and place YuMi top_cover to buffer(0)*

Y2. *move YuMi to idle_position and wait YuMi bottom_cover*

Y3. *wait YuMi assembly and pick YuMi top_cover from buffer(0) and place YuMi top_cover to pallet and move YuMi to assembly_help_pose and wait YuMi end_assembly and move YuMi to idle_position*

FIgures 4, 5 and 6 shows the automatically generated state-charts of the tasks for the RB-Kairos, Kuka, and YuMi robot, respectively.

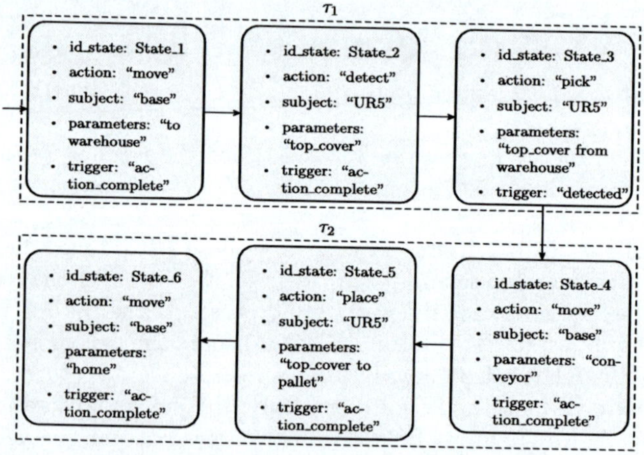

Fig. 4. RB-Kairos Tasks. τ_i encloses all the primitives composing the i-th task.

In the task definition, a particular keyword allows the generation of waiting states ("wait trigger"). The command "if trigger" is used to block the current operation, until a trigger occurs. The primitives that will be used internally implement different algorithms for the control of robots and for the detection of objects. Thanks to these primitives, the whole code necessary for controlling

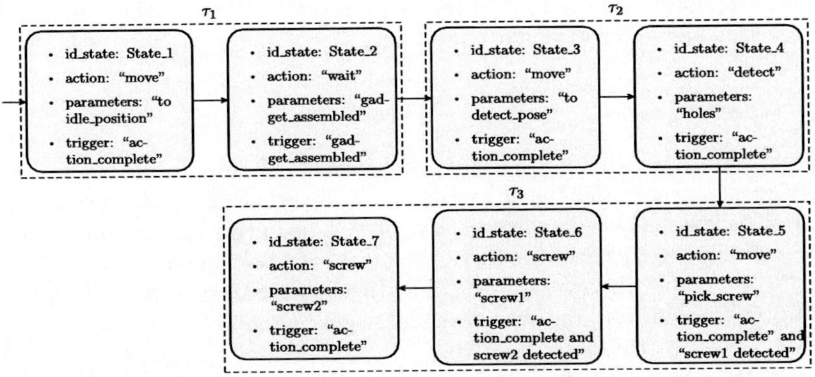

Fig. 5. Kuka Tasks. τ_i encloses all the primitives composing the i-th task. The subject of each operation is the robot itself, so it has not been specified in the image.

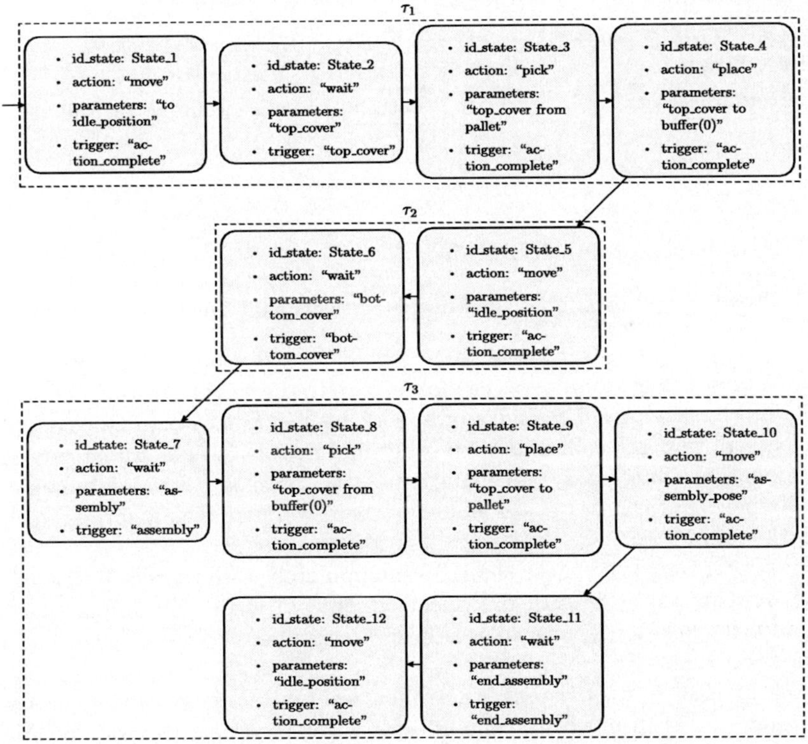

Fig. 6. Yumi Tasks. τ_i encloses all the primitives composing the i-th task. The subject of each operation is the robot itself, so it has not been specified in the image.

the robots is hidden, so this methodology can be applied transparently to all installed devices. The available primitives make available by the robots will be explained in the following subsections.

3.1 Observer

The software layer that the observer implements is responsible for generating the triggers necessary for the advancement of the previously self-generated statechart. This layer (1) knows the objective of each primitive instruction, (2) continuously checks the status of the robots (e.g. sensors, position, arm configuration), and (3) generates the events according to the execution status of the actual operation. Table 1 lists and describes the trigger actions.

Table 1. Triggers generated by the observers.

Trigger action	Description
action_complete	The operation has been completed
gadget_assembled	The gadget has been assembled
top(bottom)_cover detected	The top (bottom) cover has been detected
$screw_i$ detected	The i-th screw has been detected
top(bottom)_cover	The top (bottom) cover has come to the robotic bay

3.2 Primitives of the RB-Kairos 5

RB-Kairos is responsible for pick-and-place tasks that need the following primitives

- `move_base(target)`: moves the mobile base to the target position. The path is defined by solving a topological graph of the ICE laboratory. This commands avoids all possible collisions with other mobile robots. The target position is an identifier: in this way the definition of the task is simpler and dynamic. For example, "move base to warehouse", where warehouse is a specific location in the map.
- `move_arm(target)`: moves the UR5 manipulator to the target position.
- `detect(object)`: detects the specified object. Once the object has been found, its position frame with respect to the base frame of the manipulator is created.
- `pick_arm(object, target)`: the robot reaches the target point, closes the gripper and then moves the end-effector with the object to a safe position.
- `place_arm(object, target)`: the manipulator moves the end-effector to the target position, opens the gripper to place the object and moves to a safe position.

3.3 Primitives of the KUKA LBR Iiwa 14

In our task, the KUKA LBR iiwa 14 has to screw in two screws that will fix the top cover with the bottom cover of the gadget. This operation is supported by a computer-vision algorithm implemented in ViSP [11,12] which, thanks to the previous knowledge of the gadget's CAD description, recognizes its shape and identifies the coordinates of the two holes. The position of the camera attached to the screwdriver has been previously calibrated [13].

The robot has the following primitives:

- `move_kuka(target)`: performs a motion (in joint space) from the current position to the target one.
- `wait_kuka(synch)`: during this primitive the robot does not perform any operation and waits for the synchronization variable to change value.
- `detect_kuka(holes)`: enables the computer-vision algorithm for the detection of the holes.
- `screw_kuka(target_hole)`: the manipulator picks a screw and then moves over the target hole, enables the screwdriver and moves linearly downwards.

3.4 Primitives ABB YuMi

The ABB YuMi is in charge of picking the top and bottom covers of the gadget from the conveyor, assembles them, and finally keeps the gadget still during the screwing process. Its primitives are:

- `move_yumi(target)`: performs a motion (in the joint space) from the current position to the target position.
- `wait_yumi(synch)`: does not perform any movement, however keeps listening for synchronization.
- `pick_yumi(object, target)`: moves the left arm from the current position to the pre-pick pose that is above the target position. Then the robot moves downwards linearly until it touches the surface of the object; the vacuum gripper attached to the end-effector raises the cover; the arm returns to the pre-pick position.
- `place_yumi(object, target)`: moves the left arm from the current position to a pre-place pose that is above the target position. Then the robot moves downwards linearly until it touches the bottom cover and then disables the vacuum gripper to release the top cover; finally it returns to the pre-place position.

4 Experimental Results

The software architecture has been obtained using different frameworks and programming languages for the implementation of the methodology and primitives. ROS [14] has been used for the implementation of the observer and for the primitives of the RB-Kairos and the Kuka iiwa. The primitives of the YuMi have been implemented with the RobotStudio programming environment. The following plots will show the behavior of the robots while performing each task.

4.1 RB-Kairos

Figure 7 shows the execution of the first two tasks RB1 and RB2 that the RB-Kairos base must perform. In the first one the base moves towards the target position (in front of the vertical warehouse) following the path marked in blue, while in the second task, the base reaches the position in front of the conveyor. Figures 8 and 9 show the Cartesian position of the UR5 robotic arm during task RB1 and RB2. The green colored area defines the "home" position of the arm, while the blue area defines the gripper actions ("close" first and "open" later). In Fig. 8, the yellow area identifies the detection phase of the covers. After both pick and place motions, the manipulator returns to its home position.

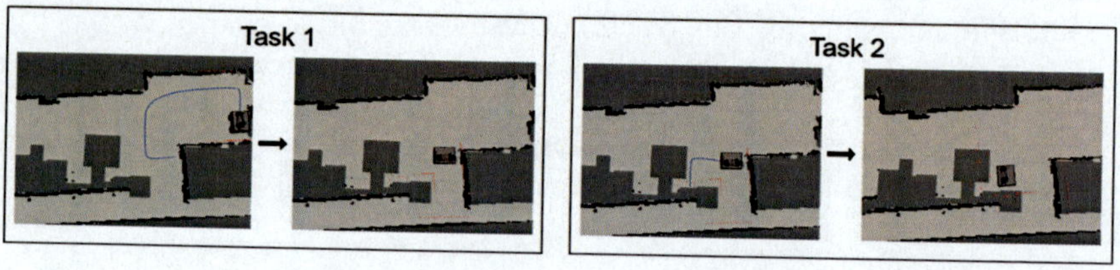

Fig. 7. RB-Kairos base motion during tasks RB1 and RB2.

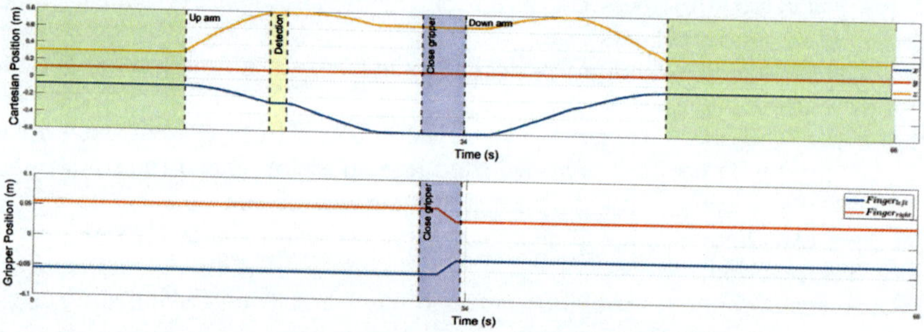

Fig. 8. Cartesian position of the UR5 during task RB1.

4.2 KUKA LBR Iiwa 14 and ABB Yumi for the Assembly Process

Figure 10 shows the Cartesian position of the KUKA LBR iiwa manipulator expressed with respect to its base frame during the task.

In the first task (K1), the robot is in idle_position and waits for a new gadget to be assembled on the pallet. The second task (K2) represents the motion of the end-effector to reach a good position for the gadget recognition task. In K3

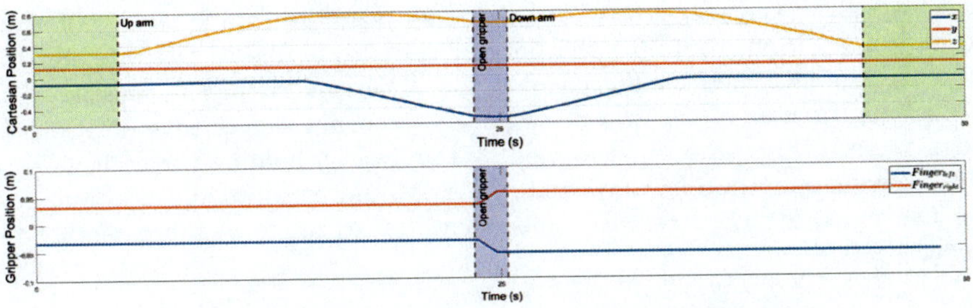

Fig. 9. Cartesian position of the UR5 during task RB2.

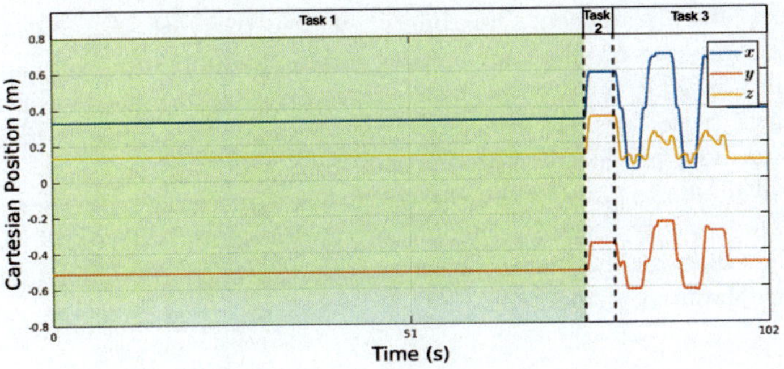

Fig. 10. Cartesian position of the KUKA LBR iiwa. The green area represents its idle_position.

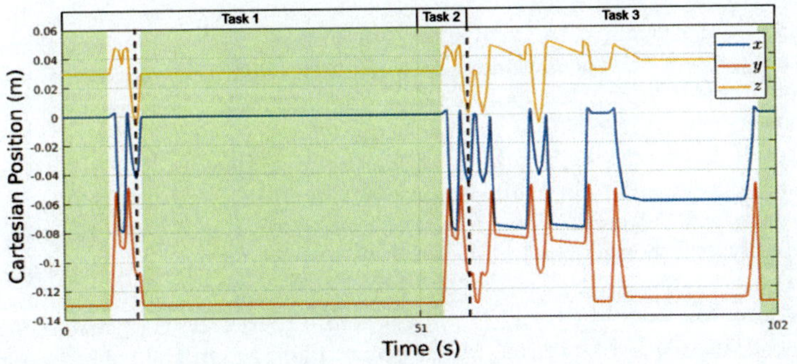

Fig. 11. Cartesian position of the ABB YuMi. The green areas represent its idle_position.

the robot performs the screwing operation and returns to a safe position for the next operations.

Figure 11 shows the Cartesian motion of the vacuum gripper attached to the Smart Gripper of the ABB YuMi cobot, during the assembly process. In the first task (Y1) the manipulator waits for the top cover to come to the robotic bay and then picks and places it to the buffering position. In Y2, the robot moves to its idle_position, waits for the bottom cover to arrive to the bay and then stores the object in the buffering position. The last task (Y3) consists of the assembling process of the two pieces, then the cobot moves to a position that helps the screwing process performed by the iiwa by holding the gadget. When the screwing operation is completed, the YuMi moves to its idle_position.

The video attached to this paper shows an example of the assembly task under study.

5 Conclusion

In this paper we proposed a methodology able to synthesize state-charts for executing complex tasks by coordinating different robots in an Industry 4.0 scenario. Automatically generated state-charts make affordable to control robots and devices from different manufacturers while maintaining a high level of abstraction. The proposed approach has been validated and tested using the facilities available in the ICE laboratory of the University of Verona. The next step will be to study the robustness of the approach and to extend the capabilities of the Observer module in order to early detect faults and malfunctioning for tasks replanning purposes.

References

1. Trstenjak, M., Cosic, P.: Process planning in industry 4.0 environment. Proced. Manuf. **11**, 1744–1750 (2017)
2. Michael, N., Zavlanos, M.M. Kumar, V., Pappas, G.J.: Distributed multi-robot task assignment and formation control. In: 2008 IEEE International Conference on Robotics and Automation, pp. 128–133, IEEE (2008)
3. Ice laboratory, universirty of verona. https://www.icelab.di.univr.it
4. Ziparo, V.A., Iocchi, L., Lima, P.U., Nardi, D., Palamara, P.F.: Petri net plans. Autonomous Agents and Multi-Agent Systems **23**(3), 344–383 (2011)
5. Palamara, P.F., Ziparo, V.A., Iocchi, L., Nardi, D., Lima, P.: Teamwork design based on petri net plans. In: Robot Soccer World Cup, pp. 200–211, Springer (2008)
6. Cashmore, M., Fox, M., Long, D., Magazzeni, D., Ridder, B., Carrera, A., Palomeras, N., Hurtos, N., Carreras, M.: Rosplan: planning in the robot operating system. In: Proceedings of the International Conference on Automated Planning and Scheduling, vol. 25 (2015)
7. Canal, G., Cashmore, M., Krivić, S., Alenyà, G., Magazzeni, D., Torras, C.: Probabilistic planning for robotics with rosplan. In: Annual Conference Towards Autonomous Robotic Systems, pp. 236–250, Springer (2019)
8. Spellini, S., Chirico, R., Lora, M., Fummi, F.: Languages and formalisms to enable eda techniques in the context of industry 4.0. In: 2019 Forum for Specification and Design Languages (FDL), pp. 1–4, IEEE (2019)
9. Galindo, C., Fernández-Madrigal, J.-A., González, J., Saffiotti, A.: Robot task planning using semantic maps. Robot. Autonom. Syst. **56**(11), 955–966 (2008)

10. Falezza, F., Piccinelli, N., De Rossi, G., Roberti, A., Kronreif, G., Setti, F., Fiorini, P., Muradore, R.: Modeling of surgical procedures using state charts for semi-autonomous robotic surgery. IEEE Trans. Med. Robot. Bion. **3**(4), 888–899 (2021)
11. Marchand, E., Spindler, F., Chaumette, F.: ViSP for visual servoing: a generic software platform with a wide class of robot control skills. IEEE Robot. Automat. Magaz. **12**, 40–52 (2005)
12. Trinh, S., Spindler, F., Marchand, E., Chaumette, F.: A modular framework for model-based visual tracking using edge, texture and depth features. In: IEEE/RSJ International Conference on Intelligent Robots and Systems, IROS'18, (Madrid, Spain), October 2018
13. Tsai, R.Y., Lenz, R.K., et al.: A new technique for fully autonomous and efficient 3 d robotics hand/eye calibration. IEEE Trans. Robot. Automat. **5**(3), 345–358 (1989)
14. Stanford Artificial Intelligence Laboratory et al.: Robotic operating system

Automated Driving and Traffic

Training Traffic Light Behavior with End-to-End Learning

Maël Wildi[1](✉), Alexandre Alahi[1], and Arnoud Visser[2]

[1] EPFL, Visual Intelligence for Transportation Laboratory, Bâtiment GC, Station 18, CH-1015 Lausanne, Switzerland
mael.wildi@alumni.epfl.ch

[2] Intelligent Robotics Lab, Universiteit van Amsterdam, Science Park 904, 1098 XH Amsterdam, The Netherlands
a.visser@uva.nl

Abstract. In this work, we study neural network architectures that will reduce the number of infractions made by autonomous-driving agents. These agents control vehicles by providing future waypoints directly from a forward-facing camera. Building on top of the teacher-student approach of *Cheating by Segmentation*, we investigate the impact of Pyramid Pooling Module and Feature Pyramid Network with the aim to learn more representative features. We run our experiment with CARLA simulator and show that pyramid perception modules have a positive impact in reducing the number of traffic light infractions and collisions.

Keywords: Conditional imitation learning · Feature pyramid network

1 Introduction

In 2021, Honda in Japan and Mercedes in Germany received the authorization to deploy a vehicle where the driver is allowed to let the car drive itself, as long as they do not exceed a speed of 60km/h and are able to take back control if needed.

Most self-driving vehicles are based on a modular approach, where perception, planning and control are separated from one another [13]. In the last decade, the use of convolutional neural networks (CNN) has exploded, thanks to the increased power of computers. It is now widely used for perception tasks, as it enables to learn a model capable of recognizing objects from a large amount of annotated images. For autonomous driving, this enabled the possibility of end-to-end learning, an alternative to the modular method [10]. The control commands for the autonomous car are directly derived from the sensory input (such as a forward-looking camera), letting the network learn the intermediate features relevant for this task.

Imitation learning is, alongside the emerging reinforcement learning approach, the main method to perform end-to-end learning for autonomous cars [9].

I. Petrovic et al. (Eds.): IAS 2022, LNNS 577, pp. 753–764, 2023.
https://doi.org/10.1007/978-3-031-22216-0_50

Fig. 1. Illustration of CARLA simulator, widely used for autonomous driving [5].

It consists in training a model to reproduce the actions an expert would have performed if presented to the same situation (Fig. 1).

Many models rely on data collected by a human driver [1], however driving simulators such as Car Learning to Act (CARLA) [6] are more and more used as they are getting very realistic. Most importantly, a simulator allows to put the autopilot in constant varying situations, which is essential for efficient learning. Secondly, it enables to have scenarios where other cars and pedestrians are disrespecting traffic rules, which is important in order to learn a robust policy trained for worst-case scenarios.

Learning by Cheating (LBC) [3], one of the best performing driving policy on CARLA simulator, was modified in *Cheating by Segmentation* (CBS) [8] to rely only on images coming from a driver perspective instead of a bird's-eye view (BEV). It led to promising results, however the number of collisions and traffic lights infractions increased due to the less informative viewpoint.

Thus, this study pursues the goal of generating a robust end-to-end imitation learning driving policy using RGB and semantically segmented images from a driver perspective as only inputs. The focus is on reducing the number of traffic light infractions and collisions. In order to do this, we rely on pyramid perception modules that are described in the Method section.

2 Related Work

Although there are several interesting developments in learning to drive based on existing datasets (see for instance [12], [4] and [9]), we concentrate for conciseness on methods that train in an end-to-end fashion an autonomous agent to drive using CARLA simulator.

2.1 Conditional Imitation Learning

Learning by Cheating *Learning by Cheating* (LBC) [3] proposes a novel way of training an agent to learn a robust driving policy. It separates the learning process in two: learning to act and learning to perceive using a teacher-student network approach.

The teacher agent is trained offline with the supervision of a dataset of expert trajectories and is privileged in the sense that it has access to a bird's-eye view (BEV) semantically segmented image. The student agent is trained under the supervision of the teacher and has only access to a standard forward-facing RGB picture.

Both the teacher and student agents have multiple prediction heads connected to the backbone, which outputs for each of the possible directional command (turn left, turn right, go straight, follow lane) heatmaps that are converted into waypoints using a soft-argmax operation. These waypoints are then converted to vehicle commands by a PID controller.

The dataset of expert trajectories is collected directly in the simulator using an agent based on CARLA autopilot. It collects both the BEV semantically segmented image and the forward-facing RGB image. Data augmentation is performed by rotating the BEV to simulate steering noise. Multiple driving episodes are gathered in different traffic and weather conditions.

Cheating by Segmentation Although the student of LBC could be transferred to the real-world without too much difficulties according to the authors, the teacher used to train it relies on the ground-truth BEV semantic segmentation image which could be difficult and expensive to gather. *Cheating by Segmentation* (CBS) [8] addresses this by replacing it by a 120deg field-of-view forward-facing camera (although still semantic segmented for the teacher). This would enable to use pre-existing large datasets to train the student.

CBS gives promising results, but has difficulty anticipating braking actions, which results in a much higher amount of collisions and traffic light infractions than LBC on the *NoCrash* [5] benchmark. According to its author, it could be a consequence of the change of perspective in the segmentation camera from BEV to forward-facing: in the perspective view, close objects appear bigger which results in a shorter-term behaviour (a strong signal at a late moment, resulting in late braking).

2.2 Reinforcement Learning

World on Rails *Learning to drive from a world on rails* (WoR) [2], published in 2021 by the same authors as LBC, proposes a model-based reinforcement learning method. It relies on a model of the ego-vehicle, which enables to simulate the outcome of the agent's actions. This model is learned by training a network to predict the next agent's state (location, orientation, speed) given its initial state and an action (steer, throttle, brake). This is done using pre-recorded driving logs from CARLA.

The main assumption of the paper is that the world is on rails, meaning that the agent has no influence on its environment. Consequently, the latter does not depend on the agent's state or actions which means the initial world state determines the entire sequence of world states. As the agent is unable to change this sequence, the state transitions for the world are simply the ordered sequence of pre-recorded world states from the driving logs.

The model of the ego-vehicle (and the known sequence of world states) are used alongside a reward function to compute the Q-value, $Q_t(s, a)$, for all possible combination of agent state s and action a at timestep t. The Q-value is the reward received for taking a given action in the current state plus the discounted estimated optimal return on the long term that it will get in the state it ends up in. The reward function is designed in a way to encourage the vehicle to stay within the target lane as well as to stop for a pedestrian or a traffic light, and is penalized otherwise. The Q-value table is used to supervise a visuomotor agent that takes a RGB image and the vehicle speed as input.

3 Method

The approach taken in this study uses as starting point *Cheating by Segmentation* (CBS): a teacher learns to drive from a dataset of forward-facing semantically segmented images and predicts waypoints in the image frame. It then supervises a student, that has the same image perspective as the teacher but with an RGB image input. Finally, the learned student model is used alone in the environment and its predicted waypoints are converted by a controller into vehicle commands.

In this study, the goal is to modify the student network architecture using two pyramid perception modules, PPM and FPN, which will be described later in this section.

3.1 Network Architecture

This section presents the network architecture on which this study relies. The learning process uses a teacher-student approach. The teacher is supervised by ground truth waypoints retrieved from expert driving logs and, once trained, is used to supervise a student. Figure 2 illustrates the whole framework. In the following sub-sections, the teacher and student are presented in detail.

Teacher The teacher architecture is unchanged from CBS. The network gets as input the forward speed of the ego-vehicle `spd` and the forward-facing semantically segmented image `SSI` representing roads, road lines, red traffic lights, cars, and pedestrians. The image is fed into a Resnet18 backbone and its output at the 4th layer, just before the fully connected layers, is retrieved. It is then concatenated with a repeated version of the forward speed `spd` and fed to one of the specialized waypoint prediction head, according to the directional command `cmd`.

These prediction heads perform deconvolution to predict heatmaps that are converted into waypoints in the image frame using a spatial arg-softmax. The model is supervised by the ground truth waypoints coming from the dataset. The loss function is the mean squared error (MSE) between the network predictions and the ground truth waypoints.

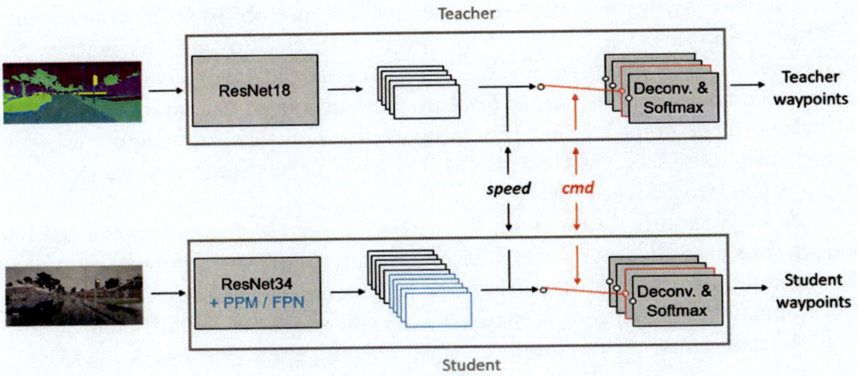

Fig. 2. Illustration of the network architecture. The teacher takes a semantically segmented image as input, while the student gets an RGB image. The teacher is supervised by the ground truth waypoints and the student by the teacher's.

Student The student uses the same framework as the teacher, except that it takes an RGB image as input and uses a ResNet34 backbone instead of a ResNet18. It is supervised by the waypoints predictions of the teacher. The student uses the same loss function as the teacher.

The main objective of this study is to improve the traffic light perception in the network, with the two modules detailed hereafter.

Pyramid Pooling Module Pyramid Scene Parsing Network (PSPNet) [14] was introduced in 2017 and aim to increase the effective receptive field and add context to the final feature maps of a feature extractor backbone such as ResNet. The heart of the PSPNet is the Pyramid Pooling Module (PPM). For different scales, adaptive average pooling is applied on these feature maps, followed by an up-sampling step to bring them back to their original dimensions. Then, the newly obtained maps are concatenated to the original ones. This provides for each pixel multiple level of context: from global context (whole image) to local context (small sub-region around the pixel). In other words, by looking at all the feature maps at a same specific location, information about features concerning the pixels outside its receptive field is also provided.

This PPM module is integrated into the architecture by placing it after the ResNet34 backbone but before the concatenation with the ego-vehicle speed. It thus takes as input the feature maps of the last ResNet34 convolutional layer.

Using PPM should provide a more robust representation of the image by fusing together features from different sub-regions. Its authors have established its effectiveness against mismatched relationships. In our case, a potential mismatched relationship would be to confound a car rear light for a red traffic light.

Feature Pyramid Network Feature Pyramid Network (FPN) [7], also introduced in 2017, is quite similar to PSPNet. However, its feature maps are coming from different layers, not only the last one. They are obtained by concatenating the intermediate feature maps of one layer with the up-sampled feature maps of the next layer. Thus, combining the good spatial resolution of the earliest layers with the higher semantic level in the features of the latest. This enables to include more complex features in spatially more accurate maps, as well as considering small objects that would not have been detected with the feature maps of the latest layer only.

This FPN module takes as input the feature maps of each layer of the backbone, in our case ResNet34. Thus, it gets feature maps with different numbers of channels and dimensions. The first step is to convolve each of these inputs with 256 kernels of size ($feature\_maps\_in \times 1 \times 1$) in order to get for each backbone layer the same amount of channels, while keeping their original dimensions.

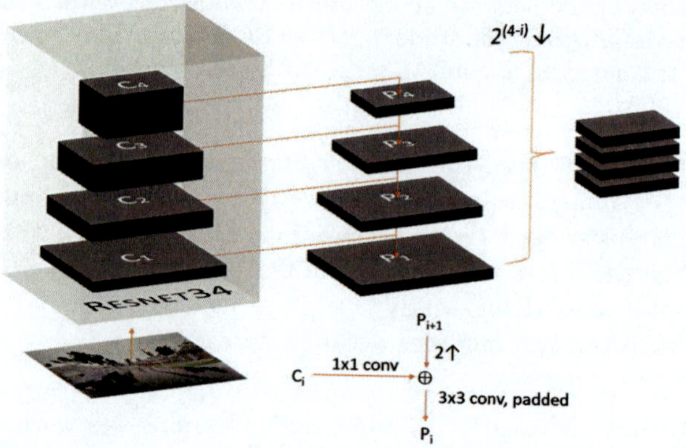

Fig. 3. Illustration of the proposed Feature Pyramid Network architecture.

To be able to integrate it to the existing student network without needing to change it, we propose to add a down-sampling step so that each set of feature maps has the same dimensions as the last backbone convolutional layer. Finally, we concatenate them to end up with a unique output of the same dimensions as PPM. The resulting architecture is illustrated in Fig. 3.

By fusing together high semantic—low resolution features of the latest layers with low semantic—high resolution of the earliest layers, a more accurate representation of the image is constructed, which should help the waypoint prediction heads in their task.

3.2 PID Controller

The controller responsible for transforming waypoints to vehicle commands is the same as the one used for LBC and CBS. The target velocity is the average speed needed to go from one waypoint to the next one. Given the current speed of the vehicle, a longitudinal PID controller must then compute the throttle and brake commands that shall make it reach its target speed. The steer command is computed by fitting an arc through the predicted waypoints. The angle formed by the current position and the projection of one of the waypoints on the arc is fed to a lateral PID controller.

Compared to LBC and CBS, the braking threshold is incremented from 2.0 m/s to 3.8 m/s in order to adapt the controller to the sensitivity our model has to obstacles. Moreover, the target speed is clipped to be in a range from 0 m/s to 5 m/s as in CBS.

4 Experimental Setup

The approach having been described, now we present the setup and dataset used to train our networks, as well as the evaluation process used to measure the performances of the different models. We compare two different modules that can be added to the existing student network architecture, in order to favour a better recognition of traffic lights. The process remains end-to-end and no classification is made. Both modules can be introduced between the backbone and the waypoint prediction heads. The architecture and code used for the study is described in more detail in the thesis [11].

4.1 Simulation Environment

We used CARLA 0.9.10.1 which allows to have reproducible runs with a fixed random seed responsible for traffic generation and an evaluator that is now used by the latest state-of-the-art implementations, allowing us to compare performances with them on common metrics. Note that an earlier version 0.9.6 of CARLA was employed to evaluate CBS algorithm (with a non-fixed random seed), which means that some of those results had to be recreated in CARLA 0.9.10.1 to get a fair comparison.

4.2 Dataset

The dataset is collected using a collector agent which follows the approach of WoR, but has been collected especially for this work. It receives a route through a given CARLA town and receives rewards for attaining intermediate waypoints, for breaking at red lights as well as for pedestrians and cars. Information about the agent's surroundings is based on the semantically segmented BEV retrieved from the simulator. For example, attributing a non-zero reward for braking for a pedestrian is possible only if the latter is visible in the BEV. The BEV is centered on the ego-vehicle and has a range of 32 m in each cardinal direction. The collected dataset is then split into two distinct datasets used respectively for training and validation, with a 80%–20% ratio.

4.3 Evaluation

Our models are evaluated on the *NoCrash* benchmark, in which a run is stopped as soon as the agent collides. It is also the case if it deviates by more than 30 m from the route or if it does not move for 180 s (timeout). *NoCrash* assesses the performances of the agent in train (*Town01*) and test (*Town02*) environments, as well as train (#1, #3, #6, #8) and test (#10, #14) weather conditions. For all of these combinations, the agent is always evaluated in three different traffic conditions: *Empty*, *Regular* and *Dense*, detailed in Table 1.

Table 1. Number of vehicles and pedestrians under the three traffic conditions, *Empty*, *Regular* and *Dense*, in Town01 (left) and Town02 (right).

Config.	Vehicles	Pedestrians
T1-Empty	0	0
T1-Regular	20	50
T1-Dense	100	250

Config.	Vehicles	Pedestrians
T2-Empty	0	0
T2-Regular	15	50
T2-Dense	70	150

There are 25 predefined routes in each town that are evaluated in every traffic and weather conditions. Each of these runs is called an episode. For instance, in the test town with test weather conditions, there are 25 routes × 4 weathers × 3 traffics resulting in 300 episodes. Therefore, with three algorithms to be evaluated, the complete benchmark consists in total in 900 episodes.

We compare the performances not on the *Success Rate*, the metric used for WoR, but zoom in on underlying metrics like the *Route Completion*, *In Lane Rate*, *No Collision Rate*, *No Block Rate* and finally the metric which is the focus of the study: *Lights ran per hour*. Note that on average in each episode (depending on the route) two traffic lights are encountered. Those performance measures are compared with WoR, the leading implementation on *NoCrash* and the former state-of-the-art, LBC.

5 Results

For the results, first the learning process is described and followed by the evaluation of the selected models on the *NoCrash* benchmark.

5.1 Learning Curves

Figure 4 shows the training and validation curves of all the models. The teacher's 50th epoch checkpoint (loss of respectively 0.0028 and 0.015) is chosen to supervise the student network. Training it for longer lead to an increased validation loss that signify an overfit of the training dataset. Three student models are trained with the supervision of the teacher previously obtained. One without additional module (Original), one with a PPM, and one relying on a FPN. The chosen model is the one with the lowest validation loss after 100 epochs at most. Their validation loss is similar, which, although it is not increasing, stays quite constant and maintain a non-negligeable gap with the training curve. This could indicate a small overfit of the dataset. To overcome this in the future, the regularization could be strengthened for instance by performing more data augmentation or by introducing a weight decay in the optimizer.

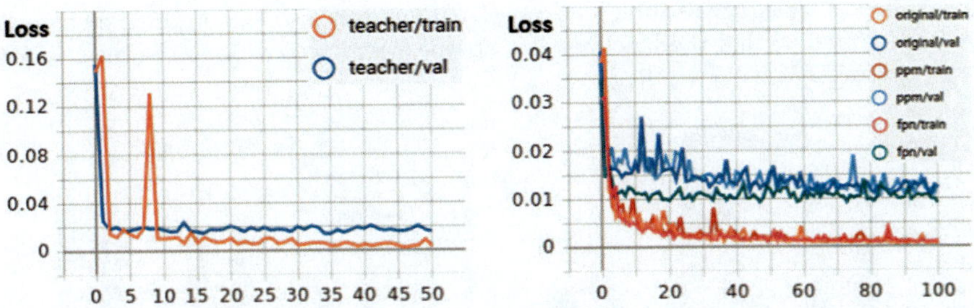

Fig. 4. Evolution of the training and validation loss of the teacher (left) and of the three students (right) with the number of epochs. The loss function is the mean L2 distance between predicted and supervision waypoints.

5.2 *NoCrash* results

We evaluate the three student models selected in the previous section on *NoCrash* and compare them with the results of WoR and LBC as published in [2], if available. For a fair comparison, the results for CBS marked as 'Original', are the results of the original CBS algorithm trained in the exact same circumstances as FPN and PPM, so these are not based on the previous reported results.

Overall Results As shown in Fig. 5 (left), the *Route Completion* results are still well below the state-of-the-art. This is not due to the vehicle leaving the road or even its lane. Our three student models have actually learned to drive and to follow a route through the city. The performances on lane keeping are excellent in every configuration, as illustrated in Fig. 5 (right).

Error Sources When staying within the lane, still three error sources for not completing the route in the *NoCrash* benchmark remain. The agent can collide with cars and pedestrians in their lane (*No Collision Rate*), stop due to a "false obstacle" or fail to restart when a traffic light becomes green (*No Block Rate*) or, finally, run a red light (*Lights ran per hour*). This latter was a major error source for CBS, but with PPM and FPN this is no longer the case, as demonstrated at the end of this section.

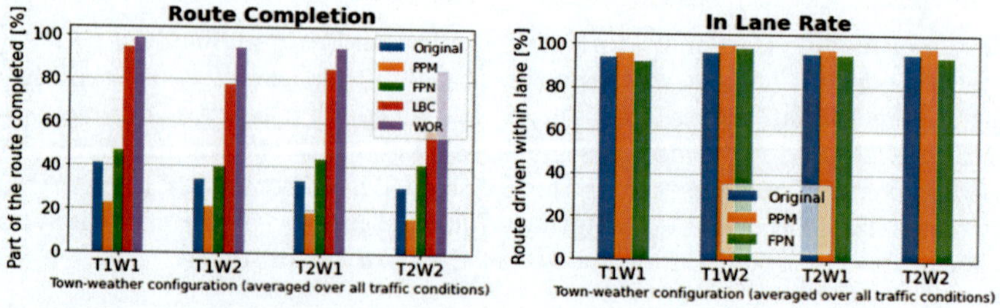

Fig. 5. Left: Percentage of the route completed without collision. Right: Percentage of the route driven within the lane. Metrics are measured according to the town-weather configuration (1:train, 2:test).

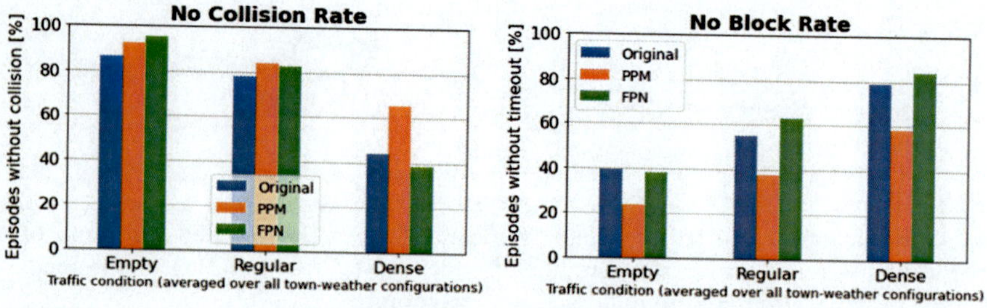

Fig. 6. Left: Percentage of episodes ended without collision according to the traffic condition. Right: Percentage of episodes ended without being blocked. This is defined to be the case when the vehicle does not move for 180 s.

Unfortunately, the *No Collision Rate* and *No Block Rate* are hard to balance. The student models' sensitivity to other vehicles is moderate. Other vehicles are detected, because the network reduces its target speed, but too often that is not sufficiently to avoid a collision. This can be repaired by adjusting the braking threshold of the PID controller, described in Sect. 3.2. Yet, this also influence the risk of getting stuck due to a "false obstacle" such as a puddle of water. This trade-off can be clearly seen in Fig. 6.

We can observe that the agent is less likely to be blocked in heavy traffic, as the resulting change in the visible environment is the only reason for it to make a change in its waypoint predictions, thereby giving it a chance to move. Moreover, it is clear that the PPM model is more often stuck. This could be explained by the fact that it is more capable than the others to differentiate objects from the road even if they have a similar texture. Nonetheless, this higher sensitivity also has the effect that PPM is better at perceiving cars. This is illustrated on the left of Fig. 6, where the percentage of episodes ended without collision is depicted.

Correct Behavior The original goal and the positive result of this study is that, with the FPN algorithm, the traffic light infractions are no longer the major source of error. The major improvement against the original CBS algorithm can be clearly seen in Table 2 and summarized in Fig. 7.

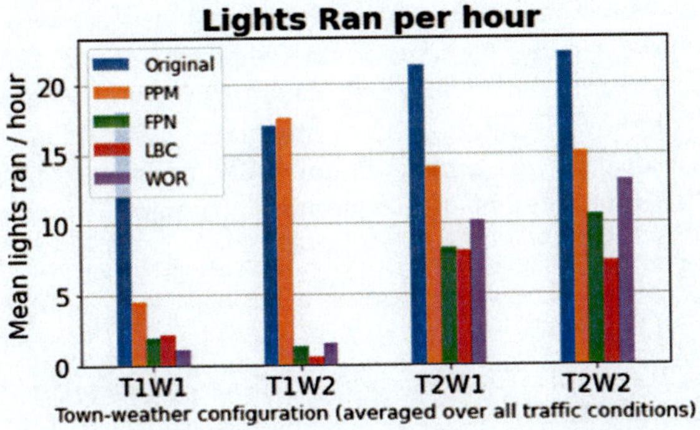

Fig. 7. Traffic light ran per hour of driving according to the town and weather.

Table 2. Number of traffic lights infractions per hour.

	Train town						Test town					
	Train weather			Test weather			Train weather			Test weather		
	Empty	Reg	Dense	Empty	Reg	Dense	Empty	Reg	Dense	Empty	Reg	Dense
LBC	1.35	1.89	3.27	0.36	0.81	**0.52**	**8.45**	8.22	**7.26**	**8.17**	**8.61**	**4.87**
WoR	**0.00**	**0.43**	**2.61**	**0.00**	**0.00**	4.29	10.68	**6.95**	12.90	14.46	11.30	13.28
Original	15.07	18.45	20.7	16.74	18.06	16.1	19.27	20.1	24.34	25.23	24.94	16.28
PPM	3.37	5.71	4.37	14.9	16.66	20.9	17.84	13.87	10.12	**2.53***	24.62	18.05
FPN	**1.15**	**2.12**	**2.51**	**0.49**	**1.65**	**1.57**	**6.70**	**9.13**	**8.69**	7.74	**10.70**	**13.27**

FPN provides a robust traffic light handling, competitive to WoR and close to LBC[1]. Fusing feature maps from different layers of the backbone enables it to recognize more precisely where and which features in the image are responsible for stopping or restarting. Also, it generalizes well to unseen environments, with a number of infractions per hour of only 10.57 in the test town—test weather configuration. Besides, it seems that the weather conditions have a very limited impact on its performances.

6 Conclusion

In this study, we trained an agent end-to-end to imitate an expert using only images taken from a driver perspective. Two different student network architectures were tested. Both the Pyramid Pooling Module and the Feature Pyramid Network resulted in an agent presenting a very good lane keeping. PPM enabled to react more strongly to cars but also to inoffensive obstacles.

FPN enabled to significantly reduce the number of traffic lights ran, reaching a number of infractions per hour of 10.6% in test conditions on *NoCrash*. This

[1] Note: the infraction score of the PPM agent in the *Empty* scenario of the test town and weather (marked with *) is underestimated due to it being significantly more stuck (72% of the episodes) compared to the other traffic scenarios.

brings it close to the state-of-the-art implementations *Learning to drive from a world on rails* and *Learning by Cheating.* Diversifying the distance at which the expert stops for a traffic light could further improve the model.

This study has shown insights on the potential of pyramid perception modules for conditional imitation learning and hopefully their perception abilities will contribute to the improvement of autonomous driving.

References

1. Bojarski, M., Testa, D.D., Dworakowski, D., Firner, B., Flepp, B., Goyal, P., Jackel, L.D., Monfort, M., Muller, U., Zhang, J., Zhang, X., Zhao, J., Zieba, K.: End to End Learning for Self-driving Cars (2016)
2. Chen, D., Koltun, V., Krähenbühl, P.: Learning to drive from a world on rails. In: Proceedings of the IEEE/CVF International Conference on Computer Vision (ICCV) (2021)
3. Chen, D., Zhou, B., Koltun, V., Krähenbühl, P.: Learning by cheating. In: Conference on Robot Learning (CoRL) (2019)
4. Chi, L., Mu, Y.: Deep steering: Learning End-to-End Driving Model from Spatial and Temporal Visual Cues. preprint arXiv:1708.03798 (2017)
5. Codevilla, F., Santana, E., López, A.M., Gaidon, A.: Exploring the limitations of behavior cloning for autonomous driving. In: Proceedings of the IEEE International Conference on Computer Vision (ICCV) (2019)
6. Dosovitskiy, A., Ros, G., Codevilla, F., Lopez, A., Koltun, V.: Carla: an open urban driving simulator. In: Proceedings of the 1st Annual Conference on Robot Learning (PMLR) (2017)
7. Lin, T.Y., Dollár, P., Girshick, R., He, K., Hariharan, B., Belongie, S.: Feature pyramid networks for object detection. In: IEEE Conference on Computer Vision and Pattern Recognition (CVPR) (2017)
8. van Orden, T.: Cheating by Segmentation. Bachelor thesis, Universiteit van Amsterdam (2021)
9. Pan, Y., Cheng, C.A., Saigol, K., Lee, K., Yan, X., Theodorou, E.A., Boots, B.: Imitation learning for agile autonomous driving. Int. J. Robot. Res. **39**(2–3), 286–302 (2020). https://doi.org/10.1177/0278364919880273
10. Tampuu, A., Matiisen, T., Semikin, M., Fishman, D., Muhammad, N.: A survey of end-to-end driving: architectures and training methods. IEEE Trans. Neural Netw. Learn. Syst. (TNNLS) (2020)
11. Wildi, M.: Conditional Imitation Learning with Pyramid Perception Modules. Master's thesis, École Polytechnique Fédérale de Lausanne (2022)
12. Xu, H., Gao, Y., Yu, F., Darrell, T.: End-to-end learning of driving models from large-scale video datasets. In: Proceedings of the IEEE Conference on Computer Vision and Pattern Recognition (CVPR) (2017)
13. Yurtsever, E., Lambert, J., Carballo, A., Takeda, K.: A survey of autonomous driving: common practices and emerging technologies. IEEE Access **8**, 58443–58469 (2020). https://doi.org/10.1109/ACCESS.2020.2983149
14. Zhao, H., Shi, J., Qi, X., Wang, X., Jia, J.: Pyramid scene parsing network. In: IEEE Conference on Computer Vision and Pattern Recognition (CVPR) (2017)

On Demand Ride Sharing: Scheduling of an Autonoumous Bus Fleet for Last Mile Travel

Jörg Husemann(✉), Simon Kunz, and Karsten Berns

Robotics Research Lab, TU Kaiserslautern, 67663 Kaiserslautern, Germany
husemann@cs.uni-kl.de, s_kunz10@cs.uni-kl.de, berns@cs.uni-kl.de
https://agrosy.cs.uni-kl.de

Abstract. Autonomous buses are expected to expand the mobility-on-demand options in cities in the next 10 years. An essential aspect of ensuring optimal usage of fleets of autonomous buses is the task of scheduling. This paper presents a scheduling approach based on the construction of all possible trips used to formulate an optimization problem. While most research is focused on the scheduling of taxi trips, there is almost no research for other applications, for instance, the scheduling of last-mile travel options. A street network has been generated based on open street map data as a basis for the scheduling. All possible combinations of buses and requests are calculated, and for each of those trips, the optimal order of requests is created. These are used to formulate the optimization problem, calculating an optimal assignment of requests on the buses. The approach has considered constraints such as maximum waiting time, travel delay, and targets to exploit shared trips for higher efficiency. Experiments have been carried out in a simulated environment of a university campus area with fleets of up to 10 vehicles. By performing various trials with changing parameters, the influence of the constraints on waiting time and travel delay to the scheduling is determined. Depending on the setup, service rates above 90%, while trails with strict constraints show that the approach can handle short-term requests. Based on the results, a use case-specific composition of the autonomous bus fleet can be done.

Keywords: Mobility-on-demand · Bus fleet scheduling · Shared mobility

1 Introduction

In recent years, the increase of mobility-on-demand services has been shaping the appearance of and mobility within cities. A large number of offers like scooter-, bike- and car-sharing concepts are available in most cities aiming to solve problems like air pollution and missing parking space while increasing the flexibility of people. The primary purpose of these mobility concepts is the first mile and

I. Petrovic et al. (Eds.): IAS 2022, LNNS 577, pp. 765–777, 2023.
https://doi.org/10.1007/978-3-031-22216-0_51

last-mile transportation, in which the passenger is traveling, for instance, from a bus stop or train station to its target location. Due to the advances in the development of autonomous vehicles, autonomous bus shuttles are an option for these trips. These shuttles can complement the current services in a way that they can be used regardless of the physical limitations of the passengers. Another advantage comes from the fact that they are allowed to drive in traffic-calmed areas, such as pedestrian zones, where conventional traffic is not allowed.

One example for such a shuttle bus is the AutoBus (Fig. 1) currently developed at the TU Kaiserslautern. The hardware and software architecture is designed to drive in pedestrian zones. The central aspect is archiving a robust control architecture, interacting with the pedestrians. The interaction is realized by recognizing pedestrians' awareness regarding the bus and avoiding critical situations with them either by targeted driving maneuvers or by interacting with them via output interfaces, for instance, speakers and displays [1]. Furthermore, using safety certificated hardware and a validated software concept, it is aimed to pass the governmental validation processes for the bus [2].

Fig. 1. Autonomous minibus (AutoBus) currently developed at TU Kaiserslautern.

For the success of a fleet of autonomous shuttles, a high degree of user acceptance is indispensable. This can be reached by achieving high reliability by minimizing waiting and driving times. Known mobility-on-demand offers are typically for single exclusive usage and follow a first come, first serve policy. Compared to these, the optimal use of a bus fleet has to include shared rides and a dynamic adaptation to new requests, leading to a more complex problem and requires a high degree of planning. [1]

In this paper, a scheduling approach is presented, which calculates close to optimal routes for a fleet of autonomous buses. The approach is adapted from a concept for scheduling of shared taxi trips [3]. The database for the street network

[1] https://agrosy.informatik.uni-kl.de/roboter/autobus.

is based on freely available data of the OpenStreetMap (OSM) project [4], by which the approach is easily transferable to other areas. Experiments are done in an area of the campus of the TU Kaiserslautern and its immediate surroundings using a bus fleet of up to 10 vehicles. In Experiments of several hours, delays and the rate of serviced requests are evaluated under varying degrees of utilization, testing the limits of the presented approach.

The paper is structured as follows. Section 2 introduces previous works in the field of scheduling for shared mobility concepts. Section 3 details the scheduling and assignment process. In Sect. 4 the evaluation of the concept is done, evaluating simulated experiments giving suggestions for the composition of the bus fleet. Finally, Sect. 5 summarizes the work showing the further way to the scheduling of a real fleet of autonomous buses.

2 Related Work

The field of mobility-on-demand concepts, especially concerning autonomous vehicles, has already been investigated in several studies. Dia et al. [5] investigated the feasibility of agent-based simulation tools to evaluate the effects of shared autonomous vehicles, stating a drastic decrease of the traffic volume with the example of Melbourne traffic. Similar studies with varying focus were evidencing positive aspects of shared-mobility concepts in terms of pollution, congestion, and the availability of parking space [6] and were also predicting benefits from an economic point of view [7]. Besides simulation scenarios, which were used to show the advantages of mobility-on-demand concepts, there were also studies about real autonomous vehicles. Kim et al. [8] presented experiments of a campus mobility service using an autonomous taxi, further targeting to extend to a fleet of vehicles in the future.

For the general categorization of vehicle fleet management systems, Hyland et al. [9] presented a taxonomy to differentiate between various aspects of them. One part of the mentioned research was predefined aspects as the vehicle fleet's composition, defined timing constraints, and the availability of information about the vehicle states. Another aspect mentioned with a massive influence on the scheduling was the knowledge about the occurrence of passenger requests. The scheduling efficiency can be increased if knowledge about the occurrence of future requests is provided, for instance, by using stochastic models. Furthermore, various categorizations regarding ride-sharing, the road network, and reservation policies were set out.

So far, the application of fleet scheduling mostly refers to the organization of taxi trips. To reduce the number of required trips, the notion of shareability networks has been introduced [10]. The application of shareability networks includes transforming the spatio-temporal sharing problem into a graph-theoretical problem targeting to take advantage of well-known methods for the solution. *Shareability networks* are hypergraphs in which trips of different sizes are saved. Those trips are created combining smaller trips, while satisfying predefined constraints for delay and seating capacity. The concept was verified on a large set of recorded taxi trips, capable of scheduling trips smaller than size four in tractable time. Based on shareability networks, an anytime-optimal algorithm for scheduling was created, enabling the calculation of trips of arbitrary size [3]. In this approach, possible trips with a dedicated vehicle were incrementally constructed. After determining the optimal route for each trip, an integer linear programming problem was formulated, calculating an optimal assignment. Later the concept was extended to take stochastic information into account to bias future requests [11]. Based on experiments using a data set of actual taxi trips, the authors state an opportunity to strongly reduce the number of required vehicles while keeping waiting times and travel delays low.

3 Bus Fleet Scheduling

The scheduling aims to create a system with a high level of user acceptance. This acceptance should be achieved by creating a distribution of all requests to the bus fleet while creating optimal routes. In the scenario, a route is optimal by minimizing the sum of delays over all requests while satisfying given constraints. Considered constraints of the scheduling are the maximum waiting delay which defines the latest time the trip is allowed to start, and the maximum travel delay, which indicates the maximal additional driving time compared to the optimal travel time of each passenger in the bus.

An overview of the presented system is shown in Fig. 2. Passengers can send requests for a ride from a specified start to a target position. The requests of the passengers are sent to the request management and are collected there until the next scheduling cycle is triggered. The scheduling is creating plans for the bus fleet and is sending them to the individual buses, which are then executing them. During execution, the buses send their current state information to the scheduling to be used in the following cycles.

3.1 Representation of Street Network

The precise representation of the road network is of high importance for a correct scheduling procedure. For a good representation of the street network, it is generated based on the free-to-use database of Openstreetmap[2] [4]. The information is saved in topological nodes, which are extracted and saved in a graph structure [12]. In addition to the location, semantic information is saved in the nodes. Using these traversable roads for the bus can be determined. Figure 3a shows the full graph generated from the OMS information while Fig. 3b shows a cleaned-up version in which prohibited and unreachable regions are removed from the network. Common path-finding algorithms as A* are implemented on the graph structure, which are later used in the scheduling process to calculate the shortest paths for the requests.

[2] https://www.openstreetmap.org.

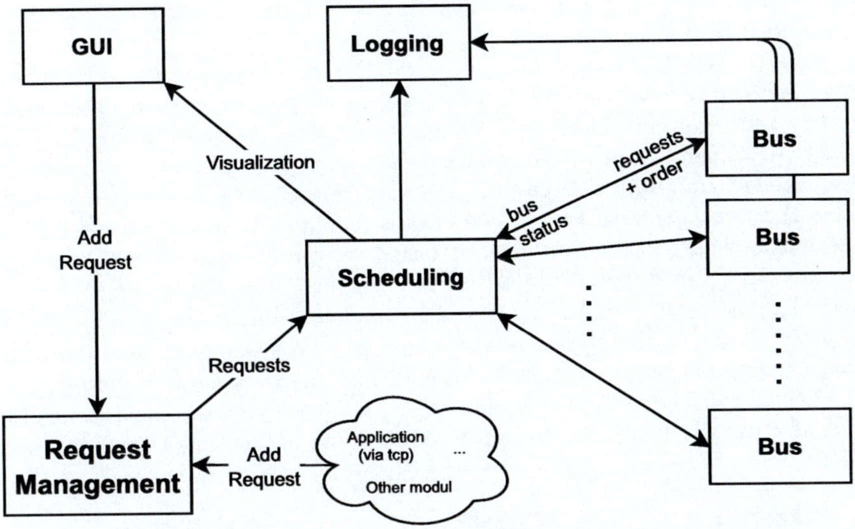

Fig. 2. Overview of the system components and their communication.

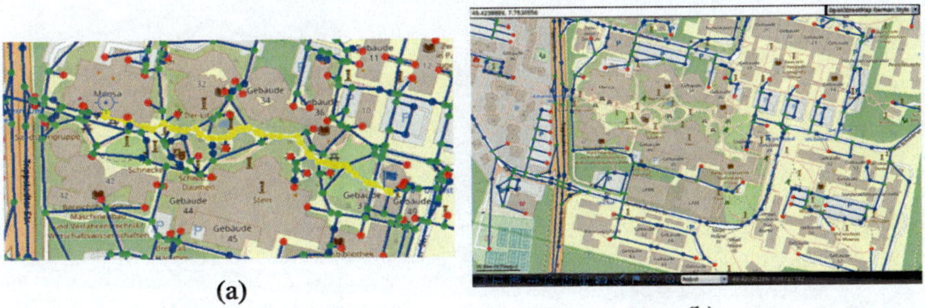

(a) (b)

Fig. 3. Figure 3a shows a zoomed in part of the complete osm graph. Nodes are visualizing topological nodes (e.g. crossings, dead-ends), edges show the paths. The yellow path is a calculated shortest path between two nodes. Figure 3b shows the reduced graph after removing edges, which are not drivable by the bus.

3.2 Graph Generation

The graph generation is performed to calculate all possible combinations of buses and requests, which are later used to formulate the *integer linear programming* (ILP) problem. A request sent by a passenger is defined by the desired pick-up and drop-off location and the time of the request. To ensure their uniqueness, the request management provides each request with an exclusive identifier. Furthermore, for simplicity, the pick-up and drop-off locations are mapped to the closed topological node in the open street map graph. Based on the parameters of the vehicle and the OSM graph, the travel duration of the request is calculated, and it is added to the request pool.

As introduced by Alonso-Mora et al. [3], the scheduling process is periodically triggered and is based on solving an *integer linear programming* problem using a greedy assignment as initial distribution to speed up the convergence. To formulate this problem, all possible trips for each bus are created using two graphs, the *Request-Vehicle-Graph* (RVGraph) and the *Request-Trip-Vehicle-Graph* (RTV-Graph) as visualized in Fig. 4. The graphs include all unassigned requests and the buses. The RVGraph is created based on the current pool of unassigned requests and the states of the buses. In the graph, two requests r_i, r_j are connected if both requests can be served by a virtual bus starting at the pick-up location of one of the requests. The edge $e(r_i, r_j)$ is weighted with the sum of travel delays of the two requests. An edge $e(r_i, b_j)$ between a request r_j and a bus b_i exists if the bus b_i is able to satisfy r_j while keeping the constraints of this and all currently picked up requests, representing all possible trip of size one.

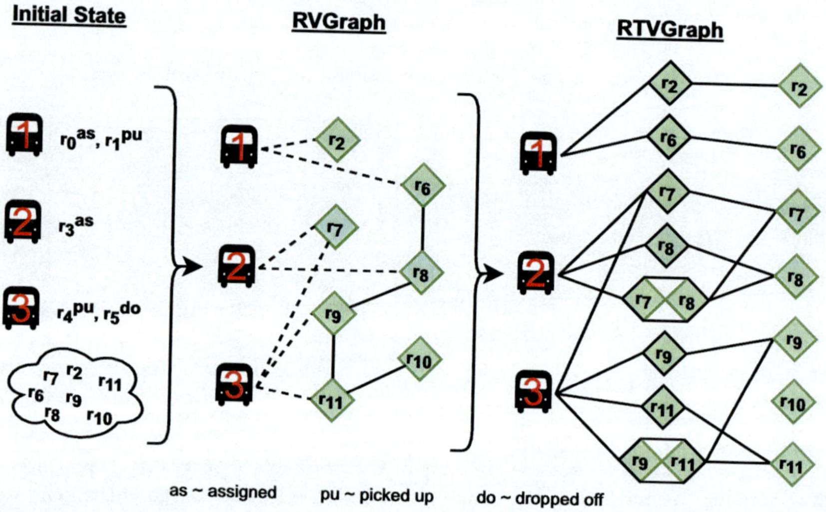

Fig. 4. Construction of the RTVGraph using the initial state and the RVGraph. RVGraph: dotted lines show which request a a bus can fulfill considering the already assigned tasks, full lines show which requests can be combined independently of the bus. RTVGraph: framed request are pick-up tasks, non-framed are drop-off tasks.

Using the RVGraph the RTVGraph is calculated, which is used to calculate trips starting with trips with two requests. A trip of size k is defined as a set of requests $T = \{r_0, \ldots, r_n\}$ with $|T| = k$. A subgraph of the RVGraph with k nodes is a trip of size k when it is complete. By that, the problem of finding all trips of a specific size corresponds to finding all cliques of the same specific size. Using the sub-feasibility lemma [13, p. 7] this problem can be solved incrementally. T_1

is created by adding all requests r servable by at least one bus and by that are part of an edge $e = (r, b)$. Next, the trips of size two up to the maximum seating capacity are calculated. For the calculation of T_k, all combinations of two trips from T_{k-1} with size k are calculated. For each valid combination, it is checked whether all subsets of the combined set of size $k-1$ are part of T_{k-1}. If this is the case, the combined set might form a feasible trip of size k. Then it is checked if there exists a valid route given the current bus states. If this route exists, the trip will be added to T_k. After calculating all T_k for all buses, they are merged into a trip set used for the assignment process.

3.3 Optimization of Bus Trips Using Simulated Annealing

To formulate the *integer linear programming* problem, the costs for each trip have to be known. Optimal solutions for each trip are calculated using simulated annealing [14]. To deliver valid solutions for the bus application, some requirements have to be defined. Firstly, the bus location has always to be the first location of the trip. Secondly, the position of the pick-up of a request r_{i_pu} has to be before the drop-off position of the same request r_{i_do}. Thirdly, if the pick-up of r_{i_pu} is not in the list of requests, the passenger is already picked up and by that r_{i_do} can be placed at an arbitrary position after the bus. The last requirement is that constraints maximum waiting time and the maximum travel delay are fulfilled.

The generator function considers the first three constraints. As an initial solution, a sorted list with the bus at the first position and each pick-up r_{i_pu} directly followed by its drop-off r_{i_do} is used. By that, the constraints 1–3 are initially met. The generator function is creating new samples based on the current solution. This is done by randomly selecting one of the request locations of the trip and placing it at a new position as shown in Fig. 5. By ensuring to satisfy the second constraint, the possible range of the request is limited by the position of the corresponding request.

Next, the energy function calculates the travel duration of the generated solution. The travel duration is calculated by adding the distances between adjacent requests, which are calculated by a path-finding algorithm on the open street map graph. In addition, the constraints for all requests are checked as visualized in Fig. 6. Using the distance between the requests, an expected timestamp for the completion of each request is calculated. Next, it is checked whether the constraints are met. For the pick-ups the maximum waiting delay is checked $t_{req} - t_{pu} \leq d_{wt\_max}$ using the starting time of the request t_{req} and the expected time for finishing the request t_{pu}. For the drop-off tasks, compliance with the maximum travel delay is checked. This delay describes the maximal time the trip is allowed to exceed the expected driving time $t_{do} - t_{pu} - d_{travel\_time} \leq d_{td\_max}$. If the delays are exceeded, a high penalty is added to the travel delay calculated in the energy function. By that, a solution satisfying the constraints is always preferred.

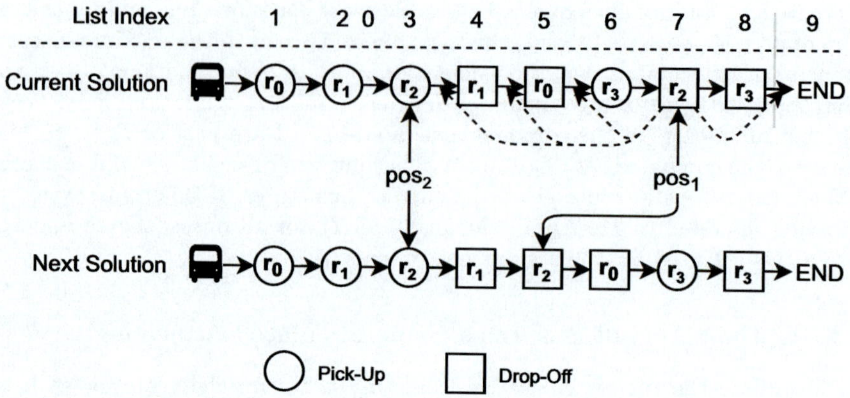

Fig. 5. Generation of the next valid solution based on the current solution. The drop-off of r_2 has been selected to change its position of the current solution, pos_2 limits the possible positions for r_2, since it is a drop-off and can not be placed in front its corresponding of the pick-up.

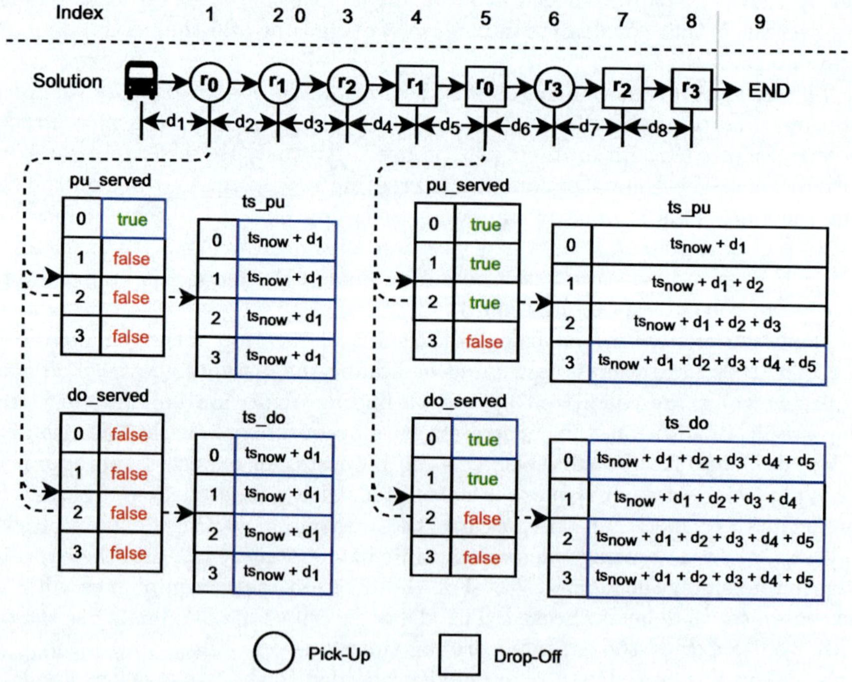

Fig. 6. Check of waiting time and travel time constraints during the optimization of each trip.

3.4 Construction and Solution of the ILP Problem

Finally, the trips and their calculated delays are used to find an optimal assignment of requests to the buses. For this, the trips represented in the RTV graph are transferred to a set of variables used in the ILP [3]. Linear programming is a method of mathematical optimization. It can be used for the maximization and minimization of linear object functions, while satisfying defined constraints [15]. Integer linear programming, as used in this approach, is a special case of linear programming, including the restriction that all variables are integer values $x \in \mathbb{Z}^n$. Each edge e between a trip (T) and a bus (b) $e(T_i, b_j)$ is represented as a binary variable $\epsilon_{i,j}$ which equals 1 if the trip T_i is assigned to the vehicle v_j. The set $\mathcal{E}_{TV}$ is defined as the set of indexes for all edged $e(T_i, v_j)$. Further each request $r \in \mathcal{R}$ a binary variable x_k is defined and set to 1 if the corresponding request r_k is not served. The cost function is defined in Eq. 1, where $c_{i,j}$ is the cost for the trip and c_{ko} is a penalty added if the trip is not assigned, which leads to avoiding unassigned trips if possible.

$$\mathcal{C}(\mathcal{X}) := \sum_{i,j \in \mathcal{E}_{TV}} c_{i,j}\epsilon_{i,j} + \sum_{k \in \{0,\ldots,n\}} c_{ko}x_k \tag{1}$$

Additionally, to get a valid solution the following constraints are defined. First each bus can only be assigned to a most one trip Eq. 2.

$$\sum_{i \in \mathcal{I}_j^V} \epsilon_{i,j} \leq 1, \quad \forall v_j \in V \tag{2}$$

Secondly, each request can either be assigned to a bus or ignored Eq. 3.

$$\sum_{i \in \mathcal{I}_k^R} \sum_{j \in \mathcal{I}_i^T} \epsilon_{i,j} + \mathcal{X}_k = 1, \quad \forall r_k \in \mathcal{R} \tag{3}$$

As an initial solution for the ILP, a greedy solution is generated by looping over the trips trying to assign the largest trips with the lowest costs to the vehicles. Then while solving the ILP problem, the initial solution is replaced as soon as a better one is calculated. The process ends if an optimal solution is calculated or if there is a timeout. The timeout is triggered when the next scheduling cycle is started. Then the currently best solution is used for the assignment. This solution is at least as good as the initial greedy solution by construction.

Finally, after the calculation of the solution, each bus is assigned with its associated trip and can start to execute it. If an old trip is assigned, it is replaced with the new one. Then the new request execution order is checked for validity since, due to the calculation time, it is possible that some of the requests assigned are already executed. This is done by erasing finished tasks from the request execution order. The execution order is processed by calculating the route to the next request based on the open street map graph, which is then followed by the bus.

4 Experiments

The experiments have been simulated on the street network of the TU Kaiserslautern and its surroundings. The buses are simulated on the OSM network traveling at a constant velocity. Kinematic constraints and delays are not considered, but non-traversable paths are removed from the OSM graph. The seating capacity of the buses is set to a maximum of six passengers per bus. Each bus is simulated self sufficiently, receiving new trips from the scheduling algorithm and sending its status information using the same interface like a real bus. Owing to this, it is possible to replace one or all buses with real vehicles without the need for adjustments to the scheduling algorithm. Besides testing the general functionality of the approach, it was also targeted to analyze the behavior under heavy load.

During the experiments, different parameters are varied to analyze their effect on the scheduling. To analyze the influence of the constraints maximum waiting time d_{wt_max} and maximum travel delay d_{td_max} used in the scheduling algorithm, they are set on values from three to seven minutes. In terms of the request, two parameters are set: their occurrence rate and their distribution. Each experiment is using a set of 500 requests, which are spawned based on the occurrence rate $r_{req} \in \{2, 4, 6\}$ $[requests/min]$. The cycle time used to pool new requests is $30s$. Since there are no real-world data to derive the locations of the request, data sets for the spawning of requests are created. One set is constructed using random positions on the campus area. For the second data set, five points of interest are defined, which mark locations in which a larger occurrence of requests is expected. Using a Gaussian distribution around these points, request are generated. In addition to these two sets, combinations of both using different distributions between them are created.

First tests with at most three buses showed that the scheduling manages to reach serviced request rates close to 100%. After these results, a more extensive test setup with a fleet of ten buses is prepared. Based on the occurrence rate of the request, the experiments took between 1.5 and 4.5 h. During the tests, timestamps for assignment, pick-up, and drop-off are logged as well as internal states of the buses. The results for the percentage of serviced requests is shown in Fig. 7. As expected, the service rate benefits from a lower occurrence of requests and more generous constraints. Especially in the experiments with the $r_{req} = 6$ the service rates are significantly less. This service rate is explainable by monitoring the utilization of the buses. Even though the buses are active the whole time, they cannot serve the majority of the requests. Primarily in the cases with strict constraints, the utilization of the buses is bad due to the reduced possibilities of shared rides, which further decreases the service rate. In the trial with $r_{req} = 6$ and high constraints, the scheduling failed due to a large number of possible bus-trip combinations. To avoid this, timeouts are required to stop the graph construction, leading to not calculating large trip sizes but enabling the scheduling to work under high load.

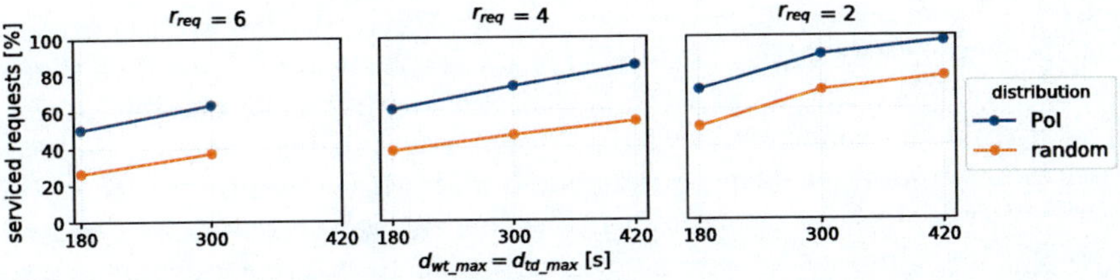

Fig. 7. Propotion of serviced request depending on request occurrence rate r_{req}, constraints d_{wt_max} and d_{td_max}, and the distribution of requests.

The utilization of the buses gives an indication of the optimal composition of the fleet. The results for a trials with $r_{req} = 4$ are shown in Fig. 8. It can be seen that except for the start-up and decay phase, the buses are active most of the time. In the case of higher delay constraints, the utilization of the seating capacity is significantly higher. Even though some buses are close to full capacity, most vehicles are underutilized. Considering the current structure of request, this leads to the result that a larger number of vehicles should be preferred over a smaller fleet with increased capacity. Nonetheless, it is also a possibility that a restructuring of the request data set would, for example, by considering group requests of multiple people, could lead to varying results. Due to that, the use of real passenger data is required to make a precise assumption about the fleet composition.

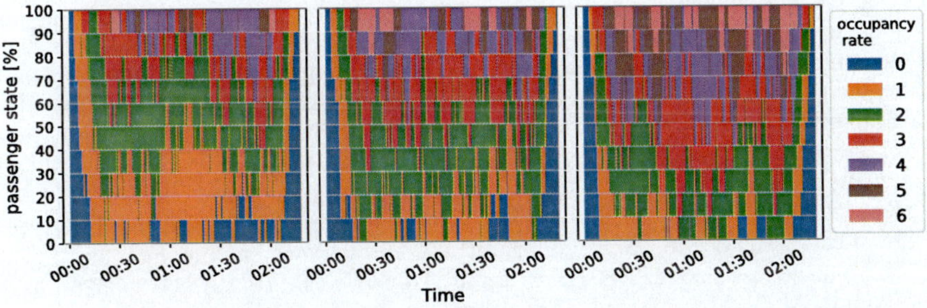

Fig. 8. Occupancy of buses over time for d_{wt_max} and d_{td_max} of 180s (left), 300s (center) and 420s (right), $r_{req} = 4$.

5 Conclusion

In this paper, a scheduling approach for a mobility-on-demand service of autonomous buses which focused on the utilization of shared rides has been demonstrated. The proposed work minimizes waiting time and travel delays while aiming to achieve high service rates. The used road network is extracted from open street map data transformed into a graph structure. Based on the pool of currently open requests and the positions of the buses, all possible combinations of trips and vehicles are calculated. After optimizing the trips, an integer linear programming problem is formulated and solved, resulting in an optimal distribution of the trips to the buses. The concept was evaluated by performing various experiments using a bus fleet up to size 10, varying different parameters for constraints and the composition of passenger requests. Experiments on data sets of different structures showed a significant impact of the distribution on the scheduling results, which leads to the conclusion that real-world data on the request occurrence is needed to model the bus fleet successfully. Furthermore, in future experiments, the combination of simulated buses with at least one real operating bus should be performed to verify the scheduling procedure and investigate the robustness against delays.

References

1. Jan, Q.H., Kleen, J.M.A., Berns, K.: Self-aware pedestrians modeling for testing autonomous vehicles in simulation. In: VEHITS, pp. 577–584 (2020)
2. Jan, Q.H., Berns, K.: Safety-configuration of autonomous bus in pedestrian zone. In: VEHITS, pp. 698–705 (2021)
3. Alonso-Mora, J., Samaranayake, S., Wallar, A., Frazzoli, E., Rus, D.: On-demand high-capacity ride-sharing via dynamic trip-vehicle assignment. Proc. Nat. Acad. Sci. **114**(3), 462–467 (2017)
4. Haklay, M., Weber, P.: Openstreetmap: user-generated street maps. IEEE Pervas. Comput. **7**(4), 12–18 (2008)
5. Dia, H., Javanshour, F.: Autonomous shared mobility-on-demand: Melbourne pilot simulation study. Transp. Res. Proced. **22**, 285–296 (2017)
6. Martinez, L., Crist, P.: Urban mobility system upgrade-how shared self-driving cars could change city traffic. In: International Transport Forum, Paris (2015)
7. Fagnant, D.J., Kockelman, K.M.: Dynamic ride-sharing and fleet sizing for a system of shared autonomous vehicles in Austin, Texas. Transportation **45**(1), 143–158 (2018)
8. Kim, S.-W., Gwon, G.-P., Hur, W.-S., Hyeon, D., Kim, D.-Y., Kim, S.-H., Kye, D.-K., Lee, S.-H., Lee, S., Shin, M.-O., et al.: Autonomous campus mobility services using driverless taxi. IEEE Trans. Intell. Transp. Syst. **18**(12), 3513–3526 (2017)
9. Hyland, M.F., Mahmassani, H.S.: Taxonomy of shared autonomous vehicle fleet management problems to inform future transportation mobility. Transp. Res. Rec. **2653**(1), 26–34 (2017)
10. Santi, P., Resta, G., Szell, M., Sobolevsky, S., Strogatz, S.H., Ratti, C.: Quantifying the benefits of vehicle pooling with shareability networks. Proc. Nat. Acad. Sci. **111**(37), 13290–13294 (2014)
11. Alonso-Mora, J., Wallar, A., Rus, D.: Predictive routing for autonomous mobility-on-demand systems with ride-sharing. In: 2017 IEEE/RSJ International Conference on Intelligent Robots and Systems (IROS), pp. 3583–3590. IEEE (2017)
12. Fleischmann, P., Pfister, T., Oswald, M., Berns, K.: Using openstreetmap for autonomous mobile robot navigation. In: Proceedings of the 14th International Conference on Intelligent Autonomous Systems (IAS-14), Shanghai, China, July 3–7 2016. Best Conference Paper Award - Final List

13. Alonso-Mora, J., Samaranayake, S., Wallar, A., Frazzoli, E., Rus, D.: On-demand high-capacity ride-sharing via dynamic trip-vehicle assignment-supplemental material (2017)
14. Aarts, E.H.L., et al.: Simulated annealing: theory and applications (1987)
15. Saigal, R.: Linear programming: a modern integrated analysis. J. Operat. Res. Soc. **48**(7), 762–762 (1997)

![Check for updates]

Lane Change Classification with Neural Networks for Automated Conversion of Logical Scenarios

Vjekoslav Diklić[1](✉) and Ivan Marković[2]

[1] dSPACE, Ltd., Zagreb, Croatia
vdiklic@dspace.hr

[2] University of Zagreb, Faculty of Electrical Engineering and Computing, Zagreb, Croatia
ivan.markovic@fer.hr

Abstract. Validation and testing based on simulated scenarios is at the heart of the automotive industry for all vehicle development phases, since it enables perfect repeatability of the experiments and control of certain parameters. However, simulations alone can hardly capture all the complexities of the real world, thus true driving scenarios also represent an indispensable part of the process. Although invaluable, they offer very little freedom in changing the parameters, which motivates approaches for automated conversion of real-world driving scenarios to so-called logical scenarios, which can offer higher abstraction level. To be able to perform the complex process of converting real-world driving data, primarily it is necessary to be able to perform vehicle motion classification. For that purpose, this paper proposes and analyzes five different neural network models. The networks were trained and evaluated on a custom generated dataset to classify lateral vehicle behaviours in three main classes with respect to road lanes: lane keep, lane change right and lane change left. The dataset represents highway driving scenarios on a road with 7 lanes in the curvilinear coordinate system. Model training and evaluation was performed on four different subsets, each of them having a different signal-to-noise ratio. In the end, the best overall result was achieved with the network model composed of a bidirectional long-short term memory and multi-scale convolutional neural network layers.

Keywords: Lane change classification · Logical scenarios · Deep learning · Highway driving scenarios

1 Introduction

Scenario based simulation testing [1] and validation [2] represent standard practice in automotive industry for almost all phases of vehicle development [3]. The main motivation behind this lies in the ability to make a large number of tests with perfect repeatability and, depending on the abstraction level, with

I. Petrovic et al. (Eds.): IAS 2022, LNNS 577, pp. 778–789, 2023.
https://doi.org/10.1007/978-3-031-22216-0_52

parameter changes to emphasize certain critical situations. There are two possible paths for generating driving scenarios: (i) manually from prior knowledge and experience, or (ii) from recorded real world driving as described in Menzel et al. [4]. The generated scenarios from real world data are much more valuable for testing and validation, but their abstraction level allows only the replay of the exact same scenario, hence the name *replay* scenarios or *concrete* scenarios [4]. In other words, it is not possible to change any parameters, like the vehicle velocity or the time instant of the started lane change. On the other hand, high level abstraction scenarios, know as *logical* scenarios [4,5] are usually the result of manually generated scenarios and as such allow most, if not all, parameters to be changed. In order to increase the abstraction level of replay scenarios, it is necessary to have the capability to classify all vehicle motions, and based on those classification results recreate scenarios from the ground up. Manual conversion from a replay scenario to a logical scenario is possible, but the process is slow and with an increased number of scenarios it becomes impractical, thus automated scenario generation becomes a necessary step in the process.

The terms *scene*, *situation* and *scenario*, were defined within the automotive field by Ulbrich et al. [6], while the aforementioned concrete and logical scenarios were introduced by Menzel et al. [4]. Additionally, authors in [7] define a *data-driven* approach for generating driving scenarios for testing and validation with ISO 26262 [3] as the reference. Abstract scenario description with sequential acts and event triggering was proposed by Bash et al. [8] and has been used widely in both academic and industry applications, while a more mathematical description was provided by Andreotti et al. [9]. As a solution to the problem of how to generate a large number of testing scenarios, Zhao et al. [10] suggested a conversion of real world driving data through automated classification algorithms based on big data and ontology [11]. Practical solutions to automated scenario generations are mostly very specialized, and depend on the used methods, type and number of vehicle motions recognizable by the method, and format of the input and output data. For example, Lages et al. [12] and Made el al. [13] base their automated conversion on laser sensors. In [14–17] vehicle motion classification has been mostly based on some form of a machine learning algorithm. Other than learning methods, some authors like Hülnhagen et al. [18] used fuzzy logic, while a hybrid method with neural networks and fuzzy logic has been presented by Lin et al. [19]. Sonka et al. [20] used probabilistic and fuzzy classification achieving combined accuracy of up to 95%.

This paper compares the performance of five different neural networks for the problem of classifying lateral vehicle motion. With the assumption that the vehicle is driving in a highway scenario and that the vehicle position is given in the curvilinear coordinate system. Such interpretation of input data allows to neglect most, if not all, dependencies on the scenario recording process, as long as the recorded data can afterwards be successfully converted to the assumed format. Using only lane change based labels for lateral behaviour allows the output data to be used in most driving simulators. The performance of the five different neural networks was analyzed on a custom generated dataset and defined three main classes with respect to road lanes: lane keep, lane change right and lane change left. Final results showed that the best overall score was achieved with the network model composed of bidirectional long-short term memory and multi-scale convolutional neural network layers.

2 Problem Description

To achieve a more generalized level of input data and an increase the ease of use, the input data format is defined in the shape of an array that consists of vehicle trajectory and road lane coordinates in the curvelinear coordinate system [21]. The curvilinear coordinate system, widely used both in related work and industry, is a road reference curve based coordinate system SD where S is position on the road curve and D is lateral deviation from the road centre curve. This coordinate system allows an easier description of vehicle movement, the ability of easy target road change and the manipulation of vehicle trajectory in the position and time domains, respectively. Conversion from and to curvilinear coordinates is done by referencing the vehicle relative position to the road centerline, which is usually stored in a separate road description file. Road lane lines can also be represented in curvilinear coordinates as an array of lateral distances D, calculated with regard to the vehicle's specific position on road S. The size of the input array $(m + 2) \times n$ depends on the number of samples n and the number of lane lines m. This approach enables importing many different types of recorded data as long as they can be converted to this specific, but simple format. An example of the data can be seen in Fig. 1a. Additionally, for faster learning and convergence, the input data is normalized so that it is in range $[0, 1]$ by the expression (1), where a_i represents a single value from the array and a represents the whole array.

$$a_i = \frac{a_i - \min(a)}{\max(a) - \min(a)}. \tag{1}$$

A prerequisite for generating logical segments from trajectory data is classified data, usually both in the lateral and the longitudinal direction and the process is illustrated in Fig. 1b. Simply put, it is necessary to have the capability to recognize the exact moment when the vehicle initiates and finishes the lane change, and this paper covers lateral classification with three possible classes: lane keep (0), lane change right (1), and lane change left (2). For simple highway driving scenarios the longitudinal classes can be described in between lateral classes with time, start velocity, and end velocity. The chosen classification is a simplification of the one described in [11,14,20], and they closely match the scenario description language of the dSPACE ASM ModelDesk [22] that was used for scenario generation in this paper.

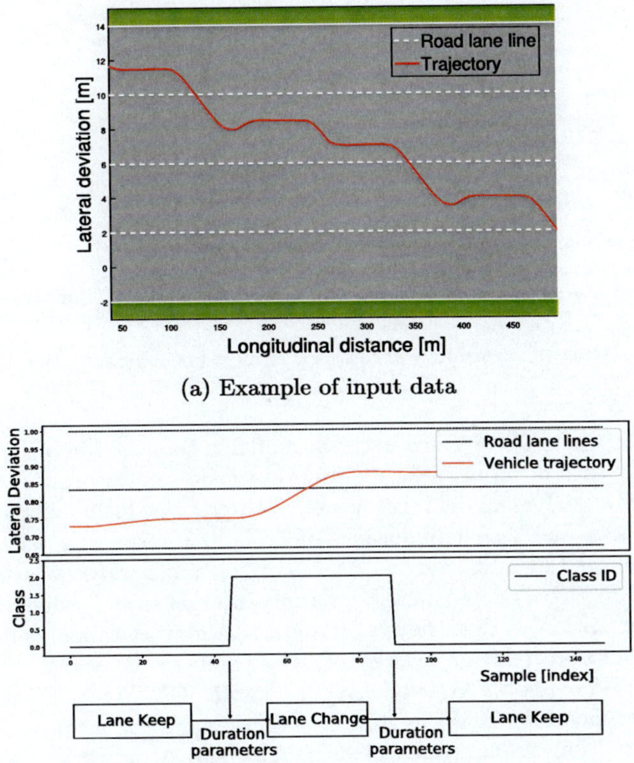

(b) Logical scenario conversion illustration

Fig. 1. Vehicle motion classification problem for automated logical scenarios.

3 Proposed Neural Network Models for Lane Change Classification

Driven by the work done in vehicle motion classification using neural networks [23], in this section five different neural network models with different levels of complexity are proposed. This was done in order to determine which model is suitable for this specific classification task. The tested models range from simple fully connected (FC) and convolutional neural networks (CNNs), to combinations of convolutional and long-short term memory (LSTM) network layers, and in the following each of them is presented.

Model 1: Simple FC. A Fully Connected (FC) network [24] is an introductory network for simple classification problems. The proposed network model is composed of 4 fully connected layers and a softmax layer, where FC layers are composed of the following neuron numbers per layer (256, 128, 64, 32) as shown in Fig. 2a. The main advantage of the FC neural networks is that no special assumptions about the structure of the input data is needed.

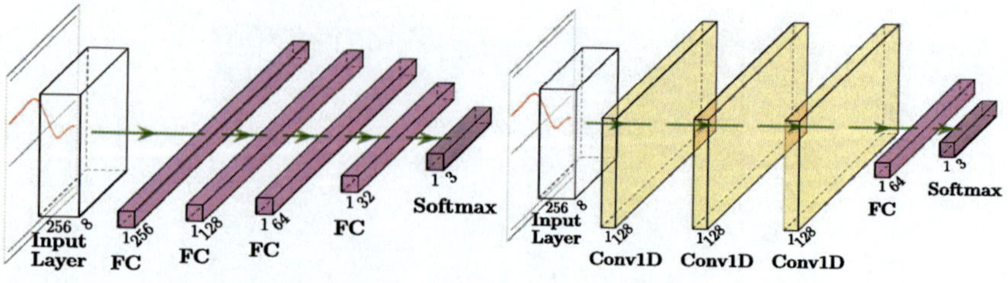

(a) Model 1: Simple FC (b) Model 2: Simple CNN

Fig. 2. Two straightforward neural network models for vehicle motion classification.

Model 2: Simple CNN. This network model is proposed in order to consider the spatial nature of input data, as they represent vehicle position in reference to road lane lines. The model is composed of three convolutional layers each containing 32 filters and kernel dimensions of 7, 5 and 3, respectively. Convolutional layers are followed by a FC layer with 64 neurons, and a softmax layer for output as shown in Fig. 2b. The advantage of convolutional layers, which are inspired by visual perception, is that they take tensors as input data and can understand spatial relations (correlation between nearby pixels in the image) [25].

Model 3: CNN + LSTM. Models that combine CNN and LSTM layers were initially developed for visual time series prediction and the purpose of generating textual descriptions from sequences of images [26]. A network model based on a combination of convolutional layers with LSTM layers is proposed because of its ability to tackle spatio-temporal information in input data. Configuration of a simple CNN model is extended with an LSTM layer with 32 neurons inserted after 3 convolutional layers and before the FC layer, and the output softmax layer, as shown in Fig. 3a.

Model 4: Simple Bi-LSTM. This model was chosen to see how a temporal model deals with structured input data and how structured input data will influence the model's classification capability [16]. Composition of the network is based on two bidirectional LSTM (Bi-LSTM) layers with 64 cells, accompanied by a FC layer with 64 neurons and a softmax layer as output, as shown in Fig. 3b. The model based on Bi-LSTM is capable of running input data in both directions, which allows to preserve information from both past and future.

Model 5: Bi-LSTM + MSCNN. This network model was taken from [16] in which it was applied for driver behaviour recognition and it achieved accuracy of 85.27%. According to [16], the advantage of the Bi-LSTM over LSTM in this application is: "Compared to vanilla LSTM, bidirectional LSTM (Bi-LSTM) architecture can use the bidirectional information of the trajectory which prompts us to introduce a Bi-LSTM module to model the dynamic evolution of trajectories.". The structure of the model is composed of two main branches which are a Bi-LSTM and multi-scale CNN (MSCNN) followed by a FC layer with 32 neurons, and a softmax output layer. The Bi-LSTM module is composed of two LSTM layers each containing 64 cells, while the MSCNN branch has three parallel convolutional layers, accompanied by a FC layer with 64 neurons. Convolutional layers each consist of 32 filters and kernel sizes of 7, 5 and 3, respectively (Fig. 4).

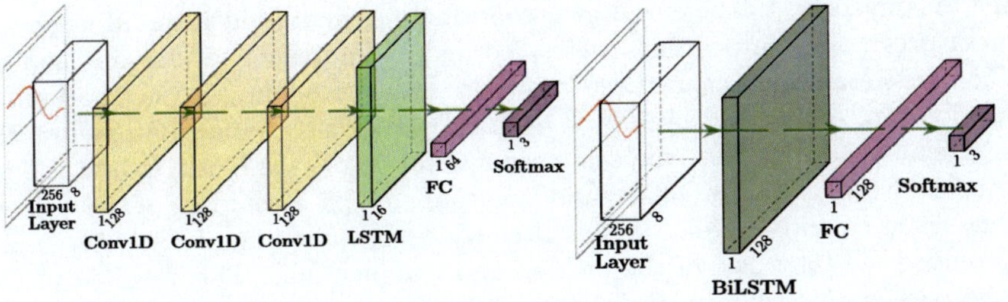

(a) Model 3: CNN + LSTM

(b) Model 5: Simple Bi-LSTM

Fig. 3. Neural network models using convolutional and LSTM layers and a bidirectional LSTM model.

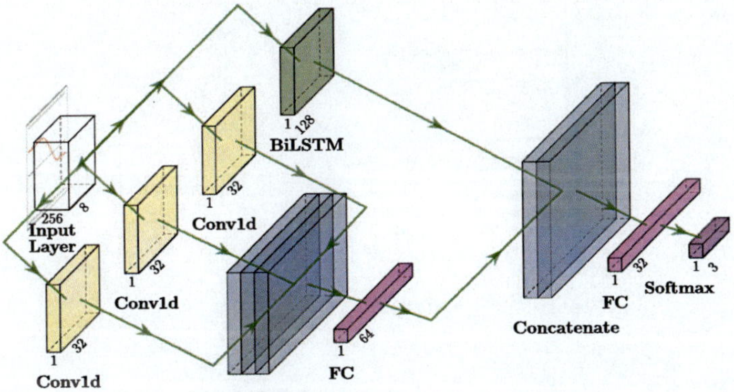

Fig. 4. Model 5: Bi-LSTM + MSCNN.

4 Training and Evaluation

4.1 Dataset Generation

In order to train and evaluate the aforementioned neural network models, a custom dataset was created using dSPACE ASM ModelDesk simulation Toolbox [22]. Although there are trajectory based datasets, such as HighD [27], they lack labels for lane changes and this work can be seen as a step in development of automated annotation of such real world data. Since the goal was to generate a large amount of diverse scenarios, simulated logic scenarios were run in the loop while changing parameters for each loop pass. Base scenarios consisted of multiple lane keep and lane change segments stacked one after the other, where the base scenario is shown in Fig. 5. The road was set as a straight highway with 7 lanes all going in the same direction and curvatures were unnecessary due to conversion to the curvilinear coordinate system (which would neglect all curvatures). Vehicle positions were recorded as lateral deviations on the road. The parameters were changed in two ways: either randomly by using a uniform distribution, or linearly by growing with a preset step value. Parameters that were changed during the simulation were: duration of each block, lateral and longitudinal velocity, lateral offset, and lane change type. Recording of the data was done with a sample rate of 0.02 s, after which subsampling of 1:40 was applied to reduce the total size of the dataset and training time. This resulted in 2912 driving scenarios where vehicles drove 1002 km, in duration of more than 20 h with 975 left lane changes and 1145 right lane changes.

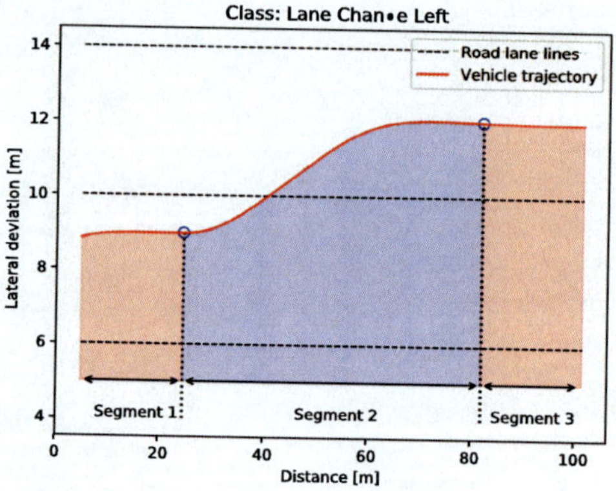

Fig. 5. Base scenario for dataset generation.

The generated dataset was divided in a 80:20 ratio, where 80% was used for learning and 20% for evaluation. To enrich the dataset, it was further augmented with white noise based on the following expression 2 and shown in Fig. 6:

$$SNR = 10\log_{10}\left(\frac{P_{signal}}{P_{noise}}\right). \tag{2}$$

White noise was applied on the training dataset in steps of signal to noise ratios of 70, 50 and 30 dB, to simulate a more realistic input data. These values ware chosen as representative examples of noise distributions. During the training of models, noise was progressively increased starting with training on data without

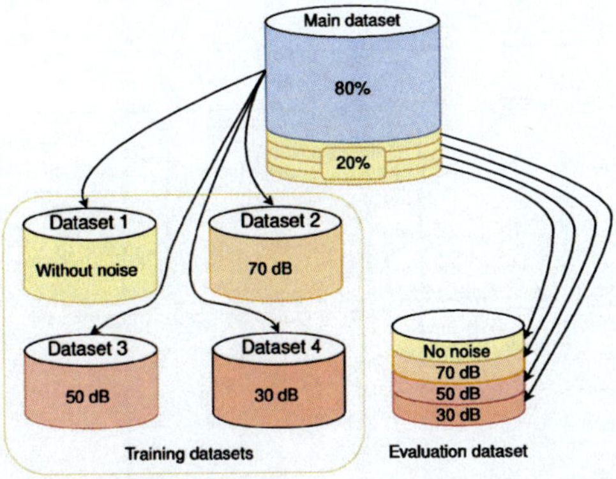

Fig. 6. Division of the dataset for training.

any noise and ending with fully noisy data at 30 dB signal-to-noise ratio as shown in Fig. 7.

Evaluation was compiled from four non-overlapping chunks of data with applied different levels of noise, starting without any noise and ending with 30 dB signal-to-noise ratio, same as the last part of training data. Adam [28] was used as the training optimizer based on stochastic gradient descent since it is well suited for problems that are both large in data and the number of parameters. The following metrics for evaluation were used: accuracy calculated as frequency of prediction matching labels, categorical cross-entropy for the loss function, and confusion matrix macro parameters: True positive (TP), True negative (TN), False positive (FP) and False negative (FN).

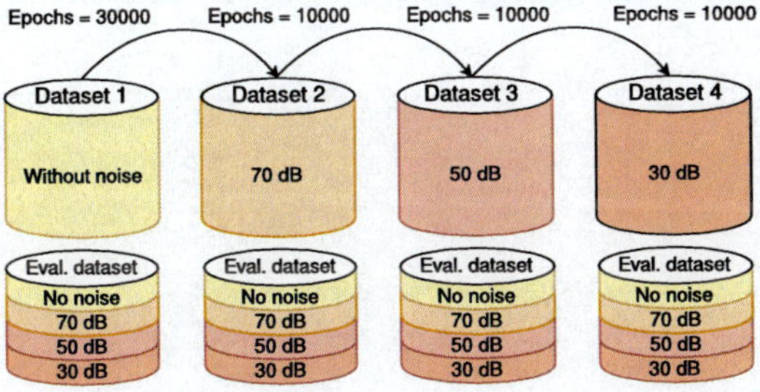

Fig. 7. The dataset in the training process.

Table 1 shows the evaluation results, from which it can be see that the best overall score was achieved by the Bi-LSTM + MSCNN model. This is most likely due to its ability to combine spatio-temporal features of backward and forward temporal looking capability of Bi-LSTM and spatial characteristics of MSCNN module. The worst performing model was the simple FC because it makes no

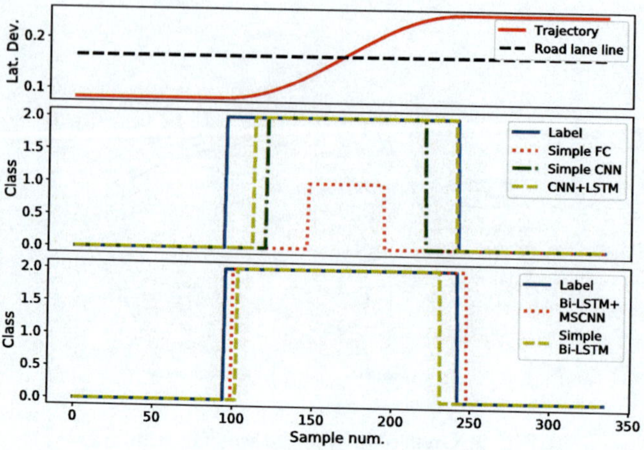

(a) Lane Change Left

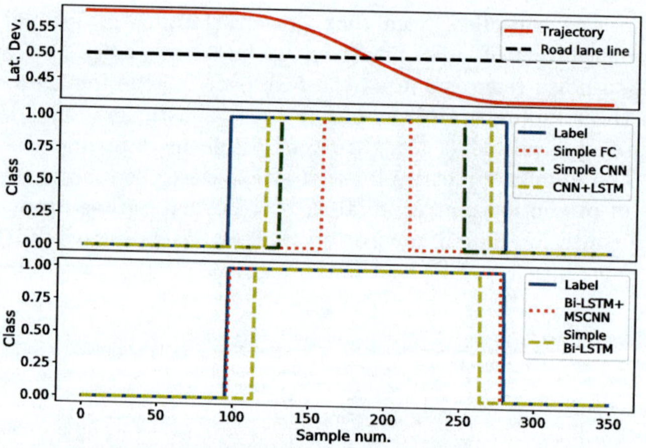

(b) Lane Change Right

Fig. 8. Evaluation results.

assumptions about the spatio-temporal correlation of the input data, followed by the simple CNN whose performance was significantly better but does not have the capability to learn temporal aspects of data. CNN + LSTM model showed initial poor performance, but gradually got better as it learned on noisy data. That is in correspondence with results from [29], which states that LSTM is less susceptible to perturbation in data if noise is present. This model could provide better results if LSTM layer is replaced by Bi-LSTM which would make it backward and forward looking capable. The simple Bi-LSTM model, that does

Table 1. Evaluation results of the analyzed network models on the custom lane change detection dataset.

Model name	Value	Without noise epoch = 30,000	Noise 70 dB epoch = 40,000	Noise 50 dB epoch = 50,000	Noise 30 dB epoch = 60,000
Simple FC	Accuracy	74.4889%	74.0898%	74.6113%	74.3476%
	Loss	0.6881	0.6923	0.6911	0.6952
	TP	0.44	0.44	0.44	0.4333
	TN	1.4366	1.44	1.4399	1.4333
	FP	0.5566	0.56	0,56	0.5667
	FN	0.5566	0.56	0,56	0.5667
Simple CNN	Accuracy	72.0325%	75.4199%	88.1666%	88.9023%
	Loss	0.6589	0.6963	0.3227	0.3106
	TP	0.38	1.1933	0.8233	0.8
	TN	1.3799	1.06	1.82	1.8
	FP	0.62	0.42	0.1733	0.2
	FN	0.62	0.3184	0.1733	0.2
CNN + LSTM	Accuracy	79.1458%	90.1328%	90.513%	90.9863%
	Loss	0.5148	0.3038	0.2712	0.2586
	TP	0.7333	0.89	0.86	0.8533
	TN	1.7333	1.8867	1.86	1.8566
	FP	0.2667	0.1066	0.1399	0.1499
	FN	0.2667	0.1066	0.14	0.1499
Bi-LSTM + MSCNN	Accuracy	**96.6529%**	94.8645%	**97.2591%**	**97.2929%**
	Loss	0.1049	0.1375	0.0852	0.0976
	TP	0.9667	0.94	0.9667	0.9667
	TN	1.9667	1.9399	1.9667	1.9667
	FP	0.0333	0.06	0.0333	0.0333
	FN	0.0333	0.0599	0.0333	0.0333
Simple Bi-LSTM	Accuracy	94.7434%	**96.9804%**	93.1367%	93.3509%
	Loss	0.1401	0.1212	0.2233	0.1458
	TP	0.9367	0.96	0.92	0.9067
	TN	1,94	1.9533	1.9167	1.9067
	FP	0.0667	0.04	0.08	0.0933
	FN	0.0666	0.04	0.2899	0.0933

not include the MSCNN part, has initially showed good accuracy until data with larger noise was introduced. Figure 8 shows the results of classification for all models, after having completed the whole learning process (60,000 epochs), on the typical examples of lane change left and lane change right without added noise. According to Table 1, which contains five different architectures of neural networks, the final architecture recommendation would be: series of convolutional layers whose outputs are combined with input and feed into LSTM and then into series of fully connected layers followed by softmax.

5 Conclusion

This paper investigated lane change classification with neural networks for automated logical scenario conversion. The lane change classification problem was categorized into three classes of lateral vehicle motion: lane keep, lane change right, and lane change left. This classification problem was tackled by five different deep learning models with varying complexity. Neural network models were trained and evaluated on a custom dataset that was generated using the dSPACE ASM ModelDesk. The resulting dataset was composed of 7 road lane lines and lateral deviation of the vehicle on the road. The road was set as a straight highway going in one direction, as curvelinear coordinates were used, which neglects all curvature features of the road. Best overall results were achieved with the network model composed of Bi-LSTM and MSCNN layers, achieving accuracy of up to 97.2%, due to its ability to learn spatio-temporal features of input data.

References

1. Berger, C., Block, D., Hons, C., Kühnel, S., Leschke, A., Plotnikov, D., Rumpe, B.: Large-scale evaluation of an active safety algorithm with EuroNCAP and US NCAP scenarios in a virtual test environment – an industrial case study. In: 2015 IEEE 18th International Conference on Intelligent Transportation Systems, pp. 2280–2286 (2015)
2. Pütz, A., Zlocki, A., Bock, J., Eckstein, L.: System validation of highly automated vehicles with a database of relevant traffic scenarios (2017)
3. ISO: 26262: Road Vehicles: Functional Safety, Geneva, Switzerland (2016)
4. Menzel, T., Bagschik, G., Maurer, M.: Scenarios for development, test and validation of automated vehicles (2018)
5. Wagatsuma, H.: Logical scenarios and coverage analyses enhanced by a representative trajectory model to reduce test cases to limited combinations. In: 2018 Joint 10th International Conference on Soft Computing and Intelligent Systems (SCIS) and 19th International Symposium on Advanced Intelligent Systems (ISIS), pp. 734–738 (2018)
6. Ulbrich, S., Menzel, T., Reschka, A., Schuldt, F., Maurer, M.: Defining and substantiating the terms scene, situation, and scenario for automated driving. In: 2015 IEEE 18th International Conference on Intelligent Transportation Systems, pp. 982–988 (2015)
7. Menzel, T., Bagschik, G., Isensee, L., Schomburg, A., Maurer, M.: From functional to logical scenarios: detailing a keyword-based scenario description for execution in a simulation environment, pp. 2383–2390 (2019)
8. Bach, J., Otten, S., Sax, E.: Model based scenario specification for development and test of automated driving functions. In: 2016 IEEE Intelligent Vehicles Symposium (IV), pp. 1149–1155 (2016)
9. Andreotti, E., Boyraz, P., Selpi, S.: Mathematical definitions of scene and scenario for analysis of automated driving systems in mixed-traffic simulations. IEEE Trans. Intell. Veh. **6**(2), 366–375 (2021)
10. Zhao, S., Wu, Y., Zhu, X., Zhou, B., Bao, H., Zhang, L.: Research on automatic classification for driving scenarios based on big data and ontology. In: 2018 IEEE 4th International Conference on Computer and Communications (ICCC), pp. 1857–1862 (2018)
11. Bagschik, G., Menzel, T., Maurer, M.: Ontology based scene creation for the development of automated vehicles (2018)
12. Lages, U., Spencer, M., Katz, R.: Automatic scenario generation based on laserscanner reference data and advanced offline processing. In: 2013 IEEE Intelligent Vehicles Symposium (IV), pp. 153–155 (2013)
13. Made, R.V.D., Tideman, M., Lages, U., Katz, R., Spencer, M.: Automated generation of virtual driving scenarios from test drive data (2015)
14. Beglerovic, H., Schloemicher, T., Metzner, S., Horn, M.: Deep learning applied to scenario classification for lane-keep-assist systems. Appl. Sci. **8**, 1 (2018)
15. Hartjen, L., Philipp, R., Schuldt, F., Howar, F., Friedrich, B.: Classification of driving maneuvers in urban traffic for parametrization of test scenarios (2019)

16. Zhang, H., Nan, Z., Yang, T., Liu, Y., Zheng, N.: A driving behavior recognition model with bi-LSTM and multi-scale CNN. In: 2020 IEEE Intelligent Vehicles Symposium (IV), pp. 284–289 (2020)
17. Woo, H., Ji, Y., Kono, H., Tamura, Y., Yamashita, A., Asama, H.: Detection method of lane change intentions in other drivers using hidden Markov models. The Abstracts of the International Conference on Advanced Mechatronics: Toward Evolutionary Fusion of IT and Mechatronics: ICAM 2015, pp. 253–254 (2015)
18. Hülnhagen, T., Dengler, I., Tamke, A., Dang, T., Breuel, G.: Maneuver recognition using probabilistic finite-state machines and fuzzy logic. In: 2010 IEEE Intelligent Vehicles Symposium, pp. 65–70 (2010)
19. Lin, Y.-C., Thi Nguyen, H.-L., Wang, C.-H.: Adaptive neuro-fuzzy predictive control for design of adaptive cruise control system. In: 2017 IEEE 14th International Conference on Networking, Sensing and Control (ICNSC), pp. 767–772 (2017)
20. Sonka, A., Krauns, F., Henze, R., Küçükay, F., Katz, R., Lages, U.: Dual approach for maneuver classification in vehicle environment data. In: 2017 IEEE Intelligent Vehicles Symposium (IV), pp. 97–102 (2017)
21. Kim, J., Jo, K., Lim, W., Lee, M., Sunwoo, M.: Curvilinear-coordinate-based object and situation assessment for highly automated vehicles. IEEE Trans. Intell. Transp. Syst. **16**(3), 1559–1575 (2015)
22. dSPACE GMbH: dspace asm. https://www.dspace.com/en/inc/home/products. Accessed 26 Jan 2021
23. Khosroshahi, A., Ohn-Bar, E., Trivedi, M.M.: Surround vehicles trajectory analysis with recurrent neural networks. In: 2016 IEEE 19th International Conference on Intelligent Transportation Systems (ITSC), pp. 2267–2272 (2016)
24. Tamura, S., Tateishi, M.: Capabilities of a four-layered feedforward neural network: four layers versus three. IEEE Trans. Neural Netw. **8**(2), 251–255 (1997)
25. Gu, J., Wang, Z., Kuen, J., Ma, L., Shahroudy, A., Shuai, B., Liu, T., Wang, X., Wang, L., Wang, G., Cai, J., Chen, T.: Recent advances in convolutional neural networks (2017)
26. Donahue, J., Hendricks, L.A., Rohrbach, M., Venugopalan, S., Guadarrama, S., Saenko, K., Darrell, T.: Long-term recurrent convolutional networks for visual recognition and description (2016)
27. Krajewski, R., Bock, J., Kloeker, L., Eckstein, L.: The highD dataset: a drone dataset of naturalistic vehicle trajectories on German highways for validation of highly automated driving systems. In: 2018 21st International Conference on Intelligent Transportation Systems (ITSC), pp. 2118–2125 (2018)
28. Kingma, D.P., Ba, J.: Adam: a method for stochastic optimization (2017)
29. Yeo, K.: Short note on the behavior of recurrent neural network for noisy dynamical system. CoRR arXiv:1904.05158 (2019)

Trajectory Analysis in a Lane-Based UAS Traffic Management System

David Sacharny, Thomas C. Henderson(✉), and Vista Marston

University of Utah, Salt Lake City, UT, USA
sacharny@cs.utah.edu, tch@cs.utah.edu, u0923256@utah.edu

Abstract. The development of Advanced Air Mobility (AAM) systems require not only a robust and safe approach to planning flights, but also a way to monitor Unmanned Aircraft Systems (UAS) flights in real-time to determine whether flights are deviating from their nominal flight paths or if there are rogue (i.e., unplanned) flights in the area. We have proposed a lane-based airways methodology for lane creation, scheduling and strategic deconfliction, and here we describe *Nominal vs. Anomalous Behavior* (NAB) an efficient and effective way to monitor flight trajectories to determine normal versus anomalous behavior.

Keywords: Air Traffic Operations Center · Utah Department of Transportation · Advanced air mobility · Nominal vs. anomalous behavior · Rogue flight detection

1 Introduction and Background

The AAM community, including UAS service providers, operators and relevant government authorities, aims to provide a wide number of services (e.g., package delivery, air taxi, etc.) by means of robust and safe UAS Traffic Management (UTM) systems which achieve large-scale (i.e., thousands per day) operations in urban areas without human control, but with reliable communications and contingency planning (see [1]). UAS Service Suppliers, for example AirMap [2] have dealt with the operational interfaces, integration of Geographic Information Systems (GIS), registration of flights, UAS communications, and monitoring UAS activity. These capabilities are all developed to operate as described in the strategic deconflicton context defined by NASA [3] which is defined in terms of a geographic grid (a set of cells). Each new flight must be deconflicted pairwise in terms of grid cells which have other scheduled flights. The trajectory of a UAS flight is a curve in 4-dimensional space (x, y, z, t). Given a set of such curves, strategic deconfliction (i.e., make sure that no two flights are ever within a specified minimum distance called *spatial headway*) necessitates a pairwise comparison of the curves with motion constrains, and this is in general PSPACE hard. Moreover the FAA and NASA have yet to specify contingency protocols, and some research suggests that all flights simply return to base in these scenarios (e.g., lost-link).

I. Petrovic et al. (Eds.): IAS 2022, LNNS 577, pp. 790–816, 2023.
https://doi.org/10.1007/978-3-031-22216-0_53

In previous research, we have proposed and studied various aspects of a lane-based approach [4–9]. This lane-based approach reduces strategic deconfliction complexity (to 1D from 4D) and makes the handling of contingencies a spatially local problem [10]. The use of lanes for commercial flights (Victor and Jet Routes) has a long-standing history [11]. However, human air traffic controllers manage commercial airway lanes, and this management function which must be automated if a large number of autonomous flights are to take place daily over major metropolitan areas.

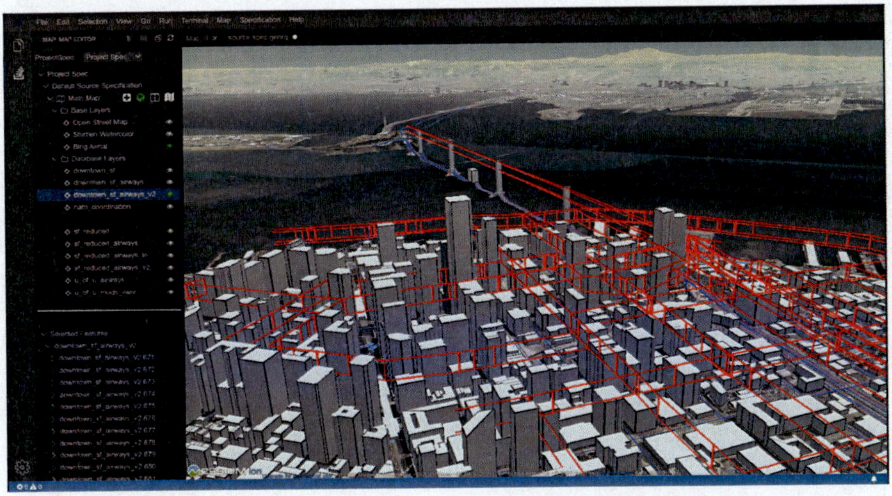

Fig. 1. Lane-based airway lanes above downtown San Francisco, CA.

Given a lane-based UTM system, lanes are created as a static structure (much like ground road networks) and all scheduled flights will follow a sequence of assigned (reserved) lanes from launch to landing. Figure 1 shows a lane-based airway over downtown San Francisco, CA. If all UAS flights performed in the lane-based system as planned, and no unexpected events occurred (e.g., unplanned flights entering), then there would be no need to monitor the airspace. Since unexpected things do happen, some sort of traffic management center is necessary to help mitigate the effects of unforeseen events. In addition, an air traffic control operations center (ATOC) can serve to apply UTM policies and enforce the rules as necessary. ATOC operators can monitor activity in the lanes and use existing conditions as part of the flight plan approval process. For example, if there are strong winds in part of the airspace, the ATOC can provide advisories and route flights away from the impacted area. In addition, the ATOC can gather airway performance data and modify policies accordingly to improve things.

It is also possible that existing ground transportation infrastructure can be leveraged to support ATOC operations. For example, if road intersections already provide power (say for cameras at traffic lights) and network access, then upward looking radars can be placed there to observe the airways. It may also be useful to have close coördination of the ATOC and ground traffic management operations, especially if UAS fly over roads. This allows rapid dissemination of disruptive events like an emergency landing on a road.

It is most likely that ATOC's will improve transportation performance. Certainly, this has been the case with ground transportation management systems (GTMS). For example, the TransStar system in Houston, Texas helped reduce traffic delay by over eleven million hours per year. In Utah, the CommuterLink system decreased intersection delay by 27% and increased freeway speed by 20% (source: Texas A&M Transportation Institute).

A full-blown GTMS for a large metropolitan area can be a costly proposition, and the cost for an ATOC would presumably be similar. A ball-park cost projection given by [12] is about $ 3M for a building, and another $ 3M annually for operation and maintenance. Of course, monitoring or provisioning hardware infrastructure imposes more cost. The same study gives a Business Plan Development Guide which shows the steps necessary to develop a Traffic Management Business Plan (see Fig. 2). Planning for an ATOC would have a similar set of considerations. However, one major additional requirement would be for some form of interaction with both the FAA and also directly with UAS flights. Working with the FAA is necessary to maintain the safety of simultaneous manned and unmanned air activity. Communication with UAS operators in necessary to receive real-time telemetry data from UAS flights.

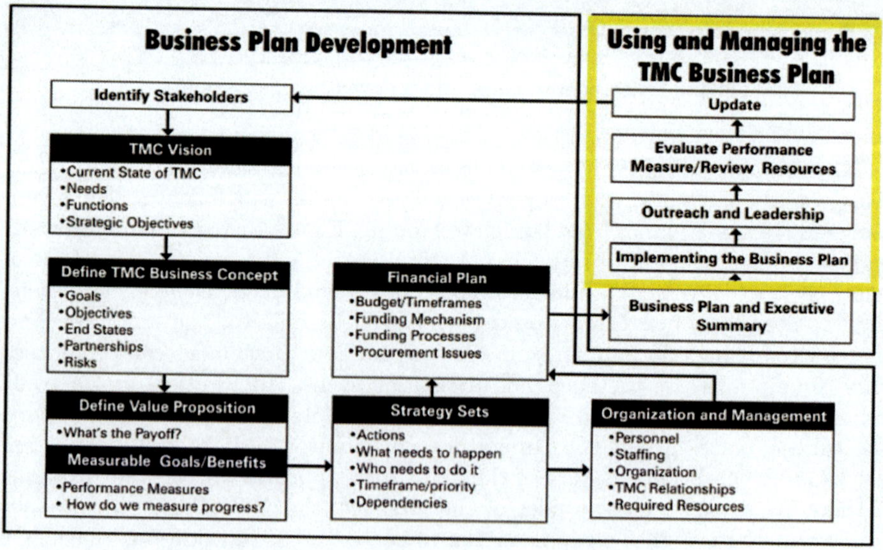

Fig. 2. ATOC business plan (from [12]).

1.1 Example ATOC Requirements by Utah Department of Transportation

The Utah Department of Transportation[1] (UDOT) has been studying the issues of Urban Air Mobility since 2019 with their prime focus on the airspace of the Provo to Ogden corridor. The main two use cases considered here are small package delivery and air taxis. A major goal is to exploit UAS to reduce ground traffic congestion. Other pertinent factors include the push by industry (e.g., Airbus, etc.) to have full integration of autonomous vertical take off and landing vehicles for human transportation.

The hardware infrastructure deemed necessary for a reliable urban airway system includes:

- Automatic Dependent Surveillance-Broadcast (ADS-B) with links to unmanned aircraft,
- GPS Real-Time Kinetics (RTK) for precision vertical operations,
- Dedicated Short-Range Communications (DSRC) for seamless connectivity, and
- Active Radar to provide detection and avoidance capabilities.

Figure 3 shows the radar and other coverage proposed by UDOT. Figure 4 shows the GPS coverage network and example hardware. Figure 5 shows the current ATMS and fiber network capable of supporting UAM in the main Salt Lake Valley corridor. Finally, UDOT proposed the acquisition of (1) several more GPS-RTK reference stations for safety redundancy, (2) Mini-radar/ADS-B with an estimated 40 stations needed covering about $1267\,\text{km}^2$ with power and network connections already available through UDOT, and (3) DSRC wireless. The estimated cost for this infrastructure system was estimated as about \$ 200K for GPS reference stations and \$ 12.7M for mini-radar/ADS-B hardware.

In terms of ATOC UTM capabilities, UDOT established goals to:

- manage low-altitude air traffic,
- provide real-time notification of flight plan changes,
- communicate hazards or priority routing, and
- account for traffic load, impact of skyport placement, weather, rogue and emergency air traffic.

UDOT also proposed that UAM traffic be routed over existing roads and along planned flight corridors with pre-approved flight plans, with assigned altitudes related to direction of flight and with speed and location constraints. Figure 6 shows the proposed air layout. Of course, complete UAM systems will most likely be built up incrementally as technology, demand and resources converge to make it possible. Current UAS platforms require a great deal of improvement and demonstrated robustness in order to meet the stringent safety requirements of air transportation. UAS management software is not yet available, the package and air taxi industries are not yet at scale, and air monitoring and enforcement mechanisms have yet to be developed and demonstrated.

[1] We would like to thank Jared Esselman for this material.

Fig. 3. Radar and other sensing capabilities proposed by UDOT.

Fig. 4. GPS coverage and example hardware.

Thus, the development of Advanced Air Mobility (AAM) systems requires not only a robust and safe approach to planning flights, but also a way to monitor UAS flights in real-time to determine whether flights are deviating from their nominal flight paths or if there are rogue (i.e., unplanned) flights in the area. We have proposed a lane-based airways methodology for lane creation, schedul-

Fig. 5. ATMS and fiber network for Salt Lake Valley airway corridor.

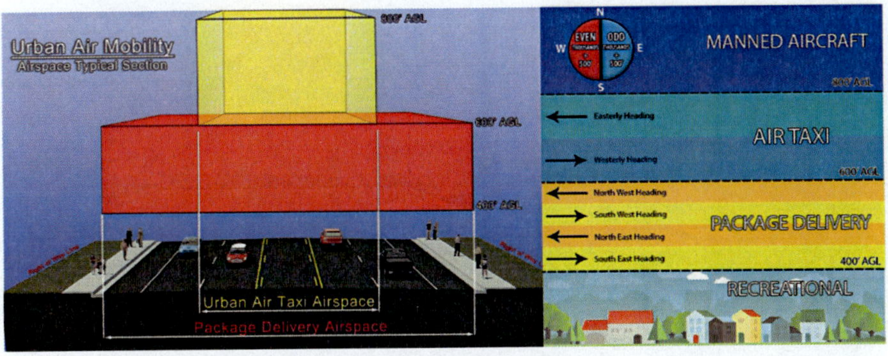

Fig. 6. Proposed airspace layout for the Salt Lake Valley airway corridor.

ing and strategic deconfliction, and here we describe *Nominal vs. Anomalous Behavior* (NAB) an efficient and effective way to monitor flight trajectories to determine normal versus anomalous behavior.

2 Lane-Based Monitoring

The basic problem addressed here is how a lane-based UTM system supports the recognition of rogue flights of a variety of sorts: amateur recreational hobbyists, UAS operators making an unscheduled up, over and down flight, malicious operators, etc. In order to detect such rogue flights, we propose the analysis of trajectories based on their deviation from the lane structure, including both location in space and direction of flight at that location. The basic idea is that a model can be produced directly from the lane structure and compared to any flight individually. The alternative FAA approach would require knowledge of all 4-dimensional flight trajectories, as well as multi-target tracking to monitor the flights along those curves and a comparison of an unidentified flight to all of those curves. Thus, the proposed lane-based method is much more efficient and effective.

2.1 NAB Modeling

Given a set of UTM lanes, a convenient model is just a set of point samples on the lanes, each with an associated direction of the travel in the lane. Figure 7 shows a lane-based airway over the East Bench area of SLC, UT. Figure 8 shows a set of sample points from the lanes. These provide a good model since any nominal flight (i.e., following its assigned lane sequence) should be near a lane and headed in the direction of the (one-way) lane. As part of the model, the direction vectors can also provide significant information about a flight. Figure 9 shows a subset of the trajectory direction vectors used in the model.

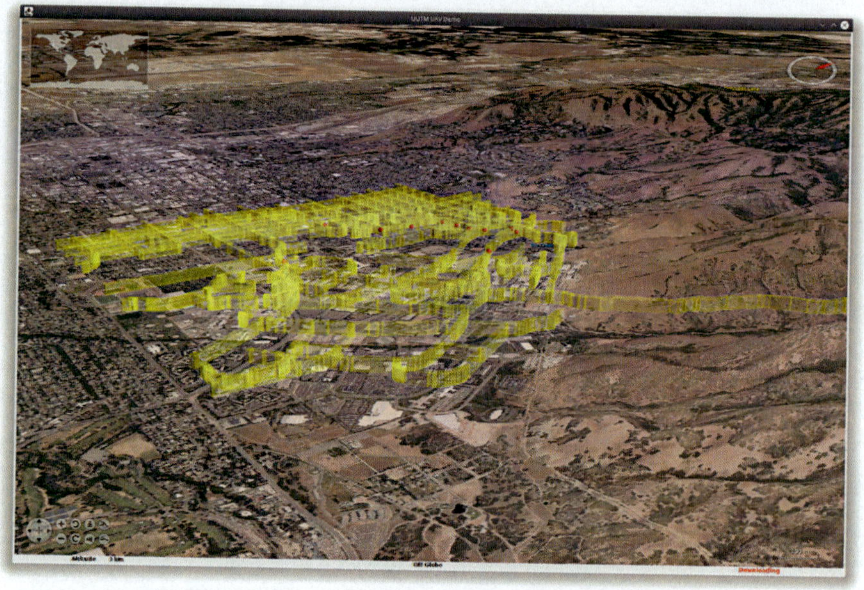

Fig. 7. Set of airway lanes created over Salt Lake City, UT.

The Lane-Based Trajectory Model A trajectory is a sequence of 4-tuples (x, y, z, t) which provides a representation of the actual UAS motion through space and time. Trajectories can be divided into two basic types: *expected* and *unexpected*. An expected trajectory is one which results from a planned flight trajectory perhaps combined with some error due to contextual influences (e.g.,

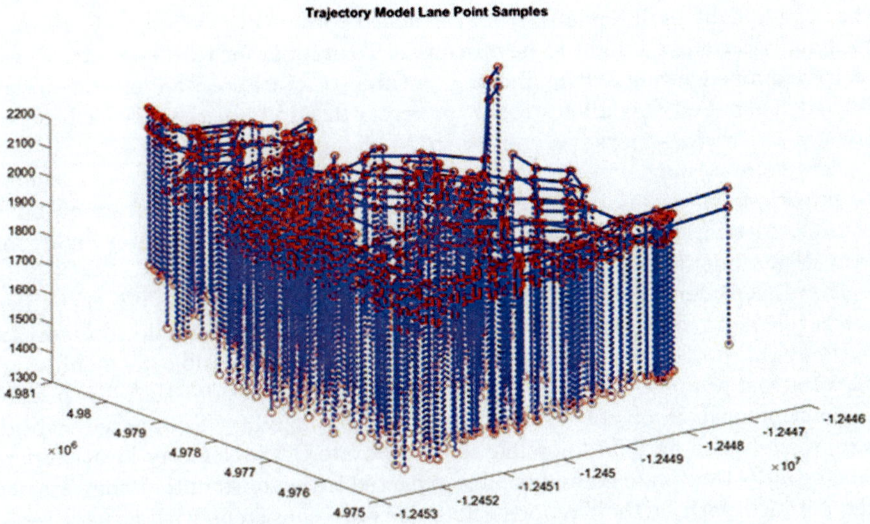

Fig. 8. Trajectory point set model of airway lanes over East Bench of Salt Lake City, UT. Red circles are lane endpoints; blue points are samples along lane.

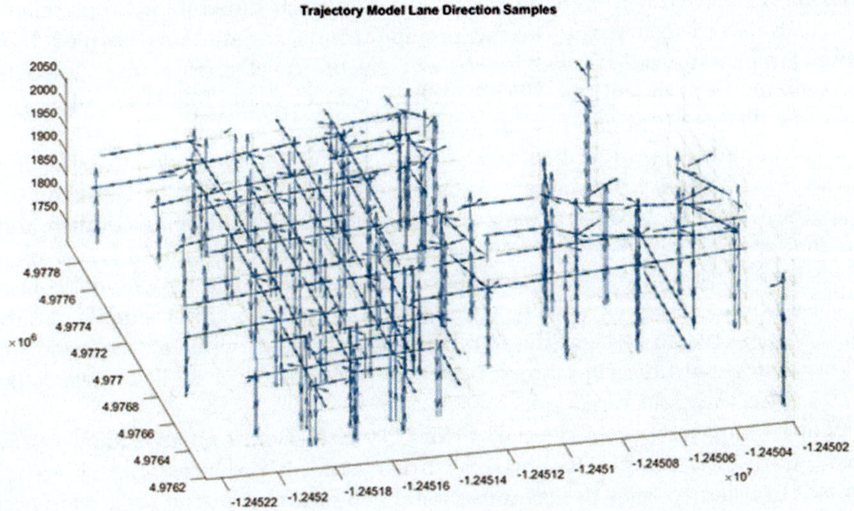

Fig. 9. Trajectory direction vector model of airway lanes over East Bench of Salt Lake City, UT.

weather, power, etc.). An assigned trajectory in terms of the lane structure is a sequence of straight line segments, but when exiting one lane to enter another, the actual flight path deviates from the sharp corner turn. Wind and precipitation may also cause a flight to be off-course. To account for this, lanes are viewed as having an enclosing 3D volume (e.g., a tube structure) containing the straight line segment, and this allows room for error. In addition, the spatial headway constraint provides more cushion for speed changes, etc.

Trajectories may arise from various flights, but a scheduled UAS is required to provide telemetry data at a set update rate; this data comes from on-board sensors. Alternatively, radar or other sensors may be used to monitor flights and provide an independent source for trajectory data.

Next, consider a planned flight and its associated trajectory. When the UAS sends telemetry data, it also sends its ID. This makes it possible to determine if the flight is off-course and by how much. It is also possible to monitor the airspace and produce locations of airborne objects. We assume that it is possible to determine which objects are UAS with high probability (as opposed to birds, etc.). Given such data it is possible to corroborate UAS telemetry location data. The result is that planned flights have expected trajectories in that they are near the planned path or their reported locations are consistent with ground sensor data.

On the other hand, a flight that has not been planned produces a trajectory that corresponds to the type of the flight, and, in general, the trajectory will not correspond to any planned flight, will not be corroborated by sensor data, and will not follow a sequence of connected lanes. This is an unexpected trajectory. Of course, there are ways that UAS can insert themselves into the lane structure and mimic a scheduled flight (e.g., by following a scheduled flight), but this can also be detected in that they do not provide telemetry data. Unscheduled flights in the airspace are called *rogue flights*, whereas unexpected trajectories are called *anomalous*. In practice, the ATOC needs to detect rogue flights as robustly as possible. The NAB method is one such approach.

NAB operates as described in Fig. 10. The lane data is made available along with the UTM policy parameters, and the set of scheduled flights. Based on this, a spatial database is constructed consisting of a set of 3D points sampled along the lane, and to each of these points is associated the direction of travel vector at each of those points (recall that lanes are one-way). The lane model consists of this data organized so as to be efficiently exploited. The inter-sample distance must be selected so to keep the number of points down while at the same time allowing adequate discriminatory power to determine if a flight is near a lane and headed the right direction.

For example, consider the Salt Lake City East Bench airways shown previously. If an inter-sample distance of 2 m is chosen, it produces a set of 454,331 points. In order to keep the computational complexity low, the points are organized as a kd-tree using the 3D points. A kd-tree divides the points at the median of each of the vector dimensions in turn (or picking the dimension with the greatest spread in values in turn), and thus achieves a relatively well-balanced tree

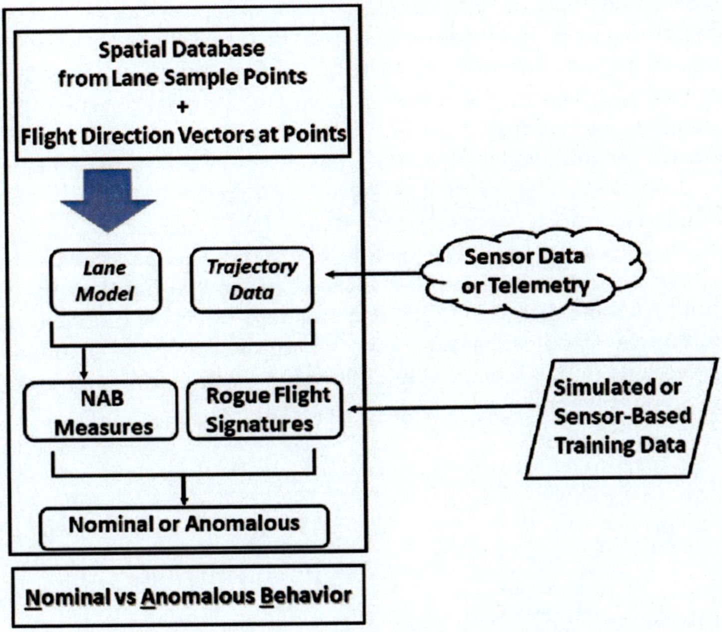

Fig. 10. The nominal vs. anomalous behavior (NAB) method.

which allows $O(log(n))$ search time to find the closest points in the spatial data to a query point.

Any nominal flight should be near one of the sample points and headed in the appropriate direction. Of course, a temporal analysis can be performed by checking the associated Space Time Lane Diagram (see [6]) which specifies the position of each flight in a lane at each time instant. Also, with the FAA-NASA approach, there is no fixed set of lanes, and therefore, every existing flight would require target tracking against the set of all flights. Since this approach requires large-scale sensing and computing resources, it is not considered further here.

Next a set of NAB measures are determined which allow the discrimination of the different types of flights, both nominal and rogue. These are computed either by comparing the UAS trajectory to the lane data, or simply in terms of the trajectory itself. For example, two lane related measures are:

1. M_{dist}: minimum distance to a lane at each time step, and
2. M_{dir}: cosine of the angle between UAS direction of travel and the lane direction of travel at each point.

These measures are applied at each point in the trajectory to produce a temporal signature to represent the flight. An example of a measure based solely on the trajectory data would be the amount of time spent hovering (i.e., staying for some minimal duration in time in one place in space). Given a characterization of the types of flights of interest, then a set of trajectory signature templates can be constructed and used as class models. Such templates can be the result of a set of simulations or produced from data sets of actual flight trajectories. Given a new trajectory, its measured features are compared to the flight signature templates and matched to the closest in order to classify the type of trajectory (i.e., nominal or anomalous).

Consider a nominal flight which does not perfectly follow the lane but rather has some noise associated with it. Figure 11 top row shows the x values of a nominal flight trajectory (with Gaussian noise of 0.16 variance), and a smoothed version of that data (in red). The middle row shows the distance to the closest lane, and the bottom row gives the cosine of the angle between the direction of flight and the lane direction. This distance and direction difference are NAB measures. For the distance measure, over 96% of the trajectory points are within 1 unit of the lane, and for the angle difference measure, 70% are within 10 °C. The large angle differences arise at lane changes.

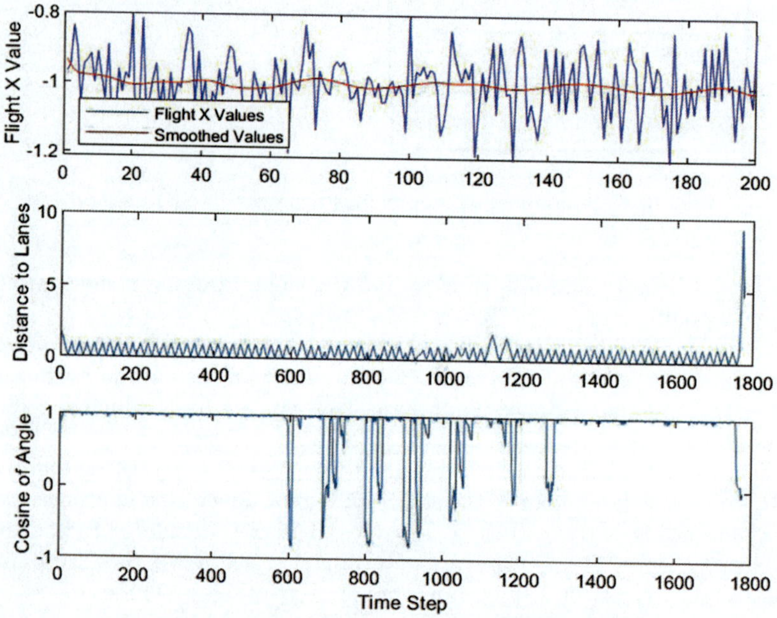

Fig. 11. NAB measures for a nominal flight.

Now consider rogue flights. Five categories are explored here:

- *Hobbyist Type I*: Flies up from one place and makes a few moves above the launch site, then eventually lands at the same site.
- *Hobbyist Type II*: Flies up from one place and makes a few moves above the launch site, hovers after each move, then eventually lands at the launch site.

– *Hobbyist Type III*: Flies up in a circular motion to some highest point then flies down in a circular motion to land.
– *Rogue Type I*: Flies up over and down as for a delivery.
– *Rogue Type II*: Flies up to a lane, flies along the lane to the end, then flies to another lane (not necessarily connected), and eventually flies down to land.

These anomalous flight patterns are representative of the types of flights to be expected. Figure 12 shows an example of these types.

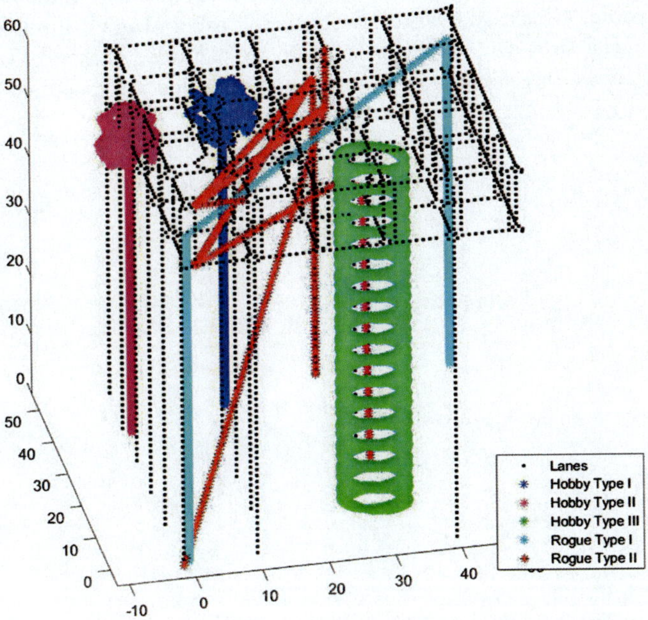

Fig. 12. Examples of the five anomalous flight trajectories.

Sensor Data to Track Rogue Flights Radar is used to scan the airspace lane network so as to make sure that scheduled flights adhere to their flight plans and to detect unscheduled flights. Here we provide an overview of the considerations necessary to employ radar sensors in airspace monitoring. Although the description addresses radar in the context of the simulation system, the actual exploitation of radar systems would require a similar analysis.

In order to analyze flights in the simulation, there first needs to be a monitoring system that can detect UAS flights and report flight information back to the Air Traffic Operations Center (ATOC). The monitoring system that was selected was Radar. The structure and performance of the radar's objects was to imitate real radar systems. One can think of the system as the radar sending signals to detect objects over certain time intervals. If an object is detected, it will be reflected through broadcasting to those that are watching the scanner. This section will discuss the mechanism used to achieve this performance and structure.

Figure 13 shows a single radar field in the grid layout lane-based simulation. One might notice that the radar's detection field is conical in shape as the field progresses through the atmosphere, this is due to the natural occurrence of refraction. In real life this radar field would not contain distinct lines as a result of attenuation, meaning that the radar's signal on the outer edges are weaker than those closer to the center. One can think of it similar to a flashlight shining in a dark room, where the center is the brightest and gets dimmer moving to the outer edges. However, in our simulation we excluded these weaker areas and created distinct edges for simpler computations which are discussed later in the chapter.

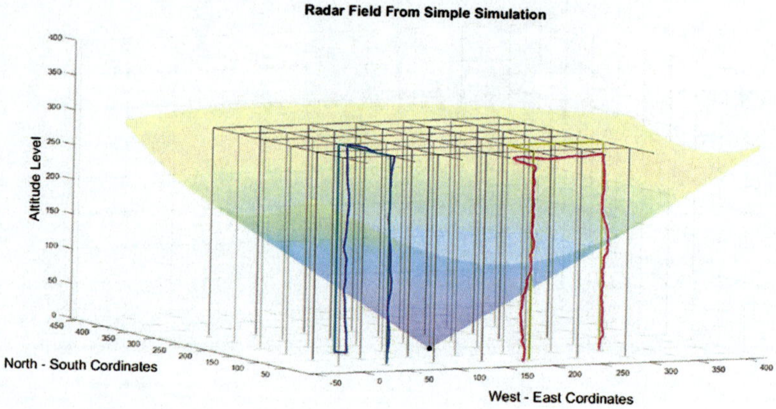

Fig. 13. Demonstrates how a vertical radar field would be placed in a grid layout lane system simulation.

Radar Placement Radar systems can be selectively placed within city coördinates in order to provide the most coverage of a given air space region that surrounds the lanes. There is a lot of variability in the creation of the radar objects that can affect the coverage (see Fig. 14). For example, some of these features are the maximum range of the radar field, the direction of the radar field, the beam angle, and the placement of the radar itself. Our goal is to find radar placements that provide the best coverage while minimizing the amount of radar field overlaps.

First let's make an assumption that the area we wish to cover is a cube that encapsulates the total area around the lane-structure. The reason we chose to maximize the radar fields' coverage of a cube instead of the actual lane structure was because we wanted to ensure additional coverage beyond the lane system that would allow for the detection of rogue flights. In addition, turning the

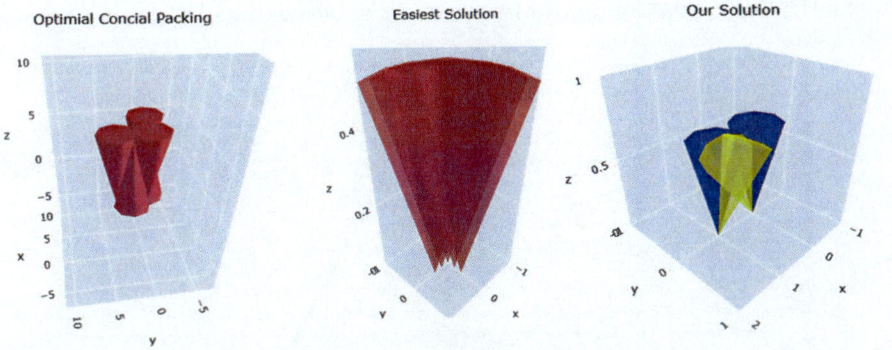

Fig. 14. Displays a couple of different packing arrangements for conical shapes. The image on the left is the optimal conical packing but not the most realistic solution. The middle image is the easiest solution to optimize coverage but it is too costly of a solution to be implemented. The rightmost image is our solution, which provides a high percentage of coverage with minimal overlaps.

region of interest into a cube allows us to solve our problem by using packing optimizations for conical structures. The best solution would be to set the length of the radar's field to the length of the cube. Then alternating the radar field direction in the vertical direction while alternating which edge of the cube to place the radar on, which is shown in the figure below. However, this solution is not really practical for a real life scenario, because it has radar systems that are floating in the air. Unfortunately, it would be impractical to suspend radar systems in midair like this. A more practical solution is to mount the radar systems to other objects that are on the ground such as on buildings, light posts, or the ground itself. Therefore, the arrangement of radar systems must also take into account their physical mounting requirements. The easiest solution would be to place radar systems away from each other at small increments. This would provide significant coverage, but would also come with a significant price tag if it were to be implemented in the real world.

Our solution takes a different approach, instead of placing several radar systems on top of one another, we spaced them based on the height and radius of the radar fields. No matter how many lanes are stacked on top of one another, we decided to use the lower lane height as the point where the radar fields are okay to overlap, this ensures that the entire lane system will be covered. To find the distance between radar systems we take the radius of the radar field at the lower lane's height and then double this radius. The final result is the distance between the radar systems. Figure 15 shows the basic parameters of a cone, and given that

$$tan(\theta) = \frac{radius}{height}$$

then we have:

$$radius = height \times tan(\theta)$$

Let's shift the attention to how to measure the coverage of a set of radar systems.

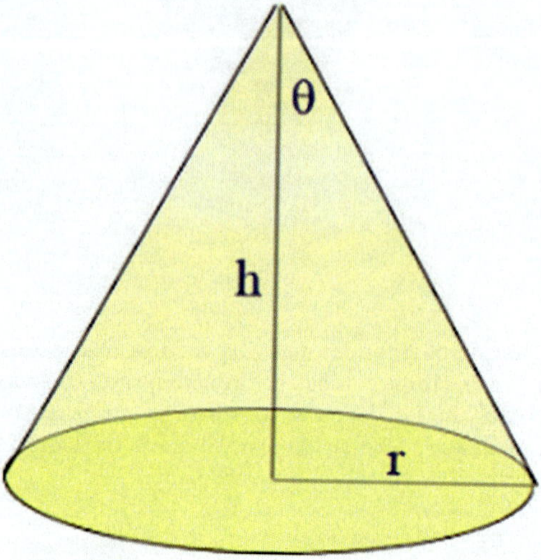

Fig. 15. The basic components of a conical shape; these are used to establish the required distance between radar placements.

Getting an exact measurement is quite difficult, therefore, an approximation of coverage is calculated using Monte Carlo experiments. In these experiments, randomly generated samples, which are bound to the cube area around the lane system, will be used to count the number of points detected by the radar systems. To approximate the volume of coverage, we can simply divide the number of detected samples over the total samples in the experiment. It is through these experiments that we can determine the differing factors that help maximize coverage. Figure 16 shows the effects of changing the beam angle through the Monte Carlo experiments.

Due to the conical shape of the radar fields, most of the coverage that is gained when using this method is in the upper altitude lane-structures; however, there are some regions that are left uncovered, which happens to correspond to areas that could contain launch and landing lanes, which must have coverage if this system is to function safely in the real world. There are two potential solutions: first, placing radar systems at every launch and land lane, or second, including more horizontally directed systems.

There will be a crossover point between placing radar systems at every launch and land site and just placing horizontally directed radars. If there are a significant number of launch and landing sites relatively close to one another, there will also be more radar systems close to each other, leading to greater overlap between radar fields. In contrast, strategic placement of a few horizontally directed radar systems could cover a larger area while also reducing the amount

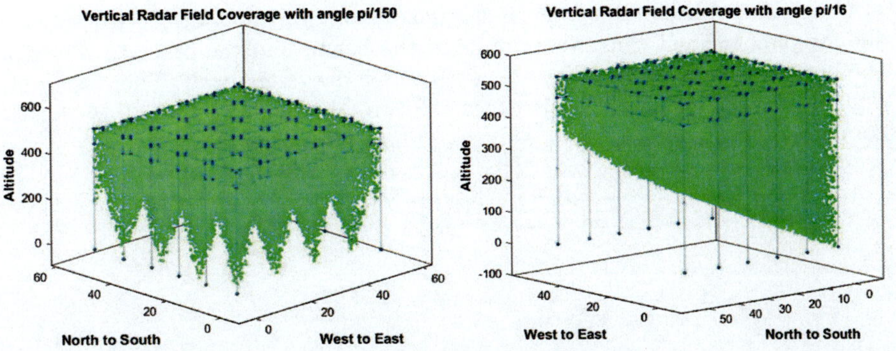

Fig. 16. Difference comparison between varying beam angles in the Monte Carlo experiment of 100,000 randomly generated sample points. This displays the inverse relationship between the angle size and the number of radar systems that would be needed for coverage. The image on the left uses a significantly smaller angle, resulting in more radars to cover the given area; whereas, the image on the right uses a larger angle of detection. The green dots represent the sample points that were within the radar fields. The empty spaces are areas in the lane-system that are not covered by the radar systems that have been placed.

of overlap between systems, meaning the number of radar systems needed will be lower to maintain the same amount of coverage. For example, consider the lane structure in the figure above with a radar angle of $\frac{\pi}{16}$, it contains 6 launch and 6 land sites for a total of 12 sites that should be covered. Therefore, 12 additional radar systems would be needed to cover all of these sites using the first approach; whereas, the second approach would only need one horizontal radar system to cover the remaining area, if it were placed in the same location as the other original radar.

Now consider how these radar systems detect objects during the simulation. When detecting objects, the main goal is to determine whether the object falls within at least one individual radar field. When determining if an object is within a radars field, one must consider a couple of factors: distance away from the radar, the relative angle from the radar, and the radar's maximum range.

To simulate a more realistic scenario, every radar and object has to be looped over to check which object falls into which radar field. Therefore, the time complexity of this check would be $O(RO)$, where R is the number of radar systems and O is the number of objects in flight during a time step. Therefore, since there could be a significant amount of radar systems and objects to detect, the calculation must be simple enough to ensure that the time complexity does not become excessive. To keep the time complexity manageable there are only two checks involved when determining if a given object is within an individual radar's field. The first check is to see if the object is within the radar's range, this is done by calculating the Euclidean distance from the object itself to the radar. The second check is to find the relative angle between the object and the middle of the radar field. For this two directional vectors are created: one to represent the radar field itself using the middle of the beam, and the other to represent the directional vector between the object and the radar itself. Then using the dot product to find the angle between the two directional vectors allows for the comparison between this angle and the radar beam angle (Fig. 17).

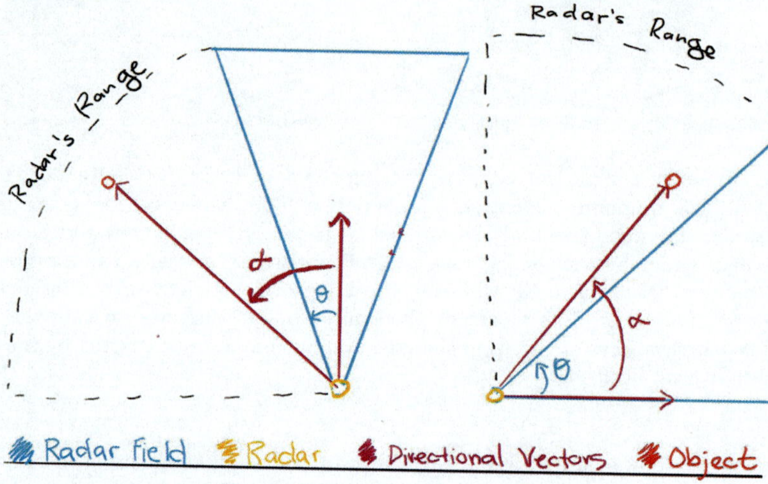

Fig. 17. Properties that are used to determine if an object is within the radar's field of view. These properties include the radar's maximum range, and the angle between the object and the radar. The left drawing shows a vertical radar field, and the right drawing shows a more horizontal radar field.

During each simulation step, each radar object will receive a list of objects that are in flight. Then each radar will perform a detection step for each individual object. Each radar object contains a list that is clear at the beginning of each simulation step, and then replenished with all of the objects that were detected. Once the radar object has gone through all of the objects in flight, it will then broadcast to any listeners that it has detected objects in the given simulation step. Any subscribers that are listening will have access to this list of objects, and they can proceed to handle the information according to their guidelines.

Each simulation will contain an individual lane-system, radar system, number of flights, and an individual ATOC. As the simulation is running, information between all of these components are hard to capture until the entire simulation is completed. Therefore, each of these classes contain functions that allow you to gain some insights while the simulation is running in order to prevent a complete black box of the simulation process. The radar systems method is to display the region it covers, along with a red dot to represent the location of any detected object at that given simulation step. Figure 18 displays the graph.

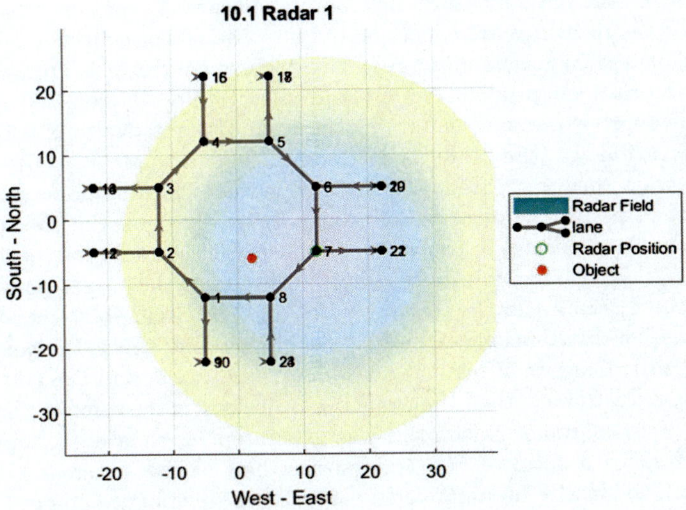

Fig. 18. Visualization of the radar detection process during a simulation. The colorful circle represents the radar field, and the red dot represent the location of the detected object within the field. The first number in the title represents the time step and is followed by the radar identification number.

This is a useful tool for many reasons, one being able to target locations that might contain a higher density, or an area that potentially contains a lot of anomaly flight behaviors. The user can then see directly what is happening, and be able to make necessary changes if possible.

In summary, the main goal for the radar systems in the simulation is to gather information that is happening in the lane system and the surrounding area through scanning and reporting any detect behaviors. The structure and performance that was designed was intended to imitate real life radar systems. In addition, these radar systems are dynamic, meaning their locations and specific features can be changed in each simulation based on their designated purpose and the goal of the simulation.

2.2 NAB Analysis

The two NAB measures given previously allow the discrimination of nominal from anomalous flight trajectories in almost all cases. This is due to the fact that anomalous flights, generally speaking, do not stay near the lanes nor do they fly in the same direction as the nearest lane. However, trajectories (i.e., x, y, z, t 4-tuple sequences) are of variable length depending on the distance of the flight and the sampling rate. Thus, in order to compare trajectories, it may be necessary to normalize the length of each trajectory to some standard length.

The nominal flights can be distinguished from the anomalous flights by means of a simple feed forward neural network. First, the trajectory lengths are normalized. Next, the NAB measures are computed at every point on the trajectory, and finally, the measures are concatenated into one vector (in this case, distance measure followed by cosine measure). A trajectory generator is created for each flight type based on random launch-land sites (uniformly selected over flight area), and appropriate parameters for the type of flight. Noise is added to the trajectory as follows (the same type of noise is added to all trajectories). First, the ideal trajectory is created. Then starting with the first point and moving to the second point, the error is defined by a circle around the goal point (the circle in the plane normal to the vector from the first point to the second point). A point in the circle (uniformly selected) is chosen as the target point. Next, a point on the line between the starting point and the circle point is chosen using a half Gaussian distribution centered at the circle point; this is the next point in the modified trajectory. When the circle has radius zero, and the Gaussian has zero mean and variance, then the resulting trajectory is the same as the original.

A set of 100 sample trajectories was generated for each flight type, including nominal, for a total of 600 trajectories; half of these were used to train the network to classify nominal versus anomalous flights (two classes), and half were used to validate the result. Figure 19 shows the training performance (from Matlab) as well as the 100% correct classification results on the test set.

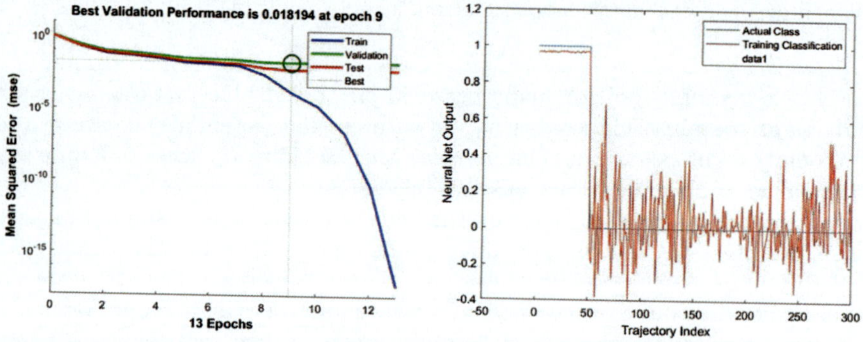

Fig. 19. Results of feed forward neural net classification of flight trajectories into two classes: nominal (first fifty) and anomalous (remaining 250). On the left is the network learning performance data, and on the right is the classification result.

Once an anomalous flight has been identified, it is possible to develop more refined and model-based techniques to distinguish between the sub-classes. Some characteristics of anomalous flights are:

- *Hobbyist Type I*: not on lanes, not in correct direction, change of altitude in non-vertical direction, launches and lands near same site.
- *Hobbyist Type II*: not on lanes, not in correct direction, change of altitude in non-vertical direction, launches and lands near same site, hovers for short periods of time.

- *Hobbyist Type III*: not on lanes, not in correct direction, change of altitude in non-vertical direction, launches and lands near same site, makes circular motion.
- *Rogue Type I*: not on lanes, only goes up, over and down, middle segment may not align with lane, may not be at normal lane altitude, launch and land sites may not be near lanes.
- *Rogue Type II*: not on lanes some of the time, not in correct direction some of the time, lanes followed may not be connected in lane network, some changes of altitude not vertical.

These characteristics are used to develop models of the various trajectories, and a classifier built based on them. Using the same set of simulated trajectories already described, the classification confusion matrix is given in Table 1 are achieved. From these results it can be seen that the trajectories of the Hobbyist Type I and the Rogue Type II are similar and require further refinement for discrimination.

Table 1. Classification results. The 6 trajectory types are (1) nominal, (2) hobbyist type I, (3) hobbyist type II, (4) hobbyist type III, (5) rogue type I, and (6) rogue type II.

	1	2	3	4	5	6
1	100	0	0	0	0	0
2	0	92	0	3	2	3
3	0	0	100	0	0	0
4	0	4	0	83	10	3
5	0	0	0	0	100	0
6	0	14	0	0	0	86

Now, suppose that there is another type of flight, for example, an amateur just flying circles above some location in the city. Figure 20 shows an example trajectory with respect to the ground, and Fig. 21 shows it with respect to the airways. Figure 22 shows the distance measure for the amateur flight compared to the noisy data from a nominal scheduled flight. It is clear that the error for such an unscheduled flight is much larger, and that flying in circles introduces a noticeable frequency component to the distance signature.

Next, consider the trajectory direction measure. Figure 23 shows the signatures for the noisy trajectories of the regular flight. Although it is difficult to see, the mean angle for these are: 26, 31, and 59 °C for variances of 0.1, 1, and 10, respectively. Now consider the signature for the amateur flight (see Fig. 24). The values are all close to 0 indicating that the direction is usually perpendicular to the lane direction. This make sense in that for the first few hundred feet, the lane is directed up while the circular flight is mainly in the x or y direction.

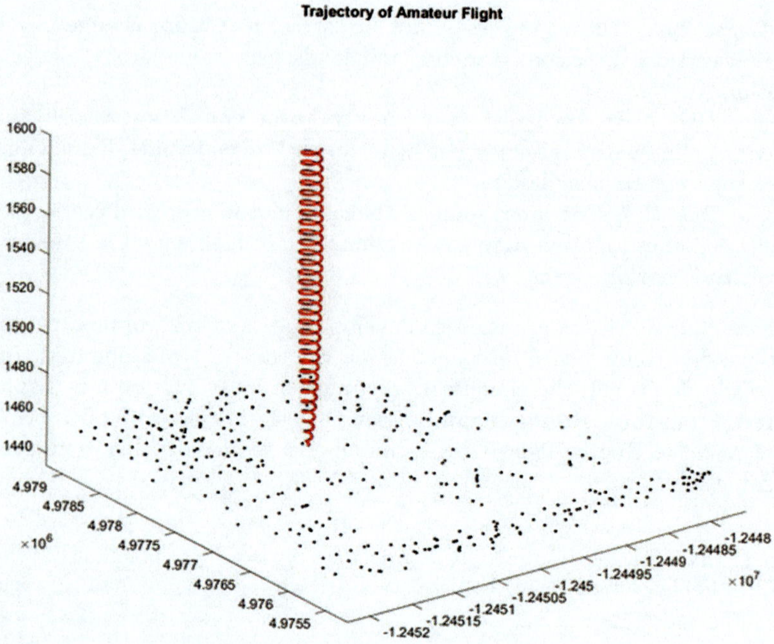

Fig. 20. Amateur trajectory flying in circles with respect to the ground.

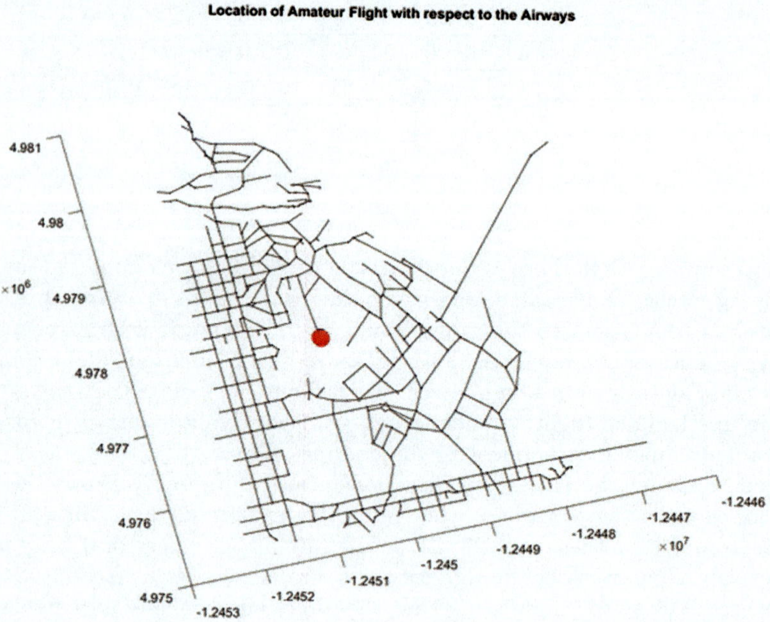

Fig. 21. Amateur flight with respect to airway network.

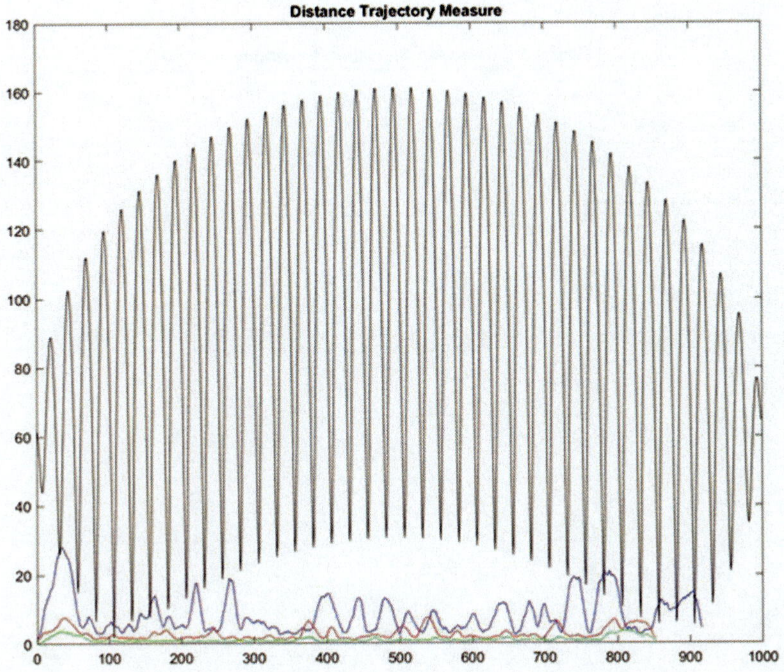

Fig. 22. Amateur flight distance signature compared to those of noisy flights.

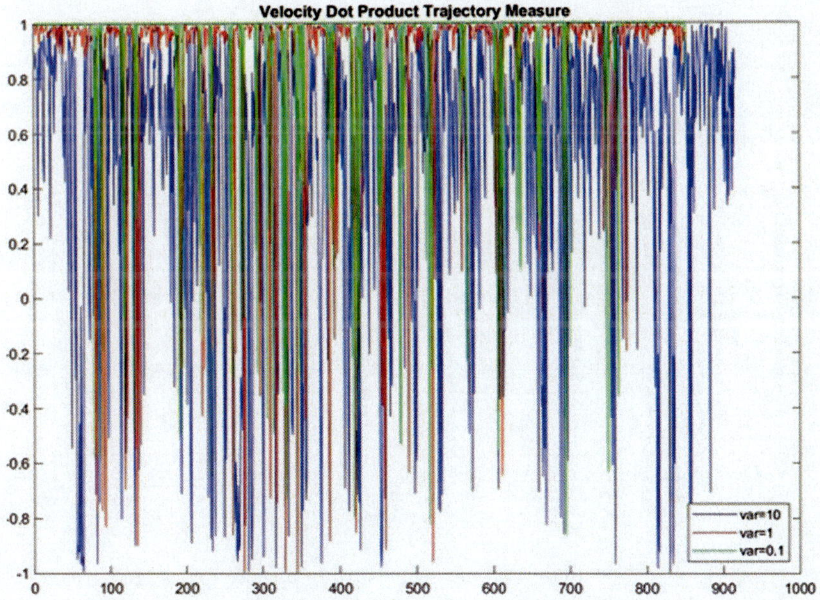

Fig. 23. Noisy flights vector cosine wrt Lanes.

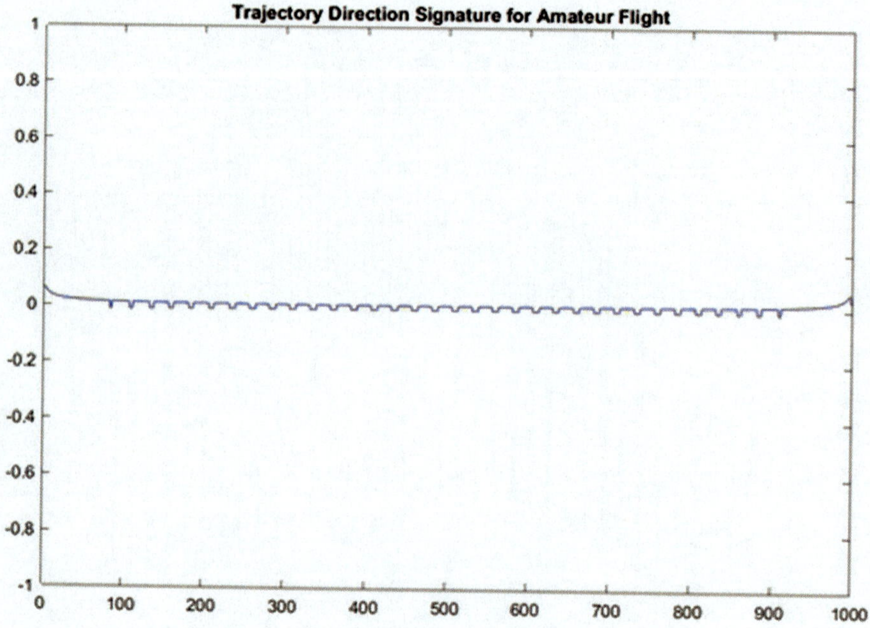

Fig. 24. Amateur flight vector cosine wrt lanes.

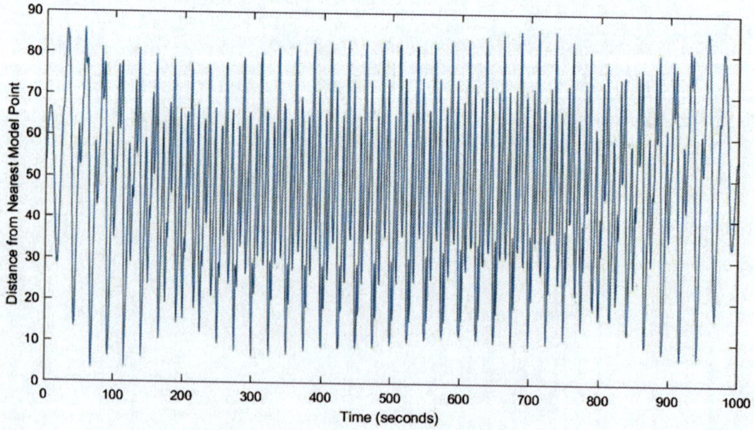

Fig. 25. Amateur flight distance signature in a more dense lane area.

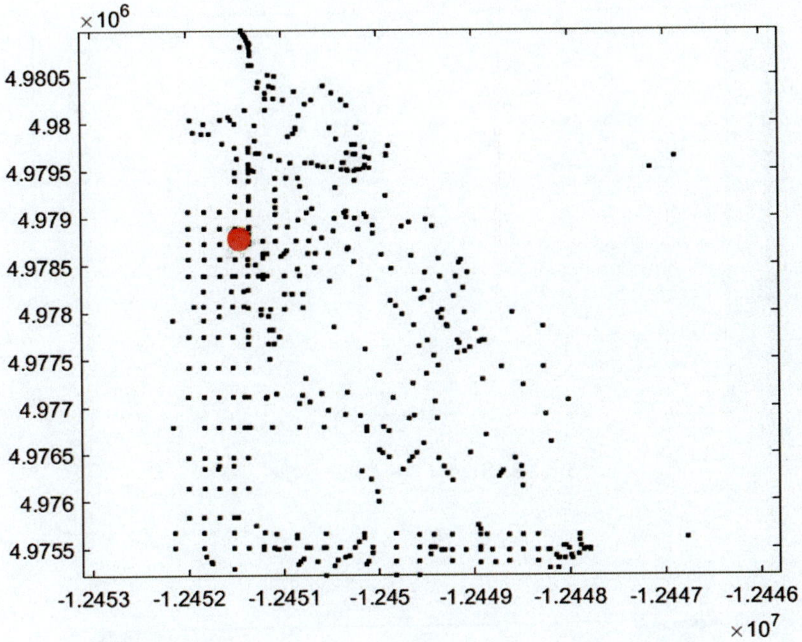

Fig. 26. Amateur flight location in a more dense lane area.

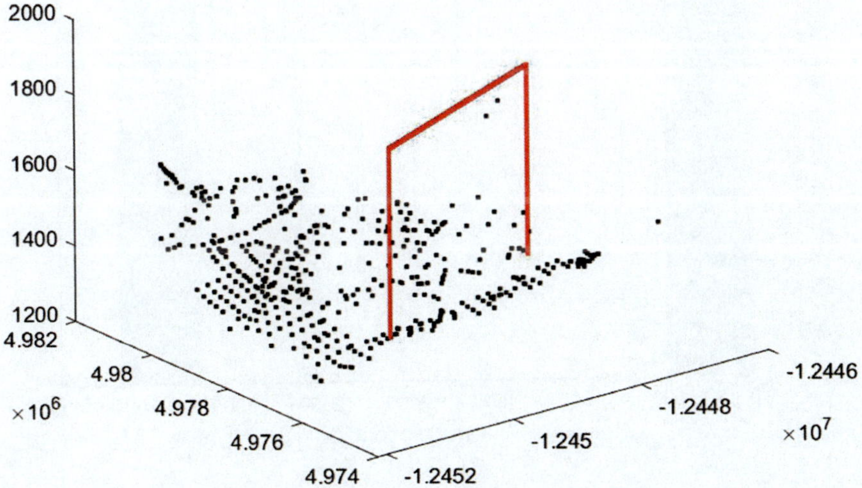

Fig. 27. Example rogue flight path.

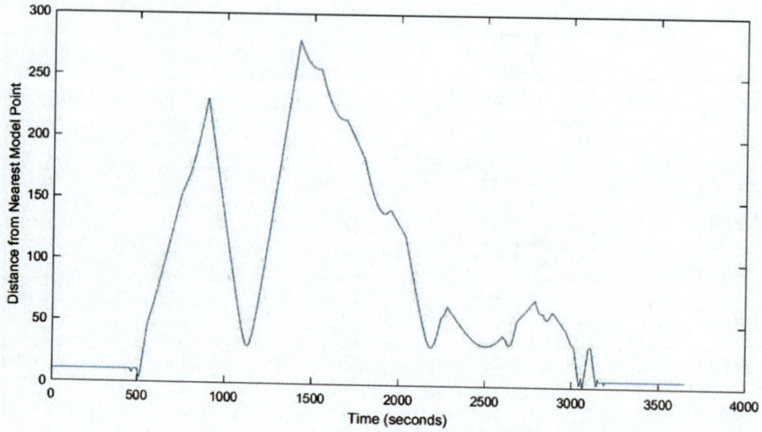

Fig. 28. Rogue flight distance error.

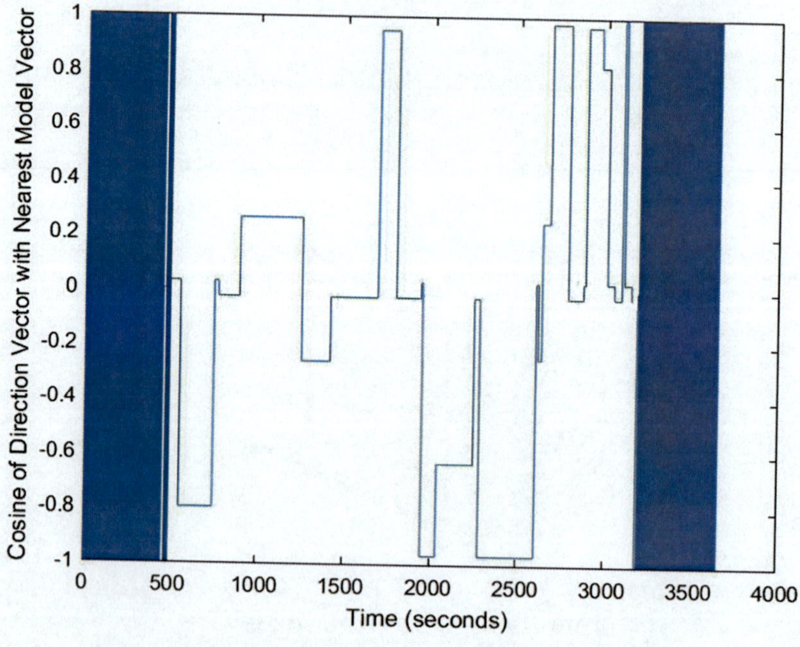

Fig. 29. Rogue flight direction error.

This particular amateur flight is always near a single lane. Figure 25 shows the distance signature from a more densely packed lane area as shown in Fig. 26, and a similar oscillating distance is observed.

Next, consider a rogue flight that simply takes the shortest route between launch and land sites (see Fig. 27). Figures 28 and 29 show the distance and direction errors signatures, respectively.

3 Next Steps

The lane-based UAS traffic management approach supports efficient and effective trajectory analysis of UAS flights in the airspace. This allows the straightforward detection of unplanned flights through the airspace without having to compare to every existing flight at the time of occurrence. In addition, it is possible to distinguish different types of rogue flights according to the trajectory distance and direction measures.

Several new avenues of investigation are under consideration:

- model updates due to dynamic lane creation and deletion,
- how to exploit knowledge of UTM parameters (e.g., UAS speed limits, lane network topology, 3D corridor constraints, etc.),
- any influence on trajectory measures due to weather, congestion, or other environmental or contingency effects,
- the constraints on sensor data to ensure effective identification of anomalous flight patterns, and
- the role of communications in UAS flight trajectory analysis.

References

1. Rios, J.: UAS Traffic Management (UTM) Project Strategic Deconfliction: System Requirements Final Report. Technical Report NASA Report, NASA, Moffet Field, CA (2018)
2. AirMap: Five Critical Enablers or Safe, Efficient, and Viable UAS Traffic Management (UTM). Technical Report, AirMap, Santa Monica, CA (2018)
3. Rios, J., Smith, I., Venkatesan, P., Smith, D., Baskaran, V., Jurack, S., Iyer, S., Verma, P.: UTM UAS Service Supplier Development, Sprint 2 Toward Technical Capability Level 4. Technical Report NASA/TM-2018-220050, NASA, Moffet Field, CA (2018)
4. Sacharny, D., Henderson, T., Cline, M., Russon, B.: Reinforcement Learning at the Cognitive Level in a Belief, Desire, Intention UAS Agent. Technical Report UUCS-20-013, University of Utah (2020)
5. Sacharny, D., Henderson, T.: A lane-based approach for large-scale strategic conflict management for UAS service suppliers. In: 2019 International Conference on Unmanned Aircraft Systems (ICUAS), Atlanta, GA, Institute of Electrical and Electronics Engineers (IEEE), pp. 937–945 (2019)
6. Sacharny, D., Henderson, T.: Optimal policies in complex large-scale UAS traffic management. In: IEEE International Conference on Industrial Cyber-Physical Systems, Taipei, Taiwan (2019)

7. Sacharny, D., Henderson, T., Cline, M.: Large-scale UAS traffic management (UTM) structure. In: IEEE Multisensor Fusion and Integration Conference, Karlsruhe, Germany (2020)
8. Sacharny, D., Henderson, T., Cline, M., Russon, B., Guo, E.: FAA-NASA vs. lane-based strategic deconfliction. In: IEEE Multisensor Fusion and Integration Conference, Karlsruhe, Germany (2020)
9. Sacharny, D., Henderson, T., Guo, E.: A DDDAS protocol for real-time UAS flight coordination. In: InfoSymbiotics/Dynamic Data Driven Applications Systems Conference, Boston, MA (2020)
10. Sacharny, D., Henderson, T.: A lane-based approach for large-scale strategic conflict management for UAS service suppliers. In: IEEE International Conference on Unmanned Aerial Systems, Atlanta, GA (2019)
11. Barbagello, J.: Instrument Procedures Handbook. Technical Report FAA-H-8083-16B, Federal Aviation Administration, Washington, DC (2017)
12. Booz Allen Hamilton Inc., Kimley-Horn and Associates Inc.: Transportation Management Center Business Planning and Plans Handbook. Technical Report TMC Pooled Fund Study, Federal Highway Administration, Washington, DC (2005)

Advanced and Intelligent Control Design for Underwater Robots

Acoustical Underwater Localization of a Remotely Operated Vehicle in Mariculture

Nadir Kapetanović(✉), Đula Nađ, Ivan Lončar, Vladimir Slošić, and Nikola Mišković

Laboratory for Underwater Systems and Technologies (LABUST), University of Zagreb Faculty of Electrical Engineering and Computing, Zagreb, Croatia
nadir.kapetanovic@fer.hr

Abstract. Localization of underwater vehicles, namely remotely operated vehicles (ROVs), used in mariculture autonomous inspection applications represents a challenging problem. The need for accurate localization of an ROV is further emphasized by often cluttered underwater environment of fisheries with many ropes and mooring around the net pens potentially causing entanglement with ROV's tether. This paper presents an overview and preliminary results of the HEKTOR (Heterogeneous Autonomous Robotic System in Viticulture and Mariculture) project regarding the ROV localization using acoustics mounted onto an autonomous surface vehicle (ASV). The ROV and the developed ASV are described, together with the hardware and software integration of a short baseline acoustic localization system . Preliminary sea trial results show promising performance of the localization system, proving that it could be used in autonomous net pen inspection missions.

Keywords: Marine robotics · Remotely operated vehicle (ROV) · Autonomous surface vehicle (ASV) · Mariculture · Autonomous inspection · Bio-fouling detection · Acoustical localization · Short baseline (SBL)

1 Introduction

For millennia, agriculture and aquaculture have been a component of human society [1,2]. They've evolved at various speeds since then. Aquaculture, as well as its component relating to sea water known as mariculture, did not begin to expand exponentially until the twentieth century, according to [2]. Globally, mariculture is the fastest growing part of food industry [3,4]. In the EU 20% of the fish stock comes from aquaculture and this industry branch accounts for around 80,000 employments [5]. Mariculture is heavily reliant on human labor, with workers generally doing arduous, repetitive, sometimes even dangerous tasks for long periods of time [6,7]. The HEKTOR (Heterogeneous Autonomous Robotic System in Viticulture and Mariculture) project [8,9] is looking for a solution

I. Petrovic et al. (Eds.): IAS 2022, LNNS 577, pp. 819–830, 2023.
https://doi.org/10.1007/978-3-031-22216-0_54

to these issues. Currently, divers must monitor fish farming net pens for long periods of time even in the worst weather conditions, even the most extreme. Furthermore, environmental laws and continuous monitoring of the condition surrounding and beneath the net pen in mariculture need the measurement of all important parameters as well as sampling from the bottom (sediment) and in a water column.

In the scope of HEKTOR project, on approach to the net pen, both LAAR and ROV are docked to the surface platform (ASV). Upon arrival, the LAAR takes off and the ROV separates from the ASV. For autonomous aerial inspection, the LAAR must be autonomously positioned over the net pen and collect video data on the structure of the net pen and the activity of fish at the sea surface. The LAAR shall be able to detect the net pen in the video image and position itself over the net pen using image processing. It must also be able to autonomously take off and land on an autonomous surface platform located near the net pen.

For underwater inspection of net pens, the ROV must be able to navigate around the entire underwater portion of the net pen and collect video. The localization of the ROV with respect to the net pen is done by an acoustic positioning system. For this purpose, there should be cooperation between the ROV and the surface platform (ASV), i.e. the surface platform must constantly monitor the position of the ROV in order to reduce the amount of cable connecting it and the surface platform (ASV) in the water, thus reducing interference caused by the sea currents (Fig. 1).

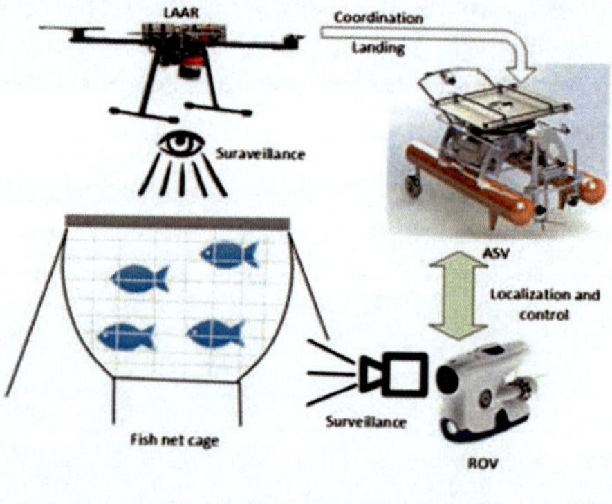

Fig. 1. Conceptual design of HEKTOR autonomous mariculture inspection system.

Main contributions of this paper are: (1) development of the mechanical and the electrical integration of the commercially available Underwater GPS (UWGPS) G2 acoustic underwater localization system with the developed ASV, (2) software integration of UWGPS API with ROS2 in a form of an open-source package, and (3) experimental results of successful field trials performed to validate the integration.

The rest of this paper is organized as follows: an overview of net pen inspection methods using robotic systems and acoustic underwater localization methods are described in Sect. 2. Specifications of the remotely operated vehicle (ROV) Blueye Pro and the description of the autonomous surface vehicle (ASV) developed in the scope of the project is given in Sect. 3. The acoustic underwater localization system, its acoustical specifications as well as hardware and software integration are given in Sect. 4. Section 5 present the preliminary field trials results. Finally, concluding remarks are given in Sect. 6.

2 Overview of Net Pen Inspection and Acoustical Underwater Localization Methods

Underwater vehicles have been used in various applications for inspection and monitoring, and some of them are related to aquaculture. In [10], a solution for navigating a net pen based on Laser-Camera Triangulation is presented, while in [11], the authors present an autonomous underwater vehicle for automatic inspection of net pens. A similar system is presented in [12], where the authors present an integrated ROV solution for underwater inspection of net pens in fish farms. A system consisting of an unmanned surface platform and an ROV for Sea Farm inspection in Norwegian aquaculture is presented in [13]. Underwater navigation presents a great challenge in marine robotics due to absence of global positioning signals that are usually available in areas reachable by satellites. Because of measurement noise and model inaccuracies, localization error is unbounded over time, so many underwater vehicles occasionally surface in order to get GNSS measurements and thereby determine their own position [14]. In order to tackle this problem, acoustic based localization techniques are used.

An Ultra Short Base-Line (USBL) system comprises two elements, a transceiver with an array of transducers that are usually less than 10cm apart, thus forming the ultra-short baseline, and a transponder. The direction of the transponder from the transceiver is measured on the transceiver with a phase-differencing within this transducer array to calculate the angle to the underwater transponder. USBL is the most popular category of underwater positioning systems due to its easy deployment, at the cost of positioning accuracy [15]. Short Base-Line systems SBL systems are mounted on floating platforms like boats or ships and are used for tracking underwater targets. They are characterized by smaller inter-baseline transponder distance of 20 to 50 m compared to LBL. The accuracy of the SBL system improves as the distance between baseline transponders increases [16]. In Long Base-Line (LBL) the distance between the baseline stations typically ranges from 50 m to more than 2000 m. LBL systems are independent of water depth and provide high accuracy of more than 1 m, with the capability of achieving accuracies of a few centimeters [16]. GNSS Intelligent Buoy (GIB) systems can be viewed as inverted LBL devices. In a typical deployment scenario, several GIBs are deployed over a given area of operation where the total number required is determined by the size of the operation area and the desired accuracy. Accuracies of centimeter to meter level can be achieved [16].

3 Vehicles

3.1 Remotely Operated Underwater Vehicle (ROV)

Blueye Pro ROV (shown in Fig. 2a) is produces by the Norwegian company Blueye Robotics. Its dimensions are 48.5 × 25.7 × 35.4 cm ($L \times W \times H$). It weights 9 kg in air, is rated for up to 300 m depth (with a 400 m tether) and has autonomy of 2 h ensured by its 96 Wh battery. The ROV has in total four 350 W thrusters. Two rear thrusters in the horizontal plane, one in the vertical plane, and one lateral thruster enable high maneuverability of the vehicle together with auto-heading and auto-depth control modes. It can reach surge speed of up to 3 knots and operate in water currents of 2 knots at most. Its main sensor is a full HD camera with a $[-30\,\text{deg}, 30\,\text{deg}]$ tilt angle mechanism working at 25–30 fps. Blueye PRO ROV has powerful 3300 lum lights. Additional payload can be mounted at the top and/or the bottom. The ROV comes with a surface unit that enables connection and control of the ROV via WiFi as well as via Ethernet in difficult wireless environments.

Blueye software development kit (SDK) that is Python3-based has been integrated into ROS2 for the operator to be aware of the situation underwater while viewing video stream in ROS2 graphical user interface (GUI). Also, this ROS2 interface enables future control of the ROV by the ASV. Code for the Blueye SDK-ROS2 interface and GUI is publicly available at [17]. Details about the Blueye Pro ROV, the developed ROS2 interface, and the biofouling and net pen rope detection can be found in [18,19].

(a) (b)

Fig. 2. (a) Blueye Pro ROV with its prototype tether management system for pool testing. (b) Testing of the ASV Korkyra at Biograd na Moru, Croatia in October 2021.

3.2 Autonomous Surface Vehicle (ASV)

Autonomous surface vehicle named Korkyra (shown in Fig. 2b) has been developed in the scope of HEKTOR project for mariculture scenarios. It's been designed as a catamaran to enable better stability and hydrodynamic properties at sea states up to state two. It is made of aluminium, 200 cm long, 100 cm wide, with hollow hulls of 24 cm in diameter. The catamaran is 80 cm high from the bottom of the hull to the top of the carbon hull and 140 cm high to the top of the rowbar. Currently, without any additional payload it weights 100 kg in air, and can withstand additional 100 kg of payload.

The autonomous catamaran can be mounted with various payloads, e.g. a multibeam sonar, a remotely operated vehicle (ROV) secured in place by a docking system (DM), tether management system (TMS) for the ROV, landing platform for the LAAR, as shown conceptually in Fig. 1 on the top and front bottom part of the ASV Korkyra. Four electrical T200 thrusters in X-configuration enable it to navigate complex marine environments omnidirectionally at lower speeds (1–2 knots), while it's booster electrical motor Minn Kota RT 55 EM enables top surge speeds of 3–4 knots. Total energy that the ASV can get from its batteries is 252 Ah or 3.73 kWh. This means that on average, in mixed consumption scenarios when the ASV would need to quickly arrive from the shore to the fishery using high-power motor and also perform slow omidirectional inspection maneuvers, the ASV would have the autonomy of 10–11 h. This is more than enough for HEKTOR project usecases.

Global Navigation Satellite System (GNSS) assisted with inertial navigation system IMU, named Applanix SurfMaster, coupled with corrections from base stations over LTE network, enables the ASV to globally localize itself with precision of up to 10 cm, which is very important in survey and inspection missions. Long-range surveillance of the catamaran via video stream from a pan-tilt-zoom (PTZ) Hikvision IP camera (mounted at the top of the rowbar) ensures that the operator can react to any risky situations that might happen at sea.

With an objective of further simplifying the integration of subsystems onto the ASV Korkyra, an extension box was made and mounted on the lower deck. An IP67 aluminium box, shown in Fig. 3 was fitted to act as power, Ethernet and serial communication (USB) hub. On its bottom side it has three cable glands for fixed inputs of ASV's battery voltage, Ethernet cable connected to the main Ethernet switch, and USB cable connected to the main NUC10 onboard computer, all placed inside the carbon hull. On the output side, it has waterproof Bulgin connectors: 6 for power, 4 for Ethernet, and 4 for USB. Inside the extension box are 8 port Gigabit Ethernet switch, 7 port USB3.0 USB hub, a 12 V voltage stabilizer for powering the switch and the USB hub, as well as 8-relay board controlled over Ethernet used for power control of the subsystems.

4 Underwater Acoustical Localization System

4.1 Acoustics Specifications

Control of the ROV by the ASV needs a position feedback. In the underwater environment this in most cases means using some of the acoustical localization systems, since electromagnetic waves are strongly attenuated. Since the ASV Korkyra's dimensions in the horizontal plane are 2×1 m, the use of range-based short baseline (SBL) system with transducers placed at four corners of the ASV

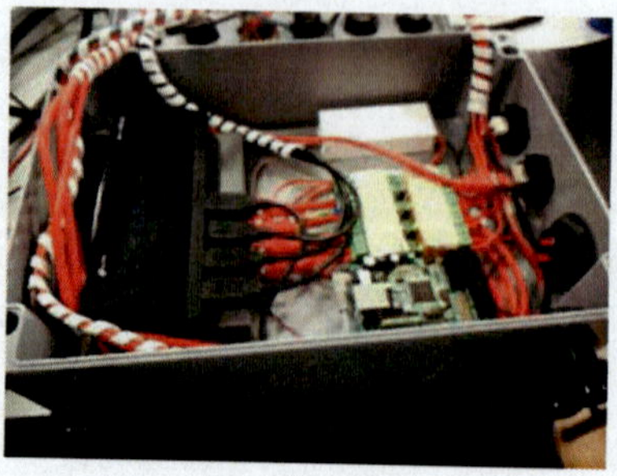

(a)

(b)

Fig. 3. (a) Interior of the subsystem extension box. (b) Exterior of the extension box.

(or even wider) is a better choice that an ultra-short baseline (USBL) system. With larger baseline, the underwater localization system is less sensitive to range measurement noise. Furthermore, since Blueye Pro ROV is a small ROV without guest ports to integrate additional sensors, a standalone lightweight battery powered bottomside unit is needed. WaterLinked Underwater GPS (UWGPS) G2 SBL system (see Fig. 4a) with a bottomside Locator U1 system met all these requirements.

The UWGPS G2 system is quite convenient since the topside comes in a Pelican case that is easy to integrate onto the ASV. It has an acoustical range of 100 m, which is enough for HEKTOR project applications. However, the range can be upgraded to 300 m just by a software update. It's omidirectional, with an option to change both the maximum range and bearing angle sector to improve the positioning accuracy. UWGPS G2 system comes with an onboard computer that takes GNSS, IMU, and acoustics measurements and reprojects them into the relative position of the ROV w.r.t. the ASV and in georeferences coordinates as well. There is also an option to provide the topside with external GNSS and IMU measurements either by NMEA messages or as HTTP requests to the local HTTP servers which runs all the computations and communication. Ping rate is 2–4 Hz.

It offers Ethernet and WiFi communication interfaces. Ethernet is better suited for HEKTOR autonomous missions since it has wider bandwidth and less delay than WiFi, and does not interfere with the WiFi that the vehicles are communicating over. WiFi option is however suitable in manual dataset collection tasks or any tasks which do not demand that the ASV is present. Also, using GNSS and heading measurements from UWGPS G2 all the datasets can be georeferenced. Setting all the parameters of the system is made easy by a web GUI that runs on topside's computer, which is accessible through any web browser. Also, this web GUI enables the user to visually track the ROV with its estimated global coordinates in a georeferenced map, as shown in Fig. 4b.

Locator U1 is a lightweight (75 g in water) battery powered bottomside transponder. Its battery can last for up to 10 h and is easily charged (~6 h)

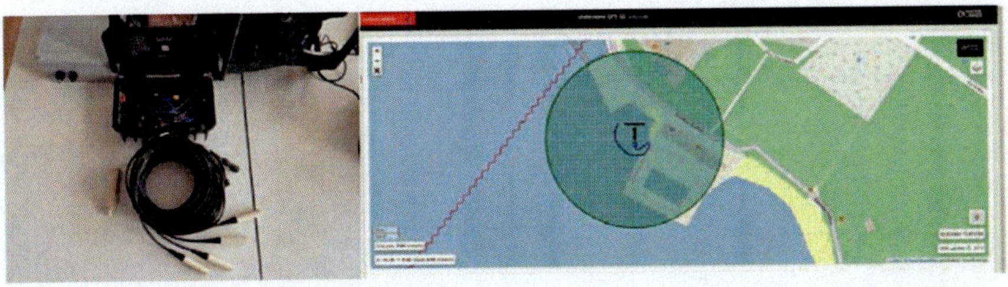

(a) (b)

Fig. 4. (a) WaterLinked Underwater GPS G2 topside with 4 transducers for underwater acoustical localization and Locator U1 (left). (b) Web GUI run on the topside computer that allows system setup and tracking of the ROV visually in a georeferenced map.

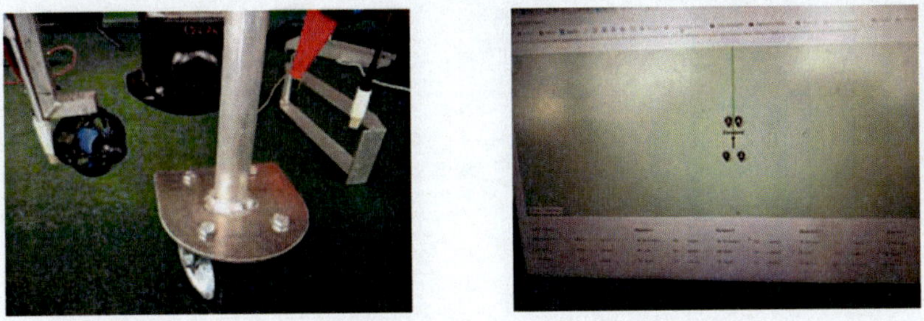

(a) (b)

Fig. 5. (a) Acoustic receivers mount on the thruster carriers in the front of the ASV Korkyra, lowered below the line of thrusters to avoid acoustic noise. (b) Receiver positions w.r.t. the topside offset setup in the web GUI.

using a USB-C port. It is rated up to 300 m. It has a GNSS chip integrated for time synchronisation with the topside. Indicator LED signals the following states: GNSS signal searching state, GNSS fix locked state, and no signal state. The bottomside unit is mounted onto the Blueye ROV by using a bracket sold by WaterLinked company. Frequency channel is chosen by a switch inside the cap of Locator U1. Of course, the same channel has to be chosen in the web GUI of the topside to enable acoustic pinging between the topside and the bottomside.

4.2 Mount Design

During the initial tests, the transponders were fixed to metal poles and mounted on thruster mounts in the front and on ASV's rear wheels' axels in the back, as shown in Fig. 5a. They were positioned below the thrusters to avoid acoustic noise. Moreover, they are fixed at different heights (15, 25, 35 and 45 cm below the thrusters, i.e. at 50, 60, 70 and 80 cm depth) to increase the positioning accuracy, as advised by the manufacturer. Configuration of transducer offsets w.r.t. the topside as shown in web GUI is given in Fig. 5b. The baseline in this configuration was 1.46×0.487 m.

A new SBL system mount with adjustable transponder depth is under development, as shown in Fig. 6a. It is designed to provide a 2×2 m baseline in more open areas, but also a 2×1 m baseline in more congested areas around the fish net pens with many buoys and ropes. The depth of each transponder can be changed separately and can be fixed at up to 1 m. Furthermore, a foldable floating version of SBL mount is developed and made from plastic tubes with Styrofoam floats, as shown in Fig. 6b. This allows testing of the acoustical localization system to be uncoupled from the necessity of using ASV Korkyra.

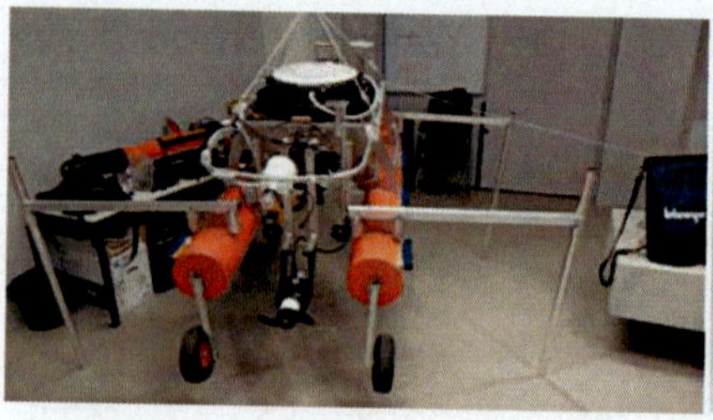

(a)

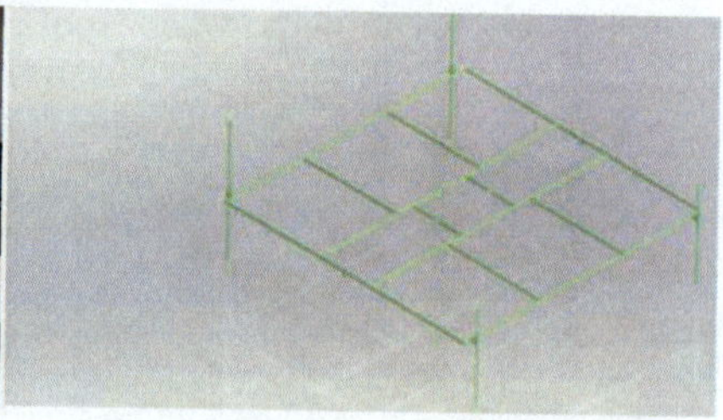

(b)

Fig. 6. (a) The finished aluminium acoustics baseline mounting mechanism. (b) 3D model of the foldable floating SBL mount.

4.3 UWGPS-ROS2 Interface

A Python v3.9 API developed by the manufacturer enabled easier software integration of the UWGPS G2 system. A ROS2 package was developed to integrate the acoustical localization measurements into the ROS framework. It sends HTTP request to the topside's HTTP server, parses the returned relative and global positions and publishes them to ROS topics. The code of the package is open-source and publicly available at [20]. Furthermore, an extension of the UWGPS-ROS2 interface, also available at [20], has been made in three modes: (1) In the first mode, the interface enables sending parametrized static GNSS and heading measurements to the UWGPS topside in a form of JSON-formatted HTTP PUT request, assuming Static setting in the web GUI of the UWGPS topside. In cooperative marine robotics it will rarely be the case that the ASV is completely static during the whole mission, even when in dynamic positioning (DP) mode. (2) The second operation mode of the UWGPS ROS2 interface allows relaying GNSS and heading measurements of the ASV (the same way as in the first mode) at parametrized rate matched with the update rate of the UWGPS system. This mode assumes that GNSS and heading data sources in UWGPS topside web GUI are set to External. The first and second mode send

GNSS and heading data statically or dynamically, and the UWGPS-RO2 interface reads the relative Locator U1 position from topside's server, but also the transformed georeferenced Locator U1 position. This works fine if the ASV is perfectly horizontal all the time, which is far from being guaranteed at sea. In the presence of waves, this assumption deteriorates the accuracy of the localization system linearly with the increase in range. (3) Therefore, frame transform between Locator U1 relative position w.r.t. the topside aboard the ASV and its georeferenced position in World Geodetic System 1984 (WGS84) must include not only the WGS84 position of the ASV and ASV's heading, but also its roll and pitch angles. Frame transform from relative to NED to WGS84 position of the ROV is given by:

$$\eta_{ROV}^{NED} = R_x(\phi_{ASV})R_Y(\theta_{ASV})R_z(\psi_{ASV})\eta_{ROV}^{ASV} + \eta_{ASV}^{NED}, \tag{1}$$

$$\eta_{ROV}^{WGS84} = ned2wgs84(\eta_{ROV}^{NED}, \eta_0^{WGS84}) \tag{2}$$

where η_{ROV}^{NED} is ROV's position in NED frame common to the ASV, R are rotation matrices around x, y, and z axes with roll, pitch and yaw angles of ASV's orientation ϕ_{ASV}, θ_{ASV}, and ψ_{ASV}, respectively. η_{ROV}^{ASV} represents ROV's position relative to the ASV, η_{ASV}^{NED} is ASV's position in NED frame, η_{ROV}^{WGS84} is ROV's georeferenced position, and η_0^{WGS84} is the georeferenced origin position of the chosen NED frame.

This is implemented as the third mode of the UWGPS-ROS2 mode in which relative Locator U1 position is fused directly in ROS2 with GNSS and orientation measurements of the ASV. Since all autonomy, navigation, guidance and control systems of the ASV Korkyra are implemented in ROS1, a ROS1-ROS2 bridge (built from source to include all custom messages from both ROS1 and ROS2) was used to relay GNSS and orientation measurements to UWGPS ROS2 interface in operation modes two and three. Graphical representation of this data relaying and processing is shown in Fig. 7. In Fig. 7, the square block in the middle with `/catamaran` namespace contains one ROS2 node named `/...localization_node` that publishes to three topics: (1) `/...position_relative_wrt_topside`, (2) `/...position_topside_ned`, and (3) `/...position_global`. Square block in the lower left corner with `/usv` namespace represents all things related to ROS1. If UWGPS ROS2 interface is in its third mode, then ROS2 node subscribes to `/usv/applanix_position` over `/ros_bridge` node to get ASV global, NED position, and orientation.

5 Results

During the field trials in Biograd na Moru, Croatia in October 2021. a dataset of acoustic localization and ASV position/orientation measurements was recorded. Usually after a successful IMU calibration the topside computer can accurately estimate the relative position of the ROV w.r.t. the topside based on trilateration and known GNSS and heading of the topside. However, due to the faulty IMU, the calibration was not successful. UWGPS ROS2 interface was thus used in

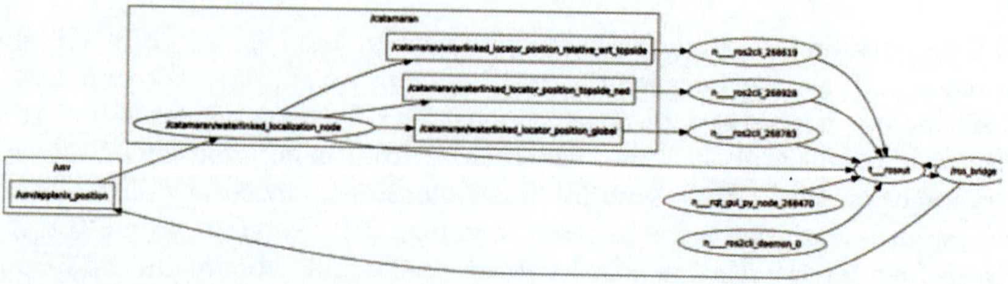

Fig. 7. Node graph of the UWGPS ROS2 interface with ROS1-ROS2 bridge for GNSS and orientation measurements.

its first mode with fixed GNSS and heading measurements, as explained in the previous section. The initial solution was to set the GNSS and IMU source as static, i.e. a fixed GNSS position close to the testing site with heading 0° looking to North, as shown in Fig. 4b. In this configuration the relative positions of the ROV were not corrupted, but its transformed global position estimates were. This issue was solved in the post-processing phase by fusing the relative ROV positions with heading and GNSS measurements from the ASV, see Fig. 8.

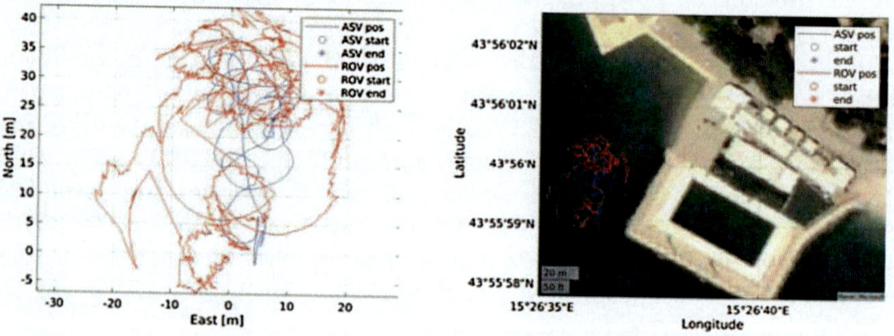

(a) (b)

Fig. 8. (a) Relative positions of the ROV fused with ASV's heading in the local NED frame. (b) Relative positions from (a) fused with ASV's GNSS and heading measurements to get the estimate of ROV's georeferenced positions.

6 Conclusion

Preliminary results of the HEKTOR project show a big potential for use of a heterogeneous robotic system in autonomous inspection and monitoring mission in mariculture. The system consists of an autonomous aerial vehicle that inspects the fish net pens from above and assists in biomass estimation as well. The ROV is integrated with the ASV. It is in a way autonomized and used as a remote camera for the ASV is envisioned to process all the visual data from ROV's camera and send the commands to the ROV to visually cover the whole net looking for biofouling and damage.

Localization of the ROV is a challenging problem. Accurate localization of an ROV is needed to autonomously inspect the whole net, but also avoid entanglements with many ropes and moorings around the fish net pens whose positions are more or less a-priori known. An overview and preliminary results of the HEKTOR project regarding the ROV localization using acoustics mounted onto an autonomous surface vehicle (ASV) are presented. After a successful integration of the ROV and the UWGPS G2 acoustic localization system with the ASV, the preliminary sea trial results performed in Biograd na Moru, Croatia show promising performance of the localization system, proving that it could be used in autonomous fish net pen inspection missions.

Acknowledgements. Research work presented in this article has been supported by the project Heterogeneous autonomous robotic system in viticulture and mariculture (HEKTOR) financed by the European Union through the European Regional Development Fund-The Competitiveness and Cohesion Operational Programme (KK.01.1.1.04.0036); the ERDF-funded project (KK.01.1.1. 07.0069)

"Multifunkcionalne pametne bove (Multifunctional smart buoys)"; the European Regional Development Fund through the Interreg Italy-Croatia InnovaMare project (Partnership ID 10248782); and by the "Razvoj autonomnog besposadnog višenamjenskog broda" project (KK.01.2.1.02.0342) co-financed by the European Union from the European Regional Development Fund within the Operational Program "Competitiveness and Cohesion 2014–2020". The content of the publication is the sole responsibility of the project partner UNIZG-FER.

References

1. Fussel, G.E., et al.: Origins of agriculture. Encyclopedia Britannica. Accessed 12 April 2021
2. White, K., O'Niell, B., Tzankova, Z.: At a Crossroads: Will Aquaculture Fulfill the Promise of the Blue Revolution? A SeaWeb Aquaculture Clearinghouse report, 17 (2004)
3. Subasinghe, R., Soto, D., Jia, J.: Global aquaculture and its role in sustainable development. Rev. Aquac. **1**, 2–9 (2009)
4. Food, Agriculture Organization of the United Nations: The State of World Fisheries and Aquaculture 2020, p. 107. FAO (2020)
5. Directorate-General for Maritime Affairs and Fisheries (European Commission): The EU Fish Market: 2020 Edition, p. 107 (2021)
6. Myers, M.L.: Review of occupational hazards associated with aquaculture. J. Agromed. **15**(4), 412–426 (2010)
7. Holen, S.M., Utne, I.B., Holmen, I.M., Aasjord, H.: Occupational safety in aquaculture - Part 1: Injuries in Norway. Mar. Policy **96**, 184–192 (2018)
8. : Heterogeneous autonomous robotic system in viticulture and mariculture (Hektor project) - 2020–2023. http://hektor.fer.hr/en/homepage/. Accessed 09 April 2020
9. Goričanec, J., et al.: Heterogeneous autonomous robotic system in viticulture and mariculture - project overview. In: Proceedings of the 16th International Conference on Telecommunications - ConTEL, Zagreb, Croatia, pp. 1–8 (2021)

10. Bjerkeng, M., et al.: ROV navigation in a fish cage with laser-camera triangulation. J. Mar. Sci. Eng. **9**(1), 1–16 (2021)
11. Livanos, G., Zervakis, M., Chalkiadakis, V., Moirogiorgou, K., Papandroulakis, N.: Intelligent navigation and control of a prototype autonomous underwater vehicle for automated inspection of aquaculture net pen cages. In: IEEE International Conference on Imaging Systems and Techniques, pp. 1–6 (2018)
12. Betancourt, J., Coral, W., Colorado, J.: An integrated ROV solution for underwater net-cage inspection in fish farms using computer vision. SN Appl. Sci. **2**(12), 1–15 (2020)
13. Osen, O.L., Leinan, P.M., Blom, M., Bakken, C., Heggen, M., Zhang, H.: A novel sea farm inspection platform for Norwegian aquaculture application. In: OCEANS 2018 MTS/IEEE Charleston, OCEAN 2018 (2019)
14. Mandić, F.: Control of autonomous surface marine vessels for underwater vehicle localization using single range acoustic measurements. PhD thesis (2019)
15. Stojanovic, M., Beaujean, P.P.J.: Acoustic communication (2013)
16. Tan, H.P., Diamant, R., Seah, W.K., Waldmeyer, M.: A survey of techniques and challenges in underwater localization. Ocean Eng. **38**(14), 1663–1676 (2011)
17. Kapetanović, N., Vuković, J.: Blueye SDK-ROS2 interface [Online]. https://github.com/labust/blueye-ros2-pkg.git (2021). Accessed 22 July 2021
18. Kapetanović, N., Nad, D., Mišković, N.: Towards a heterogeneous robotic system for autonomous inspection in mariculture. In: Proceedings of the OCEANS 2021 Conference and Exposition, San Diego - Porto (Hybrid), pp. 1–6 (2021)
19. Borković, G., et al.: Underwater ROV software for fish cage inspection. In: Proceedings of the 44th International ICT Convention MIPRO, Opatija, Croatia, pp. 1–6 (2021)
20. Kapetanović, N.: WaterLinked Underwater GPS G2 Python API-ROS2 interface [Online]. https://github.com/labust/waterlinked-uwgpsg2-ros2-pkg (2021). Accessed 05 Nov 2021

A Virtual Online Simulator Design for the Docking of Unmanned Underwater Vehicle

Fanghao Huang[1,2,3], Tuoyu Yan[3], and Zheng Chen[1,2,3](✉)

[1] Donghai Laboratory, Zhoushan 316021, People's Republic of China
zheng_chen@zju.edu.cn
[2] State Key Laboratory of Fluid Power and Mechatronic Systems, Zhejiang University, Hangzhou 310027, People's Republic of China
[3] Ocean College, Zhejiang University, Zhoushan 316021, People's Republic of China

Abstract. In this paper, a virtual online simulator is designed for the docking simulation task of Unmanned Underwater Vehicle (UUV), which provides an economic way to evaluate the reliability of related algorithms before the real underwater test. The host and lower computers are selected with several units' design to send and receive the data via TCP/IP communication in the virtual online simulator. The host computer runs the simulink control program, which calculates the mathematical model, monitors the state of UUV, and simulates or interacts with the real sensor during the docking process. The host computer runs the unity simulation software, which obtains the data from the lower computer, and implements the navigation and control of the UUV. The virtual docking scene is designed to display the simulation of UUV on the screen, and four different navigation modes are designed to implement the experiment with motion verification and underwater docking for the UUV. The experiment results show that the designed virtual online simulator can achieve the simulation of docking task in a lower cost.

Keywords: Virtual reality · Online simulator · Unmanned underwater vehicle

1 Introduction

With the development of technology [1–6] and the requirement of underwater resource exploration [7], Unmanned Underwater Vehicle (UUV) is playing an important role in the underwater monitoring due to its advantages of wide operation range, good mobility and high intelligence. However, considering the small structures of the UUV, it has the limited energy and restrictions on its working

Supported by National Natural Science Foundation of China (Nos. 52075476 and 92048302).

I. Petrovic et al. (Eds.): IAS 2022, LNNS 577, pp. 831–843, 2023.
https://doi.org/10.1007/978-3-031-22216-0_55

time and sailing distance. Therefore, the UUV needs to regularly implement the docking tasks back to docking station for the energy replenishment and the data communication. During the docking tasks, due to the underwater currents, no cable of the UUV, and no fixed base of the docking station, it's quite difficult to accurately obtain the coordinate of the docking point [8,9]. Subsequently, the researches of navigation and control design are developed for the UUV to accomplish the docking process under higher accuracy [10,11], but these researches need the highly skilled teams and high-cost devices, which may not be suitable for the long-term underwater test.

The use of simulation provides a new solution to verify the navigation and control of the UUV in an economic way. The traditional way of simulation is to use offline simulators (e.g., MATLAB/Simulink) to evaluate the feasibility of algorithm in the docking tasks, which has the advantages of high programming efficiency, efficient and convenient matrix and array operation [12–14]. However, the implement of algorithm highly depends on the information directly obtained from the actual docking process, but the offline simulator cannot interact with the real sensors efficiently due to the technical limitations. Therefore, based on the technology of virtual reality, the online simulator is developed recently, which combines the virtual scene with the real information, and has been widely used in the robot operation training, task planning and other related tasks [15–19].

Since the use of simulation provides a new solution of underwater tasks in an economic way, some researchers preliminarily investigated the online simulator for the underwater robots (e.g., AUV, ROV, UUV) [20–23]. However, most of the existing underwater online simulators have the drawbacks of low image frame rate and lack of interaction with the real sensor of underwater robots, and the complicate tasks with multi-sensor sensing and collaboration are difficult to be simulated. Besides, few researches are carried out for the specific underwater tasks (e.g., docking, navigation, motion control, etc.).

Therefore, addressed the aforementioned problems, a virtual online simulator is designed in this paper for the specific docking task of the UUV. Namely, the mathematical model and controller are designed to implement the motion control of the UUV in the virtual online simulator, and the integrated navigation modes combining four docking modes are designed. Subsequently, the navigation and control algorithms are run in the host computer with Matlab/Simulink, while the mathematical model combined with the sensor data are run in the lower computer. Through the communication of host and lower computers via TCP/IP communication, the navigation and control problems can be verified and the immersive motion of the UUV can be displayed in the virtual scene. Therefore, the proposed virtual online simulator can real time simulate or interact with the multi-sensor of the real UUV in the docking process, and provide a virtual platform and a convenient and effective experimental tool for the research of navigation and control problems for the UUV in a lower cost.

2 Overall Framework of Virtual Online Simulator

Based on the virtual reality technology, a virtual online simulator is designed for the docking task of the UUV. The flow chart of the virtual online simulator is shown in Fig. 1, where the host computer is run in the Matlab/Simulink environment, and the lower computer is run in the unity simulation software. The TCP/IP communication is selected to exchange data between the host and lower computers.

The blue line in Fig. 1 shows the virtual simulation. Based on the location of the UUV and the docking point, the control unit transmits the control inputs (e.g., propeller velocity, rudder angle of the UUV) to the motion unit, and simulate the motion of the UUV. With regard to the distance between the UUV and docking point obtained from the environmental perception and localization unit, different docking modes are selected with regard to the navigation strategy in the docking navigation unit. Therefore, by constantly updating the data (obtained from simulated sensor), the docking navigation unit can select different navigation modes, and the motion and control unit can simulate the motion of UUV to the docking point. The entire process is displayed in the virtual docking scene.

The red line in Fig. 1 shows the real platform simulation. Based on the location of the UUV and docking point obtained from sensor in the real UUV, the control unit transmits the control inputs to the real UUV control panel and drives its motion in the real underwater environment. The related data obtained from the real sensor are transmitted to the lower computer, which are used to reconstruct the motion and docking process of UUV in the virtual docking scene.

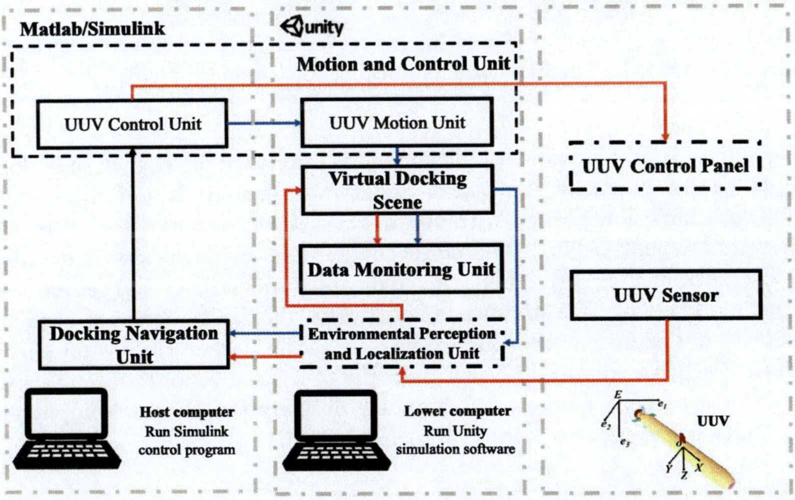

Fig. 1. The flow chart of virtual online simulator.

3 Unit Design for Virtual Online Simulator

3.1 Virtual Docking Scene Design

Based on the real underwater environment, a virtual docking scene of UUV is established in the system, which mainly includes the docking task scene and underwater environment.

Firstly, a proportional UUV 3D model is built by the SolidWorks software, as shown in Fig. 1. A tail propeller is used as the power system, and a cross rudder is used to control the direction of the UUV. Two coordinate systems are defined including the ground coordinate system and the carrier coordinate system. Also, a proportional 3D model of the docking station is built, as shown in Fig. 2. The light of the base station is simulated through the visual effect module in Unity software as the green light, which can also been seen in Fig. 2.

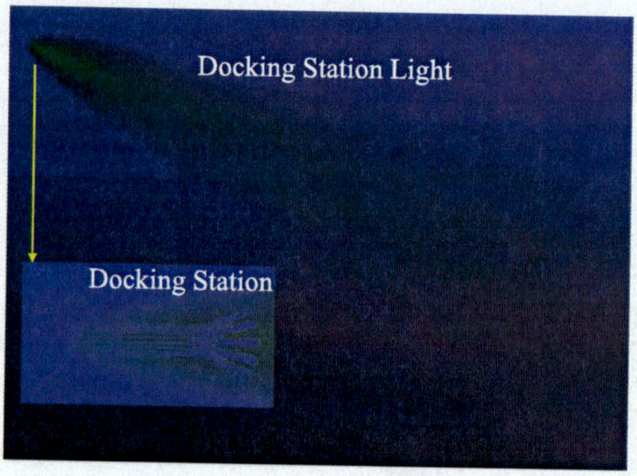

Fig. 2. The 3D model of docking station and its light.

Secondly, the underwater environment is modelled in the virtual docking scene, as shown in Fig. 3. A random underwater current is added, whose flow velocity and direction change with time. The velocity range is 0–0.25 m/s, and the direction range is 0–360°. Therefore, there exists the physical waves in the sea level, as shown in Fig. 3(a). Then, the color and transparency of the underwater area are modified by the atomization plugin in the Unity software, which can simulate that the deeper the underwater area, the darker the visual effect, as shown in Fig. 3(b). Finally, the raw data file of seabed topography in a certain underwater area near Zhoushan, China are built by the Height Mapper plugin in the Unity software, as shown in Fig. 3(c) and (d).

3.2 Motion and Control Unit Design

The motion and control unit mainly includes the mathematical modeling and motion control of UUV, who has the tail thruster and the cross rudder.

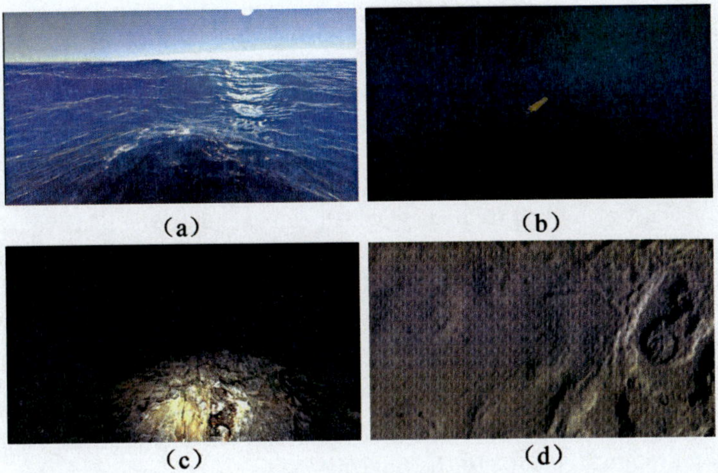

Fig. 3. The underwater environment. (a) Sea level; (b) underwater area; (c) local seabed topography; (d) global seabed topography.

Firstly, the kinematics and dynamics of the UUV can be written as:

$$\dot{\eta} = J(\eta)V \tag{1}$$

$$M\dot{V} + C(V)V + D(V)V + g(\eta) = \tau \tag{2}$$

where $\eta = [x, y, z, \varphi, \theta, \psi]^T$ are the position, roll, pitch and heading angle of UUV, $V = [u, v, w, p, q, r]^T$ are the velocity and angular velocity in X, Y and Z axis respectively, M is the mass matrix, $C(V)$ is the coriolis and centripetal matrix, $D(V)$ is the hydrodynamics matrix, $g(\eta)$ is the gravity and buoyancy matrix, τ is the control input matrix, $J(\eta)$ can be defined as:

$$\begin{aligned} J(\eta) &= \begin{bmatrix} R(\varphi,\theta,\psi) & 0_{3\times 3} \\ 0_{3\times 3} & T(\varphi,\theta) \end{bmatrix} \\ R(\varphi,\theta,\psi) &= \begin{bmatrix} c\theta c\psi & s\varphi s\theta c\psi - c\varphi s\psi & c\varphi s\theta c\psi + s\varphi s\psi \\ c\theta s\psi & s\varphi s\theta s\psi + c\varphi c\psi & c\varphi s\theta s\psi - s\varphi c\psi \\ -s\theta & s\varphi c\theta & c\varphi c\theta \end{bmatrix} \\ T(\varphi,\theta) &= \begin{bmatrix} 1 & s\varphi t\theta & c\varphi t\theta \\ 0 & c\varphi & -s\varphi \\ 0 & s\varphi/c\theta & c\varphi/c\theta \end{bmatrix} \end{aligned} \tag{3}$$

Secondly, the control unit is designed to control the rudder angle and propeller velocity based on the data obtained from the UUV sensor, which can achieve the motion of UUV to accomplish the docking task. Since it is not the main work of this paper, and due to the page limit, the main control design can refer to [24,25], which can be programmed in the Matlab/Simulink environment.

3.3 Environmental Perception and Localization Unit Design

The environmental perception and localization unit mainly includes the inertial navigation system, the Doppler speedometer and the CCD camera.

The inertial navigation system is used to obtain the velocity and acceleration based on (2). By collecting the posture and angular velocity of the UUV in three directions of roll, pitch and bow, it is packaged into a TCP/IP message and sent to the multicast at a fixed frequency, which can calculate the relative position of the UUV and docking point through the attitude angle and angular velocity, and guide the UUV to move to the docking point, thereby realizing the navigation.

The Doppler speedometer is used to obtain the echo frequencies of different beams, which are emitted downward, with regard to the principle of the acoustic reflection, and the beam collision detection method in the Unity software. Based on the echo frequencies of different beams, the velocity of the UUV relative to the ground coordinate system can be calculated, thereby assisting the inertial navigation system to complete the navigation.

The CCD (charge couple device) camera is placed in front of the UUV to simulate the shooting of underwater environment and used to obtain the image in front of the UUV. The CCD camera takes a screenshot of acquired image at a frequency 10 Hz, and packs it into a TCP/IP message and sends it to the multicast at a fixed frequency. By calculating the pixel information, the relative position of the UUV and docking point can be obtained, thereby assisting the inertial navigation system to complete the navigation.

3.4 The Docking Navigation Unit Design

In order to guarantee the high accuracy of the docking process, four different navigation modes are designed in docking navigation unit with regard to the distance between the UUV and docking station, as shown in Fig. 4.

(I) The remote docking guidance: when the UUV needs to return to the docking station but the distance between the docking station is far away and located outside the acoustic navigation range, the inertial navigation system is combined with Doppler speedometer to implement the hybrid navigation, and help navigate the UUV move to acoustic navigation range.

(II) The medium and long distance docking guidance: since the data of the inertial navigation system is generated through the integration, the position error will accumulate over time, which results in the poor accuracy. Therefore, when approaching the acoustic navigation range (at about 50–1000 m), the ultra short baseline (USBL) navigation system is used, where the underwater acoustic signals are used to determine the distance between UUV and docking station.

(III) The medium and short distance docking guidance: the accuracy of the acoustic navigation is inversely proportional to the navigation distance, that means, the closer the distance between the UUV and the docking station, the poorer the accuracy. Therefore, when approaching the medium

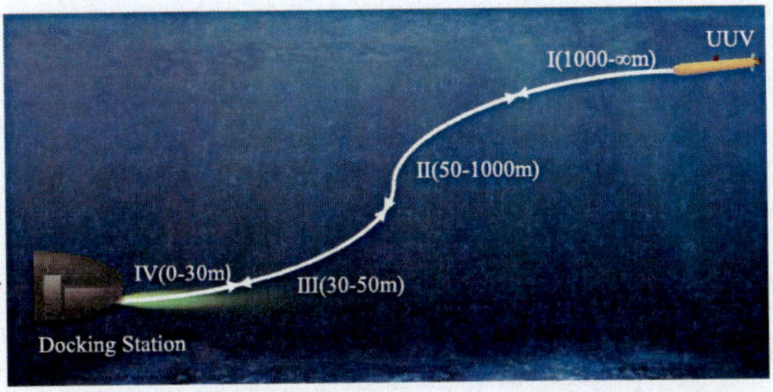

Fig. 4. The four navigation modes in the docking navigation unit.

and short distance (at about 30–50 m), an image navigation system is designed, which relies on a strong penetrating light as a guide signal, by shooting the light with a CCD camera, the relative position of UUV and light can be calculated, which can guide the UUV to approach the docking station.

(IV) The Final docking guidance: when approaching the final distance (less than 30 m) and the light is bright enough, the image navigation system (III) and acoustic navigation system (II) are turned off, and an light navigation system is designed to achieve the final docking guidance. The light navigation system is composed of a light (green) installed on the docking station, a four-quadrant photoelectric detector installed on the head of UUV as a signal detector. Firstly, a cone-shaped diverging light field is generated, after the four-quadrant photoelectric detector senses the light, the relative position between UUV and docking station can be identified through signal processing. More specifically, as shown in Fig. 5, in order to improve the sensitivity of four-quadrant photoelectric detector, a convex lens is installed ahead. When the light is condensed by convex lens, a facula will be formed on the four-quadrant photoelectric detector, whose position can be changed with regard to the posture of UUV, which can result in the change of current in diffident quadrants. Finally, the relative position between the UUV and docking station (green light) can be calculated by the measured current data.

3.5 The Data Monitoring Unit Design

The data monitoring unit is designed to monitor the UUV's state during the docking process. As shown in Fig. 6, the current longitude, latitude, velocity, roll, pitch and heading angles of UUV are presented on the interface, which make it clear for the operator to observe the process of the docking process.

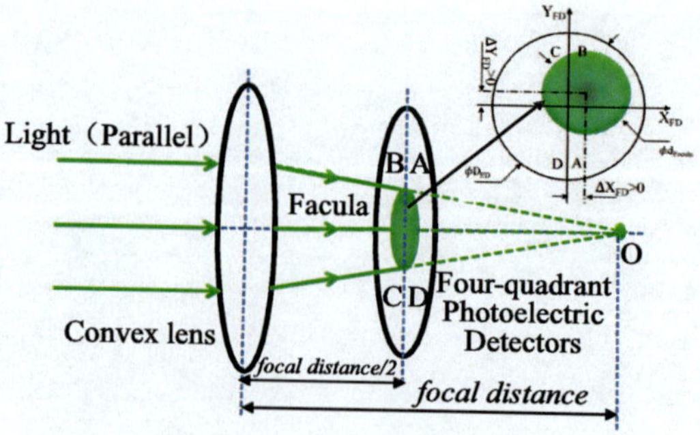

Fig. 5. The principle of light navigation system.

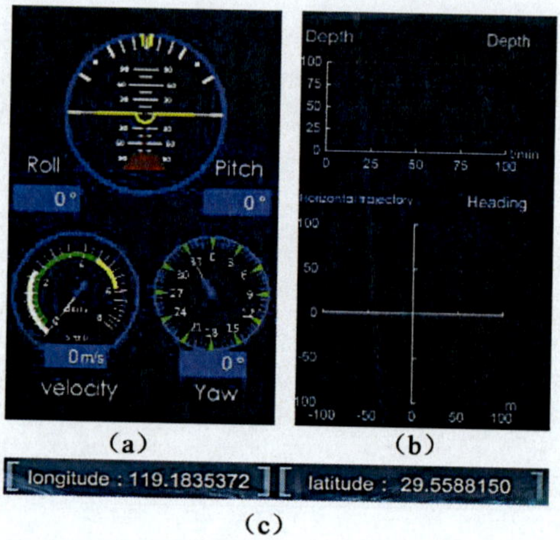

Fig. 6. The data monitoring interface. (a) State monitoring interface; (b) trajectory monitoring interface; (c) longitude and latitude monitoring interface.

4 Application for the Underwater Tasks of UUV

Considering the complexity and expensive cost of the underwater tests, based on the proposed virtual online simulator, the simulation experiment with the motion verification and the underwater docking of UUV are carried out, where the communication between the host and lower computers are achieved by the TCP/IP communication.

4.1 The Motion Verification of UUV

In order to verify the effectiveness of the motion and control unit design, the motion verification experiment is carried out in the virtual online simulator. The random underwater current is set with the velocity range of 0–0.25 m/s and the direction range of 0–360°, the initial position and attitude angle of the UUV are set as $(0, 0, 0)$ and $(0, 270°, 0)$, the desired trajectory is a spiral line and can be given as:

$$\begin{aligned} x_a &= 13.5\cos\left(\frac{\pi}{33}t\right) + 48 \\ y_a &= 13.5\sin\left(\frac{\pi}{33}t\right) + 15 \\ z_d &= -\frac{13.5}{66}t \end{aligned} \tag{4}$$

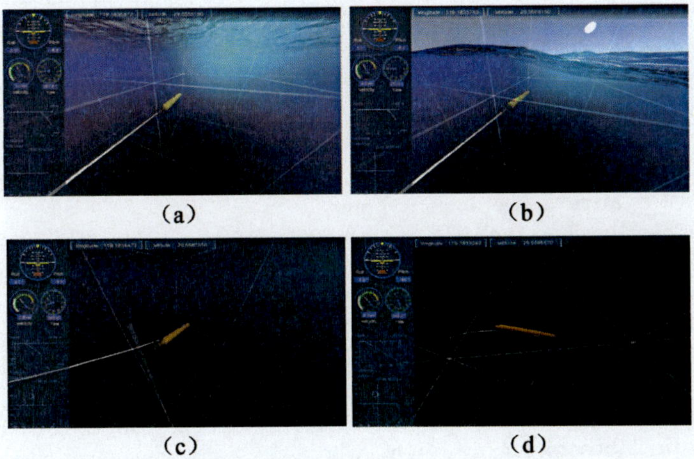

Fig. 7. The process of motion verification. (a) Rectilinear motion; (b) downward motion; (c) spiral motion; (d) motion trajectory.

Figure 7 shows the process of the motion verification. At first, the UUV moves in a rectilinear line until the velocity is stable, and then moves downward in an uniform velocity with the spiral motion. During the motion verification, the data monitoring unit presents the current state of the UUV in real time, such as the longitude, latitude and motion trajectory. To note that, when the UUV moves downward, the color of the underwater environment becomes darker, which means that the virtual online simulator can well simulate the condition of the real underwater environment. Thus, the proposed virtual online simulator can provide the immersive feeling in the motion process of the UUV.

4.2 The Underwater Docking of UUV

In order to verify the effectiveness of the simulated docking process via the whole units proposed in this paper, the underwater docking experiment is carried out in the virtual online simulator, where the relative position between the UUV and docking station is achieved by the docking navigation unit.

Figure 8 shows the process of the underwater docking experiment. Firstly, set the position of the docking point, i.e., the depth, the abscissa and the ordinate as 25 m, 1000 m, 1000 m, respectively. Besides, set the orientation of the docking station be the same as the initial orientation of the UUV. Since the position of the docking point is larger than 1000 m, which is beyond the range of the acoustic navigation system starting from the Period II, the inertial navigation system combined with the Doppler speedometer in the Period I is set as the first way to implement the docking task, as shown in Fig. 8(a). In details, the inertial navigation system and Doppler speedometer are used to calculate the position and posture of the UUV relative to the docking point, and the data are transmitted to the motion and control unit via TCP/IP communication, which achieves the motion of UUV towards the docking point, and the data are real time displayed in the virtual docking scene.

Secondly, when the UUV is within the distance of 1000 m from the docking point, which means it enters the acoustic navigation range, the USBL navigation system in the Period II is set as the second way to implement the docking task, as shown in Fig. 8(b), which can obtain the more accurate position and posture of the UUV relative to the docking point.

Subsequently, when the UUV is within the distance of $50m$ from the docking point, the green light becomes visible, where the image navigation system with a CCD camera in the Period III is set as the third way to implement the docking task, as shown in Fig. 8(c). By sending the data captured by the CCD camera and USBL navigation system in the Period II, the motion and control unit controls the UUV to adjust its posture to prepare for the final docking.

Finally, when the UUV is within the distance of $30m$ from the docking point, the green light becomes more bright, where the light navigation system in the Period IV is set as the forth way to implement the docking task, as shown in Fig. 8(d)–(f). In this period, the image and USBL navigation systems are turned off. Then, by sending the relative position between the UUV and docking station, which can be calculated by the measured current data from four-quadrant photoelectric detector, the motion and control unit can control the UUV to arrive the docking station and finish the docking task.

Based on the two experiments, it can be concluded that the virtual online simulator can simulate the whole process of the UUV to complete the underwater motion and docking tasks. Moreover, through the application for the underwater tasks of UUV in the virtual online simulator, it is possible to safely, conveniently, and low-costly provide the researchers with a platform to optimize the algorithms, and gain experience before the real underwater tests.

5 Conclusion

This paper develops a low-cost virtual online simulator for the docking tasks of UUV, where the virtual reality technology is used to simulate the real underwater environment and the motion of the UUV. In the proposed virtual online simulator, the host and lower computers are selected with several units' design to

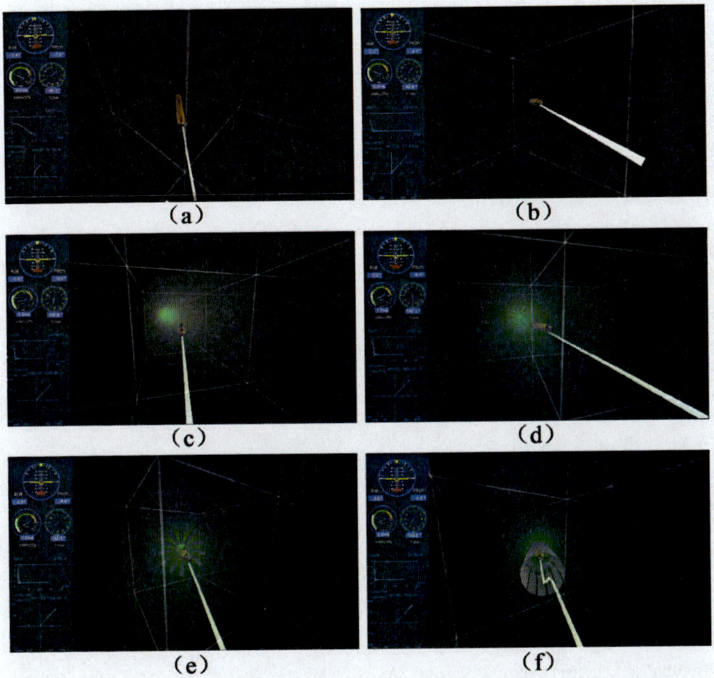

Fig. 8. The process of underwater docking. (a) Period I: remote docking guidance; (b) Period II: medium and long distance docking guidance; (c) Period III: medium and short distance docking guidance; (d) Period IV: final docking guidance; (e) docking preparation stage; (f) finish docking.

send and receive the data via TCP/IP communication in the virtual online simulator. The host computer runs the simulink control program, which calculates the mathematical model, monitors the state of UUV, and simulates or interacts with the real sensor during the docking process. The host computer runs the unity simulation software, which obtains the data from the lower computer, and implements the navigation and control of the UUV. The virtual docking scene is designed to display the simulation of UUV on the screen, and four different navigation modes are designed respectively. Finally, two different experiments are implemented with the specific motion verification and underwater docking for the UUV, and verify the effectiveness and reliability of the proposed virtual online simulator to realize the application for the underwater tasks.

Funding Information. This work is supported by Natural Science Foundation of Zhejiang Province (No. LR23E050001), National Natural Science Foundation of China (No. 52075476), Science Foundation of Donghai Laboratory (No. DH-2022KF01006), and Fundamental Research Funds for the Central Universities (2022FZZX01-06).

References

1. Wang, J., Chen, X., Yang, P.: Adaptive H-infinite Kalman filter based on multiple fading factors and its application in unmanned underwater vehicle. ISA Trans. **108**, 295–304 (2021)
2. Chen, Z., Li, C., Yao, B., Yuan, M., Yang, C.: Integrated coordinated/synchronized contouring control of a dual-linear-motor-driven gantry. IEEE Trans. Ind. Electron. **67**(5), 3944–3954 (2020)
3. Chen, Z., Huang, F., Yang, C., Yao, B.: Adaptive fuzzy backstepping control for stable nonlinear bilateral teleoperation manipulators with enhanced transparency performance. IEEE Trans. Ind. Electron. **67**(1), 746–756 (2020)
4. Sun, W., Tang, S., Gao, H., Zhao, J.: Two time-scale tracking control of nonholonomic wheeled mobile robots. IEEE Trans. Control Syst. Technol. **24**(6), 2059–2069 (2016)
5. Yao, J., Deng, W.: Active disturbance rejection adaptive control of hydraulic servo systems. IEEE Trans. Ind. Electron. **64**(10), 8023–8032 (2017)
6. Yuan, M., Chen, Z., Yao, B., Hu, J.: A general online trajectory planning framework in the case of desired function unknown in advance. IEEE Trans. Ind. Inf. **15**(5), 2753–2762 (2019)
7. Lin, C., Han, G., Du, J., Bi, Y., Shu, L., Fan, K.: A path planning scheme for AUV flock-based internet-of-underwater-things systems to enable transparent and smart ocean. IEEE Internet Things J. **7**(10), 9760–9772 (2020)
8. Mu, X., He, B., Wu, S., Zhang, X., Song, Y., Yan, T.: A practical INS/GPS/DVL/PS integrated navigation algorithm and its application on Autonomous Underwater Vehicle. Appl. Ocean Res. **106**, 102441 (2021)
9. Wang, T., Zhao, Q., Yang, C.: Visual navigation and docking for a planar type AUV docking and charging system. Ocean Eng. **224**, 108744 (2021)
10. Sahoo, A., Dwivedy, S.K., Robi, P.S.: Advancements in the field of autonomous underwater vehicle. Ocean Eng. **181**, 145–160 (2019)
11. Melo, J., Matos, A.: A survey on advances on terrain based navigation for autonomous underwater vehicles. Ocean Eng. **139**, 250–264 (2017)
12. Zhang, W., Teng, Y., Wei, S., Xiong, H., Ren, H.: The robust H-infinity control of UUV with Riccati equation solution interpolation. Ocean Eng. **156**, 252–262 (2018)
13. Qi, X., Xiang, P., Cai, Z.: Spatial target path following and coordinated control of multiple UUVs. Int. J. Naval Archit. Ocean Eng. **12**, 832–842 (2020)
14. Makavita, C.D., Jayasinghe, S.G., Nguyen, H.D., Ranmuthugala, D.: Experimental study of a command governor adaptive depth controller for an unmanned underwater vehicle. Appl. Ocean Res. **86**, 61–72 (2019)
15. Nagao, R., Matsumoto, K., Narumi, T., Tanikawa, T., Hirose, M.: Ascending and descending in virtual reality: simple and safe system using passive haptics. IEEE Trans. Visual Comput. Graphics **24**(4), 1584–1593 (2018)
16. Wang, Q., Jiao, W., Yu, R., Johnson, M.T., Zhang, Y.: Virtual reality robot-assisted welding based on human intention recognition. IEEE Trans. Autom. Sci. Eng. **17**(2), 799–808 (2020)
17. Faure, C., et al.: Adding haptic feedback to virtual environments with a cable-driven robot improves upper limb spatio-temporal parameters during a manual handling task. IEEE Trans. Neural Syst. Rehabil. Eng. **28**(10), 2246–2254 (2020)
18. Kluger, D.T., et al.: Virtual reality provides an effective platform for functional evaluations of closed-loop neuromyoelectric control. IEEE Trans. Neural Syst. Rehabil. Eng. **27**(5), 876–886 (2019)

19. Lipton, J.I., Fay, A.J., Rus, D.: Baxter's homunculus: virtual reality spaces for teleoperation in manufacturing. IEEE Robot. Autom. Lett. **3**(1), 179–186 (2018)
20. Monferrer, A., Bonyuet, D.: Cooperative robot teleoperation through virtual reality interfaces. In: Proceedings Sixth International Conference on Information Visualisation, pp. 243–248 (2002)
21. de la Cruz, M., Casan, G.A., Sanz, P.J., Marín, R.: A new virtual reality interface for underwater intervention missions. IFAC-PapersOnLine **53**(2), 14600–14607 (2020)
22. Bukhari, A.C., Kim, Y.G.: A research on an intelligent multipurpose fuzzy semantic enhanced 3D virtual reality simulator for complex maritime missions. Appl. Intell. **38**(2), 193–209 (2013)
23. Nie, Y., Luan, X., Gan, W., Ou, T., Song, D.: Design of marine virtual simulation experiment platform based on Unity3D. In: Global Oceans 2020: Singapore-U.S. Gulf Coast, pp. 1–5 (2020)
24. Wang, F., Wan, L., Su, Y., et al.: AUV modeling and motion control strategy design. J. Mar. Sci. Appl. **9**, 379–385 (2010)
25. Kong, F., Guo, Y., Lyu, W.: Dynamics modeling and motion control of an new unmanned underwater vehicle. IEEE Access **8**, 30119–30126 (2020)

Simulation Environment for Underwater Vehicles Testing and Training in Unity3D

Piotr Szlęg(✉), Paweł Barczyk, Bruno Maruszczak, Szymon Zieliński, and Emilia Szymańska

Wroclaw University of Science and Technology, Wroclaw, Poland
szlegp@gmail.com

Abstract. Design and operation of Autonomous Underwater Vehicles (AUVs) are relatively new topics, and building a functional one often requires an iterative approach. Because of the risk of damaging costly hardware, the limited availability of infrastructure suitable for testing AUVs, and the significant time investment required, it is often impractical to conduct thorough live experiments for every incremental change. In this paper, the benefits of using a simulated environment for running software tests, training neural networks, and testing the AUV's performance in arbitrary scenarios are explored. By utilizing Unity3D, a cross-platform game engine, a customized framework was developed, allowing for setting up environments for simulating various aspects of underwater operation such as buoyancy and caustics. This framework supports communication with AUV systems (collecting observations through simulated sensors and controlling simulated actuators), gathering datasets for offline training, and randomizing parts of the environment to test the system's robustness.

1 Introduction

Conquest desire is embedded in human history, being a triggering factor for continents and space exploration. However, despite technological progress, the oceans remain a vast basin of undiscovered biological secrets, unmapped ocean relief and buried constructions or wrecks. With both fresh and salt water covering over 70% of the Earth's surface [1], the importance and potential of underwater expeditions are indiminishable. The environment there is hostile – dangerous animals, high pressure and poor or no visibility in aphotic and dysphotic zones make them challenging to investigate.

Because of the invulnerability to such factors as deep-water creatures, decompression sickness and lack of breathable air, carrying out an underwater mission for a machine-composed team is an efficient and safe alternative to a human diver expedition. Implementing artificial intelligence allows for widening the range of applications of such vehicles [2], opening a path for e.g. narrow underwater cave explorations. They have already been tested for multiple possible implementations such as mapping marine wildlife [3], oceanographic sampling or under-ice

I. Petrovic et al. (Eds.): IAS 2022, LNNS 577, pp. 844–853, 2023.
https://doi.org/10.1007/978-3-031-22216-0_56

operations [4], marine archaeology [5] and ocean mining [6]. However, developing a fully functional Autonomous Underwater Vehicle (AUV) or Remotely Operated Vehicle (ROV) in an attempt to optimize the costs is a highly challenging objective. What quickly becomes apparent during the development process is that there are expenses that are avoidable and do not necessarily influence the quality of a robot. Costs associated with replacements for damaged devices, or with access to a testing infrastructure suitable for AUVs can be reduced by lowering the number of physical test iterations. Although it is not feasible in terms of hardware examination, this number is possible to be minimized when conducting experiments with control algorithms. Utilizing a simulation environment adjusted for the robot's tasks allows for evaluating the software even before the real-life vehicle is fully operative. The output of computer vision, classical path planning algorithms, or implemented neural networks can be easily and conveniently assessed. Moreover, the simulation is helpful when gathering the dataset for supervised learning or training the system based on the reinforcement learning approach.

There are various robot-specific simulators such as Gazebo [7], MuJoCo [8] and CoppeliaSim [9]. They accommodate highly-precise physics engines for the difficult task of prototyping and designing robots alongside software development. However, this paper proposes using a more general-purpose engine, namely Unity3D [10]. It offers an open platform for the creation of complex and customizable environments [11,12] and is remarkably extensible, adapted to evolving needs with C# language and providing a comprehensive API with extensive documentation. Albeit intended for game development, it allows for robot design along with simulations and modelling [13–15], assuring high visual fidelity with real-time rendering. With its practical GUI and features, the engine makes it relatively easy to launch a working prototype and avoid low-level programming or framework design decisions. These properties contributed to the selection of Unity3D for the underwater vehicle simulation. Similarly to [16–18], an attempt to implement the water-oriented simulation was made with an extension of a custom graphic user interface and reinforcement learning interface assessment.

In this paper, the implemented simulation system is analysed, including the presentation of its features:

- graphic elements resembling an underwater environment,
- collision detection and water physics inclusion,
- changeable first- and third-person view,
- light and scene randomization for reinforcement learning iterations,
- sensoric data mocking,
- connection to external robot control scripts.

As a case study, Blue Nemo AUV from PWr Diving Crew project was added to the simulator and tested. Its operation, sensor data collection and reinforcement learning training potential were explored and described. The research is complemented with conclusions and a video demonstration.

2 Approach

The main criteria considered when choosing the simulation software were the quality of physics simulation, visual fidelity, developer experience, and adjustability. One of the proposed simulators was Gazebo – a popular choice given its excellent integration with Robot Operating System. It supports Open Dynamics Engine, Bullet, Dart and Simbody engines. Being open-source makes it highly adjustable, although the effort and expertise required to bend it to individual needs are often too high for an average developer. Selected for both general and constrained motion simulation, Gazebo has a visualisation framework, but it has its limitations compared to other modern real-time 3D technologies.

Another simulation software, MuJoCo, uses a proprietary physics engine with incredible accuracy in constrained motion and contact dynamics. Unfortunately, it has limited capabilities in terms of visual fidelity – OpenGL used to power its visualisation requires an enormous amount of work to be able to match modern 3D visualisation software. MuJoCo lacks a GUI suitable for modelling tasks, but it has an XML-based human-readable serialization format. Although this software is currently closed-source, DeepMind announced it will open-source it in 2022, which would increase its adjustability.

Unity3D, on the other hand, is a general-purpose real-time 3D software. Its main physics engine is focused on performance at the expense of precision, but it exposes granular control over the simulation loop and is fairly flexible. AUVs do not require a lot of constrained motion simulation – they rather need an extendable dynamics simulation to cover domain-specific interactions like buoyancy and underwater currents. When it comes to visual fidelity Unity is among the top real-time 3D rendering software second only to Unreal Engine and dedicated artistic software, which usually does not run in real-time. Unity is a fully-featured platform, providing the user with visual modelling tools. Although Unity is closed-source, it exposes a very extensive C# API and supports plug-in architecture, integrating even ROS and MuJoCo with the Unity environment. The software has a comparatively low entry level because of the active game development community providing tutorials and examples.

After running more comparisons with other available platforms, Unity3D was eventually selected as the base of the presented system. The simulator is supposed to be an alternative solution to financially unattainable underwater vehicle development and is expected to be versatile enough to be easily adjusted to any vehicle specification. It has two modes of operation: user manual control and external script input.

In user input mode, Remotely Operated Vehicles' pilots can be trained before controlling a real robot. This possibility is critical in preparation for expeditions in hazardous and tight environments, where the recovery of the robot may be more expensive than accomplishing the original task.

The other mode allows for collecting datasets and creating virtual environments for neural network training. The property of generating randomized and distinct, yet still realistic, scenes is crucial if one wishes to employ machine learning algorithms. Due to the underlying nature of such an artificial intelligence approach, models created on a relatively small set of data have little to no use, as they would fail to carry out assignments in environments different from the samples given during learning. Even though Unity3D provides a variety of tools useful for swift environment creation, the needs arising from machine learning requirements resulted in developing a way of automating this task. This concept is implemented via a system of semi-random scenes.

Regarding the surrounding generation, the body of water was hand-modelled and textured. In designated areas of the scene, sets of props and decorations are randomly distributed each time a simulation is run. In terms of imitating buoyancy, objects can be laid on the ground, left suspended in the water, or set afloat on the surface. Moreover, each object has predefined boundaries that determine how close it can be placed in relation to the surroundings; without them, props would often unnaturally intersect each other. Besides the physical objects randomization, other environmental parameters are also different for each run – mainly watercolour, lighting conditions (Fig. 1), water current and particles.

Fig. 1. Different ambient lights comparison.

The omniscient nature of the simulator, having known the location of each object, including the robot, provides numerous opportunities to enhance machine learning processes. The generation of datasets used for training object recognition networks is trivial, as the software is able to both capture positive examples, i.e. pictures containing the target, and accurately generate a 2D bounding box around objects, a process that would normally have to be done manually. Complex phenomena responsible for e.g. sonar and hydrophone functionality, are approximated via built-in mechanisms such as raycasts, reducing the computation power required. The simulation of the propulsion system was determined not to be of vital concern. Unity3D's built-in physics system would not be sufficient to simulate intricate fluid dynamics, an element crucial to a meaningful recreation of propeller-based movement. Moreover, even slight changes in the hardware of the AUV would require at least partial remodelling of the propulsion system. However, an abstraction based on unified vectors of thrust and torque eases the process of adjusting the virtual AUV to the behaviour of the real one. This approach opens new paths for future expansion, e.g. introducing options to fine-tune the AUV properties in runtime, not only in the Unity3D editor.

The part of the codebase responsible for communication with the simulator and wrapping it in workable abstractions is built with Python. It has a structure based on component-oriented programming and Liskov's substitution principle. For each device managed by the autonomy system, there is an interface with utility methods and different implementations that can be freely modified. The AUV software model is then composed of these device implementations as in Fig. 2. The actual sensors can be connected to the personal computer and the autonomy behaviour in the simulation can be examined. The simulator also allows for testing control algorithms on simulated devices if the physical devices are not available at that particular moment.

Fig. 2. Example of robot's available sensoric data display. Zeros indicate the exclusion of the features in this specific case.

In addition to native Unity features and in-house developed features, the simulator makes use of third-party, specialized solutions. Two base packages make this level of simulation possible in real time: Aura 2 and Unity After Effects. Aura 2 adds high-performance volumetric lighting to Unity 3D, while Unity After Effects allows for miscellaneous post-processing effects like colour grading or blurs. The algorithms used in these packages are the same as the ones used in a lot of current commercial games, and most importantly work in real time. Two custom post-processing shaders were added for more realism as various existing and easily available real time solutions were too simplistic and unable to simulate physical phenomena. The first shader works by displacing pixels using 3D Perlin noise vector fields. The second one applies different amounts of blurring depending on the depth and 3D Perlin noise (Fig. 3). The three Perlin noise dimensions correspond to 2D screen coordinates and time.

The simulator implements a camera management system that offers a third-person perspective (Fig. 4). The system selects two points from a set of points on a grid or from hand-picked locations. It attaches the camera to the point

Fig. 3. Blurring process visualization: simplex noise (left), simplex noise and depth value (center), image with varying depth blur (right).

closest to the robot and observes the robot's actions. Should the vehicle come too close, the camera switches to the previously selected second closest point in order to avoid the robot dominating the viewport. When the robot moves away, the camera returns to its previous position.

Fig. 4. Camera from 1st and 3rd person perspectives.

3 Simulation Tests

Tests of the simulator applications were conducted on the example of the AUV Blue Nemo of the PWr Diving Crew project (Fig. 5).

Fig. 5. Blue Nemo of PWr Diving Crew project in simulation.

Receiving and processing sensoric data, and the use of vision to perform tasks autonomously were evaluated in this paper. The default environment for testing the robot in real life is a swimming pool, so a similar surrounding was implemented in the simulator. Additionally, a custom graphical user interface (Fig. 6) was developed to enable user input for robot control. It allows for convenient remote operation and sensoric data observation.

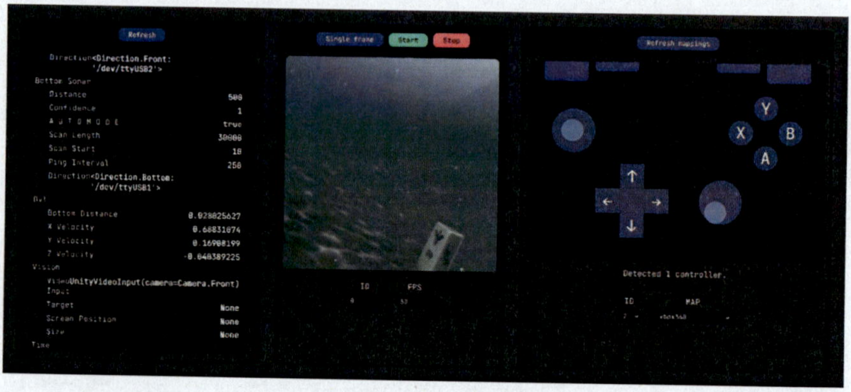

Fig. 6. Graphical user interface used for the drone operation connected to the simulator.

First, after adding the implementations of the relevant sensors (in this case, sonar, Doppler Velocity Log and accelerometer), it was possible to observe the correct change in the readings when moving the robot. These values were recorded and used to train a neural network, which enabled (in combination with computer vision) the robot to perform the assigned task. The considered objective was to navigate through a gate. Therefore, a model based on YOLOV5 [19] vision (Fig. 8) was trained on the collected dataset and used to detect the object (Fig. 7).

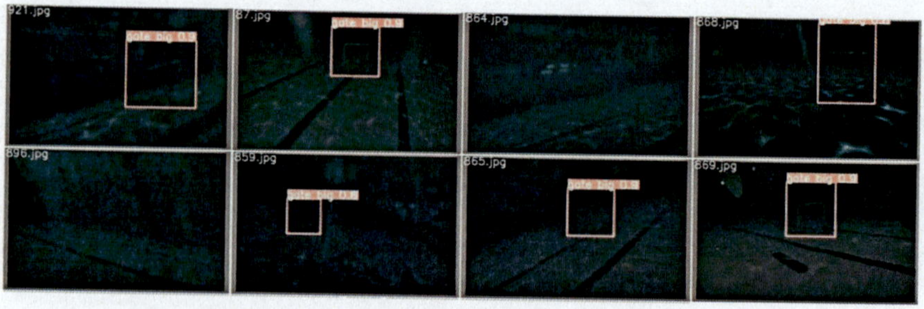

Fig. 7. YOLOv5 vision training validation for gate detection.

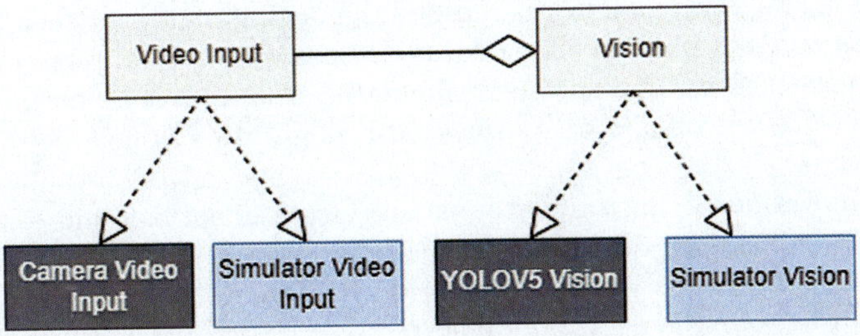

Fig. 8. A part of the virtual AUV model responsible for image object detection. YOLOV5 Vision refers to an abstract wrapper class around a popular YOLOV5 neural network architecture. The Simulator Vision component uses mesh bounding boxes available in the simulator to mimic the neural network results. Simulator Vision readings can be also used to generate labelled computer vision datasets that can be later adjusted to different neural network architectures and training methods.

Then, using this information and sensor data, the system was trained with a reinforcement learning approach for the robot to reach the goal autonomously (Fig. 9).

Fig. 9. Gate detection and navigation towards it.

4 Conclusions

In this paper, a simulation environment for underwater vehicles was developed and presented. The experimental evaluation on the example of Blue Nemo AUV of PWr Diving Crew project proved the simulator to be valuable in terms of neural network training or algorithm assessment before implementing them on a real-life robot. The simulator is a useful debugging platform when testing new devices, as one can provide the simulated robot with actual sensor readings or activate real actuators in response to simulated stimuli. It is especially important when the vehicle is off-site or its physical configuration is not ready for testing these devices.

One limitation of the simulator is the visual fidelity of the environments. It would be interesting to explore technologies used in autonomous car training environments like neural rendering and raytracing. Another improvement to consider would be to implement more nuanced physical simulations, including more realistic underwater currents or simulated vegetation. The currently developed Unity3D physics solver for joint constrained kinematics aims to match the precision of more robust physics simulators; it would cover more edge cases where the vehicle could get damaged, stuck or behave unpredictably. A lot of new small features are planned to be added, from ergonomics improvements, through devices emulation to utilizing different already existing components for neural networks training.

References

1. Karachevtseva, I.P., Kokhanov, A.A., Zubarev, A.E., Rodionova, Z.F., Matveev, E.V., Garov, A.S.: Modern methodology and new tools for planetary mapping. In: Gartner, G., Jobst, M., Huang, H. (eds.) Progress in Cartography. Lecture Notes in Geoinformation and Cartography. Springer, Cham (2016)
2. Petillot, Y.R., Antonelli, G., Casalino, G., Ferreira, F.: Underwater robots: from remotely operated vehicles to intervention-autonomous underwater vehicles. IEEE Robot. Autom. Mag. **26**(2), 94–101 (2019). https://doi.org/10.1109/MRA.2019.2908063
3. Barrett, N., Seiler, J., Anderson, T., Williams, S., Nichol, S., Nicole Hill, S.: Autonomous Underwater Vehicle (AUV) for mapping marine biodiversity in coastal and shelf waters: implications for marine management. In: OCEANS'10 IEEE SYDNEY, pp. 1–6 (2010). https://doi.org/10.1109/OCEANSSYD.2010.5603860
4. Manley, J.E.: Autonomous underwater vehicles for Ocean Exploration. In: Oceans 2003. Celebrating the Past... Teaming Toward the Future (IEEE Cat. No. 03CH37492), vol. 1, pp. 327–331 (2003). https://doi.org/10.1109/OCEANS.2003.178578
5. Mindell, D., Bingham, B.: New archaeological uses of autonomous underwater vehicles. In: MTS/IEEE Oceans 2001. An Ocean Odyssey. Conference Proceedings (IEEE Cat. No. 01CH37295), vol. 1, pp. 555–558 (2001). https://doi.org/10.1109/OCEANS.2001.968781
6. Senke, W.: Applications of autonomous underwater vehicles (AUVs) in ocean mining exploration. In: 2013 OCEANS - San Diego, pp. 1–3 (2013). https://doi.org/10.23919/OCEANS.2013.6740944
7. Koenig, N., Howard, A.: Design and use paradigms for Gazebo, an open-source multi-robot simulator. In: 2004 IEEE/RSJ International Conference on Intelligent Robots and Systems (IROS) (IEEE Cat. No. 04CH37566), vol. 3, pp. 2149–2154 (2004). https://doi.org/10.1109/IROS.2004.1389727
8. Todorov, E., Erez, T., Tassa, Y.: MuJoCo: a physics engine for model-based control. In: 2012 IEEE/RSJ International Conference on Intelligent Robots and Systems, pp. 5026–5033 (2012). https://doi.org/10.1109/IROS.2012.6386109

9. Rohmer, E., Singh, S.P.N., Freese, M.: CoppeliaSim (formerly V-REP): a Versatile and Scalable Robot Simulation Framework. In: IEEE/RSJ International Conference on Intelligent Robots and Systems, 2013. www.coppeliarobotics.com
10. Technologies, U.: Unity - Manual: Unity Manual [online]. Docs.unity3d.com (2005). Available at: http://docs.unity3d.com/Manual/index.html. Accessed 10 Feb 2022
11. Yang, K., Jie, J., Haihui, S.: Study on the virtual natural landscape walkthrough by using Unity 3D. In: 2011 IEEE International Symposium on VR Innovation, pp. 235–238 (2011). https://doi.org/10.1109/ISVRI.2011.5759642
12. Kuang, Y., Bai, X.: The research of virtual reality scene modeling based on Unity 3D. In: 2018 13th International Conference on Computer Science & Education (ICCSE), pp. 1–3 (2018). https://doi.org/10.1109/ICCSE.2018.8468687
13. Yong-kang, J., Yong, C., Daquan, T.: Design of an UAV simulation training and assessment system based on Unity3D. In: 2017 IEEE International Conference on Unmanned Systems (ICUS), pp. 163–167 (2017). https://doi.org/10.1109/ICUS.2017.8278334
14. Xueying, L., Shi, M., Li, C.: Design and implement of swarm self-organization model based on Unity3D. In: 2017 IEEE 2nd Information Technology, Networking. Electronic and Automation Control Conference (ITNEC), pp. 146–149 (2017). https://doi.org/10.1109/ITNEC.2017.8284926
15. Song, K., Xu, H., Ding, Y., Li, H.: A target tracking realization method of UAV simulation training system. In: 2016 14th International Conference on Control, Automation, Robotics and Vision (ICARCV), pp. 1–4 (2016). https://doi.org/10.1109/ICARCV.2016.7838616
16. Nie, Y., Luan, X., Gan, W., Ou, T., Song, D.: Design of marine virtual simulation experiment platform based on Unity3D. In: Global Oceans 2020: Singapore–U.S. Gulf Coast 2020, pp. 1–5 (2020). https://doi.org/10.1109/IEEECONF38699.2020.9389006
17. Li, D.: Application of 3D virtual ocean image simulation in ship vision. J. Coastal Res. **94**, 530 (2019)
18. Chen, Ge., et al.: Design and implementation of a 3D ocean virtual reality and visualization engine. J. Ocean Univ. China **11**(004), 481–487 (2012)
19. Karthi, M., Muthulakshmi, V., Priscilla, R., Praveen, P., Vanisri, K.: Evolution of YOLO-V5 algorithm for object detection: automated detection of library books and performance validation of dataset. In: 2021 International Conference on Innovative Computing, Intelligent Communication and Smart Electrical Systems (ICSES), pp. 1–6 (2021). https://doi.org/10.1109/ICSES52305.2021.9633834

Sliding Mode Control for Underwater Multi-DoF Hydraulic Manipulator

Shizhao Zhou[1], Zheng Chen[1,2](✉), and Shiqiang Zhu[1,3]

[1] Ocean College, Zhejiang University, Zhoushan 316021, China
[2] State Key Laboratory of Fluid Power and Mechatronic Systems, Zhejiang University, Hangzhou 310000, China
zheng_chen@zju.edu.cn
[3] Zhejiang Lab, Hangzhou 310000, China

Abstract. Underwater hydraulic manipulator is a typical underwater operation tool. With the progress of industrial automation, the application occasions of underwater hydraulic manipulator are increasing, and some of occasions put higher demand on operation accuracy. However, due to the negative influence caused by joint coupling, underwater dynamics, model nonlinearity and unknown external disturbance, the control performance of underwater hydraulic manipulators is limited, which results that underwater hydraulic manipulators are gradually restricted in the high-precision operations. In this paper, for achieving good control performance of multi-DoF underwater hydraulic manipulator, the comprehensive dynamic model of multi-DoF underwater hydraulic manipulator considering joint coupling and underwater dynamics is established, and the sliding mode controller is designed to cope with the negative influence caused by model nonlinearity and unknown external disturbance, and also, guarantees stability of the system. Finally, the experiment is carried out to verify the effectiveness and reliability of the designed controller.

Keywords: Sliding mode control · Multi-DoF hydraulic manipulator · Model nonlinearity · Underwater dynamics

1 Introduction

With the development of society, the demand for automation and intelligence in the industrial field is increasing, and industrial automation equipment represented by multi-degree-of-freedom(multi-DoF) manipulator is more and more widely applied in various industrial occasions [1,2]. Commonly used manipulators are divided into electric manipulators and hydraulic manipulators. Due to the larger output torque and higher power-to-weight ratio, hydraulic manipulators are more widely used in practical applications such as underwater operation [3]. However, in the control of traditional underwater hydraulic manipulator, whether the operation behavior is completed or not is more concerned

I. Petrovic et al. (Eds.): IAS 2022, LNNS 577, pp. 854–866, 2023.
https://doi.org/10.1007/978-3-031-22216-0_57

about, while the operational accuracy has received little attention [4]. However, with the continuous expansion of the application range, underwater hydraulic manipulators are gradually restricted in certain specific occasions that require high-precision operations, which is due to the low operation accuracy compared to electric manipulators [5,6].

The poor operation accuracy of the underwater hydraulic manipulator at this stage is mainly because that it usually adopts open-loop control or closed-loop PID control [7]. Due to that these methods do not consider the dynamic characteristics of the hydraulic manipulator, it is impossible to obtain better control effect [8]. To achieve better control performance, the nonlinear controller needs to be designed.

The main difficulties in the design of the nonlinear controller of the underwater hydraulic manipulator can be summarized as follow:

Firstly, the rectilinear hydraulic cylinder is used to drive the joint rotation, so that there is nonlinear mapping between the joint angle and the displacement of the underwater hydraulic cylinder rod [9,10]. Besides, factors such as hydraulic transmission components and environmental changes will introduce negative influence such as strong nonlinearity, unknown external disturbance, and joint motion coupling [11]. The adaptive backstepping control method for overcoming the influence of nonlinear characteristics was proposed based on the hydraulic servo positioning system in [12–14]. The negative influence of joint coupling is considered and a damping decoupling compensation method is proposed in [15]. However, most of these methods are solutions to a specific problem mentioned above but do not comprehensively consider the impact of all aspects.

Secondly, in the underwater operation occasion, there is extra resistance and lift force generated by the fluid, which resulting in extra stronger external disturbance. To solve this problem, an observer based PD inversion robust nonlinear controller is proposed in [16]. However, the control effect is not idea because the hydraulic dynamic characteristics are not considered in detail.

In this paper, for realizing the precise movement control of underwater hydraulic manipulator, the comprehensive dynamic model of multi-DoF hydraulic manipulator considering multi-joint coupling and underwater dynamics is established, and the nonlinear sliding mode control law is designed to overcome the problem caused by model nonlinearity and unknown external disturbance, and also, guarantees stability of the system. Finally, serious of experiments were carried out based on a two-Dof hydraulic manipulator platform to verify the effectiveness and reliability of the designed controller.

The remainder of this paper is organized as follows. Section 2 describes the dynamic model of the underwater hydraulic manipulator. The sliding mode controller is designed in Sect. 3. Section 4 proposes the experiment to test the performance of the designed controller. Conclusion is involved in Sect. 5.

2 Problem Formulation and Dynamic Models

Most of the general hydraulic manipulators are link mechanisms, and the joints of two adjacent links constitute joints, so the n-joint hydraulic manipulator is the same as the n-link hydraulic manipulator. The typical n-link underwater hydraulic manipulator is shown in Fig. 1.

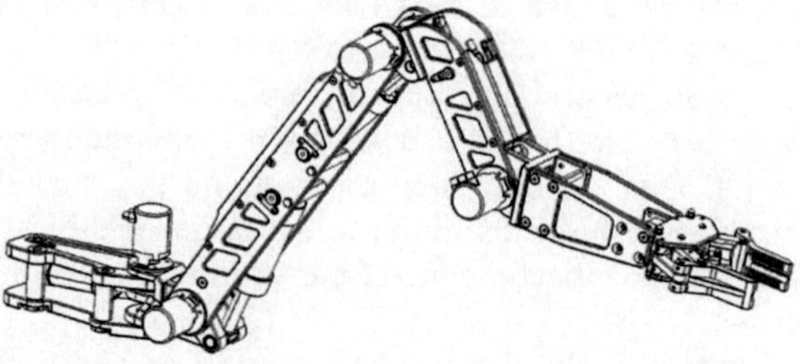

Fig. 1. Typical hydraulic manipulator model.

The rotation angle value of each joint is defined as $q = [q_1, q_2, q_3, \cdots\cdots, q_n]^T$. The general dynamic model of the n-link manipulator can be expressed as Eq. 1:

$$M(q)\,\ddot{q} + C(q,\ \dot{q})\,\dot{q} + G(q) + F(\dot{q}) = \tau + D + \Delta \tag{1}$$

where M, C, G is the inertia matrix, the Coriolis and centrifugal force matrix, the gravity matrix of n-link manipulation. D represents the calculable disturbance such as friction. Δ represents the incalculable unknown external disturbance. τ is joint rotation torque. $F(\dot{q})$ represents the underwater dynamic including viscous drag and added mass effect. Because that the added mass effect in underwater environment is several magnitudes smaller than hydraulic driving force, only the viscous drag term is modeled as follows by referring to the relative research:

$$F(\dot{q}) = \frac{1}{2}\rho C_d A_c L^2 \dot{q} \tag{2}$$

where ρ is water density, C_d is water resistance coefficient, A_c represents contact area of manipulator while moving relative to water and L represents the length of manipulator.

Considering that each joint of the manipulator is driven by the corresponding hydraulic cylinder, the joint rotation torque can be expressed as:

$$\tau = J(d, q)(P_i\ A_i - P_o\ A_o) \tag{3}$$

where $d = [d_1, d_2, d_3, \cdots\cdots, d_n]^T$ represents the displacement of joint cylinder rod. $J(d, q)$ represents the conversion relationship between the displacement of the joint hydraulic cylinder rod and the joint rotation angle value which can be expressed as:

$$\frac{\partial d}{\partial q} = \begin{bmatrix} \frac{\partial d_1}{\partial q_1} & \cdots & \frac{\partial d_1}{\partial q_n} \\ \vdots & \ddots & \vdots \\ \frac{\partial d_n}{\partial q_1} & \cdots & \frac{\partial d_n}{\partial q_n} \end{bmatrix}. \tag{4}$$

$P_i = [P_{i_1}, P_{i_2}, \cdots\cdot, P_{i_n}]^T$ and $P_o = [P_{o_1}, P_{o_2}, \cdots\cdot, P_{o_n}]^T$ represent the pressure in the oil inlet chamber and oil return chamber of the joint hydraulic cylinder respectively. $A_i = [A_{i_1}, A_{i_2}, \cdots\cdot, A_{i_n}]^T$ and $A_o = [A_{o_1}, A_{o_2}, \cdots\cdot, A_{o_n}]^T$ the area of the oil inlet chamber and oil return chamber of the joint hydraulic cylinder rod respectively.

In addition to the general link mechanisms dynamic equation, the dynamic model of the hydraulic manipulator also includes the dynamic model of the hydraulic drive system. In this paper, the dynamic model of the hydraulic drive system is established as follows:

$$\frac{V_i}{\beta_e}\dot{P_i} = -J(d,q)A_i\dot{q} + Q_i \tag{5}$$

$$\frac{V_o}{\beta_e}\dot{P_o} = J(d,q)A_o\dot{q} - Q_o \tag{6}$$

where $V_i = V_{hi} + A_i diag[d]$ and $V_o = V_{ho} + A_o diag[d]$ represents the volume of oil inlet chamber and oil return chamber of hydraulic cylinder respectively. V_{hi} and V_{ho} represents the initial volume of the inlet chamber and oil return chamber respectively. β_e is the effective bulk modulus.

In order to describe the hydraulic system modeling more intuitively, the diagram of n-joint manipulator hydraulic drive system is shown in Fig. 2. The power source of the hydraulic drive system is a hydraulic pump, and the flow of driving each joint hydraulic cylinder is controlled by the corresponding joint hydraulic servo valve.

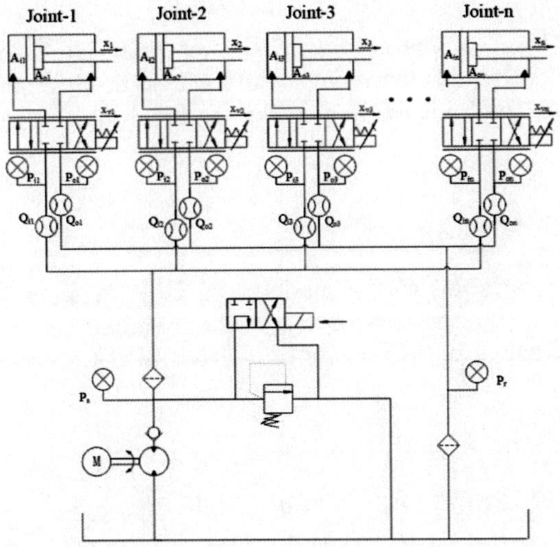

Fig. 2. Schematic of hydraulic system.

Assuming no leakage of hydraulic cylinder, the flow rate of hydraulic cylinder inlet chamber $\mathrm{Q_i} = [\mathrm{Q}_{i1}, \mathrm{Q}_{i2}, \cdots\cdot, \mathrm{Q}_{in}]^{\mathrm{T}}$ and the flow rate of hydraulic cylinder return chamber $\mathrm{Q_o} = [\mathrm{Q}_{o1}, \mathrm{Q}_{o2}, \cdots\cdot, \mathrm{Q}_{on}]^{\mathrm{T}}$ can be determined according to [9]:

$$Q_i = k_{qi} g_i (P_i, x_v) x_v \tag{7}$$

$$Q_o = k_{qo} g_o (P_o, x_v) x_v \tag{8}$$

where $x_v = [x_{v1}, x_{v2}, \ldots\ldots, x_{vn}]^T$ is the hydraulic valve displacement of each joint. $k_{qi} = diag\,[k_{qi1}, \ldots\ldots, k_{qin}]$ and $k_{qo} = diag\,[k_{qo1}, \ldots\ldots, k_{qon}]$ represents the flow gain constants of the oil inlet chamber and oil return chamber respectively. $g_i (P_i, x_v)$ and $g_o (P_o, x_v)$ is the function of P and x_v which is written as follows:

$$g_i (P_i, x_v) = diag\,[g_{i_1} (P_{i_1}, x_{v_1}), g_{i_2} (P_{i_2}, x_{v_2}), \ldots\ldots, g_{i_n} (P_{i_n}, x_{v_n})] \tag{9}$$

$$g_o (P_o, x_v) = diag\,[g_{o_1} (P_{o_1}, x_{v_1}), g_{o_2} (P_{o_2}, x_{v_2}), \ldots\ldots, g_{o_n} (P_{o_n}, x_{v_n})] \tag{10}$$

$$g_{i_n} (P_{i_n}, x_{v_n}) = \begin{cases} \sqrt{P_s - P_{i_n}} & x_{v_n} \geq 0 \\ \sqrt{P_{i_n} - P_r} & x_{v_n} < 0 \end{cases} \tag{11}$$

$$g_{o_n} (P_{o_n}, x_{v_n}) = \begin{cases} \sqrt{P_{o_n} - P_r} & x_{v_n} \geq 0 \\ \sqrt{P_s - P_{o_n}} & x_{v_n} < 0 \end{cases} \tag{12}$$

where P_s and P_r represents is the supply pressure of the hydraulic pump and the reference pressure of the whole hydraulic system.

Besides, in the actual control process, the hydraulic valve displacement can be directly controlled by the hydraulic valve input voltage V:

$$x_v = f(V) \tag{13}$$

The control target of the actual multi-joint hydraulic manipulator system is controlling the valve displacement x_v of each joint through the valve input voltage V to make the joint angle q tracks the ideal joint angle q_d as closely as possible.

3 Controller Design

Defined the state variables of the model x as $x = [x_1, x_2, x_3]^T = [q, \dot{q}, \ddot{q}]$ and the output of control system u as $u = x_v$. The dynamic model of the hydraulic manipulator established in Sect. 2 can be written in the form of the state space shown below:

$$\begin{cases} \dot{x}_1 = x_2 \\ \dot{x}_2 = x_3 \\ \dot{x}_3 = M^{-1} \left[J_t \beta_e \left(A_i V_i{}^{-1} k_{qi} g_i + A_o V_o{}^{-1} k_{q_o} g_o \right) u \right. \\ \qquad \left. - J_t \beta_e J_t \left(A_i V_i{}^{-1} + A_o V_o{}^{-1} \right) x_2 - C x_3 + \dot{D} - \dot{F} - \Delta \right] \end{cases} \tag{14}$$

It can be seen from the state space that the dynamic model of the multi-joint hydraulic manipulator is a typical high-order nonlinear dynamic model.

Therefore, in this paper, the backstepping method is used to cope with the higher-order characteristics of the dynamic model, and the high-order sliding mode controller is designed to achieve good tracking control performance of the nonlinear system.

Defined the tracking error: $z_1 = x_1 - x_{1d}$. x_{1d} is the desired joint angle.

Defined the first visual input x_{2d} for x_2 as: $x_{2d} = \dot{x}_{1d} - k_1 z_1$, in where k_1 is the controller gain. And the error of x_2 can be written as: $z_2 = x_2 - x_{2d}$.

Defined the second visual input x_{3d} for x_3 as: $x_{3d} = \dot{x}_{2d} - z_1 - k_2 z_2$, in where k_2 is the controller gain. And the error of x_3 can be written as: $z_3 = x_3 - x_{3d}$.

Defined the sliding manifold as: $s = \lambda_1 z_1 + \lambda_2 z_2 + z_3$, in where λ_1 and λ_2 is the sliding surface gain. The differential of s can be written as:

$$\dot{s} = \lambda_1 \dot{z}_1 + \lambda_2 \dot{z}_2 + \dot{z}_3 \tag{15}$$

The differential of z_1, z_2, z_3 can be written respectively as:

$$\begin{aligned}\dot{z}_1 =& x_2 - \dot{x}_{1d} \\ =& x_2 - x_{2d} - k_1 z_1\end{aligned} \tag{16}$$

$$\begin{aligned}\dot{z}_2 =& x_3 - \dot{x}_{2d} \\ =& x_3 - x_{3d} - k_2 z_2 - z_1\end{aligned} \tag{17}$$

$$\begin{aligned}\dot{z}_3 =& \dot{x}_3 - \dot{x}_{3d} \\ =& M^{-1}\left[J_t\beta_e\left(A_i{V_i}^{-1}k_{qi}g_i + A_o{V_o}^{-1}k_{q_o}g_o\right)u\right. \\ & \left.-J_t\beta_e J_t\left(A_i{V_i}^{-1} + A_o{V_o}^{-1}\right)x_2 - Cx_3 + \dot{D} - \dot{F} - \Delta\right] - \dot{x}_{3d}\end{aligned} \tag{18}$$

Combined with Eqs. (15)–(18), The differential of s can be written as:

$$\begin{aligned}\dot{s} =& M^{-1}\left[J_t\beta_e\left(A_i{V_i}^{-1}k_{qi}g_i + A_o{V_o}^{-1}k_{q_o}g_o\right)u - J_t\beta_e J_t\left(A_i{V_i}^{-1} + A_o{V_o}^{-1}\right)x_2\right. \\ & \left.-Cx_3 + \dot{D} - \dot{F} - \Delta\right] + \lambda_1(z_2 - k_1 z_1) + \lambda_2(z_3 - z_1 - k_2 z_2) - \dot{x}_{3d}\end{aligned} \tag{19}$$

According to Eq. (19), the sliding mode control law can be obtain as:

$$\begin{cases} u = u_c + u_r \\ u_c = {\rho_1}^{-1}[-\lambda_1(z_2 - k_1 z_1) - \lambda_2(z_3 - z_1 - k_2 z_2) \\ \qquad +\rho_2 x_2 + \rho_3 x_3 + \rho_4 + \dot{x}_{3d}] \\ u_r = -k_3 s + \varepsilon sgn(s) \end{cases} \tag{20}$$

where k_3 is the controller gain. ρ_1, ρ_2, ρ_3 and ρ_4 is the controller characteristic quantity which can be written as:

$$\begin{cases} \rho_1 = M^{-1}J_t\beta_e\left(A_i{V_i}^{-1}k_{qi}g_i + A_o{V_o}^{-1}k_{q_o}g_o\right) \\ \rho_2 = J_t\beta_e J_t\left({A_i}^2{V_i}^{-1} + {A_o}^2{V_o}^{-1}\right) \\ \rho_3 = C \\ \rho_4 = -\dot{D} + \dot{F} \end{cases} \tag{21}$$

$sgn(s)$ is the switching function which can be given as

$$sgn(s) = \begin{cases} 1 & \text{if } s > 0 \\ 0 & \text{if } s = 0 \\ -1 & \text{if } s < 0 \end{cases} \tag{22}$$

Theorem 1. *If after finite time, when the unknown external disturbance is bounded and the boundaries are known ($|\Delta| \leq \delta$), and the parameters of the designed SMC ($K = [k_1, k_2, k_3]^T$, $\lambda = [\lambda_1, \lambda_2]^T$) meet Eq. (23), the hydraulic manipulator system can converge to the sliding manifold $s = 0$, and the asymptotic output tracking can be achieved.* ■

$$\begin{cases} K > 0 \\ \lambda > 0 \\ k_1 k_3 \left(k_2 + \lambda_2\right) - \left(k_1 + k_3 {\lambda_1}^2/4\right) > 0 \end{cases} \tag{23}$$

Proof. Defined the Lyapunov function as: $V = \frac{{z_1}^2}{2} + \frac{{z_2}^2}{2} + \frac{s^2}{2}$. According to Eqs. (16), (17) and (19), the differential of V can be written as:

$$\begin{aligned} \dot{V} =& - k_1 {z_1}^2 - k_2 {z_2}^2 - z_2 z_3 + s\,[\lambda_1 (z_2 - k_1 z_1) \\ & + \lambda_2 (z_3 - z_1 - k_2 z_2) + \rho_1 u - \rho_2 x_2 - \rho_3 x_3 - \rho_4 - \Delta - \dot{x_{3d}}] - \varepsilon |s| \end{aligned} \tag{24}$$

Substituting Eq. (20) into Eq. (24):

$$\dot{V} = -k_1 {z_1}^2 - k_2 {z_2}^2 - k_3 s^2 - z_2 z_3 - \Delta s - \varepsilon |s| \tag{25}$$

Combined the definition of sliding surface and Eq. (25), the differential of the Lyapunov function can be rewritten as:

$$\dot{V} = -Z^T H Z - \Delta s - \varepsilon |s| \tag{26}$$

$$\leq -Z^T H Z \tag{27}$$

where $Z = [z_1, z_2, z_3]^T$ represents the vector of error and H is the error coefficient matrix, which can be expressed as:

$$H = \begin{bmatrix} k_1 + k_3 \lambda_1^2 & \lambda_1 \lambda_2 k_3 & \lambda_1 k_3 \\ \lambda_1 \lambda_2 k_3 & k_2 + k_3 \lambda_2^2 & \lambda_2 k_3 - \frac{1}{2} \\ \lambda_1 k_3 & \lambda_2 k_3 - \frac{1}{2} & k_3 \end{bmatrix}. \tag{28}$$

When the condition shown as Eq. (23) is satisfied, the error coefficient matrix H remains positive definite. And Eq. (27) can meet the following condition:

$$\dot{V} \leq -Z^T H Z < 0 \tag{29}$$

Since then, the asymptotic output tracking in the process of control law design is achieved.

The proof for Theorem 1 is completed. □

Remark I. Due to the existence of unknown external disturbance, the existence of $sgn(s)$ can cause system chattering. In order to reduce chattering, the switching function $sgn(s)$ designed in the controller is replaced by the saturation function $sat(s)$, which can be written as:

$$sat(s) = \begin{cases} sgn(s), & |s| > \varphi \\ s/\varphi, & |s| \le \varphi \end{cases} \tag{30}$$

where φ is designed boundary thickness. ◇

Remark II. Theorem 1 shows that the proposed controller can guaranteed stability in the manipulator system, and also achieve good control performance eventually ($z_1 \to 0$ as $t \to \infty$). ◇

4 Experiment Verification

In order to verify the control performance of the designed sliding mode controller, the experiments based on a specific hardware platform are implemented. At the same time, the traditional PID controller is selected as the comparison reference, so as to reflect the performance of the designed controller more intuitively.

4.1 Experimental Setup

The experimental platform is a typical underwater hydraulic manipulator (shown in Fig. 3). In order to verify the control effect of the proposed SMC in each DoF, and also to avoid the negative influence for control performance evaluation while too many joints moving simultaneously, only two joints are moved in the experiment in this paper, as marked with the red box in Fig. 3. The input voltage of hydraulic servo valve is mapping the valve opening (1 V mapping 10%) and the hydraulic pump can provide 6 MPa supply pressure in the whole hydraulic circuit.

The hardware platform can be simplified as the linkage mechanism such as Fig. 4. The system includes a lateral DoF (in the XOY plane of Cartesian coordinate) which called joint-1 in following content and a vertical DoF (in the XOZ plane of Cartesian coordinate) which called joint-2 in following content.

In the two-DoF hydraulic manipulator system, the coefficient matrices in the dynamic model established in Sect. 2 can be written in the following form:

$$\begin{cases} M(q) = \begin{bmatrix} (m_1+m_2){l_1}^2 + m_2{l_2}^2 + 2m_2l_1l_2\cos(q_2) & m_2{l_2}^2 + 2m_2l_1l_2\cos(q_2) \\ m_2{l_2}^2 + 2m_2l_1l_2\cos(q_2) & m_2{l_2}^2 \end{bmatrix} \\ C(q,\dot{q}) = \begin{bmatrix} -m_2l_1l_2\sin(q_2)\,\dot{q}_2 & -m_2l_1l_2\sin(q_2)\,(\dot{q}_1+\dot{q}_2) \\ -m_2l_1l_2\sin(q_2)\,(\dot{q}_1+\dot{q}_2) & 0 \end{bmatrix} \\ G(q) = \begin{bmatrix} (m_1+m_2)\,l_1g\cos(q_2) + m_2l_2g\cos(q_1+q_2) \\ m_2l_2g\cos(q_1+q_2) \end{bmatrix} \end{cases} \tag{31}$$

Fig. 3. Experimental platform.

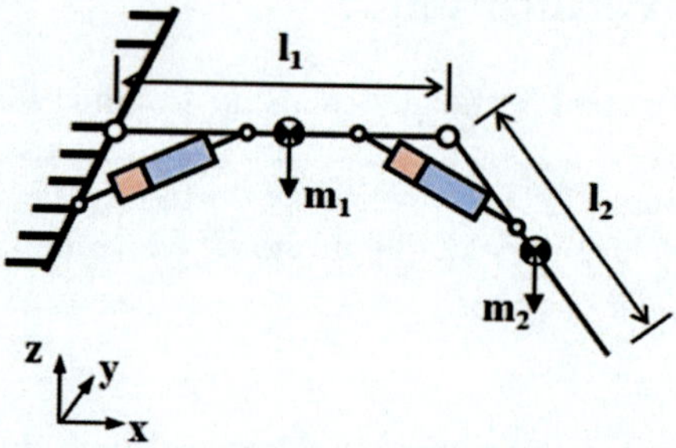

Fig. 4. Brief diagram of two DoF manipulator system.

The joint control target trajectory for experiment is shown as Fig. 5. The trajectory includes a transient phase (variable-speed motion and uniform motion) and a steady-state phase. The transient phase can evaluate the dynamic response performance of the controllers, and the steady-state phase can verify the steady-state positioning accuracy of controllers.

The PID parameters used in this paper are obtained by the following PID parameter determined method: first of all, the parameters of I and D are adjusted to 0, and the parameters are gradually reduced from large to small, and P is adjusted according to the characteristics of the oscillation value until the constant amplitude oscillation is achieved. Then, the duration (Tk) from the first vertex to the second vertex of the oscillation waveform is obtained from the oscillation waveform. Finally, the integral time and differential time values (Ti is generally 0.5Tk, Td is generally 0.125Tk) are obtained from the duration. On this basis, we also used many groups of similar parameters to carry out practical experiments and compare the result to get the most ideal PID parameters finally.

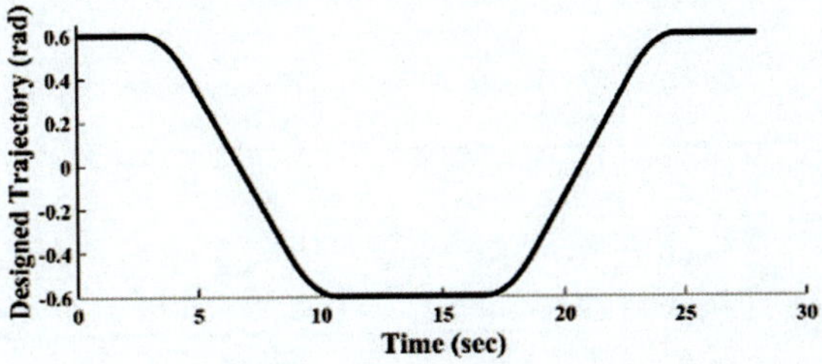

Fig. 5. Joint control target trajectory.

4.2 Experimental Result

The experimental result of SMC and PID are shown in Fig. 6.

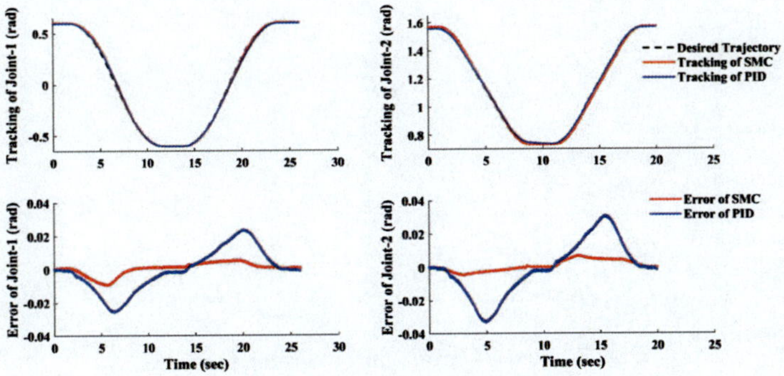

Fig. 6. Trajectory tracking result.

It can be seen as Fig. 6 that since the SMC designed in this paper take the influence of the hydraulic manipulator dynamics into account and compensates some dynamic model items, it can achieve better control effects compared with PID during trajectory tracking. Specifically, compared with PID, the trajectory tracking curve under SMC control can respond quickly to joint transient changes due to the robust control item, thereby reducing the maximum error. In the steady-state stage, since the model compensation is included in the SMC, the steady-state error is smaller. The specific control performance parameters is shown as Table 1.

Through the analysis of tracking result of SMC, the maximum tracking error during the process occurs at the moment of sudden change in joint angular velocity and angular acceleration, which is resulted by that the offline estimated value, rather than the exact value or the online adaptive value of the model parameter matrix are adopted in SMC designed, which makes the controller very sensitive to changes of joint angular velocity and acceleration, resulting in the maximum tracking error at the sudden change of those.

Table 1. Performance parameters of different controller in experiment.

Controller	Joint	Maximum error (rad)	Steady-state error (rad)	RMSE
Sliding mode controller	Joint-1	1.27×10^{-2}	2.75×10^{-3}	4.4949×10^{-4}
	Joint-2	1.09×10^{-2}	2.79×10^{-3}	5.4419×10^{-4}
PID controller	Joint-1	2.57×10^{-2}	3.07×10^{-3}	6.0731×10^{-4}
	Joint-2	3.34×10^{-2}	3.95×10^{-3}	8.1946×10^{-4}

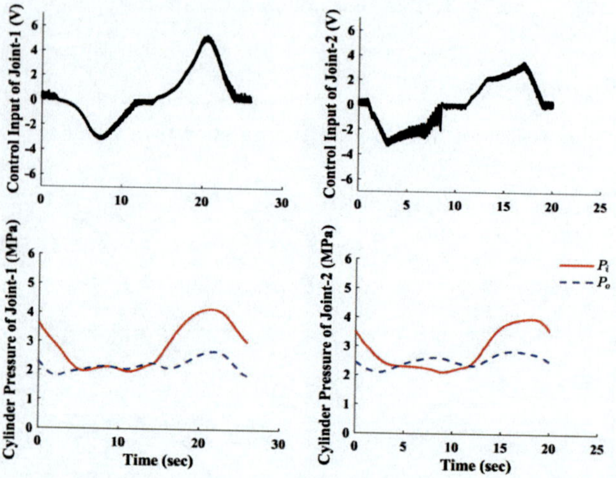

Fig. 7. Control input and cylinder pressure.

The control input result and the hydraulic cylinder pressure result is shown as Fig. 7. It can be seen from the control input result that the input voltage was always within −10 V during the whole experiment. According to the characteristics of hydraulic servo valve (mentioned in Sect. 4.1), the opening of the joint hydraulic valve was within 100%, indicating that the hydraulic manipulator exist no input saturation during the movement, which ensures that the trajectory tracking result can accurately reflect the control performance of the SMC designed.

And the cylinder pressure result show that the maximum input pressure was within the supply pressure of the hydraulic pump (P_s), and the oil pressure in the low-pressure chamber was always positive, which indicated that there was no back-sucking phenomenon and can verify the reliability of the SMC designed.

5 Conclusion

In this paper, for achieving good control performance of multi-DoF underwater hydraulic manipulator, the comprehensive dynamic model of multi-DoF underwater hydraulic manipulator considering joint coupling and underwater dynamics is established, and the sliding mode controller is designed to cope with the negative influence caused by model nonlinearity and unknown external disturbance, and also, guarantees stability of the system. Finally, the experiment with designed SMC and PID controller for comparison is carried out based on a two-Dof hydraulic manipulator platform to verify the effectiveness and reliability of the designed controller. In the future work, our team will take the uncertain parameters of the hydraulic manipulator dynamics into account, and also conduct experiments in actual underwater environment to verify the reliability of the designed controller.

Acknowledgment. This work is supported by National Natural Science Foundation of China (No. 52075476), and Natural Science Foundation of Zhejiang Province (No. LR23E050001).

References

1. Koivumki, J., Mattila, J.: Stability-guaranteed impedance control of hydraulic robotic manipulators. IEEE/ASME Trans. Mechatron. **22**, 601–612 (2017). https://doi.org/10.1109/TMECH.2016.2618912
2. Yao, J., Deng, W.: Active disturbance rejection adaptive control of hydraulic servo systems. IEEE Trans. Ind. Electron. **64**(8023–675), 8032 (2017). https://doi.org/10.1109/TIE.2017.2694382
3. Wei, S., Jiehao, W.: An integral terminal sliding mode control scheme for speed control system using a double-variable hydraulic transformer. ISA Trans. (2019). https://doi.org/10.1016/j.isatra.2019.08.068
4. Bartlett, H.L., Lawson, B.E., Goldfarb, M.: Design, control, and preliminary assessment of a multifunctional semipowered ankle prosthesis. IEEE/ASME Trans. Mechatron. **24**, 1532–1540 (2019). https://doi.org/10.1109/TMECH.2019.2918685
5. Chen, Z., Huang, F., Sun, W., Gu, J., Yao, B.: Rbf neural network based adaptive robust control for nonlinear bilateral teleoperation manipulators with uncertainty and time delay. IEEE/ASME Trans. Mechatron. **25**, 906–908 (2019). https://doi.org/10.1109/TMECH.2019.2962081
6. Helian, B., Chen, Z., Yao, B.: Precision motion control of a servomotor pump direct-drive electrohydraulic system with a nonlinear pump ow mapping. IEEE Trans. Ind. Electron. **67**, 8638–8648 (2019). https://doi.org/10.1109/TIE.2019.2947803
7. Zhou, S., Shen, C., Xia, Y., Chen, Z., Zhu, S.: Adaptive robust control design for underwater multi-DoF hydraulic manipulator. Ocean Eng. **248**, 110822 (2022). https://doi.org/10.1016/j.oceaneng.2022.110822
8. Mattila, J., Koivumki, J., Caldwell, D.G., Semini, C.: A survey on control of hydraulic robotic manipulators with projection to future trends. IEEE/ASME Trans. Mechatron. **22**, 669–680 (2017). https://doi.org/10.1109/TMECH.2017.2668604
9. Yao, Bin, Fanping, Bu., Reedy, J., Chiu, G.T.: Adaptive robust motion control of single-rod hydraulic actuators: theory and experiments. IEEE/ASME Trans. Mechatron. **5**, 79–91 (2000). https://doi.org/10.1109/3516.828592

10. Bu, F., Yuo, B.: Desired compensation adaptive robust control of single-rod electro-hydraulic actuator. In: Proceedings of the 2001 American Control Conference (Cat. No. 01CH37148), vol. 5, pp. 3926–3931(2001). https://doi.org/10.1109/ACC.2001.946267
11. Angue Mintsa, H., Venugopal, R., Kenne, J., Belleau, C.: Feedback linearization-based position control of an electrohydraulic servo system with supply pressure uncertainty. IEEE Trans. Control Syst. Technol. **20**, 1092–1099 (2012). https://doi.org/10.1109/TCST.2011.2158101
12. Helian, B., Chen, Z., Yao, B., Lyu, L., Li, C.: Accurate motion control of a direct-drive hydraulic system with an adaptive nonlinear pump flow compensation. IEEE/ASME Trans. Mechatron. **26**, 2593–2603 (2020). https://doi.org/10.1109/TMECH.2020.3043576
13. Lyu, L., Chen, Z., Yao, B.: Advanced valves and pump coordinated hydraulic control design to simultaneously achieve high accuracy and high enerciency. IEEE Trans. Control Syst. Technol. 1–13 (2020). https://doi.org/10.1109/TCST.2020.2974180
14. Lyu, L., Chen, Z., Yao, B.: Development of pump and valves combined hydraulic system for both high tracking precision and high energy efficiency. IEEE Trans. Ind. Electron. **66**, 7189–7198 (2018). https://doi.org/10.1109/TIE.2018.2875666
15. Cheng, M., Zhang, J., Xu, B., Ding, R., Wei, J.: Decoupling compensation for damping improvement of the electrohydraulic control system with multiple actuators. IEEE/ASME Trans. Mechatron. **23**, 1383–1392 (2018). https://doi.org/10.1109/TMECH.2018.2834936
16. Won, D., Kim, W., Tomizuka, M.: High-gain-observer-based integral sliding mode control for position tracking of electrohydraulic servo systems. IEEE/ASME Trans. Mechatron. **22**, 2695–2704 (2017). https://doi.org/10.1109/TMECH.2017.2764110

Path Following for Underwater Inspection Allowing Manoeuvring Constraints

Alexandre Cardaillac[1(✉)] and Martin Ludvigsen[1,2,3]

[1] Department of Marine Technology, Norwegian University of Science and Technology, Trondheim, Norway
{alexandre.cardaillac,martin.ludvigsen}@ntnu.no

[2] Department of Arctic Technology, University Centre in Svalbard, Longyearbyen, Norway

[3] Department for Arctic and Marine Biology, The Arctic University of Norway, Tromsø, Norway

Abstract. A guidance system is proposed for underwater navigation and inspection of structures to enable path-following control objectives with manoeuvring constraints such as velocity and orientation instructions. To document a vertical surface like a ship hull, a submerged drone will take benefit of maneuvering with the heading perpendicular object, while during transit the most efficient would be to align the heading towards the next way point. The proposed system is simulated using a small underactuated Remotely Operated underwater Vehicle (ROV) with control in surge, sway, heave and yaw (4 DOF). It is based on the Line Of Sight (LOS) steering laws and PID controllers for the 4 DOF motion control. The waypoints are generated together with a list of instructions for orientation and velocity for the ROV using the Parametrised Rapidly exploring Random Graph (PRRG). The LOS vector is used for heading control during transit whereas during inspection, it is used for course control. The proposed framework is tested in simulation to follow 3D straight lines in a lawnmower pattern and a typical path for ship hull inspection. Simulations shows that the paths generated using the proposed solution are viable for inspection tasks taking into account the manoeuvring constraints posed by the inspection mission and the properties of the vehicle.

Keywords: Line of sight guidance · 3D path-following · Inspection · Remotely operated underwater vehicle · Control system

1 Introduction

In the maritime industry, inspections of ship hulls and submerged installations are often done with divers. Some structures require regular supervision and can be difficult to access. Inspections are often required to guaranty the safety and integrity of the structure. This is a repeated task that often require human intervention.

I. Petrovic et al. (Eds.): IAS 2022, LNNS 577, pp. 867–880, 2023.
https://doi.org/10.1007/978-3-031-22216-0_58

Using a Remotely Operated underwater Vehicle (ROV) can help reduce risks for human operators and divers. ROVs can be deployed with a simple set up. They can also be used for underwater inspection of ship hulls, and in some cases to avoid expensive dry docking. The vehicle can fly around the hull for visual inspection and determine whether maintenance is required. Such inspection tasks can be automated, reducing human interventions. To enable autonomous ROV-based ship hull inspections, multiple dedicated systems need to be developed including mission planning, path planning and control.

This paper presents a method for path following with constraints to enable visual inspection of underwater structures. The paths are planned based on the Parametrised Rapidly exploring Random Graph (PRRG) presented in [4] with two types of paths according to the steps of the mission: transit and inspection. For inspection missions, transit paths are defined to bring the vehicle in position for the observation, while inspection paths define the maneuver during the actual survey. The optimal and possible velocity will depend on the quality of the acquired optical or acoustical data and the hydrodynamic parameters of the vehicle and vary between transit lines and inspection lines. This online 3D path-planning module generates a set of waypoints which contains instructions for orientation and velocity for the ROV.

The proposed system is tested in a simulation environment with two missions designs. The first one is based on a lawnmower pattern composed of four rows and the second one is a path following simulation for the inspection of one side of a ship.

The combination of the proposed path following system and the previously developed path planning framework makes it possible to perform efficiently inspection missions which may contain constraints and requirements.

For path following, straight-lines are considered, and the Line-Of-Sight (LOS) steering law is applied for both course control and heading control according to the type of path and the constraints in a 3-D decoupled motions design inspired by [10]. When following transit paths, the LOS law is extended by adding integral action of a function of the cross track error as modelled in [11]. The surge, sway and yaw motions is controlled based on the instructions received from the horizontal LOS model whereas the heave motion is based on a desired underwater depth at which the trajectory should be followed. The LOS is a popular method which is known to be stable and effective with a simple design. It is largely used in underwater applications [8,12,13].

2 Modelling of the ROV

For the numerical simulation, the drone is modelled based on the 6 DOF model for ROV provided in [7]. The 6 DOF nonlinear equations of motion can be written as,

$$\dot{\eta} = J_{\Theta}(\eta)\nu \tag{1}$$

$$M\dot{\nu} + C(\nu)\nu + D(\nu)\nu + g(\eta) = \tau \tag{2}$$

where $M \in \mathcal{R}^{6\times6}$ is the system inertia matrix from the addition of the rigid body and added mass inertia matrix $M = M_{RB} + M_A$. $C \in \mathcal{R}^{6\times6}$ represents Coriolis and centrifugal forces similarly based of the rigid body and added mass. $D \in \mathcal{R}^{6\times6}$ is the damping matrix, $g(\eta) \in \mathcal{R}^6$, the vector of restoring forces and moments and $\tau \in \mathcal{R}^6$, the vector of generalized forces. The velocity vector $\nu = [u, v, w, p, q, r]^T$ is in the drone body frame and its position $\eta = [x, y, z, \phi, \theta, \psi]^T$ is in the NED frame based on the rotation matrix $J_\Theta(\eta)$ from body-fixed to NED frame.

3 Path Planning

Planning a path is primordial to execute a mission. If it has online capabilities, it makes it more efficient in order to plan safe path that can adapt to the changes in the environment. For underwater drones, the planned paths need to be in the 3D space.

The proposed guidance system utilizes the Parametrised Rapidly exploring Random Graph (PRRG) [4] which is a sampling based planner with efficient real time capabilities. The large set of parameters proposed makes it possible to easily adapt it to any mission. It includes parameters such as the minimum and maximum distance of a path segment, obstacle inflation radius, about node neighboring management, and a set of rules for specific customisation of the planning graph. The Dijkstra search algorithm [5] is employed in this paper to find a path in the graph. This framework also proposes a rule system for dynamic node management. It is employed to generate two types of paths: transit paths, which correspond to the paths the drone will have to follow to reach the target of the mission, and inspection paths, to carry out the inspection mission.

This path planning framework allows an easier and more adaptable creation of inspection routes as it can take into account the vehicle parameters and inspection criteria.

When the final path is computed, each attached node receives a list of guidance instructions that will be shared with the drone when it reaches the point. All of them contain at least the guidance mode. Transit paths can be free of constraints, but still receive instruction about the LOS vector and waypoint management that will be explained in more details in their dedicated sections. On the contrary, for inspection paths, constraints are frequently included. The most common constraints are related to the drone's velocity and heading because they can have an important impact in the inspection results. The 3D waypoints of the planned path are then used as reference positions that the ROV will have to go through. The waypoint set is defined as,

$$\mathrm{WP} = [\mathrm{WP}_1, \mathrm{WP}_2, \mathrm{WP}_3, \ldots, \mathrm{WP}_n] \tag{3}$$

$$\mathrm{WP}_k = [x_k, y_k, z_k]^T \tag{4}$$

where WP_k is the last waypoint reached by the vehicle and WP_{k+1} the target. The list of instructions that the waypoint contains is of variable size, and a typical set can be defined as,

$$S_k = (\Delta_k, \psi_k, d_{hk}, d_{vk}) \tag{5}$$

with Δ_k, ψ_k, d_{hk} and d_{vk} respectively the lookahead distance, desired heading, horizontal acceptance distance and vertical acceptance distance.

4 Path Following

4.1 Objective

The path following objective is to converge to and follow the desired path which is represented as a set of 3D straight lines.

Given the vehicle position $P = [x, y, z]^T$ and WP_k the last waypoint it has reached, the along-track error x_e and the cross-track error y_e can be calculated as,

$$\begin{bmatrix} x_e \\ y_e \end{bmatrix} = R^T(\gamma_p) \begin{bmatrix} x - x_k \\ y - y_k \end{bmatrix} \tag{6}$$

where $R^T(\gamma_p)$ is defined as,

$$R^T(\gamma_p) = \begin{bmatrix} \cos(\gamma_p) & -\sin(\gamma_p) \\ \sin(\gamma_p) & \cos(\gamma_p) \end{bmatrix} \tag{7}$$

and γ_p is the horizontal path-tangential angle,

$$\gamma_p = \mathrm{atan2}(y_{k+1} - y_k, x_{k+1} - x_k) \tag{8}$$

Regarding the error in depth, the direct depth measurements are used and is therefore defined as,

$$z_e = z - z_{k+1} \tag{9}$$

The path tracking control objective is therefore to minimize the error $e(t) = [x_e(t), y_e(t), z_e(t)]^T$ over time (10) by applying guidance laws.

$$\lim_{t \to \infty} e(t) = 0 \tag{10}$$

Figure 1 represents the geometry used for the path following system.

4.2 Guidance Law

For path following in the horizontal plane, the LOS guidance law is applied using the lookahead-based version. According to the scenario, it is used for course control or augmented with integral action for heading control. In both cases, the lookahead distance Δ is provided by the mission handler which extract it from the waypoint properties. Based on [2,3], the lookahead-based guidance law can be given by,

$$\psi_d = \gamma_p + \arctan\left(\frac{-y_e}{\Delta}\right) \tag{11}$$

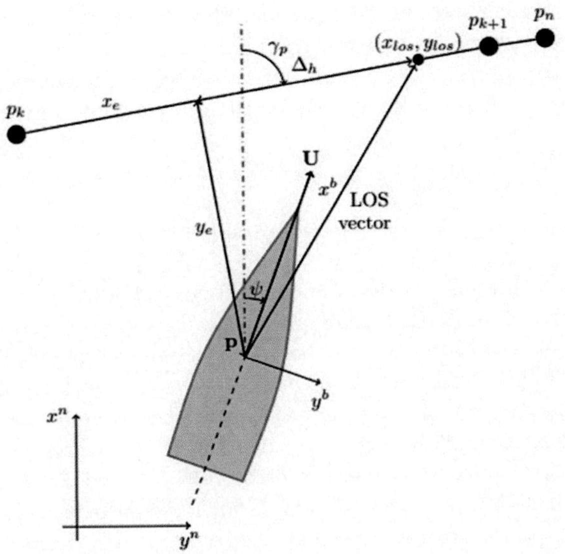

Fig. 1. Horizontal line of sight guidance for straight lines [10].

$$\chi_d = \psi_d \tag{12}$$

It defines the desired heading angle ψ_d and desired course angle χ_d with γ_p the horizontal path-tangential angle and y_e the cross-track error which needs to be minimized. However, the two angles are not always aligned, when the drone is turning, or because of the conditions in water, therefore,

$$\chi_d = \psi_d + \beta \tag{13}$$

where ψ_d and χ_d are related by the sideslip angle β that can be viewed as the angle between the orientation of the global velocity of the vehicle and its heading. It is defined as,

$$\beta = \arcsin\left(\frac{v}{U}\right) \tag{14}$$

where v is the sway velocity and U the total velocity of the ROV. In the presence of external forces, such as currents, the relative velocities need to be considered. However, calculating this drift angle based on direct measurements will results in small offsets and singularities at zero speed. Therefore, for transit paths, even though the sideslip angle can be assumed to be small, the Integral LOS guidance is applied, defined as in [11]:

$$\psi_d = \gamma_p - \arctan\left(\frac{y_e + \kappa y_{int}}{\Delta}\right) \tag{15}$$

$$\dot{y_{int}} = \frac{U y_e}{\sqrt{\Delta^2 + (y_e + \kappa y_{int})^2}} \tag{16}$$

with $\kappa > 0$, a tuning parameter. It makes the system capable of removing the cross-track error caused by external disturbances.

For inspection paths, conventional LOS is applied for course control and the desired heading is set according to the constraint contained in the last waypoint the drone has reached.

$$\psi_d = \psi_k \tag{17}$$

$$\chi_d = \gamma_p + \arctan\left(\frac{-y_e}{\Delta}\right) \tag{18}$$

which means the difference between the heading constraint and the desired course angle that points towards the LOS point should match the orientation difference β.

The choice of the lookahead distance Δ is very important as it can impact significantly the maneuvering characteristics of the drone. A low distance value can result in more aggressive steering compared to a larger value [10] which can be useful for a large cross-track error. On the contrary a high distance value will smooth out the steering and reduce oscillatory movements which is convenient when the error is small. For this reason, it can be practical to have a distance value Δ that varies. If the lookahead distance is defined in the waypoint properties as a set of two different values, the Time-Varying Lookahead Distance $\Delta(y_e)$ [9] is utilized. It is defined as,

$$\Delta(y_e) = (\Delta_{max} - \Delta_{min})e^{\gamma_r |y_e|} + \Delta_{min} \tag{19}$$

where Δ_{max} and Δ_{min} are the minimum and maximum values that Δ can adopt, and $\gamma_r > 0$ is the convergence rate. It allows to adopt a more flexible behavior which contributes to faster convergence to the desired path and reduced the oscillations around the path.

For vertical guidance, the desired depth z_d is the corresponding depth of the closest point to the vehicle on the reference line such that,

$$P_{proj} = \mathrm{WP}_k + ((P - \mathrm{WP}_k) \cdot n) \circ n \tag{20}$$

$$z_d = P_{proj\_z} \tag{21}$$

with P_{proj} the projected position on the reference line and n the normalized vector of the reference line:

$$n = \frac{WP_{k+1} - WP_k}{||WP_{k+1} - WP_k||} \tag{22}$$

The vehicle is considered to have reached the waypoint when it has entered the acceptance area which is defined according to an horizontal and vertical distance, d_{hk} and d_{vk}. If a heading constraint is present, it also need to be satisfied before moving to the next waypoint.

5 Control System

Surge, sway, heave and yaw, are controlled using Proportional Integral Derivative (PID) controllers. However, for the surge and sway, additional processing is performed to ensure the safety and successful completion of the operation. The resulting vector of desired forces and moments $\tau = [\tau_X, \tau_Y, \tau_Z, \tau_N]$ is based on the vector of error $\epsilon = [\epsilon_x, \epsilon_y, \epsilon_z, \epsilon_\psi]^T$ where the error components are defined as,

$$\epsilon = \begin{bmatrix} \epsilon_x \\ \epsilon_y \\ \epsilon_z \\ \epsilon_\psi \end{bmatrix} = \begin{bmatrix} x_d - x(t) \\ y_d - y(t) \\ z_d - z(t) \\ \psi_d - \psi(t) \end{bmatrix} \tag{23}$$

All controllers have their own set of gains including the proportional, integral and derivative gains, K_p, K_i and K_d. They were experimentally tuned.

The heading controller which provides the desired yaw moment to reach the desired heading is formulated as,

$$\tau_N = K_p\epsilon_\psi + K_i \int \epsilon_\psi dt + K_d\dot{\epsilon_\psi} \tag{24}$$

Similarly for the depth controller but with desired heave force,

$$\tau_Z = K_p\epsilon_z + K_i \int \epsilon_z dt + K_d\dot{\epsilon_z} \tag{25}$$

Regarding, the surge and sway controllers, a dynamic force limit is considered. For this, multiple conditions need to be taken into account. First, if the heading and depth errors are below the accepted error $\delta \in \mathcal{R}^2_{>0}, \delta = [\delta_z, \delta_\psi]^T$, the commanded force can be up to the maximum allowed force τ_{max}. It is important to consider the error in depth and heading since they are both important criteria for the inspection. Also, too large errors may result in a 3D drift from the reference trajectory if not compensated. To avoid situations were high forces for sway and surge are combined with large deviation terms for depth and heading a reduction term is introduced defined by λ.

$$|\epsilon_z| \leq \delta_z \tag{26}$$

$$|\epsilon_\psi| \leq \delta_\psi \tag{27}$$

If at least one of these values is exceeded, an unique ratio λ is calculated so that the allowed maximum forces for surge and sway decrease as the error increases. It is formulated as an exponential function as follows:

$$\lambda = e^{\delta_r \ln(\gamma_e)} \tag{28}$$

where δ_r is the total residual error defined as,

$$\delta_r = \begin{cases} |\epsilon_z| - \delta_z & \text{if } |\epsilon_z| > \delta_z \land |\epsilon_\psi| \leq \delta_\psi \\ |\epsilon_\psi| - \delta_\psi & \text{if } |\epsilon_z| \leq \delta_z \land |\epsilon_\psi| > \delta_\psi \\ |\epsilon_z| - \delta_z + |\epsilon_\psi| - \delta_\psi & \text{if } |\epsilon_z| > \delta_z \land |\epsilon_\psi| > \delta_\psi \end{cases} \tag{29}$$

γ_e, with $0 < \gamma_e < 1$, is the convergence rate. A γ_e close to 1 will result in a slow convergence towards 0 whereas a γ_e close to 0 will result in a fast convergence. If the tuning parameter $K_\gamma = 0.01$ is considered as a value close enough to the exponential plateau of (28), a good value of γ_e can be approximated as,

$$\gamma_e = e^{\frac{\ln(K_\gamma)}{\delta_z + \delta_\psi}} \tag{30}$$

It is important to note that if the vehicle is slow to reduce the error whether it is because of external forces or how the vehicle functions or other reasons, the approximation of the plateau should be decreased. On the contrary, if the vehicle is fast to minimize the error, it can be increased.

The resulting maximum allowed surge and sway forces can therefore be defined as follows:

$$\tau_{max\_capped} = \begin{cases} \tau_{max} & \text{if } |\epsilon_z| \leq \delta_z \wedge |\epsilon_\psi| \leq \delta_\psi \\ \lambda(\delta_r)\tau_{max} & \text{otherwise} \end{cases} \tag{31}$$

with τ_{max} the vector containing the original maximum force allowed as a percentage in surge and sway such that,

$$\tau_{max} = \begin{bmatrix} \tau_{max\_X} \\ \tau_{max\_Y} \end{bmatrix} \tag{32}$$

Both surge and sway controllers are PIDs that include the previous characteristics. By default, they are based on the horizontal errors ϵ_x and ϵ_y,

$$\tau_X = K_p\epsilon_x + K_i \int \epsilon_x dt + K_d\dot{\epsilon_x} \tag{33}$$

$$\tau_Y = K_p\epsilon_y + K_i \int \epsilon_y dt + K_d\dot{\epsilon_y} \tag{34}$$

However, in the presence of velocity instructions and/or constraints, the PIDs are based on desired velocities instead of desired positions.

6 Simulation

6.1 Setup

The simulation is based on the Robot Operating System (ROS) [14] using a python simulation node provided by Blueye Robotics [1] to simulate their underwater drones such as the one presented in Fig. 2. The simulated vehicle is a Pioneer X3 developed by Blueye Robotics. It is a 4 DOF vehicle, actuated for surge, sway, heave and yaw. There are two thrusters in the surge direction, one lateral and one vertical. The vehicle is equipped with pressure sensor for depth estimation, and IMU for heading. The vehicle also supports USBL and DVL navigation systems providing measurements of positions and velocities. For roll and pitch angle, the drone is kept passively stable by the separation of centre of gravity and centre of buoyancy. The two different patterns, both constructed from straight lines, was tested, a lawnmower pattern and an inspection pattern. The former includes six rows of 10 m each, separated from each other by a distance of 2 m. The first half of the pattern, i.e. the first three rows, are free of constraints and are considered as transit path. The last three are considered as inspection path with heading constraints. The latter pattern resembles an inspection of the side a small ship hull, considering also the navigation to the beginning of the inspection route.

Fig. 2. Blueye X3 ROV [1].

For both simulations, the same variables defining the guidance and control systems are used and presented in Table 1.

Table 1. Table of variables used during the simulations.

Name	Parameter	Value
Lookahead distance (m)	$(\Delta_{min}, \Delta_{max})$	(0.8, 2.0)
LOS convergence rate	γ_r	1.3
Integral gain parameter	κ	0.1
Horizontal acceptance distance (m)	d_{hk}	0.2
Vertical acceptance distance (m)	d_{vk}	0.35
Accepted heading error (deg)	δ_ψ	4.0
Accepted depth error (m)	δ_z	0.5
Exponential convergence rate	γ_e	0.77
Starting position (m, deg)	$(x, y, z, \phi, \theta, \psi)$	(0, 0, 2, 0, 0, 0)

6.2 Lawnmower Pattern

With this pattern the goal is to test a typical path often used for surveys of areas. To test the proposed system, it is divided in two parts, using line of sight for heading control for the first part, and course control for the second part.

The constraints requested by the waypoints of this last part are related to the heading, the underwater drone has to follow the line while being perpendicular to it, therefore moving sideways.

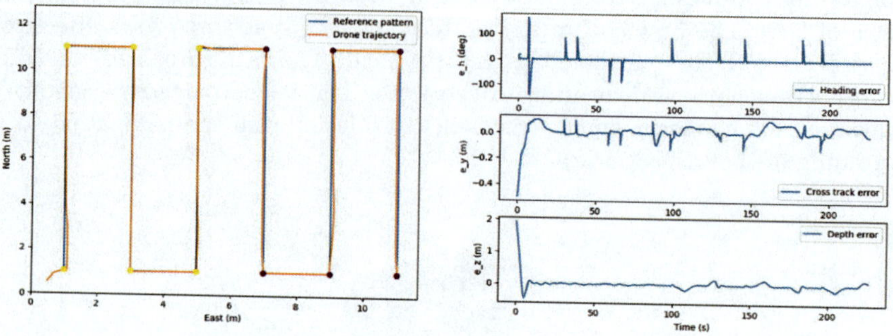

(a) (b)

Fig. 3. Results of the framework with the lawnmower pattern. (a) shows the trajectory of the underwater drone following a lawnmower pattern in the horizontal plane. The dots represent the waypoints and their color, the guidance mode they contain. Yellow for transit, black for inspection. (b) shows the errors recorded over time while the drone is following the path. From top to bottom, the heading error ϵ_{ψ}, cross track error y_e and depth error ϵ_z.

With this pattern, the desired depth is constant at 4 m deep. So only at the beginning of the operation a notable change in depth is expected as the drone starts at 2 m depth.

The results of the mission are presented in Fig. 3a and b. In the first Figure, the dots represent the waypoints, and their color represents their status, i.e. transit (yellow) or inspection (black) path.

It is possible to observe that the mission is carried out successfully. The desired heading and depth are quickly reached, the errors grow larger only during the transitions, i.e. change of targeted waypoint and/or constraints, which is expected. Over the complete duration of the operation, the system successfully maintained the cross track error close to 0.

It took the drone more time to complete the second part that the first, about 80 s for the first and about 180 s for the second part. This was expected since the drone is moving slower sideways due to larger drag forces and lower thrust force in the lateral direction.

6.3 Inspection Pattern

In this case, the goal is to test a simplified version of an inspection route for ship hull inspection. Only one side of the hull is considered. The path is divided again into two parts, first navigating to the beginning of the inspection route and then completing the inspection path.

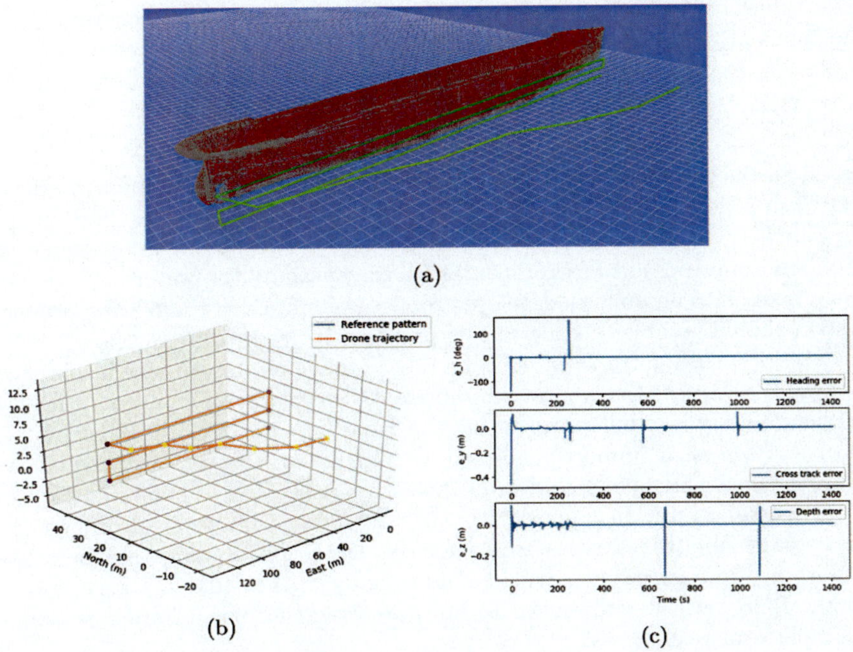

Fig. 4. Results of the simulation of an underwater ship hull inspection. (a) shows the planned path using PRRG to inspect the ship. (b) shows the 3D trajectory of the underwater drone, moving to the beginning of an inspection route and following it. The spheres represent the waypoints and their color, the guidance mode they contain. Yellow for transit, black for inspection. And (c) shows the errors recorded over time while the drone is following the path. From top to bottom, the heading error ϵ_{ψ}, cross track error y_e and depth error ϵ_z.

The second part is a vertical lawnmower pattern composed of three rows of 132 m separated from each other by a distance of 3 m. While the drone is following this pattern, the heading is kept perpendicular to the lateral direction of motion, simulating the drone facing the ship hull for visual inspection.

This simulation is based on the scenario developed in [4], using PRRG to create a planning graph around the ship and calculate a collision-free path under constraints to safely execute the path. This framework enables fast online re-planning for collision avoidance and/or in case there is a change in the mission plan.

For this simulation the depth is slowly increasing from the beginning of the transit path until its end. The next section of the simulation has also varying depth from 0.6 m depth to 6.6 m. The inspection lines are either vertical or horizontal.

The 3D trajectory of the underwater drone can be observed in Fig. 4b and the errors related to the heading, cross track error and depth are presented in Fig. 4c.

The system performed as good as in the previous simulation. More variations in the depth error were recorded but still remain in an acceptable range.

7 Discussion

Based on the two simulations experimented, the proposed system for the Blueye ROV has proven its efficiency, and made it possible to exploit the high maneuverability capabilities offered by this drone. It was however tested considering a calm environment, but environmental loads, especially the water current, need to be taken into account to develop a robust system. Even though the proposed system should be able to scale to this problem, it needs to be tested.

The proposed force limit has its role to play in this situation and helps attenuate and potentially prevents horizontal drift when the error in heading and/or depth grows. This ensures that the drone does not deviate too far from the initial trajectory by proportionally reducing its potential speed in surge and sway while the errors are being corrected. This makes it possible to maintain the drone in its last valid position. Being able to perform sway motions is advantageous in that regard. Higher priorities are set for the depth and heading because of their importance in inspection missions. Additionally, with a guidance system based on the LOS laws, it is essential to have a correct heading in order to achieve good performance.

Despite the fact that features requiring additional processing were included, such as the time varying lookahead distance, integral action for the desired heading and the force limit ratio, the processing complexity and time remain low which is important for real time applications and onboard processing with low computational capacity vehicles.

The time varying lookahead distance allowed a more stable system and reduced significantly the oscillations around the path while keeping the possibility for a short lookahead distance.

Thanks to the proposed combination of the guidance system and the path planning framework, it was possible to successfully perform an inspection mission with heading constraints to visually inspect the ship's hull. However, the possibilities offered by the rule system of the planner and the guidance instructions system were not fully exploited here as only the line of sight instructions and heading constraints were considered during the simulations.

Conventional PID controllers were implemented and could benefit some improvement, especially for the depth controller which can have difficulties to maintain the desired depth in some conditions.

A specific reference model such as [6] could also help improve the solution, especially when taking into account multiple constraints.

8 Conclusion

In this paper, a 3D path following system of straight line was presented. It is based on the horizontal lookahead based line of sight guidance law which is utilized for heading control and course control according to the presence of heading constraints. For heading control, Integral LOS was applied to improve the stability of the trajectory of the drone. To further improve it, the time varying lookahead distance was included together with a dynamic surge and sway force limit ratio.

The simulations proved the potential of this method when combined with the PRRG for inspection tasks which require for example visual feedback. And will be developed further for that purpose.

For future work, environmental loads will be considered and the system will be tested in real world conditions in the context of ship hull inspection.

The guidance system will be augmented with a constant jerk reference model [6], which is an alternative to filter based reference models. It provides a desired velocity and position by integrating of a constant jerk which is on or off. Improvements of the PID controllers will also be considered, including the addition of acceleration feedbacks.

Acknowledgment. The authors wish to thank Blueye Robotics for their help by providing the simulator of the drone and relevant data.

This work was supported by the BugWright2 EU H2020-Project under the Grant agreement No. 871260.

References

1. Blueye robotics. https://www.blueyerobotics.com/
2. Breivik, M., Fossen, T.: Principles of guidance-based path following in 2D and 3D. In: Proceedings of the 44th IEEE Conference on Decision and Control, pp. 627–634 (2005)
3. Breivik, M., Fossen, T.: Guidance Laws for Autonomous Underwater Vehicles (2009)
4. Cardaillac, A., Ludvigsen, M.: Ruled path planning framework for safe and dynamic navigation. In: Global Oceans 2021, San Diego, pp. 1–7 (2021)
5. Dijkstra, E.W.: A note on two problems in connexion with graphs. Numer. Math. **1**(1), 269–271 (1959)
6. Dukan, F.: ROV Motion Control Systems (2014)
7. Fossen, T.: Handbook of Marine Craft Hydrodynamics and Motion Control, Chap. 7, pp. 133–186. Wiley (2011)
8. Haugaløkken, B.O., Lekkas, A., Schjølberg, I.: 3D Path Following and Tracking for an Inspection Class ROV, p. V07AT06A019 (2017)
9. Lekkas, A., Fossen, T.: A time-varying lookahead distance guidance law for path following. IFAC Proc. Vol. **45**, 398–403 (2012)
10. Lekkas, A., Fossen, T.: Line-of-Sight Guidance for Path Following of Marine Vehicles (2013)
11. Lekkas, A.M., Fossen, T.I.: Integral LOS path following for curved paths based on a monotone cubic hermite spline parametrization. IEEE Trans. Control Syst. Technol. **22**(6), 2287–2301 (2014)
12. Molnar, L., Omerdic, E., Toal, D.: Guidance, navigation and control system for the tethra unmanned underwater vehicle. Int. J. Control **80**(7), 1050–1076 (2007)

13. Nie, W., Feng, S.: Planar path-following tracking control for an autonomous underwater vehicle in the horizontal plane. Optik **127**(24), 11607–11616 (2016)
14. Quigley, M., Conley, K., Gerkey, B., Faust, J., Foote, T., Leibs, J., Wheeler, R., Ng, A.: ROS: an open-source robot operating system, vol. 3 (2009)

Autonomous Robots in Warehouse Automation

Path Planning of Multiple Automatic Guided Vehicles with Tricycle Kinematics Considering Priorities and Occupancy Time Windows

Matic Ljubi, Gregor Klančar, and Andrej Zdešar(✉)

Faculty of Electrical Engineering, University of Ljubljana, Tržaška 25, 1000 Ljubljana, Slovenia
andrej.zdesar@fe.uni-lj.si

Abstract. This paper addresses the problem of path planning for multiple automated guided vehicles that have a kinematic model of a tricycle and drive in a network of one-way roads. The objective of path planning is that all AGVs reach their destination in the shortest possible time without collisions. The proposed path planning algorithm is an extension of the A* algorithm and takes into account the priorities of the AGVs. An occupancy time window approach is used to detect and resolve predicted collisions. The algorithm determines waiting times and waiting locations that allow collision-free driving. The node occupancy algorithm also considers the shape of the vehicle and its speed. The result of the path search algorithm is the path for each vehicle and the waiting times on each road. The resulting paths are converted into corresponding action plans that are sent to all AGVs for execution. The applicability of the proposed path planning algorithm was evaluated in a simulation environment and on real small-scale AGVs.

Keywords: Automatic guided vehicles · Tricycle kinematics · Path planning · Time windows · Road and node occupancy · Priority planning

1 Introduction

The demand for Automatic Guided Vehicles (AGVs) in industrial applications is increasing. AGVs are most commonly used to transport various materials between stations in industrial warehouses, large logistics centres and lean manufacturing facilities. A fleet of AGVs in modern automated warehouses can consist of thousands of AGVs that can interact with each other, with people and with other infrastructure. AGVs typically drive autonomously along a predefined route layout and use a set of sensors (e.g. magnetic line sensors, NFC readers, proximity sensors, LIDAR) to localize and navigate between obstacles. To achieve optimal, robust and collision-free performance of such a multi-agent system, optimisation of task scheduling and path planning is required [1,2].

I. Petrovic et al. (Eds.): IAS 2022, LNNS 577, pp. 883–896, 2023.
https://doi.org/10.1007/978-3-031-22216-0_59

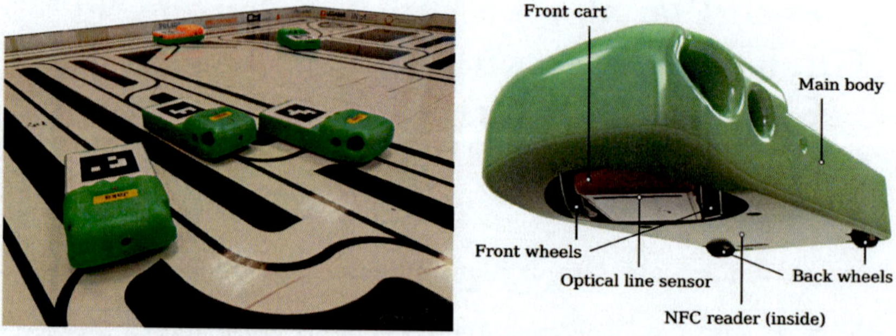

Fig. 1. Multiple small-scale AGVs (left) and structure of the AGV (right)

In this paper, we consider the problem of path planning of multiple AGVs (Fig. 1) that have the kinematics of a front wheel drive tricycle. This is a non-holonomic system. Each AGV can automatically localise itself, navigate and follow the desired path in a network of directed (one-way) roads that intersect at crossroads. AGVs can only move in the forward direction and detect obstacles in front of them. The goal of path planning is to determine the optimal paths of all AGVs so that they reach their destinations in the shortest possible time and avoid possible collisions. This is a challenging task that requires optimising the paths and schedules of all AGVs that comprise the multiple agents, considering various constraints imposed by the AGVs, the environment and the tasks.

In 1968, P. Hart et al. developed the A* algorithm [3] as an improvement of Dijkstra's algorithm [4,5]. The A* algorithm is an optimal informed search algorithm that finds the optimal path from a given starting node to a given destination node [3,6]. Many multi-agent path planning algorithms use graphs [7] and some are also based on the A* algorithm [8]. To cope with the complexity of planning, coordinating, controlling and localising multiple robotic vehicles, both AGVs and autonomous mobile robots are in practice constrained to predefined paths with fixed delivery locations along the path [9–11]. Optimally coordinated planning with A*-based approaches [12,13] is only possible for a small number of vehicles, as the complexity grows exponentially with the number of vehicles. Therefore, computational complexity must be reduced in practice by introducing suboptimal approaches that first decompile the problem to find individual plans and then resolve conflicts by introducing traffic rules, local multi-agent coordination/negotiation or priorities [9,14,15]. This work builds on a path planning algorithm for multiple robotic vehicles that considers priorities and occupancy time windows [16].

The rest of the paper is structured as follows: Sect. 2 presents the proposed path planning algorithm for multiple AGVs, Sect. 3 describes the experimental setup for simulations and real experiments, Sect. 4 discusses the results of different test case scenarios and in Sect. 5 some conclusions are drawn.

2 Path Planning of Multiple AGVs

The proposed algorithm enables path planning of multiple AGVs on the same map, while considering the priorities of the AGVs. For the AGV with the highest priority, we want it to reach the destination in the shortest possible time. At the beginning, we therefore determine the optimal path of the AGV with the highest priority. Then we determine the paths of the other AGVs, taking into account the paths of all vehicles with higher priorities. The proposed algorithm is derived from the well-known A* search algorithm.

In our case, the cost of the connections in the A* algorithm is determined by the time it takes an AGV to get from one node to another. The algorithm also allows for on-site waiting to ensure conflict-free driving. Considering the waiting time, the cost of the entire vehicle path increases. Thus, the total cost of the path is the sum of the driving time and the waiting times [16]. The modified A* algorithm has an open and a closed list, just like the basic A* algorithm. The individual lists contain information that is displayed in Table 1.

Table 1. Information about the nodes in open and closed list

Symbol	Description
A	Current node ID
B	Previous node ID
C_{HWW}	Cost from the start to the current node, excluding the waiting cost
C_{WCR}	Cost of waiting on the current road
C_{WPR}	Cost of waiting on the previous road
C_{HE}	Heuristic estimate from the current to the destination node
C_{SUM}	Sum of the costs C_{HWW}, C_{WCR}, C_{WPR} and C_{HE}
R	Current road ID

When the algorithm is initialised, the node with the lowest total cost C_{SUM} in the open list is moved to the closed list. If the moved node is the target node, the optimal path is found and the main part of the search algorithm can be terminated. In the final step, the algorithm then generates the path and determines the occupancy time of each road based on the information stored in the closed list. In addition to the list of roads that comprise the path, the algorithm provides the waiting times for each road. The road occupancy data is [16]: time of arrival on the road, time of departure from the road, waiting time on the road and direction of travel.

When adding nodes to the open list, there may occur a conflict that needs to be resolved. The algorithm selects one of the neighbours, adds it to the open list and calculates the connection cost. This means that it sets the costs of the following variables C_{HWW} (excluding the waiting cost) as the sum of the cost from the start to the current node and the cost between the current node and the neighbour node. It sets the cost of waiting on the previous and current road to zero ($C_{WCR} = C_{WPR} = 0$). It determines the cost to the destination C_{HE} with a heuristic estimate and then adds all the costs together to obtain the total cost C_{SUM} [16]. Before adding a new node to the open list, the algorithm checks whether the node and the road are not already occupied by other AGVs. If this is the case, a waiting time must be set.

2.1 Road Occupancy Check

The occupancy of the road is checked against the estimated arrival and departure time of the AGV and the estimated waiting time on the road. The vehicle's departure time from the road is equal to the total cost of the node, which already includes all waiting times on the roads. The algorithm checks the occupancy of the road by comparing the predicted time window with the occupancy time windows previously set for this road by higher priority vehicles. The direction of movement of the AGV must also be taken into account [16], so we consider the cases of driving in the opposite and same direction separately.

Driving in Opposite Directions When the time window of the current AGV and the new AGV overlap, a conflict occurs. When determining road occupancy, we also need to take into account the safety time that allows adequate spacing between AGVs. The road occupancy time is determined

$$t_{TRC} = t_{out} + \Delta t_{safe}. \tag{1}$$

In (1), t_{out} represents the time of departure of the previous vehicle from the road and Δt_{safe} is the safety time. If the road is not occupied in the estimated time window of the current vehicle, then t_{TRC} is not defined.

Driving in Same Direction Even if the AGVs are driving in the same direction, there may be a conflict. If two AGVs have the same speed, the problem is greatly simplified. Under this assumption, we can only check the time of arrival of the current AGV and the previous AGV on the road. If the difference in the arrival time of the AGVs on the road is greater than the safety time, there is no conflict between the AGVs. In the second case, the end time of node occupancy t_{TRC} is calculated as the sum of the arrival time of the first vehicle t_{in}, the safety time Δt_{safe} and the waiting time of the first vehicle t_{wait}:

$$t_{TRC} = t_{in} + \Delta t_{safe} + t_{wait}. \tag{2}$$

If the speeds of the AGVs are not equal, it is not sufficient to compare only the arrival times of the two AGVs on the road. If a faster AGV arrives after a slower one, it may catch up, causing a conflict. In this case, in addition to the arrival time of the AGV on the road, the time interval of the departure of the AGV from the road must also be considered. When checking the departure time interval, we compare the arrival time at the waiting place of a vehicle that entered the road later with the departure time of a vehicle that arrived earlier on the road [16].

2.2 Node Occupancy Check

The occupancy of the node t_{TNC} is checked by the algorithm based on the time of arrival at the node of the current AGV. A node is occupied at a time when the difference in arrival times of both AGVs at the node is less than the safety time. The end time of the node occupation is determined by adding the safety time Δt_{safe} to the time of arrival at the node t_{inn}:

$$t_{TNC} = t_{inn} + \Delta t_{safe}. \tag{3}$$

2.3 Road and Node Occupancy Times for Tricycle Kinematics

The occupancy times of the road and the node are described in Subsects. 2.1 and 2.2 and apply to a dimensionless (point body) AGV. In our case, the AGVs have a non-circular shape and cannot be considered as points, so we need to consider this in the arrival and departure times of the AGV for the road and the node.

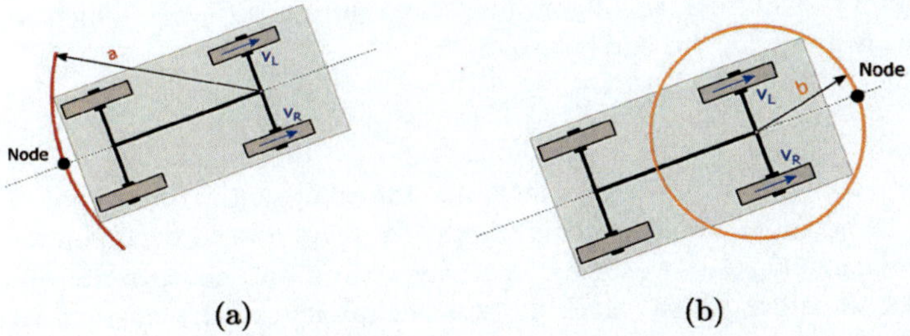

Fig. 2. AGV (a) departing from a node and (b) arriving to a node

To calculate the node arrival and departure time, we need to know the shape of the vehicle. The distances a and b (shown in Fig. 2b) represent the distance of the circular radius from the centre to the back and front of the robot respectively. The new time of entry of the vehicle to the node $t_{in_{new}}$ is obtained by subtracting the old entry time t_{in} and the quotient of the distance b and the velocity of the vehicle v_v:

$$t_{in_{new}} = t_{in} - b/v_v. \tag{4}$$

Similarly, the old departure time t_{out} is updated to the new departure time of the AGV from the node $t_{out_{new}}$ as:

$$t_{out_{new}} = t_{out} + a/v_v. \tag{5}$$

From the updated AGV arrival and departure times, we can now determine the actual occupancy time of the node.

2.4 Waiting Time Due to Occupancy

When adding nodes to the open list, if the algorithm encounters an occupied node, it proposes to wait for the AGV on the road in front of the occupied node to release it. However, if the road to the node is also occupied, the algorithm suggests waiting on the previous road. In the open list, the cost C_{WCR} represents the waiting time on the current road due to the occupancy of the current node, and the cost C_{WPR} represents the waiting time on the previous road due to the occupancy of the current road.

After the algorithm has checked the occupancy of the node and the road, it determines the waiting time, if any. The cost of waiting on the previous road C_{WPR} is calculated as

$$C_{WPR} = t_{TRC} - (C_{HWW} - p_{PT}). \tag{6}$$

In (6), t_{TRC} represents the end time of the road occupancy, C_{HWW} represents the cost to get here without waiting for the current node, and p_{PT} represents the cost of the connection between the previous node and the current node. If the current node is occupied, the algorithm determines the C_{TNC}, which represents the cost of waiting on the current road, as

$$C_{TNC} = t_{TRC} - (C_{HWW} - C_{WPR}). \tag{7}$$

In (7), the second term on the right-hand side represents the estimated time of arrival at the current node, which we want to delay due to occupancy.

By updating C_{WPR}, we delay the travel time on the current road, which we did not consider when checking the occupancy of the current road. The occupancy of the current node and the current road need to be checked again, so we repeat the process of adding the node taking into account the new cost of C_{WCR} and C_{WPR}. If waiting on the crossing road is not possible, the algorithm continues to search for a free place in front of this road. The next option would be to retreat to the side road of the previous node [16].

3 Experimental Set-up

The map of the environment used in the experiments is shown in Fig. 3a. The outer dimensions of the map are 220 cm by 180 cm. The basic structure of the AGV is shown in Fig. 1. The AGV has a width of about SI 10 cm and a length of about SI 20 cm. Although the AGV has four wheels, it has the same kinematic model as a tricycle. Each of the two front wheels is equipped with a motor and an incremental encoder. The front wheels are attached to the front cart, which is connected to the main body via a passive rotational joint that is also equipped with an encoder. The two rear wheels are passive.

The front cart is equipped with an optical line sensor that can detect black guide lines on the white road surface and enable line following control of the AGV. The control enables tracking of the left or right guide-line edge, so that different roads in the crossroads can be selected if the AGV follows the appropriate line edge when passing through the crossroad. The encoders on the AGV enable implementation of wheel-based odometry. In addition, under the ground surface is a constellation of NFC tags at known locations and unique IDs that the AGV can detect when it passes over them, hence the AGV can also localize itself on the map as well.

3.1 Control of Multiple AGVs Along the Desired Paths

Based on the desired AGV path, we create a list of actions (see Table 5 for an example) with the desired AGV space-time positions, with information about:

- R, road ID,
- L, road length,
- r, desired relative location on the road ($r \in [0, 1]$, distance from the start node of the road is rL),
- t, desired time at desired location on the road,
- m, desired control mode on the first part of the road ($m \in \{\mathrm{L}, \mathrm{R}\}$),
- M, desired control mode on the second part of the road ($M \in \{\mathrm{L}, \mathrm{R}\}$).

The duration of the AGV on a given road section is obtained by calculating the quotient of the length of the road section and the desired speed of the AGV.

The AGVs have synchronised clocks so that each AGV can perform the desired action at the desired absolute time. The AGV calculates the required forward speed based on the remaining distance and the time needed to complete the current action. Depending on the location on the road, the appropriate control mode is determined. The control algorithm goes through the list of actions to make the AGV move along the specified path until it reaches the destination.

4 Results

To determine the paths of the AGVs, the algorithm needs data about the starts and destinations of all AGVs, the speeds of the AGVs, the safety time and the dimensions of the AGVs. We tested the path planning algorithm under different conditions. The map considered in the experiments consists of many intersections that could potentially cause conflicts between AGVs. In this study, we analysed only one critical area, located in the upper right part of the map (Fig. 3a). We chose this area mainly because the intersection from 15 to 1 is followed by a long straight line where a potentially faster vehicle could catch up with a slower one.

4.1 Test Case 1: All the AGVs with the Same Speeds (Simulation)

In this test case, all AGVs had the same speed. The start of AGV1 was node 21 and the goal was node 9, while the start of AGV2 was node 23 and the goal was node 1. The algorithm determined that AGV2 should wait at node 19 (Fig. 3a), so that AGV1 (with higher priority) could pass node 15 before AGV2.

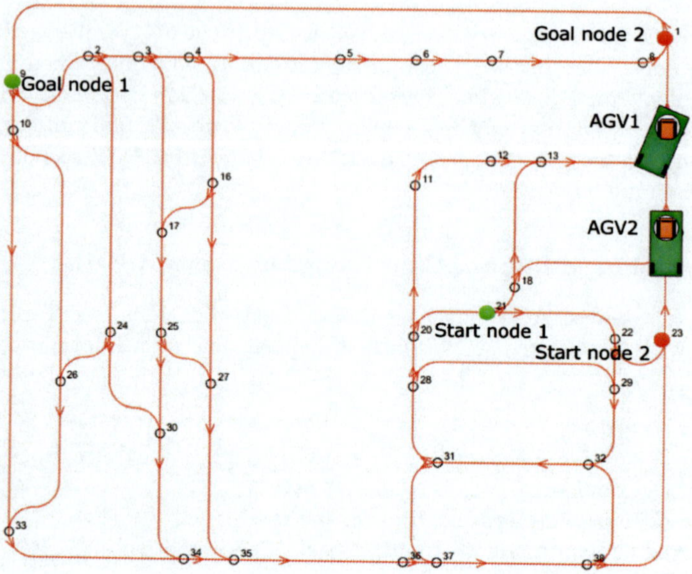

(a)

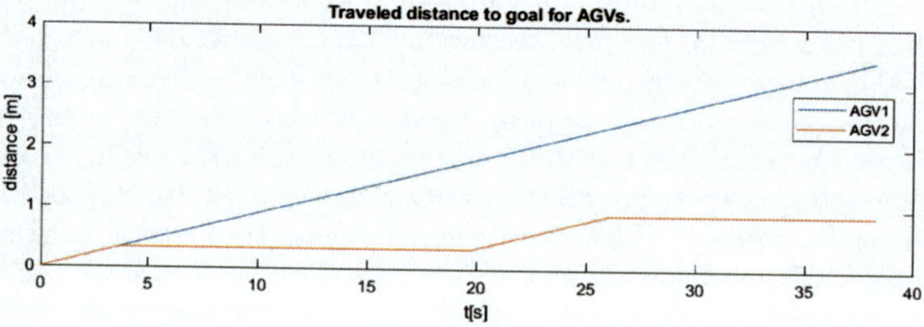

(b)

Fig. 3. Test case 1, (a) map with AGVs at time $t = 20$ s, and (b) travel distances of the AGVs

In the graph in Fig. 3b we see the distance travelled in time. If the distance travelled is constant, it means that the AGV is waiting or has already reached its destination. In this case, the algorithm finds the optimal solution because all AGVs have reached their targets in the shortest possible time.

4.2 Test Case 2: AGVs with Different Speeds (Simulation)

In this test case, the AGVs had different speeds. The AGV1 with the highest priority had the start at node 31, a goal at node 3 and a speed of 0.11 m/s. The AGV2 had a start at node 23, goal at node 1 and a speed of 0.033 m/s.

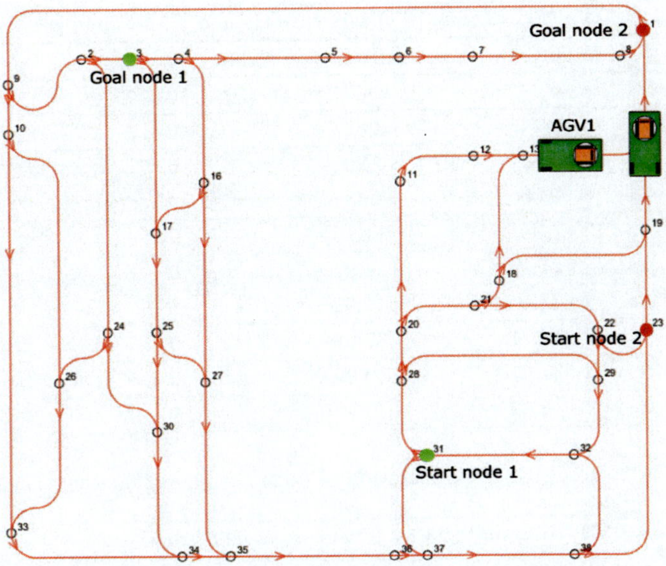

(a)

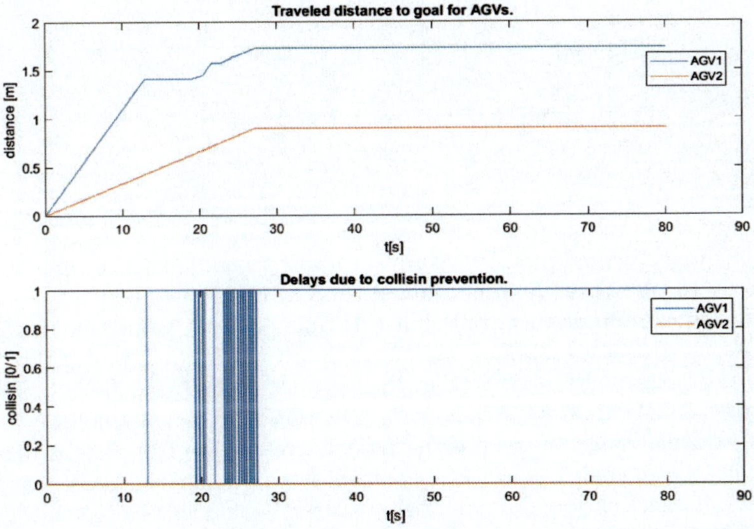

(b)

Fig. 4. Test case 2, (a) map with AGVs at time $t = 20$ s, and (b) travel distances and collisions of the AGVs

Table 2. Test case 2 – calculation of node occupancy for the AGV1

AGV1, from node 31 to 3, velocity is 0.11 m/s				
A	*B*	Arrival time [s]	Departure time [s]	Wait time [s]
0	31	0	0	0
31	28	0	2.44	0
28	20	2.44	3.82	0
20	21	3.82	6.26	0
21	18	2.26	7.33	0
18	13	7.33	11.15	0
13	15	11.15	14.97	0
15	*1*	*14.97*	*17.71*	*0*
1	9	17.71	37.67	0
9	2	37.67	40.89	0
2	3	40.89	42.26	0

Table 3. Test case 2 – calculation of node occupancy for the AGV2

AGV2, from node 23 to 1, velocity is 0.033 m/s				
A	*B*	Arrival time [s]	Departure time [s]	Wait time [s]
0	23	0	0	0
23	19	0	9.09	0
19	*15*	*9.09*	*19.17*	*0.99*
15	1	19.17	28.25	0

The algorithm results of the node occupancy calculations are shown in Tables 2 and 3. From both tables we can see that the intersection of AGVs occurs again at the node 15. To ensure the appropriate departure time of AGV2 from node 15, the algorithm determined the waiting time of 0.99 s.

In the simulation, it turns out that this time does not guarantee that AGV1 will cross the intersection before AGV2, as shown in Fig. 4a. The distance sensor of the AGV1 detects the AGV2 (Fig. 4b shows when the AGVs detect collisions) and causes AGV1 to stop. As a result, the AGV2 overtakes the AGV1 at the intersection, blocking the path of the AGV1. We can see from Fig. 4b that the AGV1 stops to avoid a collision, but then stays behind the AGV2 and does not reach its destination. The algorithm correctly calculates the waiting times of AGV2, since AGV1 reaches the node 15 before AGV2. The problem arises from the fact that we have considered AGVs as points (dimensionless objects). Therefore the AGV1 detected the AGV2 with a proximity sensor and stopped for safety reasons.

4.3 Test Case 3: AGVs with Different Speeds Considering Tricycle Kinematics (Simulation)

In this test case, the AGVs had the same input data as in the test case 2, with the difference that the path search algorithm now considers the vehicles as objects with tricycle kinematics.

Table 4. Test case 3 – calculation of node occupancy for the AGV2

AGV2, from node 23 to 1, velocity is 0.033 m/s				
A	*B*	Arrival time [s]	Departure time [s]	Wait time [s]
0	23	0	0	0
23	19	0	9.09	0
19	*15*	*9.09*	*25.10*	*6.92*
15	1	25.10	34.19	0

The algorithm results of the node occupancy calculations for AGV2 are shown in Table 4 and the results for AGV1 are the same as shown in Table 2. From the tables we can see that the intersection of the AGVs occurs at node 15. In order to ensure the appropriate departure time from node 15, the algorithm determined the waiting time of 6.92 s, with the difference that in the previous simulation (test case 2) this time was 0.99 s. The simulation shows that the algorithm has correctly determined the new waiting time, which also considers the shape of the vehicles. Thus, the AGV1 leaves the node 15 before AGV2 and continues its way to the destination as shown in Fig. 5a. Also, in Fig. 5b we can see that AGV2 waited before node 15.

4.4 Test Case 4: AGVs with Different Speeds Considering Tricycle Kinematics (Real AGVs)

The multi-vehicle path planning algorithm was also tested on small scale models of real AGVs with tricycle kinematics. The input data for the path calculation was the same as in test case 3. The algorithm calculates the individual AGV paths and waiting times. The algorithm then uses this information to determine a list of actions. The list of actions for the second vehicle (data from Table 4) is shown in Table 5. In Table 5 of AGV2 we can see in the rows 2, 3 and 4 that AGV2 is on the same road. This is because the algorithm has calculated a waiting time of 6.92 s on the road 1915. So the AGV2 in the third row is in the same place as in the second row, the only difference is the time, which is increased by the waiting time. The time in the second row is calculated as the quotient of the product of the length of the road and the relative location on the road and the speed of the vehicle. In Table 5 we can also see that the AGV2 must constantly follow the right edge of the line if it wants to reach its

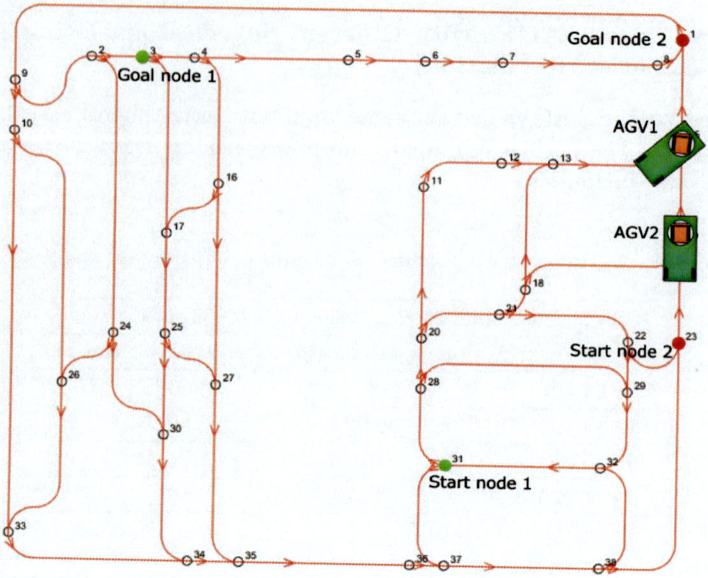

(a)

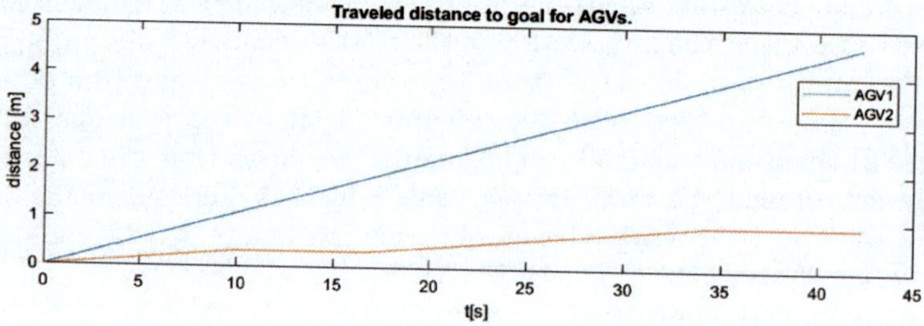

(b)

Fig. 5. Test case 3, (a) map with AGVs at time $t = 17$ s, and (b) travel distances of the AGVs

destination. Figure 6 shows the AGVs on the roads at different time steps. The AGV1 has crossed the intersection 15 first, while the AGV2 has waited in front of this intersection.

5 Conclusions

This paper presents a multi-robot path planning algorithm suitable for non-holonomic AGVs with tricycle kinematics. The algorithm is based on the A* algorithm and uses priorities and occupancy time windows. The applicability of the proposed algorithm was confirmed through a series of test cases in a simulation environment and also on real small-scale AGVs. The presented experiments

Table 5. List of actions for the AGV2

Time t [s]	Road length L [m]	Relative location r	First mode m	Second mode M	Road ID R
9.09	0.3	1	R	R	2319
9.85	0.3	0.083	R	R	1915
16.77	0.3	0.083	R	R	1915
25.10	0.3	1	R	R	1915
34.19	0.3	1	R	R	1501

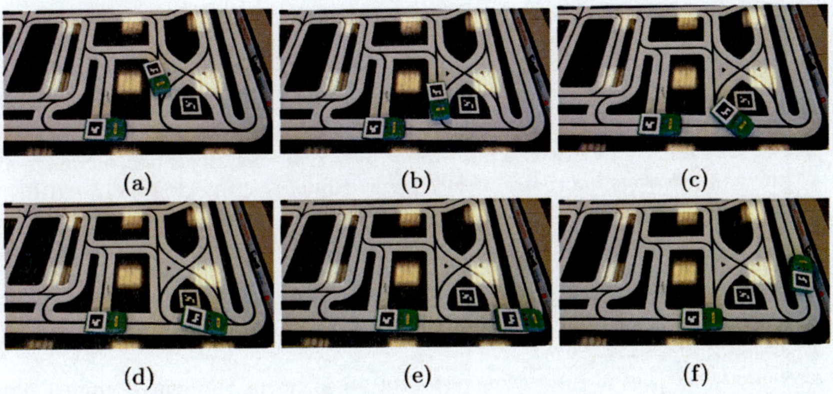

Fig. 6. Captured sequence of AGVs during execution of the planned actions at time instants (a) $t = 17$ s, (b) $t = 18$ s, (c) $t = 20$ s, (d) $t = 21$ s, (e) $t = 23$ s and (f) $t = 27$ s—see also the supplementary video

included only two AGVs, but the number of vehicles does not affect the algorithm, so the algorithm also works for multiple AGVs. The results show that the proposed algorithm can plan a collision safe path, taking into account the priorities of the AGVs and their dimension. The AGVs in the current implementation can only drive forward along one-way roads, since the line following control is only stable in the forward direction. In the future, we plan to evaluate the performance for multiple AGVs, extend the algorithm for two-way roads and with an optional reverse driving for situations where turning is not possible. This will allow for additional path optimization options, such as waiting for AGVs in side roads to allow driving of AGVs in an appropriate sequence that avoids dead locks.

Acknowledgment. The authors acknowledge the financial support from the Slovenian Research Agency and Epilog d.o.o. (research core funding No. L2-3168 and No. P2-0219).

References

1. Costa, M.M., Silva, M.F.: A survey on path planning algorithms for mobile robots. In: 2019 IEEE International Conference on Autonomous Robot Systems and Competitions (ICARSC), pp. 1–7 (2019)
2. Bagade, A.A.K., Nagare, P., Shinde, P., Gudadhe, S.: Path planning and error correction algorithm for autonomous guided vehicles: a survey. J. Crit. Rev. **7**(19), 1237–1246 (2020)
3. Hart, P.E., Nilsson, N.J., Raphael, B.: A formal basis for the heuristic determination of minimum cost paths. IEEE Trans. Syst. Sci. Cybern. **4**(2), 100–107 (1968)
4. Dijkstra, E.W.: A note on two problems in connexion with graphs. Numer. Math. **1**, 269–271 (1959)
5. Lixiao, G., Qiang, Y., Wenjun, Y.: Intelligent path planning for automated guided vehicles system based on topological map. In: IEEE Conference on Control, Systems & Industrial Informatics, vol. 1, pp. 69–74 (2012)
6. Klančar, G., Zdešar, A., Blažič, S., Škrjanc, I.: Path planning, Chap. 4. In: Wheeled Mobile Robotics, pp. 161–206. Butterworth-Heinemann (2017)
7. Yu, J., LaValle, S.M.: Optimal multirobot path planning on graphs: complete algorithms and effective heuristics. IEEE Trans. Robot. **32**(5), 1163–1177 (2016)
8. Wagner, G., Choset, H.: M*: a complete multirobot path planning algorithm with performance bounds. In: 2011 IEEE/RSJ International Conference on Intelligent Robots and Systems, pp. 3260–3267 (2011)
9. Digani, V., Sabattini, L., Secchi, C., Fantuzzi, C.: Ensemble coordination approach in multi-AGV systems applied to industrial warehouses. IEEE Trans. Autom. Sci. Eng. **12**(3), 922–934 (2015)
10. Oyekanlu, E.A., et al.: A review of recent advances in automated guided vehicle technologies: integration challenges and research areas for 5G-based smart manufacturing applications. IEEE Access **8**, 202312–202353 (2020)
11. Fragapane, G., de Koster, R., Sgarbossa, F., Strandhagen, J.O.: Planning and control of autonomous mobile robots for intralogistics: literature review and research agenda. Eur. J. Oper. Res. **294**(2), 405–426 (2021)
12. Sharon, G., Stern, R., Felner, A., Sturtevant, N.R.: Conflict-based search for optimal multi-agent pathfinding. Artif. Intell. **219**, 40–66 (2015)
13. Standley, T.: Finding optimal solutions to cooperative pathfinding problems. In: Proceedings of the AAAI Conference on Artificial Intelligence, vol. 24, no. 1, pp. 173–178 (2010)
14. Zhang, Z., Guo, Q., Chen, J., Yuan, P.: Collision-free route planning for multiple AGVs in an automated warehouse based on collision classification. IEEE Access **6**, 26022–26035 (2018)
15. Pallottino, L., Scordio, V.G., Bicchi, A., Frazzoli, E.: Decentralized cooperative policy for conflict resolution in multivehicle systems. IEEE Trans. Robot. **23**(6), 1170–1183 (2007)
16. Presečnik, N.: Path planning of multiple robotic vehicles considering priorities and occupancy time windows for the map segments. Master's thesis, University of Ljubljana, Faculty of Electrical Engineering (2021)

Cyber-Physical Platform with Miniature Robotic Vehicles for Research and Development of Autonomous Mobile Systems

Andrej Zdešar(✉), Matevž Bošnak, and Gregor Klančar

Faculty of Electrical Engineering, University of Ljubljana, Tržaška 25, 1000 Ljubljana, Slovenia
{andrej.zdesar,matevz.bosnak,gregor.klancar}@fe.uni-lj.si

Abstract. The paper presents a cyber-physical platform that enables development and evaluation of algorithms for autonomous driving of mobile robots. The platform consists of three main parts: a vision system for object tracking, a small-scale physical model of an environment with miniature mobile robots and a system for simulating soft sensors. The vision system consists of a single camera above the platform, that enables global tracking of all objects in real time. The small-scale models of the environment are presented as drawings on a flat surface on which the miniature mobile robots with wheels can drive. Two physical models developed are presented: a town with miniature wheeled mobile robots and an industrial hall with miniature automated guided vehicles. Since not all essential sensors are available in a suitable, small form, an approach that enables soft sensors is presented. Different modular systems are integrated together using the framework provided by the Robot Operating System. The proposed platform can be deployed quickly and is easy to use. It is also very affordable, customisable and expandable.

Keywords: Autonomous mobile robotics · Automated guided vehicles · Multi agent system · Cyber-physical model · Research and development

1 Introduction

Simulation environments are valuable or even indispensable in computer-aided design and development of autonomous mobile systems. The capabilities of contemporary computers with powerful hardware components enable (near) real-time simulation of accurate system dynamics and also photorealistic 3D visualisation of complex scenes under different environmental conditions. This narrows the gap between simulation and real environments and promises a seamless transition of solutions developed in simulation to real applications. Simulations in digital computers are based on mathematical models, but models can also be

I. Petrovic et al. (Eds.): IAS 2022, LNNS 577, pp. 897–908, 2023.
https://doi.org/10.1007/978-3-031-22216-0_60

physical. Each model needs to suit its purpose, i.e. it must represent a certain behaviour of the system that is relevant to its usage. In this paper, we present a cyber-physical platform that enables development and evaluation of algorithms for autonomous driving. We have created two different small-scale physical models (with two different types of miniature robotic vehicles). These models do not attempt to accurately reflect reality, but they are designed in such a way that all major problems in autonomous mobile systems can still be addressed.

The cyber-physical platform is used in the laboratory practice of the Autonomous Mobile Systems graduate course at the University of Ljubljana, Faculty of Electrical Engineering. In the past we have used various teaching approaches, such as simulation challenges, individual or multi-team projects and competitions [1], but the proposed platform has proven to be very suitable for the educational purposes in the field of wheeled mobile robotics. During the course, which lasts one semester, students use the platform to implement various algorithms that address the problems of signal processing, control, path planning and localisation so that the mobile robots become autonomous [2]. Nevertheless, the platform is also useful for research and development activities and enables evaluation of many advanced algorithms for autonomous multi-agent systems [3–5]. There are similar platforms for education and research, like *Duckietown* [6], *Robotarium* [7] and many others [8–10] that can address various problems in mobile robotics. Our system is relatively small and very affordable, but also very modular and expandable. We developed our first version of the platform in 2017 as a small-scale town model with mobile robots. A year later, we developed a small-scale model of an industrial hall with automated guided vehicles. We have gradually upgraded the platform with many features. The Robot Operating System (ROS) is used to achieve a modular design and to allow the use of standard tools to interact with the platform. Our overall motivation for implementing this cyber-physical system is to research and develop multi-agent wheeled mobile robots for industrial automation.

The paper presents all the essential building blocks that comprise the cyber-physical platform. Section 2 describes physical models of two different environments, and Sect. 3 presents two types of miniature robotic vehicles. Section 4 introduces the vision system for global tracking of multiple objects. Section 5 introduces a feature of soft sensors. Section 6 summarises the capabilities of the platform and outlines plans for future work.

2 Small-Scale Physical Model of the Environment

2.1 Model of a Town

In our first version, we created a physical model of a small-scale town that is depicted in the left Fig. 1. The model is fairly simple. It consists consist of a flat area with one-way and two-way roads that intersect at crossroads, and some green areas where vehicles are not allowed. The map consists of a rectangular grid of 14 by 12 square tiles that are 15 cm wide. Therefore, it is easy to design

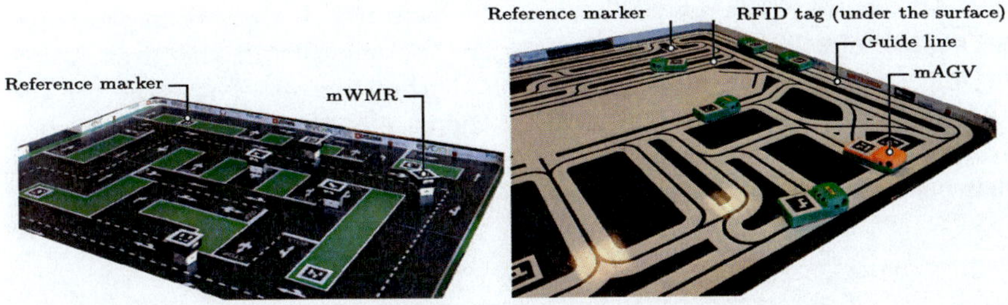

Fig. 1. Physical model of a small-scale town with mWMRs (left) and a small-scale industrial hall with mAGVs (right)

a custom map (e.g. with a tool like [11]), and it is also easy to represent the map in computer memory for the purpose of algorithm development.

Miniature Wheeled Mobile Robots (mWMRs) with differential drive have been developed for use in this model of a town (see Sect. 3.1). The mWMRs (and the map) are equipped with markers that allow their poses to be tracked with a camera above the platform (see Sect. 4).

2.2 Model of an Industrial Hall

In our second version, we have created a small-scale physical model of an industrial hall with Automated Guided Vehicles (AGVs). For the purposes of AGV navigation and autonomous driving in industrial halls, magnetic tapes are normally laid along drivable paths, and RFID tags are embedded in the floor to mark important locations such as intersections and (loading/charging) stations. In our model we also use RFID tags embedded in the ground surface, but to mark the passable paths (instead of magnetic tape) we have drawn black guide lines on a white background that can be detected with an optical line sensor. The dimensions of the floor area are 220 cm by 180 cm.

The complete map of the environment in the right Fig. 1 with all paths and locations of the RFID tags is depicted in Fig. 2. The paths consist of an arbitrary continuous spline of three basic primitives: straight segments, circular arcs and B'ezier curves. All paths in the map are directed, so the miniature AGV (mAGV) is only allowed to travel in the prescribed direction. RFID tags enable the mAGV to determine its location and take appropriate paths in the crossroads, where paths may normally split into two directions or merge into a single path. Thus, for the mAGV to take the desired path in the crossroad, it only needs to follow either the left or the right edge of the guide line. However, since the RFID tags are not located along all paths and in front of every crossroad, some additional virtual tags are introduced to allow the whole map to be displayed in the form of a graph. The nodes in the graph represent the RFID tags and the virtual tags, and the edges are the paths between the tags. In Fig. 2, blue (red) lines represent paths taken when the left (right) edge of the line is followed (green lines represent some special cases).

In the centre of the model shown in the right Fig. 1, there is an area without the guide lines. The mAGV therefore cannot navigate in this area based on the line tracking, but it needs to rely on other sensors. The structure of the mAGV is presented in Sect. 3.2. As in the case of the model of the town, the flat ground surface and the mAGVs are again equipped with markers that allow their position to be tracked with an overhead camera (see Sect. 4).

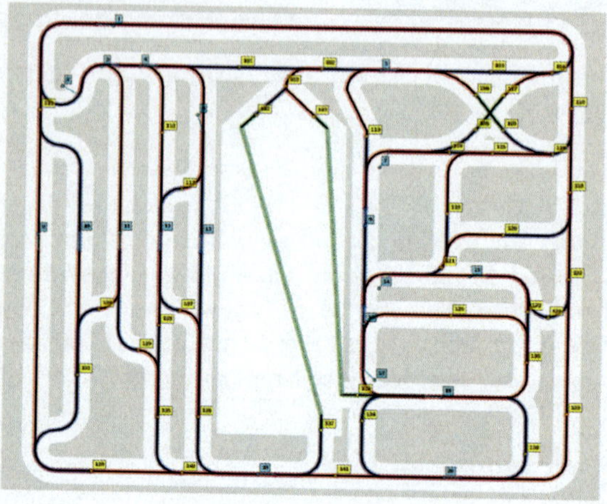

Fig. 2. Map of the small-scale industrial hall with locations of RFID (light blue) and virtual (yellow) tags

3 Miniature Robotic Vehicles

3.1 Miniature Wheeled Mobile Robot with Differential Drive

The basic parts of the first generation of the miniature WMR (mWMR) with differential drive are shown in Fig. 3. The two wheels of the mWMR are driven by two motors equipped with incremental encoders. And these are also the only two sensors on the mWMR, there are not even proximity sensors for obstacle detection and collision avoidance. Inside the mWMR is a microcontroller that processes the data from the encoders and drives the wheels (the wheels are speed-controlled). The mWMR is equipped with an embedded computer *Raspberry Pi Zero W* that enables signal processing, implementation of algorithms for autonomous driving and communication with external systems. The marker on the top is used to track the mWMR with the vision system (see Sect. 4).

The mWMR with differential drive (Fig. 3) has a non-holonomic constraint:

$$\dot{x} \sin \varphi - \dot{y} \cos \varphi = 0 \tag{1}$$

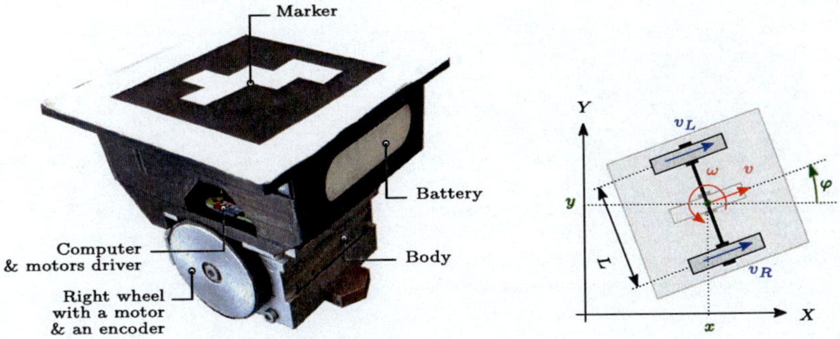

Fig. 3. Structure of the mWMR (left) and its kinematic model (right)

since the wheels cannot move in the lateral direction. From the kinematics point of view, the model of a differential drive is equivalent to a unicycle. The pose of the mWMR can be given in the generalised coordinates $\boldsymbol{q}^T = [x\ y\ \varphi]$. The constraint (1) can then be written as $\boldsymbol{A(q)}\dot{\boldsymbol{q}} = \boldsymbol{0}$, where $\boldsymbol{A(q)} = \begin{bmatrix}\sin\varphi & \cos\varphi & 0\end{bmatrix}$ is the constraint matrix with a single row. The kinematic model of the mWMR results from the null space of the constraint matrix:

$$\dot{\boldsymbol{q}} = \begin{bmatrix}\cos\varphi & \sin\varphi & 0 \\ 0 & 0 & 1\end{bmatrix}^T \boldsymbol{u} \tag{2}$$

where $\boldsymbol{u}^T = [v\ \omega]$ are the control inputs. The linear velocity v and the angular velocity ω are directly related to the velocities of the front wheels:

$$v_L = v_s - \frac{L\omega_s}{2} \qquad v_R = v_s + \frac{L\omega_s}{2} \tag{3}$$

where L is the distance between the wheels. The velocities of the wheels v_L and v_R are limited to

$$v_{MIN} \leq |v_{L,R}| \leq v_{MAX} \tag{4}$$

3.2 Miniature Automated Guided Vehicle

The structure of the miniature AGV (mAGV) we have designed is shown in Fig. 4. Originally, we wanted to create a mAGV with tricycle drive powered by a single front wheel, with one motor turning the wheel and the other motor determining the direction of the wheel with respect to the main body. However, it turned out that we can achieve a similar kinematic model if we instead use a differential drive attached to the main body with a passive bearing for the front part. In this way, it is much easier to implement the mechanics in a small and compact form. Only the two wheels on the front cart are actuated, the rear wheels are passive.

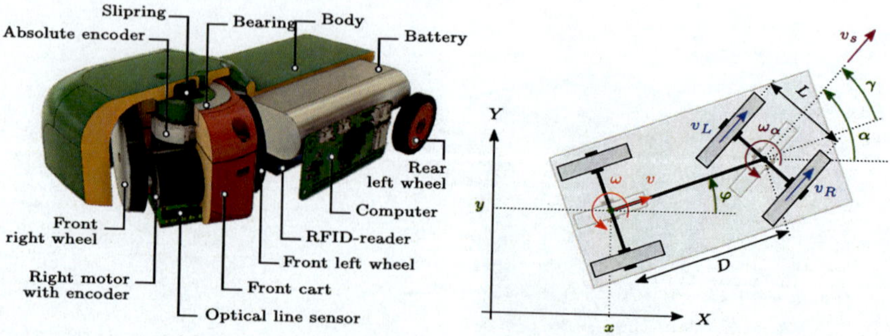

Fig. 4. Structure of the mAGV (left) and its kinematic model (right)

The mAGV is equipped with several sensors that enable implementation of algorithms for autonomous driving in the small-scale model of an industrial hall presented in Sect. 2.2. Both motors driving the front wheels are equipped with incremental encoders. The orientation of the front cart with respect to the main body is measured with an additional absolute encoder. On the underside of the front cart is a custom designed circuit board with a microcontroller responsible for real-time signal processing of the encoder data and speed control of the front drive wheels. The circuit board also contains a seven-segment linear optical reflective sensor that can be used to detect a line on the floor under the mAGV. To detect RFID tags on the floor, the main body is also equipped with an RFID tag reader on the bottom. The main body contains an embedded computer *Raspberry Pi Zero W*, which serves the same purpose as in the case of the mWMR (Sect. 3.1). All on-board systems are powered by a battery integrated into the 3D-printed main body.

Our four-wheeled implementation of a three-wheel drive has the same kinematic constraints as the bicycle drive (see 4). In this case, in addition to the (1), there is also the constraint $\dot{x}\sin\alpha - \dot{y}\cos\alpha - D\dot{\varphi}\cos\gamma = 0$, where D is the distance between the rear axle and the pivot point of the front wheels. Let us introduce the states of the system in the generalised form $\boldsymbol{q}^T = [x \ y \ \varphi \ \gamma]$. The constraints can then be given as rows in the constraint matrix $\boldsymbol{A}(\boldsymbol{q})$:

$$\boldsymbol{A}(\boldsymbol{q})\dot{\boldsymbol{q}} = . \begin{bmatrix} \sin\varphi & -\cos\varphi & 0 & 0 \\ .\sin\alpha & -\cos\alpha & -D\cos\gamma & 0 \end{bmatrix} \dot{\boldsymbol{q}} = \boldsymbol{0} \tag{5}$$

The kinematic model of the system with respect to the inputs $\boldsymbol{u}^T = [v_s \ \omega_\alpha]$ is

$$\dot{\boldsymbol{q}} = \boldsymbol{S}\boldsymbol{u} = \left[\cos\gamma\sin\varphi \ \ \cos\gamma\sin\varphi \ \ \tfrac{\sin\gamma}{D} \ \ -\tfrac{\sin\gamma}{D} \ 0 \ 0 \ 0 \ 1\right]^T \left[v_b \gamma \omega_\alpha\right] \tag{6}$$

where $\boldsymbol{S}$ is the null space of the constraint matrix $\boldsymbol{A}(\boldsymbol{q})$. The linear velocity v_s and the angular velocity $\omega_\alpha = \dot{\alpha}$ are related to the velocities of the front wheels:

$$v_L = v_s - \frac{L\omega_\alpha}{2} \qquad v_R = v_s + \frac{L\omega_\alpha}{2} \tag{7}$$

where L is the distance between the front wheels. The speeds of the wheels are again constrained by (4).

Table 1. Bicycle kinematic models $\dot{\boldsymbol{q}} = \boldsymbol{S}\boldsymbol{u}$ for common velocity inputs ($\boldsymbol{q}^T = [x\ y\ \varphi\ \gamma]$)

	Global angular velocity ω_α	Local angular velocity ω_γ
Rear drive v	$\dot{\boldsymbol{q}} = \begin{bmatrix} \cos\varphi & 0 \\ \sin\varphi & 0 \\ -\frac{\tan\gamma}{D} & 0 \\ \frac{\tan\gamma}{D} & 1 \end{bmatrix} \begin{bmatrix} v \\ \omega_\alpha \end{bmatrix}$	$\dot{\boldsymbol{q}} = \begin{bmatrix} \cos\varphi & 0 \\ \sin\varphi & 0 \\ -\frac{\tan\gamma}{D} & 0 \\ 0 & 1 \end{bmatrix} \begin{bmatrix} v \\ \omega_\gamma \end{bmatrix}$
Front drive v_s	$\dot{\boldsymbol{q}} = \begin{bmatrix} \cos\gamma\cos\varphi & 0 \\ \cos\gamma\sin\varphi & 0 \\ -\frac{\sin\gamma}{D} & 0 \\ \frac{\sin\gamma}{D} & 1 \end{bmatrix} \begin{bmatrix} v_s \\ \omega_\alpha \end{bmatrix}$	$\dot{\boldsymbol{q}} = \begin{bmatrix} \cos\gamma\cos\varphi & 0 \\ \cos\gamma\sin\varphi & 0 \\ -\frac{\sin\gamma}{D} & 0 \\ 0 & 1 \end{bmatrix} \begin{bmatrix} v_s \\ \omega_\gamma \end{bmatrix}$

Based on the control inputs $\boldsymbol{u}^T = [u_1\ u_2]$ (Table 1), several different kinematic models of a bicycle can be defined: Rear drive $u_1 = v$ or front drive $u_1 = v_s$, angular velocity in global frame ($u_2 = \omega_\alpha = \dot{\alpha}$) or local frame ($u_2 = \omega_\gamma = \dot{\gamma}$). All these kinematic models satisfy the constraints defined in (5). Often an auxiliary controller is used that sets the desired angle γ based on the input ω_γ or ω_α. In such a case, the desired angle γ becomes the control input instead of the angular velocity.

4 Global Vision-Based Object Tracking

Above the physical model is a single overhead camera that enables tracking of all objects in its field of view moving on a flat ground plane. The pose of the camera with respect to the ground plane is arbitrary. Let us assume that the camera frame with respect to the world frame is given by the rotation matrix $\boldsymbol{R} = [\boldsymbol{r}_1\ \boldsymbol{r}_2\ \boldsymbol{r}_3]$ and the translation vector $\boldsymbol{t}$. Considering the pinhole camera model, the relation between the point given in homogeneous coordinates $\boldsymbol{p}_W^T = [x_W\ y_W\ 1]$ in an arbitrary world plane and the homogeneous point in the image frame $\boldsymbol{p}_P^T = [x_P\ y_P\ 1]$ is

$$\boldsymbol{p}_P \propto \boldsymbol{S}\left[\boldsymbol{r}_1\ \boldsymbol{r}_2\ \boldsymbol{t}\right]\boldsymbol{p}_W = \boldsymbol{H}\boldsymbol{p}_W. \tag{8}$$

The matrix $\boldsymbol{S}$ in (8) contains intrinsic parameters of the camera model. The matrix $\boldsymbol{S}$ can be determined with calibration, e.g. with the well-known checkerboard calibration approach [12]. The transformation matrix $\boldsymbol{H}$ in (8) is known as *homography*—it represents the transformation between two planes.

The homography $\boldsymbol{H}$ can be estimated from a set of at least four corresponding points in the image and the world plane. For this reason, the ground plane in the

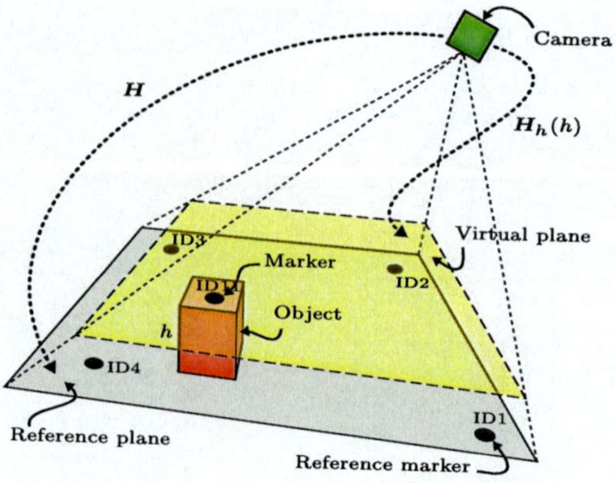

Fig. 5. Vision-based object tracking

physical model has four reference markers (see Fig. 1)—the positions of these markers with respect to the world frame are known. Each WMR (or any other object we wish to track) must also be equipped with a unique marker. As the WMR travels on the flat ground, the marker moves in the virtual plane that is parallel to the ground plane. The situation is shown in Fig. 5. Therefore, the pose of the WMR cannot be estimated directly based on the homography $\boldsymbol{H}$. Since the height h at which the marker is mounted on the WMR is assumed to be known, it is possible to estimate the $\boldsymbol{H}_h(h)$ for this parallel plane based on the homography $\boldsymbol{H}$.

The image position $\boldsymbol{p}_P$ of the marker, which lies in the plane parallel to the ground plane at height h, can be transformed (with an orthogonal projection) into a point $\boldsymbol{p}_W$ in the ground plane: $\boldsymbol{p}_W \propto \boldsymbol{H}_h^{-1}(h)\boldsymbol{p}_P$. In order to estimate the homography $\boldsymbol{H}_h(h)$, the intrinsic camera parameters $\boldsymbol{S}$ must be known:

$$\boldsymbol{H}_h(h) = \boldsymbol{S}\left[\boldsymbol{q}_1\ \boldsymbol{q}_2\ (\boldsymbol{q}_1 \times \boldsymbol{q}_2)\tfrac{h}{n} + \boldsymbol{q}_3\right] \tag{9}$$

where the normalisation coefficient n is determined as follows

$$n = (\|\boldsymbol{q}_1\| + \|\boldsymbol{q}_2\|)/2 \tag{10}$$

The vectors $\boldsymbol{q}_1$ to $\boldsymbol{q}_3$ in (9) and (10) are $\boldsymbol{S}^{-1}\boldsymbol{H} = \left[\boldsymbol{q}_1\ \boldsymbol{q}_2\ \boldsymbol{q}_3\right]$.

Since the markers are also not rotation invariant, it is also possible to determine the orientation of the markers and thus the complete pose (position and orientation) of the object in the world plane. The proposed approach enables measurement of the object's pose for arbitrary configuration of the camera with respect to the ground plane, as long as all markers are in the camera's field of view, but the accuracy of the measurements may vary. If the reference markers are never occluded, object tracking can be made even if the camera is moving. Otherwise, the homography $\boldsymbol{H}$ is estimated when all reference markers are visible. As long as the WMR is only moving on the flat ground, the proposed position estimation approach allows for very accurate object tracking since the system constraints are implicitly taken into account. The depth of the markers could also be estimated from their size, since we assume that we have a calibrated camera. In our case, however, the depth is implicitly estimated from the four reference markers on the ground plane. Since these are not colinear and are far apart, the estimate of the homography $\boldsymbol{H}$ can be very accurate.

The camera may contain large lens distortions that can be given in a non-linear model $\boldsymbol{p}_P = \boldsymbol{f}_{lens}(\boldsymbol{p}_P^*)$, where $\boldsymbol{p}_P^*$ represents a point in the distorted image and $\boldsymbol{p}_P$ is its projection into the rectified image. The distortion model can be estimated during camera calibration (e.g. [13]).

The overall tracking system is computationally very efficient and allows implementation on low-cost hardware. We implemented the complete object tracking system on an embedded computer *Raspberry Pi 3 B+* with the *Raspberry Pi Camera Board v2*. For image-based marker detection and tracking, we used the ArUco library [14,15], which provides robust and fast tracking of artificial markers with unique patterns. We achieved tracking of multiple objects while estimating their poses with respect to the world frame with a frequency of SI 15 Hz. The standard deviation of the pose estimate for a stationary object was measured to be less than 0.001 m in position and 0.01 rad in orientation. The data about the current poses of the tracked objects (WMRs) are published on the ROS topic `tf`, which collects information about the transformations between the coordinate frames. The proposed vision system for object tracking enables fast and simple deployment. Once the intrinsic camera parameters are obtained (e.g. using a standard chessboard calibration procedure), only the positions of the four reference markers on the ground plane and the heights of all other markers above the ground plane need to be known.

5 Soft Sensors

The vision-based object tracking system provides the pose of each object with respect to the world frame. In essence, this system can already be considered as Global Navigation Satellite System (GNSS) or some other (indoor) positioning system if only the position part of the data (without orientation) is considered. However, this represents an almost ideal measurement system, but a user-defined model can be used that resembles more realistic data from the GNSS, which may include noise, bias, delay etc. Therefore, the miniature robotic vehicle can obtain its global position even though it is not equipped with a GNSS receiver.

The mobile robots presented in Sect. 3 can be equipped with many additional sensors, such as internal measurement sensors (e.g. accelerometers, giroscopes and additional encoders) and proximity sensors (e.g. optical and ultrasonic sensors). In robotic applications, cameras are emerging as the most versatile sensor for advanced environment perception. Cameras are small enough to be fitted also onboard the miniature mobile robots (e.g. for visual line following [16]).

However, due to the small size of the physical model, the choice of sensors that can be mounted on the miniature robotic vehicles is limited, as not all sensors normally used in mobile robotics can be miniaturised or are not available in such a small form. For example, LIDAR is often used in mobile robotics, and may even be a required safety feature in industrial applications of AGVs. Some mobile robots may even be equipped with more than one such sensor to cover the entire working area around the robot and prevent blind spots. LIDARs are very useful for solving localisation and mapping problems [17].

Although there is no small and affordable LIDAR that would fit on our miniature robotic vehicle, we can implement the measurements of this sensor in software. The vision-based global object tracking system allows us to measure the poses of all miniature robotic vehicles, and the shapes of these vehicles and the physical model are also known. If the boundaries of all these objects are described with line segments, we can use an algorithm to calculate the intersection points of the laser rays with the segments of the objects. In this way we can obtain the measurements of the planar laser range scanner in absence of the physical sensor.

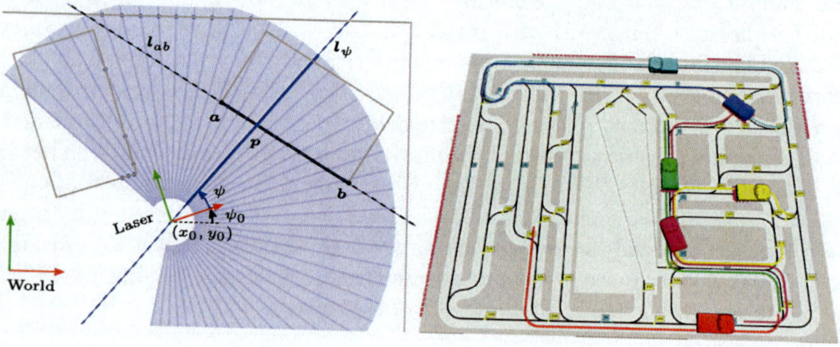

Fig. 6. Modeling of the laser range scanner (left) and visualization of its measurements (violet points) in RViz (right)

The algorithm for implementation of the planar laser range scanner as soft sensor needs to go through all the segments of all the objects that represent static and dynamic obstacles, and do the following for each of the line segments (Fig. 6):

1. If $\boldsymbol{a}$ and $\boldsymbol{b}$ are the endpoints of the segment given in homogeneous coordinates, then the line through these two points is $\boldsymbol{l}_{ab} = \boldsymbol{a} \times \boldsymbol{b}$ and the unit vector of the segment is $\boldsymbol{e}_{ab} = \frac{\boldsymbol{b}-\boldsymbol{a}}{\|\boldsymbol{b}-\boldsymbol{a}\|}$.
2. For each ray $\boldsymbol{l}_\psi = [-\sin(\psi)\ \cos(\psi)\ x_0 \sin(\psi) - y_0 \cos(\psi)]$s do:
   (a) Find the intersection $\boldsymbol{p} = [\kappa x\ \ \kappa y\ \ \kappa]^T = \boldsymbol{l}_\psi \times \boldsymbol{l}_{ab}$ of the ray with the segment line and calculate the value of the factor $q = \boldsymbol{e}_{ab}^T(\frac{\boldsymbol{p}}{\kappa} - \boldsymbol{a})$.
   (b) If $0 \leq q \leq 1$ holds, then the intersection $\boldsymbol{p}$ lies on the segment. In this case calculate the distance r from the object to the origin of the laser range scanner (in the direction of the ray): $r = \left(\frac{\boldsymbol{p}^T}{\kappa} - \boldsymbol{p}_0^T\right)[\cos\psi\ \sin\psi\ 0]^T$. If

$0 \leq r \leq r_{max}$ and r is less than the current value of r_ψ, then update the value of $r_\psi = r$.

One or more soft laser range scanners can be placed anywhere (e.g. on the front of the robot with respect to its base frame). Various properties of the soft sensor can be configured, e.g. maximum and minimum range, precision, field of view, angular resolution, etc. The measurements of the soft sensor are published as ROS message of type `sensor_msgs/LaserScan`. Due to the sensor abstraction, we cannot distinguish between physical and soft sensor measurements. It is obvious that the soft sensor cannot detect objects with unknown shapes and poses as well as objects that are not detected with the global object tracking system.

In Fig. 6 the measurements of the soft laser scanner mounted on the purple mAGV are visualised (purple points) in RViz (a 3D data visualisation tool from ROS). The laser reflections can be seen on the green, blue and yellow mAGV, but not on the cyan mAGV as it is obscured by the green one; the red mAGV is not in the sensor's field of view. Due to the limited range, distant objects are not detected.

6 Conclusions

The paper presented basic components and support systems of the cyber-physical platform with miniature robotic vehicles that enable research and development of essential problems in wheeled mobile robotics, even thought the physical model of the environment does not accurately represent reality. Some of the problems that can be evaluated on the proposed platform are: control (line following, control to a point/pose, control along a path/trajectory), path planning, localisation and mapping (based on encoders, GNSS, laser range scanner and/or RFID tags), multi-agent navigation and control and many more. The platform consists of many modular systems that are highly adaptable and expandable. The modules can therefore be reused and reconfigured to accommodate for new scenarios. All systems are designed to be robust and easy to use. A key component of the platform is the vision system for object tracking, which can provide very accurate measurements as movement constraints are implicitly taken into account. The vision system enables introduction of various soft sensors, such as GNSS and laser range scanner, but many other sensors could also be simulated in a similar way. In the near future, we plan to upgrade the miniature robotic vehicles with some proximity sensors and an onboard camera that will allow the implementation of vision-based algorithms. We have also already prepared some additional support systems that enable remote interaction with the platform, and we are also considering releasing the whole system as open source. As part of our current research and development activities, we are using the presented platform to evaluate advanced multi-agent path scheduling and planning algorithms.

Acknowledgment. The authors acknowledge the financial support from the Slovenian Research Agency and Epilog d.o.o. (research core funding No. L2-3168 and No. P2-0219).

References

1. Zdešar, A., Blažič, S., Klančar, G.: Engineering education in wheeled mobile robotics. IFAC-PapersOnLine **50**(1), 12173–12178 (2017)
2. Tavčar, M., Skalar, U., Perko, J., Zdešar, A.: Implementation of model-based control systems for autonomous driving of miniature robotic vehicles. In: 27th International Electrotechnical & Computer Science Conference (ERK), pp. 131–134. IEEE (2018)
3. Klančar, G., Škrjanc, I.: Tracking-error model-based predictive control for mobile robots in real time. Robot. Auton. Syst. **55**(6), 460–469 (2007)
4. Klančar, G., Matko, D., Blažič, S.: A control strategy for platoons of differential drive wheeled mobile robot. Robot. Auton. Syst. **59**(2), 57–64 (2011)
5. Klančar, G., Zdešar, A., Blažič, S., Škrjanc, I.: Wheeled Mobile Robotics: From Fundamentals Towards Autonomous Systems. Butterworth-Heinemann (2017)
6. Paull, L., et al.: Duckietown: an open, inexpensive and flexible platform for autonomy education and research. In: International Conference on Robotics and Automation (ICRA), pp. 1497–1504. IEEE (2017)
7. Wilson, S., et al.: The robotarium: globally impactful opportunities, challenges, and lessons learned in remote-access, distributed control of multirobot systems. IEEE Control Syst. Mag. **40**(1), 26–44 (2020)
8. Hyldmar, N., He, Y., Prorok, A.: A fleet of miniature cars for experiments in cooperative driving. In: International Conference on Robotics and Automation (ICRA), pp. 3238–3244. IEEE (2019)
9. Balaji, B., et al.: DeepRacer: autonomous racing platform for experimentation with Sim2Real reinforcement learning. In: International Conference on Robotics and Automation (ICRA), pp. 2746–2754. IEEE (2020)
10. Guadagnino, T., Corte, B.D., Grisetti, G.: Low-cost sonar navigation system. In: International Conference on Intelligent Robots and Systems (IROS), pp. 692–697. IEEE (2019)
11. Lindeijer, T., et al.: Tiled map editor. https://www.mapeditor.org. 22 Feb 2021
12. Zhang, Z.: A flexible new technique for camera calibration. IEEE Trans. Pattern Anal. Mach. Intell. **22**(11), 1330–1334 (2000)
13. Perš, J., Kovačič, S.: Nonparametric, model-based radial lens distortion correction using tilted camera assumption. In: Computer Vision Winter Workshop, pp. 286–295 (2002)
14. Garrido-Jurado, S., Muñoz-Salinas, R., Madrid-Cuevas, F.J., Medina-Carnicer, R.: Generation of fiducial marker dictionaries using mixed integer linear programming. Pattern Recognit. **51**, 481–491 (2016)
15. Romero-Ramirez, F.J., Muñoz-Salinas, R., Medina-Carnicer, R.: Speeded up detection of squared fiducial markers. Image Vis. Comput. **76**, 38–47 (2018)
16. Castro, O., Céspedes, A., Ubaldo, R., Ramos, O.E.: Lane following with a duckiebot vehicle using visual feedback. In: Sciences and Humanities International Research Conference (SHIRCON), pp. 1–4. IEEE (2019)
17. Teslić, L., Škrjanc, I., Klančar, G.: Using a LRF sensor in the Kalman-filtering-based localization of a mobile robot. ISA Trans. **49**(1), 145–153 (2010)

Autonomous Hierarchy Creation for Path Planning of Mobile Robots in Large Environments

Jelena Gregorić(✉), Marija Seder, and Ivan Petrović

Laboratory for Autonomous Systems and Mobile Robotics (LAMOR), Faculty of Electrical Engineering and Computing, University of Zagreb, Zagreb, Croatia
jelena.gregoric@fer.hr , marija.seder@fer.hr , ivan.petrovic@fer.hr

Abstract. The main task of mobile robots in large environments such as factories, warehouses, and open spaces is to transport goods and people. Planning the path in large environments using classical graph-based search is computationally too intensive. A representation by a hierarchical graph (H-graph) facilitates graph creation and reduces the complexity of path planning. In this paper, we present an algorithm for autonomously generating hierarchy of the environment from floor plans. The hierarchical abstraction depicts the environment in levels, from the most detailed to the most abstract representation of the environment, where pre-computed partial paths at the most detailed level are graph edges in a higher level. We use the E* algorithm to find partial paths in the most detailed abstraction level, and we propose the extraction of higher levels automatically from lower levels. We verified the proposed H-graph creation on our University premises resulting in five abstract levels.

1 Introduction

Mobile robots have increasingly been used in industry and households. Their main task is to move through environment avoiding obstacles and living beings. Until now, mobile robots have mostly performed their tasks in small environments that can be easily described using classical representations of space such as occupancy grid maps [12] and topological graphs [2,11]. It is obvious that the workspaces of mobile robots expand and encompass large, complex outdoor and indoor environments. Spaces like multi-room floors, multi-floor buildings, or a group of buildings are difficult to represent using classical approaches. Also, path planning in a large graph becomes very inefficient. To solve the mentioned problem of describing large environments, it is necessary to use a different approach and organize the space into simpler graphs to reduce the complexity and increase the efficiency of path planning. A suitable choice is the hierarchical decomposition of the environment based on hierarchical graphs (H-graphs) [3,4,8].

The H-graph is a combination of classical metric and topological maps at different levels of abstraction (see Fig. 1). The H-graph uses a skeleton of precalculated paths (partial paths) between the key points called the bridge nodes.

I. Petrovic et al. (Eds.): IAS 2022, LNNS 577, pp. 909–922, 2023.
https://doi.org/10.1007/978-3-031-22216-0_61

This approach to graph creation represents large environments simultaneously at different resolutions, i.e., at different abstract levels. A lot of research on creating the hierarchical graph exist in the literature [1,3–5,8,9,13]. The creation of an H-graph is usually done manually, which is a time-consuming and complex process. In order to speed up the process of graph creation and path planning, it is necessary to aim for autonomous hierarchy creation. The algorithm for the autonomous creation of an H-graph presented in [8] maintains consistency across the hierarchical levels and keeps the model consistent even after changes in the environment. However, the algorithm creates only three levels of the hierarchy.

This paper proposes autonomous hierarchy creation with extension to more than three levels of abstraction. The proposed algorithm receives the floor plans as input data and creates an H-graph as output. We follow the approach of room extraction algorithm presented in [8] and expand it to create higher levels automatically from lower levels. We also performed an autonomous selection of nodes representing elevators inside the building and the passages between the buildings. We apply the E* algorithm [10] to find partial paths at the most detailed level represented by the occupancy grid map. Partial paths at higher levels are formed by merging partial paths from lower levels. The E* algorithm extends the standard D* algorithm by using interpolation based on the Level Set Method [6] and provides a path that is very close to the optimal solution [7]. Using E*, we avoid the resulting partial paths to be a zigzag line with sharp turns, angles of which are limited to an increment of 45° [7,9]. The process of graph creation is done offline and without any human intervention. The generated H-graph is a good basis for online and inter-building path planning. We showed the efficiency of the H-graph creation of a multi-building environment at the Faculty of Electrical Engineering and Computing (FER). The algorithm creates a graph consisting of five abstract levels and can be easily extended to more if needed.

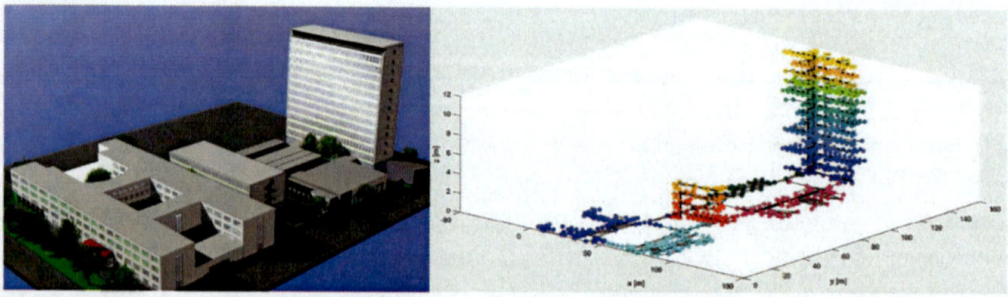

Fig. 1. Representation of FER buildings in reality and as an H-graph.

In this paper, we highlight three main contributions: i) autonomous creation of an H-graph with five or more levels; ii) generation of nodes in H-graph that represent the elevators and the passages between the buildings; and iii) autonomous creation of an H-Graph with natural close-to-optimal paths and autonomous identification of bridge nodes.

The rest of the paper is organized as follows. Section 2 describes the concept of the H-graph. Section 3 describes the autonomous hierarchy creation. Section 4 shows the algorithm validation on FER example concluded by Sect. 5.

2 Modeling of the Large Environment Using H-Graph

Hierarchical decomposition is necessary to achieve feasible path planning in large environments. In the following, the definition of the H-graph is given [9], together with the classification of nodes and optimality conditions for graph creation.

2.1 H-Graph Definition

The H-Graph is a sequence of hierarchical levels $L = L_0, L_1, \ldots, L_D$, where D is the depth of the hierarchy. L_0 is the 'root level', which represents the most abstract description of the environment, i.e., the highest level of abstraction. Analogously, the L_D level is the lowest level of hierarchical abstraction and the most detailed description of environment. Each level $L_i(0 \leqq i \leqq D)$ is represented by graph $G_i = (N_i, A_i, C_i, W_i, T_i)$, where N_i is a set of nodes, A_i is a set of edges, C_i is a set of Cartesian coordinates for N_i, W_i is a set of weights for A_i and T_i is a set of partial paths. The union of graphs $G_0 \cup G_1 \cup \cdots \cup G_D$ represents the graph $G = (N, A, C, W, T)$, where $N = \cup_{i=1}^{D} N_i$, $A = \cup_{i=1}^{D} A_i$, $C = \cup_{i=1}^{D} C_i$, $W = \cup_{i=1}^{D} W_i$, $T = \cup_{i=1}^{D} T_i$. An egde $a(n_J, n_K, w_H) \in A$ is defined by three elements n_J, n_K, w_H, where $n_J, n_K \in N$, $n_J \neq n_K$ and $w_H \in W$. Edges are non-directed: a mobile robot can navigate between two nodes in both ways. Cartesian coordinates $c_I \in C$ are defined by (x, y, z), where $x, y, z \in \mathbb{N}$. A weight $w_I \in W$ is a positive real number ($w_I \in \mathbb{R}$).

Nodes of the H-graph are classified into three groups: submap, end and bridge nodes. Submap nodes, noted by $S \subset N$, represent a subset of nodes in a deeper level of abstraction. Submap nodes can be found at all levels of hierarchical abstraction except at the deepest level L_D. A node $s \in N_i$ is a *parent* submap node from the level L_i, if it is composed of $K > 1$ *child* nodes from the level L_{i+1}. End nodes are start or goal points which can be selected from a robot path planner. Bridge nodes, noted by $B \subset N$, were divided in two classes in [3]: vertical bridge nodes are the vertical connections, i.e., elevators, that connect two floors (submap nodes) in a building, and horizontal bridge nodes connect two submap nodes as the horizontal connections, i.e., doorways. The concept of vertical bridge nodes does not imply a connection between hierarchical levels but a connection of the environment parts within the same level and allow the robot movement in the vertical direction through environment.

Each submap node has its set of partial paths defined between bridge nodes. The partial path $T(a, b, s)$ is defined by bridge nodes $a, b \in B$ and submap node $s \in S$ as follows:

$$
\begin{aligned}
T[1] &= a, & T[|T|] &= b, \\
T[i] &\in s, & i &= 1, \ldots, |T|, \\
a(T[j], T[j+1], w_{j,j+1}) &\in A, & j &= 1, \ldots, |T| - 1 \ ,
\end{aligned}
\tag{1}
$$

where $|T|$ denotes the number of nodes on the partial path T. The cost of the partial path $c(T)$ is defined as the sum of the weights of the edges on the partial path. Partial paths are optimal-length paths calculated offline. Partial paths are introduced to avoid time-consuming path corrections at deeper levels of abstraction in the case of path replanning. On the other hand, partial paths and their costs require extra memory space. However, achieving a short online computational time is extremely important for path planning in large dynamic environments.

2.2 Conditions for Optimality

To ensure optimal global path and reduce the computational complexity of path replanning in real-time, we follow [9] and extend it for a higher number of hierarchical levels. First, we put bridge nodes at the narrow passages to ensure path optimality, and second we duplicated bridge nodes to maintain consistency through hierarchy.

The indoor environment is geometrically divided into rooms and halls. At level L_D, bridge nodes are located on the border between the two rooms (submap nodes) and belong to both rooms, i.e., they have two parent submap nodes. To ensure path optimality, bridge nodes must be placed in a way that the union of the optimal partial paths between the bridge nodes gives the optimal global path. Also, the positions of bridge nodes play an important role in solving the problem of robots passing through narrow passages. The safest path in narrow passages is the one in the middle of the passage, because in this way the robot has the minimized risk of colliding with an obstacle. Therefore, we placed the bridge nodes exactly in the middle of the passage, at the door that separates the rooms. In this way, path optimality is achieved, and safety is increased.

We organize the higher levels of the hierarchy to allow optimal horizontal and vertical hierarchical path planning. To that goal, we defined a new division of bridge nodes: *regular* and *marginal*. Marginal bridge nodes represent connections to a higher level of the hierarchy, and regular bridge nodes represent connections between two submap nodes at the same level. We duplicate the marginal bridge nodes to the higher level. A duplicated bridge node $m^i \in N_i$ is defined as a newly created node at level L_i, which coordinates are equal to the coordinates of the marginal bridge node $m^{i+1} \in N_{i+1}$ at level L_{i+1}. This two nodes are connected with an edge $a(m^i, m^{i+1}, w)$, where $w = 0$. At level L_i, these duplicated bridge nodes are again divided into regular and marginal bridge nodes, where marginal bridge nodes are at the borders of two submap nodes of level above and regular bridge nodes are at the borders of two submap nodes of that level. At level L_D, all bridge nodes are marginal and represent a connection with the level L_{D-1}. End nodes appear only at the lowest level, bridge nodes can be found at all levels except the root level L_0, and submap nodes are at all levels except the lowest level L_D. For each submap node, partial paths between all possible pairs of marginal bridge nodes are determined. Once partial paths are determined at one level, they form the basis for calculating new partial paths at a higher level. In addition to the above conditions for optimality, it must be considered that consistency must be maintained through the hierarchical levels of the model, and the model must remain consistent after changes in the environment. This is achieved by creating abstract levels from the most detailed level.

3 Autonomous H-Graph Creation

In this work, the autonomous hierarchy creation is based on the principles of consistency and optimality, as described in the previous section. An H-graph is created only once and offline. The H-graph is organized into a data structure which is shown in Fig. 2(a) and (b). In this structure is a list of *nodes* at each level. The nodes are described by the corresponding (x, y, z) coordinates. The variables n and m indicate which nodes are children at the level below. The variable b is the index of each node at some level. At each level is a structure called *edges* that describes all partial paths of that level. The structure *edges* consists of the matrix *path* that stores the path described in the coordinate space and the real number *cost* that stores the price of the path. At each level, there is also a substructure that leads to a lower level. At the lowest level, we have a set of room matrices that contains a room occupancy grid maps. The occupancy grid map is continuously updated as the robot moves in the environment according to the detected obstacles by the robot sensors.

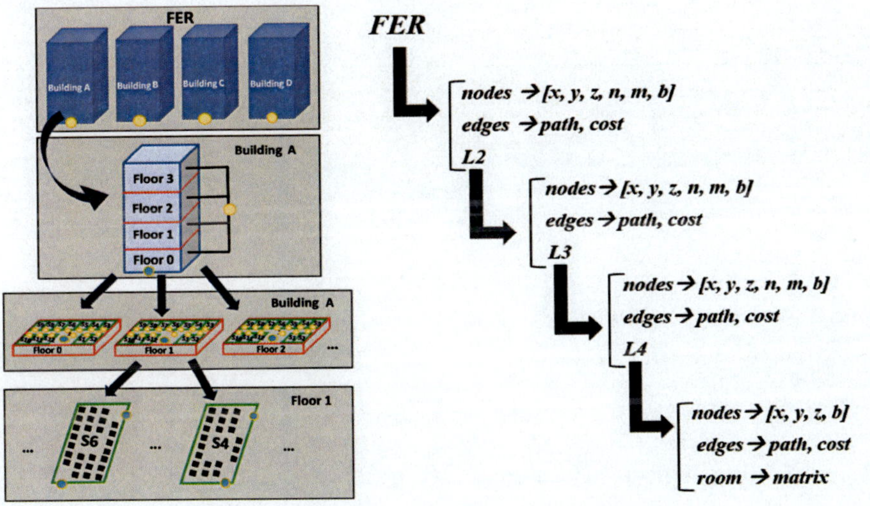

(a) (b)

Fig. 2. (a) An illustration of the hierarchical division of FER; (b) structure as output data from the autonomous hierarchy creation algorithm.

3.1 Creating Occupancy Grid Map from the Floor Plan

The input data given to the algorithm for the autonomous hierarchy creation are floor plans of all floors in *.png* format and the corresponding parameter files in *.yaml* format. The floor plan images could be obtained from CAD program or automatically via SLAM, and must not contain details such as door markings, room descriptions, elevations, etc. Each *yaml* file contains characteristic parameters that describe each floor plan: the name of image which is paired with *yaml* file, map resolution and origin in meters. Additional parameters specify the creation of the occupancy grid map and the hierarchy: the radius of the described circle around the robot's footprint noted as the radius of the robot in the following, the size of the cell, and the maximal width of the narrow passage. The algorithm read the list of *yaml* files. Loading the floor plan and its parameters is followed by the creation of the grid map with the cell occupancy function. Obstacles are extended by the radius of the robot so that the robot could be considered as a point in space. The safety cost mask defines the values of the cell occupancy function depending on the proximity of the obstacle. The cell occupancy function is represented as a function of two variables and denoted as $z(x, y)$. The lowest value of the occupancy function is in cells outside the safety cost mask and is equal to 1, while the highest value is in occupied cells, as shown in Fig. 3. The safety cost mask is equal to half of the maximal width of the narrow passage.

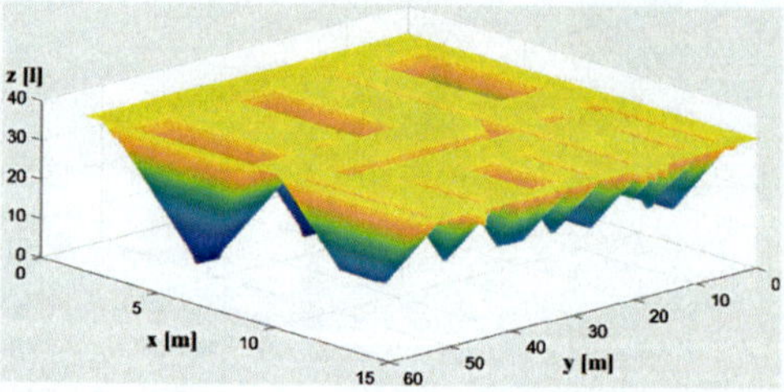

Fig. 3. Occupancy grid map with a safety cost mask represented as cell occupancy function in 3D.

3.2 Finding Narrow Passages, Bridge Nodes and Rooms

The method for determining narrow passages, bridge nodes, and rooms is performed on the previously created occupancy grid map with the safety cost mask. We adapt the room extraction algorithm from [8] and extend the algorithm to autonomously find locations of elevators. A narrow passage is defined as a set of cells where the local minimum of the function z is located and the value of the function is greater than 1. Therefore, the maximal width of the narrow passage that can be detected is equal to two widths of the safety cost mask. Each group of cells which are defined as narrow passage is assigned a central node with coordinates at the center of the narrow passage.

Bridge nodes must be placed in the middle of narrow passages and at the boundary between two rooms. The position of the bridge nodes is not necessarily determined at the door location, but it is certainly in the middle of a narrow passage. Cells belonging to a room are separated from other rooms by walls and narrow passages. The task is to find which narrow passages separate the rooms at doors and not only between obstacles in the room. For this purpose, we find a closed loop that passes through the boundaries of the room. If such closed loop exists, the examined coordinates are declared as the bridge node. Otherwise, the narrow passage lies inside the room and not at the border of the two rooms. In the case when a room is filled with obstacles that form a closed area, that area will be declared as a room, although in reality, it is not. But this case benefits hierarchical planning since it has been shown that the size of the submap node, i.e., the number of cells within one room, is closely related to the optimality and computational efficiency of the path planning algorithm [13]. Furthermore, the higher the parameter for the maximal width of the narrow passage is, i.e., the larger safety cost mask, the more found rooms will exist. The procedure of determining the bridge nodes is described in more details in [8].

The described method for finding bridge nodes is also the method for finding rooms. Rooms are rounded by the loop. Figure 4(a) shows the found loop across the boundaries of the rooms and the narrow passages. Nodes inside the room represent one submap node at the higher level of the hierarchy. A room consists of a set of cells with the same associated number. Bridge nodes are assigned to the two rooms at whose boundaries they are located. Figure 4(b) shows the narrow passage (black line), bridge nodes (diamonds), and the rooms with their assigned numbers and colors (filled color).

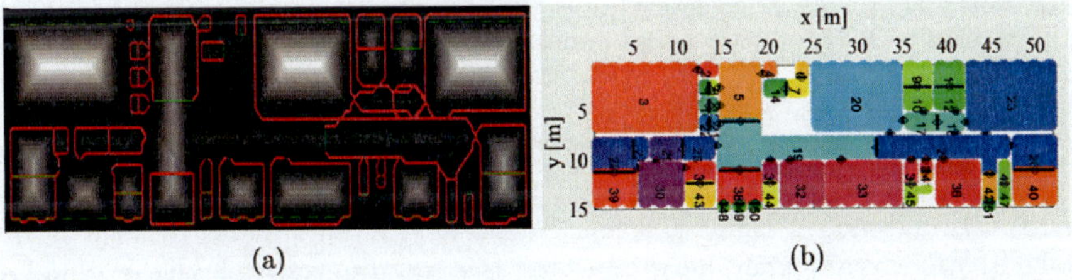

Fig. 4. (a) Loops across the occupied cells that surround the rooms (red) and narrow passages (green); (b) map of the floor with division to rooms: narrow passage (black line), bridge nodes (diamonds), and the rooms with their assigned numbers and colors (filled color).

3.3 Creation of Higher Level Nodes

The described H-graph of FER has five hierarchical levels ($D = 4$). The most abstract level L_0 contains only one node. At level L_D is a graph created from the occupancy grid map. The occupancy grid map at level L_4 is shown in Fig. 5(a) along with the marginal bridge nodes at that level (black x). Bridge nodes are located at each passage between two rooms.

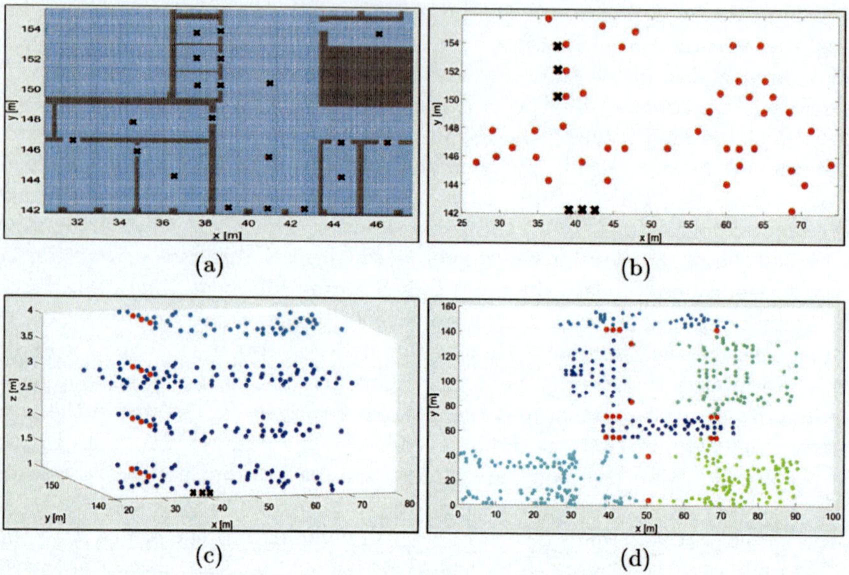

Fig. 5. Nodes of: (a) level L_4; (b) level L_3; (c) level L_2; (d) level L_1. For clarity all bridge nodes are drawn at each level, but only the red ones are actually nodes of the graph at that level. The black x sign indicates marginal bridge nodes.

The marginal bridge nodes (all bridge nodes) from level L_4 are duplicated to the level above and they appear as nodes of the graph at level L_3. In that way, the bridge nodes connect levels L_3 and L_4. Bridge nodes in level L_3 are shown abstractly and they have inherited the coordinates from the level below. The bridge nodes from the most detailed level are stored in the structure with their corresponding (x, y, z) coordinates and used to create a graph on level L_3. The nodes at this level are the parent nodes for nodes of both rooms between which they are located. The connections between nodes in the graph at level L_3 are partial paths determined between all bridge nodes at the level below.

The marginal bridge nodes at level L_3 are located at places of elevators and at the ground floor intersections of two adjacent buildings. The bridge nodes at the elevator locations play the role of vertical bridge nodes. Their determination is done automatically under three conditions. First, comparing the positions of the nodes on level L_3, second, the size of the rooms to which these nodes belong is taken into account, and third, the number of room doors is considered. Figure 5(b) shows the nodes of level L_3 composed of regular (red dots) and marginal (black x) bridge nodes.

At the next level L_2, the graph consists of marginal bridge nodes from level L_3. Thus, the marginal bridge nodes from level L_3 with the corresponding (x, y, z) coordinates, are duplicated in level L_2, and used to create a graph on that level. The marginal bridge nodes at level L_2 are determined as the bridge nodes at the transition points from one building to another and they are located on the ground floor where the two buildings meet. The nodes colored red in Fig. 5(c) represent all bridge nodes at level L_2, while the marginal bridge nodes (three black x) are duplicated to a higher level where they form a new, more abstract graph. Blue dots are shown only for a reference to have a view of other floors in a building, and they represent bridge nodes from lower level L_3.

At level L_1, the graph is composed of marginal bridge nodes from level L_2. Figure 5(d) shows the nodes of level L_1 colored in red, while all other dots are the bridge nodes from level L_3 colored in different color representing each building. We can notice that the bridge nodes of level L_1 are at connections of buildings.

3.4 Finding Partial Paths

To complete the abstract graphs at higher levels, it is necessary to connect selected nodes with edges. Connecting nodes is also performed from the most detailed level. In this work, we use partial paths that connect two marginal bridge nodes within a common submap node. Partial paths are determined within each room submap node between all possible pairs of marginal bridge nodes that belong to that room, shown in Fig. 6(a). Once the partial paths are determined at the most detailed level, they are the basis for computing the new partial paths at the higher level. The process of determining partial paths ends when the paths are determined in the last level which contains bridge nodes.

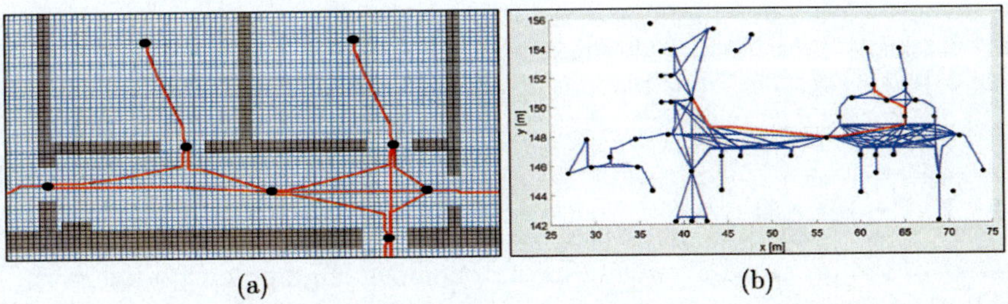

Fig. 6. Partial paths: (a) determined at the most detailed level within each room submap node between all possible pairs of marginal bridge nodes of the room; (b) at level L_3 determined as the union of the partial paths from the level below.

Partial paths between the marginal bridge nodes at level L_4 are determined by an E* algorithm. The E* algorithm search the graph from the goal node and finds all possible paths in the graph. For this purpose, we single out an occupancy grid map of each room. If the selected room has more than one marginal bridge node, an algorithm performs a search of the entire room in order to find the paths between all marginal bridge nodes within that room, storing the costs of the found partial paths. If a room has N bridge nodes, there must be found $\frac{N(N-1)}{2}$ partial paths within the room and E* must be called $(N-1)$ times because one search connects the goal node with all other bridge nodes in the same room.

At the level above, L_3, the abstract graph consists of nodes and edges which build a simple one-level graph. The nodes are marginal bridge nodes from level L_4, and the edges are partial paths determined at level L_4. At all higher levels, the partial paths between two marginal bridge nodes are simply determined by the union of the corresponding partial paths from the level below, as shown in Fig. 6(b). In other words, a simple search using the A* algorithm finds the partial paths between all marginal bridge nodes by concatenating the partial paths from the level below. Partial paths from all levels and their costs are stored in memory.

4 Simulation Results

In this section we analyse the proposed algorithm according to the reduction of nodes comparing to the plain graph search, and partial paths quality compared to the ones obtained by the A* graph search. FER has four buildings marked with letters: A, B, C, and D (see Fig. 2). The floor plan of each floor in each building is individually conducted through the described algorithm. Each floor plan is the basis for creating all abstract levels using the previously described parameters: we selected a Pioneer-like robot with the radius 0.25 m, we selected the size of the cell 0.1 m to ensure passable doorways in the grid representation after enlargement of obstacles for the radius of the robot, and we selected the maximal width of the narrow passage as 6 m according to the average width of the room to create more submap nodes in the hierarchy. The lowest level contains the occupancy grid map of all rooms, and the highest level contains only one node called FER. The floor plans of buildings B and D were divided into two parts due to their large dimensions and therefore considered as four buildings.

4.1 H-Graph Analysis

Results of the proposed algorithm for autonomous H-graph creation is given in Table 1 in terms of the number of submap or end nodes K, the number of bridge nodes B_{N} and the number of marginal bridge nodes $B_{\mathrm{Nmarginal}}$. Expectingly, the number of nodes decreases at higher levels of the hierarchy. At the most detailed level, all bridge nodes are also marginal bridge nodes. Also, at each higher level, bridge nodes are marginal bridge nodes from a lower level. The created H-graph is shown in Fig. 1. The graph is completely connected by bridge nodes and the corresponding partial paths, so it is possible to plan the robot's path through the whole FER.

Table 1. Review of nodes at each level.

Level	*K*	B_N	$B_{Nmarginal}$
L_0	1 environment	0	0
L_1	6 buildings	18	0
L_2	27 floors	92	18
L_3	1924 rooms	1882	92
L_4	1,992,344 nodes	1882	1882

4.2 H-Graph and Plain Graph Paths

We tested the proposed hierarchy in the path planning example where the start is on the third floor of building A and the goal is on the seventh floor of building C. First, the rooms where the start and goal are located are searched by the E* algorithm to find the path to the bridge nodes (doors). The rest of the path is found by searching the marginal bridge nodes at each level. On the other hand, path planning within the plain graph searches all nodes on each floor through which the path passes. The results of the comparison are shown in Table 2 and the path comparison is shown in Fig. 7. The results show that about 50 times fewer nodes need to be searched using the H-graph compared to the plain graph. Searching fewer nodes significantly saves computational time. Path planning within an H-graph results in a 2% longer path compared to a plain graph because somewhere bridge nodes are not on the doors but in the middle of the passage to form rooms with fewer end nodes. This proves that position of the bridge nodes is indeed at optimal place because they do not significantly affect the length of the path.

Table 2. Comparison of path planning in H-graph and plain graph.

Graph	*N*	l (m)
H-graph	5497	189.4575
Plain graph	289,199	185.3772

4.3 Comparison of Partial Paths by E* and A*

We compared the partial paths creation by the E* algorithm to the standard A* algorithm according to the following path characteristics: the length of the

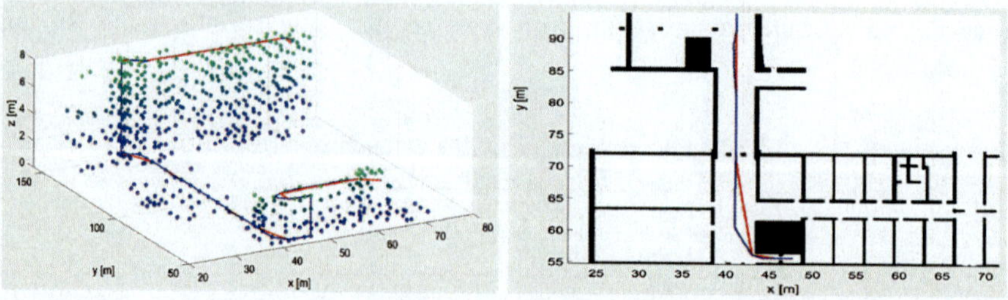

(a) (b)

Fig. 7. Path found within H-graph (blue) and within plain graph (red): (a) set in the context of the H-graph; (b) zoomed in order to detect deviations.

path l, the number of points in which path changes direction $n\alpha$, the sum of the all direction change angles $\sum\alpha$ and the number of points in which the change of direction is greater than or equal to 45° $n\alpha_{\geqq 45^\circ}$. Algorithms E* and A* are executed at occupancy grid map of one room. The start is chosen within the room, and the goal is set at the door of the room. The numerical results for the initial planning are shown in Table 3 and the path comparison is shown in Fig. 8. The E* algorithm outperforms the A* algorithm in path length and direction changes. On average, for the found paths from each node in room shown in Fig. 8 to the door, A* has about 3% higher path cost and the path has significant directional changes at several path points.

Table 3. Comparison of E* and A* algorithm in map of room.

Algorithm	l (m)	$n\alpha$	$\sum\alpha$ (°)	$n\alpha_{\geqq 45^\circ}$
E*	7.2504	4	93.74	1
A*	7.6569	7	315.00	7

5 Conclusion and Future Work

We presented the autonomous hierarchy creation for path planning of mobile robots in large environments. We demonstrate the necessary conditions of optimality during the creation of H-graph. The test of the algorithm on FER buildings demonstrated successful construction of the graph. We have shown that the path planning algorithm in the H-graph explored 50 times fewer nodes than in the classical non-hierarchical graph. Furthermore, testing shows that our algorithm supports modular graph upgrades when it is needed. Future work will focus on fast path replanning within created H-graph.

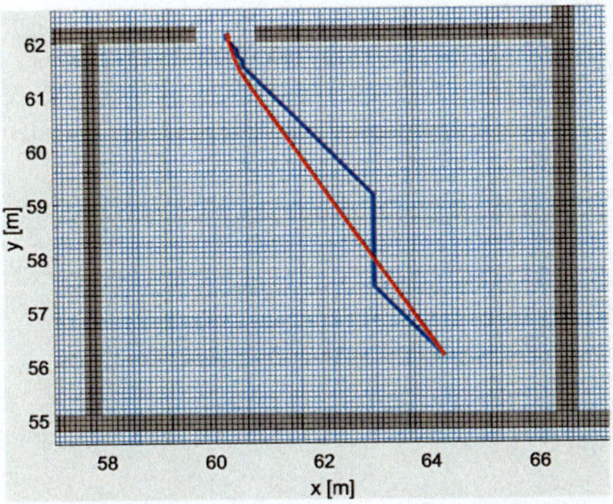

Fig. 8. Path found in the occupancy grid map with E* algorithm (red) and A* algorithm (blue).

Acknowledgments. This work has been supported by the European Regional Development Fund under the grant KK.01.2.1.02.0119 – Research and development of an advanced unit for autonomous control of mobile vehicles in logistics (A-UNIT).

References

1. Galindo, C., Fernandez-Madrigal, J.A., Gonzalez, J.: Improving efficiency in mobile robot task planning through world abstraction. IEEE Trans. Robot. **20**(4), 677–690 (2004)
2. Bonnington, C.P., Little, C.H.C.: The Foundations of Topological Graph Theory. Springer, New York (1995)
3. Cagigas, D.: Hierarchical D* algorithm with materialization of costs for robot path planning. Robot. Auton. Syst. **52**, 190–208 (2005)
4. Ryu, H.: Hierarchical path-planning for mobile robots using a Skeletonization-informed rapidly exploring random tree*. Appl. Sci. **10**(21) (2020)
5. Chen, J., Liu, C., Wu, J., Furukawa, Y.: Floor-SP: inverse CAD for floorplans by sequential room-wise shortest path. In: 2019 IEEE/CVF International Conference on Computer Vision (ICCV), pp. 2661–2670 (2019)
6. Sethian, J.: A fast marching level set method for monotonically advancing fronts. Proc. Natl. Acad. Sci. **93**, 1591–1595 (1996)
7. Čikeš, M., Đakulović, M., Petrović, I.: The path planning algorithms for a mobile robot based on the occupancy grid map of the environment—A comparative study. In: 2011 XXIII International Symposium on Information, Communication and Automation Technologies, pp. 1–8 (2011)
8. Seder, M., Juric-Kavelj, K., Petrović, I.: Automatic creation of hierarchical maps for indoor environments. In: Proceedings of the Fifth International Conference on Computational Intelligence, Robotics and Autonomous Systems, pp. 19–24 (2008)
9. Seder, M., Mostarac, P., Petrović, I.: Hierarchical path planning of mobile robots in complex indoor environments. Trans. Inst. Meas. Control **33**(3–4), 332–358 (2011)

10. Philippsen, R.: A Light Formulation of the E* Interpolated Path Replanner. Autonomous Systems Lab, Ecole Polytechnique Federale de Lausanne (2006)
11. Sam Ge, S., Zhang, Q., Thomas Abraham, A., Rebsamen, B.: Simultaneous path planning and topological mapping (SP2ATM) for environment exploration and goal oriented navigation. Robot. Auton. Syst. **59** (2011)
12. Thrun, S., Burgard, W., Fox, D.: Probabilistic Robotics. MIT Press, Cambridge, Massachusetts (2005)
13. Huang, Y.W., Jing, N., Rundensteiner, E.: A hierarchical path view model for path finding in intelligent transportation systems. GeoInformatica **1**, 125–159 (1997)

Improving the Flow in Multi-robot Logistic Systems Through Optimization of Layout Roadmaps

Rok Vrabič[1(✉)], Tena Žužek[1], Gašper Škulj[1], Igor Banfi[2], and Viktor Zaletelj[2]

[1] Faculty of Mechanical Engineering, University of Ljubljana, Ljubljana, Slovenia
rok.vrabic@fs.uni-lj.si
[2] Epilog d.o.o., Ljubljana, Slovenia

Abstract. In intralogistic systems, set-up is extremely important and largely determines their performance. In systems that use autonomous guided vehicles (AGVs), planning the roadmaps is a complex problem that is usually solved by human experts. While roadmaps in AGV systems are fixed using magnetic tape, autonomous mobile robots (AMRs) provide additional flexibility because they are able to move around freely. However, to maintain system performance, free movement needs to be constrained, e.g., to avoid deadlocks in narrow aisles. A method is presented that suggests the movement constraints in terms of preferred directions of motion. The free space is modeled as a grid graph with directed, weighted edges. The weights are then optimized by an algorithm inspired by ant colony optimization (ACO) that aims to reduce the amount of conflicting situations, such as collisions, in a given intralogistic problem. The approach is illustrated in a case study which shows that the proposed method can inform a feasible roadmap plan that leads to fewer conflict situations.

Keywords: Mobile robots · Intralogistics · Ant colony optimization

1 Introduction

Determining roadmaps on a given layout is an important problem in setting up autonomous guided vehicle (AGV) intralogistic systems. A good roadmap can lead to better logistic performance and fewer conflicts such as evasive maneuvers, grid-locks, and forced waiting times. However, the set-up problem is very difficult because it is not only an optimization challenge, but also usually requires the knowledge of human experts who understand additional context and have an overview of how the system will operate.

Several approaches to the AGV system set-up can be found in the literature. For the definition of roadmaps, Digani et al. [1] propose an automatic algorithm that maximizes the efficiency of the overall system by maximizing algebraic connectivity, coverage, and redundancy. Beinshob et al. [2] further integrate the

I. Petrovic et al. (Eds.): IAS 2022, LNNS 577, pp. 923–934, 2023.
https://doi.org/10.1007/978-3-031-22216-0_62

proposed automatic roadmap creation algorithm with a semi-automated mapping methodology. Uttendorf et al. [3] propose an expert system that combines traditional path finding algorithms (A* and Bellman-Ford algorithm) with a fuzzy inference system that incorporates the human knowledge of AGV system planners. Xu et al. [4] use Voronoi diagrams to establish the road model in a static environment. The optimal path of AGV is then found using a path planning algorithm based on a high fitness genetic algorithm. Yu et al. [5] formulate a mixed-integer programming model of a multi-AGV system for path planning and propose an improved routing method with a penalty item to generate the optimal paths for AGVs.

Recently, an alternative to AGV systems has emerged in the form of autonomous mobile robots (AMRs) that are fully autonomous in performing navigation tasks. AMRs can move freely without being bound to a magnetic tape and therefore offer more flexibility. This, however, does not mean that they automatically provide higher logistic performance. To improve the performance of AMR systems, additional rules and constraints need to be defined. This is best illustrated by the example of a narrow aisle, where it makes sense to enforce a one-way rule to avoid potential deadlocks. By extension, rules that constrain the movement in aisles, corridors, intersections, etc., can have a major impact on the performance of the entire system. The goal, then, is to maintain the flexibility offered by AMRs while achieving logistic performance comparable to that of AGV systems. Setting up the roadmaps for AMR systems on a given layout has received little attention in the existing literature. Kleiner et al. [6] present an adaptive roadmap optimization based on a linear programming formulation. The proposed approach is capable of adapting the roadmap in a dynamically changing environment. Lee et al. [7] propose an improved A* algorithm that finds all possible AMR paths. Comparing the time window of predetermined paths, it finds different types of collisions (stay-on, head-on, cross conflict) and solves them using different strategies (go-away, detour, wait-before-startup strategy). Henkel and Toussaint [8] present Optimized Directed Roadmap Graph approach that builds a directed roadmap graph that allows for collision avoidance in multi-robot navigation and can be used both in centralized and decentralized planning setup. The core idea of the approach is to optimize directed roadmaps for minimal expected path length. Kozjek et al. [9] propose an automated AMR route generation approach based on reinforcement learning.

In contrast to the research mentioned above, this paper focuses on the set-up problem in AMR systems by abstracting the free space in the form of a grid map with directed weighted costs of movement. To determine the costs, an ant-colony-optimization-type algorithm is proposed. The resulting weighted graph can then be used by the designer of the system when designing the set-up. Several case studies are used to illustrate the results and confirm the feasibility of the approach.

2 Problem Formulation

This section formalizes the roadmap set-up problem for AMR systems. The problem can be stated as follows and is elaborated further below.
Given an intralogistic problem definition, determine the costs of a weighted graph, corresponding to the layout, so that the flow of multiple agents traversing the graph is maximized.

An intralogistic problem is described by a floor plan, the corresponding layout, and a statistical description of the tasks. The floor plan specifies the coordinates of the pick-up and drop-off points. Each task type is specified by a set of pick-up points, a set of drop-off points, an arrival rate, and a due-time. The arrival rate is exponentially distributed, while the due-time is normally distributed. The description of the tasks is then expressed as a set of task types.

Given the layout, an occupancy grid map can be constructed. The cell size is assumed to be just large enough to accommodate two AMRs traversing in opposite directions, i.e., two robots on a single cell would be hindered, but not incapacitated. A weighted grid graph is then laid over the grid map, with weighted edges connecting the free cells. The weights are used for pathfinding by each agent traversing the graph, e.g., using Dijkstra's algorithm or A*.

AMRs are represented by agents. When tasks are assigned to a vehicle, the corresponding agent creates an execution plan in the form of a path connecting pick-up and drop-off points. As the agents execute their plans in time, conflicts can occur, such as deadlocks or multiple AMRs in a confined space, resulting in collision avoidance maneuvers, and disrupting the overall flow. The frequency of these types of conflicts needs to be minimized by optimizing the costs of the weighted graph used for pathfinding. Thus, the goal is to find the preferred directions of movement for each cell. This is illustrated in Fig. 1.

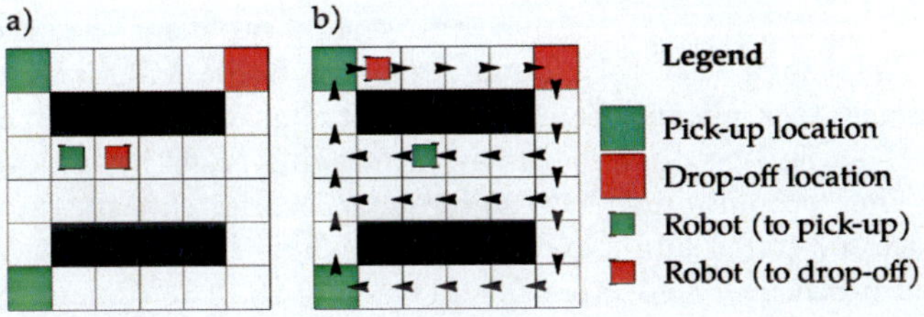

Fig. 1. Illustration of a) an intralogistic problem, and b) an example solution.

The illustration shows a simple problem in which two pick-up points and one drop-off point are specified within a given layout. The layout consists of narrow corridors on the sides and a wider corridor in the middle. If the movement were not constrained, agents could end up in scenarios where they have to avoid collisions (Fig. 1a), while with predefined preferred movement directions(Fig. 1b), a better flow and consequently a higher throughput can be achieved. The problem can be interpreted as one of defining roadmaps through free space. To illustrate with a simple example that uses the same concepts: the throughput of car traffic can be improved by building highways. That is, as long as cars move along the lanes in the correct, predefined directions.

The difficulty with this problem is that it is hard to identify what exactly the optimization criterion should be. While human experts can quickly identify important features of the layout, optimization techniques require well-defined criteria. However, the throughput of the system depends largely on the parameters of the problem. If there are relatively few agents relative to the layout size the movement constraints should be relaxed, while in heavy-traffic systems, roadmaps must to be well-defined and strictly enforced to avoid congestion.

Therefore, the paper suggests a method inspired by ant colony optimization (ACO) as these types of methods can intrinsically generate sensible pathways that result in improved flow. The method, presented in the next section, is then evaluated from the perspectives of conflict minimization, increase of path lengths arising due to the movement constraints, and from the perspective of common sense.

3 Roadmap Generation Inspired by Ant Colony Optimization

Ant colony optimization is a probabilistic technique for solving combinatorial optimization problems, most commonly applied to path-finding problems in graphs. The technique is inspired by the real behavior of ants, which use pheromones as a communication medium [10]. The pheromone trails in ACO serve as distributed information that guides the agents (ants) in optimizing the pathways. Notably, the optimization is not explicit, i.e., the optimization criterion is not explicitly used to evaluate the solutions, rather, the optimization is an implicit consequence of the pheromone update function.

ACO techniques have been employed in several problems that arise in AGV systems. In Kulatunga et al. [11], a meta-heuristic based ACO technique is used for simultaneous task allocation and path planning of AGVs in material handling. Bedi et al. [12] propose the usage of foraging trails by ants to eliminate congestion. They present an ACO-inspired approach for choosing an alternative optimum path for traffic jam avoidance. In Fuellerer et al. [13], an ACO-based heuristic has proven very effective for addressing the two-dimensional loading vehicle routing problem. An ACO-based algorithm has also been proposed for finding the shortest bidirectional path for AGVs [14]. Compared to other existing algorithms (i.e. cutting-plane algorithm), the algorithm provides improved results in terms of both time and quality of the solutions. Yi et al. [15] propose a two-stage improved ACO algorithm for solving job-shop scheduling and collision-free routing problem in multi-AGV systems. The algorithm is improved based on multi-object programming to optimize the pheromone matrix.

Furthermore, ACO algorithms have been employed to improve the behavior of AMRs and corresponding systems. Porta Garcia et al. [16] combine a simple ACO heuristic with fuzzy cost function evaluation for AMR path planning.

The proposed approach supports both static and dynamic obstacle avoidance. A combination of ACO and fuzzy logic for AMR planning in an unknown dynamic environment has also been proposed by Purian and Sadeghian [17]. ACO has also been successfully applied for path planning of multiple autonomous marine vehicles. Xiong et al. [18] present a hybrid Voronoi-based ACO technique that is efficient and effective in generating near optimal and collision-free trajectory. Compared to other frequently used algorithms (i.e. RRT*, Dijkstra's algorithm, conventional ACO), the proposed technique shows better performance in terms of searching ability, accuracy and convergence rate.

However, there is no literature that addresses the roadmap generation problem using ACO. An algorithm is thereby proposed that works as follows. First, a directed graph is created based on the layout. Each edge is assigned a cost c and a movement count m, initialized to $c_{max}/2$ and 0, respectively. In each iteration, N agents are spawned on random nodes and J jobs are created connecting the pick-up and the drop-off points. The dispatching is randomized, i.e., jobs are assigned to agents at random. Then, for each agent, based on its initial position, the least-cost paths to each pickup and drop-off point are created for each job assigned to the agent. In this way, N paths are generated. In this way, N paths are generated.

In each iteration, for every agent, and for every edge on the agent's path, the movement count of the edge is decremented (thereby decreasing the cost), while the movement count for every opposing directed edge is incremented. The cost of the edge is set using the logistic function (Eq. 1), to prevent runaway costs and to ensure smooth updates.

$$c = \frac{c_{max}}{1 + e^{-k \cdot m}} \tag{1}$$

In Eq. 1, k is the logistic growth rate, i.e., the steepness of the logistic curve.

Head-on collisions, movement steps, and total path costs are counted for each agent. A head-on collision occurs 1) when two or more agents moving in opposite directions plan to occupy the same grid cell at the same time, or 2) when two agents traverse the edges in opposite directions between the same two cells at the same time. Figure 2 illustrates collision scenarios on a grid map.

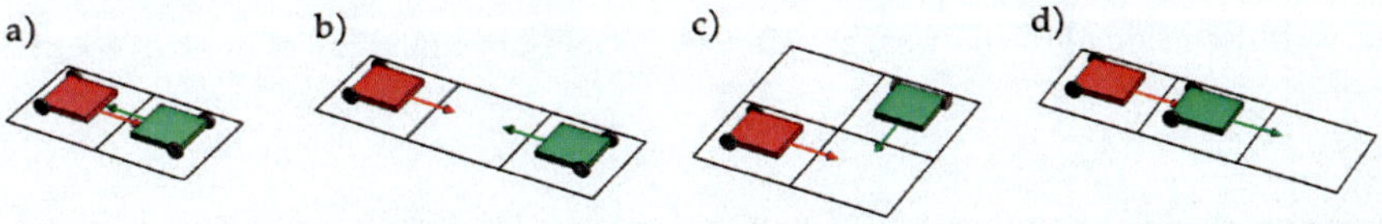

Fig. 2. a) Head-on edge collision, b) head-on node collision, c) side collision, and d) no collision, as the AMRs are moving in the same direction.

The pseudocode of the algorithm is shown in Algorithm 1.

```
load a layout and create the corresponding directed graph;
set all world graph edges movement counts m to 0;
set all world graph edges costs c to c_max/2;
foreach iteration do
    generate N agents on random nodes;
    generate J jobs connecting random pick-up and drop-off points;
    foreach job do
        pair job with a random agent;
    end
    foreach agent do
        generate a least-cost (c) path for the agent over assigned jobs;
    end
    foreach path do
        foreach move in path do
            decrease m in direction of move by 1;
            increase m in opposite direction of move by 1;
            set c according to Eq. 1;
        end
    end
end
```

Algorithm 1: ACO-inspired edge cost optimization algorithm.

Based on the resulting weighted graph, the preferred movement directions on the given layout could be visualized, which can then be used by an expert to determine the roadmaps. The steps of the approach are shown in Fig. 3.

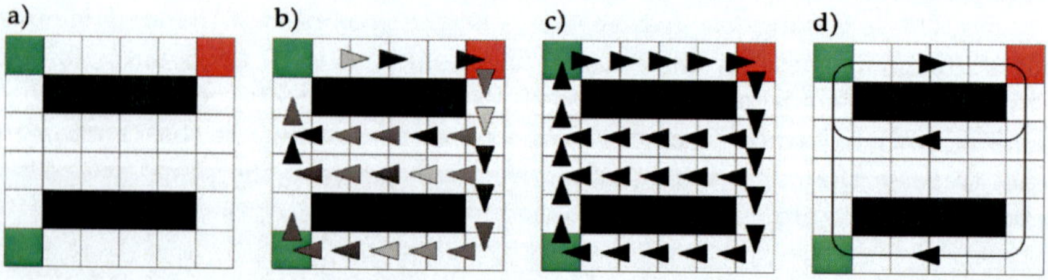

Fig. 3. Illustration of the approach: a) layout, b) mid-optimization, c) final, d) roadmap graph.

Figure 3a shows a layout with marked pick-up (green) and drop-off (red) points. Over the course of the iterations, the algorithm determines the costs and the corresponding preferential movement directions for each cell, as shown in Fig. 3b. After the costs converge (Fig. 3c), the system designer can use this information to determine the roadmaps that make up a directed graph (Fig. 3d).

4 Approach Validation

The approach is validated using four illustrative examples. Three of the examples are based on the multi-agent pathfinding benchmark (MAPF) [19], while the fourth one mimics a realistic warehouse set-up. All layouts use a 32×32 grid and are shown in Fig. 4.

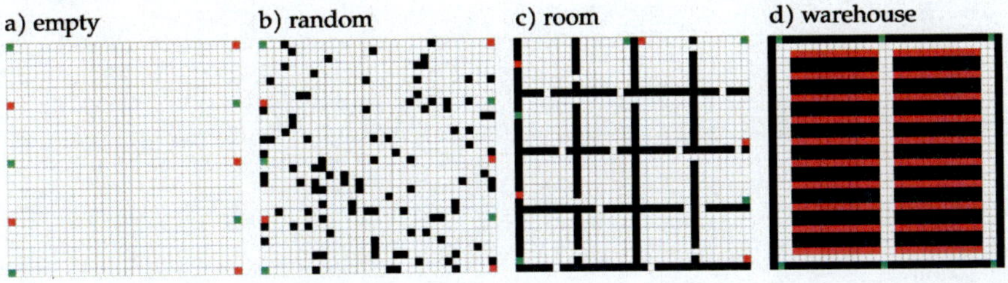

Fig. 4. Layouts for validation: a) empty, b) random obstacles, 10% fill, c) room, and d) warehouse.

Scenarios a) empty and b) random are used to illustrate the behavior of the algorithm and to obtain working parameters (N, J), while c) room, and d) warehouse are used to showcase the behavior of the algorithm for more realistic scenarios. The algorithm parameters are set to $N = 32$ agents, $J = 1024$ jobs. The parameters of the logistic function (Eq. 1) are set to $c_{max} = 100$ and $k = 0.5$. The choice of parameters is based on the size of the layout and mainly affects the speed of convergence. The algorithm is run for 8 iterations and with 8 repetitions.

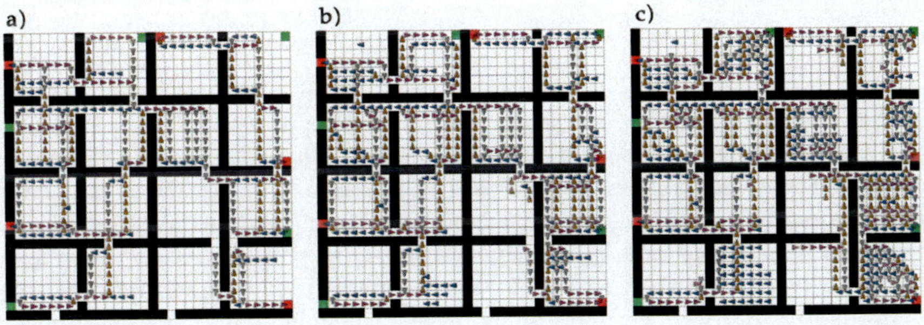

Fig. 5. Example progression over iterations, room layout: after a) 1, b) 2, and c) 8 iterations.

Figure 5 shows the progression of the costs over 1, 2, and 8 iterations of the algorithm for the room scenario. The arrows show the preferred, i.e. least-cost directions of the movement. While the algorithm is able to find a sensible solution after 1 iteration (Fig. 5a), paths are reinforced and new paths are added in subsequent ones (Fig. 5b and c). The final result in Fig. 5c shows that roadmaps should be placed to connect the rooms, while movements within the rooms should be either circular or can be of any kind that connects the passages between the rooms, e.g. any diagonal.

To analyze how the algorithm improves the system flow, collisions between the agents are analyzed and the ratio between head-on collisions and all collisions r is plotted in Fig. 6.

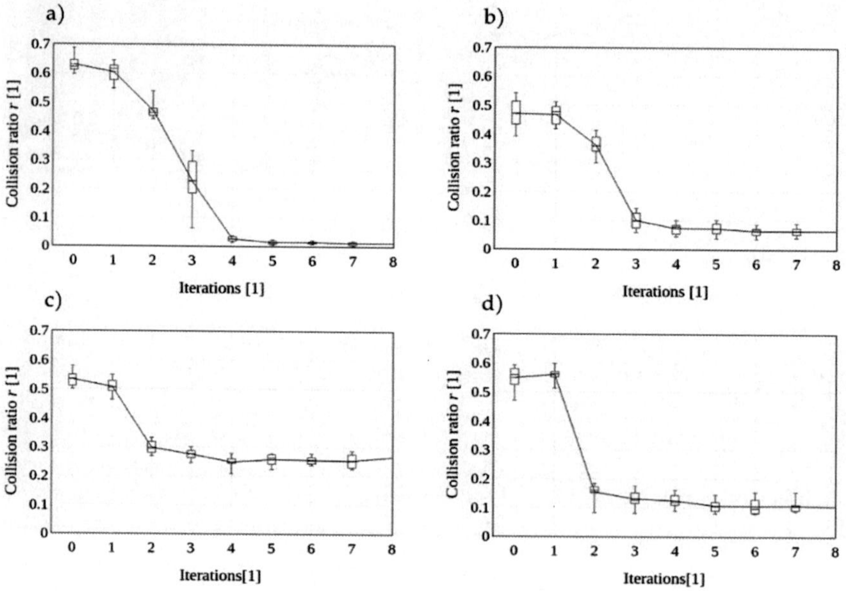

Fig. 6. The ratio of head-on collisions in a) empty, b) random, c) room, and d) warehouse scenarios.

In the empty space scenario (Fig. 6a), the collision ratio drops to almost 0, signifying that when the costs obtained by the algorithm are used, the agents almost never move in opposite directions, thereby improving the overall system flow. According to the definition used, side collisions still occur, but they are not considered to disrupt the flow (in the same way as e.g. merging onto highways in automobile traffic does not disrupt the flow, while driving in the opposite direction does). The absolute number of head-on collisions decreases in the simulations with the given algorithm parameters from approx. 3200 to 0 (empty), 1800 to 200 (random), 4200 to 1600 (room), and 2300 to 300 (warehouse).

The algorithm converges in 2–4 iterations in all of the presented cases. However, the final ratio depends on the case and is approx. 7%, 25% and 10% in random, room, and warehouse scenarios, respectively. This means that, especially in the room scenario, it sometimes still makes sense for the agents to go in the opposite direction of the flow, as going all the way around would increase the cost too much. This relates to the distance traveled. The ratio of the distance traveled to the shortest distance, i.e. without any cost optimization, is shown in Fig. 7.

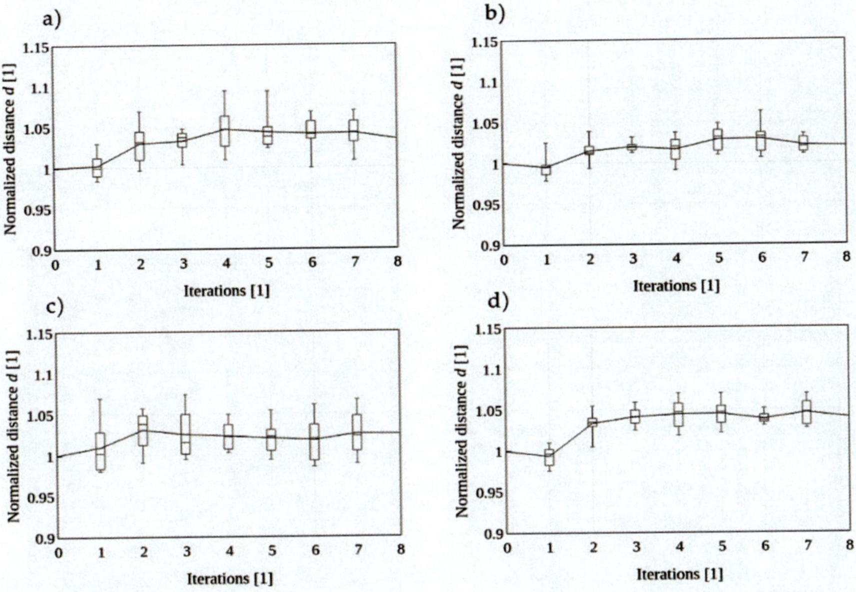

Fig. 7. Normalized distances in a) empty, b) random, c) room, and d) warehouse scenarios.

Since the costs obtained by the algorithm guide the agents away from the shortest path in order to improve the flow, the normalized distances increase in all scenarios, but for less than 5%, on average. The smallest increases are observed in random (Fig. 7b) and room (Fig. 7c) scenarios in which the layout itself is the main factor that determines the distance lengths. In the warehouse scenario (Fig. 7d), the distance increases the most, as one-way traffic is necessary through the narrow corridors to improve the flow.

Finally, the resulting preferential movement directions after 8 iterations are shown in Fig. 8.

In all scenarios, the solutions obtained present a sensible ground for determining the roadmaps. For example, in the warehouse scenario (Fig. 8d), outward directions are suggested between the aisles with inward traffic at the top and bottom sides, while roundabouts are suggested in the vertical part in the middle of the layout.

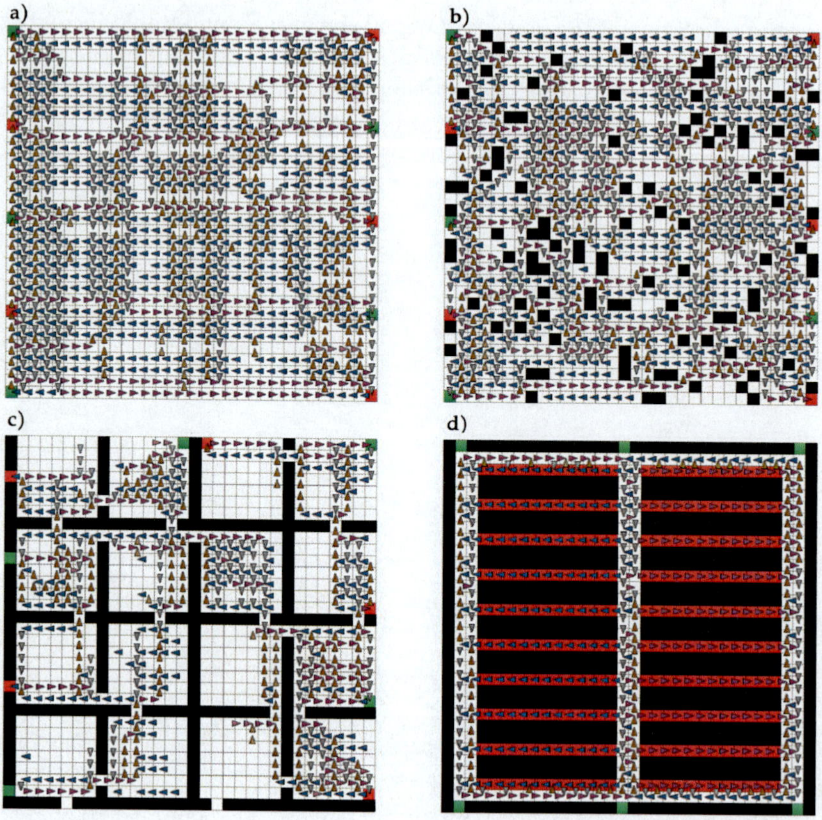

Fig. 8. a) Empty, b) random, c) rooms, and d) warehouse layout cost maps after 8 iterations.

5 Discussion, Conclusions, and Future Work

The paper addresses the problem of set-up in intralogistic systems that use AMRs. While using AMRs instead of AGVs improves the system flexibility, constraining AMR movements in free space can lead to improved intralogistic flow.

An algorithm inspired by ACO is proposed for determining the preferred directions of movement. The algorithm is tested in four scenarios. The results show that head-on collisions are reduced significantly at the expense of prolonging the routes by less than 5%, thereby creating useful suggestions for roadmap planning.

This work presents initial research with the ultimate goal of creating a suggestion system that will help experts improve the overall set-up of intralogistic systems in warehouses and factories. An illustrative example of how an expert can use the results of the proposed approach is shown in Fig. 9 for the warehouse scenario.

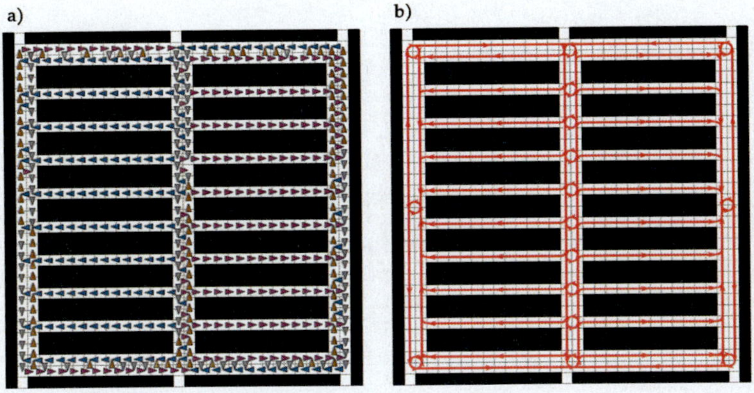

Fig. 9. a) The result of the proposed algorithm, serving as a basis for determining b) roadmaps.

Although the results presented are promising, further work is needed and planned. This includes building a discrete-time, grid-based simulation and applying other optimization methods, particularly exploring multi-objective optimization, integrating the simulation with route planning and dispatching systems, evaluating the benefits through multi-agent simulations [19], as well as validating the approach in realistic simulations and real-world scenarios.

Acknowledgments. This work and future work is supported by The Slovenian Research Agency ARRS, grants nb. L2-3168 and P2-0270, and by Epilog d.o.o., Ljubljana, Slovenia.

References

1. Digani, V., Sabattini, L., Secchi, C., Fantuzzi, C.: Ensemble coordination approach in multi-AGV systems applied to industrial warehouses. IEEE Trans. Autom. Sci. Eng. **12**, 922–934 (2015)
2. Beinschob, P., Meyer, M., Reinke, C., Digani, V., Secchi, C., Sabattini, L.: Semi-automated map creation for fast deployment of AGV fleets in modern logistics. Robot. Auton. Syst. **87**, 281–295 (2017)
3. Uttendorf, S., Eilert, B., Overmeyer, L.: A fuzzy logic expert system for the automated generation of roadmaps for automated guided vehicle systems. In: 2016 IEEE International Conference on Industrial Engineering and Engineering Management (IEEM), vol. 87, pp. 977–981 (2016)
4. Xu, W., Wang, Q., Yu, M., Zhao, D.: Path planning for multi-AGV systems based on two-stage scheduling. Int. J. Perform. Eng. **13**, 1347–1357 (2017)

5. Yu, D., Hu, X., Liang, K., Yong, J.: A parallel algorithm for multi-AGV systems. J. Ambient Intell. Human. Comput. (2021)
6. Kleiner, A.. Sun, D., Meyer-Delius, D.: ARMO: adaptive road map optimization for large robot teams. In: 2011 IEEE/RSJ International Conference on Intelligent Robots and Systems, pp. 3276–3282 (2011)
7. Lee, C.K.M., Lin, B., Ng, K.K.H., Lv, Y., Tai, W.C.: Smart robotic mobile fulfillment system with dynamic conflict-free strategies considering cyber-physical integration. Adv. Eng. Inform. **42**, 100998 (2019)
8. Henkel, C., Toussaint, M.: Optimized directed roadmap graph for multi-agent path finding using stochastic gradient descent. In: Proceedings of the 35th Annual ACM Symposium on Applied Computing, pp. 776–783 (2020)
9. Kozjek, D., Malus, A., Vrabič, R.: Reinforcement-learning-based route generation for heavy-traffic autonomous mobile robot systems. Sensors **21**, 4809 (2021)
10. Dorigo, M., Stützle, T.: Ant colony optimization: overview and recent advances. In: Handbook of Metaheuristics, 3rd edn., pp. 311–351 (2019)
11. Kulatunga, A., Liu, D., Dissanayake, G., Siyambalapitiya, S.B.: Ant colony optimization based simultaneous task allocation and path planning of autonomous vehicles. In: 2006 IEEE Conference on Cybernetics and Intelligent Systems, pp. 1–6 (2016)
12. Bedi, P., Mediratta, N., Dhand, S., Sharma, R., Singhal, A.: Avoiding traffic Jam using ant colony optimization—a novel approach. In: International Conference on Computational Intelligence and Multimedia Applications (ICCIMA 2007), pp. 61–67 (2007)
13. Fuellerer, G., Doerner, K.F., Hartl, R.F., Iori, M.: Ant colony optimization for the two-dimensional loading vehicle routing problem. Comput. Oper. Res. **36**, 655–673 (2009)
14. Hamzheei, M., Farahani, R.Z., Rashidi-Bajgan, H.: An ant colony-based algorithm for finding the shortest bidirectional path for automated guided vehicles in a block layout. Int. J. Adv. Manuf. Technol. **64**, 399–409 (2013)
15. Yi, G., Feng, Z., Mei, T., Li, P., Jin, W., Chen, S.: Multi-AGVs path planning based on improved ant colony algorithm. J. Supercomput. **75**, 5313–5898 (2019)
16. Porta Garcia, M.A., Montiel, O., Castillo, O., Sepúlveda, R., Melin, P.: Path planning for autonomous mobile robot navigation with ant colony optimization and fuzzy cost function evaluation. Appl. Soft Comput. **9**, 1102–1110 (2009)
17. Purian, F.K., Sadeghian, E.: Mobile robots path planning using ant colony optimization and Fuzzy Logic algorithms in unknown dynamic environments. In: 2013 International Conference on Control, Automation, Robotics and Embedded Systems (CARE), pp. 1–6 (2013)
18. Xiong, C., Chen, D., Lu, D., Zeng, Z., Lian, L.: Path planning of multiple autonomous marine vehicles for adaptive sampling using Voronoi-based ant colony optimization. Robot. Auton. Syst. **115**, 90–103 (2019)
19. Stern, R., Sturtevant, N., Felner, A., Koenig, S., Ma, H., Walker, T., Li, J., Atzmon, D., Cohen, L., Kumar, T.K.S., Boyarski, E., Bartak, R.: Multi-agent pathfinding: definitions, variants, and benchmarks. In: Twelfth Annual Symposium on Combinatorial Search (SoCS 2019), pp. 151–158 (2019)

Author Index

I. Petrovic et al. (Eds.): IAS 2022, LNNS 577, pp. 935–937, 2023.
https://doi.org/10.1007/978-3-031-22216-0